MOLECULAR DYNAMICS AND SPECTROSCOPY BY STIMULATED EMISSION PUMPING

Advanced Series in Physical Chemistry

Editor-in-Charge

Cheuk-Yiu Ng, *Ames Laboratory USDOE, and Department of Chemistry, Iowa State University, USA*

Associate Editors

Paul F. Barbara, *Department of Chemistry, University of Minnesota, USA*

Sylvia T. Ceyer, *Department of Chemistry, Massachusetts Institute of Technology, USA*

Hai-Lung Dai, *Department of Chemistry, University of Pennsylvania, USA*

Benny Gerber, *The Fritz Haber Research Center and Department of Chemistry, The Hebrew University of Jerusalem, Israel, and Department of Chemistry, University of California at Irvine, USA*

James J. Valentini, *Department of Chemistry, Columbia University, USA*

Published:

Vol. 1: Physical Chemistry of Solids: Basic Principles of Symmetry and Stability of Crystalline Solids
H. F. Franzen

Forthcoming:

Vol. 2: Modern Electronic Structure Theory
ed. D. R. Yarkony

Vol. 3: Progress and Problems in Atmospheric Chemistry
ed. J. R. Barker

Vol. 5: Laser Spectroscopy and Photochemistry on Metal Surfaces
eds. H.-L. Dai and W. Ho

Vol. 6: The Chemical Dynamics and Kinetics of Small Radicals
eds. K. Liu and A. Wagner

Vol. 7: New Developments in Theoretical Studies of Proteins
ed. R. Elber

Advanced Series in Physical Chemistry — Vol. 4

MOLECULAR DYNAMICS AND SPECTROSCOPY BY STIMULATED EMISSION PUMPING

Editors

Hai–Lung Dai
Univ. Pennsylvania

Robert W. Field
MIT

World Scientific
Singapore • New Jersey • London • Hong Kong

Published by

World Scientific Publishing Co. Pte. Ltd.
P O Box 128, Farrer Road, Singapore 9128
USA office: Suite 1B, 1060 Main Street, River Edge, NJ 07661
UK office: 57 Shelton Street, Covent Garden, London WC2H 9HE

Library of Congress Cataloging-in-Publication Data

Molecular Dynamics and Spectroscopy by Stimulated Emission Pumping /
   editor[s], Hai-Lung Dai, Robert W. Field.
      p.   cm. -- (Advanced Series in Physical Chemistry ; Vol. 4)
   Includes bibliographical references and index.
   ISBN 9810217498. -- ISBN 9810221118 (pbk.)
    1. Vibrational spectra.   2. Molecular dynamics.
 I. Dai, Hai-Lung.   II. Field, Robert W.   III. Series.
 QD96.V53M63   1995
 543'.0858--dc20   95-5003
                                        CIP

Printed in Singapore by Uto-Print

INTRODUCTION

Many of us who are involved in teaching a special-topic graduate course may have the experience that it is difficult to find suitable references, especially reference materials put together in a suitable text format. Presently, several excellent book series exists and they have served the scientific community well in reviewing new developments in physical chemistry and chemical physics. However, these existing series publish mostly monographs consisting of review chapters of unrelated subjects. The modern development of theoretical and experimental research has become highly specialized. Even in a small subfield, experimental or theoretical, few reviewers are capable of giving an in-depth review with good balance in various new developments. A thorough and more useful review should consist of chapters written by specialists covering all aspects of the field. This book series is established with these needs in mind. That is, the goal of this series is to publish selected graduate texts and stand-alone review monographs with specific themes, focusing on modern topics and new developments in experimental and theoretical physical chemistry. In review chapters, the authors are encouraged to provide a section on future developments and needs. We hope that the texts and review monographs of this series will be more useful to new researchers about to enter the field. In order to serve a wider graduate student body, the publisher is committed to making available the monographs of the series in a paperbound version as well as the normal hardcover copy.

Cheuk-Yiu Ng

PREFACE

What is Intramolecular Vibrational Redistribution? Where does the energy flow? How? Why? How fast? How is classical ball-and-spring vibrational motion encoded in quantum eigenstate spectra? Eigenstates are stationary yet atom move and energy flows. For decades Physical Chemists and Molecular Physicists have fantasized about external control over unimolecular dynamics: bond specific chemistry! The Born–Oppenheimer approximation works well and serves to provide a rigorous definition of potential energy surfaces. Nevertheless, traditional high resolution vibration–rotation spectra tell us mostly only about the near equilibrium region of the potential. At low excitation energy, eigenstates are assignable to $3N - 6$ vibrational normal mode quantum numbers, yet at chemically significant levels of excitation the success of statistical reaction rate models seems to imply that the quantum number names of eigenstates are either irrelevant or nonexistent. Unimolecular isomerization occurs and Organic Chemists have assembled a large and reliable body of information about the heights of barriers to isomerization and reaction mechanisms involving unstable isomers. Nevertheless, even the simplest isomerization, a transfer of a Hydrogen atom from one Carbon atom to another has been extremely resistant to detection in rotation–vibration spectrum.

The quality of the questions being asked is intimately related to the capabilities of the available experimental and computational techniques. The first hint that the evolution from assignable eigenstates to statistical dynamics contained some surprises and mysteries was revealed by high overtone spectroscopy. But overtone of R–H stretching vibrations provide a decidedly nondemocratic sample of initially localized excitations. When the first Stimulated Emission Pumping (SEP) experiments were reported in 1981, a period of explosive growth in both theoretical and experimental studies of intra- and inter-molecular dynamics was initiated.

SEP grew out of the Optical Optical Double Resonance (OODR) and Electronic Transition Optically Pumped Laser (OPL) experiments of one of us (RWF). In fact, as a postdoc in the UCSB Laboratory of the late H.P. Broida, the seeds were sown for the universally adopted but curiously crude "PUMP–DUMP" nomenclature of SEP when Broida refused to consider the use of the descriptive but less poetic alternative of DUMP-ed: "de-pumped."

The motivations for the initial development of SEP by James L. Kinsey and Robert W. Field were hardly prescient! PUMP–DUMP–PROBE studies of high vibrational levels of I_2 $X^1\Sigma_g^+$ were designed to characterize the mechanism of an atomic Iodine chemical laser. Then there was the idea that small polyatomic molecules near a dissociation limit might exhibit a simplified asymptotic energy level pattern similar to the long range $(v - v_D)^\alpha$ expressions then being developed for diatomic molecules. Nevertheless, when the initial MIT and Griffith University SEP experiments on I_2, H_2CO, HCCH, and *para*-difluorobenzene were successful, it was clear we had a tiger by the tail.

SEP is a folded variant of Optical Optical Double Resonance. The target excited vibrational levels of the electronic ground state $(e''vJ)$ are reached by two linked sequential electronic transitions originating from an initial, thermally populated rotation–vibration level $(e''v''J'')$: PUMP $e'v'J' \leftarrow e''v''J''$, DUMP $e'v'J' \rightarrow e''vJ$. Some of the advantages of SEP over other forms of vibrational spectroscopy that utilize zero or one laser are: reduced rotational congestion, ability to detect transitions spanning an intensity range of over 10^5, systematic exploitation of Franck–Condon access/discrimination via selection of a suitable $e'v'J'$ intermediate level, high sensitivity at high resolution (limited only by the laser linewidth), ability to span a wide energy region, and time resolution limited directly by the $\sim 10\,\mathrm{ns}$ duration of typical pulsed dye lasers on indirectly by coverage of $\sim 10\,000\,\mathrm{cm}^{-1}$ (i.e., $3\,\mathrm{fs}$) frequency regions. These advantages have been exploited in the study of stable molecules (ranging in size from diatomic molecules to substituted benzenes) and transient species (radicals, ions, atomic and molecular clusters, and weakly bound complexes). Many unimaginably difficult types of spectra have become almost routine; among these are fully resolved vibrational "resonances" above the dissociation threshold and systematic examination of extremely low frequency inter-molecular van der Waals vibrational modes.

Large, systematic, high quality vibration–rotation data sets have forced a reexamination of the question: How can the classical ball-and-spring and quantum tunneling motions encoded in such data sets best be extracted? This question has affected the experimentalists' choices of tactics and forced theorists to strive for more powerful computational methods, new statistical measures of complexity, and new kinds of spectral patterns or dynamical classifications. Naive ideas such as "intrinsically unassignable spectra" and "statistical spectroscopy" for placing a system on a simple, single continuum between regular and chaotic dynamics have evolved into robust and generally applicable methods for describing early time dynamics from assignable "feature states" (rather than eigenstates) and a powerful arsenal of "spectroscopic statistics" designed to extract specific types of patterns from a spectrum.

SEP has been used to explore the effects of initial evJ state on the rate and products of unimolecular reactions, both isomerization and dissociation. SEP or other more efficient variants such as STIRAP could also be used to populate specific highly excited vibrational levels for studies of bimolecular reactions. PUMP–DUMP–PROBE schemes have already provided unique information about collision induced rotation–vibration energy transfer processes.

The twenty-eight contributed articles in this book include discussions of a wide range of experimental techniques and results and theoretical methods, measures, and models. We have divided the book into four parts in order to emphasize the inter-relationships between techniques and to clarify the key issues. Part I, *Experimental Methods and Spectroscopy of Vibrationally Excited Molecules*, is focused on experimental techniques and spectroscopic results. Techniques range from the original fluorescence dip SEP detection scheme to the Resonant Four Wave Mixing, STIRAP, and Dispersed Fluorescence schemes which some have claimed will make SEP obsolete. Part II, *Intramolecular Vibrational Redistribution and Unimolecular Dissociation*, is a sampling of experimental studies of unimolecular dynamics, ranging from diatomic half-collisions, IVR in triatomics such as HCN and SO_2, tetra-atomics [HFCO and HCCH], and larger molecules [CH_3O and substituted benzenes], and the possibility of mode-specific unimolecular dissociation. In Part III, *Intermolecular Interactions*, There are studies of collision induced vibrational relaxation in small [NO] and large molecules as well as energy transfer within a van der Waals complex

[ArOH] and molecular clusters. Part IV, *Theoretical Methods for Extracting Vibrational Dynamics*, includes diverse topics such as *ab initio* computation of realistic SEP spectra, group theoretical classification of spectra of nonrigid molecules, pattern recognition and "spectroscopic statistics", and multi resonance superpolyad models and local/normal classifications of eigenstates.

We wish to thank Dr. Stephen Coy (MIT), Dr. Michael Davis (Argonne), Dr. Gregory Hartland (University of Pennsylvania), Dr. David Jonas (University of Chicago), Dr. Trevor Sears (Brookhaven), and Professor Patrick Vaccaro (Yale) for their assistance in the reviewing and editing of the contributed chapters.

Hai-Lung Dai
Robert W. Field

CONTENTS

Part II. Intramolecular Vibrational Redistribution and Unimolecular Dissociation

Part I. Experimental Methods and Spectroscopy of Vibrationally Excited Molecules

CHAPTER 1

RESONANT FOUR-WAVE MIXING SPECTROSCOPY: A NEW PROBE FOR VIBRATIONALLY-EXCITED SPECIES

Patrick H. Vaccaro

Department of Chemistry, Yale University

225 Prospect Street, New Haven, CT 06511, USA

Contents

1. Introduction

Potential energy hypersurfaces, describing the internal and external forces that couple the constituent atoms of a molecular ensemble, play a pivotal role in our present-day understanding of molecules and their behavior. In low energy regimes, theoretical descriptions for the structural properties of molecules are based upon the premise of adiabatic Born–Oppenheimer surfaces that serve to constrain and correlate the motions of individual nuclei.[1,2] For molecular systems displaced from their equilibrium configurations, the inelastic scattering processes responsible for the phenomenon of collisional relaxation are critically dependent upon the nature of intermolecular interactions (viz., potentials) and their relationship to internal degrees of freedom.[3,4] In a similar manner, modern treatments of reaction dynamics rely upon potential energy surfaces to mediate the subtle interplay between energy and matter that constitutes a chemical transformation.[5] Clearly, any detailed interpretation of molecular behavior is predicated upon the availability of suitable potential energy surfaces, with highly-refined information required over a broad range of both excitation energies and nuclear configurations.[6] The expedient extraction of such quantum state-specific data from experimental measurements demands the development and implementation of spectroscopic probes which are not encumbered by the complex, multidimensional character of polyatomic systems.

Since its inception over a decade ago at the M.I.T. laboratories of Field and Kinsey,[7] the method of Stimulated Emission Pumping (SEP) has emerged as a powerful tool for exploring the topography of potential energy hypersurfaces, with special utility for elucidating the dynamical processes which occur in regimes of extreme vibrational excitation.[8] Essentially a folded variant of optical–optical double resonance, SEP employs two narrow-band, tunable lasers, known as the PUMP and the DUMP, to execute a sequential transfer of molecules from an initial, thermally-populated level, via an electronically-excited intermediate state, to pre-selected rotation–vibration levels of the ground electronic state. While the magnitude of energy deposited into a target molecule is governed primarily by the frequency offset between the PUMP and DUMP beams (i.e., $\omega_{\text{PUMP}} > \omega_{\text{DUMP}}$), the type of excitation created within the molecular framework depends upon the nature of the spectroscopic transitions utilized for the doubly-resonant SEP scheme. In particular, by exploiting the substantial changes in molecular structure which often accompany electronic excitation, SEP enables the state-specific preparation and characterization of molecules having vibrational properties radically different from those encountered in close proximity to the equilibrium configuration of nuclei. Such chemically-relevant regions of the ground state potential surface are not readily accessible to other experimental techniques.

The canonical implementation of SEP entails tuning of the PUMP laser frequency into resonance with a preselected rovibronic transition, thereby forming a state-selected ensemble of electronically-excited molecules. The transient population produced in this intermediate state can be sensitively monitored by observing the spontaneous emission of photons which accompanies radiative relaxation processes. Now, as the frequency of the spatially-overlapping DUMP laser is scanned, coincidences occur with rovibronic transitions originating from the single rotation–vibration level prepared by the PUMP and terminating on vibrationally-excited eigenstates in the ground electronic potential surface. At these points of resonance the DUMP stimulates molecules to undergo "downward" transitions, resulting in a concomitant decrease in the fluorescence intensity emerging from the intermediate state. It is this DUMP-induced depletion of spontaneous emission (viz., a fluorescence-dip) that constitutes the signal measured in a conventional SEP experiment. When recorded as a function of DUMP frequency, these variations in fluorescence magnitude provide a detailed glimpse of the rovibrational structure supported by the ground state potential surface, with the electronically-excited intermediate state serving as a window that mediates the visibility (i.e., the intensity) of individual spectroscopic features. The doubly-resonant nature of the SEP scheme, coupled with the high spectral and temporal resolution afforded by modern dye laser technology, essentially eliminates the spectral complexity and rovibronic congestion that are normally associated with the optical transitions of polyatomic species.

As demonstrated by the many contributions to the present monograph, the substantial advantages afforded by SEP have rapidly transformed it into a mainstream methodology of modern-day physical chemistry. For nearly all studies performed to date, the fluorescence-dip detection scheme, as originally proposed by Kittrell *et al.*,[7] has remained the predominant means of observing resonant DUMP transitions. While the utility and versatility of this technique have been established firmly by a variety of research efforts, the restrictions that it imposes on the design and execution of experiments must not be discounted. In particular, since the fluorescence depletion method is based upon a competition between spontaneous and stimulated emission processes, it can successfully be applied only to systems that exhibit reasonable fluorescence quantum yields and relatively long fluorescence lifetimes. This dismisses from consideration the large and important class of molecules which have excited electronic states that are either predissociative or dissociative in nature. Furthermore, the conventional implementation of fluorescence-dip SEP can achieve the highest sensitivity and resolution only through use of dual-beam, null-detection configurations[7,9,10] that are, at best, awkward to integrate into existing experimental apparatus (e.g., molecular beams). Even under the most favorable of circumstances, high DUMP laser intensities, inducing significant saturation broadening of resonant transitions,[9] often are required for the effective discrimination of fluorescence depletion features from baseline noise. Single beam analogs of fluorescence-dip SEP, including transient gain,[11-13] polarization,[14-16] photoacoustic,[17] and ion-dip[18-20] variants, have been proposed. However, the experimental realization of these schemes is usually accompanied by a considerable degradation in attainable peak-to-noise ratios and/or detection limits owing to the necessity of extracting weak signals from large, fluctuating backgrounds.

Recent work in our laboratory[21,22] has demonstrated the feasibility of using resonant four-wave mixing spectroscopy as a background-free means of monitoring the SEP process. Similar experiments, based on a two-color transient grating scheme, have been reported by Hayden and co-workers,[23] with additional refinements introduced by Butenhoff and Rohlfing.[24] All of these techniques exploit the nonlinear behavior of an absorbing or amplifying medium so as to couple and redirect various optical fields impinging upon a gas-phase ensemble of target molecules. In particular, through judicious choice of experimental geometry, a coherent and spatially distinct beam of light can be made to emanate from the molecular sample whenever a resonant DUMP transition is encountered. For SEP spectroscopic measurements, the direct observation of this emerging signal wave provides a viable alternative to the detection of fluorescence depletion from an electronically-excited intermediate state.

Since the beam-like response inherent to a four-wave mixing process stems from the phase-locked emission of radiation by an induced array of target molecules, it is quite evident that such nonlinear optical probes can never hope to attain the "single-particle" detection limits afforded by their linear counterparts. However, in situations where the inherent sensitivity of a linear technique is compromised by competing effects, methods based upon the coherent interaction of light with matter can provide substantial advantages. This is the case in conventional fluorescence-dip SEP, where the zero-background capabilities of Laser-Induced Fluorescence (LIF) spectroscopy are abrogated by the need to discern small depletion signals on a large baseline of spontaneous emission. It is here that a nonlinear detection scheme can significantly enhance the ability to extract and discriminate spectroscopic resonances from background fluctuations.

This chapter focuses primarily on a recently-developed, zero-background variant of SEP which is especially well-suited for the investigation of polyatomic species containing chemically-significant quantities of vibrational energy. Based upon a novel implementation of phase-conjugate Degenerate Four-Wave Mixing (DFWM) spectroscopy,[25,26] this method provides a quantum state-specific probe of molecular structure and dynamics that offers substantial advantages over more conventional techniques. In particular, our exploitation of the DFWM detection scheme builds upon the experimental simplicity inherent to the doubly-resonant SEP process while eliminating any need to rely upon spontaneous emission as a means of identifying DUMP transitions.

The remainder of this chapter is organized as follows. The conceptual and theoretical foundation of DFWM spectroscopy are introduced in Sec. 2, with special emphasis on the unique characteristics exhibited by this nonlinear optical process. After demonstrating how the DFWM interaction can be utilized for the observation of resonant DUMP transitions, the details of a spectrometer based upon the so-called SEP–DFWM detection scheme are presented in Sec. 3. Experimental results contained in Sec. 4 allow the capabilities of this new spectroscopic tool to be compared and contrasted with those afforded by the ubiquitous fluorescence-dip technique. Following a brief description of analogous transient grating methods (cf. Sec. 5), the future prospects of resonant four-wave mixing spectroscopy as a probe of potential hypersurfaces in polyatomic species are discussed with particular attention directed toward the strengths and limitations revealed by ongoing studies.

2. Methodology

2.1. *Introduction*

Most of the difficulties associated with previously-developed SEP detection schemes stem from their dependence upon secondary processes (e.g., a decrease in fluorescence intensity from the intermediate state) to monitor the transfer of molecular population which accompanies a resonant DUMP interaction. Consequently, the sensitivity and dynamic range of these methods are often constrained by the need to discern weak depletion features on a large and noisy baseline of superfluous signal. This situation, completely analogous to that encountered in conventional (linear) absorption spectroscopy, can be improved significantly through use of "dual-beam" experimental configurations where the response of a sample to the PUMP and DUMP radiation is compared to that induced through PUMP excitation alone. While such null-detection techniques substantially enhance the ability to extract and discriminate SEP transitions, practical considerations usually preclude the realization of true background-free sensitivity under all but the most ideal of experimental conditions.

Two complementary schemes have been devised for the observation of SEP without reliance upon DUMP-induced modifications to a secondary process. On the one hand, a third optical beam can be introduced to directly interrogate the small subset of molecules that are transferred to target rotation–vibration levels of the ground electronic state as a result of the sequential PUMP and DUMP interactions.[27–29] While this method has the potential to yield sensitivities approaching the best attainable through background-free detection, its implementation requires a substantially more complicated experimental configuration, including the availability of a third laser source. Furthermore, the use of such a technique for *resonant* spectroscopic studies is virtually impossible owing to the nontrivial tuning of the probe beam frequency which must accompany scanning of the DUMP laser.

Alternative methods for detecting SEP without recourse to secondary effects rely upon direct observation of changes in the DUMP radiation which accompany its resonant interaction with the molecular sample. In particular, the inverted and spatially-anisotropic medium resulting from PUMP excitation can lead to simultaneous amplification and depolarization of the incident DUMP beam. Transient gain techniques, which monitor the stimulated emission of DUMP photons produced during the SEP process, have been employed for the time-resolved investigation of collision dynamics in both ground[12,13] and excited[11,30] electronic states. By eliminating the need to detect spontaneous emission, transient gain spectroscopy and other "absorption-based" schemes should be applicable to a much broader class of target species. However, owing to their extraction of weak gain signals from a large background of incident DUMP radiation, such techniques are subject to many of the same limitations imposed upon the fluorescence-dip methodology.

Polarization-based forms of SEP, which exploit the optical anisotropy (viz., birefringence and/or dichroism) created by polarized PUMP excitation so as to label and subsequently detect molecules involved in a double resonance process, have been demonstrated in fully-pulsed[15] and hybrid-pulsed continuous-wave[14,16] configurations. More recently, Chen and co-workers have employed a multiplexed variant of this scheme for the time-resolved study of photodissociation dynamics.[31] The crossed-polarization geometry inherent to such experiments effectively eliminates a substantial fraction of the superfluous background light which compromises the sensitivity and resolution of transient gain techniques. Unfortunately, the greater-than-linear dependence of polarization spectroscopy on sample number density,[32,33] as well as the limited extinction coefficients ($\sim 10^{-7}$) afforded by even the best polarization optics, usually restrict application of these methods to the investigation of stable species that can be maintained at relatively high sample pressures under bulk gas conditions.

Ideal detection of SEP transitions would entail use of a direct, "absorption-based" scheme which incorporates the high sensitivity and dynamic range afforded by a background-free response. Many of these features are embodied in various forms of nonlinear spectroscopy built upon the methods of resonant four-wave mixing.[34,35] In particular, the previously mentioned technique of degenerate four-wave mixing (DFWM) has been shown to provide a convenient and powerful probe of vibrationally-excited polyatomic species prepared under both bulk-gas[21] and free-jet[22] conditions.

The primary applications of DFWM have traditionally involved the study of aberration correction through optical phase-conjugation[36–41] and the investigation of ultrafast relaxation phenomena by means of time-resolved measurements.[42–44] The Doppler-free nature of this coherent interaction has also motivated considerable interest in its utility as a tool for high-resolution spectroscopy.[25,45–48] While DFWM does exhibit a greater-than-linear dependence on sample number density, the enormous resonant enhancement inherent to this parametric process offers the possibility of achieving sensitivities that far exceed those attainable by other nonlinear techniques. These unique capabilities, combined with a relatively simple experimental configuration that results in formation of a highly-collimated signal beam, have led to numerous applications of DFWM, with recent efforts focussing on trace species detection in hostile combustion environments.[49]

Before introducing the SEP–DFWM scheme in Sec. 2.3, the ensuing discussion will address various aspects of four-wave mixing spectroscopy, including theoretical models required for the description of both resonant and nonresonant interactions. More specifically, DFWM will be shown to exhibit behavior substantially different from that encountered with more conventional single-photon techniques. This results in advantages as well as disadvantages, both of which must be considered in any practical application of the nonlinear DFWM process.

2.2. Degenerate Four-Wave Mixing

2.2.1. General description

In brief, degenerate four-wave mixing is a coherent optical process whereby three input beams of identical (i.e., degenerate) frequency, ω, interact with a nonlinear medium so as to produce a fourth output beam also having frequency ω. Figure 1 schematically illustrates the basic geometry for phase-conjugate DFWM[50,51] in which two strong pump waves, with electric field vectors $\mathbf{E_f}$ and $\mathbf{E_b}$, are directed through an isotropic molecular sample in a coaxial and counterpropagating fashion. In this configuration, the wavevectors for the forward-going (viz., $\mathbf{E_f}$) and backward-going (viz., $\mathbf{E_b}$) pump beams are related by:

$$\mathbf{k_f} = -\mathbf{k_b} \ . \tag{1}$$

A probe beam, having electric vector $\mathbf{E_p}$, intersects the pump fields under a small crossing angle θ. The nonlinearity of the medium couples these three input beams to form a signal wave, $\mathbf{E_s}$, that is exactly collinear and counterpropagating with respect to the incident probe radiation. The directional properties of this signal beam follow from conditions of phase-matching or momentum conservation[26] which demand

$$k_s = k_f - k_p + k_b \tag{2}$$

or, by substitution of expression (1),

$$k_s = -k_p . \tag{3}$$

In a similar manner, the frequency of the emerging signal wave stems from restrictions imposed upon the system by energy conservation:

$$\omega_s = \omega_f - \omega_p + \omega_b \tag{4}$$

or, recalling that the pump and probe beams are degenerate in frequency, $\omega_s = \omega$.

The experimental configuration depicted in Fig. 1 satisfies momentum conservation criteria for all angles θ, thereby resulting in a nonlinear process that maintains perfect phase-matching regardless of any variations in the common frequency of the incident fields. In addition, the counterpropagating geometry has the ability to generate a signal wave that is the time-reversed or phase-conjugated image of the probe beam.[52,53] Indeed, a DFWM interaction often is described in terms of a phase-conjugate mirror that precisely retroreflects a time-reversed replica of the incoming probe radiation.[54] This unique feature has been exploited extensively for the real-time correction of wavefront aberrations which accrue as a result of light propagation through inhomogeneous optical media.[38–40,55]

The bottom panel of Fig. 1 displays the rather simplistic energy level diagram for a near-resonant DFWM interaction, where coupling of the pump, probe, and signal beams is dominated by the nonlinear effects associated with a particular spectroscopic transition (viz., $|2\rangle \leftrightarrow |1\rangle$ having resonance frequency ω_{21}). The illustrated four-wave mixing scheme clearly belongs to the broader class of elastic photon scattering phenomena in which the molecular sample serves as a frequency-dependent "catalyst" designed to mediate the direction and magnitude of energy flow among various optical fields.[56] This description, strictly valid only in the case of a purely reactive (i.e., transparent or nonabsorbing) medium, demands that individual target molecules be returned to their initial quantum states upon

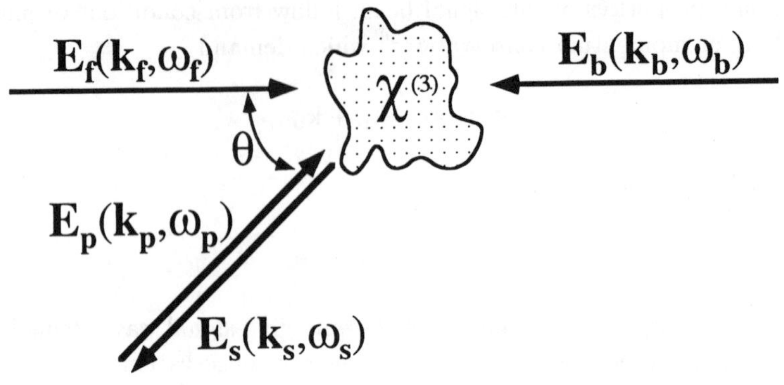

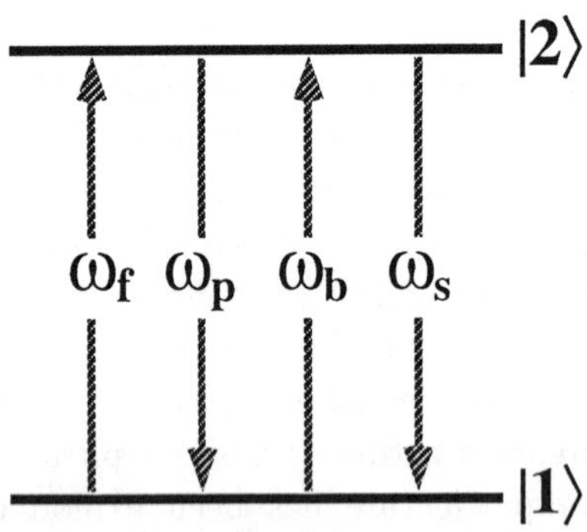

Fig. 1. Schematic diagram of the DFWM process. The upper panel depicts implementation of the backwards-going, phase-conjugate geometry for DFWM. Two pump beams ($\mathbf{E_f}$ and $\mathbf{E_b}$) are directed through a molecular sample in a coaxial and counterpropagating fashion with a probe wave ($\mathbf{E_p}$) intersecting them under a small crossing angle θ. Momentum conservation demands that the signal beam ($\mathbf{E_s}$) emerge in a direction that is exactly collinear and counterpropagating with respect to the incident probe radiation. Wavevectors and frequencies for each electric field, $\mathbf{E_m}$, are denoted by $\mathbf{k_m}$ and ω_m, respectively (where $\mathbf{m} = \mathbf{f}, \mathbf{b}, \mathbf{p}$, or $\mathbf{s}$). The symbol $\chi^{(3)}$ represents the third-order susceptibility tensor which mediates the nonlinear wave coupling phenomenon.

The lower panel depicts the energy level structure for a resonant DFWM interaction. The nonlinear effects arising from a specific one-photon absorption process (viz. $|1\rangle \leftrightarrow |2\rangle$ with rest frequency ω_{21}) couple three input beams of common frequency $\omega \approx \omega_{21}$ so as to generate an outgoing signal wave.

completion of the DFWM process (e.g., $|1\rangle \xrightarrow{\hbar\omega_f} |2\rangle \xrightarrow{\hbar\omega_p} |1\rangle \xrightarrow{\hbar\omega_b} |2\rangle \xrightarrow{\hbar\omega_s} |1\rangle$). Such parametric interactions[57] must be distinguished from their inelastic counterparts (e.g., stimulated Brillouin and Raman scattering) which, by definition, leave the system in a different final state as a result of the photon coupling. Therefore, in contrast to conventional SEP and the more sophisticated technique of <u>STI</u>mulated <u>R</u>aman <u>A</u>diabatic <u>P</u>assage (STIRAP),[58–61] the methods presented in this chapter do not provide a viable means for the efficient preparation of vibrationally-excited molecules. Instead, these nonlinear detection schemes are aimed specifically at the spectroscopic investigation of polyatomic potential energy surfaces.

The conventional framework of nonlinear optics provides a convenient means for demonstrating the general characteristics of DFWM spectroscopy.[53,62,63] By neglecting effects arising from magnetic interactions, the local response of a medium to incident electromagnetic radiation can be described in terms of the induced electric polarization vector, $\mathbf{P}(\mathbf{r}, t)$:

$$\mathbf{P}(\mathbf{r}, t) = \sum_{n=1}^{\infty} \mathbf{P}^{(n)}(\mathbf{r}, t), \tag{5}$$

where $\mathbf{P}^{(n)}(\mathbf{r}, t)$ is proportional to the total electric field at point $\mathbf{r}$ and time t raised to the nth power. Provided that rapid convergence is achieved, this perturbation expansion of the induced polarization permits specific molecule-field processes to be identified with individual $\mathbf{P}^{(n)}(\mathbf{r}, t)$ terms. However, in the presence of strong resonant interactions, the response of a medium can have contributions of comparable magnitude from several orders of nonlinearity,[57] thereby diminishing the utility of any expansion in powers of the optical field. While later sections will introduce models designed to address such situations, the well-established formalism embodied in Eq. (5) enables the characteristic features of four-wave mixing spectroscopy to be correlated immediately with the intrinsic properties of a molecular sample.

For isotropic media (e.g., liquids and gases), the lowest order term of Eq. (5) that is both nonzero and capable of supporting nonlinear behavior is found to be $\mathbf{P}^{(3)}(\mathbf{r}, t)$.[56] This polarization, which describes effects induced by the coupling of three electromagnetic waves, is responsible for generation of the signal field, $\mathbf{E_s}(\mathbf{r}, t)$, during a DFWM interaction. In its most general form, the signal polarization, $\mathbf{P}_s^{(3)}(\mathbf{r}, t)$, entails a multidimensional frequency integration over all optical waves present within a sample of target molecules.[62] However, in keeping with the pedagogical nature of this discussion, the ensuing analysis will assume that all electromagnetic fields can be expressed as simple monochromatic plane waves:

$$\mathbf{E_q}(\mathbf{r}, t) = \frac{1}{2}\left(E_{\omega_q}e^{-i\omega_q t} + E^*_{\omega_q}e^{i\omega_q t}\right)$$

$$= \frac{1}{2}\left(\mathbf{A_q}e^{i(\mathbf{k_q}\cdot\mathbf{r}-\omega_q t)} + \mathbf{A^*_q}e^{-i(\mathbf{k_q}\cdot\mathbf{r}-\omega_q t)}\right), \tag{6}$$

where the vector $\mathbf{A_q} = E_{\omega_q}e^{-i\mathbf{k_q}\cdot\mathbf{r}}$ (with $\mathbf{q} = \mathbf{f}, \mathbf{p}, \mathbf{b}$, or $\mathbf{s}$) specifies the amplitude and polarization properties for the corresponding electromagnetic wave. As is the case for all classical fields,[64,65] the sum of complex conjugates contained in Eq. (6) serves to define an explicitly real vector quantity. Under this plane wave approximation, the integral form for the DFWM signal polarization collapses to yield:

$$\mathbf{P_s^{(3)}}(\mathbf{r}, t) = \frac{1}{2}\left(\mathbf{P}_{\omega_s}^{(3)}e^{-i\omega_s t} + \mathbf{P}_{\omega_s}^{(3)*}e^{i\omega_s t}\right), \tag{7}$$

with the amplitude vector for the induced polarization, $\mathbf{P}_{\omega_s}^{(3)}$, expressed in terms of the incident electromagnetic fields (viz., $\mathbf{E_f}$, $\mathbf{E_p}$, and $\mathbf{E_b}$) and the third-order susceptibility tensor, $\chi^{(3)}(-\omega_s; \omega_f, -\omega_p, \omega_b)$, which mediates the nonlinear wave coupling phenomenon.[66] In particular, the inherent symmetries of a DFWM interaction enable $\mathbf{P}_{\omega_s}^{(3)}$ to be formulated as:[62]

$$\mathbf{P}_{\omega_s}^{(3)} = \frac{3}{4}\varepsilon_0\chi^{(3)}(-\omega_s; \omega_f, -\omega_p, \omega_b) \vdots \mathbf{E}_{\omega_f}\mathbf{E}^*_{\omega_p}\mathbf{E}_{\omega_b}$$

$$= \frac{3}{4}\varepsilon_0\chi^{(3)}(-\omega_s; \omega_f, -\omega_p, \omega_b) \vdots \mathbf{A_f}\mathbf{A^*_p}\mathbf{A_b}e^{i(\mathbf{k_f}-\mathbf{k_p}+\mathbf{k_b})\cdot\mathbf{r}}, \tag{8}$$

where ε_0 denotes the permittivity of free space and the column of three dots, $\vdots$, represents a tensor contraction. Taking into account the phase matching criteria embodied in Eq. (2), expressions (7) and (8) show that $\mathbf{P_s^{(3)}}(\mathbf{r}, t)$ has the same spatial and temporal characteristics (viz., $\mathbf{k_s} = -\mathbf{k_p}$ and $\omega_s = \omega$) as those expected for the generated signal field, $\mathbf{E_s}(\mathbf{r}, t)$. The induced nonlinear polarization is also found to be proportional to the complex conjugate (or phase-conjugate) of the amplitude vector for the incident probe radiation, $\mathbf{E}^*_{\omega_p}$.

The intensity of the signal beam generated during a DFWM interaction, I_s, can be related to the corresponding electric field vector, $\mathbf{E_s}(\mathbf{r}, t)$, by means of a time or cycle average:[65]

$$I_s = \frac{c\varepsilon_0}{2}\,\mathrm{Re}\,\left(\mathbf{E^*_s}(\mathbf{r}, t)\cdot\mathbf{E_s}(\mathbf{r}, t)\right) = \frac{c\varepsilon_0}{2}|\mathbf{A_s}|^2, \tag{9}$$

where the latter expression stems from the plane wave definition for $\mathbf{E_s}(\mathbf{r}, t)$. In turn, the value of $\mathbf{A_s}$ follows from solution of a differential wave equation which contains the induced nonlinear polarization as the source term for creation of the signal field:[62]

$$\frac{\partial \mathbf{A_s}}{\partial \mathbf{r}} = \frac{i\omega_s}{2c\varepsilon_0}\mathbf{P}_{\omega_s}^{(3)}e^{i\mathbf{k_s}\cdot\mathbf{r}}$$

$$= \frac{3i\omega_s}{8c}\chi^{(3)}(-\omega_s; \omega_f, -\omega_p, \omega_b) \mathbin{\vdots} \mathbf{A_f}\mathbf{A_p^*}\mathbf{A_b}e^{i\Delta\mathbf{k}\cdot\mathbf{r}} , \qquad (10)$$

where the polarization amplitude, $\mathbf{P}_{\omega_s}^{(3)}$, has been recast as a contraction between the susceptibility tensor and the incident electromagnetic fields. The derivation of expression (10) is based upon the slowly-varying envelope approximation[26,62] which requires that changes in the amplitude and phase of the generated signal wave (viz., $\mathbf{A_s}$) be small over the distance scale of an optical wavelength. The final term of this differential equation contains the so-called wavevector mismatch:

$$\Delta\mathbf{k} = (\mathbf{k_f} - \mathbf{k_p} + \mathbf{k_b}) - \mathbf{k_s} , \qquad (11)$$

which is rigorously equal to zero for the counterpropagating experimental configuration of Fig. 1 [cf. expression (2)]. A finite value for this quantity would imply constructive and destructive interference of the nascent signal wave as it propagates through the nonlinear medium, thereby decreasing the overall efficiency of the DFWM process.

Expression (10), with its explicit dependence upon field amplitudes, is but one of four coupled equations which describe the evolution of incident and generated electromagnetic waves within the molecular medium. The resulting system of differential equations can effectively be decoupled by assuming a small signal limit,[41] where the pump and probe beams propagate through an optically thin sample of length ℓ without experiencing significant modifications in their corresponding amplitudes (viz., $\mathbf{A_f}$, $\mathbf{A_p}$, and $\mathbf{A_b}$ are constant). This approximation, in conjunction with the self-phase-matched nature of a counterpropagating experimental configuration (viz., $\Delta\mathbf{k} = 0$), enables a direct integration of expression (10) to be performed so as to yield the following DFWM signal intensity:

$$I_s \propto \ell^2 |\chi^{(3)}(-\omega_s; \omega_f, -\omega_p, \omega_b) \mathbin{\vdots} \hat{\mathbf{e}}_\mathbf{f}\,\hat{\mathbf{e}}_\mathbf{p}^*\,\hat{\mathbf{e}}_\mathbf{b}|^2 I_f I_p I_b$$

$$= R I_p , \qquad (12)$$

where $I_q (q = f, p, b,$ or $s)$ represents the time averaged intensity of $\mathbf{E_q}(\mathbf{r}, t)$ and R denotes the phase-conjugate reflectivity induced by the four-wave mixing process. For derivation of this expression, the field amplitude vectors, $\mathbf{A_q}$, have been recast in terms of their magnitudes, A_q, and polarization unit vectors, $\hat{\mathbf{e}}_\mathbf{q}$, such that, $\mathbf{A_q} = A_q \hat{\mathbf{e}}_\mathbf{q}$.

Equation (12) predicts that the DFWM signal strength will scale as the square of the interaction length, ℓ, and the product of intensities for all three input beams, $I_f I_p I_b$. The pump and probe waves, being of identical frequency, typically are derived from a common source of tunable light. The magnitude of I_s can therefore be expected to exhibit a cubic dependence upon applied power, with small fluctuations in the incident radiation translating into large variations of observed signal intensity. Consequently, successful exploitation of DFWM as a sensitive tool for spectroscopic measurements demands the availability of lasers that have superior amplitude stability and coherence lengths, ℓ_c, capable of supporting phase-matched nonlinear interactions over considerable distances. In addition, a significant enhancement in detection limits can be realized through use of experimental configurations that maximize effective sample lengths (e.g., the use of slit nozzle sources[67,68] for target species entrained within a supersonic expansion).

The magnitude of the DFWM signal also depends on the third-order susceptibility, a complex, frequency-dependent quantity which contains the spectroscopic response of the nonlinear medium.[66] In the absence of strong resonant interactions, this fourth-rank tensor is independent of the applied field amplitudes. However, the polarization characteristics of the incident radiation determine which elements of $\chi^{(3)}(-\omega_s; \omega_f, -\omega_p, \omega_b)$ are required for the description of a specific four-wave mixing process.[69–71] In particular, the constitutive relationship in expression (8) enables the Cartesian components of the induced polarization vector, $(\mathbf{P}^{(3)}_{\omega_s})_j$ with $j = x, y,$ or z, to be formulated in terms of the directional properties for the pump and probe waves:

$$(\mathbf{P}^{(3)}_{\omega_s})_j = \frac{3}{4}\varepsilon_0 \sum_{k,l,m} \chi^{(3)}_{jklm}(-\omega_s; \omega_f, -\omega_p, \omega_b)(\mathbf{E}_{\omega_f})_k (\mathbf{E}^*_{\omega_p})_l (\mathbf{E}_{\omega_b})_m , \quad (13)$$

with $(\mathbf{E}_{\omega_q})_k$ denoting the k-component ($k = x, y,$ or z as does l and m) of the electric field vector $\mathbf{E}_{\omega_q}$ ($\mathbf{q} = \mathbf{f}, \mathbf{p},$ or $\mathbf{b}$).

For an isotropic medium, symmetry arguments can be used to show that the 81 elements of a general fourth-rank susceptibility tensor reduce to 21 nonzero terms of which only three are linearly independent.[62] The identical nature of the frequencies employed in a DFWM interaction places additional restrictions on the form of $\chi^{(3)}(-\omega_s; \omega_f, -\omega_p, \omega_b)$. More specifically, all Cartesian components of the third-order susceptibility can be reformulated in terms of two tensor elements:[63]

$$\chi^{(3)}_{jklm}(-\omega_s; \omega_f, -\omega_p, \omega_b) = \chi^{(3)}_{\eta\eta\zeta\zeta}(-\omega_s; \omega_f, -\omega_p, \omega_b)[\delta_{jk}\delta_{lm} + \delta_{jm}\delta_{kl}]$$
$$+ \chi^{(3)}_{\eta\zeta\eta\zeta}(-\omega_s; \omega_f, -\omega_p, \omega_b)\delta_{jl}\delta_{km} , \qquad (14)$$

where the Kronecker delta symbol, δ_{jk}, has the usual definition of being equal to unity if $j = k$ and zero otherwise. The subscripts η and ζ can be x, y, or z subject to the constraint that $\eta \neq \zeta$. By substituting this expression into Eq. (13), the vector amplitude of the signal polarization induced during a DFWM process can be written as a sum of three terms,[26] each of which contains the scalar product of electric fields for two of the incident light waves:

$$\mathbf{P}^{(3)}_{\omega_s} = A(\mathbf{E}_{\omega_b} \cdot \mathbf{E}^*_{\omega_p})\mathbf{E}_{\omega_f} + B(\mathbf{E}_{\omega_f} \cdot \mathbf{E}^*_{\omega_p})\mathbf{E}_{\omega_b} + C(\mathbf{E}_{\omega_f} \cdot \mathbf{E}_{\omega_b})\mathbf{E}^*_{\omega_p} , \qquad (15)$$

where the factors A, B, and C are proportional to the independent susceptibility components introduced in Eq. (14) (viz., $\chi^{(3)}_{\eta\eta\zeta\zeta}$ and $\chi^{(3)}_{\eta\zeta\eta\zeta}$) with $A = B$ for a purely reactive (i.e., absorptionless) medium.[57] As will be shown in Sec. 2.2.2, this definition of $\mathbf{P}^{(3)}_{\omega_s}$ leads to a convenient transient grating interpretation for resonant DFWM processes. In addition, expression (15) enables the polarization of a DFWM signal to be predicted from the vectorial properties of the applied electromagnetic fields.[72] For example, consider the experimental configuration usually employed for SEP–DFWM spectroscopy, where the counterpropagating pump waves are cross-polarized ($\mathbf{E_f} \perp \mathbf{E_b}$) while the nearly-collinear probe beam has the same polarization as $\mathbf{E_f}(\mathbf{E_p}\|\mathbf{E_f})$. This geometry, often referred to as a cross-population scheme,[25,73,74] demands that only the second term in Eq. (15) be nonzero. Consequently, the direction of polarization for the emerging DFWM signal will be determined by that of the backward-going pump field (viz., $\mathbf{P}^{(3)}_{\omega_s} \propto \mathbf{E}_{\omega_b}$). The shortcomings of this simple scalar theory have been documented by Ducloy and Bloch,[75] with more sophisticated models required to account for polarization-specific effects that arise from coherences created between the degenerate magnetic sublevels of gas-phase target molecules.

The spectroscopic capabilities of DFWM stem from variations in the susceptibility tensor which accompany tuning of the common input frequency, ω [cf. Eq. (12)].[35] The functional form for $\chi^{(3)}(-\omega_s; \omega_f, -\omega_p, \omega_b)$, which reflects the intrinsic properties (e.g., transitions frequencies and decay rates) of a molecular sample, can be derived from a third-order perturbation expansion of the polarization vector, $\mathbf{P}_s^{(3)}(\mathbf{r}, t)$, induced by the incident electromagnetic waves. This can most readily be accomplished through use of a density operator formalism,[76,77] with various diagrammatic techniques[78-80] serving to both organize and interpret individual molecule-field interactions. The general scheme for performing such calculations has been well documented[26,62,63] and explicit susceptibility expressions for various nonlinear processes, including several variants of resonant four-wave mixing,[34,71] have been reported.

In order to demonstrate the basic features of $\chi^{(3)}(-\omega_s; \omega_f, -\omega_p, \omega_b)$, consider an isotropic, gas-phase sample of target molecules that can be treated as an isolated ensemble of independent and distinguishable particles. For a (one-photon) near-resonant DFWM process, two distinct electronic manifolds (e.g., $|1\rangle$ and $|2\rangle$ in Fig. 1) are involved, each of which supports a discrete pattern of rovibrational structure symbolically labelled by $\{|\alpha\rangle\}$ in the ground state and $\{|\beta\rangle\}$ in the excited state. In a symmetric rotor basis, a more precise definition for eigenstates might entail $|\alpha\rangle$ and $|\beta\rangle$ expressed as $|vJKM\rangle$, where K and M signify projections of the total angular momentum J on the molecule-fixed and space-fixed quantization axes, respectively, and v represents all other quantum numbers (viz., electronic and vibrational) required to fully describe the system. The unperturbed energies and population decay rates for individual levels are represented by E_q and Γ_{qq} ($q \in \{\alpha\}$ or $q \in \{\beta\}$), with the latter quantity equal to the inverse of the longitudinal relaxation time, T_1^q, for state $|q\rangle$ (viz., $T_1^q = 1/\Gamma_{qq}$). The resonant portion of the third-order susceptibility tensor is found to consist of eight distinct terms which reflect various time-orderings for the incident pump and probe waves.[25] Assuming that molecules interact with the applied electromagnetic fields under a dipole approximation, each of these contributions to $\chi^{(3)}(-\omega_s; \omega_f, -\omega_p, \omega_b)$ can be shown to have the general form:[62]

$$\frac{N}{\varepsilon_0 \hbar^3} \sum_{\alpha,\alpha',\beta,\beta'} \rho_{\alpha\alpha}^{(0)}$$

$$\times \frac{\langle\alpha|\boldsymbol{\mu}\cdot\hat{\mathbf{e}}_s^*|\beta'\rangle\langle\beta'|\boldsymbol{\mu}\cdot\hat{\mathbf{e}}_b|\alpha'\rangle\langle\alpha'|\boldsymbol{\mu}\cdot\hat{\mathbf{e}}_p^*|\beta\rangle\langle\beta|\boldsymbol{\mu}\cdot\hat{\mathbf{e}}_f|\alpha\rangle}{[\omega_{\beta\alpha}-\omega_f-i\Gamma_{\beta\alpha}][\omega_{\alpha'\alpha}-\omega_f+\omega_p-i\Gamma_{\alpha'\alpha}][\omega_{\beta'\alpha}-\omega_s-i\Gamma_{\beta'\alpha}]}, \quad (16)$$

where N is the number density of target species, $\boldsymbol{\mu}$ is the transition dipole moment operator, $\omega_{\beta\alpha} = (E_\beta - E_\alpha)/\hbar$ is the resonant angular frequency for a particular rovibronic transition (viz., $|\beta\rangle \leftrightarrow |\alpha\rangle$), and the summation is performed over all quantum numbers used to describe fine structure within the two electronic manifolds. The molecular population is assumed to reside initially only in the ground electronic state, with a distribution specified by the diagonal elements of the zero-order density matrix, $\rho_{\alpha\alpha}^{(0)}$. Equation (16) also contains phenomenological damping constants, such as $\Gamma_{\beta\alpha}$, which ensure that the susceptibility does not diverge in an unphysical manner at resonance.[62] These parameters correspond to the decay of molecular coherences, as described by off-diagonal elements of the density matrix,[77] and under quite general conditions can be related to the depopulation rates for individual rovibronic levels:[81]

$$\Gamma_{\beta\alpha} = \frac{1}{2}\left(\Gamma_{\beta\beta} + \Gamma_{\alpha\alpha}\right) + \Gamma_{\beta\alpha}^{\varphi} = \frac{1}{T_2^{\beta\alpha}}, \quad (17)$$

where $T_2^{\beta\alpha}$ denotes the transverse relaxation time for the $|\beta\rangle \leftrightarrow |\alpha\rangle$ rovibronic transition. The quantity $\Gamma_{\beta\alpha}^{\varphi}$ is the pure dipole dephasing rate that results from processes which destroy molecular coherence without disrupting the corresponding populations (e.g., elastic scattering events).[82] In a weak collision limit $2\Gamma_{\beta\alpha}$ can be identified as the so-called natural (i.e., homogeneous) linewidth for the spectroscopic resonance that occurs at angular frequency $\omega_{\beta\alpha}$.[33]

The third-order susceptibility tensor is found to be proportional to the number density of target species, N, and the product of four electric dipole matrix elements, $\mu_{\alpha\beta'}^s \mu_{\beta'\alpha'}^b \mu_{\alpha'\beta}^p \mu_{\beta\alpha}^f$, where, for example, $\mu_{\beta\alpha}^f$ denotes $\langle\beta|\boldsymbol{\mu}\cdot\hat{\mathbf{e}}_f|\alpha\rangle$. Since the DFWM signal intensity is related to the square modulus of $\chi^{(3)}(-\omega_s;\omega_f,-\omega_p,\omega_b)$ [cf. Eq. (8)], I_s can be expected to scale quadratically with sample pressure and as the 8th power of the rovibronic transition moment. This behavior must be contrasted with the $N|\mu|^2$ response which is typical for a linear, one-photon technique.[83,84] Consequently, molecular spectra recorded through the use of four-wave mixing schemes should exhibit radically different intensity patterns from those obtained by means of conventional spectroscopic probes. In addition, the N^2 dependence of DFWM implies a reduced sensitivity for the detection of target species in rarified environments.

The frequency dependence of the susceptibility tensor and, therefore, the spectroscopic response of the corresponding four-wave mixing process is contained in the three resonance denominators of Eq. (16). Since the square modulus of $\chi^{(3)}(-\omega_s; \omega_f, -\omega_p, \omega_b)$ is proportional to the observed signal strength, these terms, when taken separately, lead to Lorentzian-shaped functions centered at particular rovibronic transitions of the molecular sample. For example, the first denominator yields a spectral lineshape of the form:

$$\left| \frac{1}{\omega_{\beta\alpha} - \omega_f - i\Gamma_{\beta\alpha}} \right|^2 = \frac{1}{\Gamma_{\beta\alpha}^2 + (\omega_{\beta\alpha} - \omega_f)^2} = \frac{1/\Gamma_{\beta\alpha}^2}{1 + \delta^2} , \qquad (18)$$

where $\delta = (\omega_{\beta\alpha} - \omega_f)/\Gamma_{\beta\alpha}$ is the dimensionless detuning parameter for a Lorentzian of full-width at half-maximum (FWHM) $2\Gamma_{\beta\alpha}$ centered at $\omega_{\beta\alpha}$. In particular, a tremendous increase in the amplitude of the susceptibility tensor can be realized whenever the combination of incident frequencies in one or more of the denominators approaches resonance with a rovibronic transition [e.g., $\omega_{\beta\alpha} \approx \omega_f$ in expression (18)]. While this generally can be expected to enhance the four-wave mixing output, direct interpretation of such resonant interactions is complicated by the presence of the corresponding phenomenological decay rates and of additional cross terms that have been neglected in the presentation of Eq. (18). Since the expression for I_s involves a coherent summation of several $\chi^{(3)}(-\omega_s; \omega_f, -\omega_p, \omega_b)$ elements, constructive and destructive interference effects can also modulate observed signal intensities.[85–88]

The single optical frequency, ω, employed for a DFWM interaction suggests that the greatest enhancement in $\chi^{(3)}(-\omega_s; \omega_f, -\omega_p, \omega_b)$ will occur for those terms of Eq. (16) where $E_\alpha = E_{\alpha'}$ and $E_\beta = E_{\beta'}$ such that $\omega_{\beta\alpha} = \omega_{\beta'\alpha}$ and $\omega_{\alpha'\alpha} = 0$. For each distinct transition of the molecular sample (e.g., $|\beta\rangle \leftrightarrow |\alpha\rangle$) this results in a fully-resonant denominator which, in turn, yields an expression for I_s that is roughly proportional to $[\Gamma_{\alpha\alpha}^2 \Gamma_{\beta\alpha}^4 (1 + \delta^2)^2]^{-1}$. Consequently, as ω is tuned through individual rovibronic features (i.e., as $\delta \to 0$), the generated four-wave mixing signal can be augmented by many orders of magnitude, thereby providing an essentially background-free means for probing the energy level structure of target molecules. As will be shown in Sec. 2.2.3, the unusual frequency dependence of the lineshapes expected under these conditions has implications for the effective spectral resolution attained in DFWM measurements.

2.2.2. *Molecular holography*

The physical manifestation of DFWM can be rationalized in terms of dynamic light scattering or laser-induced grating spectroscopy.[26,89] Here, two coherent beams of light (e.g., $\mathbf{E_f}$ and $\mathbf{E_p}$) spatially interfere so as to create a modulation in the sample's complex index of refraction. The remaining incident beam (e.g., $\mathbf{E_b}$) subsequently Bragg scatters off of this transient diffraction grating so as to produce the observed signal wave (viz., $\mathbf{E_s}$). In many respects, this overall process strongly resembles the simultaneous formation and recreation of a volume hologram,[36,90] with the spatial distribution of molecules replacing the more conventional pattern of fringes recorded on a photographic plate.

The interpretation of DFWM in terms of transient diffraction gratings stems from the well-known classical phenomenon of interference, where a periodic modulation in the intensity and/or polarization of electromagnetic radiation can be created by the spatial overlap of two coherent light sources. This description, which forms the basis for an entire class of powerful spectroscopic techniques,[89,91] will prove to be useful for characterizing many (but not all) aspects of four-wave mixing interactions. More specifically, the behavior of DFWM spectroscopy in gas-phase samples can be explained readily through the temporal evolution of spatial structures that are formed upon resonant molecular excitation.

In places where the forward-going pump and probe waves spatially overlap, the total electric field, $\mathbf{E_{tot}}$, is given by the vector sum of $\mathbf{E_f}$ with $\mathbf{E_p}$. The corresponding intensity of the electromagnetic radiation, I_{tot}, can be obtained by means of a time or cycle average such as that defined in Eq. (9). Therefore, the intersection of $\mathbf{E_f}$ and $\mathbf{E_p}$ in the DFWM interaction region should result in an intensity distribution of the form:

$$I_{\text{tot}} = I_f + I_p + \Delta I_{fb} \,, \tag{19}$$

where

$$\Delta I_{fb} = c\varepsilon_0 \, \mathrm{Re}\,(\mathbf{E}_{\omega_f} \cdot \mathbf{E}^*_{\omega_p}) = c\varepsilon_0 \, \mathrm{Re}\,(\mathbf{A_f} \cdot \mathbf{A}^*_{\mathbf{p}} e^{i(\mathbf{k_f}-\mathbf{k_p})\cdot\mathbf{r}}) \,, \tag{20}$$

and I_q denotes the independent intensity for electric field $\mathbf{E_q}$ (with $q = f$ or p). Provided that the polarizations for the forward-going pump and probe beams are not orthogonal (i.e., $\mathbf{A_f} \cdot \mathbf{A_p^*} \neq 0$), the final term in expression (19), ΔI_{fb}, leads to a position-dependent modulation of light intensity which remains stationary in time. The structure of this interference phenomenon, schematically illustrated in the upper portion of Fig. 2, strongly resembles the repetitive pattern of grooves associated with a conventional diffraction grating. Its spatial characteristics can conveniently be described by means of the grating wavevector, $\mathbf{k_g}$:

$$\mathbf{k_g} = \mathbf{k_f} - \mathbf{k_p} \, , \tag{21}$$

a quantity which appears in the defining expression for ΔI_{fb}. Moreover, the scalar product of $\mathbf{E}_{\omega_f}$ and $\mathbf{E}_{\omega_p}^*$ that generates this interference pattern is identical in form to the second term of the nonlinear DFWM signal polarization, $\mathbf{P}_{\omega_s}^{(3)}$, presented in Eq. (15).

As shown by the coupling diagram in the lower portion of Fig. 2, the grating wavevector points in a direction which is orthogonal to the planes of constant light intensity (i.e., across the pattern of grooves) and has a magnitude that is inversely proportional to the spatial period, Λ, of the interference pattern:

$$|\mathbf{k_g}| = \frac{2\pi}{\Lambda} \, . \tag{22}$$

The value of Λ, which corresponds to the groove or fringe spacing of a canonical diffraction grating, can be obtained from Eqs. (21) and (22):

$$\Lambda = \frac{\lambda}{2 \sin\left(\frac{\theta}{2}\right)} \, , \tag{23}$$

where θ and λ denote the crossing angle and common wavelength for the forward-going pump and probe beams. Using values of $\theta = 1.5°$ and $\lambda = 500$ nm, which are quite typical for experiments based on the SEP–DFWM scheme, the periodicity of the spatial modulation resulting from the intersection of $\mathbf{E_f}$ with $\mathbf{E_p}$ is characterized by $\Lambda \approx 20$ μm.

If a material substance is introduced into the region where $\mathbf{E_f}$ and $\mathbf{E_p}$ overlap, then matter–radiation interactions (e.g., absorption) can provide a mechanism for transforming the subtle phenomenon of interference into a tangible entity that can be probed and measured. In short, the optical properties of the medium will be altered in a manner that reflects the spatial

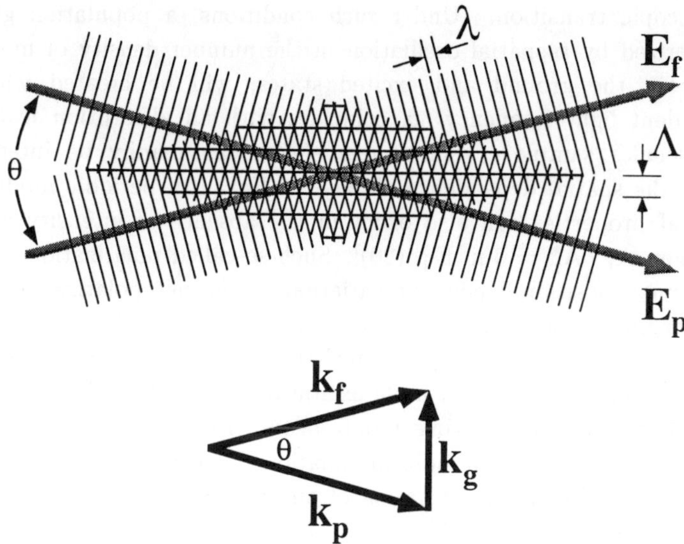

Fig. 2. Description of the DFWM process in terms of transient diffraction gratings. The upper panel illustrates the modulation of optical properties produced by overlap of the forward-going pump (E_f) and probe (E_p) waves. The resulting interference pattern has a spatial periodicity, Λ, that can be related to the common wavelength, λ, and the crossing angle, θ, for the intersecting beams. The interaction of the backward-going pump (E_b) and probe (E_p) fields can be described by means of an analogous figure. The lower panel depicts the coupling of wavevectors for E_f and E_p to produce the so-called grating vector, k_g, (i.e., $k_g = k_f - k_p$) which describes the spatial properties of the interference pattern.

modulation of intensity and/or polarization in the incident electromagnetic field. These variations can manifest themselves as changes in either the absorption coefficient or the index of refraction, thereby resulting in the formation of macroscopic amplitude or phase gratings that have the same periodicity as the interference pattern responsible for their creation (cf. Fig. 2). Obviously, the magnitude of these effects will be a strong function of the initial light frequency, with the most pronounced modification to the optical properties of a molecular sample occurring upon resonant excitation.[92–94]

For rarified media (e.g., low pressure bulk gases), where the collisional transport of heat and mass need not be considered, the modulation of optical properties resulting from the overlap of E_f with E_p derives primarily from the different nature of the upper and lower levels involved in a resonant spectroscopic transition. Under such conditions, a population grating, characterized by a spatial oscillation in the number density of molecules residing in the ground and excited states, will be created whenever the incident light frequency, ω, coincides with an absorption feature of the molecular sample. While these effects can readily be interpreted through the spatial dependence of ΔI_{fb} in Eq. (19), periodic modulation of optical properties can also occur in the absence of intensity changes [e.g., when $\mathbf{A_f} \cdot \mathbf{A_p^*} = 0$ in Eq. (20)]. Such so-called polarization gratings stem from a position-dependent variation in the net polarization of the electromagnetic field, a phenomenon which manifests itself as an anisotropic orientational distribution (e.g., nonuniform magnetic sublevel distribution) for the molecules taking part in an excitation process.[89,92–94]

When the backward-going pump beam, $\mathbf{E_b}$, traverses the DFWM interaction region, the periodic modulation of optical properties that it encounters results in the formation of an induced electromagnetic polarization which exhibits identical spatial characteristics. This phenomenon is described by the second term of Eq. (15), where the DFWM signal polarization is found to be proportional to $(\mathbf{E}_{\omega_f} \cdot \mathbf{E}_{\omega_p}^*)\, \mathbf{E}_{\omega_b}$. From a more physical perspective, a three-dimensional array of phase-locked antennas is created upon interaction of the backward-going pump beam with the anisotropic target medium. The radiation emitted from various portions of this structure can constructively interfere only in certain directions with respect to the orientation of the material grating (viz., $\mathbf{k_g}$) and the path of the incident electromagnetic wave (viz., $\mathbf{k_b}$). If the dimensions of the material grating are substantially larger than its periodicity, the scattered light resulting from this diffraction process emerges with a wavevector, $\mathbf{k_s}$, that is precisely determined by the Bragg condition:[64]

$$\mathbf{k_s} - \mathbf{k_b} = m\mathbf{k_g} \, , \tag{24}$$

where $m = \pm 1, \pm 2, \dots$. Taking into account the defining expression for the grating wavevector, the first-order Bragg scattering condition (i.e., $m = +1$) for the backward-going pump wave reproduces the phase-matching criterion associated with the nonlinear optical description of DFWM [cf. Eqs. (2) and (3)]. For sufficiently thin samples, the restrictions imposed by Eq. (24) will be somewhat relaxed thus allowing for the observation of scattered signal photons even in situations where phase-matching requirements are not fully satisfied.[89]

In the experimental studies that form the basis of this chapter, a four-wave mixing process is induced by the interaction of short (≤ 6 ns) light pulses with a gaseous sample of target molecules. While the near-Gaussian spatial profiles of most laser sources would seem to invalidate a simple plane wave description, the principal results derived from the above analysis, including the direction and magnitude of the grating wavevector $\mathbf{k_g}$, can still be shown to provide a reasonable interpretation for the underlying interference phenomena.[95,96] However, since the modulation of optical properties is created through the abrupt excitation of a *gas-phase* ensemble of molecules, the random translation motion (of individual molecules) that subsequently occurs will lead to a temporal dissipation of any spatially anisotropic features. Consequently, the macroscopic grating-like structure formed through the mutual action of $\mathbf{E_f}$ and $\mathbf{E_p}$ on the material medium will be transient in nature, with its central position or phase remaining constant while its amplitude decays in time. Steel and co-workers[72] have incorporated such effects into the scalar expression for the DFWM signal polarization by defining the A and B coefficients of Eq. (5) to be explicit functions of $\mathbf{k_g}$, T_0/T_1, and T_0/T_2, where T_1 and T_2 are the longitudinal and transverse relaxation times for a given spectroscopic transition. The quantity T_0 is a characteristic "motional" time that can be related to the inverse of the Doppler width for an inhomogeneously-broadened, gas-phase sample.

Under low pressure conditions, a reasonable estimate for the lifetime, τ_y, of a transient gas-phase grating is given by the average time required for a molecule to traverse the distance between two fringes of the periodic structure:

$$\tau_g = \frac{\Lambda}{\bar{v}} \propto T_0 \; , \tag{25}$$

where $\bar{v}$ denotes the mean velocity for the ensemble of target molecules. Using values of $\Lambda \approx 20$ μm and $\bar{v} \approx 288$ m/s, which are appropriate for SEP–DFWM measurements on room-temperature CS_2 samples, a lifetime of $\tau_g \approx 70$ ns is determined for the transient grating formed through the overlap of $\mathbf{E_f}$ with $\mathbf{E_p}$. The observation of large diffraction signals therefore demands that the backward-going pump beam, $\mathbf{E_b}$, enter and traverse the DFWM interaction region on a timescale that is short compared to 70 ns. The validity of this result and of expression (25) in general is contingent upon having the mean free path for a molecule exceed the characteristic dimensions (viz., Λ) of the initial interference pattern, with the longitudinal and transverse decay rates for the molecular system being sufficiently slow so as not to introduce additional constraints. For environments dominated by collisional processes, such as high-pressure gases or condensed media, other dissipative phenomena which critically depend on the macroscopic thermodynamic properties of the sample (e.g., diffusive transport of heat and mass) must be taken into account. These new effects can lead to significant complications, including the formation of secondary grating structures (e.g., thermal phase gratings) that exhibit their own unique features for diffractive scattering and temporal relaxation.[89]

While the above discussion has focussed on the interference of $\mathbf{E_f}$ with $\mathbf{E_p}$ so as to form a volume diffraction grating or hologram that subsequently scatters $\mathbf{E_b}$ into $\mathbf{E_s}$, similar considerations can be applied for other groupings of the three incident beams involved in a DFWM process. In particular, spatial overlap of the backward-going pump and probe waves can create a modulation of the sample's optical properties that leads to the Bragg scattering of $\mathbf{E_f}$ into the signal field. This phenomenon is described by the first term of Eq. (15), where the vector amplitude of the induced polarization, $\mathbf{P}_{\omega_s}^{(3)}$, is proportional to $(\mathbf{E}_{\omega_b} \cdot \mathbf{E}_{\omega_p}^*)\mathbf{E}_{\omega_f}$. However, since the angular deviation of the grating-forming beams is now on the order of 180°, the corresponding fringe spacing will be quite small, with $\Lambda \approx \lambda/2$. Assuming that $\lambda = 500$ nm, and taking into account the mean velocity for a room temperature ensemble of CS_2 molecules, the periodic structure established by $\mathbf{E_b}$ and $\mathbf{E_p}$ is found to have a lifetime of roughly 0.8 ns. This value of τ_g is considerably shorter than the duration of laser pulses utilized for SEP–DFWM studies, strongly suggesting that the transient grating formed through interaction of the backward-going pump and probe waves will have a negligible effect on observed diffraction signals.

As demonstrated by the final term in Eq. (15), a DFWM signal polarization can also be generated through interference of the counterpropagating pump beams. However, in contrast to the energy level diagram depicted in Fig. 1, this process can be resonantly enhanced only by a two-photon absorption transition (e.g., $|1\rangle \xrightarrow{\hbar\omega_f} |2\rangle \xrightarrow{\hbar\omega_b} |3\rangle \xrightarrow{\hbar\omega_p} |2\rangle \xrightarrow{\hbar\omega_s} |1\rangle$, where $E_3 > E_2 > E_1$) which, if present, leads to the formation of a spatially invariant grating that oscillates in time at twice the optical input frequency (viz., 2ω).[26] The Doppler-free response afforded by such DFWM processes has been exploited for high-resolution measurements of two-photon spectroscopy in both atoms and molecules.[25]

2.2.3. *Resonant DFWM theory*

While the conventional formalism of nonlinear optics offers useful guidelines for characterizing the behavior of DFWM, it fails to provide an adequate description for the response of a system undergoing strong molecule-field interactions. In particular, resonant excitation of individual molecular transitions can easily destroy convergence of the series expansion usually employed for the induced signal polarization [cf. expression (5)].[57,62] Under such conditions, however, it is usually appropriate to consider only the two energy levels coupled by the applied electromagnetic field, with the increase in complexity required for a nonperturbative treatment partially offset by the ability to neglect nonresonant processes associated with other quantum states of the target molecule. Such two-level models have been applied to a variety of nonlinear processes, especially those that demand explicit incorporation of dynamical effects (e.g., saturation) for their successful interpretation.[97] The present section will utilize this approach in order to examine the behavior of DFWM spectroscopy when incident light intensities approach or exceed those needed for the saturation of rovibronic transitions. The previous susceptibility analysis of four-wave mixing will be shown to represent a low-power, limiting case of the results obtained below.

The near-resonant behavior of DFWM can be demonstrated succinctly through use of a simple analytical theory based upon the interaction of a homogeneously-broadened medium with three continuous-wave, monochromatic fields of identical frequency ω. Here, the molecular sample is treated as an ensemble of *stationary*, two-level systems (cf. lower portion of Fig. 1), each characterized by a common resonance frequency, ω_{21}, and transition dipole moment, μ_{21}, with relaxation rates for the decay of population and coherence (viz., longitudinal and transverse relaxation) denoted by Γ_0 and Γ_{21}, respectively. While Γ_{21} can be defined in a manner analogous to that of Eq. (17), Γ_0 is equal to the inverse of the average lifetime for the ground and excited states. As first derived by Abrams and Lind,[25,98,99] this saturable absorber model considers the counterpropagating pump beams to be of equal intensity, I, with no significant modifications in their amplitudes resulting from propagation through the interaction region of length ℓ. In contrast, the probe and signal waves, having intensities I_p and I_s, are assumed to be considerably weaker than the pump fields (viz., $I_s, I_p \ll I$) and are subject to depletion or amplification as a consequence of their coupling with the nonlinear medium. Under these conditions, the frequency-dependent magnitude of the four-wave mixing signal can be expressed in terms of the phase-conjugate reflectivity, R, for the incident probe radiation:

$$I_s = RI_p = \left| \frac{\beta \sin \gamma \ell}{\gamma \cos \gamma \ell + \alpha \sin \gamma \ell} \right|^2 I_p \; , \tag{26}$$

where α is the attenuation or gain coefficient:

$$\alpha = \alpha_0 \frac{1}{1+\delta^2} \frac{1 + \frac{2I}{I_{\text{sat}}}}{\left(1 + \frac{4I}{I_{\text{sat}}}\right)^{3/2}} \; , \tag{27}$$

β is the nonlinear coupling coefficient:

$$\beta = \alpha_0 \frac{\delta + i}{1+\delta^2} \frac{\frac{2I}{I_{\text{sat}}}}{\left(1 + \frac{4I}{I_{\text{sat}}}\right)^{3/2}} \; , \tag{28}$$

and γ provides a measure for the relative strengths of absorption and nonlinearity in the medium:

$$\gamma^2 = |\beta|^2 - \alpha^2 \; . \tag{29}$$

The magnitudes of α and β are directly proportional to the small-signal, *field* absorption coefficient at line-center, α_0:

$$\alpha_0 = \frac{\omega |\mu_{21}|^2}{2\Gamma_{21} c \hbar \varepsilon_0} \Delta N_0 \; , \tag{30}$$

where ΔN_0 denotes the population difference (per unit volume) between the ground and excited states of the two-level system in the absence of applied fields.[a] While ΔN_0 is positive for samples exhibiting absorption, a negative value for this parameter implies an inverted or gain medium such as that prepared between the upper and lower levels of the DUMP transition during

---

[a]As originally derived by Abrams and Lind, the definition for the line-center absorption coefficient contains an intensity and frequency dependent index of refraction, n, which is rigorously equal to unity at resonance ($\delta = 0$). For the present analysis, it is assumed that this parameter can be well approximated by $n \approx 1$ under all conditions of frequency detuning and light intensity.

an SEP process. The frequency dependence of the DFWM signal intensity is described in terms of the normalized detuning, $\delta = (\omega - \omega_{21})/\Gamma_{21}$, of the incident radiation from the resonance condition (viz., $\omega = \omega_{21}$). This dimensionless quantity also appears in the defining expression for the saturation intensity, I_{sat}:

$$I_{\text{sat}} = I_{\text{sat}}^0 (1 + \delta^2) \, , \tag{31}$$

where the saturation intensity at line-center, I_{sat}^0, is given by:

$$I_{\text{sat}}^0 = \frac{\Gamma_{21} \Gamma_0 c \hbar^2 \varepsilon_0}{2|\mu_{21}|^2} \, . \tag{32}$$

Saturation phenomena, such as that embodied in Eqs. (31) and (32), characterize the balance reached between optical pumping and dissipative processes.[33,100,101] This fact can best be appreciated by recasting I_{sat}^0 in terms of the line-center absorption cross-section for the incident electromagnetic field, $\sigma_0 = \alpha_0 (\Delta N_0)^{-1}$, and the longitudinal or population relaxation time for the two-level medium, $T_1 = (\Gamma_0)^{-1}$.

The phase-conjugate reflectivity, R, is a function of three dimensionless parameters: the product of the line-center absorption (or gain) coefficient with the sample interaction length, $\alpha_0 \ell$; the ratio of the pump and saturation intensities, I/I_{sat}^0; and the normalized detuning of the incident radiation from resonance, δ. In the limit of small absorption (or gain), where $|\alpha_0 \ell| \ll 1$ and $|\gamma \ell| \ll 1$, the expression for R can be greatly simplified by substituting $\sin \gamma \ell \approx \gamma \ell$ and $\cos \gamma \ell \approx 1$. As shown by Farrow, Rakestraw, and Dreier,[102] this approximation yields a four-wave mixing signal intensity of the form:

$$I_s \approx (\alpha_0 \ell)^2 \frac{1}{1 + \delta^2} \frac{\left(\frac{2I}{I_{\text{sat}}}\right)^2}{\left(1 + \frac{4I}{I_{\text{sat}}}\right)^3} I_p \, . \tag{33}$$

Under these conditions, the signal strength varies as the square of both the sample interaction length, ℓ, and the line-center absorption or gain coefficient, α_0, with the latter also implying a quadratic dependence on the unperturbed population difference, ΔN_0. This behavior is illustrated graphically in the upper portion of Fig. 3, where the phase-conjugate reflectivity, as given by expression (26), is plotted as a function of $\alpha_0 \ell$ and normalized detuning, δ, for the case of $I/I_{\text{sat}}^0 = 1/3$. At line-center (viz., $\delta = 0$), the calculated values of R trace out a parabolic curve, in keeping with the predictions of Eq. (33). However, for comparable values of $|\alpha_0 \ell|$, gain media (i.e., $\alpha_0 < 0$) can yield much higher reflectivities than those attained under conditions of pure absorption (i.e., $\alpha_0 > 0$). This phenomenon becomes more pronounced with increasing magnitude of $|\alpha_0 \ell|$ and can be rationalized in terms of the additional amplification experienced by the emerging signal wave as it propagates through a region of inverted population. The SEP–DFWM technique, with its incorporation of four-wave mixing spectroscopy into the (inverted) DUMP transition, exploits such effects in order to enhance the overall sensitivity achieved during the double resonance detection process.

The dependence of the DFWM signal amplitude on the intensities of the pump and probe beams can be demonstrated by considering the limiting forms of Eq. (33) when $I \ll I_{\text{sat}}$ and $I \gg I_{\text{sat}}$:

$$I \ll I_{\text{sat}}; \qquad I_s \approx (\alpha_0 \ell)^2 \left(\frac{1}{1 + \delta^2} \right)^3 \left(\frac{2I}{I_{\text{sat}}^0} \right)^2 I_p \,, \tag{34}$$

$$I \gg I_{\text{sat}}; \qquad I_s \approx (\alpha_0 \ell)^2 \frac{I_{\text{sat}}^0}{16I} I_p \,, \tag{35}$$

where the validity of the latter expression hinges upon small detuning from resonance [viz., $I \gg I_{\text{sat}}^0(1 + \delta^2)$]. Since the pump and probe waves in a typical DFWM experiment are derived from a common laser source, it is reasonable to examine the behavior of I_s when $I_p \propto I$. For the case $I \ll I_{\text{sat}}$, the signal amplitude is found to scale as the cube of the incident intensity, a result in keeping with that derived from the previous nonlinear susceptibility analysis [cf. Eq. (12)]. In contrast, the high-power limit ($I \gg I_{\text{sat}}$) suggests a near constant (i.e., saturated) value of I_s which is essentially independent of both I and I_p.

The asymptotic predictions of Eqs. (34) and (35) are illustrated in Fig. 4, where the resonant response (viz., $\delta = 0$) induced by a four-wave mixing process is plotted as a function of $\alpha_0 \ell$ and $\log(I/I_{\text{sat}}^0)$. Two contour maps are presented; the bottom panel depicting calculated values for the phase-conjugate reflectivity, R, and the top panel representing the corresponding signal strength, $I_s = RI_p \approx RI$. For a given combination of α_0 and ℓ, the amplitude of I_s rises rapidly ($\propto I^3$) and ultimately reaches a plateau ($\propto I^0$) when $I \gg I_{\text{sat}}^0$. In contrast, the reflectivity associated with a particular $\alpha_0 \ell$ product increases abruptly until a maximum is attained, whereupon

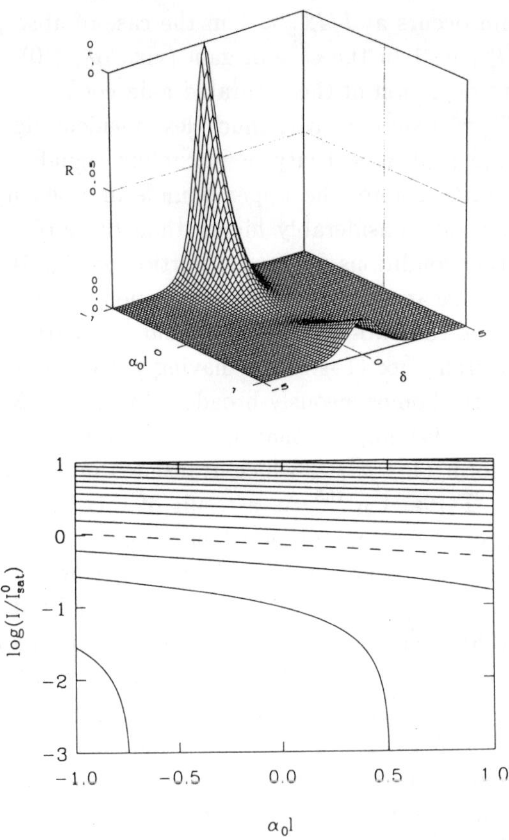

Fig. 3. DFWM frequency dependence for a homogeneously-broadened, two-level system. The top panel displays the calculated phase-conjugate reflectivity (R) as a function of two dimensionless parameters: the line-center absorption or gain coefficient times sample interaction length ($\alpha_0\ell$) and the normalized frequency detuning (δ). These results, obtained for a $I/I_{\rm sat}^0$ intensity ratio of 1/3, demonstrate the much higher R values and, therefore, signal strengths afforded by a gain medium ($\alpha_0\ell < 0$). The resonant transition clearly exhibits a spectral profile which is Lorentzian in nature. The bottom panel depicts the widths of spectral features predicted for a resonant DFWM interaction. Contours of constant linewidth, measured in units of the natural FWHM for the homogeneously-broadened transition ($\Delta\omega_{21}^0$), are displayed as a function of $\alpha_0\ell$ and $\log(I/I_{\rm sat}^0)$. The narrowest linewidths are found in the lower left corner of the plot, with a minimum contour level corresponding to $0.4\Delta\omega_{21}^0$. Successive contours represent increments of $0.2\Delta\omega_{21}^0$, with the dashed curve denoting the natural linewidth, $\Delta\omega_{21}^0$.

additional intensity results in the eventual decrease of R back towards zero. This extremum occurs at $I/I_{\text{sat}}^0 \approx 1$ in the case of absorption (viz., $\alpha_0 > 0$) and at $I/I_{\text{sat}}^0 \ll 1$ in the case of gain (viz., $\alpha_0 < 0$), a fact that can be attributed to depletion of the saturated gain coefficient that takes place when $I \geq I_{\text{sat}}^0$.[25] Consequently, much less incident light intensity is required to achieve peak reflectivity or maximum signal output in an inverted medium. Furthermore, the upper bounds imposed upon R and I_s in a gain situation are considerably higher than those obtained under comparable absorption conditions (cf. upper portion of Fig. 3).

In the case of a linear (unsaturated) absorption measurement, the resonant transition for the two-level model would be characterized by a Lorentzian spectral profile, $\propto (1 + \delta^2)^{-1}$, having a FWHM equal to the natural linewidth for the homogeneously-broadened medium, $\Delta\omega_{21}^0 = 2\Gamma_{21}$. When $I \ll I_{\text{sat}}$, Eq. (34) implies that the DFWM signal will exhibit a lineshape of the form $(1 + \delta^2)^{-3}$ which translates into a linewidth of $\Delta\omega_{21} = 2[\sqrt[3]{2} - 1]^{1/2}\Gamma_{21} \approx \Gamma_{21}$.[103] As a result of this Lorentzian-cubed dependence, the spectral features observed in a four-wave mixing study can actually be narrower in appearance than homogeneous broadening parameters would indicate. In contrast, the high-power limit (viz., $I \gg I_{\text{sat}}^0$) of expression (35) suggests no explicit dependence on frequency and, therefore, a near-infinite spectral bandwidth. The lower portion of Fig. 3 displays a contour map of calculated DFWM linewidths, $\Delta\omega_{21}$, as a function of both $\alpha_0\ell$ and $\log(I/I_{\text{sat}}^0)$. For a given combination of α_0 with ℓ, the magnitude of $\Delta\omega_{21}$ remains approximately constant when $I \ll I_{\text{sat}}^0$ but grows rapidly with increasing light intensity after attaining the natural value of $\Delta\omega_{21}^0$ at $I \approx I_{\text{sat}}^0$. Furthermore, the linewidths associated with a particular I/I_{sat}^0 ratio are always smaller in the case of an inverted medium, with low power excitation yielding spectral features that are substantially narrower than the homogeneous bandwidth (viz., $2\Gamma_{21}$). The enhancement in resolution attained under gain conditions can be attributed to the difference in amplification experienced by the peak and the wings of the signal beam as it traverses the interaction region.

In order for DFWM to provide a viable probe of molecular structure and dynamics, the dependence of the signal amplitude on the intrinsic properties of the nonlinear medium (e.g., magnitude of transition dipole moments and rates of various relaxation processes) must be understood at a level that permits the interpretation of observed spectral features. The limiting forms assumed by I_s under conditions of low and high power excitation can furnish some insight into the response variations that accompany changes in molecular parameters. In particular, substitution of the defining expressions for α_0 and I_{sat}^0 into Eqs. (34) and (35) yields:

$$I \ll I_{\text{sat}}; \qquad I_s \approx \left(\frac{2\omega}{\hbar^3 c^2 \varepsilon_0^2}\right)^2 (\Delta N_0 \ell)^2 \frac{|\mu_{21}|^8}{\Gamma_0^2 \Gamma_{21}^4} I^2 I_p , \qquad (36)$$

$$I \gg I_{\text{sat}} ; \qquad I_s \approx \frac{\omega^2}{128 c \varepsilon_0} (\Delta N_0 \ell)^2 \frac{\Gamma_0 |\mu_{21}|^2}{\Gamma_{21}} \frac{I_p}{I} . \qquad (37)$$

For either case, the amplitude of the emerging DFWM signal wave is found to scale quadratically with both the interaction length ℓ and the population difference ΔN_0, the latter also implying a square dependence upon the number density or pressure of target species. From the viewpoint of spectroscopic measurements, where the intensities of transitions can provide as much information as their location in frequency, the influence of the transition dipole moment, μ_{21}, upon the magnitude of I_s is of crucial importance. As a result of saturation effects, Eqs. (36) and (37) display pronounced differences in their dependence on transition dipole, with the low-power limit being characterized by the same $|\mu_{21}|^8$ functional form suggested by the previous susceptibility analysis, while the high-power regime scales as $|\mu_{21}|^2$. Since one-photon linestrengths,[84] S_{21}, are proportional to $|\mu_{21}|^2$, this behavior translates into signal amplitudes that increase as $(S_{21})^4$ for $I \ll I_{\text{sat}}$ and $(S_{21})^1$ for $I \gg I_{\text{sat}}$.

Farrow and co-workers have made an extensive study of the influence which transition dipole moments exert upon a DFWM interaction.[102] Despite the use of a Doppler-broadened medium having intrinsic magnetic sublevel degeneracy (viz., bulk-gas NO samples), their experimental observations were in reasonable agreement with predictions derived from the stationary, two-level model of Abrams and Lind. In particular, the signal amplitude at line-center was found to be well described by a power law of the form, $I_s \propto (\mu_{21})^m$, where the numerical value of the exponent, m, initially exhibited a rapid decrease with increasing light intensity followed by a plateau which asymptotically approached a high-power limit. It is quite evident that this unusual dependence upon transition moment will complicate direct interpretation of spectral line intensities, especially when considering closely-spaced features of vastly different character (e.g., different ground state vibrational levels accessed via SEP). These deleterious effects, as well as those incurred from power fluctuations in the incident laser source, can be partially mitigated by operating in a near-saturation regime (viz., $I \gg I_{\text{sat}}$). Unfortunately, such conditions will also result in a decrease of effective spectral resolution (cf. Fig. 4).

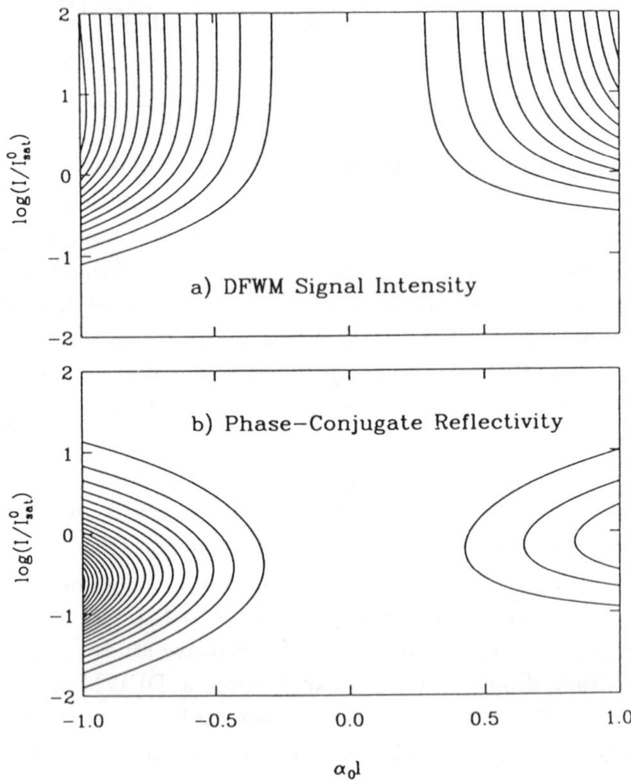

Fig. 4. DFWM response for a homogeneously-broadened, two-level system. The resonant ($\delta = 0$) response predicted for a DFWM interaction is displayed as a function of two dimensionless parameters: the line-center absorption or gain coefficient times sample interaction length ($\alpha_0 \ell$) and the logarithmic ratio of incident to saturation intensities $[\log(I/I_{\text{sat}}^0)]$. The bottom panel depicts the calculated phase-conjugate reflectivity (R) while the top panel presents the corresponding DFWM signal intensity ($I_s = RI$). Each contour represents an increment of 0.005 units in R or I_s with a minimum value of zero located at $\alpha_0 \ell \approx 0$. The peak reflectivities and signal amplitudes in a gain medium ($\alpha_0 \ell < 0$) are found to be substantially higher and to occur at significantly lower incident powers than under comparable conditions of absorption ($\alpha_0 \ell > 0$).

2.2.4. *Refinements for gas-phase spectroscopy*

Various extensions of the saturable absorber model (Sec. 2.2.3) have been reported, including analyses designed to incorporate the consequences of pump absorption/depletion,[104] unequal pump fields,[105,106] multiple-level behavior,[104,107] and pump–probe detuning.[25,108,109] However, for the interpretation of results derived from a gas-phase ensemble of target molecules, it is most important to consider the modifications introduced by motional effects[72,110] and level degeneracies.[73,74] While the random thermal motion of target species can be expected to diminish the effective nonlinearity of the medium,[25] the presence of degenerate sublevels, such as those associated with the magnetic quantum number of an angular momentum eigenstate,[111] can lead to new and potentially useful four-wave mixing interactions.[70]

The stationary (i.e., zero velocity) ensemble of two-level systems employed for the saturable absorber model provides no mechanism to distinguish scattering processes that result from the various holographic gratings created within a nonlinear medium. In particular, the phase-conjugate reflectivity of expression (26) implicitly contains a coherent summation of two *equivalent* resonant signal amplitudes, the origins of which can be traced to the spatial modulation of optical properties that accompanies pair-wise interference of the forward-going and backward-going pump beams with the nearly-collinear probe wave. Translational motion of the target molecules will lead to a differential dissipation for these periodic structures on timescales that can be correlated to their individual fringe spacings, Λ, and the mean molecular speed, $\bar{v}$. Since Λ is a function of the angular separation, θ, between the corresponding pump and probe beams [cf., Eq. (23)], four-wave mixing in a nonstationary medium can, therefore, be expected to exhibit a strong dependence upon the pump–probe crossing angle.[110,112] In contrast, the stationary perturber analysis suggests a DFWM signal that is completely independent of θ. The rectilinear nature of free molecular motion will also influence both the efficacy of grating formation and the rate of grating destruction, with the former following from the frequency shifts inherent to the Doppler effect[33,113] (cf. discussion below), while the latter stems from the more pronounced attenuation of fringe contrast obtained for molecules moving parallel to the grating wavevector (i.e., across the pattern of fringes in Fig. 2). Consequently, the spectral, temporal, and angular response observed in a gas-phase DFWM measurement can differ significantly from predictions of the saturable absorber model.

The modulation of optical properties and the Bragg diffractive scattering that are responsible for the creation of a DFWM signal follow from the basic dynamics of resonant absorption and emission.[89] In order to participate in this process, individual molecules within an inhomogeneously-broadened medium must be able to couple simultaneously with both the incident and the generated electromagnetic fields. As shown in Sec. 2.2.2, formation of a transient spatial grating requires interference of the probe wave, $\mathbf{E_p}$, with one of the two pump beams, for example, $\mathbf{E_f}$. Matter-field interactions provide the vehicle for transferring the resulting variations in light intensity and/or polarization into the nonlinear medium. In the case of a gas-phase sample, the first-order Doppler effect[113] will place restrictions on the velocity vector, $\mathbf{v}$, of a molecule that can successfully contribute to the preparation of this volume hologram. While $\mathbf{E_f}$ and $\mathbf{E_p}$ have degenerate frequencies in the stationary laboratory-fixed frame (viz., for $\mathbf{v} = 0$, $\omega_f = \omega$ and $\omega_p = \omega$), translational motion of the target species will lead to velocity-dependent frequency shifts that can be related to the propagation direction for the corresponding electromagnetic wave:[33]

$$\omega_f = \omega - \mathbf{k_f} \cdot \mathbf{v} \ , \tag{38}$$

$$\omega_p = \omega - \mathbf{k_p} \cdot \mathbf{v} \ , \tag{39}$$

where ω_f and ω_p now denote apparent frequencies for the forward-going pump and probe beams in the moving molecule-fixed frame of reference. For a rovibronic transition having rest frequency ω_{21}, simultaneous interaction of $\mathbf{E_f}$ and $\mathbf{E_p}$ with an individual target molecule demands that both $\omega_f \approx \omega_{21}$ and $\omega_p \approx \omega_{21}$. Similarly, coherent re-emission of radiation via Bragg diffractive scattering of the remaining pump wave, $\mathbf{E_b}$, requires that $\omega_b \approx \omega_{21}$ and $\omega_s \approx \omega_{21}$ with:

$$\omega_b = \omega - \mathbf{k_b} \cdot \mathbf{v} = \omega + \mathbf{k_f} \cdot \mathbf{v} \ , \tag{40}$$

$$\omega_s = \omega - \mathbf{k_s} \cdot \mathbf{v} = \omega + \mathbf{k_p} \cdot \mathbf{v} \ , \tag{41}$$

where the latter portion of these expressions follows from the counterpropagating geometry of the phase-conjugate experimental configuration (viz., $\mathbf{k_b} = -\mathbf{k_f}$ and $\mathbf{k_s} = -\mathbf{k_p}$). The identical frequencies employed in DFWM imply that the resonance conditions embodied in Eqs. (38) through (41) can be satisfied concurrently only for molecules having a zero net velocity along the nearly-collinear axis of wave propagation. This selection of individual velocity groups by the radiation field results in a sub-Doppler response[25,46] which manifests itself as a spectral feature located exactly at ω_{21} with a linewidth determined either by the effects of homogeneous broadening or by the bandwidth of the excitation source.

While a detailed theory of DFWM that simultaneously incorporates saturation, polarization, and motional effects still remains to be formulated,[114–116] some insight into the angular dependences introduced by a nonstationary medium can be gleaned from the susceptibility analysis of conventional nonlinear optics. For a gas-phase ensemble of target molecules having a velocity distribution specified by $f(\mathbf{v})$, the same approximations that lead to the small signal response of Eq. (12) now suggest a signal intensity of the form:[25,116]

$$I_s \propto \ell^2 \left| \int_{\mathbf{v}} f(\mathbf{v}) \chi^{(3)}(-\omega_s; \omega_f, -\omega_p, \omega_b) \vdots \hat{\mathbf{e}}_\mathbf{f}\, \hat{\mathbf{e}}_\mathbf{p}^*\, \hat{\mathbf{e}}_\mathbf{b}\, d\mathbf{v} \right|^2 I_f I_p I_b \,, \qquad (42)$$

where the susceptibility tensor elements, as defined by expression (16), must be rewritten in terms of the velocity-dependent frequencies [cf. Eqs. (38) through (41)] experienced by the moving target molecules. Figure 5 presents the results of numerical calculations performed on a model two-level system having a Maxwell–Boltzmann distribution of velocities characterized by $\bar{v} = 288$ m/s. Assuming that a resonant spectroscopic transition occurs at $\lambda = 500$ nm (viz., $\omega_{21} \approx 3.8 \times 10^{15}$ Hz) with the lifetimes for both longitudinal and transverse relaxation fixed at 10 ns (viz., $\Gamma_{22} = \Gamma_{11} = \Gamma_{21} = 1 \times 10^8$ Hz), the Doppler width for this inhomogeneously-broadened ensemble, $\Delta\omega_{21}^D = 2\sqrt{\pi \ln 2}\, \omega_{21} \bar{v}/c \approx \bar{v}/\lambda$,[113] is expected to be roughly fifty times larger than the natural FWHM of $\Delta\omega_{21}^0 = 2\Gamma_{21}$.

The upper panel of Fig. 5 displays the line-center DFWM intensity calculated as a function of the pump–probe crossing angle θ. In addition to showing the total signal amplitude I_s, this plot also depicts the relative contributions made by scattering processes that originate from interference of the probe beam with the forward-going (I_{fp}) and backward-going (I_{bp}) pump waves. As demonstrated by the logarithmic nature of the ordinate scale, the four-wave mixing output drops precipitously with increasing θ, ultimately attaining a value some five orders of magnitude smaller in the case of perpendicular pump and probe beams ($\theta = 90°$). Provided that

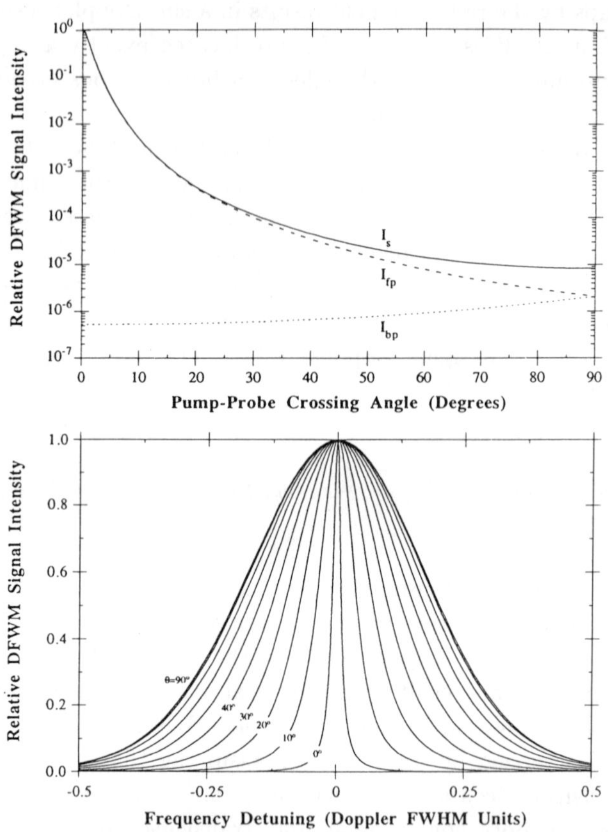

Fig. 5. Angular dependence of DFWM in a Doppler-broadened medium. Numerical calculations were performed on a Doppler-broadened ensemble ($\bar{v} = 288$ m/s) of two-level systems having a resonant transition at $\omega_{21} \approx 3.8 \times 10^{15}$ Hz with $\Gamma_{22} = \Gamma_{11} = \Gamma_{21} = 1 \times 10^8$ Hz. The top panel depicts the angular dependence of the line-center DFWM signal amplitude I_s (solid curve) with separate contributions arising from interference of the probe wave with the forward-going and backward-going pump beams explicitly denoted by I_{fp} (dashed curve) and I_{bp} (dotted curve), respectively. For orthogonal pump and probe waves ($\theta = 90°$), the two transient gratings formed within the nonlinear medium become equivalent and $I_{fp} = I_{bp}$. Note that $I_s \neq I_{fp} + I_{bp}$ since the actual DFWM response entails a coherent summation of the electromagnetic fields generated by the individual scattering processes. The bottom panel displays variations in the shape of the DFWM spectral profile as a function of the pump–probe crossing angle. Each curve represents a $10°$ increment in θ with the abscissa scaled in units of the Doppler width (FWHM). The form of the resonant response is found to evolve from a Doppler-free Lorentzian (FWHM equal to the natural linewidth) at $\theta = 0°$ to a Doppler-broadened Gaussian (FWHM roughly 20 times the natural linewidth) at $\theta = 90°$.

saturation and optical pumping effects can be neglected,[25,117] Wandzura has suggested that $I_s(\theta) \propto (\sin\theta)^{-4}$ where the apparent singularity at $\theta = 0°$ disappears when the full expression for the nonlinear response is considered.[110] By using a treatment similar to that of Eq. (42), Ducloy and Bloch have confirmed this pronounced angular dependence and have derived an analytical form for the ratio of DFWM intensities measured in orthogonal ($\theta = 90°$) and collinear ($\theta = 0°$) experimental geometries:[112]

$$\frac{I_s(\theta = 90°)}{I_s(\theta = 0°)} = 4\theta_0^4 = 4\left(\frac{\lambda\Gamma_{21}}{\pi\bar{v}}\right)^4 , \tag{43}$$

where θ_0 denotes a characteristic angle for which the mean free path ($d = \bar{v}\tau \approx \bar{v}/\Gamma_{21}$) and grating periodicity ($\Lambda = \lambda/2\sin(\theta/2) \approx \lambda/\theta$) are approximately equal. For a Doppler-broadened medium, maximum DFWM response clearly demands use of the smallest pump–probe crossing angle that is consistent with convenient extraction of the emerging signal beam and efficient rejection of any incoherently-scattered background light (e.g., from window substrates).[b] The distinct angular behavior exhibited by the two contributions to the total signal amplitude (viz., I_{fp} and I_{bp}) reflects expected variations in grating lifetimes, with $I_{fp} = I_{bp}$ for the orthogonal configuration of pump and probe waves ($\theta = 90°$) where both spatial modulation patterns have the same fringe spacing. The infinite plane wave approximation employed for derivation of these results assumes that the interaction length, ℓ, remains constant as the crossing angle is varied. Since practical laboratory experiments usually entail the use of laser beams having finite transverse extent, an even more pronounced decline in the magnitude of I_s can be expected to arise from the decrease in effective sample volume that must occur as θ is increased from $0°$ to $90°$.

The lower panel of Fig. 5 shows resonant DFWM lineshapes calculated for various pump–probe crossing angles, with each curve normalized to an amplitude of unity at line-center. For small values of θ, the spectral profile is found to be essentially Lorentzian with a FWHM roughly equal to the natural linewidth of $\Delta\omega_{21}^0 = 2\Gamma_{21}$. This characteristic Doppler-free shape

---

[b]From the viewpoint of transient grating spectroscopy, the large spatial periodicities obtained for near-collinear experimental configurations (viz., $\Lambda \to \infty$ as $\theta \to 0$) would tend to imply a decrease in the overall efficiency of a DFWM scattering process. This behavior contradicts the predictions of nonlinear optical theory (cf. Fig. 5), as well as the results of laboratory measurements, and suggests one of the limitations imposed upon a "holographic" description of the four-wave mixing interaction.

broadens and evolves into a Gaussian-like feature as θ moves away from the collinear geometry. Ducloy and Bloch have predicted a residual Doppler broadening that scales approximately as $\Delta\omega(\theta) \propto \sin\theta$ with a linewidth for the perpendicular configuration of pump and probe beams given by:[112]

$$\Delta\omega(\theta = 90°) = 1.17\frac{\Delta\omega_{21}^0}{\theta_0}. \tag{44}$$

For the model two-level system used to generate the results presented in Fig. 5, expression (44) implies a linewidth that increases by more than a factor of twenty as θ varies from $0°$ to $90°$. While orthogonal pump and probe beams still provide an effective spectral resolution that is at least two times narrower than the full Doppler width, maximum exploitation of the Doppler-free capabilities inherent to DFWM spectroscopy requires the use of nearly-collinear experimental geometries. Saturation effects have been shown to modify significantly the spectral response of four-wave mixing measurements performed on inhomogeneously-broadened media, with broadening, as well as bifurcation, of resonant lineshapes predicted theoretically[114,115,118] and observed experimentally.[116,119,120]

Figure 6 compares DFWM and LIF spectra recorded simultaneously in a low pressure (~ 0.020 Torr) bulk-gas sample of NO. For these measurements, Farrow and Rakestraw[48,49] utilized a single-frequency (≤ 0.004 cm^{-1} bandwidth) source of tunable ultraviolet light in order to probe the O_{12} bandhead region in the NO $A^2\Sigma^+ - X^2\Pi$ $(0,0)$ system. The LIF data display linewidths of roughly 0.1 cm^{-1} which are consistent with the Doppler broadening expected for a room temperature ensemble of NO molecules. In contrast, the Doppler-free response afforded by the DFWM detection scheme ($\theta = 2.5°$) leads to laser-limited spectral profiles that are located precisely at the rest frequencies for individual molecular transitions. The small depletion features observed on the top of individual fluorescence peaks are sub-Doppler Lamb dips[33] which arise from nonlinear saturation effects induced by the counterpropagating pump waves.

As in the related techniques of saturation spectroscopy[33] and polarization spectroscopy,[32,121] the standing-wave nature of the counterpropagating pump beams enables the DFWM interaction to be exploited for elimination of inhomogeneous Doppler broadening. While this feature can prove useful for analysis of dense spectral regions containing many closely spaced transitions, the enhancement in resolution is accompanied by a decrease in overall sensitivity owing to the small subset of target species that can participate

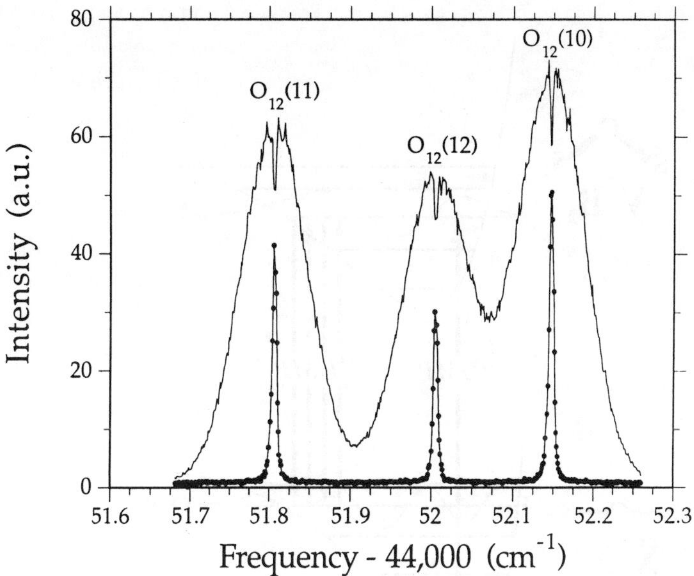

Fig. 6. Example of the spectral resolution afforded by DFWM detection. Rovibronic features in the O_{12} bandhead region of the NO $A^2\Sigma^+$—$X^2\Pi$ (0-0) system are simultaneously examined by means of DFWM spectroscopy (connected dots) and fluorescence excitation spectroscopy (solid curve). The sub-Doppler nature of lineshapes measured in the phase-conjugate four-wave mixing geometry is clearly evident. Both data sets were acquired at a total sample pressure of 0.02 Torr through use of a pulse-amplified laser system having a Fourier Transform-limited bandwidth of < 0.004 cm^{-1}. Small depletion features appearing on the tops of individual fluorescence peaks are Lamb Dips induced by the counterpropagating pump waves of a phase conjugate DFWM geometry. (courtesy of R. L. Farrow and D. J. Rakestraw)

in the signal generation process. For example, less than 5% of the NO molecules residing in a given ground state rovibrational level have velocities that enable them to contribute to the formation of DFWM spectral features depicted in Fig. 6.[122] Improved detection limits can be realized either through use of excitation sources having bandwidths more comparable to the Doppler-width[123,124] or by implementing experimental geometries other than the counterpropagating phase-conjugate configuration.[24,26,125]

2.3. *SEP–DFWM Spectroscopy*

The discussion above has demonstrated many of the unique features that distinguish resonant four-wave mixing spectroscopy from other linear and

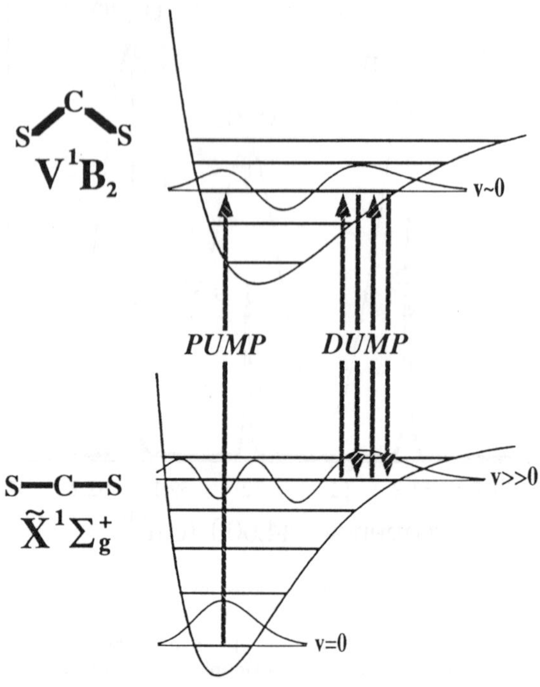

Fig. 7. Energy level diagram for SEP–DFWM spectroscopy. The PUMP laser is used to transfer population to an electronically-excited intermediate state with DUMP resonances detected through a nearly-instantaneous DFWM interaction. Implementation of the four-wave mixing scheme in an inverted gain medium (e.g., the DUMP transition) offers significant advantages in terms of both sensitivity and resolution. Structural formulas illustrate the change in nuclear configuration which accompanies the bent-from-linear V^1B_2–$\tilde{X}^1\Sigma_g^+(\pi^* \leftarrow \pi)$ electronic transition for CS_2.

nonlinear techniques. The inherent flexibility of this detection scheme suggests several possibilities for investigating the structural and dynamical effects of vibrational excitation in polyatomic species. In order to combine the widely recognized advantages afforded by the double resonance nature of SEP with the zero-background capabilities of DFWM, recent studies performed in our laboratory have retained the pulsed PUMP, so as to create a state-selected ensemble of electronically excited molecules, but have replaced the simple DUMP beam with a four-wave mixing probe. As illustrated in Fig. 7, this SEP–DFWM configuration basically utilizes the PUMP laser as an "incoherent" means of preparing molecules within a preselected rotation–vibration level of the intermediate state. Before this transient molecular population can decay through either radiative or nonradiative processes, a nearly-instantaneous DFWM interaction, driven by the coherent DUMP radiation, is used to record an "absorption-like" spectrum terminating on highly excited eigenstates within the ground electronic potential surface.

SEP–DFWM spectroscopy provides an essentially background-free means for elucidating the nature of vibrationally-excited polyatomic species. In particular, this scheme retains much of the experimental simplicity inherent to the original fluorescence-dip methodology while simultaneously eliminating any need for cumbersome dual-beam, null-detection configurations. True Doppler-free response can be realized by exploiting the counterpropagating phase-conjugate geometry of Fig. 1. Furthermore, the population inversion created by the PUMP between the upper and lower levels of the DUMP transition implies that the four-wave mixing process will take place in a transitory gain medium. As demonstrated in Sec. 2.2.3, this situation can greatly augment both the sensitivity and the resolution attained in a DFWM interaction, with optimum signal generation demanding the use of DUMP laser powers substantially below those required for optical saturation of resonant transitions. Indeed, SEP–DFWM measurements performed under both static bulb[21] and molecular beam[22] conditions have demonstrated detection limits and spectral linewidths that corroborate many of these expectations.

3. Experimental Implementation

Figure 8 schematically illustrates the basic experimental configuration utilized for SEP–DFWM studies in bulk-gas samples. Two Nd:YAG-pumped dye lasers, each containing an intracavity etalon designed to reduce its spectral bandwidth to < 0.04 cm^{-1}, provide the pair of tunable optical pulses (< 10 ns duration) required for the doubly-resonant SEP process. While this single etalon mode performance is far removed from ideal, single-frequency (i.e., single longitudinal mode) operation, the effective coherence length, ℓ_c, for these commercial laser sources probably lies somewhere between the bandwidth-limited and transform-limited values of ~ 10 cm and ~ 1 m, respectively. As in conventional holography, ℓ_c provides an upper bound on the path length mismatch that can be

     Molecular Dynamics and Spectroscopy

tolerated between two beams involved in a spatial interference process. Since the transient grating response of DFWM derives from completely analogous interference phenomena (cf. Fig. 2), experimental realization of the SEP–DFWM detection scheme demands careful consideration of such optical path constraints. More importantly, the value of ℓ_c places a limit on the length of sample over which coherent nonlinear interactions (e.g., DFWM signal generation) can occur.

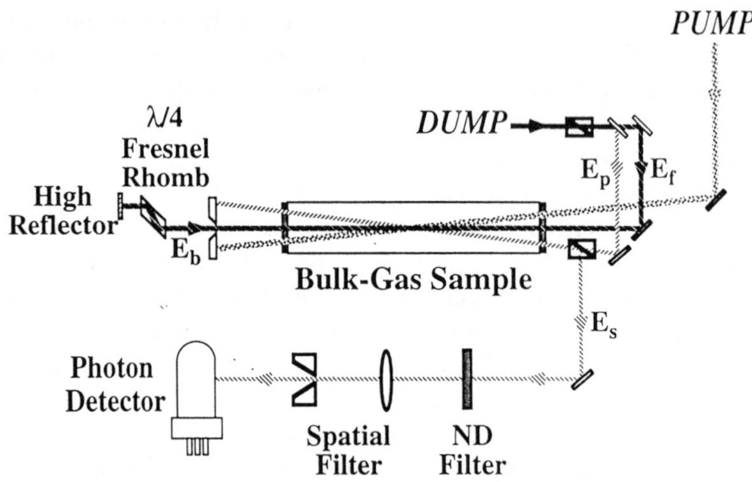

Fig. 8. Schematic diagram of the SEP–DFWM experimental apparatus. The phase-conjugate DFWM geometry is implemented with a cross-population detection scheme in which the forward-going pump (E_f) and probe (E_p) waves are orthogonally-polarized with respect to the backward-going pump beam (E_b). As a result, the signal field (E_s) is generated with a polarization normal to that of the incident probe radiation. This permits facile extraction and discrimination of signal photons through the use of polarization optics.

The output of one dye laser, designated as the PUMP, is frequency doubled to obtain the ultraviolet wavelengths required for population of an electronically-excited intermediate state in the target molecules (cf. Fig. 7). Following separation of residual fundamental light, this ultraviolet beam is directed through a Keplerian telescope where it is spatially filtered by means of a diamond pinhole and recollimated to a diameter of < 5 mm. The visible output of the second (DUMP) dye laser propagates through an analogous set of optical components. The use of beams possessing high quality wavefronts can greatly enhance both the efficacy of the four-wave mixing interaction and the sensitivity of the signal detection process, the latter following from the elimination of incoherently-scattered background light. Given the poor mode quality associated with most pulsed dye lasers, spatial filtering of the DUMP radiation is essential for successful implementation of the SEP–DFWM scheme in rarified environments such as low pressure bulk-gases and free-jet expansions.

After traversing a polarization filter oriented to transmit vertically-polarized light (i.e., linearly-polarized light with its electric vector normal to the horizontal plane of the optical table) and an optical delay line designed to ensure temporal overlap with the DUMP radiation, the collimated PUMP beam propagates through the target medium at a slight angle ($\leq 1.5°$) with respect to the central axis of a cylindrical containment vessel (cf. Fig. 8). For bulk-gas studies, the sample cell consists of a simple glass tube of ~ 1 m total length with high quality fused silica windows mounted upon o-ring seals at both ends. A factory-calibrated capacitance manometer provided an accurate indication for the absolute pressure of target species. In order to minimize the polarization scrambling effects associated with stress-induced birefringence, the windows are held in place only by the pressure differential between the interior and exterior of the cell. The unusual size of this apparatus stems not from the desire to increase the DFWM interaction length, but rather from the need to avoid overlap of the four-wave mixing (DUMP) beams at the window substrates, a situation that has been found to result in the generation of large, nonresonant background signals. Similar considerations must be applied in the design of other experiments (e.g., molecular beam studies) based upon the DFWM detection scheme.

The spatially-filtered and collimated DUMP radiation passes through a polarization filter and subsequently enters an achromatic 30–50% beam-splitter designed to separate it into two vertically-polarized beams destined to become the pump and probe waves in a resonant DFWM process (cf. Fig. 1). For the DFWM probe, $\mathbf{E_p}$, the weaker of these DUMP beams is directed through the sample vessel at a slight angle ($\leq 1.5°$) with respect to its central axis. The remaining DUMP light propagates along the central axis of the cell, thereby forming the forward-going pump, $\mathbf{E_f}$. After traversing the sample, this beam passes through a quarter-wave Fresnel rhomb and impinges upon a mirror that precisely retroreflects it back along its initial path so as to generate the backward-going pump, $\mathbf{E_b}$. In this manner, collinear and counterpropagating pump fields of orthogonal polarization (viz., $\mathbf{E_f} \perp \mathbf{E_b}$) can readily be established. While this method is convenient and requires a minimum number of components, it does result in an intrinsic delay between the arrival times for $\mathbf{E_f}$ and $\mathbf{E_b}$ at the SEP–DFWM interaction region. This temporal mismatch can prove to be detrimental for spectroscopic investigations aimed at short-lived or transient species. In such cases, the various beams involved in the DFWM interaction should be produced and controlled by means of independent optical trains.

For the phase-conjugate DFWM geometry depicted in Fig. 1, the signal wave emerges in a direction that is exactly collinear and counterpropagating with respect to the incident probe radiation. Consequently, appropriate optical elements (e.g., a beamsplitter) must be inserted into the path of the probe beam in order to extract the signal photons. The present implementation of SEP–DFWM spectroscopy makes use of a cross-population scheme[25,74] in which the polarization properties of light are exploited for the sensitive detection of resonant DUMP transitions. In particular, by arranging for the forward-going pump and probe waves to have a common linear polarization which is orthogonal to that of the backward-going pump (viz., $\mathbf{E_p} \parallel \mathbf{E_f} \perp \mathbf{E_b}$), the generated signal field, $\mathbf{E_s}$, will be cross-polarized with respect to the incoming probe beam (i.e., $\mathbf{E_s} \perp \mathbf{E_p}$). As shown in Sec. 2.2.2, this phenomenon can be rationalized through the Bragg scattering of $\mathbf{E_b}$ from a transient grating formed by the interaction of $\mathbf{E_f}$ with $\mathbf{E_p}$. A polarization optic incorporated into the probe beam can therefore be used to extract and discriminate the orthogonally-polarized signal wave. In SEP–DFWM spectroscopy, this is accomplished through use of an achromatic polarizer (i.e., a specially fabricated calcite prism) which provides a wavelength-independent means of deviating counterpropagating signal photons out of the path followed by the probe light.

After passing through neutral density filters to attenuate its amplitude and a spatial filter to reject stray room light, the DFWM signal beam impinges upon the photocathode of a simple photomultiplier tube. The highly polarized nature of the signal wave permits the use of polarization filters as an additional means of discriminating signal photons from background scatter. The resulting photocurrent is amplified and directed to a CAMAC gated integrator system that enables the DFWM signal intensity to be recorded as a function of DUMP wavelength. Typical spectra are obtained by averaging the signals derived from 10–20 laser pulses for each increment of the DUMP frequency.

Initial alignment of the signal collection optics is greatly facilitated by the counterpropagating geometry exploited in phase-conjugate DFWM. In particular, since the signal is expected to emerge in a direction that exactly retraces the optical path of the probe radiation, its presence can be simulated (i.e., an artificial "signal" can be created) by using a simple mirror to precisely retroreflect the incoming probe beam. This procedure enables a crude adjustment of the detection train to be performed in an expedient and reliable manner that is usually more than sufficient to ensure the observation of resonant SEP–DFWM transitions. Once located, such spectral features can be used for final alignment of the apparatus through real-time maximization of the measured signal intensity.

When carefully aligned and optimized, the major source of baseline noise in the SEP–DFWM apparatus appears to be residual DUMP light scattered from the optical elements used to extract signal photons from the counterpropagating probe beam. While not a significant limitation for most studies, this background level can be minimized through the judicious selection of optical components fabricated from high quality, scatter-free materials. Recently, Farrow and Rakestraw have demonstrated the direct detection of DFWM signals in a manner that eliminates the need to use any beamsplitting or polarization optics.[48] This was accomplished by slight deviation of the backward-going pump wave out of the plane formed by $\mathbf{E_f}$ and $\mathbf{E_p}$, thereby resulting in a pyramidal geometry that spatially isolates the signal beam from scattering sources. However, implementation of this scheme requires a much more critical alignment of the signal detector (i.e., the spatial location of the signal beam is no longer a trivial matter) and might be expected to result in some reduction of signal strength owing to a decrease in the effective interaction length.

4. Experimental Results and Discussion

4.1. *Comparison to Fluorescence-Dip Detection*

Figure 9 compares SEP spectra obtained from a bulk-gas sample (i.e., ~ 0.08 Torr) of CS_2 target molecules through use of (a) dual-beam fluorescence-dip detection and (b) DFWM detection. For each of these measurements, the frequency of the PUMP laser was tuned to coincide

with a "single" rovibronic transition [viz., $P(12)$] in the 320 nm V^1B_2-
$\tilde{X}^1\Sigma_g^+(\pi^* \leftarrow \pi)$ system of CS_2, thereby forming a state-selected ensemble
of electronically-excited molecules in a $J = 11$ rotational level. While
spectral features in this region had previously been assigned to the
$(0\,0\,0)_{K=0} - (0\,0^0\,0)$ 10V origin band,[126] double-resonance and molecular
beam studies performed in our laboratory have indicated the presence of
at least five distinct vibronic transitions.[127] The bent $^1B_2$ intermediate
state utilized for these measurements has been the subject of several
spectroscopic investigations[126,128–130] and has been shown to correlate with
the upper Renner–Teller component of a linear $^1\Delta_u$ electronic manifold.[126]
The forbidden nature of a $^1\Delta_u -^1\Sigma_g^+$ transition accounts for the relatively
small integrated oscillator strength of 2.7×10^{-4} associated with the
bent-from-linear $V-\tilde{X}$ system.[132]

Scanning of the DUMP laser frequency in the vicinity of 447 nm
reveals two strong SEP resonances that can be assigned to the $P(12)$ and
$R(10)$ rotational lines for the parallel $(0\,0\,0)_{K=0} \to (2\,18^0\,0)$ vibronic band.
The enormous spectral simplification afforded by the double-resonance
nature of SEP is clearly demonstrated by the limited number of downward
transitions, permitted by rotational selection rules, between the single
rovibrational level prepared by the PUMP in the excited electronic state and
the target vibrational level in the ground electronic state. The lower level
for these DUMP transitions nominally contains 2 quanta of excitation in the
ν_1 symmetric stretching mode combined with 18 quanta of excitation in the
ν_2 degenerate bending mode [viz., $(2\,18^0\,0)$] and is located some 8500 cm^{-1}
above the vibrationless zero-point energy of the $\tilde{X}^1\Sigma_g^+$ electronic potential
surface. This amount of internal excitation constitutes approximately 25%
of that required for the dissociation of CS_2 molecules into their $CS(X^1\Sigma^+)$
and $S(^3P)$ ground state fragments.[132,133]

A definitive normal mode assignment for the ground state vibrational
level depicted in Fig. 9 is hampered by the pervasive anharmonic interactions that characterize the highly excited CS_2 system.[134,135] While
extrapolation of previous spectroscopic work strongly supports the $(2\,18^0\,0)$
designation,[136] this state is but one member of a twelve-dimensional
Fermi polyad that entails the successive coupling of levels differing by
one quantum of ν_1 and two quanta of ν_2. Numerical diagonalization of
model Hamiltonians expressed in a normal mode set of basis vectors shows
that an extensive mixing of vibrational character occurs among all of the

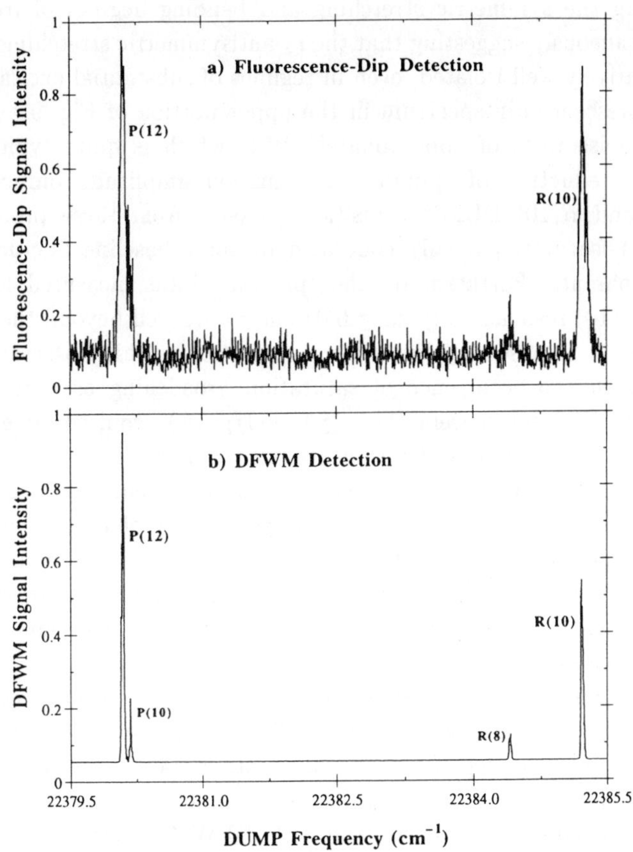

Fig. 9. Comparison between fluorescence-dip and DFWM detection schemes for SEP. SEP spectra recorded under bulk-gas conditions (0.08 Torr) are presented for the $(2\,18^0\,0)$ vibrational level of CS_2. The upper panel (a) depicts results obtained from dual-beam fluorescence-dip detection while the lower panel (b) displays identical measurements performed with DFWM detection. In each case, the frequency of the PUMP laser was tuned to coincide with a $P(12)$ rovibronic transition (at $30\,893.10$ cm^{-1}) in the 10V band of the CS_2 V–$\tilde{X}$ system. Individual DUMP resonances are labeled by the rotational quantum numbers of the $\tilde{X}$ state, with the extra features revealed by the DFWM scheme attributable to overlapping PUMP transitions.

states involved in this Fermi interaction.[137] Vibrational wavefunctions, obtained for both empirical and *ab initio* CS_2 potential surfaces through direct variational calculations,[138,139] confirm a loss of distinct nuclear motion along the symmetric stretching and bending degrees of freedom while simultaneously suggesting that the ν_3 antisymmetric stretching mode remains relatively well isolated, even in regimes of substantial excitation.

The fluorescence-dip spectrum in the upper portion of Fig. 9 exhibits a signal-to-noise ratio of approximately 10:1, which is quite typical for the 10–15% reduction of spontaneous emission amplitude induced by the $P(12)$ and $R(10)$ DUMP transitions. Aside from these prominent peaks, small features, partially concealed by large baseline fluctuations, are also apparent. Furthermore, the spectral widths measured for the $P(12)$ and $R(10)$ resonances (viz., ≥ 0.08 cm^{-1}) are well beyond the value of ~ 0.03 cm^{-1} imposed by instrument resolution. This observation is consistent with the occurrence of saturation broadening as a result of the high DUMP laser powers (i.e., ≥ 10 mJ/pulse) required to extract depletion signals from a noisy fluorescence background. While not the best achieved in our laboratory, the quality of the data presented in this figure is representative of that obtained for a large number of fluorescence-dip measurements performed on the CS_2 system. The conventional dual-beam, null-detection techniques employed for this study were found to greatly enhance the ability to discriminate resonant SEP processes from superfluous spontaneous emission.

The lower panel of Fig. 9 displays one of the first SEP spectra recorded through the use of DFWM detection. This result must be compared with the fluorescence depletion measurements presented in the upper panel, with both data sets being acquired under essentially identical experimental conditions. Clearly, the SEP–DFWM spectrum, obtained with an unfocussed DUMP beam having a total energy of less than 1 mJ/pulse, exhibits a superior signal-to-noise ratio as expected for a true zero-background technique. The spectral widths for the $P(12)$ and $R(10)$ resonances (viz., ≤ 0.035 cm^{-1}) reflect the sub-Doppler, laser-limited resolution afforded by the four-wave mixing scheme. Additional peaks, previously hidden by the saturation broadening and baseline fluctuations inherent to canonical fluorescence-dip SEP, are now quite evident. These new features have been attributed to an overlap in the PUMP transition that results in the simultaneous preparation of a previously unobserved $J = 9$ level in the electronically excited intermediate state. This assignment has been corroborated by subsequent double-resonance studies designed to systematically examine rovibronic structure within the CS_2 10V band.[127]

4.2. *Dynamic Range and Sensitivity*

It is quite evident from the results presented in Fig. 9 that the SEP–DFWM scheme has an enormous potential for exploring the structural and dynamical features of vibrationally-excited polyatomic molecules. However, extension of SEP–DFWM measurements to species other than CS_2 and to environments more rarified than those encountered under typical bulk-gas conditions (e.g., molecular beams) clearly depends on the overall sensitivity and versatility afforded by this new spectroscopic tool. It is here that the ubiquitous fluorescence-dip methodology excels since, in principle, it can be utilized for any molecule that is amenable to detection through the exceptionally sensitive technique of LIF spectroscopy. Consequently, depletion schemes based on a DUMP-induced competition between spontaneous and stimulated emission have been applied successfully to the investigation of a wide variety of molecular species, including ionic systems (e.g., $H–(C{\equiv}C)_2–H^+)^{140}$ where space–charge effects constrain the total number density of target molecules to extremely low values ($\ll 10^8 cm^{-3}$). In contrast, four-wave mixing is a parametric optical process that generates signals having a greater-than-linear dependence on both sample concentration and incident light intensity. Based on experience with more prevalent forms of nonlinear spectroscopy (e.g., CARS[71]), one might therefore be misled into believing that application of the SEP–DFWM technique will be limited to the study of stable species that can be maintained at relatively high pressures in a static bulb environment.

The enormous dynamic range inherent to the four-wave mixing detection scheme is demonstrated by the SEP–DFWM spectrum presented in Fig. 10. Here, the frequency of the PUMP was tuned to coincide with that of a $P(16)$ rovibronic transition in the CS_2 10V system. The DUMP laser frequency was scanned over a 45 cm^{-1} range designed to probe regions of the CS_2 potential surface located $\sim$ 8500 cm^{-1} above the zero-point energy. Two pairs of spectral features, which can be attributed to the $P(16)$ and $R(14)$ rotational lines for two different vibronic bands, are readily resolved. The strong resonances found at lower DUMP frequencies belong to transitions that terminate on the same $(2\,18^0\,0)$ vibrational level that was depicted in Fig. 9. At higher DUMP frequencies, corresponding to deeper portions of the CS_2 potential well, a much less intense pair of transitions is observed. These peaks can nominally be assigned to DUMP resonances that involve a $(0\,20^0\,0)$ ground state vibrational level.

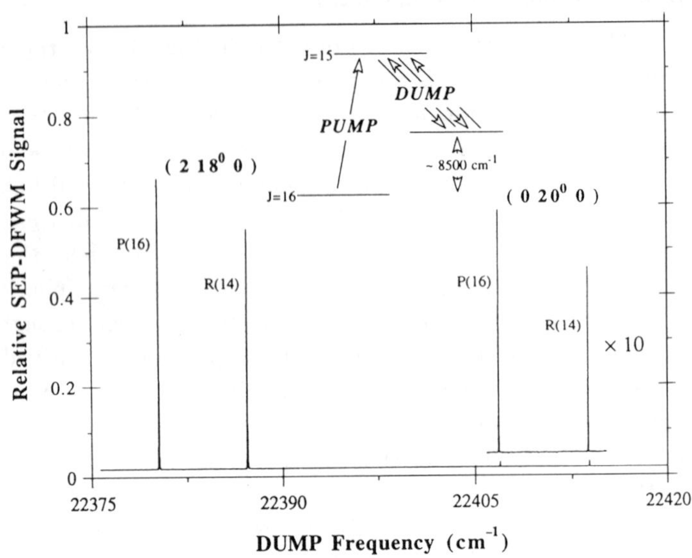

Fig. 10. Example of SEP–DFWM spectroscopy under bulk-gas experimental conditions. A segment of the SEP–DFWM spectrum recorded for CS_2 molecules containing ~ 8500 cm^{-1} of internal excitation is presented. Two pairs of rovibronic features are observed and nominally assigned to the $P(16)$ and $R(14)$ rotational lines for DUMP resonances terminating on the $(2\,18^0\,0)$ and $(0\,20^0\,0)$ vibrational bands. This data set was obtained at a sample pressure of ~ 0.03 Torr with total DUMP powers of < 500 μJ/pulse.

The primary spectroscopic advantage of DFWM detection stems from the lack of any appreciable signal in the absence of a resonant response from the molecular sample under investigation. The exceptionally large peak-to-noise ratios ($> 1000 : 1$) exhibited by the strong resonances of Fig. 10 clearly demonstrate this unique property. More importantly, without the presence of significant background fluctuations to compete with the discrimination of signals, the gain of a multiplicative detector (e.g., a photomultiplier tube) can be increased until the theoretical limit of single photon sensitivity is achieved. In this manner, weak transitions, such as those associated with the $(0\,20^0\,0)$ vibronic band of Fig. 10, can be uncovered. Thus, while the dynamic range for a given DUMP scan will be limited by the resolution of the apparatus used to record it (e.g., the number of bits in an A/D converter), the inherent dynamic range for the SEP–DFWM technique spans many additional orders of magnitude.

The unusual dependence of four-wave mixing interactions upon incident laser power and target transition moment [cf., Eqs. (36) and (37)] suggests yet another means for uncovering exceptionally weak spectral features. While SEP–DFWM signals obtained for strong downward transitions will saturate at relatively low values of the DUMP intensity, I_{DUMP}, those derived from resonances having substantially smaller oscillator strengths will continue to grow in size as roughly I^3_{DUMP}. Since the magnitude of any incoherently-scattered background light scales linearly with I_{DUMP}, an increase in the DUMP laser power can be exploited in order to extract and discriminate weak transitions. This method has enabled the observation of SEP features having effective intensities roughly two orders of magnitude smaller that those associated with the $(0\,20^0\,0)$ band of Fig. 10. A direct comparison of these exceptionally weak signals, as derived through the nonlinear SEP–DFWM technique, to those expected from conventional fluorescence depletion measurements is, by no means, a trivial matter. However, at least in the case of CS_2, there is no doubt that the DFWM detection scheme provides substantial advantages, in both sensitivity and resolution, over the fluorescence-dip methodology.

Figure 11 presents a plot of SEP–DFWM signal strength, I_s, as a function of CS_2 sample pressure, p, with the previously mentioned $(2\,18^0\,0)$ vibrational state serving as the terminal level of the DUMP transition. The solid curve represents the result of fitting these data to a power law expression of the form $I_s \propto p^m$, where the value of the exponent derived from least squares regression is $m = 1.975 \pm 0.025$ (one standard deviation uncertainty). This essentially quadratic response is in keeping with the N^2 behavior predicted both by the conventional nonlinear optical description of four-wave mixing [cf. Eqs. (12) and (16)] and by the saturable absorber model for a resonant DFWM interaction [cf. Eq. (33)].

The quadratic dependence of signal strength upon sample number density places the SEP–DFWM scheme into the same category as a variety of well-established laser techniques, including the related Doppler-free methods of polarization spectroscopy[32,121] and saturation spectroscopy.[33] As suggested by Fig. 11, this nonlinear behavior will lead to a rapid reduction in apparent sensitivity with decreasing concentration of target molecules. Nevertheless, experiments performed on bulk-gas samples of CS_2 have been able to observe SEP–DFWM resonances at total sample pressures of less than 1 mTorr, while still retaining peak-to-noise ratios which ex-

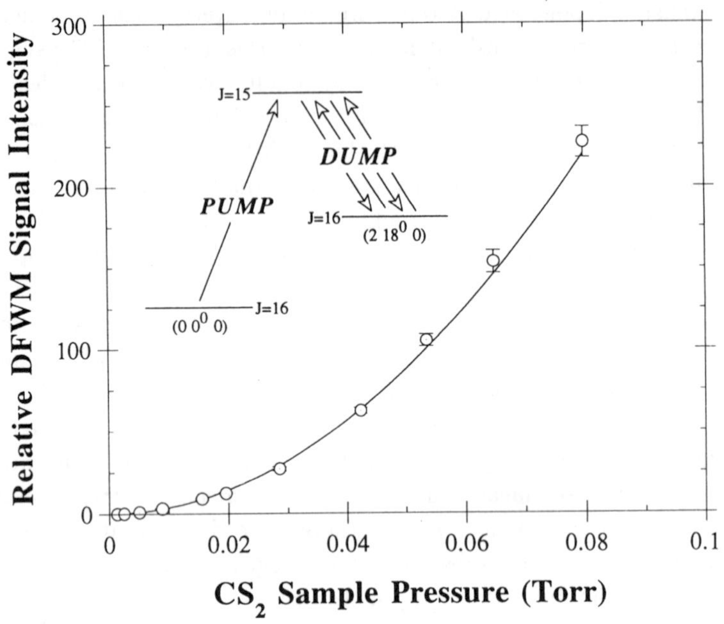

Fig. 11. Pressure dependence of the SEP–DFWM signal intensity. The indicated PUMP and DUMP transitions were used to record bulk-gas SEP–DFWM spectra for CS_2 as a function of sample pressure, p. Measured signal intensities are found to scale as p^m, where $m = 1.975 \pm 0.025$ (1σ uncertainty) as derived by least squares regression. This behavior, denoted by the solid curve, confirms the quadratic dependence on number density expected for the nonlinear DFWM process.

ceed those attainable through conventional fluorescence-dip measurements. Taking into account the distribution of rovibrational levels populated in a room temperature ensemble of CS_2 molecules and the effective volume of the DFWM interaction region (≤ 1 cm^3), these results translate into an SEP–DFWM detection limit for CS_2 of less than 10^{12} molecules per cm^3 per quantum state. Optimization of various parameters (e.g., the size and collimation of the PUMP and DUMP beams) should enable double-resonance signals to be acquired under even more rarified conditions.

A further indication for the ultimate sensitivity of the SEP–DFWM scheme can be found in the nature of the intermediate state utilized for the CS_2 studies. Since the efficiency of a resonant four-wave mixing process should scale roughly as the square of the corresponding absorption or gain coefficient [cf. Eq. (33)], the high quality of spectra obtained through use of the nominally-forbidden $V\,^1B_2$–$\tilde{X}\,^1\Sigma_g^+$ system of CS_2 implies that even greater sensitivities can be achieved for species having larger electronic transition moments.[c] This conjecture is supported by recent DFWM "absorption" measurements performed on diatomic radicals (e.g., OH,[116,141,142] NH,[143,144] and NO[48,102]) and polyatomic transients (e.g., C_3[24] and S_2O[127]), where substantial oscillator strengths have permitted the demonstration of detection limits approaching those of laser-induced fluorescence spectroscopy.

4.3. *Molecular Beam Studies*

The high spectral resolution that characterizes the SEP–DFWM technique stems from both the double-resonance nature of stimulated emission pumping and the sub-Doppler response of degenerate four-wave mixing. In theory, these combined capabilities yield a spectroscopic probe that is constrained only by the degree of state selectivity achieved in the PUMP transition and the overall bandwidth of the DUMP laser source. However, in a static bulb environment several factors conspire to diminish the practical limits of sensitivity and resolution achieved by the DFWM detection scheme. This effective reduction in capabilities is especially apparent in the case of massive polyatomic species, where small rotational and vibrational constants can lead to optical transitions that exhibit extreme spectral congestion. For the most part, the complexity of such spectra derives from the multitude of internal quantum states that is populated thermally in a room temperature ensemble of molecules. Aside from the potential loss of intermediate state selectivity that arises from overlapping PUMP transitions (cf. Fig. 9), this dilution of population over a large number of rovibrational levels decreases the state-specific concentration of molecules available for detection. In the case of exceptionally monochromatic DUMP sources (e.g., single frequency lasers), the finite Doppler width associated with the random translational motion of gas-phase molecules

---

[c]While the bent-from-linear V–$\tilde{X}$ system of CS_2 has a reported oscillator strength of 2.7×10^{-4}, *ab initio* calculations performed in our laboratory (Q. Zhang and P. H. Vaccaro, in preparation) suggest that a pronounced increase in electronic transition moment accompanies distortion of the ground state molecular framework along the ν_2 bending coordinate. Consequently, oscillator strengths for DUMP transitions terminating on vibrationally excited eigenstates of CS_2 can be substantially larger than that ascribed to the $(0\,0\,0)_{K=0} \leftarrow (0\,0^0\,0)$ origin band.

further restricts the DFWM interaction to only a small fraction of the total population residing within a given rotation–vibration eigenstate (cf. Figs. 5 and 6).

Most of the difficulties encountered under static bulb conditions can be alleviated by incorporating a molecular beam source into the SEP–DFWM apparatus. An impressive body of previous work has documented the utility of supersonic beams, with their narrow and controllable distribution of particle velocities, as facile probes of the translational dynamics that govern inelastic and reactive scattering.[5] The compression of velocities obtained in a molecular beam environment can be exploited to substantially decrease the Doppler width exhibited by entrained species,[145] a featattained in a Doppler-free four-wave mixing interaction. Of greater importance to polyatomic spectroscopy, however, is the extensive cooling of internal degrees of freedom that accompanies the supersonic expansion process.[146,147] In addition to providing for an increase in the state-specific number density of target molecules, the tremendous reduction in spectral congestion afforded by this freezing of rotational and vibrational motion significantly enhances the selectivity achieved for the first step (viz., the PUMP) of the doubly-resonant SEP scheme.

The application of SEP to molecular beams is not a novel concept, with several studies already documented in the chemical literature. While the majority of these efforts have been motivated by the reduction in spectral congestion which accompanies supersonic expansion,[148–151] SEP has also been used to explore vibrational predissociation in weakly bound van der Waals complexes that can only be formed and maintained under molecular beam conditions.[152–154] Nearly all of these previous experiments have relied on the canonical fluorescence depletion technique which, in turn, builds upon the exceptional sensitivity afforded by LIF spectroscopy. An alternate scheme, which is especially well suited to the low density conditions of a supersonic expansion, can be found in various methods based upon the observation of DUMP-induced changes in resonant ionization processes.[18–20,155–158] When applicable, such ion-dip variants of SEP offer the possibility of achieving state-specific detection limits that are well beyond those attainable through optical techniques. This capability stems from the near-perfect efficiency with which ionization signals can be collected and measured.

In contrast to more conventional (linear) techniques, the application of nonlinear optical probes to the low pressure environment of a molecular beam has been quite limited owing to the expected greater-than-linear dependence of signal intensity on sample number density (cf. Fig. 11). Nevertheless, both coherent anti-Stokes Raman spectroscopy[159,160] and stimulated Raman spectroscopy[161] have successfully been exploited for the characterization of small polyatomic species entrained within a free-jet expansion. Variations of these schemes, based upon the observation of secondary ionization processes rather than the primary photon response,[162,163] have demonstrated greatly enhanced sensitivities that enable even the minor constituents of a supersonic beam (e.g., van der Waals complexes) to be spectroscopically interrogated.[164,165] More recently, Butenhoff and Rohlfing have shown the feasibility of implementing optically-detected DFWM as a probe of transient species prepared under free-jet conditions.[24]

When directly applied to the rarified environment of a molecular beam, fluorescence-dip and ion-dip methods can be expected to yield SEP spectra having only limited peak-to-noise ratios. This situation can be improved dramatically through the use of various null-detection techniques designed to enhance the discrimination of weak depletion features from an unavoidable background of either spontaneous emission or ionization current. In particular, dual-beam sample geometries (viz., using two matched supersonic nozzles[149]) and/or pulse-by-pulse subtraction procedures[150,151] have been implemented successfully for the efficient extraction of DUMP-induced signals. SEP–DFWM spectroscopy, with its essentially background-free response, provides a viable alternative to such experimental configurations. In addition, the coherent, "beam-like" nature of the signal generated during a four-wave mixing interaction should enable the observation of resonant DUMP transitions with minimal optical access to the supersonic source, thereby permitting rapid incorporation of the DFWM detection scheme into existing apparatus.

Figure 12 presents an example of the results that can be obtained by applying SEP–DFWM spectroscopy in a molecular beam environment. For this measurement the static cell utilized in previous bulk-gas studies (cf. Fig. 8) was replaced by a pulsed supersonic nozzle of the current-loop actuated design.[166] Typical experiments entailed the use of a room temperature stagnation mixture consisting of $\leq 5\%$ CS_2 seeded into 2–3 atmospheres of helium. Following expansion through an aperture of circular cross-section (0.5 mm diameter), this sample entered a diffusion-pumped vacuum chamber that was capable of maintaining operating pressures of $< 5 \times 10^{-5}$ Torr, as measured by an uncalibrated ionization gauge. In order to achieve maximum compression of the observed Doppler width, interacting laser beams were directed through the supersonic expansion at an angle of roughly $90°$ with respect to the axis of molecular flow and at a distance ~ 10 cm downstream from the nozzle orifice. Fluorescence excitation spectra obtained under these experimental conditions have demonstrated a rotational temperature of < 1.5 K for the entrained CS_2 molecules. While not measured, the corresponding vibrational temperature should be substantially higher owing to the inability of the helium carrier gas to quench vibrational degrees of freedom efficiently.[130]

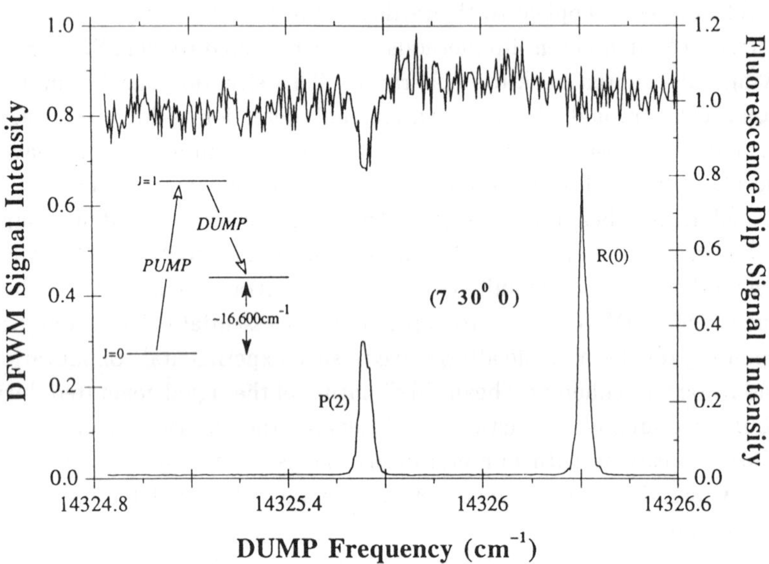

Fig. 12. Example of SEP–DFWM spectroscopy under supersonic expansion conditions. Fluorescence-dip (upper trace) and DFWM (lower trace) signals are presented for regions of the CS_2 potential energy surface corresponding to $\sim 16\,600$ cm^{-1} (~ 2 eV) of internal excitation. Two DUMP resonances are observed and nominally assigned to the $P(2)$ and $R(0)$ rotational lines for the $(7\,30^0\,0)$ vibrational band. These measurements utilized a supersonic free-jet expansion of 5% CS_2 in helium with an interaction region located roughly 10 cm downstream from the orifice of the pulsed nozzle source.

In order to acquire the data displayed in Fig. 12, the PUMP laser frequency was tuned to coincide with the $R(0)$ rovibronic transition in the $(0\,0\,0)_{K=0} - (0\,0^0\,0)$ 10V origin band of the CS_2 $V\text{–}\tilde{X}$ system. A definitive assignment for this spectral feature was made possible by the substantial reduction in rotational and vibrational congestion that accompanies the process of supersonic expansion. The upper trace represents the intensity of undispersed fluorescence emerging from the vibrationless $J = 1$ rotational level that has been selectively prepared in the electronically-excited intermediate state while the lower curve depicts the amplitude of the corresponding DFWM signal. The baseline fluctuations illustrated for each of these data sets are typical for SEP measurements performed under molecular beam conditions.

When the frequency of the DUMP laser is scanned in the vicinity of 698 nm, two strong resonances that can be attributed to the $P(2)$ and $R(0)$ rotational lines for a single vibronic band are observed. While a vibrational assignment for the terminal levels of these DUMP transitions is far from definitive, extrapolation of results derived from previous spectroscopic studies[136] suggests the normal mode designation $(7\,30^0\,0)$. This state is located some $16\,600$ cm^{-1} above the vibrationless origin of the ground electronic potential surface and represents a degree of internal excitation that corresponds to roughly 50% of the CS_2 bond dissociation energy.[132,133]

For the investigation of CS_2 molecules entrained in a supersonic expansion, it is quite evident from Fig. 12 that the DFWM detection scheme provides significant advantages over the conventional fluorescence-dip methodology. The SEP–DFWM spectrum, acquired with a DUMP energy of ≤ 100 μJ/pulse, exhibits a peak-to-noise ratio in excess of 2000:1. In contrast, the simultaneously recorded fluorescence-dip data show a peak-to-noise ratio that barely approaches 3:1. While higher dump energies might be expected to improve the quality of the fluorescence results, the substantial magnitude of the observed depletion feature (viz., $\sim 15\%$) and the saturation broadening already apparent on the $P(2)$ transition argue against this possibility. In short, the background of superfluous signal inherent to measurements based upon a competition between spontaneous and stimulated emission effectively limits the ability to discriminate DUMP resonances from baseline fluctuations. Incorporation of null-detection techniques into the experimental apparatus might enhance the obtained signal-to-noise ratios by as much as a factor of ten. However, this still leaves a considerable disparity between capabilities afforded by the fluorescence-dip and four-wave mixing schemes.

When the relative intensities for the rotational lines of Fig. 12 are compared, there appears to be a discrepancy between the results derived by DFWM detection and those obtained through fluorescence-depletion measurements, with the latter barely indicating the presence of an R(0) resonance. This reproducible effect is most pronounced for studies that involve low rotational quantum numbers and can be correlated with the nature of the PUMP excitation process. Figure 13 displays SEP–DFWM spectra recorded through the use of two different PUMP transitions [viz., $P(2)$ and $R(0)$ in the CS_2 10V origin band] that prepare the same $J = 1$ rotational level in the electronically-excited intermediate state. The experimental data, denoted by solid black dots, clearly show a complete reversal in the amplitudes of DUMP resonances as the PUMP transition is switched between $P(2)$ and $R(0)$.

The intensity variations in Figs. 12 and 13 stem from the polarization specificity of the cross-population detection scheme utilized for the present implementation of DFWM. In essence, the linearly polarized PUMP radiation leads to the creation of an electronically-excited population that is nonuniformly distributed among the degenerate magnetic sublevels of the $J = 1$ intermediate state.[111] As shown by the inserts of Fig. 13, the nature of this spatial anisotropy depends upon the spectroscopic details of the PUMP transition. Incorporation of these effects into the conventional description of four-wave mixing interactions [viz., $\rho_{\alpha\alpha}^{(0)}$ factors in Eq. (16)] permits the calculation of relative signal strengths that are in close agreement with experimental measurements (cf. solid curve in Fig. 13). The polarization characteristics exhibited by the SEP–DFWM technique are analogous to those encountered in conventional polarization spectroscopy.[14,32] As such, they can be exploited as a means of identifying rotational quantum numbers[167–169] for individual DUMP resonances, a feature of some utility for unravelling the spectral congestion that would accompany a high density of ground state vibrational levels. The pronounced dependence of four-wave mixing signals on the degree of alignment and/or orientation exhibited by the target species also provides a viable method for experimentally distinguishing the contributions of individual transient gratings to the overall DFWM process (cf. Sec. 2.2.2).

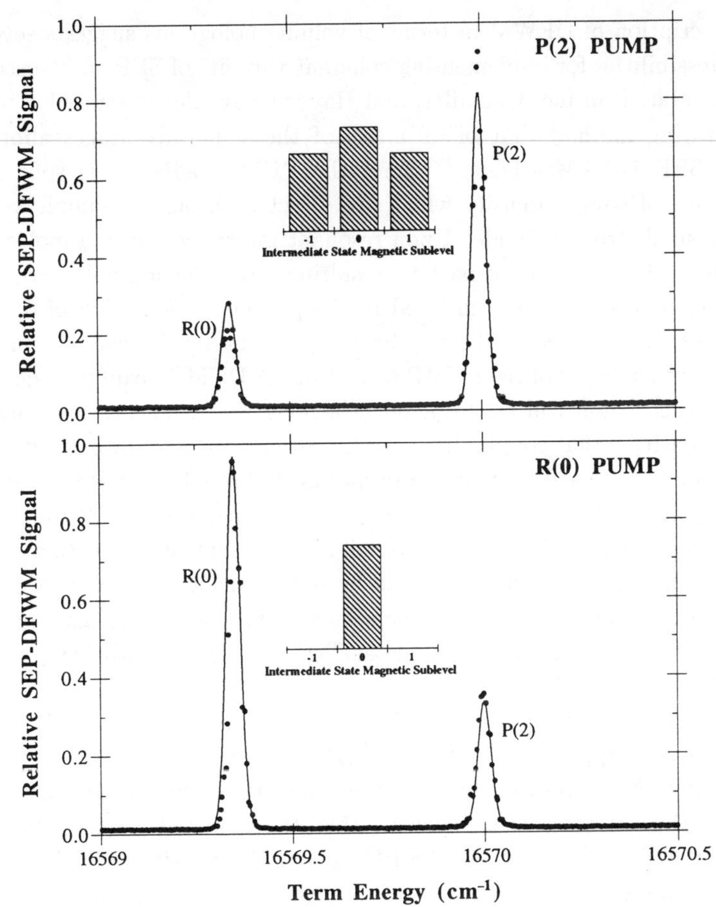

Fig. 13. Polarization dependence of SEP–DFWM spectroscopy. Supersonic expansion conditions were used to record two CS_2 data sets, each corresponding to DUMP resonances that originate from the same $J = 1$ level in the electronically-excited intermediate state and terminate on the same rovibronic levels of the vibrationally-excited ground state. Solid dots in the upper panel depict experimental results obtained through use of a $P(2)$ PUMP transition while those in the lower panel represent identical measurements performed with an $R(0)$ PUMP transition. The obvious variations in line intensities can be attributed to different magnetic sublevel distributions created in the intermediate state by the linearly-polarized PUMP radiation (cf. inserts). The solid curves are spectral simulations resulting from incorporation of these anisotropic population effects into the expression for DFWM signal intensities.

5. Related Techniques

The description of DFWM in terms of volume holograms suggests several other possibilities for implementing coherent variants of SEP spectroscopy. In particular, Buntine, Chandler, and Hayden have demonstrated a transient grating method that offers many of the same advantages afforded by the SEP–DFWM scheme.[23] Here, the PUMP radiation is split into two beams of equal intensity which are directed through a sample vessel under a small crossing angle. Upon resonant interaction with a molecular transition, the interference pattern resulting from the mutual overlap of these coherent sources is translated into a periodic modulation of optical properties. More specifically, population gratings are formed within the initial and final levels of the PUMP transition. A DUMP beam propagating through this region will be diffracted whenever its frequency, ω, coincides with a transition that couples to *either* the ground state or the excited state populations that have been spatially modified through PUMP excitation. Thus, by fixing the PUMP laser on a known spectroscopic feature and monitoring the magnitude of scattered DUMP light as a function of ω, a background-free SEP spectrum can be obtained for the molecule under investigation. Hayden and co-workers have appropriately designated this resonant four-wave mixing technique as $\underline{T}$wo-$\underline{C}$olor $\underline{L}$aser-$\underline{I}$nduced $\underline{G}$rating $\underline{S}$pectroscopy or TC-LIGS.

Figure 14 presents an example of the initial results obtained by Buntine *et al.* through application of the TC–LIGS scheme to the $B^3\Pi_{0u}^+$–$X^1\Sigma_g^+$ system of I_2.[23] As previously mentioned, diffractive signals are obtained for DUMP resonances that involve either the depleted ground state or the populated excited state of the PUMP transition, with only the latter corresponding to spectral signatures for conventional SEP. Consequently, in spite of the large number of intense rovibronic absorption features traversed by the DUMP laser frequency, only those levels of I_2 that have been labelled through PUMP interaction are observed. This true double-resonance capability provides a distinct advantage over the fluorescence-dip and SEP–DFWM techniques, both of which respond to any DUMP-induced transfer of population between the ground and excited electronic states. While only moderate peak-to-noise ratios were achieved in the case of I_2, Butenhoff and Rohlfing have extended and refined the TC–LIGS methodology so as to probe transient species entrained in a supersonic free-jet expansion.[24] Figure 15 illustrates their results for the SiC_2 system, a reactive molecule

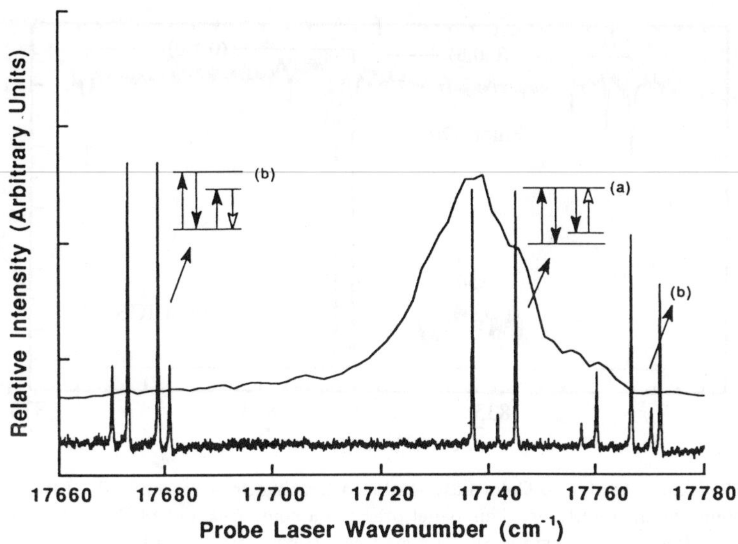

Fig. 14. Examples of Two-Color Laser-Induced Grating Spectroscopy (TC–LIGS) under bulk gas conditions. This panel compares the laser-induced grating (bottom trace) and dispersed emission (top trace) spectra recorded for a bulk-gas sample of I_2. The pump beam was provided by the second harmonic of an injection-seeded Nd:YAG laser which coincides in frequency with specific rovibronic features contained within the (32–0) and (34–0) bands of the I_2 $B^3\Pi_{0u}^+ - X^1\Sigma_g^+$ system. As demonstrated by the inserted energy level diagrams, TC–LIGS detects resonances that are due to both (a) stimulated emission from the populated intermediate state (i.e., terminating on $v = 5$ rovibronic levels in the $X^1\Sigma_g^+$ potential) and (b) absorption from the depleted ground state [i.e., involving the (18–0), (19–0), and (20–0) bands of the B–X transition]. The true double resonance capabilities of this technique are clearly illustrated; although the DUMP frequency scans through many I_2 transitions, TC–LIGS signals are observed only from levels labelled through interaction with the PUMP radiation. (Courtesy of C. C. Hayden)

that was prepared *in situ* with state-specific number densities of roughly 10^{12}cm^{-3}.

To some extent, the advantages of TC–LIGS are offset by the more complicated nature of the phase-matching constraints imposed upon this detection scheme. As shown in Sec. 2.2.2, optimum diffraction efficiency demands that the incoming DUMP beam and the emerging signal wave satisfy the Bragg scattering condition associated with the PUMP-induced transient grating [cf. Eq. (24)]. For a given TC–LIGS experimental configuration, exact phase-matching (or first-order Bragg scattering) can be obtained for only one set of PUMP and DUMP frequencies with other

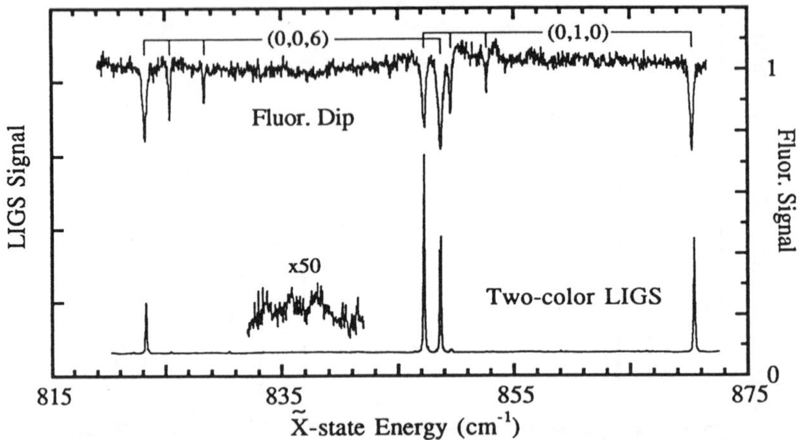

Fig. 15. Examples of Two-Color Laser-Induced Grating Spectroscopy (TC–LIGS) under supersonic beam conditions. This panel presents a comparison of SEP spectra recorded for the $\tilde{A}^1B_2$–$\tilde{X}^1A_2$ system if SiC_2 under free-jet expansion conditions. The top trace represents the fluorescence-dip signal while the bottom curve denotes the corresponding laser-induced grating results [i.e., all transitions correspond to energy level diagram (a) in Fig. 14]. Both data sets were obtained with the PUMP frequency tuned to the $^rR_2(2)_{e,0}$ rovibronic transition. Ground state vibrational assignments are shown with rotational lines terminating on the 2_2, 3_2, 4_2, and 4_4 asymmetry doublets, in order of increasing energy. The expanded region illustrates the magnitude of baseline fluctuations in the TC–LIGS spectrum. (Courtesy of E. A. Rohlfing)

combinations resulting in a reduction of maximum signal amplitude due to a finite wavevector mismatch. In contrast, the counterpropagating DFWM geometry of Fig. 1 satisfies phase-matching criteria for all incident wavelengths and for all crossing angles, θ.

A judicious choice of experimental configuration can significantly enhance the range of wavelengths over which the two-color LIGS scheme remains approximately phase-matched. In particular, Trebino and Siegman have demonstrated that the widest bandwidths for Bragg scattering are obtained for geometries in which the beams of varying frequency nearly copropagate.[170] By assuming that the signal wave emerges in a direction which tends to minimize any overall phase mismatch, Rohlfing and co-workers[24,171] have found that the bandwidth for efficient TC–LIGS signal generation scales as roughly $\{\ell\sin^2(\theta/2)\}^{-1}$, where ℓ and θ denote the sample interaction length and PUMP crossing angle, respectively. This increase in phase-matching bandwidth with decreasing ℓ is consistent with the changes in diffractive scattering that are expected to accompany the contraction of a thick transient grating.[89] The limited sample length afforded by a molecular beam environment would thus appear to be particularly advantageous for the TC–LIGS technique. Under such conditions, small values of θ and ℓ can be maintained without incurring the deleterious effects (viz., nonresonant background signals) associated with overlap of the PUMP and DUMP beams on the window substrates of a bulk-gas sample vessel.

While a definitive comparison remains to be performed, at this juncture it appears that the SEP–DFWM and TC–LIGS techniques are quite complementary in nature. Each scheme has its own unique capabilities that afford a background-free means for the detection of DUMP resonances with high overall sensitivity (viz., $\leq 10^{12}$ molecules per quantum state per cm^{-3}). SEP–DFWM spectroscopy seems especially well-suited for the investigation of ground state molecules in regimes of extreme vibrational excitation, where the self-phase-matched nature and Doppler-free resolution of this technique can prove to be of significant advantage. TC–LIGS, by virtue of its flexibility and true double-resonance response, provides a powerful tool for probing vibrational states that might be populated under ambient conditions (i.e., either thermally or as a result of chemical/photochemical processes utilized for the preparation of target species).

6. Conclusions and Future Directions

This chapter has demonstrated the feasibility and utility of exploiting resonant four-wave mixing spectroscopy as a state-specific probe of vibrationally-excited polyatomic molecules. In contrast to the dual-beam, null-detection geometries required for canonical fluorescence depletion measurements, SEP studies can now be accomplished through use of a relatively simple, "single-beam" experimental configuration that combines an essentially background-free response with laser-limited spectral resolution. The coherent, highly-collimated signal wave generated during such nonlinear interactions also promises to enable the spectroscopic investigation of species entrained in luminous and/or hostile environments (e.g., plasma discharges and photolysis sources), conditions that are not conducive to the application of more conventional techniques based upon the detection of either fluorescence intensity or ionization current.[49]

By eliminating the need to observe DUMP-induced modifications in the spontaneous emission emerging from an intermediate state, the "absorption-based" SEP–DFWM scheme should be applicable to a variety of molecular systems that have escaped previous study owing to either low fluorescence quantum yields or short radiative lifetimes. This includes a large and important class of polyatomic species which support electronically-excited manifolds that are subject to pronounced nonradiative decay via predissociation or other dissipative processes. The nearly-instantaneous response afforded by the four-wave mixing interaction also suggests extension of SEP-like measurements to isotropic dense media, such as high-pressure gases or liquids, where the influence of intermolecular forces upon vibrational dynamics could systematically be explored.

The major limitations for SEP–DFWM spectroscopy stem from its unusual dependence upon both the intensity of the DUMP laser and the oscillator strength of the target molecules. In the pre-saturation regime, where the conventional susceptibility theory of nonlinear optics is approximately valid, the magnitude of the signal field generated in a four-wave mixing interaction is expected to scale roughly as the cube of the incident light intensity. Consequently, even small pulse-to-pulse power fluctuations in the DUMP source will translate into large deviations in observed signal amplitude, thereby imposing practical limitations on the sensitivity and resolution that can be achieved in a particular measurement. The use of active laser stabilization schemes,[102,172] in conjunction with standard signal averaging techniques, should minimize these detrimental effects. The influence of DUMP instabilities can also be suppressed by operating under near-saturation conditions; however, this is usually accompanied by an unacceptable broadening of spectroscopic transitions.

The amplitude of the signal beam emerging from a resonant DFWM interaction is expected to be a strong function of the transition dipole moment, $|\mu|$, for the target molecules, with the precise form of this dependence contingent upon the incident light intensity. More specifically, the saturable absorber model introduced in Sec. 2.2.3 predicts a response that ranges from $|\mu|^8$ to $|\mu|^2$, where the former corresponds to a low-power limit and the latter pertains to situations of extreme optical saturation. This unusual behavior will have a profound influence on the appearance of SEP–DFWM spectra recorded under a given set of experimental conditions (viz., with a fixed DUMP energy and detector amplification). In particular,

DUMP resonances having sufficient oscillator strength so as to approach saturation will result in intense spectral features while weaker transitions, being far removed from the saturation regime, may barely be visible.

By repeating individual measurements with increasing amounts of DUMP energy, the four-wave mixing scheme should theoretically be able to uncover all SEP resonances that are permitted by spectroscopic selection rules. For rarified media, where the effects of collisional processes need not be considered, the intensity pattern exhibited by the fine structure within a given vibronic band can be predicted from a straightforward calculation involving rotational linestrengths and the distribution of molecular population prepared by the PUMP among the degenerate magnetic sublevels of the electronically-excited intermediate state (cf., Fig. 13). However, the direct comparison of peak heights for SEP–DFWM features that terminate on different ground state vibrational levels is not a trivial matter. The fluorescence-dip technique, with its quadratic dependence on transition moment over a broad range of incident DUMP energies, provides a distinct advantage for the quantitative interpretation of such intensity information. As described in a later chapter of this monograph, the ion-dip detection scheme offers additional benefits for the extraction of meaningful relative intensity data. This unique capability stems from use of the same laser source to induce both stimulated emission and resonant ionization processes.

While sensitivity and resolution play pivotal roles in determining the ultimate utility of a spectroscopic tool, the simplicity and flexibility associated with the implementation of a technique are also of primary importance. At present, SEP–DFWM spectroscopy and its transient grating analogs have been applied to only a handful of molecular systems, a situation much different from that associated with the ubiquitous fluorescence-dip methodology. For all of these initial studies, coherent light scattering techniques have demonstrated detection capabilities and resolving powers that rival or exceed those attainable through more conventional means. The limitations imposed upon these methods, both in terms of practical considerations (e.g., species applicability) and theoretical interpretations (e.g., intensity measurements), still remain to be fully characterized. However, given the relative ease with which they can be incorporated into existing experimental apparatus, there is no doubt that the future will witness a rapid growth in the application of resonant four-wave mixing probes to problems of molecular spectroscopy and dynamics.

Acknowledgments

The author wishes to thank T. A. W. Wasserman, Q. Zhang, S. A. Kandel, and A. A. Arias for their assistance in the collection and interpretation of results presented in this manuscript. He is especially indebted to Drs. R. L. Farrow, D. J. Rakestraw, E. A. Rohlfing, and C. C. Hayden of Sandia National Laboratories for providing the material contained in Figs. 6 and 14. Acknowledgement is made to the donors of the Petroleum Research Fund, administered by The American Chemical Society, The Camille and Henry Dreyfus Foundation, The National Science Foundation, and The David and Lucile Packard Foundation for support of this research effort.

References

1. M. Born and R. Oppenheimer, *Ann. Phys. (Leipzig)* **84**, 457 (1927).
2. M. Born and K. Huang, *Dynamical Theory of Crystal Lattices* (Oxford University Press, New York, 1956).
3. A. D. Buckingham, *Adv. Chem. Phys.* **12**, 107 (1967).
4. J. T. Yardley, *Introduction to Molecular Energy Transfer* (Academic Press, New York, 1980).
5. R. D. Levine and R. B. Bernstein, *Molecular Reaction Dynamics and Chemical Reactivity* (Oxford University Press, Oxford, 1987).
6. J. N. Murrell, S. Carter, S. C. Farantos, P. Huxley, and A. J. C. Varandas, *Molecular Potential Energy Functions* (John Wiley and Sons, New York, 1984).
7. C. Kittrell, E. Abramson, J. L. Kinsey, S. A. McDonald, D. E. Reisner, R. W. Field, and D. H. Katayama, *J. Chem. Phys.* **75**, 2056 (1981).
8. C. E. Hamilton, J. L. Kinsey, and R. W. Field, *Ann. Rev. Phys. Chem.* **37**, 493 (1986).
9. D. E. Reisner, P. H. Vaccaro, C. Kittrell, R. W. Field, and J. L. Kinsey, *J. Chem. Phys.* **77**, 573 (1982).
10. E. Abramson, R. W. Field, D. Imre, K. K. Innes, and J. L. Kinsey, *J. Chem. Phys.* **80**, 2298 (1984).
11. P. H. Vaccaro, R. L. Redington, J. Schmidt, J. L. Kinsey, and R. W. Field, *J. Chem. Phys.* **82**, 5755 (1985).
12. F. Temps, S. Halle, P. H. Vaccaro, R. W. Field, and J. L. Kinsey, *J. Chem. Phys.* **87**, 1895 (1987).
13. F. Temps, S. Halle, P. H. Vaccaro, R. W. Field, and J. L. Kinsey, *J. Chem. Soc., Farad. Trans. 2* **84**, 1457 (1988).
14. P. H. Vaccaro, *Ph. D. Thesis*, Massachusetts Institute of Technology, 1986.
15. D. Frye, H. T. Liou, and H.-L. Dai, *Chem. Phys. Lett.* **133**, 249 (1987).
16. P. H. Vaccaro, F. Temps, S. Halle, J. L. Kinsey, and R. W. Field, *J. Chem. Phys.* **88**, 4819 (1988).

17. D. J. Moll, G. R. J. Parker, and A. Kuppermann, *J. Chem. Phys.* **80**, 4800 (1984).

18. D. E. Cooper, C. M. Klimcak, and J. E. Wessel, *Phys. Rev. Lett.* **46**, 324 (1981).

19. D. E. Cooper and J. E. Wessel, *J. Chem. Phys.* **76**, 2155 (1982).

20. T. Suzuki, N. Mikami, and M. Ito, *J. Phys. Chem.* **90**, 6431 (1986).

21. Q. Zhang, S. A. Kandel, T. A. W. Wasserman, and P. H. Vaccaro, *J. Chem. Phys.* **96**, 1640 (1992).

22. S. A. Kandel, T. A. W. Wasserman, Q. Zhang, H. Wang, A. A. Arias, and P. H. Vaccaro, in *Proceedings of SPIE Conference on Laser Techniques for State-Selected and State-to-State Chemistry*, Los Angeles, 1993, ed. C.-Y. Ng (SPIE, Bellingham, 1993), p. 126.

23. M. A. Buntine, D. W. Chandler, and C. C. Hayden, *J. Chem. Phys.* **97**, 707 (1992).

24. T. J. Butenhoff and E. A. Rohlfing, *J. Chem. Phys.* **97**, 1595 (1992).

25. R. L. Abrams, J. F. Lam, R. C. Lind, D. G. Steel, and P. F. Liao, "Phase Conjugation and High-Resolution Spectroscopy by Resonant Degenerate Four-Wave Mixing," in *Optical Phase Conjugation*, ed. R. A. Fisher (Academic Press, San Diego, 1983), p. 211.

26. Y. R. Shen, *The Principles of Nonlinear Optics* (John Wiley and Sons, New York, 1984).

27. W. D. Lawrance and A. E. W. Knight, *J. Chem. Phys.* **77**, 570 (1982).

28. W. D. Lawrance and A. E. W. Knight, *J. Phys. Chem.* **87**, 389 (1983).

29. W. D. Lawrance and A. E. W. Knight, *J. Chem. Phys.* **79**, 6030 (1983).

30. F. Temps, S. Halle, P. H. Vaccaro, R. W. Field, and J. L. Kinsey, *J. Chem. Phys.* **91**, 1008 (1989).

31. Y. Chen, L. Hunziker, P. Ludowise, and M. Morgen, *J. Chem. Phys.* **97**, 2149 (1992).

32. R. E. Teets, F. V. Kowalski, W. T. Hill, N. Carlson, and T. W. Hänsch, in *Proceedings of SPIE Conference on Advances in Laser Spectroscopy I*, San Diego, CA, 1977, ed. A. H. Zewail (SPIE, Bellingham, 1977), p. 88.

33. M. D. Levenson, *Introduction to Nonlinear Laser Spectroscopy* (Academic Press, New York, 1982).

34. J.-L. Oudar and Y. R. Shen, *Phys. Rev.* **A22**, 1141 (1980).

35. R. M. Hochstrasser and H. P. Trommsdorff, *Acc. Chem. Res.* **16**, 376 (1983).

36. A. Yariv, *IEEE J. Quantum Electron.* **14**, 650 (1978).

37. R. C. Lind, D. G. Steel, and G. J. Dunning, *Opt. Eng.* **21**, 190 (1982).

38. T. R. O'Meara, *Opt. Eng.* **21**, 231 (1982).

39. J. O. White and A. Yariv, *Opt. Eng.* **21**, 224 (1982).

40. T. R. O'Meara, D. M. Pepper, and J. O. White, "Applications of Nonlinear Optical Phase Conjugation," in *Optical Phase Conjugation*, ed. R. A. Fisher (Academic Press, San Diego, 1983), p. 537.

41. R. W. Boyd and G. Grynberg, "Optical Phase Conjugation," in *Contemporary Nonlinear Optics*, eds. G. P. Agrawal and R. W. Boyd (Academic Press, Boston, 1992), p. 85.
42. J. O. Tocho, W. Sibbett, and D. J. Bradley, *Opt. Comm.* **34**, 122 (1980).
43. R. K. Jain, *Opt. Eng.* **21**, 199 (1982).
44. G. A. Kenny-Wallace and S. C. Wallace, *IEEE J. Quantum Electron.* **19**, 719 (1983).
45. D. Bloch, R. K. Raj, and M. Ducloy, *Opt. Comm.* **37**, 183 (1981).
46. J. F. Lam, *Opt. Eng.* **21**, 219 (1982).
47. D. A. Chen and W. G. Tong, *J. Anal. Atom. Spect.* **3**, 531 (1988).
48. R. L. Vander Wal, R. L. Farrow, and D. J. Rakestraw, in *Proceedings of Twenty-Fourth (International) Symposium on Combustion*, 1992, (Combustion Institute, Pittsburgh, 1992), p. 1653.
49. R. L. Farrow and D. J. Rakestraw, *Science* **257**, 1894 (1992).
50. R. W. Hellwarth, *J. Opt. Soc. Am.* **67**, 1 (1977).
51. A. Yariv and D. M. Pepper, *Opt. Lett.* **1**, 16 (1977).
52. D. M. Pepper, *Opt. Eng.* **21**, 156 (1982).
53. D. M. Pepper and A. Yariv, "Optical Phase Conjugation using Three-Wave and Four-Wave Mixing via Elastic Photon Scattering in Transparent Media," in *Optical Phase Conjugation*, ed. R. A. Fisher (Academic Press, San Diego, 1983), p. 23.
54. A. Yariv and R. A. Fisher, "Introduction to Optical Phase Conjugation," in *Optical Phase Conjugation*, ed. R. A. Fisher (Academic Press, San Diego, 1983), p. 1.
55. T. R. O'Meara and A. Yariv, *Opt. Eng.* **21**, 237 (1982).
56. D. C. Hanna, M. A. Yuratich, and D. Cotter, *Nonlinear Optics of Free Atoms and Molecules* (Springer–Verlag, Berlin, 1979).
57. J. F. Reintjes, *Nonlinear Optical Parametric Processes in Liquids and Gases* (Academic Press, New York, 1984).
58. U. Gaubatz, P. Rudecki, M. Becker, S. Schiemann, M. Külz, and K. Bergmann, *Chem. Phys. Lett.* **149**, 463 (1988).
59. J. R. Kuklinski, U. Gaubatz, F. T. Hioe, and K. Bergmann, *Phys. Rev.* **A40**, 6741 (1989).
60. U. Gaubatz, P. Rudecki, S. Schiemann, and K. Bergmann, *J. Chem. Phys.* **92**, 5363 (1990).
61. G.-Z. He, A. Kuhn, S. Schiemann, and K. Bergmann, *J. Opt. Soc. Am.* **B7**, 1960 (1990).
62. P. N. Butcher and D. Cotter, *The Elements of Nonlinear Optics* (Cambridge University Press, Cambridge, 1990).
63. R. W. Boyd, *Nonlinear Optics* (Academic Press, Boston, 1992).
64. J. D. Jackson, *Classical Electrodynamics* (John Wiley and Sons, New York, 1975).
65. A. Yariv, *Introduction to Optical Electronics* (Holt, Rinehart, and Winston, New York, 1976).

66. R. W. Hellwarth, *Prog. Quant. Electr.* **5**, 1 (1977).

67. C. M. Lovejoy and D. J. Nesbitt, *J. Chem. Phys.* **86**, 3151 (1987).

68. S. W. Sharpe, R. Sheeks, C. Wittig, and R. A. Beaudet, *Chem. Phys. Lett.* **151**, 267 (1988).

69. I. Aben, W. Ubachs, P. Levelt, G. van der Zwan, and W. Hogervorst, *Phys. Rev.* **A44**, 5881 (1991).

70. I. Aben, W. Ubachs, G. van der Zwan, and W. Hogervorst, *Mol. Phys.* **76**, 591 (1992).

71. S. A. J. Druet and J.-P. Taran, *Prog. Quant. Electr.* **7**, 1 (1981).

72. D. G. Steel, R. C. Lind, J. F. Lam, and C. R. Giuliano, *Appl. Phys. Lett.* **35**, 376 (1979).

73. J. F. Lam, D. G. Steel, R. A. McFarlane, and R. C. Lind, *Appl. Phys. Lett.* **38**, 977 (1981).

74. J. F. Lam and R. L. Abrams, *Phys. Rev.* **A26**, 1539 (1982).

75. M. Ducloy and D. Bloch, *Phys. Rev.* **A30**, 3107 (1984).

76. U. Fano, *Rev. Mod. Phys.* **29**, 74 (1957).

77. K. Blum, *Density Matrix Theory and Applications* (Plenum Press, New York, 1981).

78. T. K. Yee, T. K. Gustafson, S. A. J. Druet, and J.-P. E. Taran, *Opt. Comm.* **23**, 1 (1977).

79. T. K. Yee and T. K. Gustafson, *Phys. Rev.* **A18**, 1597 (1978).

80. Y. Prior, *IEEE J. Quantum Electron.* **20**, 37 (1984).

81. S. Stenholm, *Foundations of Laser Spectroscopy* (John Wiley and Sons, New York, 1984).

82. P. R. Berman, *Phys. Rev.* **A5**, 927 (1972).

83. M. Weissbluth, *Atoms and Molecules* (Academic Press, San Diego, 1978).

84. R. C. Hilborn, *Am. J. Phys.* **50**, 982 (1982).

85. D. Cotter, D. C. Hanna, and R. Wyatt, *Opt. Comm.* **16**, 256 (1976).

86. D. S. Bethune, J. R. Lankard, and P. P. Sorokin, *Opt. Lett.* **4**, 103 (1979).

87. J. E. Bjorkholm and P. F. Liao, *Phys. Rev. Lett.* **33**, 128 (1974).

88. G. S. Agarwal, *Adv. At. Mol. Opt. Phys.* **29**, 113 (1992).

89. H. J. Eichler, P. Günter, and D. W. Pohl, *Laser-Induced Dynamic Gratings* (Springer–Verlag, Berlin, 1986).

90. R. J. Collier, C. B. Burkhardt, and L. H. Lin, *Optical Holography* (Academic Press, New York, 1971).

91. J. T. Fourkas and M. D. Fayer, *Acc. Chem. Res.* **25**, 227 (1992).

92. J. T. Fourkas, T. R. Brewer, H. Kim, and M. D. Fayer, *J. Chem. Phys.* **95**, 5775 (1991).

93. J. T. Fourkas, R. Trebino, and M. D. Fayer, *J. Chem. Phys.* **97**, 78 (1992).

94. J. T. Fourkas, R. Trebino, and M. D. Fayer, *J. Chem. Phys.* **97**, 69 (1992).

95. D. B. Brayton, *Appl. Opt.* **13**, 2346 (1974).

96. A. E. Siegman, *J. Opt. Soc. Am.* **67**, 545 (1977).

97. L. Allen and J. H. Eberly, *Optical Resonance and Two-Level Atoms* (John Wiley and Sons, New York, 1975).

98. R. L. Abrams and R. C. Lind, *Opt. Lett.* **2**, 94 (1978).

99. R. L. Abrams and R. C. Lind, *Opt. Lett.* **3**, 205 (1978).

100. M. S. Feld, M. M. Burns, T. U. Khl, P. G. Pappas, and D. E. Murnick, *Opt. Lett.* **5**, 79 (1980).

101. P. G. Pappas, M. M. Burns, D. D. Hinshelwood, M. S. Feld, and D. E. Murnick, *Phys. Rev.* **A21**, 1955 (1980).

102. R. L. Farrow, T. Dreier, and D. J. Rakestraw, *J. Opt. Soc. Am.* **B9**, 1770 (1992).

103. M. S. Brown, L. A. Rahn, and T. Dreier, *Opt. Lett.* **17**, 76 (1992).

104. D. G. Steel, R. C. Lind, and J. F. Lam, *Phys. Rev.* **A23**, 2513 (1981).

105. G. J. Dunning and D. G. Steel, *IEEE J. Quantum Electron.* **18**, 3 (1982).

106. G. P. Agrawal, A. Van Lerberghe, P. Aubourg, and J. L. Boulnois, *Opt. Lett.* **7**, 540 (1982).

107. W. P. Brown, *J. Opt. Soc. Am.* **73**, 629 (1983).

108. T.-Y. Fu and M. Sargent, *Opt. Lett.* **4**, 366 (1979).

109. J. Nilsen and A. Yariv, *J. Opt. Soc. Am.* **71**, 180 (1981).

110. S. M. Wandzura, *Opt. Lett.* **4**, 208 (1979).

111. R. N. Zare, *Angular Momentum: Understanding Spatial Aspects in Chemistry and Physics* (John Wiley and Sons, New York, 1988).

112. M. Ducloy and D. Bloch, *J. Physique* **42**, 711 (1981).

113. W. Demtröder, *Laser Spectroscopy* (Springer–Verlag, Berlin, 1981).

114. D. Bloch and M. Ducloy, *J. Opt. Soc. Am.* **73**, 635 (1983).

115. M. Pinard, B. Kleinmann, G. Grynberg, D. Bloch, and M. Ducloy, *J. Physique* **46**, 149 (1985).

116. H. Bervas, B. Attal-Trétout, S. Le Boiteux, and J. P. Taran, *J. Phys. B: At. Mol. Opt. Phys.* **25**, 949 (1992).

117. L. M. Humphrey, J. P. Gordon, and P. F. Liao, *Opt. Lett.* **5**, 56 (1980).

118. D. Bloch, R. K. Raj, K. S. Peng, and M. Ducloy, *Phys. Rev. Lett.* **49**, 719 (1983).

119. P. F. Liao, D. M. Bloom, and N. P. Economou, *Appl. Phys. Lett.* **32**, 813 (1978).

120. J. P. Woerdman and M. F. H. Schuurmans, *Opt. Lett.* **6**, 239 (1981).

121. C. Wieman and T. W. Hänsch, *Phys. Rev. Lett.* **36**, 1170 (1976).

122. D. J. Rakestraw (personal communication).

123. J. Cooper, A. Charlton, D. R. Meacher, P. Ewart, and G. Alber, *Phys. Rev.* **A40**, 5705 (1989).

124. M. Kaczmarek, D. R. Meacher, and P. Ewart, *J. Mod. Opt.* **37**, 1561 (1990).

125. G. Meijer and D. W. Chandler, *Chem. Phys. Lett.* **192**, 1 (1992).

126. Ch. Jungen, D. N. Malm, and A. J. Merer, *Can. J. Phys.* **51**, 1471 (1973).

127. Q. Zhang, T. A. W. Wasserman, S. A. Kandel, and P. H. Vaccaro, in preparation.

128. B. Kleman, *Can. J. Phys.* **41**, 2034 (1963).

129. H. Kasahara, N. Mikami, M. Ito, S. Iwata, and I. Suzuki, *Chem. Phys.* **86**, 173 (1984).

130. N. Ochi, H. Watanabe, S. Tsuchiya, and S. Koda, *Chem. Phys.* **113**, 271 (1987).

131. J. W. Rabalais, J. M. McDonald, V. Scherr, and S. P. McGlynn, *Chem. Rev.* **71**, 73 (1971).

132. H. Okabe, *J. Chem. Phys.* **56**, 4381 (1972).

133. H. Okabe, *Photochemistry of Small Molecules* (John Wiley and Sons, New York, 1978).

134. D. F. Smith Jr. and J. Overend, *J. Chem. Phys.* **54**, 3632 (1971).

135. I. Suzuki, *Bull. Chem. Soc. Jpn* **48**, 1685 (1975).

136. P. F. Bernath, M. Dulick, R. W. Field, and J. L. Hardwick, *J. Mol. Spectrosc.* **86**, 275 (1981).

137. Q. Zhang and P. H. Vaccaro, in preparation.

138. B. R. Johnson and W. P. Reinhardt, *J. Chem. Phys.* **85**, 4538 (1986).

139. B. R. Johnson, *J. Chem. Phys.* **92**, 574 (1990).

140. F. G. Celii, J. P. Maier, and M. Ochsner, *J. Chem. Phys.* **85**, 6230 (1986).

141. T. Dreier and D. J. Rakestraw, *Opt. Lett.* **15**, 72 (1990).

142. D. J. Rakestraw, R. L. Farrow, and T. Dreier, *Opt. Lett.* **15**, 709 (1990).

143. T. Dreier and D. J. Rakestraw, *Appl. Phys.* **B50**, 479 (1990).

144. D. J. Rakestraw, T. Dreier, and L. R. Thorne, in *Proceedings of Twenty-Third (International) Symposium on Combustion*, 1991, (Combustion Institute, Pittsburgh, 1991), p. 1901.

145. D. R. Miller, "Free Jet Sources," in *Atomic and Molecular Beam Methods: Volume 1*, ed. G. Scoles (Oxford University Press, New York, 1988), p. 14.

146. R. E. Smalley, L. Wharton, and D. H. Levy, *Acc. Chem. Res.* **10**, 139 (1977).

147. A. R. Skinner and D. W. Chandler, *Am. J. Phys.* **48**, 8 (1980).

148. K. Yamanouchi, H. Yamada, and S. Tsuchiya, *Chem. Phys. Lett.* **132**, 361 (1986).

149. D. Frye, L. Lapierre, and H.-L. Dai, *J. Chem. Phys.* **89**, 2609 (1988).

150. F. J. Northrup and T. J. Sears, *Chem. Phys. Lett.* **159**, 421 (1989).

151. E. A. Rohlfing and J. E. M. Goldsmith, *J. Chem. Phys.* **90**, 6804 (1989).

152. D. Frye, P. Arias, and H.-L. Dai, *J. Chem. Phys.* **88**, 7240 (1988).

153. M. T. Berry, M. R. Brustein, M. I. Lester, C. Chakravarty, and D. C. Clary, *Chem. Phys. Lett.* **178**, 301 (1991).

154. M. T. Berry, R. A. Loomis, L. C. Giancarlo, and D. C. Clary, *J. Chem. Phys.* **96**, 7890 (1992).

155. A. Goto, M. Fujii, and M. Ito, *Chem. Phys. Lett.* **135**, 407 (1987).

156. T. Kakinuma, M. Fujii, and M. Ito, *Chem. Phys. Lett.* **140**, 427 (1987).

157. T. Suzuki, M. Hiroi, and M. Ito, *J. Chem. Phys.* **92**, 3774 (1988).

158. T. Ebata and M. Ito, *J. Chem. Phys.* **96**, 3224 (1992).

159. P. Huber-Wälchli, D. M. Guthals, and J. W. Nibler, *Chem. Phys. Lett.* **67**, 233 (1979).

160. M. D. Duncan, P. Oesterlin, and R. L. Byer, *Opt. Lett.* **6**, 90 (1981).

161. P. Esherick and A. Owyoung, "High Resolution Stimulated Raman Spectroscopy," in *Advances in Infrared and Raman Spectroscopy*, eds. R. J. H. Clark and R. E. Hester (Heyden, London, 1982), p. 130.

162. P. Esherick and A. Owyoung, *Chem. Phys. Lett.* **103**, 235 (1983).

163. G. V. Hartland, B. F. Henson, L. L. Connell, T. C. Corcoran, and P. M. Felker, *J. Chem. Phys.* **92**, 6877 (1988).

164. G. V. Hartland, B. F. Henson, V. A. Venturo, and P. M. Felker, *J. Chem. Phys.* **96**, 1164 (1992).

165. B. F. Henson, G. V. Hartland, V. A. Venturo, and P. M. Felker, *J. Chem. Phys.* **97**, 2189 (1992).

166. W. R. Gentry, "Low-Energy Pulsed Beam Sources," in *Atomic and Molecular Beam Methods: Volume 1*, ed. G. Scoles (Oxford University Press, New York, 1988), p. 54.

167. V. Stert and R. Fisher, *Appl. Phys.* **17**, 151 (1978).

168. N. W. Carlson, F. V. Kowalski, R. E. Teets, and A. L. Schawlow, *Opt. Comm.* **29**, 302 (1979).

169. M. Raab, G. Höning, R. Castell, and W. Demtröder, *Chem. Phys. Lett.* **66**, 307 (1979).

170. R. Trebino and A. E. Siegman, *Opt. Comm.* **56**, 297 (1985).

171. E. A. Rohlfing (personal communication).

172. R. L. Vander Wal, B. E. Holmes, J. B. Jeffries, P. M. Danehy, R. L. Farrow, and D. J. Rakestraw, *Chem. Phys. Lett.* **191**, 251 (1992).

CHAPTER 2

FEMTOSECOND TRANSIENT STIMULATED EMISSION PUMPING: THEORY AND EXPERIMENT

L. Hunziker, P. Ludowise, W. Price,
M. Morgen, M. Blackwell and Y. Chen*
*Department of Chemistry, University of California
Berkeley, CA 94720, USA*

Contents

---

*Alfred P. Sloan Research Fellow

1. Introduction

The last ten years has witnessed tremendous progresses in femtosecond transition state spectroscopy (FTS) by which chemical reactions are studied in real time.[1,2] The generic experimental scheme in the time resolved studies of isolated chemical reactions can be described as follows: the reaction is initiated at a well defined t_0 through the absorption of a femtosecond laser pulse. Its time evolution is then monitored by the absorption of a second laser pulse as a function of time delay between the two pulses. Pioneered by Zewail's group at Caltech,[1] these FTS experiments allow us for the first time to take 'snapshots' of the chemical events that occur while molecules are transforming from the reactants to the products. Recently, we have successfully developed a new variant of the femtosecond pump-probe technique — transient stimulated emission pumping (TSEP).[3] The key to TSEP is in probing the reaction dynamics through stimulated emissions to the ground electronic state. Owing to generations of work on molecular spectroscopy and *ab initio* quantum chemistry calculations, the ground state PES's of many small polyatomic molecules have been well characterized. Intuitively, the spectral profile of the transient stimulated emission, induced by a femtosecond laser pulse, is directly related to the projection of the dissociating wave function onto the ground state. The spectrum can, therefore, provide an 'instantaneous' image of the molecule undergoing dissociation.

In this paper, we will give a full account of the physical principle of the TSEP experiment. The general theoretical foundation of the TSEP will be laid out in Sec. 2. In Sec. 2.1, we show that the TSEP signal is directly related to the time- dependent nonlinear polarization, $\mathbf{P}^{NL}(t)$, of the molecular system. In Sec. 2.2, the $\mathbf{P}^{NL}(t)$ is evaluated explicitly through its correlation with the third order density matrix. It is also shown in Sec. 2.2 that, in the impulsive limit, the nonlinear polarization generated in the TSEP experiment is proportional to the instant projection of the dissociating wavepacket onto the ground state potential surface. In Sec. 2.3, we will discuss some of the physical implications of the TSEP process. In particular, comparisons between the femtosecond TSEP and the frequency domain resonance Raman experiment will be made. Section 3 will discuss our recent TSEP experiments on ozone visible photodissociation. By photodissociating the molecule with a Fourier-transform limited femtosecond PUMP pulse and probing the stimulated emission with a delayed white-light continuum DUMP pulse, we have obtained transient stimulated emission spectrum from the dissociating ozone molecule. The time evolution of the TSEP spectrum can be directly correlated to the motion of the nuclei when the molecule falls apart. Recurrence of up to about 600 fs has also been observed in the TSEP spectrum. This is tentatively attributed to the bifurcation of the wavepacket near the conical intersection between the two excited electronic states. Our result is in qualitative agreement with the recent computational studies by Stock *et al.*,[4] who investigated the ultrafast molecular dynamics of the ozone Chappuis band, and observed quasi-periodic oscillations in their simulated TSEP spectrum.

2. Theory

The energy level diagram for the femtosecond TSEP is shown in Fig. 1. In the experiment, a femtosecond pulse excites the molecule from the ground state to a dissociative surface (PUMP). The subsequent dynamics of the molecular system are then probed at various delays, Δt, by a femtosecond 'supercontinuum'[5] through stimulated emission (DUMP).

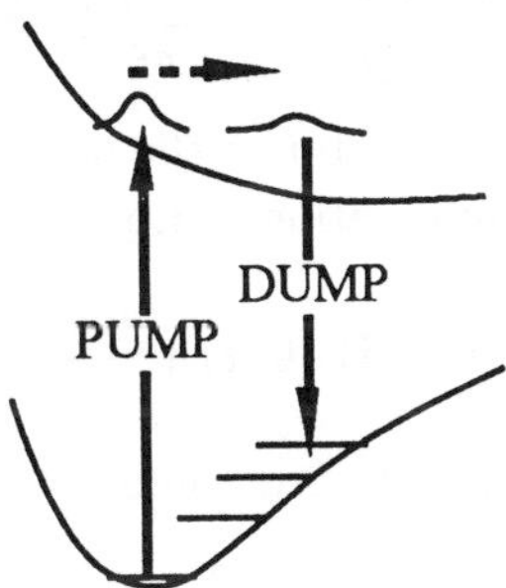

Fig. 1. Energy level diagram for the TSEP. In the experiment, the molecule is excited into a dissociative electronic state through a polarized fs pump pulse. The subsequent dynamics are probed by a DUMP pulse of different polarization at later time.

The TSEP is based on transient coherent polarization spectroscopy as a mean of observing resonant DUMP transitions.[6] In brief, when molecules are excited onto the repulsive PES by a linearly polarized PUMP pulse at wavelength λ_1 and later stimulated back to the ground PES by a linearly polarized DUMP pulse at λ_2 with its polarization at an angle of 45° with respect to the PUMP field, a component of the DUMP othorgonal to its incident polarization is produced. This component can then be detected with high sensitivity by a pair of crossed polarizers.

2.1. *Relationship between the Nonlinear Polarization and the TSEP Signal*

Let both the PUMP and the DUMP beams propagate in the y- direction, the input polarizer be aligned in the z-direction, and the analyzing polarizer in the x-direction. The TSEP signal, I_{TSEP}, is then proportional to the total number of DUMP photons passing through the analyzing polarizer. Assume that $E_d(y, t)$ is the amplitude of the DUMP field at the space-time point (y, t) and the origin $y = 0$ is at the position right before the DUMP pulse interacts with the sample; we can then write

$$I_{\text{TSEP}} \propto \int_{-\infty}^{\infty} |\mathbf{E}_d(L, t)\hat{\mathbf{x}}|^2 dt \tag{2.1}$$

where L is the optical path-length of the sample. In an optically thin sample, the propagation of the DUMP field in the frame moving with the pulse is governed by

$$\frac{\partial \mathbf{E}_d(y, t)}{\partial y} = \frac{i2\pi k}{n^2}\mathbf{P}^{\text{NL}}(t) \tag{2.2}$$

where k is the DUMP wave vector, n the index of refraction, and $\mathbf{P}^{\text{NL}}$ the nonlinear polarization.[6] Integration of Eq. (2.2) leads to

$$\mathbf{E}_d(L, t) = \mathbf{E}_d(0, t) + \frac{i2\pi kL}{n^2}\mathbf{P}^{\text{NL}}(t). \tag{2.3}$$

Inserting Eq. (2.3) into Eq. (2.1), we obtain

$$I_{\text{TSEP}} \propto \omega_d^2 \int_{-\infty}^{\infty} |\mathbf{P}_x^{\text{NL}}(t)|^2 dt. \tag{2.4}$$

The experimental configuration that monitors a signal proportional to the square of the nonlinear polarization is generally referred to as homodyne detection. Often, optical heterodyne detection (OHD) is also used in the coherent polarization experiments. In the latter case, the electromagnetic field impinged upon the detector is a superposition of the signal field and the DUMP field, i.e.,

$$\mathbf{E}_d(L, t)\hat{\mathbf{x}} = \frac{i2\pi kL}{n^2}\mathbf{P}_x^{\text{NL}}(t) + \beta\mathbf{E}_d(t),$$

where β is a proportionality constant. When the second term on the right hand side is much larger than the first term, the experimental observable can be approximated by

$$I_{\text{TSEP}} \propto i\omega_d \int_{-\infty}^{\infty} \mathbf{E}_d \mathbf{P}_x^{\text{NL}}(t)dt + \text{ c.c.} \qquad (2.5)$$

where c.c. stands for the complex conjugate. In deriving Eq. (2.5), we have neglected the smaller term proportional to P^2 and the delay-independent term proportional to the total energy of the DUMP pulse.

When a stable DUMP is available, OHD is often the favored detection scheme. Not only can one obtain the relative phase of the induced nonlinear polarization, but also the signal is amplified by the DUMP field. Equation (2.5) also bears many similarities to the expressions for the dynamic absorption line shape derived previously by other groups.[7-9] On the other hand, the 'zero-background' homodyne spectroscopy is preferred if the pulse-to-pulse fluctuation is large. Over the years, various other configurations have also been developed in the coherent polarization spectroscopic experiments.[10-22] Although the observables in these experiments appear in different forms, the dynamic information is contained exclusively in the nonlinear polarization $\mathbf{P}^{\text{NL}}(t)$. In the discussion below, we will therefore focus attention on the evaluation of $\mathbf{P}^{\text{NL}}(t)$ generated from a TSEP experiment.

2.2. *Correlation of the TSEP Nolinear Polarization to the Molecular Dynamics*

Let H^0 be the molecular Hamiltonian, and $H'(t)$ the external perturbation. Under the dipole approximation, the full Hamiltonian is then

$$H = H^0 + H'(t) = H^0 - \mu\mathbf{E}(t) \qquad (2.6)$$

where μ is the dipole operator and $\mathbf{E}(t)$ the external field

$$\mathbf{E}(t) = \mathbf{E}_p(t) + \mathbf{E}_d(t)$$

with

$$\mathbf{E}_p(t) = \mathbf{E}_p(t) + \text{ c.c.} = \mathbf{e}_p\varepsilon_p(t)e^{-i\omega_p t} + \text{ c.c.} \qquad (2.7a)$$

and

$$\mathbf{E}_d(t) = \mathbf{E}_d(t) + \text{c.c.} = \mathbf{e}_d \varepsilon_d(t - \Delta t)e^{-i\omega_d t} + \text{c.c.}, \qquad (2.7\text{b})$$

the transient amplitudes of the PUMP and the DUMP laser pulses.

As has been shown in Appendix A, in the interaction representation, the resonant component of the third-order density matrix corresponding to the stimulated emission process is

$$\rho^{(3)}(t) = \rho_a^{(3)} + \rho_b^{(3)} + \text{c.c.}$$

$$= \left(\frac{-1}{i\hbar}\right)^3 \int_{-\infty}^{t} dt_3 \int_{-\infty}^{t_3} dt_2 \int_{-\infty}^{t_2} dt_1$$

$$\times \mu(t_3)\mathbf{E}_d^*(t_3)\mu(t_2)\mathbf{E}_p(t_2)\rho^{(0)}\mu(t_1)\mathbf{E}_p^*(t_1) \qquad (2.8)$$

$$+ \left(\frac{-1}{i\hbar}\right)^3 \int_{-\infty}^{t} dt_3 \int_{-\infty}^{t_3} dt_2 \int_{-\infty}^{t_2} dt_1$$

$$\times \mu(t_3)\mathbf{E}_d^*(t_3)\mu(t_1)\mathbf{E}_p(t_1)\rho^{(0)}\mu(t_2)\mathbf{E}_p^*(t_2) + \text{c.c.} . \qquad (2.9)$$

For an isotropic medium, the second order nonlinear polarization disappears due to the inversion symmetry of the sample, i.e.,

$$\mathbf{P}^{(2)}(t) = N \, \text{Tr}\left[\mu(t)\rho^{(2)}(t)\right]$$

$$= N \left(\frac{-1}{i\hbar}\right)^2 \int_{-\infty}^{t} dt_2 \int_{-\infty}^{t_2} dt_1 \, \text{Tr}\{\mu(t)[\mu\mathbf{E}(t_2), \, [\mu\mathbf{E}(t_1), \, \rho^{(0)}]]\}$$

$$= 0 \qquad (2.10)$$

where N is the number density of the molecules, and $\text{Tr}(O)$ the trace of the operator O. The lowest order nonlinear polarization is therefore $\mathbf{P}^{(3)}(t)$. From Eqs. (2.7) to (2.9), we have

$$\mathbf{P}^{\mathrm{TSEP}}(t) = \mathbf{P}^{(3)}(t) = N \, \mathrm{Tr}[\mu(t)\rho^{(3)}(t)]$$

$$= -N \left(\frac{-1}{i\hbar}\right)^3 \int_{-\infty}^{t} dt_3 \int_{-\infty}^{t_3} dt_2 \int_{-\infty}^{t_2} dt_1 \varepsilon_d(t_3 - \Delta t)\varepsilon_p(t_2)\varepsilon_p(t_1)$$

$$\times \left\{ \mathrm{Tr}\{\mu(t)[\mathbf{e}_d^* \cdot \mu(t_3)][\mathbf{e}_p \cdot \mu(t_2)]\rho^{(0)}[\mathbf{e}_p^* \cdot \mu(t_1)]\} \right.$$

$$\times \, e^{i(\omega_d t_3 - \omega_p t_2 + \omega_p t_1)}$$

$$+ \mathrm{Tr}\{\mu(t)[\mathbf{e}_d^* \cdot \mu(t_3)][\mathbf{e}_p \cdot \mu(t_1)]\rho^{(0)}[\mathbf{e}_p^* \cdot \mu(t_2)]\}$$

$$\left. \times \, e^{i(\omega_d t_3 - \omega_p t_1 + \omega_p t_2)} \right\} + \text{c.c.} . \tag{2.12}$$

In deriving Eq. (2.12), the usual rotating wave approximation has been adopted.[6] According to the discussion in Appendix B, the projection of $\mathbf{P}^{\mathrm{TSEP}}(t)$ along the polarization direction of the analyzing polarizer, $\mathbf{e}_s$, is

$$\mathbf{e}_s \cdot \mathbf{P}^{\mathrm{TSEP}}(t) = -N \left(\frac{-1}{i\hbar}\right)^3 \int_{-\infty}^{t} dt_3 \int_{-\infty}^{t_3} dt_2 \int_{-\infty}^{t_2} dt_1$$

$$\times \, \varepsilon_d(t_3 - \Delta t)\varepsilon_p(t_2)\varepsilon_p(t_1)[R^{(1)}(t_3, t_2, t_1)$$

$$+ R^{(2)}(t_3, t_2, t_1)] + \text{c.c.} \tag{2.13}$$

where

$$R^{(1)}(t_3, t_2, t_1) = e^{i(\omega_d t_3 - \omega_p t_2 + \omega_p t_1)}$$

$$\times \, \mathrm{Tr}\{[\mu(t) \otimes \mu(t_3)]_0^{(2)}[\mu(t_2) \otimes \rho^{(0)}\mu(t_1)]_0^{(2)}\} , \tag{2.14}$$

$$R^{(2)}(t_3, t_2, t_1) = e^{i(\omega_d t_3 - \omega_p t_1 + \omega_p t_2)}$$

$$\times \, \mathrm{Tr}\{[\mu(t) \otimes \mu(t_3)]_0^{(2)}[\mu(t_1)\rho^{(0)} \otimes \mu(t_2)]_0^{(2)}\} \tag{2.15}$$

are the components of the response function tensor and

$$[\mathbf{A} \otimes \mathbf{B}]_0^{(2)} = \frac{1}{\sqrt{6}}(3A_z B_z - \mathbf{A} \cdot \mathbf{B}) \tag{2.16}$$

is the zeroth component of the second-rank irreducible tensor constructed from the vectors $\mathbf{A}$ and $\mathbf{B}$. $R^{(1)}$ and $R^{(2)}$ can be expressed explicitly in the Schrödinger representation:

$$R^{(1)}(t_3,\, t_2,\, t_1) = e^{i(\omega_d t_3 - \omega_p t_2 + \omega_p t_1)}$$

$$\times \, \mathrm{Tr}\{[\mu \otimes U(t - t_3)\mu]_0^{(2)} U(t_3 - t_2)$$

$$\times \, [\mu \otimes U(t_2 - t_1)\rho^{(0)}\mu U^\dagger(t_2 - t_1)]_0^{(2)}$$

$$\times \, U^\dagger(t_3 - t_2)U^\dagger(t - t_3)\} \tag{2.17}$$

$$R^{(2)}(t_3,\, t_2,\, t_1) = e^{i(\omega_d t_3 - \omega_p t_1 + \omega_p t_2)}$$

$$\times \, \mathrm{Tr}\{[\mu \otimes U(t - t_3)\mu]_0^{(2)} U(t_3 - t_2)$$

$$\times \, [U(t_2 - t_1)\mu\rho^{(0)} U^\dagger(t_2 - t_1) \otimes \mu]_0^{(2)}$$

$$\times \, U^\dagger(t_3 - t_2)U^\dagger(t - t_3)\}\,, \tag{2.18}$$

where $U(t) = \exp\left(\frac{H^0 t}{i\hbar}\right)$ is the time-evolution operator. In (2.17) and (2.18), we have regrouped the evolution operators to keep all the time variables non-negative. Although the rearrangement is trivial for an isolated molecular system, it is important when there is dissipation, since in the latter case $U(t)$ is defined only for $t \geq 0$.

Equations (2.13) to (2.18) directly correlate the microscopic dynamic properties of the molecular system, $U(t)$, μ and ρ, to the macroscopic nonlinear polarization, $\mathbf{P}(t)$, from which the experimental observable can be trivially obtained (e.g., from Eqs. (2.4) and (2.5)). These equations are, therefore, of fundamental importance to our understanding of the molecular dynamics via the TSEP. For those who are more familiar with the theoretical treatment for the transient gain (loss) spectroscopy,[7-9] the similarity between the two can be easily recognized. As have been shown in Appendix B, the two types of spectroscopies are sensitive to the different components of the same nonlinear response function tensor. Consequently, they are complementary to each other in revealing molecular dynamic information. In particular, if the rotational motion of the molecules can be ignored, these techniques probe the same molecular dynamic properties.

Some general physical insight can be obtained from Eqs. (2.13) to (2.18). In a typical TSEP experiment, the PUMP pulse duration is about 50–100 fs. This corresponds to a Fourier transformed spectral width of $\sim$ 150 cm^{-1}, much less than the typical vibrational frequency for a stable molecule. As we have mentioned earlier, the optical transitions in the TSEP experiment involve a bound ground electronic state $|g\rangle$ and a dissociative excited state $|e\rangle$. In the time scale of interest, the transitions can, therefore, be considered as being between a discrete ground state vibrational manifold and a moving wavepacket on the excited state potential surface. The corresponding molecular Hamiltonian for such a system is

$$H^0 = |g\rangle \left[\sum_n \hbar\omega_n |v_n\rangle\langle v_n| \right] \langle g| + |e\rangle [\hbar\omega_e + H_e(\mathbf{Q})]\langle e| \tag{2.19}$$

and

$$\rho(0) = |g\rangle|v\rangle\langle v_0|\langle g|, \quad \mu = |e\rangle\mu_{eg}(\mathbf{Q})\langle g| + \text{ c.c.} \tag{2.20}$$

Here, $H_e(\mathbf{Q})$ is the excited state nuclear Hamiltonian, $\mathbf{Q}$ the nuclear coordinates and $\mu_{eg}(\mathbf{Q})$ the transition dipole between the two electronic states $|g\rangle$ and $|e\rangle$.

Inserting Eqs. (2.19) and (2.20) into (2.17) and making the usual Condon approximation that $\mu_{eg}(\mathbf{Q})$ is independent of $\mathbf{Q}$, i.e., $\mu_{eg}(\mathbf{Q}) = \mu_{eg}$, we obtain

$$R^{(1)}(t_3, t_2, t_1) = e^{i(\omega_d t_3 - \omega_p t_2 + \omega_p t_1)} |[\mu_{eg} \otimes \mu_{ge}]_0^{(2)}|^2$$

$$\times \sum_n \langle v_n|U_e(t_3 - t_2)|v_0\rangle\langle v_0|U_e^\dagger(t - t_1)|v_n\rangle$$

$$\times e^{-i(\omega_e + i\Gamma_0)(t_2 - t_1)} e^{i(\omega_{en} + i\Gamma_n)(t - t_3)}. \tag{2.21}$$

Similarly,

$$R^{(2)}(t_3, t_2, t_1) = e^{i(\omega_d t_3 - \omega_p t_1 + \omega_p t_2)} |[\mu_{eg} \otimes \mu_{ge}]_0^{(2)}|^2$$

$$\times \sum_n \langle v_n|U_e(t_3 - t_1)|v_0\rangle\langle v_0|U_e^\dagger(t - t_2)|v_n\rangle$$

$$\times e^{-i(\omega_e + i\Gamma_0)(t_2 - t_1)} e^{i(\omega_{en} + i\Gamma_n)(t - t_3)}, \tag{2.22}$$

where Γ_n is the electronic dephasing rate, $\omega_{en} \equiv \omega_e - \omega_n$, and $U_e(t)$ the propagator on the excited state potential surface

$$U_e(t) \equiv \exp[-iH_e(\mathbf{Q})/\hbar]. \tag{2.23}$$

The introduction of dephasing in (2.21) and (2.22) is performed by considering the ensemble averaged time-evolution of the density matrix

$$\langle U(t)|g, v_n\rangle\langle e, \mathbf{Q}|U^\dagger(t)\rangle_{av} = e^{i\omega_{en} t} e^{-\Gamma_n t} |v_n\rangle\langle \mathbf{Q}|U_e^\dagger(t) \quad (t \geq 0). \tag{2.24}$$

It can be easily verified that this is equivalent to the usual approach where a damping matrix is explicitly introduced into the quantum Liouville equation.

Inserting Eqs. (2.21) and (2.22) into (2.13), we finally obtain

$$\mathbf{e}_s \cdot \mathbf{P}^{\text{TSEP}}(t)$$

$$= -N\left(\frac{-1}{i\hbar}\right)^3 |[\mu_{eg} \otimes \mu_{ge}]_0^{(2)}|^2 e^{i\omega_d t} \sum_n \int_{-\infty}^t dt_2 \int_{-\infty}^{t_3} dt_2 \int_{-\infty}^{t_2} dt_1$$

$$\times \, \varepsilon_d(t_3 - \Delta t)\varepsilon_p(t_2)\varepsilon_p(t_1) e^{-i(\omega_d - \omega_{en} - i\Gamma_n)(t-t_3)}$$

$$\times \left[\langle v_n|U_e(t_3 - t_2)|v_0\rangle\langle v_0|U_e^\dagger(t - t_1)|v_n\rangle e^{-i(\omega_p - \omega_e - i\Gamma_0)(t_2 - t_1)} \right.$$

$$\left. + \langle v_n|U_e(t_3 - t_1)|v_0\rangle\langle v_0|U_e^\dagger(t - t_2)|v_n\rangle e^{i(\omega_p - \omega_e + i\Gamma_0)(t_2 - t_1)} \right] + \text{ c.c..}$$

$$(2.25)$$

Equation (2.25) is the most important result in this section. It represents the nonlinear polarization response of the molecular system in a TSEP experiment. The physical significance of the equation can be more appreciated by examining the case in the impulsive limit, i.e., when the PUMP pulse duration is much shorter than the molecular dynamics of interest. Under such circumstance, we can approximate the PUMP pulse profile by a δ-function. Consequently, Eq. (2.25) reduces to

$$\mathbf{e}_s \cdot \mathbf{P}^{\text{PTSEP}}(t) = -2N\left(\frac{-1}{i\hbar}\right)^3 |[\mu_{eg} \otimes \mu_{ge}]_0^{(2)}|^2 |\varepsilon_p(0)|^2 e^{i\omega_d t}$$

$$\times \sum_n \int_{-\infty}^t dt_3 e^{-i(\omega_d - \omega_{en} - i\Gamma_n)(t-t_3)}$$

$$\times \, \varepsilon_d(t_3 - \Delta t)\langle v_n|U_e(t_3)|v_0\rangle\langle v_0|U_e^\dagger(t)|v_n\rangle$$

$$+ \text{ c.c.} \qquad (2.26)$$

Integrating by parts, and making the fast dephasing approximation[7,23]

$$\left| \frac{\partial \varepsilon_d(\tau - \Delta t)\langle v_n|U_e(\tau)|v_0\rangle\langle v_0|U_e^\dagger(t)|v_n\rangle}{\partial \tau} \right| \ll |\omega_d - \omega_{en} - i\Gamma_n|, \qquad (2.27)$$

Eq. (2.26) reduces to

$$
\mathbf{e}_s \cdot \mathbf{P}^{\text{TSEP}}(t) = -2N \left(\frac{-1}{i\hbar} \right)^3 |[\mu_{eg} \otimes \mu_{ge}]_0^{(2)}|^2 |\varepsilon_p(0)|^2 e^{i\omega_d t}
$$

$$
\times \left[\sum_n \frac{\varepsilon_d(t - \Delta t)|\langle v_n|U_e(t)|v_0\rangle|^2}{-i(\omega_d - \omega_{en} - i\Gamma_n)} \right] + \text{c.c.} .
\tag{2.28}
$$

When the DUMP laser is in resonance with the transition $|e\mathbf{Q}\rangle \rightarrow |gv_f\rangle$, we can neglect the rest of the terms in (2.28), i.e.,

$$
\mathbf{e}_s \cdot \mathbf{P}^{\text{TSEP}}(t) = -2N \left(\frac{-1}{i\hbar} \right)^3 |[\mu_{eg} \otimes \mu_{ge}]_0^{(2)}|^2 |\varepsilon_p(0)|^2
$$

$$
\times \varepsilon_d(t - \Delta t)e^{i\omega_d t}|\langle v_f|U_e(t)|v_0\rangle|^2 \frac{1}{-i(\omega_d - \omega_{ef} - i\Gamma_f)}
$$

$$
\propto |\varepsilon_p(0)|^2 \mathbf{E}_d^*(t)|\langle v_f|U_e(t)|v_0\rangle|^2 \frac{1}{(\omega_d - \omega_{ef} - i\Gamma_f)}
$$

$$
+ \text{c.c.} .
\tag{2.29}
$$

From (2.29), we see that, in the impulsive limit, the nonlinear polarization in the TSEP is proportional to the PUMP intensity, $|\varepsilon_p(0)|^2$, and follows the temporal profile of the DUMP pulse, $\mathbf{E}_d^*(t)$. The amplitude of the nonlinear polarization is directly proportional to the transient projection of the dissociating wavepacket $U_e(t)|v_0\rangle$ onto the ground state vibrational wave function $|v_f\rangle$. This instantaneous Franck–Condon (FC) factor, $|\langle v_f|U_e(t)|v_0\rangle|^2$, approaches $|\langle v_f|U_e(\Delta t)|v_0\rangle|^2$ for a δ-function DUMP pulse. Equation (2.29) is, therefore, a direct mathematical manifestation of the following physical process. At $t = 0$, the $|v_0\rangle$ wavepacket is transferred onto the excited state potential surface by a femtosecond PUMP pulse. It then evolves under the influence of the Hamiltonian $H_e(\mathbf{Q})$ for a period Δt, and then DUMPed back to the ground PES.

2.3. *Discussions*

It is worthwhile to point out that, according to Eq. (2.29), if we scan through the PUMP–DUMP delay, Δt, at fixed DUMP frequency, the nonlinear polarization will map out the time dependent overlap of the dissociating wave function with $|v_f\rangle$. On the other hand, the spectral line width of the frequency resolved TSEP spectrum at a fixed Δt is determined by the electronic dephasing rate Γ_f, even if both the PUMP and the DUMP approach $\delta(t)$-pulses! This 'paradox' of finite frequency resolution in the impulsive limit has been discussed previously by many authors.[8,9] Physically, it is because the light-matter interaction modifies not only the amplitude but also the phase of the ultrashort probe pulse. The experiments designed to measure either the time (the amplitude) response or the frequency (the phase) response of the molecular system with ultrashort light pulses will, therefore, provide complementary dynamic information about the system.

It is easy to show from Eq. (2.13) that the impulsive response of the molecular system to the δ-function excitations approaches the response function. In general, the polarization generated in a laboratory experiment involving finite-duration pulses is a convolution of the response function with the pulse envelopes. Consequently, the time resolution in a TSEP experiment (or any time-domain experiment) is ultimately limited by the pulse widths. At this point, a comparison of the TSEP with the CW resonance Raman experiment (RR)[24-27] seems appropriate. In the RR, the signal is proportional to the square of the molecular polarizability. Within the Condon approximation, it can be expressed as[27]

$$I_{0,f}(\omega) \propto |\alpha_{0,n}(\omega)|^2 \propto \left| \int_0^\infty e^{i\omega t - \Gamma_f t} \langle v_f | U_e(t) | v_0 \rangle dt \right|^2 . \tag{2.30}$$

Both (2.30) and (2.25) are related to $\langle v_f | U_e(t) | v_0 \rangle$. However, neither provide a complete dynamic picture about the time-dependent Franck–Condon overlap. We, therefore, return to the decades-old question: Is there any *unique* dynamic information contained in the gas-phase ultrafast time-resolved experiments?

Experimental observables always involve time-average and ensemble-average. Physically, the time-domain experiment performs the ensemble average of a system at a fixed t, then studies its time evolution. On the other hand, the frequency-domain experiment takes a weighted time-average of the molecular dynamics (the Fourier transform), then performs the ensemble average. Only for an isolated molecule with *completely* specified Hamiltonian can the two types of experiments provide the same amount of information. For a dynamic system, the order of the two averages is not necessarily exchangeable. It is not difficult to show that there is no one-to-one correspondence between Eqs. (2.29) and (2.30). The nature of the time-resolved pump-probe spectroscopy dictates that it is much less sensitive to the electronic dephasing. As is shown in Eq. (2.29), the dephasing parameter (which is a result of ensemble averaging) appears as a constant decoupled from the time-dependent observable. Over the years, various time-resolved spectroscopic techniques have been vital to our understanding of the condensed phase molecular dynamics. This is due mainly to the omnipresent fast dephasing time (T_2) in the condensed phase. It should be emphasized, however, that there is dephasing whenever the subject of interest is divided into 'system' and 'bath'. As a result, it could also play an important role in the dynamics of isolated gas phase molecules. The well-known dephasing mechanisms are intramolecular vibrational-energy redistribution (IVR) in large polyatomic,[28,29] and electronic interactions, such as conical intersections/intersystem crossings, in medium or even small polyatomic systems.[30,31]

It is well established that stochastic dynamics or chaos is ubiquitous in classical systems with finite degrees of freedom ($\geq$ 2).[32] Although the corresponding quantum manifestation is still not well understood,[33,34] numerous experimental evidences show that the reaction rates and product state distributions can often be described by statistical theories, such as RRKM and phase space theory. The basic physical assumption of the statistical theories is the complete mixing of all available phase space. From a dynamic point of view, the reaction mechanism depends critically on the relative time scales of energy transfer (or mixing) and bond breaking/rearrangement. Detailed knowledge about these time scales and the interconnections between them is, therefore, vital to our understanding of chemical reactions. While frequency domain techniques, such as resonance Raman, are at their best for revealing dynamic information faster than the electronic dephasing, time domain spectroscopies are capable of studying molecular dynamics at 'long' times (currently $\geq$ 50 fs due to technical reasons, though there are exceptions[35]). After $\sim$ 50 fs, the vibrational characters of those motions orthogonal to the reaction coordinate are reasonably well defined. Chemically important questions, such as the time scale of energy-flow between different vibrational modes and its impact on the reaction rates/paths can, therefore, be addressed. *It is the capability of obtaining real-time dynamic information beyond the initial dephasing, which enables the time-resolved techniques such as TSEP to establish a unique position in the study of gas-phase chemical reaction dynamics.*

The capabilities of real-time spectroscopy in the study of 'long' time molecular dynamics are demonstrated clearly in our recent TSEP studies of ozone visible photodissociation.

3. Ozone Visible Photodissociation Dynamics

Due to its role in atmospheric chemistry, the photodissociation of ozone has been the subject of many laboratory and theoretical studies.[24,36–40] The visible–UV spectrum of ozone consists of several diffusive band systems:[41] Chappuis ($\sim 600\,\mathrm{nm}$), Huggins ($\sim 340\,\mathrm{nm}$), and Hartley ($\sim 260\,\mathrm{nm}$). The UV dissociation dynamics have been studied extensively through resonance Raman spectroscopy.[24] However, efforts in the Chappuis band

$$O_3(^1A_1) + h\nu(\sim 600\,\mathrm{nm}) \to O_3^* \to O_2(^3\Sigma_g^-) + O(^3P) \qquad (3.1)$$

have not been very successful due to the small absorption cross-section ($\sim 5 \times 10^{-21}\,\mathrm{cm}^2$). The only experimental work reported so far on reaction (3.1) is the product state distribution studies of Valentini *et al.*[37] The results of this study are taken to imply rotational impulsive and vibrational adiabatic dissociation on a single electronic surface. However, based on the measurement of the isotope shifts between $^{16}O_3$ and $^{18}O_3$ in absorption, Anderson *et al.* concluded that the Chappuis band consists of transitions from the ground electronic state ($^1A_1$) to a $^1B_1$ state and a still unassigned 'X' state.[42] Recently, several large scale *ab initio* calculations related to the Chappuis band have appeared in literature.[39,40] These studies suggest that the oscillator strength in the Chappuis band comes from the dipole allowed transition $^1B_1 \leftarrow {}^1A_1$ (or $2^1A'' \leftarrow {}^1A'$ in the C_s symmetry notation). A conical intersection that couples the bounded $^1B_1$ ($2^1A''$) to a 'dark' dissociative $^1A_2$ ($1A''$) state is responsible for the diffusive nature of the absorption band. However, calculations based on the adiabatic potential surfaces failed to provide an absorption spectrum in satisfactory agreement with the experiments.[39] Only when the conical intersection is accounted for explicitly, is the theoretical model able to reproduce the essential features of the experimental absorption spectrum.[4] It is interesting to point out that in the calculation, the dissociation is modeled phenomenologically with a very fast optical dephasing constant ($T_2 = 18\,\mathrm{fs}$) and a relatively slow population decay constant ($T_1 = 300\,\mathrm{fs}$). The fast T_2 also agrees well with the decay of the spectral autocorrelation function (i.e., the Fourier transform of the absorption spectrum).[42] As has been shown in the model calculations for other polyatomic systems with similar PES topology, conical intersections can often induce ultrafast electronic relaxation.[30,31] A TSEP study of the relative time scales involved in the visible photodissociation of ozone, in particular the nuclear dynamics after the initial dephasing, should, therefore, shed light on our general understanding of chemical reactions occurred on multiple potential surfaces.

3.1. *Experimental*

Compared with many other femtosecond pump-probe spectrometers reported in the literature, there are two salient features about our TSEP setup. First, we adopted the coherent polarization detection technique. This provides $\sim 10^6$ discrimination against the background noise and renders TSEP a background-free environment. Second, we used the supercontinuum as the DUMP pulse. This not only allows us to bypass the experimental difficulty associated with the generation and synchronization of two tunable femtosecond pulses, but also offers 'single shot' collection of many different spectral features.

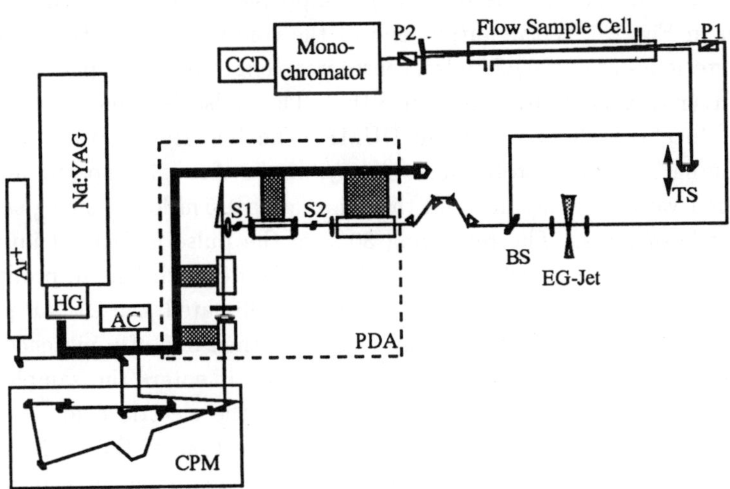

Fig. 2. TSEP schematic. HG: harmonic generator; AC: zero-background autocorrelator; S1, S2: saturable absorber jets; BS: beam splitter; EG-Jet: ethylene glycol jet; TS: precision translation stage; P1, P2: matched polarizers; PDA: pulsed dye amplifier ; CCD: CCD detector.

The heart of our femtosecond TSEP spectrometer is the home-built amplified colliding-pulse mode-locked (CPM) laser system (Fig. 2). Since similar systems have been around for quite a while, only a brief description will be given here. The seven-mirror CPM, after the original design of Valdmanis *et al.*,[43] is pumped by a CW Ar^+ laser (Coherent 318AS). Running at the repetition rate of 100 MHz, the CPM delivers typically 50 fs pulses at ~ 625 nm. The use of an actively stabilized Ar^+ pump laser results in extremely reliable CPM performance. Consequently, very little adjustment is needed during the daily operation. The output of the CPM is further amplified by a pulsed-dye-amplifier (PDA). The 4-stage PDA uses prism dye cells, and is transversely pumped by the 532 nm output from a Nd:YAG laser (Spectral Physics, GCR4-30) running at 30 Hz. In a typical configuration, the pump energy partition among the four stages is 7:14:30:120 mJ. To minimize the amplified spontaneous emission, a 200 μm pinhole is placed between the first two dye cells and ethylene glycol jets with Malachite–Green as saturable absorber between the later cells. After the amplification, the pulse duration is typically 600 fs due to the group velocity dispersion induced by the various optical components along the amplifier beam path. The pulse is recompressed using a 4-prism setup[44] shown in Fig. 2. At the output end of the compressor, the typical pulse characteristics are: 500 μJ/pulse, center wavelength 624 nm, spectral width 10 nm (FWHM), pulse duration 80 fs. The pulse is then split in two to produce the PUMP and the DUMP. The linearly polarized PUMP, containing 70 % of the total pulse energy, passes through a spatial filter, a precision variable delay line and then through a 1.5 m long flowing-sample cell made of glass. The remaining 30 % of the pulse is focused into a 1 mm thick ethylene glycol jet using a 50 mm focal-length lens to produce a white-light continuum DUMP. After being recollimated, the DUMP passes through a long-pass filter (Corning 2-58) and then one of a matched pair of polarizers (extinction ratio 5×10^{-7}) before it enters the sample cell. This 'polarizing' polarizer is adjusted so that the electric field vector of the continuum DUMP is at 45° with respect to that of the PUMP. The DUMP beam then propagates through the cell at a crossing angle of $\sim 0.7°$ with respect to the PUMP so that there is no overlap on the sample cell windows. After the cell, the PUMP is blocked by an iris pinhole and the DUMP signal passes through an 'analyzing' polarizer, which is adjusted to minimize the DUMP leakage when $\Delta t \ll 0$. The signal is then focused into an imaging monochromator (Spex 270M, spectral resolution $\sim 2\,\text{nm}$) and analyzed by a CCD detector (Princeton Instruments TE512TKB). Because of the broad spectral range of the continuum, many resonant components can be simultaneously recorded. Varying the delay between the PUMP and the DUMP pulses thus generates a two-dimensional (intensity vs. frequency-time) TSEP spectrum. The entire TSEP spectrometer is controlled by a PC 386 computer.

The spectral width of the DUMP also leads to temporal dispersion of the pulse. However, the dispersion is readily corrected in our experiment by measuring the cross-correlation between PUMP and DUMP pulses at various wavelengths. In our polarization detected TSEP experiment, the cross-correlation is conveniently obtained via overlapping the two pulses spatially and temporally on the exit window of the sample cell. The cross-correlation, due to the instantaneous Raman induced Kerr effect (RIKES) of quartz,[6] also provides an accurate measure of the instrument response. The typical cross-correlation width (FWHM) is $\sim 120\,\text{fs}$ at any wavelength of interest.

Our TSEP experiment of ozone is carried out in a room temperature sample cell. Because ozone dissociates in less than a picosecond, collisions and molecular rotations should not play a significant role in the reaction dynamics. Ozone is prepared using a commercial ozonator and stored in a silica-gel trap at 195 K. Before use, the residual oxygen in the trap is pumped off until the gas flow contains only a few percentages of oxygen. During the experiment, the ambient-temperature sample cell is filled with ~ 80 Torr of ozone. Due to its small absorption cross-section, the depletion of ozone from photolysis is insignificant. Diffusion is, therefore, sufficient to replenish the ozone in the interaction zone after each laser shot. With the current setup, no changes of signal intensities were observed from the beginning to the end of the experiment.

In a typical data collection cycle, the PUMP–DUMP delay is swept from $-500\,\text{fs}$ to $5,000\,\text{fs}$ with a step size of $10\,\text{fs}$. At each fixed delay, the signal is integrated on the CCD chip for 60 shots and then transferred onto the computer. The resulted two-dimensional TSEP spectrum is then normalized to the DUMP spectral distribution, and dechirped numerically using the dispersion curve obtained experimentally from the RIKE signal of the quartz window.

3.2. *Results and Discussion*

Figure 3 presents the ozone TSEP spectrum we reported earlier.[3] As in our original report, the dispersion of the DUMP pulse has already been corrected. The whole spectrum covers from $900\,\mathrm{cm}^{-1}$ to $5000\,\mathrm{cm}^{-1}$ in the Raman frequency $\Delta\omega$ ($\equiv \omega_{\mathrm{PUMP}} - \omega_{\mathrm{DUMP}}$), and from $-200\,\mathrm{fs}$ to $600\,\mathrm{fs}$ in the PUMP–DUMP delay, Δt. The curve in the background is the time integrated TSEP intensity during this period. Two 'wavepackets' are clearly observable from the spectrum. One peaks at $\Delta t = 0$ with its frequency centered at $\Delta\omega = 1110\,\mathrm{cm}^{-1}$ and the other at $\Delta t = 90\,\mathrm{fs}$ with $\Delta\omega = 2700\,\mathrm{cm}^{-1}$. From the known vibrational spectroscopy of O_3,[45] the two peaks can be attributed to the stimulated emission from the dissociating state to the $(1,0,0)$ and $(0,1,2)$ vibrational levels of the ground state, respectively. The observed TSEP spectral activity can be reconciled with the following dynamic process. Upon the absorption of the PUMP photon, the motion of the ozone wavepacket is dominantly symmetric-stretching (ν_1) like, resulting in the ν_1 mode as the most prominent feature in the earlier time TSEP spectrum. After $90\,\mathrm{fs}$, the energy flows into the reaction coordinate, which is a superposition of bending (ν_2) and antisymmetric stretch (ν_3) normal modes. Such a picture is consistent with the recently published *ab initio* potential surfaces of ozone.[39,40] According to the calculations of Braunstein *et al.*[39] and Banichevich *et al.*,[40] the O–O bond length in both the 'bright' $^1B_1$ and the 'dark' $^1A_2$ states are significantly different from that in the ground state. As a result, when the molecule absorbs a photon from its C_{2v} ground state, it will land on the $^1B_1$ surface at a point with steepest descent along the symmetric stretch (q_1) direction. (At the center of the Franck–Condon region, the symmetry dictates that the gradient be zero along the q_3-antisymmetric stretch coordinate.) Consequently, the early time movement of the ozone wavepacket will be dominated by motions along the q_1 axis. Eventually, the molecule will fall apart, therefore, TSEP activity containing ν_3 excitation is expected. However, since the $^1B_1$ is bound, the only passage to dissociation is through the $^1B_1$–$^1A_2$ conical intersection. The time delay between the appearance of ν_1 and the $\nu_2 + 2\nu_3$ modes is, therefore, due to the bottleneck on the reaction path.

It is interesting to compare the time scale in the observed TSEP spectrum with the electronic dephasing time T_2 ($\equiv 1/\Gamma$). The latter can be readily estimated from the decay in the Fourier transform of

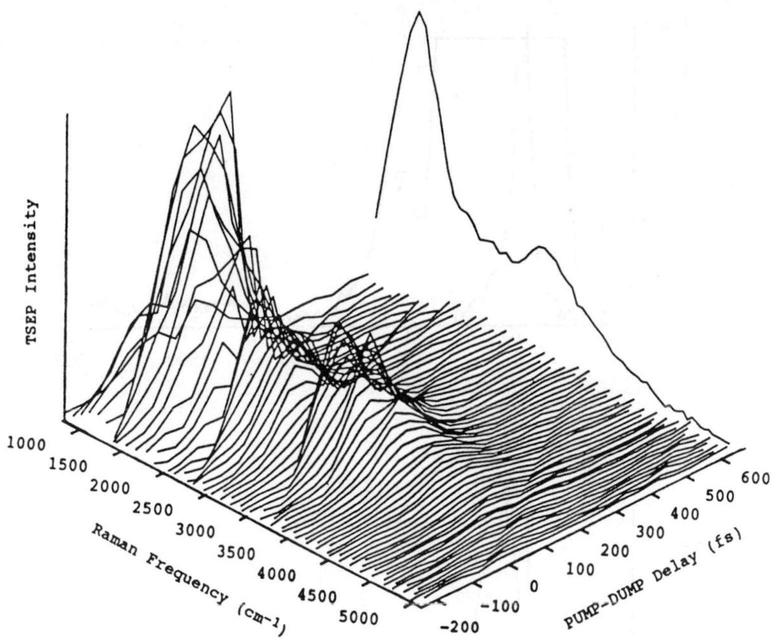

Fig. 3. TSEP spectrum of ozone at 624 nm photolysis. The Raman frequency $\Delta\omega \equiv \omega_{\mathrm{PUMP}} - \omega_{\mathrm{DUMP}}$. The curve in the background is the time-integrated TSEP intensity. The two prominent features in the spectrum correspond to emissions to $(1,0,0)$ and $(0,1,2)$ of the ground state, respectively. Note that while the $(1,0,0)$ peak appears promptly, the $(0,1,2)$ is delayed by ~ 90 fs.

the absorption spectrum.[27] This was obtained recently by Anderson *et al.* for the ozone Chappuis band.[42] The estimated T_2 from their data is approximately 20 fs. The TSEP spectral activity in the $(\nu_2 + 2\nu_3)$ mode could, therefore, reflect the molecular dynamics after the initial dephasing. To confirm the result, we have recently repeated the ozone experiment with improved signal-to-noise ratio. Figure 4 displays a cut from the TSEP spectrum at fixed DUMP frequency. ($\Delta\omega = 1750\,\mathrm{cm}^{-1}$, i.e., in the valley between the two peaks of Fig. 3.) To our surprise, there is not only a 90 fs delay for the main feature, but also a weak recurrence at ~ 550 fs. The global structure of the recurrence can be visualized more clearly in the contour plot (Fig. 5). To highlight the recurrence, the main feature shown here in Fig. 3 is truncated. However, we can still see vividly that

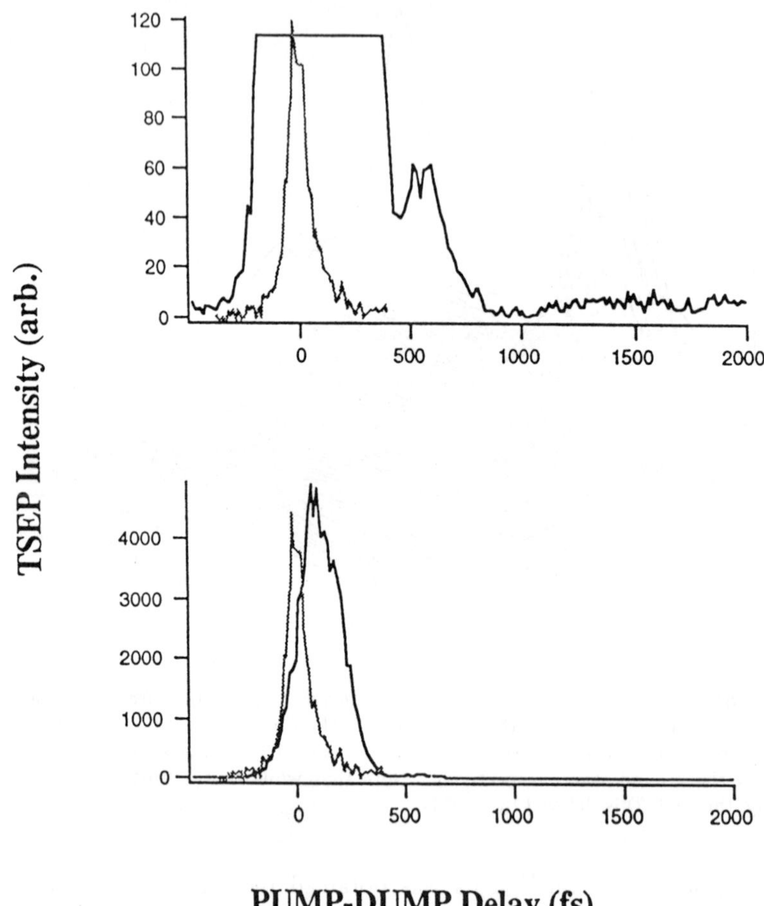

Fig. 4. Ozone TSEP spectrum at fixed DUMP frequency. The lighter line is the instrumental response obtained via cross correlation of PUMP and DUMP on the quartz window. The appearance of the main TSEP feature is delayed by $\sim 90\,$fs (lower figure). The zoom-in also indicates a recurrence at $\sim 550\,$fs (top figure).

the main body of the TSEP spectrum is spreading towards red in $\sim 200\,$fs. On the other hand, the weak recurrence observed at $\sim 500\,$fs after the photodissociation moves in an opposite direction. It is blue shifting towards the Franck–Condon region as the PUMP-DUMP delay increases. The time evolution of the TSEP spectrum can be explained qualitatively in a zeroth

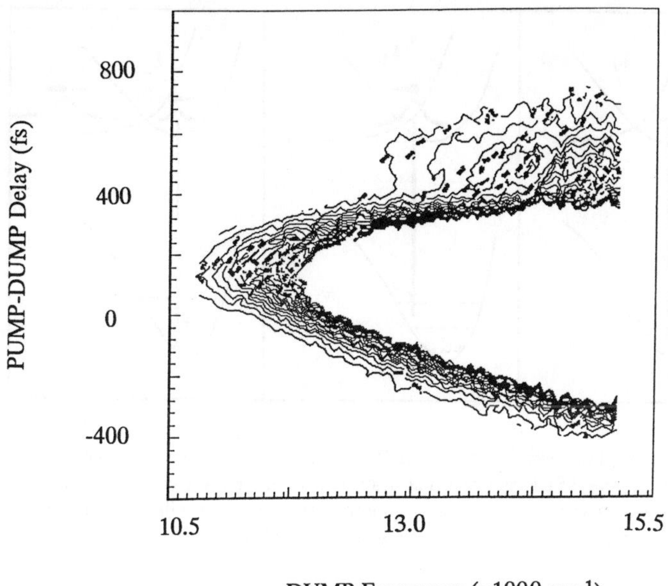

Fig. 5. Contour plot of the ozone TSEP spectrum. Only the ones with intensities between 0 to 600 counts are shown. The main spectral feature spreads toward red at earlier t. A weaker recurrence appears at $\sim 500\,\mathrm{fs}$ and moves towards blue.

order picture using simple Franck–Condon argument (Fig. 6). When the wavepacket moves out toward dissociation on the excited PES, it overlaps with higher energy vibrational levels of the ground electronic state, leading to the red shift of the emission spectrum. At the conical intersection region, the wavepacket encounters a discontinuity on the potential surface. Partial reflection can, therefore, split the wavepacket, sending a portion back towards the Franck–Condon region. Such a tentative dynamic picture is very appealing due to its simplicity. It is worthwhile to point out, however, that the time scales associated with the experiment are much longer than what one would expect from a simple bond–rupture reaction on a smooth PES. The localization time for a wavepacket (i.e., the time during which the wavepacket can be approximated by a Gaussian) can be estimated from the characteristic vibrational period, T_0, of the molecule. In ozone, the lowest vibrational frequency is c.a. $500\,\mathrm{cm}^{-1}$, corresponding to T_0 of c.a. $10\,\mathrm{fs}$. Therefore, by the time the TSEP spectral features

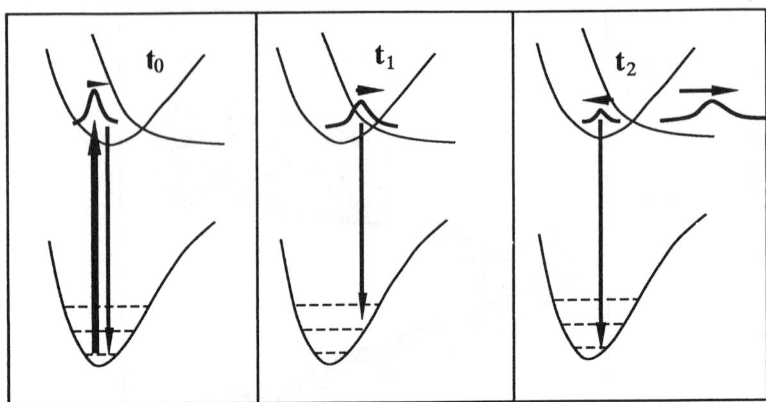

Fig. 6. Cartoon explanation for the TSEP recurrence. When the molecule moves toward dissociation, the TSEP spectrum shifts toward longer wavelength. At the conical intersection, the wavepacket splits. The reflection (recurrence) will move towards the Franck–Condon region, causing the TSEP spectrum to blue-shift.

appear (at 90 or 500 fs after the PUMP pulse), the nuclear motions orthogonal to the reaction coordinate have obtained well-defined vibrational characters. Consequently, the time evolution of the TSEP spectrum should be dominated by the molecular dynamics along the reaction coordinate (soft mode), or energy transfer between various vibrational degrees of freedom. The 'slow' time scale in the TSEP spectrum has also been observed in a recent model study of ozone dissociation by Stock *et al.*[4] In their simulation, the nonadiabatic vibronic coupling between the $^1A_2$ and the $^1B_1$ states (conical intersection) is accounted for explicitly, but the dissociation is modeled by a phenomenological T_1 of 300 fs and a T_2 of 18 fs. In spite of the approximation, they were able to reproduce semiquantitatively the ozone visible absorption spectrum. In addition, they have also observed quasi-periodic oscillations (recurrence) in the simulated TSEP spectrum with period of about 100 fs. The time scale of the spectral motion can not be associated with any vibrational mode of the ozone molecule. It is attributed by the authors as due to the dynamics across the conical intersection between the two coupled electronic surfaces.

Both the experiment and the theoretical simulation on the visible photodissociation of ozone, therefore, seem to indicate a process with $T_2 \ll T_1$. In the absence of pure dephasing, it is well known that T_2 should be equal to $2T_1$. For a gas phase molecule, there should be no intermolecular interactions in the femtosecond time scale. What is then the mechanism for the ultrafast pure dephasing? The T_2 obtained from the absorption spectrum simply reflects the spectral width. A trivial explanation is, therefore, the inhomogeneous broadening. However, previous work showed that there is almost no temperature dependence of the absorption cross-section in the Chappuis band.[46] Recent effort to study the absorption in the molecular beam also found no additional structure in the spectrum.[47] The linewidth, therefore, more likely reflects the excited state properties. As has been demonstrated in many large scale quantum mechanical calculations, conical intersection will often increase the density of optically accessible vibronic states by many orders of magnitude.[30,31] In the work by Stock *et al.*, it is also shown that the $^1B_1$–$^1A_2$ intersection causes similar type of strong vibronic mixing.[4] In analogy with the intramolecular vibrational energy redistribution,[28,29] a dephasing mechanism in addition to the dissociation can, therefore, be expected. It should be pointed out that the dephasing mechanism is different from that due to IVR. Conical intersection essentially causes an electronic energy transfer. The time scales involved can be much shorter than that for the nuclear motions.

4. Summary

A theoretical description of the femtosecond transient stimulated emission pumping process has been formulated. Within the perturbation-theory framework, the TSEP signal is directly correlated to the third-order nonlinear polarization arising from the interaction of the molecular system with the PUMP and the DUMP pulses. The theory explicitly takes into account the polarizations of the laser pulses. It shows that when rotational motion can be ignored, the coherent polarization spectroscopic version of the TSEP probes the same molecular dynamics as the more familiar gain/loss spectroscopies. In particular, within the impulsive limit and the Condon approximation, the polarization created by TSEP is proportional to the transient overlap of the dissociating wavepacket with the ground state vibrational wave functions. In general, however, different configurations of PUMP–DUMP polarizations will access different components of the same nonlinear response function tensor. These techniques are, therefore, complementary to each other in revealing the molecular dynamics. A comparison of TSEP with the frequency-resolved Resonance Raman spectroscopy has also been made. While the latter is a powerful technique in studying dynamics faster than the electronic dephasing time, TSEP is more advantageous for processes after the initial dephasing. We have shown analytically that in the ultrafast TSEP spectrum, dephasing and population dynamics naturally separate into two parts. Unique information about the underlying molecular dynamics can, therefore, be obtained from the time and frequency resolved femtosecond TSEP experiment.[48] The current theoretical treatment, however, neglects all the nonresonant terms in the third-order nonlinear density matrix, and in particular the stimulated Raman contributions to the TSEP signal. It has been shown that when the PUMP pulse width is shorter than the ground state vibrational period, ground state transients will also appear in the time-resolved pump-probe spectra.[7-9] In addition, like many others,[7-9] we have also neglected the rotational motions in our description for the TSEP. These approximations greatly reduce the complexity of the problem and, therefore, offer clearer conceptual understanding about the underlying dynamics. However, it is our belief that a quantitative agreement between theory and experiment is only possible when all these effects are taken care of.

Recent TSEP experiments on the visible photodissociation of ozone have also been discussed. Two salient features of our TSEP apparatus are (1) the coherent polarization detection and (2) white-light supercontinuum DUMP. The former offers a sensitive way to extract the transient stimulated emission signal, and the latter allows multichannel detection. Like its frequency domain counterpart, the resonance Raman spectroscopy, the main strength of TSEP is in the study of *polyatomic* photoreactions. The combination of unequivocal spectral assignment with real-time pump-probe technique, can provide some unique information about the energy-transfer and bond rearrangement process in polyatomic systems. In ozone, the TSEP spectrum at 625 nm photolysis is found to contain mainly two spectral features. The emission to the symmetric stretching mode ($1\nu_1$) appears promptly, and the one to the combination mode of bending + antisymmetric stretching ($\nu_2 + 2\nu_3$) is delayed by $\sim 90\,$fs. The spectral movement is tentatively attributed to the nuclear dynamics on the excited state potential surfaces: i.e., the early time motion is predominantly symmetric stretching like, and the later motion along the reaction coordinate is more like a combination of bending and antisymmetric stretch. Recurrence in the TSEP spectrum is also observed at $\sim 600\,$fs after the photolysis. This is probably due to the reflection of the wavepacket near the discontinuity on the excited state potential surfaces (conical intersection). The picture on the ozone visible photodissociation, however, is built entirely on qualitative descriptions. Clearly, more experiments at various excitation wavelengths in the Chappuis band will be very helpful to confirm the proposed dynamic picture. Quantitative comparison of theories with experiments will also play an important role in our understanding of this interesting system. Work towards this direction is in progress.

Acknowledgments

The authors wish to thank Dr. G. Stock of Technical University of Munich for very fruitful discussions. YC wishes to thank the Camille and Henry Dreyfus Foundation and the Alfred P. Sloan Foundation. This work was also made possible by a Junior Faculty Research Grant from Committee on Research, University of California, Berkeley.

Appendix A. Third-Order Nonlinear Density Matrix for TSEP

The connection between the microscopic molecular dynamics and the macroscopic observables in the ultrafast pump-probe spectroscopies can be established through the solution of the Schrödinger equation,[47]

$$i\hbar \frac{\partial \Psi}{\partial t} = H\Psi \tag{A.1}$$

or the quantum Liouville equation.[6–9,48]

$$i\hbar \frac{\partial \rho}{\partial t} = [H, \rho]. \tag{A.2}$$

The former often provides an intuitive physical picture about the dynamics. The latter is more commonly used by the nonlinear optics community[6] due to the straightforward relationship between the density matrix ρ and the experimental observables. In this appendix, both approaches will be adopted to obtain the mathematical expression for the TSEP process. We show that under certain approximations, the second-order perturbation solution of the Schrödinger equation is equivalent to the third-order perturbation solution of the Liouville equation.

We will use the usual dipole approximation and assume that the pump laser field $\mathbf{E}_p(t)$ connects the initial state $|g\rangle$ to the intermediate state $|e\rangle$

and the probe field $\mathbf{E}_d(t)$ connects the intermediate state $|e\rangle$ to the final state $|f\rangle$. In the interaction representation, Eq. (A.1) can be written as

$$i\hbar\frac{\partial\Psi}{\partial t} = H'\Psi. \tag{A.3}$$

Let

$$\mathbf{E}(t) = \mathbf{E}_p(t) + \mathbf{E}_d(t) \tag{A.4}$$

with

$$\mathbf{E}_p(t) = \mathbf{E}_p(t) + \text{ c.c.} = \mathbf{e}_p\varepsilon_p(t)e^{-i\omega_p t} + \text{ c.c.}$$

and

$$\mathbf{E}_d(t) = \mathbf{E}_d(t) + \text{ c.c.} = \mathbf{e}_d\varepsilon_d(t - \Delta t)e^{-i\omega_d t} + \text{ c.c.}.$$

The second-order perturbation solution of (A.3) can be obtained easily using iteration method:

$$|\Psi\rangle = |\Psi^{(0)}\rangle + |\Psi^{(1)}\rangle + |\Psi^{(2)}\rangle$$

$$= \left\{ 1 - \frac{1}{i\hbar}\int_{-\infty}^{t} dt_1\mu(t_1)\mathbf{E}_p(t_1)\right.$$

$$\left. + \left(\frac{1}{i\hbar}\right)^2 \int_{-\infty}^{t_3} dt_2 \int_{-\infty}^{t_2} dt_1\mu(t_2)\mathbf{E}_d(t_2)\mu(t_1)\mathbf{E}_p(t_1) \right\}|g\rangle \tag{A.5}$$

where we have assumed that $H'(-\infty) = 0$. From (A.5), we can write the third-order density matrix

$$\rho^{(3)}(t) = |\Psi^{(2)}(t)\rangle\langle\Psi^{(1)}(t)| + \text{ c.c.}$$

$$= \left(\frac{-1}{i\hbar}\right)^3 \int_{-\infty}^{t} dt_3 \int_{-\infty}^{t_3} dt_2 \int_{-\infty}^{t} dt_1$$

$$\times \mu(t_3)\mathbf{E}_d(t_3)\mu(t_2)\mathbf{E}_p(t_2)|g\rangle\langle g|\mu(t_1)\mathbf{E}_p^*(t_1) + \text{ c.c.} \tag{A.6a}$$

$$= \left(\frac{-1}{i\hbar}\right)^3 \int_{-\infty}^{t} dt_3 \int_{-\infty}^{t_3} dt_2 \left(\int_{-\infty}^{t_2} + \int_{t_2}^{t_3} + \int_{t_3}^{t}\right) dt_1$$

$$\times \mu(t_3)\mathbf{E}_d(t_3)\mu(t_2)\mathbf{E}_p(t_2)|g\rangle\langle g|\mu(t_1)\mathbf{E}_p^*(t_1) + \text{ c.c.}. \tag{A.6b}$$

Since

$$\int_{-\infty}^{t} dt_3 \int_{t_3}^{t} dt_1 f(t_3, t_1)\mathbf{E}_d(t_3)\mathbf{E}_p^*(t_1) = 0 \quad \text{(pump after the probe)}$$

and

$$\int_{-\infty}^{t_3} dt_2 \int_{t2}^{t_3} dt_1 f(t_2, t_1) = \int_{-\infty}^{t_3} dt_1 \int_{-\infty}^{t_1} dt_2 f(t_2, t_1)$$

$$= \int_{-\infty}^{t_3} dt_2 \int_{-\infty}^{t_2} dt_1 f(t_1, t_2)$$

hold for an arbitrary function $f(t_2, t_1)$, Eq. (A.6b) becomes

$$\rho^{(3)}(t) = \left(\frac{-1}{i\hbar}\right)^3 \int_{-\infty}^{t} dt_3 \int_{-\infty}^{t_3} dt_2 \int_{-\infty}^{t_2} dt_1$$

$$\times \mu(t_3)\mathbf{E}_d(t_3)\mu(t_2)\mathbf{E}_p(t_2)|g\rangle\langle g|\mu(t_1)\mathbf{E}_p^*(t_1)$$

$$+ \left(\frac{-1}{i\hbar}\right)^3 \int_{-\infty}^{t} dt^3 \int_{-\infty}^{t_3} dt_2 \int_{-\infty}^{t_2} dt_1$$

$$\times \mu(t_3)\mathbf{E}_d(t_3)\mu(t_1)\mathbf{E}_p(t_1)|g\rangle\langle g|\mu(t_2)\mathbf{E}_p^*(t_2) + \text{ c.c.}$$

$$\equiv \rho_a^{(3)} + \rho_b^{(3)} + \text{ c.c..} \tag{A.6c}$$

The same solution can also be obtained from the perturbative solution of the interaction representation Liouville equation

$$i\hbar\frac{\partial\rho}{\partial t} = [H', \rho]. \tag{A.7}$$

(A.7) can again be solved using the iteration method. To the third order, the result is:

$$\rho(t) = \rho^{(0)} + \rho^{(1)}(t) + \rho^{(2)}(t) + \rho^{(3)}(t)$$

with

$$\rho^{(1)}(t) = \frac{-1}{i\hbar} \int_{-\infty}^{t} dt_1 [\mu(t_1)\mathbf{E}(t_1), \rho^{(0)}]$$

$$\rho^{(2)}(t) = \left(\frac{-1}{i\hbar}\right)^2 \int_{-\infty}^{t_3} dt_2 \int_{-\infty}^{t_2} dt_1 [\mu(t_2)\mathbf{E}(t_2), [\mu(t_1)\mathbf{E}(t_1), \rho^{(0)}]]$$

and

$$\rho^{(3)}(t) = \left(\frac{-1}{i\hbar}\right)^3 \int_{-\infty}^{t} dt_3 \int_{-\infty}^{t_3} dt_2 \int_{-\infty}^{t_2} dt_1$$

$$\times [\mu(t_3)\mathbf{E}(t_3), [\mu(t_2)\mathbf{E}(t_2), [\mu(t_2)\mathbf{E}(t_2), \rho^{(0)}]]]. \tag{A.8}$$

Expanding the Possion bracket in (A.8) and writing $\mathbf{E}(t)$ explicitly according to (A.4), we will obtain 48 terms containing the product of both the pump and the probe fields. However, it can be shown that only 4 terms satisfy the resonant conditions mentioned earlier. Let

$$\rho^{(0)} = |g\rangle\langle g| \,.$$

Under the rotating wave approximation (RWA), the 4 resonant terms are the same as in (A.6c)

$$\rho^{(3)}(t) = \rho_a^{(3)} + \rho_b^{(3)} + \text{ c.c. } = |\mathbf{\Psi}^{(2)}(t)\rangle\langle\mathbf{\Psi}^{(1)}(t)| + \text{ c.c.}\,. \tag{A.9}$$

As have been shown by many authors,[6–9] in a pump-probe experiment, the nonlinear transient absorption (emission) of the probe pulse at frequency ω_d is

$$I = 2\omega_d \text{Im} \int_{-\infty}^{\infty} dt\, \mathbf{E}_d(t)\mathbf{P}^{(3)}(t) \tag{A.10}$$

where

$$\mathbf{P}^{(3)}(\omega_d,\, t) = N\, \text{Tr}[\mu\rho^{(3)}(t)]$$

is the third-order nonlinear polarization, N the number density and $\text{Tr}(O)$ the trace for the operator O. From (A.9) and (A.10) we then obtain (RWA)

$$I = -iN\omega_d \int_{-\infty}^{\infty} dt\Big\{\text{Tr}[|\mathbf{\Psi}^{(2)}(t)\rangle\langle\mathbf{\Psi}^{(1)}(t)|\mu(t)\mathbf{E}_d^*(\omega_d,\, t)$$

$$- \mu(t)\mathbf{E}_d(\omega_d,\, t)|\mathbf{\Psi}^{(1)}(t)\rangle\langle\mathbf{\Psi}^{(2)}(t)|]\Big\}\,. \tag{A.11}$$

Since $i\hbar\frac{\partial\Psi^{(2)}}{\partial t} = -\mu(t)\mathbf{E}_d(\omega_d,\, t)\Psi^{(1)}$, (A.11) can be written as

$$I = -iN\omega_d \int_{-\infty}^{\infty} dt\left\{\text{Tr}\left[\left|\Psi^{(2)}(t)\right\rangle\left\langle i\hbar\frac{\partial}{\partial t}\Psi^{(2)}(t)\right|\right.\right.$$

$$\left.\left. + \left|i\hbar\frac{\partial}{\partial t}\Psi^{(2)}\right\rangle\left\langle\Psi^{(2)}(t)\right|\right]\right\}$$

$$= -iN\omega_d \int_{-\infty}^{\infty} dt\left[\left\langle f\left|\Psi^{(2)}(t)\right\rangle\left\langle i\hbar\frac{\partial}{\partial t}\Psi^{(2)}(t)\right|f\right\rangle\right.$$

$$\left. + \left\langle f\left|i\hbar\frac{\partial}{\partial t}\Psi^{(2)}(t)\right\rangle\left\langle\Psi^{(2)}(t)\right|f\right\rangle\right]\,. \tag{A.12}$$

Because $|f\rangle$ is independent of the time in the interaction representation, Eq. (A.11) becomes

$$I = N\hbar\omega_d \int_{-\infty}^{\infty} dt \frac{\partial}{\partial t}\left|\left\langle \Psi^{(2)}(t) \middle| f \right\rangle\right|^2 = N\hbar\omega_d \left|\left\langle \Psi^{(2)}(\infty) \middle| f \right\rangle\right|^2 . \quad (A.13)$$

(A.13) is also consistent with the law of the conservation of energy.

In the above treatment, the transition $|e\rangle \to |f\rangle$ is assumed to be an absorption. If $\mathbf{E}_d$ is replaced by $\mathbf{E}_d^*$ and *vice versa*, the entire discussion can be applied to stimulated emission as well. i.e.,

$$\rho^{(3)}{}_{\mathrm{TSEP}}(t) = \rho_a^{(3)} + \rho_b^{(3)} + \text{c.c.}$$

$$= \left(\frac{-1}{i\hbar}\right)^3 \int_{-\infty}^{t} dt_3 \int_{-\infty}^{t_3} dt_2 \int_{-\infty}^{t_2} dt_1$$

$$\times \mu(t_3)\mathbf{E}_d^*(t_3)\mu(t_2)\mathbf{E}_p(t_2)\rho^{(0)}\mu(t_1)\mathbf{E}_p^*(t)$$

$$+ \left(\frac{-1}{i\hbar}\right)^3 \int_{-\infty}^{t} dt_3 \int_{-\infty}^{t_3} dt_2 \int_{-\infty}^{t_2} dt_1$$

$$\times \mu(t_3)\mathbf{E}_d^*(t_3)\mu(t_1)\mathbf{E}_p(t_1)\rho^{(0)}\mu(t_2)\mathbf{E}_p^*(t_2) + \text{c.c.} . \quad (A.14)$$

In keeping only the resonant terms in the third order nonlinear density matrix (A.8), however, the present treatment neglects the transient stimulated Raman (TSR) contributions to the observed signal. As discussed by many authors,[7-9] if the pump pulse is sufficiently short, transient population/coherence in the *ground* electronic state will be created due to the stimulated Raman excitation by the pump pulse. The resulted transients can interact with the probe pulse, generating additional signal at the probe wavelength. In our TSEP experiment of ozone, the pump pulse width is much longer than the vibrational period of the ground state molecule, and the TSR contributions due to ground-state vibrational motions can, therefore, be neglected.

Appendix B. Tensorial Analysis of the TSEP Nonlinear Polarization

In this appendix, we consider the dependence of the TSEP signal on the polarization configurations of the PUMP and the DUMP lasers. The discussion will be restricted to isotropic samples only. Using spherical tensor algebra, we explore explicitly the tensorial properties of the nonlinear response function. We show that, different components of the response function tensor can be extracted from the pump-probe experiment using different polarization-configurations and detection methods. In particular, the gain/loss spectroscopy is sensitive to the zeroth and second rank tensor-products. The coherent polarization spectroscopy on the other hand is sensitive to the first and second rank tensor products. From Eq. (2.12), the projection of nonlinear polarization along the direction of $\mathbf{e}_s$ is

$$\mathbf{e}_s \cdot \mathbf{P}^{\text{TSEP}}(t) = S_1(t) + S_2(t) \tag{B.1}$$

where

$$S_1(t) = -N \left(\frac{-1}{i\hbar}\right)^3 \int_{-\infty}^{t} dt_3 \int_{-\infty}^{t_3} dt_2 \int_{-\infty}^{t_2} dt_1$$

$$\times \, \varepsilon_d(t_3 - \Delta t)\varepsilon_p(t_2)\varepsilon_p(t_1)e^{i(\omega_d t_3 - \omega_p t_2 + \omega_p t_1)}$$

$$\times \, \text{Tr}\left\{[\mathbf{e}_s \cdot \mu(t)][\mathbf{e}_d^* \cdot \mu(t_3)][\mathbf{e}_p \cdot \mu(t_2)]\rho^{(0)}[\mathbf{e}_p^* \cdot \mu(t_1)]\right\} + \text{ c.c.} \tag{B.2}$$

$$S_2(t) = -N \left(\frac{-1}{i\hbar}\right)^3 \int_{-\infty}^{t} dt_3 \int_{-\infty}^{t_3} dt_2 \int_{-\infty}^{t_2} dt_1$$

$$\times \, \varepsilon_d(t_3 - \Delta t)\varepsilon_p(t_2)\varepsilon_p(t_1)e^{i(\omega_d t_3 - \omega_p t_2 + \omega_p t_1)}$$

$$\times \, \text{Tr}\left\{[\mathbf{e}_s \cdot \mu(t)][\mathbf{e}_d^* \cdot \mu(t_3)][\mathbf{e}_p \cdot \mu(t_1)]\rho^{(0)}[\mathbf{e}_p^* \cdot \mu(t_2)]\right\} + \text{ c.c.} \tag{B.3}$$

We can express the product of the two scalars, $\mathbf{e}_s \cdot \mu(t)$ and $\mathbf{e}_d^* \cdot \mu(t_3)$, in the form of spherical tensors,[49]

$$[\mathbf{e}_s \cdot \mu(t)][\mathbf{e}_d^* \cdot \mu(t_3)] = \sum_{kq}(-1)^{k-q}(\mathbf{e}_s \otimes \mathbf{e}_d^*)_q^{(k)}[\mu(t) \otimes \mu(t_3)]_{-q}^{(k)} \tag{B.4}$$

where

$$(\mathbf{a} \otimes \mathbf{b})_q^{(k)} \equiv \sum_m (-1)^q \sqrt{2k+1} \begin{pmatrix} 1 & 1 & k \\ m & q-m & q \end{pmatrix} \mathbf{a}(1, m)\mathbf{b}(1, q-m) \tag{B.5}$$

is the qth component of the kth rank compound irreducible tensor formed by contracting the vectors $\mathbf{a}$ and $\mathbf{b}$, and $\mathbf{a}(1, m)$ the mth component of the vector $\mathbf{a}$ in spherical tensor notation. Similarly,

$$[\mathbf{e}_p \cdot \mu(t_2)]\rho^{(0)}[\mathbf{e}_p^* \cdot \mu(t_1)] = \sum_{k'q'}(-1)^{k'-q'}(\mathbf{e}_p \otimes \mathbf{e}_p^*)_{q'}^{(k')}[\mu(t_2) \otimes \rho^{(0)}\mu(t_1)]_{-q'}^{(k')}.$$

$$\tag{B.6}$$

Consequently, the trace in (B.2) can be expressed as

$$
\begin{aligned}
T_1 &\equiv \mathrm{Tr}\left\{[\mathbf{e}_s \cdot \mu(t)][\mathbf{e}_d^* \cdot \mu(t_3)][\mathbf{e}_p^* \cdot \mu(t_2)]\rho^{(0)}[\mathbf{e}_p^* \cdot \mu(t_1)]\right\} \\[2mm]
&= \mathrm{Tr}\left\{\sum_{kq}(-1)^{k-q}(\mathbf{e}_s \otimes \mathbf{e}_d^*)_q^{(k)}[\mu(t) \otimes \mu(t_3)]_{-q}^{(k)} \right. \\[2mm]
&\qquad \left. \times \sum_{k'q'}(-1)^{k'-q'}(\mathbf{e}_p \otimes \mathbf{e}_p^*)_{q'}^{(k')}[\mu(t_2) \otimes \rho^{(0)}\mu(t_1)]_{-q'}^{(k')}\right\} \\[2mm]
&= \sum_{kk'qq'}(-1)^{k+k'-q-q'}(\mathbf{e}_s \otimes \mathbf{e}_d^*)_d^{(k)}(\mathbf{e}_p \otimes \mathbf{e}_p^*)_{q'}^{(k')} \\[2mm]
&\qquad \times \mathrm{Tr}\left\{[\mu(t) \otimes \mu(t_3)]_{-q}^{(k)}[\mu(t_2) \otimes \rho^{(0)}\mu(t_1)]_{-q'}^{(k')}\right\}. \qquad (\text{B.7})
\end{aligned}
$$

We can again construct an Lth rank irreducible compound tensor, $M_m^L(k, k')$, from the two tensors, $[\mu(t) \otimes \mu(t_3)]^{(k)}$ and $[\mu(t_2) \otimes \rho^{(0)}\mu(t_1)]^{(k')}$:

$$
\begin{aligned}
M_m^L(k, k') &= \sum_{qq'}(-1)^{k-k'+m}\sqrt{2L+1}\begin{pmatrix} k & k' & L \\ -q & -q' & -m \end{pmatrix} \\[2mm]
&\qquad \times [\mu(t) \otimes \mu(t_3)]_{-q}^{(k)}[\mu(t_2) \otimes \rho^{(0)}\mu(t_1)]_{-q'}^{(k')}. \qquad (\text{B.8})
\end{aligned}
$$

Using the orthogonality relations of the 3-J symbols, it can be shown that

$$
\begin{aligned}
&[\mu(t) \otimes \mu(t_3)]_{-q}^{(k)}[\mu(t_2) \otimes \rho^{(0)}\mu(t_1)]_{-q'}^{(k')} \\[2mm]
&= \sum_{Lm}(-1)^{k-k'+m}\sqrt{2L+1}\begin{pmatrix} k & k' & L \\ -q & -q' & -m \end{pmatrix}M_m^L(k, k'). \\
&\hspace{11cm}(\text{B.9})
\end{aligned}
$$

Let $q = q' = 0$ in (B.9) and taking traces on both sides of (B.8) and (B.9), one obtains

$$
\begin{aligned}
&\mathrm{Tr}\left\{[\mu(t) \otimes \mu(t_3)]_{-q}^{(k)}[\mu(t_2) \otimes \rho^{(0)}\mu(t_1)_{-q'}^{(k')}\right\} \\[2mm]
&= (-1)^q\delta_{kk'}\delta_{-qq'}\mathrm{Tr}\left\{[\mu(t) \otimes \mu(t_3)]_0^{(k)}[\mu(t_2) \otimes \rho^{(0)}\mu(t_1)]_0^{(k')}\right\}. \\
&\hspace{11cm}(\text{B.10})
\end{aligned}
$$

Eq. (B.7) thus reduces to

$$T_1 = \sum_k \left[\sum_q (-1)^q (\mathbf{e}_s \otimes \mathbf{e}_d^*)_q^{(k)} (\mathbf{e}_p \otimes \mathbf{e}_p^*)_{-q}^{(k)} \right]$$

$$\times \, \mathrm{Tr} \left\{ [\mu(t) \otimes \mu(t_3)]_0^{(k)} \left[\mu(t_2) \otimes \rho^{(0)} \mu(t_1) \right]_0^{(k)} \right\} . \tag{B.11}$$

Similarly, it can be shown that

$$T_2 \equiv \mathrm{Tr} \left\{ [\mathbf{e}_s \cdot \mu(t)] \, [\mathbf{e}_d^* \cdot \mu(t_3)] \, [\mathbf{e}_p \cdot \mu(t_1)] \, \rho^{(0)} \, [\mathbf{e}_p^* \cdot \mu(t_2)] \right\}$$

$$= \sum_k \left[\sum_q (-1)^q (\mathbf{e}_s \otimes \mathbf{e}_d^*)_q^{(k)} (\mathbf{e}_p \otimes \mathbf{e}_p^*)_{-q}^{(k)} \right]$$

$$\times \, [\mu(t) \otimes \mu(t_3)]_0^{(k)} [\mu(t_1) \rho^{(0)} \otimes \mu(t_2)]_0^{(k)} . \tag{B.12}$$

In the TSEP experiment, $\mathbf{e}_p = \sqrt{\frac{1}{2}}(\hat{\mathbf{x}} + \hat{\mathbf{z}})$, $\mathbf{e}_d = \hat{\mathbf{z}}$ and $\mathbf{e}_s = \hat{\mathbf{x}}$. As a result,

$$\left[\sum_q (-1)^q (\mathbf{e}_s \otimes \mathbf{e}_d^*)_q^{(k)} (\mathbf{e}_p \otimes \mathbf{e}_p^*)_{-q}^{(k)} \right] = \delta_{2,k} \qquad \text{(linear)} \tag{B.13}$$

and

$$\left[\sum_q (-1)^q (\mathbf{e}_s \otimes \mathbf{e}_d^*)_q^{(k)} (\mathbf{e}_p \otimes \mathbf{e}_p^*)_{-q}^{(k)} \right] = i\delta_{1,k} \qquad \text{(circular)} . \tag{B.14}$$

Therefore, for linearly polarized PUMP,

$$T_1 = \mathrm{Tr} \left\{ [\mu(t) \otimes \mu(t_3)]_0^{(2)} [\mu(t_2) \otimes \rho^{(0)} \mu(t_1)]_0^{(2)} \right\} , \tag{B.15}$$

$$T_2 = \mathrm{Tr} \left\{ [\mu(t) \otimes \mu(t_3)]_0^{(2)} [\mu(t_1) \rho^{(0)} \otimes \mu(t_2)]_0^{(2)} \right\} . \tag{B.16}$$

For circularly polarized PUMP

$$T_1 = i \, \mathrm{Tr} \left\{ [\mu(t) \otimes \mu(t_3)]_0^{(1)} [\mu(t_2) \otimes \rho^{(0)} \mu(t_1)]_0^{(1)} \right\} , \tag{B.17}$$

$$T_2 = i \, \mathrm{Tr} \left\{ [\mu(t) \otimes \mu(t_3)]_0^{(1)} [\mu(t_1) \rho^{(0)} \otimes \mu(t_2)]_0^{(1)} \right\} . \tag{B.18}$$

It is worthwhile to compare the above results with the more familiar pump-probe spectroscopy that measures the gain (or loss) of the probe pulse. For the latter, the corresponding nonlinear polarization is $\mathbf{e}_d \cdot \mathbf{P}(t)$ and the relevant trace of the tensorial product is

$$T = \sum_k \left[\sum_q (-1)^q (\mathbf{e}_d \otimes \mathbf{e}_d^*)_q^{(k)} (\mathbf{e}_p \otimes \mathbf{e}_p^*)_{-q}^{(k)} \right]$$
$$\times \operatorname{Tr} \left\{ [\mu(t) \otimes \mu(t_3)]_0^{(k)} [\mu(t_2) \otimes \rho^{(0)} \mu(t_1)]_0^{(k)} \right\} . \tag{B.19}$$

Let θ be the angle between the polarization planes of the linearly polarized pump and the probe fields, then it is easy to show that

$$\left[\sum_q (-1)^q (\mathbf{e}_d \otimes \mathbf{e}_d^*)_q^{(k)} (\mathbf{e}_p \otimes \mathbf{e}_p^*)_{-q}^{(k)} \right] = \frac{1}{3} \delta_{0,\,k} + \frac{2}{3} P_2(\cos\theta) \delta_{2,\,k} , \tag{B.20}$$

where $P_2(\cos\theta)$ is the second Legendre polynomial. Consequently,

$$T = \frac{1}{3} \operatorname{Tr} \left\{ [\mu(t) \otimes \mu(t_3)]_0^{(0)} [\mu(t_2) \otimes \rho^{(0)} \mu(t_1)]_0^{(0)} \right\}$$
$$+ \frac{2}{3} P_2(\cos\theta) \operatorname{Tr} \left\{ [\mu(t) \otimes \mu(t_3)]_0^{(2)} [\mu(t_2) \otimes \rho^{(0)} \mu(t_1)]_0^{(2)} \right\} . \tag{B.21}$$

From (B.21), we see that the difference absorption (emission) signal in the linearly polarized pump-probe spectroscopy contains contributions from two terms. One is the product of the scalars and the other the product of the second-rank irreducible tensors. In general, the dependence of the third order nonlinear polarization, $\mathbf{P}^{(3)}(t)$, on the experimental configurations can be easily obtained from equations similar to (B.11) and (B.12). In this paper, we will focus on the TSEP in which

$$S_1(t) = -N \left(\frac{-1}{i\hbar} \right)^3 \int_{-\infty}^t dt_3 \int_{-\infty}^{t_3} dt_2 \int_{-\infty}^{t_2} dt_1$$
$$\times \varepsilon_d(t_3 - \Delta t) \varepsilon_p(t_2) \varepsilon_p(t_1) e^{i(\omega_d t_3 - \omega_p t_2 + \omega_p t_1)}$$
$$\times \operatorname{Tr} \left\{ [\mu(t) \otimes \mu(t_3)]_0^{(2)} [\mu(t_2) \otimes \rho^{(0)} \mu(t_1)]_0^{(2)} \right\} + \text{c.c.} \tag{B.22}$$

$$S_2(t) = -N \left(\frac{-1}{i\hbar} \right)^3 \int_{-\infty}^{t} dt_3 \int_{-\infty}^{t_3} dt_2 \int_{-\infty}^{t_2} dt_1$$

$$\times \, \varepsilon_d(t_3 - \Delta t)\varepsilon_p(t_2)\varepsilon_p(t_1)e^{i(\omega_d t_3 - \omega_p t_1 + \omega_p t_2)}$$

$$\times \text{Tr} \left\{ [\mu(t) \otimes \mu(t_3)]_0^{(2)} [\mu(t_1) \otimes \rho^{(0)}\mu(t_2)]_0^{(2)} \right\} + \text{ c.c.} \qquad \text{(B.23)}$$

When the pulse duration is much shorter than the time scale of the molecular dynamics, the pulse envelops, $\varepsilon_p(t)$ and $\varepsilon_d(t)$, can be approximated by δ-functions. Equations (B.22) and (B.23), therefore, converge into one:

$$\mathbf{e}_s \cdot \mathbf{P}^{\text{TSEP}}(t)$$

$$= -2N \left(\frac{-1}{i\hbar} \right)^3 \text{Tr} \left\{ [\mu(t) \otimes \mu(\Delta t)]_0^{(2)} [\mu(0)\rho^{(0)} \otimes \mu(0)]_0^{(2)} \right\} e^{i\omega_d \Delta t} .$$
$$\text{(B.24)}$$

References

1. M. Gruebele, and A. H. Zewail, *Physics Today* **43**, 24 (1990); A. H. Zewail, *Science* **242**, 1645 (1988).
2. J. H. Glownia, J. A. Misewich, and P. P. Sorokin, *J. Chem. Phys.* **92**, 3335 (1990).
3. Y. Chen, L. Hunziker, P. Ludowise, and M. Morgen, *J. Chem. Phys.* **97**, 2149 (1992).
4. G. Stock, C. Woywod and W. Domcke, *Chem. Phys. Lett.* **200**, 163 (1992).
5. R. R. Alfano, *The Ultrafast Supercontinuum Laser Source* (Springer, Berlin, 1989).
6. Y. R. Shen, *The Principles of Nonlinear Optics* (John Wiley & Sons, New York, 1984).
7. Y. J. Yan, L. E. Fried, and S. Mukamel, *J. Phys. Chem.* **93**, 8149 (1989).
8. W. T. Pollard, S-Y. Lee, and R. A. Mathies, *J. Chem. Phys.* **92**, 4012 (1990); W. T. Pollard and R. A. Mathies, *Ann. Rev. Phys. Chem.* **43**, 497 (1992).
9. G. Stock and W. Domcke, *J. Opt. Soc. Am.* **B7**, 1970 (1990).
10. C. Wieman and T. W. Hänsch, *Phys. Rev. Lett.* **36**, 1170 (1976).
11. M. D. Levenson and G. L. Eesley, *Appl. Phys.* **19**, 1 (1979).
12. M. D. Levenson, *J. Raman Spec.* **10**, 9 (1981).
13. G. L. Eesley, *Coherent Raman Spectroscopy* (Pergamon Press, New York, 1981).
14. Q. Zhang, S. A. Kandel, T. A. W. Wasserman, and P. H. Vaccaro, *J. Chem. Phys.* **96**, 16407 (1992).
15. J-L. Oudar, R. W. Smith, and Y. R. Shen, *Appl. Phys. Lett.* **34**, 758 (1978).
16. C. V. Shank, E. P. Ippen, O. Teschke, and K. B. Eisenthal, *J. Chem. Phys.* **67**, 5547 (1978).

17. J. J. Song, J. H. Lee, and M. D. Levenson, *Phys. Rev.* **A17**, 1439 (1978).

18. D. Waldeck, A. J. Cross, Jr., D. B. McDonald, and G. R. Fleming, *J. Chem. Phys.* **74**, 3381 (1981).

19. N. Kohles and A. Laubereau, *Chem. Phys. Lett.* **138**, 365 (1987).

20. J. Chesnoy and A. Mokhtari, *Phys. Rev.* **A38**, 3566 (1988).

21. D. McMorrow, W. T. Lotshaw, and G. A. K-Wallace, *Chem. Phys. Lett.* **145**, 309 (1988).

22. N. F. Scherer, L. D. Ziegler, and G. R. Fleming, *J. Chem. Phys.* **96**, 5544 (1992).

23. T. K. Yee and T. K. Gustafson, *Phys. Rev.* **A18**, 1597 (1978).

24. D. Imre, J. L. Kinsey, A. Sinha, and J. Krenos, *J. Phys. Chem.* **88**, 3956 (1984).

25. A. B. Myers, *J. Opt. Soc. Am.* **B7**, 1665 (1990); K. Q. Lao, M. D. Person, P. Xayariboun and L. J. Butler, *J. Chem. Phys.* **82**, 823 (1990).

26. L. D. Ziegler, *J. Chem. Phys.* **84**, 6013 (1986).

27. E. J. Heller, *Acc. Chem. Res.* **14**, 368 (1981), E. J. Heller, R. Sundberg, and D. Tannor, *J. Phys. Chem.* **86**, 1822 (1982).

28. W. M. Gelbart, S. A. Rice, and K. F. Freed, *J. Chem. Phys.* **57**, 4699 (1972).

29. K. F. Freed, in *Radiationless Processes*, ed. F. K. Fong, *Topics in Applied Physics*, **15** (Springer, Berlin, 1976), p. 23.

30. H. Köppel, L. S. Cederbaum, and W. Domcke, *J. Chem. Phys.* **77**, 2014 (1982).

31. R. Schneider and W. Domcke, *Chem. Phys. Lett.* **150**, 235 (1988).

32. V. I. Arnold and A. Avez, *Ergodic Problems of Classical Mechanics* (W. A. Benjamin, Inc., New York, 1968).

33. M. C. Gutzwiller, *Chaos in Classical and Quantum Mechnics* (Springer, New York, 1990).

34. F. Haake, *Quantum Signatures of Chaos* (Springer, Berlin-Heidelberg, 1991).

35. C. V. Shank, in *Ultrashort Laser Pulses and Applications* ed. W. Kaiser (Springer–Verlag, Berlin, 1988).

36. J. I. Steinfeld, S. M. A. Golden, and J. W. Gallagher, *J. Phys. Chem.* Ref. Data, (1987).

37. H. B. Levene, J. C. Nieh, and J. J. Valentini, *J. Chem. Phys.* **87**, 2583 (1987).

38. P. J. Hay and T. H. Dunning, Jr., *J. Chem. Phys.* **67**, 2290 (1977).

39. M. Braunstein, P. J. Hay, R. L. Martin, and R. T. Pack, *J. Chem. Phys.* **95**, 8239 (1991); M. Braunstein, and R. T. Pack, *J. Chem. Phys.* **96**, 891 (1992).

40. A. Banichevich, S. D Peyerimhoff, J. A. Beswick, and O. Atabek, *J. Chem. Phys.* **96**, 6580 (1992).

41. G. Herzberg, *Molecular Spectra and Molecular Structure, III* (van Nostrand Reinhold Co., New York, 1966).

42. S. M. Anderson, J. Maeder, and K. Mausersberger, *J. Chem. Phys.* **94**, 6351 (1991).

43. J. A. Valdmanis, R. L. Fork, and J. P. Gordon, *Opt. Lett.* **10**, 131 (1985).

44. R. L. Fork, O. E. Martinez, and J. P. Gordon, *Opt. Lett.* **9**, 150 (1984).

45. A. Barbe, C. Secroun, and P. Jouve, *J. Mol. Spectrosc.* **49**, 171 (1974).

46. E. Vigroux, *Ann. Phys. (Paris)* **8**, 709 (1953).

47. J. J. Scherer and R. Saykally, private commuication.

48. Similar conclusions have been arrived in the context of dynamic absorption spectroscopy by Pollard *et al* and Stock *et al.* ee Ref. 8 and 9.

49. D. J. Tannor and S. A. Rice, *Adv. Chem. Phys.* **70**, 441 (1987); S. O. Williams and D. G. Imre, *J. Phys. Chem.* **92**, 6648 (1988); V. Engel and H. Metiu, *J. Chem. Phys.* **91**, 1596 (1989).

50. B. Fain, S. H. Lin, and N. Hamer, *J. Chem. Phys.* **91**, 4485 (1989).

51. R. N. Zare, Eq. (4.15), *Angular Momentum* (John Wiley & Sons, New York, 1988).

CHAPTER 3

STIMULATED EMISSION PUMPING
BY FLUORESCENCE DIP: EXPERIMENTAL METHODS

Carter Kittrell

Department of Chemistry and The Rice Quantum Institute
Rice University, Houston, Texas, USA
Mailing address: Office of the Dean of Natural Science
P. O. Box 1892 Houston, TX, 77251, USA

Contents

109

1. Introduction: Creating and Detecting Molecular Vibrational–Rotational Levels

Stimulated emission pumping (SEP) is a method to excite molecules from thermally populated rotational–vibrational levels in the ground electronic state to selected rotational–vibrational levels, usually also in the ground electronic state. Since this method uses allowed transitions to and from an intermediate electronically-excited state, the transfer can be quite efficient. There are two parts to the successful SEP experiment. The first is to transfer population to the selected level. The second is to determine that such a transfer has occurred, preferably in a quantitative manner. This distinction is not trivial: Consider a gas phase species in which a small number of molecules in a single highly excited rotational–vibrational level have been instantly generated by an unspecified means. How would this excited level population manifest its presence? There would be spontaneous infrared emission, but this is much slower than electronic emission and, along with the difficulty of monitoring small numbers of low energy infrared photons, very low sample pressures would be needed to prevent collisional relaxation which would, in turn, alter the detected population. To use a third probe laser back to the electronically excited state would add complexity to the experiment and there is no guarantee that it would always be at the right frequency to observe the newly excited rovibrational level. The best and simplest case would be that the SEP process be one which also simultaneously provides a signal that it has succeeded.

SEP is a folded variant of optical–optical double resonance (OODR).[1] For the latter case, the molecule is excited into a high-lying electronic state by utilizing two upward electronic transitions, with a bound intermediate state in between. Barring dissociation or rapid internal conversion, and depending on selection rules, fluorescence will occur back to the ground state. Or, in a two step cascade, it will also occur back through the intermediate electronic state. By this means, the molecules provide a spontaneous, quantitative report of their state of excitation. But for SEP, the second laser stimulates emission rather than absorption, usually returning population to the ground electronic state; and there is lack of a convenient spontaneous emission means for monitoring the final populated level. The experiment needs to be designed so that it will both move the population and simultaneously generate a quantitative report of the successful transfer to the final level. If this can be accomplished, the technique will be suitable for both spectroscopy and dynamics of highly excited vibrational levels. In addition, since detector sensitivity currently favors observation of visible or ultraviolet photons over mid-infrared photons, it would be preferable to take advantage of this factor.

It quickly becomes apparent that the electronically excited intermediate state is going to play the major role in providing the optical information for an SEP experiment. The generally used techniques of transient gain, polarization rotation[2,3] and fluorescence from the intermediate state all fit this approach. All of these have proven successful. To launch the SEP effort at MIT, one method needed to be selected. Transient gain, like absorption, is generally less sensitive than fluorescence detection, since it has a small signal riding on top of the large amplitude of the downward stimulating laser beam. Polarization rotation requires some rather careful interpretation to obtain a quantitative measure of the population that is transferred. For both of these techniques, use of increased laser power to force a transition to very high vibrational levels with poor Franck–Condon overlap means looking for small changes in ever larger laser pulse energies. Both of these techniques, however, have proved very useful for monitoring collisionally induced population loss in the excited state,[4] and the SEP-pumped vibrationally excited ground state level.[5]

Fluorescence is a very sensitive and direct measure of an excited state population and was chosen because it is straightforward and of general applicability. The approach used in this chapter is to provide to a researcher the information and special considerations which will lead to a reliable SEP experimental apparatus based on detection of fluorescence from the intermediate state. A basic familiarity with pulsed laser spectroscopy will be assumed.

2. Laser Light for Fluorescence Dip SEP

The desired transfer of population from the intermediate electronically excited state to the ground electronic state can be readily monitored by measuring the effect of the downward stimulated transition on the fluorescence from this intermediate state. The PUMP laser populates the excited intermediate state. The DUMP laser follows immediately after the PUMP laser and causes stimulated emission. This removes population from the intermediate state, and thereby causes a decrease in the fluorescence emission from this state compared to what it would be if there were no DUMP laser beam or if the DUMP laser were tuned off resonance. Since this depletion of fluorescence provides a quantitative measure of the population transfer, a means to observe it is needed. The nature of pulsed laser sources play a major role in the design of the experiment; a robust system, which is relatively insensitive to the fluctuations in the laser light is desirable. A balanced, dual beam null detection system is designed to observe this "fluorescence dip" with good sensitivity, even with the irregular fluctuations typically occurring in the output of pulsed dye lasers.

2.1. *Pulse Dye Laser Sources*

Pulsed laser pumped dye lasers of about 10 ns duration are an excellent light source for rapid pumping of molecules in large numbers. These lasers are available commercially and are sufficiently narrowband (with some care) to roughly match the Doppler width of small gas-phase molecules. The pulse is short enough to transfer population to the final state with a minimal number of gas-kinetic collisions when working in the sub-Torr pressure range. Such rapid, level-specific transfer of population is particularly attractive for relaxation, reaction, and dynamics studies. A large selection of dyes and efficient nonlinear frequency conversion methods provide continuously tunable laser radiation which covers all wavelengths from the deep ultraviolet to the near infrared. However, the output characteristics of this laser significantly impact the choice of method used to detect the fluorescence dip signal. If all of the laser pulses were truly identical, simply blocking the "DUMP" laser on alternate pulses and comparing the two signal levels would yield the desired result, with the shot-to-shot difference in fluorescence intensity providing a measure of population transfer.

Such an approach had previously been tried on an OODR experiment which had the alternate diagnostic of monitoring final level fluorescence.

Emission from the final level in high-lying electronic state of Li_2 yielded excellent signals,[6] using broadband detection. But the fluorescence dip from the intermediate state was not a useful monitor of the upward transfer of population in this single detection channel pulse experiment using alternate beam blocking. The problem lay not in the first part of the experiment, which is creating the population transfer, and it was very successful. It lay in the second part, which is in determining that population has been transferred out of the intermediate level. Pulse-to-pulse fluctuations in the laser output obscured the fluorescence decrease signal.

This problem is inherent in the general nature of 10 ns pulse dye lasers. The laser light characteristics interfere with a single channel method of monitoring population removal. The design of a reliable SEP fluorescence dip experiment requires a more detailed understanding of the characteristics of the pulsed dye laser.

2.2. *Characteristics of Commercial Pulse Dye Lasers*

The various commercial dye lasers, pumped by excimer (actually exciplex) or harmonics of the Q-switched Nd:YAG lasers, have considerable pulse-to-pulse variations. This manifests itself in many ways, three of which will be considered in relation to the fluorescence dip experiment. The first, pulse energy amplitude jitter, is the most easily observed of these fluctuations and will be directly added to the amplitude of the fluorescence signal. It is hard to observe a 1% fluorescence decrease on alternate pulses when there is a 20% laser amplitude jitter. This is not a fatal flaw, as the laser pulse amplitude can be measured and the signal normalized to total laser energy, compensating for pulse-to-pulse amplitude fluctuations. For accurate normalization, this needs an intermediate level fluorescence signal which is precisely linear in laser pulse energy.

Spatial beam jitter is the second problem. Deviation of laser beams entering the SEP cell varies the overlap and causes the SEP fluorescence dip signal to fluctuate. This is aggravated even further if the fluorescent light passes through a spectrometer slit or other aperture. Wide aperture or open detection systems will reduce this problem, but at the cost of collecting more scattered light from the cell. Furthermore, the scattered light may itself vary with beam position.

The third and perhaps the most pernicious problem is the frequency structure in the dye laser output and the pulse-to-pulse jitter in this

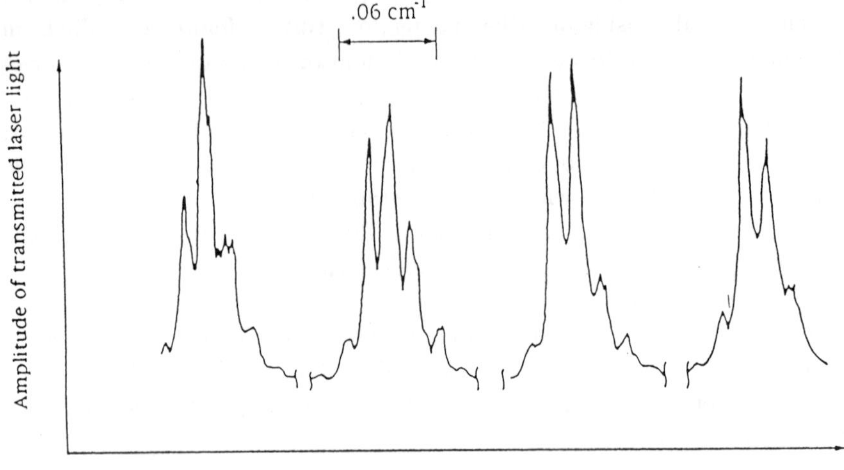

Fig. 1. Interferometric spectral analysis of the Quanta–Ray PDL-1 dye laser output operating on a fixed wavelength. A Molectron 1 cm^{-1} dye laser etalon, with a finesse of about 25, was used intracavity. The spectrum was recorded with a scanning Fabry-Perot interferometer. The cavity mode spacing is 0.013 cm^{-1}. Four or five cavity modes are observable. The four sets of scans were each taken a couple of minutes apart to show the gradual drift in relative amplitude of the modes. The dark space between the modes is considerably more pronounced than shown here but is partially filled in by pulse-to-pulse jitter. About 300 laser shots at 10 pps were needed to generate one scan. Higher dye flow rates and stronger pumping by 532 nm from the Nd:YAG laser increases the frequency jitter of the modes, and the time averaging of many jittering pulses will appear to wash out mode structure in the observed laser bandshape. Detecting only one frequency at a time may therefore lead to the impression that mode structure is much less pronounced than it actually is for a single laser shot.

structure. Of these three problems, it is the most difficult to observe and the most difficult for which to compensate. Although the output is usually characterized by an overall spectral bandwidth, this is really an envelope, and not a spectral profile; the laser output of most commercial dye lasers is highly structured (Fig. 1). Even if the average of many pulses becomes spectrally smooth, the problem is not solved. Each laser shot should be viewed as a separate experiment. Longitudinal cavity modes dominate the spectral structure within the envelope. These modes shift more or less with every laser pulse, especially if the laser is being scanned. For commercial lasers using a beam expander and grating only, there are many

~ 0.013 cm^{-1} spaced modes in the typical 1/3 cm^{-1} bandwidth envelope. So one may ask: "What is worse than a lot of cavity modes in the spectral envelope of the laser output?" The answer, which is far from obvious, is: "Few cavity modes in the spectral envelope."

It is very fortunate that insertion of an intracavity etalon next to the grating of a pulsed dye laser provides a spectral envelope which is approximately equal to room temperature Doppler width of small gas phase molecules. This makes it easy to do the first part of the SEP experiment, which is level specific population transfer with minimal overlap from adjacent rotational levels. It is very unfortunate that the laser will now be operating on only a few longitudinal modes (typically 2–4) and, often, with a dominant single mode. This makes it hard to do the second part of the experiment, which is to quantitatively observe the fluorescence dip by comparing alternately blocked DUMP laser pulses.

Longitudinal mode structure was examined for a Quanta–Ray PDL-1, a Lambda Physik FL2002E, and a Molectron DL 16P dye laser,[7] all equipped with intracavity etalons. The TecOptics[8] etalon used in the Quanta–Ray dye laser was purchased separately by the author. It is important for intracavity etalons to meet stringent specifications on mirror flatness and parallelism so that the finesse is primarily dictated by reflectivity. If not, the etalon will have diminished throughput even at the peak transmission wavelength. The spectral analysis was made using a scanning Fabry–Perot interferometer.[8,9] (A second non-scanning interferometer was used as a spectral pre-filter.) The lasers show sharp modes with a bandwidth typically less than 1/3 of the intermode spacing. The old Molectron DL300 used by the author[10] has closely spaced longitudinal modes. Using high intensity laser light more or less saturates the excitation, and tends to interact with all velocity groups. However, saturated PUMPing and DUMPing makes the spectra sensitive to more subtle structure in the dye laser, such as spectral hole burning in the gain medium, and leads to nonlinear response and spectral broadening; when making linewidth measurements, a study of fluence dependence is particularly important.

The dye laser mode structure shifts with every pulse. The Quanta–Ray PDL-1 mode structure can be fairly well stabilized using reduced pump power to the oscillator section and by varying the dye flow rate; spectral mode patterns tend to persist for about one minute. The Molectron DL-II dye laser is less readily stabilized, and the Lambda Physik FL2002E proved to be the least amenable to mode stabilization. (Mode randomization was generally considered beneficial to SEP results, so that stabilization techniques were not employed.) The FL2002E laser has been recently analyzed.[11] Mode spacing for this laser is $\sim$ 500 MHz with mode widths $\sim$ 80 MHz. This observed longitudinal mode width is roughly the same bandwidth typically achieved from pulse amplification of a single frequency continuous wave dye laser by a multi-mode Nd:YAG laser.

The above Quanta–Ray dye laser often operates with a dominant single longitudinal mode. While the near single mode operation may appear appealing from a narrow linewidth standpoint, it can be hazardous for accurate energy level measurements. If the near single mode PUMP laser tends to select, for example, "positive" velocity groups more often than "negative" ones, the linewidth may be quite narrow, but the line center of the SEP signal will be shifted away from the true value. Since studies of dye laser mode structure by the author showed that such structure may persist for 30 seconds or more, this is generally a greater length of time than it takes to scan over an SEP spectral feature, and the effect of nonrandom velocity selection will leave its imprint on the frequency of the SEP transition, giving an incorrect value.

2.3. *PUMPing and DUMPing: Too much for Some and Not Enough for Others*

Assume that the PUMP laser is tuned to a single rovibronic transition in a small gas phase species and that the laser linewidth is roughly matched to the Doppler width of the selected transition in the molecule to be studied. The result of this mode structure is that the dye laser, or frequency doubled output thereof, selects only some of the velocity groups of the target molecule. For modest fluences, some groups of molecules which are just in the right place in space-velocity coordinates will be strongly PUMPed, and maybe these will have a transition which is "saturated". But the majority of the molecules in the selected initial state are subject to moderate or minimal PUMPing on any given laser pulse. This is too much for some and too little for others, and it changes with every laser pulse. The transition may be heavily saturated by an intense fluence, if enough laser energy is available. But, saturation causes the fluorescence to no longer be proportional to laser pulse energy, hence reducing the effectiveness of ratioing the fluorescence signal to the pulse energy. To eliminate saturation and regain linearity, PUMP laser power would need to be made quite small, thereby reducing overall population transfer.

A series of "fingers" are cut into the ground state Doppler profile by the PUMP laser, and this part of the population is transferred to the excited state. These features, one to four in number, form a very coarse grid upon the Doppler profile. The interfinger spacing is constant; and is a characteristic of the dye laser design. It is not difficult to imagine that a frequency shift of the grid would not only change the velocity groups selected, but also the total number of molecules available for interaction. For higher fluences, the amount of saturation will vary with the molecular velocity, as does the amount of Rabi cycling. The width and depth of "spectral dark space" between modes may vary. The FL2002E has an even more peculiar problem in that most of the time the mode pattern is pronounced, but sometimes it is washed out. The PUMPing rate changes with the mode pattern, even if the pulse energies remain identical, so that ratioing to pulse energy would be ineffective for this problem. Furthermore, the pattern shifts with nearly every pulse, a fact that is reflected in the fluorescence amplitude.

The DUMP laser has a similar set of shifting spectral fingers, which may or may not overlap with those of the PUMP laser. So even if saturation effects are avoided by using a weak PUMP, there will still be a considerable fluctuation in the fluorescence dip signal due to drifting and random fluctuations in the interweaving of the two mode structures with the interacting velocity groups of the target molecules. Driving the system harder to alleviate this problem exacts a price of nonlinear dependence of the fluorescence on PUMP laser energy. Some possibilities for improved mode structure are discussed in a later section.

An etalon narrowed dye laser can have extra etalon modes which generate artificial spectra corresponding to the etalon free spectral range, typically 1 cm^{-1}. These can be troublesome in regions of spectral congestion. When transitions are saturated by high laser fluence, these false spectra become more pronounced. The use of a second harmonic generation crystal does not mitigate the problem; although the weak extra mode will be considerably lower in amplitude after frequency doubling, it will mix with the fundamental and produce spurious ultraviolet output which is shifted by one mode spacing, rather than two, e.g., 1 cm^{-1} and not 2 cm^{-1} from the main ultraviolet output. The summation conversion is linear in the intensity of the weak mode. An extra mode monitor, consisting of an extracavity etalon with a viewing screen, is quite important.

A more sophisticated monitor can be constructed using a Fizeau interferometer and a photodiode array.[11] The laser mode structure could then be viewed on an oscilloscope or digitized for computer acquisition. Such an electronic monitor could be used to generate a warning flag if the extra mode amplitude exceeds a certain fraction of the amplitude of the main laser mode. Unacceptably distorted laser mode structure may also be cause for selective rejection of those laser pulses.

The laser frequency is determined by passing the beam through an iodine, tellurium, or optogalvanic cell. The absorption of the laser is detected by a photodiode as the laser is scanned. The vapor pressure in the cell must be low enough to avoid condensation on the cell windows; a cold finger is one way to do this. The iodine cell may also use fluorescence detection instead of absorption. It is helpful for the high temperature tellurium cell to have tubular extensions beyond the windows and the furnace area to minimize air turbulence which distorts the transmitted laser beam.

Due to the multitude of beams and the large number of optical components utilized in an SEP experiment, there are some special considerations with regard to laser safety, and should be included in the design of the apparatus. This is in addition to, and not in lieu of, general laser safety requirements which are covered in Ref. 12. The control of stray beams is particularly important, as even a few microjoules of a 10 ns laser pulse reaching the retina can cause eye damage. The $\sim 4\%$ reflection which occurs for near-normal incidence light impinging on uncoated glass or fused silica surfaces is a common source of this problem. The best approach is that all specular reflections, no matter how small, be appropriately terminated. Stray beams should not be permitted to leave the optical table. The optical layout should be, as much as possible, configured in a horizontal plane well below eye level for the researcher. When the laser beam passes to apparatus off of the table, it is necessary to prevent personnel from crossing this beam path.

There are a variety of beam shields and beam stops which have been utilized in the laboratory. For low energy beams, an "L" shape of 1/16″ aluminum which has been sandblasted and black anodized is suitable for attaching to the optical table via a 1/4″ slot in the base. An extra fold in the "L", such as illustrated in Fig. 2, item 22, will provide better balance for a freestanding beam stop. The aluminum surface is roughened, as a shiny surface will cause dangerous specular reflection even if it is jet black. For moderate energy beams, the black beam stop is folded like a letter "C", but with the top sloping upward. The laser light hits the inside of the top and is deflected downward towards the back of the "C", further diffusing the beam.

For somewhat higher energy beams, a long stainless steel tube (or copper, which absorbs ultraviolet light rather well) with a tapered, pinched off end is useful. The tube axis is slightly at an angle to the entering laser beam, which undergoes multiple scattering reflections inside, diffusing and absorbing the beam. The high average power beams of currently available Nd:YAG and excimer lasers will need a carefully designed beam DUMP with convective/radiative cooling fins or water cooling. The black anodized aluminum beam stops are not suitable for high energy beams, as the beam will bleach and ablate the black coating. The exposed metallic aluminum will scatter the light. It is important to recognize that even diffuse light from an intense laser beam scattering from a rough surface can be harmful. When a "snapping" sound is heard on any item blocking a laser beam, this is an indication of material being ablated. Not only will the target be damaged, but optics in the vicinity may become coated with debris! The resulting plasma can itself be a source of intense scattering.

Single element lenses have two back reflections. For the lenses most commonly used, a converging, collimated, or weakly diverging beam entering the lens will generate at least one converging back reflection. The typical 4% reflection of an uncoated glass surface will generated a focal spot which is considerably more intense than the incident beam and can damage the laser, optics, and mirrors, and possibly cause injury. Many a graduate student independently rediscovers this unanticipated source of optics damage. The back reflections of lenses must be located and blocked. The incident beam is passed through a hole in a beam stop at an appropriate distance from the lens; this stop will catch the diverging reflection. Tilting the lens slightly off axis will direct the converging beam away from the incoming beam path so that it can be blocked. (Excessive tilting of the lens will introduce off-axis aberrations into the beam.)

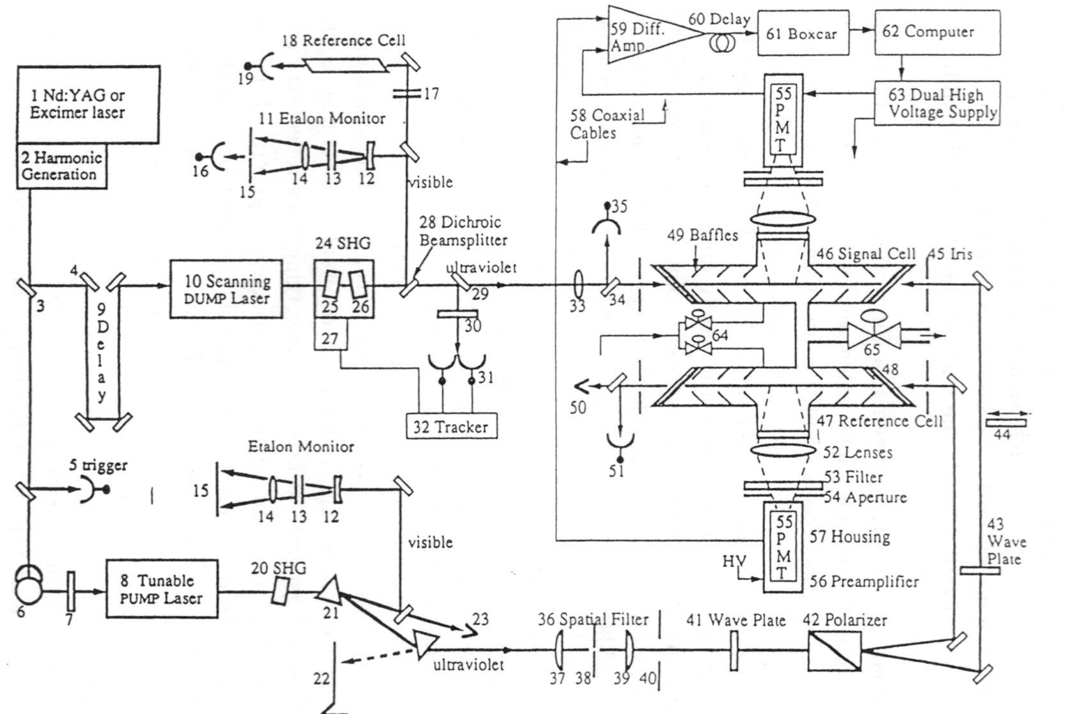

Fig. 2. Generalized layout of the fluorescence dip SEP experiment using pulsed lasers. All light propagation is in the plane of the page which is assumed to be horizontal. This system is designed for an excellent null signal to provide high sensitivity to small dip signals and a large number of features have been incorporated, some of which may be omitted if more robust signals are expected. Tests with iodine have shown that the null balance can come within a factor of two of the shot noise limit for fluorescence signals of about 10 000 detected photons per laser pulse.

1. Q-switched Nd:YAG or excimer laser source. It typically has a 6–20 ns pulse duration.
2. Harmonic generation and beam separation for the Nd:YAG. This is not needed for the excimer.
3. Beamsplitter to divide main laser. It is dielectric coated and anti-reflection (AR) coated on back. A polarizing beamsplitter for the excimer source is preferable, as both of the tunable lasers can then be PUMPed with same polarization after rotating one component. When using both 532 nm and 355 nm from the Nd:YAG, the beamsplitter isn't necessary, but the polarizations from the mixing crystal are orthogonal, and one polarization must be rotated.
4. Mirror or beamsteering 45-45-90 prism. The hypotenuse of the prism should be mounted parallel to the face of the mirror mount holder so that the mirror adjustments provide orthogonal beam steering and minimize beam translation.
5. Photodiode trigger. An optically driven trigger pulse has less jitter than that from laser electronic trigger circuits. Operation in the photoconductive mode with applied voltage at a substantial fraction of the breakdown voltage causes charge depletion and low junction capacitance for fastest response.
6. Polarization rotator. A pair of $90°$ turning laser mirrors arranged with a $90°$ dihedral angle between them makes an economical polarization rotator, although it changes the beam height. Polarization must be precisely in the plane of incidence ("P"), or precisely perpendicular to it ("S") to avoid introducing ellipticity.
7. Waveplate. (Used instead of item 6.) Horizontal polarization is considerably less efficient than vertical for PUMPing a dye laser and may not work at all.
8. Tunable PUMP laser. There should be an optical delay of about 3 ns in the Nd:YAG or excimer PUMP beam between the oscillator and the first amplifier to suppress amplified spontaneous emission (ASE) in a dye laser; this is especially important if an intracavity etalon is used, which increases the oscillator "wake-up time". An optical parametric oscillator/amplifier (OPO/OPA) or Ti:Sapphire laser provides a solid state gain medium instead of the dye.
9. Variable DUMP laser delay. It may be before or after the laser.
10. Scanning DUMP laser. Older dye lasers used pressure scanning for high resolution. Computer synchronization of scanning the grating and etalon may also be used. Dye lasers with only a grating can be designed to be quite narrowband, but reproducibility of the scan must be considered. Solid state tunable sources may be used here.
11. Etalon fringe monitor. This is for observing laser bandwidth and the presence of extra modes originating from the intracavity etalon.
12. Diverging lens for etalon monitor. Increased divergence increases the number of fringes seen, but does not change the diameter of the individual fringes.
13. Etalon. An aluminum coated etalon on fused silica is very broadband and has a finesse up to about 10. The delicate aluminum coating is easily damaged by an intense laser beam.
14. Imaging lens. This will help sharpen fringes (optional). The diameter of the fringes is proportional to the focal length
15. Etalon fringes displayed on screen.
16. Photodiode fringe monitor. Placed behind a center pinhole in screen, this provides a regular wavenumber marker for the scanning DUMP laser.
17. Variable attenuator. A pair of film polarizers with one that is rotatable works well.
18. Heated fused silica tellurium absorption cell. This provides calibration in the

blue part of the spectrum. A cool finger (and hot windows) eliminates window condensation and regulates vapor pressure. Sleeves extending beyond the heated region reduces thermal turbulence in the beam path. Alternately, an iodine absorption cell (or fluorescence cell) provides calibration in the red-yellow part of the spectrum.

19. Photodiode. A 9 Volt battery and an RC circuit of 200k ohms and a 1 nF capacitor provides a large signal. An inexpensive sample and hold chip or peak detector may be used in place of a gated integrator to provide the signal for the computer to read.

20. Second harmonic generation (SHG). This SHG crystal is usually angle tuned and so is placed in a gimbal mount. KDP or b-barium borate are suitable.

21. Prism pair for second harmonic separation. The 60° prisms are close to Brewster's angle for horizontal polarization and little ultraviolet is lost.

22. Beam stops. Sandblasted or chemically etched black anodized aluminum 1/16″ sheet has been used in the laboratory to block low energy stray beams. A variety of sizes should be available in the laboratory in considerable abundance. A simple folded single piece beam stop is illustrated in profile . These are not suitable for use as a high pulse energy beam DUMP, as the black dye used with the aluminum will be bleached.

23. Fundamental beam DUMP. A deep "V" or pinched off metal tube should be used instead of a flat plate for moderately high pulse energy.

24. Dual gimbal mount for angle tuned SHG crystal. The axis of rotation is vertical to accept vertically polarized visible input. It provides horizontally polarized ultraviolet output.

25. SHG crystal.

26. Fused silica compensator. This counter-rotates in same dual gimbal mount and eliminates the translation of the laser beam caused by angle tuning of the SHG crystal.

27. Servo motor or stepper motor. This may be driven by a tracker circuit or by a computer.

28. Dichroic beamsplitter for second harmonic separation. This has a limited wavelength range. Alternately, a set of four 60° prisms arranged in complementary pairs will cancel beam walk and has very broad band acceptance.

29. Pickoff for second harmonic servo tracking.

30. Ultraviolet band pass filter and attenuator.

31. Dual ultraviolet photodiode. This monitors the angular shift of the ultraviolet beam from the SHG crystal which indicates detuning from an optimum crystal angle.

32. Tracker circuit.

33. Beamshaping lens. Use of a lens, a telescope, or a spatial filter with a lens pair adjusts the DUMP beam size to overlap the PUMP beam in the cell. If the DUMP beam is slightly larger than the PUMP beam the system is more tolerant of slight spatial jitter or misalignment.

34. Pickoff beamsplitter (BS). A good beam sampling BS is thick, slightly wedged, and has an AR coating on the back. Used at 45° incidence, a single uncoated fused silica surface favors "S" (vertical) polarization, $\sim 10\%$, to "P" (horizontal) polarization, $\sim 1\%$. These numbers vary rapidly with angle. If it is desirable that the sample beam retain the polarization characteristics of the main beam, the BS should be used near normal incidence; this approach should be used if the polarization of the beam is not clean. Be careful of "accidentalon" effects[22] if both sides are uncoated.

35. Ultraviolet photodiode monitor. A filter and attenuator will suppress room light pickup and provide appropriate signal levels. When using a photodoide to normalize signals to laser pulse energy, check for linearity. Several photodiodes operated in the photoconductive mode (9V, 200 k ohm load) have become quite nonlinear with pulsed ultraviolet irradiation. The cause of the damage is not known. DC leakage current which increases over a period of several hours in the dark, considerably in excess of the manufacturer's specifications, appears to be an indicator of such damage.

36. Spatial filter. This provides a spatially clean PUMP beam. The lenses also function as a telescope to control the beam geometry in the cell. Longer focal length lenses increase the waist size of the beam at the pinhole and minimize damage to it.

37. Focusing lens for spatial filter.

38. Sapphire or diamond spatial filter pinhole on 3-axis translator.

39. Imaging lens. This projects a magnified image of the pinhole into the sample cell. The Gaussian beam minimum waist may be separate from the geometric image point.

40. Iris. This cleans up diffraction rings from the spatial filter and is an alignment aid

41. Half wave plate on rotation stage. This determines the ratio of the signal and reference PUMP beams.

42. Polarizer. If calcite is used, it should be air spaced and selected for ultraviolet transmission. Use a crystal quartz Wollaston for shorter wavelengths. (See item 43 about optical contacted surfaces.) Alternatively, a dielectric coated beamsplitter may be used, but is not readily adjustable to optically balance the amplitude of the two beams. This may be compensated by adjusting the photomultiplier gains, but unless the detector response is quite linear in pulse energy, the null may not be as good for unequal signal and reference pulse energies.

43. Half waveplate or optically active rotator. A pair of mirrors may also be used to rotate the polarization. Moderate intensity ultraviolet light may cause an optically contacted "zero-order" waveplate to separate, even though there is no damage to the optical surfaces.

44. Signal amplitude calibrator. This is an attenuator which is inserted into the signal PUMP beam path to simulate the loss of fluorescence by the DUMP laser beam. It should be coated for the desired level of attenuation on one side and anti- reflection coated on the other side with a wedge angle of about 10 arc seconds. Do not use uncoated fused silica microscope slide, as it is highly "accidentalon-prone". (See the text, Sec. 3.3.)

45. Iris. This reduces scattered light and is a valuable alignment tool. Good visual alignment is achieved when the counterpropagating beams overlap along their entire paths; additional irises are helpful in observing this.

46. SEP signal fluorescence cell. A common outlet connection assures equal pressure.

47. SEP reference fluorescence cell.

48. Brewster angle windows. An inert gas purge is needed to protect the windows from samples which photolyze. The reflection from a Brewster window should not be used to monitor the pulse laser energy, as the measurement will be extremely biased toward the pulse energy contained in the "wrong" polarization!

49. Conical baffles. These are quite effective for suppressing scattered light.

50. Beam DUMP for Pump laser.

51. Photodiode reference monitor.

52. Collection optics. A set of 3 or 4 fused silica lenses can be designed for f/1 collection without excessive spherical aberration.
53. Filters. A laser mirror coating on a $2'' \times 2''$ fused silica substrate is quite effective and is generally non-fluorescent. Many glass absorption-type filters will fluoresce.
54. Image aperture. This is rectangular and independently adjustable in length and width.
55. Photomultiplier. It should be suitable for pulsed detection and have a large pulse current capability without saturation. Both photomultipliers should be the same.
56. Preamplifier. This should have 50 ohm or 25 ohm output impedance and be the same for both photomultipliers.
57. Housing. A metal housing will reduce electrical noise pickup, but the photocathode should not be close to grounded conducting surfaces. An internal magnetic shield at cathode potential is best (Use a multimegohm resistor to connect the shield to the high voltage!).
58. 50 ohm coaxial cables. A double braid or foil shield is more resistant to noise pickup. The cable lengths should be matched so that the signals arrive simultaneously.
59. Differential amplifier. This provides the null signal for the balanced dual beam detection
60. Delay line. This delays the differential signal until the trigger pulse can activate the boxcar integrator.
61. Gated integrator. "Boxcar" averaging normalizes the integrated signal to the gate width.
62. Computer. This can be used both to record data and control the scanning of the laser.
63. High voltage power supply. A stable supply is important, as the gain of the photomultiplier strongly depends on voltage; for example, a 12 dynode PMT has a gain which scales roughly as the 9th power of the voltage. A single potentiometer may be used to share the voltage between the two PMT's, thereby permitting simultaneous adjustment of the two outputs in opposite directions. (Caution: an insulated extension handle should be added to the metal shaft of the pot.)
64. Two equal metering valves. These provide balanced flow to the two cells. A single valve followed by a tee is not suitable since, with virtually zero flow resistance in both directions, flow may not be equal. An equivalent pair of needle valves or capillary flow restrictions on the inlets will provide equal replenishment of a sample if photolysis is a problem.
65. Vacuum metering valve. A single large tube connecting both cells assures equal pressure in both cells.

A large area shield has been used which consists of a series of posts with 2 slots cut vertically along the length on opposite sides, and are bolted into place at regular intervals. Removable black aluminum sheets are inserted into the slots. Such a barrier can be removed in seconds for access, then replaced for laser operation. Inverted "L" or "U" shaped tops on these shields cover the beam path. Shields should not be used as beam stops; rather each beam should be terminated by an individual stop appropriately designed to handle the energy level. Pipes will provide total beam enclosure, but there is a special problem with metal pipes. They are usually smooth inside and if the beam is slightly misaligned, the grazing incidence reflection(s) will throw nearly the full power of the laser beam in unpredictable directions. Machining circular grooves or installing a series of washer-like apertures will mitigate this problem. Appropriate beam stops and shielding materials should be abundantly available in the laser laboratory. These items are neither expensive or burdensome to use. The large number of optical components utilized in an SEP setup makes control of stray beams particularly important.

When considering laser safety, the hazards posed by the laser power supply cannot be overlooked. Before working on a high voltage supply, always discharge the capacitors with a chassis-grounded cable attached to the end of a long insulating rod after disconnecting the power. Use the "one hand rule": Only one hand is used for work, the other is not in contact with any circuitry, the power supply case, laser table, or any other grounded conductor. Finally, never work alone on a laser power supply. Pulling the plug does not make it "safe". See Ref. 12 for more details.

3. The Fluorescence Dip SEP Experimental Configuration

In constructing the first fluorescence dip SEP experiment, there appeared a choice between designing a near perfect tunable laser or building a detection system which was relatively insensitive to the all of the pulse-to-pulse fluctuations described above. The latter course was chosen, as it would minimize laser engineering and allow the focus to be on molecular science. Also, if commercial pulse dye lasers could be used, with all of their flaws, the technique would be more widely available to other researchers. The single detector approach was not attempted, until some time after the initial SEP success and the signal-to-noise ratio was considerably inferior. Single beam systems do work under the right conditions, as discussed in a later section.

3.1. *An Overview of the Null Experiment*

The dual beam null detection system is schematically illustrated in Fig 2. Two identical PUMP laser beams passing through two matched cells, or through two equivalent locations in one cell, should give off equal fluorescence. A matched pair of collection optics and photomultipliers detect the fluorescence with equal efficiency. A high speed differential amplifier subtracts the signals from the signal and reference cells. The amplified output can be viewed on a fast oscilloscope. If everything is carefully balanced, there is an average null output, subject of course, to photoelectron shot noise. All of the PUMP laser amplitude jitter, beam pointing jitter, frequency and mode jitter, and saturation effects vanish into the null signal.

High detection sensitivity of the SEP system now centers around obtaining a good null balance. Once this is achieved, the sample is ready for a DUMP laser beam to remove population from the intermediate state. This unbalances the null signal and thereby provides an SEP signal. The DUMP beam both transfers the population and causes a quantitative fluorescence decrease signal proportional to this population transfer. The dual beam, null balance method was used to obtain the first SEP signal by pulsed laser fluorescence dip on December 22, 1979. The original data from this experiment, is reproduced in Fig. 3. A single rotational–vibrational level, $v'' = 10$ and $J'' = 38$, was selected and populated in iodine vapor. The laser fluences were chosen to be approximately those necessary to approximately "saturate" the transitions, assuming uniform spectral and spatial beam quality. As previously indicated, this is not a very good assumption for a pulsed dye laser output, and further tests showed that the signal was nonlinear in DUMP laser energy, with attenuation of the DUMP beam by a factor of 3 diminishing the fluorescence dip amplitude to about 2/3 of its previous value. Attenuation of the DUMP laser by another factor of 3 further diminished the fluorescence dip amplitude by another factor of 2.5 and the signal response approached linear behavior. The fluorescence dip, not the DUMP laser pulse energy, provides the correct information about population transfer. Only about six seconds were needed to pressure scan the peak width of 0.05 cm^{-1}, at 10 pulses per second. The fact that the author obtained such a robust SEP signal on the very first attempt was a harbinger that fluorescence dip SEP would provide a new and reliable source of data for spectroscopy and dynamics of highly excited vibrational levels.

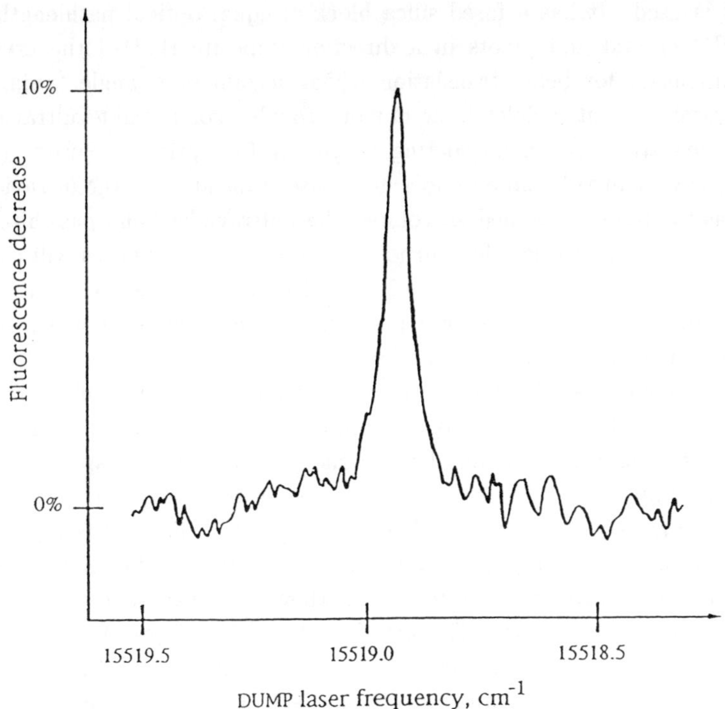

Fig. 3. This is the first observed pulsed laser SEP fluorescence dip signal. The experimental conditions are described in more detail in Ref. 7. A 3 mm diameter PUMP beam with pulse energy of 30 microjoules was used to excite the B–X (17–0), $P(38)$ transition in I_2, at $17\,594.0$ cm^{-1}. A 5 mm DUMP beam of 50 microjoules was pressure scanned over the (17–10), $P(38)$ transition at $15\,518.9$ cm^{-1}. A fluorescence decrease of about 10% was observed, and the FWHM linewidth was about .05 cm^{-1}. Excellent signal-to-noise was provided by the dual beam null detection scheme in spite of the 10%–20% pulse-to-pulse fluctuations which occur for single channel fluorescence detection.

3.2. *The Laser Beams and SHG*

A pulsed dye laser beam, with the myriad of flaws described in the previous section, is directed through a second harmonic generation (SHG) crystal to generate ultraviolet light. After the SHG crystal, separation of the ultraviolet PUMP beam from the fundamental may be accomplished by means of a laser mirror on a fused silica substrate or a pair of dispersing prisms. If SHG is needed for the DUMP laser, a home built double gimbal mount is used. It has a fused silica block of equal optical pathlength to the SHG crystal and pivots in a direction opposite that of the crystal to compensate for beam translation which accompanies angle tuning of the crystal. An ultraviolet laser mirror provides good visible–ultraviolet beam separation within its coating range. A four prism separator (not shown) is very broadband and does not cause translation of the ultraviolet beam as the laser is scanned. A colored glass ultraviolet band pass filter is generally not suitable for this purpose since high pulse fluences will cause it to solarize[13] and become opaque in the ultraviolet. Nevertheless, it may successfully be used in front of a photodiode energy monitor which does not need high fluence pulses.

An optional spatial filter will improve beam pointing stability in the sample cell. The pinhole serves to improve beam quality to a TEM_{00} mode at the expense of reduced pulse energy, and converts spatial jitter into amplitude jitter. Since amplitude jitter is readily quantifiable, this trade-off may well be justified. The spatial filtering apparatus is placed before the beamsplitting optics and is described in Ref. 14. Sapphire or diamond pinholes are the most durable, as they are transparent, minimizing the probably of breakdown. Be sure that the two surfaces of the pinhole substrate are not parallel, so that the unwanted light will be refracted away from the cleaned up beam and onto the surface of a downstream aperture. There is a band gap for diamond, about 220 nm, and hence is not as suitable for these shorter wavelengths. Care must be taken in the optical layout; if the lens which focuses the laser light into the sample cell is in the near field of either the spatial filter or the image point in the cell, the geometric image point of the pinhole will separate from that of the Gaussian beam waist (i.e., the minimum imaged beam diameter). Since light passing through the spatial filter is restricted in position, but not angle, only the geometric image point is translationally stable. A separated Gaussian waist position will be translated if the angle of the laser beam entering the spatial filter changes. Separation of the waist from the geometric focus is described in Ref. 15. It is normally minimal except when the Rayleigh range is quite long. But, this is often the case for the SEP setup, as a long collimated beam is commonly used in the cell.

The geometric image point is located by illuminating the pinhole with an extended source, such as a light bulb filament. The image of the pinhole formed from this incoherent source will be the geometric image point. If the beam entering the pinhole changes position, including angular deviations, it will still pass through this image point.

The Gaussian waist may be located by the knife-edge test; a sharp edge is placed on a precision translator and moved across the beam while a distant image of the beam is viewed. For a good TEM_{00} Gaussian beam such as that from a HeNe laser, a shadow passes across the round image in the same direction as the motion of the knife edge when it is past the focal point. If the knife edge is between the focusing lens and the waist, the shadow will traverse the image in the opposite direction. When the knife edge is at the focal point, the image fades uniformly. Surprising as it may seem, the far field image of this beam with the waist half blocked by a straight edge is still round. A negative lens may be used to increase the divergence of beam and enlarge the spot on the screen. Recording the transmitted energy measured with a photodiode can yield a beam profile.[16]

For the less than perfect beams emitted from many pulse lasers, the distant image will fade with varying degrees of irregularity depending on the beam quality. Lack of shading from one direction or the other still indicates that the knife edge is passing through the waist of the beam. A rather robust edge may be needed, as razor blade edges are ablated with even modest pulse energies. Fused silica or sapphire polished to an acute angle will handle considerable fluence as the beam is refracted (into a beam stop) rather than absorbed. This technique is similar to the standard Foucault knife-edge test for optics described in Ref. 17. However, the observer should not look directly into the laser beam! While direct viewing is the usual approach for an incandescent source, the image is projected onto a screen when testing a laser source.

The "acoustic method" is a quick way to locate the approximate waist and collimated region of a pulsed beam. A black target is placed in the beam path. The ablation of the surface will produce an audible "snap" when the intensity is sufficiently high. The more intense the beam, the louder the snap; the maximum is in the vicinity of the waist. For higher intensity beams a diffuse white target should be used, as the absorbing target will snap everywhere. (White paper is not "white" for ultraviolet light.) Wear eye protection! The ablation plasma strongly scatters the focused light.

3.3. *The Optically Balanced Dual Beam*

The PUMP laser, after SHG and spatial filtering, passes through a 1/2 λ plate and polarizer to separate the beam into two equal components. Rotation of the waveplate provides a continuously adjustable ratio without introducing absorption losses in either beam path. A calcite polarizer works well for longer wavelengths. Since optical quality calcite is still obtained from natural sources,[18] an ultraviolet grade must be selected for shorter wavelengths. A Wollaston polarizer made of synthetic crystal quartz works for shorter wavelengths down to about 190 nm. (Again, be sure to specify that it will be suitable for such short wavelengths when purchasing the device.) Optically contacting of Wollastons present a problem in that high intensity tends to separate the contact. Expanded beams or an air-spaced Wollaston will mitigate the problem. Unfortunately, an air spaced quartz Wollaston separates the two polarizations by only about 1/2 degree. Dielectric coated beamsplitters[19,20] may also be used. These optical components tolerate intense ultraviolet laser light well but relative intensities are not so easily adjusted. The Glan-laser calcite polarizer is normally air spaced and also handles high intensity well if it does not absorb the PUMP wavelength.

The polarization of one of the beams is reversed with a second waveplate or by two 90° mirror or prism reflections with a 90° dihedral angle between them. A zero-order two plate crystal quartz waveplate has a considerably wider spectral tuning range than a multi-order single plate quartz retarder, but if it is optically contacted the same decontacting problem arises as with the Wollaston. A double Fresnel rhomb will provide a broadband half-wave plate. Using one lens before the polarizer, or matched lenses afterwards, the laser fluence in the two beams may now be readily balanced.

If any saturation is present, unequal PUMP fluence will translate into imbalance of the null as the laser pulse energy fluctuates. Saturation first occurs for the strongest modes and in the center of the beam at fluences considerably less than would be calculated for an "average" fluence. The DUMP beam is usually a little larger to encompass all of the molecules excited to the intermediate state by the PUMP laser and allow for diffusion. This is more critical for long delays, especially for a delayed probe.[21] A two lens telescope allows for beam adjustment to virtually any desired size. Tight focusing is generally avoided in fluorescence dip experiments to minimize competing multiphoton processes.

An attenuator may be inserted into the signal PUMP beam to provide amplitude calibration. A fused silica window reflects about 4% from a single surface at normal incidence in the ultraviolet, and the variation in index of refraction with wavelength is well known. But accurate attenuation is not as trivial as it might seem as a single surface cannot be inserted. A plane parallel window at normal incidence is an etalon, or "accidentalon"[22] with attenuation ranging from $\sim$ 0% to 16%. An uncoated fused silica microscope slide is an accidentalon waiting to happen. It also can introduce beam distortion and deviation of the laser beam.

A tiny wedge angle doesn't solve the problem because of the small laser beam diameter; it becomes a position sensitive accidentalon. A wedge angle large enough to cause the two reflected beams to be separate will also cause angular displacement of the PUMP beam and put a secondary weak beam into the cell due to double reflection. A thick plane parallel substrate inserted off normal incidence has similar problems, except that the PUMP beam undergoes translation rather than an angle change.

An antireflective (AR) coating on one side of the substrate, and a reflective coating on the other side, if desired, will reduce the contrast in the etalon fringes. For a fixed wavelength, the AR coating may be less than 0.1% reflective, mitigating the problem. Broadband AR coatings are not as good. A wedge angle of 10 arc seconds will provide a few fringes across a 25 mm diameter, so that possible etalon effects may be observed by translating the flat across the beam; yet this will not significantly deviate the beam path.

More involved approaches are also available. A pair of matched wedged prisms will solve this problem and are available as prism beamsteerers from some optics suppliers. There are enough degrees of freedom in the pair that angular and translational shifts are canceled, and no two surfaces parallel to each other. It is probably best that three of the four surfaces be AR coated. An alternative is a "dummy" wedged attenuator with AR coatings on both sides which stays in the beam, only to be replaced with an identical substrate with an attenuating reflective surface for a measurement. A dual coating on a single substrate would serve the same purpose. The beam alignment is done with the dummy wedge in place. All of this assumes that if a meaningful comparison between PUMP attenuation and DUMP laser induced fluorescence dip is to be obtained, the PUMP laser fluence must be low enough that the fluorescence is linear in pulse energy.

3.4. *Sample Cell and Collection Optics*

Sharp-edged cone shaped baffles effectively clean up scattered laser light as the beam passes through the cell. Two matched cells, or equivalent paths through a single cell, help to achieve a better balance in null signal. Compared to metal sample vessels, glass cells are clean and easy to make but are prone to fluorescence under illumination with ultraviolet laser light. Furthermore, "light-piping" of longer wavelengths can originate from the cell windows . A ring of black Pyrex fused into the cell arms just past the laser beam windows eliminates the latter problem.

Matched collection optics and image restricting apertures in front of the photomultipliers help to ensure that the same length of excited sample is viewed in both cells. Camera optics are a bargain if the emission is in the visible part of the spectrum. For the ultraviolet, multi-element f/1 collection optics assure efficient light collection. Spherical aberration of an all-fused silica lens train can be minimized with careful design and moderate imaging quality can be achieved. However, chromatic aberration remains quite significant. The system should be aligned with a wavelength roughly corresponding to that of the strongest fluorescent emission. Alternatively, an image is formed with a source of a convenient wavelength, with the difference in optimum imaging position being computed from fused silica dispersion curves.

A quartz crystal from an old-style radio transmitter oscillator has a finely ground surface and makes an excellent non-fluorescent ultraviolet scatterer. Alignment using a fluorescent card in the laser beam path will cause the collection optics to be severely out of focus for ultraviolet light! Focal lengths will change 10–15% from visible to deep ultraviolet. It is better to use the fluorescent card to locate the image of the ultraviolet light after it has passed through the collection optics. Reflective collection optics are more cumbersome, but have no chromatic aberration.

Collection lenses are not essential, but quite often a 2″ diameter lens can be placed closer to the sample than can a 2″ diameter photomultiplier, thereby collecting a larger solid angle of emission. Also, since the viewed area is not so readily controlled with a wide open no-lens system, scattered laser light is of greater significance. The viewed area may be fine tuned with apertures at the lens image point in front of the PMT. A width restricting aperture reduces scattered light detected from the sample vessels, but should not be too small or else the signal will be quite sensitive to small amounts of spatial jitter in the PUMP beam. Laser mirror coatings on synthetic fused silica substrates make effective laser line blocking filters[20,23] and many manufacturers will put such a coating on a thin $2'' \times 2''$ square substrate that is interchangeable with a standard glass filter. Since the coatings may be relatively narrowband, such filters may be used to block scattered light from both PUMP and DUMP laser beams without severely depleting the fluorescence signal. The laser mirror coatings generally show little fluorescence under ultraviolet illumination. In contrast, colored glass filters may exhibit considerable emission under similar conditions.

3.5. *Balanced Detection and Electronics*

The two photomultipliers should have similar photocathode quantum efficiencies, as this property is not user adjustable. PMT gain is adjustable. A single high voltage power supply may be used, with a single ten turn potentiometer acting to simultaneously increase the voltage on one PMT and decrease it on the other. This balances the gain of the two detectors. (Caution: the potentiometer should have a non-conducting extension shaft to the control knob as it is not normally designed to provide several thousand volts isolation.) The PMT gain is very sensitive to voltage fluctuations. The gain is approximately proportional to the ratio of two voltages, raised to the power ~ 0.7 times the number of dynodes. Typically, the gain varies as about the 8th to the 10th power of the applied voltage for a $2''$ end-on 12 dynode PMT. The single power supply means that the gain of both PMT's fluctuates simultaneously with power supply drift, minimizing adverse effects of such drift on the stability of the null signal.

Differences in transit time for the electrons in the PMT may be compensated by using different lengths of 50 ohm terminated coaxial cable (e.g., RG223/U) to the differential amplifier. Signal cables should be new or in excellent condition. Kinked or damaged cables degrade pulse transmission; don't step on or roll carts over signal cables! The total PMT transit time is significantly affected by voltage changes and can range over many nanoseconds. A difference of one or two nanoseconds in signal arrival time at this differential amplifier will be observed on a fast oscilloscope as a large derivative-like output. This indicates that the leading edges of the fluorescence signals from the two PMT's are arriving at different times. When everything is properly balanced, e.g., to $< 1/2\%$, then an imbalance in the null signal caused by the DUMP laser removing population from the intermediate state of this amount or even less will be observable. Scattered DUMP laser light will provide an erroneous positive imbalance, or "negative" SEP signal. A well baffled cell is needed to minimize scattered laser light, and filters such as the previously mentioned laser mirror coatings are quite useful in this regard.

While high speed balancing isn't vital to the overall goal of achieving a pulse-by-pulse null signal, this approach does allow the system's temporal response to be followed in detail on an oscilloscope. One advantage of sub-nanosecond nulling is that the differential signal may be considerably amplified without worrying about clipping, thereby making the signal less vulnerable to electrical interference such as that generated by noisy excimer lasers. The high speed null technique does not require demanding tolerances on the gated integrator. However, the leading edge of the differential pulse is prone to large fluctuations for even small differences in transit time. The gated integrator which accepts the differential signal should have a 1–2 ns rise time for its gate, and jitter will be comparable.

A 100 MHz oscilloscope is adequate to align the gate and signal pulses, but care should be taken with cable lengths. Electrical pulses travels about 2/3 c in a high quality 50 ohm coaxial cable. The gate onset should not be too close to the leading edge of the signal. PMT electron transit times very considerably (tens of nanoseconds) when the high voltage is changed substantially and the gate delay will need to be adjusted accordingly. This phenomena is often overlooked and even catches experienced researchers off guard. If the trigger signal to the boxcar arrives a bit late, the differential amplifier output signal can be kept occupied running around 50 feet of coaxial cable without serious degradation while waiting for the tardy trigger pulse to provide the gate wake-up call.

4. Some Experimental Considerations.

4.1. *Sensitivity Considerations for Fluorescence Dip SEP*

One of the attractive features of the fluorescence dip approach is that when the Franck–Condon overlap factors to a high-lying vibrational state are very weak, one can drive the transition with all the DUMP power that can be mustered, short of damaging the cell windows. The dynamic range of observable SEP intensities in the fluorescence dip detection scheme has been estimated to be as large as 1000.[24] Unlike an experiment using transient gain, the detector does not look directly at the laser. Consequently, the increased pulse energy does not decrease the sensitivity of detection. Instead, sensitivity is based on population loss from the intermediate state.

A caveat is that competing processes, such as transition to higher electronic states or multiphoton ionization, can give false SEP signals. The transitions to higher states are often diffuse or continuous such as in HCCH[25] (dubbed "potato" spectra by Y. Chen, as the amorphous lumpy continuum resembled a potato underlying the sharp SEP peaks sprouting above it.) These features can be distinguished readily from spectrally narrow transitions back to the ground state. Transitions into continuum states do not saturate easily, so once the transitions into discreet levels become saturated, increasing of the laser fluence will increase which only serves to augment the unwanted the continuum amplitude relative to the desired SEP transition. This competition can become the principle source of noise in the fluorescence dip experiment,[24] with the large dynamic range being somewhat dependent on the fortuitous presence or absence of a strong high-lying continuum at the sum of the PUMP and DUMP laser energies. A rotational analysis can distinguish upward transitions from downward ones when the upper state transitions are discreet. A photoacoustic SEP scheme provides for such a distinction by measuring the signal size.[26]

An additional feature is the great sensitivity of fluorescence dip to small amounts of sample. Basically, any molecule for which fluorescence can be observed from a bound excited state is a candidate for fluorescence dip SEP described here. In the early stages of SEP development at MIT, an iodine cell was evacuated with a diffusion pump for 15 minutes, and then an SEP experiment was performed on the "empty" cell which was still open to the pump. The residual outgassing provided 10–20 detectable photons (reference channel) per laser pulse and a good iodine SEP signal was obtained when the DUMP laser was turned on. This sensitivity also shows that considerable care must be given to the sample handling system, as the observation of SEP signals from trace impurities is a real possibility. Reducing the PUMP laser energy by 5 orders of magnitude also did not inhibit the observation of a fluorescence dip.[10] Fluorescence detection shows remarkable sensitivity for tiny amounts of sample; this sensitivity is likewise available to SEP when utilizing fluorescence dip detection. If the fluorescence lifetime is much shorter than the duration of the laser pulse, e.g., due to predissociation, non-radiative decay provides a competing channel which diminishes the fluorescence signal strength. This loss can be somewhat mitigated by using shorter duration, closely spaced PUMP and DUMP laser pulses.

4.2. *Dump Laser Delay*

The DUMP laser is usually delayed relative to the PUMP laser. The DUMP laser may immediately follow the falling edge of the PUMP laser pulse, which is the usual case, but in principle it may be incident anytime later during the observable fluorescence emission. The total interaction time from the beginning of the PUMP pulse to the end of the DUMP pulse should be short compared to collisional relaxation times for the intermediate electronic state so as to avoid the appearance of extra SEP spectra from collisionally populated levels. Rotational relaxation cross-sections are often quite large[1] and those for vibrational relaxation are usually smaller. Low pressures mitigate this problem and take advantage of one of the virtues of the fluorescence dip approach to SEP: The signal amplitude is rather insensitive to a large decrease in sample concentration. The delayed DUMP can, of course, become a monitor for relaxation of the electronically excited population. Diffusion of the excited sample out of the laser beam region is a significant factor for longer delays, especially in PUMP–DUMP–PROBE experiments.[21]

Complete pulse overlap with intense, spectrally structured 10 ns pulses can produce strange results. In an early test utilizing the iodine *B*-state, occasional *negative* SEP signals were observed; i.e., the fluorescence from the excited state sometimes was *increased* by the presence of the DUMP laser! This was observable only when the data were recorded on a shot-by-shot basis, as the average still showed the strong positive SEP signal of fluorescence dip. The best guess for this effect is that the DUMP laser, with a mildly saturating intensity, transferred population to the final level, creating more room for the saturating PUMP laser to further deplete the initial ground state population. Then, in the closing nanoseconds of the DUMP laser pulse, the population from the final level was cycled back into the intermediate state, with the result being that the intermediate level was receiving population from both the initial and final levels of the SEP scheme. With the two dye lasers undergoing separate frequency and amplitude jitter, it is not surprising that the right circumstances for this negative SEP signal occurred randomly and only occasionally. Likewise, with two strong fields, the SEP signal was sometimes seen to be unusually large.

The above-mentioned behavior suggests that with excellent pulse dye laser mode control, and perhaps specially shaped pulses, it might be possible to apply π-like pulses and transfer a much larger fraction of the population to the final state than would be calculated using simple rate models. Overall, the appearance of a fluorescence dip signal utilizing precisely simultaneous laser pulses on iodine produced an unaveraged signal for which a peak looked like a lump of excessively large noise, mostly but not entirely above the baseline. The noise envelope constitutes the SEP signal for the simultaneous lasers. More recently, some very interesting results have been achieved utilizing continuous wave single frequency dye lasers and jet cooling of the sample to provide precise pumping conditions and controlled overlap.[27] Here, the coherent interaction of all three levels leads to the "counterintuitive" result that the largest signal is achieved with the Stokes (DUMP) laser being slightly ahead of the PUMP laser. It is essential in this case to follow the time evolution of Rabi frequencies. This fascinating new approach will be presented elsewhere in this book in Chapter 8.

4.3. *Variations: Single PUMP Beam, Jet-Cooled Samples, ps Pumping*

Two sample cells may also be placed end-to-end, with the same PUMP beam passing through both. However, in this case, the pulse energies can no longer be balanced precisely. There will be some transmission losses, including losses in the beam combiner (or separator) where the DUMP beam enters (or exits). Care must be taken to achieve roughly equal fluences in both cells, otherwise velocity group saturation will be unequal and consequent variations in the nonlinear response will make null balancing more troublesome. Relay imaging optics may be required between the cells.

The dual beam approach to fluorescence dip need not be used, provided that some other reference method is employed to compensate for fluctuations in the single PUMP beam. An example is time-resolved monitoring of the single photomultiplier output before and after the DUMP beam passes through the sample cell. Our experience with this delayed gate method indicates that the nulling is not as good as that obtained via the dual-beam scheme. For some experiments, where adding a second reference beam makes it considerably more complex, other methods become attractive. A delayed gate has been used for the jet-cooled radical C_3, and the effect of the delayed DUMP on temporal profile is presented.[28] This was also used with excellent results in a jet-cooled experiment on HFCO.[29] The authors point out the additional sources of noise which occur with this approach: scattered laser light increases the detected light in the prompt gate channel, time jitter of the two boxcar gates will change the part of the decay curve sampled, and there is nonlinear response of the PMT. This method clearly works best for long fluorescent decay times. A short decay time would require high gain on the PMT for the rapidly decaying signal in the delayed gate, a situation which would also increase PMT nonlinearity for the large initial signal entering the prompt gate. A PMT with a high saturation current was used for PUMP–DUMP–PROBE experiments where a tiny signal must be observed after a large one.[29] A gated PMT[21] may also prove to be useful here. Boxcar gate timing jitter would be much more critical for a short decay time. Use of a pulse amplified continuous dye laser for the PUMP in fluorescence dip spectra on jet-cooled benzene[30] is a good approximation to the "near perfect" dye laser and single channel detection was used with good success in this case.

When two single frequency continuous wave dye lasers can be used, excellent results may be obtained without the need for the dual beam null detection. SEP studies on the HCF radical clearly demonstrated this capability. The DUMP dye laser beam was alternately blocked by a chopper and lock-in detection provided excellent signal-to-noise ratios. Very narrow linewidths were also observed.[31] A caveat for using narrow linewidths is that such sub-Doppler SEP spectra do not have the same spectral characteristics as those obtained by Doppler-free coherent two photon excitation using counterpropagating beams. The SEP spectra are not automatically centered at the zero velocity transition frequency; the PUMP laser selects a velocity group which may or may not have nonzero velocity and this, in turn, offsets the DUMP signal frequency.

Collisional losses of the fluorescing population, which would reduce the signal passing through the delayed gate, are minimized by using a free jet expansion instead of a cell. For jet-cooled species, this single beam approach can work quite well; and a similar approach has been used for OH–Ar van der Waals complexes which have a ~ 700 ns fluorescence decay time.[32] Dual jets and dual beam differential nulling have also been used on glyoxal with excellent signal-to-noise; it is estimated that only 10^{-4} Torr of cold molecules are needed for SEP.[33,34] This may be compared with earlier single beam work of the same type.[35] Twin flow reactors were used in a dual beam SEP study of the radical CH_3O.[36] A dual flash photolysis system has also been used for SEP on the methylene radical.[37,38] The C_3 has also been studied in a jet-cooled beam.[28,39]

The creation of the final level may be detected by a third probe laser rather than observing fluorescence decrease. Three laser beams and a spectrometer were used to study p-difluorobenzene. The added complexity has the benefit of allowing for the study of relaxation processed in the vibrationally excited ground state.[40,41] An excellent new SEP detection scheme with an astonishing signal-to-noise ratio based on degenerate four-wave mixing has recently been developed[42] and will be discussed elsewhere in this book, in Chapter 1. Additional techniques are in Ref. 43.

Picosecond SEP experiments using mode-locked dye lasers do not have the multiple longitudinal mode problem of the nanosecond dye lasers. Alternate blocking and unblocking of the DUMP laser beam provided a reference signal of maximum fluorescence to be compared to the fluorescence dip induced in a free-jet expansion of p-cyclohexylaniline.[44] Delay of the ps DUMP laser provides information on the intramolecular vibrational energy redistribution in the excited intermediate state.

4.4. *SEP Using Intermediate Predissociating States*

When laser wavelengths correspond to excitation into electronic states which predissociate, dissociate, or decay through other non-radiative processes, the use of SEP is often inhibited. In a rough approximation of a definition presented in Ref. 45, a distinction will be made here between resonance Raman scattering and laser induced fluorescence. It is considered a scattering process when the temporal coherence length of the laser is considerably longer (here ~ 10 ns) than the excited state lifetime of the molecule. The excited state is a dressed molecular state in a strong laser field, and emission ceases when the laser is turned off. The term fluorescence is used when the excited state has a lifetime comparable to or longer than the temporal coherence length of the excitation source. Photons are absorbed and Stokes-shifted emission occurs even after the laser pulse is terminated. The delayed DUMP laser stimulates emission and causes fluorescence dip near the end or even after the end of the PUMPing process.

However, for resonance Raman excitation, the DUMP laser must temporally coincide with the PUMP laser, as there will be no population to "DUMP" after the PUMP laser ceases. The DUMP laser competes with non-radiative depopulation mechanisms to remove the dressed states before they scatter. For a short lived excited state, a very intense DUMP laser field is needed to deplete the steady state dressed population significantly faster than the non-radiative rate. But, if the DUMP laser is intense enough to move the population quickly down to the final state, to some extent it will cycle it back to the excited level, thereby reducing the net population transfer into the ground electronic state. If the DUMP laser fluence is substantially reduced, then it will fail to compete with the non-radiative process and cause little change in the steady-state excited population. The SEP signal will be correspondingly weakened. For SEP involving an electronically excited final level, the final level may be predissociated, e.g., the A-state of ammonia.[46] Then there is a route for population removal, mitigating the above problem. Recently, success has been reported in performing SEP on oxygen even though it is predissociated.[47] Clearly, this problem can be handled if the experimental conditions are carefully controlled. SEP is quite sensitive and even a weak signal may be observed.

Early awareness of the experimental difficulties associated with performing SEP on molecules which have only short lived excited electronic states led to discussions in R. W. Field group meetings proposing the use of short laser pulses as a general way to mitigate this problem. If the laser pulses were on the order of the duration of the excited state lifetime, they may effectively compete with the non-radiative processes, thus giving strong signals, and providing for large population transfers. This would broaden the applicability of the SEP technique. A successful ps SEP experiment was recently carried out[48] and is discussed in this book in Chapter 2.

5. Future Directions

Several possible experimental developments are examined in this section. Regarding the short-lived intermediate state problem above, another approach should be considered. If a means for rapid removal of the final level can be achieved, such as by an intense laser pulse connecting this ground state vibrationally excited bound level to a short lived dissociating excited state, then it should be feasible to use nanosecond pulses for SEP on a molecule such as ozone, while retaining the relatively high resolution common to SEP experiments.

5.1. *Rydberg Intermediate States*

To generalize the above approach using predissociated intermediate levels, consider that most molecules have Rydberg states with pronounced vibrational structure. The Rydberg state itself does not usually lead to direct dissociation, so that it can serve as intermediate for SEP spectroscopy. The molecule could be PUMPed to a Rydberg level, then DUMPed back to a high-lying ground state. A longer wavelength fixed laser then connects this high rotational–vibrational state to a continuum state. Most molecules have some continuum states lying at lower energies than the Rydberg states. This third laser provides a competing channel, mitigating the cycling of the population to the intermediate state that would otherwise nullify the SEP experiment. If the Rydberg state is not well characterized, or there are overlapping Rydberg states, all is not lost. In the absence of collisions, the energy difference between the initial level and final SEP pumped level is precisely determined by the difference in the laser pulse frequencies and does not require identifying the excited state.

Laser sources for populating Rydberg states are not as problematic as at one time. Stimulated Raman shifting[49] (SRS) can convert Nd:YAG 266 nm radiation to 140 nm and ArF excimer radiation to 120 nm. Anti-Stokes shifting in atomic vapor shows promise for highly efficient conversion. The recent availability of narrowband tunable 193 lasers for SEP[19,50] should also be of utility. The F_2 laser is commercially available for 157 nm and SRS will also extend its wavelength range. Third harmonic generation in gases can achieve tunability for much shorter wavelengths. The tunable DUMP laser also has several interesting possibilities. High pulse energies from the newer Nd:YAG photon cannons are suitable for mixing down to ~ 180 nm and SRS can achieve tunability for much shorter wavelengths. While these short wavelength pulse energies will not be large, Rydberg transitions are often orders of magnitude stronger than the valance transitions in classic SEP molecules like formaldehyde. Development of such a system would make SEP a universally applicable technique for preparing high vibrational levels, even in the absence of a convenient bound valence excited state in the chosen molecule, radical, or ion.

5.2. *Laser Mode Cleanup*

The value of laser mode reduction or elimination has already been seen in the single mode cw and ps SEP experiments. Some possibilities for reducing the mode problem include the use of stimulated Raman shifting (SRS) and self phase modulation. This will provide some broadening of the spectral structure and overlap of the cavity modes. The SRS gain medium needs to be carefully chosen; most SRS diatomics, gas or liquid, have relatively small broadening and SRS linewidths are (gas) pressure sensitive. These media are probably most suitable for blending the longitudinal modes of SEP lasers. Polyatomics and especially liquids or solids provide more broadening and may go well beyond blending the modes and increase the spectral bandwidth considerably. Of course, the dye laser must operate at a correspondingly shorter wavelength to allow for the SRS frequency shift. Such stimulated processes generally increase the pulse-to-pulse amplitude fluctuations, but these are usually easier to handle than complex frequency structure.

Other options for better spectral characteristics include use of a distributed feedback dye laser[51] which, lacking a cavity, minimizes logitudinal mode structure. Intracavity insertion of an electro-optic device could be used to "chirp" the laser frequency to sweep across all molecular velocity groups. A stretched cavity dye laser decreases the mode spacing. Since a dye laser typically needs a PUMP pulse duration which is twice the cavity round trip time, pulse stretching techniques may be needed. The stretched cavity Molectron DL300 dye laser used in the original iodine SEP experiment needed a long PUMP pulse. Part of the 532 nm light from an Nd:YAG laser was split off, delayed, and then recombined with an undelayed beam before entering the oscillator dye cell. Although some laser light is wasted in the process, this effectively stretched the pulse length by a factor of two. While it may appear that the fraction of velocity groups prepared by the more densely spaced but spectrally narrower cavity modes would not be improved, the amplifier stages can still use the regular short PUMP and will transform-broaden the cavity modes so that the "spectral dark space" between modes is reduced.

A smoother PUMP beam can improve laser performance. Prominent temporal structure on the PUMP laser, such as from mode beating, will prevent the dye laser cavity modes from achieving the transform limited bandwidth which is calculated from the overall pulse duration. A single longitudinal mode (SLM) Nd:YAG laser (some designs use injection seeding) as the PUMP source will narrow the dye laser bandwidth and reduce the background, including that in between the cavity modes.[52] Recent developments in tunable lasers based on the SLM PUMPed optical parametric oscillator or solid state gain media show considerable promise for cleaner spectral output, and smoothly tunable single mode operation.

5.3. *Temporal Tailoring*

The push deep into the ultraviolet is one future direction. Another direction is the tailoring of laser pulses to suit specific needs. Pulse durations can be shortened to compete with non-radiative processes in the intermediate state, yet be transform-limited to maintain as much resolution as possible. Femtosecond pulses are useful for short-lived intermediate states as shown in the previous chapter. A theoretical analysis has been done for use of fs pulses in SEP experiments on very short-lived electronic excited states of polyatomic molecules.[53]

To efficiently move large populations, temporally shaped pulses should be able to do more than provide simple saturation. A π-like excitation pulse[27] would move nearly all of the initial population. This would be particularly attractive where there is a predissociating intermediate state that precludes simple saturation, as the population is lost. But a coherent sequence of laser pulses should be able to move the population through the intermediate state into the final state with reasonable efficiency. Another very interesting approach for achieving highly efficient population transfers involves adiabatic passage with short, chirped laser pulses.[54] Perhaps this can eventually be applied to SEP experiments to create population transfers which are considerably larger than those for simple saturation of the PUMP and DUMP transitions.

6. Summary

The apparatus for generic fluorescence dip stimulated emission PUMPing experiments has been described. The experiment provides both the population transfer to a selected rotational–vibrational level of the ground state and the quantitative diagnostic signal to indicate that the transfer has occurred. This makes it very useful for studies of molecular spectroscopy and dynamics in molecules having a high degree of vibrational excitation.

To help SEP gain widespread acceptance, it was considered quite important that the basic experiment be simple and reliable enough for an unassisted researcher to maintain and operate the apparatus so as to generate consistent, high quality data. The dual beam null detection approach appears to have lived up to that promise, as the first and many subsequent SEP experiments at MIT have been operated by a single experimentalist. Knowledge of the choices made in originally designing the experiment and the rationale behind those choices will help a researcher engaged in constructing a new apparatus. Many modifications of the basic configuration can and have been made. An understanding of the basic principles will help a researcher determine the impact that such modifications will have on the quality of the spectral data. This, in turn, alludes to new experiments which can improve the quality and extend the capabilities of SEP.

Acknowledgments

The author appreciates the encouragement and suggestions for the manuscript by Patrick H. Vaccaro. The author is deeply indebted to James L. Kinsey for his guidance and support, and especially to Robert W. Field, whose vision and enthusiasm made it possible for the author to build the SEP pulse laser laboratories at MIT and develop the fluorescence dip experimental technique. Acknowledgment is made to the National Science Foundation and the Air Force office of Scientific Research for providing the initial support for the SEP project, and to the Department of Energy for providing subsequent funding; and for the use of the facilities of the NSF Laser Research Center of the George R. Harrison Spectroscopy Laboratory.

References

1. C. E. Hamilton, J. L. Kinsey, and R. W. Field, *Ann. Rev. Phys. Chem.* **37**, 493–524, (1986).
2. R. Teets, R. Feinberg, T. W. Hänsch, and A. L. Schawlow, *Phys. Rev. Lett.* **37**, 683, (1976).
3. J .C. D. Brand, K. J. Cross, and R. J. Hayward, *Can. J. Phys.* **57**, 1455, (1979).
4. P. H. Vaccaro, F. Temps, S. Halle, J. L. Kinsey, and R. W. Field, *J. Chem. Phys.* **88**, 4819, (1988).
5. F. Temps, S. Halle, P. H. Vaccaro, R. W. Field, and J. L. Kinsey, *J. Chem. Phys.* **87**, 1895, (1987).

6. R. A. Bernheim, L. P. Gold, P. B. Kelly, C. Kittrell, and D. K. Veirs, *Phys. Rev. Lett.* **43**, 123, (1979).

7. Lambda Physik, Acton, MA 01720, USA, or Lambda Physik GmbH, D-3400 Goettingen, Germany; Quanta-Ray lasers are made by Spectra Physics, Inc., Mountain View, CA 94039; the laser division of Molectron Corp. was acquired by Laser Photonics, Orlando, FL 32826.

8. TecOptics, 1760 Grand Ave., Merrick, NY, 11566, distributor for Technical Optics Ltd., 2nd Ave., Onchan, Isle of Man, 1M3 4PA, British Isles; This vendor was also the supplier of intracavity etalons for Molectron dye lasers.

9. J. M. Vaughan, "The Fabry-Perot Interferometer", Chapter 4, (Adam Hilger, Philadelphia, 1989), pp 135–183.

10. C. Kittrell, E. Abramson, J. L. Kinsey, S. A. McDonald, D. E. Reisner, R. W. Field, and D. H. Katayama, *J. Chem. Phys.* **75**, 2056, (1981).

11. T. T. Kajava, H. M. Lauranto, and R. R. E. Salomaa, *Appl. Opt.* **31**, 6987, (1992), and references therein.

12. D. C. Winburn, "Practical Laser Safety", (Marcel Dekker, Inc., New York, 1989); D. H. Sliney and M. L. Wolbarsht, "Safety with Lasers and Other Optical Sources: A Comprehensive Handbook", (Plenum, New York, 1980). A recent review is by D. H. Sliney, in "Laser Safety, Eyesafe Laser Systems, and Laser Eye Protection", *Society of Photo-Optical Instrumentation Engineers (SPIE) Proceedings*, **1207**, 2, (1990); Retinal lesions from only 3 microjoules exposure are reported by R. G. Allen, J. A. Labo, and M. W. Mayo, in "Laser Safety, Eyesafe Laser Systems, and Laser Eye Protection", *SPIE Proceedings*, **1207**, 34, (1990).

13. "Solarization" is a term used in glass technology to describe permanent reduction in transmission of filter glass in the normally spectrally transparent region, induced by intense ultraviolet exposure. "Optical Glass Filters", Schott Glass Technologies, Inc., Duryea, PA 18642.

14. P. H. Vaccaro, Ph. D. thesis, Massachusetts Institute of Technology, 1986. Page 219–222.

15. A. E. Siegman, "Lasers", (University Science Books, Mill Valley, CA 94941, 1986), p. 678.

16. T. D. Milster and J. P. Treptau, *Society of Photo-Optical Instrumentation Engineers (SPIE) Proceedings*, **1414**, 91, (1991).

17. J. Ojeda-Castañeda, "Foucault, Wire, and Phase Modulation Tests", in *Optical Shop Testing*, D. Malacara ed., (Wiley, New York, 1978), p. 231.

18. Vino Vats, Karl Lambrecht Corp., Chicago, IL 60618, private communication.

19. X. Yang, C. A. Rogaski, and A. M. Wodtke, *J. Opt. Soc. Am.* **B7**, 1835, (1990).

20. Y. Chen, Ph. D. thesis, Massachusetts Institute of Technology, 1988, pp. 74,75.

21. X. Yang, E. H. Kim, and A. M. Wodtke, *J. Chem. Phys.* **96**, 5111, (1992).

22. R. W. Field coined the term "accidentalon" from accident + etalon, which occurs when reflections from two nearby reflective parallel surfaces overlap.

For an uncoated glass substrate ($n = 1.5$), this imposes a wavelength dependent modulation up to 16% on the transmitted beam and up to 100% on the reflected beam.

23. Using a laser mirror as a blocking filter convincingly demonstrated that some unexpected laser induced emission from ozone was in fact Stokes shifted emission and was not stray laser light which had defied all attempts to banish it (C. Kittrell and D. Imre, unpublished results). A glass absorption filter often exhibits fluorescence, especially with ultraviolet light. The non-absorbing dielectric coating exhibits very little fluorescence by comparison. A follow-up experiment (D. G. Imre, J. L. Kinsey, R. W. Field, and D. H. Katayama, *J. Phys. Chem.* **86**, 2564, (1982)) with a spectrometer confirmed that the observed Stokes emission was from the molecule, not from the filter, cell, or optics.

24. D. M. Jonas, S. A. B. Solina, B. Rajaram, R. J. Silbey, R. W. Field, K. Yamanouchi, and S. Tsuchiya, *J. Chem. Phys.* **97**, 2813, (1992).

25. Y. Chen, D. M. Jonas, J. L. Kinsey, and R. W. Field, *J. Chem. Phys.* **91**, 3976, (1989)

26. D. J. Moll, G. R. Parker, Jr., and A. Kuppermann, *J. Chem. Phys.* **80**, 4800, (1984).

27. U. Gaubatz, P. Rudecki, S. Schiemann, and K. Bergmann, *J. Chem. Phys.* **92**, 5363, (1990). See also Chapter 8.

28. E. A. Rohlfing and J. E. M. Goldsmith, *J. Opt. Soc. Am.* **B7**, 1915, (1990); F. J. Northrup and T. J. Sears, *J. Opt. Soc. Am.* **B7**, 1924, (1990).

29. Y. S. Choi and C. B. Moore, *J. Chem. Phys.* **94**, 5414, (1991); Y. S. Choi and C. B. Moore, *J. Chem. Phys.* **97**, 1010, (1992).

30. Th. Weber, E. Riedle, and H. J. Neusser, *J. Opt. Soc. Am.* **B7**, 1875, (1990).

31. T. Suzuki and E. Hirota, *J. Chem. Phys.* **88**, 6778, (1988).

32. M. T. Berry, R. A. Loomis, L. C. Giancarlo, and M. I. Lester, *J. Chem. Phys.* **96**, 7890 (1992).

33. D. Frye, L. Lapierre, and H. L. Dai, *J. Chem. Phys.* **89**, 2609, (1988).

34. D. Frye, L. Lapierre, and H. L. Dai, *J. Chem. Phys.* **88**, 7240, (1988).

35. K. Yamanouchi, H. Yamada, and S. Tsuchiya, *Chem. Phys. Lett.* **132**, 361, (1986).

36. A. Geers, J .Kappert, F. Temps, and J. W. Wiebrecht, *J. Opt. Soc. Am.* **B7**, 1935, (1990).

37. W. Xie, A. Ritter, C. Harkin, K. Kasturi, and H. L. Dai, *J. Chem. Phys.* **89**, 7033, (1988).

38. W. Xie, C. Harkin, and H. L. Dai, *J. Chem. Phys.* **93**, 4615, (1990).

39. E. A. Rohlfing and J. E. M. Goldsmith, *J. Chem. Phys.* **90**, 6804, (1989).

40. S. H. Kable, J. W. Thoman, Jr., and A. E. W. Knight, *J. Chem. Phys.* **88**, 4748, (1988).

41. W. D. Lawrance and A. E. W. Knight, *J. Chem. Phys.* **76**, 5637, (1982); E. Abramson, H. L. Dai, R. W. Field, D. G. Imre, J. L. Kinsey, C. Kittrell, D. E. Reisner, and P. H. Vaccaro, in *Lasers as Reactants and Probes in*

Chemistry, W. M. Jackson and A. B. Harvey, eds., (Howard University Press, Washington, D.C., 1985), p. 393.

42. Q. Zhang, S. A. Kandel, T. A. W. Wasserman, and P. H. Vaccaro, *J. Chem. Phys.* **96**, 1640, (1992).

43. See the collection of papers in the special issue on Stimulated Emission Pumping in *J. Opt. Soc. Am.* **B7**, 1802–1980, (1990).

44. P. G. Smith and J. D. McDonald, *J. Chem. Phys.* **96**, 7344, (1992).

45. J. M. Friedman and R. H. Hochstrasser, *Chem. Phys.* **6**, 155–165, (1974).

46. X. Li, B. Jiang, X. Xie, and C. Zhang, *J. Opt. Soc. Am.* **B7**, 1884, (1990).

47. X. Yang, *et. al.*, *J. Phys. Chem.* **97**, 3944, (1993).

48. Y. Chen, L. Hunziker, P. Ludowise, and M. Morgen, *J. Chem. Phys.* **97**, 2149, (1992); also see A. Gragowska, J. Sepiol, and C. Rulliere, *J. Chem. Phys.* **95**, 10493, (1991).

49. J. C. White, "Stimulated Raman Scattering", in *Tunable Lasers*, L. F. Mollenauer and J. C. White, eds., (Springer-Verlag, Berlin, 1987) pp. 146–207.

50. X. Yang, C. A. Rogaski, and A. M. Wodtke, *J. Chem. Phys.* **92**, 2111, (1990).

51. I. A. McIntyre and M. H. Dunn, *Opt. Comm.*, **55**, 45, (1985); A. N. Rubinov and T. Sh. Efendiev, *Optica Acta*, **32**, 1291, (1985).

52. Patrick H. Vaccaro, private communication.

53. G. Stock and W. Domcke, *J. Opt. Soc. Am.* **B7**, 1970, (1990).

54. D. Goswami and W. S. Warren, *J. Chem. Phys.* **99**, 4509, (1993).

CHAPTER 4

LASER INDUCED DISPERSED FLUORESCENCE SPECTROSCOPY (LIDFS) OF CS_2 AND NO_2

Rémy Jost and Antoine Delon

Laboratoire des Champs Magnétiques Intenses de Grenoble (CNRS)
B.P. 166 X - 38042 Grenoble Cedex - France

Jean-Paul Pique and Georges Sitja

Laboratoire de Spectrométrie Physique (UA 08)
Université Joseph Fourier de Grenoble
B.P. 87 - 38402 Saint-Martin-d'Hères Cedex - France

Contents

1. Introduction

The increasing interest in highly excited vibrational and sometimes vibronic levels of polyatomic molecules has led to the development or adaptation of new experimental laser techniques, like Stimulated Emission Pumping (SEP), Degenerate Four Wave Mixing (DFWM), etc.

A rather conventional and simple technique, Laser Induced Dispersed Fluorescence Spectroscopy (LIDFS), has, however, given numerous and extensive results for many molecules such as NO_2,[1] CS_2,[2] C_3,[3] SiC_2,[4] SO_2,[5] C_2H_2,[6] $C_2H_2O_2$,[7] and azabenzenes.[8]

The simplest spectroscopic technique, IR absorption, gives access only to a few of the lowest vibrational levels, mainly because the transition probabilities are governed by the Δv propensity rule due to the usual harmonic character of the bottom of the Potential Energy Surfaces (PES) of stable polyatomic molecules. The transition from the vibrationless level to the $v = 1$ (fundamental IR) transition is usually strong, when allowed by symmetry, and the intensities of overtones ($0 \rightarrow v$, $v \geq 2$) decrease by typically one order of magnitude for each unit increase in v. Consequently, high vibrational levels are usually not observed by absorption. A well-known counter-example is that of the overtones of the CH stretch, for which levels up to $v = 7$ have been observed in several molecules of the methane family. In order to overcome the above mentioned $\Delta v = 1$ propensity rule, it is necessary to take advantage of two facts:
- The strong oscillator strength between different electronic states.
- The possible difference in nuclear equilibrium position and/or vibrational frequencies between the ground electronic state and excited electronic state.

These two properties are used in the well-known Laser Induced Fluorescence (LIF) excitation technique, which is used to characterize ground to excited state electronic transitions.

Thus, in order to observe the high vibrational levels of the ground state, two (consecutive) optical transitions are necessary. Figure 1 illustrates this in the case of a diatomic molecule for which the PES is reduced to one dimension: the $|0\rangle \to |10\rangle$ transition is very weak but the two-step process $|0\rangle \to |3\rangle$ and $|3'\rangle \to |10\rangle$ is highly favoured.

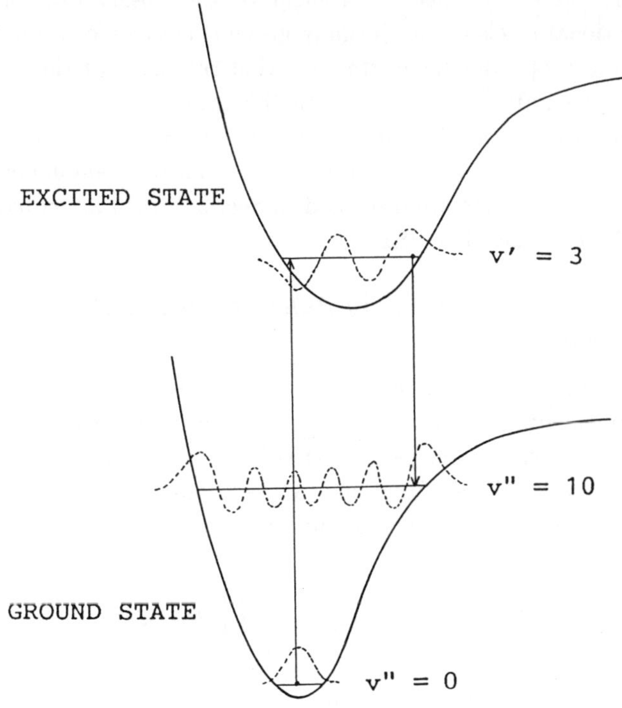

Fig. 1. Schematic representation of Franck–Condon access to highly excited vibrational levels.

In an N atom polyatomic molecule with a PES of dimension $3N - 6$, there are many possible situations concerning the relative location of the minimum and the shape of the PES along each vibrational degree of freedom. For example, in a triatomic molecule the ground state may be linear (or bent) and an excited state (optically connected to the ground state) may be bent (or linear). We will see the very important role of the bending angle in the LIF and LIDFS spectra of CS_2 and NO_2 detailed in Secs. 3 and 4.

In Sec. 2, we introduce the general features of the LIDFS technique and compare the advantages and disadvantages of this technique with those of other techniques, like SEP.

2. LIDFS of Polyatomic Molecules

We emphasize here the specific problem of the observation of highly excited vibrational levels of one (usually ground) electronic state. We will not discuss the properties (strength, selection rules,...) of the electronic transition(s) which are involved in the LIDFS technique.

The crucial role played by the shape of the PES of the two electronic states involved in LIDFS is discussed in Sec. 2.1. The relevant experimental parameters and apparatus are discussed in Sec. 2.2, with a comparison to other related techniques, like SEP.

2.1. *The Vibrational Propensity Rules Related to Potential Energy Surfaces*

We consider now an N atom molecule and describe its wavefunction in the Born–Oppenheimer approximation. The vibrational wavefunction, in the harmonic approximation, can be written as a product of $3N - 6$ one dimensional vibrational wavefunctions ($3N - 5$ for a linear molecule), the variables being the normal mode coordinates, Q_i.

$$\psi_{v_1...v_{3N-6}}(Q_1 \cdots Q_{3N-6}) = \prod_{i=1}^{3N-6} \phi_{v_1}(Q_i) . \tag{1}$$

The set of normal mode coordinates $Q_1 \ldots Q_{3N-6}$ is specific for each PES and there is generally no one-to-one correspondence between the two sets of coordinates. There is, instead, a linear transformation from one set to the other. We consider first the case of two electronic states having the same PES. We consider that the probability of a transition can be written as the product of an electronic part, $|\langle \psi^{\text{upper}}|\mathbf{E} \cdot \mathbf{d}|\psi^{\text{lower}}\rangle|^2$, times the square of the vibrational overlap, $\langle \psi^{\text{upper}}_{v_1'...v_{3N-6}'} |\psi^{\text{lower}}_{v_i''...v_{3N-6}''} \rangle^2$: this last term is the Franck–Condon factor. Thus, the optical transitions from one vibrational level of the upper (lower) electronic state to the set of vibrational levels of the lower (upper) electronic state follow the selection rule $\Delta v_i = v_i' - v_i'' = 0$ for $1 \leq i \leq 3N - 6$. This is an extreme case of the Franck–Condon principle. In real molecules the PES are never identical and, in addition, there are anharmonicities which relax the above-mentioned selection rules into propensity rules. The $A^1A_u \to X^1A_g$ electronic transition of glyoxal, $C_2H_2O_2$, is one good example of the $\Delta v = 0$ propensity rule. The LIDFS emitted from any vibrationally excited level of the A^1A_u state has its strongest line which follows the $\Delta v = 0$ propensity rule.[7] Often, in triatomic molecules there is a good correspondence between the stretching normal mode coordinates (symmetric and antisymmetric) for different electronic states, but a significant difference in the equilibrium bending angle. Thus, long progressions in the bending mode are observed in absorption and in LIDFS, but very short progression in the two stretching modes (i.e., a $\Delta v = 0$ propensity rule for these modes). The Duschinsky effect[9] corresponds to the next step of complexity: one normal coordinate of an electronic state is a linear combination of two (or more) normal coordinates of another electronic state. Thus, the LIDFS spectrum displays progressions in the two modes involved. Ultimately, any normal coordinate of one electronic state may be a combination of all the normal coordinate of the other electronic state (but there are symmetry restrictions), and then there are no propensity rules. Other subsidiary effects, like anharmonicities, vibronically induced transitions, etc., should also be taken into account in order to explain quantitatively LIF and LIDF spectra of polyatomics. In addition to this, the vibrational couplings between the various vibrational degrees of freedom can break the separability of the various vibrational coordinates in the total vibrational wavefunction (see Eq. 1), and at the limit of complete chaos, can destroy any propensity rules. These vibrational and/or vibronic couplings are discussed in Secs. 3 and 4 and in many contributions to this book. However, when the vibrational energy levels are "regular", (i.e., not strongly coupled) the dominant feature of each spectra can usually be explained by the Franck–Condon principle, i.e., by the correspondence between normal coordinates of the two PES involved in an electronic transition.

2.2. *LIDFS: Towards High Resolution and High Sensitivity*

The LIDFS experimental setup is simple but the observation of weak transitions requires the optimisation of: (i) the shape of the fluorescence source, (ii) the spectrometer, and (iii) the detector. We will discuss here mainly the case of observations using a monochromator. Fourier Transform (FT) spectrometers[10] have, however, two advantages: the "multiplex" nature which is essential in the IR range (where detectors are noisy) and that they have a large throughput (compared with a monochromator slit).

(i) The optimum imaging of an extended incoherent fluorescence source onto the entrance slit of the spectrometer is achieved when the solid angle of the incident light matches the acceptance angle of the spectrometer. The high resolution spectrometers used for the CS_2 and NO_2 studies (see Secs. 3 and 4) have an acceptance angle of about $5°$. A good compromise between the resolution and the luminosity of the spectrometer leads to an entrance slit width on the order of 100 μm. A typical laser waist is of the same order so that an optical system with a magnification of one is usually enough. Diffusion phenomena can, however, broaden the fluorescence source. In LIDFS the pressure has to be sufficiently low that collisional quenching and rovibrational energy transfer do not occur within the lifetime of the emitting rovibronic level. In this case, thermal gas diffusion spreads the fluorescence source. The lifetime of the excited level has to be compared with the time of flight $\tau = d/\bar{v}$, where $\bar{v}$ is the mean thermal velocity and d the laser waist. $\bar{v}$ is of the order of 0.3 mm/μs in a gas cell at room temperature and about 0.7 mm/μs in an argon, supersonic, free jet. For a lifetime shorter than 1 μs a gas phase experiment is adequate and easy to use: the laser beam is then parallel to the entrance slit of the spectrometer and the diffusion of the fluorescence light source in the perpendicular direction is of the order of the slit width. For longer lifetimes a supersonic jet experiment is preferable: the laser beam is then perpendicular to the slit which is parallel to the flame of fluorescence (see Fig. 8). But, in this case, a single mode laser frequency ($\sim$ 10 MHz) is necessary in order to obtain a narrow fluorescence source of the order of 100 μm (a free jet expands over about one steradian with a Doppler distribution of about 1 GHz).

(ii) The holographic production of high-quality gratings has now reached a high technological standard. High resolution spectrometers are now equipped with large gratings ($\sim$ 20 $\times$ 20 cm), with a large number of grooves (1200 to 3600 per mm), and have high efficiency ($\sim$ 70%). A typical dispersion of 1 Å/mm allows a resolution of 0.1 Å with a 100 μm slit. The best spectrometers in the world now reach a resolution of 1 million. For small triatomic molecules such resolution is not necessary.

(iii) The sensitivity of detectors has also increased during recent years, especially in the near IR. Behind the output slit of a spectrometer one conventionally uses a photomultiplier (PMT). Now the output slit can be replaced by an optical multichannel analyser (OMA). The first OMA's were composed of a photocathode amplifier in front of a linear array of photodiodes. Their intrinsic noise did not, however, allow the observation of very weak transitions, especially in the near IR. The liquid nitrogen cooled CCD camera technology provides a dramatic improvement in detection sensitivity and resolution, enabling experiments that would not otherwise have been possible. Section 3 will show LIDFS in CS_2 performed with such a camera and Sec. 4 LIDFS in NO_2 performed mainly with a PMT.

The combination of lasers and detectors of high performance makes the LIDFS technique a high resolution, high sensitivity, reproducible, and fast technique for spectroscopists. A comparison with the SEP technique was done using the example[2] of CS_2 (see Sec. 3). The advantages and disadvantages of the two techniques are the following: i) Since SEP is a double optical laser resonance it is theoretically sub-Doppler. Unfortunately, high peak power lasers are needed, which usually have a linewidth of the order of the Doppler width. ii) SEP is a collision-free technique. The delay between pump and dump lasers can be small, avoiding quenching even at rather high pressures. Low pressure gas cells or a supersonic free jet can also avoid quenching problems in the LIDFS technique. iii) LIDFS is a zero background technique while conventional SEP is not. Several variant techniques can increase the sensitivity of SEP, but usually lead to experimental complications. iv) Probably the best advantage of the LIDFS versus the SEP is its multichannel aspect, which considerably reduces the experimental time. v) Finally, the intensities in LIDFS are linear whereas strong nonlinearities have been observed in SEP experiments.[11]

3. LIDFS of CS_2

3.1. *Introduction*

The behaviour of the CS_2 molecule should be analogous to that of the CO_2 molecule. Since they have the same number of valence electrons, Mulliken predicted that their ground states are similar. The 320 nm excitation of CS_2 is experimentally more attractive than the 210 nm excitation of CO_2, which indeed leads to a predissociated excited state. The numerous theoretical studies of the CO_2 molecule can help to understand the experimental results of the CS_2 molecule. Our motivation is the study of the transition to vibrational chaos in an electronic state which does not interact with another excited electronic state. The ground state $\tilde{X}\Sigma_g^+$ of CS_2 is well adapted and

CS_2 appears to be a very good prototype for such a study. The first excited triplet state $^3A_2$ lies above 26 000 cm^{-1}.[12] Theoretical classical calculations have predicted a chaotic situation for CS_2 at about 15 000 cm^{-1} in the ground electronic state.[13] Before this work only a few levels (~ 50) were observed up to 12 000 cm^{-1} with low resolution experiments.[14] Previous work has shown that extrapolation of the potential or *ab initio* calculation to high vibrational energy usually leads to contradictory conclusions because classical calculations can be very sensitive to details of the potential surface. Experiment, therefore, becomes necessary.

As it was already mentioned the spectroscopy of high vibrational levels is not easy, particularly when one needs a "pure" spectrum (levels belonging to a given symmetry). In fact, high vibrational level spectroscopy does not mean high absolute energy but rather numerous quanta in a given mode: for example, at 20 000 cm^{-1} a CH overtone has only 6 quanta of vibration and is rather harmonic, whereas a bending mode of CS_2 is highly anharmonic with 50 quanta of vibration. Up until now the sensitivity of LIDFS was not enough to obtain spectra of CS_2 in its chaotic regime. The new technologies of high power and high repetition rate copper vapour lasers (Cu laser) and very high sensitivity liquid nitrogen-cooled CCD cameras (LN-CCD) have allowed us to push the sensitivity of the LIDFS technique further out, so that we have already recorded a continuous spectrum up to 19 200 cm^{-1} (above the ground state) with an intensity dynamic range of 5000 and an absolute precision of 0.01 Å. We have recorded 964 vibrational levels of the same symmetry.[15] This spectrum allows us to observe experimentally and continuously for the first time, the transition from a regular spectrum to a "chaotic" one in CS_2.[15]

In the following section, we will discuss the LIDFS of CS_2 versus SEP; we will then show some results of the LIF excitation spectra of CS_2, and discuss the interpretation of the LIDFS spectra below 12 000 cm^{-1} and the transition to chaos.

3.2. *The Cu Laser System and the LN-CCD Camera*

The experimental arrangement is shown in Fig. 2 and is based on a 60 W Oxford Lasers copper vapour laser (Cu laser). The high repetition rate of the Cu laser (5 to 30 kHz) with its, nevertheless, reasonably high peak power (250 kW for a mean power of 50 W, a repetition rate of 6.5 kHz, and a pulse duration of 40 ns) and high efficiency ($> 0.5\%$ directly in the visible) make it, in our view, a useful instrument for spectroscopy. The quality of the beam is excellent when an unstable cavity is used and when convection currents around the windows are suppressed. A home made dye laser consists of an oscillator and two amplifiers. The oscillator is composed of a special holographic grazing incidence grating (Jobin Yvon, 2400 lines/mm), a high refractive index expander consisting of two prisms and a high flow dye cell. The resolution we obtained is 0.02 cm^{-1} without an etalon. Continuous tuning over the whole gain of the dye is achieved by a high precision (10^{-4} degrees) rotation of the end miror. The efficiency of our dye laser (using DCM) is on the order of 15%. A BBO crystal allows us to obtain UV output power between 0.5 and 1 W with a pumping power of 45 W. Because of the high repetition rate, the peak power at 320 nm is low enough to prevent multiphoton dissociation and excitation of CS_2. This is a significant point compared to YAG or Eximer lasers systems. Conversely it is usually difficult to obtain high mean UV power and large continuous tuning spectral range with CW lasers, whereas with the Cu laser system this is rather simple.

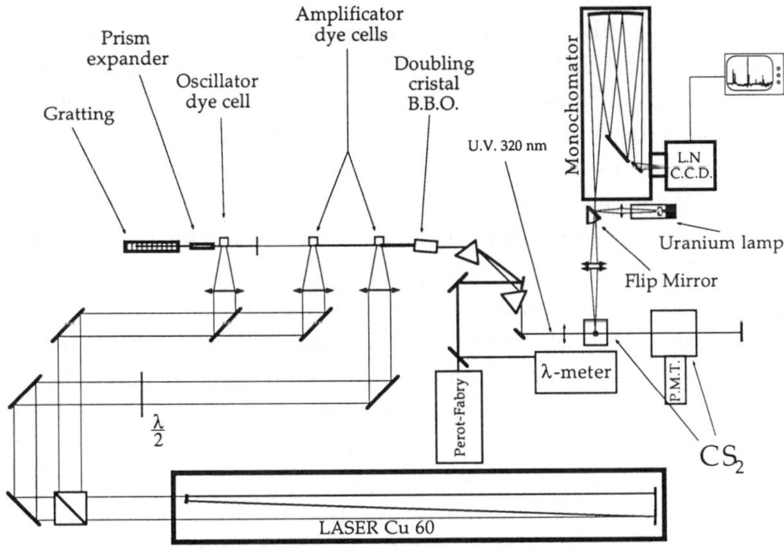

Fig. 2. CS_2 LIDFS experimental setup.

In the next section, we will discuss a particular single rovibrational excitation of CS_2 in the singlet V^1B_2 electronic state. The main advantage of high resolution double resonance spectroscopy of the CS_2 molecule is that the first step (pump) alleviates the need for a supersonic jet. The excitation of a single rovibrational level in the intermediate state can be seen as a zero rovibrational temperature preparation of a nonequilibrium state out of the thermal bath. From this intermediate state it is possible to stimulate emission to rovibrational states of the ground electronic state using a second laser (SEP), or to analyse the LIDFS through a high resolution monochromator. The first technique has been extensively used to reach high vibrational states. Here, we again list briefly the limitations of SEP in the case of the CS_2 molecule: i) The SEP is a single channel technique and laser intensity fluctuations necessitate long time averaging of the signal, so that the experiments often become very long. The SEP experiments we performed on the CS_2 molecule have shown that to record a continuous 20 000 cm^{-1} long spectrum several thousand hours would have been necessary! ii) Large nonlinear effects render the interpretation of SEP intensity difficult. iii) Because of laser intensity fluctuations and the photon noise of fluorescence intensity from the excited state, the signal to noise ratio of the SEP is limited. As was described in the introduction, different variants of SEP technique significantly increase its sensitivity. In CS_2, Vaccaro *et al.*[16] observed a vibrational state at 8500 cm^{-1} above the ground state with excellent signal to noise. The fineness of the four wave mixing technique used limits this experiment, however, to short spectral excursions. We have shown that it is possible to observe the SEP spectrum of a molecule inside the cavity of a broad band Ti/Sa laser pumped by the Cu laser.[17] The interesting advantages of the intracavity SEP technique are its zero background and multichannel averaging. The cross-section of the transitions involved of the molecule must, however, be enough so that the additional gain can compete with the Ti/Sa gain.

LIDFS of CS_2 can be performed in a gas cell since we can resolve the rotational structure with our high resolution laser, and also because the optimum imaging discussed in Sec. 2.2 can be achieved. The self-quenching of near UV-excited CS_2 emission is very fast. It occurs at almost every gas-kinetic collision and involves electronic quenching to the ground state. The pressure must be adjusted in order to avoid collisions during the electronic lifetime of CS_2, which is about 2 μs at 320 nm of excitation.

The number of collisions which a single CS_2 molecule makes per unit time is about $5\times10^5 s^{-1}$, taking a collision diameter of 4 Å at room temperature and for a pressure of 1 Torr. Moreover, CS_2 LIDFS spectra show an apparent absence of collision-induced vibrational and rotational relaxation effects.

For a long time high resolution LIDFS was limited to the study of low energy vibrational levels of the CS_2 molecule. We have increased the sensitivity of this technique using the Cu laser system described above and a liquid nitrogen CCD camera. Several types of detectors can be employed at the output of the monochromator: a PMT behind the output slit is sensitive but is a mono spectral detector; an intensified OMA is a multi spectral detector but is very noisy. The new technology of liquid nitrogen silicon CCD cameras have both avantages. The noise of each spectral element is ±10 photons, which is very close to the noise of a PMT. The camera we used (Princeton LN/CCD-1152) has 1152 elements. Another favorable point of a LN-CCD camera is that the resolution is not effected by the smear of the intensity of one 23 μm spectral element, which is generally produced by an intensifier. Moreover, a LN-CCD camera allows long exposure times and a large spectral range (200 nm to 1100 nm). The only limitation is that cosmic rays can overlap the spectrum. To correctly use the high resolution of the monochromator (SOPRA UHRS 2000, 2 meters, grating 1200 lines/mm, dispersion 0.8 Å/mm, aperture: f/11) a homemade UV cylindrical doublet expands by a factor of 6 the output image on the camera. The emission spectrum of a uranium hollow cathode discharge can be used via a flip mirror to precisely calibrate each spectral window. The uranium atlas spreads from the UV to the IR and is sufficiently dense that several emission lines cover a spectral window. The precise fit of the line positions, the identification in the uranium atlas, and the fit of the spectral dispersion and planar distortions are done on a 486 PC with a program written using Spectra Calc. The absolute line position precision is then 0.01 Å, which is close to the precision of the uranium atlas. Under these conditions, LIDFS is a high resolution and high sensitivity spectroscopy technique. Figure 3 shows the LIDFS spectra of CS_2.

3.3. *LIF Excitation Spectrum of CS_2 Close to the Barrier to Linearity in the V^1B_2 State*

Particular attention to the excitation has been made in order to: i) avoid "impure" spectra, and ii) to give the best Franck–Condon access to high

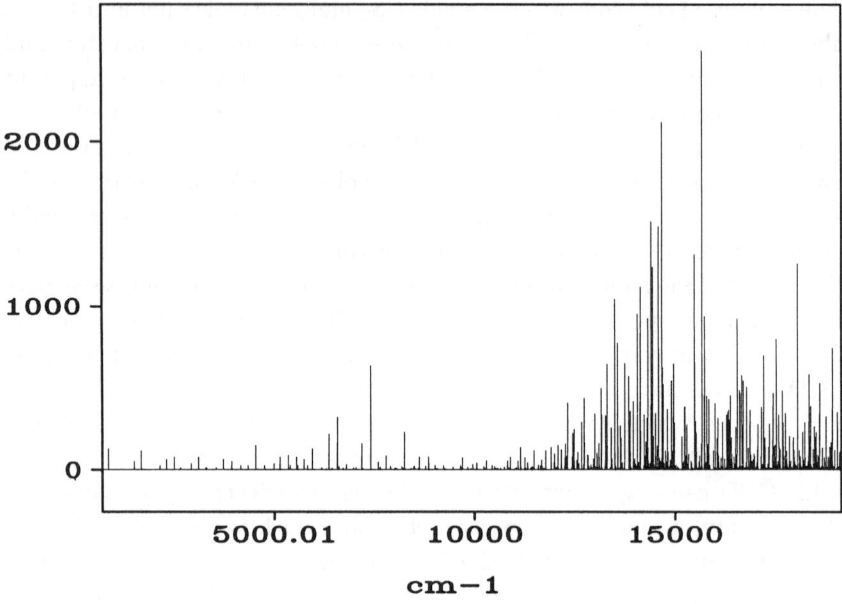

Fig. 3. LIDFS spectrum of CS_2 from the $|0,3,0\rangle^{K'=0}|J' = 1, K' = 0\rangle$ level. The absolute precision is 0.01 Å. We have observed 964 levels of Σ_g^+ symmetry.[15]

vibrational states.

The excitation spectrum of CS_2 at 320 nm corresponds to an electronic transition from the ground state $\tilde{X}\Sigma_g^+$ to the singlet excited state V^1B_2 (V is a nomenclature due to Kleman[12]). The observation of magnetic rotation spectra[18] and of a weak magnetic moment, detected using the Hanle effect, shows that V is a singlet state coupled with a lower triplet state, R^3A_2, by a strong spin–orbit interaction. The V^1B_2 state is one of the two Renner–Teller bent component states which correlate with a linear $^1\Delta_u$ state. Because of strong interactions between the V^1B_2 state and several other electronic states (Fig. 4) which intersect around 31 000 cm^{-1} the rotational structure of the vibronic bands of CS_2 at 320 nm is rather complicated. A supersonic jet is, however, not necessary. We have shown[2] that in the CS_2 vapour phase one can excite a single rotational level with a high resolution laser. The precise identification of the excited rotational quantum J' and possible overlapping rovibrational transitions were analysed from LIDFS spectra to known levels of the ground electronic state (low energy levels).

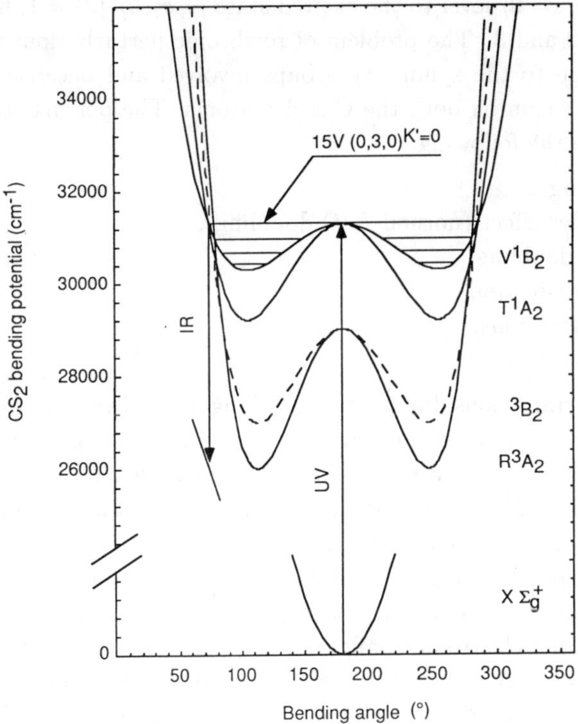

Fig. 4. Electronic potentials of CS_2 along the bending coordinate.

Because the study of vibrational chaos necessarily needs "pure" spectra, the allowed rovibrational perturbations in the excited state must be carefully examined and corresponding forbidden transitions should be avoided. The involved symmetry groups of CS_2 are, respectively, $D_{\infty h}$ and C_{2v} for the ground and excited electronic states. The stretching v_1 and bent v_2 normal modes belong to the a_1 species of C_{2v} and the stretching v_3 mode to the b_2 species. The rovibrational normalized basis states of the linear ground $\tilde{X}\Sigma_g^+$ and bent excited V^1B_2 states will be labelled $|v_1, v_2^{\ell''}, v_3\rangle|J'', \pm\ell''\rangle$ and $|v_1, v_2, v_3\rangle^{K'}|J', \pm K'\rangle$, respectively. ℓ'' is the vibrational angular momentum of the linear molecule and K' is the projection of the angular momentum onto the axis of the maximum inertia of the bent molecule. The vibrational bands we discuss here are all of parallel type (i.e., $\Delta K = K' - \ell'' = 0$) from the ground state

$|0, 0^0, 0\rangle|J'' = 2, \ell'' = 0\rangle$ level to the excited state $|0, v_2, 0\rangle^0|J' = 1, K' = 0\rangle$ levels with $v_2 = 2$ and 3. The problem of rovibronic perturbations in CS_2 is very simple due to the symmetry groups involved and because of the absence of nuclear spin on both the C and S atoms. The perturbations to be considered are the following:

(i)   axis switching,
(ii)   Renner–Teller effect (normal $K(\ell)$ doubling),
(iii) asymmetry doubling,
(iv) centrifugal distortion,
(v)   Coriolis interactions,
(vi) Fermi resonances.

(i) Forbidden transitions due to axis switching arise physically from the change in Eckart axes between the two electronic states. Transition between groups of symmetry $D_{\infty h}$ and C_{2v} is a particular case where no change of the axes are expected.[19] As a consequence, no axis switching transitions ($\Delta K \neq 0$) have been observed in the $\tilde{X} - V$ band system of CS_2.

(ii) Renner–Teller splittings arise from the interaction of the electronic and vibrational angular momenta, which produces the energy differences of states of the same electronic configuration. This effect is absent in the ground state for which $\Lambda = 0$. In the excited state $\Lambda = 2$, the Renner–Teller interaction coupling $2A\Lambda K'$ ($A = 88$ cm^{-1}) does not affect $K' = 0$ levels.

(iii) CS_2 is only a near prolate symmetric top molecule in its excited state. Usually a bent-to-linear transition shows in addition to the ΔD effect a K-type doubling. However, since CS_2 has no nuclear spin, J is a good quantum number ($F \equiv J$) and the antisymmetric rovibronic levels are missing. Specifically, the $\Sigma_g^-(-a)$ and $\Sigma_u^+(+a)$ rotational levels of the ground state and the b_1 and b_2 rotational levels of the excited states are missing. This leads to a double consequence: first for all $\ell' = 0$ (and $K' = 0$) levels, J'' is always an even number (J' is odd); secondly for $\ell' \neq 0$ (and $K' \neq 0$) there is no ℓ doubling (or K doubling); alternately, the upper or lower components of k signed type doublet levels are missing. Since we excite $K' = 0$ and because of the above remarks asymmetry doubling is not expected.

(iv) The coupling operators for centrifugal distortion are such as $Q_r J_\alpha J_\beta$. In the excited state of CS_2, the operator $Q_2 J_c^2$ can couple $K' = 0$ and $K' = 2$ levels of the same vibrational band. But as we excite $J' = 1$, $K' = 2$ does not exist ($K' \leq J'$).

(v) It is well-known that in the CS_2 molecule Coriolis interactions can occur because of accidental level crossings. In the V state we have observed a Coriolis perturbation[2] between the quasi resonant vibrational levels $|0,3,0\rangle^{K'=0}$ and $|0,0,1\rangle^{K'=1}$ for relatively large values of J'. The Coriolis operator involved is of the type $Q_2{}^3 Q_3 J_c$. This perturbation has been seen in the LIDFS of the CS_2 molecule. When the $|0,3,0\rangle^{K'=0}$ state was excited, in addition to even v_3'' levels of the ground state, odd v_3'' levels were observed through weak transitions, whereas, when the $|0,2,0\rangle^{K'=0}$ state was excited they were unobservable. Vibrational levels with odd v_3'' and even v_3'' belong to different symmetries and must be separated. However, for the lowest value of $J' = 1$, the corresponding forbidden transitions have not been detected.

(vi) Fermi resonances in the ground electronic state will be discussed in detail in Sec. 3.4. In the excited state no Fermi resonances occur since none of the sublevels of the $|0,3,0\rangle$ state have the same species as the $|0,0,1\rangle$ state. Vibrational states of the same species are far from the $|0,3,0\rangle$ state and could not cause significant Fermi interactions.

In conclusion, the excitation of the $V^1 B_2 |0,3,0\rangle^0 |J' = 1, K' = 0\rangle$ rovibrational state is well adapted to minimize the effect of "impure" transitions due to perturbations in the excited state. Moreover, because this state lies close to the barrier to linearity (Fig. 4), its vibrational Franck–Condon overlap with the $|0,0^0,0\rangle$ ground state is very good leading to efficient excitation. Finally, as was discussed in detail in the introduction, the large difference in the shape of the ground and excited state potential energy surfaces gives good Franck–Condon access to high vibrational levels of the $\tilde{X}^1 \Sigma_g^+$ state, connected to transitions which are emitted from the turning points of the $V^1 B_2$ potential (see the intensity envelop in Fig. 3).

3.4. *LIDFS of CS_2 below $12\,000$ cm^{-1}: Fermi Resonances*

Because of the particular situation of the excitation described above, all breaking of the optical selections rules observed in the LIDFS spectra are due to interactions within the set of ground electronic state vibrational levels. Only two sets of independant levels are observed in the LIDFS. They correspond, respectively, to $J'' = 0$ (R(0) lines) and $J'' = 2$ (P(2) lines). Only optical double resonance spectroscopy techniques allow such simplification. The $J'' = 0$ set of levels is free of rovibrational coupling. Only pure vibrational interactions (Fermi resonances) can mix $J'' = 0$ levels amongst themselves. On the other hand, the $J'' = 2$ levels can present vibrational and rovibrational couplings. In the absence of rovibrational couplings the intensity ratio of the R(0) versus the P(2) lines is 1:2. Rovibrational couplings of $J'' = 2$ levels should modify this ratio. The high resolution spectra show us that the 1:2 ratio is very well conserved. So in a first approximation we can neglect rovibrational couplings.

The vibrational Hamiltonian can be written:

$$H = H_0 + H_F \, ,$$

where H_0 is the effective diagonal part of a given basis and corresponds to the Dunham expansion in the harmonic basis. The Fermi term, H_F, contributes to non-diagonal couplings. In a three-dimensional system the creation and annihilation operators C_i and A_i of the (sometimes called) vibron model[20] can be constructed using tensor algebra on the individual oscillator (harmonic or Morse) operators c_i and a_i which act on the $|v_i\rangle$ states, whereas C_i and A_i act on $|v_1, v_2, v_3\rangle$ states. On this basis we have:

$$H_0 = \sum_i \omega_i \left(C_i A_i + \frac{d_i}{2} I \right)$$

$$+ \sum_{ij} X_{ij} \left(C_i A_i + \frac{d_i}{2} I \right) \cdot \left(C_j A_j + \frac{d_j}{2} I \right)$$

$$+ \sum_{ijk} Y_{ijk} \left(C_i A_i + \frac{d_i}{2} I \right) \left(C_j A_j + \frac{d_j}{2} I \right) \left(C_k A_k + \frac{d_k}{2} I \right) + \cdots \, ,$$

$$H_F = \sum_{ijk} k_{ijk} (A_i C_j C_k + C_i A_j A_k + \cdots)$$

$$+ \sum_{ijk\ell} k_{ijk\ell} (A_i A_j C_k C_\ell + C_i C_j A_k A_\ell + \cdots) \, ,$$

where I is the identity operator, d_i the multiplicity of the vibrational modes, and ω, X, Y, $k \ldots$ are the molecular constants. The operators C_i and A_i can be constructed from the harmonic normal mode basis. According to our excitation of CS_2 the vibrational ground states we observed are linear combinations of the totally symmetric a_i basis states $|v_1, v_2^0 (\text{even}) v_3 (\text{even})\rangle$ where the number of quanta in the v_2 and v_3 modes is even (since $\ell'' = 0$ and $b_2 b_2 = a_1$). As a consequence, and because each term of H must be totally symmetric, several terms of H vanish. For example, k_{112} will vanish because the corresponding terms transform even v_2 quanta to odd ones. The k_{113} parameter also vanishes because the corresponding term belongs to b_2 symmetry.

The above expansion of H is very convenient for writing a general fit program using tensor algebra. The technique of our program will be discussed in a forthcoming paper.[21]

The Dunham expansion H_0 is not sufficient to interpret the CS_2 spectra. A strong 1:2 Fermi resonance couples the bending mode ν_2 and the symmetric stretching mode ν_1. The off-diagonal term $k_{122}(A_1C_2C_2 + C_1A_2A_2)$ must be added. Below $12\,000$ cm^{-1} the following parameters $\omega_1, \omega_2, \omega_3, X_{11}, X_{22}, X_{33}, X_{12}, X_{13}, X_{23}, Y_{222}, Y_{122}$ describe well the LIDFS spectra of CS_2.[2,22] But as most of the levels observed in this energy range correspond to $v_3 = 0$, the parameters ω_3, X_{33}, X_{13}, and X_{23} cannot be determined. They have to be fixed to values obtained by high resolution IR experiments[22] or can be integrated to effective parameters ω_1', ω_2', and ω_3':

$$\omega_1' = \omega_1 + \frac{X_{13}}{2} \, ,$$

$$\omega_2' = \omega_2 + \frac{X_{23}}{2} \, ,$$

$$\omega_3' = \omega_3 + \frac{X_{33}}{2} \, .$$

In fact, by using higher excited J' levels we observed several odd v_3 levels[2] and performed a global fit up to $12\,000$ cm^{-1}. Table V of Ref. 2 shows the corresponding parameters. The 1:2 Fermi resonance with $k_{122} \sim 40$ cm^{-1} is large and implies that the normal mode numbers v_1 and v_2 are no longer good quantum numbers. The polyad number $N = v_1 + v_2/2$, however, remains a good quantum number. Below $12\,000$ cm^{-1} the energy E, N, and v_3 are the three good quantum numbers, which implies that the vibration of CS_2 is nonchaotic. Only local resonances between levels of adjacent polyads were detected in this energy range.[2] The detailed study of the nature of such accidental interactions is very interesting since they spoil the quantum number N, and are responsible for a chaotic situation at higher energies[23] where many vibrational anticrossing levels are observed.

3.5. *LIDFS of CS_2 up to $19\,200$ cm^{-1}: Transition to Chaos*

Between $12\,000$ cm^{-1} and $19\,200$ cm^{-1} we have observed 833 vibrational levels of the same symmetry. The fit using the model described above no longer works: the deviation between experimental and calculated level positions rapidly becomes of the order of the mean level spacing. In addition to the 1:2 Fermi resonances, higher order terms of H_F must be considered. We have shown[23] that any additional terms couple levels of different polyads and, thus, lead to a chaotic situation. Traditional spectroscopic assignment with quantum numbers or pseudo quantum numbers is no longer possible. Three ways may be investigated in order to analyse the molecular dynamics: i) quantum calculations, ii) semi-classical calculations, and iii) statistical calculations. The fit of chaotic spectra is theoretically possible for a small molecule like CS_2. The only problems are technical problems of the diagonalisation of a huge and sparse matrix constructed with a known basis. CS_2 is also an interesting molecule for a semi-classical calculation. Because it is a heavy molecule, CS_2 rapidly becomes a semi-classical object: for example, at $15\,000$ cm^{-1} the number of quanta in the bending mode is already equal to 40. Quantum and semi-classical calculations are in progress and will not be discussed here.

Several statistical functions can test the rigidity and correlations of a spectrum. Any statistical test requires many levels.[24] Once again, CS_2 has a large density of states compared to other triatomic molecules. For example, CS_2 has a vibrational density in the Σ_g^+ symmetry which is four times that of NO_2 and five times that of CO_2. A very simple model gives the number of levels $N(E)$ from 0 to the energy E as is approximately:

$$N(E) = \frac{E^n}{n! \prod_i \omega_i} \, ,$$

where n is the degrees of freedom. The number of levels can also be evaluated using the Fermi Hamiltonian described in the previous section, which predicts that up to $19\,200$ cm^{-1} about 1150 levels are expected in the Σ_g^+ symmetry states ($\equiv |v_1, v_2^0(\text{even}), v_3(\text{even})\rangle$). In the same spectral range we observed 964 vibrational levels in the LIDFS spectra (Fig. 3). Figure 5 represents the experimental evolution of $N(E)$. There are clearly two domains which correspond to a different power law: $n = 2$ below $12\,000$ cm^{-1} and $n = 3$ above $14\,000$ cm^{-1}. The low energy part of $N(E)$ is well described with $\alpha_2 E^2$ and the high energy part with $N_0 + \alpha_3 E^3$.

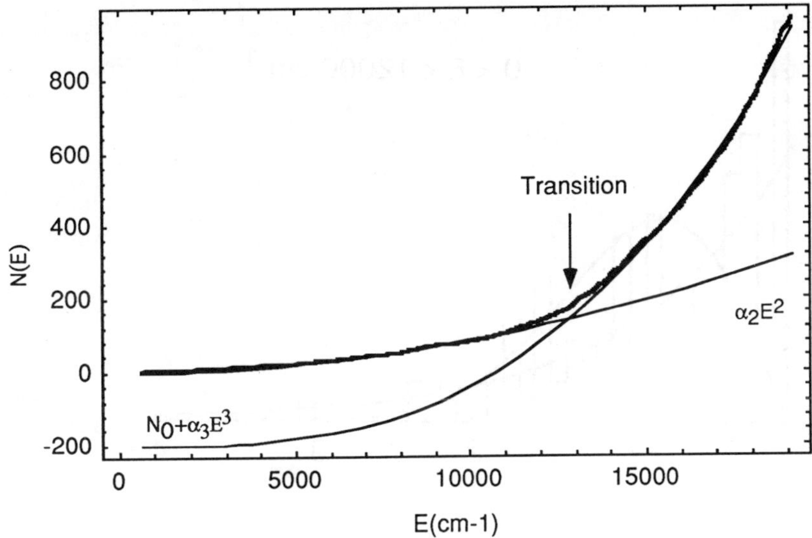

Fig. 5. Number of observed levels $N(E)$ from the $(0,0,0)$ level to the energy E. The evolution of $N(E)$ is quadratic up to $12\,000$ cm^{-1} and cubic above. The solid curves correspond to the fitted function, ($\alpha_2 E^2$ and $N_0 + \alpha_3 E^3$).

The fitted parameters α_2 and α_3 corresponds roughly to the calculated parameters using, respectively, the formula $\frac{1}{2} \cdot \frac{E^2}{2!\omega_1\omega_2}$ and $\frac{1}{4} \cdot \frac{E^3}{3!\omega_1\omega_2\omega_3}$ ($\frac{1}{2}$ and $\frac{1}{4}$ are to take account of the fact the v_2 and v_3 are even). We have effectively shown in the preceding paragraph that because of the bending excitation very few $v_3 \neq 0$ levels have been observed below $12\,000$ cm^{-1} and that most of the levels correspond to linear combinations of ν_1 and ν_2 modes due to a 1:2 Fermi interaction. In other words, below $12\,000$ cm^{-1} the antisymmetric motion (v_3) is separable and is not coupled to the two others. A better evaluation of α_2 and α_3 is obtained using a direct counting of levels generated with the diagonalisation of the Fermi Hamiltonian described above. α_2 corresponds to $v_3 = 0$ levels and α_3 to the total number of levels in the Σ_g^+ symmetry. N_0 corresponds to the $v_3 \neq 0$ missing levels below $14\,000$ cm^{-1}. This more accurate evaluation does, however, give the number of levels actually observed above $14\,000$ cm^{-1}. We conclude that, above $14\,000$ cm^{-1}, we have observed nearly the complete set of the Σ_g^+ vibrational levels.

                                    Molecular Dynamics and Spectroscopy

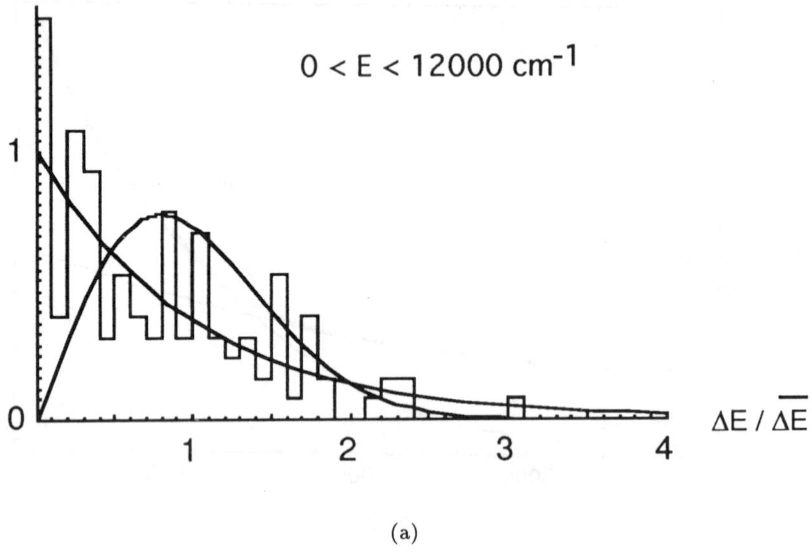

(a)

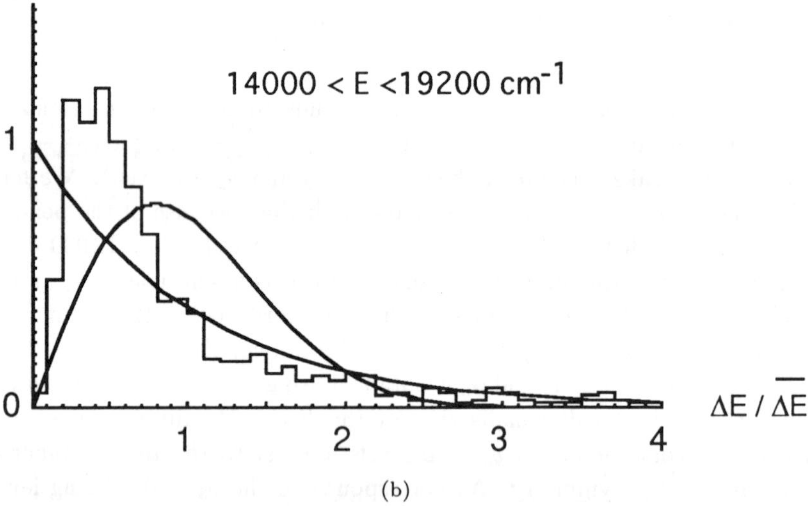

(b)

Fig. 6. Next Neighbour Distribution (NND) for different energy ranges. Continuous and dashed curves correspond to the Poisson and Wigner distribution, respectively. (a) below 12 000 cm^{-1} the distribution is Poisson. (b) above 14 000 cm^{-1} level repulsions are evidenced.

The above results suggest that when the vibrational energy of the ground state increases above $12\,000$ cm^{-1} the even quanta of the antisymmetric mode gradually couples to the two others and become observable due to the Dunshinsky effect, or to a breaking of optical selection rules. This result alone does not necessarily imply that the molecule will become chaotic. Effectively, a simple integrable (not necessarilly separable) change of the ground state basis can explain the observation of ν_3 modes above $12\,000$ cm^{-1}. Below $12\,000$ cm^{-1} we already know that the ν_1 and ν_2 normal modes are strongly mixed by a 1:2 Fermi resonance. We can easily imagine that above $12\,000$ cm^{-1} the two new modes (described by N and M quantum numbers in Ref. 22) and the ν_3 mode change to a new modes description similar to the normal to local mode transition. In the absence of a precise model which could fit the experimental spectra, one can check the spectral correlations in order to decide if the system becomes chaotic or not. Figure 6 shows the next neighbour distribution (NND) of levels for the two energy domains. Theoretical and numerical calculations predicted that the NND passes continuously from a Poisson to a Wigner distribution (respectively continuous and dashed lines in Fig. 6) when a system becomes more and more chaotic. Figure 6 shows that the experimental distribution is Poisson below $12\,000$ cm^{-1} and that levels repulsions are evidenced above this energy. The width of the first bin in Fig. 6b is equal to 0.7 cm^{-1} which is close to the LIDFS resolution (0.5 cm^{-1}) of the present experiment. So, even if the first bin is slightly distorted, a repulsion of levels is clearly demonstrated. The existence of spectral correlations prove that the transition observed in CS_2 is not a simple basis change, but that the couplings involved in the transition are non-integrable and the good quantum numbers gradually disappear. The transition to chaos in CS_2 is slow and at $19\,000$ cm^{-1} the system is still not completely chaotic. The LIDFS spectra of CS_2 is the first example which shows continuously the evolution to chaos in a single electronic state. The challenge is to build the corresponding Hamiltonian in order to study the molecular dynamics in more detail.

4. LIDFS of NO_2

4.1. *Vibronic Coupling in NO_2*

The four lowest electronic states of NO_2 are doublets (NO_2 has one unpaired electron) with various geometries as shown on Fig. 7. The relevant symmetry group is C_{2v} (4 species A_1, A_2, B_1, B_2), the $\tilde{X}^2 A_1$ ground state being bent with an equilibrium angle of 134°, the two bonds being equal. This $\tilde{X}^2 A_1$ state is a lower Renner–Teller component and the second component, the $\tilde{B}^2 B_1$ state, can also be described as a $^2\Pi$ state in the $D_{\infty h}$ group. The two other states $\tilde{A}^2 B_2$ and $\tilde{C}^2 A_2$ have an equilibrium angle evaluated by *ab initio* calculation to be of about 102° and 112°, respectively.[25] The three vibrational normal mode coordinates of NO_2 in the C_{2v} symmetry group are the symmetric stretch Q_1 which belongs to the a_1 species, the angle Q_2 which belongs to the a_1 species, and the antisymmetric stretch Q_3 which belongs to the b_2 species. Thus, the 3D vibrational wavefunctions $|v_1, v_2, v_3\rangle$ belong to a_1 or b_2 species according to the parity of n_3. The relative location and shape of the PES of the ground state $\tilde{X}^2 A_1$ and the $\tilde{A}^2 B_2$ state are such that the Q_1 and Q_3 normal coordinates are almost the same for the two electronic states, but $Q_2(A^2 B_2) = Q_2(X^2 A_1) + \Delta Q_2$. ΔQ_2 corresponds to the large difference in the 134° and 102° equilibrium angles. The optically allowed transitions are $\tilde{X}^2 A_1 \leftrightarrow \tilde{A}^2 B_2$ and $\tilde{X}^2 A_1 \leftrightarrow \tilde{B}^2 B_1$. The $\tilde{C}^2 A_2$ is then a "dark" state which has only been observed indirectly. The well-known complexity of the visible NO_2 spectrum is due to strong vibronic couplings between the high vibrational levels of the ground state $\tilde{X}^2 A_1$ and the isoenergetic vibrational levels of the first excited electronic state $\tilde{A}^2 B_2$ whose origin probably lies at 9737 cm^{-1}.[1] The vibronic coupling can only mix levels having the same vibronic symmetry. The vibrationless level $|\tilde{A}^2 B_2; 0, 0, 0\rangle$ of the $\tilde{A}^2 B_2$ electronic state is a vibronic level which belongs to the B_2 species ($B_2 \otimes a_1 = B_2$). This vibronic level, therefore, interacts mainly with (almost isoenergetic) high vibrational levels of the $\tilde{X}^2 A_1$ ground electronic state having an odd n_3 quantum number. As a result we have observed the lowest significantly shifted level of the ground state, the $|\tilde{X}^2 A_1; 0, 11, 1\rangle$ level whose matrix element with $|\tilde{A}^2 B_2; 0, 0, 0\rangle$ is about 26 cm^{-1}. Here, we do not want to analyze the mechanism of the vibronic interactions. We admit that these vibronic interactions can be described by a conical intersection of the $\tilde{X}^2 A_1$ and $\tilde{A}^2 B_2$ electronic states PES. Such a problem can be analyzed in two steps: First we consider the eigenstates of H_0 (corresponding to the diabatic approximation) which are the products of an electronic part ($\tilde{X}^2 A_1$ or $\tilde{A}^2 B_2$) and a vibrational part ($|v_1, v_2, v_3\rangle$). These zero order states form a basis for the second step which takes into account the vibronic interactions represented by H_1. At low energy, around

10 000 cm^{-1}, the vibronic interactions are weak enough to be treated perturbatively and, moreover, numerous vibrational levels of the $\tilde{X}^2A_1$ state are probably not effected by the vibronic interactions. In contrast, at higher energies around and above 17 000 cm^{-1}, the vibronic interactions induce a vibronic chaos[26] as demonstrated by the strong correlations within energy levels which are close to those predicted by the G.O.E. model.[27] We have used the LIDFS technique in order to observe the vibrational levels of the $\tilde{X}^2A_1$ state in the 0 to 12 000 cm^{-1} energy range, from the vibrationless level of the ground state up to and above the energy of the vibrationless level of the $\tilde{A}^2B_2$ state. About 190 vibrational levels of the $\tilde{X}^2A_1$ state are located from zero to 10 000 cm^{-1} but only 20 had been observed previously by infra-red absorption. Two LIDFS experiments had been performed previously[27,28] and globally about 40 levels had been observed with this technique. It should be mentioned that these two previous experiments were done in a cell.

Any LIDFS experiment requires a pertinent choice of the excited level(s), from which the fluorescence is emitted. This is discussed in the following section.

4.2. *LIF Excitation Spectrum of Jet Cooled* NO$_2$

The NO$_2$ excitation spectrum spreads continuously from about 10 000 cm^{-1} up to 25 130 cm^{-1} where dissociation occurs.[30] At room temperature this spectrum is dense and irregular for two reasons:

- First, the above mentioned vibronic chaos mixes the dense set of "dark" vibrational levels of the ground state with the sparse set of "bright" vibrational levels of the optically active $\tilde{A}^2B_2$.
- Second, numerous rovibronic interactions (due to spin–orbit interaction) smear the usually regular rotational sequence of each vibronic level.

These two points explain the high complexity of the visible NO$_2$ spectrum. However, in 1975 Smalley *et al.*[31] made a breakthrough by using a supersonic jet: the rotational temperature is then reduced to a few kelvin and the spectrum is considerably simplified. Since that time, most of the laser experiments on NO$_2$ have been performed with a supersonic jet.

Concerning LIDFS experiments, we have selected several vibronic upper levels around 23 000 cm^{-1}: The selected levels give the strongest

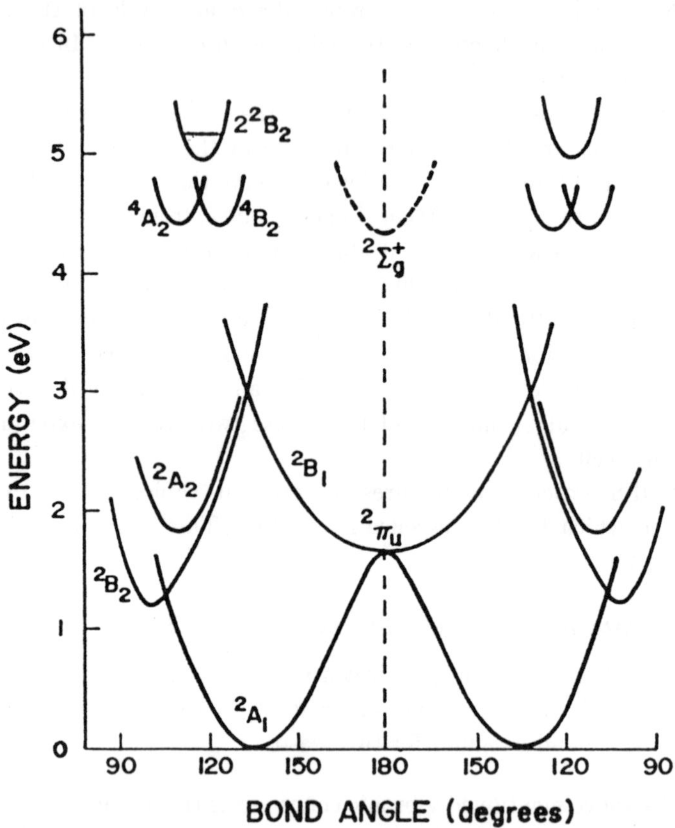

Fig. 7. Electronic energy as a function of bond angle for the low-lying electronic states of NO_2 (from Gillispie *et al.*, *J. Chem. Phys.* **63**, 3425 (1975)).

(unresolved) fluorescence (excitation spectrum) and the fluorescence occurs at wavelengths where the detectors (PMT, CCD, etc) are very sensitive. We will see later that the irregular (i.e., perturbed) behaviour of the *rovibronic* wavefunctions allows one to observe vibronically forbidden transitions (to the set of b_2 vibrational levels of the ground state).

It should be noted that the LIF excitation spectrum of NO_2 in the blue region ($23\,000$ cm^{-1}) is almost inextricable even with a supersonic jet and with a single mode narrow laser: The rotational levels are almost "randomly" distributed everywhere in the 20 000 to 25 000 cm^{-1} energy ranges. However, we have been able to assign numerous R_0 and P_2 lines ($K = 0$ manifold) and the strongest ones have been used as the upper level of our LIDFS experiments.

4.3. *Experimental Considerations*

We have used a supersonic free jet (50 microns pinhole) without any skimmer combined with a single mode cw ring dye laser (stilbene 3) as shown on Fig. 8. The experimental setup is described in detail in Refs. 1 and 7. The fluorescence, emitted perpendicularly to the jet axis and to the laser beam, is focused onto the entrance slit (50 to 100 microns) of a Jobin–Yvon T.H.R. monochromator (1.5 meter, grating 2400 lines/mm, dispersion: 2.6 Å/mm, aperture: f/12). The dispersed fluorescence is detected by photon counting with a low dark current PMT. The monochromator scan is linear and the signal is recorded on a PC. The wavelength calibration is obtained by superposition of an atomic spectrum (He, Ar, or Ne according to the range of wavelength).

When the single mode laser frequency is tuned to the center of a given optical transition, the excited molecules are moving perpendicularly to the laser beam and form (when observed perpendicularly to the laser beam and to the jet axis) a very bright and narrow flame of fluorescence (we excite selectively a narrow slice within the broad angular distribution which comes from the pinhole). This flame has an exponentially decreasing intensity with a typical length of $v\tau$, where v is the molecule speed and τ the lifetime of the level excited by the laser. When NO_2 (1%) is seeded in Ar the speed is about 700 m/sec and the lifetime is of the order of few μsec, leading to a typical length of a few mm. This bright and narrow source of fluorescence can be efficiently focused onto the entrance slit of a monochromator. The scan speed of the monochromator ranges between 50 Å/min down to 1 Å/min according to the scanned energy range. The complete set of vibrational level have been easily observed below 8000 cm^{-1}, but as the spectrum becomes weaker at higher vibrational energy (longer wavelength of fluorescence) we have scanned slowly over some reduced energy ranges in order to observe levels which were missing in the first rapid scans.

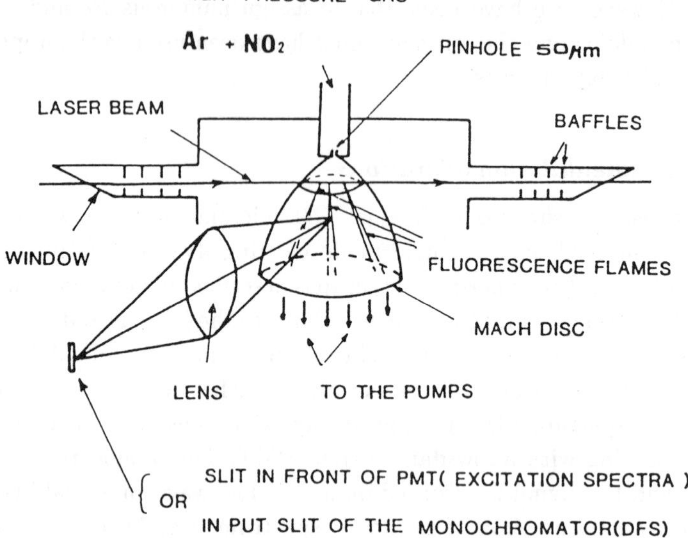

Fig. 8. Schematic view of the jet experiments. NO_2 is cooled in a supersonic jet and excited by a cw dye laser. The laser frequency is locked by collecting the unresolved fluorescence emitted perpendicular to the laser on a PMT. Simultaneously, on the opposite side, the fluorescence is imaged onto the entrance slit of a monochromator.

4.4. *LIDFS from Ten Rovibronic Levels Located Around* $23\,000$ *cm*$^{-1}$

The efficient rotational cooling effect of the supersonic jet ($T_{\rm rot} \cong 3$ kelvins) allows us to excite efficiently the lowest $N = 1$, $K = 0$ rotational level ($J = 3/2$) of several upper B_2 vibronic levels around $23\,000$ cm^{-1}. Then the fluorescence, according to the selection rules of parallel transitions, contains, for each lower a_1 vibrational level, only one R_0 and one P_2 line splitted by $6\bar{B} \cong 2.53$ cm^{-1}, which are in a constant 1 to 2 intensity ratio. This typical pattern can be easily seen in Fig. 6 of Ref. 1. However, in this figure there are also transitions to the $K = 1$, $N = 1$ and/or 2 and/or 3 manifold of vibrational levels of b_2 symmetry in the ground state (these rotational levels have an A_1 rovibronic symmetry). These optically forbidden transitions (vibronic $B_2 \rightarrow$ vibronic B_2) are observed because the excited upper levels around $23\,000$ cm^{-1} are a *rovibronic* mixture of B_2 vibronic level(s) ($K = 0$, $N = 1$) with A_1 vibronic level(s) ($K = 1$, $N = 1$ or 2).

A detailed explanation of this kind of *ro*vibronic mixings is out of the field of this paper and will be analysed elsewhere.[32] The result of this is our ability to observe, by LIDFS, vibrational levels of a_1 and b_2 symmetry, and to assign, with the rotational pattern, this symmetry. It is very important to note that, in each LIDFS spectrum, the intensity ratio of various lines is specific to the upper level excited by the laser. By playing with this, when several upper levels (vibrationally and/or electronically different) can be efficiently excited, it is possible to have various (and then complementary) Franck–Condon access to the set of vibrational levels of the lowest (usually the ground) electronic level. Conversely, the analysis of the rotational structure (selection rules) and of the vibrational propensity rules, i.e., of the Franck–Condon envelope, gives information on the nature of the upper state excited with a laser. This approach has been used on NO_2 and is presented in Sec. 3.5.

Our ability to observe the *complete* set of vibrational levels of the ground state (see below) is mainly due to the "chaotic" behaviour of the B_2 vibronic levels excited around $23\,000$ cm^{-1}. As a matter of fact, the relative intensities in our LIDFS are distributed according to a Porter Thomas law (in agreement with the GOE model[27]) which is expected for a set of transition from a given "chaotic" level to a set of "regular" levels. (It should be noted that the reverse situation, i.e., a "regular" initial level and a set of "chaotic" final levels, is more often encountered). The Porter Thomas Law is highly favorable when we want to observe many levels in *one* fluorescence spectrum. In contrast, the LIDF spectrum emitted from one "regular" level (of the upper electronic state) to a set of "regular" levels (of the lower electronic state) usually displays only a subset of levels because there are propensity rules and there are numerous unobservable very-weak transitions. A good example of propensity rules is the LIDFS of CS_2 presented in Sec. 3: Most of the vibrational levels having a non-zero antisymmetric stretch quantum number are not seen (or very weak) in the LIDFS below $12\,000$ cm^{-1}. This propensity rule disappears progressively above $12\,000$ cm^{-1}. The absence of any propensity rule (i.e., Porter Thomas intensity distribution) in the ten LIDFS spectra observed from NO_2 vibronic levels located around $23\,000$ cm^{-1} is a strong indication of the chaotic behaviour of these levels. On the other hand, we take advantage of the absence of propensity rules to observe numerous vibrational levels in each LIDFS spectrum: typically 60 to 70% of the possible levels below

10 000 cm^{-1}. Then, by concatenation of the measurements obtained from the 10 LIDFS spectra corresponding to various upper levels, we have been able to observe all the vibrational levels located below 10 000 cm^{-1}. Obviously this statement requires an unambiguous vibrational assignment of each level. The procedure was the following: We have first assigned the sub-set of levels located below about 6000 cm^{-1} and we have fitted these $|v_1, v_2, v_3\rangle$ levels with a Dunham expansion of 12 terms. Then, we have calculated (i.e., predicted) the levels at higher energy and compared these energies with our observations. The assignment has been straightforward with few exceptions of pairs of levels which are in resonance. Progressively, we have increased the number of fitted levels and the number of terms in the Dunham expansion. At the end of the procedure, all the observed transitions have been assigned (except one, see below) and all the 191 predicted levels above 10 000 cm^{-1} have been observed. The standard deviation between observed and calculated energy is lower than 1 cm^{-1}, compared with a mean spacing of about 20 cm^{-1} around 10 000 cm^{-1}. The set of 24 Dunham coefficients contains three w_i, six x_{ij}, but only eight y_{ijk} out of ten (two are insignificant, i.e. $\approx$ zero) and some higher order $z_{ijk\ell}$ terms. The a_1 and b_2 vibrational levels have been considered both globally and separately and the two fits give very close results. Around and above 10 000 cm^{-1} the assignments are more difficult, not only because the density of states increases, but also because of vibronic interactions between the high vibrational levels of $\tilde{X}^2 A_1$ and low vibrational levels of $\tilde{A}^2 B_2$.

Figure 9 displays the differences between observed and calculated values. The assignments become doubtful near the expected locations of the three lower bending levels of $\tilde{A}^2 B_2$ which are separated by about 750 cm^{-1}. We have not been able to analyse quantitatively the above-mentioned vibronic interactions except for the $|\tilde{X}^2 A_1; 0, 11, 1\rangle - |\tilde{A}^2 B_2; 0, 0, 0\rangle$ pair of levels, for which a matrix element of 26 cm^{-1} and an energy shift of about 13 cm^{-1} have been determined. The energy shift of about 13 cm^{-1} was observed for the $|\tilde{X}^2 A_1; 0, 11, 1\rangle$ level. We are now using Intracavity Laser Absorption Spectroscopy (ICLAS), combined with a 24 cm slit jet,[33] to observe the vibronic levels of NO_2 between 10 000 and 16 000 cm^{-1}. With this data we expect to understand quantitatively the $\tilde{X}^2 A_1 - \tilde{A}^2 B_2$ vibronic interactions as a function of energy and, thus, the transition to vibronic chaos.[26]

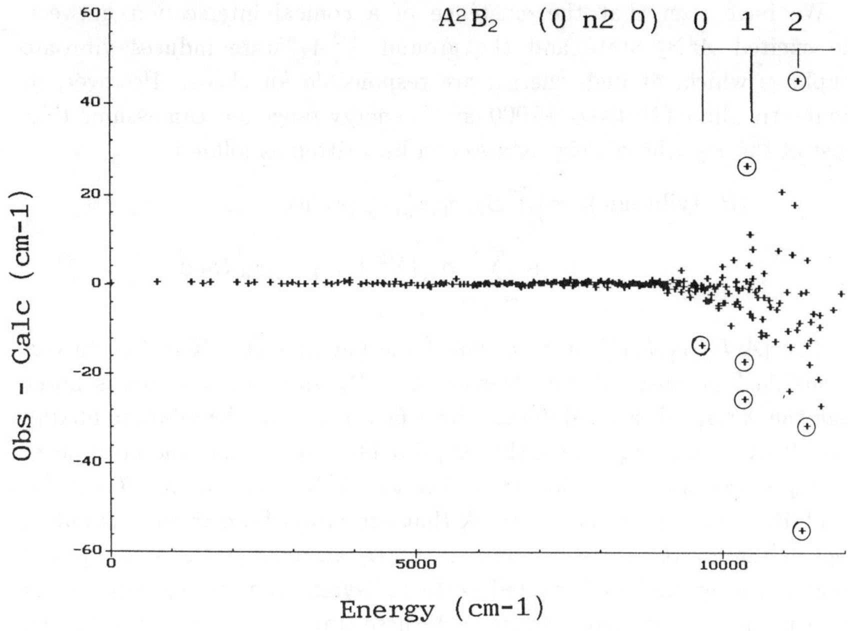

Fig. 9. Observed minus calculated vibrational energies of NO_2. The calculated values have been obtained by fitting the observed values in the 0 to 9600 cm^{-1} energy range. The large discrepancies around 9750, 10 500 and 11 250 cm^{-1} are attributed to the perturbating presence of the $0, 0, 0$, $0, 1, 0$, and $0, 2, 0$ bending levels of the $\tilde{A}^2 B_2$ electronic state.

4.5. *LIDFS from Levels Located Around 17 000 cm^{-1}*

We have seen in the previous section that LIDFS spectra obtained after excitation of levels located around 23 000 cm^{-1} do not display any propensity rules. This is due to the chaotic nature of the laser excited levels. On the other hand, when one excites B_2 vibronic levels around 17 000 cm^{-1} propensity rules are observed in LIDFS spectra: (i) only transitions to a_1 vibrational levels of the $\tilde{X}^2 A_1$ state are observed (this means that rovibronic mixing is weak); (ii) among this a_1 subset only the $|v_1'', v_2'', 0\rangle$ levels are observed (a similar situation has been observed in the CS_2 molecule below 12 000 cm^{-1}; see Sec. 3.4). In addition to this, we will see that it is possible to assign vibrationally some excited levels by analysing the intensity ratio of transitions in terms of Franck–Condon factors. We will first discuss theoretical aspects and then present preliminary experimental results.

We have seen that the existence of a conical intersection between the excited $\tilde{A}^2 B_2$ state and the ground $\tilde{X}^2 A_2$ state induces vibronic couplings which, at high energy, are responsible for chaos. However, in the intermediate ($10\,000 \sim 16\,000$ cm^{-1}) energy range one can assume that most of the B_2 vibronic eigenstates can be written as follows:

$$|B_2 \text{ (vibronic)}\rangle \propto |\tilde{A}^2 B_2; v_1', v_2', v_3' \text{ (even)}\rangle$$

$$+ \sum_{v_1 v_2 v_3} \alpha_{v_i} |\tilde{X}^2 A_1; v_1, v_2, v_3 \text{ (odd)}\rangle \,.$$

The $|\tilde{A}^2 B_2; v_1' v_2', v_3'\rangle$ state is called the parent state. This notion can be justified because: i) the density of $\tilde{A}^2 B_2$ vibrational states is lower than the $\tilde{X}^2 A_1$ vibrational density by a factor of 6; ii) the vibronic mixing extends over an energy range that is probably smaller than the mean level spacing between a_1 vibrational levels of the $\tilde{A}^2 B_2$ state (about 70 cm^{-1}). In addition to this we must remark that excitation from the vibrationless level of the ground state occurs only to B_2 vibronic states with $v_3' = 0$ because the ground and excited state antisymmetric stretch vibrational frequencies are close and because both these states have symmetric bonds. Therefore, the Franck–Condon factors involved in the transitions between the B_2 vibronic excited states and $\tilde{X}^2 A_1$ vibrational states ($v_1'', v_2'', v_3'' = 0$) are: $|\langle v_1'', v_2'', 0| \rangle v_1', v_2', 0\rangle|^2$. Taking into account the Duschinsky effect,[9] one can identify the parent state $|v_1', v_2', 0\rangle$ of each excited level by visual comparison between the calculated and observed intensities $I(v_1'', v_2'', 0)$.

We have performed LIDFS experiments from 29 B_2 vibronic states excited between $16\,728$ and $17\,627$ cm^{-1}. These levels have been chosen among the set of 68 observed levels in this energy range because of their strong unresolved fluorescence intensities. The fluorescence spectra of 22 excited levels displays intensity ratios which allow a level assignment of the dominant parent state (v_1', v_2') (see, for example, Fig. 10). Unfortunately, in many cases, calculated Franck–Condon factors differing by only one quantum number in the upper state are often similar, preventing a unique assignment of the parent state. Nevertheless, as a primary conclusion, one can say that the possible identifications are compatible with an excited electronic state lying at 9737 cm^{-1}, and that the notion of a parent state is relevant up to at least $17\,500$ cm^{-1} for the more intense levels.

Many questions can be raised: how is the existence of a parent state around $17\,000$ cm^{-1} compatible with the chaotic statistical properties

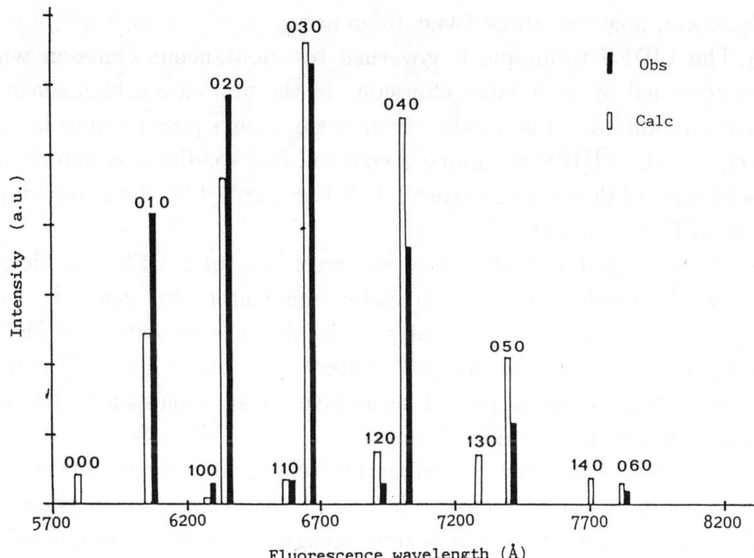

Fig. 10. Comparison between experimental intensities obtained after excitation of a B_2 vibronic level located at $17\,280.76$ cm^{-1} and the coresponding Franck– Condon factors calculated by assuming that the parent state is $|\tilde{A}^2B_2; 3, 5, 0\rangle$. The numbers indicated on the figure are the vibrational quantum numbers $v_1'', v_2'', v_3'' = 0$ of the ground state.

observed in the same energy range? How does the Franck–Condon signature evolve between consecutive vibronic upper levels? If the notion of a parent state is relevant then the character of the dispersed fluorescence of the intermediate low intensity levels should evolve progressively from one Franck–Condon signature (corresponding to a $\tilde{A}^2B_2$ vibrational level) to the following one (corresponding to the next $\tilde{A}^2B_2$ vibrational level). We plan to extend this kind of experiment to lower energy ranges where the Franck–Condon signature must become more and more obvious.

5. Conclusion

When one wants to observe highly excited vibrational levels of small polyatomic molecules, the LIDFS technique is simple, efficient and quite general, even if each molecule has it own spectroscopic character. The LIDFS technique uses the same optical transitions as the SEP technique, but is simpler and so should be used first. General limitations like the Franck–Condon principle and related propensity rules are similar for both SEP and LIDFS.

There are, however, three basic differences:

(i) The LIDFS technique is governed by spontaneous emission while SEP is governed by stimulated emission. In the first case a high sensitive detector is required, while in the second case a high power dump laser is required. In the LIDFS technique a collision free condition is determined by the lifetime of the excited state while it is controlled by the pump–dump delay in SEP experiments.

(ii) For moderate size molecules the resolution of LIDFS is sufficient. Typically, the resolution of a two meter monochromator can vary from 1 cm^{-1} down to 0.1 cm^{-1}. For large molecules, the resolution of SEP is well adapted. LIDFS is a Doppler limited technique while SEP can be sub-Doppler. Conventional pulsed lasers give a resolution of few GHz and injected lasers can be single mode.

(iii) LIDFS is a zero background method which can make use of the new, low noise, multichannel detectors, like a LN-CCD camera. In contrast, the sensitivity of SEP, which is not a zero background method, is limited by laser intensity fluctuations.

Our experiences show that it is possible to obtain very good LIDF spectra over a range of several thousand cm^{-1} in a day. Such a large range scan necessitates months of SEP experiments. In any case, LIDFS should be used first so that high resolution SEP, or other techniques, like DFWM, can then be performed locally.

Acknowledgments

This work was supported by the contract SADOVEM n° SCI-CT 91-0711 (TSTS) with the European Economic Community. We thank Graham Naylor for helpful reading and Martine Aumont for the preparation of this section.

References

1. A. Delon and R. Jost, *J. Chem. Phys.* **95**, 5686 (1991).
2. J.P. Pique, J. Manners, G. Sitja, and M. Joyeux, *J. Chem. Phys.* **96**, 6495 (1992).
3. E.A. Rohlfing, *J. Chem. Phys.* **91**, 4531 (1989).
4. T.J. Butenhoff and E.A. Rohlfing, *J. Chem. Phys.* **95**, 1 (1991).
5. K. Yamanouchi, S. Takenchi, and S. Tsuchiya, *J. Chem. Phys.* **92**, 4044 (1990).
6. K. Yamanouchi, N. Ikeda, S. Tsuchiya, D.M. Jonas, J.K. Lundberg, G.W. Adamson, and R.W. Field, *J. Chem. Phys.* **95**, 6330 (1991).

7. E. Pebay-Peyroula, A. Delon, and R. Jost, *J. Mol. Spectros.* **132**, 123 (1988).

8. J.J. O'Brien, G. Fisher, and B.K. Selinger, *Chem. Phys.* **117**, 275 (1987).

9. F. Duschinsky, *Acta Physicochim., URSS*, **7**, 551 (1937)

10. G.V. Hartland, D. Qin, and H.L. Dai, *J. Chem. Phys.* **98**, 2469 (1993); G.V. Hartland and H.L. Dai, in "Molecular Spectroscopy and Dynamics by Stimulated Emission Pumping", eds., H.L. Dai and R.W. Field (World Scientific, 1995).

11. D.M. Jonas, PhD Thesis, Massachussetts Institute of Technology, 1991.

12. B. Kleman, *Can. J. Phys.* **41**, 2034 (1963).

13. S. Kato, *J. Chem. Phys.* **82**, 3020 (1985).

14. H. Kasahara, N. Mikami, M. Ito, S. Iwata, and I. Susuki, *Chem. Phys.* **86**, 173 (1984); S.J. Silvers and M.R. Mc Keever, *Chem. Phys.* **18**, 333 (1976); P.F. Bernath, M. Dulick, R.W. Field, and J.L. Hardwick, *J. Mol. Spectrosc.* **86**, 275 (1981); R. Vasurdev, *Chem. Phys.* **64**, 167 (1982); A. Matsuzaki, Y. Nakamura, and T. Itah, *Chem. Phys. Lett.* **87**, 389 (1982).

15. G. Sitja and J.P. Pique, *Phys. Rev. Lett.* **73**, 232 (1994).

16. Q. Zhang, S.A. Kandet, T.A. Wasserman, and P.H. Vaccaro, *J. Chem. Phys.* **92**, 1640 (1992).

17. F.J. Olawsky, M. Ziadé, G. Sitja, and J.P. Pique, *J. de Physique* **4**, C7 (1991).

18. J.W. Mills and R.N. Zare, *Chem. Phys. Lett.* **5**, 37 (1970).

19. G. Herzberg, *Molecular Spectra and Molecular Structure,* III (Van Nostrand Reinhold, New York, 1966).

20. C. Jaffe, *J. Chem. Phys.* **85**, 2885 (1986); Z. Li, L. Xiao, and M.E. Kellman, *J. Chem. Phys.* **92**, 2251 (1990).

21. G. Sitja, in preparation.

22. A.G. Maki and R.L. Sams, *J. Mol. Spectrosc.* **52**, 233 (1974).

23. J.P. Pique, M. Joyeux, J. Manners, and G. Sitja, *J. Chem. Phys.* **95**, 8744 (1991).

24. J.P. Pique, *J. Opt. Soc. Am.* **B7**, 1816 (1990).

25. G.D. Gillispie, A.V. Khan, A.C. Wahl, R.P. Hosteny, and M. Krauss *J. Chem. Phys.* **63**, 3425 (1975); C.F. Jackels and E.R. Davidson, *J. Chem. Phys.* **64**, 2908 (1976).

26. A. Delon, R. Jost, and M. Lombardi, *J. Chem. Phys.* **95**, 5701 (1991).

27. T.A. Brody, J. Flores, J.B. French, P.A. Mello, A. Pandey, and S.S.M. Wong, *Rev. Mod. Phys.* **53**, 385 (1981).

28. J.L. Hardwick, *J. Mol. Spectrosc.* **66**, 248 (1977).

29. H.D. Bist and J.C.D. Brand, *J. Mol. Spectrosc.* **62**, 60 (1976).

30. J. Miyawaki, K. Yamanouchi, and S. Tsuchiya, *Chem. Phys. Lett.* **180**, 287 (1991).

31. R.E. Smalley, L. Wharton, and D.M. Levy, *J. Chem. Phys.* **63**, 4977 (1975).

32. A. Delon, P. Dupré, and R. Jost, *J. Chem. Phys.* **99**, 9482 (1993); A. Delon, R. George, and R. Jost, submitted to *J. Chem. Phys.*

33. R. George, A. Delon, F. Bylicki, R. Jost, A. Campargue, A. Chawat, M. Chenevier, and F. Stoeckel, accepted in *Chem. Phys.*.

CHAPTER 5

DISPERSED AND STIMULATED EMISSION STUDIES OF THE EXCITED VIBRATIONAL LEVELS OF A TRANSIENT MOLECULE: SINGLET METHYLENE

Gregory V. Hartland and Hai-Lung Dai

Department of Chemistry
University of Pennsylvania
Philadelphia, PA 19104-6323, USA

Contents

1. Introduction — Challenges for Vibration–Rotational Studies of Transient Molecules

The vibration–rotational spectroscopy of transient molecules — such as radicals, reactive intermediates, and molecular ions — is a subject of much interest in physical chemistry.[1] The molecular constants derived from these studies are useful for structural determination, modelling the potential energy surfaces, and deducing structure related properties of these species. However, such studies can be a difficult experimental undertaking. The low concentrations and short lifetimes of transient molecules means that conventional IR absorbance techniques — such as using a FT–IR spectrometer — are not sensitive or fast enough for their study. Tunable cw IR laser transient absorption spectroscopy has sufficient sensitivity.[2] However, the limited tuning range of these laser sources means that only vibrational levels in a small frequency interval can be examined. In addition, the small cross-sections of vibrational overtone and combination band transitions make it nearly impossible to study highly excited vibrational levels by direct absorption techniques.

Another problem in transient molecule spectroscopy is spectral congestion. The methods used to produce transient molecules, such as photolysis, reaction, or microwave discharge of stable precursor molecules,[3] invariably create rotationally and vibrationally hot samples. The unrelaxed initial quantum state distribution often leads to congested rovibrational spectra. When this is coupled to the extensive perturbations that often occur in transient molecule spectra — for example, Fermi or Coriolis interactions — the analysis of the direct absorption spectrum can be difficult and ambiguous. Cooling the sample in a supersonic beam[4,5] or employing threshold photodissociation techniques[6] can reduce spectral congestion. However, these techniques also limit the amount of spectroscopic information that can be acquired.

As will be discussed in this chapter, the ground vibrational levels of transient molecules can be studied with high sensitivity and resolution by recording dispersed fluorescence spectra after pulsed laser excitation, or by using a second laser pulse to achieve stimulated emission spectroscopy. The double resonance nature of these techniques also significantly reduces spectral congestion. Because pulsed lasers are used, spectra can be recorded as long as the electronic state lifetime of the transient species is comparable to the laser pulse duration. In particular, results from our studies of the excited vibrational levels of $\tilde{a}^1 A_1$ CH_2 will be presented. Singlet methylene is the prototypical example of a transient molecule: it can only be produced in small concentrations from photolysis and it is extremely reactive.[7-9] In our experiments, ketene is used as the precursor; at 800 mTorr of ketene approximately 1 mTorr of $\tilde{a}^1 A_1$ CH_2 is produced by XeCl laser photolysis. Under such conditions, the lifetime of $\tilde{a}^1 A_1$ CH_2 is < 100 ns.[10]

2. Vibrational Spectroscopy Through Examining the Downward Transitions From an Excited Electronic State

Dispersed fluorescence spectroscopy (DFS) from an excited electronic state prepared by laser excitation has been widely used for studying vibrational bands in the electronic ground state. Its high sensitivity, compared to direct IR absorption, arises from several factors. First, two relatively strong electronic transitions are used, rather than a much weaker vibrational transition. These transitions can also be accessed with tunable visible or UV laser sources, which usually have higher powers than tunable IR sources. Second, for molecules that undergo a geometry change upon electronic excitation, the Franck–Condon factors for absorption and emission allow observation of long vibrational progressions in the mode that corresponds to the geometry change. Thus, vibrational levels with energies from zero to tens-of-thousands of wavenumbers energy can be studied by DFS. And third, for transient species pulsed dye lasers can be used to efficiently transfer the population to an excited electronic state. By using DFS, species with lifetimes as short as 10^{-8}s can be studied. An additional advantage to recording DFS spectra is that spectral congestion can be reduced because only a single, or a few, rotational levels in the upper electronic state are populated by the excitation laser. This will greatly simplify the analysis of the spectra, especially for species with complicated band structures, even for small molecules like CH_2. Initial selection of the rovibronic level for study also gives a high degree of species specificity to DFS, i.e., only transitions due to the laser excited species will appear in the spectra. This is not true for direct absorption techniques, where all species in the sample can contribute. This is a nontrivial consideration for studying minority species in a sample — such as molecules produced from photolysis.

The traditional method for recording dispersed fluorescence spectra is to use a grating spectrometer for wavelength resolution. Grating spectrometers can be used to obtain spectra over extremely wide spectral ranges to very high levels of vibrational excitation. Some recent examples are the studies of the $\tilde{X}$ state vibrational structure of C_3,[4] glyoxal,[11] SO_2,[12] NO_2,[13] and C_2H_2.[14] However, the resolution that can be obtained with a grating spectrometer is usually low, at typically 4–20 cm^{-1}. For most polyatomic molecules this means that individual rotational transitions cannot be resolved and, hence, the ground state rovibrational constants cannot be accurately determined. Higher resolution can be obtained, but at the cost of sensitivity; this compromise between resolution and sensitivity makes grating spectrometers inadequate for vibrational studies of transient species at low concentrations. Alternative methods for recording dispersed fluorescence spectra with high resolution *and* sensitivity, that we have demonstrated in our studies of singlet CH_2, are stimulated emission pumping spectroscopy (SEP)[1,10,15,16] and the newly developed time-resolved Fourier transform emission spectroscopy (TR–FTES) technique.[17] The advantages and drawbacks of these two techniques are discussed in the following sub-sections.

2.1. *Flash Photolysis Stimulated Emission Pumping (SEP)*

SEP has been the subject of several reviews[15,16] and will be discussed in detail in this volume by other authors, hence, only a brief overview is given here. A diagram of the experimental apparatus used for SEP on singlet methylene is shown in Fig. 1.[10] In our experiments a three laser pulse sequence was used. First, $\tilde{a}^1A_1$ CH_2 was produced by photolysis of ketene at 308 nm with 20% of the output of a Lambda Physik 201 MSC excimer laser (unstable resonator optics, 300 mj/pulse, 30 Hz repetition rate). Ketene was flowed continuously through the gas cell to avoid build up of reaction products, where the typical ketene pressure used was 0.5–0.8 Torr. The rest of the excimer laser output was used to pump two dye lasers (Lambda Physik 2002E and 3002E) to provide the pump and dump laser pulses. The pump laser was delayed from the photolysis pulse by ca. 15 ns; this allows partial thermalization of the vibrationally and rotationally hot $\tilde{a}^1A_1$ CH_2 produced from ketene photodissociation. The dump laser was delayed by a further 15 ns from the pump laser. The pump laser was tuned to a single rotational level of the CH_2 $\tilde{b}^1B_1$ state and total fluorescence was

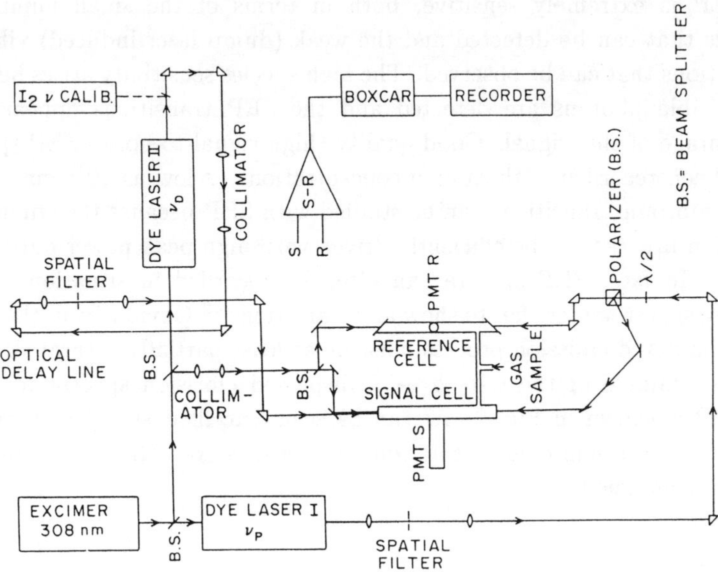

Fig. 1. Experimental apparatus for flash-photolysis stimulated emission pumping of singlet CH_2.

monitored, while the dump laser was scanned over the spectral region of interest. When the dump laser is resonant to a $\tilde{b} \to \tilde{a}$ transition, stimulated emission occurs and there is a decrease in the fluorescence signal.

A two cell null reference scheme was used to increase the signal-to-noise in the SEP spectra.[10,15] The pump and photolysis lasers were split into two equal intensity beams and passed through both the signal and reference cells. The dump laser was passed through the signal cell only, co-linear and counterpropagating to the pump laser. The fluorescence signal from the reference cell was subtracted from that of the signal cell by a fast differential amplifier, and the output was averaged by a boxcar integrator. This reference scheme greatly reduces noise due to pump laser fluctuations.

The resolution of SEP is limited by a combination of laser bandwidth and the scanning rate; for our experiments the resolution was ~ 0.07 cm^{-1}. The dump laser frequency was calibrated by recording an I_2 absorption spectrum concomitantly with the SEP scan, giving absolute frequencies accurate to ≤ 0.04 cm^{-1}.[18]

SEP is extremely sensitive, both in terms of the small number of species that can be detected and the weak (dump laser induced) vibronic transitions that can be observed. The high species sensitivity arises because UV–visible photons are detected and the SEP transitions appear as a percentage of this signal. Good quality (high signal-to-noise) SEP spectra have been recorded with sample concentrations as low as 10^{12} cm^{-3}.[16,19] Weak vibronic transitions can be studied with SEP because the stimulated emission process can be efficiently driven with high peak power pulsed dye lasers. In fact, SEP spectra can often be recorded in situations where grating spectrometers fail to show any transitions.[12] Obviously, in this case, the stimulated emission process must be at least partially saturated!

An example of the CH_2 $\tilde{b} \rightarrow \tilde{a}$ dispersed emission spectra recorded by SEP is shown in Fig. 2. As can be seen, excellent signal-to-noise was obtained and the individual rotational transitions are well resolved and can be easily assigned.

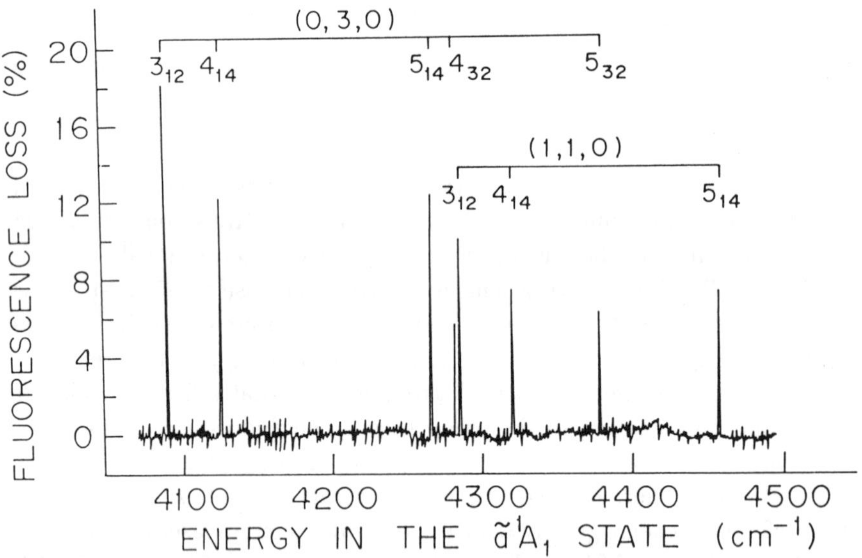

Fig. 2. Example of SEP spectra of $\tilde{a}^1A_1$ CH_2. The pump laser was set to the $\tilde{b}^1B_1$ $(0, 16^0, 0)$ 4_{04} level. The $\tilde{a}^1A_1$ levels probed by the dump laser are labelled in the spectra. The intensity scale was calibrated by observing the $\sim 8\%$ fluorescence loss caused by inserting a glass slide into the pump laser beam.

The major disadvantage of SEP for spectroscopy studies of species like CH_2 is that only a small range of the emission spectrum can be recorded in a single experiment. For example, with our dye lasers the maximum spectral bandwidth that can be obtained in a single scan is approximately 40–60 cm^{-1} — the tuning range of the etalon. [Even if no etalon is used, which leads to a loss in resolution, the spectral range is limited by the dye gain curve.] Because CH_2 is a light molecule with large rotational constants, the rotational transitions in a given band can be separated by hundreds of wavenumbers (see Fig. 2). This situation makes for a tedious and strenuous experiment if no prior knowledge of the approximate position of these transitions is available. However, this would not be a problem if the fluorescence of transient species can be efficiently detected and dispersed with high resolution. The newly developed time-resolved Fourier transform emission spectroscopy technique can be used to achieve these goals.[17]

2.2. *Time-Resolved Fourier Transform Emission Spectroscopy (TR–FTES)*

Before discussing our TR–FTES apparatus in detail, it is relevent to briefly review the principles of Fourier transform (FT) spectroscopy and why it is advantageous to use a FT spectrometer for dispersed emission studies.[20]

In optical FT experiments spectral resolution is achieved with a Michelson interferometer, which consists of a beam splitter, a fixed mirror, and a moving mirror (see Fig. 3). Light from a suitable source (in our case fluorescence from a gas cell) is collimated and split into two equal intensity beams by the beam splitter, reflected off the fixed and moving mirrors, and recombined at the beam splitter. After recombination the two beams interfere. Because of this interference the light intensity at the detector is modulated as a function of the path difference between the two arms of the interferometer. Recording the intensity versus the path difference, determined by the moving mirror position, x, gives the "interferogram" $I(x)$. Fourier transformation of the interferogram yields the spectrum $I(v)$ of the source.[20] In the ideal case — perfectly collimated light and dispersionless optics — the resolution in the frequency domain spectrum is given by $1/\Delta$, where Δ is the maximum path difference, which is twice the maximum travelling distance of the moving mirror in the Michelson interferometer. However, in practical situations the resolution is limited by dispersion through the interferometer optics.[20] With a standard 10 cm

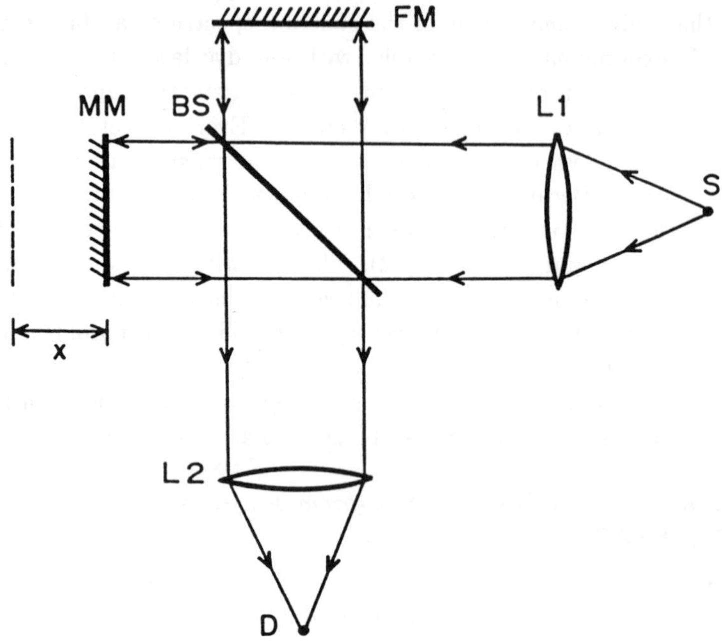

Fig. 3. Schematic diagram of a Michelson interferometer. BS is the beam splitter, FM is the fixed mirror, MM is the moving mirror, L1 and L2 are lenses, and S and D are the source and the detector, respectively.

interferometer designed for IR spectroscopy, the best resolution is 0.1 cm^{-1}. However, in the visible region the maximum resolution is 0.25 cm^{-1}.

An important consideration in FT spectroscopy is the sampling interval along the moving mirror axis. The interferogram is constructed by recording data at a series of equally spaced mirror positions. If the optical path interval between successive mirror positions is d, then only the frequencies in the range $-v_{\max} \leq v \leq v_{\max}$ can be correctly resolved, where $v_{\max}$, the Nyquist frequency, is given by $v_{\max} = 1/(2d)$.[20] Frequencies outside this range will not appear at their correct position, but will appear in the spectrum as aliased peaks. For example, if the detector sees a signal with a frequency v which is greater than $v_{\max}$, peaks will appear in the Fourier spectrum at $\pm(v - mv_{\max})$, where m is an interger that depends on how much greater v is than the Nyquist frequency. For $v_{\max} < v < 3v_{\max}$

$m = 2$, for $3v_{max} < v < 5v_{max}$ $m = 4$, etc. Aliased peaks can only be distinguished from "real" peaks by changing the sampling interval (i.e., changing v_{max}) and seeing if their position changes. The optical path difference is usually calibrated by monitoring the zero-crossings in the interference pattern of a HeNe reference laser that passes along the axis of the interferometer. The zero-crossing is defined as the position where the ac-coupled HeNe intensity $I_{NeHe} = 0$. As each HeNe wavelength has two zero-crossings, $d = n\lambda_{HeNe}/2$, where $n = 1, 2, \ldots$ depending on whether data is collected at every zero-crossing, or every second zero-crossing, etc. From these considerations we find that $v_{max} = 15\,798/n$ cm^{-1}.

Aliasing can be very helpful in Fourier transform spectroscopy. If the light reaching the detector can be limited to a known bandwidth — for example, by optical filtering — then the spectrum can be recorded as an aliased spectrum, which can be reconstructed after Fourier transformation to give the true spectrum. In this case v_{max} need only be as large as the experimental bandwidth, and not the highest frequency. Recording spectra in this fashion is known as undersampling.[20] The advantage of undersampling is that larger intervals between data sampling points can be used than if an unaliased spectrum were recorded and, thus, less data points are needed to obtain a given resolution. For example, to obtain a spectrum in the 11 849–15 798 cm^{-1} range, data can be collected at every fourth HeNe zero-crossing (a 3949 cm^{-1} bandwidth) rather than at every HeNe zero-crossing (a 15 798 cm^{-1} bandwidth). This represents a large saving in experiment time! Of course, great care must be taken to ensure that only frequencies within the desired bandwidth reach the detector.

To record Fourier transform spectra with a pulsed laser source the laser firing must be synchronized to the interferometer mirror movement.[21] There are two ways to do this. In the "record-on-the-flight" approach the mirror is scanned at a constant rate and the laser firing and data collection are triggered by monitoring the HeNe laser zero-crossings.[22–28] The second approach, which we employ, is to use a step-scan interferometer.[17,29,30] In this "arrest-and-record" scheme the interferometer mirror is sent to a set position (again referenced by the HeNe zero-crossings) and kept stationary while data is collected and averaged. The advantage of this scheme is that it allows more flexibility in the time-scale of the experiment, and the laser repetition rate does not depend on the mirror speed, i.e., spectra can be efficiently collected from low repetition rate laser sources.

TR–FTES has been extensively used for IR emission studies,[25-28] where FT spectrometer are known to out perform grating spectrometers The superior performance of Fourier transform spectrometers in the IR region arises from the throughput and multiplex advantages.[20] The multiplex advantage for FT spectrometers occurs when the detector noise is dominated by the background; in this case by detecting all the photons at once the signal-to-noise is increased. For UV–visible emission studies where a photo-multiplier tube (PMT) is used, photon shot-to-shot noise, PMT dark current, and laser source fluctuations all contribute to the noise. There is no multiplex advantage for the shot-to-shot and laser fluctuation noises, but a multiplex advantage for the PMT dark current still exists. This can be significant for low intensity fluorescence signals.

The throughput advantage of Fourier transform spectrometers over grating spectrometers still applies for UV–visible studies. This means that dispersed fluorescence spectra can be recorded with better sensitivity by FT spectrometers compared to grating spectrometers. Higher sensitivity in turn means that better resolution can be obtained. For example, we have acquired good quality, < 1 cm^{-1} resolution TR–FTES spectra with sample concentrations as low as 10^{13} cm^{-3}.

There are also some significant advantages of TR–FTES compared to SEP. First, by coupling a transient digitizer to the detector, complete temporal and spectral infomation can be obtained in a single TR–FTES experiment. Second, emission spectra can be recorded over a much wider bandwidth. We routinely record spectra with a 3949 cm^{-1} bandwidth, and the maximum bandwidth that can be obtained is 15 798 cm^{-1}. Third, by simply changing the beam splitter and the detector TR–FTES spectra can be recorded with high sensitivity from the UV to the IR. For SEP, the IR and near IR frequency regions are inaccesible due to the limited dye laser range.

In addition, accurate absolute line positions are obtained with Fourier transform techniques, without the need for calibration. This significantly simplifies the experiment, especially in spectral regions where no convenient wavelength standards are available. At 0.5 cm^{-1} resolution the absolute peak positions measured in our TR–FTES spectra are accurate to within 0.1 cm^{-1}.

Of course, there are disadvantages to TR–FTES. The most signaficant of these is that the spectral resolution of TR–FTES is not as good as that for SEP. However, all things considered, its advantages make TR–FTES an attractive technique for fast recording survey spectra of excited vibrational levels over wide spectral ranges, for species with short lifetimes and/or low concentrations, and for small molecules with large rotational constants. These advantages of TR–FTES will be clearly demonstrated in spectroscopy studies of $\tilde{a}^1 A_1$ CH_2.

A diagram of our TR–FTES apparatus is shown in Fig. 4. The laser system used is similar to that for the flash photolyis SEP experiments, except that only a single dye laser is needed. Half of the 308 nm output of a Lambda Physik LPX 210i excimer laser (running at a 30 Hz repetition rate, 360 mj/pulse) was used for ketene photoysis and the other half pumped a dye laser (Lambda Physik 3002E). The dye laser, tuned to a single $\tilde{b} \leftarrow \tilde{a}$ transition, was delayed by ca. 20 ns from the photolysis pulse to allow for $\tilde{a}^1 A_1$ CH_2 thermalization. A dye laser end mirror (DM) was used to double pass both laser pulses through the gas cell; the delay between the first and second set of pulses was ca. 2 ns. The resulting fluorescence was collected with a White cell mirror arrangement, collimated, and focussed into the FT spectrometer (FTS, Bruker IFS-88) with $f/4$ characteristics. The fluorescence was focussed through an adjustable aperture (A) to limit dispersion through the interferometer optics, recollimated, directed through the Michelson interferometer, and focussed onto a PMT (Hamamatsu R1477 or R936). The PMT signal was amplified (25 ns time response fast preamplifier) and then digitized by a transient digitizer (TD, Markenrich waveform acquisition board, 25 ns rise time, 8 bit resolution, 20 MHz maximum sampling rate). The transient digitizer was triggered by a fast photo-diode (PD) which monitored the excimer laser pulse.

The Michelson interferometer of the Bruker IFS-88 spectrometer can be run in continuous or step-scan mode. In step-scan mode the movable mirror can be positioned at optical path differences of multiples of 0.316 μm by locating the zero-crossings in the HeNe reference laser interference pattern. The positioning accuracy of this interferometer system has been measured as ± 1.1 nm.[30] The time taken for the mirror to stabilize at a zero-crossing — the "settling time" — is approximately 30 ms; however, the actual time between steps also depends on the distance travelled and, so, is slightly greater than the settling time. The operation of the IFS-88 spectrometer

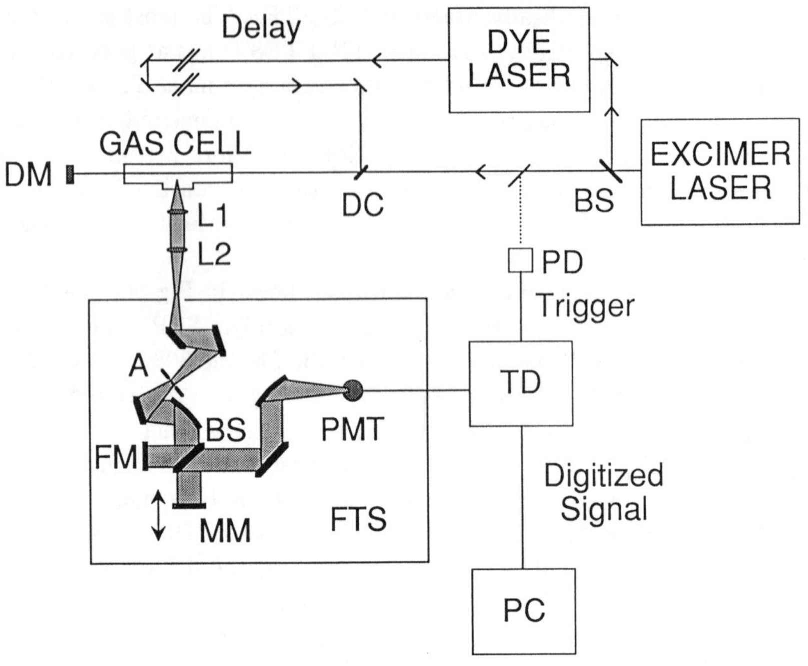

Fig. 4. Experimental apparatus for time-resolved Fourier transform emission spectroscopy; see text for details.

and the data collection are controlled by a Dell 325D, AT compatible, personal computor (PC–AT) equipped with a Bruker acquisition processor board and OPUS^{TM} software.

During a TR–FTES experiment, once the moving mirror has settled to a HeNe zero-crossing, a trigger is sent to the PC–AT to start data acquisition. The digitized signal from the transient digitizer is then collected for a given number of laser shots and stored on disk, while the mirror position is held constant. The spacing between digitization times can be set from 25 to 50 000 nsec and the number of times collected is only limited by the available RAM of the acquisition processor (4 MB). After the required number of laser shots have been averaged, the mirror is moved to the next position and the process is repeated until the complete interferogram is obtained. The excimer laser is run continuously and each laser shot is digitized; however, only those shots that occur while the mirror is being held in position are included in the average.

At the end of the experiment a two-dimmensional data array is obtained which contains the interferometric intensity as a function of time at each mirror position. A schematic diagram of the array is shown in Fig. 5. This array is subsequently rearranged and the signal as a function of mirror position, the interferogram, at each time is Fourier transformed to give the fluorescence spectrum at that time.

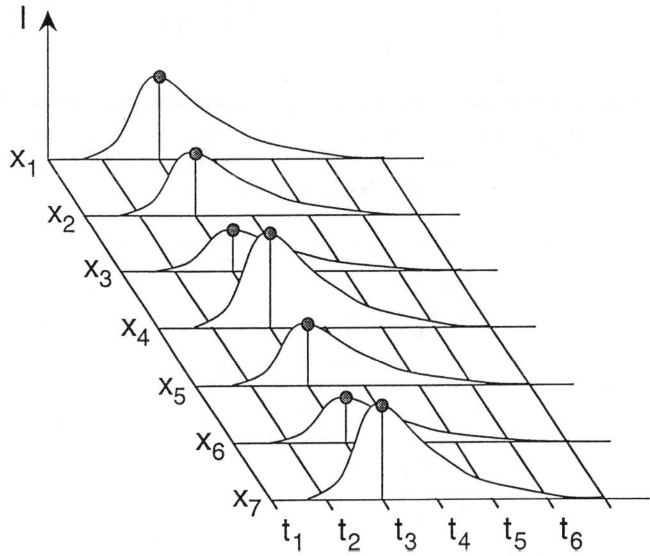

Fig. 5. Representation of the data array collected during a TR–FTES experiment. At each mirror position, x_i, the transient signal is digitized at times $t_1, t_2, t_3, \ldots$. After the scan is completed, the signal as a function of mirror position at each time, represented here by the filled circles at t_3, is Fourier transformed to give the spectrum at that time.

Examples of the CH_2 $\tilde{b} \rightarrow \tilde{a}$ fluorescence spectra obtained by TR–FTES are shown in Fig. 6. For these experiments a 665 nm long-pass filter was used to limit the fluorescence reaching the PMT to frequencies less than $15\,798$ cm^{-1} (the HeNe laser frequency) and to cut the scattered HeNe light.[31] Fluorescence to the red was limited by the drop in sensitivity of the PMT. The PMT cut-off was 900 nm for the Hamamatsu R1477 and 930 nm for the R936. In Fig. 6a emission from the CH_2 $\tilde{b}^1B_1$ $(0, 18^0, 0)$ 0_{00} level was recorded with a 3949 cm^{-1} bandwidth and 2 cm^{-1} resolution. Emission to the CH_2 $\tilde{a}^1A_1$ $(0, 4, 0)$, $(1, 2, 0)$, $(2, 0, 0)$, $(0, 5, 0)$, and $(1, 3, 0)$

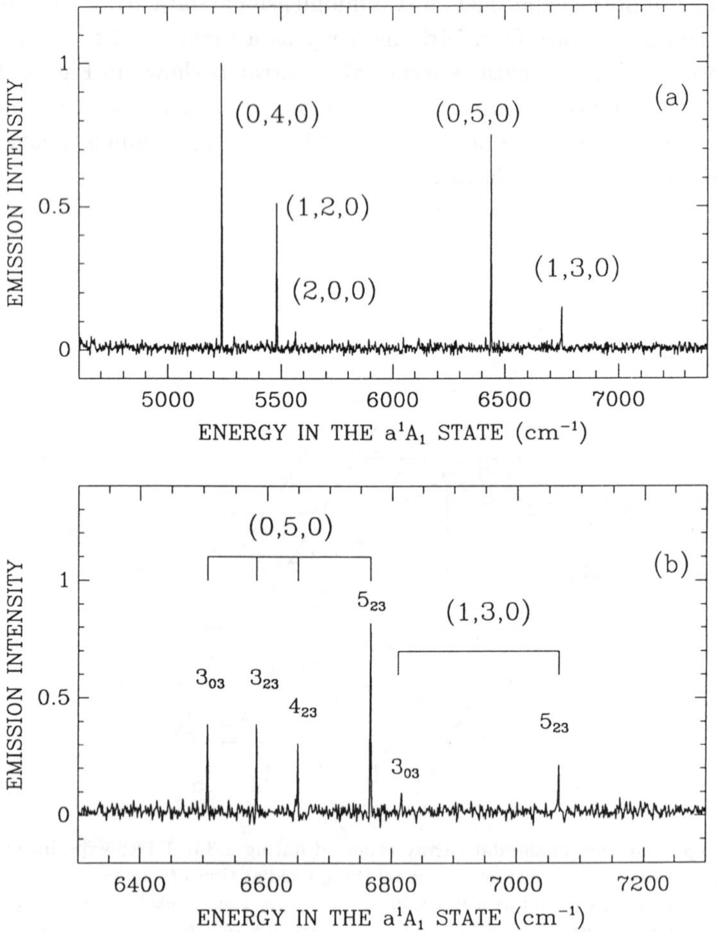

Fig. 6. Examples of TR–FTES spectra of CH_2. (a) Spectrum recorded after excitation of the CH_2 $\tilde{b}^1B_1$ $(0, 18^0, 0)$ $J_{K_aK_c} = 0_{00}$ level. (b) Spectrum recorded after excitation of the CH_2 $\tilde{b}^1B_1$ $(0, 19^1, 0)$ $J_{K_aK_c} = 4_{13}$ level. The rovibrational levels in the $\tilde{a}^1A_1$ state are labelled in the figure.

vibrational levels can be seen and are assigned in the spectra. Each of these bands consists of a single rotational transition: $0_{00} \rightarrow 1_{10}$. The peak intensities in this spectrum have been corrected for the transmission of the filter and the sensitivity of the PMT. In Fig. 6b a TR–FTES spectrum obtained after excitation of the CH_2 $\tilde{b}^1B_1$ $(0, 19^1, 0)$ 4_{13} level is shown.

This spectrum was recorded with 1 cm^{-1} resolution and a bandwidth of 1974 cm^{-1}. Emission to the CH$_2$ $\tilde{a}^1 A_1$ $(0,5,0)$ and $(1,3,0)$ vibrational levels can be seen and the rotational levels in the $\tilde{a}^1 A_1$ state are labelled in the figure. The total length of the interferogram for both these TR–FTES spectra was 2000 mirror positions. For the data in Fig. 6a, 60 laser shots per mirror position were collected, and in Fig. 6b, 90 laser shots were collected. With the present laser repetition rate, these spectra of a few thousands wavenumber and ~ 1 cm^{-1} resolution take only a couple of hours to obtain. In comparison, the SEP spectra shown in Fig. 2 was collected over a period of a few days!

3. The Excited Vibrational Levels of $\tilde{a}^1 A_1$ CH$_2$

Singlet methylene (CH$_2$) has been the subject of much experimental and theoretical interest ever since the red $\tilde{b} \leftarrow \tilde{a}$ absorption spectrum was first recorded and partially analysed by Herzberg and Johns.[32] This interest stems from the fact that CH$_2$ is the simplest carbene — and therefore the prototype for this class of reagents — and an important reactive intermediate. Because of perturbation of the $\tilde{a}^1 A_1$ levels through spin–orbit coupling with nearby $\tilde{X}^3 B_1$ levels, strong Renner–Teller coupling between the $\tilde{a}^1 A_1$ and $\tilde{b}^1 B_1$ states, Coriolis and Fermi interactions, and the fact that $\tilde{a}^1 A_1$ CH$_2$ is a highly asymmetric top, the $\tilde{b} \leftarrow \tilde{a}$ spectrum of singlet methylene is very complex. For example, over 10 000 lines have been recorded in the 15 600 to 18 650 cm^{-1} region,[32,33] but only 10% of these have been assigned.[6,10] The complete $\tilde{b} \leftarrow \tilde{a}$ absorption spectrum extends from 11 000 to over 20 000 cm^{-1}.

In addition to finding an explanation for the many unassigned lines in the CH$_2$ absorption spectrum, another question of interest is the barrier height to linearity in the $\tilde{a}^1 A_1$ state. Herzberg and Johns originally gave a barrier height of ~ 8000 cm^{-1} deduced from the $\tilde{b}^1 B_1$ state origin, which was estimated from the spacing of the higher $\tilde{b}^1 B_1$ bending levels.[32] A barrier height of 9800 cm^{-1} was obtained by fitting a bending potential function to the $\tilde{a}^1 A_1$ $(0, v_2, 0)$, $v_2 = 0$–3 levels,[34,35] observed in a low resolution dispersed fluorescence experiment.[36] Recently, an *ab initio* calculation where the $\tilde{a}^1 A_1$ state potential energy surface was empirically adjusted to the absorption spectrum around 15 000 cm^{-1} produced a barrier height of 8800 cm^{-1},[37] which disagrees with a previous *ab initio* calculation result of 11 000 cm^{-1}.[38] To more accurately determine the barrier height,

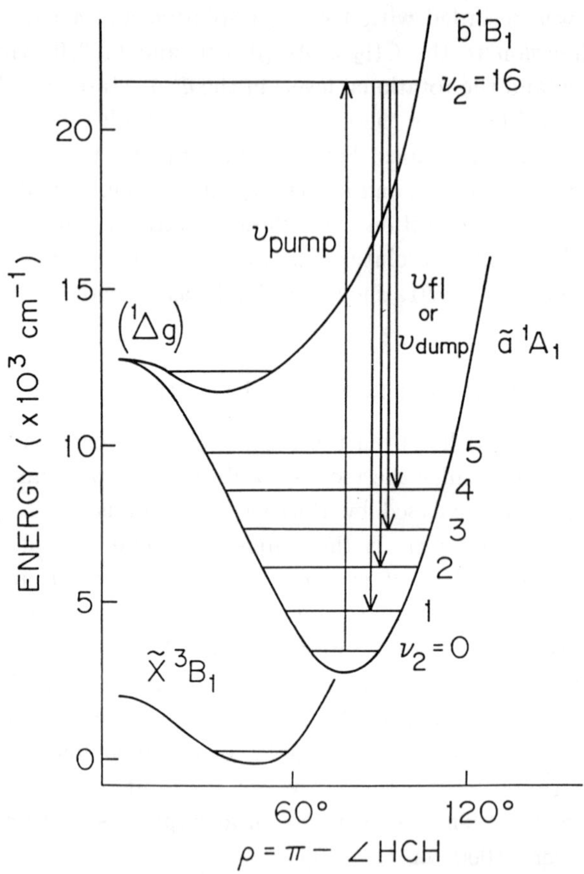

Fig. 7. Diagram of the CH_2 $\tilde{X}^3B_1$, $\tilde{a}^1A_1$, and $\tilde{b}^1B_1$ potential energy surfaces plotted against $\rho = \pi - \angle$ HCH. The $\tilde{a}^1A_1$ and $\tilde{b}^1B_1$ states become a degenerate $^1\Delta_g$ pair in the linear configuration ($\rho = 0$). Also shown is a typical dispersed fluorescence experiment: ν_{pump} is the pump laser and ν_{fl}, ν_{dump} represent spontaneous or stimulated emission, respectively, from the pump laser excited vibronic level.

term values for higher $\tilde{a}^1A_1$ vibrational levels, particularly the bending overtone levels, are required.

A schematic diagram of the CH_2 $\tilde{X}^3B_1$, $\tilde{a}^1A_1$, and $\tilde{b}^1B_1$ potential energy curves along the bending coordinate are shown in Fig. 7. Also displayed in the figure is a typical dispersed emission experimental scheme. In our experiments, CH_2 is laser excited from the $\tilde{a}^1A_1$ $(0, 0, 0)$ level to a single rovibrational level of the $\tilde{b}^1 B_1$ state, and transitions from this excited rovibronic level back down to the $\tilde{a}^1 A_1$ state are detected and analyzed. The $\tilde{b} \leftrightarrow \tilde{a}$ transition of CH_2 is a bent-to-linear transition (the bending vibrational levels excited by the pump laser are well above the barrier to linearity in the $\tilde{b}^1 B_1$ state; see Fig. 7) and thus extremely high bending overtone levels can be observed in both absorption and emission.[32]

3.1. *Rovibrational Analysis of* CH_2 $\tilde{b} \rightarrow \tilde{a}$ *Spectra*

3.1.1. *Spectroscopy and molecular constants*

The CH_2 $\tilde{b} \leftrightarrow \tilde{a}$ transition is a type-c band with the rotational selection rules $\Delta J = 0 \pm 1$; $\Delta K_a = \pm 1, \pm 3$; and $\Delta K_c = 0, \pm 2, \pm 4$. In the following discussion, all term values are refered to the CH_2 $\tilde{a}^1 A_1$ $(0,0,0), 0_{00}$ level. Molecular rotation–vibration constants are available for the $\tilde{a}^1 A_1$ $(0,0,0)$ and $(0,1,0)$ levels from the high resolution visible absorption spectrum,[33,39] and for the $\tilde{a}^1 A_1$ $(1,0,0)$ and $(0,0,1)$ levels from diode laser transient absorption spectra.[40] The $\tilde{a}^1 A_1$ $(0,2,0)$, $(0,3,0)$, and $(1,1,0)$ levels were observed at ~ 10 cm^{-1} resolution by dispersed fluorescence spectroscopy with a grating spectrometer.[36] However, because of the low resolution and the limited number of levels studied, accurate molecular constants could not be determined.

SEP was used to study the $\tilde{a}^1 A_1$ $(0,2,0)$, $(0,3,0)$, $(1,1,0)$, and $(0,4,0)$ levels with 0.07 cm^{-1} resolution.[10] For these experiments the pump laser was tuned to single rotational transitions in the $\tilde{b} \leftarrow \tilde{a}$ 2_0^{15}, 2_0^{16}, or 2_1^{17} bands. Additional transitions were measured for the $\tilde{a}^1 A_1$ $(0,2,0)$ level by hot-band laser induced fluorescence.[41] Higher vibrational levels could not be observed by SEP because a near-IR dye would be needed for the dump laser. The $\tilde{a}^1 A_1$ $(1,2,0)$, $(2,0,0)$, $(0,5,0)$, and $(1,3,0)$ levels were examined by TR–FTES. Dispersed fluorescence spectra of the $\tilde{a}^1 A_1$ $(1,2,0)$ level were recorded by tuning the pump laser to transitions in the $\tilde{b} \leftarrow \tilde{a}$ 2_0^{16}, 2_0^{17}, or 2_0^{18} bands. For the $\tilde{a}^1 A_1$ $(2,0,0)$ level, spectra were only obtained from pump laser transitions in the $\tilde{b} \leftarrow \tilde{a}$ 2_0^{16} and 2_0^{18} bands, because of the reduced Franck–Condon factors for this level (see Fig. 6a). The $\tilde{a}^1 A_1$ $(0,5,0)$ and $(1,3,0)$ levels were observed exclusively by exciting the $\tilde{b} \leftarrow \tilde{a}$ 2_0^{18} and 2_0^{19} bands. These highly excited vibrational levels could only be observed from the $\tilde{b}^1 B_1$ $(0, 18^0, 0)$ and $(0, 19^1, 0)$ states because of the drop in sensitivity of the PMT at longer wavelengths.

To set the pump laser transition for the $\tilde{b} \leftarrow \tilde{a}$ 2_0^{15}, 2_0^{16}, and 2_1^{17} bands the rotational assignments of Petek *et al.* were used.[33] However, no assignments were available for the $\tilde{b} \leftarrow \tilde{a}$ 2_0^{18} or 2_0^{19} bands. The CH_2 $\tilde{b} \leftarrow \tilde{a}$ absorption spectrum in the region of the 2_0^{18} and 2_0^{19} bands (20 000 to 21 500 cm^{-1}) appears very complicated and nontrivial for rotational assignment. We have assigned 8 rotational transitions in each of these bands from assigning the TR–FTES spectra in the 2_2^{18} or 2_2^{19} region recorded following excitation of these transitions.[42] The assignments were performed as follows: First, using the known term values for the $\tilde{a}^1 A_1$ $(0, 2, 0)$ levels[41] the peaks in the TR–FTES spectra were identified. The $\tilde{b}^1 B_1$ state rotational level can be deduced straightfowardly from the rotational selection rules. The $\tilde{b}^1 B_1$ level term values were then determined from the $\tilde{a}^1 A_1$ $(0, 2, 0)$ term values and the transition frequencies in the TR–FTES spectra. Finally, the assignment of the 2_0^{18}, 2_0^{19} pump laser transition was made by subtracting the pump laser frequency from the $\tilde{b}^1 B_1$ term value, and comparing this number to the $\tilde{a}^1 A_1$ $(0, 0, 0)$ term values[33] to identify the initial rotational level. Because combination differences are extremely hard to observe for these bands, it is almost essential to perform a double resonance experiment of this kind to make any progress in assigning the spectra.

The transition frequencies in the dispersed fluorescence spectra were converted to $\tilde{a}^1 A_1$ state term values by $E(\tilde{a}) = E(\tilde{b}) - v$; where $E(\tilde{a})$ is the $\tilde{a}^1 A_1$ level term value, $E(\tilde{b})$ the term value of the intermediate $\tilde{b}^1 B_1$ level, and v the frequency in the dispersed fluorescence spectra. The values of $E(\tilde{b})$ for the $\tilde{b}^1 B_1$ $(0, 15^1, 0)$, $(0, 16^0, 0)$, and $(0, 17^1, 0)$ levels were taken from Ref. 33, and those for the $\tilde{b}^1 B_1$ $(0, 18^0, 0)$ and $(0, 19^1, 0)$ levels were determined from the $\tilde{b} \to \tilde{a}$ 2_2^{18} and 2_2^{19} TR–FTES spectra, as described above.[42] The measured rovibrational term values were analyzed by a nonlinear least squares fitting procedure with a complete rigid asymmeteric rotor Hamiltonian (type I^r). In fitting each vibrational band, all the rotational levels were included at first, but those with large differences between their experimental and calculated term values, which indicate perturbations for these levels, were removed in the final fitting. The vibrational term values and rotational constants obtained from this analysis are given in Table 1. Also included in Table 1 are the vibration/rotation constants for the $(0, 0, 0)$, $(1, 0, 0)$, $(0, 1, 0)$, and $(0, 0, 1)$ levels determined by other workers.[39,40] Only $K_a = 1$ rotational levels were observed for the $\tilde{a}^1 A_1$ $(2, 0, 0)$ state, hence, the vibrational term value and A rotational constant could not be independently determined. Thus, for this level the A rotational constant from the *ab initio* calculation of Green *et al.*[37] was used in our analysis, and held fixed during the least squares fitting. Also included in Table 1 are the harmonic frequencies and anharmonicity constants determined from the vibrational term values.

Table 1. Experimental vibrational term values (G_v) and rotational constants (A, B, C) determined for $\tilde{a}^1 A_1$ CH_2. The numbers in parenthesis represent 1σ uncertainties. Also included are the harmonic frequencies and anharmonicity constants obtained from the vibrational term values. All values are in cm^{-1}.

level	G_v	A	B	C
$(0,0,0)$	–	20.118(2)	11.205(2)	7.069(2)
$(0,1,0)$	1352.59(9)	21.826(1)	11.329(1)	6.943(2)
$(0,2,0)$	2667.7(1)	23.62(5)	11.38(1)	6.844(9)
$(1,0,0)$	2806.07(3)	19.829(2)	11.0437(2)	6.9455(2)
$(0,0,1)$	2863.99(8)	19.324(2)	11.1478(4)	6.9855(1)
$(0,3,0)$	3950.5(1)	25.59(6)	11.46(1)	6.709(9)
$(1,1,0)$	4152.8(2)	21.53(9)	11.18(2)	6.8(1)
$(0,4,0)$	5196.57(9)	27.47(9)	11.54(1)	6.62(1)
$(1,2,0)$	5444.9(3)	23.1(3)	11.20(3)	6.72(3)
$(2,0,0)$	5531.(1)	20.1[a]	10.95(4)	6.83(2)
$(0,5,0)$	6403.1(2)	23.1(3)	11.55(3)	6.55(2)
$(1,3,0)$	6714.1(5)	24.1(5)	11.22(7)	6.6(1)

Vibrational constants for $\tilde{a}^1 A_1$ CH_2

$\omega_1^0 = 2,846 \pm 1$	$\chi_{11}^0 = -40 \pm 1$	$\chi_{12}^0 = -6 \pm 1$
$\omega_1^0 = 1371.3 \pm 0.2$	$\chi_{22}^0 = -18.7 \pm 0.2$	

[a]Taken from Ref. 37.

It was found by fitting selected subsets of the $\tilde{a}^1 A_1$ $(0,2,0)$ and $(0,3,0)$ rotational levels that the rotational constants determined are strongly J and K_a dependent.[10] For example, the A rotational constant decreases as K_a increases, with as much as a 2% change for the $K_a = 0,1$ levels compared to the $K_a = 1,2,3$ levels. This effect appears even though the asymmetric rotor Hamiltonian includes the D_K and D_{JK} constants. A possible reason for this problem is the inadequacy of the rigid rotor

Hamiltonian for describing molecules — like CH_2 — that undergo large amplitude bending motion. It seems likely that the nonrigid bending Hamiltonian of Bunker *et al.*[43] will be a better approach to analyzing the rotational structure of the CH_2 bending overtone levels.

3.1.2. *Perturbations in the $\tilde{a}^1 A_1$ rovibrational levels*

For many of the transitions observed in the SEP and TR–FTES spectra there are large differences (> 0.3 cm^{-1}) between the experimental term values and those calculated using the asymmetric rotor Hamiltonian. Perturbations due to Fermi resonance, Coriolis coupling, and singlet–triplet interactions with nearby $\tilde{X}^3 B_1$ levels are possible causes of these discrepencies.[10,33,39,40] Each of these interactions has a specific set of vibration/rotation selection rules. Whether a particular $\tilde{a}^1 A_1$ level is perturbed depends on whether there is a nearby level that satisfies the selection rules. The effect of each of the above perturbing mechanisms is discussed below.

a. Fermi Resonance: The rotational selection rules for Fermi interactions in the CH_2 $\tilde{a}^1 A_1$ state are $\Delta K_a = 0, \pm 2$, $\Delta K_c = 0, \pm 2$. Resonances between $(0, v_2, 0)$ and $(1, v_2-2, 0)$ levels are the most likely, as these are the closest spaced levels with the same vibrational symmetry. However, the difference in term values between these vibrational levels is still large enough for no $\Delta K_a = 0$ Fermi interactions to have been detected. Because CH_2 is a strongly asymmeteric top, interactions between levels with $\Delta K_a = 2$ are possible. An example of this type of interaction is between the $(0, 3, 0)$ 6_{34} and $(1, 1, 0)$ 6_{16} levels. Deperturbation analysis of this nearly resonant interaction gives a matrix element of 2.0 ± 0.3 cm^{-1} between these two levels.[10] Similar interactions have been observed for the $(0, 2, 0)$ 6_{33} and $(1, 0, 0)$ 6_{15} levels.[10,33] As v_2 increases the $(0, v_2, 0)$ and $(1, v_2-2, 0)$ vibrational levels move further apart (see Table 1) and $\Delta K_a = 2$ Fermi resonance interactions become less likely.

b. Coriolis coupling: For CH_2 c-type Coriolis interaction can occur between vibrational levels with a_1 and b_2 symmetry. The symmetric stretching mode (v_1) and the bending mode (v_2) have a_1 symmetry, while the asymmetric stretching mode (v_3) has b_2 symmetry. The rotational selection rules for c-type Coriolis coupling are $\Delta K_a = \pm 1, \pm 3$, $\Delta K_c = 0, \pm 2$. The Coriolis interaction between the $\tilde{a}^1 A_1$ $(1,0,0)$ and $(0,0,1)$ levels has been analyzed by Petek *et al.* from their diode laser transient absorption data, and the interaction matrix elements were determined.[40] None of the $\tilde{a}^1 A_1$ state bending overtone levels are close enough to levels with b_2 symmetry to be discernibly perturbed. The $\tilde{a}^1 A_1$ $(1,1,0)$, $(1,2,0)$, and $(2,0,0)$ levels are calculated to lie close (< 50 cm^{-1}) to the $(0,1,1)$, $(0,2,1)$ and $(1,0,1)$ levels, respectively.[37] However, no large deviations of the C rotational constant of these levels can be detected. To determine how Coriolis coupling affects the rotational structure of these levels, a detailed spectroscopic analysis of the b_2 symmetry vibrational levels is also required. Unfortunately, this can not be done through DFS because of the $\tilde{b}^1 B_1 \leftrightarrow \tilde{a}^1 A_1$ vibronic selection rules.

c. Singlet–triplet interactions: The rotational selection rules for spin–orbit coupling are $\Delta K_a = 0, \pm 2$, $\Delta K_c = \pm 1, \pm 3$, for $\tilde{a}^1 A_1$ and $\tilde{X}^3 B_1$ vibrational levels of the same symmetry, and $\Delta K_a = \pm 1, \pm 3$, $\Delta K_c = \pm 1, \pm 3$, for $\tilde{a}^1 A_1$ and $\tilde{X}^3 B_1$ vibrational levels of different symmetry. The calculated *ab initio* spin–orbit matrix element between the $\tilde{a}^1 A_1$ and $\tilde{X}^3 B_1$ states is 1–2 cm^{-1}.[44] This value means that the $\tilde{a}^1 A_1$ levels are only significantly perturbed by singlet–triplet coupling when an accidental degeneracy with a $\tilde{X}^3 B_1$ level occurs. Using a triplet vibrational level density of 0.005 per cm^{-1} (obtained by direct counting in the 6000–8000 cm^{-1} region of the $\tilde{X}^3 B_1$ state), the probability that an $\tilde{a}^1 A_1$ level is perturbed to generate an energy deviation > 0.2 cm^{-1} is estimated to be 20%.[10] Thus, singlet triplet interactions are very likely to be the cause of many of the discrepencies between the calculated and experimental term values. Without a knowledge of the $\tilde{X}^3 B_1$ state rotation–vibration structure, these perturbations will appear to be essentially random. For the $(0,3,0)$ 5_{32} and 6_{33} levels the perturbing triplet levels have been identified from calculations of Bunker *et al.*[45] For the singlet 5_{32} level they are the $\tilde{X}^3 B_1$ $(0,6,0)$ 4_{13} and $(0,5,0)$ 5_{33} levels, and for the singlet 6_{33} level they are the $\tilde{X}^3 B_1$ $(0,6,0)$ 5_{14} and $(0,5,0)$ 7_{34} levels. Deperturbation analysis yields a spin–orbit matrix element of 2.3 cm^{-1},[10] which is close to the *ab initio* value.

3.2. Barrier Height to Linearity in the $\tilde{a}^1 A_1$ State

Using the term values for the bending overtone levels determined by SEP and TR–FTES, a more accurate estimate of the barrier height to linearity in the $\tilde{a}^1 A_1$ state can be made. The experimental term values were fit to the bending potential function used by Bunker and Sears[34] and Duxbury and Jungen,[35]

$$V(\rho) = \frac{h_b f (\rho^2 - \rho_e^2)}{f\rho^4 + (8h_b - f\rho_e^2)\rho^2} \,, \tag{1}$$

where h_b is the barrier height, f is the bending force constant, $\rho = \pi - \angle\,\text{HCH}$, and ρ_e is the equilibrium value of ρ. The bending level energies in $V(\rho)$ were evaluated numerically using WKB theory,[10,46] i.e., the vibrational energies were determined by the quantization condition

$$\oint [2\mu(E_n - V(\rho))]^{1/2} d\rho = (n + 1/2)h, \quad n = 0, 1, 2, \ldots , \tag{2}$$

where $\mu = (f/\omega_e)^{1/2}$ is the reduced mass, and $E_n = G_v + E_{zpt}$ where E_{zpt} is the zero-point energy and G_v is the vibrational term value. ρ_e was held fixed at the experimentally determined value of $77.6°$ (Ref. 39) and h_b, f, ω_e, and E_{zpt} were optimized using a nonlinear least-squares fitting procedure. The values obtained are $h_b = 10\,000$ cm^{-1}, $f = 3.2\times10^5$ cm^{-1}, $\omega_e = 1380$ cm^{-1}, and $E_{zpt} = 690$ cm^{-1}. The vibrational term values obtained using these parameters are given in Table 2. Included in Table 2 for comparison are the experimental term values and the term values obtained from *ab initio* calculation.[37] The agreement between the WKB and experimental results is reasonable: the differences are $< 1\%$. The *ab initio* term values are consistently too high, which indicates that the potential energy surface used for these calculations needs to be adjusted — particularly along the bending coordinate.

Table 2. Experimental and calculated vibrational term values for the CH_2 $a^1 A_1$ $(0, v_2, 0)$ levels, $v_2 = 1$–5.

v_2	WKB	Experiment[a]	*ab initio*[b]
1	1350	1352.6	1356
2	2671	2667.7	2675
3	3957	3950.6	3962
4	5201	5196.7	5216
5	6395	6403.1	6430

[a]Taken from Refs. 10, 33, 40 and 41.
[b]Taken from Ref. 37.

The WKB calculation is a purely one-dimmensional model of the potential energy surface. Thus, if there is significant mixing of the bending vibrational mode with the stretching modes, the calculated barrier height will be incorrect. The most likely mixing for the $(0, v_2, 0)$ vibrational levels is Fermi resonance with the $(1, v_2\text{--}2, 0)$ levels. However, as discussed in Sec. 3.1.2.a, this interaction takes the form of isolated perturbations and is not expected to significantly shift the $(0, v_2, 0)$ vibrational term values. However, if Fermi resonance were significant between the $(0, v_2, 0)$ and $(1, v_2\text{--}2, 0)$ levels, deperturbation analysis would yield higher term values for the $(0, v_2, 0)$ levels as they lie below the $(1, v_2\text{--}2, 0)$ levels, which implies that the barrier height determined from Eq. (1) is a lower limit.

One should also bear in mind that the accuracy of the barrier height in the WKB calculation depends on how well the real potential is described by the chosen functional form. For example, the *ab initio* barrier height is lower than that deduced from Eq. (1), but the *ab initio* level energies are consistently higher than the ones generated by the WKB calculation. Thus, one can conclude that the bending function of Eq. (1) with a barrier height of $10\,000$ cm^{-1} gives a better representation of the bending level energies than the most up-to-date and precise *ab initio* calculation, but not that this represents the true bending potential and barrier height.

3.3. *The Renner–Teller Effect*

Because the $\tilde{a}^1 A_1$ and $\tilde{b}^1 B_1$ states of CH$_2$ become a degenerate $^1\Delta_g$ pair in the linear configuration,[32] they are strongly coupled by the Renner–Teller effect.[35,47] In bent CH$_2$ Renner–Teller coupling acts as an a-axis Coriolis interaction between electronic and nuclear motions. The matrix element of this coupling is proportional to K_a.[35,47–49] The $K_a = 0$ states are not affected by Renner–Teller coupling, therefore, the $\tilde{a}^1 A_1$ vibrational term values will not be affected.

The effect of Renner–Teller coupling in the CH$_2$ $\tilde{b}^1 B_1$ state (the upper component of the Renner–Teller pair) has been observed through the perturbed rotational structure of $K_a \neq 0$ levels compared to $K_a = 0$ levels, and from the observation of singlet–triplet interactions for $K_a \neq 0$ levels, but not $K_a = 0$ levels.[33] However, there has been no reported experimental observation of the effects of Renner–Teller coupling on the $\tilde{a}^1 A_1$ state, the lower Renner–Teller component.

One of the most dramatic consequences of Renner–Teller coupling in a bent molecule is the inversion of the K_a rotational structure of the high bending vibrational levels in the lower Renner–Teller component state.[48,49] Specifically, as the barrier to linearity in the lower state bending potential is approached, the Renner–Teller coupling with the upper state levels becomes stronger. The spacing between succesive K_a stacks in the bending vibrational levels decreases. For the very high bending levels the $K_a = 1$ levels may fall below the $K_a = 0$ levels and the $K_a = 2$ levels fall below the $K_a = 1$ levels, so the A rotational constant becomes negative! In contrast, when there is no Renner–Teller coupling, the A rotational constant should increase with the bending vibrational quantum number because the molecule becomes more linear.[50] The inversion of the K_a rotational structure does not occur for the upper Renner–Teller component state. This effect has been observed in the CS_2 $\tilde{a}^3 A_2 \leftarrow \tilde{X}^1 \Sigma_g^+$ absorption spectrum, where the $\tilde{a}^3 A_2$ state is the lower component of a Renner–Teller pair of states. Analysis of the spectrum shows that A starts to decrease at $v_2 = 7$ and is negative at $v_2 = 9$.[51]

The dependence of the CH_2 A rotational constant on v_2 in the $\tilde{a}^1 A_1$ state has been calculated using *ab initio* methods by Green *et al.*[37] These calculations predict that A should increase with v_2 up to $v_2 = 4$, decrease at $v_2 = 5$, and become negative at $v_2 = 6$ (i.e., the K_a rotational structure is inverted). This trend is also predicted by Alijah and Duxbury,[47] who used a least squares fitting of the $\tilde{a}^1 A_1$ and $\tilde{b}^1 B_1$ vibronic levels to obtain their potential energy curves; unfortunately, these authors only reported K_a sub-band origins and not rotational constants. The A rotational constants of Green *et al.*[37] and the experimental A constants for the $(0, v_2, 0)$ levels are plotted against v_2 in Fig. 8. There is excellent agreement between the experimental result and the *ab initio* calculations. In particular, the decrease in A from $v_2 = 4$ to $v_2 = 5$ predicted by theory is observed in the experiments.

Unfortunately, we were unable to record spectra of the $\tilde{a}^1 A_1$ $(0, 6, 0)$ level even though our detection sensitivity should allow its observation as long as the transition strength of the 2_6^n bands is greater than 10% of that for the 2_5^n bands. The transition strength is expected to decrease due to a decrease in the Franck–Condon factor from the overlap of the $\tilde{b}^1 B_1$ $(0, 18^0, 0)$ and $(0, 19^1, 0)$ vibrational wavefunctions with the $\tilde{a}^1 A_1$ $(0, 6, 0)$ vibrational wavefunction. Additionally, for the $K_a \neq 0$ levels strong

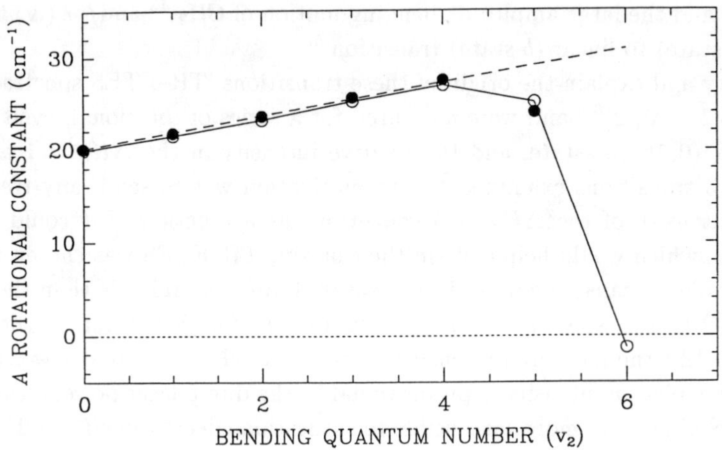

Fig. 8. Plot of the CH_2 A rotational constant versus number of quanta in the bending vibrational mode (v_2) for the $\tilde{a}^1A_1$ $(0, v_2, 0)$ levels. (o) Taken from the *ab initio* calculations of Green *et al.* (Ref. 37); (•) experimental values (Refs. 10, 33, 40, and 41); --- extrapolation of A from the $v_2 = 0$–4 levels.

Renner–Teller coupling will mix the electronic wavefunctions of the $\tilde{a}^1A_1$ $(0, 6, 0)$ and the $\tilde{b}^1B_1$ $(0, 0, 0)$ levels and, hence, dilute the oscillator strength of the $\tilde{b} \rightarrow \tilde{a}2_6^{18}$ and 2_6^{19} transitions.

3.4. *Strong $\Delta K_a = 3$ Transitions in the CH_2 $\tilde{b} \leftrightarrow \tilde{a}$ Spectrum*

One intriguing aspect of CH_2 spectroscopy is the occurance of intense $\Delta K_a = 3$ transitions — in some cases over 50% of the intensity of the corresponding $\Delta K_a = 1$ transition. These transitions were first recognized in SEP spectra of the $\tilde{b} \rightarrow \tilde{a}$ 2_2^{16} and 2_3^{16} bands.[10,41] Since these initial results, strong $\Delta K_a = 3$ transitions have been identified in the $\tilde{b} \leftarrow \tilde{a}$ 2_0^{16}, 2_1^{16}, 2_0^{14}, 2_0^{15} and 2_1^{15} bands by laser induced fluorescence (LIF),[6,33] and in the $\tilde{b} \rightarrow \tilde{a}$ 2_4^{16} band by TR–FTES.[42] Even some $\Delta K_a = 5$ transitions have been assigned in the $\tilde{b} \leftarrow \tilde{a}$ 2_0^{16} band.[52] Of the 10 000 lines in the CH_2 $\tilde{b} \leftarrow \tilde{a}$ spectrum that have been observed,[32,33] so far only 500 have been assigned to $\Delta K_a = 1$ transitions. High ΔK_a transitions can account for some of the remaining unassigned lines. However, no definative explanation has yet been given for their seemingly large intensity, although it has been speculated[6,10,41] that the following reasons may be the cause, or causes: (i) the asymmetry of CH_2, (ii) Renner–Teller coupling, (iii) interaction between rotation and the large amplitude bending motion of CH_2,[43] and/or (iv) the bent ($\tilde{a}$ state) to linear ($\tilde{b}$ state) transition.[53]

To try and explain the origin of these transitions, TR–FTES spectra of the CH_2 $\tilde{b} \rightarrow \tilde{a}$, 2_1^{16} band were measured for a series of rotational levels in the $\tilde{b}^1B_1$ $(0, 16^0, 0)$ state, and the relative intensity of the $\Delta K_a = 1$ and $\Delta K_a = 3$ transitions examined.[54] Our motivation was to see if any trend in the intensity of the $\Delta K_a = 3$ transitions as a function of J could be observed, which would help explain their origin. TR–FTES was chosen for these studies because the intensities measured are more reliable than those from SEP or LIF spectra. There are two reasons for this: First, for both SEP and LIF the measured intensity is proportional to the laser power (in SEP the depletion intensity is proportional to the dump laser power), thus, these techniques are more susceptible to saturation effects than TR–FTES, where spontaneously emitted photons are detected. Second, in the LIF $\tilde{b}$ $(0, 16^0, 0) \leftarrow \tilde{a}$ spectrum the $\Delta K_a = 1$ and $\Delta K_a = 3$ lines originate from different ground state K_a'' levels. Hence, to compare line strengths the ground state population distribution must be known. The non-Boltzmann distribution of CH_2 produced by photo-dissociation of ketene may introduce an error in the measured line strength.

FTES spectra were recorded for eight rotational levels in the $\tilde{b}^1B_1$ $(0, 16^0, 0)$ state, $J_{K_a' K_c'} = 0_{00}, 1_{01}, 2_{02}, 3_{03}, 4_{04}, 5_{05}, 6_{06}$, and 7_{07}, with a sampling bandwidth of 987 cm^{-1} and a resolution of 0.5 or 1.0 cm^{-1}. $\Delta K_a = 3$ transitions were observed from the higher J levels. In Fig. 9 the FTES spectra from the $4_{04}, 5_{05}, 6_{06}$, and 7_{07}, rotational levels are shown. From Fig. 9 it can be seen that the intensity of the $\Delta K_a = 3$ transitions relative to the $\Delta K_a = 1$ transitions increases as J increases in the $\tilde{a}^1A_1$ $(0, 1, 0)$ state, to the extent that for emission from the 7_{07} level the intensity of the P-branch $\Delta K_a = 3$ transition is equal to that of the $\Delta K_a = 1$ transition!

The $\Delta K_a = 3$ transition intensity may arise from a mixing of K and $K \pm 2$ symmetric top basis rotational wavefunctions in the $\tilde{a}^1A_1$ $(0, 1, 0)$ state. Such a mixing can originate purely from the asymmetry of $\tilde{a}^1A_1$ $(0, 1, 0)$ CH_2, defined by the value of $(B - C)/2$, which introduces off-diagonal matrix elements between the K and $K \pm 2$ symmetric top basis wavefunctions. For the $K = 1$ and $K = 3$ symmetric top rotational wavefunctions the matrix elements are proportional to $\frac{1}{2}(B - C)\sqrt{(J - 2)(J - 1)(J + 2)(J + 3)}$, i.e., the mixing between $\Delta K = 2$

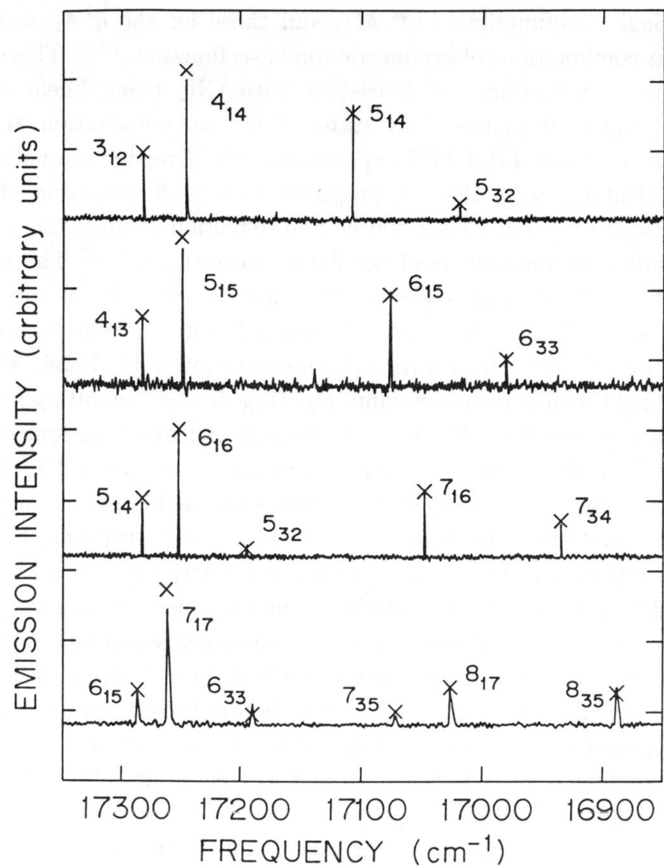

Fig. 9. TR–FTES spectra of the CH_2 $\tilde{b} \rightarrow \tilde{a}$ 2_1^{16} band. The rotational levels in the $\tilde{b}^1 B_1$ $(0, 16^0, 0)$ state are, from top-to-bottom, $J_{K_a K_c} = 4_{04}$, 5_{05}, 6_{06}, and 7_{07}. The $\tilde{a}^1 A_1$ rotational levels are labelled in the figure. The crosses represent line strengths calculated using an asymmetric rotor Hamiltonian; see text for details.

rotational wavefunctions increases with J. In contrast, in Renner–Teller coupling the mixing of $\Delta K = 2$ wavefunctions is proportional to K_a and not J.[47–49] Therefore, Renner–Teller coupling does not seem to offer a reasonable explanation for the observed intensity pattern. To quantitatively test this hypothesis a simulation of the dispersed fluorescence spectrum was performed with a rigid asymmetric rotor Hamiltonian.

The rotational line strengths were calculated by assuming that the rotational wavefunctions in the $\tilde{b}^1 B_1$ state are given by the symmetric top rotational wavefunctions $|J, 0, M\rangle$, and those for the $\tilde{a}^1 A_1$ state are Wang linear combinations of symmetric top basis functions.[55,56] This choice of rotational wavefunctions is consistent with CH_2 being linear in the $\tilde{b}^1 B_1$ state and bent in the $\tilde{a}^1 A_1$ state. Since no polarization analysis was performed in our TR–FTES experiments, the direction cosine matrix elements calculated were those appropriate to $\tilde{b} \leftarrow \tilde{a}$ excitation of CH_2 with a vertically polarized laser source and detection of both the parallel and perpendicular components of the fluorescence signal.[55–57] The mixing between the $\tilde{a}^1 A_1$ $(0,1,0)$ state $K = 1$ and $K = 3$ basis rotational wavefunctions was calculated using Watson's A-reduced Hamiltonian (I^r representation)[58] with the rotational constants given by Petek *et al.*[39] (which include the distortion constants D_K D_{JK} D_J δ_J, δ_K and ϕ_K). It was assumed that no mixing of the $\tilde{b}^1 B_1$ state rotational wavefunctions occurs.

Results from the asymmetric rotor calculation are included in Fig. 9. The agreement between the experimental and simulated intensities is within the experimental error for almost all the observed transitions, and the increase in intensity of the $\Delta K_a = 3$ transitions with increasing J is well reproduced by the calculation. Note that removing the distortion constants (which can also mix the $K = 1$ and $K = 3$ rotational wavefunctions) from the asymmetric rotor calculation does not affect the calculated intensities. The asymmetric rotor calculation also predicts that the maximum $\Delta K_a = 5$ transition intensity for the low J rotational levels studied is less than 1% of the corresponding $\Delta K_a = 1$ transition and, therefore, not observable with our present signal-to-noise.

In the SEP spectra presented, the pump and dump laser polarizations were parallel, which is equivalent to only detecting the parallel component of the dispersed fluorescence spectrum. This difference in the detection geometry causes the different intensity distribution of the ${}^P P$, ${}^P Q$, and ${}^P R$ branches observed by SEP compared to TR–FTES (see Fig. 2 compared to Fig. 9). In some cases, this pump–dump polarization configuration makes the $\Delta K_a = 3$ intensities in the SEP spectra larger than those in the corresponding TR–FTES spectra without polarization selection.

In conclusion, the enormously strong $\Delta K_a = 3$ transitions that occur in the CH_2 $\tilde{b} \to \tilde{a}$ 2_1^{16} band are attributed to the asymmetry in the $\tilde{a}^1 A_1$ $(0,1,0)$ state of CH_2. No other effects, such as interaction between rotation and vibration[43] or axis switching,[53] are needed to account for the observed intensities. The intensity of the high $\Delta K_a = 3$ transitions increases with J, to the extent of being stronger than the $\Delta K_a = 1$ transitions for $J = 7$. The high ΔK_a transitions will give strong lines from high J levels in the LIF spectra. Since the CH_2 molecules generated from photolysis of ketene may contain significant population at high J levels, these asymmetry induced high ΔK_a transitions may account for many of the previously unassigned lines observed in the $\tilde{b} \leftarrow \tilde{a}$ LIF spectra.

4. Reaction and Rotational Energy Transfer of Vibrationally Excited $\tilde{b}^1 B_1$ CH_2

In this section, the state-resolved dynamics of single rotational levels in the fifteenth bending overtone of the CH_2 $\tilde{b}^1 B_1$ state are presented and discussed. In its $\tilde{b}^1 B_1$ state CH_2 is a quasilinear molecule and the rotation and bending degrees of freedom are strongly coupled.[37] This coupling, plus the existance of numerous perturbations in the CH_2 $\tilde{b}^1 B_1$ state (singlet–triplet, Renner–Teller, and Fermi resonance interactions have all been observed[32,33]), makes the question of how energy relaxation occurs for individual $\tilde{b}^1 B_1$ levels an interesting one. Singlet methylene is also an interesting species for study because of its high reactivity. In the $\tilde{a}^1 A_1$ state the reaction products have been studied for a wide variety of partners.[7] The kinetics of the $\tilde{a}^1 A_1$[8,9] and $\tilde{b}^1 B_1$[59,60] states have also been investigated, but without quantum state resolution. To properly determine the collisional dynamics of CH_2, i.e., the extent of collision induced rotation/vibrational energy transfer versus reactive collisions, a single rovibronic state must be prepared and the population of that state monitored.

State-resolved kinetic measurements for $\tilde{b}^1 B_1$ CH_2 were performed by exciting a single rotational level of the $\tilde{b}^1 B_1$ $(0, 16^0, 0)$ state and recording the subsequent dispersed fluorescence spectrum as a function of time by TR–FTES. The fluorescence spectrum shows peaks due to the initially excited rovibronic level as well as "daughter" peaks from nearby rotational levels populated by collision induced rotational energy transfer. The rotational energy changing collisions are primarily with the bath gas, ketene.[61]

An example of the time-resolved spectra obtained is shown in Fig. 10. In this experiment the $\tilde{b}^1 B_1$ $(0, 16^0, 0)$, 0_{00} level was initially excited; transitions can be clearly seen from the 0_{00} level, and from the 2_{02} level which is populated by rotational energy transfer (RET) from the 0_{00} level. In Fig. 11, the relative intensities of the $\tilde{b} \rightarrow \tilde{a}$ 2_1^{16} $0_{00} \rightarrow 1_{10}$ and $2_{02} \rightarrow 3_{12}$

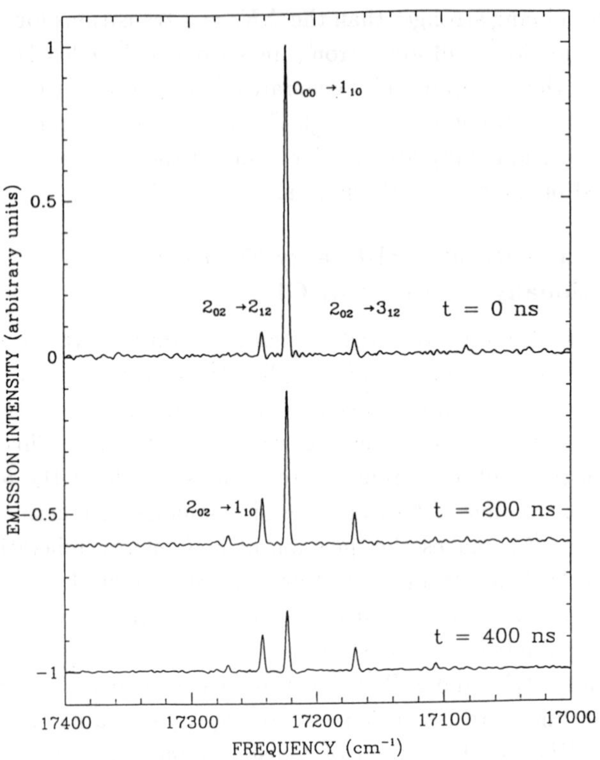

Fig. 10. TR–FTES spectra of the CH$_2$ $\tilde{b} \to \tilde{a}$ 2_1^{16} band at $t = 0$, 200, and 400 ns after excitation of the $J_{K_a K_c} = 0_{00}$ level. The 2_{02} level is populated by RET from the 0_{00} level. The ketene pressure was 200 mTorr.

transitions are plotted against time. Note that after a short instrument response rise time (ca. 70 ns) the $\tilde{b} \to \tilde{a}$ 2_1^{16} $0_{00} \to 1_{10}$ transition simply decays, indicating the decrease of the 0_{00} population in time. In contrast, the $2_{02} \to 3_{12}$ "daughter" transition first grows in and then decays. This behaviour confirms that the 2_{02} level is populated by an energy transfer process.

RET from five different $\tilde{b}^1 B_1$ $(0, 16^0, 0)$ rotational levels, $J'_{K'_a K'_c} = 4_{04}$, 3_{03}, 2_{02}, 1_{01} and 0_{00}, has been monitored. Figure 12 shows TR–FTES spectra of the CH$_2$ $\tilde{b} \to \tilde{a}$ 2_1^{16} band following excitation of the 4_{04}, 2_{02}, and 0_{00} rotational levels. The peaks labelled with an asterisk in Fig. 12 are assigned to transitions from levels populated by RET. In all cases only

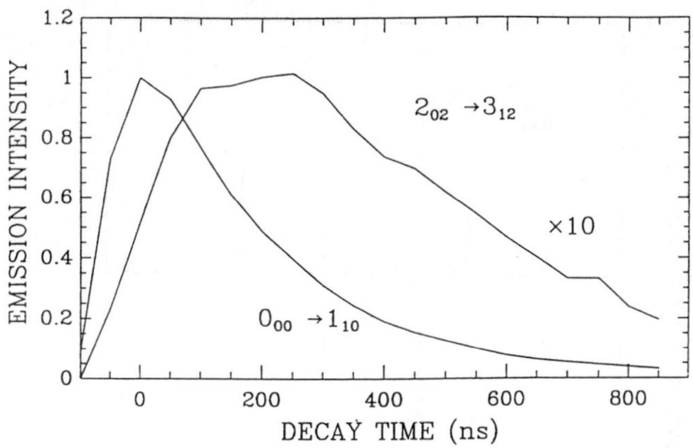

Fig. 11. Intensity of the CH_2 $\tilde{b} \rightarrow \tilde{a}$ 2_1^{16} $0_{00} \rightarrow 1_{10}$ and $2_{02} \rightarrow 3_{12}$ transitions as a function of time (ns) after excitation of the CH_2 $\tilde{b}^1 B_1$ $(0, 16^0, 0)$ 0_{00} level at 200 mTorr of ketene. The intensities have been normalized to the maximum intensity of the $0_{00} \rightarrow 1_{10}$ transition, and the $2_{02} \rightarrow 3_{12}$ transition has been multiplied by 10 for display purposes.

RET transitions with quantum number changes of $\Delta J = \pm 2$, $\Delta K_a = 0$ and $\Delta K_c = \pm 2$ were observed. The weak $\Delta J = \pm 4$ RET transitions in Fig. 12 are assigned to consecutive $\Delta J = \pm 2$ transitions from their time behaviour. No vibrational energy transfer was observed in our measurements.

The observed RET propensity rules suggest that CH_2 is quasilinear in the highly excited bending overtone levels of the $\tilde{b}^1 B_1$ state.[32] The lowest order electric multipole moment in the quasilinear CH_2 intermolecular potential (after time averaging of the nuclei positions) is a quadrupole. Collisional interactions through a quadrupole have propensity rules of $\Delta J = \pm 2, \pm 1$, and 0, and conservation of parity.[62] However, additional restrictions arise from nuclear spin statistics. CH_2 has two nuclear spin states, where rotational levels with $K_a K_c = ee$ and oo belong to one nuclear spin state and $K_a K_c = eo$ and oe belong to the other.[63] Because collisions do not interconvert nuclear spin states, and parity is conserved, only the transitions $K_a K_c = ee \leftrightarrow ee$ $oo \leftrightarrow oo$, etc. ... are allowed. Finally, as CH_2 is quasilinear in the $\tilde{b}^1 B_1$ state, K_a increases with l, the vibrational angular momentum quantum number.[32] The energy separation between states with l and $l \pm 2$ is much larger than that between rotational levels with the same l, thus transitions with $\Delta l \neq 0$, and therefore $\Delta K_a \neq 0$, do not occur for RET at room temperature. Consideration of the above points gives the

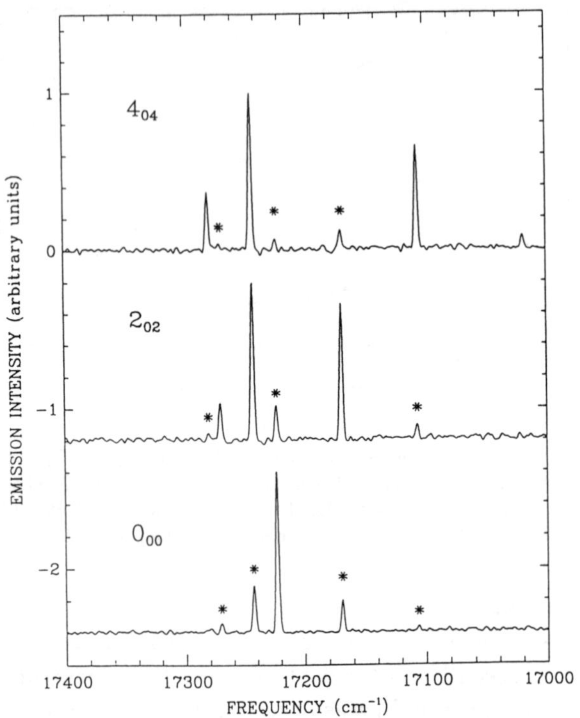

Fig. 12. TR–FTES spectra of the CH_2 $\tilde{b} \to \tilde{a}$, 2_1^{16} band after excitation of the $J_{K_a K_c} = 4_{04}$, 2_{02}, and 0_{00} levels. The transitions labelled with an asterisk occur from levels populated by RET.

propensity rules $\Delta J = 0,\ \pm2$, $\Delta K_a = 0\ (= \Delta l)$ and $\Delta K_c = 0,\ \pm2$; which agree with our observations.

These results show that RET is sensitive to vibrational dynamics. For example, if CH_2 had a small bending amplitude, so that the time averaged structure was bent, the lowest order electric multipole in the intermolecular potential would become a dipole, and the propensity rules expected for RET would be $\Delta J = 0,\ \pm1$, $\Delta K_a = \pm1$ and $\Delta K_c = \pm1$.[62] However, because the CH_2 $\tilde{b}^1 B_1$ $(0, 16^0, 0)$ level is above the barrier to linearity with a large amplitude vibrational motion — and the duration of a collision between CH_2 and ketene is longer than the period of vibration — the time-averaged CH_2 structure is linear, giving a quadrupole dominated intermolecular potential.

A quantitative study of RET was undertaken for the $\tilde{b}^1 B_1$ $(0, 16^0, 0)$ 4_{04}, 2_{02}, and 0_{00} rotational levels. The total fluorescence decay rate constant k_J^{tot} for the $\tilde{b}^1 B_1$ $(0, 16^0, 0)$ J_{0J} level is given by $k_J^{\text{tot}} = k_J^{\text{fl}} + k_J^{\text{col}}[M]$, where k_J^{fl} is the rate constant for spontaneous emission, $[M]$ is the density of the bath gas (ketene), and k_J^{col} is the rate constant for total collisional decay. k_J^{col} is given by $k_J^{\text{col}} = k_J^{\text{rxn}} + \Sigma_{J'} k(J \to J')$, where k_J^{rxn} is the state-resolved rate constant for reactive collisions with ketene, and $k(J \to J')$ is the state-to-state rate constant for RET from J_{0J} to $J'_{0J'}$. By measuring k_J^{tot} as a function of ketene pressure, the spontaneous fluorescence lifetime $\tau = 1/k_J^{\text{fl}}$ and k_J^{col} can be obtained from a Stern–Volmer analysis of the data. The results are presented in Table 3. The state-resolved cross-sections for collisional decay (σ_J^{tot}) of the $\tilde{b}^1 B_1$ $(0, 16^0, 0)$ levels are also given in Table 3.

Table 3. Total collisional decay rates and cross-sections (k_J^{col} and σ_J^{tot}) and spontaneous fluorescence lifetimes (τ) for the $J_{K_a K_c} = 0_{00}$, 2_{02}, and 4_{04} rotational levels of $\tilde{b}^1 B_1$ $(0, 16^0, 0)$ CH_2. Also given are the cross- sections for reactive collisions with ketene (σ_J^{rxn}), see text. The quoted errors are one sigma confidence limits.

J_{0J}	$k_J^{\text{col}} \times 10^{12}$ $(\text{cm}^3 \text{molec}^{-1}\text{s}^{-1})$	τ (μsec)	σ_J^{tot} (Å^2)	σ_J^{rxn} (Å^2)
0_{00}	577 ± 22	1.5 ± 0.5	54 ± 7	14 ± 6
2_{02}	491 ± 18	1.7 ± 0.7	46 ± 6	24 ± 7
4_{04}	586 ± 24	2.7 ± 1.2	56 ± 8	46 ± 7

The state-to-state rate constants for RET can be determined from the intensity of the daughter peaks in the spectrum and the total decay rates (k_J^{tot}) of the two levels involved in the rotational energy changing transition (see Ref. 61 for details of this calculation). The values of $k(J \to J')$ obtained and the corresponding state-to-state RET cross-sections $(\sigma_{J \to J'})$ are given in Table 4. Note that the RET rate constants obey microscopic reversibility.[64] The ratio of the "foward" to "backward" rate constant for RET is given by the degeneracy of the final state weighted by a Boltzmann factor of the energy difference between the two states: $k(0 \to 2)/k(2 \to 0) = 4.8 \pm 0.4$ and $k(2 \to 4)/k(4 \to 2) = 1.4 \pm 0.4$. The expected values from the M degeneracies and energies of the final rotational levels are 4.5 and 1.4, respectively.

Table 4. State-to-state rotational energy transfer rate constants and cross-sections for the $J_{K_a K_c} = 0_{00}$, 2_{02}, and 4_{04} rotational levels of $\tilde{b}^1 B_1$ $(0, 16^0, 0)$ CH_2. The quoted errors are one sigma confidence limits.

$J_{0J} \rightarrow J'_{0J'}$	$k(J \rightarrow J') \times 10^{12}$ $(cm^3 molec^{-1} s^{-1})$	$\sigma_{J \rightarrow J'}$ $(\text{\AA}^2)$
$0_{00} \rightarrow 2_{02}$	424 ± 31	40 ± 7
$2_{02} \rightarrow 0_{00}$	88.4 ± 5	8.4 ± 1.2
$2_{02} \rightarrow 4_{04}$	151 ± 28	14 ± 4
$4_{04} \rightarrow 2_{02}$	108 ± 6	10 ± 2

Subtraction of the RET contribution from the total collisional decay rate yields the rate constant for removal of $\tilde{b}^1 B_1$ CH_2 by reactive collisions with ketene.[7-9] In principle these rate constants include contributions from collision-induced intersystem crossing and internal conversion. However, these effects are expected to be small for the $(0, 16^0, 0)$ level of $\tilde{b}^1 B_1$ CH_2.[33,61] The state-resolved reactive cross-sections σ_J^{rxn} are included in Table 3.

The data in Tables 3 and 4 shows that both RET and reaction occur rapidly for $\tilde{b}^1 B_1$ CH_2 in collisions with ketene. The cross-sections for RET range approximately one to four times the hard-sphere gas kinetic cross-section, which was calculated to be 10 $\text{\AA}^2$,[65] while those for reactive scattering vary from one to five times the hard-sphere rate. Note that the reaction cross-sections increase monotonically with J. A plausible explanation is that as J increases more states become accesible at the transition state of the reaction between CH_2 and ketene. Thus, from phase space theories of reaction dynamics, the reaction cross-section will be greater.[64] The measured reactive cross-sections show that the CH_2 $\tilde{b}^1 B_1$ state is approximately twice as reactive as the $\tilde{a}^1 A_1$ state.[9,61,66] This increased reactivity may arise because $\tilde{b}^1 B_1$ CH_2 is a bi-radical (the two non-bonding valence electrons occupy different molecular orbitals), whereas in $\tilde{a}^1 A_1$ CH_2 the two non-bonding electrons occupy the same molecular orbital.[3] Finally, measurement of time-dependent polarized TR–FTES spectra shows that elastic ($\Delta J = 0$) M changing collisions between $\tilde{b}^1 B_1$ CH_2 and ketene are negligible. This is in agreement with recent measurements for dealignment collisions between $\tilde{A}^1 A_2$ $H_2 CO$ and $\tilde{X}^1 A_1$ $H_2 CO$.[67]

5. Concluding Remarks: A Comparison of the SEP and TR–FTES Techniques

We complete this chapter by comparing the attributes of SEP and TR–FTES. The advantages of SEP are: First, much higher resolution can be obtained with SEP than with TR–FTES. For SEP the resolution is only limited by the bandwidth of the laser sources and, thus, with single mode lasers can be better than 0.01 cm^{-1}. This sub-Doppler resolution arises because only a small fraction of the Doppler profile is excited by the narrow band pump laser and, therefore, the Doppler width of the dump transition depends on the pump laser linewidth. The best demonstrated resolution with TR–FTES is 0.25 cm^{-1},[17] although better resolution should be possible with a larger (> 10 cm) interferometer and specially designed optics. Second, weaker vibronic transitions can be studied with SEP because the stimulated emission process can be saturated by the dump laser.[14] And third, by using SEP significant population can be transferred to a specific ground state rovibrational level. The subsequent dynamics of this level can then be investigated. This can be done by either monitoring LIF from the SEP prepared vibrational level,[68,69] or transient absorption of a cw laser.[70] Information about the dynamics of the ground state vibrational levels cannot be obtained by TR–FTES.

In contrast, the attributes of TR–FTES are that: (i) spectra can be recorded over a much wider spectral bandwidth than is possible with SEP (i.e., thousands of wavenumbers compared to tens of wavenumbers); (ii) the intensities measured by TR–FTES are much more reliable than those recorded by SEP. In SEP the relative intensities of the peaks in the spectra are often distorted due to dump laser saturation. This is not a problem with TR–FTES because spontaneously emitted photons are detected (see Sec. 3.4); (iii) only transitions from the excited vibronic state to a lower electronic state can appear in TR–FTES spectra. Upward transitions, which may appear in an SEP spectrum and confuse the assignment of the downward transitions, are not possible; and (iv) the time-resolved capability of TR–FTES means that kinetic processes in the excited electronic state can be investigated. Examples of the processes that have been studied so far are collisional relaxation (specifically RET) and bimolecular reactions (see Sec. 4). Other processes that are amenable to study are intra-molecular vibrational energy redistribution, unimolecular reactions such as photodissociation, and vibrational predissociation. Recording time-resolved spectra is also useful for assigning peaks in the dispersed emission spectra: transitions from levels populated by energy transfer can be uniquely identified (see Figs. 10 and 12). This is not possible with SEP, because spectra are recorded at only one time, determined by the delay between the pump and dump lasers.

Both SEP and TR–FTES have been shown to be extremely sensitive, high resolution, and actually complementary techniques for transient molecule studies. Whether SEP or TR–FTES should be used depends on which of the attributes discussed above is required.

Acknowledgements

It is our pleasure to acknowledge Dr. Wei Xie and Ms. Dong Qin, without whose talent and effort this work would not have been completed. We would also like to thank Dr. Carmel Harkin, Mr. Alan Ritter, and Dr. Kamu Kasturi for their contributions in different parts of the experiments. Funding for this project was provided by a grant from the Basic Energy Sciences division of the US Department of Energy, and by a US Department of Energy University Instrumentation Grant.

References

1. For a recent review of the vibrational spectroscopy of transient molecules, see F. J. Nortrup and T. J. Sears, *Annu. Rev. Phys. Chem.* **43**, 127 (1992).
2. E. Hirota, *High-Resolution Spectroscopy of Transient Molecules* (Springer, New York, 1985).
3. G. Herzberg, *The Spectra and Structure of Simple Free Radicals: An Introduction to Molecular Spectroscopy* (Dover Publications, NY, 1988).
4. (a) E. A. Rohlfing and J. E. M. Goldsmith, *J. Chem. Phys.* **90**, 6804 (1989); (b) E. A. Rohlfing, *J. Chem. Phys.* **91**, 4531 (1989).
5. F. J. Northrup and T. J. Sears, *J. Chem. Phys.* **91**, 762 (1989).
6. W. H. Green, I.-C. Chen, H. Bitto, D. R. Guyer, and C. B. Moore, *J. Mol. Spectros.* **138**, 614 (1989).
7. For a review of CH_2 chemistry, see A. H. Laufer, *Rev. Chem. Int.* **4**, 225 (1981).
8. (a) M. N. R. Ashfold, M. A. Fullstone, G. Hancock, and G. W. Ketley, *Chem. Phys.* **55**, 245 (1981); (b) G. Hancock and M. R. Heal, *J. Phys. Chem.* **96**, 10316 (1992).
9. A. O. Langford, H. Petek, and C. B. Moore, *J. Chem. Phys.* **78**, 6650 (1983).
10. (a) W. Xie, A. Ritter, C. Harkin, K. Kasturi, and H.- L. Dai, *J Chem. Phys.* **89**, 7033 (1988); (b) W. Xie, C. Harkin, and H.-L. Dai, *ibid.* **93**, 4615 (1990).
11. E. Pebay-Peyroula, A. Delon, and R. Jost, *J. Mol. Spectrosc.* **132**, 123 (1988).

12. K. Yamanouchi, S. Takeuchi, and S. Tsuchiya, *J. Chem. Phys.* **92**, 4044 (1990).

13. A. Delon and R. Jost, *J. Chem. Phys.* **95**, 5687 (1991).

14. K. Yamanouchi, N. Ikeda, S. Tsuchiya, D. M. Jonas, J. K. Lundberg, G. W. Adamson, and R. W. Field, *J. Chem. Phys.* **95**, 6330 (1991).

15. (a) C. Kittrell, E. Abramson, J. L. Kinsey, S. A. McDonald, D. E. Reisner, and R. W. Field, *J. Chem. Phys.* **75**, 2056, (1981); (b) C. E. Hamilton, J. L. Kinsey, and R. W. Field, *Ann. Rev. Phys. Chem.* **37**, 493 (1983).

16. For a recent review of SEP, see papers in the special issue "Molecular Spectroscopy and Dynamics by Stimulated Emission Pumping." ed. H.-L. Dai *J. Opt. Soc. Am.* **B7**, (1990).

17. G. V. Hartland, W. Xie, H.-L. Dai, A. Simon, and M. J. Anderson, *Rev. Sci. Instrum.* **63**, 3261 (1992).

18. S. Gerstenkorn and P. Luc, "Atlas du spectre d'absorption de la molecule de l'iode $(14\,800\text{–}20\,000 \text{ cm}^{-1})$" (Editions du C.N.R.S., Paris, France, 1978.)

19. (a) D. Frye, P. Arias, and H. L. Dai, *J. Chem. Phys.* **88**, 7240 (1988); (b) D. Frye, L. Lappierre, and H. L. Dai, *ibid.* **89**, 2609 (1988).

20. See, for example, P. R. Griffiths, *Chemical Infrared Fourier Transform Spectroscopy* (John Wiley and Sons, New York, 1975).

21. For a recent review of time-resolved Fourier transform spectroscopy, see J. J. Sloan and E. J. Kruus in *Time-resolved Spectroscopy*, Advances in Spectroscopy, Volume 18, Eds. R. J. H. Clark and R. E. Hester, (John Wiley and Sons, 1989), 219–253.

22. (a) A. Mantz, *Appl. Spectrosc.* **30**, 459 (1976); (b) A. Mantz, *Appl. Opt.* **17**, 1347 (1978).

23. A. A. Garrison, R. A. Crocombe, G. Mamantov, and J. A. de Haseth, *Appl. Spectrosc.* **34**, 399 (1980).

24. B. D. Moore, M. Polaikoff, M. B. Simpson, and J. J. Turner, *J. Phys. Chem.* **89**, 850 (1985).

25. (a) G. E. Caledonia, B. D. Green, and R. E. Murphy, *J. Chem. Phys.* **71**, 4369 (1979); (b) B. D. Green, G. E. Caledonia, R. E. Murphy, and F. X. Robert, *J. Chem. Phys.* **76**, 2441 (1982).

26. (a) P. M. Aker and J. J. Sloan, *J. Chem. Phys.* **85**, 1412 (1986); (b) P. M. Aker, B. I. Niefer, J. J. Sloan, and H. Heydtmann, *ibid.* **87**, 203 (1987); (c) E. J. Kruus, B. I. Niefer, and J. J. Sloan, *ibid.* **88**, 985 (1988).

27. (a) T. R. Fletcher and S. R. Leone, *J. Chem. Phys.* **88**, 4720 (1988); (b) T. R. Fletcher and S. R. Leone, *ibid.* **90**, 871 (1989); (c) E. L. Woodbridge, M. N. R. Ashfold, and S. R. Leone, *ibid.* **94**, 4195 (1991).

28. G. Hancock and D. Heard, *Chem. Phys. Lett.* **158**, 167 (1989).

29. (a) R. A. Palmer, C. J. Manning, J. A. Rzepiela, J. M. Widder, and J. L. Chao, *Appl. Spectrosc.* **43**, 193 (1989); (b) V. G. Gregoriou, J. L. Chao, H. Toriami, and R. A. Palmer, *Chem. Phys. Lett.* **179**, 491 (1991); (c) C. J. Manning, R. A. Palmer, and J. L. Chao, *Rev. Sci. Instrum.* **62**, 1219 (1991).

30. W. Uhmann, A. Becker, C. Taran, and F. Siebert, *Appl. Spectrosc.* **45**, 390 (1991).

31. G. V. Hartland, D. Qin, and H. L. Dai, *J. Chem. Phys.* **98**, 2469 (1993).

32. G. Herzberg and J. W. C. Johns, *Proc. R. Soc. London Ser.* **A295**, 107 (1966).

33. H. Petek, D. J. Nesbitt, D. C. Darwin, and C. B. Moore, *J. Chem. Phys.* **86**, 1172 (1987).

34. P. R. Bunker and T. J. Sears, *J. Chem. Phys.* **83**, 4866 (1985).

35. G. Duxbury and Ch. Jungen, *Mol. Phys.* **63**, 981 (1988).

36. D. Feldmann, K. Meier, R. Schmeidl, and K. H. Welge, *Chem. Phys. Lett.* **60**, 30 (1978).

37. W. H. Green, N. C. Handy, D. J. Knowles, and S. Carter, *J. Chem. Phys.* **94**, 118 (1991).

38. C. F. Bender, H. F. Schaefer, D. R. Franceschetti, and L. C. Allen, *J. Am. Chem. Soc.* **94**, 6888 (1972).

39. H. Petek, D. J. Nesbitt, C. B. Moore, F. W. Birss, and D. A. Ramsay, *J. Chem. Phys.* **86**, 1189 (1987).

40. (a) H. Petek, D. J. Nesbitt, P. R. Ogilby, and C. B. Moore, *J. Phys. Chem.* **83**, 5367 (1983); (b) H. Petek, D. J. Nesbitt, D. C. Darwin, P. R. Ogilby, C. B. Moore, and A. Ramsay, *ibid.* **91**, 6566 (1989).

41. W. Xie, C. Harkin, H.-L. Dai, W. H. Green, Q. K. Cheng, and A. J. Mahoney, *J. Mol. Spectrosc.* **138**, 596 (1989).

42. D. Qin, G. V. Hartland, and H. L. Dai, *J. Mol. Spectrosc.*, accepted for publication.

43. P. R. Bunker, *Ann. Rev. Phys. Chem.* **34**, 59 (1983).

44. A. R. W. McKellar, P. R. Bunker, T. J. Sears, K. M. Evenson, T. J. Saykally, and S. R. Langhoff, *J. Chem. Phys.* **79**, 5251 (1983).

45. P. R. Bunker, private communication.

46. A. Messiah, *Quantum Mechanics,* Vol 1 (North-Holland, Amsterdam, 1972).

47. A. Alijah and G. Duxbury, *Mol. Phys.* **70**, 605 (1990).

48. T. Barrow, R. N. Dixon, and G. Duxbury, *Mol. Phys.* **27**, 1217 (1974).

49. (a) Ch. Jungen and A. J. Merer, in *Molecular Spectroscopy, Modern Research,* Vol II, ed. K. Narahari Rao (Academic Press, 1976), p. 127; (b) Ch. Jungen and A. J. Merer, *Mol. Phys.* **40**, 1 (1980).

50. R. N. Dixon, *Trans. Faraday Soc.* **60**, 1363 (1964).

51. Ch. Jungen, D. N. Malm, and A. J. Merer, *Can. J. Phys.* **51**, 1471 (1973).

52. W. Xie, Ph. D. dissertation, University of Pennsylvania, 1992.

53. J. T. Hougen and J. K. G. Watson, *Can. J. Phys.* **43**, 298 (1965).

54. G. V. Hartland, W. Xie, D. Qin, and H.-L. Dai, *J. Chem. Phys.* **97**, 7010 (1992).

55. R. N. Zare, *Angular Momentum* (John Wiley and Sons, New York, 1988).

56. C. H. Townes and A. L. Schawlow, *Microwave Spectroscopy* (Dover Publications, New York, 1975).

57. G. W. Loge and C. S. Parmenter, *J. Chem. Phys.* **74**, 29 (1981).

58. (a) J. K. G. Watson, *J. Chem. Phys.* **46**, 1935 (1967); (b) J. K. G. Watson, *Vib. Spectra Struct.* **6**, 1 (1977).

59. J. Danon, S. V. Filseth, D. Feldmann, H. Zacharias, C. H. Dugan, and K. H. Welge, *Chem. Phys.* **29**, 345 (1978).

60. J. M. Figuera, I. Garcia-Moreno, J. C. Rodriguez, and H. A. Zeaiter, *Chem. Phys. Lett.* **183**, 577 (1991).

61. G. V. Hartland, W. Xie, D. Qin, and H.-L. Dai, *J. Chem. Phys.* **98**, 6906 (1993).

62. T. Oka, *Advances in Atomic and Molecular Physics* (Academic, New York) **9**, 127 (1973).

63. P. R. Bunker, *Molecular Symmetry and Spectroscopy* (Academic Press, San Diego, 1979).

64. R. D. Levine and R. B. Bernstein, *Molecular Reaction Dynamics and Chemical Reactivity* (Oxford University Press, Oxford, 1987).

65. The hard-sphere cross-section for collisions between $\tilde{b}^1 B_1$ CH_2 and ketene was calculated from the data given in: J. T. Edwards, *J. Chem. Ed.* **47**, 261 (1970).

66. E. R. Lovejoy, R. A. Alvarez, and C. B. Moore, *Chem. Phys. Lett.* **174**, 151 (1990).

67. S. L. Coy, S. D. Halle, J. L. Kinsey, and R. W. Field, *J. Mol. Spectrosc.* **153**, 340 (1992).

68. (a) W. D. Lawrance and A. E. W. Knight, *J. Chem. Phys.* **79**, 6030 (1983); (b) J. W. Thoman, S. H. Kable, A. B. Rock, and A. E. W. Knight, *ibid.* **85**, 6243 (1986); (c) S. H. Kable, J. W. Thoman, and A. E. W. Knight, *ibid.* **88**, 4748 (1988); (d) S. H. Kable and A. E. W. Knight, *ibid.* **93**, 3151 (1990).

69. (a) X. Yang and A. M. Wodtke, *J. Chem. Phys* **92**, 116 (1990); (b) X. Yang, E. H. Kim, and A. M. Wodtke, *ibid.* **93**, 4483 (1990); (c) X. Yang, E. H. Kim, and A. M. Wodtke, *ibid.* **96**, 5111 (1992).

70. (a) F. Temps, S. Halle, P. H. Vaccaro, R. W. Field, and J. L. Kinsey, *J. Chem. Phys.* **87**, 1895 (1987); (b) F. Temps, S. Halle, P. H. Vaccaro, R. W. Field, and J. L. Kinsey, *J. Chem. Soc. Faraday Trans. 2* **84**, 1457 (1988).

CHAPTER 6

STUDIES OF THE RENNER–TELLER EFFECT
IN NCO BY SEP SPECTROSCOPY

Ming Wu and Trevor J. Sears

Department of Chemistry,
Brookhaven National Laboratory,
Upton, NY 11973, USA

Contents

1. Introduction

The Renner–Teller effect (RTE) is one of the best characterized breakdowns of the Born–Oppenheimer approximation in molecular spectroscopy. In its simplest manifestation, the RTE occurs in linear triatomic molecules in orbitally degenerate electronic states and it is caused by a coupling between the electronic orbital angular momentum and the nuclear vibrational angular momentum associated with the bending vibration. The effect was originally discussed theoretically in 1934[1] when Renner formulated a model based on a perturbation operator which connects bending vibrational levels

223

associated with the two components of the electronic surface. The presence of this operator means that one cannot separate bending vibrational levels into those associated with each single electronic wavefunction, as would conventionally be possible in a Born–Oppenheimer separation. Instead, the vibronic wavefunctions contain linear combinations of both electronic wavefunctions and their associated vibrational functions. Renner introduced a coupling parameter, ϵ, which measures the gap between the two electronic potential surfaces and showed how the bending vibronic states could be expressed in terms of a convergent expansion in powers of ϵ.

Interestingly, the molecular system in which the consequences of the RTE were first recognized was in the spectrum of NH_2[2,3] which is a bent free radical in which the two lowest electronic states correlate with a single Π electronic state in the linear configuration. This was a situation not forseen by Renner in his original development and a model formulated by Pople and Longuet–Higgins[4] in which the coupling was described in terms of an electrostatic interaction between a single electron and the bent nuclear framework was used to describe the effect in NH_2. As a footnote, it was only a few years following the publication of Renner's paper that an analysis of the electronic spectrum of CO_2^+ was reported[5,6]. The ground $\tilde{X}^2\Pi$ state of CO_2^+ is now known[7] to represent a classic example of the RTE, exactly of the type predicted by Renner. At the time of the original spectroscopic work, however, many bands remained unassigned and the presence of the RTE went unrecognized.

Since the original work on NH_2, there have been a vast number of experimental and theoretical studies of the RTE and its different manifestations. Most of the molecular species known to exhibit the effect are free radicals and other transient molecules. The sensitive spectroscopic techniques required to measure their spectra have provided the experimental data which have stimulated significant theoretical advances in the description of the RTE during the past 30 years. While the RTE may seem to be a spectroscopic curiosity, it is responsible for the complexity of the spectra of such well-known species as NO_2. Also, in a way completely unforeseen by spectroscopists, the mathematical formulation developed to describe the RTE in triatomic molecules has proved equivalent to that required to describe inelastic scattering between a diatomic radical in a $^2\Pi$ state and a closed shell atom.[8] The term dynamical RTE has been coined to describe the nature of the scattering process in this case.[9]

Several reviews concerned with the development of theoretical models to describe the RTE and their applications to spectroscopic examples have been published. They include a section in a wider ranging review of nuclear electronic coupling by Longuet–Higgins,[10] Jungen and Merer's discussion of the RTE[11] and, more recently, a review by Brown and Jørgensen[12] which returned to Renner's original formulation of the problem and presented a logical perturbation theory construction of the model. Other developments have included the incorporation of large amplitude bending[13,14] and the addition of some smaller correction terms.[15–18]

On the experimental side, much interest has centered on a series of 15 valence electron triatomic radicals exemplified by BO_2, $CO_2{}^+$ and NCO. These species all possess linear $^2\Pi$ ground electronic states and therefore represent classic examples of the effect. They also do not contain hydrogen atoms so that the models incorporating large amplitude motion[14] need not be used, at least for the lower bending vibronic states. At the simplest level, the presence of the RTE in these species is demonstrated by the strong lifting of the degeneracies associated with the bending vibration in the molecules due to coupling between the angular momentum associated with the bending motion and the electronic orbital angular momentum. The different vibronic states produced by this coupling and the splittings between them give information on the RT coupling in the molecule and its interaction with other couplings (for example, the spin–orbit effect) present in the molecule.

In this chapter, we describe experiments designed to probe the vibronic energy levels in radicals subject to a RTE and have chosen NCO as the primary example. The emphasis is on experimental results obtained using stimulated emission pumping (SEP) spectroscopy which is a powerful tool for accessing the higher vibronic states in these molecules, for which it has traditionally been difficult to obtain high resolution data. The reader is referred to earlier papers discussed above for rigorous discussions of the derivation of the quantum mechanical model required to analyze the data. Here, we discuss the results necessary to interpret the data and outline how the calculations are performed. In the final section, we discuss the additional experimental information that is required before the harmonic and anharmonic contributions to the bending potential of NCO can be fully characterized.

2. Experimental Details

The technique of SEP spectroscopy[19] has already been described in the literature and, during the past decade, the method has been demonstrated to be a unique tool for studies of spectroscopy and dynamics[20] of vibrational levels at energies far above the vibrational potential minimum. There are two transitions involved in the SEP process. The pump transition excites molecules out of a low lying level to a specific rotation–vibration level in an excited electronic state, while the dump transition drives some molecules from the electronically excited rotation–vibration level back down to a vibrationally excited rotational level in the ground electronic state. When a pump transition is selected, the SEP spectrum is usually obtained by monitoring depletion of the pump laser induced fluorescence (LIF) as the dump laser is scanned. The fluorescence depletion depends on the dump laser power, the vibrational Franck–Condon factors, and the rotational Hönl–London factors for the downward transition.

In terms of population transfer in an SEP process, the stronger the fluorescence depletion the more population transfer to the final SEP prepared state. The maximum population transfer in a typical SEP experiment has been estimated to be approximately 30%.[21] The shot-to-shot fluctuation in the commercial dye lasers and source instability in the production of the transient species of interest can easily reach at least one third or more of this value, and a reference signal is required to correct for these and other fluctuations in order to observe SEP signals in most cases. Several referencing schemes have been reported. In a dual cell scheme, the pump and dump laser beams propagate through two identical gas sample cells. The two beams are crossed in the signal cell, while they are arranged parallel to one another, but not overlapping in the reference cell. The ratio or difference of the signals from the two cells yields the SEP signal corrected for such factors as laser intensities and source instability. When the dump laser causes strong LIF, the dual cell scheme is limited. An alternative scheme used for extracting weak SEP signals in the case of strong dump LIF is to block the pump laser pulse on every other shot using a chopper. The two resultant florescence signals, $I_{\mathrm{pump\&dump}}$ and I_{dump} are subtracted and normalized to the average value of I_{pump} to yield the desired SEP signal.[22] Another way to reduce the effect of dump LIF is to discriminate against the unwanted signal by using bandpass filters or a small monochromator set to accept fluorescence only at wavelengths shorter than the dump laser wavelength. The SEP spectrum is recorded by monitoring the desired pump LIF feature while the dump laser is scanning.

If the fluorescence lifetime of the molecule under study is long enough, a second, time-gated reference scheme is possible. In this scheme, the early portion of the fluorescence decay, induced solely by the pump laser, is measured using a gated integrator and serves as the reference channel; the fluorescence signal arising from both the pump and dump laser is measured using a second gated integrator and is the signal channel. The typical delay between the pump and dump laser is 20–50 ns. The ratio of the two signals yields the percent of fluorescence dip when the dump laser scans through the resonance. The ratio corrects for fluctuations in laser intensity and in the source of the transient species. This scheme is suitable for transient species in a supersonic jet because the experimental apparatus can be simpler than for a dual cell scheme.

The experimental apparatus used in our NCO SEP experiments is as follows. A cylindrical stainless steel vacuum chamber which is 12 in. high with 13.25 in. top and bottom flanges was pumped by an expanded 10 in. diffusion pump (Torr Vacuum Co.) backed by a mechanical roughing pump (Varian SD-700). A water cooled baffle was used to prevent diffusion pump oil backstreaming into the chamber. The supersonic free jet expansion entered from the top of the chamber and was controlled by a pulsed valve (General Valve, Series-9). A nozzle diameter of 0.25 mm was employed with backing pressures of 60–100 psi. The steady state pressure in the experimental chamber with the pulsed valve off or on was 2×10^{-7} or 4×10^{-5} torr, respectively, at the 10 Hz operation rate used. There were two 12 in. long baffle arms with Brewster angle windows mounted on opposite sides of the chamber. Each arm contained ten slotted disks to allow passage of both the photolysis and probe lasers at the proper separation and still decrease scattering light. Each disk had a 0.25 in. slot machined from its edge terminating in a 0.25 in. diameter hole in its center. The entire arm assembly was coated with colloidal graphite to reduce reflections. A photomultiplier tube (EMI 9954QB) viewed the center of the chamber via an f/1 lens system from one side perpendicular to the baffle arms and a 0.32 m monochromator (ISA HR320) with a 2400 grooves/mm grating was mounted opposite to it on the other side of the chamber. An f/4 lens system imaged light from the center of the chamber onto the entrance slit of the monochromator and allowed the collection of dispersed fluorescence

(DF) spectra following laser excitation of single vibronic levels in addition to discriminating against dump LIF in some SEP experiments as discussed above.

The photolysis laser beam was provided by an excimer laser (Questek 2260) operating on the ArF transition (193 nm). The pump and dump laser beams came from two pulsed dye lasers (Lambda–Physik FL3002E and Lumonics HD 300), pumped by two separate XeCl excimer lasers (Questek 2260). The use of two separate pump lasers avoided the need for a long optical delay path for the dump laser and increased the pulse energies available from it. Strong dump laser intensity is required, particularly for weak Franck–Condon transitions. The linewidth of the pump light from the Lambda–Physik FL3002E is about 0.15 cm^{-1} operating without the intracavity etalon, while the dump laser linewidth from the Lumonics HD 300 is about 0.08 cm^{-1}. The photolysis and probe laser beams passed from opposite directions through the baffle arms. The 193 nm photolysis laser beam was focussed at the pinhole orifice using a 75 cm focal length lens. The pump and dump dye laser beams were collimated by telescopes and crossed at an angle of approximately 1° in the center of the free jet, approximately 1.5 cm down stream from the nozzle orifice. The rotational temperature of the NCO free radical in this region was between 10 and 25 K. Both the probe beams were about 0.3 cm in diameter; however, the dump beam was arranged to be the slightly larger of the two, and dump beam energies used were typically in the range 5–10 mJ while the pump beam was typically less than one mJ. The laser wavelength was calibrated by measuring well-known molecular spectra or recording the optogalvanic spectra of argon or neon in a hollow cathode lamp.

Total or dispersed fluorescence was monitored by imaging the fluorescence on the PMT tube or on the entrance slit of the monochromator. The signal from the PMT tube was amplified by an ORTEC preamplifier then averaged using a gated integrator or a transient recorder and stored on a PC/AT type computer. Linewidths in excitation spectra or SEP spectra were smaller than 0.2 cm^{-1}, which is a combination of laser linewidth and Doppler broadening in the jet. Sometimes laser power broadening occurred on strong lines in an SEP spectrum because higher dump laser power was required to observe the weaker transitions in the same spectral range. The observed linewidths in the dispersed fluorescence spectra were about 0.13 nm in most cases, which was just sufficient to partially resolve P- and

R-branch rotational transitions at the peak of the rotational distribution of NCO in our source.

The NCO radical was formed in the reaction $CN + O_2 \rightarrow NCO + O$, where the CN radicals were produced by photolysis of cyanogen at 193 nm. The reactants were diluted in an inert gas, usually N_2 in the present experiments. The ratios in the gas mixture were about 15% $(CN)_2$, 30% O_2, and 55% N_2.

3. Results

3.1. *Renner–Teller Effect in $^2\Pi$ Triatomics*

In this section, we qualitatively describe the pattern of vibronic levels arising due to RT coupling between the electronic orbital angular momentum and the bending vibration in the NCO radical, which we shall use as a model for the discussion. We attempt to give an experimentalist's approach to the process, indicating how one may calculate the level positions for this type of molecule. In addition to the RT coupling in molecules of this sort, complications arise due to the presence of spin–orbit coupling, which is typically of the same order of magnitude, and Fermi-resonances between levels associated with the bending vibration and the lower frequency stretching vibration, v_1. Throughout this paper we shall use the symbols v and ℓ for the bending vibrational quantum number and the associated angular momentum, and v_1 and v_3 for the two stretching vibrations. The resultant of coupling ℓ and the orbital angular momentum, Λ, we label K. Levels with $K = 0, \pm 1, \pm 2 \ldots$ are labeled Σ, Π, $\Delta \ldots$ etc.

Figure 1 shows the vibronic energy levels associated with $v = 0, 1,$ and 2 in NCO. How does this pattern of vibronic levels arise? The vibronic energy levels shown in Fig. 1 can be thought of as the eigenvalues of the matrix of an effective Hamiltonian operator expressed in a suitable basis set. The development of this operator has been the focus of much effort since the work of Pople and Longuet–Higgins.[4] A feature of many of the discussions of the RTE in the literature is the extensive use of perturbation theory in order to reduce the numerical computation. This is of course achieved at the expense of considerable algebraic complexity which can confuse the reader and cause one to lose sight of the physics of the problem. The matrix of the RT Hamiltonian is, in fact, infinite. However, its truncation, at a point where the energy levels of interest can be calculated to a precision which is adequate compared to the experimental measurement accuracy, leads to matrices that are comparatively easily handled by modern workstation computers and it would seem that future work need rely much less on extensive perturbation theory developments.

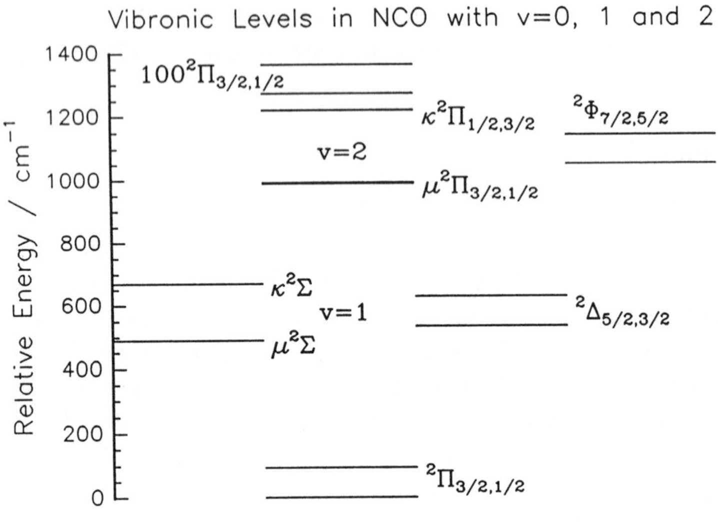

Fig. 1. Vibronic energy levels in the ground electronic state of NCO with $v = 0, 1$, and 2 and $v_1 = 1$. Their positions are taken from Refs. 24 and 33. The subscripts on the state labels are the P $(= \Lambda + \ell + \Sigma)$ values with the lowest energy component first. Note that the unique levels, with K $(= \Lambda + \ell) = v + 1$, have a spin–orbit splitting which is similar to the ground state while the $v = 2$ $^2\Pi$ levels have a much reduced splitting, which is too small to be resolved in the figure for the $(020)\kappa\,^2\Pi$ levels.

Following Jungen and Merer[11] and in the notation of Ref. 12, Table 1 gives the matrix elements of the RT and spin–orbit Hamiltonian around a block for a particular v and K value. In the absence of end-over-end rotation, K is a good quantum number. The matrix also includes a relatively small correction term first pointed out by Brown[15] and associated with the parameter g_K appearing in the matrix elements of Table 1. This arises due to coupling between the pair of nearly degenerate states and excited electronic states, and represents a Coriolis interaction between the states. The reason that the matrix becomes infinite is that the tri-diagonal blocks shown in Table 1 extend up to infinite vibrational quantum number. Diagonalization of a suitably truncated matrix for a given set of coupling parameters leads to the required energy levels as the eigenvalues of the matrix. Brown and Jørgensen[18] discussed additional contributions, such as anharmonicities associated with the RT perturbed bending vibration. We do not include the explicit matrix elements for these smaller contributions. If required, they may be included by simply incorporating the appropriate matrix elements from the general expressions given in Ref. 12.

One further interaction that cannot be neglected in NCO and related species is the Fermi resonance between the bending and lower frequency stretching vibration. This occurs because $\omega_1 \approx 2\omega_2$. Levels of the same symmetry with $v = v_1/2$ mutually perturb one another via a term in the cubic anharmonicity of the form $q_1 q^2$ where q_1 and q are the dimensionless coordinates associated with the symmetric stretching and bending vibrations, respectively. The treatment of Fermi-resonance interactions in $^2\Pi$ triatomic molecules has been considered by Hougen and Jesson[16] and Hougen.[23] For the kind of radicals considered here, all workers have adopted the approach developed originally by Hougen.[23] In this model, the effect is modeled by two parameters, W_1 and W_2 and the appropriate matrix elements are given in Table 2. W_1 is related to the average of the force constants ϕ_{122}, which are the coefficients of terms of the form $q_1 q^2$ in the potential for the two components of the Π state, while W_2 is related to their difference and is expected to be small in the present case. The effective operators whose coefficients are W_1 and W_2 also have non-zero matrix elements, not shown in Table 2, with $\Delta v_1 = \pm 1$, $\Delta v_2 = 0$ and $\Delta v_1 = \pm 1$, $\Delta v_2 = \pm 2$ that connect levels that lie far apart in energy. The effects of these matrix elements are conventionally collapsed into the diagonal anharmonicity constants, x_{11}, g_{21} and $\hat{g}_{21}$.[18] The RT, spin–orbit and Fermi resonance terms discussed above have been found by Patel–Misra *et al.*[24] to be adequate to model the vibronic energy levels in NCO up to some $2500 \ cm^{-1}$ in energy to an accuracy on the order of a few wavenumbers, which compared favorably with the experimental uncertainty in their work.

Inclusion of the rotational contributions to the energy of the RT split vibronic states was first considered in detail by Hougen[25] who derived explicit perturbation theory based expressions for the rotational energy levels in molecules of this type. More recently, Brown and Jørgensen[12] also discussed rotational contributions to the energy. If we ignore x- and y-type Coriolis contributions to the vibronic level energy, i.e., non J-dependent shifts to the level positions which can be treated as additional contributions to the anharmonicities, then the rotational contributions derive from the

Table 1. Central block of the Renner–Teller matrix for a given v, K.[a]

	$|-, v+2, K\rangle$[b]	$|+, v+2, K\rangle$	$|+, v, K\rangle$	$|-, v, K\rangle$	$|-, v-2, K\rangle$	$|+, v-2, K\rangle$
$\langle -, v+2, K|$	$\omega(\tilde{v}+2) - g_K K - AS'_z$					
$\langle +, v+2, K|$	$\varepsilon\omega[(\tilde{v}+2)^2 - K^2]^{1/2}/2$	$\omega(\tilde{v}+2) + g_K K + AS'_z$				
$\langle +, v, K|$	$\varepsilon\omega[(-\tilde{v}-K)(-\tilde{v}-2-K)]^{1/2}/4$		$\omega\tilde{v} + g_K K + AS'_z$			
$\langle -, v, K|$		$\varepsilon\omega[(-\tilde{v}+K)(-\tilde{v}-2+K)]^{1/2}/4$	$\varepsilon\omega[\tilde{v}^2 - K^2]^{1/2}/2$	$\omega\tilde{v} - g_K K - AS'_z$		
$\langle -, v-2, K|$			$\varepsilon\omega[(-\tilde{v}+2+K)(-\tilde{v}+K)]^{1/2}/4$		$\omega(\tilde{v}-2) - g_K K - AS'_z$	
$\langle +, v-2, K|$				$\varepsilon\omega[(-\tilde{v}+2-K)(-\tilde{v}-K)]^{1/2}/4$	$\varepsilon\omega[(\tilde{v}-2)^2 - K^2]^{1/2}/2$	$\omega(\tilde{v}-2) + g_K K + AS'_z$

[a]The matrix is actually doubled due to the two possible values of the expectation value of the electron spin angular momentum (S'_z) on the linear axis of the molecule. In the absence of rotation, the two matrices are not connected. Diagonal matrix elements contain only the bending vibrational energy. The symbol $\tilde{v}$ is used for $v+1$.

[b]Basis vectors are labelled by the electronic orbital angular momentum ($\Lambda = +/-1$), the bending vibrational quantum number, v, and the value of K ($= \Lambda + \ell$, where ℓ is the bending vibrational angular momentum).

Table 2. Matrix of the Fermi-resonance interaction Hamiltonian for NCO.[a]

	$\lvert +, v_1 + 1, v, K\rangle$	$\lvert -, v_1 + 1, v, K\rangle$	$\lvert +, v_1, v_2 + 2, K\rangle$	$\lvert -, v_1, v_2 + 2, K\rangle$
$\langle +, v_1 + 1, v, K\rvert$	$\omega_1(\tilde{v}_1 + 1) - \omega\tilde{v} + g_K K + AS'_z$			
$\langle -, v_1 + 1, v, K\rvert$	$\epsilon\omega[(\tilde{v}^2 - K^2]^{1/2}/2$	$\omega(\tilde{v}_1 + 1) + \omega\tilde{v} - g_K K + AS'_z$		
$\langle +, v_1, v_2 + 2, K\rvert$	$W_1[(v_1 + 1)(\tilde{v} + 2 - K)(\tilde{v} + K)]^{1/2}$	$W_2[(v_1 + 1)(\tilde{v} - K)(\tilde{v} - K + 2)]^{1/2}$	$\omega_1\tilde{v}_1 + \omega(\tilde{v} + 2) + g_K K + AS'_z$	
$\langle -, v_1, v_2 + 2, K\rvert$	$W_2[(v_1 + 1)(\tilde{v} + K)(\tilde{v} + K + 2)]^{1/2}$	$W_1[(v_1 + 1)(\tilde{v} + 2 + K)(\tilde{v} - K)]^{1/2}$	$\epsilon\omega[(\tilde{v} + 2)^2 - K^2]^{1/2}/2$	$\omega_1\tilde{v}_1 + \omega(\tilde{v} + 2) - g_K K - AS'_z$

[a]Only matrix elements connecting close lying vibrational states are given in this table. The basis vectors are labeled as in Table 1 with the addition of the stretching quantum number $v_1 \cdot \tilde{v}_1$ is equal to $v_1 + 1/2$. W_1 and W_2 are the interaction parameters defined by Hougen.[23]

matrix elements given in Table 3. The effect of adding the rotational operator is to double each matrix element of Tables 1 and 2 because we have to include the spin variable explicitly in the basis. Brown and Jørgensen[12] considered the problems associated with the introduction of the spin variable in detail. We note that basis states with different values of the quantum number $P(= \Lambda + \ell + \Sigma)$ are mixed; however, levels with $+P$ and $-P$ remain degenerate. Smaller J-dependent contributions to the energy such as those leading to K-type doubling[26] are not included in the matrix. They lead to additional non-zero matrix elements between basis states of different K and they may be included using the general matrix element expressions given in Ref. 26. The effects of these terms are only evident in the electronic spectra of NCO at high J values, and Bolman *et al.*[27] and Woodward *et al.*[28] included them in analysis of NCO spectra recorded at ambient temperatures. Their effects are expected to be too small to observe for the low rotational levels represented in the SEP spectra reported here.

3.2. *SEP Spectroscopy of NCO*

The NCO radical has been the subject of much spectroscopic research since the original work of Dixon[29] who recorded many bands in both the $\tilde{A} - \tilde{X}$ and $\tilde{B} - \tilde{X}$ band systems of the radical. More recent work on the electronic spectrum of NCO by Bolman *et al.*[27] concentrated on measurement of bending hot bands in the $\tilde{A}^2\Sigma^+ - \tilde{X}^2\Pi$ system, and this work provided a more accurate determination of the Renner parameter and firmly established the positions of the Σ and Δ states associated with $v = 1$ in the ground state. These levels are shown in Fig. 1. In an effort to obtain more information on the perturbed vibronic potentials and also determine anharmonic contributions to the vibrational energies, Woodward *et al.*[28] recorded DF spectra following excitation of rotational lines in the band origin of the $\tilde{A} - \tilde{X}$ system. The fluorescence was dispersed using a medium resolution monochromator and rotationally resolved information on vibronic levels associated with $v = 2$ and $v_1 = 1$ was obtained.

Very recently, Patel–Misra *et al.*[24] detected NCO radicals following the reaction between CN radicals, which were photolytically generated, and O_2. This reaction is highly exothermic and leads to product radicals containing a large degree of internal energy. The LIF excitation spectrum of the nascent radical product exhibited a large number of bending hot bands.

Table 3. Matrix elements of the effective rotational Hamiltonian for NCO.[a]

	$\|+, v, K, 1/2\rangle$	$\|-, v, K, 1/2\rangle$	$\|+, v, K, -1/2\rangle$	$\|-, v, K, -1/2\rangle$
$\|+, v, K, 1/2\rangle$	$\omega\tilde{v} + g_K K + A/2 + B_v(J(J+1) - P^2)$[b]			
$\|-, v, K, 1/2\rangle$	$\epsilon\omega[\tilde{v}^2 - K^2]^{1/2}/2$	$\omega\tilde{v} - g_K K - A/2 + B_v(J(J+1) - P^2)$		
$\|+, v, K, -1/2\rangle$	$B_v[J(J+1) - (K-1/2)(K+1/2)]^{1/2}$		$\omega\tilde{v} + g_K K - A/2 + B_v(J(J+1) - P^2)$	
$\|-, v, K, -1/2\rangle$		$B_v[J(J+1) - (-K - 1/2)(-K + 1/2)]^{1/2}$	$\epsilon\omega[\tilde{v}^2 - K^2]^{1/2}/2$	$\omega\tilde{v} - g_K K + A/2 + B_v(J(J+1) - P^2)$

[a]This matrix explicitly includes basis vectors with $S'_z(\Sigma) = 1/2$ and $-1/2$, i.e., $P = K + 1/2$ and $K - 1/2$, respectively. It is constructed from two 2×2 matrices that correspond to the central block of Table 1 with the two possible values of S'_z.
[b]$B_v = B_e - \alpha\tilde{v}_1 - \alpha\tilde{v} - \alpha\tilde{v}_3$. The α constants are the vibrational corrections to the rotational constant. The K-doubling contributions are not given here; they may be incorporated as described by Brown.[26] See also Refs. 28 and 33.

However, the rotational temperature of the source was similarly elevated and the positions of the different bands had to be estimated from band head information, limiting the accuracy and in some cases the certainty of the assignments. This work, together with an earlier study by Copeland and Crosley[30] who recorded LIF spectra in a flame, established positions of many excited bending levels in NCO for the first time.

Both of these studies were limited by the exceptionally complex nature of the spectrum of the radical at the high source temperatures used. SEP spectroscopy, in common with other double resonance techniques, has the great advantage that such complex spectra can be enormously simplified. In addition, the experimental techniques described in the previous section produce a sample at very low temperatures, contributing still further to the potential spectral simplification. Figure 2 shows an example of a rotationally cold but vibrationally only partially relaxed spectrum obtained in our apparatus. It is the $\tilde{A}(000)^2\Sigma^+ - \tilde{X}(010)\mu^2\Sigma^+$ band first analyzed by Bolman *et al.*[27] In contrast to the spectrum recorded in the earlier work, our spectrum shows near complete resolution at the lowest J values without the overlapping high J lines present in the ambient temperature spectrum. This figure also illustrates the fact that although the radicals are rotationally cold in our source, the vibrational cooling is much less complete and low-lying bending vibronic states especially retain significant population.

Information on low lying states can be obtained from these hot bands, but for the most part this information is available from the earlier conventional spectroscopic work on NCO. Also, the presence of hot bands can be a disadvantage when the dump laser excites hot band transitions in the same wavelength region as the expected SEP transitions. The major motivation of our work has been to obtain data on higher bending vibronic states, however. We can get broad, low resolution, information from DF spectra. Figure 3 shows examples of such spectra obtained when pumping different excited state vibrational levels in the $\tilde{A}^2\Sigma^+$ electronic state. For the strongest bands in these spectra, we were able to measure the band positions to an accuracy of between 5 and 10 cm^{-1}. However, our spectrometer does not have sufficient resolution to separate the individual rotational transitions under these band envelopes. This information is crucial to making unambiguous assignments to the vibronic levels when $v \geq 2$.

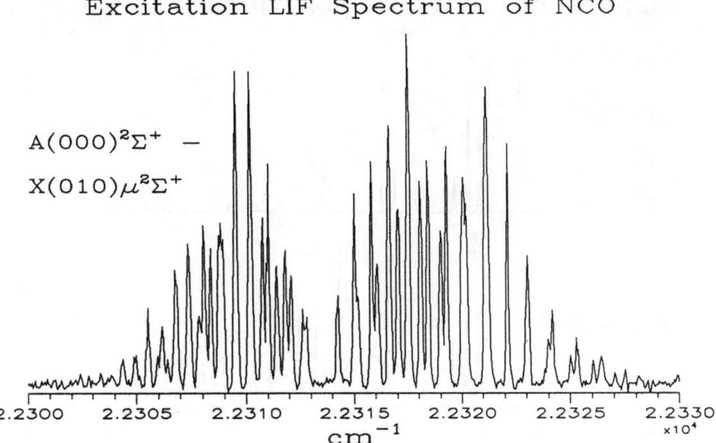

Fig. 2. Excitation spectrum of the $\tilde{A}(000)^2\Sigma^+$–$\tilde{X}(010)\mu^2\Sigma^+$ band of NCO observed in these experiments. The rotational temperature of between 10 and 20 K is much lower than the vibrational one in our source and bands originating in $v = 1$ retain significant intensity.

The lack of sufficient resolution can possibly be addressed in the future by replacing the current monochromator with a higher resolution grating or Fourier transform instrument. The experience of Woodward *et al.*,[28] however, suggests that this solution will only be possible for the strongest bands in the spectrum. It is also doubtful whether the source could be kept stable for the extended periods that high resolution scans over long wavelength intervals require. Very recently, Hartland *et al.*[31] have reported the observation of LIF DF spectra using a Fourier transform interferometer. The resolution obtained is significantly higher than can be practically expected from a dispersive instrument in this kind of experiment. However, sensitivity to weakly fluorescing transitions will remain a problem. SEP, on the other hand, can obtain high resolution information for even the weakest features in the DF spectra shown in Fig. 3.

In Fig. 3, the variation in the Franck–Condon activity obtained following excitation of the $(v_1 v v_3) = (100)$ and (120) levels of the excited electronic state can be seen. This electronic transition is predominantly diagonal with the strongest activity in the DF spectrum to vibronic bands derived from similar vibrational quantum numbers to the pumped level.   RT

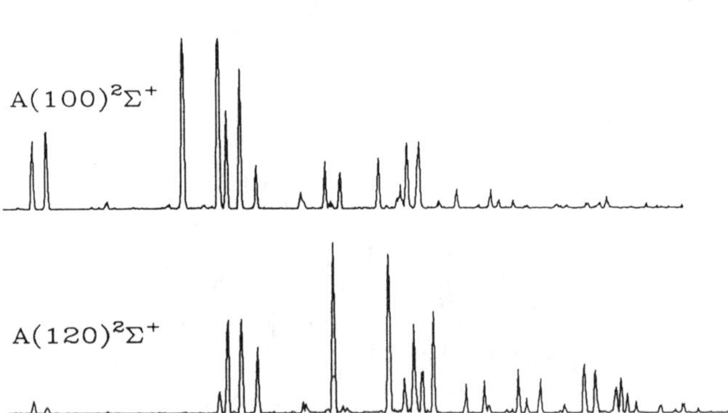

Fig. 3. Wavelength dispersed fluorescence spectra of NCO following excitation at the Q branch head of two different bands in the A–X system of NCO in our source. The different Franck–Condon distributions in the two cases are clearly seen. The x-axis scale represents energies in wavenumbers of the different vibronic levels above the lowest 000 $^2\Pi_{3/2}$ fine structure component in the ground electronic state. Bands in the 900–1200 region are due to vibronic levels associated with the 020/100 states while those between approximately 1950 and 2300 are predominantly associated with the 040/120/200 Fermi-resonant triad.

and Fermi-resonance in the ground state causes considerable mixing and complexity in the DF spectrum, particularly at higher energies. The most intense bands in these DF spectra are to vibronic levels of Π symmetry and Fig. 1 shows the positions of some of them relative to the $^2\Pi_{3/2}$ ground state. There is, however, appreciable intensity in some bands where the lower vibronic state is of $^2\Sigma$ character which are nominally forbidden in a perpendicular electronic spectrum such as this. Bolman and Brown[32] showed how the RT Hamiltonian mixes character of the vibrationless level of the $\tilde{A}$ electronic state (and excited $^2\Sigma$ states in general) into levels with one quantum of bending excitation in the ground state. This mechanism allows for only integral changes in v, however, and the extensive mixing within the ground state means many other transitions are observed. It was the observation of these nominally forbidden transitions that allowed the first reliable characterization of the RTE in NCO[27] and, in fact, Fig. 2 shows an example of such a forbidden band in excitation.

Here, we aim to use SEP spectroscopy to measure rovibronic energy levels at higher energies in order to characterize the RT perturbed bending potential surfaces more accurately. The first set of levels to which the technique was applied were those of $^2\Pi$ vibronic symmetry associated with $v = 2$. As Fig. 1 shows, these are spaced between approximately 1000 cm^{-1} and 1200 cm^{-1} above the vibronic ground state. SEP transitions to these levels were detected following excitation of single rotational lines in the $\tilde{A}(100)^2\Sigma^+$–$\tilde{X}(000)^2\Pi_{3/2}$ band around 415 nm. The DF spectrum from this level is shown in Fig. 3 and some examples of the observed SEP spectra are shown in Fig. 4.

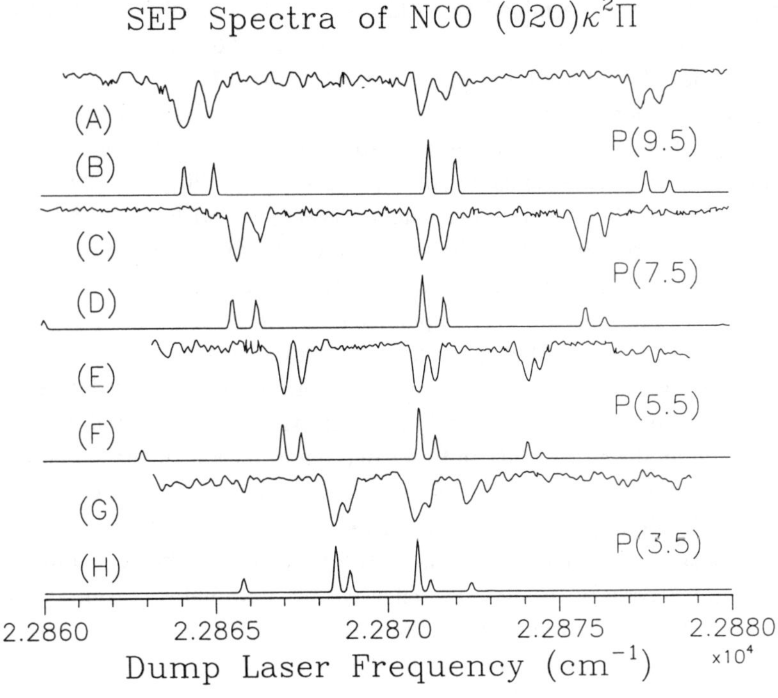

Fig. 4. Examples of SEP spectra to the $\tilde{X}$ (020) $\kappa^2\Pi$ vibronic levels. The pumped transitions are in the $\tilde{A}$ (100)$^2\Sigma^+$–$\tilde{X}^2\Pi_{3/2}$ band. The dump transitions are in the strong band at approximately 1100 cm^{-1} in the upper panel of Fig. 3. Calculated spectra are shown under each observed spectrum in the figure. The observed spectra are slightly laser power broadened.

Transitions to all four fine structure states associated with the (020) $\mu\,^2\Pi$ (lower) and $\kappa\,^2\Pi$ (upper) vibronic levels were detected.[33] These give precise information on the low J rotational structure of these vibronic states which in contrast to the unique states have case(b) rotational structure. Together with DF spectra to rotational levels in the (100) $^2\Pi$ fine structure components reported previously[28] these data allow a complete characterization of the the Fermi-resonant (020)/(100) $^2\Pi$ symmetry diad in NCO. Woodward *et al.*[28] described a fairly complete perturbation theory based effective Hamiltonian which can be used to model the rovibronic structure in these states and this Hamiltonian was used to fit all the existing data.[33] The quality of the fit was close to that expected on the basis of the measurement uncertainties and the absolute positions of the vibronic components was determined to within 0.1 cm^{-1}, a number limited by the calibration accuracy of the laser wavelengths. Table 4 shows the values obtained for the vibronic parameters from the fitting.

Similar experiments accessing $^2\Pi$ vibronic levels associated with four quanta of bending vibration were also performed. Here, there exists a Fermi-resonance triad consisting of the levels (040), (120), and (200). There are a total of 10 vibronic components associated with the triad. They lie between approximately 1950 cm^{-1} above the $^2\Pi_{3/2}$ vibrationless level, for the (040) $\mu\,^2\Pi$ levels, and 2625 cm^{-1}, for the (200) $^2\Pi_{1/2}$ spin–orbit component. Figure 5 shows an expanded energy level diagram in this region and some examples of SEP spectra to these levels are shown in Fig. 6. Wu, Northrup, and Sears[33] reported measurements that included data on 8 of the 10 vibronic components of $^2\Pi$ symmetry that are associated with these levels, and analyzed the data in terms of a perturbation theory based model that is an extension of that used for the analysis of the lower (020) and (100) diad. In the fitting, the Fermi-resonance parameters were kept fixed at the values obtained from the analysis of the lower energy data. However, it was found necessary to allow the effective RT parameter (taken as $\epsilon\omega$) to vary slightly and the final value found was some 1.7% smaller in magnitude than had been determined for the (020)/(100) levels. In retrospect, it seems likely that the model used was deficient in its neglect of the anharmonicity parameter $\hat{g}_4$, g_{21}, or $\hat{g}_{21}$, which would allow the precise change in the RT parameter that is apparently observed. It did not prove possible to determine an estimate for these parameters at the time due to the lack of sufficient data. We have recently been able to record some SEP spectra

Table 4. Molecular Parameters Obtained from the Fitting of SEP Data to $\tilde{X}$ (100)/(020) and (200)/(120)/(040) and (00v_3) $^2\Pi$ Vibronic States of NCO

Parameter[a]	cm^{-1}	Parameter[a]	cm^{-1}
$A_{(100)}$	−95.5871	$A_{(001)}$	−97.084(10)
$A_{(020)}$	−95.2609	$A_{(002)}$	−98.946(7)
$G_{(100)}$	1251.329(53)	$A_{(003)}$	−102.315(8)
$G_{(020)}$	1070.605(26)	$G_{(001)}$	1921.350(10)
$A_{(200)}$	−95.5871	$G_{(002)}$	3813.148(11)
$A_{(120)}$	−95.2609	$G_{(003)}$	5675.082(11)
$A_{(040)}$	−94.9460	$B_{(100)}$	0.38769(6)
$G_{(200)}$	2496.038(22)	$B_{(020)}$	0.39233(12)
$G_{(120)}$	2319.157(20)	$B_{(200)}$	0.38871(39)
$G_{(040)}$	2151.650(27)	$B_{(120)}$	0.39472(33)
W_1	25.951(37)	$B_{(040)}$	0.39711(40)
W_2	0.890(21)	$B_{(001)}$	0.386178[b]
$\epsilon\omega_2$[c]	−75.965(16)	$B_{(002)}$	0.382875(72)
$\epsilon\omega_2$[d]	−74.6600(63)	$B_{(003)}$	0.380075(84)

[a]The values with parenthesis were varied in the fitting.[33,34] One standard deviation of the least squares fit is given in parenthesis. The values fixed in the fit are taken from Refs. 33 and 34.
[b]The values are taken from *J. Mol. Spectrosc.* **92**, 485 (1982).
[c]The value of $\epsilon\omega_2$ was taken from the (100)/(020) fittings.
[d]The value of $\epsilon\omega_2$ was taken from the (200)/(120)/(040) fittings.

to the (040) $\mu^2\Pi$ vibronic components which completes the measurement of all the components of this symmetry in the triad. Examples of the new data are shown in Fig. 7.

In Table 5 we give the energies of the rotational levels in the (040) $\mu_2\Pi_{1/2,3/2}$ vibronic components relative to the $J = 3/2$ rotational level of the ground $^2\Pi_{3/2}$ state. These were obtained from SEP spectra such as those shown in Fig. 6 using the relation:

$$E_{J'} = E_{J''} + \nu_{\text{pump}} - \nu_{\text{dump}} \tag{1}$$

where $E_{J'}$ is the energy of the rotational level in Table 5, $E_{J''}$ is the energy of the pumped rotational level above $J = 3/2$, and ν_{pump} and ν_{dump} are the wavenumbers of the pump and dump lasers. The new (040) spectra were obtained following excitation of the $\tilde{A}(100)^2\Sigma^+$ level and the pump wavenumbers were obtained from the data reported by Dixon[29] following

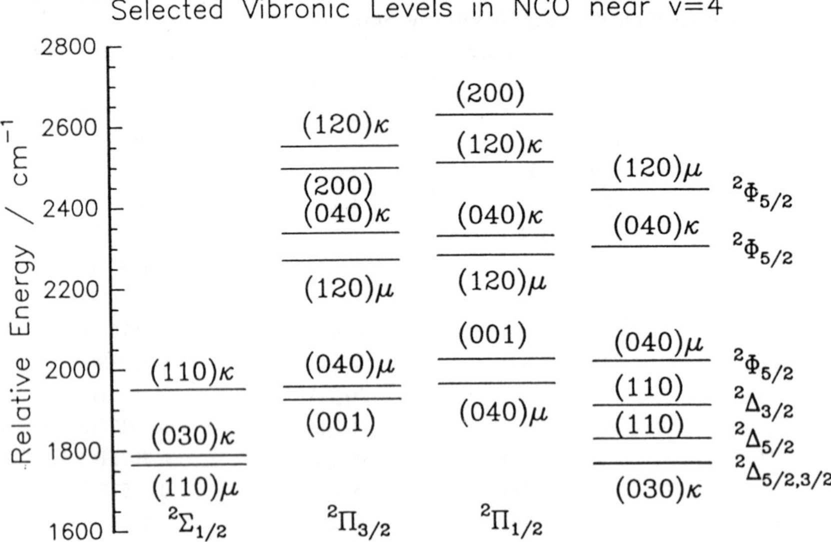

Fig. 5. Selected vibronic energy levels of NCO at energies near to the (200)/(120)/(040) states.

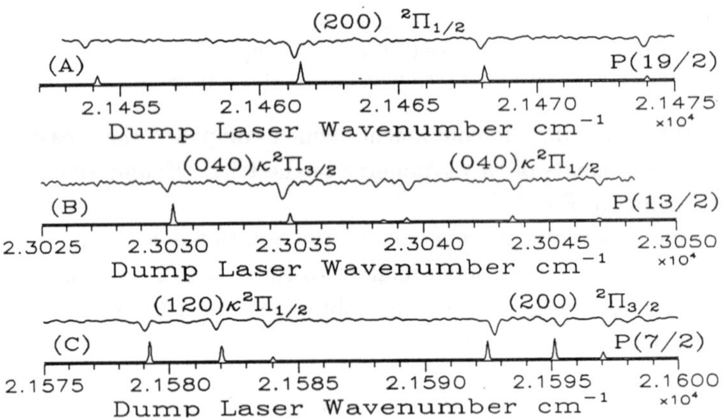

Fig. 6. Some examples of simulated SEP spectra compared to experiment. (A) The pumped transition is $P(19/2)\tilde{A}(100)^2\Sigma^+ - \tilde{X}(000)^2\Pi_{3/2}$, the dumped level is the $\tilde{X}(200)^2\Pi_{1/2}$. (B) The pumped transition is $P(13/2)\tilde{A}(200)^2\Sigma^+ - \tilde{X}(000)^2\Pi_{3/2}$, the dumped levels are the $\tilde{X}(040)\kappa^2\Pi$. (C) The pumped transition is $P(7/2)\tilde{A}(100)^2\Sigma^+ - \tilde{X}(000)^2\Pi_{3/2}$, the dumped levels are the $\tilde{X}(120)\kappa^2\Pi_{1/2}$ and $\tilde{X}(200)^2\Pi_{3/2}$. The upper trace is the experimental and the lower trace is the simulation in each pair of spectra.

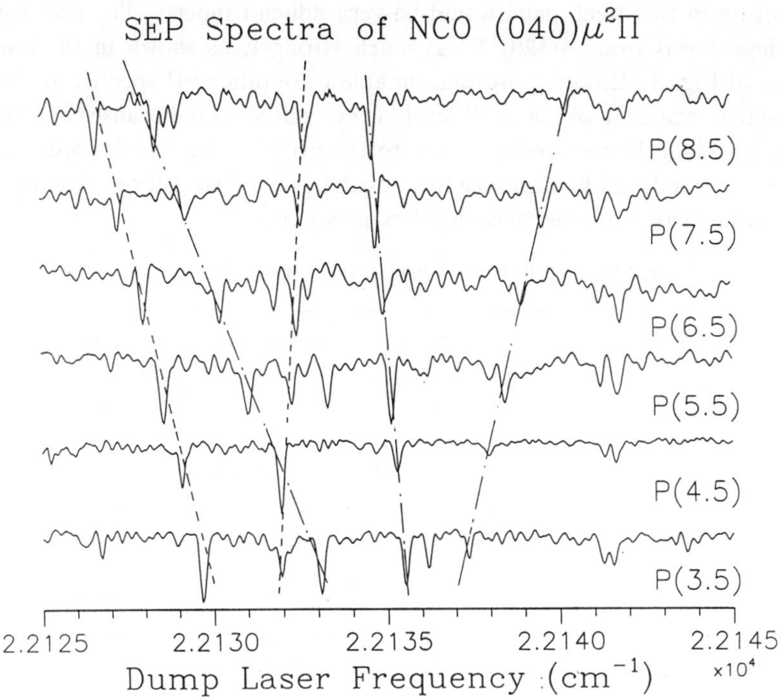

Fig. 7. SEP spectra to $\tilde{X}(040)\mu^2\Pi_{1/2}$ and $^2\Pi_{3/2}$ and $\tilde{X}(110)^2\kappa\Sigma$. The pumped transitions are in the $\tilde{A}(100)^2\Sigma^+ - \tilde{X}(000)^2\Pi_{3/2}$ band. The 040 $\mu^2\Pi$ transitions are indicated by the dotted lines except that the highest energy (smallest wavenumber) P branch transitions are at lower wavenumbers than are included in the figure. The lowest J member of this branch is discernable at approximately $22\,126.8$ cm^{-1} in the lowest trace. Other transitions not so indicated are to the (110) $\kappa^2\Sigma$ level and show small spin–rotation splittings that increase with increasing J.

a refitting to a modern effective Hamiltonian as described by Northrup *et al.*[34] Figure 3 shows that the Franck–Condon factors for transitions from (100) in the excited state back to the (040) $\mu^2\Pi$ levels are very small; the band in question lies between the spin–orbit components of the (001) level at approximately 1960 cm^{-1} from the ground state in the upper panel of Fig. 3. The observation of SEP spectra, given the weakness of the spontaneous fluorescence to these levels, is a nice demonstration of the sensitivity of the SEP technique, given the fact that it is not a zero background technique. To obtain DF spectra with sufficient resolution to separate the rotational structure in this weak band would be very difficult indeed. The transition to these levels from $\tilde{A}(120)^2\Sigma^+$ is much stronger, as shown in the lower panel of Fig. 3. However, we were unable to obtain SEP spectra to them following excitation of the (120) level in the excited state because the dump laser wavelengths required are close to the origin of the band system and dump laser induced fluorescence from weak hot band transitions dominated the, even weaker, fluorescence depeletion signals.

Table 5. SEP data of NCO in the $\tilde{X}(040)\mu^2\Pi$ States[a]

Pumped transition	Final-state rotational quantum number		
$P_1(J'')$	$J'' - 2$	$J'' - 1$	J''
	Final state $\tilde{X}(040)\mu^2\Pi_{3/2}$		
7/2	1955.34	1957.19	1959.63
9/2	1957.14	1959.91	1963.17
11/2	1959.88	1963.17	1967.32
13/2	1963.39	1967.43	1972.12
15/2	1967.38	1972.24	1977.72
17/2	1972.17	1977.81	1984.16
	Final state $\tilde{X}(040)\mu^2\Pi_{1/2}$		
7/2	1960.76	1963.00	1966.01
9/2	1963.10	1966.00	1969.84
11/2	1966.05	1969.77	1974.32
13/2	1969.92	1974.34	—
15/2	1974.39	1979.73	—
17/2	1979.72	1985.82	—

[a]The laser wavenumbers have been converted to level energies relative to the $\tilde{X}(000)$ $^2\Pi_{3/2}$ $J = 3/2$ ground state of the molecule using Eq. (1) and the ground state energies given in Table VII of Ref. 34.

The entries in Table 4 contain information on the force constants entering the expresion describing the potential energy function of the radical. Brown and Jørgensen[18] have discussed the effects of vibrational anharmonic terms on the energy levels of triatomic molecules in degenerate states and show that they depend on 14 anharmonicity parameters in addition to the three harmonic frequecies, ω_1, ω, and ω_3. These anharmonicity parameters provide the information on the shape of the RT perturbed potential whose determination is the ultimate aim of these spectroscopic measurements. We are still far from determining these parameters for NCO, but the energies $G(v_1, v, v_3)$ summarized in Table 4 allow us to go part way, although there is very little information on combinations with ν_3. From $G(001)$ and $G(002)$, we calculate $x_{33} = 14.78$ and $\omega_3 = 1950.902$ cm^{-1}. Note that an arithmetic error was made in Ref. 33 when calculating these numbers. The value of ω_3 calculated here is an effective one. Brown and Jørgensen show that we have in effect determined the quantity

$$\omega_3 + x_{13}/4 + \{g_{23} + B_e(\Omega_3^{-1} + \Omega_3)\} \, , \tag{2}$$

where $\Omega_3 = \omega_3/\omega$ and B_e is the equilibrium rotational constant. In order to extract the other anharmonicities entering this expression, information on combination bands involving ν_3 is required and we discuss potential methods for their measurement in the following section.

We have much more information on ν_1 and its combinations with the bending vibration. However, the contributions to the anharmonicity corrections are complicated in this case by the Fermi-resonance and the RT and spin–orbit splittings. The data for levels of Π vibronic symmetry is most extensive and the new data reported here completes the experimental information for the levels associated with the $v = 4$ Fermi-resonance triad. The data for the $v = 2$ and 4 Π vibronic levels was previously fit separately leading to the estimates of the depertubed energies $G_{v_1,v,0}$ given in Table 4. These quantities are related to the harmonic and anharmonic potential parameters by

$$G_{v_1,v,0} = \tilde{v}\omega + \tilde{v}_1\omega_1 + g_4\langle q^4\rangle + \hat{g}_4\langle q^4\sigma_z\rangle + \{g_{21}\langle q^2\rangle + \hat{g}_{21}\langle q^2\sigma_z\rangle\}\tilde{v}_1 \tag{3}$$

in the notation of Brown and Jørgensen.[18] We have neglected zero point energy contributions involving ν_3 and the g_K contribution in Eq. (3). The latter was explicitly included in the previous fitting[33] and held fixed at a value derived from the positions of the (010) $^2\Delta$ and $^2\Sigma$ states. Note that Eq. (3) includes only diagonal anharmonic contributions to the energies and neglects terms such as ϕ_{122} that contribute to the energy in second and higher order only. The major effects of some of these parameters are in the Fermi-resonance interactions where they are explicitly included as W_1 and W_2. To this approximation, Eq. (3) shows that there are six parameters that characterize the anharmonic energies while we have only determined the positions of five relevant levels of $^2\Pi$ symmetry. In addition, the assumed value of g_K is most likely affected by the neglect of anharmonic contributions to the (010) vibronic energies because the anharmonic constants are expected to be similar in size to g_K. In order to arrive at a consistent set of anharmonic constants, we need to fit the presently available data using a single, complete, harmonic and anharmonic Hamiltonian operator. Patel–Misra *et al.*[24] have come closest to this ideal in practice; however, the data available to them was of much lower precision than is now available and they did not include the anharmonicities g_{21} and $\hat{g}_{21}$ given in Eq. 3. This may explain the problems encountered in their analysis in that a negative value for g_K was apparently required. Such a result implies contamination of the ground state by one or more excited electronic states of $^2\Delta$ symmetry. There is no other evidence that such states in NCO perturb the ground state structure significantly and no evidence for them in any known spectra of the radical.

In addition to levels of $^2\Pi$ symmetry associated with $v_1, v, 0$ states characterized by SEP measurements, the data of Bolman *et al.*[27] fix the positions of the 010 $^2\Sigma$ and $^2\Delta$ states very precisely. We have also, very recently, obtained SEP data to levels of $^2\Sigma$ symmetry associated with the 030 and 110 Fermi-resonance diad. To our knowledge, these data represent the first accurate information on levels of this type in NCO or any of the related isoelectronic linear radicals. Rotational R-branch transitions to the 110 $\kappa^2\Sigma$ level are evident in Fig. 7 as the weak unidentified features at the highest wavenumbers in the experimental traces. The P-branch is weaker; the low J transitions start in the $P(7/2)$ spectrum at $22\,136.2$ cm^{-1} and move to longer wavelength with increasing J. SEP spectra to the 030 μ and $\kappa^2\Sigma$ states and the 110 $\mu^2\Sigma$ state were obtained following excitation of other vibronic states. The analysis of these data in conjunction with all the rotationally resolved data for the ground state of the NCO radical will form the basis of a future publication.[35]

4. Future Work

More than 30 years after Dixon's original work on the electronic spectrum of NCO,[29] the radical continues to provide new subtleties to occupy spectroscopists. SEP spectroscopy has provided a wealth of new information on higher vibronic levels in the radical which have proved to be ideal for assessing the utility and accuracy of model Hamiltonian descriptions of the RTE. As pointed out in the previous section, the currently existing data should now be sufficient to determine many of the harmonic and anharmonic force field parameters for the radical. From these quantities, a reliable bending potential for NCO may be constructed. This process is more difficult for NCO than was the case for C_3, studied previously,[36] because the combined effects of the RTE, Fermi-resonance and spin–orbit coupling in NCO conspire to make it harder to extract the effective vibrational energy level positions. Also, the SEP spectra to any given vibronic sub-level are weaker than for C_3 which has a $^1\Sigma$ ground electronic state. For C_3 it was possible to construct effective bending levels for many different stretching states and extract the equilibrium bending potential. We are still some way from this ideal for NCO and, in particular, further measurements of bending vibronic levels in combination with the ν_3 vibrational manifold are required.

Fortunately the experimental challenge is not insurmountable and SEP spectroscopy that accesses vibronic levels associated with 011 and the pair of levels 101/021 are certainly feasible. Low resolution DF spectra obtained by Northrup *et al.*[34] showed that several of the levels in question can be accessed following excitation of the $\tilde{A}(001)^2\Sigma^+$ and $\tilde{A}(002)^2\Sigma^+$ levels. It is more difficult to obtain information on vibronic levels with $K \geq 2$ in the ground state. Transitions to such levels from $\tilde{A}$ state levels of $^2\Sigma^+$ symmetry are forbidden. It is possible that excitation from hot bands, as exemplified in Fig. 2, may prove a feasible route to access $^2\Delta$ levels at least. Pumping the $\tilde{A}(010)^2\Pi_b-\tilde{X}(010)\mu^2\Sigma^+$ transition would, in principle, allow dump transitions to levels with $K = 2$ in the ground state. This possibility has yet to be tested in practice.

Acknowledgments

This work was carried out at Brookhaven National Laboratory and was performed under Contract No. DE-AC02-76CH00016 with the U. S. Department of Energy and supported by its Division of Chemical Sciences, Office of Basic Energy Sciences.

References

1. R. Renner, *Z. Phys.* **92**, 172 (1934).
2. K. Dressler and D. A. Ramsay, *J. Chem. Phys.* **27**, 971 (1957).
3. J. W. C. Johns, D. A. Ramsay, and S. C. Ross, *Can. J. Phys.* **54**, 1804 (1976).
4. J. A. Pople and H. C. Longuet-Higgins, *Mol. Phys.* **1**, 372 (1958).
5. S. Mrozowski, *Phys. Rev.* **60**, 730 (1941).
6. F. Bueso-Sanllehi, *Phys. Rev.* **60**, 556 (1941).
7. J. M. Frye and T. J. Sears, *Mol. Phys.* **62**, 919 (1987).
8. K. Liu, R. G. MacDonald, and A. F. Wagner, *Int. Rev. Phys. Chem.* **9**, 187 (1990).
9. R. G. MacDonald and K. Liu, *J. Chem. Phys.* **97**, 978 (1992).
10. H. C. Longuet-Higgins, *Adv. Spectrosc.* **2**, 429 (1961).
11. Ch. Jungen and A. J. Merer in "Molecular Spectroscopy: Modern Research" Vol II, ed. K. N. Rao, (Academic Press, NY, 1976) 127.
12. J. M. Brown and F. Jørgensen, *Adv. Chem. Phys.* **52**, 117 (1983), eds. I. Prigogine and S.A. Rice, (Wiley, NY).
13. T. Barrow, R. N. Dixon, and G. Duxbury, *Mol. Phys.* **27**, 1217 (1974).
14. Ch. Jungen and A. J. Merer, *Mol. Phys.* **40**, 1 (1980).
15. J. M. Brown, *J. Mol. Spectrosc.* **68**, 712 (1977).
16. J. T. Hougen and J. P. Jesson, *J. Chem. Phys.* **38**, 1524 (1963).
17. D. Gauyacq and Ch. Jungen, *Mol. Phys.* **41**, 383 (1980).
18. J. M. Brown and F. Jørgensen, *Mol. Phys.* **47**, 1065 (1982).
19. C. Kittrell, E. Abramson, J. L. Kinsey, S. A. McDonald, and R. W. Field, *J. Chem. Phys.* **75**, 2056 (1981).
20. See, for example, papers in *J. Opt. Soc. Amer.* **B7(9)**, 1802–1970 (1990).
21. G.-Z. He, A. Kuhn, S. Schiemann, and K. Bergmann, *J. Opt. Soc.* **B7**, 1960 (1990).
22. F. G. Celii and J. P. Maier, *Chem. Phys. Letts.* **166**, 517 (1990).
23. J. T. Hougen, *J. Chem. Phys.* **37**, 403 (1962).
24. D. Patel-Misra, D. G. Sauder, and P. J. Dagdigian, *J. Chem. Phys.* **93**, 5448 (1990).
25. J. T. Hougen, *J. Chem. Phys.* **36**, 519 (1962).
26. J. M. Brown, *J. Mol. Spectrosc.* **56**, 159 (1975).
27. P. S. H. Bolman, J. M. Brown, A. Carrington, I. Kopp, and D. A. Ramsay, *Proc. R. Soc. London Ser.* **A343**, 17 (1975).
28. D. R. Woodward, D. A. Fletcher, and J. M. Brown, *Mol. Phys.* **62**, 517 (1987); *Mol. Phys.* **68**, 261 (1989).
29. R. N. Dixon, *Philos. Trans. R. Soc. London Ser.* **A252**, 165 (1960).
30. R. A. Copeland and D. R. Crosley, *Can. J. Phys.* **62**, 1488 (1984).
31. G. V. Hartland, D. Qin, and H. L. Dai, *J. Chem. Phys.* **98**, 2469 (1993), G. V. Hartland and H. L. Dai, in "Molecular Spectrsocopy and Dynamics by Stimulated Emission Pumping", eds., H. L. Dai and R. W. Field (World Scientific, 1995).

32. P. S. H. Bolman and J. M. Brown, *Chem. Phys. Letts.* **21**, 213 (1973).

33. M. Wu, F. J. Northrup, and T. J. Sears, *J. Chem. Phys.* **97**, 4583 (1992).

34. F. J. Northrup, M. Wu, and T. J. Sears, *J. Chem. Phys.* **96**, 7218 (1992).

35. M. Wu and T. J. Sears, *Mol. Phys.* **82**, 503 (1994).

36. F. J. Northrup, T. J. Sears, and E. A. Rohlfing, *J. Mol. Spectrosc.* **145**, 74 (1991).

CHAPTER 7

CONTINUOUS WAVE PERTURBATION-FACILITATED OPTICAL–OPTICAL DOUBLE RESONANCE SPECTROSCOPY OF Na$_2$ and Li$_2$

Li Li

Department of Modern Applied Physics
Tsinghua University
Beijing 100084, China

and

Robert W. Field

Department of Chemistry
Massachusetts Institute of Technology
Cambridge, MA 02139, USA

Contents

1. Introduction

Stimulated Emission Pumping is a special case of Optical–Optical Double Resonance (OODR) spectroscopy.[1] The PUMP laser is tuned to selectively excite the target molecule from the electronic ground state (a Boltzmann distribution of e'', v'', J'' states) to a single, predetermined intermediate rovibronic state (e', v', J'). The frequency of a second laser, called the PROBE in OODR or the DUMP in SEP, is tuned to stimulate transitions from the selected e', v', J' intermediate to usually well-resolved and easily assignable final rovibronic states (e, v, J). Whether the transition stimulated by the second laser is upward, terminating in a highly excited (usually Rydberg) electronic state, or downward (SEP), terminating in highly excited vibrational levels of the electronic ground state (or other low-lying electronic states), is a matter of considerable practical but minimal fundamental importance. In this paper, we will be primarily concerned with the upward PROBE form of OODR spectroscopy, as recorded using two continuous wave (cw), single mode dye lasers.

We will discuss two types of cw OODR spectra: (i) sub–Doppler OODR fluorescence excitation spectra of high-lying electronic states, and (ii) OODR dispersed fluorescence spectra, which provide monochromator-limited resolution information about all of the lower-lying rovibronic levels sampled as terminal levels of spontaneous fluorescence transitions from the single OODR-selected (e, v, J) level. These two forms of OODR spectroscopy are very powerful complementary techniques for characterizing both high- and low-lying molecular electronic states.[1] The sub-Doppler resolution arises because the PUMP is a single mode, frequency stabilized (<1 MHz bandwidth) laser which excites the $e', v', J' \leftarrow e'', v'', J''$ transition only for molecules within a narrow range of the velocity projection on the PUMP propagation direction, v_{PUMP}. The co- or counter-propagating PROBE laser then excites the velocity-selected $e', v', J', v_{\text{PUMP}}$ intermediate state molecules to the final state without any change in longitudinal velocity. OODR spectra are, therefore, "Doppler-free" in terms of resolution but do contain residual Doppler shifts, which depend in a trivially calculable manner on the offset of the PUMP frequency from the center $(v_{\text{PUMP}} = 0)$ of the $e', v', J' \leftarrow e'', v'', J''$ transition. These Doppler shifts are often of diagnostic or spectroscopic value. When recorded in the forward or backward direction, OODR dispersed fluorescence spectra are also potentially "Doppler-free" in the same sense as OODR fluorescence spectra.

In the mid-1970's laser spectroscopic techniques began to be used to study the electronic spectra of Li_2 and Na_2. By the early 1980's, the low-lying $X^1\Sigma_g^+$, $A^1\Sigma_u^+$, and $B^1\Pi_u$ states of Na_2[2,3] and Li_2[4,5] had been well characterized by single-laser and modulated population techniques, and the singlet–gerade Rydberg states were being rapidly characterized by two-photon and OODR techniques.[6–10] The explosive growth of experimental information stimulated accurate *ab initio* electronic structure calculations.[11–15] A nearly complete set of *ab initio* potential energy curves for the valence and low-Rydberg states of Li_2 and Na_2 became available. Essentially, nothing was known from laser experiments about triplet electronic states, because the electronic ground state $(X^1\Sigma_g^+)$ is a singlet state and triplet $\leftarrow$ singlet transitions are spin-forbidden.

Triplet states are not merely three times more complicated than singlet states. They contain a great deal of diagnostically valuable information about the electronic structure which is not available from the study of singlet states alone: spin–orbit, spin–spin, spin–rotation, and electron spin–nuclear spin hyperfine splittings. Perturbation-Facilitated Optical–Optical Double Resonance (PFOODR) spectroscopy was developed in order to gain access to the triplet states while retaining the extreme spectral simplification typical of OODR. This spectral simplification is more valuable for triplet states than singlet states because of the vastly more complex nuclear spin, electron spin, and rovibronic structure of triplet states. The first PFOODR experiments were performed on Li_2 and Na_2.

More than 80 years ago the magnetic rotation spectrum of Na_2[16] was observed to contain anomalies that were later shown to arise from spin–orbit perturbations of the $A^1\Sigma_u^+$ state by the $b^3\Pi_u$ state.[17] Many $A^1\Sigma_u^+ \sim b^3\Pi_u$ mixed levels of Na_2 were observed and characterized by several groups in the early 1980's.[18–21] The key feature of PFOODR spectroscopy is that, by (asymptotically equal to) selecting a singlet$\sim$triplet mixed rovibronic level as the intermediate in an OODR scheme, *both* triplet and singlet e, v, J final levels become accessible. In the case of Na_2, it is possible to distinguish between transitions in the OODR excitation spectrum which terminate on $^3\Lambda_g$ versus $^1\Lambda_g$ states. The $^3\Lambda_g$ states decay radiatively into the $a^3\Sigma_u^+$ and $b^3\Pi_u$ states whereas the $^1\Lambda_g$ states decay into the higher lying $A^1\Sigma_u^+$ and $B^1\Pi_u$ states. Thus, by recording an OODR fluorescence excitation spectrum through a simple short-λ passing filter one obtains a high degree of discrimination against detection of transitions terminating in $^1\Lambda_g$ states and in favor of those terminating in $^3\Lambda_g$ states.

Figure 1 is a schematic energy level diagram which illustrates PFOODR fluorescence excitation and dispersed fluorescence spectroscopy of Na_2 and Li_2. Laser-1, the PUMP, excites from the $X^1\Sigma_g^+$ ground state to an $A^1\Sigma_u^+ \sim b^3\Pi_u$ mixed intermediate level. Laser-2, the PROBE, further excites to triplet–gerade states. The PFOODR excitation spectrum of $^3\Lambda_g$ ($\Lambda = 0, 1$) states is recorded by selectively detecting $^3\Lambda_g \rightarrow a^3\Sigma_u^+$ fluorescence. $^1\Lambda_g$ Rydberg states cannot decay radiatively into the $a^3\Sigma_u^+$ state (they can only decay into singlet–ungerade states, the lowest of which ($A^1\Sigma_u^+$) lies at considerably higher energy than the $a^3\Sigma_u^+$ state, thus transitions into $^3\Lambda_g$ states will be mostly eliminated from the PFOODR fluorescence excitation spectrum.

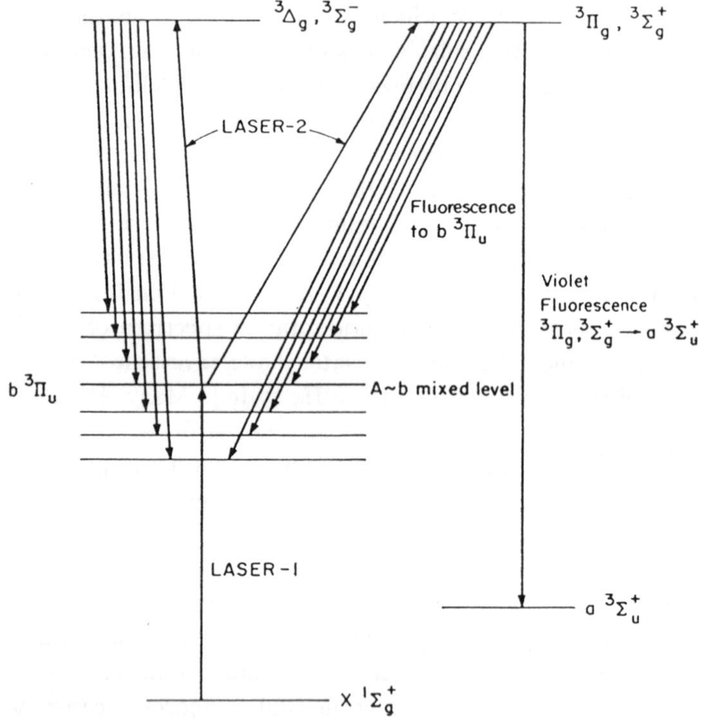

Fig. 1. Schematic diagram of PFOODR fluorescence excitation and resolved fluorescence spectroscopy of Na_2 and Li_2 triplet states.

PFOODR fluorescence excitation spectroscopy is both highly selective and sensitive. It is essentially a zero-background technique because the detected $^3\Lambda_g \rightarrow a^3\Sigma_u^+$ fluorescence is well to the blue of both laser frequencies. PFOODR excitation transitions via intermediate levels containing only 0.01% $A^1\Sigma_u^+$ character are readily detectable. The resultant factor of 10^{-4} in the PUMP transition probability is partially offset by the effect of the $\sim 10^4$ times longer radiative lifetime on the steady state population of the triplet intermediate level. However, the ability to use nearly unperturbed $b^3\Pi_u$ levels as PFOODR intermediates stems primarily from the strong discrimination against detection of fluorescence from $^1\Lambda_g$ upper states.

Figure 1 also illustrates how both perturbed and unperturbed levels of the $b^3\Pi_u$ state, and both bound and continuum regions of the $a^3\Sigma_u^+$ state can be studied by PFOODR dispersed fluorescence spectroscopy.

PFOODR spectroscopy has provided a wealth of information (in fact, almost everything that is known) about the triplet states of Na₂,[22–28] Li₂,[29–31] and K₂.[36] Nine triplet states of Na₂ ($a^3\Sigma_u^+$, $b^3\Pi_u$, $2,3,4\,^3\Sigma_g^+$, $2,3\,^3\Pi_g$, $1,2\,^3\Delta_g$), five triplet states of Li₂ ($a^3\Sigma_u^+$, $b^3\Pi_u$, $3\,^3\Sigma_g^+$, $2\,^3\Pi_g$, $1\,^3\Delta_g$) and five triplet states of K₂ ($a^3\Sigma_u^+$, $b^3\Pi_u$, and a number of rovibronic levels of the $4\,^3\Sigma_g^+$, $1\,^3\Sigma_g^-$, and $4\,^3\Pi_g$ states) have been observed and well characterized during the last decade. In the case of Li₂, owing to the extreme weakness of the spin–orbit interaction ($\zeta_{2p} = 0.2$ cm^{-1}), only two and three $A^1\Sigma_u^+$ rovibronic levels of ^{6}Li₂ and ^{7}Li₂, respectively, have been found simultaneously to have appreciable $b^3\Pi_u$ character, and to be free of accidental predissociation. (The accidental predissociation arises from the interaction of the $b^3\Pi_u$ state with the continuum of the $a^3\Sigma_u^+$ state. The lifetimes of predissociated levels are usually too short for these levels to be usable as PFOODR intermediates.) The PFOODR technique has been applied to many other diatomic and polyatomic molecules.

2. Experimental Details

Figure 2 illustrates the setup of a typical cw OODR experiment. Sodium (or lithium) vapor is generated in a heatpipe oven. 1–2 Torr of Ar or He buffer gas is used to maintain the temperature in the metal vapor region at ~ 700 K (or 1000 K for Li) and to protect the windows.

Two single-mode, frequency stabilized (<1 MHz spectral linewidth) dye lasers are used as PUMP and PROBE. The output power of the dye lasers is in the 10–700 mW range. After passing through a mechanical chopper,

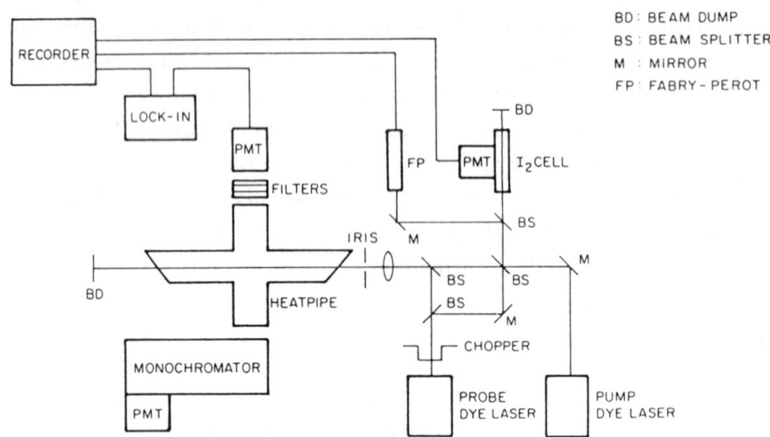

Fig. 2. A typical experimental setup for a cw PFOODR experiment on Na$_2$ and Li$_2$. If a linear heatpipe is used, an aluminum-coated mirror with a 4 mm diameter 45° hole through its center is placed in front of the entrance window of the heatpipe at a 45° angle in order to collect fluorescence (in the backward direction) and to transmit the co-propagating PUMP and PROBE beams. In the case of counter-propagating PUMP and PROBE beams, the two laser beams cross at an angle of $\sim$ 3 mrad near the center of the heatpipe.

the PUMP beam is combined with the PROBE beam using either a 50–50% beam splitter (if the frequencies of the two lasers are similar) or a dichroic mirror. The copropagating PUMP and PROBE beams are then focused into the center of the heat pipe. For some experiments, the counter-propagating PUMP and PROBE geometry is preferable. In that case, no beamsplitter or dichroic mirror is needed and the two beams are arranged to cross at a < 3 mrad angle in the center of the heatpipe.

The PUMP laser frequency is held fixed at the center of an $A^1\Sigma_u^+ \sim b^3\Pi_{\Omega u} J' \leftarrow X^1\Sigma_g^+ v'', J''$ transition while the PROBE laser frequency is scanned. In this way PFOODR transitions into high-lying $^3\Lambda_g$ states are detected by monitoring the $^3\Lambda_g$ ($\Lambda = 0$ or 1 only) $\rightarrow a^3\Sigma_u^+$ violet fluorescence (near 436 nm) using a filter/PMT/lock-in system. $^3\Sigma_g^-$ and $^3\Delta_g$ states cannot radiate directly to a $^3\Sigma_u^+$ state and the detected violet fluorescence must be induced by collision-induced transfer into $^3\Sigma_g^+$ or $^3\Pi_g$ states. As we will discuss in Sec. 5, for Na$_2$ and Li$_2$ the efficiency of collision-induced triplet $\leftrightarrow$ triplet transitions is typically much higher than that for singlet $\leftrightarrow$ triplet transitions. However, the most sensitive and selective method for detecting $^3\Sigma_g^-$ and $^3\Delta_g$ states in a PFOODR excitation spectrum is to monitor collisionally unrelaxed $^3\Lambda_g \rightarrow b^3\Pi_u$ fluorescence. This scheme also works well using only a short-λ passing filter because in Na₂ (Li₂) the $b^3\Pi_u$ $v = 0 - 13$ ($v = 0 - 6$) levels lie below $A^1\Sigma_u^+$ $v = 0$.

A PFOODR dispersed fluorescence spectrum is recorded by holding the PUMP and PROBE laser frequencies fixed on both steps of a PFOODR $^3\Lambda_g \leftarrow b^3\Pi_u \sim A^1\Sigma_u^+ \leftarrow X^1\Sigma_g^+$ excitation and directing the induced fluorescence into a scanning monochromator. The strongest fluorescence signals are obtained by collecting fluorescence in the backward direction using a concave mirror with a ~ 4 mm diameter 45° hole through its center. PFOODR dispersed fluorescence spectra are potentially of sub-Doppler resolution, but are typically recorded at monochromator-limited ~ 1 cm^{-1} resolution.

The PUMP and PROBE laser frequencies are calibrated (to ± 0.005 cm^{-1}) by recording simultaneously the PFOODR fluorescence excitation spectrum, the excitation spectrum of I₂, and transmission fringes from a 300 MHz free spectral range half-confocal etalon. The PFOODR dispersed fluorescence spectrum is calibrated by being recording simultaneously with atomic emission lines from a hollow cathode lamp.

3. Fine and Hyperfine Structure of Triplet Rydberg States

The PFOODR fluorescence excitation technique has been used to characterize the following states of Na₂ and Li₂ [where the adiabatic states are designated by a number, which specifies the energy rank at dissociation within the Λ–S–g/u symmetry class (1 means the lowest energy), and (redundantly) by the atomic dissociation asymptote]: Na₂ $2(3s + 4s)^3\Sigma_g^+,$[28] $3(3s + 3d)^3\Sigma_g^+,$[27] $4(3s + 4p)^3\Sigma_g^+,$[25,32] $2(3s + 3d)^3\Pi_g,$[22,25,33] $3(3s + 4p)^3\Pi_g,$[22,25] $1(3s + 3d)^3\Delta_g,$[25,34,35] and $2(3s + 4d)^3\Delta_g;$[26] [6,7]Li₂ $3(2s + 3p)^3\Sigma_g^+,$ $2(2p + 2p)^3\Pi_g,$ and $1(2s + 3d)^3\Delta_g.$[30,31]

The Na₂ $(\sigma_g 3s)(\delta_g 3d)$ $1^3\Delta_g$ state dissociates adiabatically to 3s + 3d separated atoms. The atomic Na 3d orbital has a very small spin–orbit splitting ($E_{5/2} - E_{3/2} = -0.0494$ cm^{-1}) and, therefore, the $1^3\Delta_g$ state is expected to be at the Hund's case (b) limit. The singly occupied $\sigma_g 3s$ valence orbital contributes zero to any electronically diagonal spin–orbit matrix element and its nominal character implies that it will also make a negligible contribution to any electronically off-diagonal spin–orbit matrix element. This simple atom-in-molecule based argument implies that the $1^3\Delta_g$ state will exhibit negligibly small electronic triplet splittings of either first-order $\mathbf{H}^{SO}$ type (A_Δ) or from second-order $\mathbf{H}^{SO} \times \mathbf{H}^{SO}$ and $\mathbf{H}^{SO} \times \mathbf{H}^{rot}$ contributions, respectively, to the effective spin–spin (λ_Δ) and spin–rotation (γ_Δ) constants. Every rotational level of the $1^3\Delta_g$ state will be well described by the pattern-forming N quantum number ($\mathbf{N} = \mathbf{J} - \mathbf{S}$) and will (except for the $N < 3$ levels) be sixfold degenerate. The six exactly degenerate fine structure components will each be described by a $\pm$ parity and $J = N, N \pm 1$ quantum number. These $J, \pm$ quantum numbers are rigorously good in the absence of external fields or hyperfine effects, thus the six fine structure components belong to six distinct, non-interacting symmetry classes.

Even in the absence of hyperfine effects, one expects to observe O,P,Q,R, and S form (respectively $N - J' = -2, -1, 0, 1, 2$) rotational transitions in a case (b) $^3\Delta_g \leftarrow$ case (a) $^3\Pi_{\Omega u}$ OODR spectrum if the PUMP populates a single Ω, J, parity component of the $^3\Pi$ intermediate state.[25] The presence of 5 groups of rotational transitions is consistent with the rigorous $\Delta J = 0, \pm 1, + \leftrightarrow -$ electric dipole selection rules. In the Na$_2$ $1^3\Delta_g \leftarrow b^3\Pi_{0u}$ and $1^3\Delta_g \leftarrow b^3\Pi_{1u}$ OODR spectra, respectively, O, P, Q, R and O, P, Q, R, S form transitions are observed. All of the $1^3\Delta_g \leftarrow b^3\Pi_u \Omega' = 0, 1$ rotational transitions exhibit resolvable fine/hyperfine structure, despite the expectation that A_Δ, λ_Δ, and γ_Δ are all too small to give rise to observable splittings. The hyperfine splittings in the $b^3\Pi_{0u}$ and $b^3\Pi_{1u}$ sub-states are known to be, respectively, non-negligible and negligible. Thus, it is difficult to interpret the observed structure on the $1^3\Delta_g \leftarrow b^3\Pi_{0u}$ rotational transitions as originating either in the $1^3\Delta_g$ or the $b^3\Pi_{0u}$ state.[25] However, owing to the absence of detectable hfs (< 15 MHz) in the $b^3\Pi_{1u}$ substate,[21] the observed structure on the $1^3\Delta_g \leftarrow b^3\Pi_{1u}$ transitions belongs entirely to the $1^3\Delta_g$ state and can be readily assigned.

Due to the nonzero electronic orbital ($\Lambda = 2$) and spin ($S = 1$) angular momenta in the $1^3\Delta_g$ state, electron–nuclear magnetic dipole interactions are expected to be the dominant source of hyperfine splittings. Figure 3 illustrates the usual magnetic hyperfine coupling schemes appropriate for the case (b) limit.[37] I is the total nuclear spin angular momentum of the molecule. The subscript β indicates that the nuclear spin, I, is not coupled to the internuclear axis but rather to N (case $b_{\beta N}$) or S (case $b_{\beta S}$) or J (case $b_{\beta J}$).

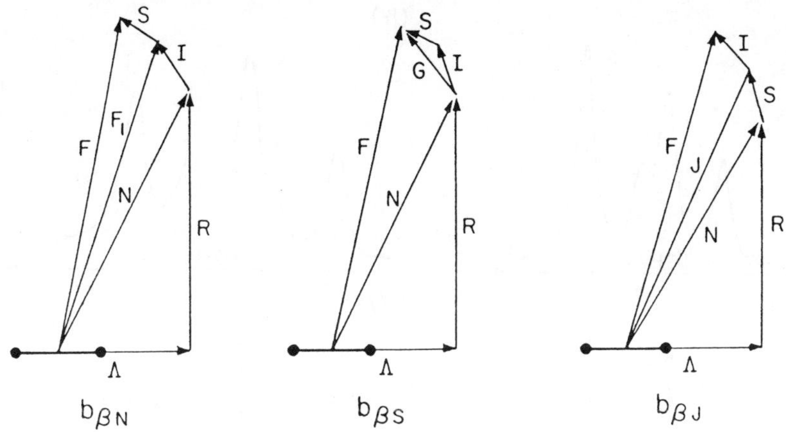

Fig. 3. Hyperfine coupling schemes for Hund's case b.

PFOODR spectra, such as those shown in Fig. 4, prove that the Na$_2$ $1^3\Delta_g$ state is a textbook example of case $b_{\beta S}$ coupling. In this scheme, I first couples to S to yield $\mathbf{G} = \mathbf{I} + \mathbf{S}$, and then G couples to N to yield $\mathbf{F} = \mathbf{N} + \mathbf{G}$. (Fig. 3b). The J and e/f quantum numbers are completely destroyed (by $\Delta N = 0$, $\Delta J = \pm 1$ matrix elements of the Fermi contact $bI\bullet S$ term within the two groups of 3 exactly degenerate same-N, same-parity $J = N, N - 1, N + 1$ fine structure components). Each rotational level, N, splits into components described by

$$E_{G,I,S} = (b/2)[G(G + 1) - I(I + 1) - S(S + 1)]$$

where b is independent of N and arises almost entirely from the singly occupied $\sigma_g 3s$ valence orbital common to all $(\sigma_g 3s)(n\ell\lambda_g)$ $^3\Lambda_g$ ($\ell = $ even) and $(\sigma_g 3s)(n\ell\lambda_g)$ $^3\Lambda_u$ ($\ell = $ odd) Rydberg states. The Fermi contact constants, b, for all Na$_2$ and Li$_2$ Rydberg states built on the $M_2^+ X^2\Sigma_g^+$ ion-core ground state are predicted to be equal to $\sim 1/4$ the value of the Fermi contact constant of the atomic M ^2S ground state[34] (for Li 2^2S $b = 152$ MHz and 402 MHz for ^{6}Li and ^{7}Li, respectively, for Na 3^2S $b = 886$ MHz).

Figures 4(a) and (b) show the observed OODR excitation spectra of the Na$_2$ $1^3\Delta_g$ $v = 53$, $N = 17 \leftarrow b^3\Pi_{1u}$ $v' = 21$, $J' = 17e,$s $[Q_s(17)]$ and $N = 15 \leftarrow J' = 16e,$a $[P_a(16)]$ transitions. The Na atom has $I = 3/2$

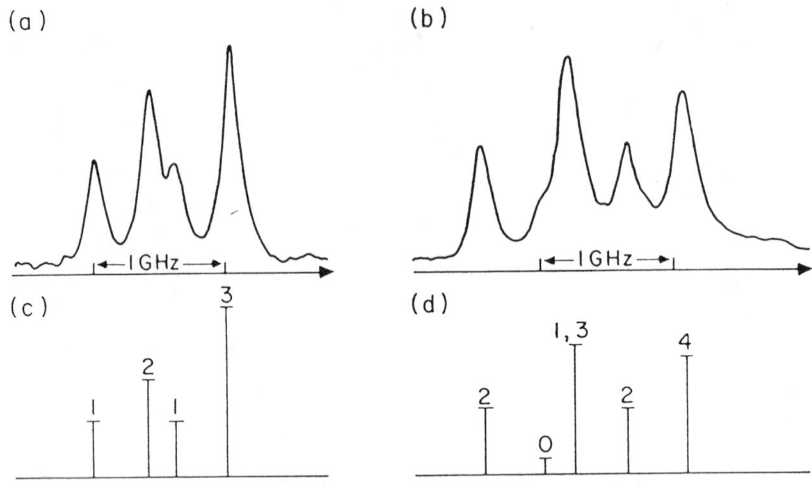

Fig. 4. Observed and computed relative intensities for subfeatures of various $1^3\Delta g \leftarrow b^3\Pi_{1u}$ $(N \leftarrow J')$ transitions: (a) observed $1^3\Delta_g$ $v = 53$, $N = 17 \leftarrow b^3\Pi_{1u}$ $v' = 21$, $J' = 17\,e,\mathbf{s}$ hypermultiplet; (b) observed $1^3\Delta_g v = 53, N = 15 \leftarrow b^3\Pi_{1u}v' = 21, J' = 16\,e,\mathbf{a}$ hypermultiplet; (c) calculated G component relative intensities (summed over N–$G \leq F \leq N + G$) and frequency intervals for spectrum (a); (d) calculated G component relative intensities and frequency intervals for spectrum (b). In (c) and (d) the hyperfine components are labeled by the G quantum numbers in the $1^3\Delta_g$ state. For an odd $J'\,e,\mathbf{s}$ level, the possible I values are $I = 0, 2$, thus $G = 1, 1, 2, 3$; for an even $J'\,e,\mathbf{a}$ level, $I = 1, 3$, thus $G = 0, 1, 2, 2, 3, 4$.

(a Fermion) and therefore $I = 0, 2$ for symmetric rotational levels (**s**) and $I = 1, 3$ for antisymmetric rotational levels (**a**). The sub-features in the OODR excitation rotational features are upper state G components with unresolved F splittings. Figures 4(c) and (d) show the computed relative intensities and positions of the G components and their agreement with the observed spectra in Figs. 4(a) and (b) is quite satisfactory. The predicted b value, $b = 886/4 = 222$ MHz for all $X^2\Sigma g^+$-core $^3\Lambda$ Rydberg states of Na_2, is in excellent agreement with the observed value, $b = 210.3 \pm 7.8$ MHz.

The predicted $b_{\beta S}$ coupling scheme and atom-in-molecule Fermi contact b-values have been verified by PFOODR spectroscopy for the Na_2 $4^3\Sigma_g^+$ and $2^3\Delta_g$ states and the $^7Li_2$ $1^3\Delta_g$ state. All vibrational levels of the Na_2 $2^3\Delta_g$ state follow $b_{\beta S}$ coupling and have $\langle b \rangle = 221$ MHz. The $4^3\Sigma_g^+$ state follows $b_{\beta S}$ coupling when it is not perturbed by the $2^3\Pi_g$ and/or $3^3\Pi_g$ states and $\langle b \rangle = 214 \pm 50$ MHz.[32] However, the $4^3\Sigma_g^+$ levels perturbed by the $2,3^3\Pi_g$

states exhibit $b_{\beta J}$ coupling.[32] The spin–orbit coupling constants for the $2^3\Pi_g$ and $3^3\Pi_g$ states are $A_\Pi \approx 4$ and 3 cm^{-1}, respectively.[25,33] Since the spin–orbit splitting in the $2,3^3\Pi_g$ states is large relative to the $\Delta J = \pm 1$ off-diagonal Fermi-contact matrix elements, J remains a good quantum number and these $^3\Pi_g$ states follow case $a_{\beta J}$ coupling. The $^3\Pi_g \sim 4^3\Sigma_g^+$ perturbations lift the otherwise perfect threefold degeneracy within each $^3\Sigma_g^+ N$ level and, therefore, prevent the destruction of J within each N by the $b\,I\bullet S$ term and the coupling case is $b_{\beta J}$ rather than $b_{\beta S}$. No hfs has been observed for ^{6}Li$_2$ triplet states.[30] However, the hfs in the ^{7}Li$_2$ $1^3\Delta_g$ and $2^3\Pi_g$ states has, recently, been resolved by cw PFOODR spectroscopy.[31,39] The $1^3\Delta_g$ state dissociates adiabatically to 2s and 3d atoms and is expected to exhibit case $b_{\beta S}$ coupling with $b \approx 100$ MHz. These expectations are confirmed by experiment, with $b = 98.6 \pm 4$ MHz.[31] The ^{7}Li$_2$ $2^3\Pi_g$ state dissociates adiabatically to 2p + 2p atoms, therefore, it is not built on the Li$_2^+$ $X^2\Sigma_g^+$ ion-core and is expected to follow a different hfs coupling scheme. Recent PFOODR experiments[39] show that the ^{7}Li$_2$ $2^3\Pi_g$ state follows $b_{\beta J}$ coupling.

The rovibronic level densities of Li$_2$ or Na$_2$ in the 3–5 eV energy region are very high. Two-state and multi-state perturbations are frequent and reduction of spectra to traditional molecular constants (deperturbed potential energy curves) can be difficult. However, since *a priori* known intermediate levels are selectively excited, the rotational and parity quantum numbers of the upper states observed in PFOODR excitation spectra may be unambiguously assigned. Often, vibrational assignments may also be secured by observing the upper state vibrational nodal structure, as reflected in the Franck–Condon distribution of PFOODR dispersed fluorescence transitions into the $b^3\Pi_u$ and $a^3\Sigma_u^+$ states.

Many different kinds of perturbations have been observed and characterized in the PFOODR spectrum of Na$_2$. The $2^3\Pi_g \sim 3^3\Pi_g$ interaction (due to the electrostatic e^2/r_{ij} perturbation term in the diabatic basis) affects every level of the $3^3\Pi_g$ state and deperturbation will require diagonalization of $a > 100 \times 100$ vibronic interaction matrix.[22,25] The $2^1\Pi_g \sim 2^3\Pi_{2g}$ perturbation involves spin–orbit interaction of a $^1\Pi$ state with a partially S-uncoupled $^3\Pi$ state.[40] The $4^1\Sigma_g^+ \sim 2^3\Pi_g \sim 1^3\Delta_g$ triple perturbation involves $^1\Sigma^+ \sim {}^3\Pi_0$ and $^3\Pi_\Omega \sim {}^3\Delta_\Omega$ spin–orbit interactions as well as $^3\Pi_\Omega \sim {}^3\Delta_{\Omega\pm1}$ L-uncoupling interactions.[41] The $4^3\Sigma_g^+ \sim 3^3\Pi_g$ perturbation involves both spin–orbit and L-uncoupling interactions.[42]

PFOODR fluorescence excitation spectra have been observed extending all the way up the dissociation limits of the $2^3\Pi_g$ and $1^3\Delta_g$ states.

4. PFOODR Dispersed Fluorescence Spectroscopy: Bound $\to$ Bound, Bound $\to$ Quasibound, and Bound $\to$ Free Transitions

The lowest energy triplet state of all alkali dimer molecules is of $^3\Sigma_u^+$ symmetry and is well-known, at the internuclear distance of the $X^1\Sigma_g^+$ molecular ground state, to be repulsive relative to two ground state ^2S atoms. However, at large internuclear distance this $a^3\Sigma_u^+$ state is weakly bound, owing to an attractive $C_6 R^{-6}$ induced dipole-induced dipole interaction. The bound and quasibound levels and the repulsive continuum of this $a^3\Sigma_u^+$ state have been directly observed for the first time by $^3\Lambda_g$, $\Lambda = 0,1 \to a^3\Sigma_u^+$ PFOODR dispersed fluorescence spectroscopy.[24,30,36] The $^6Li_2$ $a^3\Sigma_u^+$ potential well supports only nine ($v = 0 - 8$) bound or quasibound vibrational levels and only 193 (239) bound (quasibound) rotation–vibration levels.[30] The Na_2 $a^3\Sigma_u^+$ well supports 17 bound vibrational levels.[24] The K_2 $a^3\Sigma_u^+$ state has also been observed recently by PFOODR dispersed fluorescence spectroscopy.[36] Molecular constants and RKR potential curves for the bound part of the $a^3\Sigma_u^+$ states of Li_2,[30] Na_2,[24] and K_2[36] have been reported.

A portion of the PFOODR dispersed fluorescence spectrum originating from the $^6Li_2$ $3^3\Sigma_g^+$ $v = 3$, $J = 33$, $N = 32$ level at $T_{vJ} = 32\,646.517$ cm^{-1} term value[30] is shown in Fig. 5. The oscillatory continuum in the 415–485 nm region provides the absolute vibrational numbering of the $3^3\Sigma_g^+$ state as well as information about the shape of the repulsive wall of the $a^3\Sigma_u^+$ state. The presence of four distinct maxima in the short-λ oscillatory continuum indicates that the upper state vibrational quantum number is $v = 3$. Absolute vibrational assignments of the Na_2 $2^3\Sigma_g^+$,[28] $4^3\Sigma_g^+$,[25] $2^3\Pi_g$,[33] $3^3\Pi_g$,[25] and $^6Li_2$ $2^3\Pi_g$[30] states have also been determined via direct node-counts in PFOODR dispersed-fluorescence spectra into the $a^3\Sigma_u^+$ state, and later confirmed by comparing observed PFOODR $^3\Lambda_g \to b^3\Pi_u$ v relative intensities to computed Franck–Condon factors.

Figure 6 is a PFOODR dispersed fluorescence survey spectrum which originates from the $^6Li_2$ $2^3\Pi_g$ $v = 13$, $J = 34$, e, $N = 33$ level at $T_{vJ} = 32\,788.963$ cm^{-1}. This spectrum can be divided into three distinct regions: (i) in the 410–420nm region there are sharp lines assigned as

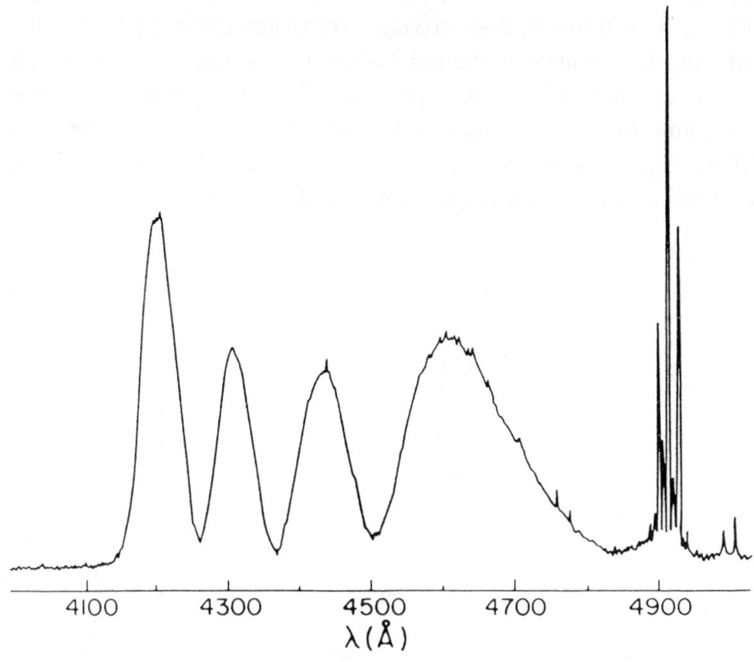

Fig. 5. Resolved fluorescence spectrum from $^6\mathrm{Li}_2$ $3^3\Sigma_g^+$ ($v = 3$, $J = 33$, $N = 32$) level at $T_{vJ} = 32\,646.517$ cm^{-1}. The $3^3\Sigma_g^+ - a^3\Sigma_u^+$ oscillatory continuum at $415 < \lambda < 485$ nm provides the absolute vibrational numbering of the $3^3\Sigma_g^+$ state as well as information about the inner wall of the $a^3\Sigma_u^+$ state. Similar oscillatory continua are observed from the $^6\mathrm{Li}_2$ $2^3\Pi_g$ (Fig. 6),[30a] Na$_2$ $2^3\Pi_g$,[25,33] $3^3\Pi_g$,[25] $4^3\Sigma_g^+$[25,32] and $2^3\Sigma_g^+$[28] states. The sharp features at $\lambda > 488$ nm include a strong Q_1 transition terminating on $b^3\Pi_u$ ($v = 0$, $J = 33$, e, $N = 32$) as well as unassigned collision-induced rotational satellite lines.

$2^3\Pi_g \rightarrow a^3\Sigma_u^+$ bound $\rightarrow$ bound Q-branch transitions (see detail in Fig. 7); (ii) several broad features appear between 420 and 480nm which correspond to bound $\rightarrow$ free $2^3\Pi_g \rightarrow a^3\Sigma_u^+$ transitions (notice that the fluorescence pattern near 450 nm in this spectrum is very different from that in Fig. 5, because in the present spectrum [as well as spectra of the Na$_2$ $2^3\Pi_g \rightarrow a^3\Sigma_u^+$ system[24,25,33,43]] the structure arises from quantum mechanical interference structure rather than the nodal structure of the bound vibrational level); and (iii) at $\lambda > 490$ nm there is a series of pairs of R- and P-branch $2^3\Pi_g \rightarrow b^3\Pi_u$ bound $\rightarrow$ bound transitions. Figures 7 (top) and 7 (bottom) illustrate bound $\rightarrow$ bound and bound $\rightarrow$ quasibound $2^3\Pi_g \rightarrow a^3\Sigma_u^+$ transitions originating from the $N = 33$ and 31 rotational levels of the $^6\mathrm{Li}_2$ $2^3\Pi_g$ $v = 13$ and $v = 9$ levels, respectively. Note that the $a^3\Sigma_u^+$ $v = 3$ level is present (and broadened) at the red end of the bottom ($N = 31$) spectrum, but absent in the top ($N = 33$) spectrum. The $N = 31$ level of the $a^3\Sigma_u^+$ state is bound only in $v = 0$ and quasibound behind the centrifugal barrier $[\hbar^2 J(J+1)/2\mu R^2]$ only in $v = 1-3$; in contrast $N = 33$ is rigorously bound in no vibrational level and is quasibound only in $v = 0 - 2$.

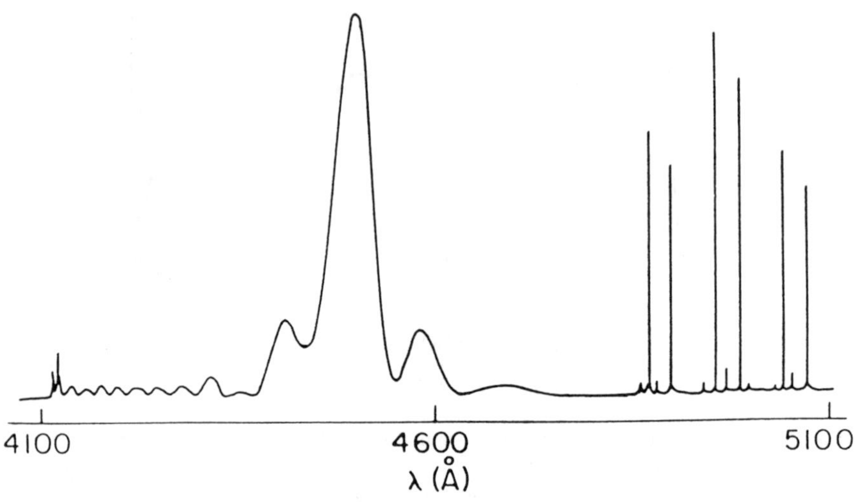

Fig. 6. Survey spectrum from the $^6\mathrm{Li}_2$ $2^3\Pi_g$ ($v = 13$, $J = 34$, $N = 33$) level at $T_{v,J} = 32\,788.963$ cm^{-1}. The sharp lines at the short-λ end of the spectrum correspond to transitions into the $v' = 0-2$ levels of the $a^3\Sigma_u^+$ state. The $2^3\Pi_g \to a^3\Sigma_u^+$ oscillatory continuum extends from 411.5 to 475 nm. The sharp features at long-λ are P, R branch rotational doublets belonging to the $2^3\Pi_g \to b^3\Pi_u$ ($v' = 0-2$) progression. The weaker features are unassigned collision-induced rotational satellite lines.

The PFOODR dispersed fluorescence spectra of $^6\mathrm{Li}_2$ contain transitions terminating in another type of quasibound level: (i) levels of the $b^3\Pi_u$ state that are *predissociated* by L-uncoupling ($-\mathrm{BN}\bullet\mathrm{L}$) interaction with the continuum of the $a^3\Sigma_u^+$ state; (ii) levels of the $A^1\Sigma_u^+$ state that are *accidentally predissociated* owing to $A^1\Sigma_u^+ \sim b^3\Pi_u$ spin–orbit perturbations by predissociated $b^3\Pi_u$ levels.[44] Radiative lifetimes in the $A^1\Sigma_u^+$ state and homogeneous linewidths in the $b^3\Pi_u$ state (as directly measured in PFOODR dispersed fluorescence spectra) exhibit an extraordinary range of

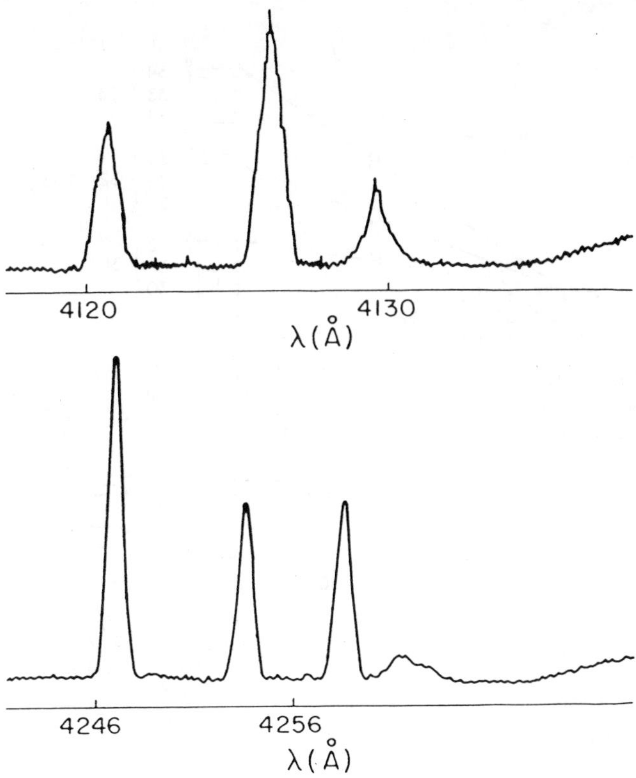

Fig. 7. High-resolution PFOODR dispersed fluorescence spectra of the $2^3\Pi_g \to a^3\Sigma_u^+$ bound → bound transition. (Top) Fluorescence from $2^3\Pi_g$ $v = 13$, $N = 33$, $J = 34$ at $T_{vJ} = 32\,788.963$ cm^{-1} into $a^3\Sigma_u^+$ $v' = 0 - 2$, $N' = 33$, $J' = 34$. There is no trace of a transition into $v' = 3$. (Bottom) Fluorescence from $2^3\Pi_g$ $v = 9$, $N = 31$, $J = 32$ at $T_{vJ} - 32\,036.281$ cm^{-1} into $a^3\Sigma_u^+$ $v' = 0 - 3$, $N' = 31$, $J' = 32$. The $v' = 3$ level is present but broadened, possibly asymmetrically.

v, J, N, and parity dependent variation, all of which has been quantitatively modeled by two electronic perturbation parameters (one for the $b \sim A$ spin–orbit interaction, and one for the $b \sim a$ L-uncoupling interaction) and the shape of the inner wall of the $a^3\Sigma_u^+$ potential curve.[44]

Figure 8 illustrates the PFOODR dispersed fluorescence scheme used to characterize the $a^3\Sigma_u^+$ state through its perturbations of the $b^3\Pi_u$ state.[44] Perturbation and optical selection rules are indicated on the figure. The

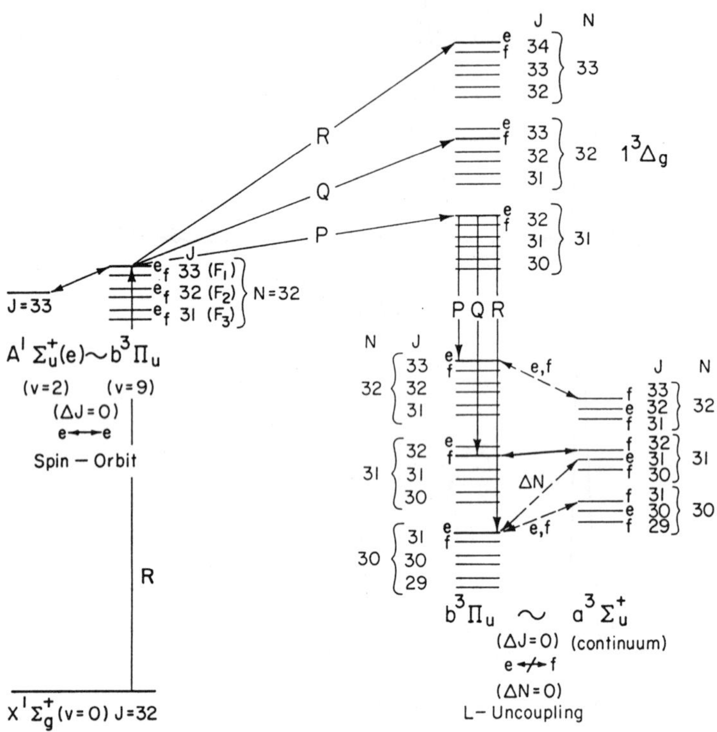

Fig. 8. PFOODR fluorescence scheme used to study the predissociation of the $^6Li_2$ $b^3\Pi_u$ state by the $a^3\Sigma_u^+$ state. Perturbation and optical selection rules are indicated. Dashed lines between $b^3\Pi_u$ and $a^3\Sigma_u^+$ levels depict interactions which are forbidden by the selection rules cited adjacent to them.

$b^3\Pi_u$ $v = 9$, $J = 33$, $N = 32$, e-parity intermediate level (perturbed by $A^1\Sigma^+$ $v = 2$, $J = N = 33$) is populated by the PUMP. The PROBE excites to $1^3\Delta_g$ $v = 4$, ($J = 34$, $N = 33$, e-parity), ($J = 33$, $N = 32$, f-parity), or ($J = 32$, $N = 31$, e-parity) upper levels via R, Q, or P transitions. $1^3\Delta_g \to b^3\Pi_u$ dispersed fluorescence spectra originating from either the ($J = 34$, $N = 33$, e) or ($J = 32$, $N = 31$, e) level are expected to contain sharp R and P lines and a predissociatively broadened Q line; in contrast, spectra originating from the ($J = 33$, $N = 32$, f) level are expected to contain broadened R and P lines and a sharp Q line. If the only

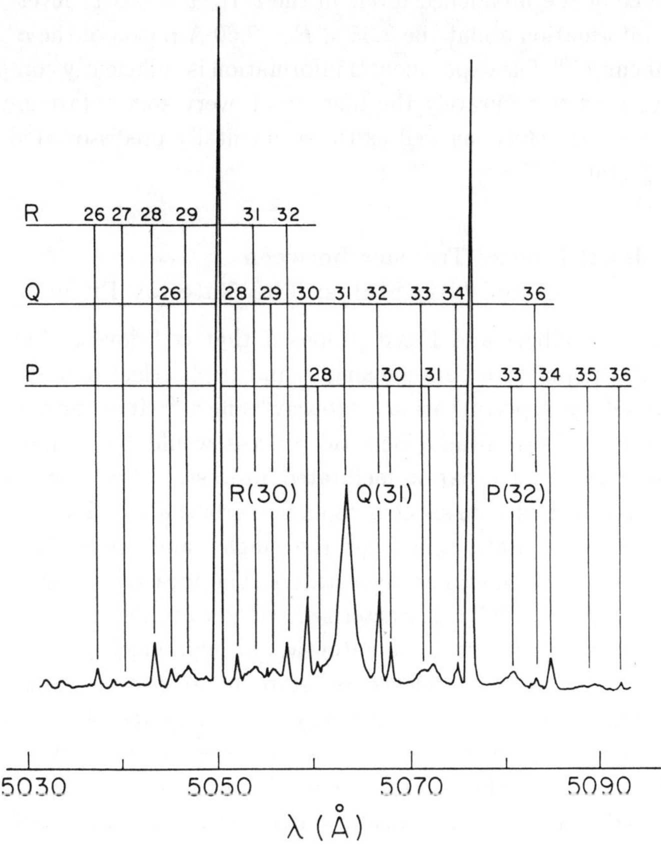

Fig. 9. PFOODR fluorescence spectrum from $1^3\Delta g$ ($v = 4$, $J = 32$, $N = 31, e$) to $b^3\Pi_u$ ($v' = 0$). The broadening of the $Q(31)$ line reflects the predissociation of the $b^3\Pi_u$ state by $a^3\Sigma_u^+$. The weak features which surround the main lines originate from collisional relaxation within the $1^3\Delta_g$ state.

$b^3\Pi_u \sim a^3\Sigma_u^+$ perturbation interaction is $-BN{\bullet}L$ (spin–orbit interaction is negligible), the $\Delta N = 0$, $e \leftrightarrow f$ perturbation selection rules restrict the predissociation to only 3 of the 6 fine structure components of each N in the $^3\Pi$ state [e.g., for even N, odd-J, f and even-J, e levels are predissociated; for odd N, odd-J, e and even-J, f levels are predissociated]. Figure 9 shows the expected sharp (R, P), broad Q pattern of rotational lines in the $1^3\Delta_g$ $v = 4$, $J = 32$, $N = 31$ $e \to b^3\Pi_u$ $v' = 0$ transition.[44] The v and N

dependence of the broadened levels in the $b^3\Pi_u$ $v = 0 - 11$ levels provides detailed information about the $2.35 < R < 2.60$ Å region of the $a^3\Sigma_u^+$ state potential curve.[30] The experimental information is sufficiently complete and precise to predict accurately the lifetime of every spin–rotation–vibration level of the $b^3\Pi_u$ state, as well as the accidentally predissociated levels of the $A^1\Sigma_u^+$ state.[29,30]

5. Collisional Energy Transfer between the Na$_2$ $A^1\Sigma_u^+$ and $b^3\Pi_u$ States: The Gateway Effect

20 years ago Gelbart and Freed proposed that collision-induced transfer of molecular population between singlet and triplet electronic states might be described by a perturbation "gateway" effect.[46] It seemed reasonable that a direct, electronically inelastic process would be significantly less probable than a perturbation-facilitated process. (The "perturbations" here are isolated molecule spectroscopic perturbations, and a spectroscopic analysis yields accurate values for the singlet and triplet state mixing coefficients in each molecular eigenstate.) The idea that a small number of spectroscopically, fully characterized, singlet~triplet mixed rovibronic levels would act as a restrictive gateway, through which population must funnel *en route* from one electronic state to another is both plausible and appealing. Moreover, the gateway idea suggested the possibility of both predictive modeling (based on spectroscopic mixing coefficients) and external control over electronically inelastic processes.

PFOODR spectroscopy is ideally suited to the study of the relevance of spectroscopic perturbations to the propensity rules for collisionally-induced transfer of population between electronic states of different spin multiplicities. Taking the Na$_2$ $A^1\Sigma_u^+ \sim b^3\Pi_u$ perturbation as an example,[45,47] the PUMP laser can directly populate three types of **parent** rovibronic levels: unperturbed $A^1\Sigma_u^+$ singlets, $A \sim b$ mixed levels of dominant singlet character, and $A \sim b$ mixed levels of dominant triplet character. The PROBE laser can monitor , with near perfect singlet versus triplet discrimination (by choosing an appropriate filter to select the wavelength region of detected OODR fluorescence) the population in **daughter** levels (populated by a *single* collision step from the parent level) or **granddaughter** levels (populated via multiple collision steps) of four types: pure singlets, pure triplets, a mixed state of dominant singlet, or dominant triplet character.

Although the PFOODR scheme is extremely sensitive and versatile for monitoring the population flow pathways between two electronic states, there is one serious deficiency. Owing to the usually enormous difference between the radiative lifetimes (and hyperfine splittings) of isoenergetic singlet and triplet states and to the rapid J-variation of lifetimes and PUMP and PROBE transition oscillator strengths (and hfs) near perturbations, it can be prohibitively difficult to relate the PFOODR signal to the population in the PROBEd level. The cw PFOODR experiments described here were capable of demonstrating qualitatively the gateway effect but not measuring quantitative state-to-state rate coefficients.

The Na$_2$ $A^1\Sigma_u^+$ $v' = 26 \sim b^3\Pi_u$ $v' = 28$ perturbation has been examined by Doppler-free polarization spectroscopy.[47] The $b^3\Pi_{\Omega u}$ $v' = 28$, $\Omega' = 2, 1$, and 0 substates cross $A^1\Sigma_u^+$ $v' = 26$, respectively, at $J' = 16, 22$, and 27. Because of the extremely small size of the $A^1\Sigma_u^+ \sim b^3\Pi_{2u}$ $(J \approx 16)$ spin–orbit matrix element, only the $J' = 16$ levels are appreciably mixed (the $J = 12$–17 levels and their mixing coefficients are illustrated in Fig. 10). The pair of $J = 16\,e, \mathbf{a}$ levels are expected to act as gateways through which all population must flow on its way between the $A^1\Sigma_u^+$ and the $b^3\Pi_{2u}$ states. The perturbation matrix elements are larger near the $J = 22$ and 27 levels where the $A^1\Sigma_u^+$ state crosses the $b^3\Pi_{1u}$ and $b^3\Pi_{0u}$ substates and the $A \sim b$ mixing is not restricted to a single J value. The gateway effect will be more complex at the $b^3\Pi_{1u}$ and $b^3\Pi_{0u}$ perturbations, so the PFOODR experiments were focused on the $J = 16$ $A^1\Sigma_u^+ \sim b^3\Pi_{2u}$ perturbation.

The PUMP prepares a specific $A^1\Sigma_u^+$ $v' = 26$, J', $e, \mathbf{s}/\mathbf{a}$ parent level. The PROBE interrogates $b^3\Pi_{2u}$ $v' = 28$ daughter and granddaughter rotational levels via fully resolved and assigned rotational lines in the $2^3\Pi_{2g}$ $v = 28 \leftarrow b^3\Pi_{2u}$ $v' = 28$ vibrational band. PFOODR resonances were detected by monitoring $2^3\Pi_g \rightarrow a^3\Sigma_u^+$ fluorescence.

The relative intensities of parent and daughter/granddaughter OODR PROBE transitions, which were observed when each of the $A^1\Sigma_u^+$ $v' = 26$, $J' = 12 - 16$ levels was selectively populated by the PUMP, are summarized in Table 1. The columns and rows are, respectively, labeled by the parent level populated initially by the PUMP and the daughter/granddaughter level sampled by the PROBE. The % $^3\Pi_2$ and $^1\Sigma^+$ characters (mixing coefficient squared times 100), and e/f, $\mathbf{s}/\mathbf{a}$ symmetries are also given for every level. Notice that the $A^1\Sigma_u^+$ J levels all have e-parity and that the permutation symmetry alternates between $\mathbf{s}$ for odd-J and $\mathbf{a}$ for even J. These facts are of central importance in view of the rigorous $\Delta J = 0$, $e \leftrightarrow f$

Molecular Dynamics and Spectroscopy

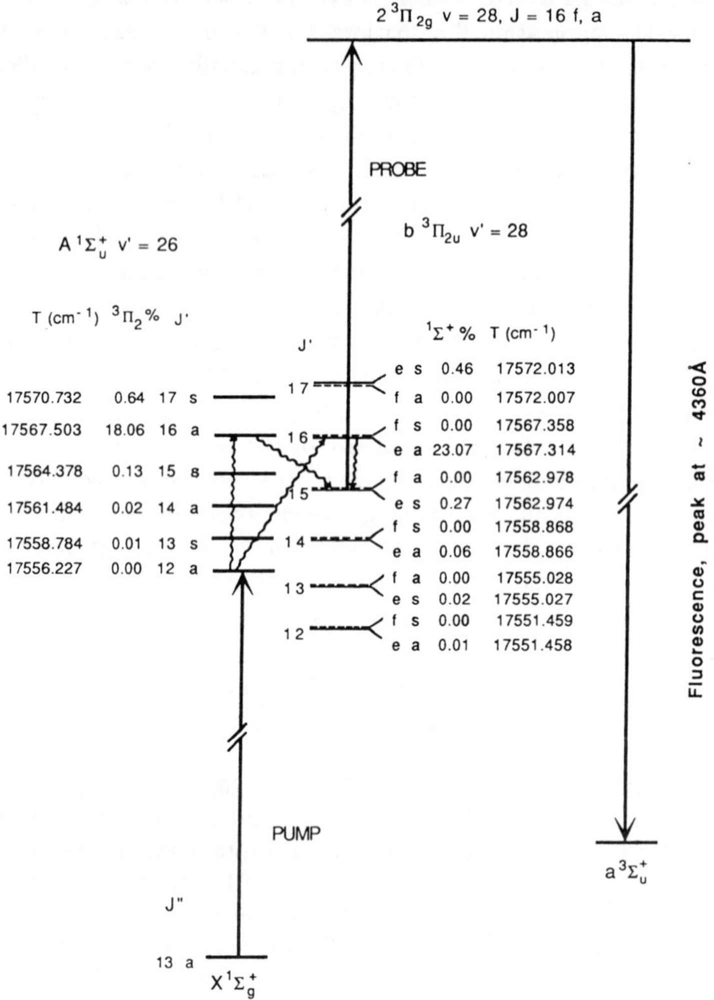

Fig. 10. Level diagram for the $A^1\Sigma_u^+ \sim b^3\Pi_{2u}$ perturbation utilized in the Na$_2$ PFOODR experiment. Term values, mixing % (mixing coefficeint squared times 100), and e/f, **a/s** symmetry labels are given for parent, daughter, and granddaughter rotational levels. For the scheme actually illustrated, $A^1\Sigma_u^+$ $v' = 26$, $J' = 12\,e,$**a** is the parent, the two gateway levels ($J' = 16\,e,$**a** for both $A^1\Sigma_u^+$ $v' = 26$ and $b^3\Pi_{2u}$ $v' = 28$) are the daughters, and $b^3\Pi_{2u}$ $J' = 15\,f,$**a** is the granddaughter level being probed. Note the near perfect energy match between the $A^1\Sigma_u^+$ $J' = 13\,e,$**s** and $b^3\Pi_{2u}$ $J' = 14\,f,$**s** pair.

perturbation selection rules and the extremely strong $\mathbf{s} \nleftrightarrow \mathbf{a}$ collisional transfer propensity rule.

Three features of the present Na_2 $A^1\Sigma_u^+ \sim b^3\Pi_{2u}$ PFOODR experiments demonstrate the existence of a highly selective and efficient collisional gateway between the $A^1\Sigma_u^+$ and $b^3\Pi_{2u}$ states: (i) efficient transfer from one, dominantly A-state, gateway level to the other, dominantly b-state, gateway level when either gateway level is selected as the parent level directly populated by the PUMP; (ii) efficient transfer to $J = 1\,6e,\mathbf{a}$ gateway levels and to a-symmetry levels of the $b^3\Pi_{2u}$ substate near $J = 16$ whenever an a-symmetry rotational level of the A-state is selected, the b-state satellite line rotational population distribution being essentially independent of A-state parent J level; (iii) essentially no $A^1\Sigma_u^+ \rightarrow b^3\Pi_{2u}$ collisional transfer whenever an s-symmetry rotational level of the A-state is selected as the parent.

(i) When the $A^1\Sigma_u^+$ $J' = 16\,e,\mathbf{a}$ level is selected as the parent level, the parent $2^3\Pi_{2g} \leftarrow A^1\Sigma_u^+$ $R(16)$ and $P(16)$ lines are extremely strong because this nominally $A^1\Sigma_u^+$ $J' = 16\,e,\mathbf{a}$ level contains 18.06% $^3\Pi_{2u}$ character. The daughter lines associated with the other gateway level $(b^3\Pi_{2u}$ $J' = 16\,e,\mathbf{a})$ are 1/3 as strong as the parent lines, which implies that the population in this daughter level is $\sim 10\%$ that of the parent level. This sort of extremely efficient population transfer between mutually perturbing rotational levels has been observed previously in CN,[48] BaO,[49] Na_2,[25] Li_2,[29] CO^+,[50] CH_2,[51,52] and glyoxal.[53] In addition to the efficient gateway$\leftrightarrow$gateway $\Delta J = 0$ $e \leftrightarrow e$ collision induced transitions, less intense $A^1\Sigma_u^+$ $J' = 16e,\mathbf{a} \rightarrow b^3\Pi_{2u}$ $\Delta J = \pm 1, \pm 3e, \mathbf{a} \rightarrow f, \mathbf{a}$ and $\Delta J = \pm 2\,e, \mathbf{a} \rightarrow e, \mathbf{a}$ satellite lines are also observed. These observations are summarized in Table 1.

(ii) When the unperturbed or weakly perturbed $A^1\Sigma_u^+$ $J = 12e,\mathbf{a}$ or $14e,\mathbf{a}$ level is selected as the parent, unlike the result for the $A^1\Sigma_u^+$ $J = 16\,e,\mathbf{a}$ parent case, the parent lines and the $b^3\Pi_{2u}$ $J' = 12e,\mathbf{a}$ or $14e,\mathbf{a}$ $\Delta J = 0$ collisional satellite lines are very weak. The strongest lines in the OODR spectrum are daughter transitions originating from the two $J = 16e,\mathbf{a}$ gateway levels, followed in relative intensity by the granddaughter transitions originating from the $b^3\Pi_{2u}$ $J' = 15f,\mathbf{a}$, $17f,\mathbf{a}$, $14e,\mathbf{a}$, and $18e,\mathbf{a}$ levels which received their population primarily by collisional transfer from the two gateway levels. That the $J' = 15f, 17f, 14e$, and $18e$ levels are not daughter but granddaughter levels populated via the $J' = 16e$ gateway levels can be proved clearly by the absence of detectable OODR lines originating from the $b^3\Pi_{2u}J' = 13f,\mathbf{a}$ level when the $A^1\Sigma_u^+$ $J' =

Table.1. Relative OODR Parent and Satellite Intensities. The parent level selectively populated by the PUMP is specified at the top of each column and the PROBEd level is specified to the left of each row. The intensities are tabulated as the ratio of the specified parent/PROBEd level combination to the intensity observed in the $b^3\Pi_{2u}$ $J' = 16\,e,$**a** level when the $A^1\Sigma_u^+$ $J' = 12\,e,$**a** parent level is selected. No entry means that the PFOODR intensity was undetectable for that parent/PROBEd level combination.

PROBEd Level	Parent Level	$A^1\Sigma_u^+$ $12\,e,$**a**	$A^1\Sigma_u^+$ $13\,e,$**s**	$A^1\Sigma_u^+$ $14\,e,$**a**	$A^1\Sigma_u^+$ $15\,e,$**s**	$A^1\Sigma_u^+$ $16\,e,$**a**
$A^1\Sigma_u^+$ 17556.227	$12\,e,$**a** $^3\Pi_2$ 0.00%					
$b^3\Pi_{2u}$ 17551.458	$12\,e,$**a** $^1\Sigma^+$ 0.01%					0.5
$b^3\Pi_{2u}$ 17551.459	$12\,f,$**s** $^1\Sigma^+$ —					
$A^1\Sigma_u^+$ 17558.784	$13\,e,$**s** $^3\Pi_2$ 0.01%					
$b^3\Pi_{2u}$ 17555.027	$13\,e,$**s** $^1\Sigma^+$ 0.02%		0.05			
$b^3\Pi_{2u}$ 17555.028	$13\,f,$**a** $^1\Sigma^+$ —			0.05		0.6
$A^1\Sigma_u^+$ 17561.484	$14\,e,$**a** $^3\Pi_2$ 0.02%			0.1		
$b^3\Pi_{2u}$ 17558.866	$14\,e,$**a** $^1\Sigma^+$ 0.06%	0.1		0.4		1.2
$b^3\Pi_{2u}$ 17558.868	$14\,f,$**s** $^1\Sigma^+$ —				0.05	
$A^1\Sigma_u^+$ 17564.378	$15\,e,$**s** $^3\Pi_2$ 0.13%				0.3	
$b^3\Pi_{2u}$ 17562.974	$15\,e,$**s** $^1\Sigma^+$ 0.27%		0.05		0.3	
$b^3\Pi_{2u}$ 17562.978	$15\,f,$**a** $^1\Sigma^+$ —	0.1		0.4		1.3
$A^1\Sigma_u^+$ 17567.503	$16\,e,$**a** $^3\Pi_2$ 18.06%	1.0		25		440

Table 1 (continue)

PROBEd Level	Parent Level	$A^1\Sigma_u^+$ $12\,e,\mathbf{a}$	$A^1\Sigma_u^+$ $13\,e,\mathbf{s}$	$A^1\Sigma_u^+$ $14\,e,\mathbf{a}$	$A^1\Sigma_u^+$ $15\,e,\mathbf{s}$	$A^1\Sigma_u^+$ $16\,e,\mathbf{a}$
$b^3\Pi_{2u}$ 17567.314	$16\,e,\mathbf{a}$ $^1\Sigma^+$ 23.07%	1.4		35		140
$b^3\Pi_{2u}$ 17567.358	$16\,f,\mathbf{s}$ $^1\Sigma^+$ —				0.05	
$A^1\Sigma_u^+$ 17570.732	$17\,e,\mathbf{s}$ $^3\Pi_2$ 0.64%					
$b^3\Pi_{2u}$ 17572.013	$17\,e,\mathbf{s}$ $^1\Sigma^+$ 0.46%					
$b^3\Pi_{2u}$ 17572.007	$17\,f,\mathbf{a}$ $^1\Sigma^+$ —	0.1		0.3		1.3
$A^1\Sigma_u^+$ 17574.208	$18\,e,\mathbf{a}$ $^3\Pi_2$ 0.32%					
$b^3\Pi_{2u}$ 17576.928	$18\,e,\mathbf{a}$ $^1\Sigma^+$ 0.12%	0.1		0.2		1.2
$b^3\Pi_{2u}$ 17576.925	$18\,f,\mathbf{s}$ $^1\Sigma^+$ —					

$12\,e,\mathbf{a}$ level is selected as the parent. The $b^3\Pi_{2u}J' = 13\,f,\mathbf{a}$ level is closest both in J' and energy to the $A^1\Sigma_u^+ J' = 12\,e,\mathbf{a}$ parent. It is also more distant in J' and energy from the $J' = 16\,e,\mathbf{a}$ gateway levels than the observably populated $J' = 15\,f, 17\,f, 14\,e$, and $18\,e$ granddaughter levels. The absence of $b^3\Pi_{2u}J' = 13\,f,\mathbf{a}$ satellite lines (in the same $A^1\Sigma_u^+ J' = 12\,e,\mathbf{a}$ parent spectrum in which $b^3\Pi_u$ $J = 15\,f, 17\,f, 14\,e$, and $18\,e$ granddaughter transitions are observable) is the strongest quantitative evidence ever obtained for any molecule that a gateway mechanism is operative.

(iii). The $b^3\Pi_{2u} \sim A^1\Sigma_u^+$ perturbation is so weak that significant mixing is confined to the $J' = 16\,e,\mathbf{a}$ levels. This means that there is essentially no perturbation at all, hence no gateway, for the **s**-symmetry rotational levels. Since the **s** $\leftrightarrow$ **a** collisional propensity rule is very strong,[54,55] there is no way that an **s**-symmetry parent level can take advantage of an **a**-symmetry gateway! The $A^1\Sigma_u^+ J' = 13\,e,\mathrm{s}$ and $15\,e,\mathrm{s}$ levels are actually very weakly mixed with the $b^3\Pi_{2u}$ $J = 13\,e,\mathrm{s}$ and $15\,e,\mathrm{s}$ levels (respectively 0.01% and 0.13%). The $A^1\Sigma_u^+$ $J' = 13\,e,\mathrm{s}$ and $b^3\Pi_{2u}$ $J' = 14\,f,\mathrm{s}$ levels are nearly degenerate, but no $b^3\Pi_{2u} J' = 14\,f,\mathrm{s}$ satellite line is detectable in an $A^1\Sigma_u^+ J' = 13\,e,\mathrm{s}$ parent OODR spectrum. This absence of $b^3\Pi_{2u} J' = 14\,f,\mathrm{s}$ satellite lines is strong qualitative evidence for the gateway mechanism: no perturbation $\to$ no gateway $\to$ no transfer!

When more than a single pair of rotational levels is appreciably mixed, interference effects can occur and perturbation facilitated energy transfer processes can become more complicated than the Gelbart-Freed[46] gateway model.[55]

The $\mathrm{Na_2}$ $b^3\Pi_{2u} \sim A^1\Sigma_u^+$ perturbation is ideal for producing a gateway effect. There have been several well documented examples of efficient electronically inelastic collision-induced processes which are not perturbation facilitated. The three best reasons for non-gateway behavior are: (i) the existence of a strong electric dipole electronic transition that can be induced by the time-dependent interaction with a permanent electric monopole or dipole or an induced electric dipole on the collision partner; (ii) Förster type exchange of excitation between two collision partners; and (iii) $M_2^* + M' \leftrightarrow M + MM'^*$ atom exchange. Process (i) is ruled out by the forbidden electric dipole character of the $A^1\Sigma_u^+ \leftrightarrow b^3\Pi_u$ transition. Process (ii) is ruled out by the dipole forbidden character of the $b^3\Pi_u \leftrightarrow X^1\Sigma_g^+$ transition. Process (iii) is ruled out by our failure to detect any breakdown of the $\mathrm{s} \nleftrightarrow \mathrm{a}$ collisional propensity rule.

To summarize, cw PFOODR spectroscopy has been used to demonstrate the perturbation gateway effect proposed by Gelbart and Freed.[46] Interelectronic state perturbations do act, under special well defined conditions, as a bottleneck through which all population must funnel as it is collisionally transferred from one electronic state to another. The gateway effect offers some interesting possibilities for external manipulation of plasmas and chemical laser gain media. It is also a very satisfying situation when information (perturbation matrix elements, mixing coefficients), derived from the analysis of the spectrum of an *isolated* molecule, has qualitative and quantitative relevance to the dynamical behavior of a *non-isolated* molecule.

6. Pulsed PFOODR Spectroscopy of Na_2 and Li_2

Pulsed PFOODR experiments on Na_2 have been performed in Schawlow's group.[56,57] Although the resolution of pulsed laser OODR spectroscopy (~ 0.2 cm^{-1}) is usually inferior to that of cw OODR, with pulsed laser OODR a much wider energy region can be explored rapidly. The Na_2 $2^3\Pi_g$, $3^3\Pi_g$, $1^3\Delta_g$, and $2^3\Delta_g$ states (also observed by cw PFOODR spectroscopy) as well as two highly excited $^3\Pi_g$ states have been observed by two-step polarization labeling spectroscopy via $A^1\Sigma_u^+ \sim b^3\Pi_u$ mixed intermediate levels.[56] Pulsed laser PFOODR experiments on $^7Li_2$ are currently in progress at Tsinghua University and several $^3\Lambda_g$ Rydberg states have been observed.[58]

Acknowledgments

We thank Professors W.C. Stwalley, A.M. Lyyra, and X. Xie, Dr. S.F. Rice and Dr. T.-J. Whang for their contributions to the acquisition and analysis of cw PFOODR spectra of Na_2 and Li_2 at both MIT and the University of Iowa. We thank the Journal of Molecular Spectroscopy for permission to reproduce Figures 1 and 2, 3, and 4 from *J. Mol. Spectrosc.* **117**, 245 and 228 (1986) and Figures 2 and 3 from *J. Mol. Spectrosc.* **134**, 50 (1989), the Journal of Chemical Physics for permission to reproduce Figure 1 from *J. Chem. Phys.* **97**, 8836 (1992), and Chemical Physics for permission to reproduce Figures 1 and 2 from *Chem. Phys.* **104**, 161 (1986). The MIT portion of this work was supported by grants from the AFOSR (F49620-83-C-0010) and the NSF (CHE81-12966, PHY83-20098, PHY87-09759).

References

1a. R.W. Field, *Faraday Disc. Roy. Soc. Chem.* **71**, 111 (1981).

1b. R.W. Field, M. Revelli, and G.A. Capelle, *J. Chem. Phys.* **63**, 3228 (1975).

2. P. Kusch and M.M. Hessel, *J. Chem. Phys.* **68**, 2591 (1978).

3. M.E. Kaminsky, *J. Chem. Phys.* **66**, 4951 (1977).

4. P. Kusch and M.M. Hessel, *J. Chem. Phys.* **67**, 586 (1977).

5. M.M. Hessel and C.R. Vidal, *J. Chem. Phys.* **70**, 4439 (1979).

6. J.P. Woerdman, *Chem. Phys. Lett.* **43**, 279 (1976).

7. N.W. Carlson, A.J. Taylor, and A.L. Schawlow, *Phys. Rev. Lett.* **45**, 18 (1980).

8. A.J. Taylor, K.M. Jones, and A.L. Schawlow, *J. Opt. Soc. Amer.* **73**, 994 (1983).

9. R.A. Bernheim, L.P. Gold, P.B. Kelley, C. Kittrell, and D.K. Veirs, *Phys. Rev. Lett.* **43**, 123 (1979).

10. R.A. Bernheim, L.P. Gold, P.B. Kelly, C. Tomczyk, and D.K. Veirs, *J. Chem. Phys.* **74**, 3249 (1981).

11. D.D. Konowalow, M.E. Rosenkrantz, and M.L. Olson, *J. Chem. Phys.* **72**, 2612 (1980).

12. D.D. Konowalow and M.E. Rosenkrantz, *J. Chem. Phys.* **86**, 1099 (1982).

13. G. Jeung, *J. Phys. B: At. Mol. Phys.* **16**, 4289 (1983).

14. D.D. Konowalow and P.S. Julienne, *J. Chem. Phys.* **72**, 5815 (1980).

15. D.D. Konowalow and J.L. Fish, *Chem. Phys.* **77**, 435 (1983); **94**, 463 (1984).

16. R.W. Wood and F.E. Hackett, *Astrophys. J.* **30**, 339 (1909).

17a. R.S. Mulliken, *Rev. Mod. Phys.* **4**, 15 (1932).

17b. W.R. Fredrickson and C.R. Stannard, *Phys. Rev.* **44**, 632 (1933).

17c. T. Carroll, *Phys. Rev.* **52**, 822 (1937).

18. P. Kusch and M.M. Hessel, *J. Chem. Phys.* **63**, 4087 (1975).

19. M.E. Kaminsky, R.T. Hawkins, F.V. Kowalski, and A.L. Schawlow, *Phys. Rev. Lett.* **36**, 671 (1976).

20. F. Engelke, H. Hage, and C.D. Caldwell, *Chem. Phys.* **64**, 221 (1982).

21. J.B. Atkinson, J. Becker, and W. Demtrder, *Chem. Phys. Lett.* **87**, 92 (1982); 87, 128 (1982).

22. Li Li and R.W. Field, *J. Phys. Chem.* **87**, 3020 (1983).

23. Li Li, S.F. Rice, and R.W. Field, *J. Mol. Spectrosc.* **105**, 344 (1984).

24a. Li Li, S.F. Rice, and R.W. Field, *J. Chem. Phys.* **82**, 1178 (1985).

24b. E. J. Friedman-Hill and R.W. Field, *J. Chem. Phys.* **96**, 2444 (1992).

25. Li Li and R.W. Field, *J. Mol. Spectrosc.* **117**, 245 (1986).

26. T.J. Whang, A. M. Lyyra, W.C. Stwalley, and Li Li, *J. Mol. Spectrosc.* **149**, 505 (1991).

27. T.J. Whang, W.C. Stwalley, Li Li, and A.M. Lyyra, *J. Mol. Spectrosc.* **155**, 184 (1992).

28. T.J. Whang, C.C. Tsai, W.C. Stwalley, A.M. Lyyra, and Li Li, *J. Mol. Spectrosc.* **160**, 411 (1993).

29. X. Xie and R.W. Field, *Chem. Phys.* **99**, 337 (1985).

30a. X. Xie and R.W. Field, *J. Mol. Spectrosc.* **117**, 228 (1986).

30b. X. Xie and R.W. Field, *J. Chem. Phys.* **83**, 6193 (1985).

31. Li Li, T. An, T.J. Whang, A.M. Lyyra, W.C. Stwalley, R.W. Field, and R.A. Bernheim, *J. Chem. Phys.* **96**, 3342 (1992).

32. Li Li, Q. Zhu, and R.W. Field, *Mol. Phys.* **66**, 685 (1989).

33. X. Xie, R.W. Field, Li Li, A.M. Lyyra, J.T. Bahns, and W.C. Stwalley, *J. Mol. Spectrosc.* **134**, 119 (1989).

34. Li Li, Q. Zhu, and R.W. Field, *J. Mol. Spectrosc.* **134**, 50 (1989).

35. Li Li, A.M. Lyyra, and W.C. Stwalley, *J. Mol. Spectrosc.* **134**, 113 (1989).

36. Li Li, A.M. Lyyra, W.T. Luh, and W.C. Stwalley, *J. Chem. Phys.* **93**, 8452 (1990).

37. C.H. Townes and A.L. Schawlow, "Microwave Spectroscopy", (McGraw-Hill, New York, 1955.) 194–199.
38. R.A. Frosch and H.M. Foley, *Phys. Rev.* **88**, 1337 (1952).
39. Li Li, A. Yiannopoulou, K. Urbanski, A.M. Lyyra, B. Ji, T. An, T.-J. Whang, and W. C. Stwalley, to be published.
40. Li Li and R.W. Field, *J. Mol. Spectrosc.* **123**, 237 (1987).
41. Li Li, A.M. Lyyra, W.C. Stwalley, M. Li, and R.W. Field, *J. Mol. Spectrosc.* **147**, 215 (1991).
42. Li Li and M. Li, to be published.
43. G. Pichler, J.T. Bahns, K.M. Sando, W.C. Stwalley, D.D. Konowalow, Li Li, R.W. Field, and W. Muller, *Chem. Phys. Lett.* **129**, 425 (1987).
44. S.F. Rice, X. Xie, and R.W. Field, *Chem. Phys.* **104**, 161 (1986).
45. Li Li, Q. Zhu, A.M. Lyyra, T.-J. Whang, W.C. Stwalley, R.W. Field, and M.H. Alexander, *J. Chem. Phys.* **97**, 8835 (1992).
46. W.M. Gelbart and K.F. Freed, *Chem. Phys. Lett.* **18**, 470 (1973).
47. M. Li, C. Wang, Y. Wang, and Li Li, *J. Mol. Spectrosc.* **123**, 161 (1987).
48. D.W. Pratt and H.P. Broida, *J. Chem. Phys.* **50**, 2181 (1969).
49. R.W. Field, C.R. Jones, and H.P. Broida, *J. Chem. Phys.* **60**, 4377 (1974).
50. A.V. Dentamaro and D.H. Katayama, *J. Chem. Phys.* **90**, 91 (1989).
51. A.R.W. McKellar, P.R. Bunker, T.J. Sears, K.M. Evenson, R.J. Saykally, and S.R. Langhoff, *J. Chem. Phys.* **79**, 5251 (1983).
52. U. Bley and F. Temps, *J. Chem. Phys.* **98**, 1058 (1993).
53. C. Michel, M. Lombardi, and R. Jost, *Chem. Phys.* **109**, 357 (1986).
54. G. Herzberg, "Molecular Spectra and Molecular Structure. I" (van Nostrand Reinhold, New York, 1950).
55. M.H. Alexander, *J. Chem. Phys.* **76**, 429 (1982).
56. K.M. Jones, "Rydberg States in Diatomic Sodium", Ph.D. Thesis, (Stanford University, December 1983.)
57. K.M. Jones, E. Breford, and A.L. Schawlow, private communication.
58. D. Chen, L. Li, X. Wang, Li Li, Q. Hui, H. Ma, L.Q. Li, X.Y. Xu, and D.Y. Chen, *J. Mol. Spectrosc.* **161**, 7 (1993).

CHAPTER 8

APPLICATIONS OF MASS-SELECTIVE IONIZATION-LOSS STIMULATED RAMAN SPECTROSCOPY IN STUDIES OF MOLECULAR COMPLEXES AND CLUSTERS

Peter M. Felker

Department of Chemistry and Biochemistry
University of California, Los Angeles, CA 90024-1569, USA

Contents

1. Introduction

Molecular complexes and clusters represent species whose study is relevant to the characterization of intermolecular interactions, many-body dynamics, surface adsorption, local structure in solutions, chemistry in condensed phases, photodissociation, and numerous other issues (for books and reviews see Refs. 1–12). In studying such species one desires an experimental approach capable of high sensitivity and high species-selectivity. The former is necessary because the samples one generally deals with are seeded, supersonic molecular beams having densities characteristic of sparse gases. The latter is desirable because these samples tend to be composed of a distribution of cluster sizes and, moreover, may contain a distribution of cluster isomers at any one size: ultimately, one wants information specific to particular species.

Given the twin desirables of sensitivity and species-selectivity, it is natural that spectroscopic methods have played an important role in cluster studies. Indeed, the number and variety of spectroscopic schemes that have been used in such studies is considerable—so considerable that an attempt to list them all here would invariably lead to significant omissions. One area of particular interest has been the ground-state vibrational spectroscopy of molecular clusters. Such studies, making use of direct infrared[10,13] and far-infrared[11] absorption, infrared predissociation spectroscopy,[7] optothermal infrared methods,[9,14,15] infrared–vibronic double-resonance,[16] dispersed fluorescence spectroscopy,[3–5,12,17] nonlinear Raman methods,[18] and others, are valuable because they provide information pertaining to ground electronic state surfaces. The characterization of ground-state properties is highly desirable because such tend to be much more accessible to interpretation and theoretical modeling than excited-state properties and because the majority of chemical systems of interest involve species in their ground electronic states.

Notwithstanding the high level of effort pertaining to the vibrational spectroscopy of clusters, there are some capabilities in this area that have only recently become available. One such capability pertains to the measurement of species-specific cluster vibrational spectra throughout the vibrational fundamental region at subwavenumber resolution. This capability derives from the development of double-resonance methods that involve a stimulated Raman transition followed by some probe process sensitive to that transition. This class of methods includes the fluorescence-dip[19] and ion-dip[20,21] variants of stimulated emission pumping spectroscopy. It also comprises the methods that we call mass-selective, ionization-detected stimulated Raman spectroscopy (IDSRS).[22,23]

IDSRS methods involve the use of resonantly enhanced multiphoton ionization (REMPI)[24] to probe the rovibrational population shifts that are induced by stimulated Raman transitions. There are two general variants of IDSRS. In ionization-gain stimulated Raman spectroscopy (IGSRS),[25] the population gain of the final state in a Raman transition is monitored by REMPI. In ionization-loss stimulated Raman spectroscopy (ILSRS)[26] the population loss of the initial state in the Raman transition is so monitored. Coupled with mass-selective detection of the photoions these two schemes have proved to be powerful means by which to obtain vibrational spectroscopic information on weakly bound molecular complexes and clusters.[22,23,27−32] Mass-selective ILSRS, in particular, has been shown to have especial promise in this regard.

In this chapter we focus on the characteristics of mass-selective ILSRS and its application in the vibrational spectroscopy of molecular complexes and clusters. The chapter is outlined as follows. In the section following this one (Sec. 2) we consider the basics of the mass-selective ILSRS method, its implementation, and the advantages it has in studies of clusters. In Sec. 3 we outline results from this laboratory on benzene dimer,[27,29] benzene–$(Ar)_n$ clusters,[31] and benzene–$(N_2)_n$ clusters.[32] These results demonstrate the capabilities of the technique and what can be learned about clusters by using it. In the final section we speculate about some future applications.

2. Fundamentals of Mass-Selective ILSRS

2.1. *The Method*

ILSRS is a Raman/vibronic double-resonance spectroscopy.[26] The basic scheme is depicted schematically in the level diagram of Fig. 1. A two-color (ω_1, ω_2) pulse induces a two-photon-resonant stimulated Raman transition from initial state $|a\rangle$ to excited vibrational state $|b\rangle$. The loss in population of state $|a\rangle$ in the sample is then probed by a second pulse whose frequency (ω_3) is tuned to be resonant with a vibronic transition originating in $|a\rangle$. This second pulse photoionizes by a REMPI process those species that remain in $|a\rangle$ subsequent to the stimulated Raman pulse. (In IGSRS, ω_3 is tuned to a vibronic transition originating in $|b\rangle$ and photoionizes those species that have been Raman-excited.[25]) An ILSRS experiment consists of measuring photoion signal as a function of $\omega_1 - \omega_2$. When $\omega_1 - \omega_2$ does not correspond to the frequency of a Raman transition originating in $|a\rangle$, then the photoion signal level is unaffected by the stimulated Raman pulses. However, when $\omega_1 - \omega_2$ does match such a Raman resonance, the photoion signal decreases owing to the Raman-induced transfer of population out of $|a\rangle$ into $|b\rangle$. (Photoion depletion mechanisms in mass-selective ILSRS are discussed in more detail in Sec. 2.3.)

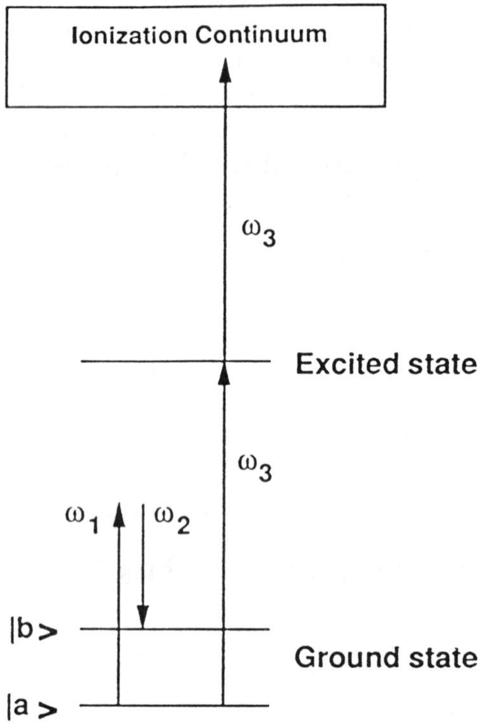

Fig. 1. Level diagram depicting ILSRS. ω_1 and ω_2 excite a stimulated Raman transition and ω_3 probes the corresponding population loss of the initial state $|a\rangle$ by resonantly enhanced multiphoton ionization. Reprinted with permission from Ref. 29. Copyright 1992, American Institute of Physics.

Several points should be made about the ILSRS scheme. First, like the most common variants of SEP,[19-21] it is a depletion spectroscopy—it is not a zero-background method. Its success in an application relies on the efficiency of Raman-induced population transfer relative to the noise on the photoion signal. Second, the photoionization process depicted in Fig. 1 (i.e., a single-color, "1 + 1" process[24]) is not the only type that can be employed in ILSRS. For example, two-color schemes, schemes involving multiphoton resonances, and even schemes involving single-photon ionization, can all be used, in principle, as the probe process in ILSRS. Third, and most important, the photoions created in ILSRS can be readily subjected to mass analysis such that the signal monitored in an experiment is not due to all the photoions but only to those of a particular mass. This *mass-selective* ILSRS method[22,23] has considerably greater capabilities than ILSRS without mass selectivity. This is particularly true in studies of weakly bound complexes and clusters. This chapter pertains entirely to the mass-selective method.

2.2. *Implementation*

Figure 2 shows a schematic diagram of the apparatus used for mass-selective ILSRS experiments in this laboratory. The ω_1 field is provided by part of the frequency-doubled output of an injection-seeded Nd:YAG laser operating at a repetition rate of 30 Hz. The temporally coincident ω_2 field is obtained from the output of a dye laser that is pumped by the remainder of the 532 nm light from the injection-seeded Nd:YAG. The ω_1 and ω_2 pulse trains are combined on a beam splitter. Ultimately, about 30 mJ per pulse at 532 nm and 25 mJ per pulse at the dye-laser fundamental impinge on the sample. The Raman resolution (i.e., effective bandwidth of $\omega_1 - \omega_2$) provided by these lasers is about 0.1 cm^{-1}. ω_3 is obtained from a second Nd:YAG laser and second dye laser. The second Nd:YAG laser is triggered synchronously at a $\sim$ 10 ns delay from the firing of the first Nd:YAG. Either the 532 nm or 355 nm output of the second Nd:YAG laser pumps the second dye laser. The output of this dye laser is frequency-doubled in β-barium borate, and the doubled output serves as the ω_3 pulse. The ω_3 pulse train, containing pulses with typical energies of $\sim$ 0.2 mJ, is combined with the two-color ω_1, ω_2 pulse train on a dichroic beam splitter. The resulting three-color pulse train is focused into the sample with a 25 cm focal-length lens.

It is important to note that the ω_3 pulse train must not overlap the ω_1, ω_2 pulses temporally. When there is such overlap, the high intensity of the latter pulses very efficiently destroys the species that are vibronically excited or photoionized by the former pulses. Evidently, this is due to

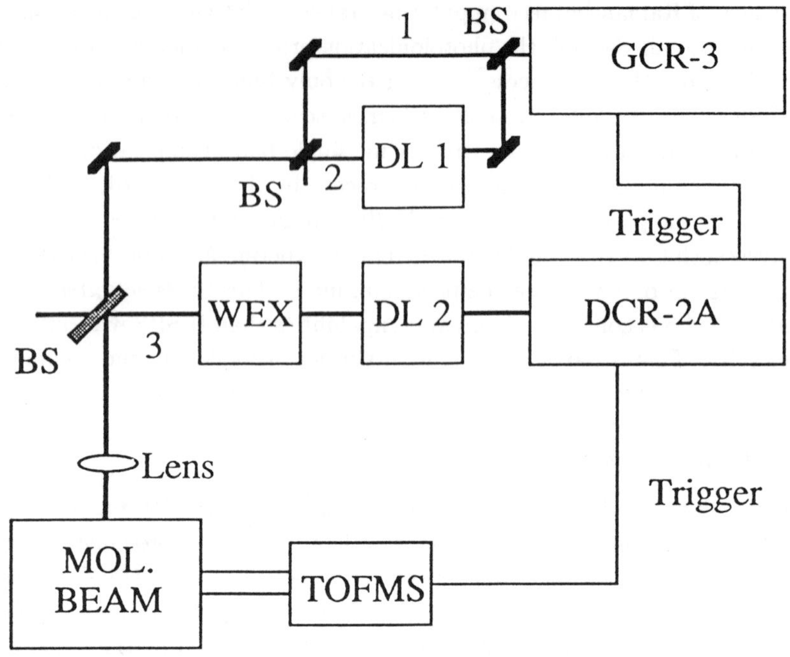

Fig. 2. Schematic diagram of the experimental apparatus used for mass-selective ILSRS in this laboratory. BS = beam splitter; MOL. BEAM = molecular beam apparatus; TOFMS = time-of-flight mass spectrometer; DL1 and DL2 denote Spectra-Physics PDL-2 (or PDL-3) dye lasers; GCR-3 and DCR-2A denote Spectra-Physics Nd:YAG lasers. The pump laser field—"1"—was supplied by 20% of the 2nd harmonic of the GCR-3. The Stokes laser field—"2"—was supplied by the output of DL1. The probe laser field—"3"—was supplied by the output of DL2 frequency-doubled by a Spectra-Physics Wavelength Extender (WEX). Reprinted with permission from Ref. 28. Copyright 1992, American Chemical Society.

efficient absorption of the visible photons by ω_3-excited neutrals or parent ions. In any case, for ILSRS to be successful, the ω_3 pulse train must be delayed from the ω_1, ω_2 pulse train by a time on the order of the pulse widths involved. In fact, the striking reduction in parent-ion signal that occurs when the ω_3 pulse is temporally and spatially overlapped with the two-color stimulated Raman pulse provides one with a convenient means by which to align the spatial overlap of ω_1 and ω_2 with ω_3—one simply reduces to zero the temporal delay between ω_3 and $\omega_1(\omega_2)$ and adjusts the alignment of ω_1 (or ω_2) relative to that of ω_3 so as to minimize parent-ion signal. The minimal parent-ion signal corresponds to the maximal overlap of the ω_1 (ω_2) focal volumes. To do an ILSRS experiment one then increases the delay of the ω_3 pulse train such that there is no temporal overlap with the ω_1, ω_2 pulses. The usefulness of this procedure is the prime reason why two Nd:YAG lasers, one of which can be fired at an electronically variable delay with respect to the firing of the other, are employed in the ILSRS laser system. The two pump-laser system allows for trivial alternation between temporal overlap and nonoverlap of the pulses.

The pulsed molecular beam apparatus consists of a source chamber and a differentially pumped, Wiley–McLaren-type,[33] time-of-flight mass spectrometer (TOFMS). The pulsed molecular beam is generated by a commercial solenoid-actuated valve (General Valve Series 9) that is fired in synchrony with the firing of the lasers. Typically, 200 to 300 μs-duration pulses of helium carrier gas seeded with the species of interest expand through a circular orifice of 0.1 to 0.5 mm in diameter. The expansion is skimmed several cm downstream from the valve orifice, after which the skimmed beam enters the ionization region of the TOFMS. Here, the molecular beam intersects the laser beams at a 90° angle. Photoions created by interaction of the molecular beam with the ω_3 pulse are accelerated in a direction at right angles to both the molecular-beam and laser-beam propagation directions. After passing through an ion lens, the photoions are focused onto a dual microchannel plate detector. The output of the detector is amplified by a fast video amplifier, the output of which is directed to a 400 MHz-bandwidth oscilloscope and to a boxcar integrator. The oscilloscope serves as a monitor of the mass spectrum of the sample. It is also used to set the boxcar gate to the ion mass peak of interest. With the boxcar gate set to an ion mass peak, the averaged output of the boxcar is fed to a computer in synchrony with the scanning of the ω_2 dye laser. In this way mass selected ion signal is monitored as a function of $\omega_1 - \omega_2$. To obtain a useful mass-selective ILSRS spectrum, it is generally necessary to signal average by scanning the ω_2 laser repetitively and co-adding the spectra so measured.

2.3. *Depletion Mechanisms*

The description of the ILSRS process presented in Sec. 2.1. is somewhat simplistic in that it makes the assumption that no Raman-excited species contribute to the measured photoion signal. That is, it assumes that photoionization is a perfectly selective monitor of the population of initial state $|a\rangle$. In practice, such perfect selectivity need not obtain. Raman-excited species may absorb ω_3 photons and be photoionized. Species other than the one of interest may absorb ω_3 photons and produce daughter ions in the mass channel being monitored. Obviously, these channels for ion production can affect mass-selective ILSRS spectra. Thus, it is important to understand the general mechanisms by which Raman transitions can give rise to photoion depletions and the ways in which spectral artifacts can occur.

2.3.1. *Desirable Raman-induced Depletions*

To produce a Raman-dependent depletion in a mass-selective ILSRS spectrum, Raman-excited species must have a smaller probability for producing photoions in the mass channel of interest than that of species that remain unexcited by the stimulated Raman pulse. There are four major mechanisms by which this situation can arise. The first ("mechanism #1") is one that most often obtains in bare molecules or molecular clusters composed of just a few moieties. In such species, the vibronic resonance frequencies of Raman-excited vibrational states tend to be different than those associated with the vibrational ground state. If ω_3 is tuned to one of the latter vibronic resonances, then Raman-excited species will not be photoionized efficiently. This situation is depicted in Fig. 3a.

A second spectral mechanism ("#2") of Raman-induced photoion depletion is depicted in Fig. 3b. It corresponds to cases where vibrational energy redistribution ("IVR") occurs in Raman-excited species during the time between the stimulated Raman excitation pulse and the ω_3 pulse. Such redistribution processes tend to populate a wide range of "bath" vibrational states.[34] The resulting vibronic spectrum of these vibrationally "relaxed" species tends to consist of broad features which reflect this range of bath states. (This broadening phenomenon is well known in the fluorescence spectra of molecules that undergo IVR.[34]) The upshot is that although the vibronic spectrum of the Raman-excited species may overlap that of the species unexcited by the stimulated Raman pulse, there is a Raman- and IVR-induced dilution of vibronic spectral intensity. Thus, with ω_3 tuned to a resonance of the vibrationally unexcited species, the probability for photoionizing such species will be greater than that for photoionizing Raman-excited species, and the Raman resonance will give rise to a photoion depletion.

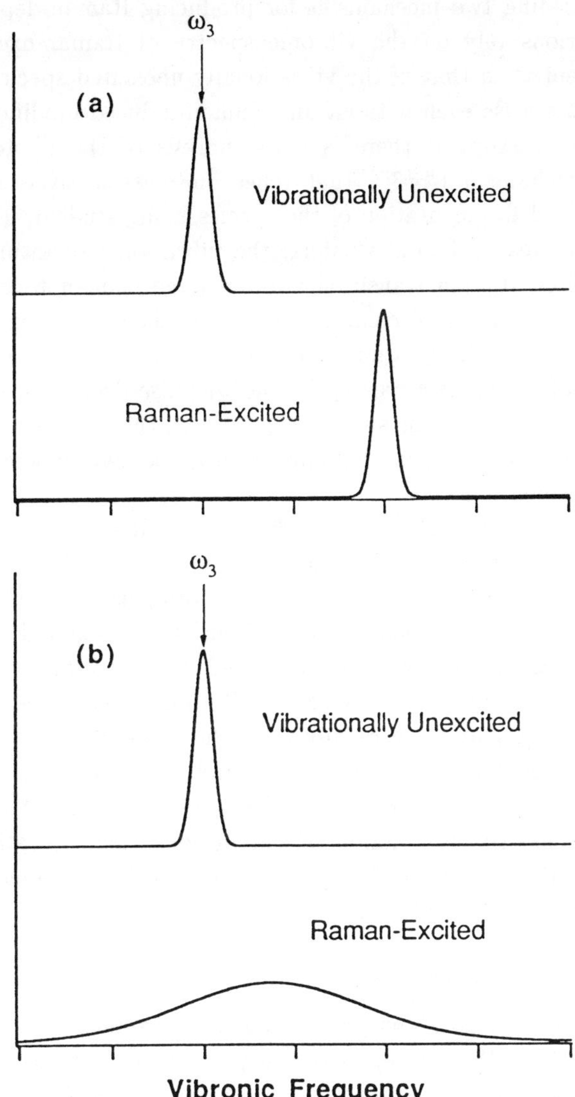

Fig. 3. Schematic depiction of vibronic spectra for vibrationally-unexcited and Raman-excited species, two situations that can lead to depletions in ILSRS. (a) The vibronic resonance of the Raman-excited species shifts completely away from that of the vibrationally unexcited species on which the REMPI probe (ω_3) is set. (b) The vibronic spectrum of the Raman-excited species broadens and decreases in maximal intensity due to IVR processes subsequent to Raman excitation.

The preceding two mechanisms for producing Raman-dependent photoion depletions rely on the vibronic spectra of Raman-excited species being different than that of the vibrationally unexcited species. However, depletions can arise even without any Raman-induced modification of the vibronic spectroscopy if there is mass-analysis of the photoions. One depletion mechanism ("#3") that relies on mass analysis involves the Raman-induced fragmentation of the species being studied. For example, in studies of weakly bound clusters, the vibrational quantum deposited in a stimulated Raman transition is often large enough to break one or more of the weak intermolecular bonds holding the cluster together. When such vibrational predissociation processes occur in a species the number of parent ions of the species created by the ω_3 pulse decreases, regardless of any features of the vibronic spectroscopy. Thus, there is a Raman-induced depletion in parent ion signal that appears as a resonance in a mass-selective ILSRS experiment.

A second depletion mechanism ("#4") that relies on mass analysis can arise even in situations where Raman-excited and vibrationally unexcited species have identical chances of being photoionized by the ω_3 pulse. In general, because photoions created from Raman-excited species have greater internal energy (by an amount equal to the vibrational quantum) than photoions produced from vibrationally unexcited species, the former will tend to fragment more readily than the latter. Such fragmentation reduces the signal in the parent ion channel. Thus, by virtue of ion fragmentation processes Raman-excited species will tend to have a smaller cross-section for the production of parent ions than vibrationally cold species. Selective detection of the parent-ion signal will therefore manifest Raman transitions as depletions.

2.3.2. *Other Spectral Features*

There are two types of processes that can produce complications in mass-selective ILSRS spectra, particularly in studies of molecular clusters. One type of complication arises when photoions created from vibrationally cold species undergo prompt fragmentation in the acceleration region of the TOFMS. Such fragmentation contaminates the signal in lower mass channels with contributions from higher clusters. If stimulated Raman transitions in the higher clusters give rise to modifications in the photoionization (and/or fragmentation) cross-sections of those clusters, then

Raman transitions in the higher clusters will be manifest in the signal corresponding to lower mass channels. This can reduce the size-specificity of a mass-selective ILSRS spectrum. As an example of this type of effect consider the series of clusters $A\text{–}(B)_n$, $n = 1, 2, \ldots$. Suppose the signal in the mass channel corresponding to $A\text{–}(B)_{n-1}^+$ has contributions from parent ions of $A\text{–}(B)_{n-1}$ and from fragments of the $A\text{–}(B)_n^+$ parent ion. Suppose further that stimulated Raman transitions in the $A\text{–}(B)_{n-1}$ and $A\text{–}(B)_n$ clusters reduce the efficiency with which these neutrals can be photoionized. In this situation there will be depletions in the $A\text{–}(B)_{n-1}^+$ signal that correspond to Raman resonances in both the $A\text{–}(B)_{n-1}$ and $A\text{–}(B)_n$ clusters. Generally, this loss of size selectivity is undesirable.

In regard to the above, it should be pointed out that ion fragmentation *per se* is not necessarily undesirable. For example, if the parent ions of both $A\text{–}(B)_n$ and $A\text{–}(B)_{n-1}$ fragment quantitatively by loss of a single B moiety (e.g., see Ref. 35), then one can measure the Raman spectrum of $A\text{–}(B)_n$ without contamination by monitoring the signal in the $A\text{–}(B)_{n-1}^+$ mass channel. It is only when two or more neutrals of different size contribute significantly to the same mass channel that the size-selectivity of mass-selective ILSRS suffers. It should also be pointed out that Raman contamination from higher clusters due to ion fragmentation processes need not only produce depletions in ILSRS spectra. Raman-dependent *gains* in ion signal at lower mass channels can occur if Raman-excited species have a greater propensity for producing fragment ions than do vibrationally unexcited species of the same size.

A second source of complication in mass-selective ILSRS spectra can occur due to Raman-induced fragmentation of neutrals. If such fragmented neutrals are subsequently photoionized by ω_3, they will contribute to a Raman-dependent gain in ion signal at masses smaller than the mass of the original cluster. So, for example, by this mechanism positive peaks may occur in the ILSRS spectrum associated with detection of the $A\text{–}(B)_n^+$ mass,[30] which peaks reflect Raman resonances in the clusters $A\text{–}(B)_m$, where $m > n$.

Given that fragmentation processes, whether in the neutrals or the ions, are the prime source for complications in mass-selective ILSRS spectra, one strategy to minimize those complications is to manipulate the size distribution of the species in the molecular beam. In particular, one seeks to produce a beam in which the species of interest is considerably more abundant than all those species larger than it. That is, one tries to minimize the number of species that can contaminate the mass channel of interest via fragmentation relative to the number of those species that one wants to study. In studies of molecular clusters the use of different expansion orifice sizes and shapes, different carrier gas pressures and mixtures, different beam-valve temperatures, different valve-to-skimmer distances, different valve pulse widths, etc. provide one with some power to manipulate the size distribution of species in the beam.

2.4. *Desirable Features of Mass-Selective ILSRS*

Having outlined the basics of mass-selective ILSRS and the way in which it is implemented, we now consider explicitly those features that render the method valuable in the spectroscopy of complexes and clusters. First, the technique is very sensitive. This is due to the high efficiency of the REMPI probe process and the ability of stimulated Raman scattering to effect substantial population transfer between vibrational levels. The sensitivity achievable by mass-selective ILSRS is clearly demonstrated by the results presented in Sec. 3.

Second, the technique can have considerable species-selectivity. Such selectivity derives from the size selectivity associated with the mass analysis of the photoions and from the vibronic resonance condition associated with the REMPI probe process. In practice, the degree of species-selectivity characterizing a given study depends in part on experimental parameters, particularly those pertaining to the molecular beam. As described in Sec. 2.3.2 these parameters influence the degree to which Raman spectra are contaminated with features due to larger clusters. Also important is the degree to which different species have different vibronic spectra. Indeed, the measurement of isomer-specific Raman spectra by mass-selective ILSRS can be achieved only if such spectral differences between isomers exist. An experimental example[28] of such selectivity is given in Fig. 4, which shows isomer-specific spectra of m-chlorophenol obtained with mass-selective ILSRS by tuning ω_3 to vibronic bands of each species.

A third important feature of mass-selective ILSRS is that it can be implemented with visible lasers as the sources of the ω_1 and ω_2 fields. Such sources permit wide spectral coverage and relatively easy generation of high-energy narrow bandwidth pulses with a single laser system.

Fourth, mass-selective ILSRS gives one access to all the vibrational

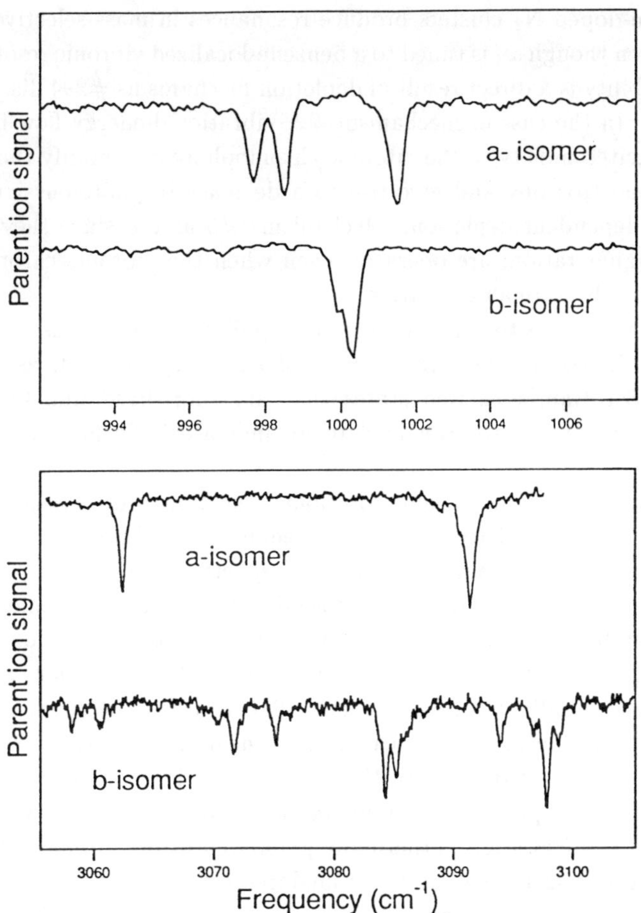

Fig. 4. Isomer-specific ILSRS spectra of the *syn* and *anti* isomers of the *m*-chlorophenol monomer in the 1000 cm^{-1} (top) and the 3080 cm^{-1} (bottom) spectral regions. The assignment of which pair of spectra belongs to the *syn* and which to the *anti* isomer is not known. Hence, we label them as *a*- and *b*-isomer. The spectra were recorded by monitoring the *m*-chlorophenol parent ion signal and tuning the probe laser to either the *a* *m*-chlorophenol 0^0_0 transition or the *b* *m*-chlorophenol 0^0_0 transition. See Ref. 28 for further details. Reprinted with permission from Ref. 28. Copyright 1992, American Chemical Society.

chromophores within a molecular complex or cluster, regardless of whether that chromphore is the same one through which the REMPI process proceeds. As shown in Sec. 3.3., for example, the N–N stretch fundamentals in benzene-doped N_2 clusters produce resonances in mass-selective ILSRS spectra even though ω_3 is tuned to a benzene-localized vibronic resonance.[32] This capability is a direct result of depletion mechanisms #2–4 discussed in Sec. 2.3.1. In the case of mechanism #2, vibrational energy flow from the Raman-excited moiety to the vibronic chromophore can modify the latter's vibronic spectroscopy and give rise to a decrease in photoion signal, i.e., a Raman-dependent depletion. Mechanisms #3 and 4, since they rely on cluster fragmentation, are operative even when the vibronic chromophore is unaffected by Raman excitation.

Fifth, mass-selective ILSRS can be applied in situations where the final state in the Raman transition is short-lived.[23] Indeed, as outlined in Sec. 2.3.1 Raman-excited states that are short-lived due to IVR or predissociation processes can lead to an increase in Raman-induced ion depletion.

Finally, the vibrational and rotational selection rules associated with the Raman effect lead to desirable consequences. Vibrational transitions different than those available to infrared spectroscopy can be probed by Raman spectroscopy. Thus, mass-selective ILSRS is complementary to infrared studies of clusters. In addition, the rotational selection rules in a polarized Raman transition favor a dominant Q-branch (when ω_1 and ω_2 are polarized parallel to one another).[36] This causes most of the depletion intensity to overlap in a single feature at or near the vibrational transition frequency. The consequences of this are several. First, it allows for large Raman-induced depletions in the photoion signal, even when a wide distribution of rotational levels is thermally populated in the sample. Second, it facilitates the measurement of vibrational frequencies and splittings. Third, it makes it feasible to relate the measured widths of vibrational resonances to the rates of dynamical processes in the vibrationally excited state, even when individual rovibrational resonances are not resolved.

2.5. *Comparison with Other Methods*

Having discussed the characteristics of mass-selective ILSRS it is pertinent to compare the method with other options for the subwavenumber-resolution, ground-state vibrational spectroscopy of species in seeded supersonic molecular beams. We consider, in particular, infrared methods, stimulated emission pumping techniques, and other nonlinear Raman schemes.

There is no question that infrared methods have been extremely valuable in the molecular beam spectroscopy of complexes and clusters[7-11,13-16] and will continue to be so. Nevertheless, there are some limitations of such methods that can be substantially overcome by using mass-selective ILSRS. One limitation pertains to spectral range. Access to infrared transitions throughout the vibrational fundamental region requires several light sources and, indeed, several significantly different technologies. The same is not true of stimulated Raman schemes. Second, those infrared schemes that rely on direct absorption[10,11,13] or optothermal detection[9,14,15] produce spectra to which all of the species in the sample contribute. The spectra are species-selective only to the extent that resolved features can be confidently attributed to specific species, something which often requires the observation and analysis of rotational structure. This places limits on the size of the species that can be studied with any sort of size- or species-selectivity. These limits are well below those characterizing mass-selective ILSRS. Third, of course, there are vibrational chromophores and vibrational transitions whose study by infrared methods is effectively precluded by symmetry. Such chromophores and transitions can often be accessed by stimulated Raman methods (e.g., see Refs. 18 and 32). Finally, the rotational band structure of an infrared resonance tends to be at least several cm^{-1} in breadth. Therefore, without rotational resolution it can be difficult to extract linewidths, vibrational splittings, and/or vibrational frequency shifts precisely. The dominance of Q-branch structure in the resonances that appear in mass-selective ILSRS spectra renders the extraction of such parameters considerably easier (e.g., see Ref. 29).

One major drawback of mass-selective ILSRS in comparison to infrared methods is that the former requires access to a vibronic chromophore while the latter do not necessarily. This places restrictions on the type of species that can be studied conveniently by ILSRS. Second, the resolution that has been achieved with CW infrared methods[9-11,13-15] is considerably greater than that which is feasible in stimulated Raman studies of clusters. The high-intensity requirements placed on the Raman excitation fields necessitate pulsed light sources that have Fourier-transform-limited bandwidths of about 100 MHz or greater. Third, the fact that Q-branches dominate the band structure of those Raman resonances that appear with any intensity in ILSRS effectively precludes the determination of rotational constants from the spectra. Such constants and the structural information they contain are more readily available from the rotational structure associated with an infrared transition.

In comparison with stimulated emission pumping methods[19-21] mass-selective ILSRS has several advantages in molecular beam studies. First, light-source requirements are less stringent in ILSRS since only one vibronic resonance need be accessed and the vibrational transition can be driven by two visible lasers, only one of which need be tunable. In contrast, SEP techniques require access to two vibronic resonances. Generally, this means that two tunable ultraviolet light sources are required and one of these must be scanned to obtain a spectrum. Second, mass-selective ILSRS provides access to all of the vibrational moieties in a cluster. The vibrational transitions associated with SEP processes will tend to be localized in the chromophore associated with the pump–pulse-induced vibronic transition. Third, SEP depletions are dependent on the lifetime of the excited vibronic intermediate state relative to the pulsewidths of the light sources employed.[37] Short excited-state lifetimes require short-pulse lasers to achieve appreciable depletions. This requirement does not obtain in ILSRS. Finally, the rotational band structure associated with a depletion resonance in a frequency-domain SEP experiment depends on the bandwidth and position of the pump laser.[19] Moreover, the structure is that associated with a dipole-allowed transition. The point is that the rotational structure associated with a SEP resonance of a reasonably large species will generally be considerably broader than the Q-branch associated with the same resonance in an ILSRS spectrum.

There are also disadvantages of mass-selective ILSRS relative to SEP techniques. First, SEP methods are better suited to the study of in-termolecular vibrational motions (e.g., see Refs. 38–40) than is ILSRS. This is because SEP methods are effectively stimulated *resonance*-Raman techniques. Thus, one can choose the excited intermediate state in SEP so as to maximize the Franck–Condon factors associated with transitions between intermolecular vibrational levels. Furthermore, even though pump and probe transitions may be weak in a SEP process, the fact that they are single-photon resonances means that they can generally be driven close to saturation without much difficulty. Neither of these capabilities obtains in stimulated Raman methods that are not resonantly enhanced. Second, the presence of significant rotational structure on SEP resonances allows for analysis leading to the rotational constants of the species (assuming the rotational structure can be resolved). As we have pointed out above, this is generally not feasible in ILSRS.

Finally, mass-selective ILSRS has some significant advantages relative to other nonlinear Raman methods. Consider first a comparison with mass-selective IGSRS.[22,23] Unlike ILSRS, IGSRS cannot be applied in situations where the final state in the Raman transition is short-lived relative to the pulsewidths of the lasers. In such cases, the excited vibrational state decays before the ω_3 pulse can photoionize the Raman-excited species.[23] (One way around this is to tune ω_3 to ionize from those states that are populated via relaxation from the Raman-excited state.[29] Clearly, however, this adds complexity that does not exist in ILSRS.) Second, IGSRS is not generally applicable to the characterization of vibrational transitions localized in moieties other than the one through which the REMPI process proceeds.[23] (Once again, this limitation can be overcome, in principle, if vibrational energy flow occurs subsequent to Raman excitation and ω_3 is tuned to photoionize the vibrationally "relaxed" species that result from this energy flow process.[29]) Finally, IGSRS is not as useful as ILSRS as a survey spectroscopy because to cover a broad spectral range in the former ω_3 has to be tuned as $\omega_1 - \omega_2$ is tuned, whereas in the latter ω_3 remains fixed. IGSRS also has advantages over ILSRS, however. In particular, in situations where IGSRS can be applied, it produces spectra with much better signal-to-noise ratios than ILSRS because, in principle, it is a zero-background method. In addition, IGSRS can be readily implemented as a pump–probe spectroscopy capable of monitoring the dynamics of Raman-excited states in real time.[22] ILSRS does not have this capability. Finally, IGSRS allows one to measure the vibronic spectra of Raman-excited states by tuning ω_3 while $\omega_1 - \omega_2$ is fixed to the Raman resonance of interest.[25] Again, there is no such capability in ILSRS.

The principal advantage of mass-selective ILSRS relative to the other stimulated Raman method that has proven utility in cluster studies[18]—stimulated Raman loss spectroscopy—is that the latter is not very species- or size-selective. On the other hand, stimulated Raman loss does not require the presence of a convenient vibronic chromophore in the species of interest. Thus, it is possible to study species such as neat $(N_2)_n$ clusters,[18] for example. In comparison to coherent anti-Stokes (or Stokes) Raman spectroscopy,[41] which has also been applied in limited fashion to molecular beam samples,[42,43] mass-selective ILSRS has an overwhelming advantage in sensitivity. This is due to the fact that the former methods have signals that scale with the square of the density of the sample[41] unlike the linear density dependence of ILSRS. This disparity in sensitivity can be mitigated to some extent if the coherent Raman scheme is implemented with resonance enhancement.[44] However, this entails a considerable increase in experimental complexity and, moreover, requires one or more of the three excitation fields in the coherent Raman scheme to be resonant with a vibronic transition.

3. Selected Results

To date, we have applied mass-selective ILSRS to several molecular cluster systems. These include species of the form benzene$_n$,[22,27,29,45] benzene–$(Ar)_n$,[31] benzene–$(N_2)_n$,[32] benzene–$(water)_n$,[46] carbazole–$(Ar)_n$,[30] and phenol–X complexes[28] (X = H_2O, NH_3, methanol, ethanol, phenol, benzene, diethylether, CH_4, and Ar). In this section we review some of the results for benzene dimer, benzene–$(Ar)_n$, and benzene–$(N_2)_n$ clusters. These results are representative of the kind of information that is currently available from the technique.

3.1. *Benzene Dimer*

We have reported an extensive nonlinear Raman study of benzene dimer isotopomers.[29] The study, which employed both mass-selective ILSRS and IGSRS methods, was aimed primarily at learning about the geometry of this prototypical aromatic–aromatic dimer. In this regard, the results from both of the IDSRS methods proved to be informative. However, in several respects the ILSRS method was uniquely valuable.

One of our principal concerns was to ascertain whether the benzene moieties in the dimer are symmetrically equivalent or inequivalent. To address this question, mass-selective ILSRS spectra in the region of benzene's ν_1 (totally symmetric ring-breathing mode) and ν_2 (totally symmetric C–H stretch) fundamentals[47] were measured. Figure 5 shows spectra in the ν_2 region of the fully protonated and fully deuterated dimers. In both cases, one observes two Raman bands. As we have argued,[29] the very presence of two ν_2 bands in these species is a strong indication that the benzene moieties comprising the dimer are inequivalent. Similarly, two ν_1 bands were observed by mass-selective ILSRS for these same two dimer isotopomers. For both species, the higher frequency resonance of the ν_1 doublet has significantly less intensity than the lower frequency member. These observations also strongly indicate a dimer composed of benzenes in inequivalent sites.

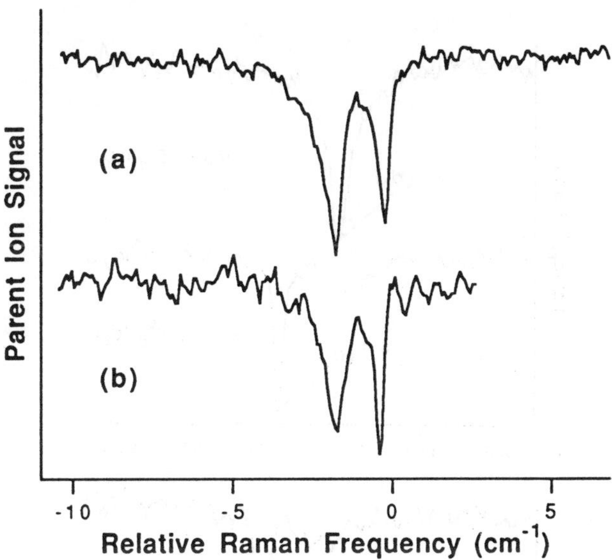

Fig. 5. Mass-selective ILSRS spectra of (a) the fully protonated benzene dimer in the region of the ν_2 fundamental of C_6H_6 and (b) the fully deuterated dimer in the region of the ν_2 fundamental of C_6D_6. Reprinted with permission from Ref. 29. Copyright 1992, American Institute of Physics.

Mass-selective ILSRS experiments on isotopically mixed species [i.e., (h_6–benzene)–(d_6–benzene), (h_6–benzene)–(d_1–benzene), and (d_6–benzene)–(d_1–benzene)] yielded even more information on the dimer's structure. First, the results of these experiments demonstrate that two different isomers of each of these isotopomers exists. Figure 6 shows spectra relevant to this issue. The spectra pertain to the h_6-localized and d_6-localized ν_2 fundamentals of the h_6–d_6 isotopomer. The top spectra in Figs. 6a and 6b were obtained by setting ω_3 to the $S_1 \leftrightarrow S_0$ 0_0^0 band localized in the h_6 benzene moiety. The bottom spectra were measured by tuning ω_3 to the d_6–localized 0_0^0 band. One notes that there are two Raman resonances per nondegenerate monomer fundamental. (The analogous behavior was observed for the ν_1 and ν_2 fundamentals of all mixed isotopomers.) This can only occur if the benzene sites in the dimer are inequivalent, in which case two different h_6–d_6 isomers exist—ones corresponding to the interchange of the h_6 and d_6 moieties between the inequivalent sites.

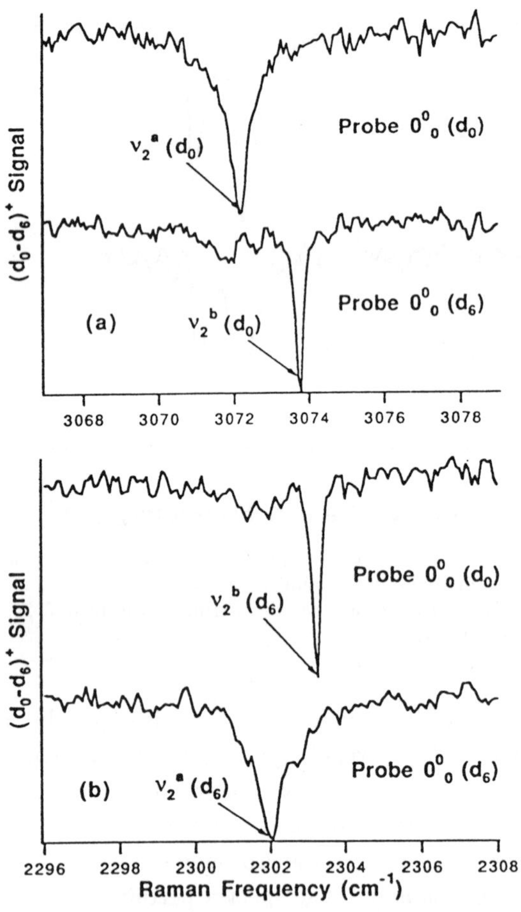

Fig. 6. Mass-selective ILSRS spectra of the $d_0 - d_6$ benzene heterodimer. (a) Top—Spectrum in the $\nu_2(h_6)$ region obtained by tuning ω_3 to the h_6-localized 0^0_0 band of the heterodimer. Bottom—Spectrum in the same region obtained by tuning ω_3 to the d_6-localized 0^0_0 band of the heterodimer. (b) Top—Spectrum in the $\nu_2(d_6)$ region obtained by tuning ω_3 to the h_6-localized 0^0_0 band of the heterodimer. Bottom—Spectrum in the same region obtained by tuning ω_3 to the d_6-localized 0^0_0 band of the heterodimer.

The second point to be noted from the spectra of Fig. 6 pertains to the correlation between the frequency of the Raman resonances and the identity of the moiety through which the ionization process proceeds. In particular, the red-most h_6-localized Raman band (which we call $\nu_2^a(h_6)$) is associated with a probe process that proceeds through the 0_0^0 band localized in the h_6 moiety. Similarly, the red-most d_6-localized Raman band ($\nu_2^a(d_6)$) is associated with probing through the d_6-localized 0_0^0 band. Now, the similarity of the shifts of $\nu_2^a(h_6)$ and $\nu_2^a(d_6)$ away from their respective monomer fundamental frequencies strongly indicates that the two bands arise from benzene moieties in the same site—"site a." That is, $\nu_2^a(h_6)$ comes from the h_6–d_6 isomer in which the h_6 moiety is bound in site a, and $\nu_2^a(d_6)$ comes from the other isomer, in which the d_6 moiety is bound in site a. Given this, the double-resonance data of Fig. 6 show that the two 0_0^0 bands observed for the h_6–d_6 isotopomer are transitions localized in site-a benzene moieties. This brings up the question—where are the two site-b-localized 0_0^0 bands that should also be present? Only two, not four 0_0^0 bands have ever been observed for this dimer isotopomer.[48,49] A plausible answer to this question is that the b-site 0_0^0 bands have no intensity because site b is of high symmetry—the $S_1 \leftrightarrow S_0$ 0_0^0 band of a benzene moiety is known to have no intensity if the moiety resides in a high-symmetry site (e.g., see Ref. 35b).

The upshot of all of the above is that the ILSRS results not only inform one as to the inequivalence of the sites in the benzene dimer, they also provide information about the nature of those sites. Taken together, mass-selective ILSRS and IGSRS go a long way toward characterizing the geometry of the dimer. In Ref. 29 detailed arguments supporting specific structural forms are presented. The interested reader is referred to this reference for further information.

3.2. *Benzene–(Ar)$_n$ Clusters*

Benzene–(Ar)$_n$ clusters are species that have received considerable attention in the past few years.[50–54] Part of the motivation for studying these species arises from the mass-selective REMPI spectra of Hahn and Whetten and the interpretation thereof in terms of liquid- and solid-like clusters.[51] These vibronic spectra, which pertain to the benzene-localized 6_0^1 vibronic band, evolve in the manner depicted schematically in Fig. 7. The broad features in the $n \simeq 6$–16 size range were assigned to liquid-like clusters.[51] The simultaneous appearance of broad and sharp spectral features in the $n \simeq 17$–21 range was interpreted in terms of the coexistence of liquid-like (broad feature) and solid-like (sharp features) clusters.[51] Finally, the exclusive appearance of sharp structure for $n > 21$ was attributed to solid-like species.[51] We undertook mass-selective ILSRS studies of these clusters in the hope of shedding more light on their nature.

Figure 8 shows mass-selective ILSRS spectra pertaining to the benzene–localized ν_1 fundamental of benzene–$(Ar)_n$ species.[31] These spectra were obtained by setting ω_3 to the peak of the 6_0^1 vibronic structure in the $n \leq 14$ size range and to the positions marked by arrows in the schematic spectra of Fig. 7 for the $n = 15$ to 22 sizes. One notes the following points about the Raman spectra. First, for $n < 16$ they consist of a single peak (to within our resolution) which shifts steadily to the blue with increasing n. Second, at $n = 16$ a second band appears to the red to produce a resolved doublet. Third, as n increases from 16 to 21 this doublet structure remains. However, the blue member of the doublet, which correlates with the single band observed at lower cluster sizes, decreases in intensity relative to the redder band. By $n = 22$, the redder Raman band is the only one observed. Experiments were also carried out on the ν_2 fundamental. The results of these experiments exhibit the same qualitative trends as those exhibited by the ν_1 spectra.

Mass-selective ILSRS spectra were also measured as a function of ω_3 for particular cluster sizes. Figure 9 shows selected results. In Fig. 9a, ω_3 was tuned to the red-most sharp vibronic band associated with the $n = 18$ cluster size. Figure 9b shows the spectrum obtained by tuning ω_3 to a position that overlaps both the blue-most sharp vibronic band and the broad band associated with the clusters of this size. One clearly sees that only one red-shifted ν_1 band is present in Fig. 9a, whereas a doublet appears in Fig. 9b. Analogous behavior was observed for other sizes in the $n = 16$ to 20 range and for the ν_2 fundamental as well.

The ILSRS results on benzene–$(Ar)_n$ clusters allow one to make several statements regarding the clusters in the $n = 16$ to 21 size range.[31] First, the very presence of two ν_1 and two ν_2 bands for a given cluster size means that at least two grossly different cluster types are present in the sample in this size range. This statement follows from the fact that the ν_1 and ν_2 modes of benzene are nondegenerate; a splitting of the fundamentals

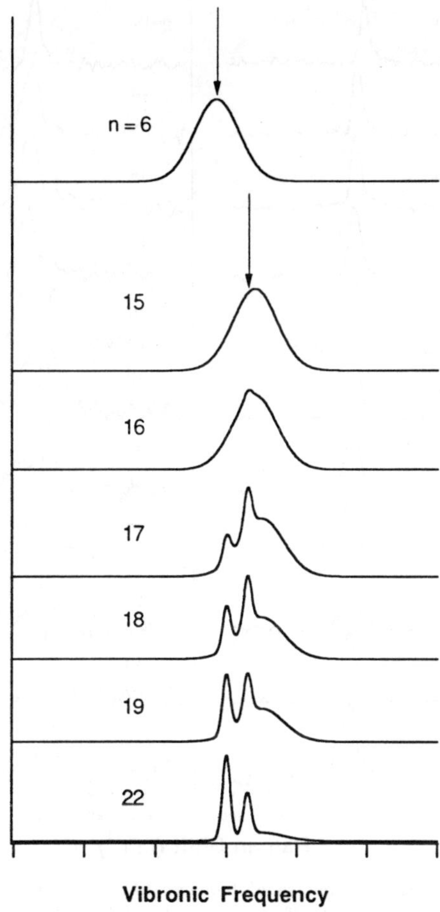

Fig. 7. **Schematic** vibronic spectra of benzene–$(Ar)_n$ clusters in the region of the $S_1 \leftarrow S_0$ 6_0^1 benzene-localized transition. The spectra were constructed by reference to the results of Ref. 51. The arrows indicate the positions of ω_3 used to obtain the mass-selective ILSRS spectra of Fig. 8. (Identical ω_3 values were used for $n = 15$ to 22.)

can only arise from different cluster isomers. Second, the evolution of the Raman spectra from $n = 15$ to 22 shows that there is a changeover from one of these gross cluster types at small n to the other type at large n. Notably, this changeover correlates with the evolution of the vibronic spectra of the species. Third, the results of Fig. 9 show that the member of the Raman doublet that grows in at $n \simeq 16$ and persists at larger sizes arises from

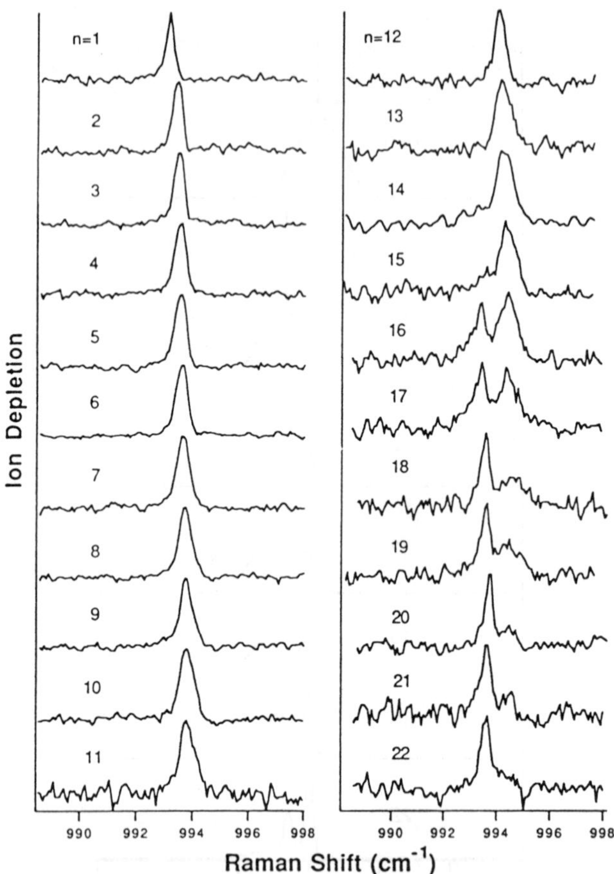

Fig. 8. Mass-selective ILSRS spectra of benzene–$(Ar)_n$ clusters in the region of the benzene-localized ν_1 fundamental. The number that labels each spectrum corresponds to the number of Ar atoms in the detected benzene–$(Ar)_n^+$ ion. The spectra are plotted such that the positive ordinate direction corresponds to increasing depletion. Also, the baselines of the spectra have been removed, and they have been scaled such that the largest depletion in any one spectrum has the same height as that in the others.

the same cluster isomers that produce the sharp vibronic bands. The bluer member of the Raman doublet arises from the species that produce the broad vibronic feature. Thus, the assignment[51] of sharp vibronic features to one cluster type and broad vibronic features to another type is substantiated by the ILSRS results.

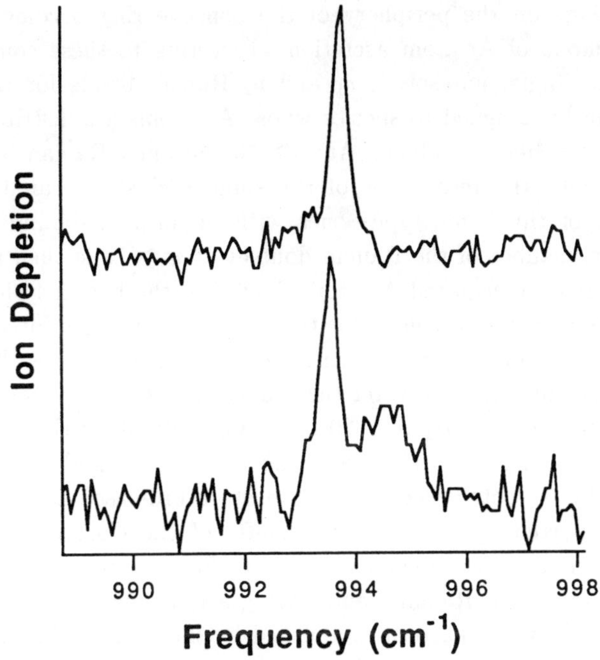

Fig. 9. Mass-selective ILSRS spectra pertaining to the benzene-localized ν_1 fundamental of benzene–$(Ar)_{18}$. The top spectrum was obtained by setting the probe frequency (ω_3) to the position of the red-most sharp vibronic band reported in Ref. 51 (see also Fig. 7). The bottom spectrum was obtained by setting ω_3 to the peak of the blue-most sharp vibronic band (Ref. 51), a frequency that also significantly overlaps the broad vibronic feature observed for this cluster size (the position of the lower arrow in Fig. 7).

Just what distinguishes the different benzene–$(Ar)_n$ cluster types that appear in the $n \simeq 16$ to 21 size range? Our ILSRS results do not rule out an interpretation based on the coexistence of liquid-like and solid-like species. On the other hand, an interpretation based on structure—the building up of the Ar solvent shell about the benzene molecule—is a plausible alternative that fits the data very well.[31] Computational (Monte Carlo) studies by Hahn[51b] and by Fried and Mukamel[54] have addressed the issue of structure in the benzene–$(Ar)_n$ species. By using a Lennard–Jones, atom–atom potential they found that the most favorable structures for $n = 1$ to 14 involve Ar atoms above and below the benzene plane. (These results are supported by experiments on the smallest clusters, which clearly indicate out-of-plane binding.[50,52]) Beyond $n = 14$, however, Ar binding in the benzene plane on the periphery of the benzene ring becomes the most favorable mode of Ar-atom accretion. Referring to these computational results, the single, unresolved ν_1 and ν_2 Raman bands for the $n < 16$ clusters can be assigned to species whose Ar atoms are distributed above and below the benzene plane. At $n \geq 16$ the bluer Raman band, which correlates with the resonances of the smaller clusters, can be assigned to clusters of this same type—ones with no in-plane-bound Ar atoms. The redder member of the Raman doublet at $n \geq 16$ is then assigned to clusters that have acquired Ar atoms bound in the benzene plane.[31] This interpretation readily accounts for the increase in relative intensity of the redder Raman band as cluster size increases from $n = 16$ — the more Ar atoms in the cluster, the greater the chance that some Ar atoms will bind in the benzene plane. By $n = 20$ or 21 virtually all of the clusters have such Ar atoms present.[52] Thus, only the red Raman band remains.

It should be noted that the above structure-based interpretation is further supported by the frequency shifts of the cluster Raman bands relative to the benzene monomer ν_1 and ν_2 fundamentals. At $n \geq 16$, the presence of a second Raman band having a frequency shift that deviates significantly from the pattern established by the smaller clusters signals a qualitative change in the perturbation of these in-plane vibrational modes. Furthermore, the ν_2 results indicate that the change is in the direction of an increase in that perturbation.[31] The natural, if somewhat simple-minded conclusion, is that such perturbation is the consequence of in-plane binding by solvent species.

Additional evidence in support of this structural interpretation was obtained by mass-selective ILSRS experiments on smaller benzene–$(Ar)_n$ clusters.[31] In particular, ν_1 and ν_2 Raman spectra were measured for the $n = 2$ to 4 clusters. Even at these cluster sizes different isomers are present. These isomers are distinguished by different distributions of Ar atoms above and below the benzene plane and by different vibronic resonance frequencies. Isomer-specific ILSRS spectra of the clusters show that the different distributions of Ar atoms above and below the benzene do not appreciably affect the frequencies of the ν_1 and ν_2 fundamentals. The implication is that the two cluster types that give rise to the Raman doublets in the $n \simeq 16$ to 20 size range are isomers that differ by more than just the distribution of out-of-plane solvent species. The natural conclusion is that they differ in the number of in-plane species.

3.3. *Benzene–$(N_2)_n$ Clusters*

In contrast to benzene–$(Ar)_n$ species, benzene–$(N_2)_n$ clusters are solute–(solvent)$_n$ species in which the individual solvent moieties have vibrational structure. This provides one with the opportunity to monitor solvent structure and dynamics by direct probing of the vibrational spectroscopy of the solvent species. Mass-selective ILSRS is particularly suited to this task given its ability to probe the vibrational structure of all the moieties in a molecular cluster, not just the one through which the REMPI process proceeds (i.e., the benzene).

Figure 10 shows mass-selective ILSRS spectra in the region of the N–N stretch fundamental (~ 2330 cm^{-1}) of benzene–$(N_2)_n$ clusters for $n = 1$ to 16.[32] (Spectra were measured for n as large as 32.) The spectra were obtained by setting ω_3 to the maximum of the benzene-localized 6_0^1 band[55,56] associated with the cluster size of interest. The spectra exhibit several notable features. First, the $n = 1$ complex has only a single resonance, which is shifted to the red of the bare N_2 fundamental by 2 cm^{-1}. Second, the ILSRS spectra corresponding to the detection of benzene–$(N_2)_n^+$, $n = 2 - 10$, exhibit doublets, with one member of the doublet (band "a") correlating with the $n = 1$ resonance and the other (band "b") shifted to the blue by about 1 cm^{-1}. Third, the intensity of band a relative to band b decreases as n increases. Finally, at $n = 11$ a third band (band "c") grows in to the blue of both band a and band b. The intensity of this band increases relative to that of the latter two as the cluster size increases. Near $n = 20$ (not shown), band c becomes unresolvable from band b.

The results of Fig. 10 can be interpreted in terms of the building up of an N_2 solvent shell about the benzene solute. The $n = 1$ species is known to have a geometry in which the N_2 moiety is centrally bound above the benzene ring.[57] Thus, the N–N band in the $n = 1$ spectrum is due to such a moiety. Given this, and the fact that the a bands in the higher-n spectra have frequencies similar to the $n = 1$ band, it is reasonable to assign the latter to centrally bound N_2 moieties. The shift of band b away from band a and towards the free N_2 resonance frequency suggests that it arises from N_2 moieties bound in sites different from the two available central binding sites. Such sites will open up subsequent to the filling of one or both of the central sites. In particular, binding sites out of the benzene plane and in a circle about a centrally bound N_2 might

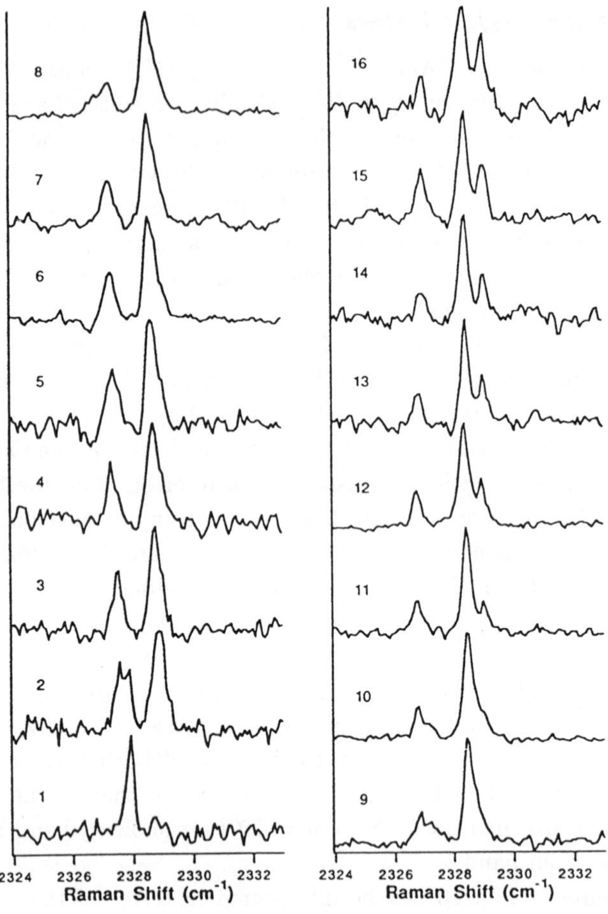

Fig. 10. Mass-selective ILSRS spectra of benzene–$(N_2)_n$ clusters. The number that labels each spectrum corresponds to the number of N_2 moieties in the detected benzene–$(N_2)_n^+$ ion. The spectra are plotted such that the positive ordinate direction corresponds to increasing depletion. Also, the baselines of the spectra have been removed, and they have been scaled such that the largest depletion in any one spectrum has the same height as that in the others.

be expected to be energetically favorable.[55,58] We assign the b bands to N_2 moieties bound in such sites. Notably, the number of these sites is greater than the number of central binding sites. Thus, as cluster size increases from $n = 2$, the fraction of N_2 species in the latter sites will decrease relative to the fraction in the former. This accounts for the trend in relative intensities of the a and b bands in going from $n = 2$ to 10. Finally, we assign the c bands to N_2 moieties bound in a third kind of site. This interpretation is consistent with the growth of band c relative to bands a and b as n increases from 11 to 20. It is probable that this third site corresponds to positions in or near the plane of the benzene moiety at a distance from the center of the benzene ring. One expects such sites to become the energetically most favorable ones after the sites corresponding to bands a and b are filled.[58] The assignment of the c bands to in-plane-bound N_2 moieties is supported by mass-selective ILSRS results pertaining to the benzene-localized ν_1 fundamental. In short, a break in the steady, size-dependent frequency shift of this in-plane fundamental has been observed[46] over the $n = 9 - 11$ size range, the same size range in which band c first appears. A reasonable interpretation is that this qualitative change in the perturbation of the in-plane vibrational motion arises from the in-plane binding of N_2 species.

4. Future Prospects

4.1. *Neutrals*

The results presented above make it clear that size-selective, and indeed species-selective, vibrational spectroscopy of neutral molecules, molecular complexes, and molecular clusters is possible with mass-selective ILSRS. There is little doubt that the technique will be applied to many more species of these types. Several classes of studies, in particular, suggest themselves. First, ILSRS provides one with the capability to examine systematically, at subwavenumber resolution, and throughout the fundamental region, the vibrational structure of large molecules. By such studies one can expand the knowledge pertaining to vibrational couplings and vibrational energy flow in molecular systems. One can also characterize the vibrational structure of individual conformers of the same compound. In this regard, the coupling of mass-selective ILSRS with laser-desorption vaporization of the sample[59] promises to be a powerful probe of how the conformational structure of biomolecules, such as di- and tri-peptides, is reflected in their vibrational spectroscopy. Finally, the combination of ILSRS with laser photolysis of radical precursors in a supersonic expansion[60] offers one a promising approach for the study of the ground-state vibrational spectroscopy of large organic and organometallic radicals.

The size-selective vibrational spectroscopy of molecular clusters like the benzene–$(Ar)_n$ and benzene–$(N_2)_n$ series described above has only just become possible by virtue of the development of mass-selective ILSRS. Given this, even straightforward extensions of these studies, such as experiments on substituted-benzene–$(N_2)_n$ clusters or benzene–$(CO)_n$ clusters, have the potential to be rich sources of information about intermolecular interactions, local structure in solutions, solute and solvent vibrational dynamics, etc. A more significant extension to studies of clusters formed in a "pick-up" source[15b] is also a promising direction to follow. In pick-up experiments one can deposit a probe species onto the surface of an intact homogeneous cluster. With mass-selective ILSRS one can then probe the vibrational spectroscopy of the adsorbate, or one can use the adsorbed moiety as the photoionization chromophore in ILSRS and measure the Raman spectrum of the rest of the cluster. In either case, one has access to information that generally is not available in experiments on clusters formed by co-expanded species.

Used in conjunction with a laser vaporization molecular beam source,[61] mass-selective ILSRS also has considerable promise in the study of atomic clusters. For example, the sharp, accessible vibronic structure of many small metal clusters[61,62] make them ideal candidates for study by ILSRS. The ability to perform ground-state vibrational spectroscopy on such species at subwavenumber resolution should add considerably to one's knowledge of metal–metal bonding and the size evolution thereof. ILSRS studies of these clusters in complexation with other species[63] have the potential to inform one about surface adsorption on metals and to connect experiment directly with theory in this area. The applicability of mass-selective ILSRS to larger metal clusters and to other types of atomic clusters (e.g., C_n, Si_n, alkali halides) is not clear at this point because one is not certain that efficient, Raman-dependent photoion depletion mechanisms will obtain. Nevertheless, it is probable that at least for some such species, mass-selective ILSRS will be possible. Alternatively, the approach of using a weakly bound "messenger" species[64] that dissociates from the cluster upon Raman excitation could prove to be a useful way to get at the vibrational spectroscopy of these kinds of clusters.

4.2. *Ions*

The ILSRS method involves photoionization subsequent to stimulated Raman excitation. It is not a scheme that one would think to apply to species that are already ions. However, the general idea of a stimulated-Raman action spectroscopy based on charged-particle detection is one that has potential in the study of ions. There are at least two motivations for the development of such spectroscopies. First, of course, ionic species are important ones for which one would like to have vibrational information. Stimulated Raman spectroscopies might be expected to be particularly valuable in obtaining such information on large ions. Second, with cluster ions one can size-select the species prior to any spectroscopic measurement. With such size selection one need not worry about the degradation of size-specificity due to light-induced ion fragmentation processes, something that must be worried about in mass-selective ILSRS experiments on neutrals.

The major issue in performing stimulated Raman experiments on ions is how one detects the occurrence of the stimulated Raman transitions. There are several possible approaches. First, one can use a Raman scheme analogous to the infrared photodissociation-induced ion depletion method.[65] Such a scheme would rely on stimulated Raman-induced fragmentation of size-selected ions to produce a depletion in the number of those ions. The major potential drawback to this approach derives from the fact that fairly tight focusing of the Raman excitation fields is required to effect substantial fractional population transfers. If the pertinent focal volume does not overlap a significant portion of the volume in which the ions reside, then large Raman-induced depletions will not be possible. Thus, some spatial focusing of the size-selected ions may be necessary for this depletion scheme to work. An alternative approach to monitoring the depletion of size-selected ions is to look for fragment ions that result from the Raman excitation of the size-selected species.[66] This scheme, in principle, is a zero-background one. It should be much less dependent on achieving a close match between the sample volume and the Raman excitation volume and should have greater sensitivity than the depletion scheme.

In the case of negative ions, there are some additional feasible approaches besides those that depend on Raman-induced fragmentation.[67] These involve probing the stimulated Raman transitions by photodetachment and electron detection. One scheme, which is analogous to IGSRS, involves setting the frequency of a probe laser just below the photodetachment threshold of a size-selected negative ion in such a way that stimulated-Raman excitation of the ion meets the energy condition for photodetachment. One then measures the photoelectron signal as a function $\omega_1 - \omega_2$. Raman resonances, in principle, will appear as gain signals in the resulting spectrum. A similar scheme involves photodetachment probing with time-of-flight analysis of the photoelectrons. The idea behind this approach is that photodetachment from Raman-excited ions should produce photoelectrons with a different energy distribution and/or a different electron-emission rate than photodetachment from vibrationally unexcited ions. Thus, detection of photoelectrons within a selected range of arrival times, as a function of $\omega_1 - \omega_2$, can reflect the stimulated Raman spectrum of the negative ion of interest. This scheme allows for the use of a fixed-frequency photodetachment light source that produces photoelectrons from both vibrationally excited and unexcited species. Raman resonances may appear as either gain or depletion signals, respectively, depending on whether the photoelectron time gate selects electrons preferentially from vibrationally excited or unexcited species. Finally, a negative-ion method that is effectively another depletion version of stimulated emission pumping is also possible. In this method the ω_1 frequency is chosen to be above the photodetachment threshold of the vibrationally cold negative ion. The photoelectron signal is then monitored as a function of ω_2. When $\omega_1 - \omega_2$ is resonant with a Raman transition of the ion, the photoelectron signal produced by ω_1 will, in principle, decrease due to downward transitions induced by ω_2.

4.3. Conclusion

The use of mass-selective, ionization-detected stimulated Raman spectroscopies in molecular beam studies is a relatively recent development. Nevertheless, the results already obtained encourage one that such methods will be a fruitful source of ground-state spectroscopic information on molecules, complexes, and clusters in the years to come. Such information will not only be complementary to that obtained by other ground-state vibrational spectroscopies; for some classes of important species it may well be the only ground-state spectroscopic information realistically available from experiment. The continued development and application of mass-selective IDSRS methods is currently in progress in this laboratory.

Acknowledgments

It is a pleasure to acknowledge past and present members of my group for their contributions to the research described herein. I am especially grateful to Dr. Gregory V. Hartland, Dr. Bryan F. Henson, and Vincent A. Venturo. I also thank Robert Hertz, Dr. Patrick Maxton, Mark Schaeffer, and Wousik Kim for their contributions. Financial support for this research has been provided by the Department of Energy, Department of Basic Energy Sciences (grant no. DE-FG03-89-ER 14066).

References

1. *Structure and Dynamics of Weakly Bound Complexes*, ed. A. Weber (Reidel, Dortrecht, 1987).
2. *Large Finite Systems. Proceedings of the 20th Jerusalem Symposium on Quantum Chemistry and Biochemistry*, eds. J. Jortner and B. Pullman (Reidel, Dortrecht, 1987).
3. *Studies in Physical and Theoretical Chemistry, Vol. 68. Atomic and Molecular Clusters*, ed. E. R. Bernstein (Elsevier, Amsterdam, 1990).
4. D. H. Levy, *Adv. Chem. Phys.* **47**, 3742 (1981).
5. (a) U. Even, A. Amirav, S. Leutwyler, M. J. Ondrechen, Z. Berkovitch-Yellin, and J. Jortner *Faraday Discuss. Chem. Soc.* **73**, 153 (1982); (b) S. Leutwyler and J. Jortner, *J. Phys. Chem.* **91**, 5558 (1987).
6. K. I. Peterson, G. T. Fraser, D. D. Nelson, and W. Klemperer, in *Comparison of Ab Initio Quantum Chemistry with Experiment for Small Molecules: The State of the Art*, ed. R. J. Bartlett (Reidel, Dortrecht, 1985).
7. F. G. Celii and K. C. Janda, *Chem. Rev.* **86**, 507 (1986).
8. A. C. Legon and D. J. Millen, *Chem. Rev.* **86**, 635 (1986).
9. (a) R. E. Miller, *Acc. Chem. Res.* **23**, 10 (1990); (b) R. E. Miller *Science* **240**, 447 (1988); (c) R. E. Miller, *J. Phys. Chem.* **90**, 3301 (1986).
10. D. J. Nesbitt, *Chem. Rev.* **88**, 843 (1988).
11. R. C. Cohen and R. J. Saykally, *J. Phys. Chem.* **96**, 1024 (1992).
12. M. Ito, *J. Mol. Struct.* **177**, 173 (1988).
13. For example, see (a) A. R. W. McKellar and H. L. Welsh, *J. Chem. Phys.* **55**, 595 (1971); (b) A. S. Pine, W. J. Lafferty, and B. J. Howard, *J. Chem. Phys.* **81**, 2939 (1984); (c) G. D. Hayman, J. Hodge, B. J. Howard, J. S. Muenter, and T. R. Dyke, *J. Chem. Phys.* **86**, 1670 (1987).
14. T. E. Gough, R. E. Miller, and G. Scoles, *Appl. Phys. Lett.* **30**, 338 (1977).
15. See also (a) T. E. Gough, D. G. Knight, and G. Scoles, *Chem. Phys. Lett.* **97**, 155 (1983); (b) T. E. Gough, M. Mengel, P. A. Rowntree, and G. Scoles, *J. Chem. Phys.* **83**, 4958 (1985); (c) X. J. Gu, D. J. Levandier, B. Zhang, G. Scoles, and D. Zhuang, *J. Chem. Phys.* **93**, 4898 (1990).
16. R. H. Page, Y. R. Shen, and Y. T. Lee, *J. Chem. Phys.* **88** 4621 (1988).

17. For example, M. M. Doxtader and M. R. Topp, *J. Phys. Chem.* **89**, 4291 (1985).

18. For example, R. D. Beck, M. F. Hineman, and J. W. Nibler, *J. Chem. Phys.* **92**, 7068 (1990).

19. C. E. Hamilton, J. L. Kinsey, and R. W. Field, *Ann. Rev. Phys. Chem.* **37**, 493 (1986).

20. D. E. Cooper, C. M. Klimcak, and J. E. Wessel, *Phys. Rev. Lett.* **46**, 324 (1981); D. E. Cooper and J. E. Wessel, *J. Chem. Phys.* **76**, 2155 (1982).

21. (a) T. Suzuki, N. Mikami, and M. Ito, *Chem. Phys. Lett.* **120**, 333 (1985); (b) T. Suzuki, N. Mikami, and M. Ito, *J. Phys. Chem.* **90**, 6431 (1986); (c) T. Suzuki, M. Hiroi, and M. Ito, *J. Phys. Chem.* **92**, 3774 (1988).

22. B. F. Henson, G. V. Hartland, V. A. Venturo, and P. M. Felker, *J. Chem. Phys.* **91**, 2751 (1989).

23. G. V. Hartland, B. F. Henson, V. A. Venturo, R. A. Hertz, and P. M. Felker, *J. Opt. Soc. Am.* **B7**, 1950 (1990).

24. For example, see D. H. Parker in *Ultrasensitive Laser Spectroscopy*, ed. D. S. Kliger (Academic Press, New York, 1983), p. 233.

25. P. Esherick and A. Owyoung, *Chem. Phys. Lett.* **103**, 235 (1983); P. Esherick, A. Owyoung, and J. Pliva, *J. Chem. Phys.* **83**, 3311 (1985).

26. W. Bronner, P. Oesterlin, and M. Schellhorn, *Appl. Phys.* **B34**, 11 (1984).

27. B. F. Henson, G. V. Hartland, V. A. Venturo, R. A. Hertz, and P. M. Felker, *Chem. Phys. Lett.* **176**, 91 (1991).

28. G. V. Hartland, B. F. Henson, V. A. Venturo, and P. M. Felker, *J. Phys. Chem.* **96**, 1164 (1992).

29. B. F. Henson, G. V. Hartland, V. A. Venturo, and P. M. Felker, *J. Chem. Phys.* **97**, 2189 (1992).

30. V. A. Venturo, P. M. Maxton, B. F. Henson, and P. M. Felker, *J. Chem. Phys.* **96**, 7855 (1992).

31. V. A. Venturo, P. M. Maxton, and P. M. Felker, *Chem. Phys. Lett.* **198**, 628 (1992).

32. V. A. Venturo, P. M. Maxton, and P. M. Felker, *J. Phys. Chem.* **96**, 5234 (1992).

33. W. C. Wiley and I. H. McLaren, *Rev. Sci. Instrum.* **26**, 1150 (1955).

34. For example, (a) P. S. H. Fitch, L. Wharton, and D. Levy, *J. Chem. Phys.* **70**, 2018 (1979); (b) S. M. Beck, J. B. Hopkins, D. E. Powers, and R. E. Smalley, *J. Chem. Phys.* **74**, 43 (1981); (c) P. M. Felker and A. H. Zewail, *Adv. Chem. Phys.* **LXX**, 265 (1988).

35. For example, see (a) A. J. Gotch and T. S. Zwier, *J. Chem. Phys.* **93**, 6977 (1990); (b) A. J. Gotch, A. W. Garrett, D. L. Severance, and T. S. Zwier, *Chem. Phys. Lett.* **178**, 121 (1991).

36. For example, see D. A. Long, *Raman Spectroscopy* (McGraw–Hill, New York, 1977).

37. For example, see (a) Y. Chen, Ph.D. thesis, Massachusetts Institute of Technology, 1988; (b) G. V. Hartland, P. W. Joireman, L. L. Connell, and P. M. Felker, *J. Chem. Phys.* **96**, 179 (1992).
38. (a) D. Frye, L. Lapierre, and H.-L.Dai, *J. Chem. Phys.* **89**, 2609 (1988); (b) D. Frye, P. Arias, and H.-L.Dai, *J. Chem. Phys.* **88**, 7240 (1988).
39. M. T. Berry, M. R. Brustein, M. I. Lester, C. Chakravarty, and D. C. Clary, *Chem. Phys. Lett.* **178**, 301 (1991).
40. R. J. Stanley and A. W. Castleman, *J. Chem. Phys.* **94**, 7744 (1991).
41. Y. R. Shen, *The Principles of Nonlinear Optics*, (Wiley, New York, 1984).
42. (a) P. Huber-Walchli and J. W. Nibler, *J. Chem. Phys.* **76**, 273 (1982); (b) M. Maroncelli, G. A. Hopkins, J. W. Nibler, and T. R. Dyke, *J. Chem. Phys.* **83**, 2129 (1985); (c) J. W. Nibler and J. W. Yang, *Ann. Rev. Phys. Chem.* **38**, 349 (1987).
43. F. Kronig, P. Oesterlin, and R. L. Byer, *Chem. Phys. Lett.* **88**, 477 (1982).
44. P. M. Weber and S. A. Rice, *J. Chem. Phys.* **88**, 6120 (1988).
45. B. F. Henson, Ph.D. Thesis, Department of Chemistry and Biochemistry, UCLA, 1991.
46. V. A. Venturo and P. M. Felker, in preparation.
47. We use Wilson's notation for labeling benzene's normal modes. See E. B. Wilson, *Phys. Rev.* **45**, 706 (1934).
48. K. O. Börnsen, H. L. Selzle, and E. W. Schlag, *J. Chem. Phys.* **85**, 1726 (1986).
49. K. Law, M. Schauer, and E. R. Bernstein, *J. Chem. Phys.* **81**, 4871 (1984).
50. (a) Th. Weber, E. Reidle, H. J. Neusser and E. W. Schlag, *Chem. Phys. Lett.* **183**, 77 (1991); (b) Th. Weber and H. J. Neusser, *J. Chem. Phys.* **94**, 7689 (1991).
51. (a) M. Y. Hahn and R. L. Whetten, *Phys. Rev. Lett.* **61**, 1190 (1988); (b) M. Y. Hahn, Ph.D. Dissertation, Department of Chemistry and Biochemistry, UCLA, 1989.
52. (a) M. Schmidt, M. Mons, and J. Le Calvé, *Chem. Phys. Lett.* **177**, 371 (1991); (b) M. Schmidt, M. Mons, J. Le Calvé, P. Millié, and C. Cossart-Magos, *Chem. Phys. Lett.* **183**, 69 (1991); (c) M. Schmidt, M. Mons, and J. Le Calvé, *J. Phys. Chem.* **96**, 2404 (1992).
53. J. E. Adams and R. M. Stratt, *J. Chem. Phys.* **93**, 1358 (1990).
54. (a) L. E. Fried and S. Mukamel, *Phys. Rev. Lett.* **66**, 2340 (1991); (b) L. E. Fried and S. Mukamel, *J. Chem. Phys.* **96**, 116 (1992).
55. R. Nowak, J. A. Menapace, and E. R. Bernstein, *J. Chem. Phys.* **89**, 1309 (1988).
56. X. Li, M. Y. Hahn, M. S. El-Shall, and R. L. Whetten, *J. Phys. Chem.* **95**, 8524 (1991).
57. Th. Weber, A. M. Smith, E. Riedle, H. J. Neusser, and E. W. Schlag, *Chem. Phys. Lett.* **175**, 79 (1990).
58. This structural form is also suggested by the results calculated for benzene–$(Ar)_n$ in Ref. 51b.

59. For example, J. R. Cable, M. J. Tubergen, and D. H. Levy, *J. Am. Chem. Soc.* **111**, 9032 (1989).

60. For example, see (a) T. D. Lin, C. P. Damo, J. R. Dunlop, and T. A. Miller, *Chem. Phys. Lett.* **168**, 349 (1990); A. M. Ellis, E. S. J. Robles, and T. A. Miller, *J. Chem. Phys.* **94**, 1752 (1991).

61. T. G. Dietz, M. A. Duncan, D.E. Powers, and R. E. Smalley, *J. Chem. Phys.* **74**, 6511 (1981).

62. For example, M. D. Morse, J. B. Hopkins, P. R. R. Langridge–Smith, and R. E. Smalley, *J. Chem. Phys.* **79**, 5316 (1983).

63. P. Y. Cheng, K. F. Willey, and M. A. Duncan, *Chem. Phys. Lett.* **163**, 469 (1989).

64. M. Okumura, L. I. Yeh, J. D. Myers, and Y. T. Lee, *J. Chem. Phys.* **85**, 2328 (1986).

65. For example, J. A. Draves, Z. Luthey–Schulten, W.-L. Liu, and J. M. Lisy, *J. Chem. Phys.* **93**, 4589 (1990).

66. An infrared version of this scheme has been employed. See, for example, L. I. Yeh, M. Okumura, J. D. Myers, J. M. Price, and Y. T. Lee, *J. Phys. Chem.* **91**, 7319 (1989).

67. I am grateful to Profs. D. M. Neumark and O. Cheshnovsky for discussions on this point.

CHAPTER 9

COHERENT POPULATION TRANSFER*

K. Bergmann

Fachbereich Physik, Universität Kaiserslautern
D-67653 Kaiserslautern, Germany

B. W. Shore

Lawrence Livermore National Laboratory, Livermore, CA 94550, USA

Contents

---

*Manuscript submitted in November 1992.

6. Pulsed Three-State Excitation   344

7. Summary and Conclusions   360

1. Introduction

Reactions between atoms or molecules at high temperature, or in a nonequilibrium environment, involve excited energy levels. The presence of such levels, often with appreciable population, may have a significant effect upon reaction rates because the additional internal energy may increase the cross-sections for inelastic or reactive processes. The excitation may make possible reaction channels that would be inaccessible to unexcited species. To study reactions in detail one must ensure single-collision conditions, such as those that occur within molecular beams. However, most such beam experiments involve supersonic expansion, and this cools the internal degrees of freedom. Thus, most experiments have dealt with molecules in their vibrational ground state. New techniques are becoming available that permit efficient and selective preparation of atoms and molecules in specified excited levels. With such techniques, all involving laser-induced transitions, it becomes possible to explore in detail the spectroscopy and collision dynamics of excited species. (For a comprehensive discussion of molecular beam methods, with detailed description of various techniques for producing selective excitation, see the book edited by Scoles.[1])

The first techniques for state selection, developed some 20 years ago, exploited the high spectral purity of the laser to populate individual levels.[2]

The earliest work produced electronic excitation, via allowed (electric dipole) transitions. Soon afterward infrared laser radiation was used to excite the $v = 1$ (and, in a few cases, the $v = 2$) vibrational levels of a molecule.[3] More recently it has become possible, with more powerful lasers, to excite vibrational overtones and thereby place population into specific excited vibrational levels of the ground electronic state.[4]

Electronic excitation will be followed by fluorescence (spontaneous emission), leading to additional possibilities for excitation. For atoms these tertiary levels have the same parity as the ground state, and so they are typically metastable. For molecules the distribution of final vibrational levels follows a pattern predicted by Franck–Condon factors, hence the term Franck–Condon pumping (FCP) has been used to describe preparation of excited vibrational levels via optical pumping.[5] The FCP technique is easy to implement, but it suffers from a lack of selectivity: it produces a distribution of vibrational levels.

A second generation of techniques, in use for some 10 years, uses pairs of laser beams to enact various types of double-resonance transitions.[6] The selectivity provided by such techniques offer a significant advance in gaining complete control over excited populations. Stimulated emission pumping (SEP),[7] the main theme of this book, is one of these techniques. In SEP one laser moves population out of a thermally populated level into an electronically excited state. A second laser stimulates emission from this state into a specific final level. The technique is relatively easy to implement with the power and spectral purity of commercially available pulsed lasers. The selectivity of SEP can be much greater than that of FCP. However, spontaneous emission from the intermediate level is not eliminated, and so some population reaches levels other than the desired final one. This lack of complete selectivity can be tolerated in most spectroscopic experiments and in some collision dynamics studies, but it can constitute a severe problem in other cases.

This chapter describes a third generation of laser-induced state-selection techniques. For these we invoke another important property of a laser, namely its coherence[8]. Very recently, several techniques have been developed that rely on radiation coherence, and some of these have already been applied to studies of collision dynamics. These methods hold great promise to provide the experimentalist complete control over the population of excited level, including the control of alignment and orientation of angular momenta. The price one has to pay for such coherence-based methods is an increased effort to assure high quality of the laser radiation, particularly its coherence. Most pulsed lasers presently in the research laboratories of the chemical physics community do not meet the stringent requirements for coherent excitation. However, single-mode pulsed laser systems based exclusively on solid-state devices are now becoming available. The successful and widespread use of the techniques discussed in this chapter depends on further progress in the development of solid-state and short-pulse laser systems anticipated for the near future.

This chapter is meant to serve as an introduction to the more specialized literature and the recent papers on the subject, with particular emphasis on adiabatic population transfer in molecules. We note a recent topical review of this subject by Reuss and co-workers.[9] For a more thorough discussion of coherent excitation, consult the text by Shore.[10]

2. Rate Equations

To appreciate the features of coherent excitation we should begin by reviewing the kinetics of excitation by incoherent light, which is describable by sets of rate equations. Our concern will be with samples of molecules (or atoms) maintained within a volume or traveling as a beam. We are interested in predicting the molecular population that exists in a particular discrete energy level. We define the probability $P_n(t)$ to be the fraction of the molecules that are in level n at time t. Changes in molecular populations may be caused by a variety of processes, including collisions with other molecules. We are particularly concerned with changes produced by radiation, either spontaneous emission or the directed radiation from a laser. It is customary to employ radiation irradiance $I(t)$, power per unit area, as a descriptor of this radiation (for more detailed definitions see Ref. 11). We require equations of motion for the excitation probabilities, subject to specified radiation irradiance, starting from the instant $t = 0$ at which the molecules first encounter the radiation.

The earliest quantitative description of molecular population dynamics, predating the introduction of the laser, was provided by Einstein. He postulated that the rate at which a population will change due to a particular process is directly proportional to the present value of the population. As an example of the Einstein radiative rate equations,[12] consider an idealized two-level molecule free from collisions with other particles. According to Einstein, the change of population in level 1 (assumed to be the ground level) is caused by absorption of radiation (producing population in level 2), by stimulated emission from level 2, and by spontaneous emission from level 2. Both the absorption and the stimulated emission rates are directly proportional to the laser intensity. We express these statements by means of the differential equation

$$\frac{d}{dt}P_1(t) = -B_{12}I(t)P_1(t) + B_{21}I(t)P_2(t) + A_{21}P_2(t)\,. \tag{1}$$

Here, the fixed parameters B_{12} and B_{21} (sometimes termed Milne coefficients[13]) as well as the parameter A_{21} (the Einstein A coefficient[13]) are to be determined either by direct measurement or by computations based on models of molecular structure. The change in excited level population follows a similar pattern, except that we should include the possibility that spontaneous emission from this level need not necessarily place population back into the ground level. The resulting rate equation is

$$\frac{d}{dt}P_2(t) = +B_{12}I(t)P_1(t) - B_{21}I(t)P_2(t) - A_2P_2(t)\,. \tag{2}$$

The total population of these two levels will remain constant if all spontaneous emission replenishes the ground level, so that $A_{21} = A_2$ (the fraction A_{21}/A_2 is known as the branching ratio for the $2 \to 1$ transition). More generally, the total population in the two levels 1 and 2 diminishes (it is optically pumped into other levels by decay from level 2), as expressed by the equation

$$\frac{d}{dt}[P_1(t) + P_2(t)] = -(A_2 - A_{21})P_2(t)\,. \tag{3}$$

2.1. *Solutions for Two-Level Systems*

To simplify the equations for purposes of demonstrating solutions, we assume constant irradiance and equal statistical weights in the two levels, so $B_{12} = B_{21} = B$. Then, for population initially placed into the ground level, the excited population at time t is expressible as

$$P_2(t) = \frac{BI}{2\sqrt{(BI)^2 + BIA_{21} + \left(\frac{1}{2}A_2\right)^2}}[\exp(R_+t) - \exp(R_-t)]\,, \tag{4}$$

where the rates $R_\pm$ are

$$R_\pm = -\left(BI + \frac{1}{2}A_2\right) \pm \sqrt{(BI)^2 + BIA_{21} + \left(\frac{1}{2}A_2\right)^2}\,. \qquad (5)$$

In the special case $A_{21} = A_2$, so that total population remains constant, the rate $R_+ = 0$ and the population equation takes the simpler form

$$P_2(t) = \frac{BI}{(2BI + A_2)}[1 - \exp(-(2BI + A_2)t)]\,. \qquad (6)$$

At long times this excitation probability approaches an equilibrium value $P_2(\infty)$; this is the transfer efficiency of the rate equations. The time scale for approaching this equilibrium value is $1/(2BI + A_2)$. When the irradiance is low, this time is the spontaneous emission lifetime. When the irradiance is sufficiently high that spontaneous emission is much less important than stimulated emission, $2BI \gg A_2$, then the time scale is fixed by the stimulated rate and the transfer efficiency is $P_2(\infty) = 0.5$. The transition is said to be saturated. It can readily be appreciated that at most half of the population can be transferred from one level to another. This maximum transfer occurs only when spontaneous emission is negligible.

2.2. *Sequences of Transitions: Stimulated Emission Pumping*

Evidently one can proceed through a succession of two-level population transfers. One first transfers half the population from level 1 to level 2. Then, assuming that spontaneous emission causes no appreciable loss, one transfers half of this population into another level, say 3. The result of these two steps is the population distribution

$$P_1(\infty) = 0.5\,, \quad P_2(\infty) = 0.25\,, \quad P_3(\infty) = 0.25\,.$$

These fractions are upper bounds for population transferred in separate steps. Spontaneous emission into other levels that are not in this chain will diminish the transfer yield.

In principle one can improve this yield by exposing the molecules simultaneously to the two lasers. The theory for such excitation involves three coupled rate equations, each involving absorption, stimulated emission, and spontaneous emission, as well as rates for collision-induced changes. Expressions for solutions are more complicated than for two levels, and these permit a variety of steady state results. Such rate equations are used to model three-level lasers, in which population inversion occurs.[14] One possible steady state, achievable for suitable choices of irradiance, places equal population in each level, thereby improving the population transfer to $P_3(\infty) = 0.33$. Figure 1 illustrates these properties of radiative rate equations.

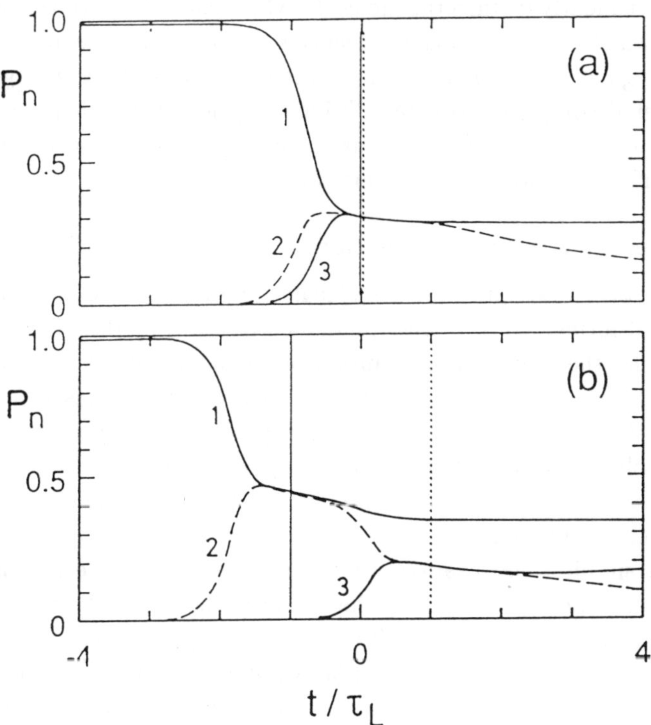

Fig. 1. Evolution of the population of levels 1, 2, and 3 in a three-level system driven by two pulsed lasers (a pump and a Stokes pulse, each with pulse width τ_L), as obtained by rate equations. The vertical solid lines mark the peak of the pump pulse, which couples states 1 and 2, while the dashed line marks the peak of the Stokes pulse, which couples states 2 and 3. Initial population resides in state 1. The peak of the pulses overlap in time in frame (a), while in frame (b) the Stokes pulse is delayed. The irradiance of the lasers is chosen sufficiently large to saturate fully both transitions. Decay of the intermediate state 2 by spontaneous emission is included. Adapted from He *et al.*[70]

3. Coherent Excitation

It is natural to ask whether it is possible to improve these population transfer probabilities. We shall see that it is possible to achieve complete population transfer. To do so, however, we must abandon the regime described by rate equations and examine coherent excitation. Rather than accept the postulates of Einstein, we examine the fundamental quantum mechanics of a molecule interacting with radiation.[15] Rate equations emerge from this theory only in limits wherein the radiation is very broadband or collisions frequently disturb the molecule. Our concern is with the opposite limit, in which the radiation is idealized as nearly monochromatic and lacking irregular fluctuations in amplitude or phase, so that the light may be considered completely coherent.[16] Laser induced excitation is coherent when the time intervals of interest are much shorter than the spontaneous emission lifetimes and the mean times between collisions.

3.1. *Coherent Excitation Equations*

The theoretical description of a molecule interacting with radiation can take several forms. On the most fundamental level, we describe both the molecule and the radiation as quantum mechanical systems, coupled by a suitable interaction.[17] However, when the field is that of a laser, it suffices to treat this field as classical, i.e., as a prescribed time varying interaction. We shall follow here this "semiclassical" approach, in which the molecule is a quantum mechanical system and the radiation field appears only in a time dependent interaction.

In the simplest idealization we consider a single molecule, isolated from its surroundings except for the radiation field, which we regard as nearly monochromatic light. We further assume spontaneous emission has negligible effect during the course of laser-induced excitation. Under these conditions the molecule can be described by a time dependent statevector, $|\Psi(t)\rangle$, whose behavior is controlled by the time dependent Schrödinger equation

$$\hbar\frac{\partial}{\partial t}|\Psi(t)\rangle = -iH(t)|\Psi(t)\rangle. \tag{7}$$

The Hamiltonian $H(t)$ appearing here is the sum of an unperturbed term H_0 for the free molecule and a time varying interaction $V(t)$:

$$H(t) = H_0 + V(t). \tag{8}$$

For present purposes we take $V(t)$ to be the usual electric dipole interaction, namely, the scalar product of a molecular dipole moment $\mathbf{d}$ (the sum of individual electron dipole moment operators $e\mathbf{r}$) and a prescribed electric field $\mathbf{E}(t)$, to be evaluated at the center of mass of the molecule:

$$V(t) = -\mathbf{d} \cdot \mathbf{E}(t) \,. \tag{9}$$

As noted in various texts,[18] this is the first term of a multipole expansion that includes magnetic dipole and electric quadrupole terms. Although an observer in a laboratory frame will recognize the electric field as that of either a standing wave or a traveling wave, the field which affects the atom is that seen by an observer on the molecule (for discussion, see Ref. 9). Thus a purely monochromatic cw field, varying with angular frequency ω_L in the laboratory frame, will appear to a moving molecule as a Doppler shifted frequency ω. Furthermore, motion will carry the molecule into and out of a beam of radiation, so that the apparent time dependence of the field acquires an additional pulse envelope. Thus we are led to consider, in the time dependence of $V(t)$, a sinusoidal carrier field, at frequency ω, modulated by a slowly varying envelope.

It is customary to extract the vector properties of the field by means of a unit vector $\mathbf{e}$ and to extract the carrier frequency as a sinusoid. For linearly polarized light this construction can be written (see Ref. 20 for a discussion of the appropriate expressions for elliptically polarized light)

$$\mathbf{E}(t) = \mathbf{e}\mathcal{E}(t) \cos(\omega t + \varphi) \,, \tag{10}$$

where φ is a phase and $\mathcal{E}(t)$ is a slowly varying envelope with dimensions of an electric field. In general, the phase will vary with time, perhaps randomly, but for simplicity we consider here constant phase. Section 6.6 presents a brief discussion of the effects of phase fluctuations.

3.2. *The Density Matrix*

Because a molecule is a bound quantum system, the free-molecule Hamiltonian H^0 has discrete internal-energy eigenstates, identifiable, say, as $|1\rangle$, $|2\rangle$, $\ldots$. If a molecule is known to be described initially by the statevector $|\Psi(0)\rangle$, then at later times it will be found in the statevector $|\Psi(t)\rangle$ of Eq. (7), and the probability of finding the molecule in state $|n\rangle$ at that time is the square of the projection of this statevector onto basis vector $|n\rangle$,

$$P_n(t) = |\langle n|\Psi(t)\rangle|^2 \,. \tag{11}$$

The use of a single statevector is too restrictive for all but the most ideal cases. Typically, we deal with a mixture of molecules, each characterized by values α of some suitable set of parameters such as initial excitation condition, the molecular trajectory, the molecular orientation or the Doppler shift. We then require an average over results obtained with statevectors $\Psi(\alpha; t)$ and weighted by the probability p_α,

$$P_n(t) = \sum_\alpha p_\alpha |\langle n|\Psi(\alpha; t)\rangle|^2 \,. \tag{12}$$

When the Hamiltonian does not depend on the parameters (e.g., when α denotes the initially populated level of a mixture) a simple means of treating such averages is by means of the density matrix,[21]

$$\rho(t) = \sum_\alpha |\Psi(\alpha; t)\rangle p_\alpha \langle \Psi(\alpha; t)|, \quad \rho_{nm}(t) = \langle n|\rho(t)|m\rangle \,. \tag{13}$$

Diagonal elements of the density matrix are various molecular excitation probabilities,

$$P_n(t) = \rho_{nn}(t) \,,$$

while off-diagonal elements ($n \neq m$) are termed coherences. The density matrix obeys the quantum Liouville equation,

$$\hbar\frac{\partial}{\partial t}\rho(t) = -iH(t)\rho(t) + i\rho(t)H(t) \equiv -i[H(t), \rho(t)] \,. \tag{14}$$

A single statevector can describe a coherent superposition of basis states, in which the amplitudes of individual basis vectors have definite phase relationships. By contrast, to describe an incoherent superposition (a mixture), one must either add probabilities computed with separate statevectors, or one may employ a density matrix. However, it should be kept in mind that to treat molecular mixtures involving different electric fields or different velocities, each case requires a different Hamiltonian, and it will be necessary to average over different density matrices.

The great power of the density matrix comes from extending the Liouville equation to include homogeneous relaxation processes such as spontaneous emission. These not only transfer population, as does a rate equation, but they diminish coherences (the off-diagonal elements of ρ). The relevant equation adds a term linear in ρ involving a relaxation operator Γ, as in the equation[22]

$$\hbar\frac{\partial}{\partial t}\rho(t) = -i[H(t), \rho(t)] + \hbar\Gamma\rho(t) \,. \tag{15}$$

When homogeneous relaxation is absent the equation is equivalent to the Schrödinger equation. When relaxation dominates coherence, thereby producing rapid damping of coherences, the equations reduce to rate equations for populations.

We shall be concerned with coherent excitation. To present the theory in the simplest form we shall work with the Schrödinger equation and a single statevector. It should be understood that application to experiment will often require an average over mixtures of such statevectors. In later sections, we shall comment on work that uses the density matrix.

4. The Two-Level System

The simplest molecular system is one in which only two discrete energy states, $|1\rangle$ and $|2\rangle$ with energies E_1 and E_2, participate in the dynamics. This is to be expected whenever the carrier frequency of the radiation, ω, nearly matches the molecular Bohr frequency $(E_2 - E_1)/\hbar$ of two nondegenerate energy levels. We assume that the dipole moment has no diagonal matrix elements, so that the Hamiltonian has matrix elements

$$H_{11} = E_1 \qquad H_{21} = \hbar\Omega(t)e^{i\varphi} \cos(\omega t + \varphi)$$
$$H_{22} = E_2 \qquad H_{12} = H_{21}^* \,. \tag{16}$$

This parametrization introduces the Rabi frequency $\Omega(t)$ as a measure of the strength of the interaction $V(t)$. For the field defined above, it has the expression

$$\hbar\Omega(t) = -\langle 2|\mathbf{d} \cdot \mathbf{e}|1\rangle \mathcal{E}(t) \exp(-i\varphi)\,. \tag{17}$$

That is, the Rabi frequency is proportional to a component of the dipole moment and to the square root of the irradiance. (We have incorporated the phase of the electric field into the definition of the Rabi frequency to facilitate application of the rotating wave approximation below). The appendix presents background needed for estimating values for the Rabi frequency.

The Rabi frequency is constant in time for a purely monochromatic field (as seen by the molecule). More generally, it has a slow variation expressive of an electric-field pulse envelope. The phase φ may fluctuate during a pulse and may also exhibit shot-to-shot variation.

With the restriction to a two-state system we know that at any time the molecular statevector must be some superposition of these two quantum states, a constraint that we use to express the statevector as

$$|\Psi(t)\rangle = |1\rangle C_1(t)\,\exp[-i\zeta_1(t)] + |2\rangle C_2(t)\,\exp[-i\zeta_2(t)]\,. \tag{18}$$

In forming this expression we have introduced explicit time-dependent phases $\zeta_n(t)$ that permit useful simplification of the equations of motion for the probability amplitudes $C_n(t)$, and whose specification defines what is usually termed a *picture:* the choice $\zeta_n = 0$ defines the Schrödinger picture, whereas the choice $\hbar(\zeta_2 - \zeta_1) = (E_2 - E1)t$ defines the Heisenberg picture.[23] For any choice of phases, the probability for finding the molecule in state $|n\rangle$ at time t is

$$P_n(t) = |\langle n|\Psi(t)\rangle|^2 = |C_n(t)|^2\,. \tag{19}$$

We typically assume that population initially resides entirely in state 1, and refer to the long-time probability $P_2(\infty)$ as the population transfer efficiency. As suits the description of near-resonant excitation by fields of moderate intensity we shall choose the rotating wave picture, for which $(\zeta_2 - \zeta_1) = \omega t$. This choice, together with the choice $\hbar\zeta_1 = E_1 t$ and $E_1 = 0$, leads to the expansion

$$|\Psi(t)\rangle = |1\rangle C_1(t) + |2\rangle C_2(t)\,\exp[-i\omega t]\,. \tag{20}$$

We substitute this expansion into the Schrödinger equation and collect separately the coefficients of $|1\rangle$ and $|2\rangle$. We thereby obtain a pair of coupled linear ordinary differential equations

$$\frac{d}{dt}\begin{bmatrix} C_1(t) \\ C_2(t) \end{bmatrix} = -\frac{i}{2}\begin{bmatrix} 0 & \Omega(t)^*f(t)^* \\ \Omega(t)f(t) & 2\Delta(t) \end{bmatrix}\begin{bmatrix} C_1(t) \\ C_2(t) \end{bmatrix}, \tag{21}$$

where $f(t)$ is a dimensionless time-dependent function

$$f(t) = 1 + \exp[2i\omega t + 2i\varphi] \tag{22}$$

and the frequency $\Delta(t)$ is the detuning between the field frequency ω (as seen by the molecule) and the Bohr frequency,

$$\hbar\Delta(t) = E_2 + E_1 - \hbar\omega\,. \tag{23}$$

We have introduced here the possibility of time variation in the detuning, as could be caused either by a variation in the Bohr frequency (e.g., a varying Stark or Zeeman shift from quasistatic electric or magnetic fields) or by variation in the carrier frequency at the molecular center of mass (as might be introduced by a deliberate sweep of the laser phase or frequency).

It should be recognized that these equations neglect spontaneous emission: the only field modes present are those that comprise the laser. It is possible to incorporate those emission events that do not move population into the ground state (but instead move it into states that are not included in our two-state approximation), by introducing imaginary parts to the unperturbed energies: the replacement $E_2 \rightarrow E_2 - iA/2$ yields an equation that correctly describes probability loss of state 2 at rate A.[24]

Furthermore, the equations take no account of collisions or other mechanisms that are responsible for equilibrating molecular populations in a thermal environment. Within those limitations the equations are an exact transcription of the Schrödinger equation for excitation of a two-state molecule exposed to a time-varying electric field. Taken with initial conditions, they provide a complete description of the two-state system.

It should also be recognized that actual molecules (or atoms) have rotational degeneracy. A rotational level having angular momentum J has $2J + 1$ degenerate sublevels, each differing by the mean value of the angle between the rotation axis and a reference axis. The Schrödinger equation applies to individual quantum states, i.e., to rotational sublevels. Because the sublevel populations differ in orientation with respect to the electric field vector, they have different Rabi frequencies. The desired molecular excitation probabilities must sum over probabilities for transitions between sublevels,[25] a particular illustration of Eq. (12), as is the average over a distribution of Doppler shifts.

4.1. *The Rotating Wave Approximation (RWA)*

The traditional methods of time-dependent perturbation theory are not adequate for dealing with these equations, because the action of the perturbation $V(t)$ is too strong — we cannot assume that the initial state remains only slightly perturbed by this interaction.

In principle, there is no insuperable difficulty in obtaining numerical solutions to these coupled equations for any reasonably well behaved form assumed for the time dependence $V(t)$. In this sense, the problem may be considered solved. However, there are practical difficulties in constructing general solutions. These arise because the equations involve three distinct time scales, the inverses of the three frequencies ω, Ω and Δ. For optical excitation at modest intensity, there is a very great difference between the time scale $1/\omega$ of electric field oscillations and the time scale $1/\sqrt{\Delta^2 + \Omega^2}$ of molecular response.

Our interest is with behavior that occurs over a time scale much longer than an optical cycle. We therefore replace the rapid variation of $f(t)$ at frequency 2ω with an average of this function over an optical cycle. That is, we make the approximation

$$f(t) \approx 1 \,. \tag{24}$$

This rotating wave approximation (RWA)[26] leads to much simpler equations,

$$\frac{d}{dt}\begin{bmatrix} C_1(t) \\ C_2(t) \end{bmatrix} = -\frac{i}{2}\begin{bmatrix} 0 & \Omega(t)^* \\ \Omega(t) & 2\Delta(t) \end{bmatrix}\begin{bmatrix} C_1(t) \\ C_2(t) \end{bmatrix} \tag{25}$$

or, in vector form,

$$\frac{d}{dt}\mathbf{C}(t) = -iW(t)\mathbf{C}(t) \,, \tag{26}$$

where $\mathbf{C}(t)$ is a two-component column vector and $W(t)$ is a 2×2 matrix,

$$\mathbf{C}(t) = \begin{bmatrix} C_1(t) \\ C_2(t) \end{bmatrix}, \quad W(t) = \frac{1}{2}\begin{bmatrix} 0 & \Omega(t)^* \\ \Omega(t) & 2\Delta(t) \end{bmatrix}. \tag{27}$$

Only the slow variation of the pulse envelope contributes to the time dependence of the RWA Hamiltonian. The rapid variation at carrier frequency ω has been incorporated into the solutions via the phase $\zeta_n(t)$ introduced explicitly in our definition of probability amplitudes $C_n(t)$. The present choice of the phase $\zeta_1(t)$ gives a constant null value to W_{11}. Other choices will add, to each diagonal element, a common term possibly time dependent. This phase has no effect on probabilities. Figure 2 presents a schematic picture of the energies and detunings.

4.2. *Resonant Pulses*

It is worth noting that when the excitation is resonant, $\Delta = 0$, then a simple transformation of independent variable from time to (dimensionless) pulse area,

$$dA = \Omega(t)\,dt \quad \text{or} \quad A = \int_0^t dt'\,\Omega(t') \tag{28}$$

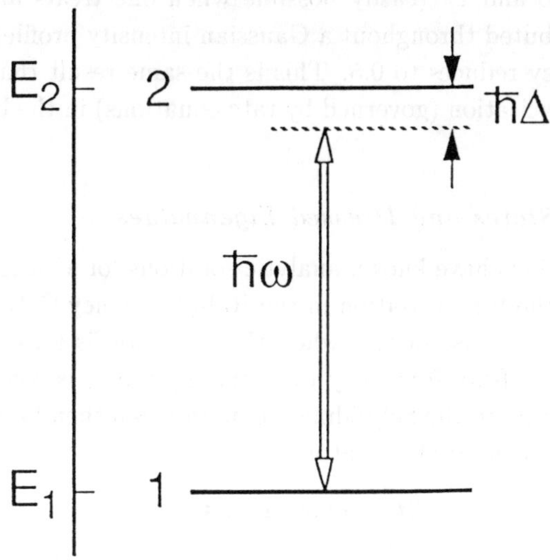

Fig. 2. The energy levels of a two-level system, showing relationship between energies E_1 and E_2, carrier frequency ω and detuning Δ.

leads to the equations

$$\frac{d}{dA}\begin{bmatrix} C_1(t) \\ C_2(t) \end{bmatrix} = -\frac{i}{2}\begin{bmatrix} 0 & 1 \\ 1 & 0 \end{bmatrix}\begin{bmatrix} C_1(t) \\ C_2(t) \end{bmatrix}. \tag{29}$$

The solution to these equations, starting from the initial condition of an unexcited atom, $C_1(0) = 1$, gives the excitation probability

$$P_2(t) = \frac{1}{2}[1 - \cos(A)]. \tag{30}$$

Complete population transfer, $P_2 = 1$, occurs whenever the pulse area A is an odd-integer multiple of π. This transfer is independent of the functional form of the pulse envelope. It might seem that such oscillations provide a convenient method for producing complete transfer of population from level 1 to level 2. Even if the Rabi frequency changes with time, complete population inversion occurs for $A = \pi, 3\pi, \ldots$. However, the success of this method for population transfer depends both on the requirement of exact resonance and on the exact value of the pulse area. Any variation of pulse fluence will diminish the transfer efficiency. In particular, if the pulse area varies between 0 and 2π (easily possible when one treats an ensemble of molecules distributed throughout a Gaussian intensity profile), the average transfer efficiency reduces to 0.5. This is the same result that one obtains for incoherent excitation (governed by rate equations) in the limit of strong saturation.

4.3. *Dressed States and Dressed Eigenvalues*

The RWA equations have known analytic solutions for a variety of analytic expressions for the time variation in the Rabi frequency $\Omega(t)$ and detuning $\Delta(t)$. The simplest case occurs when the radiation is strictly monochromatic, so that the Rabi frequency (and the detuning) is constant in time. The formal solution to the Schrödinger equation can then be written as the action of an exponentiated operator,

$$\mathbf{C}(t) = \exp(-iWt)\mathbf{C}(0)\,, \tag{31}$$

where $\hbar W$ is the constant RWA Hamiltonian matrix. To find expressions for the probability amplitudes it is useful to first find the (two) eigenvalues $\hbar\omega_\nu$ and the (two-component) eigenstates $|\Phi_\nu\rangle$ of the matrix W. These dressed states obey the matrix equation

$$W|\Phi_\nu\rangle = \omega_\nu|\Phi_\nu\rangle \quad \text{with} \quad \nu = +,- \tag{32}$$

and, hence, the action of the time evolution exponential upon such a state is simply to multiply the vector by a phase factor:

$$\exp(-iWt)|\Phi_\nu\rangle = \exp(-i\omega_\nu t)|\Phi_\nu\rangle\,. \tag{33}$$

The eigenvectors $|\Phi_+\rangle$ and $|\Phi_-\rangle$ can be obtained, following Cohen–Tannoudji *et al.*,[27] by introducing the mixing angle Θ through the definitions

$$\cos(2\Theta) = \frac{\Delta}{\sqrt{\Delta^2 + |\Omega|^2}}\,, \quad \sin(2\Theta) = \frac{|\Omega|}{\sqrt{\Delta^2 + |\Omega|^2}}\,. \tag{34}$$

Using trigonometric double-angle identities, and taking the phase of Ω to be that of the carrier, $\Omega = |\Omega|e^{-i\varphi}$, we can write the RWA Hamiltonian as

$$W = \sqrt{\Delta^2 + |\Omega|^2}\begin{bmatrix} 1 - 2\cos^2\Theta & 2\cos\Theta\sin\Theta\exp(i\varphi) \\ 2\cos\Theta\sin\Theta\exp(-i\varphi) & -1 + 2\cos^2\Theta \end{bmatrix}$$

$$+ \frac{1}{2}\begin{bmatrix} \Delta & 0 \\ 0 & \Delta \end{bmatrix}\,. \tag{35}$$

We use this form to verify that the vectors

$$|\Phi_-\rangle = \begin{bmatrix} \cos\Theta \\ -\exp(-i\varphi)\sin\Theta \end{bmatrix} = \cos\Theta\,|1\rangle - \exp(-i\varphi)\sin\Theta\,|2\rangle \tag{36a}$$

$$|\Phi_+\rangle = \begin{bmatrix} \exp(-i\varphi)\sin\Theta \\ \cos\Theta \end{bmatrix} = \exp(i\varphi)\sin\Theta\,|1\rangle + \cos\Theta\,|2\rangle \tag{36b}$$

which are demonstrably orthonormal,

$$\langle\Phi_-|\Phi_+\rangle = 0, \quad \langle\Phi_-|\Phi_-\rangle = \langle\Phi_+|\Phi_+\rangle = 1, \tag{37}$$

are eigenvectors of the two-state RWA Hamiltonian matrix W. (This choice is not unique: each of these vectors may be multiplied by an arbitrary phase without changing their status as orthonormal eigenvectors.) The corresponding eigenvalues, obtainable either with the use of these eigenvectors or by solving the quadratic determinantal equation, are

$$\omega_\pm = \frac{1}{2}\left[\Delta \pm \sqrt{|\Omega|^2 + \Delta^2}\right]. \tag{38}$$

Figure 3, discussed in the following paragraphs, illustrates some of the relationships of dressed eigenvalues. By inverting the transformation that defines the eigenstates we obtain the expression for individual molecular states as combinations of dressed states:

$$|1\rangle = \cos\Theta|\Phi_-\rangle + \exp(-i\varphi)\sin\Theta|\Phi_+\rangle \tag{39a}$$

$$|2\rangle = \cos\Theta|\Phi_+\rangle - \exp(+i\varphi)\sin\Theta|\Phi_-\rangle. \tag{39b}$$

The mixing angle Θ describes the difference in orientation of two coordinate systems in a two-dimensional Hilbert space: one is fixed with the unperturbed basis states and the other is fixed upon dressed states.

These formulas, together with the simple action of the time evolution operator upon dressed states, allow us to evaluate the time dependence of any initial state. For example, a molecule prepared in state $|1\rangle$ will at subsequent times be in the state

$$|\Psi(t)\rangle = \exp(-i\omega_- t)\cos\Theta|\Phi_-\rangle + \exp(-i\omega_+ t)\exp(-i\varphi)\sin\Theta|\Phi_+\rangle. \tag{40}$$

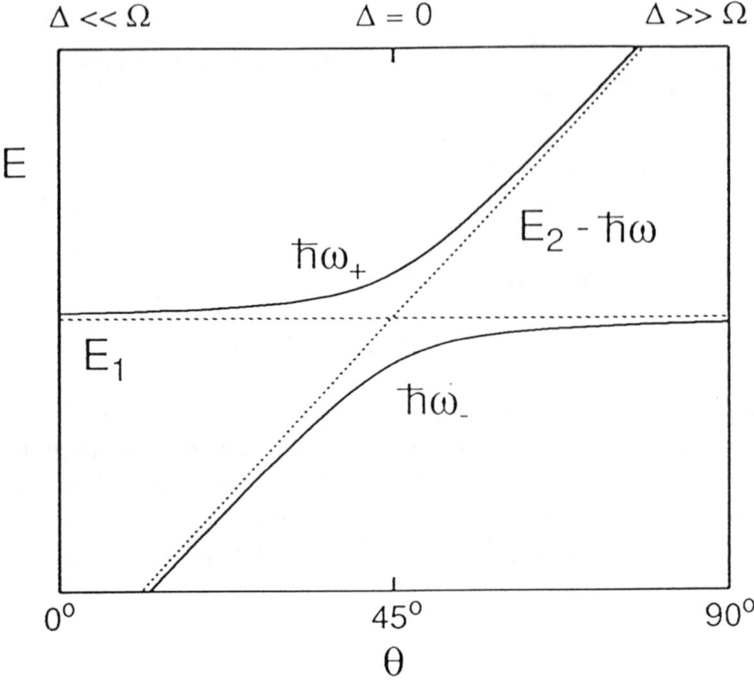

Fig. 3. Variation of the dressed-state eigenvalues $\hbar\omega_+$ and $\hbar\omega_-$ with mixing angle Θ [and hence, by Eq. (34), with detuning Δ] for the two-level system of Fig. 2. Dotted lines are diagonal elements of RWA Hamiltonian, full lines are eigenvalues $\hbar\omega_\pm$. (Often E_1 is set to zero, as in this chapter.)

In particular, the probability of population reappearing in the initial state is

$$P_1(t) = |\langle 1|\Psi(t)\rangle|^2$$

$$= |\cos\Theta\langle 1|\Phi_-\rangle + \exp[-i(\omega_+ - \omega_-)t]\exp(-i\varphi)\sin\Theta\langle 1|\Phi_+\rangle|^2 .$$

$$(41)$$

A superposition of dressed states leads to sinusoidally varying interference, observable as periodic oscillations of the populations. The frequency of these oscillations is the difference between the two dressed eigenfrequencies, the so-called flopping frequency

$$\omega_+ - \omega_- = \sqrt{|\Omega|^2 + \Delta^2} . \qquad (42)$$

It is this frequency that sets the scale for atomic response, and which must therefore be much smaller than the optical frequency for the RWA to be valid.

Unlike populations governed by rate equations, the populations presented here do not approach a definite limit at large time. When the RWA Hamiltonian is constant, as it is here, one must consider population transfer efficiency $P_2(T)$ at some fixed time T.

4.4. *Special Cases*

For exact resonance, $\Delta = 0$, the eigenvalues are

$$\omega_+ = +\frac{1}{2}\Omega, \quad \omega_- = -\frac{1}{2}\Omega \tag{43}$$

and the mixing angle is $\Theta = 45°$, so that the eigenvectors are coherent superpositions of equal probabilities for the two molecular states,

$$|\Phi_-\rangle = \frac{1}{\sqrt{2}}[|1\rangle - \exp(-i\varphi)|2\rangle] \tag{44a}$$

$$|\Phi_+\rangle = \frac{1}{\sqrt{2}}[\exp(-i\varphi)|1\rangle + |2\rangle]. \tag{44b}$$

From the formula above we find that the probability for finding the molecule in state 1, after starting in that state, is

$$P_1(t) = 1 - P_2(t) = \frac{1}{4}|1 + \exp(-i\Omega t)|^2 = \frac{1}{2}[1 + \cos(\Omega t)]. \tag{45}$$

The population oscillates at the Rabi frequency, defined above as the product of a dipole moment and an electric field magnitude, divided by $\hbar$. When detuning is present the frequency of population oscillation is the root mean square of the detuning Δ and the Rabi frequency Ω.

For large detuning the eigenvalues become

$$\omega_+ \to \frac{1}{2}[\Delta + |\Delta|], \quad \omega_- \to \frac{1}{2}[\Delta - |\Delta|] \quad \text{for} \quad |\Omega| \ll |\Delta|. \tag{46}$$

In this latter regime we have, for positive detuning, the eigenvalues

$$\omega_+ \to |\Delta|, \quad \omega_- \to 0 \quad \text{for} \quad \Delta > 0, \tag{47}$$

and, because the mixing angle $\Theta = 0$, the eigenvectors become identified with individual molecular states,

$$|\Phi_+\rangle \rightarrow |2\rangle , \quad |\Phi_-\rangle \rightarrow |1\rangle . \tag{48}$$

For large negative detuning the eigenvalues become

$$\omega_+ \rightarrow 0 , \quad \omega_- \rightarrow -|\Delta| \quad \text{for} \quad \Delta < 0 , \tag{49}$$

and, because $\Theta = 90°$, the eigenvectors again become pure molecular states, but differing from those for positive detuning:

$$|\Phi_+\rangle \rightarrow \exp(-i\varphi)|1\rangle , \quad |\Phi_-\rangle \rightarrow \exp(-i\varphi)|2\rangle . \tag{50}$$

The important aspect of these relationships is that the composition of each dressed state changes dramatically as the detuning changes. The dressed state $|\Phi_+\rangle$ changes from $|2\rangle$ to $|1\rangle$ as detuning changes from large positive to large negative values.

5. Adiabatic Passage

There are two limits that have special significance in descriptions of time dependence in quantum mechanics. First, when there is an abrupt change in an otherwise constant Hamiltonian, it is possible to find simple analytic solutions. These typically exhibit nonmonotonic variations, generalizatons of the Rabi oscillations that typify two-level atoms. Analytic solutions are available for a variety of cases and numbers of levels[28] and efficient numerical methods exist for solving systems of linear ordinary differential equations with constant or slowly varying coefficients.[29]

Alternatively, when the changes occur sufficiently slowly, the system tends to remain in a fixed combination of eigenstates of the (instantaneous) Hamiltonian. These eigenstates,

$$W(t)|\Phi_n(t)\rangle = \omega_n(t)|\Phi_n(t)\rangle , \tag{51}$$

are known variously as perturbed, dressed, or adiabatic states. Typically it is possible to associate each adiabatic state $|\Phi_n(t)\rangle$ with a single basis state in the remote past and in the remote future. These two basis states are usually not the same: by remaining in an adiabatic state at all times the system actually undergoes a transition between two basis states, a process often termed adiabatic following or adiabatic passage. The process must, of course, be completed in a time shorter than any relaxation time that would destroy coherence.

One can employ the (orthonormal) adiabatic states to express the statevector, via the expansion

$$|\Psi(t)\rangle = \sum_{n=1}^{N} a_n(t)|\Phi_n(t)\rangle \, \exp\left[-i\int_0^t dt'\omega_n(t')\right] \qquad (52)$$

subject to the normalization

$$\sum_{n=1}^{N} |a_n(t)|^2 = 1. \qquad (53)$$

There must be as many terms N as there are molecular basis states $|n\rangle$. We start from the initial condition

$$a_n(0) = \langle\Phi_n(0)|\Psi(0)\rangle. \qquad (54)$$

Because the adiabatic states are as complete as the original basis states, we can certainly express the statevector at any time as a superposition of these states by incorporating suitable time variation into the expansion coefficients $a_n(t)$. The value of such an expansion will depend on how slowly the coefficients vary and on how many terms are required in the expansion. Unless the statevector is a single adiabatic state, it will have oscillatory time dependence from interference between time evolving adiabatic states.

5.1. *The Adiabatic Condition*

In general, the Hamiltonian $H(t)$ at time t will not commute with the Hamiltonian at another time, and so the expansion coefficients $a_n(t)$ will vary with time. However, when changes in $H(t)$ are sufficiently slow, then these coefficients remain constant. Such evolution is termed adiabatic. The condition for this adiabatic evolution[30] is that the dressed eigenstates should vary slowly, on a time scale fixed by the various differences of dressed eigenvalues:

$$\left|\left\langle\Phi_n\left|\frac{d}{dt}\Phi_m\right\rangle\right| \ll |\omega_n - \omega_m|. \qquad (55)$$

When, in addition to maintaining adiabatic evolution, the initial condition corresponds to a single basis state, then the system remains in a single state at subsequent times. Such evolution is known as adiabatic following. By remaining an adiabatic state, the system undergoes a transition between basis states.

5.2. *The Bloch Vector and Adiabatic Passage*

Discussions of coherent excitation and adiabatic passage in two-level systems are often presented within the framework of the density matrix. Following the approach of Feynman, Vernon, and Hellwarth,[31] one defines a Bloch vector $\mathbf{R}(t)$ with components constructed from the density matrix

$$R_1(t) = 2\operatorname{Re}\rho_{12}(t)\,, \quad R_2(t) = -2\operatorname{Im}\rho_{12}(t)\,, \quad R_3(t) = \rho_{22}(t) - \rho_{11}(t)$$

$$(56)$$

and an angular velocity vector $\tilde{\Omega}(t)$ with components constructed from the RWA Hamiltonian

$$\tilde{\Omega}_1(t) = \operatorname{Re}\Omega(t)\,, \quad \tilde{\Omega}_2(t) = -\operatorname{Im}\Omega(t)\,, \quad \tilde{\Omega}_3(t) = \Delta(t)\,. \qquad (57)$$

The Liouville equation of motion for the density matrix (equivalent to the Schrödinger equation) then translates into a torque equation in an abstract space of three dimensions,

$$\frac{d}{dt}\mathbf{R}(t) = \tilde{\Omega}(t) \times \mathbf{R}(t)\,. \qquad (58)$$

The Bloch vector maintains constant length (unity), and so with passing time the tip of the Bloch vector traces out a curve on a spherical surface (the Bloch sphere).

The usual initial condition of population residing entirely in state $|1\rangle$ corresponds to a Bloch vector directed toward the south pole of the Bloch sphere ($R_3 = -1$) whereas population inversion, placing all population into state $|2\rangle$, corresponds to the north pole ($R_3 = +1$). Schemes for producing population transfer may be viewed as designs for the time dependent angular velocity vector that will move the Bloch vector from the south to the north pole.

When the RWA Hamiltonian is constant, the curve traced by the moving Bloch vector is a great circle: the angular velocity vector forms the axis for steady rotation of the Bloch vector at the flopping frequency. When the Hamiltonian changes adiabatically, starting from large detuning (so the angular velocity vector initially is along the polar axis, parallel to the initial Bloch vector) the Bloch vector tends to precess about the slowly changing axis of the angular velocity vector $\tilde{\Omega}(t)$.

5.3. *The Landau–Zener–Stueckelberg Model*

The simplest examples of transitions occur in two-state systems. Consider two (diabatic) quantum states, say 1 and 2, subject to a Hamiltonian that shifts their relative energies linearly with time, $\Delta(t) = -St$, while maintaining a constant interaction between them:

$$W(t) = \begin{bmatrix} 0 & \frac{1}{2}\Omega \\ \frac{1}{2}\Omega & -St \end{bmatrix}. \tag{59}$$

This model problem involves only two parameters: the (constant) Rabi frequency Ω and the rate of frequency change, $S = d\Delta/dt$. This Hamiltonian, first studied by Landau, Zener, and Stueckelberg[32] has exact analytic solutions, expressible in terms of parabolic cylinder functions,[33] from which the asymptotic transition probabilities can be deduced. It is a simple task to find the (adiabatic) eigenstates $|\Phi_+(t)\rangle$ and $|\Phi_-(t)\rangle$ that diagonalize this Hamiltonian at each instant, together with the (adiabatic) eigenvalues $\omega_+(t)$ and $\omega_-(t)$. The eigenstates are those of Eq. (33) with $\tan(2\Theta) = -St/\Omega$, and the eigenvalues of Eq. (38) become

$$\omega_\pm(t) = \frac{1}{2}\left[-St \pm \sqrt{(St)^2 + \Omega^2}\right]. \tag{60}$$

When the time variation of $W(t)$ is suitably rapid (large S or small Ω), then the interaction has insufficient opportunity to affect behavior, and the system remains in the initial unperturbed (bare) basis state (no transition). When the variation is suitably slow (small S or large Ω), then the system remains at all times in a dressed state and complete population transfer occurs. To visualize these limiting cases it is useful to plot the two (nonintersecting) adiabatic eigenvalues versus time, along with the diabatic energies (the diagonal elements of the original Hamiltonian), as was done in Fig. 3. Then we deduce the behavior by tracking one of the curves from the past (when the adiabatic and diabatic states coincide) into the future (when again there occurs a one-to-one connection between adiabatic and diabatic states). Between these two extremes of null or complete population transfer lie other possibilities. The standard expression for the probability of remaining in the initial state 1, i.e., of switching adiabatic states, the so-called Landau–Zener–Stueckelberg (LZS) probability,[32,34] is

$$P(1 \rightarrow 1) = \exp\left[-\frac{\pi\Omega^2}{2S}\right] \rightarrow \begin{cases} 0 & \text{if adiabatic (large } \Omega \text{ or small } S) \\ 1 & \text{if diabatic (small } \Omega \text{ or large } S). \end{cases} \tag{61}$$

The probability for making the (diabatic) transition $1 \rightarrow 2$ between basis states, i.e., the transfer efficiency $P_2(\infty) = P(1 \rightarrow 2) = 1 - P(1 \rightarrow 1)$, is unity for an adiabatic transition.

These formulas can be obtained in a variety of ways, such as by examining the known asymptotic properties of analytic functions that satisfy the differential equation. They apply, exactly, to the problem posed above, of an unperturbed energy difference that sweeps linearly during an infinite interval, while the interaction remains constant. That is, the exact time dependent probabilities $P_n(t)$ approach their final values asymptotically, as the time interval becomes infinitely long. (At finite times the probabilities typically oscillate about these asymptotic values.)

It should be noted that for the two-state model adiabatic population transfer will occur for either direction of frequency chirp: either from blue to red or *vice versa*. In either case population will be transferred from the initial state to the other state.

From the argument of the exponential we deduce the conditions needed to produce complete population transfer between basis states (i.e., for the system to remain adiabatic):

$$\pi\Omega^2 \gg 2S. \tag{62}$$

In practice, it is never possible to maintain a constant interaction for an infinite interval. Often the deviation from linear time dependence becomes appreciable, and there may be participation by more than two levels. Nevertheless, we can use this constraint to derive a useful rule. Let $\Delta\tau$ be the time required to shift the frequency from $-\Omega$ to $+\Omega$, so that the chirp rate is expressible as $S = 2\Omega/\Delta\tau$. Then the adiabatic condition becomes

$$\Omega\Delta\tau \gg 1. \tag{63}$$

In other words, the pulse area must be much larger than 1 over the time interval required for a frequency sweep of 2Ω. This constraint also emerges in more general models directly from the evaluation of the adiabatic condition given in Eq. (55).

5.4. *Experimental Demonstration, Two-Level System*

The phenomenon of adiabatic following is well known from nuclear magnetic resonance.[35] There, inversion in a two level system is created by slowly changing either the frequency or the direction of a magnetic field. This is done in order to study relaxation processes by monitoring the return to equilibrium. The first observation of adiabatic following in the infrared was reported by Loy,[36] who used an electric field to Stark-shift the level separation in NH_3 through resonance with a fixed frequency laser. Adiabatic following in NH_3 was also observed by Javan and co-workers.[37] They swept the frequency of a strong pump laser through resonance and monitored the population transfer by observing the time evolution of the absorption or amplification of a weak probe laser.

Of more relevance to the topics of the present chapter are the demonstrations of adiabatic passage produced by near infrared radiation, reported by Avrillier *et al.*[38] and Adam *et al.*[39] (see also the review by Liedenbaum *et al.*[9]) and by visible light, reported by Kroon *et al.*[40] and by Lorent *et al.*[41] In this work, infrared or visible laser radiation crosses a molecular or atomic beam at a right angle. If the molecules cross the laser very close to its waist, and if only those particles traversing through the center of the laser beam are detected, then Rabi oscillations are seen. These are detected by observing the level population as the laser irradiance is changed[39] or by dispersing the velocity distribution by the time-of-flight technique.[40] Each technique changes the area of the pulse [see Eq. (28)] to which the molecules are exposed, upon which the population depends. However, if the focal point of the laser is not at the molecular beam axis, then the molecules experience curved wave fronts. These appear to the moving molecules as time varying doppler shifts[40]

$$\Delta\omega(t) = \frac{v^2\omega}{cr}t, \tag{64}$$

where v is the velocity of the molecule and r is the radius of curvature of the wavefront where the molecule and laser beam axes cross. This frequency chirp, obtained without external frequency control, induces adiabatic population inversion.

In more recent work[42] the frequency sweep is imposed on the laser pulses, of a few ps duration, by reflecting them off a suitably arranged pair of gratings. It was shown there that the chirped pulse leads to more efficient population transfer into electronically excited levels of I_2 than does a pulse without chirp.

5.5. *Multilevel Adiabatic Passage: Generalized LZS*

For all but the simplest processes the energy curves are numerous and the pattern of crossings becomes complicated. Nevertheless, it is often possible to induce complete population transfer through a succession of curve crossings. Excellent examples may be found in descriptions of controlled Stark shifts of Rydberg levels.[43]

Some guidance to the behavior under such conditions comes from considering a single level whose linear change in energy causes it to cross successively several levels whose energy remains constant. Such a situation can be realized by steadily sweeping (chirping) the frequency of a laser that has dipole connections with a number of excited states. This situation is an extension of LZS to a single curve linearly crossing $N - 1$ parallel curves,[44] as shown in Fig. 4. The relevant LZS formula, the product of a succession of exponential LZS factors,[44] generalizes to treat the intersection of a single curve with a quasicontinuum of levels and then, by extrapolation, a continuum.[45] It should be recognized that when the process is completely adiabatic, all population is transferred (between bare, basis states) at the first curve crossing, and the remaining levels have no effect on the dynamics. Traditional examinations of adiabatic passage rely on sweeping the diagonal elements of the (diabatic) Hamiltonian through a degeneracy (as occurs with a frequency chirp), perhaps in combination with variation in the off-diagonal elements (as caused by pulse amplitude variation); see Fig. 4. Whereas a two-state system does not have population transfer in the absence of detuning variations, three-state (and more general multilevel) systems may have population transfer induced solely by amplitude variations[46] (cf. Sec. 6 below). Examples are known in which slowly increasing pulses, abruptly terminated, can produce population transfers.[47] However, it is difficult to satisfy the adiabatic conditions for such pulses.

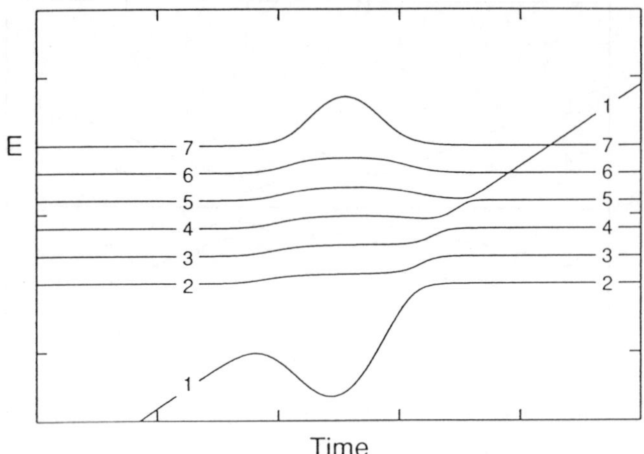

Fig. 4. Variation of the dressed state eigenvalues with time for a chirped laser pulse when level 2 (see Fig. 2) is replaced by a group of 6 evenly spaced levels. Labels at left and right identify the unperturbed states associated with these eigenvalues; adiabatic following will transfer all population from state 1 to state 2. The variation of the eigenvalues due to the pulse envelope produces the bulge. The frequency chirping is evidenced by the line crossing the group of levels. Early there occurs a pronounced avoided crossing between states 1 and 2. The interaction at the subsequent crossings is smaller because the Rabi frequency is smaller.

A different pattern of curve crossings occurs when we consider a set of N energy levels connected in a ladder linkage. Such a system has a tri-diagonal RWA Hamiltonian matrix. Let us assume that a single laser produces the excitation, so that the detunings are

$$\hbar\Delta_n = E_{n+1} - E_1 - n\hbar\omega. \tag{65}$$

For simplicity let the energy levels be evenly spaced and let the frequency vary linearly with time, so that the diagonal elements become 0, $-St$, $-2St$, $-3St$, ... , where S is the rate of frequency sweep. An alternative choice for the phase $\xi_1(t)$ will yield the more symmetric diagonal elements ... , $-2St$, $-St$, 0, $+St$, $+2St$, ... A plot of these unperturbed (diabatic) energy values as a function of time will appear as a "bow tie" of intersecting lines (cf. Harmin[43]). The dressed energies will not intersect; they will appear as separate threads, as shown in Fig. 5. In this case, the sweep of frequency through resonance will, if carried out adiabatically, move all population from the lowest energy state to the most excited state[9].

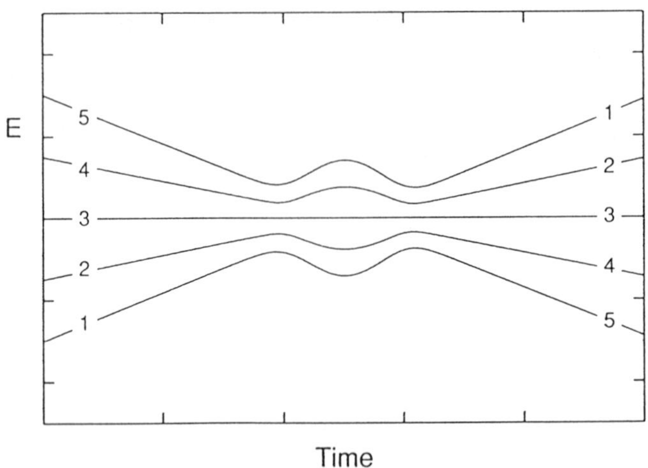

Fig. 5. Variation of the dressed state eigenvalues with time produced by a chirped laser pulse acting on a chain of 5 equidistant energy levels, coupled to nearest neighbors. Labels at left and right identify the unperturbed states associated with these eigenvalues; adiabatic following will transfer all population from state 1 to state 5.

5.6. *Experimental Demonstration, Multilevel LZS*

An instructive demonstration of selective coherent population transfer in a multilevel system, of the type shown in Fig. 4, has been presented by Warren and co-workers.[48] They exploit the dispersion in an optical fibre to induce a chirp (red to blue on a laser pulse of 3 ps duration. The process stretches the pulse to 15 ps. To reverse the chirp (blue to red) they pass the pulse twice through a pair of gratings. For the purpose of the demonstration they tune the laser to the transition $3s \rightarrow 3p$ transition of Na, near 589 nm. They detect the population in the fine-structure levels $^2P_{1/2}$ and $^2P_{3/2}$ by laser-induced fluorescence following $3p \rightarrow 5s$ excitation. The spectral width of the fluorescence-inducing probe pulse, which is delayed by about 2 ns from that of the primary excitation pulse, is sufficiently small to resolve the fine structure. As shown in Fig. 6, they found that the lower fine structure level of Na ($3p$) is predominantly populated when the frequency chirp goes from red to blue, whereas a blue to red chirp places population into the upper fine structure level. This is consistent with predictions based on adiabatic passage. The small population seen in the other fine-structure level indicates that the adiabaticity condition is not entirely fulfilled for all

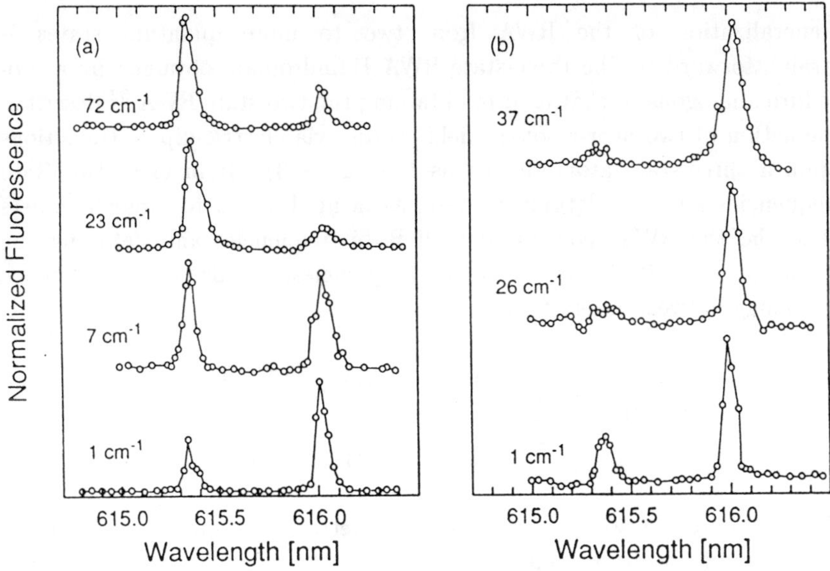

Fig. 6. Normalized relative intensities of the fluorescence from the $5s$ level of Na populated by the probe laser from the $3p$ $(^2P_{1/2})$ at 615.42 nm or $3p$ $(^2P_{3/2})$ at 616.07 nm. In frame (a) the chirp goes from red to blue and produces predominantly population of the lower fine structure level. In frame (b) the chirp goes from blue to red and leads to predominant population of the upper fine structure level. The numbers in cm^{-1} give the approximate peak Rabi frequencies. Adapted from Melinger *et al.*[42] and reproduced with permission.

atoms detected by the probe pulse (for instance those in the spatial wings of the picosecond laser). Further discussion will be found in the work of Band and Julienne.[49]

It is worth noting that very efficient photodissociation of a molecule can be achieved by short, chirped laser pulses. Chelkowski *et al.*[50] show by numerical studies that a sequence of chirp-induced population transfers to adjacent vibrational energy levels can result in high excitation or dissociation when the chirp rate matches the anharmonicity of the level spacing.

6. Pulsed Three-State Excitation

Generalization of the RWA from two to more quantum states is straightforward.[51] The three-state RWA Hamiltonian, obtained by a procedure analogous to that used in obtaining the two-state RWA,[52] describes the action of two near-resonant fields acting via electric-dipole transitions upon a three-state atom (linked as $1 \leftrightarrow 2 \leftrightarrow 3$). It involves two Rabi frequencies and two detunings (here, as in application to larger numbers of levels, the RWA requires that all Rabi frequencies and detunings be small not only with respect to carrier frequencies, but also to the difference frequency between lasers[52]):

$$W(t) = \frac{1}{2} \begin{bmatrix} 0 & \Omega_P(t) & 0 \\ \Omega_P(t) & 2\Delta_2(t) & \Omega_S(t) \\ 0 & \Omega_S(t) & 2\Delta_3(t) \end{bmatrix}. \tag{66}$$

We label here the two Rabi frequencies by letters appropriate to the pump and Stokes field of a Raman process. The detuning Δ_2 vanishes when the first-step field is resonant (with transition $1 \leftrightarrow 2$). The meaning of the third detuning depends on the relative ordering of the molecular energy levels (see Fig. 7). When $E_1 \leq E_2 \leq E_3$ (a ladder configuration), then $\hbar\Delta_3$ is the difference between $E_3 - E_1$ and the *sum* of the two-photon energies. When E_2 lies above both E_1 and E_3 (a lambda configuration), then $\hbar\Delta_3$ is the difference between $E_3 - E_1$ and the *difference* of the two-photon energies. This is the configuration for a Raman process, wherein the Raman excitation energy $E_3 - E_1$ is close to resonance with the difference $\hbar(\omega_P - \omega_S)$ between pump and Stokes carriers.

As with other uses of the Schrödinger equation, there is no account here of spontaneous emission or relaxation. These represent losses from excited states, as shown in Fig. 7. To the extent that spontaneous emission removes population without returning it to one of the 3 states of the model, radiative decay can be modeled by introducing imaginary parts to the molecular energies. We shall comment later on the effects of spontaneous emission.

Numerous analytic solutions for the three-state RWA equation are available. Simplest are those for constant Hamiltonian, expressible in terms of trigonometric functions and roots of a cubic equation.[53] Other solutions, requiring less familiar special functions, are known for particular analytic expressions for time variation of Rabi frequencies and detunings.[54]

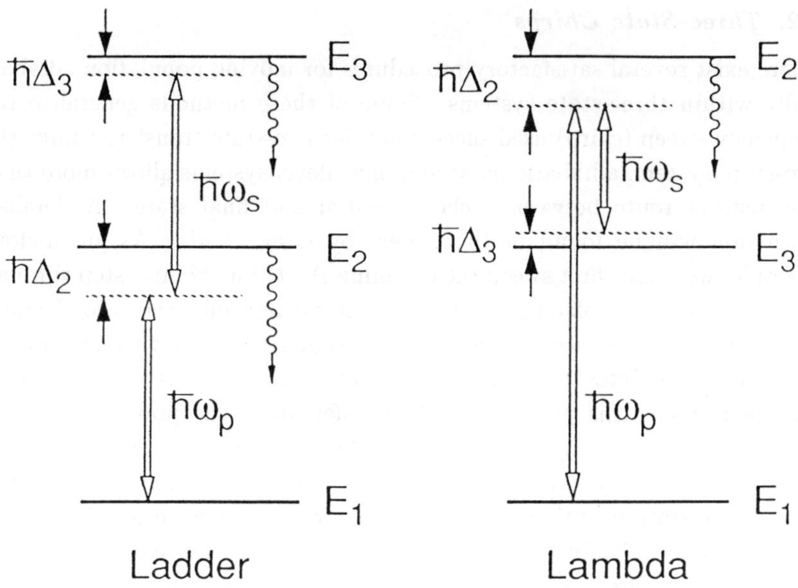

Fig. 7. Two configurations of the three-level system, showing relationships of energies, pump, and Stokes carrier frequencies, and detunings, for the ladder and the lambda configurations. Wiggly vertical lines indicate possibility of spontaneous emission, possibly also to levels other than those shown.

6.1. *Two-Photon Excitation*

When the single photon detuning $|\Delta_2(t)|$ is very much larger than either of the two Rabi frequencies or the two photon detuning $|\Delta_3(t)|$, then the first excited state $|2\rangle$ has negligible population, and the dynamics become that of a two-state atom. The effective (two-photon) Rabi frequency and detuning become[55]

$$\Omega(t) = \frac{\Omega_P(t)\Omega_S(t)}{2\Delta_2(t)}, \quad \Delta(t) = \Delta_3(t) + \frac{|\Omega_P(t)|^2 - |\Omega_S(t)|^2}{4\Delta_2(t)}. \quad (67)$$

The two-photon transition can produce adiabatic passage between states 1 and 3 by sweeping either frequency. Interestingly, the required sign-changing detuning required for two-state adiabatic passage can also be produced by amplitude variation alone, either by first establishing a steady pump field, followed by a Stokes pulse (as demonstrated by Grischkowsky and Loy[55]) or by applying first the Stokes and then the pump pulse. In each case the two carrier frequencies remain constant.

6.2. *Three-State Chirps*

There exist several satisfactory procedures for moving population adiabatically within three-state systems. Some of these methods generalize the frequency sweep (chirp) used successfully for two-state transfer. Unlike the two-state system, adiabatic passage in multilevel systems allows more than one distinct route between a chosen initial and final state. A detailed discussion will be found in the papers by Oreg *et al.*[56] As an obvious example, we might first sweep the detuning $\Delta_2(t)$ for the first step through resonance, thereby placing the population entirely into the second state. Then we sweep the second detuning through resonance, thereby moving the population into the desired final state. When adiabatic conditions are maintained, this procedure will transfer all of the population. It is surprising at first that this intuitively obvious procedure of independent population transfers can be improved by sweeping the frequencies in the opposite, counterintuitive order. We start with resonance for the second step, and adiabatically sweep the detunings until we achieve resonance with the first step. Oreg *et al.*[56] point out that this counterintuitive ordering requires less stringent control of the lasers and makes more efficient use of the laser power. However, this counterintuitive chirp is not applicable with a single laser in a lambda configuration.[57]

6.3. *Experimental Demonstration, Three-Level Chirp*

Very recently Broers *et al.*[58] demonstrated that in a three-level system population can be transferred by laser pulses that are chirped either in an intuitive or counterintuitive sequence of resonance encounters. In fact, they verify the prediction of Oreg *et al.*[56] that the counterintuitive sequence is more efficient and that little population ever resides in the intermediate level.

Broers *et al.*[58] use mode-locked laser pulses to generate a wavelength continuum which is subsequently passed through a pulse shaper for amplitude and frequency control before being amplified to a pulse energy of about 10 μJ. The chirp was large enough to cover the 4.2 nm difference between the $5s \rightarrow 5p$ and the $5p \rightarrow 5d$ transition frequency in Rb near 778 nm. They detected the upper-level population by photoionization at 532 nm.

Figure 8 shows their experimental results, together with their calculations based on numerical solution of the Schrödinger equation. The computations model the actual amplitude and frequency variation of the pulse. Figure 8 shows that both the intuitive direction of the chirp (left) as well as the counterintuitive direction (right) lead to complete population transfer into the upper level. The electronic lifetimes of the excited levels are much longer than the pulse duration. Therefore, radiative loss is negligible during the transfer process. From the parameters given by Boers et al.[58] (Rabi frequency 2.1 THz and pulse duration of 10 ps) we find a product $\Omega\Delta\tau$ of around 20, so that the criterion for adiabatic evolution, Eq. (63), is well satisfied.

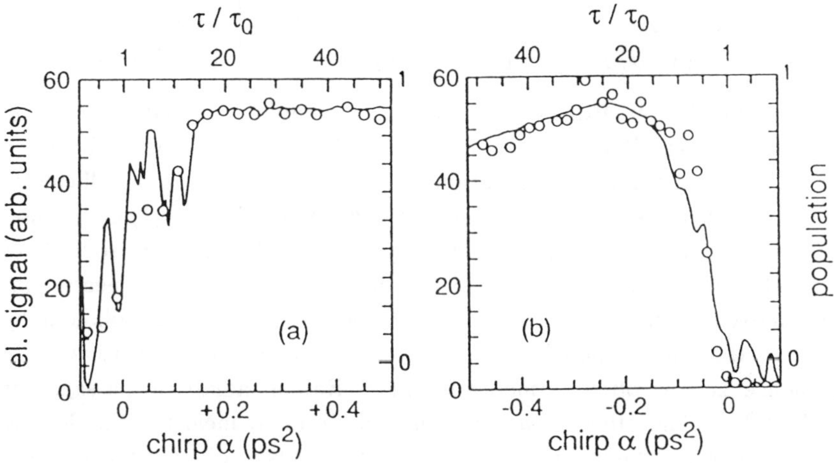

Fig. 8. Population transfer in Rb ($5s \rightarrow 5p \rightarrow 5d$) by either (a) intuitive red to blue chirp, or (b) counterintuitive blue to red chirp, for detuning range of 5.8 nm around 779 nm and pulse energy of 500 mJ/cm^2. The population of the $5d$ level is monitored by photoelectrons from a probe laser. The solid line is a theoretical result using the experimentally determined properties of the laser pulse. The frequency sweep α (lower scale) parameterizes the frequency domain of the field as $E(\omega) = |E(\omega)| \exp[i\alpha(\omega-\omega_0)^2]$. The ratio τ/τ_0 (upper scale) expresses the chirp range in units of the bandwidth of a chirp-free pulse of duration τ_0 (here 0.3 ps). [For details on the relation of chirp, bandwidth and pulse duration see A. E. Siegman, *Lasers* (University Science Books, Mill Valley, Calif., 1986) Chp. 9]. Adapted from Boers *et al.*[58]; reproduced with permission.

6.4. *Delayed Pulses: STIRAP*

Traditional processes for producing adiabatic population transfer rely on changes in the diagonal elements of the (diabatic) Hamiltonian. These may accompany pulsed variation of the off-diagonal elements, but the inversion occurs because changes in energies pass through a degeneracy. However, not all inversions require that frequencies be swept through resonance. It is possible to achieve inversion by modulating only the field amplitudes, while maintaining constant carrier frequencies. Consider resonant excitation of a three-state system by two time-dependent pulses, as embodied in the two real-valued Rabi frequencies. We term these the pump $\Omega_P(t)$ and the Stokes $\Omega_S(t)$ pulses. Such excitation involves the RWA equation

$$\frac{d}{dt}\begin{bmatrix} C_1(t) \\ C_2(t) \\ C_3(t) \end{bmatrix} = -\frac{i}{2}\begin{bmatrix} 0 & \Omega_P(t) & 0 \\ \Omega_P(t) & 0 & \Omega_S(t) \\ 0 & \Omega_S(t) & 0 \end{bmatrix}\begin{bmatrix} C_1(t) \\ C_2(t) \\ C_3(t) \end{bmatrix}. \tag{68}$$

Let the population initially reside in state 1; complete population inversion occurs whenever $|C_3(t)| = 1$. A remarkable result, first noted by Oreg *et al.*,[56] is that complete population inversion often occurs more efficiently when the second-step Stokes pulse arrives *before* the first-step pump pulse. As in the methods relying on frequency chirp, such pulses are sometimes termed counterintuitive, by contrast with an intuitive pulse sequence that first moves population from the initial state 1 into state 2, followed by a pulse that moves population into state 3. The counterintuitive pulse sequence is now often termed stimulated Raman adiabatic passage (or STIRAP).[59] The process can be examined either by means of the density matrix, as was done by Oreg *et al.* or, because it is a coherent phenomena, it can be described with a statevector and the Schrödinger equation, as was first done by Kuklinski *et al.*[60] The latter approach is slightly simpler, so we shall follow it here.

At any instant of time t the resonant RWA Hamiltonian has the (adiabatic) eigenstates $|\Phi_\nu(t)\rangle$ and eigenvalues $\omega_\nu(t)$:

$$|\Phi_0(t)\rangle = \cos\Theta|1\rangle - \sin\Theta|3\rangle \qquad\qquad \omega_0 = 0$$

$$|\Phi_+(t)\rangle = \frac{1}{\sqrt{2}}[\sin\Theta|1\rangle + |2\rangle + \cos\Theta|3\rangle] \qquad \omega_+ = +\Omega_T(t) \tag{69}$$

$$|\Phi_-(t)\rangle = \frac{1}{\sqrt{2}}[\sin\Theta|1\rangle - |2\rangle + \cos\Theta|3\rangle] \qquad \omega_- = -\Omega_T(t)$$

where $\Omega_T(t)$ is the mean square Rabi frequency

$$\Omega_T(t) = \sqrt{(\Omega_P)^2 + (\Omega_S)^2} \tag{70}$$

and Θ is a time varying angle expressing the relative strength of the two fields,

$$\tan \Theta = \frac{\Omega_P(t)}{\Omega_S(t)} . \tag{71}$$

We can express the state vector $|\Psi(t)\rangle$ at any time as a superposition of these three dressed states, as shown in the construction of Eq. (52). If the field varies sufficiently slowly then the coefficients $a_n(t)$ will remain constant. The requirement for this adiabatic approximation, derivable from Eq. (55) using Eq. (69), is[59]

$$\left| \frac{d\Theta}{dt} \right| \ll |\omega_0 - \omega_\pm| = |\Omega_T(t)| . \tag{72}$$

The significant point to notice about the structure of the eigenvectors is that the dressed state $|\Phi_0\rangle$ has no component of the intermediate state 2. This "trapped state" has been discussed extensively in other contexts.[61] The construction can be made more explicit by rewriting it as

$$|\Phi_0(t)\rangle = [\Omega_S(t)|1\rangle - \Omega_P(t)|3\rangle][\Omega_T(t)]^{-1} . \tag{73}$$

If we initially prepare the system to be in this dressed state, and maintain the adiabatic condition at subsequent times, then the system will remain in this state. As is the case with conventional stimulated Raman scattering, no population will ever reside in the intermediate state.

Now consider the initial condition

$$|\Psi(-\infty)\rangle = |1\rangle . \tag{74}$$

This is a special case of the state $|\Phi_0\rangle$ in which $\cos \Theta = 1$. We can approximate this condition by requiring that initially, for $t \to -\infty$, the pump pulse is very weak,

$$\text{initial:} \quad \Omega_P \ll \Omega_S , \quad \cos \Theta \approx 1 , \quad |\Psi\rangle \approx |\Phi_0\rangle \approx |1\rangle . \tag{75}$$

We now increase the pump intensity and decrease the Stokes, while maintaining the adiabatic condition. Eventually we approach the limit, as $t \to \infty$,

$$\text{final:} \quad \Omega_P \gg \Omega_S , \quad \cos \Theta \approx 0 , \quad |\Psi\rangle \approx |\Phi_0\rangle \approx |3\rangle . \tag{76}$$

This pulse sequence has accomplished the desired objective of transferring all population from the initial state into the final state. The procedure is robust: it is insensitive to the exact values of pulse areas (they need only be larger than about 5π) and pulse shapes.

For Gaussian pulses, the mean Rabi frequency on the right hand side of Eq. (72) approaches zero faster than does the time derivative on the left hand side. Therefore, adiabatic conditions are *not* met at very early (or very late) times. However, this is not detrimental because the coupling is weak then. At slightly later times the adiabatic criterion holds, and the system locks onto whatever adiabatic state (or superposition of states) then applies. We conclude that the pump laser should start moving population only *after* there is a sufficiently large separation of dressed eigenvalues. Once the system becomes adiabatically locked, the most critical times are during the overlap of the two pulses, when the change in Θ is largest. Evaluation of Eq. (72) when the two Rabi frequencies are equal leads to the expression $\Omega \Delta \tau \gg 1$, as in Eq. (63). More generally, when detuning is present, the condition beomes

$$\Omega_T \Delta\tau \gg \sqrt{1 + \Delta_2 \Delta\tau}. \tag{77}$$

When this condition holds, the system remains locked to the same adiabatic state. Figure 9 illustrates the relationship of various values. Inspection of Fig. 9 reveals again the need for suitably delaying the pump laser pulse. If the two pulses overlap in time, then the mixing angle Θ is near $45°$ with negligible time derivative, and Eq. (72) is fulfilled trivially. However, from Eq. (69) we realize that population is also present in the states $|\Phi_{\pm}\rangle$, from which population is subsequently lost. On the other hand, if the delay exceeds the pulse duration, then the time interval of overlapping pulses shrinks, and the mixing angle changes rapidly near $t = 0$. Under these circumstances, even though the system is neatly prepared in state $|\Phi_0\rangle$ as desired, nonadiabatic coupling puts population into state 2 at times near $t = 0$, and again population may be lost. The most efficient transfer occurs when the delay is about equal to half the pulse width. With increasing Rabi frequency the transfer becomes insensitive to the exact choice of delay.

The STIRAP process of counterintuitive pulses is perhaps most simply understood as the preparation of a coherent superposition state whose components vary with time. It may also be viewed as a limiting case of LZS processes in a 4-state linkage.[62]

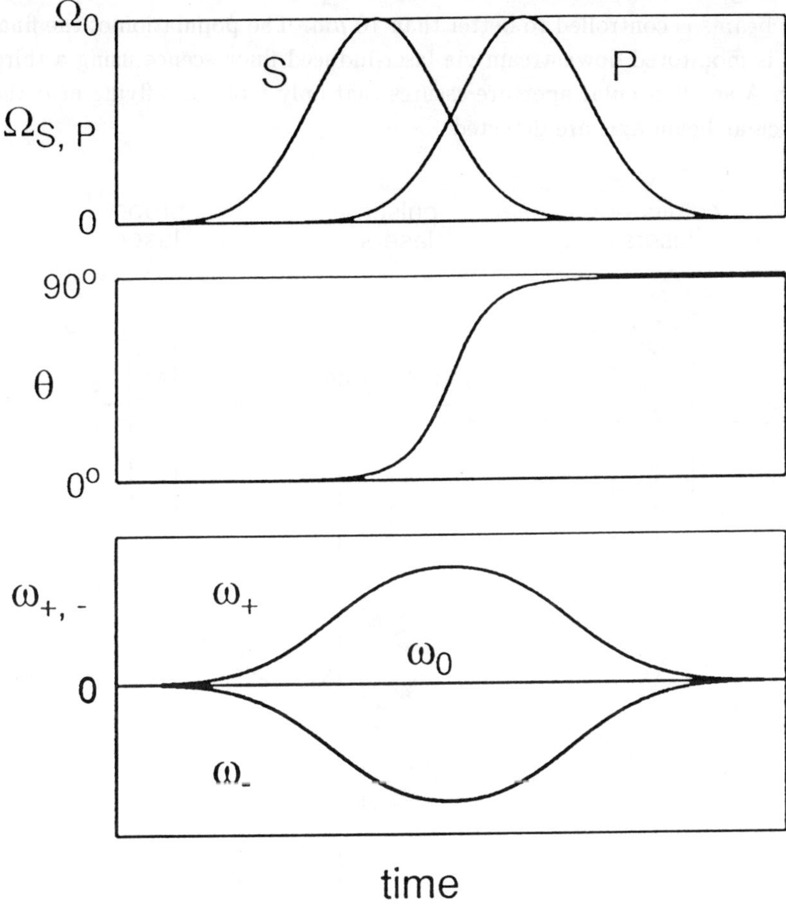

Fig. 9. Variation of the Rabi frequency Ω, the mixing angle Θ, and the dressed state eigenfrequencies ω_+ and ω_- for a three-level system such as diagrammed in Fig. 7.

6.5. *Experimental STIRAP with cw Lasers*

The first experimental indication that STIRAP works was presented by Gaubatz *et al.*[63] Figure 10 shows their experimental setup for implementing the STIRAP method with cw lasers.[59] Two single mode dye lasers are used to transfer population from the level ($v = 0$, $j = 5$) to ($v = 5$, $j = 5$) of Na_2 molecules in their electronic ground state $X\,^1\Sigma_g^+$ via an intermediate level

$(v = 7, j = 6)$ of the electronic state $A\,^1\Sigma_u^-$. The spatial overlap of the laser beams is controlled to better than 10 μm. The population of the final level is monitored downstream via laser-induced fluorescence using a third laser. A small circular aperture assures that only molecules flying near the molecular beam axis are detected.

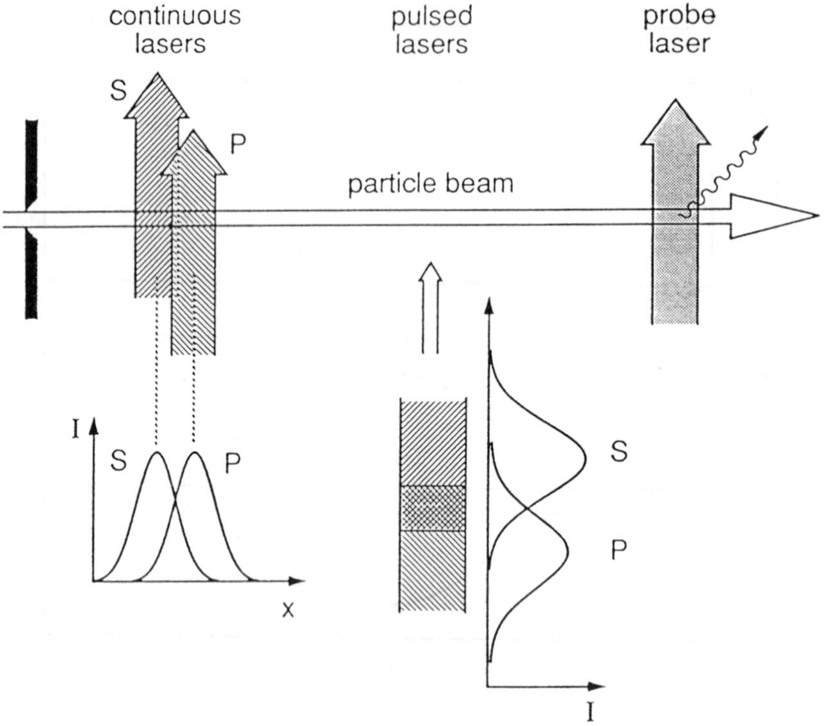

Fig. 10. Implementation of the STIRAP technique for molecular beam experiments for cw and pulsed lasers.

Figure 11 shows examples of experimental results demonstrating the STIRAP process.[59] Frame (a) shows the variation of the fluorescence induced by the probe laser as the overlap D between pump and Stokes laser (given in units of the laser beam waist) is changed. Negative values correspond to the counterintuitive sequence: the molecules interact with the Stokes laser first. Absolute transfer efficiencies can be evaluated for the following reason. The overlap of the Stokes laser with the pump

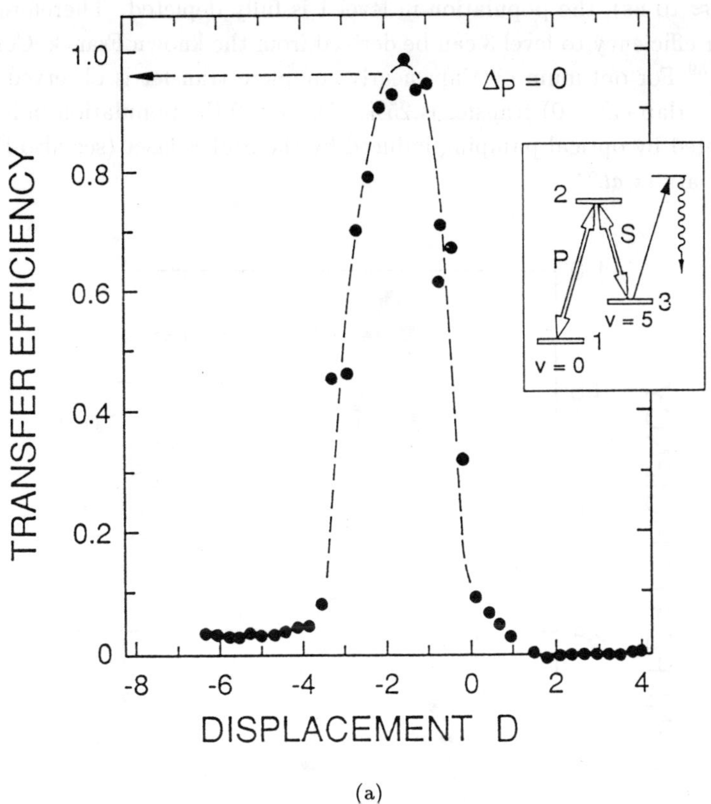

(a)

Fig. 11. Experimental demonstration of STIRAP, showing the efficiency of population transfer in Na_2 from the level 1 ($v'' = 0$) to level 3 ($v'' = 5$) produced by two lasers (from Gaubatz *et al.*[59]). (a) Population transfer as a function of the spatial separation D between the laser-beam axes. D is expressed in units of the beam waist; negative values correspond to the STIRAP sequence (Stokes before pump). Both lasers are tuned to resonance with their respective transitions (see insert showing excitation linkage and probe laser to fluorescing level). (b) Population transfer as a function of detunings Δ of the Stokes laser from the two-photon resonance. The spatial overlap is optimised in the STIRAP sequence and the pump laser is tuned to near resonance with the 1–2 transition. The maximum is reached when $\Delta = 0$.

laser in position A (where $D = -6$) is negligibly small. The (small) population transfer at that point is achieved exclusively by Franck–Condon pumping due to excitation by the pump laser. Because the interaction time is long compared to the electronic lifetime in the intermediate level

($\tau_{\mathrm{spont}} \approx 10$ ns), the population in level 1 is fully depleted. Therefore, the transfer efficiency to level 3 can be derived from the known Franck–Condon factors.[59] For optimum overlap, nearly complete transfer is observed. For exact overlap ($D = 0$) transfer is 25%. For $D > 0$ the population in level 3 is depleted by optical pumping induced by the Stokes laser (see also Fig. 9 of Gaubatz *et al.*[59]).

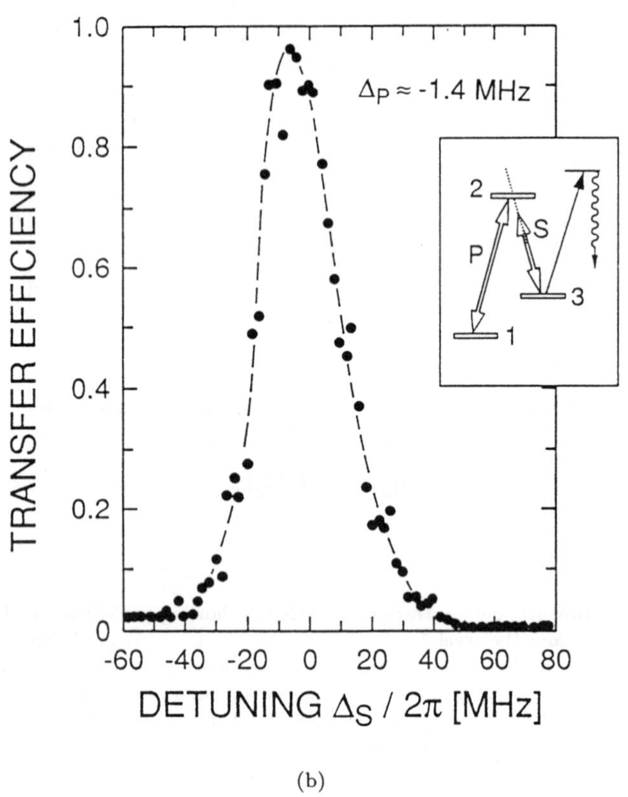

(b)

Fig. 11. (*Continued*)

Frame (b) of Fig. 11 shows the transfer efficiency for optimum overlap, with the pump laser tuned to near resonance ($\Delta_2 \approx 0$) while the Stokes laser frequency is tuned across the two-photon resonance ($\Delta_3 = 0$). It is evident from the data that the difference frequency of the pump and Stokes lasers needs to be controlled to within a few MHz to achieve nearly complete population transfer. A novel scheme for difference frequency stabilization, based on a birefringent cavity has been successfully developed and applied.[64]

Experimental results similar to those of Fig. 11 have also been obtained for population transfer between the $^3P_0$ and $^3P_2$ metastable levels of Ne.[65]

6.6. *Effect of Phase Fluctuations on STIRAP*

So far, we have not discussed the effect of phase fluctuations on the transfer efficiency. This neglect is justified if the variation of the phase of the lasers is small during the interaction time. For molecules crossing a laser beam of width 0.1 mm at a velocity of 1000 m/s the interaction time is $\Delta\tau_{\text{int}} = 100$ ns. The characteristic time scale for phase fluctuations is $\Delta\tau_{\text{phase}} > 1$ μs for a laser with a bandwidth of $\Delta\nu < 1$ MHz, as is typical of a single-mode cw laser. Under such circumstances only small or smooth changes of the phase are to be expected during 100 ns. However, quantitative evaluation of the adiabaticity criterion of Eq. (55) or (72) reveals that the intensity typically available from commercial tunable cw dye lasers (≤ 1 W) is sufficient only for atomic or molecular systems with a large electronic transition dipole moment. Many molecular systems of interest require laser power that is available only from pulsed sources. This is particularly true when radiation in the UV region is needed, and for which frequency doubling is required. The efficiency of frequency doubling is proportional to the peak irradiance, so that for a given pulse energy the efficiency of frequency doubling increases with decreasing pulse width $\Delta\tau$. From the adiabatic condition $\Omega\Delta\tau \gg 1$ we conclude that the required pulse energy, which is proportional to $\Omega^2\Delta\tau$, also increases as pulses become shorter. A pulse width of the order of 10 ns (typical for excimer or Nd-YAG pumped systems) is a good compromise.

Unlike radiation from cw sources in the visible and infrared region that have been used for coherent population transfer, the radiation from nanosecond pulsed lasers inevitably exhibits phase fluctuations. The amplitude and phase may be strongly oscillating or even be chaotic (in the case of multimode pulsed dye lasers[66]). For pulse-amplified cw radiation or in single-mode pulsed oscillators the phase may fluctuate to some extent even if the amplitude changes smoothly. The effect of field fluctuations on the population in laser-driven two- or three-level systems has been studied both theoretically and experimentally.[67]

The effect of amplitude fluctuations on the transfer efficiency in a three-level system induced by a STIRAP process can be estimated based on results given by the adiabatic condition.[68] Complete population transfer will occur if, despite fluctuations, we ensure that at early times $\Omega_S \gg \Omega_P$ and at late times $\Omega_P \gg \Omega_S$ and that condition $\Delta\Omega\Delta\tau_{\text{fluc}} \gg 1$ is fulfilled, with $\Delta\tau_{\text{fluc}}$ being the characteristic time scale of the fastest fluctuations during the pulse. If one of these conditions is not satisfied, then the transfer efficiency will be small.

It is less straightforward to determine the effect of phase fluctuations. It is important, however, to establish a generalization of Eq. (77) which includes the effect of such fluctuations. This implies that we infer the consequences of frequency detuning from the one- or two-photon resonance for the transfer efficiency, because a varying change in phase produces a shift in frequency. In this context, an important quantity is the ratio of the bandwidth imposed by phase fluctuations to the inherent Fourier limited bandwidth. For Gaussian pulses this ratio is

$$r_{\text{BW}} = \frac{\Delta\omega_{\text{phase}}}{\Delta\omega_{\text{Fourier}}}. \tag{78}$$

Kuhn *et al.*[68] have modeled phase fluctuations by Monte–Carlo simulation, in which there occurs exponentially correlated colored noise, equivalent to a phase-diffusion model with non-Lorentzian line shape. Their numerical evaluation of the effect of phase fluctuation assumed that pump and Stokes radiation showed the same fluctuation characteristics, but were independent processes. These studies revealed a strong dependence of the minimum pulse energy not only on the bandwidth of the phase fluctuations but also on the autocorrelation time. Kuhn *et al.*[68] found that for a Gaussian amplitude envelope (width $\Delta\tau$) and no static detuning of the mean carrier frequency from the one-photon and two-photon resonance the adiabaticity criterion reads

$$\Omega^2\Delta\tau \gg \frac{[1 + (r_{\text{BW}})^2]}{\Delta\tau}g. \tag{79}$$

The factor g depends on the pulse shape, the spontaneous decay rate of the intermediate level, and the autocorrelation time of the fluctuations. For details, see Ref. 68, Eq. (39).

It is clear from these studies that population transfer is best accomplished with nearly transform-limited pulses. In fact, a dramatic enhancement of the population transfer into the $v = 6$ level of NO, over what can be established by Franck–Condon pumpings, has recently been observed in the Kaiserslautern laboratory.[69] In that experiment population transfer was achieved in a pulsed supersonic beam of NO via the $A\,^2\Sigma(v = 0)$ state. Quantitative analysis of these data is currently in progress.

6.7. *STIRAP and the Density Matrix*

Although the STIRAP procedure has been studied and applied to the collisionless environment of molecular beams, it is important to study the sensitivity of the population transfer efficiency to collisions. The theory for such homogeneous relaxation effects is best treated using the density matrix. Using this approach it has been found[70] that collisions reduce the STIRAP transfer efficiency appreciably only if the collision rate exceeds the peak Rabi frequency. Of course any process that requires coherent excitation will fail if the collisions are too frequent, such that the inverse of the collision rate is less than the pulse duration. For further examples of applications of the density matrix to STIRAP, see the work by Glushko[71] and by Band.[72]

The density matrix facilitates introduction of initial states that are not single basis states. It is needed for a full treatment of excitation with allowance for spontaneous emission between states of the excitation chain. It can also provide useful insight into the theory of adiabatic passage. The density matrix for an N-state system involves N^2 elements, some of which may be complex-valued, but not all of these are independent. The diagonal elements must be real and sum to unity, and the off-diagonal elements are Hermitian adjoints. Thus a specification of the density matrix at any single time requires only $N^2 - 1$ real numbers. As emphasized by Oreg *et al.*,[56] these can be regarded as the coordinates of a generalized Bloch vector (or coherence vector) of unit length in an abstract space of dimension $N^2 - 1$. Appropriate components of this vector provide population differences as well as coherences. For the two-state atom this construction gives the three-dimensional vector model of Feynman, Vernon, and Hellwarth,[31] whereas for the three-state atom the generalized vector moves in an 8-dimensional space.

As Hioe and Eberly noted,[73] there exist a sequence of conservation laws which further constrains the motion. Oreg *et al.*[56] show that there exists an $(N - 1)$ dimensional adiabatic subspace within which the generalized

Bloch vector must remain. For a two-state system this subspace is one-dimensional, so it permits no additional freedom. But for three-level systems this subspace is two-dimensional, and there exist two independent adiabatic processes that accomplish the same population transfer.

6.8. *STIRAP with Multilevel Systems*

The elementary theory of STIRAP presented above treats nondegenerate levels. The inclusion of magnetic sublevel degeneracy is necessary for realistic modeling of rotational levels. If the transfer occurs through a sequence of rotational levels $j \to j + 1 \to j$ using linearly polarized light, then the optical selection rule $\Delta m = 0$ applies and the system decomposes into $2j + 1$ independent 3-state systems, each labeled by a quantum number m and each associated with a distinct set of Rabi frequencies. Complete transfer of population of all magnetic sublevels will occur if the adiabatic condition is satisfied for the weakest of these three-state subsystems. When optical selection rules are exploited or the direction of the linear polarization of the pump and Stokes laser is not parallel, an alignment or even orientation of the angular momentum of the final level can be achieved.[74]

For application to polyatomic molecules, particularly, it is desirable to know the effect upon the STIRAP process when there exists a high density of levels in the region of either the intermediate or the final state, so that there may be multiple interfering channels of excitation. Theoretical studies by Coulston and Bergmann[75] have shown that splitting of the intermediate level to produce two paths has no detrimental effect. In an extension of this model to a quasicontinuum of levels, Carroll and Hioe[76] have suggested that STIRAP remains possible even when the intermediate states form a continuum.*

It was shown in Ref. 75 that population transfer can occur to a single final state of a pair of possible final states when the energy difference of those states exceeds the two-photon linewidth. In the opposite extreme, when many levels fall within this linewidth, transfer can be complete into

---

*Note added in proof: very recently Lambropoulos and co-workers (T. Nakajima, M. Elk, J. Zhang, and P. Lambropoulos, *Phys. Rev.* **A50**, R913 (1994)) have shown that the model of the continuum used in Ref. 74 is not adequate to treat the process of coherent population transfer. They show that transfer is likely to fail when the coupling between initial and final state is through a continuum.

the totality of levels. It should be noted that when oscillator strengths are diluted among several levels the dipole moments are correspondingly reduced, and so it is necessary to increase the pulse intensity in order to maintain the same Rabi frequency.

The STIRAP process can be generalized to treat transfer of population in an N-state chain, using at most $N-1$ lasers. Shore *et al.*[77] have presented an analysis applicable when N is an odd integer, and have described conditions needed for complete population transfer with as few as 2 lasers, using a counterintuitive sequence for pulses. (As with the 3-level case, the pulses must have some overlap.) In these cases, some population is placed briefly into the states, 3, 5, ... , $N-2$, although no population ever resides in states 2, 4, ... , $N-1$. Therefore, the maximum transfer efficiency may be less than unity when radiative decay from intermediate levels becomes significant.

Transfer of population within an even-N chain behaves differently than that for an odd-N chain and requires slightly different procedures. An example of population transfer in a 4-level chain by a sequence of 3 pulses has been presented by Band and Julienne.[78] The cause of the difference between even- and odd-N systems can be seen in the eigenvalues: when there is no detuning there exists a null eigenvalue (a population-trapping state) for odd-N but not for even-N.[79] Consequently, it is not possible to accomplish robust adiabatic passage of the STIRAP type when all carrier frequencies are tuned to the corresponding resonances. Oreg *et al.*,[80] examining the 4-state system, have found useful analytic expressions for the adiabatic eigenstates and from these have shown that the desired complete transfer can be accomplished with either 3 or 2 lasers. Unlike the odd-N systems, here the intermediate detuning should not vanish; they find that for optimal transfer the detuning should be comparable to the peak Rabi frequency. If the detuning from intermediate resonance is smaller than this, the population transfer proceeds through more than one adiabatic state, and interference between these paths will render the final transfer probability very sensitive to small variations of laser beam parameters.

6.9. *Other Applications of STIRAP*

There are many interesting and important aspects of coherent population transfer. Space limitations permit us to mention only two of these.

A brief discussion of a first implementation of the STIRAP process for a cross-beam reactive scattering experiment is given in Ref. 81. That paper describes an investigation of the effect of vibrational excitation on the cross-section for the chemiluminescent channel in the process $Na_2(v)$ + Cl $\rightarrow$ NaCl + Na*. In combination with results using Franck–Condon pumping, it is found that the cross-sections increased by about 0.75% per vibration level in the range $3 < v < 19$. These experiments are currently being extended to even higher vibrational levels.

Zoller and co-workers[82] have proposed an interesting scheme, exploiting the absence of transient population in the intermediate level of STIRAP, to produce atomic deflection by photon momentum transfer during multiple STIRAP transitions. Because population is not lost from the 3-level system by decay to other states, even if the intermediate level has a short lifetime, repetitive momentum transfers can result in a large deflection angle. Furthermore, the deflected beams are prepared coherently, an advantage for use with atomic interferometers. In principle, such procedures could also be used to produce molecular deflection, leading to possible construction of molecular interferometers. Very recently we became aware of experimental demonstrations of population transfer in such a multilevel system.[83]

7. Summary and Conclusions

Methods of laser-induced molecular excitation based on coherence and adiabatic changes offer opportunities for efficient and selective population transfer, as a means of preparing molecules for studies of state-selective reactions. We have pointed out ways in which controlled variation of either the amplitude or the frequency of the laser (or a combination of each) can induce such changes. The variation may be produced by passing a molecular beam through cw laser beams, or by interaction with pulsed lasers. In principle, the techniques may be applied to two-state or multistate systems. Often they are robust, i.e., they are insensitive to small changes in laser characteristics.

We have given special attention to techniques that transfer population efficiently in a three-level system, either by appropriate frequency sweep of the laser (possibly including simultaneous intensity variation) or by suitable sequencing of two pulses whose carrier frequencies remain constant (as in STIRAP). In both cases the intuitive sequence of interactions (whereby the pump laser moves population into the intermediate level, from which the

Stokes laser subsequently induces transfer to the final level) is successful if the lifetime of the intermediate level is much longer than the interaction duration. The counterintuitive sequence of interactions (i.e., in a three-level system the final and intermediate levels become coupled before any population transfer occurs) does not suffer from this restriction, because the intermediate level is never appreciably populated. In fact, it has been demonstrated that complete population transfer can be achieved by the STIRAP method even if the interaction time exceeds the lifetime of the intermediate electronic level by an order of magnitude. Therefore, the counterintuitive sequence is to be preferred whenever it is practical.

To conclude, we list several experimental conditions which should be met for the success of the STIRAP method.

1. The adiabaticity criterion of Eq. (55) or (77) must be satisfied. For pulsed lasers with transform limited spectral width this criterion implies that the required pulse energy increases proportional to the inverse of the pulse length. Thus, if for a given system a pulse energy of 1 mJ is adequate for a pulse of 5 ns duration, then a 5 ps pulse will require a pulse energy of 1 J.

2. The pulses should be nearly transform limited, although an increase of the bandwidth due to phase fluctuation can be compensated by larger pulse energy. The required minimum pulse energy increases proportional to the square of the phase fluctuation bandwidth.

3. The deviation from two-photon resonance of either the frequency difference (for the lambda configuration) or the frequency sum (for the ladder configuration) of pump and Stokes lasers must be smaller than the two-photon linewidth. For cw lasers it is advisable to impose active stabilization of the difference frequencies of the otherwise uncorrelated lasers.

4. The pulse envelope must vary smoothly. For STIRAP produced with suitable spatial overlap of cw lasers, this constraint means one should avoid diffraction-induced interference fringes (as might be introduced by collimation apertures). For STIRAP produced with pulsed lasers the needed amplitude stability can be achieved with injection-seeded pump lasers.

5. The optimum pulse delay depends (but not sensitively) on the maximum Rabi frequency. However, the Rabi frequencies for the pump and Stokes transitions should be approximately equal. If they

differ greatly then the condition for efficient transfer ($\Omega_S \gg \Omega_P$ at early times but $\Omega_S \ll \Omega_P$ at late times) cannot be satisfied simultaneously.

6. The pulse envelope should be well localized and be free of long range wings that extend well beyond the main pulse. If the envelope of the pulse (spatially or in the time domain) shows an extended pedestal underneath a narrower structure, then one can expect detrimental effects. The optimum delay is given by the pulse width. Optimum delay assures that the initially populated bare state $|1\rangle$ will become identified with the non-decaying dressed state $|\Phi_0\rangle$. Deviation from optimum delay will result in population transfer into other, decaying, dressed states. If the pulses consists of two components of different width, then the optimum delay can only be adjusted for one of them.

Most of these conditions are not very stringent. When they hold then the STIRAP process is very robust: small changes of irradiance, frequency or overlap do not reduce the efficiency.

Acknowledgments

We are indebted to A. Kuhn and S. Schiemann, of the University of Kaiserslautern and Dr. J. Oreg of the Israel Nuclear Research Center of the Negev, for numerous discussions.

This work was completed during a visit of BWS to Kaiserslautern, supported by the Deutsche Forschungsgemeinschaft through SFB 91 "Energy Transfer in Atomic and Molecular Collisions". This work was also supported by a NATO collaborative research grant. The work of BWS at LLNL is furthermore supported under the auspices of the U.S. Department of Energy at Lawrence National Laboratory under contract W-7405-Eng-48.

Appendix: Numerical Estimates of the Rabi Frequencies

In this appendix, we provide the information needed to estimate Rabi frequencies from the power, beam size, and pulse duration, along with the usually known electronic lifetimes or transition strengths.

The Rabi frequency between two nondegenerate atomic or molecular energy states 1 and 2 is defined by the formula

$$\hbar\Omega(t) = -\langle 2|\mathbf{d} \cdot \mathbf{e}_q|1\rangle \mathcal{E}(t)e^{-\varphi} \equiv -d_{21}\mathcal{E}(t)e^{-\varphi}\,, \tag{A.1}$$

where $\mathbf{d}$ is the dipole moment operator, $\mathbf{e}_q$ is a unit vector describing the polarization direction of the electric field (the subscript $q = +1, 0, -1$ identifies one of the spherical components[84]), $\mathcal{E}(t)$ is a slowly varying electric-field envelope, and φ is the phase of the carrier. (Note that Ω is an angular frequency, not a circular frequency).

Molecular spectroscopists commonly express dipole moments in units of the Debye, originally defined as 10^{-18} electrostatic units, or, in SI units,

$$1 \text{ Debye} = 3.3356 \times 10^{-30} \text{ A s m}. \tag{A.2}$$

Atomic spectroscopists typically express dipole moments in atomic units, defined by taking the electron charge e, the Dirac constant $\hbar$, and the electron mass as unity. The corresponding unit of length is the first Bohr radius of hydrogen, a_0, and the unit of dipole moment is ca_0,

$$ea_0 = 2.542 \text{ Debye} = 8.478 \times 10^{-31} \text{ C m}. \tag{A.3}$$

It will be appreciated that, because typical atomic dimensions are a few atomic units, typical electronic transitions involving valence electrons have dipole transition moments of a few Debye. Rydberg atoms, with their larger orbits, have correspondingly larger moments. As an example, the Rabi frequency for a dipole moment of 1 Debye in an electric field of 1 V cm^{-1} is 3.163×10^6 s^{-1}.

To identify the relevant quantities for a particular experiment we require the electric field appropriate the particular laser geometry, and we require the dipole transition moment associated with a particular transition.

Plane Waves

When dealing with the electric field of laser radiation it is common to regard the radiation as a pulsed plane wave, whose time varying irradiance $I(t)$ is the power at time t divided by the area over which this power is spread. With SI units the needed connection between this uniform intensity and the electric field magnitude, obtained from the time averaged Poynting vector for a plane wave, is

$$I(t) = \frac{1}{2}c\varepsilon_0 |\mathcal{E}(t)|^2 \tag{A.4}$$

where ε_0 is the permittivity of free space and c is the speed of light in vacuum. An electric field of 1 V cm^{-1} requires an irradiance of 1.327 mW cm^{-2}, and leads to a Rabi frequency of 2.7457×10^7 s^{-1}. Thus, we have the scaling relationship

$$|\Omega[\text{s}^{-1}]| = 27.457 \times 10^6 \times |d_{12}[\text{Debye}]| \times \sqrt{I[\text{mW mm}^{-2}]}. \qquad (\text{A.5})$$

Focused Beams

In a typical application of a cw laser we deal with a focused Gaussian beam of known waist size, maintained at a steady known power. Molecules travel across this beam with constant velocity. We wish to know the time varying electric field experienced by a typical molecule.

For pulsed laser operations the molecular motion is typically negligible, and the time dependence of the electric field at any molecule comes directly from the time variation of the irradiance. From measured values of the energy per pulse, the beam size, and the time variation of the power during a pulse we wish to determine the time varying electric field seen by a molecule.

Consider the cw case. We have a beam directed along a propagation axis (say the z axis) which maintains constant intensity at any given point. The radiation intensity may vary along the z axis if the beam is focused, and typically has a Gaussian irradiance variation in the transverse direction (the x, y plane). The steady irradiance at position x, y, z we denote

$$\overline{I}(x,\, y,\, z) = \overline{I}^{\text{max}} g(x,\, y,\, z). \qquad (\text{A.6})$$

The distribution function $g(x,\, y,\, z)$ has unit peak value, and has area

$$\int dx \int dy\, g(x,\, y,\, z) = \pi (r_z)^2 g_1, \qquad (\text{A.7})$$

where r_z is the radius of the beam at location z and g_1 is a dimensionless factor that depends on the functional form of the distribution function. (For a Gaussian beam whose FWHM is $2r_z$ the factor is $g_1 = 1/\ln 2$). The power in the beam is the irradiance integrated over the beam cross-section, and is independent of z,

$$\overline{P} = \int dx \int dy\, \overline{I}(x,\, y,\, z) = \overline{I}^{\text{max}} \int dx \int dy\, g(x,\, y,\, z)$$

$$= \overline{I}^{\text{max}} \pi (r_z)^2 g_1. \qquad (\text{A.8})$$

We consider a typical molecule moving at right angles across this beam, at constant velocity v along the x axis. At any time this molecule experiences the irradiance

$$I(t) = \overline{I}^{\text{max}} g(vt,\, y,\, z) = \frac{\overline{P}}{\pi (r_z)^2} \frac{g(vt,\, y,\, z)}{g_1}. \qquad (\text{A.9})$$

This time variation will depend on the laser spot size at position z and on the distance of closest approach y between molecular trajectory and the laser axis. The peak value of this irradiance will depend on the pulse power and the spot size at the position of the molecules.

Because we treat coherent excitation, each distinct value for the impact parameter y and velocity v must be treated by a separate Schrödinger equation; the experiments measure an average over a distribution of such trajectories. In general, this average of solutions is not the same as the solution of a single average-molecule equation in which we use an average Rabi frequency.

Pulsed Beams

For a pulsed system the irradiance depends on time as well as on position.

$$I(t,\, x,\, y,\, z) = \overline{I}^{\text{max}} f(t) g(x,\, y,\, z)\,. \tag{A.10}$$

The energy in a pulse is measurable as the average power $\overline{P}$ multiplied by the time between pulses T_{pulse} (or divided by the pulse frequency ν_{pulse})

$$\overline{E}^{\text{pulse}} = \overline{P} T_{\text{pulse}} = \frac{\overline{P}}{\nu_{\text{pulse}}}\,. \tag{A.11}$$

This energy is the integral of the time varying power for one pulse,

$$\overline{E}^{\text{pulse}} = \overline{I}^{\text{max}} \int_{\text{pulse}} dt\, f(t) \int dx \int dy\, g(x,\, y,\, z)\,. \tag{A.12}$$

The precise connection between energy and irradiance depends upon the temporal profile $f(t)$ and the spatial profile $g(x,\, y,\, z)$ but may be written

$$\overline{E}^{\text{pulse}} = \overline{I}^{\text{max}} \tau_{\text{pulse}}\, f_1\, \pi(r_z)^2 g_1\,, \tag{A.13}$$

0where τ_{pulse} is the duration of one pulse and f_1, like g_1, is a dimensionless factor expressing the pulse area. (For a Gaussian pulse whose FWHM is τ_{pulse} the factor is $f_1 = \sqrt{\pi}/\sqrt{\ln 2}$.) Thus the irradiance experienced by a molecule at position $x,\, y,\, z$ at time t is

$$I(t) = \frac{\overline{E}^{\text{pulse}}}{\pi(r_z)^2} \frac{f(t)}{f_1} \frac{g(x,\, y,\, z)}{g_1}\,. \tag{A.14}$$

This irradiance varies with position as does the intensity of cw radiation.

Relative Line Strengths

When degeneracy is present we supplement the labels 1 and 2 with additional quantum numbers, say m, so that the dipole moment matrix elements take the form

$$\langle 2|\mathbf{d}\cdot\mathbf{e}_q|1\rangle \to \langle 2,\, m_2|\mathbf{d}\cdot\mathbf{e}_q|1,\, m_1\rangle\,. \tag{A.15}$$

One can use standard techniques of angular momentum theory[85–87] to extract the dependence upon these new quantum numbers as separate factors, in the form

$$|\langle 2,\, m_2|\mathbf{d}\cdot\mathbf{e}_q|1,\, m_1\rangle|^2 = S_q(2,\, m_2;\, 1,\, m_1)|\langle 2\|d\|1\rangle|^2\,, \tag{A.16}$$

where the reduced matrix element $\langle 2\|d\|1\rangle$ is defined, apart from a phase, by the relationship

$$|\langle 2\|d\|1\rangle|^2 = \sum_{qm_1m_2} |\langle 2,\, m_2|\mathbf{d}\cdot\mathbf{e}_q|1,\, m_1\rangle|^2\,, \tag{A.17}$$

and the strength factor used here has the normalization

$$\sum_{qm_1m_2} S_q(2,\, m_2;\, 1,\, m_1) = 1\,. \tag{A.18}$$

Usually it is possible to introduce further factorization to separate the dependence on various sets of quantum numbers. When treating atomic spectra these produce the so-called line and multiplet factors.[88,89]

The Geometric Factor

The simplest example of such factorization (the Wigner–Eckart theorem[87]) extracts the dependence on magnetic quantum numbers M from a transition between rotational levels,

$$|\langle 2,\, J_2 M_2|\mathbf{d}\cdot\mathbf{e}_q|1,\, J_1 M_1\rangle|^2 = S_q^{\text{geom}}(J_2 M_2;\, J_1 M_1)|\langle 2J_2\|d\|1J_1\rangle|^2\,. \tag{A.19}$$

The geometric factor S^{geom} contains all the information about molecular orientation in a laboratory reference frame, through the dipole component q and the magnetic quantum numbers. This factor is expressible as the square of a three-j symbol[87]

$$S_q^{\text{geom}}(J_2 M_2;\, J_1 M_1) = \begin{pmatrix} J_2 & 1 & J_1 \\ -M_2 & q & M_1 \end{pmatrix}^2\,. \tag{A.20}$$

The Molecular Line Strengths

Apart from the simple dependence on magnetic quantum numbers invoked by the Wigner–Eckart theorem, further expressions for dipole transition strengths depend on the molecular symmetry. In the simple case of an idealized vibrating symmetric top, whose energy states are labeled by vibrational quantum number v and projection K of angular momentum along the molecular symmetry axis, the dipole transition moment can be expressed in the form[87]

$$|\langle e_2 v_2 J_2 K_2 M_2 | \mathbf{d} \cdot \mathbf{e}_q | e_1 v_1 J_1 K_1 M_1 \rangle|^2$$

$$= S_q^{\text{geom}}(J_2 M_2; J_1 M_1) \qquad \text{geometric factor}$$

$$\times\, S^{\text{HL}}(J_2 K_2; J_1 K_1) \qquad \text{Hönl–London factor}$$

$$\times\, S^{\text{FC}}(e_2 v_2; e_1 v_1) \qquad \text{Franck–Condon factor}$$

$$\times\, |\langle e_2 J_2 \| d \| e_1 J_1 \rangle|^2 \qquad \text{Electronic transition strength}$$

$$\text{(A.21)}$$

The Hönl–London factor, like the geometric factor, can be expressed as the square of a three-j symbol (see Zare[87]),

$$S^{\text{HL}}(J_2 K_2; J_1 K_1) = (2J_2 + 1)(2J_1 + 1) \begin{pmatrix} J_2 & 1 & J_1 \\ -K_2 & q & K_1 \end{pmatrix}^2$$

$$\text{with} \quad q = K_2 - K_1 . \quad \text{(A.22)}$$

Expressions for Hönl–London factors will be found in several references.[90,91] The Franck–Condon factor S^{FC} involves the overlap of vibrational wavefunctions.[89,91]

When the molecule has nonzero electronic spin (e.g., a doublet or triplet electronic state), the formulas for transition moments become more complicated. Often they may be treated as one of Hund's four cases of angular momentum coupling[91] which makes it possible to deduce the various strength factors on the basis of angular momentum theory. In any case, the needed expression for the dipole moment to be used in calculating the Rabi frequency has the form

$$|d_{21}| = \sqrt{S^{\text{geom}}} \sqrt{S^{\text{HL}}} \sqrt{S^{\text{FC}}} |\langle e_2 J_2 \| d \| e_1 J_1 \rangle| . \qquad \text{(A.23)}$$

Spontaneous Emission

The Rabi frequency parametrizes the field-induced coupling between two distinct quantum states. Specifically, the Rabi frequency refers to a single molecular orientation and a single pair of magnetic sublevels. Properly speaking, there is no Rabi frequency for transitions between degenerate sublevels: we require the average of results from several Schrödinger equations, not the result of a single "average molecule" Schrödinger equation. By contrast, traditional spectroscopic measurements yield values of transition moments that are averaged over orientation or even over rotational levels within a vibrational band.

As an example, we consider the spontaneous emission rate (the Einstein A coefficient). For a transition between two *nondegenerate* levels the relevant formula is,[86] in SI unit,

$$A_{21} = \frac{1}{3} \frac{|\omega_{21}|^3}{\pi \varepsilon_0 \hbar c^3} |\langle 2|\mathbf{d}|1\rangle|^2 \quad \text{for} \quad \hbar \omega_{21} \equiv E_2 - E_1 \,. \tag{A.24}$$

The lifetime of the excited state is the inverse of the total spontaneous emission rate to lower lying energy levels

$$\frac{1}{\tau_2} = \sum_f A_{2f} \,. \tag{A.25}$$

Therefore, for the special case of nondegenerate levels and spontaneous emission in a single transition, the dipole moment is calculable as

$$|d_{21}| = 6.312 \text{ Debye} \times \left(\frac{\lambda_{21}}{500 \text{ nm}}\right)^{3/2} \times \left(\frac{\tau_2}{10 \text{ ns}}\right)^{1/2} , \tag{A.26}$$

where λ_{21} is the wavelength of the transition, in nanometers, and τ_2 is the lifetime, in nanoseconds.

When the levels are *degenerate*, and have equal sublevel population (as occurs in the absence of optical pumping), then the relevant transition probability is

$$A_{21} = \frac{1}{3} \frac{|\omega_{21}|^3}{\pi \varepsilon_0 \hbar c^3} \frac{1}{g_2} \sum_{m_1 m_2} |\langle 2m_2|\mathbf{d}|1m_1\rangle|^2 , \quad g_2 = \sum_{m_2} 1 \,. \tag{A.27}$$

For the idealized molecular transition discussed above the transition between two rotational sublevels of a vibrational band has the form

$$A_{21} = \frac{1}{3} \frac{|\omega_{21}|^3}{\pi \varepsilon_0 \hbar c^3} \frac{1}{g_2} S^{\text{HL}} S^{\text{FC}} |\langle e_2 \| d \| e_1 \rangle|^2 \,. \tag{A.28}$$

When spontaneous emission occurs only between two electronic states the lifetime of a particular rotation–vibration level is obtained as the sum over Hönl–London and Franck–Condon factors,

$$\frac{1}{\tau_2} = \frac{1}{3}\frac{|\bar{\omega}|^3}{\pi\varepsilon_0\hbar c^3}\frac{1}{g_2}|\langle e_2\|d\|e_1\rangle|^2\sum_f\frac{|\omega_{2f}|^3}{|\bar{\omega}|^3}S^{\mathrm{HL}}S^{\mathrm{FC}}\,,\qquad (A.29)$$

where $\bar{\omega}$ is an average frequency. From approximations to the sum and a known lifetime we can estimate the electronic dipole moment. We then multiply this by the various S factors of Eq. (A.21) or (A.23) to obtain the dipole transition moment portion of the Rabi frequency. We combine this, as prescribed by Eq. (A.5), with the irradiance determined from Eq. (A.9) or (A.14) to obtain the Rabi frequency.

Using those numbers and the formulae given above the Rabi frequencies for other atoms or molecules with different transition moments or electronic lifetimes and Franck–Condon factors and for other laser intensities can be estimated by scaling.

Sodium dimers have an electronic transition moment $|\langle e_2\|d\|e_1\rangle|$ of around 10 Debye (equivalent to an electronic lifetime of about 10 ns for emission centered around 700 nm). A typical magnitude of the Franck–Condon factor is $S^{\mathrm{FC}} = 0.05$. To estimate the Hönl–London factor for a transition between vibrational levels v'' and v' starting from a particular electronic j-state we assume that each level is coupled to six m-levels (P- and R-branch with $\Delta m = 0, \pm 1$ transitions). This gives the mean value $S^{\mathrm{HL}} = 0.16$. The peak irradiance of a cw laser with power $P = 100$ mw in a gaussian beam profile of $w = 100$ μm (1/e-radius of the electric field strength) is $I_0 = 6.4$.

References

1. G. Scoles, ed., *Atomic and Molecular Beam Methods* (Oxford Univ. Press, Oxford, 1988).
2. R. B. Kurzel and J. I. Steinfeld, *J. Chem. Phys.* **53**, 3293 (1970); K. Bergmann and W. Demtröder, *Z. Phys.* **243**, 1 (1971).
3. K. Bergmann, in *Atomic and Molecular Beam Methods*, ed. G. Scoles (Oxford Univ. Press, Oxford, 1988) p. 293.
4. F. F. Crim, *Ann. Rev. Phys. Chem.* **35**, 657 (1984).
5. K. Bergmann, U. Hefter, and J. Witt, *J. Chem. Phys.* **72**, 4777 (1980); M. Fuchs and J. P. Toennies, *J. Chem. Phys.* **85**, 7062 (1986).
6. K. Bergmann, R. Engelhardt, U. Hefter, P. Hering, and J. Witt, *Phys. Rev. Lett.* **40**, 1446 (1978).

7. C. H. Hamilton, J. L. Kinsey, and R. W. Field, *Ann. Rev. Phys. Chem.* **37**, 493 (1986).

8. R. Loudon, *The Quantum Theory of Light* (Clarendon, Oxford, 1973); P. L. Knight and L. Allen, *Concepts of Quantum Optics* (Pergamon, N.Y., 1983).

9. C. Liedenbaum, S. Stolte, and J. Reuss, *Phys. Rep.* **178**, 1 (1989).

10. B. W. Shore, *The Theory of Coherent Atomic Excitation* (Wiley, N.Y., 1990).

11. See Ref. 10, Sec. 1.4.

12. See Ref. 10, Sec. 2.2.

13. See Ref. 10, Sec. 2.2.

14. P. W. Milonni and J. H. Eberly, *Lasers* (Wiley, N.Y., 1988); K. Smith and R. M. Thomson, *Computer Modeling of Gas Lasers* (Plenum, N.Y., 1978).

15. See Ref. 10, Sec. 3.3; C. Cohen–Tannoudji, J. Dupont–Roc, and G. Grynberg, *Atom–Photon Interactions* (Wiley, N.Y., 1992); S. Stenholm, *Foundations of Laser Spectroscopy* (Wiley, N.Y., 1984); P. Meystre and M. Sargent III, *Elements of Quantum Optics* (Springer-Verlag, N.Y., 1990).

16. See Ref. 10, Sec. 1.9; P. L. Knight and L. Allen, Ref. 8.

17. See Ref. 10, Chap. 10; C. Cohen–Tannoudji, J. Dupont–Roc, and G. Grynberg, *Photons and Atoms: Introduction to Quantum Electrodynamics* (Wiley, N.Y., 1989).

18. See Ref. 10, Sec. 9.15; D. P. Craig and T. Thirunamachandran, *Molecular Quantum Electrodynamics* (Academic, N.Y., 1984).

19. See Ref. 10, Sec. 1.8.

20. See Ref. 10, Sec. 1.6.

21. See Ref. 10, Chap. 6; K. Blum, *Density Matrix Theory and Applications* (Plenum, N.Y., 1981); U. Fano, *Rev. Mod. Phys.* **29**, 74 (1957); A. Omont, *Prog. Quant. Electron.* **5**, 69 (1977); F. T. Hioe, *Phys. Rev.* **A30**, 3097 (1984).

22. See Ref. 10, Sec. 6.4, 21.9; W. Happer, *Rev. Mod. Phys.* **44**, 169 (1972); V. A. Alekseev, T. L. Andreeva, and I. I. Sobel'man, *Sov. Phys. JETP* **35**, 325 (1972); I. M. Beterov and V. P. Chebotaev, *Three-Level Gas Systems and Their Interaction with Radiation* (Pergamon, N.Y., 1974); K. Burnett, J. Cooper, R. J. Ballagh, and E. W. Smith, *Phys. Rev.* **A22**, 2005 (1980); C. Cohen–Tannoudji, in *Frontiers in Laser Spectroscopy*, eds. R. Balian, S. Haroche, and S. Liberman (North-Holland, N.Y., 1977); V. M. Fain and Y. I. Khanin, *Quantum Electronics. 1: Basic Theory* (M.I.T. Press, Cambridge, Mass., 1969).

23. See Ref. 10, Sec. 3.1.

24. See Ref. 10, Sec. 3.10.

25. See Ref. 10, Sec. 22.4.

26. See Ref. 10, Sec. 3.9; I. I. Rabi, N. F. Ramsey, and J. Schwinger, *Rev. Mod. Phys.* **62**, 167 (1954); J. H. Shirley, *Phys. Rev.* **B138**, 979 (1965).

27. See Ref. 10, Secs. 3.5–3.6; C. Cohen–Tannoudji, B. Diu, and F. Lalo, *Quantum Mechanics* (Wiley-Interscience, N.Y., 1977) p. 421.

28. See Ref. 10, Secs. 3.9–3.10; Z. Bialynicka–Birula, I. Bialynicki–Birula, J. H. Eberly, and B. W. Shore, *Phys. Rev.* **A16**, 2048–2054 (1977).

29. See Ref. 10, Chap. 14; G. F. Thomas and W. J. Meath, *J. Phys.* **B16**, 951 (1983); J. D. Dollard and C. N. Friedman, *J. Math. Phys.* **18**, 1598 (1977); S. Klarsfeld and J. O. Oteo, *Phys. Rev.* **A45**, 3329 (1992); C. Cerjan, *JOSA.* **B7**, 680 (1990).

30. See Ref. 10, Sec. 3.6.

31. See Ref. 10, Sec. 8.5; R. P. Feynman, F. L. Vernon Jr., and R. W. Hellwarth, *J. Appl. Phys.* **28**, 49 (1957).

32. L. D. Landau, *Phys. Z. Sowjet* **2**, 46 (1932); C. Zener, *Proc. Roy. Soc. (Lond.)* **A137**, 696 (1932); C. Stueckelberg, *Helv. Phys. Acta* **5**, 369 (1932).

33. See Ref. 10, Sec. 5.6; S. Geltman, *Topics in Atomic Collision Theory* (Academic, N.Y., 1969).

34. See Ref. 10, Sec. 5.6; Y. N. Demkov, *Sov. Phys. Dokl.* **11**, 138 (1966); E. E. Nikitin, *Theory of Elementary Atomic and Molecular Processes in Gases* (Clarendon Press, Oxford, 1974).

35. A. Abragam, *The Principles of Nuclear Magnetism* (Oxford University Press, London, 1970); C. P. Slichter, *Principles of Magnetic Resonance*, 3rd ed. (Springer, Berlin, 1990).

36. M. M. T. Loy, *Phys. Rev. Lett.* **32**, 814 (1974).

37. S. M. Hamadani, A. T. Mattick, N. A. Kurnit, and A. Javan, *Appl. Phys. Lett.* **27**, 21 (1975).

38. S. Avrillier, J.-M. Raimond, C. J. Bordé, D. Bassi, and G. Scoles, *Opt. Comm.* **39**, 311 (1981).

39. A. G. Adam, T. E. Gough, N. R. Isenor, and G. Scoles, *Phys. Rev.* **A32**, 1451 (1985).

40. J. P. C. Kroon, H. A. J. Senhorst, H. C. W. Beijerinck, B. J. Verhaar, and N. F. Verster, *Phys. Rev.* **A21**, 3724 (1985).

41. V. Lorent, W. Claeys, A. Cornet, and X. Urbain, *Opt. Comm.* **64**, 41 (1987).

42. J. S. Melinger, A. Hariharan, S. R. Gandhi, and W. S. Warren, *J. Chem. Phys.* **95**, 2210 (1991).

43. J. R. Rubbmark, M. M. Cash, M. G. Littman, and D. Kleppner, *Phys. Rev.* **A23**, 3107 (1981); D. A. Harmin, *Phys. Rev.* **A44**, 433 (1991).

44. Y. N. Demkov, *Sov. Phys. Dokl.* **11**, 138 (1966); Y. N. Demkov and V. I. Osherov, *Sov. Phys. JETP* **26**, 916 (1967); E. E. Nikitin, *Theory of Elementary Atomic and Molecular Processes in Gases* (Clarendon Press, Oxford, 1974); Y. Kayanuma and S. Fukuchi, *J. Phys.* **B18**, 4089 (1985); C. E. Carroll and F. T. Hioe, *JOSA* **B2**, 1355 (1985).

45. Y. N. Demkov and V. I. Osherov, *Sov. Phys. JETP* **26**, 916 (1967).

46. M. V. Kuz'min and V. N. Sazanov, *Sov. Phys. JETP* **52**, 889 (1980); M. V. Kuz'min, *Sov. J. Quant. Electron.* **11**, 9 (1981).

47. G. L. Peterson and C. D. Cantrell, *Phys. Rev.* **A31**, 807 (1985).

48. J. S. Melinger, S. R. Gandhi, A. Hariharan, J. X. Tull, and W. S. Warren, *Phys. Rev. Lett.* **68**, 2000 (1992).

49. Y. Band and P. Julienne, *J. Chem. Phys.* **97**, 9107 (1992).

50. S. Chelkowski, A. D. Bandrauk, and P. B. Corkum, *Phys. Rev. Lett.* **65**, 2355 (1990).

51. See Ref. 10, Sec. 14.2; T. H. Einwohner, J. Wong, and J. C. Garrison, *Phys. Rev.* **A14**, 1452 (1976).

52. See Ref. 10, Sec. 13.1.

53. See Ref. 10, Sec. 13.2; P. M. Radmore and P. L. Knight, *J. Phys.* **B15**, 561 (1982).

54. C. E. Carroll and F. T. Hioe, *Phys. Rev.* **A36**, 724 (1987); C. E. Carroll and F. T. Hioe, *J. Math. Phys.* **29**, 487 (1988); C. E. Carroll and F. T. Hioe, *JOSA* **B5**, 1335 (1988); C. E. Carroll and F. T. Hioe, *J. Phys.* **B22**, 2633 (1989); C. E. Carroll and F. T. Hioe, *Phys. Rev.* **42**, 1522 (1990).

55. See Ref. 10, Sec. 13.4; D. Grischkowsky, *Phys. Rev.* **A7**, 2096 (1973); D. Grischkowsky and M. M. T. Loy, *Phys. Rev.* **A12**, 1117 (1975); D. Grischkowsky, M. M. T. Loy, and P. F. Liao, *Phys. Rev.* **A12**, 2514 (1975); M. M. T. Loy, *Phys. Rev. Lett.* **41**, 473 (1978).

56. J. Oreg, F. T. Hioe, and J. H. Eberly, *Phys. Rev.* **A29**, 690 (1984); J. Oreg, G. Hazak, and J. H. Eberly, *Phys. Rev.* **A23**, 2776 (1985).

57. B. W. Shore, K. Bergmann, A. Kuhn, S. Schiemann, J. Oreg, and J. H. Eberly, *Phys. Rev.* **A45**, 5297 (1992).

58. B. Broers, H. B. van Linden van den Heuvell, and L. D. Noordam, *Phys. Rev. Lett.* **69**, 2062 (1992).

59. U. Gaubatz, P. Rudecki, S. Schiemann, and K. Bergmann, *J. Chem. Phys.* **92**, 5363 (1990).

60. J. R. Kuklinski, U. Gaubatz, F. T. Hioe, and K. Bergmann, *Phys. Rev.* **A40**, 6741 (1989).

61. See Ref. 10, Sec. 13.7; E. Arimondo and G. Orriols, *Nuov. Cim. Letts.* **17**, 333 (1976); H. R. Gray, R. M. Whitley, and C. R. Stroud Jr., *Opt. Letts.* **3**, 218 (1978); P. E. Coleman and P. L. Knight, *J. Phys.* **B15**, L235 (1982); P. M. Radmore and P. L. Knight, *J. Phys.* **B15**, 561 (1982).

62. B. W. Shore, K. Bergmann, and J. Oreg, *Z. Phys.* **D23**, 33 (1992).

63. U. Gaubatz, P. Rudecki, M. Becker, S. Schiemann, M. Külz, and K. Bergmann, *Chem. Phys. Lett.* **149**, 463 (1988).

64. I. C. M. Littler, P. Jung, and K. Bergmann, *Opt. Commun.* **87**, 61 (1992).

65. H.-G. Rubahn, E. Konz, S. Schiemann, and K. Bergmann, *Z. Phys.* **D22**, 401 (1991).

66. L. A. Westling, M. G. Raymer, and J. J. Snyder, *JOSA* **B1**, 150 (1984).

67. V. Buzek, P. L. Knight, and I. K. Kudryavtsev, *Phys. Rev.* **A44**, 1931 (1991); S. N. Dixit, P. Zoller, and P. Lambropoulos, *Phys. Rev.* **A21**, 1289 (1980).

68. A. Kuhn, S. Schiemann, G. Z. He, G. Coulston, W. S. Warren, and K. Bergmann, *J. Chem. Phys.* **96**, 4215 (1992).

69. S. Schiemann, A. Kuhn, S. Steuerwald, and K. Bergmann, *Phys. Rev. Lett.* **71**, 3637 (1993).

70. G. Z. He, A. Kuhn, S. Schiemann, and K. Bergmann, *JOSA* **B7**, 1960 (1990).

71. B. Glushko and B. Kryzhanovsky, *Phys. Rev.* **A46**, 2823 (1992).

72. Y. B. Band and P. S. Julienne, *J. Chem. Phys.* **94**, 5291 (1991); Y. B. Band, *Phys. Rev.* **A45**, 6643 (1992).

73. F. T. Hioe and J. H. Eberly, *Phys. Rev.* **A25**, 2168 (1982).

74. Y. Band and P. S. Julienne, *J. Chem. Phys.* **96**, 3339 (1992).

75. G. Coulston and K. Bergmann, *J. Chem. Phys.* **96**, 3467 (1992).

76. C. E. Carroll and F. T. Hioe, *Phys. Rev. Lett.* **68**, 3523 (1992).

77. B. W. Shore, K. Bergmann, J. Oreg, and S. Rosenwaks, *Phys. Rev.* **A44**, 7442 (1991).

78. Y. Band and P. S. Julienne, *J. Chem. Phys.* **95**, 5681 (1991).

79. See Ref. 10, Sec. 21.4; B. W. Shore, *Phys. Rev.* **A29**, 1578 (1984).

80. J. Oreg, K. Bergmann, B. W. Shore, and S. Rosenwaks, *Phys. Rev.* **A45**, 4888 (1992).

81. P. Dittmann, F. P. Pesl, J. Martin, G. W. Coulston, G. Z. He, and K. Bergmann, *J. Chem. Phys.* **97**, 9472 (1992).

82. P. Marte, P. Zoller, and J. L. Hall, *Phys. Rev.* **A44**, 4118 (1991).

83. P. Pillet, C. Valentin, R.-L. Yuan, and J. Yu, *Phys. Rev.* **A48**, 845 (1993); L. S. Goldner, C. Gerz, R. J. C. Spreeuw, S. L. Rolston, C. I. Westbrook, W. D. Phillips, P. Marte, and P. Zoller, *Phys. Rev. Lett.* **72**, 997 (1994); J. Lawall and M. Prentiss, *Phys. Rev. Lett.* **72**, 993 (1994).

84. See Ref. 10, Sec. 1.5.

85. See Ref. 10, Chap. 20; A. R. Edmonds, *Angular Momentum in Quantum Mechanics* (Princeton Univ. Press, Princeton, 1957).

86. E. U. Condon and G. H. Shortley, *The Theory of Atomic Spectra* (Cambridge Univ. Press, Cambridge, 1953); I. I. Sobelman, *Atomic Spectra and Radiative Transitions* (Springer, N.Y., 1992).

87. B. R. Judd, *Angular Momentum Theory for Diatomic Molecules* (Academic, N.Y., 1975); L. C. Biedenhard and J. D. Louck, *Angular Momentum in Quantum Physics* (Addison-Wesley, Reading, Mass., 1981); R. N. Zare, *Angular Momentum: Understanding Spatial Aspects in Chemistry and Physics* (Wiley, N.Y., 1988).

88. B. W. Shore and D. H. Menzel, *Appl. J. Supp.* **12**, 187 (1965); B. W. Shore and D. H. Menzel, *Principles of Atomic Spectra* (Wiley, N.Y., 1968).

89. R. W. Nicholls and A. L. Stewart, in *Atomic and Molecular Processes*, ed. D. R. Bates (Academic, N.Y., 1962).

90. H. Hönl and F. London, *Z. Phys.* **33**, 803 (1925); H. C. Allen, Jr., and P. C. Cross, *Molecular Vib-Rotors* (Wiley, N.Y., 1963); J. B. Tatum, *Appl. J. Suppl.* **14**, 21 (1967); I. Kovacs, *Rotational Structure in the Spectra of Diatomic Molecules* (American Elsevier, N.Y., 1969); see also Zare, Ref. 87.

91. G. Herzberg, *Spectra of Diatomic Molecules* (Van Nostrand, N.Y., 1950).

92. W. T. Zemke, K. K. Verma, T. Vu, and W. C. Stwalley, *J. Mol. Spectrosc.* **85**, 150 (1981).

Part II. Intramolecular Vibrational Redistribution and Unimolecular Dissociation

CHAPTER 10

**VIBRATIONAL SPECTROSCOPY,
INTRAMOLECULAR DYNAMICS AND STATE SPECIFIC
UNIMOLECULAR DISSOCIATION OF CH_3O $(\tilde{X}^2E)$**

F. Temps

Max-Planck-Institut für Strömungsforschung
Bunsenstrasse 10, 37073 Göttingen, Germany

Contents

375

1. Introduction

The elucidation of the detailed relationship between the mechanisms of unimolecular chemical reactions[1] and the underlying potential energy hypersurfaces of the species is a central issue which is behind much of the ongoing work pertaining to the spectroscopy and dynamics of highly vibrationally excited polyatomic molecules. In view of this task, great interest exists in direct investigations of the dynamics of highly vibrationally excited molecules at chemically significant energies, i.e., at a level of excitation that is comparable to the potential barriers which must be overcome for a reaction to occur. To this end, current research activities focus on a number of intimately related topics, including, e.g., the spectroscopy of highly excited molecules (see, e.g., Refs. 2–7), the extraction of the dynamical information which is encoded in the spectra, in particular information concerning intramolecular vibrational energy redistribution processes (IVR, see, e.g., Refs. 8–13), measurements of specific unimolecular reaction rate constants (e.g., Refs. 14–18) and product distributions (e.g., Refs. 19–21), and investigations of collisional energy transfer processes (e.g., Refs. 22–24). However, despite much progress, which has been made over the last few years with the arrival of new experimental and theoretical techniques, many fundamental, and often controversial, questions have remained. At least for selected examples, many of these points ask for direct and fully rotation–vibration quantum state resolved investigations of the properties of the energized reactive molecules with specific excitation comparable to the respective potential threshold energies.

Different important questions regarding the spectroscopic characterization and intramolecular dynamics of highly excited molecules are addressed in the various other articles in this monograph. In principle, precise measurements of the vibrational term values of the excited molecules allow the potentials of the species to be mapped with spectroscopic accuracy. However, it is precisely the important high energy part of the potential that is least accessible to conventional high resolution spectroscopies. A number of methods have been developed over the last years which allow investigations of the individual high-lying vibrational levels of small polyatomic molecules. These techniques, which include, e.g., vibrational overtone excitation (see, e.g., Ref. 2), Raman spectroscopy of dissociating molecules (Ref. 3), level crossing spectroscopy (Ref. 6), IR–UV double resonance spectroscopy (e.g., Ref. 4), and, last but not least, stimulated emission pumping spectroscopy (SEP, Refs. 5–6), are complementary in that they are sensitive to different regions of the potential.

As evident from the contributions to this monograph, since its development only a decade ago,[25] SEP spectroscopy has gained particular importance.[26] It stands out for its high sensitivity and unique ability to probe the skeletal vibrational modes of a molecule which are especially germane to unimolecular processes. Because of the double resonance nature of the technique, spectra can be obtained which are free of inhomogeneous rotational structure compared to the congestion encountered with thermally populated states. These spectra encode detailed and direct information on the intramolecular and dissociative dynamics which cannot easily be obtained by other methods. Moreover, the SEP technique can be employed to prepare highly excited species in well defined states in order to study the dynamics directly in the time domain.[23,24]

The present chapter is concerned with SEP studies of one special molecule, namely, the methoxy radical, CH_3O ($\tilde{X}^2E$), which can be investigated in much greater detail than other species. Being a prototype C_{3v} symmetric top molecule with an orbitally degenerate electronic ground state (^2E), CH_3O is an elementary example of a molecule that is subject to the Jahn–Teller (JT) effect.[27] At the same time, CH_3O presents an ideal and, in many respects, unprecedented model for investigating the vibrational level structure and intramolecular dynamics of a small polyatomic molecule. Namely, the phenomenon of intramolecular vibrational energy redistribution (IVR), and, eventually, the dissociation of the molecule can be studied at a fully quantum state resolved level of detail. To these ends, the spectra and the dynamics of individual CH_3O $(\tilde{X})$ rotation–vibration states with precisely defined total excitation energy E, angular momentum J, rovibronic symmetry species Γ, and, as far as possible, other specific degrees of freedom α, etc., can be probed by using SEP spectroscopy[28] in the complete range from the bottom of the $\tilde{X}$ potential well up to energies of at least $E_v \leq 10\,000\,\mathrm{cm}^{-1}$. Since the highest energies which can be reached are considerably higher than the H–H_2CO bond dissociation energy, so that the excited molecules can undergo unimolecular decomposition according to

$$CH_3O\ (\tilde{X};\ E,\ J,\ \Gamma;\ \alpha,\ \dots) \rightarrow H + H_2CO\,, \quad \Delta H_0^0 \approx 6900\,\mathrm{cm}^{-1}\,, \qquad (1)$$

the SEP spectra of CH_3O $(\tilde{X})$ can provide highly detailed and unique information on the dissociative dynamics of the molecule.

Figure 1 shows a schematic energy diagram of the CH_3O system which illustrates the experiment.

In principle, highly vibrationally excited CH_3O $(\tilde{X})$ molecules may undergo three unimolecular reaction channels, namely, dissociation to $H + H_2CO$ (Eq. (1)), isomerization to CH_2OH, or decomposition to $H_2 + HCO$. With a value for the H–H_2CO bond energy of $\Delta H_0^0 \approx 6900\,\mathrm{cm}^{-1}$ (Refs. 29–36) and an estimate for the height of the barrier for the addition of H to H_2CO of $\approx 1500\,\mathrm{cm}^{-1}$, reaction (-1), which is obtained from an investigation of the rate constant for the isotope exchange reaction $D + H_2CO \rightarrow H + HDCO$ (Ref. 37), the classical threshold energy for reaction (1) is found to be $E_0(H$–$H_2CO) \approx 8400\,\mathrm{cm}^{-1}$. Concerning the other channels, according to results of a number of *ab initio* quantum chemical calculations,[33–36] the potential barrier for the possible isomerization to CH_2OH is some $2000\,\mathrm{cm}^{-1}$ higher than that for the $H + H_2CO$ channel, and the threshold energy for decomposition to $H_2 + HCO$ is considerably higher again, Thus, reaction (1) is the lowest energy decomposition channel.

The excited vibrational levels of CH_3O in the $\tilde{X}^2E$ ground state which are of interest are probed by SEP spectroscopy through optical double resonance using the $\tilde{A}^2A_1$ first excited electronic state as intermediate state. The $\tilde{A}$ state has been characterized at rotational resolution using

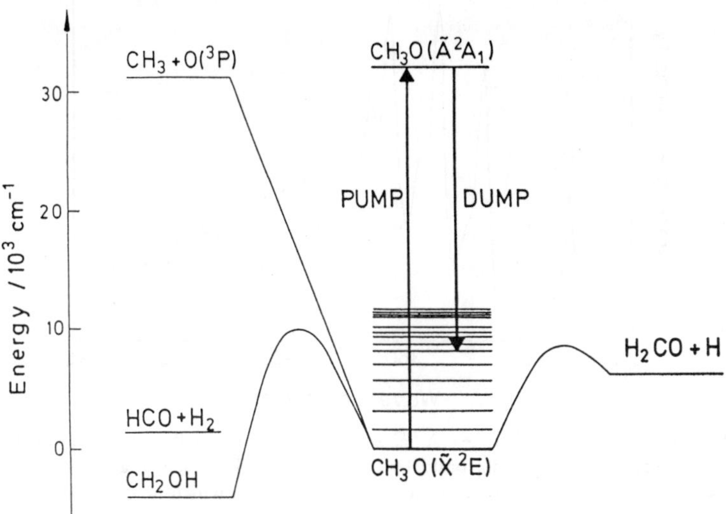

Fig. 1. Schematic potential energy for the CH$_3$O system. Highly excited vibrational levels in the $\tilde{X}^2E$ ground electronic state of CH$_3$O are probed by SEP through optical double resonance (pump and dump) via the $\tilde{A}^2A_1$ excited state. The pump laser is set to excite a precisely defined $\tilde{A}$ rovibrational level. The transition is monitored by observing the LIF. The dump laser is then employed to stimulate emission to the target $\tilde{X}$ vibrational levels which leads to a dip in the fluorescence intensity from $\tilde{A}$.

laser induced fluorescence (LIF) excitation spectroscopy.[38,39] As can be seen from a low resolution dispersed fluorescence (DF) spectrum recorded after excitation to the $\tilde{A}$ vibrational origin at $\lambda = 316$ nm (Fig. 2 and Ref. 40), the spontaneous emission to the ground state is distributed over a broad Franck–Condon range and extends to $\tilde{X}$ vibrational energies far above the H H$_2$CO dissociation limit of the molecule. However, at these high energies, DF spectra cannot be obtained with the necessary resolution to observe discrete structure. Using a strong and narrow bandwidth "dump" laser pulse, on the other hand, stimulated emission (SEP) transitions can be induced from the $\tilde{A}$ state to very high-lying vibrational levels at the interesting energies in the $\tilde{X}^2E$ state with laser-limited resolution.

We have been studying SEP spectra of CH$_3$O over a range of $\tilde{X}$ excitation energies to address different questions of both spectroscopic and dynamical interest. First, high resolution SEP spectra that were observed indicate that, even for many low-lying vibrational states, distinctively differ-

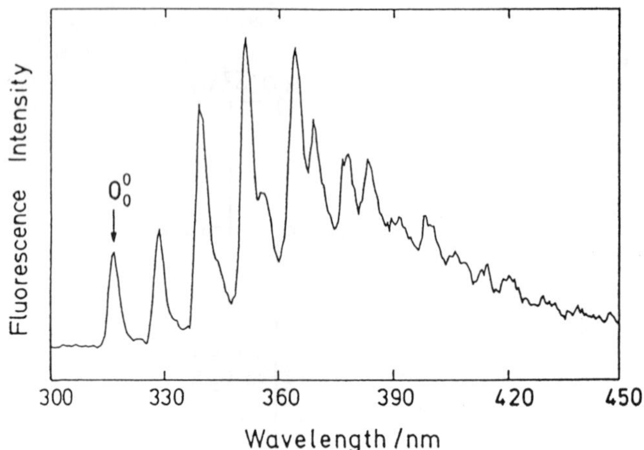

Fig. 2. Low resolution $\tilde{A}\,^2A_1 \rightarrow \tilde{X}^2E$ spectrally dispersed fluorescence spectrum of CH_3O after excitation in the 0_0^0 band to the $\tilde{A}$ origin. The resulting $\tilde{A} \rightarrow \tilde{X}$ emission can be seen to be dominated by a pronounced progression which is assigned to the C–O stretch vibration ν_3.

ent vibronic assignments must be made than were deduced previously from low resolution DF spectra.[41] These results lead to new information on the JT effect in the molecule.[42] Second, towards higher excitation, systematic series of spectra were observed which could be assigned in zero-order to the Franck–Condon active ν_3 (C–O stretch) normal mode vibration. However, at increasing densities of states, the normal mode vibrations become subject to anharmonic vibration–vibration couplings and Coriolis or centrifugal vibration–rotation interactions with the large number of Franck–Condon inactive background states.[28,43–45] The resulting extensive level mixing that can be observed in the spectra is the spectroscopic signature of IVR processes. Here, considering the calculated total density of vibrational states at, e.g., the $H–H_2CO$ bond dissociation energy of $\rho_v(6900\,\mathrm{cm}^{-1}) \approx 0.4/\mathrm{cm}^{-1}$, the energy range that is investigated corresponds precisely to the interesting transition regime between vibrational mode specific chemistry at low energies towards the expected statistical properties of large molecules at high energies. Last but not least, third, at high resolution, significantly broadened linewidths were observed in SEP spectra up to $E_v < 10\,000\,\mathrm{cm}^{-1}$, i.e., above the classical dissociation threshold.[43,45] These spectra establish vibrational "resonances" which are embedded in the dissociation quasicontinuum of the molecule. Observed homogeneous line broadenings give measures of the specific unimolecular rate constants $k(E, J, \Gamma; \alpha, \ldots)$ for the decomposition of highly energized CH$_3$O ($\tilde{X}$) with well known values of E, J, Γ, etc.

This article summarizes selected interesting results on these topics. Following this introduction, Sec. 2 will describe the experimental setup and procedures. Section 3 will be concerned with aspects of the spectroscopy of low-lying vibrational levels of CH$_3$O, namely, the fundamental vibrational modes and selected low energy combination bands. Section 4 will report on results concerning the intramolecular dynamics, namely IVR, encoded in the spectra of CH$_3$O ($\tilde{X}$) after excitation to specific vibration states between $2000 \, \text{cm}^{-1} \leq E \leq 9000 \, \text{cm}^{-1}$. Section 5 will be concerned with the evidence for unimolecular dissociation of CH$_3$O. Eventually, Sec. 6 will conclude with a discussion of some implications of the results in their own right and in relation to other work and some considerations of future perspectives.

2. Experimental

The SEP studies of CH$_3$O which are reported here were carried out using the fluorescence dip method[5] with conventional pulsed dye lasers as the "pump" and "dump" lasers (cf. Fig. 1). Fluorescence detection offers considerable advantages for investigations of short-lived free radical species which can only be produced in very low concentrations. Especially, it is a highly sensitive and well documented method for monitoring free radicals. Furthermore, apart from saturation at high laser pulse energies, it is a linear technique, in contrast, e.g., to SEP detection schemes based on four-wave mixing spectroscopy[46] which exhibit a complicated dependence on species concentration that hampers applications to dynamics studies.

Two experimental systems are used to generate CH$_3$O, a room temperature discharge flow reaction (Fig. 3) and a pulsed supersonic free jet apparatus (Fig. 4). They are complementary in that they allow a wide range of rotational levels to be covered. At room temperature, one can start with relatively high initial rotational energies. Because of specific selection rules, this can actually lead to a simplification of the interpretation of the SEP spectra. On the other hand, rotational congestion of the LIF excitation spectra, which imposes restrictions on the pump transitions, can be avoided by working in the cold jet environment.

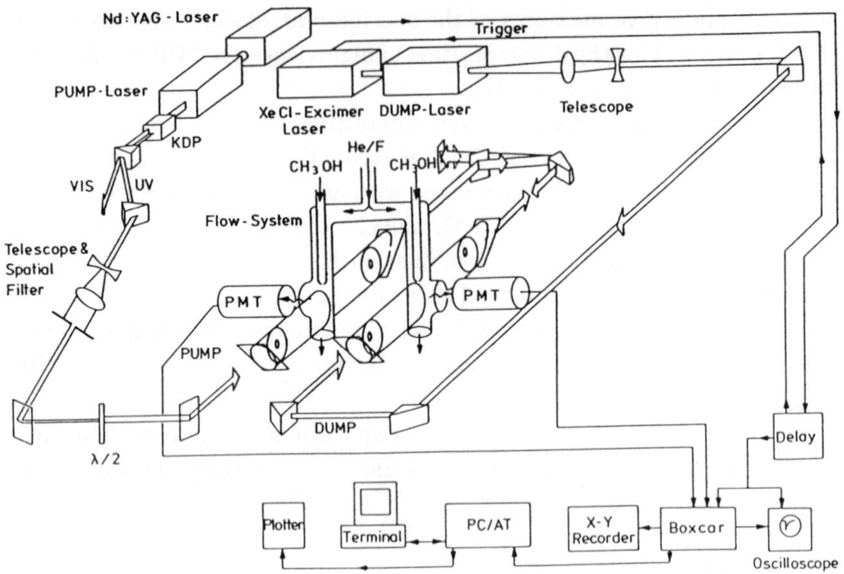

Fig. 3. Discharge flow apparatus.

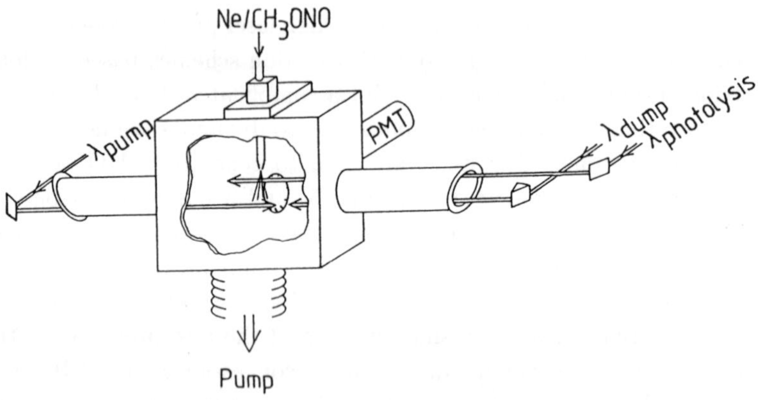

Fig. 4. Pulsed free jet apparatus.

2.1. *Discharge Flow Apparatus*

A suitable discharge flow setup for SEP studies of free radicals at ambient temperatures consists of a twin flow tube made of quartz glass for producing radicals in chemical reactions of reactive atomic species with corresponding stable precursor molecules. The flow reactors are equipped with laser input windows, fluorescence collection optics, and detection electronics for monitoring laser induced fluorescence (LIF) excited by the pump laser. The twin arrangement provides for "signal" and "reference" arms which allow compensation for pulse-to-pulse fluctuations of the pump laser intensity. A schematic diagram of the setup is shown in Fig. 3. Details are described elsewhere.[43,47]

CH_3O radicals are generated in the flow system via the reaction of F atoms with a large excess of CH_3O

$$F + CH_3OH \rightarrow CH_3O + HF. \qquad (2)$$

The reaction between F and CH_3OH produces nearly equal yields of CH_3O and CH_2OH (Ref. 48). However, CH_2OH does not have any known spectra at the wavelengths of interest which could interfere. The F atoms are produced in a 2.45 GHz microwave discharge of traces of F_2 in He inside an alumina tube. CH_3OH, which is taken from a temperature controlled reservoir, is admixed through injectors a short distance above the two detection cells. Helium at a pressure of 0.2 Torr $\leq p \leq$ 0.4 Torr serves as an inert carrier gas. The low pressure ensures absence of collisional scrambling of the rotational population of the excited molecules in the intermediate $\tilde{A}^2A_1$ state before interaction with the dump pulse. On the other hand, the carrier gas guarantees a thermal equilibrium population for the initial CH_3O vibrational and rotational levels. Flow rates are regulated using calibrated mass flow controllers and needle valves. High flow velocities ($v \approx 50\,\text{m/s}$) and short residence times (1–$2\,\text{ms}$) are used to reduce depletion of the radicals by homogeneous recombination or disproportionation reactions or heterogeneous loss reactions at the reactor walls prior to the spectroscopic excitation. Typical reactant concentrations for producing CH_3O are $[F]_0 \approx 5 \times 10^{12}$ molecules/cm^3 and $[CH_3OH] \approx 3 \times 10^{13}$ molecules/cm^3. Depending on the conditions, the concentrations of CH_3O in the detection volumes are in the range $[CH_3O] \approx 1$–2×10^{12} molecules/cm^3. In practice, this concentration is limited mainly by the fast mutual disproportionation reaction of the radicals.

The pump laser beam is obtained from a Nd:YAG pumped frequency doubled pulsed dye laser (Spectra Physics DCR-3 & Lambda Physik FL

3002E). The required wavelengths for excitation of CH_3O are between 280–320 nm. The transverse mode structure of the UV output beam is cleaned in a spatial filter consisting of a Galileian telescope and several apertures. The homogeneous central part of the beam is recollimated through an aperture to obtain a well defined near-Gaussian beam of $\approx$ 2–3 mm diameter which is steered through the two fluorescence cells (signal and reference) of the flow reactor using dielectric mirrors or 90° quartz prisms.

The dump dye laser (Lambda Physik FL 3002E) is optically pumped by a XeCl excimer laser (Lambda Physik EMG 201). The excimer laser is fired with a delay of $\approx$ 30 ns with respect to the Nd:YAG laser. The dump beam is steered colinear but counterpropagating to the pump into the signal fluorescence cell. A Pellin–Broca prism in combination with a dielectric mirror serves as a retroreflector to block the dump from entering the reference arm. Several dyes are used to cover the wavelength range between $350\,\text{nm} \leq \lambda \leq 470\,\text{nm}$. Both lasers can be equipped with intracavity etalons for narrowing the excitation bandwidths from 0.15–$0.4\,\text{cm}^{-1}$ in the normal grating controlled mode to $\Delta\nu \approx 0.02$–$0.04\,\text{cm}^{-1}$ in the etalon mode. Absolute pump and dump wavenumbers are obtained by recording I_2 absorption spectra[49] or optogalvanic spectra from Ne and Ar hollow cathode discharge lamps.[50]

Fluorescence excited by the pump laser in the signal and reference cells is monitored through $f/1.5$ quartz lenses and, optionally, suitable filters to reduce scattered light using two matched photomultipliers (Hamamatsu R928). The preamplified output signals are fed to a gated integrator (Stanford Research) interfaced to a PC/AT microcomputer for data storage and analysis. The LIF signals from the signal and reference sides are subtracted in an analog preprocessor. Thus, in the absence of a dump transition, one has a balanced null signal. SEP transitions induced by the dump laser result in a dip in the fluorescence intensity and a corresponding deflection of the balance which is recorded.

Figure 5 shows a portion of the CH_3O ($\tilde{A} \leftarrow \tilde{X}$) LIF excitation spectrum[38] at room temperature which shows well resolved individual rotational lines that can be pumped for the SEP experiment.

2.2. *Pulsed Supersonic Free Jet Apparatus*

A schematic diagram of the pulsed molecular beam apparatus is shown in Fig. 4. The pulsed source is preferable for work with free radicals to a continuous one because of the higher number densities of the molecules which can be reached. For studying CH_3O a free jet expansion in Ne inert gas is used. The Ne expansion results in a far more efficient rotational

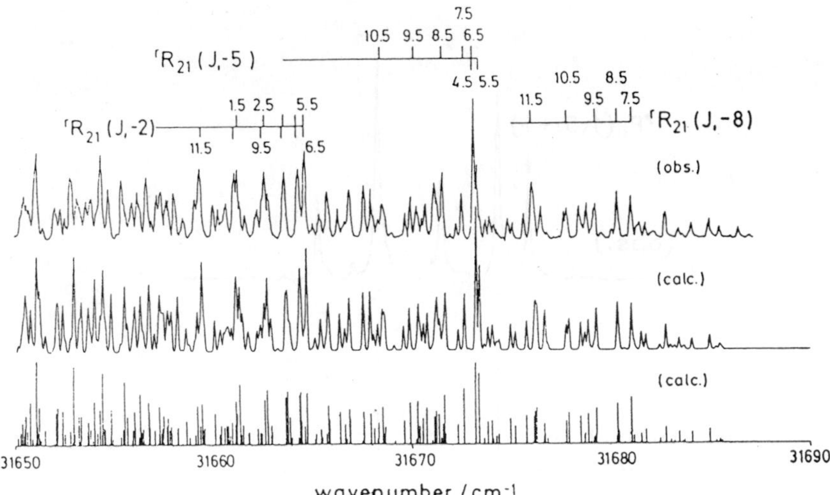

Fig. 5. LIF excitation spectrum of CH$_3$O recorded using the discharge flow apparatus at room temperature. The diagram shows a rotationally resolved portion of the $\tilde{A} \leftarrow \tilde{X}$, 0_0^0 band. Top: observed spectrum. Middle: Calculated $\tilde{A} \leftarrow \tilde{X}$ absorption spectrum assuming a Gaussian lineshape function with $\Delta\nu_D \approx 0.07\,\text{cm}^{-1}$. Bottom: Calculated stick spectrum which shows the line intensities. Nuclear spin statistics dictates weights of $g_I = 4$ each for the A_1, A_2 ($K = 3n+1$) and $E(K \neq 3n+1)$ rovibronic symmetry species. Since A_1/A_2 levels in the ground vibrational state occur as energetically degenerate pairs, these levels stand out for their higher intensities. See Ref. 38 for details.

cooling of the molecules than a comparable expansion in He without formation of clusters between CH$_3$O and the noble gas atoms as observed for Ar. The molecular beam is controlled by a solenoid actuated valve (General Valve) with a 0.8 nm diameter pinhole operating at a 20 Hz repetition rate. The radicals are generated by photodissociation of CH$_3$ONO in the $S_0 \rightarrow S_2$ system at $\lambda = 248\,\text{nm}$ (Ref. 51). The photolysis laser is focused into the free expansion at a line approximately 1 mm below the orifice of the pinhole using an $f = 1\,\text{m}$ lens. The same laser systems and electronics are employed as described above. The pump laser crosses the "cold" region of the supersonic jet up to 30 mm downstream from the valve orifice, depending on the required rotational cooling. Fluorescence is collected at right angles to both the laser and molecular beams. In view of the long radiative lifetime of the $\tilde{A}$ state of CH$_3$O (see below), the dump laser can be delayed relative to the pump by 500 ns. In order to improve the signal-to-noise limit, the ratios of the fluorescence intensities are taken for two gates of 400 ns width before and after the dump pulse, respectively. Collisional scrambling of the

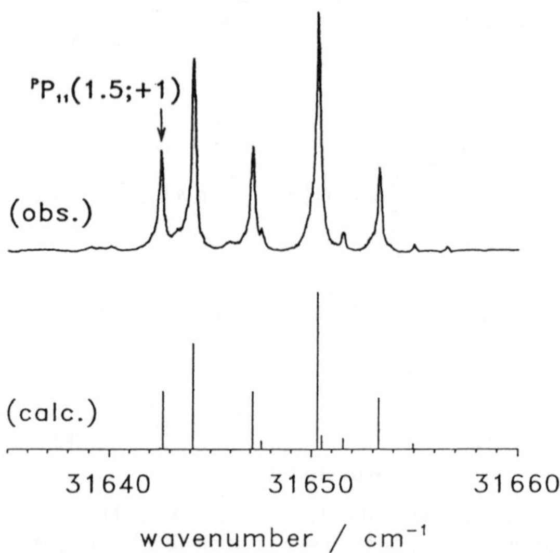

Fig. 6. $\tilde{A} \leftarrow \tilde{X}$, 0_0^0 LIF excitation spectrum of CH_3O recorded using the pulsed jet apparatus ($T_{\mathrm{rot}} \approx 1.5\,\mathrm{K}$). Top: observed spectrum. Bottom: Calculated stick spectrum. The $^PP_{11}$ ($J'' = 1.5$, $K'' = +1$) line which pumps $\tilde{A}$, 0^0, $N' = 0$, $J' = 0.5$, $K' = 0$ is pointed out by the arrow.

$\tilde{A}$ state rotational population in the jet expansion is negligible for these time intervals under the conditions used.

The conditions for generating CH_3O are optimized by recording LIF excitation spectra of the CH_3O ($\tilde{A} \leftarrow \tilde{X}$) vibrational origin band. Best signals are obtained at a Ne backing pressure of 10 bar and with photolysis pulse energies of 40 mJ. For the quantum states of interest, the number density of the CH_3O molecules in the jet were about the orders of magnitude higher than in the flow apparatus. An LIF excitation spectrum[39] is depicted in Fig. 6. Judged from a simulation, the rotational distribution of the CH_3O in the coldest expansion can be characterized by a "temperature" of $T_{\mathrm{rot}} \approx 1.5\,\mathrm{K}$. Vibrational relaxation of the CH_3O is relatively slower under the conditions used. This permits observation of numerous "hot" LIF excitation bands which furnish valuable additional information for low-lying $\tilde{X}$ vibrational states. To distinguish weak SEP lines masked by more intense hot band LIF excitation transitions, fluorescence dip spectra were recorded by scanning the dump laser with the pump laser on and off. The resulting two spectra were then subtracted on the computer to obtain the unmasked SEP spectrum.

3. High Resolution Spectroscopy of Low-Lying Vibrational States of CH_3O ($\tilde{X}^2E$)

Considerable attention has been directed to the spectroscopy of the CH_3O molecule for quite some time.

Wendt and Hunziker[52] were the first to observe the $\tilde{A}^2A_1$–$\tilde{X}^2E$ electronic transition in absorption with resolved vibrational structure. Subsequently, Inoue *et al.* recorded low resolution LIF excitation spectra of the $\tilde{A}$–$\tilde{X}$ system.[40] They could also obtain some information on the $\tilde{X}$ vibrational levels from $\tilde{A} \rightarrow \tilde{X}$ dispersed fluorescence (DF) spectra. Several other LIF studies followed.[53] Besides providing additional information on the $\tilde{A}$ vibrational energy levels, these studies also established a value of the radiative lifetime for the $\tilde{A}$ state of CH_3O of ≈ 2 μs (Refs. 40, 53). Furthermore, Brossard *et al.* observed $\tilde{A} \rightarrow \tilde{X}$ emission spectra after excitation of CH_3O in a corona discharge.[54] However, the complexity of these spectra, with emission from many excited $\tilde{A}$ vibrational states, resulted in an erroneous assignment of the exact value of the electronic origin of the $\tilde{A}$ state.

Concerning the rotational level structure, Russell and Radford obtained high resolution far infrared laser magnetic resonance (LMR) spectra of CH_3O in the $\tilde{X}$ state.[55] Later on, Hirota and coworkers analyzed the $\tilde{X}$ rotational manifold by microwave spectroscopy.[56]

With this information, rotationally resolved LIF excitation spectra could be assigned independently in this laboratory[38] and by Miller and coworkers.[39] These studies also settled earlier conflicting views[54] on the precise value of the electronic origin of the $\tilde{A}$ state.[38]

However, despite these studies, the vibrational structure of CH_3O in the $\tilde{X}$ state has remained relatively poorly understood. Most of the available information on the $\tilde{X}$ vibrational modes stems from the $\tilde{A} \rightarrow \tilde{X}$ DF spectra by Foster *et al*,[41] but this work left many questions. Weak $10\,\mu$m LMR[57] and diode laser absorption spectra in the $15\,\mu$m region[58] are still awaiting analysis.[57] The $\tilde{X}$ vibrational level structure is complicated by the vibronic coupling induced by the Jahn–Teller (JT) effect[27] in the molecule, and, in fact, the present understanding of the JT effect in CH_3O is far from complete. In particular, the previous work on the CH_3O molecular vibrations in the $\tilde{X}$ state suffers from a lack of rotational resolution. It is precisely the rotational level structure in an excited vibrational state, however, that gives a signature of the vibronic state symmetry.

With these premises, we have performed careful investigations of rotationally resolved SEP and "hot band" $\tilde{A} \leftarrow \tilde{X}$ fluorescence excitation spectra of CH_3O ($\tilde{X}$) with the aim to unravel the vibronic coupling situation for this molecule. The vibronic symmetries which could be deduced from the rotational structures led to revised assignments compared to what was inferred previously from DF spectra,[42] even for some of the $\tilde{X}$ fundamental vibrational states. The present section contains an overview of some of the main results which have been obtained to date.

Concerning the notations which will be used in the following, unprimed labels are used to denote the final excited $\tilde{X}$ rotation–vibration levels accessed by SEP. Double and single primes refer to the initial $\tilde{X}$ and intermediate $\tilde{A}$ levels of the pump transition. The notations ν_v and $\nu^{v'}$ are used to denote vibrational states of mode ν with v and v' quanta of excitation in the $\tilde{X}$ and $\tilde{A}$ electronic states, respectively; $\nu_v^{v'}$ then stands for a corresponding vibrational band in the $\tilde{A}$–$\tilde{X}$ spectrum.

3.1. *Normal Mode Vibrations and Survey SEP Spectra*

The electronic transition between the CH_3O $\tilde{A}^2A_1$ and $\tilde{X}^2E$ states is electric dipole allowed with the perpendicular component $\mu_\perp$ of the electronic transition moment. The transition arises from excitation of a C–O σ-bonding electron to a nonbonding p-orbital at the O atom. Thus, the emission from the $\tilde{A}$ state is dominated by a pronounced progression in the C–O stretch vibration. This progression is the most prominent feature in the CH_3O DF spectrum in Fig. 2.

A survey SEP spectrum with laser limited resolution, recorded with the pump laser tuned to the lowest energy rotational level ($K' = 0$, $N' = 0$, $J' = 0.5$) in the $\tilde{A}$ 0^0 vibrational origin, is shown in Fig. 7. This spectrum gives an overview of the $\tilde{X}$ vibrational structure up to $E < 3500\,cm^{-1}$. It is plotted directly versus the final $\tilde{X}$ rovibrational energy, $E_{vr}(\tilde{X}) = \nu_{pump} - \nu_{dump} + E_r''(\tilde{X})$, where the last term is the rotational energy of the initial $\tilde{X}$ level. The spectrum shows immediately the vast improvement of the SEP technique compared to DF studies in both resolution and sensitivity. The high resolution reveals a much more detailed vibrational or, actually, vibronic structure than in the DF spectrum.

The term values of the observed vibronic states of CH_3O ($\tilde{X}$) up to $E_v \leq 3000\,cm^{-1}$ are compiled in Table 1. The data (see below) were inferred from SEP spectra obtained by using a number of intermediate $\tilde{A}$ vibrational levels.

In order to rationalize the observed spectra, one must first take a look at the normal mode vibrations of the molecule. The molecular vibrations of CH_3O are described in the C_{3v} molecular symmetry group by the three totally symmetric ($\Gamma_v = a_1$) modes $\nu_s = \nu_1$ through ν_3 and the three doubly degenerate ($\Gamma_v = e$) modes $\nu_t = \nu_4$ through ν_6 with their respective

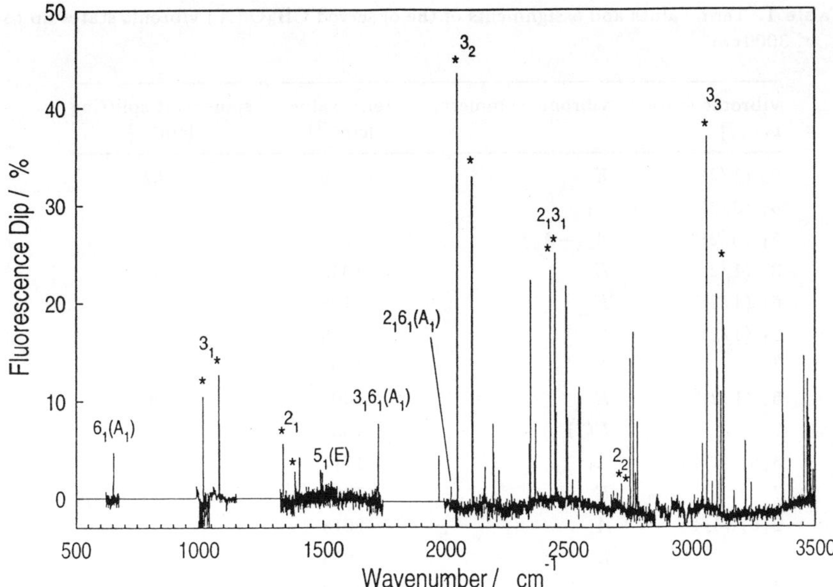

Fig. 7. Survey SEP spectrum obtained using the $^PP_{11}(J'' = 1.5,\ K'' = +1)$ pump line (upper state: 0^0, $N' = 0$, $J' = 0.5$, $K' = 0$). The asterisks denote the spin–orbit components of the respective vibrational states. See the text below for comments on the vibronic assignments.

vibrational angular momenta l_t and the resulting total vibrational angular momentum $l = \Sigma_t(l_t)$. Figure 8 shows a schematic representation of these modes.

Considering now the different types of bands that can be observed in the DF or SEP spectra of CH$_3$O from the $\tilde{A}$ origin or excited symmetric ($l = 0$) $\tilde{A}$ vibrational states, it is noted that transitions to totally symmetric ($\Gamma_v = A_1$) $\tilde{X}$ vibrational states are fully Franck–Condon allowed. This is true especially for the $\tilde{X}$ vibrational states involving ν_1 through ν_3. Corresponding assignments are fairly straightforward if SEP spectra are recorded via a number of adjacent $\tilde{A}$ rotational levels. Stimulated emission to $\tilde{X}$ vibrational states arising from the degenerate (e) models ν_4 through ν_6 with $\Delta l \neq 0$, on the other hand, is forbidden by the normal $\Delta l = 0$ vibrational selection rule and the Franck–Condon symmetry restriction, $\Gamma_v = A_1 \leftrightarrow A_1$, $A_2 \leftrightarrow A_2$, and $E \leftrightarrow E$. However, the JT effect and anharmonicities provide for interaction mechanisms that can permit the observation of $\Delta l \neq 0$ transitions.[59,60] The importance of the JT active modes is obvious in Fig. 7.

Table 1. Term values and assignments of the observed CH_3O ($\tilde{X}$) vibronic states up to $E_v \leq 3000\,cm^{-1}$

vibronic state ν_v (j)	vibronic symmetry	term value $[cm^{-1}]$	spin–orbit splitting $[cm^{-1}]$
0_0 $(1/2)$	E	0	−62
6_1 $(3/2)$	A_1	652	−
5_1 $(3/2)^a$	$A_1(+A_2)$	914	−
3_1 $(1/2)$	E	1045	−63
6_1 $(1/2)^a$	E	1198	−7
2_1 $(1/2)$	E	1359	−47
?	E	1403	?
5_1 $(1/2)^a$	E	1490	−5
?	$E(?)$	1639	?
$3_1 6_1$ $(1/2)$	A_1	1717	−
?	E	1964	?
$2_1 6_1$ $(1/2)$	A_1	2018	−
3_2 $(1/2)$	E	2073	−59
?	$A_1(?)$	2157	?
?	E	2197	−24
?	E	2338	?
?	$A_1(?)$	2363	?
$2_1 3_1$ $(1/2)$	E	2432	−24
?	$A_1(?)$	2444	?
?	E	2488	?
?	E	2550	?
?	E	2650	?
2_2 $(1/2)$	E	2728	−31
?	A_1	2748	−
?	A_1	2761	−
4_1 $(3/2)^a$	A_1	2778	−
4_1 $(1/2)^a$	E	2848	−20
3_3 $(1/2)$	E	3091	−66

[a]Vibronic assignments for the degenerate modes based on separate one dimensional Jahn–Teller calculations of the vibronic energy levels for the three modes (Ref. 66a). A full three dimensional model may require some refinements. $j = 1 - \frac{1}{2}\Lambda$.

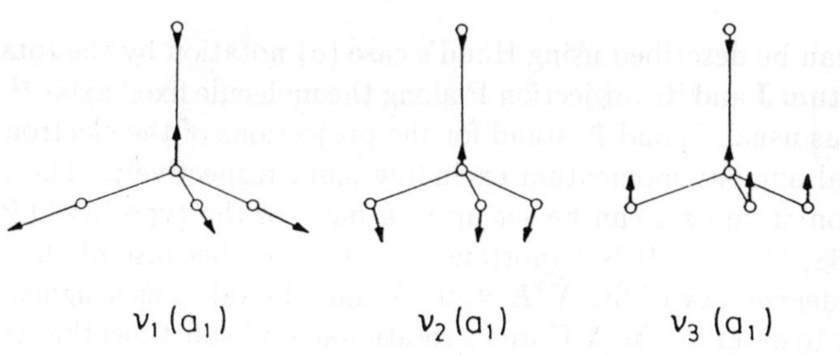

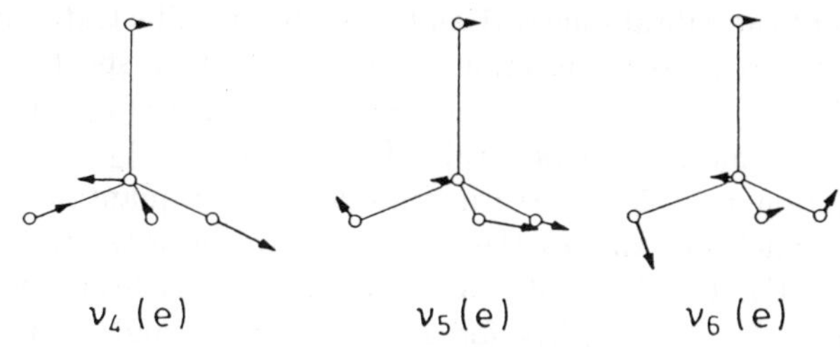

Fig. 8. Schematic representation of the normal mode vibrations of CH_3O.

3.2. *Spectra Involving Symmetric (a$_1$) Vibrational Modes: ν_1 through ν_3*

3.2.1. *Rotational line structure and selection rules*

The general form of the Hamiltonian operator for C_{3v} symmetry ^2E molecules has been discussed by Brown,[61] Hougen,[62] Watson,[63] and Liu et al.[64]

Because of its orbital degeneracy and its unpaired electron, CH_3O $(\tilde{X})$ is subject to spin–orbit interaction. This is manifest in the $\tilde{X}$ vibrational ground state by the splitting into $^2E_{3/2}$ and $^2E_{1/2}$ components. Both are separated by $\Delta E = a\xi_e d \approx -62\,\mathrm{cm}^{-1}$ (Refs. 38–39, 56). Since excitation in the symmetric modes ν_1 through ν_3 should not change the picture fundamentally, spin–orbit splittings are expected as signatures of these modes. Vibrational angular momentum is absent in these states, apart from the possible small admixtures of some character of the degenerate modes by the JT effect.

The rotational levels of the symmetric $\tilde{X}$ vibrational states ($\Gamma_v = A_1$, i.e. $\Gamma_{ev} = E$ states), which arise from excitation of modes ν_1 through ν_3

alone, can be described using Hund's case (a) notation by the total angular momentum $\mathbf{J}$ and its projection $\mathbf{P}$ along the molecule fixed axis. $\mathbf{P} = \mathbf{\Sigma} + \mathbf{K}$, where, as usual, Σ and $\mathbf{K}$ stand for the projections of the electron spin and the total angular momentum excluding spin, respectively. The rotational Hamiltonian matrix can be set up in a basis of the type $|\Lambda\rangle \, |JPM\rangle \, |S\Sigma\rangle$ (Refs. 38, 56, 62). It is important to note that, because of the electronic orbital degeneracy of the $\tilde{X}^2 E$ state, K must be taken as a signed quantum number to describe the $\tilde{X}$ Coriolis rotational level splittings that result from the couplings of the different angular momenta (see Fig. 9).

The rotational levels in the $\tilde{A}$ electronic state are described using Hund's case (b) notation by N, K, S, and $J = N + S$. Spin–rotation splittings (Ref. 65) in the $\tilde{A}$ state are too small to be observed here. The energy level structures of the vibrationless $\tilde{A}$ and $\tilde{X}$ states are illustrated in Fig. 9.

The usual rotational selection rules for the $\tilde{A}$–$\tilde{X}$ spectrum are $\Delta J = 0$, ± 1 and $\Delta K = \pm 1$. However, these rules are supplemented by the rovibronic symmetry selection rules $\Gamma_{evr} = A_1 \leftrightarrow A_2$, $E \leftrightarrow E$ and the strict selection rule $\Delta G = 0$ (modulo 3) for the quantum number $G = G_{ev} - K$, which determines the rovibronic symmetries ($G_{ev} = \Lambda + \Sigma_t l_t$; see below). For $l = 0$ vibrational states, the $\tilde{X}$ rovibronic wavefunctions transform as $\Gamma_{evr} = A_1/A_2$ for $K = +1 \pm 3n$ and as $\Gamma_{evr} = E$ for $K \neq +1 \pm 3n$. Different rotational lines are denoted in the following using the convention $^{\Delta K}\Delta J_{F'F''}(J'', K'')$, where F' and F'' ($= 1$ or 2) indicate the $\tilde{A}$ spin–rotation and $\tilde{X}$ spin–orbit components[61] ($^2E_{3/2}$ and $^2E_{1/2}$ levels are referred to as F_1 and F_2, respectively[38]). Term energies are conveniently referred to the center of the ground state spin–orbit components after removal of the spin–orbit splitting.

3.2.2. *Observed SEP spectra*

As indicated above, the ν_3 (C–O stretch) mode dominates in the $\tilde{A}$–$\tilde{X}$ emission spectrum (cf. Fig. 2) because of the difference of the C–O bond strengths in the two electronic states. Indeed, in the survey SEP spectrum in Fig. 7, the observed fluorescence dip signals to the $\tilde{X}$ ν_3 states induced by the dump laser can exceed 40 %. This means that the transitions are strongly saturated.

As an example which illustrates the rotational line structure for the symmetric modes, Fig. 10 depicts three expanded scans for the 3_2 vibrational state which were obtained with the pump laser turned to adjacent $\tilde{A}$ rotational levels with different values of N' in the $K' = 0$ stack. The spectra yield direct values of the rotational level energies. The term values are described, in the simplest case, by the conventional $B \cdot J \cdot (J + 1)$

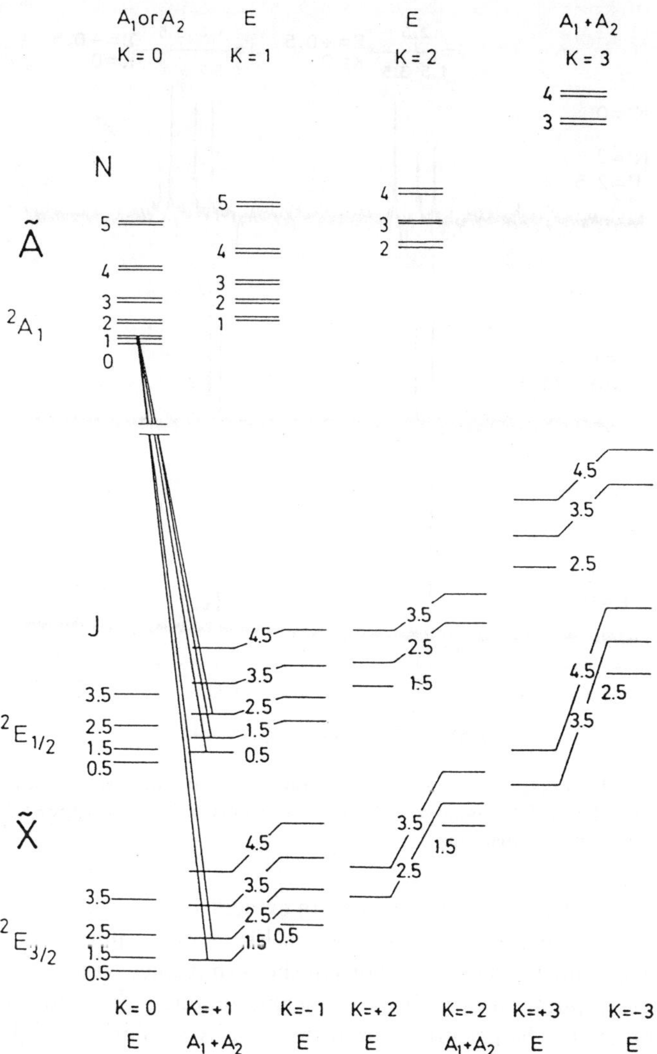

Fig. 9. Schematic rotational energy level diagram for the vibrationless levels of the $\tilde{X}$ and $\tilde{A}$ electronic states of CH_3O. The vertical lines indicate the allowed rotational transitions which could be observed in an SEP spectrum from the $\tilde{A}$, $K' = 0$, $N' = 1$, $J' = 1.5$ rotational level.

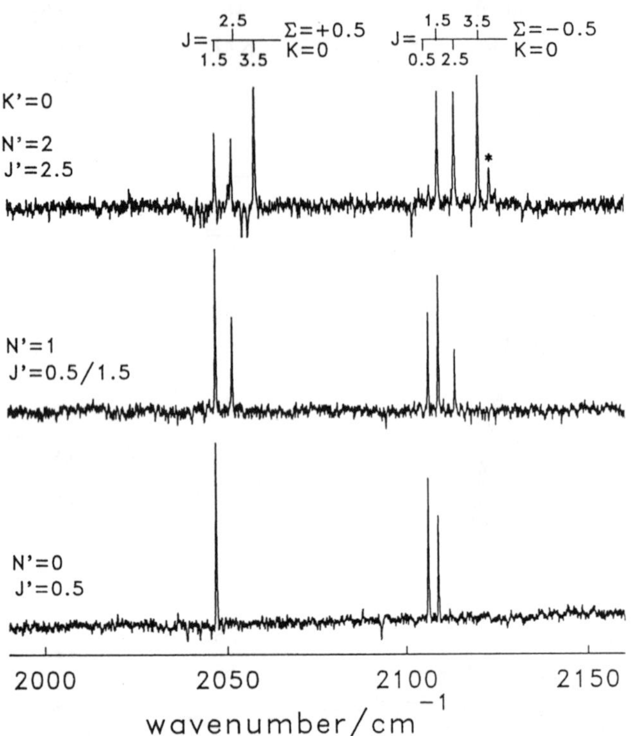

Fig. 10. SEP spectra for the $\tilde{X}$, 3_2 vibrational state of CH_3O obtained using different pump lines. The asterisk indicates a spurious line which appears because of an overlapping pump transition.

formula. The rotational assignments are evident considering the rotational energy diagram for the $\tilde{X}$ and $\tilde{A}$ states in Fig. 9. The spin–orbit splittings are obvious. Multiple determinations of the term values which are obtained using adjacent pump lines allow checking and unambiguous confirmations of the assignments. The picture is complemented by similarly straightforward spectra for higher values of K.

A corresponding analysis can be carried out from the ν_3 fundamental (3_1) up to the third overtone (vibrational state 3_4) to obtain a listing of spectroscopic constants. The observed vibrational origins are compiled in Table 2. The spin–orbit splittings for ν_3 are included in the table. It can be seen that the splittings change little upon excitation up to the third overtone. The table is completed by the vibrational term values for the

Table 2. Vibrational term values of the observed ν_3 overtone states.

vibronic state	term value $[\mathrm{cm}^{-1}]$	spin–orbit splitting $[\mathrm{cm}^{-1}]$
0_0	0	-62
3_1	1045	-63
3_2	2073	-59
3_3	3091	-66
3_4	4083	-50
3_5	5070	-36
3_6	6043	-38
3_7	7006	-36
3_8	7965	-33
3_9	8913	-36
3_{10}	9862	-30

higher ν_3 overtone states up to $v_3 = 10$, which can be inferred from the band centers of SEP spectra for the high energy regime (see Sec. 4). After removal of the spin–orbit splitting, from a Birge–Sponer plot, one obtains a value of $1050.5\,\mathrm{cm}^{-1}$ for the ν_3 fundamental frequency and a value of $-6.9\,\mathrm{cm}^{-1}$ for the diagonal anharmonicity constant.

Apart from the changes of the fundamental frequencies, quite analogous SEP spectra may be expected for the ν_2 (CH_3 umbrella) and ν_1 (symmetric C–H stretch) modes. However, the spectra are less intense than for ν_3 due to less favourable Franck–Condon factors. The ν_2 fundamental is indicated in Fig. 7 as a weaker transition. From spectra recorded via several intermediate $\tilde{A}$ rotational levels, the frequency of ν_2 could be determined to be $1359\,\mathrm{cm}^{-1}$. The spin–orbit splitting is found to be reduced slightly upon excitation of ν_2 to a value of $\approx -47\,\mathrm{cm}^{-1}$. The small difference compared to the value of the spin–orbit splitting in the vibrational ground state and in the excited ν_3 states does seem to indicate a weak perturbation of the ν_2 mode by a vibronic level of one of the degenerate vibrations (see below) at almost the same energy. An alternative assignment of ν_2 to $1412\,\mathrm{cm}^{-1}$, which could not be excluded *a priori*, would imply a spin–orbit splitting of $\approx -156\,\mathrm{cm}^{-1}$ which appears unreasonably large. This possibility was, therefore, rejected. The first overtone (2_2) and combinations with ν_3 (e.g., $2_1 3_1$) are also contained in Fig. 7 (see Table 1). Unambiguous assignments of these and further vibrational states, which are in progress, depend on the identification of excited states of the degenerate modes below.

No spectra have yet been identified with certainty involving the ν_1 (C–H stretch) mode. A reported assignment of ν_1 in the DF spectrum from the $\tilde{A}$ state[41] has to be revised in view of a reassignment[66] of the pump transition used in Ref. 41.

3.3. *Spectra Involving Degenerate (e) Vibrational Modes:* ν_4 *through* ν_6

The JT effect brings the require vibronic interaction mechanism[60] that permits observations of $\Delta l \neq 0$ transition (cf. Fig. 7) in the $\tilde{A}^2 E - \tilde{X}^2 A_1$ electronic spectrum of the CH_3O molecule, which would normally be forbidden.

3.3.1. *Jahn–Teller effect, vibronic states, and vibronic selection rules*

The JT effect is a consequence of non-zero terms in the vibrational Hamiltonian that connect the two components ($\Lambda = \pm 1$) of the degenerate electronic state.[59] Taking the electron orbital and vibrational basis functions for the $\tilde{X}^2 E$ state in the form $|\Lambda\rangle = \psi_e \cdot e^{+i\Lambda\theta}$ and $|v, l\rangle = \psi_v \cdot e^{+il\varphi}$ and adopting the notation and phase convention of Hougen,[62,67,68] the two lowest order JT interaction terms in the ^2E state can be written, for each degenerate mode t, as

$$\mathbf{H_{JT}^l} = k_1 \cdot [(e^{+2i\theta}) \cdot (r \cdot e^{+i\varphi}) + (e^{-2i\theta}) \cdot (r \cdot e^{-i\varphi})]$$

$$= k_1 \cdot [L_+^2 Q_+ + L_-^2 Q_-] \tag{3a}$$

$$\mathbf{H_{JT}^q} = k_2 \cdot [(e^{+2i\theta}) \cdot (r^2 \cdot e^{-2i\varphi}) + (e^{-2i\theta})(r^2 \cdot e^{+2i\varphi})]$$

$$= k_2 \cdot [L_+^2 Q_-^2 + L_-^2 Q_+^2], \tag{3b}$$

where $Q_\pm$ are the usual dimensionless normal coordinate shift operators for the JT active modes and $L_\pm$ are the corresponding operators for the electronic orbital angular momentum. $\mathbf{H_{JT}^l}$ and $\mathbf{H_{JT}^q}$ are referred to usually as the linear and quadratic JT interaction terms. The effect of $\mathbf{H_{JT}^l}$ leads to a moat in the effective potential around the unperturbed equilibrium positions for the respective normal coordinates Q_t, while $\mathbf{H_{JT}^q}$ introduces local minima and maxima in the moat. The JT interaction energy for CH_3O has, however, been estimated to be below the zero-point vibrational energy.[42,56,69] Therefore, there is no static distortion from C_{3v} symmetry in the $\tilde{X}^2 E$ vibrationless level.

The JT interaction terms lead to a splitting of the energy levels of the degenerate vibrational modes. The matrix elements of $\mathbf{H_{JT}^l}$ and $\mathbf{H_{JT}^q}$

satisfy the selection rules $\Delta v = \pm 1$, $\Delta \Lambda = +2\Delta l = \pm 2$, and $\Delta v = 0$, ± 2, $\Delta \Lambda = -\Delta l = \pm 2$, respectively. Thus, it can be seen that the linear JT operator couples only those vibronic levels which have the same value of $l - \frac{1}{2}\Lambda$ ($l = \Sigma_t l_t$). Following Longuet–Higgins[59], but using the phase convention as given above,[62] the vibronic states can therefore be labeled by a "JT quantum number" j, which is defined as $j = l - \frac{1}{2}\Lambda$. Considering, e.g., $\tilde{X}^2E$ vibrational states with one quantum of excitation in a degenerate mode (i.e. $l = \pm 1$), one obtains $j = \pm 1/2$ or $\pm 3/2$. Note that Λ, l, and j are taken as signed quantum numbers here. The signs may be dropped in the following unless explicitly needed.[60]

The transformation properties of the resulting $\tilde{X}$ vibronic states are important. They are determined easily by Hougen's quantum number[67] $G_{ev} = \Lambda + l$, which is defined modulo 3, with $l = \Sigma_t l_t$. States with $G_{ev} = 3n$ (n integer) have species $\Gamma_{ev} = A_1/A_2$; those with $G_{ev} \neq 3n$ have species $\Gamma_{ev} = E$. Thus, the vibronic components with $j = \pm 1/2$ and $\pm 3/2$, etc., have different vibronic symmetries. Namely, for the $j = \pm 1/2$ pair, one has $\Gamma_{ev} = E$, i.e., these levels remain degenerate. On the other hand, the wave functions of the $j = \pm 3/2$ pair of levels can be written in the form of two symmetrized linear combinations, for which $\Gamma_{ev} = A_1$ or A_2. This A_1/A_2 pair may be split by the quadratic JT interaction term.

The linear JT operator (3a) can be seen to mix the characters of the vibrationless level of the $\tilde{X}^2E$ state and the $v_t = 1$ and higher levels of the degenerate vibrations. This mechanism has been invoked to rationalize the observation of the JT active modes in the spectra of, for example, the benzenoid cations.[60] However, considering the CH_3O $\tilde{A}$–$\tilde{X}$ spectrum, it can explain only the observation of the $j = 1/2$ (i.e., $\Gamma_{ev} = E$) components of the JT active vibrations of the $\tilde{X}$ state. The transitions to these levels obey a $\Delta j = \pm 1/2$ selection rule since $j' = 0$ for the $\tilde{A}$ 0^0 level. The strict $\Delta j = \pm 1/2$ rule breaks down on addition of the quadratic JT interaction term (3b) which would allow other vibronic components to be observed. Nevertheless, it is emphasized that, even then, the $j = 3/2$ ($\Gamma_{ev} = A_1/A_2$) vibronic levels cannot mix with the vibrationless ($\Gamma_{ev} = E$) level of the $\tilde{X}$ electronic state because of the different vibronic symmetries. On the other hand, any linear JT term would still lead to observation of the $j = 1/2$ levels.

It is important to realize that, with these premises, the different types of vibronic bands that could be observed in the CH_3O $\tilde{A}$–$\tilde{X}$ spectrum are governed by different transition moments, and, therefore, exhibit different rotational selection rules. In particular, $A_1 \leftrightarrow A_1$ or $A_2 \leftrightarrow A_2$ vibronic transitions involve $\mu_\parallel$, $E \leftrightarrow A_1/A_2$ bands are governed by $\mu_\perp$, while $E \leftrightarrow E$

combinations are allowed with both $\mu_\parallel$ and $\mu_\perp$, and $A_1 \leftrightarrow A_2$ remains forbidden by the vibronic selection rules. The corresponding selection rules for K are $\Delta K = 0$ versus $\Delta K = \pm 1$ for $\parallel$ and $\perp$ transitions, respectively. Therefore, the rotational structures of the vibronic bands unveil the coupling situations and, in fact, provide indispensable clues for the vibronic assignments.

3.3.2. *Observed SEP and LIF excitation spectra*

Possible vibronic assignments of the low-lying levels resulting from excitation of the three degenerate (e symmetry) vibrational modes, ν_4 through ν_6, in the observed spectra can be made at the present stage only somewhat tentatively under the condition that the vibronic level energies are affected by the JT coupling terms of the respective modes more or less independently without significant multimode interactions. This assumption can be justified for a weak JT effect.[60d] However, already for moderate values of only the linear JT coupling parameters, the characters of the different vibronic states with a given value of the JT quantum number j are scrambled rapidly, such that assignments to the different vibrations do become ambiguous. For instance, when two degenerate modes are JT active, the vibronic states of these two modes (with a given j) can be strongly mixed.[60c] Thus, some of the vibronic assignments given in the following have to be considered carefully. Ultimately, one would want to carry out a full three dimensional JT analysis, considering the simultaneous effect of at least the linear (and, as far as possible, even quadratic) JT interaction terms of all three degenerate vibrations. Nevertheless, even at the present stage, an analysis of the rovibronic structures of the different observed states provides considerable insight into the vibronic coupling situation for the molecule.

Of the three degenerate vibrational modes of CH_3O, the ν_6 (CH_3 "rocking") mode had been identified as the lowest frequency vibration in the DF spectrum from the $\tilde{A}$ state.[41] In view of the conventional understanding of the JT effect,[60] the observed transition had been attributed to the 6_1 ($j = 1/2, \Gamma_{ev} = E$) vibronic component.[41] A corresponding signal can be seen in the survey SEP spectrum in Fig. 7 at $652\,\mathrm{cm}^{-1}$ as the first member of a progression in ν_3 built upon one quantum in ν_6. In addition, under conditions of less efficient vibrational cooling of the CH_3O in the Ne free jet expansion, a hot band LIF excitation spectrum can be observed which has to be attributed to the same $\tilde{X}$ vibration. The vibronic character of this band becomes apparent from SEP scans with a series of adjacent $\tilde{A}$ rotational levels[42] (Fig. 11a). The observed structure is seen to be that of

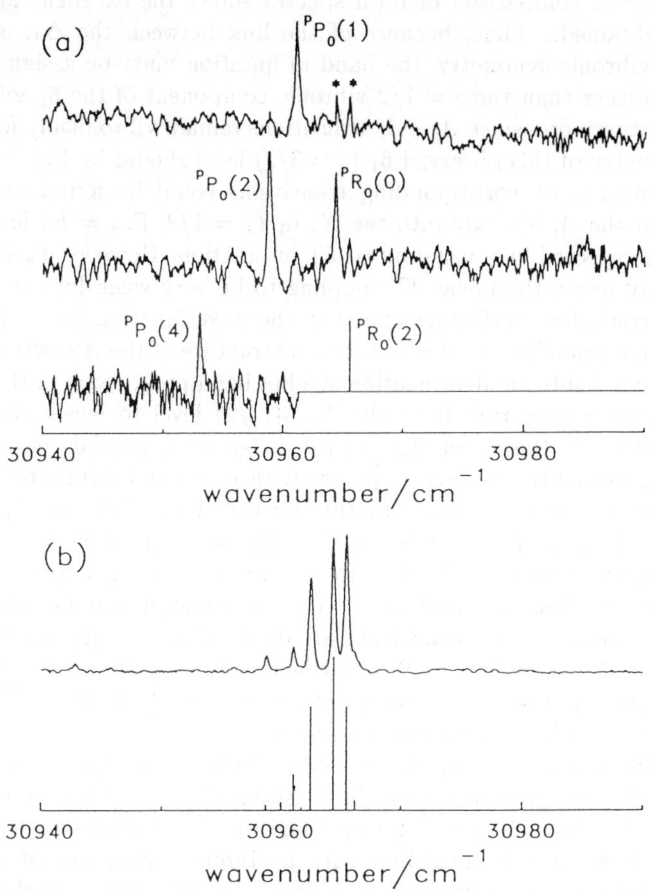

Fig. 11. (a) SEP spectra for the $\tilde{X}$, $6_1(\Gamma_{ev} = A_1$, $j = 3/2)$ vibronic state of CH_3O obtained using different pump transitions. The weak $^P R_0(2)$ line could not be observed because of interference by hot band excitation transition. (b) Observed hot band LIF excitation spectrum of the $\tilde{A}$, $0_0 \leftarrow \tilde{X}$, 6_1 ($\Gamma_{ev} = A_1$, $j = 3/2$) band together with a calculated stick spectrum for this band. See Ref. 42 for further details.

a typical parallel band ($\Delta K = 0$) of a symmetric top, where $\Delta N = 0$ is forbidden for $K = 0$. The measured rotational term values are represented by a simple formula, $F(N, K) = B \cdot N \cdot (N + 1) + (A - B) \cdot K^2$, with, for this vibronic state, $A = 5.27\,\mathrm{cm}^{-1}$ and $B = 0.937\,\mathrm{cm}^{-1}$. The observed LIF excitation spectrum is shown in Fig. 11b. The figure also depicts a simulated spectrum with line intensities calculated from the Hönl–London formulas.[70] A comparison of both spectra shows the excellent agreement that is obtained. Thus, because of the link between the ΔK structure and the vibronic symmetry, the band in question must be assigned to the $j = 3/2$ rather than the $j = 1/2$ vibronic component of the 6_1 vibrational level. Furthermore, since $A_1 - A_2$ transitions remain vibronically forbidden, the symmetry of this observed 6_1 ($j = 3/2$) level should be $\Gamma_{ev} = A_1$.

In contrast, no corresponding transitions could be found at least to date from the $\tilde{A}$, 0^0 state into the $\tilde{X}$, 6_1 ($j = 1/2$, $\Gamma_{ev} = E$) level which would be expected by a usual linear JT interaction. However, this does not necessarily imply the linear JT coupling to be very weak for the ν_6 mode as there could be interference effects in the wave function (see below).

A recent reanalysis of the vibrational structure of the $\tilde{A}$ electronic state of CH_3O has led to an identification of a hot band progression in the $\tilde{A} \leftarrow \tilde{X}$ LIF excitation spectrum from the $\tilde{X}$, $6_1(3/2)$ level to levels of the type $3^n 6^1$ upstairs.[66] Based on this secure assignment, careful searches have produced weak LIF excitation spectra with rotational structures that are consistent with the expected structure for transitions from an $\tilde{X}$, $\Gamma_{ev} = E$ symmetry level at $E_{ev} \approx 1198\,\mathrm{cm}^{-1}$ to the same $\tilde{A}$, $3^n 6^1$ progression.[66] To the extent that the JT effect for ν_6 can be separated from the other degenerate modes, the $1198\,\mathrm{cm}^{-1}$ ($\Gamma_{ev} = E$) level can be assigned as $6_1(1/2)$. The respective transitions are diagonal in v_6; they are thus fully allowed by the normal $\Delta l = 0$ vibrational selection rule, as $l = 1$ in both cases. The rotational assignment of the supposed $\tilde{A}$, $6^1 - \tilde{X}$, $6_1(1/2)$ transition could be substantiated by SEP.

The frequency of the ν_5 (C–H "scissors") mode is expected to be very close to that of ν_2. Apart from the 2_1 state, Fig. 7 shows only two other levels in the 1300–$1500\,\mathrm{cm}^{-1}$ region. The rotational level structure of the state at $1490\,\mathrm{cm}^{-1}$ is consistent with a vibronic symmetry of $\Gamma_{ev} = E$ and, under the aforementioned conditions, an assignment to the $5_1(1/2)$ vibronic component of the ν_5 vibration. The assignment is supported again by SEP spectra from neighboring $\tilde{A}$ rotational levels. The $5_1(1/2)$ component seems to exhibit a weak spin–orbit interaction, but the splitting of the two substates is reduced to $\approx -5\,\mathrm{cm}^{-1}$. Therefore, the rotational pattern changes from Hund's case (a) for low J to case (b) for high J. The transitions in Fig. 7 arise from $K' = 0$, $N' = 0$, $J' = 0.5$ in the $\tilde{A}$, 0^0 state to $K = -1$, $J = 0.5$ and 1.5 in the $^2E_{3/2}$, and $K = -1$, $J = 1.5$ in the $^2E_{1/2}$ spin–orbit substates of the $\tilde{X}$, $5_1(1/2)$ state. (The $K = -1$, $J = 0.5$ level does not exist in the $^2E_{1/2}$ spin– orbit component.) Thus, in contrast to the results for ν_6 one observes the component of ν_5 which is allowed by the linear JT operator.

Pertaining to the $5_1(3/2, \Gamma_{ev} = A_1$ and $A_2)$ vibronic components, a hot band has been observed recently in the LIF excitation spectrum which can be assigned tentatively to the $\tilde{A}\ 5^1 \leftarrow \tilde{X}\ 5_1(3/2)$ transition.[66] According to this assignment, the frequency of the $5_1(3/2)$ level is at $914\,\mathrm{cm}^{-1}$. The observed spectrum is consistent with a practically negligible splitting of the vibronic A_1/A_2 pair. In addition, initial results have been obtained recently also for the vibronic components of the ν_4 vibration.[66]

The simultaneous excitation of one quantum each in a degenerate and a symmetric mode is unlikely to lead to a fundamental change of the vibronic coupling. Accordingly, a number of combination states of ν_5 and ν_6 with ν_2 and ν_3 have been identified. As expected, progressions in ν_3 stand out because of their higher intensities. The vibronic structure of states with two quanta of excitation in a degenerate mode becomes significantly more complicated. Considering, e.g., the 6_2 level, one can have $l = 0$ and 2. Accordingly, there will be 4 vibronic sublevels, which have $j = 1/2$ $(l = 0, \Gamma_{ev} = E)$, $j = 3/2$ $(l = 2, \Gamma_{ev} = A_1)$, $j = 3/2$ $(l = 2, \Gamma_{ev} = A_2)$, and $j = 5/2$ $(l = 2, \Gamma_{ev} = E)$. If two different degenerate modes are excited simultaneously with one quantum each, as, e.g., in the $5_1 6_1$ combination state, one will have an additional $j = 1/2$ $(l = 0, \Gamma_{ev} = E)$ vibronic component.

The different vibronic assignments as they stand at this time[66] are given in Table 1.

3.4. *Conclusions on the Jahn–Teller Mechanism*

These result shed new light on the JT effect in CH_3O which may be of general significance for the interpretation of spectra of other molecules that are subject to a JT interaction.

Considering the ν_5 vibration, the observed picture seems to be consistent with activity of the linear JT interaction term for this mode. However, the observations for ν_6 show that the vibronic coupling in CH_3O cannot be described by a simple linear JT effect alone. The vibronic assignments of the JT quantum number j for a number of the observed transitions are different from those proposed previously for the same bands in a low resolution DF spectrum.[41] These new assignments outlined above are supported, however, by the corresponding rotational analysis.

The absence of the $\tilde{A}\ 0^0 - \tilde{X}\ 6_1(1/2)$ transition could be due to an interference effect in the wave function of the lower level which can occur for certain values of the JT coupling parameters. Very incommensurate magnitudes of the "vibronic transition moments" $\mu_{\parallel}$ and $\mu_{\perp}$ are another possibility, but this can only be investigated with knowledge of the exact wave functions. It is emphasized again that the present assignment for the $6_1(1/2)$ and the $5_1(1/2)$ levels are based on the assumption that the JT splittings for both modes can be described independently. However, the results outlined above would imply quite sizeable vibronic JT splittings, $\Delta E_{ev} = 546$ and $576 \, \mathrm{cm}^{-1}$ for ν_6 and ν_5, respectively. In a three dimensional JT picture, which allows for JT activity of all three degenerate modes, one could find significant interference effects which may require refined assignments.

The striking and, indeed, puzzling point is the unexpected occurrence of the $j = 3/2$ ($\Gamma_{ev} = A_1$) vibronic component of the fundamental of the ν_6 mode in the SEP spectrum from the $\tilde{A}$, 0^0 vibrational origin compared to the apparent absence, or at least much lower intensity, of transitions between the $\tilde{A}$, 0^0 level and the $j = 1/2$ ($\Gamma_{ev} = E$) vibronic component of the ν_6 mode. The "JT allowed" vibronic transition involving $6_1(1/2)$ must be at least an order of magnitude weaker than the "JT forbidden" ones to $6_1(3/2)$. The question arises as to why this could be so.

It is noted first that, in the framework of the Herzberg–Teller picture,[70] the notion of a "forbidden" transition refers to the transition moment being zero for the molecule in its undistorted equilibrium structure. In the case of CH_3O, this would mean the true C_{3v} molecule. However, even for such a "forbidden" transition, the transition moment may well have one or more non-zero components for certain molecular configurations away from equilibrium. This situation may arise because of the vibrational motion of the atoms. It may be encountered especially in the case of a JT active molecule because of the distortion of the effective potential along the coordinates of the degenerate vibrations.

A normal non-zero quadratic JT term alone does not suffice to explain the observation of the $\tilde{A}\,0^0 - \tilde{X}\,6_1(3/2)$ transition, as it could mix $v = 1$ only with an equally forbidden state with $v = 3$, but not with the vibrationless state. Alternatively, a plausible explanation which has been suggested[42] assumes a pseudo Jahn–Teller interaction operator of the form

$$\mathrm{H_{pJT}} = k_{pJT} \cdot (L_+ Q_- + L_- Q_+). \tag{4}$$

This operator has non-zero matrix elements between states with $\Delta v = \pm 1$ and $\Delta l = -\Delta \Lambda = \pm 1$ and mixes components of the ground electronic state ($\Lambda = +1$ or -1) with excited electronic states with $\Lambda = 0$. In the case of interest here, the vibronic character of excited $^2A_1$ states, the vibrationless level of the $\tilde{A}^2 A_1$ state for instance, would be admixed to the $\tilde{X}$, $6_1(3/2, A_1)$ symmetry level such that the observed transition would become allowed. In addition, there may also be other mechanisms such as "nonadiabatic" Fermi resonances[71] or Coriolis interaction within the $\tilde{X}$ electronic state between higher vibrational levels of ν_6 and other modes, which could then be mediated along the ν_6 vibrational ladder by the conventional JT coupling mechanism. The latter mechanism appears plausible in view of the extensive vibration–rotation couplings that are observed in SEP spectra for the higher vibrational energy regime (see Sec. 4). An interaction of the $\tilde{A}^2A_1$ excited electronic state with a higher electronic state of E symmetry would provide another explanation. In any case, further quantitative data on higher vibrational levels are highly desirable to unravel the mechanism.

Qualitatively, on the other hand, the strong JT activity of the ν_6 mode is well understandable by considering the electronic and vibrational structures of the CH_3O $(\tilde{X}^2E)$ molecule. The JT effect arises because of the singly occupied p orbital of the O atom. A normal mode analysis (cf. Fig. 8) shows that, for the ν_6 vibration, the motion of one of the H atoms is directed directly towards the O atom. Thus, the electronic orbital degeneracy of the $\tilde{X}^2E$ electronic wave function is lifted. Indeed, the ν_6 mode can be seen to resemble the reaction coordinate for the isomerization reaction from the CH_3O to the CH_2OH configuration, which would be possible at very high vibrational excitation. For this process, one can obviously expect significant energy differences depending on the orientation of the p orbital of the O atom. The p lobe can be either in the plane of the motion of the nuclei or perpendicular to it. These two configurations correspond to an A' or an A'' state. Of these, only the A' state correlates with the electronic ground state of CH_2OH. Thus, the A' state should be at lower energy. It is conceivable that an analysis of the mechanism which gives rise to the vibronic splittings of the ν_6 states opens a new route for investigating the isomerization of CH_3O to CH_2OH.

Undoubtedly, with future measurements, the power of the SEP technique to yield "self-assigned" rotational spectra will result in much additional information. An analysis would be of interest in particular for the region of the ν_4 (C–H stretch) mode at $\approx 3000\,cm^{-1}$. With further experimental confirmation of the preliminary vibronic assignments for these and the other observed levels (see Table 1) and additional observations for the isotopomer CD_3O, a full three dimensional JT analysis should become possible in order to determine the unperturbed vibrational frequencies of the degenerate modes and the values of the JT interaction parameters. In principle, then, the measured vibrational term values provide a map of the potential surface of the molecule which could be used to "calibrate" a theoretical potential surface from *ab initio* quantum chemical calculations.

4. Intramolecular Vibrational Energy Redistribution in Highly Excited CH_3O $(\tilde{X}^2 E)$

Renewed interest is being directed to the basic postulates of the statistical rate theories which have been so successful for modeling unimolecular reactions.[1] Much discussion is fueled by the old debate about state-specific versus statistical chemistry. In essence, the fundamental assumption of the statistical models is that, in highly excited molecules, intramolecular vibrational energy redistribution (IVR) among the different "chemical" internal degrees of freedom occurs on a much shorter time scale compared to a possible real chemical reaction, so that any local excitation, e.g., in a particular chemical bond that is formed in a recombination reaction, will be randomized very rapidly. However, the mechanisms that govern the interesting transition between state specific chemistry and statistical molecular properties at low versus high energies, especially regarding the extent of rotation–vibration interactions, are still a matter of controversy. The rates of IVR processes in different molecules, in particular the possible occurrence of stepwise IVR with different time scales, the accessible phase space volume, and the often discussed existence of periodic orbits[72] at high energies, have direct relevance for the unimolecular rate models.

The general importance of IVR processes has been demonstrated elegantly by fluorescence excitation[8,9] and chemical timing[10] experiments, multiphoton dissociation,[73] and, of course, ultra-short laser techniques.[74]

Over the last few years, considerable progress has been made regarding the connection of dynamical processes in highly excited molecules, such as IVR, with the high resolution eigenstate resolved molecular spectra of the species.[12,13,75–77]. However, the vast majority of the previous studies in this field were concerned with X–H overtone excitations[12,13] (X = C, N, O) which, in fact, contribute the least to the molecular phase space because of their high frequencies. As many observed overtone spectra show, apart from the characteristic "stretch–bend" resonances, typical X–H bonds couple only weakly with the rest of the molecular frame. In comparison, SEP excitation experiments, which were carried out as described below for CH_3O, stand out because of the unique opportunity to study directly the much more important skeletal vibrational modes.

In this section, first, some theoretical aspects are reviewed within the framework of a simple physical picture which relates directly to the experiment. The reader is referred to the original literature for details.[13,75] Second, selected interesting results are discussed concerning the onset of vibration–vibration and rotation–vibration level mixing and IVR in the ground electronic state of CH_3O at energies approaching the unimolecular decomposition threshold.

A full discussion of the available data, including the energy and angular momentum dependence, is beyond the scope of this article and is deferred to a separate publication.[45]

4.1. *Rotation–Vibration Level Mixing and IVR*

The conceptual framework for the interpretation of spectra of highly excited molecules is illustrated in Fig. 12.

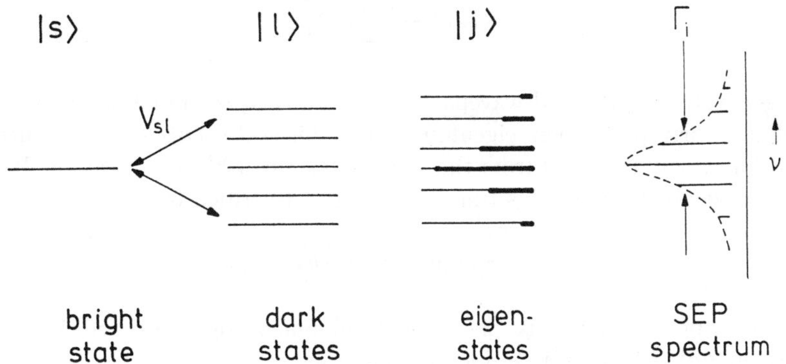

Fig. 12. Schematic level coupling diagram for a single Franck–Condon active bright zero order state with a variety of inactive dark background states.

As a starting point, it is noted that, by their nature, the high resolution molecular spectra which are observed refer to the stationary eigenstates of the molecule as the solutions of the time-independent Schrödinger equation (dissociation of the molecules is ignored for the moment). The vibrational eigenstates ψ_j of the molecule are usually expanded using a zero order basis of harmonic normal mode vibration states ϕ_i. Suppose that one of these zero order states, ϕ_s, is a Franck–Condon active state which can be accessed directly, for example, by SEP from the $\tilde{A}$ state, with a high Franck–Condon allowed transition moment. This state is frequently called a "bright" state. The other zero order states ϕ_l, called "dark" states, shall be Franck–Condon inactive. However, at the high excitation energies of interest, these dark states may couple to the bright through higher order anharmonic, Coriolis, or centrifugal interactions. Including these interactions, the wave function of an exact eigenstate ψ_j can be written as

$$\psi_j = a_{js}\phi_s + \sum_l b_{jl}\phi_l . \tag{5}$$

The intensity of the state j in the spectrum is then simply proportional to the square of the coefficient of the bright state, $I_j \propto |a_{js}|^2$. Thus, in essence, the bright state opens a "Franck–Condon window" which permits detailed observation of the "active" levels of the highly excited molecules.

In order to relate observations of eigenstate resolved spectra to the phenomenon of IVR, consider the effect of an ultrashort dump laser pulse which projects a wavepacket from the excited to the ground electronic state. The initial state that is created can be written as a coherent superposition of eigenstates,

$$\phi(0) = \sum_j c_j \psi_1 . \tag{6}$$

In general, the prepared wavepacket $\phi(0)$ on the ground electronic surface will not correspond to an eigenfunction of this state. Often, for instance, it will find itself displaced from the ground state equilibrium position. It will thus develop in time. This time evolution is described by[78]

$$\phi(t) = \exp(-2\pi i \mathbf{H} t / h) \cdot \phi(0) , \tag{7}$$

where $\exp(-2\pi i \mathbf{H} t / h)$ is the time development operator and $\mathbf{H}$ is the corresponding Hamiltonian for the molecule.

The quantity of interest is the "survival probability density" at time t of the initially prepared state. In the absence of other decay processes, such as radiative transitions or unimolecular decomposition, this probability, which one may write as $P_{ss}(t) = |\langle\phi(0)|\phi(t)\rangle|^2$, keeping the simple physical picture, can be expressed in terms of the eigenstate resolved spectrum by

$$P_{ss}(t) = \sum_j c_j^4 + 2\sum_j \sum_{i>j} c_i^2 \cdot c_j^2 \cdot \cos[(E_i - E_j)t/\hbar] . \tag{8}$$

The time dependent term on the right in Eq. (8) is the cosine Fourier transform of the autocorrelation function of the spectrum (this assumes the coefficients c_i, c_j to be real). The related quantity $C_{ss}(t) = \langle\phi(0)|\phi(t)\rangle$ is often termed the autocorrelation function of the prepared wavepacket.

In the limiting case of an infinitely short weak pulse, the initial state $\phi(0)$ will correspond to the "pure" bright state. Then, the coefficients c_i^2 are simply proportional the measured intensities, and Eq. (8) describes the "dephasing" of the initially prepared coherent superposition. The initial decay is characterized by a dephasing time τ which is a measure of the IVR lifetime. This dephasing time is related to the width $\Delta\tilde{\nu}$ (full width at half maximum) of the Lorentzian envelope function over the spectral intensity distribution $I(\tilde{\nu})$, which is approximated in the intermediate and, ultimately, large molecule coupling case, by $\tau = (2\pi c\Delta\tilde{\nu})^{-1}$. Fermi's golden rule, $\tau \approx (2\pi c|V_{sl}|^2 \cdot \rho_l)^{-1}$, relates the dephasing time to the average squared coupling matrix element and the density of states of the coupled levels.

The inverse of the time average of P_{ss},

$$\overline{P_{ss}} = \lim_{T \to 0} \frac{1}{T} \int_0^T P_{ss}(t)dt, \tag{9}$$

$(\overline{P_{ss}})^{-1}$, is interpreted as the effective number of coupled states and thus give a measured of the volume of the accessible phase space.

Equations (8) and (9) describe the route for extracting the desired dynamical information about the rate and extent of IVR processes from high resolution eigenstate spectra. It is very important to realize the inherent restriction, however. Namely, application of these equations is allowed only in the case that the spectrum shows purely homogeneous structure. The presumption which needs to be verified separately is that the eigenstate spectrum arises from coupling to a single bright state with, e.g., specific symmetry, distinctive zero-order vibrational character, and known total angular momentum J. This condition is not easily met by ordinary spectroscopies, as inhomogeneous contributions, e.g., congestion because of many different rotational states, may hamper an unambiguous analysis.

4.2. *"Franck–Condon Windows"*

The ability to probe the highly excited vibrational levels of CH_3O can be seen from the portion of a low resolution overview SEP scan[47] shown in Fig. 13a. The spectrum exhibits clear vibrational structure up to $E_v \geq 8000\,cm^{-1}$, which is above the asymptotic $H-H_2CO$ dissociation limit. The most intense peaks which occur at approximately $1000\,cm^{-1}$ intervals can easily be identified by comparison with Figs. 2 and 7 as members of the progression in ν_3 with, nominally, $6 \leq v_3 \leq 8$. Thus, the ν_3 (C–O stretch) mode of CH_3O is seen to be an optically bright mode; high overtones of ν_3 can be accessed by SEP spectroscopy via the $\tilde{A}$ state with high Franck–Condon allowed intensity. The vibrational term values for the zero order ν_3 overtone states which can be inferred from the intensity weighted centers of the observed bands are summarized in Table 2.

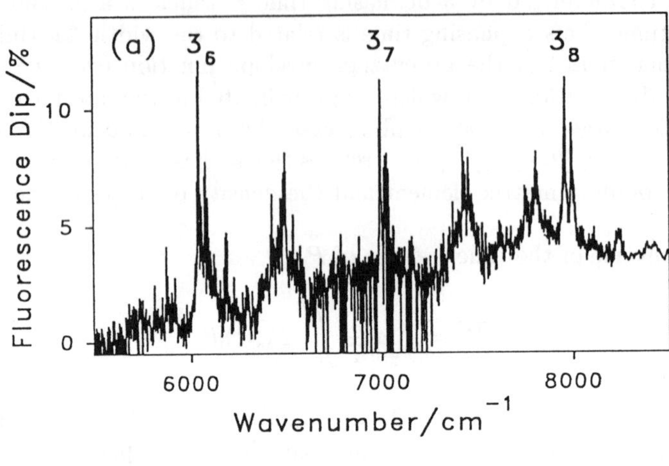

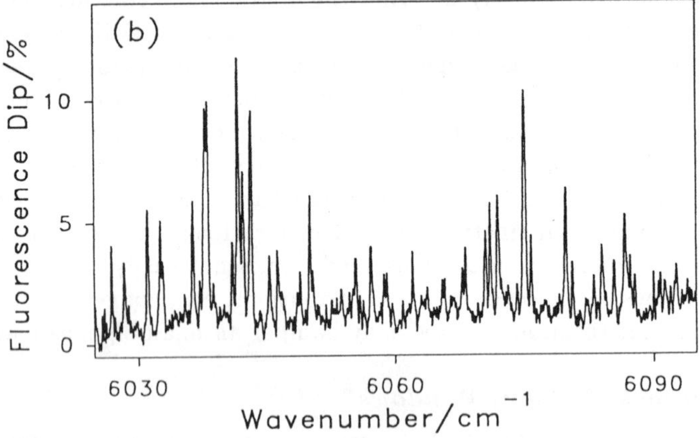

Fig. 13. (a) Low resolution SEP spectrum of CH_3O for $5000\,\mathrm{cm}^{-1} \leq E(\tilde{X}) \leq 8000\,\mathrm{cm}^{-1}$ which demonstrates the possibility for probing the vibrational structure at these chemically significant energies. Selected pump transition: 3_0^1, $^P P_{11}(J'' = 2.5, K'' = +1)$. The resolution was degraded to $\Delta\nu \approx 2\,\mathrm{cm}^{-1}$ by sampling the fluorescence dip signals over a corresponding number of laser shots. At this resolution, the different rotational lines for a given $\tilde{X}$ vibrational state coincide. However, distinctive "Franck–Condon windows" become discernible which allow the detailed structure to be probed at high energies. (b) High resolution scan for the $6000\,\mathrm{cm}^{-1}$ Franck–Condon window ($v_3 = 6$) which shows strong rotation–vibration level mixing.

A higher resolution ($\Delta\nu \approx 0.25\,\mathrm{cm}^{-1}$) scan of the $6000\,\mathrm{cm}^{-1}$ band from the 3_6 state is displayed in Fig. 13b. Considering the rotational selection rules, from the pumped level, one would expect a total of only five rotational SEP lines to the ν_3 overtone vibrational states (see Figs. 7 and 9. Figure 7 depicts the corresponding SEP spectrum for the 3_2 state, and Fig. 9 shows the relevant term diagram). However, the fine structure in Fig. 13b is much more complicated than in Figs. 7 and 9; it exhibits a very large number of lines without easily discernible pattern. This reflects the high density of vibrational states of the molecule at the high energy close to the dissociation limit.

4.3. *Rotationally "Assigned" Zero Order States and Spectra*

The aforementioned results give qualitative information on the onset of vibrational level mixing and IVR in CH_3O. However, as pointed out, a quantitative analysis requires unambiguous selection of the angular momentum (J) values of the $\tilde{X}$ rovibrational levels that are accessed.

As an illustrative example which demonstrates how J specific information can be extracted, consider the set of spectra for increasing CH_3O ($\tilde{X}$) excitation energies shown in Fig. 14. Scans are depicted over the regions of the 3_2 to 3_6 zero order vibrational states. The employed pump line, 3_0^0 $^rR_{21}$ ($J'' = 7.5$, $K'' = -8$), leads to the intermediate $\tilde{A}$ level with $N' = 9$, $J' = 8.5$, $K' = 9$ in the 0^0 vibrational origin. It is important that from this $\tilde{A}$ level, because of the rotational selection rules, there are only three allowed dump transitions to each of the target $\tilde{X}$ vibrational states of the 3_n series. Moreover, of these, because of the perpendicular transition ($\Delta K = +1$ or -1), the single line to the final level $K = +10$, $J = 9.5$ in the upper ($\Sigma = -0.5$) spin–orbit component is separated by as much as $200\,\mathrm{cm}^{-1}$ from the lines to the $K = -8$, $J = 7.5$ and 8.5 levels in the lower ($\Sigma = +0.5$) spin–orbit substate. (As outlined above, K is taken as a signed quantum number in a degenerate vibronic state of a symmetric top molecule.)

Figure 14 depicts in the bottom trace a calculated SEP spectrum from the pumped $\tilde{A}$ level which shows the expected rotational line structure and calculated line intensities for a transition to the $\tilde{X}$, 0_0 vibrational state in order to illustrate the resulting SEP spectra.

The rotationally resolved spectra for the 3_2 and 3_3 vibrational states in Figs. 14 (second and third from bottom) can be seen to show precisely the expected structure. This is true even for the relative intensities, if the saturation of the strongest SEP lines is taken into account. However, the fine structure for the 3_4 state (Fig. 14, fourth trace from bottom) starts to

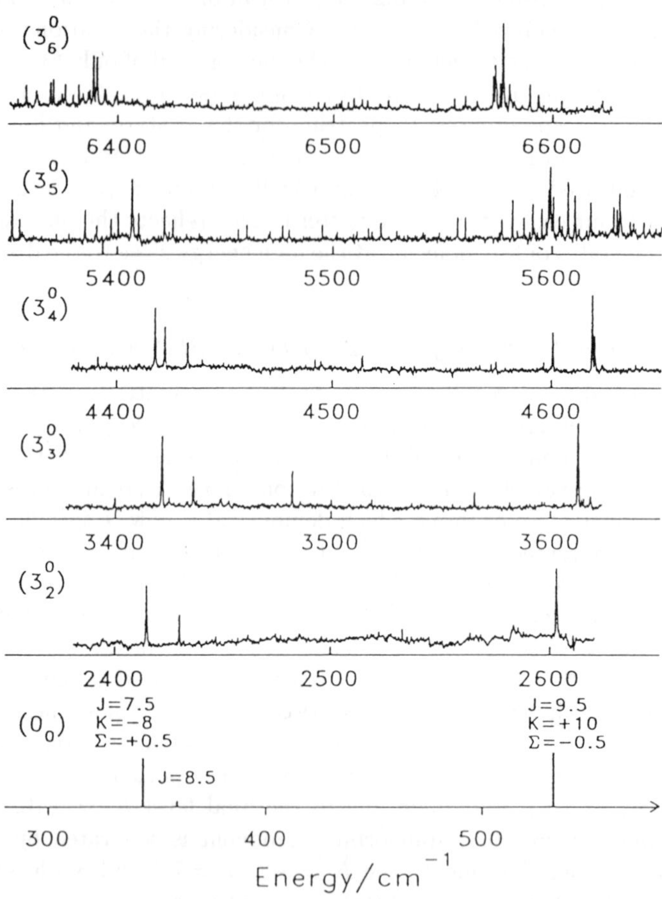

Fig. 14. SEP spectra for the regions of the $v_3 = 2$ to $v_3 = 6$ states. Pump line: 0_0^0, $^rR_{12}(J'' = 7.5, K'' - 8)$. See text for details.

exhibit unexpected satellite lines which are indicative of the beginning of level couplings. Complicated clusters or "clumps" of eigenstates are then becoming obvious for the regions of the 3_5 and 3_6 zero order vibration states (Fig. 14, top two traces). These clumps are centered just where the allowed rotational lines would occur. This picture is precisely the one that is expected for the coupling of a number of nearby dark background states with the bright $J = 7.5$ and 8.5 ($K = -8$) and $J = 9.5$ ($K = +10$) levels of the 3_5 and 3_6 vibrational states. With these premises, since the total angular momentum must be conserved and thus J remains a good quantum number, all levels within one clump feature around a single bright state may be assigned *identical* values of J. As determined by the bright states, we have either $J = 7.5$ and 8.5, respectively, or $J = 9.5$. However, the $J = 9.5$ levels are well separated from the others. Furthermore, the $J = 8.5$ features, which coincide within their widths with the $J = 7.5$ clumps, are expected to be much less intense and thus are less likely to be observed.

Similar spectra starting, however, from different $\tilde{A}$ rotational levels, were observed for higher excitation energies up to $E_v < 9500\,\mathrm{cm}^{-1}$ (Ref. 28). The results thus demonstrate the exciting possibility to probe *single rotation–vibration states* of CH_3O with *defined energy and total angular momentum* in the quasicontinuum above the H–H_2CO bond dissociation energy of the molecule.

4.4. *IVR Time Scales, Coupling Matrix Elements, and Rovibronic State Densities*

The observed clump structures are the spectroscopic signature of IVR process which would occur in the time domain. Here, CH_3O presents a striking example of a molecule which conforms to the so-called "intermediate" coupling case.[75] Following the aforementioned concept, using an ultrashort (femtosecond) laser pulse with frequency corresponding to the center of one of the clump features, one would excite a coherent superposition of the eigenstates within the spectral envelope. Having identified a specific rotation–vibration overtone state of the ν_3 mode (e.g., $v_3 = 6$, $K = 10$, $J = 9.5$, etc.) as the single active basis state which carries oscillator strength, the initial state that would be prepared would be one with corresponding initially localized motion in the C–O stretch coordinate and specific rotational motion as determined by J and K of the bright state. (It is noted in passing that the assumption of an ultrashort pulse laser excitation may seem artificial, but one might think of analogous chemical routes, for instance, formation of a C–O bond during a collision between initially separate O and CH_3 moieties or a local excitation of the C–O bond by some inelastic collision. The latter bimolecular processes do not allow the fine experimental control, however, compared to ultrashort optical excitation[74]).

As described by Eq. 8, the survival probability of a corresponding hypothetical localized initial excitation may be evaluated from the cosine transform of the autocorrelation function of the high resolution clump spectrum. These survival probabilities have been computed for a number of rotational states of different ν_3 overtone vibration states.[44,45] The results for

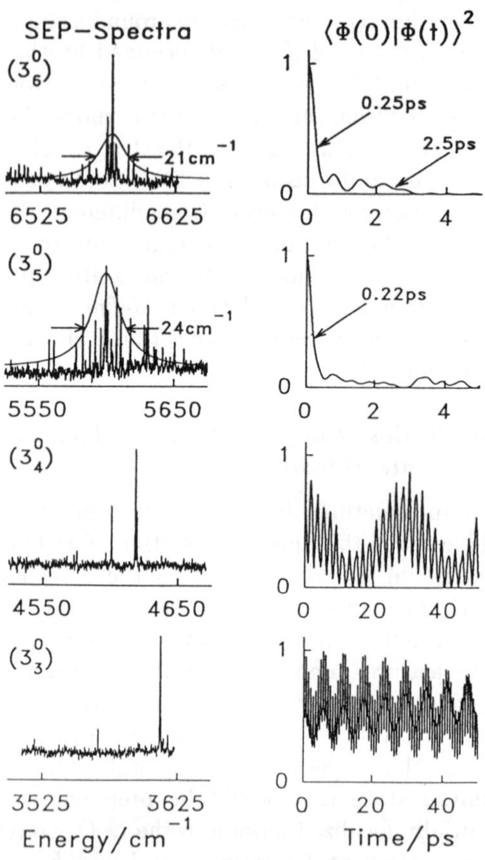

Fig. 15. SEP spectra of single "clump" features assigned to v_3, $J = 9.5$ and computed survival probability densities for the excited states according to Eq. (8).

the observed v_3 ($J = 9.5$, $K = +10$) features are depicted in Fig. 15. The diagrams show, on the left, portions of the spectra from Fig. 14 at different energies and, on the right, the computed time evolution functions. For excitation of $v_3 = 3$ and 4, the computed time evolution is clearly periodic for times longer than 40 ps. Hence, as expected for this low energy region, the excitation remains localized for at least this time interval. However, for $v_3 = 5$ and higher, the survival probability of the initial excitation falls off sharply within ≈ 0.25 ps. In addition, we note the indication of a second, somewhat longer time scale of ≈ 2.5 ps for $v_3 = 6$!

For this analysis, it is necessary because of signal-to-noise limitations to convert the observed spectra to "stick spectra". It is emphasized, however, that it can be verified that inclusion of a few spurious extra lines in the computation of the spectral autocorrelation function cannot alter the result for the IVR times very significantly. It is pointed out, also, that the computed CH_3O IVR rates can be converted to Lorentzian widths for the spectral envelope functions over the clump features. As seen from Fig. 15, the agreement that is obtained is very satisfactory.

Having determined the rate of IVR in CH_3O, the next point of interest is the question of the underlying coupling mechanism. Purely anharmonic interaction leads to vibration–vibration mixing which preserves the rotational motion, whereas Coriolis and centrifugal vibration–rotation mechanisms diminish the "goodness" of the K quantum number for the rotation of the molecule around the top axis. From high resolution scans over the J-assigned clump features, one can derive an observed density of states of the order of $\rho_{\rm obs}$ $(6500\,{\rm cm}^{-1},\ J = 9.5) \approx 4/{\rm cm}^{-1}$ to $6/{\rm cm}^{-1}$ (Ref. 28). This value may be compared with the computed symmetry-sorted density of A_1 and E symmetry vibrational states of $\rho_v(A_1 + E) \approx 0.3/{\rm cm}^{-1}$ obtained from a direct harmonic state count (allowing for vibronic coupling by the JT effect, the relevant selection rules would permit observation of both A_1 and E symmetry vibrational levels). Thus, the observed state density exceeds the computed symmetry-sorted vibrational one by more than a factor of ten. Moreover, line densities in spectra for constant vibrational energy but for different J show a pronounced J dependence. Both observations provide a strong case for the importance of vibration–rotation interactions. Indeed, the observed state densities are only slightly lower than computed full rovibronic state densities.[28]

The number of coupled states, as obtained from the inverse of the time average of P_{ss} (Eq. 7), is of the order of $\overline{P_{ss}}^{-1} \approx 25\text{--}30$ for $v_3 = 5$ and 6. This number is somewhat uncertain, however, and should be considered as a lower limit, as the spectrum contains unresolved lines and weak lines can be lost in the noise. Nevertheless, in combination with the widths of the spectral envelope functions, one can derive a density of "active" states of $\rho \approx 1/{\rm cm}^{-1}$ to $1.5/{\rm cm}^{-1}$. This value is below the calculated full rovibronic one and thus indicates some restrictions for the fast IVR within the initial $0.25\,{\rm ps}$.

Some insight into the interaction strengths can be obtained using the formalism of Lawrence and Knight[79] for extracting values of coupling matrix elements from the observed spectra. To this end, it is assumed again that the clump features arise from single bright states which are coupled to a "prediagonalized" set of dark background levels. In this case, an inversion of the spectrum can be carried out to obtain the interaction matrix elements between the bright zero order state and the different background states.[79] For the features in Fig. 15, the average root mean square coupling matrix elements are of the order of $\langle V^2 \rangle^{1/2} \approx 2\,\mathrm{cm}^{-1}$. A caveat is in order here, though, as any interaction between the background levels is excluded in the treatment. The prediagonalization of the dark states thus spoils an interpretation of the results for $\langle V^2 \rangle^{1/2}$ in terms of a physical mechanism. Nevertheless, the value of $\langle V^2 \rangle^{1/2}$ may be useful for checking internal consistency and estimating the effective density of states using the golden rule formula.

These results, especially on the IVR rates, ought to be compared to the unimolecular dissociation rates of CH_3O.

5. Unimolecular Dissociation of CH_3O $(\tilde{X}^2 E)$

For a critical comparison to predictions by statistical unimolecular rate theory, one would like direct information on the dissociation lifetimes of the highly excited molecules at and above their dissociation limit. To this end, the rotation–vibration quantum state specific lifetimes or the corresponding rate constants, $k(E, J, \Gamma; \alpha, \dots)$, for molecules with known total internal excitation E, defined angular momentum J, symmetry Γ, and other (not necessarily conserved) degrees of freedom α (e.g., v_i, K) are especially germane for developing detailed pictures of the dynamics.

The onset of the unimolecular decomposition of CH_3O is expected to be observable as homogeneous line broadenings in high resolution SEP scans for individual rovibrational states of the molecule with excitation energies of $E_v > E_0$ (H–H_2CO). With this premise, this section discusses related initial observations, which give unambiguous evidence for the decomposition reaction. Results of detailed experiments, which are currently being evaluated, will be reported in a later publication.[80]

Interesting SEP spectra obtained a high resolution with dump laser bandwidths of $\Delta\tilde{\nu}_{\mathrm{dump}} \leq 0.04\,\mathrm{cm}^{-1}$, which illustrate the evolution of the linewidths at energies above $E_0(\mathrm{H}\text{–}H_2CO)$, are shown in Fig. 16. The diagram shows three scans over $15\,\mathrm{cm}^{-1}$ regions for the $v_3 = 6$, 8, and 9 Franck–Condon windows at $E \approx 6500$, 8500, and $9500\,\mathrm{cm}^{-1}$, respectively. According to the arguments presented in Sec. 4, the value of the angular momentum quantum number of the prepared states is $J = 9.5$.

At $E = 6500\,\mathrm{cm}^{-1}$, still below the dissociation threshold, the observed spectrum exhibits exclusively sharp lines. The narrow widths reflect the convolution of the pump and dump laser bandwidths. The first evidence for

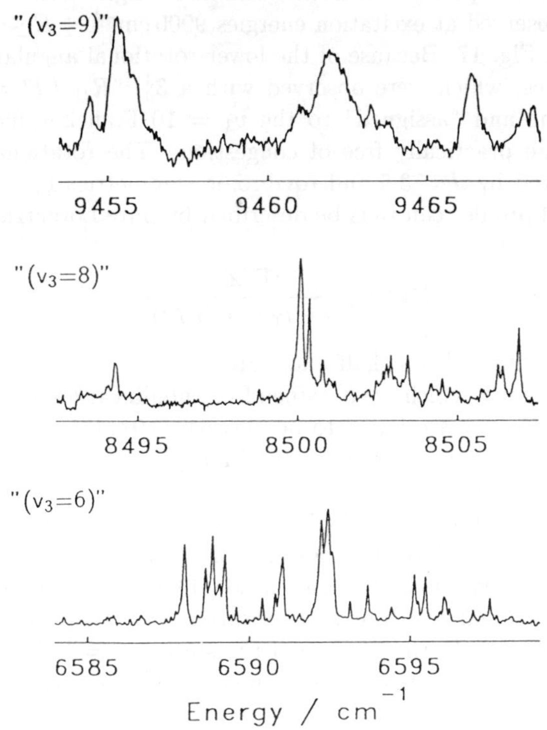

Fig. 16. Evolution of the linewidths in high resolution spectra sampling CH_3O $(\tilde{X})$ excitation energies above the H–H_2CO dissociation limit. Pump line: 0^0_0, $^rR_{21}(J'' = 7.5, K'' = -8)$.

the onset of broadening can be seen on careful inspection of the $8500\,\mathrm{cm}^{-1}$ scan. However, at higher energies, in the scan at $9500\,\mathrm{cm}^{-1}$, the linewidths can be seen to increase dramatically. In the language of scattering theory, the observed states correspond to vibrational "resonances" which are embedded in the quasicontinuum above the dissociation barrier. The total excitation energy of $\approx 9500\,\mathrm{cm}^{-1}$ exceeds the classical H–H_2CO dissociation threshold by some $1100\,\mathrm{cm}^{-1}$! Hence, although in view of the high rovibronic state density some inhomogeneous contributions cannot be fully excluded, the observed line broadening must be attributed primarily to the rapid dissociation of the excited states according to

$$CH_3O\ (\tilde{X}^2E;\ E,\ J,\ \Gamma) \rightarrow H + H_2CO\,.$$

Individual line profiles of transitions into single rovibrational levels, which were observed at excitation energies $9900\,\text{cm}^{-1} \leq E \leq 10\,000\,\text{cm}^{-1}$, are plotted in Fig. 17. Because of the lower rotational angular momentum for these states, which were observed with a 3_0^3, $^rR_{21}$ ($J'' = 1.5$, $K'' = -2$) pump line and "assigned" to the $v_3 = 10$ Franck–Condon window, these scans are practically free of congestion. The rotational states can be characterized by $J = 3.5$ and rovibronic symmetries $\Gamma_{evr} = A_1$ or A_2. The measured profiles can thus be described by pure Lorentzian line shape functions,

$$I(\tilde{\nu}) \propto \frac{(\Gamma/2)^2}{(\tilde{\nu} - \tilde{\nu}_0)^2 + (\Gamma/2)^2}\,, \tag{10}$$

where Γ is the full width at half maximum.

The results for Γ which are obtained from nonlinear least squares fits for the signals in Fig. 17 are found to be between $0.9\,\text{cm}^{-1} \leq \Gamma \leq 1.35\,\text{cm}^{-1}$. The experimental and computed Lorentzian profiles can be seen to show very good agreement.

Under the condition of negligible inhomogeneous contributions to the broadening, one can relate the linewidth of a transition into a single state directly to the corresponding dissociation lifetime τ. The dissociation rate of an individual highly excited rotation–vibration state is obtained according to $k = 2\pi c\Gamma$; the lifetime of the state is $\tau = k^{-1}$. The observed values for $k(E, J)$ are found to correspond to state-specific unimolecular dissociation rates of $1.7 \times 10^{11}\,\text{s}^{-1} \leq k$ ($E \approx 9950\,\text{cm}^{-1}$, $J = 3.5$) $\leq 2.6 \times 10^{11}\,\text{s}^{-1}$. Thus, the lifetimes of the excited states are of the order of $\tau \approx 5\,\text{ps}$, which is very short indeed!

These initial results, which are confirmed by recent results of more detailed ongoing measurements[80] for a range of energies and different J values, and their implications are discussed in Sec. 6 in the light of the observations on IVR and compared to predictions from unimolecular rate theory.

6. Discussion and Conclusions

6.1. *Spectroscopy of Low-Lying Vibrational States of* CH_3O $(\tilde{X}^2E)$

The SEP results for the low-lying vibronic states of the degenerate vibrational modes shed new light on the JT effect in CH_3O which may be of general interest for the interpretation of spectra of other molecules subject to the JT effect. It is apparent that the vibronic coupling in CH_3O cannot be described by a simple linear JT effect. Quadratic and

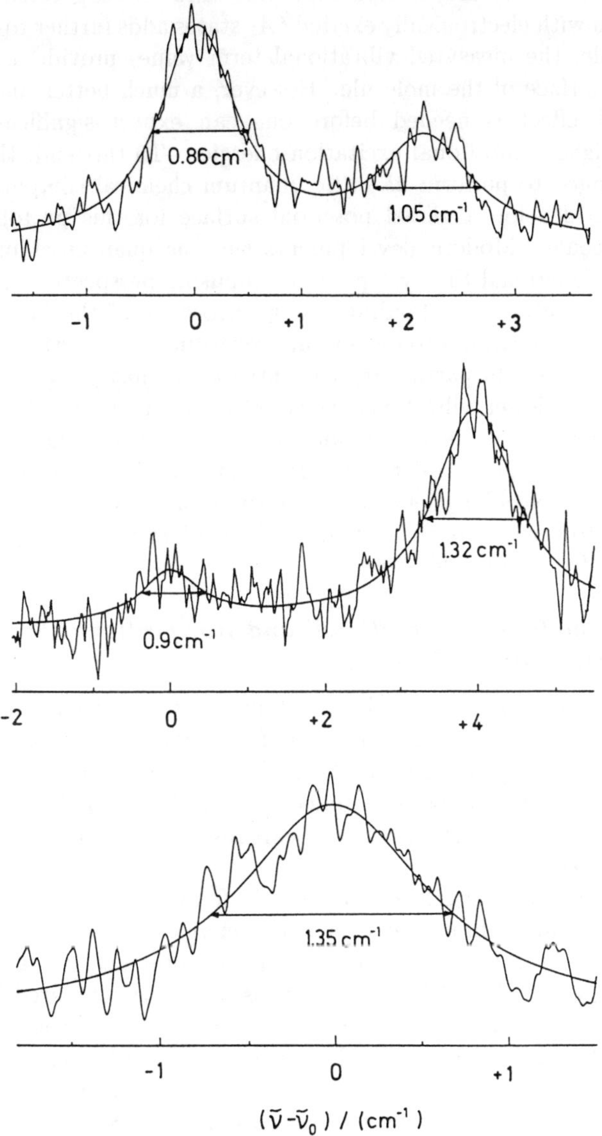

Fig. 17. Observed line profiles and fitted Lorentzian functions for individual "resonance" states between $9920\,\mathrm{cm}^{-1} \leq E \leq 9970\,\mathrm{cm}^{-1}$ of excitation energy. Pump line: $3_0^3\ ^rR_{21}(J'' = 1.5,\ K'' = -2)$.

higher order terms are coming into play, and the suggested pseudo-JT interaction with electronically excited $^2A_1$ states adds further to the picture. In principle, the measured vibrational term values provide a map of the potential surface of the molecule. However, a much better understanding of the JT effect is needed before one can expect significant progress towards higher vibrational excitation energies. To this end, theoreticians are challenged to perform *ab initio* quantum chemical computations with the aim to develop a global potential surface for this prototypical five-atom molecule. Modern developments such as quantum Monte–Carlo[81] or density functional theory[82] provide promising perspectives towards this goal. Ultimately, understanding the spectroscopy of the low-lying states of CH_3O will contribute to our picture regarding the reactive dynamics at higher excitation. In particular, it is anticipated that the vibrational term values for the lower vibrational levels can be represented by some form of a Dunham expansion which can then be scaled into the higher energy region, for example, in order to improve density of states calculations and explore intramolecular energy redistribution pathways. Furthermore, it is of interest to investigate the relevance of the JT effect for the unimolecular dynamics of the molecule.

6.2. *Intramolecular Vibrational and Rovibrational Energy Redistribution (IVR)*

The observation of extensive level couplings in the eigenstate spectra of highly excited CH_3O demonstrates the feasibility of intramolecular energy randomization as postulated by unimolecular rate theory. CH_3O is the only five-atom radical species studied in such detail all the way up to its dissociation energy. Furthermore, it bears resemblance to larger, chemically relevant molecules.

The onset of strong level coupling is found to occur rather sharply at a "threshold" vibrational excitation energy of approximately $E_v \approx 5000\,cm^{-1}$. At this energy, the calculated total density of vibrational states is $\rho_{\mathrm{vib}} \approx 0.1/cm^{-1}$. This value is very low compared to relevant state densities in larger molecules. Related values have been indicated for H_2CO (Refs. 24a, 83) or C_2H_2 (Ref. 84). On the other hand, there have been reports of remarkably stable vibrational motion in larger molecules at higher energies. SEP spectra for HFCO (Ref. 85) give evidence for quasistable extreme motion states which may be related to "periodic orbits" (Ref. 72). In some respect, the familiar local mode description of highly excited overtones of C–H or O–H stretch vibrations also has a bearing in this context.[13] However, as noted, C–H or O–H stretch vibrations are special high frequency modes with limited relevance to the postulated "statistical" molecular properties assumed in unimolecular rate theory.

The observed IVR rates in CH_3O must now be compared to the specific unimolecular dissociation rates of the molecule as discussed in Sec. 5. For the molecule with almost $10\,000\,\mathrm{cm}^{-1}$ of excitation energy, dissociation lifetimes around $\tau \approx 5\,\mathrm{ps}$ had been inferred. For the ν_3 (C–O stretch) mode in the specific rotational states that were investigated, an initial excitation would remain localized for approximately $0.25\,\mathrm{ps}$ before it spreads out over other degrees of freedom with the same J. Such rapid intramolecular randomization is known for the special case of C–H stretch/bend resonances.[13] The $0.25\,\mathrm{ps}$ time interval corresponds to only 7 C–O vibrational periods. Similar magnitudes of IVR times were obtained for a number of higher ν_3 overtone states with $v_3 \leq 5$ and other initial rotational quantum numbers ($3.5 \leq J \leq 11.5$). These observations imply, to a large extent, "statistical" reactivity at longer times. Note that the excited states that have been studied here are unique in some sense, however, as they derive from pure high overtones in the C–O stretch coordinate. In the aforementioned conceptual framework, the initial coherent excitation by a short pulse would thus be a very localized one, in contrast to the case that would be encountered for excitation of more delocalized combination states. Excited states that derive from combinations of different vibrational modes may behave differently. Experimental work in this direction is in progress.

However, the number of coupled states derived from $\overline{P_{ss}}^{-1}$ can vary drastically[45] which seems to indicate that only a fraction of the total energetically and symmetry accessible phase space is accessed within the $\approx 0.25\,\mathrm{ps}$ of the initial decay. It is important that, for $v_3 = 6$ ($J = 9.5$), the computed survival probability P_{ss} (Fig. 15) seems to indicate a second, somewhat slower time scale of $\approx 2.5\,\mathrm{ps}$. This observation would be consistent with a fast initial randomization of the energy into a limited region of the overall phase space, followed by a subsequent slower distribution over a larger region. Such a finding of an IVR tier structure is of course well in accordance with chemical intuition.[12e,76] Related observations for C_2H_2 could be reproduced by trajectory calculations on a model potential energy surface.[86] Here, for CH_3O, it is important that the longer IVR times ($\tau_{\mathrm{IVR}} \geq 2.5\,\mathrm{ps}$) are comparable (!) to the lifetimes with respect to dissociation at energies only slightly above the classical reaction threshold energy ($\tau_{\mathrm{diss}} \approx 5\,\mathrm{ps}$ at $E \approx 10\,000\,\mathrm{cm}^{-1}$). Hence, one might speculate about the possibility of some mode specificity of the reaction.

The volume of the region of phase space that the molecule explores within the first picosecond is a related central topic. The observed state densities provide a strong case for the importance of rotation–vibration interactions. Accordingly, concerning the energy available for the unimolecular reaction, the rotation about the top axis is an "active" degree of freedom, i.e., the rotational energy around the top axis contributes to overcoming the reaction barrier. This is assumed in most statistical rate calculations.[1] However, again, the above results and further J dependent investigations which cannot be detailed here[44,45] do indicate restrictions on the couplings for the specific levels that were described in Sec. 4. A subpicosecond IVR lifetime for rotation–vibration interactions appears surprisingly short, but in CH_3O, with its three degenerate vibrational modes which can carry vibrational angular momentum, one can envision particular efficient Coriolis coupling mechanisms which would explain the observations. Further studies for different excited modes, also as a function of rotational state, and perhaps even directly in the time domain, would be of great interest.

6.3. *Unimolecular Dissociation Rate Constants*

With the development of high resolution spectroscopic techniques to prepare reactive molecules in "resonance" states in the quasicontinuum above the reaction threshold energies, investigations are starting to become available of state specific unimolecular reaction rates $k(E, J, \Gamma; v_i, K)$ and product distributions for molecules with known total energy E, defined angular momentum J, symmetry Γ, and, as far as they can be specified, other possible degrees of freedom (v_i, K).[2,6-7,9a-9b,17,19]

The linewidth broadening measurements for CH_3O demonstrate first state specific lifetime information for a five-atom molecule. The results described above can be compared to predictions obtained from unimolecular rate theory. The observations show that the dissociation of CH_3O, at energies only slightly above the classical threshold, proceeds extraordinarily rapidly compared to related reactions of larger molecules.[1] The high rate can be traced to the still relatively small state density of CH_3O at E_0, which in turn is due to the low value of the $H–H_2CO$ bond dissociation energy. The short dissociation lifetime of $\approx 5\,\mathrm{ps}$ at $E \approx 10\,000\,\mathrm{cm}^{-1}$ imposes particularly rigorous tests for the application of statistical rate theory.

6.3.1. *RRKM model*

Statistical (RRKM) rate theory[1] gives the specific rate constant for a normal unimolecular dissociation or isomerization reaction of a molecule as a function of total energy E and angular momentum J according to

$$k(E, J) = \frac{W^{\neq}(E, J)}{h \cdot \rho(E, J)}, \tag{11}$$

where $\rho(E, J)$ is the active density of rovibronic states of the molecules at E and J, and $W^{\neq}(E, J)$ is the corresponding number of states of the transition state, often referred to as the number of open reaction channels. The rotational motion of the molecule around the top axis (the "K rotor") is assumed to be an active degree of freedom here, in agreement with the evidence for effective rovibrational coupling.

Equation (11) presumes a well characterized and "tight" transition state which determines the bottleneck for the reaction. Knowing the vibrational frequencies and rotational constants of the molecule and transition state, $\rho(E, J)$ and $W^{\neq}(E, J)$ can be computed by direct state counts. For larger molecules, approximate formulas have been developed which permit a convolution of the density of vibrational states with that of the K rotor.[1]

For the dissociation reaction of CH_3O to $H + H_2CO$, the transition state can be localized at the top of the $H-H_2CO$ potential barrier. However, considering the light H atom, dissociation can occur at energies below the classical barrier by quantum mechanical tunneling. Thus, $W^{\neq}(E, J)$ must be replaced by summing over the tunneling probabilities of the reaction channels,

$$W^{\neq}(E, J) = \sum_a T_a^{\neq}(E - E_a^{\neq}, J), \tag{12}$$

where $T_a^{\neq}(E - E_a^{\neq}, J)$ is the transmission coefficient for the ath reaction channel from reactant to product states. The transmission coefficients can be calculated analytically with the assumption that the reaction coordinate and potential barrier can be approximated by a one dimensional Eckart potential.[89] To this end, adiabatic channel potential curves can be constructed which describe the "transition" of the eigenvalues of the vibrational and rotational degrees of freedom from the reactant to product vibrational states. $E - E_a^{\neq}$ is then the energy in excess of the amount that must remain fixed in the adiabatic channels.[90]

The molecular parameters of CH_3O, the $H-H_2CO^{\neq}$ transition state, and the products $H + H_2CO$ which are needed for the calculations are collected in Table 3. Unfortunately, the $H-H_2CO$ bond dissociation energy ($\Delta_r H_0^0$)

Table 3. Parameters of CH_3O, $H-H_2CO^{\neq}$, and $H_2CO + H$ for the RRKM calculations.

	CH_3O	$H-H_2CO^{\neq}$	$H_2CO + H$
$\Delta H_0^0/cm^{-1}$	0.0	8400	6900
ν_i/cm^{-1} [a]			
C–H s stretch	2840	2798	2811
C–H a stretch	2800	2892	2861
C–H$'$ stretch	2800	850 i	–
CHHH$'$ deform	1220	740	–
OCH$_2$ bend	1220	1173	1170
CHH bend	1364	1593	1500
C–O stretch	1045	1414	1756
CHHH$'$ rock	980	1232	1251
OCHH$'$ bend	980	581	–
A/cm^{-1}	5.21	3.93	9.41
B/cm^{-1}	0.93	0.99	1.21
g_s	2	2	2
g_e	2	1	1
σ_{rot}	3	1	2

[a]The parameters for CH_3O have been taken either directly from the observed spectra or estimated from a preliminary JT analysis for the ν_5 and ν_6 modes, which were treated independently. Parameters for $H-H_2CO^{\neq}$ have been scaled from Ref. 36 (see text). The parameters for H_2CO were taken as given in Ref. 5a.

and the barrier height (E_0) are not known very accurately. Values for $\Delta_r H_0^0$ which are quoted are in the range $6160\,cm^{-1} \leq \Delta H_0^0 \leq 7250\,cm^{-1}$ (Refs. 29–36). From preliminary results of an investigation of the rate constant for the D/H atom isotope exchange reaction $D + H_2CO \rightarrow H + HDCO$ (Ref. 37), which is assumed to proceed via an intermediate CH_2DO^* complex, the height of the potential barrier for the addition of H to H_2CO (i.e., the reverse of the CH_3O dissociation reaction) is estimated to be $\approx 1500\,cm^{-1}$. Thus, adopting a value of $\Delta_r H_0^0 \approx 6900\,cm^{-1}$, one obtains $E_0 \approx 8400\,cm^{-1}$.

The thermal unimolecular decomposition of CH_3O has been investigated by Lin and coworkers[35d,87] and by Wantuck *et al.*[88] The latter authors were able to monitor the decay of the CH_3O directly by LIF detection. Unfortunately, these measurements refer to the low pressure regime of thermal unimolecular reactions[1]; the rates are governed by collisional processes and cannot give direct information on the potential. However, the measured apparent thermal activation energy E_A (Refs. 87–88) is in accordance with the present value of $E_0 \approx 8400\,\mathrm{cm}^{-1}$, if the required corrections (Ref. 1) are applied for the deviation of E_A from E_0.

The imaginary tunneling frequency is adjusted from the *ab initio* value of Refs. 35d and 36 by scaling the computed potentials and their curvatures to the present values for $\Delta_r H_0^0$ and E_0. This yields $\nu_i \approx 850\ (i)\ \mathrm{cm}^{-1}$. Other transition state parameters are taken from the high level *ab initio* results of Ref. 36. Only one (A_1) of the two potential surfaces that result from the ^2E ground state of CH$_3$O correlates through to products H + H$_2$CO in their ground states.

Because of the small size for the CH$_2$O molecule, all computations of $\rho(E, J)$ and $\Sigma_a T_a^{\neq}(E - E_a, J)$ can be performed by direct state counting with approximate expressions for the rotation–vibration energy levels. However, anharmonicity has been neglected (see below). The symmetry correlations from reactant to product states within the complete nuclear permutation inversion (CNPI) symmetry group,[91] on the other hand, can be taken into account.

6.3.2. *Discussion and further developments*

The calculated $k(E, J)$ curves and the inferred experimental values are shown in Fig. 18. The theoretical curve can be seen to predict the correct order of magnitude, but especially the data points at $E \approx 10\,000\,\mathrm{cm}^{-1}$ are significantly higher than the predicted values. Considering the present uncertainties, primarily regarding the exact value of the reaction threshold energy E_0 and the imaginary tunneling frequency, more detailed data are needed before one can really assess the agreement. Nevertheless, it is noted that the likely errors in the statistical calculations, primarily the neglection of anharmonicities, would lead to an overestimate of the calculated rate constants. More refined model calculations would increase the differences. On the other hand, the data points at $10\,000\,\mathrm{cm}^{-1}$ ($J = 3.5$) are unlikely to be affected strongly by inhomogeneous broadening, considering the quality of the Lorentzian line fits in Fig. 17 (this may not be quite true for the $J = 9.5$ data). It has been mentioned that the "active" state density may not quite reach the calculated full rovibronic value because of some restrictions on the extent of intramolecular energy randomization. Adopting a somewhat lower value for the density of states in the statistical rate calculations, the rate constants inferred from the linewidth measurements could indeed be reproduced. However, a detailed

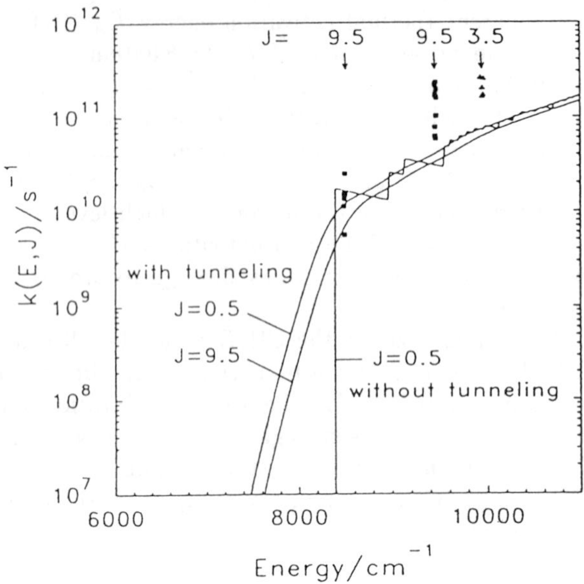

Fig. 18. Comparison of CH_3O unimolecular decay rates deduced from SEP linewidths and statistical (RRKM) calculations.

discussion of the deviations and other explanation is deferred to a special publication until after completion of further experimental investigations.

More detailed linewidth measurement using the pulsed jet apparatus for a large number of low J levels ($J = 0.5$ and 1.5) between $8000\,\mathrm{cm}^{-1} \leq E \leq 10\,000\,\mathrm{cm}^{-1}$ are currently being evaluated.[66,80] In addition, at lower energies ($E \geq \Delta H_0^0$, but $\leq E_0$), where the decay of the CH_3O proceeds by quantum mechanical tunneling of the H atom through the barrier, direct lifetime measurements are being carried out in the time domain using a pump–dump-probe scheme.[44] First results reveal strong fluctuations of single resonance dissociation lifetimes by more than an order of magnitude within narrow energy intervals both below and also *above* the top of the classical barrier.[66] These statistical fluctuations of the state specific dissociation lifetimes (or rate constants) reflect the *random* coupling situation between the individual molecular rotation–vibration states at the high excitation energies! Nevertheless, it was found that the average values of the state specific rate constants can be reproduced nicely by the RRKM model.[66]

Similar strong state specific lifetime fluctuations have been observed by Moore and coworkers for the dissociation of H_2CO to $H_2 + CO$, even though by various measures the H_2CO molecule exhibits stochastic behaviour (Refs. 9a–9b, 92–93). However, the results for H_2CO had been obtained using the less direct technique of Stark level crossing spectroscopy, and some assumptions on Stark tuning rates had to be made for the evaluation of the results. The general significance of these lifetime fluctuations and their distribution function[94] should be investigated in more detail. Obviously, they may have a pronounced effect in the low pressure regime and in the so-called "fall-off" regime at intermediate pressures of thermal unimolecular reactions.

6.4. Conclusion

The common standard analysis[1] of unimolecular reactions using statistical rate models, which has been so successful, is limited by the necessity to use a number of estimated to adjustable parameters. However, accurate global potential surfaces are starting to become available for some small polyatomic molecules (see, e.g., Refs. 95–99). Improved statistical rate theories, e.g., variational RRKM theory[100] or the statistical adiabatic channel model[101] (SACM), have been proposed which, at least in principle, allow implementation of exact potentials.[102] Complementary direct theoretical dynamics studies are in their infancy[103–106] and will undoubtedly gain much more importance. Trajectory calculations on global *ab initio* potential surfaces provide one route.[103] Important modern theoretical developments are quantum reactive scattering calculations[104] of the energies and width of metastable "resonance" states[105] and wave packet propagation methods[78,106] for evaluating short time dynamics on realistic potentials.

With its intermediate size, the CH_3O molecule presents the important link between the statistical treatment, which is the only possibility for large molecules, and direct dynamics calculations.

Acknowledgments

This research was support by the Max–Planck–Gesellschaft and the Deutsche Forschungsgemeinschaft within SFB 93. The author is greatly indebted to Prof. H. Gg. Wagner for his enthusiasm, many stimulating discussions, and continuous support and encouragement. Thanks are due to a number of research students who have been and are eagerly involved with different aspects of the work and to collegues in Göttingen and other laboratories for discussions and helpful comments.

References

1. J. Troe, *J. Phys. Chem.* **83**, 114 (1979); J. Troe, *J. Chem. Phys.* **75**, 226 (1981); J. Troe, *ibid.* **79**, 6017 (1983); J. Troe, *Z. Phys. Chem. NF* **161**, 209 (1989); J. Troe and H. Gg. Wagner, *Ber. Bunsenges. Phys. Chem.* **71**, 937 (1967); P. J. Robinson and K. A. Holbrook, *Unimolecular Reactions* (Wiley, London, 1972); R. G. Gilbert and S. C. Smith, *Theory of Unimolecular and Recombination Reactions* (Blackwell, Oxford, 1990).

2. R. G. Bray and M. S. Berry, *J. Chem. Phys.* **71**, 4909 (1979); H. R. Dübal and M. Quack, *ibid.* **81**, 3779 (1984); K. K. Lehmann, G. H. Scherer, and W. Klemperer, *ibid.* **78**, 2817 (1983); M. Child, *Chem. Soc. Rev.* **17**, 31 (1988); F. F. Crim, *Ann. Rev. Phys. Chem.* **35**, 657 (1982).

3. D. G. Imre, J. L. Kinsey, R. W. Field, and D. H. Katayama, *J. Phys. Chem.* **86**, 2564 (1982).

4. R. H. Page, Y. R. Shen, and Y. T. Lee, *J. Chem. Phys.* **88**, 4621 (1988); P. R. Fleming, M. Li, and T. Rizzo, *ibid.* **95**, 865 (1991).

5. D. E. Reisner, R. W. Field, J. L. Kinsey, and H.-L. Dai, *J. Chem. Phys.* **80**, 5968 (1984); E. Abramson, R. W. Field, D. Imre, K. K. Innes, and J. L. Kinsey, *ibid.* **80**, 2298 (1984).

6. C. E. Hamilton, J. L. Kinsey, and R. W. Field, *Ann. Rev. Phys. Chem.* **37**, 493 (1986); F. J. Northrup and T. J. Sears, *Ann. Rev. Phys. Chem.* **43**, 127 (1992).

7. W. F. Polik, C. B. Moore, and W. H. Miller, *J. Chem. Phys.* **89**, 3584 (1988); W. F. Polik, D. R. Guyer, and C. B. Moore, *ibid.* **92**, 3453 (1990); M. Lombardi, in *Excited States*, eds. E. C. Lim and K. K. Innes (Academic Press, San Diego, 1988), Vol. 7; P. Dupré, R. Jost, M. Lombardi, P. G. Green, E. Abramson, and R. W. Field, *Chem. Phys.* **152**, 293 (1991).

8. J. B. Hopkins, D. E. Powers, and R. E. Smalley, *J. Chem. Phys.* **72**, 5039 (1980); J. B. Hopkins, D. E. Powers, S. Mukamel, and R. E. Smalley, *ibid.* **72**, 5039 (1980); R. E. Smalley, *J. Phys. Chem.* **86**, 3504 (1982).

9. G. M. Stewart and J. D. MacDonald, *J. Chem. Phys.* **75**, 5949 (1981); G. M. Stewart and J. D. MacDonald, *ibid.* **78**, 3907 (1983); H. L. Lim, T. J. Kulp, and J. D. MacDonald, *ibid.* **87**, 4376 (1987).

10. C. S. Parameter, *Faraday Discuss. Chem. Soc.* **75**, 7 (1983); K. W. Holtzclaw and C. S. Parmenter, *J. Chem. Phys.* **84**, 1099 (1986).

11. A. E. W. Knight, in *Excited States*, eds. E. C. Lim and K. K. Innes (Academic Press, San Diego, 1988).

12. H. R. Dübal and M. Quack, *Chem. Phys. Lett.* **72**, 342 (1980); K. K. Lehmann, B. H. Pate, and G. Scoles, *J. Chem. Phys.* **93**, 2152 (1990); B. H. Pate, K. K. Lehmann, and G. Scoles, *ibid.* **95**, 3891 (1991); X. Luo and T. Rizzo, *ibid.* **93**, 8620 (1990); X. Luo and T. Rizzo, *ibid.* **94**, 889 (1991).

13. M. Quack, *Ann. Rev. Phys. Chem.* **41**, 839 (1990).

14. H. Hippler, K. Luther, J. Troe, and H. J. Wendelken, *J. Chem. Phys.* **79**, 239 (1983); U. Brand, H. Hippler, L. Lindemann, and J. Troe, *J. Phys. Chem.*

94, 6305 (1990); T. Shimada, Y. Ojima, N. Nakashima, Y. Izawa, and C. Yamanaka, *ibid.* **96**, 6298 (1992); H. Hippler, V. Schubert, J. Troe, and H. J. Wendelken, *Chem. Phys. Lett.* **84**, 253 (1981); K. Luther, J. Troe, and K.-M. Weitzel, *J. Phys. Chem.* **94**, 6316 (1990); R. Fröchtenicht, H.-G. Rubahn, and J. P. Toennies, *Chem. Phys. Lett.* **162**, 269 (1989).

15. L. R. Khundkar, J. L. Knee, and A. H. Zewail, *J. Chem. Phys.* **87**, 77 (1987); N. F. Scherer and A. H. Zewail, *ibid.* **87**, 97 (1987); E. D. Potter, M. Gruebele, L. R. Khundkar, and A. H. Zewail, *Chem. Phys. Lett.* **164**, 1989 (1989); I.-C. Chen and C. B. Moore, *J. Phys. Chem.* **94**, 263 (1990).

16. A. Kiermeier, H. J. Neusser, and E. W. Schlag, *Z. Naturforsch.* **42a**, 1399 (1987); A. Kiermeier, H. Kühlewind, H. J. Neusser, E. W. Schlag, and S. H. Lin, *J. Chem. Phys.* **88**, 6182 (1988); B. Abel, B. Herzog, H. Hippler, and J. Troe, *J. Chem. Phys.* **91**, 890 (1989).

17. B. R. Foy, M. P. Casassa, J. C. Stephenson, and D. S. King, *J. Chem. Phys.* **90**, 7037 (1989); B. R. Foy, M. P. Casassa, J. C. Stephenson, and D. S. King, *ibid.* **92**, 2782 (1990); Y. S. Choi and C. B. Moore, *J. Chem. Phys.* **97**, 1010 (1992).

18. G. A. Brucker, S. I. Ionov, Y. Chen, and C. Wittig, *Chem. Phys. Lett.* **194**, 301 (1992).

19. T. R. Rizzo, C. C. Hayden, and F. F. Crim, *J. Chem. Phys.* **81**, 4501 (1984); L. Brower, C. J. Cobos, J. Troe, H.-R. Dübal, and F. F. Crim, *ibid.* **86**, 6171 (1987); A. Sinha, R. L. Vander Wal, and F. F. Crim, *ibid.* **92**, 401 (1990).

20. C. X. W. Qian, A. Ogai, H. Reisler, and C. Wittig, *J. Chem. Phys.* **90**, 209 (1989).

21. I.-C. Chen, W. H. Green Jr., and C. B. Moore, *J. Chem. Phys.* **89**, 314 (1988); I.-C. Chen and C. B. Moore, *J. Phys. Chem.* **94**, 269 (1990); S. K. Kim, Y. S. Choi, C. D. Pibel, Q.-K. Zheng, and C. B. Moore, *J. Chem. Phys.* **94**, 1954 (1991); W. H. Green, A. J. Mahoney, Q.-K. Zheng, and C. B. Moore, *ibid.* **94**, 1961 (1991).

22. H. Hippler, J. Troe, and H. J. Wendelken, *Chem. Phys. Lett.* **84**, 257 (1981); H. Hippler, *Ber. Bunsenges, Phys. Chem.* **89**, 303 (1985); M. Damm, H. Hippler, and J. Troe, *J. Chem. Phys.* **88**, 3564 (1988); J. R. Barker and R. E. Golden, *J. Phys. Chem.* **88**, 1012 (1984); M. L. Yerram, J. D. Brenner, K. D. King, and J. R. Barker, *J. Phys. Chem.* **94**, 6341 (1990).

23. S. H. Kable, J. W. Thoman, and A. E. W. Knight, *J. Chem. Phys.* **88**, 4748 (1988).

24. F. Temps. S. Halle, P. H. Vaccaro, R. W. Field, and J. L. Kinsey, *J. Chem. Soc., Faraday Trans.* 2, **84**, 1457 (1988); A. L. Utz, J. D. Tobiason, E. Carrasquillo M., M. D. Fritz, and F. F. Crim, *J. Chem. Phys.* **97**, 389 (1992); X. Yang, E. H. Kim, and A. M. Wodtke, *ibid.* **96**, 5111 (1992).

25. C. Kittrell, E. Abramson, J. L. Kinsey, S. A. MacDonald, D. E. Reisner, and R. W. Field, *J. Chem. Phys.* **75**, 2056 (1991).

26. See the collection of papers in the issue on *Stimulated Emission Pumping* in *J. Opt. Soc.* **B7**(9), 1802–1980 (1990).

27. H. A. Jahn and E. Teller, *Proc. Roy. Soc. (Lond.)* **A161**, 220 (1937).

28. A. Geers, J. Kappert, F. Temps, and J. W. Wiebrecht, *J. Chem. Phys.* **93**, 1472 (1990); A. Geers, J. Kappert, F. Temps, and J. W. Wiebrecht, *Ber. Bunsenges. Phys. Chem.* **94**, 1219 (1990).

29. D. F. McMillen and D. M. Golden, *Ann. Rev. Phys. Chem.* **33**, 493 (1982); W. Tsang and R. F. Hampson, *J. Phys. Chem. Ref. Data* **15**, 1987 (1986); K. M. Ervin, private communication.

30. L. Batt, K. Christie, R. T. Milne, and J. A. Summers, *Int. J. Chem. Kin.* **6**, 877 (1974); L. D. Batt and R. T. Milne, *Int. J. Chem. Kin.* **6**, 945 (1974); L. Batt and R. D. McCulloch, *Int. J. Chem. Kin.* **8**, 491 (1976).

31. J. O. Holmes and F. P. Lossing, *Int. J. Mass Spectrosc. Ion Proc.* **85**, 113 (1984).

32. B. Ruscic, E. H. Appelman, and J. Berkowitz, *J. Chem. Phys.* **95**, 7957 (1991); R. Ruscic and J. Berkowitz, *J. Chem. Phys.* **95**, 4033 (1991).

33. J. A. Seetula and D. Gutman, *J. Phys. Chem.* **96**, 5401 (1992); S. Dóbé, *Z. Phys. Chem.*, **175**, 123 (1992).

34. C. Melius, *BAC–MP4 Heats of Formation and Free Energies*, Sandia National Laboratory, 1990.

35. S. Saebo, L. Random, and H. F. Schaefer III, *J. Chem. Phys.* **89**, 845 (1983); C. Sosa and H. B. Schlegel, *Int. J. Quant. Chem.* **29**, 1001 (1986); L. A. Curtiss, L. D. Kock, and J. A. Pople, *J. Chem. Phys.* **95**, 4040 (1991); M. Page, M. C. Lin, Y. He, and T. K. Choudhury, *J. Phys. Chem.* **93**, 4404 (1989).

36. S. P. Walch, *J. Chem. Phys.* **98**, 3076 (1993).

37. S. Dóbé, C. Oehlers, F. Temps, H. Gg. Wagner, and H. Ziemer, *Ber. Bunsenges. Phys. Chem.* **98**, 754 (1994).

38. J. Kappert and F. Temps, *Chem. Phys.* **132**, 197 (1989).

39. X. Liu, C. P. Damo, T.-Y. Lin, S. C. Foster, P. Misra, L. Yu, and T. A. Miller, *J. Phys. Chem.* **93**, 2266 (1989).

40. G. Inoue, H. Akimoto, and M. Okuda, *J. Chem. Phys.* **72**, 1769 (1980).

41. S. C. Foster, P. Misra, T.-Y. Lin, C. P. Damo, C. C. Carter, and T. A. Miller, *J. Phys. Chem.* **92**, 5914 (1988).

42. A. Geers, J. Kappert, F. Temps, and T. J. Sears, *J. Chem. Phys.* **98**, 4297 (1993).

43. J. Kappert, Ph. D. Thesis, Göttingen, 1992.

44. J. W. Wiebrecht, Ph. Thesis, Göttingen, 1992.

45. A. Geers, J. Kappert, F. Temps, and J. W. Wiebrecht, *J. Chem. Phys.* **101**, 3618 (1994); *ibid.* **101**, 3634 (1994).

46. Q. Zhang, S. A. Kandel, T. A. W. Wasserman, and P. H. Vaccaro, *J. Phys. Chem.* **96**, 1640 (1992); T. J. Butenhoff and E. A. Rohlfing, *J. Chem. Phys.* **97**, 1595 (1992).

47. A. Geers, J. Kappert, F. Temps, and J. W. Wiebrecht, *J. Opt. Soc. Am.* **B7**, 1935 (1990).

48. K. Hoyermann, N. S. Loftfield, R. Sievert, and H. Gg. Wagner, *18th Sympo-*

sium (International) on Combustion (The Combustion Institute, Pittsburgh, 1981), p. 831; J. L. Durant, *J. Phys. Chem.* **95**, 10701 (1991).

49. S . Gerstenkorn, J. Verges, and J. Chevillard, Atlas du Spectre de la Molecule de L'Iode, Laboratoire Aimé Cotton, CNRS, 91405 Orsay, France.

50. F. M. Phelps III, *M.I.T. Wavelength Tables*, Vol. 2 (M.I.T. Press, Cambridge, 1982).

51. J. G. Calvert and J. N. Pitts Jr., *Photochemistry* (Wiley, New York, 1966); P. Felder, B. A. Keller, and J. R. Huber, *Z. Phys.* **D6**, 185 (1987); P. Felder, B. A. Keller, and J. R. Huber, *J. Phys. Chem.* **91**, 1114 (1987).

52. H. R. Wendt and H. E. Hunziker, *J. Chem. Phys.* **71**, 5202 (1979).

53. D. E. Powers, J. B. Hopkins, and R. E. Smalley, *J. Phys. Chem.* **85**, 2711 (1981); D. S. King and R. F. Wormsbecher, *Proc. Soc. Phot. Opt. Instrum. Eng.* **286**, 111 (1981); K. Fuke, K. Ozawa, and K. Kaya, *Chem. Phys. Lett.* **126**, 119 (1986); N. L. Garland and D. R. Crosley, *J. Phys. Chem.* **92**, 5322 (1988); S.-Y. Chiang, Y.-C. Hsu, and Y.-P. Lee, *J. Chem. Phys.* **90**, 81 (1989).

54. S. D. Brossard, P. G. Carrick, E. L. Chapell, S. C. Hulegaard, and P. C. Engelking, *J. Chem. Phys.* **84**, 2459 (1986).

55. D. K. Russell and H. E. Radford, *J. Chem. Phys.* **72**, 2750 (1980).

56. Y. Endo, S. Saito, and E. Hirota, *J. Chem. Phys.* **81**, 122 (1984); T. Momose, Y. Endo, E. Hirota, and S. Saito, *J. Chem. Phys.* **88**, 5338 (1988).

57. A. R. W. McKellar, *Faraday Disc. Roy. Soc. (Lond.)* **71**, 63 (1981).

58. T. J. Sears, private communication.

59. H. C. Longuet-Higgins, U. Öpik, M. H. L. Pryce, and R. A. Sack, *Proc. Roy. Soc.* **A244**, 1 (1958); H. C. Longuet-Higgins, *Adv. Spectrosc.* **2**, 429 (1961).

60. T. A. Miller and V. E. Bondybey, in *Molecular Ions: Spectroscopy, Structure, and Chemistry*, eds. T. A. Miller, V. E. Bondybey (North-Holland, Amsterdam, 1983); T. J. Sears, T. A. Miller, and V. E. Bondybey, *J. Chem. Phys.* **72**, 6070 (1980); T. J. Sears, T. A. Miller, and V. E. Bondybey, *ibid.* **74**, 3240 (1981); C. Cossart-Magos and S. Leach, *Chem. Phys.* **48**, 329 (1980).

61. J. M. Brown, *Mol. Phys.* **20**, 817 (1971).

62. J. T. Hougen, *J. Mol. Spectrosc.* **81**, 73 (1980).

63. J. K. G. Watson *J. Mol. Spectrosc.* **103**, 125 (1984).

64. X. Liu, L. Yu, and T. A. Miller, *J. Mol. Spectrosc.* **140**, 112 (1990).

65. X. Liu, S. C. Foster, J. M. Wiliamson, L. Yu, and T. A. Miller, *Mol. Phys.* **69**, 357 (1990).

66. A. Geers, Ph. Thesis, Göttingen, 1993; C. Stöck, Diploma Thesis, Göttingen, 1993; A. Geers, J. Kappert, C. Stöck, and F. Temps, to be published.

67. J. T. Hougen, *J. Chem. Phys.* **37**, 1433 (1962).

68. Note that this form differs from that of Longuet-Higgins *et al.*[59] by an opposite sign for φ. This results in different definitions for the JT quantum number j. Longuet-Higgins *et al.* have $j = l + \frac{1}{2}\Lambda$, whereas Hougen[62] has $j = l - \frac{1}{2}\Lambda$. The form of Eq. (1) is necessary here for consistency with Hougen's[67] definitions of $G_{ev} = \Lambda + \Sigma_t l_t$ and $G = G_{ev} - K$ (modulo 3) and with the standard labels of the $\tilde{X}$ rotational levels.[38,56]

69. G. D. Bent, G. F. Adams, R. H. Bartram, G. D. Purvis, and R. J. Bartlett, *J. Chem. Phys.* **76**, 4144 (1982); G. D. Bent, *J. Chem. Phys.* **89**, 7298 (1988).

70. G. Herzberg, *Molecular Spectra and Molecular Structure, Vol III, Electronic Spectra and Electronic Structure of Polyatomic Molecules*, (Van Nostrand Reinhold, New York, 1966).

71. X. Li, X. Xie, L. Li, X. Wang, and C. Zhang, *J. Chem. Phys.* **97**, 128 (1992).

72. G. Hose and H. S. Taylor, *Chem. Phys.* **84**, 375 (1984).

73. M. Quack, in *Dynamics of the Excited State*, ed. K. P. Lawley (Wiley, New York, 1982).

74. L. R. Khundkar and A. H. Zewail, *Ann. Rev. Phys. Chem.* **41**, 15 (1990); W. Kaiser, J. P. Maier, and A. Seilmeier, in *Laser Spectroscopy*, eds. H. Walther and K. W. Rothe (Springer, Berlin, 1979), Vol. IV; J. F. Kauffman, M. J. Cote, P. G. Smith, and J. D. MacDonald, *J. Chem. Phys.* **90**, 2874 (1989); P. G. Smith and J. D. MacDonald, *ibid.* **96**, 7344 (1992).

75. K. F. Freed and A. Nitzan, *J. Chem. Phys.* **73**, 4765 (1980); E. Heller, *Acc. Chem. Res.* **14**, 368 (1981); E. J. Heller, *Phys. Rev.* **A35**, 1360 (1987).

76. L. Leviandier, M. Lombardi, R. Jost, and J. P. Pique, *Phys. Rev. Lett.* **56**, 2449 (1986); J. P. Pique, Y. Chen, R. W. Field, and J. L. Kinsey, *ibid.* **58**, 475 (1987); J. P. Pique, *J. Opt. Soc. Am.* **B7**, 1816 (1990).

77. Y. Chen, D. M. Jonas, J. L. Kinsey, and R. W. Field, *J. Chem. Phys.* **91**, 3976 (1989).

78. E. J. Heller, *Acc. Chem. Res.* **14**, 368 (1981); D. Imre, J. L. Kinsey, A. Sinha, and J. Krenos, *J. Phys. Chem.* **88**, 3956 (1984).

79. W. D. Lawrance and A. E. W. Knight, *J. Phys. Chem.* **89**, 917 (1985).

80. A. Geers, J. Kappert, F. Temps, and J. W. Wiebrecht, *J. Chem. Phys.* **99**, 2271 (1993); A. Geers, J. Kappert, F. Temps, and J. W. Wiebrecht, to be published.

81. W. A. Lester and B. L. Hammond, *Ann. Rev. Phys. Chem.* **41**, 283 (1991).

82. R. G. Parr and W. Yang, *Density-Functional Theory of Atoms and Molecules* (Oxford, New York, 1989); R. O. Jones and O. Gunnarsson, *Rev. Mod. Phys.* **61**, 689 (1989); B. G. Johnson, P. M. W. Gill, and J. A. Pople, *J. Chem. Phys.* **97**, 7846 (1992).

83. H. L. Dai, C. L. Korpa, J. L. Kinsey, and R. W. Field, *J. Chem. Phys.* **82**, 1688 (1985).

84. Y. Chen, S. Halle, D. M. Jonas, J. L. Kinsey, and R. W. Field, *J. Opt. Soc. Am.* **B7**, 1805 (1990).

85. Y. S. Choi and C. B. Moore, *J. Chem. Phys.* **94**, 5414 (1991).

86. T. A. Holme and R. D. Levine, *Chem. Phys. Lett.* **150**, 393 (1988); T. A. Holme and R. D. Levine, *J. Chem. Phys.* **89**, 3379 (1988); T. A. Holme and R. D. Levine, *Chem. Phys.* **131**, 169 (1989).

87. See, e.g., B. C. Garrett and D. G. Truhlar, *J. Phys. Chem.* **83**, 2921 (1979), and references therein.

88. J. Troe, in *Tunneling, Proceedings of the Jerusalem Symposia on Quantum Chemistry and Biochemistry*, Vol. 19, Eds. J. Jortner and B. Pullman, (Reidel, Dordrecht, 1986).

89. T. K. Choudhury, Y. He, W. A. Sanders, and M. C. Lin, *J. Phys. Chem.* **94**, 2394 (1990).

90. P. J. Wantuck, R. C. Oldenborg, S. L. Baughcum, and K. R. Winn, *22nd Symposium (International) on Combustion* (The Combustion Institute, Pittsburgh, 1988), p. 973.

91. P. R. Bunker, *Molecular Symmetry and Spectroscopy* (Academic Press, New York, 1979); M. Quack, *Mol. Phys.* **34**, 477 (1977).

92. W. F. Polik, D. R. Guyer, W. H. Miller, and C. B. Moore, *J. Chem. Phys.* **92**, 3471 (1990); W. H. Miller, R. Hernandez, C. B. Moore, and W. F. Polik, *J. Chem. Phys.* **93**, 5657 (1990).

93. H. L. Dai, R. W. Field, and J. L. Kinsey, *J. Chem. Phys.* **82**, 1606 (1985).

94. R. D. Levine, *Adv. Chem. Phys.* **70**, 53 (1987); R. D. Levine, *Ber. Bunsenges. Phys. Chem.* **92**, 222 (1988).

95. A. R. Hoy, I. M. Mills, and G. Strey, *Mol. Phys.* **24**, 1265 (1972); S. Carter and N. Handy, *Mol. Phys.* **57**, 175 (1986); P. Jensen, *J. Mol. Spectrosc.* **133**, 438 (1989).

96. J. M. Bowman, J. S. Bittman, and L. D. Harding, *J. Chem. Phys.* **85**, 911 (1986); J. M. Bowman and B. Gazdy, *ibid.* **94**, 816 (1991).

97. S. Carter, N. C. Handy, and I. M. Mills, *Philos. Trans. R. Soc. London, Ser. A* **332**, 309 (1990); A. B. McCoy and E. L. Sibert III, *J. Chem. Phys.* **97**, 2938 (1992).

98. L. Halonen, T. Carrington Jr., and M. Quack, *J. Chem. Soc., Faraday Trans. 2*, **84**, 1371 (1988); M. Lewerenz and M. Quack, *J. Chem. Phys.* **88**, 5408 (1988); H. R. Dübal, T. K. Ha, M. Lewerenz, and M. Quack, *ibid.* **91**, 6698 (1989).

99. L. B. Harding, *J. Phys. Chem.* **93**, 8004 (1989).

100. S. J. Klippenstein and R. A. Marcus, *J. Phys. Chem.* **92**, 3105 (1988); S. J. Klippenstein and R. A. Marcus, *ibid.* **92**, 5412 (1988); S. J. Klippenstein and R. A. Marcus, *J. Chem. Phys.* **91**, 2280 (1989).

101. M. Quack and J. Troe, *Ber. Bunsenges. Phys. Chem.* **78**, 240 (1974).

102. J. Troe, *Chem. Phys. Lett.* **122**, 3485 (1985); A. I. Maergoiz, E. E. Nikitin, and J. Troe, *J. Chem. Phys.* **95**, 5117 (1991).

103. K. Kamiya and K. Morokuma, *J. Chem. Phys.* **94**, 7287 (1991).

104. J. M. Bowman, *J. Phys. Chem.* **95**, 4960 (1991); S.-W. Cho, A. F. Wagner, B. Gazdy, and J. M. Bowman, *J. Chem. Phys.* **96**, 2799 (1992).

105. A. D. Sappey and D. R. Crosley, *J. Chem. Phys.* **93**, 7601 (1990); G. W. Adamson, X. Zhao, and R. W. Field, *J. Mol. Spectrosc.* **160**, 11 (1993).

106. R. N. Dixon, *J. Chem. Soc., Faraday Trans.* **88**, 2575 (1992); S. Henning, V. Engel, R. Schinke, M. Nonella, and J. R. Huber, *J. Chem. Phys.* **87**, 3522 (1987).

CHAPTER 11

STATE-SPECIFIC INTRAMOLECULAR AND DISSOCIATION DYNAMICS OF HFCO

Young S. Choi

Department of Chemistry
Inha University
Incheon, Korea

C. Bradley Moore

Department of Chemistry
University of California
Berkeley, CA 94720, USA

Contents

1. Introduction

Unimolecular reactions occur when sufficient energy to surmount the barrier between reactant and products is accumulated in the reaction coordinate.

Since energy is initially deposited in vibrational modes other than the reaction coordinate in nearly all activation processes, intramolecular vibrational energy redistribution (IVR) must precede reaction. In statistical rate theories, such as the Rice–Ramsperger–Kassel–Marcus (RRKM)[1] theory, it is assumed that IVR is rapid compared to reaction. The reaction rate is thus controlled by the flux through the transition state. Although RRKM theory has been successful in fitting and extrapolating thermal unimolecular reaction rate data[2,3] and is found to be quantitatively correct for several small molecules,[4] examples of 'nonstatistical' behavior have been reported in unimolecular reactions of chemically activated molecules.[5,6] This raises questions about the universality of complete IVR prior to reaction and suggests the possibility of mode- or bond-selective chemistry.

The dynamics of unimolecular processes are best studied when the molecule can be prepared in single well-defined states with energies near the reaction threshold.[4] For reactions with barriers above the zero-point energy of the products, the deposition of energy in the products is controlled by the geometry and momenta at the transition state as well as by the shape of the repulsive exit valley of the potential surface. Thus, product quantum state and velocity resolution provide information about dynamics at the transition state. Thermal distributions of initial states or even ensembles of a modest number of initial states can give averages which approach statistical results much more closely than do the state-specific properties. However, the energy levels of polyatomic molecules at energies for which reaction occurs are often so dense that it is impossible to excite a single level. Rovibrational cooling of molecules in supersonic expansions and use of narrow bandwidth lasers can significantly alleviate the difficulties.

SEP is an ideal technique[7–9] in that it selects a single quantum state from a simple spectrum in the first step and delivers the molecule to the target level in the second step. The Franck–Condon factors for the electronic transitions of SEP permit a large range of excitation energies in modes which change frequency or geometry substantially between ground and excited electronic states. Vibrational overtones may also be excited by a sequence of vibrational transitions with the same state selectivity. The accessible range of states is usually not as large as in SEP, but the modes accesible are often different from those available via SEP. These double-resonance techniques not only make it possible to reach the energy region where reactions occur but simplify the spectra of excited molecules

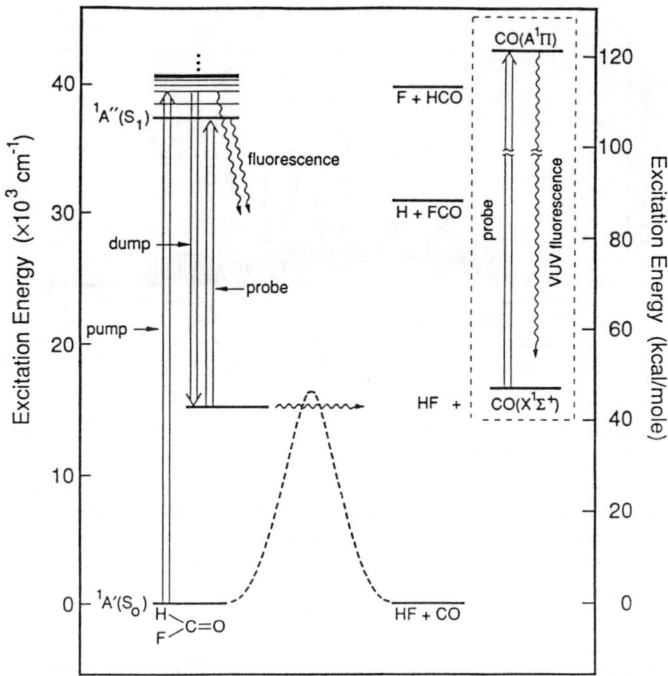

Fig. 1. Energy level diagram and experimental schematic for HFCO. The *ab initio* dissociation threshold for HFCO $\rightarrow$ HF + CO on the ground electronic potential energy surface is $\approx$ 16 500 cm^{-1}, and the origin of S$_1$ is 37 503 cm^{-1} above S$_0$. S$_0$ and S$_1$ states have planar and nonplanar structures, respectively. In SEP, HFCO molecules are excited to a single rovibrational level of S$_1$, and then dumped down to rovibrational levels of S$_0$. Depletion of S$_1$ fluorescence signals resonances for the dump laser. The number of excited S$_0$ molecules can be probed by reexciting to S$_1$ and detecting the consequent fluorescence. The CO dissociation fragment may be detected by laser-induced fluorescence in the vacuum uv.

to such an extent that each molecular eigenstate can be resolved in small molecules. In addition to the work on HFCO reviewed here, SEP has been successfully applied to spectroscopic studies of IVR and/or unimolecular reaction in C$_2$H$_2$, HCO, HCN, HCP, H$_2$CO, SO$_2$, CH$_3$O, etc.[8-9]

HFCO is an ideal molecule for the application of SEP to both IVR and unimolecular dissociation dynamics. A simple diagram of the energy levels and SEP scheme related to this work is shown in Fig. 1. The electronically excited HFCO(S$_1$) internally converts to highly excited vibrational levels of the ground electronic state, from which it dissociates mainly to H +

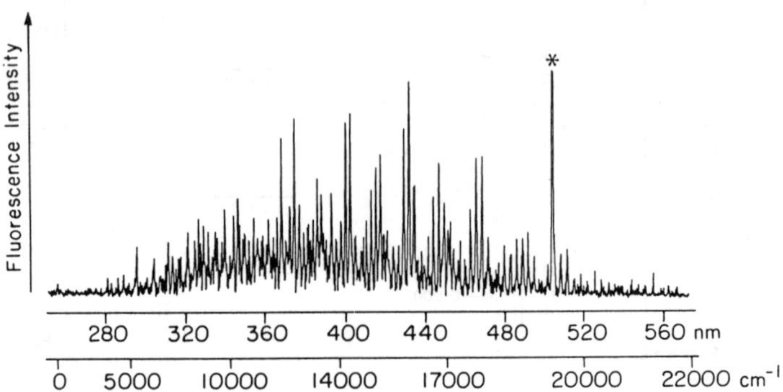

Fig. 2. Dispersed fluorescence spectra of S_1 HFCO for 6^2 excitation. The excitation frequency is fixed at $39\,194.0$ cm^{-1}, which corresponds to the overlapped transitions of $1_{11} \leftarrow 1_{01}$, $2_{12} \leftarrow 2_{02}$, and $3_{13} \leftarrow 3_{03}$ of the 6^2 band (rotational levels designated by $J_{K_a K_c}$). The wavelengths (nm) of emission lines and the energy (cm^{-1}) of vibrational levels of S_0 are given on the abscissa. The peak marked with an asterisk is scattered light from the excitation laser. The broad and red-shifted distribution of Franck–Condon factors, which makes this work possible, is easily noticed.

FCO (with a small amount of HF + CO).[10] The fluorescence quantum yield is known to be low (1% for the 6^2 level of S_1).[11] The downward transition is easily saturated with a nsec pulse of a few mJ.[12] The emission spectrum from S_1 to S_0 (Fig. 2) covers a very broad range; hence, S_0 vibrational energy levels from the vibrational fundamentals to combinations and overtones thousands of cm^{-1} above the dissociation barrier can be excited. SEP spectra of H_2CO^7 show that IVR is not complete at $15\,000$ cm^{-1} and suggest that the same may be true for HFCO; if so, IVR would be incomplete for HFCO at energies around the dissociation threshold. By using SEP as an excitation method the IVR and dissociation dynamics of HFCO are studied in a state-selective manner.

In this chapter recent SEP studies of IVR and dissociation dynamics of HFCO are reviewed. In Sec. 2 resolved spectra of the vibrational eigenstates of HFCO exhibit (a) level densities of about 10 per cm^{-1}, (b) IVR couplings which depend on the vibrational mode excited, and (c) quasistable extreme motion for levels excited with 15 to 23 quanta in the out-of-plane bending mode. These states are coupled anharmonically to a background of levels with excitation in other modes (Fig. 3). Section 3 includes the state-selectively measured dissociation rates and CO product J distributions which exhibit large variations with rotational quantum state. Rotational excitation is found to greatly speed up the reaction, presumably by Coriolis coupling of background states to states with extreme motion in the reaction coordinate. In Sec. 4 the current status of the work and possible future directions are discussed.

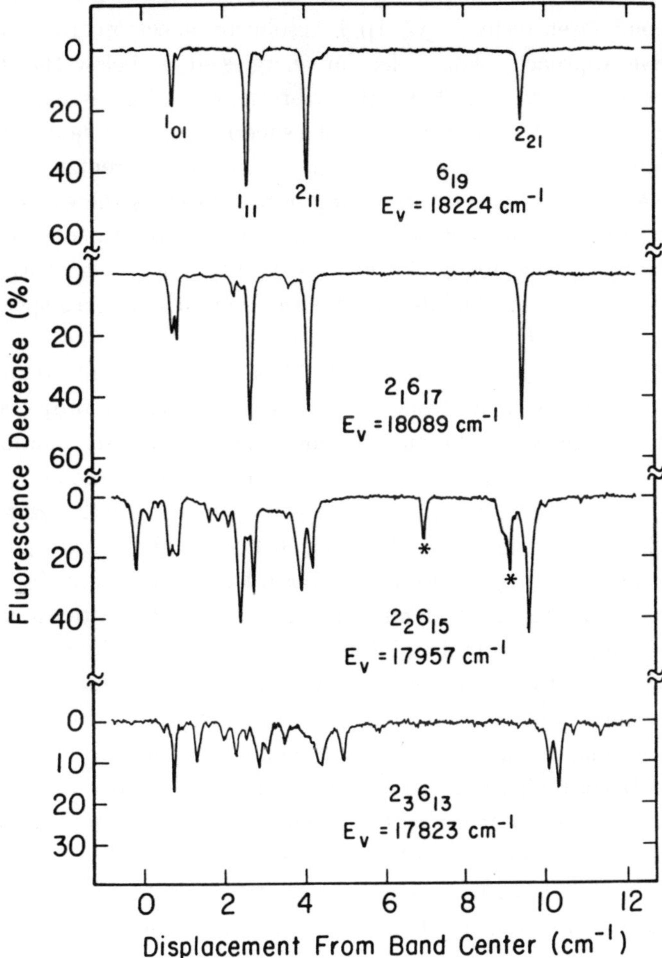

Fig. 3. State-dependent vibrational state mixing near $18\,000$ cm^{-1}. For the four vibrational states observed, the extent of state mixing decreases as the number of quanta in ν_6 increases. Most of the lines observed at this energy are broadened by rapid dissociation to HF + CO (reproduced from Ref. 12).

2. Mode-Specific IVR

IVR is one of the fastest and most complicated processes in polyatomic molecules.[13] Ideally, time-resolved experiments could be performed in which a single degree of freedom is excited in a short period of time and all other degrees of freedom are observed as they subsequently become excited. Such experiments are just beginning to be approachable with femtosecond and picosecond laser technology. High resolution spectroscopy provides an alternative approach. For molecular energies at or below the lowest threshold for dissociation, individual rovibrational molecular eigenstates may be resolved. The Fourier transform of a spectrum gives time-dependent dynamical information. For example, the width of a Lorentzian-shaped clump of lines (such as an OH stretching overtone) gives the exponential decay rate for the vibrational excitation responsible (transfer of one or more OH quanta to any other mode(s) in the molecule) for the oscillator strength of the observed spectrum. An intensity measurement of a line in a spectrum gives the projection of the wave function of the final level onto a single basis state determined by the initial level. Thus, in the HFCO SEP spectra presented here, the spectrum from a single S_1 level gives the projection of S_0 levels onto the zero-order, Franck–Condon-allowed S_0 basis function for each observed S_0 level. The width of a clump of lines gives the decay rate of that zero-order excitation if that clump were to be coherently excited by SEP through the same S_1 level. By doing SEP through several S_1 levels, several wave function projections may be observed. If these same S_0 states were accessed through CH stretching transitions, different intensities and projections would be measured and different clump widths observed. From such a spectroscopy, the relaxation time for a particular coherent excitation of CH stretching could be predicted. However, the SEP spectrum which projects out-of-plane bending and CO stretching excitations could never permit the CH stretching dynamics to be predicted. In order to calculate the full dynamics for an arbitrary excitation, the complete wave function of every molecular eigenstate must be known.

In practice, it is hard to develop one or two different high resolution spectroscopies for a group of molecular eigenstates, and unthinkable to measure the amplitude and phase of all significant basis states contributing to each of many eigenstates in even the simplest of highly excited poly-atomic molecules. Nevertheless, high and medium resolution spectroscopy does give a great deal of valuable information on anharmonic and Coriolis coupling strengths and on the extent to which coupling approaches the chaotic or rapid IVR limit required for statistical theories of unimolecular reactions. For the forseeable future, it will require judiciously selected combinations of spectroscopic, dynamical and theoretical studies to develop a predictive understanding of the basic phenomena of IVR.

Vibrational state mixing of highly excited rovibrational levels of HFCO in its ground electronic state is studied for vibrational energies up to $23\,000$ cm^{-1} with SEP spectroscopy in a highly state-selective manner. HFCO molecules cooled in a pulsed jet are excited to a single rovibrational level of S_1 and the SEP spectrum is obtained by measuring the depletion of total fluorescence as a function of the dump laser wavelength.[12]

Medium-resolution SEP spectra permit the zero-order vibrational quantum numbers to be assigned and term values to be determined for the observed bands.[12] As expected, from the fact that S_1 and S_0 differ greatly in C–O bond length and have nonplanar and planar structures, respectively, most of the peaks are assigned to the combined progressions of Franck–Condon active ν_2 (CO stretch) and ν_6 (out-of-plane bend). One notable feature of the low-resolution SEP spectra is that the linewidths of levels above the dissociation threshold are state-dependent, i.e., depend upon the rovibrational quantum numbers.[14] For a given total energy, levels with more ν_2 and fewer ν_6 quanta are usually broader.

There are two possible broadening mechanisms for the transition lines observed above threshold for dissociation, vibrational state mixing (IVR), and dissociation. In order to determine the dynamical process responsible for the state-specific broadening, higher resolution study is required. In molecules whose density of states, ρ, is so high that each molecular eigenstate cannot be resolved, it is impossible to distinguish these two broadening mechanisms by higher resolution spectroscopy alone. When dissociative line broadening, Γ, is less than the inverse level density, molecular eigenstates coupled with a zero-order state can be resolved giving information on the extent of vibrational state mixing and demonstrating IVR as the cause of state-dependent broadening in low resolution. In the statistical limit of strong vibrational coupling, transition state theory gives the dissociation rate constant as[1]

$$k = N(E, J)/h\rho(E, J) \,, \qquad (1)$$

where $N(E, J)$ is the number of energy levels for vibrations orthogonal to the reaction coordinate at the transition state and h is Planck's constant.

The product of line width and level density is thus[15]

$$\Gamma\rho = N(E, J)/2\pi \, , \tag{2}$$

and, consequently, spectral resolution of lines within a clump becomes impossible for energies above the first five or ten vibrational levels at the transition state. Of course, sharp spectra may be observed at higher energies if they correspond to states which are not strongly mixed.

High-resolution SEP spectra of HFCO in the range of $13\,000$ cm^{-1}–$23\,000$ cm^{-1} have been recorded at 0.05 cm^{-1} resolution.[12] For energies less than $18\,000$ cm^{-1}, broad features observed in low-resolution spectra were resolved into many peaks, each corresponding to a single molecular eigenstate. Polarization dependence studies show that all the lines around a single rovibronic Franck–Condon-allowed transition have the same rotational quantum numbers as the zero-order level. This result indicates that these fine structures come from the coupling (state mixing) of the zero-order states with dark states that have negligible Franck–Condon overlap with the intermediate level. The well-resolved level structures exhibit mixing with 1–10 levels per cm^{-1}, five times the $\rho(E)$ calculated for A$'$ or A$''$ vibrational levels by direct anharmonic count. The extent of mixing with background states (clump width) depends upon the zero-order vibrational state. This observation confirms that the IVR of highly vibrationally excited HFCO is nonstatistical.

Figure 3 shows the vibrational level dependence of mixing within a narrow energy range around $18\,000$ cm^{-1}. Mixing increases as energy is shifted from ν_6 into ν_2. For a given total energy, the vibrational states with the largest number of quanta in ν_6, the 6_{19} band in Fig. 3, for example, show the simplest spectra. These states are particularly stable against IVR(vibrational state mixing). Figure 4 demonstrates the effects of mode-specific vibrational state mixing in an even more striking way. The spectra exhibit increasing mixing with the first increase in out-of-plane bending quanta from $2_1 6_{13}$ to $2_1 6_{15}$, as expected from the increased importance of anharmonicity as vibrational energy increases. The spectra then sharpen as further ν_6 quanta are added even though the total vibrational energy increases. For example, the $2_1 6_{19}$ level at $19\,886$ cm^{-1} shows no mixing at all while the $2_1 6_{15}$ level at $16\,267$ cm^{-1} is mixed significantly, showing complicated structure for all four rotational levels. The extent of state mixing of all vibrational states studied by high-resolution SEP spectroscopy

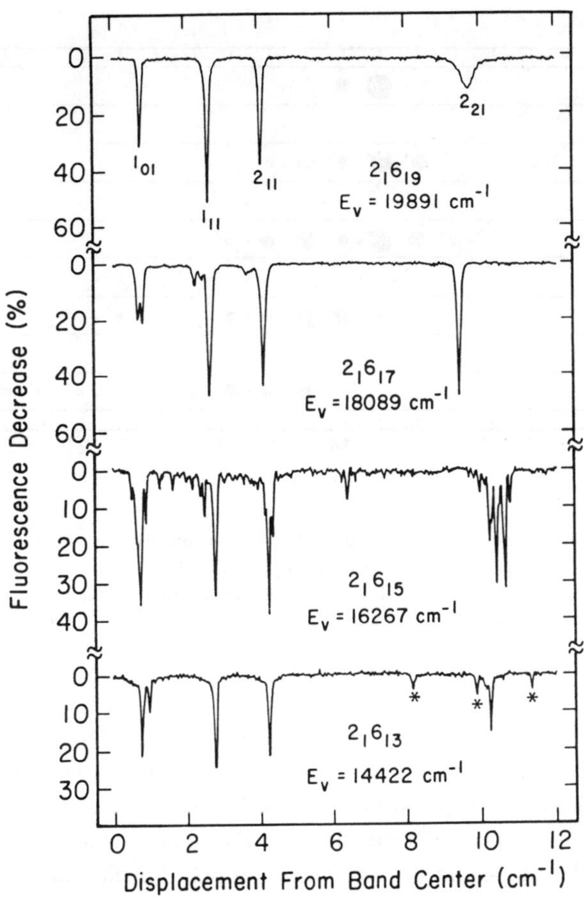

Fig. 4. ν_6 dependence of state mixing for A$''$ vibrational states. The addition of two quanta in ν_6 to the 2_16_{13} state increases mixing significantly. However, further excitation in ν_6 decreases the coupling and sharpens the spectra as energy and ν_6 increase. As a result, the 2_16_{19} state, which has the highest energy of all, is extremely narrow without any indication of mixing. The 2_16_{19} and 2_16_{17} states are homogeneously broadened by dissociation (reproduced from Ref. 12).

is summarized in Fig. 5. In this figure, it is easily noticeable that adding ν_2 quanta with ν_6 constant does not increase mixing. In summary, *IVR in the HFCO molecule excited at energies near and higher than the threshold for dissociation to HF + CO is mode-specific.* The existence of sharp, unmixed vibrational states at energies far above the dissociation threshold is striking.

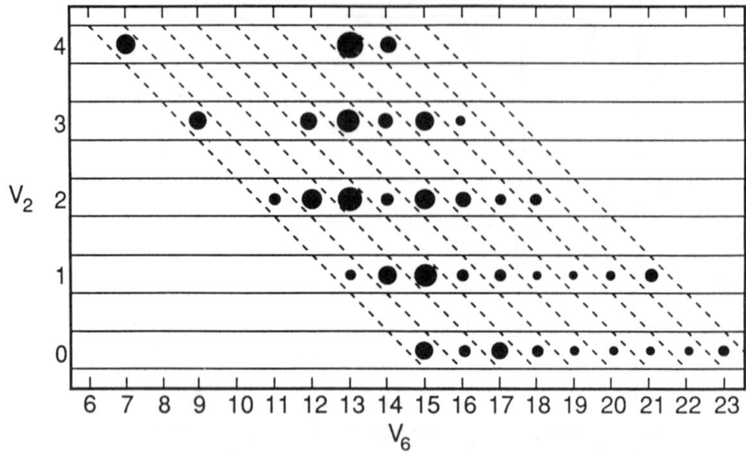

Fig. 5. The dependence of vibrational state mixing on v_2, v_6, and total energy determined by high-resolution SEP spectroscopy. The vibrational states between two dashed lines are nearly isoenergetic. The diameter of the filled circles is proportional to the extent of mixing, which is proportional to the width of a clump (reproduced from Ref. 12).

Recent theoretical studies[16–20] have suggested that a few regular states, states with well-defined periodic motion, may exist in high-energy regions where the dynamics are primarily chaotic. One class of these states is known as "extreme motion" states since nearly all of the energy is localized in a single mode. Hose and Taylor[19] have proposed a physical mechanism for the survival of extreme motion states against strong mixing at high vibrational energy and predicted the properties of such states. Sufficiently extreme motion in a mode induces dynamic potential barriers that localize the mode and decouple states with large amplitudes in a single mode from other zero-order states. Viewed from a quantum mechanical perspective, the coupling matrix elements of an extreme-motion state with other states nearby are very small since most of those nearby states have little or no excitation in the extreme-motion mode and coupling involves a large change of the extreme-motion mode quantum number and correspondingly the quantum numbers for other modes. Thus these extreme-motion states could be extremely stable against state mixing and coexist with a uniform mode-mixed quasicontinuum of states for highly excited polyatomic molecules.

Based upon the idea of adiabatic separability of fast and slow motions, Brummer and Shapiro[20] have also reached the conclusion that extreme motion states should be regular.

The experimental results show that the unmixed states in HFCO have extremely high excitation in the out-of-plane bend (ν_6) mode, and almost no excitation in other modes and that the states become more stable with increasing excitation in the extreme motion mode, ν_6. These features are those expected for the extreme-motion states suggested in theoretical studies.[19,20] It is concluded, therefore, that extreme motion in ν_6 is substantially decoupled from other vibrational modes and that the strength of coupling, or extent of IVR, varies systematically with vibrational quantum numbers.[12]

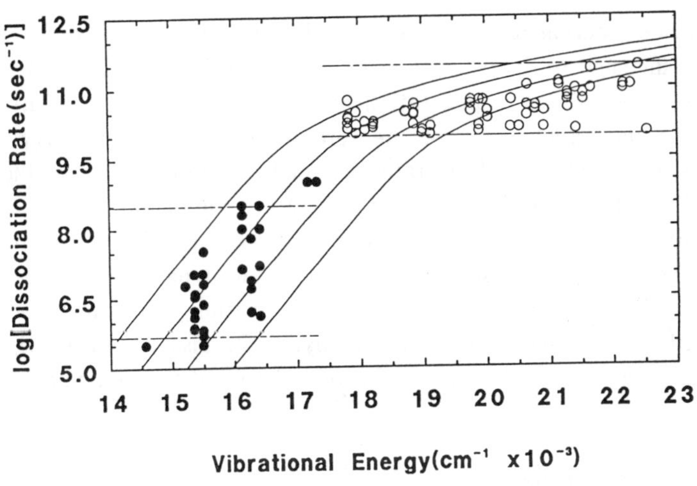

Fig. 6. Unimolecular dissociation rates of highly vibrationally excited HFCO (S_0). Filled and empty circles are the experimental values measured directly from linewidths, respectively. The dashed lines show the ranges within which accurate measurements were possible. The solid lines are RRKM rate constants calculated for four different barrier heights, 16 500, 17 200, 17 900, and 18 600 cm^{-1} using *ab initio* transition state parameters (reproduced from Ref. 21).

Although some states with very high excitation in ν_6 appear to be completely isolated from the background states, most of the observed vibrational states show some degree of mixing. There are two state mixing mechanisms at high vibrational energy — anharmonic resonance and Coriolis coupling. For states observed in this SEP spectroscopy, the extent of state mixing observed does not depend upon rotational levels (see Figs. 3 and 4); this suggests that anharmonic coupling is the dominant mechanism in the mixing of states, at least for $J \leq 2$. However, the apparent rotational constants[12] of many vibrational states indicate that they are mixed with relatively distant states through Coriolis couplings too weak to give extra lines in the SEP spectra but strong enough to shift the positions of the rotational levels of the interacting states (Fig. 6).[12] This Coriolis coupling is weak because the two interacting levels are far apart but may play an important role in slow molecular processes such as dissociation because these states can acquire a small but non-zero character of vibration along the reaction coordinate.

3. State-Specific Unimolecular Dissociation Dynamics

In the previous section it was shown that HFCO can be excited by SEP with high efficiency to dissociative levels which yield HF + CO and that the IVR of these states is nonstatistical or mode-specific. In addition, the coupling matrix elements of the zero-order states which carry the oscillator strength with the dark states are smaller than the rotational level spacing. These favorable properties of HFCO provide an excellent chance to study dissociation in a state-selective manner and to experimentally observe the state-specific dynamics. Two important dynamical observables in the unimolecular dissociation process, dissociation rates and product state distributions, are measured and analyzed for dissociation of highly vibrationally excited HFCO to HF + CO on the ground electronic potential surface.

3.1. *Dissociation Rates*

The HFCO molecules are excited to a dissociative level of S_0 by SEP. The decay of population of the level is monitored by measuring the fluorescence intensity from HFCO molecules returned to S_1 by the probe laser as a function of time delay between the dump and probe pulses.[21] Since the fluorescence intensity from the upper probe level is proportional to the number of molecules remaining in the dumped (dissociative) level, the fluorescence intensity versus delay represents the temporal evolution of the population of the dumped level under study. Because of the limit imposed by the ≈ 5 ns pulse duration of the lasers, this method is applied in the tunneling region where dissociation lifetimes are a few ns or longer. At higher vibrational energies dissociation occurs on a time scale shorter than the laser pulse duration. In this energy region dissociation rates are deduced from the linewidths of dissociative levels measured by high resolution SEP spectroscopy.[21,22]

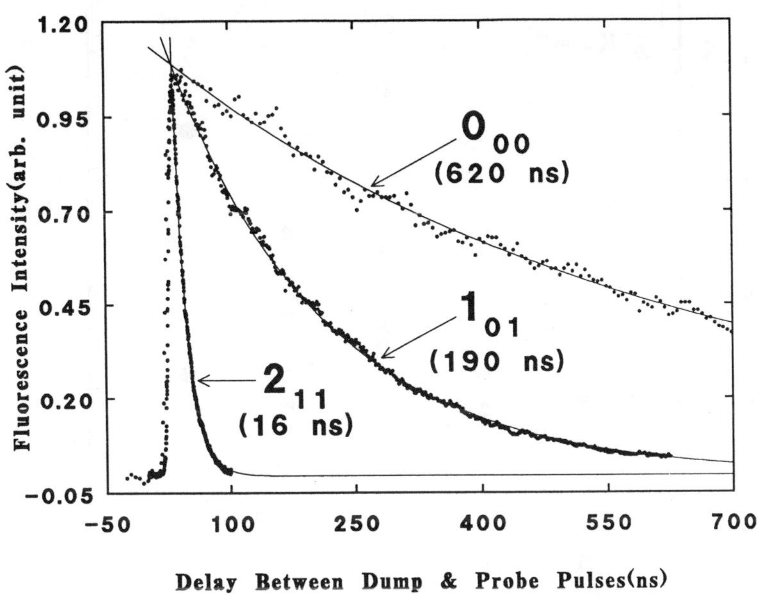

Fig. 7. Fluorescence decay curves for three rotational levels of the $2_1 6_{15}$ vibrational state (S_0). Note the drastic decrease of lifetime for a small rotational excitation. The solid lines are the least-square fitted decay curves with the τ values in parentheses under the rotational assignments (reproduced from Ref. 21).

All of the dissociation rates measured by the pump–dump–probe method are summarized in Fig. 6. The most striking feature in Fig. 6 is the fact that the dissociation rates of HFCO are not a simple function of total energy but depend strongly on the specific level and especially on rotation. Even a small rotational excitation increases the dissociation rate dramatically. An example of this strong rotational level dependence is shown in Fig. 7 for the 0_{00}, 1_{01}, and 2_{11} rotational levels of the $2_1 6_{15}$ vibrational state. Figure 8 shows the general features of the rotational level dependence of dissociation rates in the tunneling region. It is clear

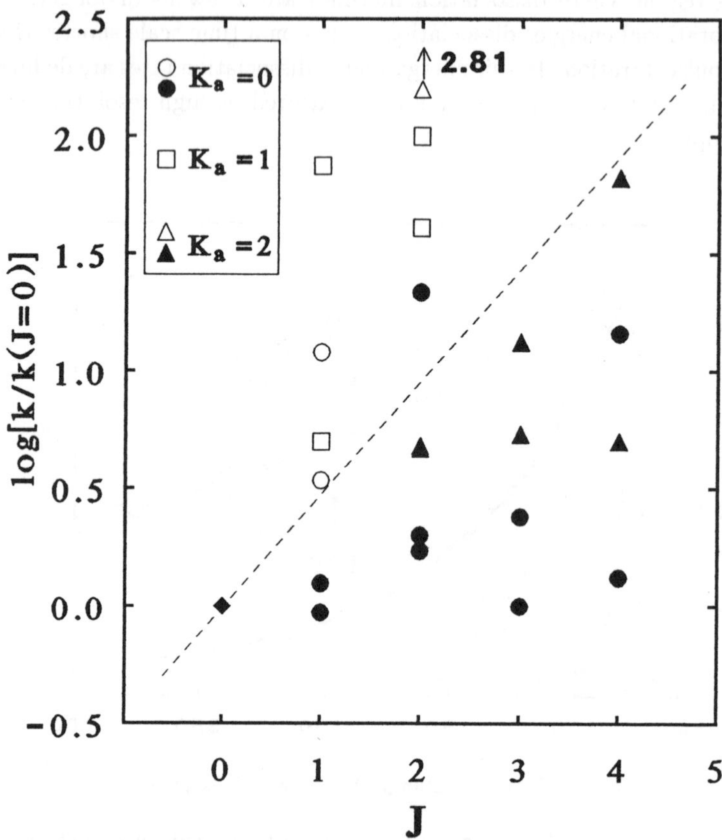

Fig. 8. Rotational level dependence of unimolecular dissociation rates of HFCO in the tunneling region. The dissociation rates of all rotational levels are normalized to that for the $J = 0$ level in the same vibrational state. The filled diamond represents the $J = 0$ reference point. The circles are for $K_a = 0$, the squares for $K_a = 1$, and the triangles for $K_a = 2$. The empty symbols represent the A'' symmetry vibrational states and the filled ones A'. Note the log base 10 scale of the vertical axis. The broken line is drawn arbitrarily to emphasize the different extent of dependence for A' and A'' symmetry states (reproduced from Ref. 21).

from Fig. 8 that the rotational level dependence is much stronger for the A'' symmetry states (ν_6 =odd) than for the A' states. Dissociation rates depend most strongly upon K_a of the three rotational quantum numbers, J, K_a, and K_c. In the tunneling region vibrational level dependence of dissociation rates is not discernable because of strong rotational dependence. At higher energies ($17\,500$ cm^{-1}–$22\,500$ cm^{-1}) where the dissociation rates are indirectly obtained from the linewidths, the effect of total energy and of rotational motion on dissociation rates is much weaker than in the tunneling region (Fig. 6), and the dependence of dissociation rates upon the vibrational states is more apparent. Figure 9 shows that dissociation rates generally increase with increasing ν_2 (decreasing ν_6) for states in a given energy range, although there are a few states which deviate from this general trend.

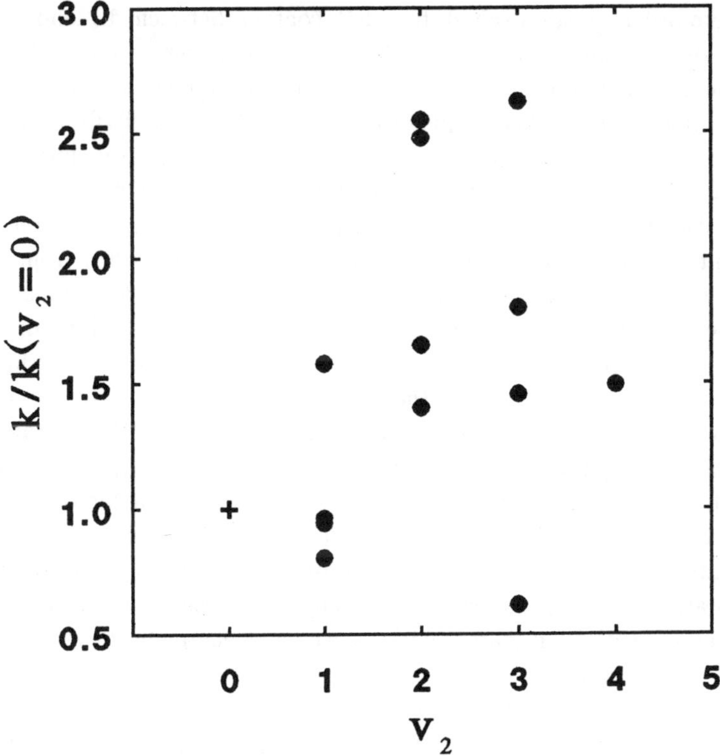

Fig. 9. Vibrational state dependence of dissociation rates of vibrational states above the dissociation threshold. The dissociation rate of each vibrational state is averaged over the observed rotational levels and divided by the average rate of the $v_2 = 0$ state in the polyad for which $2v_2 + v_6 = $ constant. The cross is the reference point for $v_2 = 0$ states (reproduced from Ref. 21).

In the tunneling region, observed rates increase by as much as 2.5 orders of magnitude as rotational quantum numbers increase up to $J = 4$ and $K_a = 2$. According to RRKM theory such a change would require increasing the total energy available or lowering the barrier by some 1,500 cm^{-1}. Above the threshold, the range of rates is only about half an order of magnitude but corresponds to roughly the same change in available energy since the slope of the RRKM curve decreases greatly at the threshold. Rotation about the a-axis has the largest effect on rate constants and the effect is much larger for A'' than A' symmetry levels. These observations are clearly not consistent with the strong coupling of vibrational coordinates presumed by statistical theories for unimolecular reactions.[2,3]

The zero-order states excited by SEP contain between 12 and 22 quanta of out-of-plane bending excitation, an excitation orthogonal to the in-plane reaction coordinate.[23,24] These states have been shown to be extreme-motion states weakly coupled to a dense background of vibrational levels. The coupling is enhanced by shifting energy from ν_6 to ν_2 but the increase of dissociation rates due to the shift of energy is not nearly as strong as the observed rotational level dependence. Thus the observed increases in rate with rotational quantum number *do not* result from a rotationally enhanced *coupling to the background levels.* Where clumps of molecular eigenstates are observed each has a wave function which is some 10–20% from the zero-order state. If all levels of excitation in the reaction coordinate were strongly coupled to the background states, the latter would exhibit dissociation rates statistically distributed about the RRKM rate. The excessive contribution to the wave function from the high-ν_6, zero-order state would decrease the rate by 10–20%, or conceivably a factor of 2 compared to RRKM, and only this small factor could be eliminated by rotation-induced vibrational mixing. The orders of magnitude increase in rate as rotational energy increases from zero must result from strongly increased mixing with basis states which have enough energy in the reaction coordinate to dissociate.

The vibrational states of HFCO with an even number of ν_6 quanta belong to the A' symmetry representation and those with an odd number belong to A''. In the absence of rotation, these are rigorous symmetries of the molecule. Hence, an A'' state with $J = 0$ must have at least one quantum of energy tied up in the out-of-plane bending mode. Since the reaction coordinate for dissociation of HFCO has A' symmetry, this energy is not available to the reaction coordinate. The A'' states cannot dissociate through the zero-point vibrational level of the transition state but only through the $v_6^\dagger = $ odd levels. Thus, the A'' states must experience a higher barrier height than the A' states by $\nu_6^\dagger$ and dissociate correspondingly less rapidly than the A' states of comparable energy. The $J = 0$ levels of the 6_{16} and 2_16_{14} (A') states near 15 400 cm^{-1} and of the 6_{17} and 2_16_{15} (A'') states near 16 350 cm^{-1} show the symmetry specificity most clearly. The 6_{17} and 2_16_{15} (A'') states dissociate only two or three times faster than the 6_{16} and 2_16_{14} (A') states. The expected ratio of rates is about 30 by RRKM theory. This suggests that only 200 cm^{-1} of the added 900 cm^{-1} of energy is available to the reaction coordinate. The difference of 700 cm^{-1} is within experimental scatter of the *ab initio* value of 805 cm^{-1} for the out-of-plane bend of the transition state.[23] The dissociation rates for $J = 0$ of the next *higher* A' levels, 6_{18} and 2_16_{16}, are at least 300 times larger than for the 6_{17} and 2_16_{15} levels, 15 times *larger* than from RRKM. These discrepancies between the expected and observed values are strong evidence for symmetry specificity. This symmetry specificity[25] is destroyed by Coriolis coupling of out-of-plane to in-plane vibration. This is observed as a larger rotational enhancement of dissociation rates for A'' than for A' symmetry levels. The effect was first predicted by Miller[25] for formaldehyde but not exhibited by the data for that molecule,[15] all of which were for $J > 0$.

Since the high-resolution SEP spectra of highly vibrationally excited HFCO exhibit mode specificity in the anharmonic coupling of the zero-order states to the background (dark) states, mode specificity in the dissociation rates is naturally anticipated. Of the vibrational states in an energy region, states with more energy in ν_2 (less energy in ν_6) usually decay more rapidly. Such states are also more strongly coupled to the background states as revealed by their more extensive clump structures. Kamiya and Morokuma[24] noted in their *ab initio* study that many forms of intramolecular coupling vanish by symmetry for the lone out-of-plane (ν_6) mode. Of course anharmonic terms in the potential and kinetic energy which contain even powers of the out-of-plane bending coordinate do mix in- and out-of-plane vibrations. Thus at a given total energy, states with larger ν_6 quantum numbers would have smaller rate constants for dissociation, as observed in this work. Therefore, the vibrational mode specificity in dissociation rates of HFCO seems to be the result of decoupling of ν_6 from other modes including the reaction coordinate.

3.2. *Rotational Distribution and Doppler Widths of CO Fragments*

HFCO molecules cooled in a pulsed jet are excited to dissociative levels by SEP and CO fragments are probed by laser-induced fluorescence(LIF) employing the $A^1\Pi \leftrightarrow X^1\Sigma^+$ transition of CO. The needed vacuum UV (VUV) wavelengths are produced in a Xe gas cell by frequency tripling (third-harmonic generation)[26] of fundamental output from a dye laser pumped with a Q-switched Nd:YAG laser. For Doppler width measurements, the bandwidth of the probe dye laser was narrowed to 0.04 cm^{-1} with an intracavity etalon in the oscillator. The experimental details are described elsewhere.[22,27]

CO fragments are observed to be rotationally hot and distributed over $J \leq 15$ up to 63, (Fig. 10). The overall distributions of CO fragments are strongly inverted and asymmetric, rising slowly between $J = 30$ and 45 and dropping steeply for $J \geq 55$. It is interesting to notice in Fig. 10 that the CO rotational distributions for the 0_{00} and 2_{20} levels of the $2_1 6_{18}$ vibrational state are significantly different. Even though the peak values of J are close, the distribution for the 2_{20} level is much broader. Its FWHM (full-width-at-half-maximum) of ≈ 30 compares to ≈ 20 for 0_{00}, and there is more population in J levels between 20 and 40 and higher than 55 levels for 2_{20} than for the 0_{00} level.

It is surprising to note that the CO J distributions are very different for the 0_{00} and 2_{20} rotational levels of the $2_1 6_{18}$ state (Fig. 10). For molecules with a barrier to dissociation, the product state distribution is dominated by the shape of the repulsive exit channel potential. The angular momentum of the parent molecule may add to or subtract from the angular momentum imparted to the CO by the repulsive forces. This systematically increases the width of the distribution by only one or two units of J.[28,29] However, this effect does not change the overall shape or the peak position of the CO rotational distribution and cannot be responsible for the significantly different shapes of the CO rotational distributions for the change of only 2 units of J between the 0_{00} and 2_{20} levels of the $2_1 6_{18}$ vibrational state.

The high-resolution SEP spectrum of HFCO in Sec. 2 shows that the $2_1 6_{18}$ vibrational state like other states with extreme excitation in ν_6 is barely mixed with the background states and thus the state forms a nearly pure molecular eigenstate by itself. Since the energy flow rate from this 'isolated' ν_6 mode to other modes is quite slow, while energy in other modes

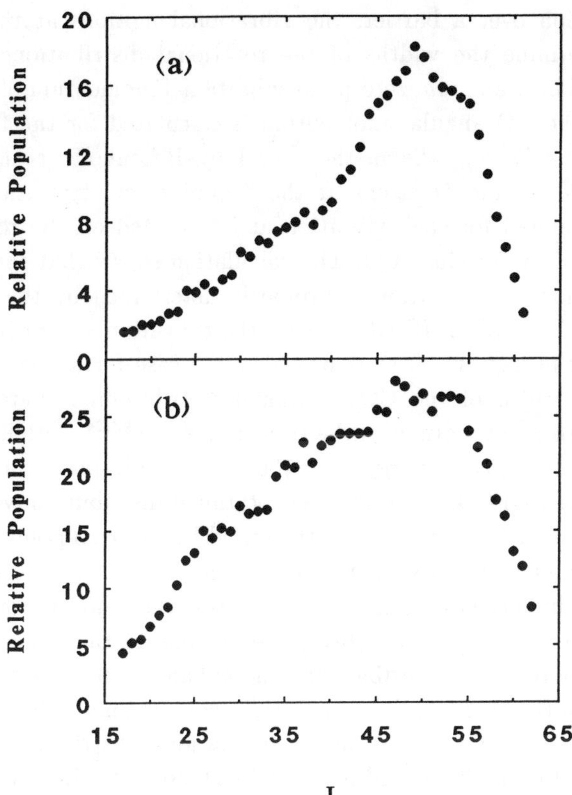

Fig. 10. The CO rotational distributions for the (a) 0_{00} and (b) 2_{20} rotational levels of $2_1 6_{18}$ state. The distributions are significantly different although the two levels differ in total energy by only 9 cm^{-1}, (Ref. 27).

redistributes much more rapidly, a non-rotating HFCO molecule excited to a non-mixed state dissociates with almost no delay once energy comparable to the barrier height is released from the localized mode, and thus the excess energy over the threshold is left mostly in the ν_6 mode when the molecule reaches the transition state. When the molecule is rotationally excited, however, Coriolis coupling provides an additional IVR path enhancing the energy flow out of ν_6, and thus causes the observed increase in dissociation rate. This increased IVR rate also causes the residual energy at the transition state to be distributed over more vibrational modes, all of which but ν_6 are entirely in-plane motions.

For dissociation over a barrier, the vibrational motions at the transition state determine the widths of the rotational distributions of the fragments.[28-30] The maximum zero-point vibrational momentum of HFCO transferable to the CO angular momentum is calculated for the five real frequency modes following the method used by Butenhoff *et al.*[30] The momentum of the C and O atoms at the "equilibrium" transition-state geometry is calculated for each vibration and projected on the rotations and translation of the product CO. The calculation shows that the width of the CO rotational distribution is primarily determined by the atomic motions of ν_3 and ν_4. When HFCO reaches the transition state with most of its vibrational energy in ν_6, as is probably the case for a non-rotating $2_1 6_{18}$ state, the width of the CO rotational distribution is determined mainly by the zero-point motions of the transition state.[23,24] Coriolis forces may redistribute the excess energy of ≈ 1800 cm^{-1} to in-plane modes for rotationally excited HFCO. If so, the CO rotational distribution would be significantly broadened beyond the width caused by the zero-point motion since mostly ν_3 and ν_4 will be excited due to their low frequencies. This argument rationalizes the experimental observation that the CO rotational distribution for the 2_{20} level of $2_1 6_{18}$ state is broader than that of the 0_{00} level. Therefore, it is plausible that the enhancement of vibrational energy flow out of the "localized" ν_6 by Coriolis coupling not only increases the dissociation rate but also broadens the rotational distribution of CO product for the rotationally excited parent HFCO molecule. However, until a statistically more significant set of data is available fluctuations in the overlap of molecular eigenstate wave functions with different transition state level wave functions[31] must also be considered.

Figure 11 shows the Doppler profiles of Q(29) lines of CO ($v = 0$) fragments formed from the 2_{02} and 2_{20} levels of the $2_1 6_{18}$ vibrational state. Strong dependence of the shape of the CO Doppler profiles upon the rotational levels of the HFCO molecule is quite striking. Similar dependences have been observed previously in the dissociation of H_2CO and successfully interpreted in terms of spatial anisotropy of the rotational wave functions of the parent molecule and the dynamics of dissociation.[32] This effect is more pronounced for relatively low rotational levels, as are studied in this work, than for higher rotational levels. An extensive study of the various vector correlations and Doppler shifts[33] involved in HFCO dissociation should reveal a great deal more about HFCO fragmentation dynamics.

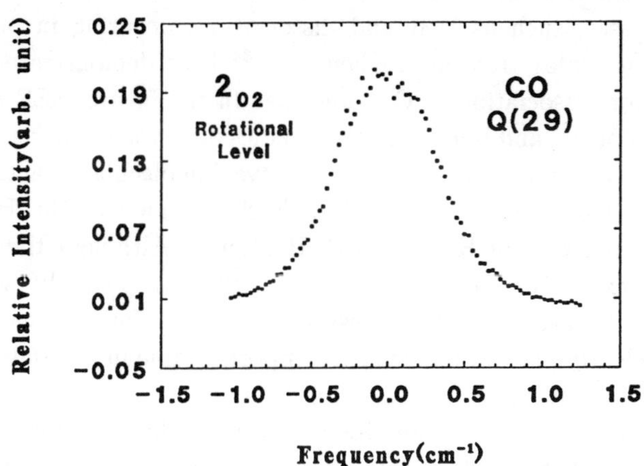

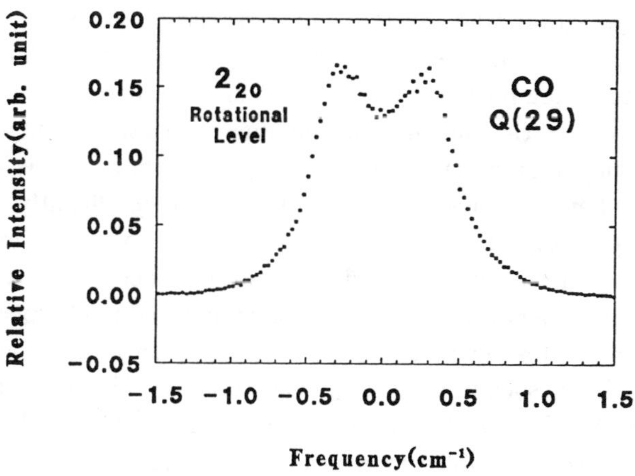

Fig. 11. CO Doppler profiles of Q(29) of CO fragments formed from the 2_{02} and 2_{20} levels of $2_1 6_{18}$ state. Two rotational levels of HFCO molecules show significantly different Doppler profiles (Ref. 27).

4. Summary

This work eloquently demonstrates the power of SEP as a tool to investigate the detailed dynamics of unimolecular processes in a state-selective manner. It is easily imaginable that the use of short-pulse lasers will extend the region of application of SEP to direct time-resolved studies of ultrafast molecular processes, such as IVR and dissociations occuring in pico- or femtosecond time scales. Recently, Chen *et al.*[34] have demonstrated such an application for dissociation of O_3 molecule. In the future, SEP should also be useful for state- and bond-selective studies of *bimolecular* reactions. Crim *et al.* have demonstrated bond-selective bimolecular reaction by exciting the third overtone of the H–O stretching mode of the H–O–D molecule in the reaction of H–O–D with H atom.[35] Although they used direct overtone excitation in that experiment, SEP may equally well be employed for molecules with the necessary properties for application of SEP. SEP will clearly play an increasingly important role in expanding our understanding of many aspects of reaction dynamics.

In this chapter, high-resolution SEP spectra of highly vibrationally excited HFCO reveal that IVR in this molecule at energies near and above the threshold for dissociation to HF + CO is mode-specific, and that states with extremely high excitation in the out-of-plane bending mode are well isolated from the background levels. The dissociation rates for HFCO in extreme motion levels are seen to increase by orders of magnitude as rotational quantum numbers are increased from $J = 0$ by a few units. Rates also increase as the out-of-plane bending excitation is made less extreme by moving quanta into CO stretching, but the effect is only a factor of 2 or 3 when the total energy is not changed significantly. These observations suggest that Coriolis forces are required to couple extreme motion in the reaction coordinate (i.e., dissociation) to background states or to out-of-plane extreme motion states. Symmetry-induced mode specificity, in which A'', $J = 0$ states must carry a quantum of ν_6 across the transition state, decreases the dissociation rates of such states by about a factor of 20 compared to A', $J = 0$ states. Two rotational levels of a localized vibrational state give very different rotational distributions and Doppler profiles of CO fragments, illustrating again the roll of parent rotational motion in dissociation of HFCO. In summary the IVR, dissociation rates, product state distributions, and Doppler profiles of dissociation fragments are strongly state-specific in HFCO and the state specificity originates from the decoupling of extremely highly excited out-of-plane bending states from other vibrational modes including the reaction coordinate.

These results describe a qualitatively new dynamics for intramolecular vibrational coupling and dissociation presented in Fig. 12. The isolation of extreme motion in the out-of-plane bending and reaction coordinates leads to highly nonstatistical behavior. While the observed reaction rates are clearly "non-RRKM", it is perhaps most remarkable that the deviations from statistical behavior occur only for the very lowest rotational states. It will be of great interest to establish the conditions under which extreme motion decouples one vibration from all of the others. At the energetic threshold for a reaction, all of the energy above the zero point must be in the reaction coordinate for reaction to occur. Under what circumstances will this extreme motion be weakly coupled to other molecular vibrations and the rate decreased below the statistical value to an extent determined by the weakness of this vibrational coupling?

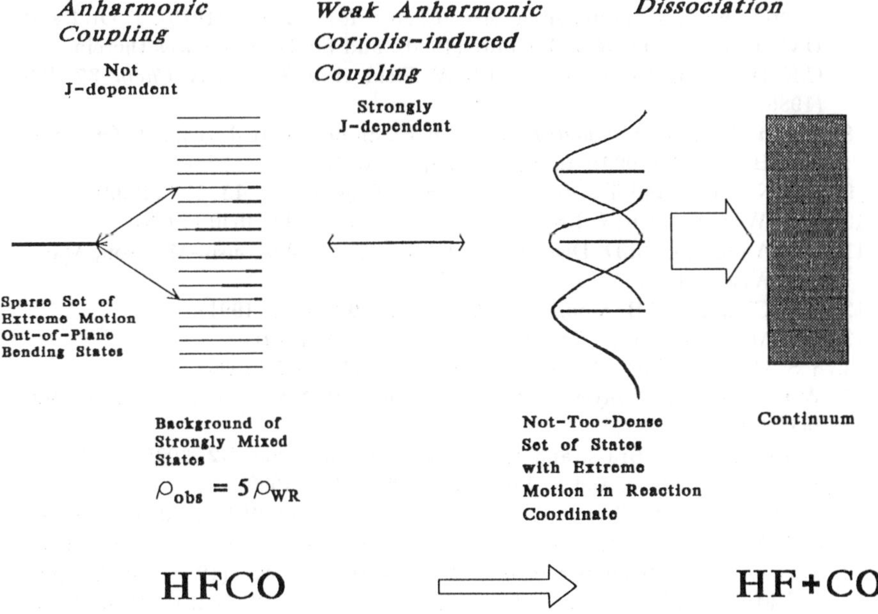

$$\rho_{obs} = 5\,\rho_{WR}$$

Fig. 12. Vibrational coupling scheme for IVR and dissociation of HFCO (reproduced from Ref. 21).

Acknowledgments

This work was supported by U.S. National Science Foundation Grant No. CHE93-16640. Y.S.C. gratefully acknowledges support by the Inha University Research Fund and the Center for Molecular Science at Korea Advanced Institute of Science and Technology.

References

1. R.A. Marcus, *J. Chem. Phys.* **20**, 359 (1952).
2. P.J. Robinson and K.A. Holbrook, *Unimolecular Reactions* (Wiley-Interscience, London, 1972); W. Forst, *Theory of Unimolecular Reactions* (Academic Press, New York, 1973).
3. R.G. Gilbert and S.C. Smith, *Theory of Unimolecular and Recombination Reactions* (Blackwell Scientific Publications, Oxford, 1990).
4. W.H. Green, Jr., C.B. Moore, and W.F. Polik, *Ann. Rev. Phys. Chem.* **43**, 591 (1992).
5. J.D. Rynbrandt and B.S. Rabinovitch, *J. Chem. Phys.* **54**, 2275 (1971); J.D. Rynbrandt and B.S. Rabinovitch, *J. Phys. Chem.* **75**, 2164 (1971).
6. I. Oref and B.S. Rabinovitch, *Acc. Chem. Res.* **12**, 166 (1979); I. Oref and D.C. Tardy, *Chem. Rev.* **90**, 1407 (1990), and further references therein.
7. C.E. Hamilton, J.L. Kinsey, and R.W. Field, *Ann. Rev. Phys. Chem.* **37**, 493 (1986).
8. *Special Issue on Stimulated Emission Pumping*, ed. H.-L. Dai, *J. Opt. Soc. Am.* **B7**, No. 9 (1990); and many chapters of this present book.
9. F.J. Northrup and T.J. Sears, *Ann. Rev. Phys. Chem.* **43**, 127 (1992).
10. B.R. Weiner and R.N. Rosenfeld, *J. Phys. Chem.* **92**, 4640 (1988).
11. S.T. Amimoto, Ph.D. Dissertation (University of Wisconsin-Madison, Madison, Wisc., 1979).
12. Y.S. Choi and C.B. Moore, *J. Chem. Phys.* **94**, 5414 (1991).
13. J.D. McDonald, *Ann. Rev. Phys. Chem.* **30**, 29 (1979).
14. Y.S. Choi and C.B. Moore, *J. Chem. Phys.* **90**, 3875 (1989).
15. W.F. Polik, D.R. Guyer, W.H. Miller, and C.B. Moore, *J. Chem. Phys.* **92**, 3471 (1990).
16. J. Tennyson, O. Brass, and E. Pollak, *J. Chem. Phys.* **92**, 3005 (1990); O. Brass, J. Tennyson, and E. Pollak, *ibid.* **92**, 3377 (1990).
17. G. Hose and H.S. Taylor, *J. Chem. Phys.* **76**, 5356 (1982); R.M. Hedges, Jr. and W.P. Reinhardt, *Chem. Phys. Lett.* **91**, 241 (1982); M. Shapiro, R.D. Taylor, and P. Brummer, *Chem. Phys. Lett.* **106**, 325 (1984); M. Shapiro and M.S. Child, *J. Chem. Phys.* **76**, 6176 (1982); Y.Y. Bai, G. Hose, C.W. McCurdy, and H.S. Taylor, *Chem. Phys. Lett.* **99**, 342 (1983); K.N. Swamy, W.L. Hase, B.C. Garrett, C.W. McCurdy, and J.F. McNutt, *J. Phys. Chem.* **90**, 3517 (1986).

18. R.J. Wolf and W.L. Hase, *J. Chem. Phys.* **73**, 3779 (1980); W.L. Hase, R.J. Duchovic, K.N. Swamy, and R.J. Wolf, *ibid.* **80**, 714 (1984); R.B. Shirts and W.P. Reinhardt, *J. Chem. Phys.* **77**, 5204 (1982).

19. G. Hose and H.S. Taylor, *Chem. Phys.* **84**, 375 (1984).

20. P. Brumer and M. Shapiro, *Adv. Chem. Phys.* **70**, 365 (1988).

21. Y.S. Choi and C.B. Moore, *J. Chem. Phys.* **97**, 1010 (1992).

22. Y.S. Choi, P. Teal, and C.B. Moore, *J. Opt. Soc. Am.* **B7**, 1829 (1990).

23. J.D. Goddard and H.F. Schaefer III, *J. Chem. Phys.* **93**, 4907 (1990).

24. K. Kamiya and K. Morokuma, *J. Chem. Phys.* **94**, 7287 (1991).

25. W.H. Miller, *J. Am. Chem. Soc.* **105**, 216 (1983).

26. R. Hilbig and R. Wallenstein, *IEEE J. Quantum Electron.* **QE-17**, 1566 (1981).

27. Y.S. Choi and C.B. Moore, in preparation; Y.S. Choi, Ph.D. Thesis, Chemistry, University of California, Berkeley (1991).

28. I-C. Chen and C.B. Moore, *J. Phys. Chem.* **94**, 269 (1990).

29. H.B. Levene and J.J. Valentini, *J. Chem. Phys.* **87**, 2594 (1987).

30. T.J. Butenhoff, K.L. Carleton, and C.B. Moore, *J. Chem. Phys.* **92**, 377 (1990).

31. R.D. van Zee, C.D. Pibel, T.J. Butenhoff, and C.B. Moore, *J. Chem. Phys.* **97**, 3235 (1992).

32. K.L. Carleton, T.J. Butenhoff, and C.B. Moore, *J. Chem. Phys.* **93**, 3907 (1990).

33. P.L. Houston, *J. Phys. Chem.* **91**, 5388 (1987); G.E. Hall and P.L. Houston, *Ann. Rev. Phys. Chem.* **40**, 375 (1989).

34. Y. Chen, L. Hunziker, P. Ludowise, and M. Morgen, *J. Chem. Phys.* **97**, 2149 (1992).

35. A. Sinha, M.C. Hsiao, and F.F. Crim, *J. Chem. Phys.* **92**, 6333 (1990).

CHAPTER 12

ALL-OPTICAL TRIPLE RESONANCE: SPECTROSCOPY AND STATE-SELECTED PHOTODISSOCIATION DYNAMICS

A. Marjatta Lyyra

Physics Department
Temple University
Philadelphia, PA 19122, USA

and

Paul D. Kleiber

Department of Physics and Astronomy
University of Iowa, Iowa City, IA 52242, USA

and

W. C. Stwalley

Department of Physics
University of Connecticut, Storrs, CT 06269, USA

Contents

459

1. Introduction

1.1. *Selection Rule Constraints*

Spectroscopic studies of molecular structure and dynamics, both intra-
and intermolecular (e.g., state-selected photodissociation, predissociation,
associative ionization and energy transfer), are severely constrained by the
selection rules for single photon transitions. Not only are the symmetries
of the accessible electronic states limited by electronic selection rules, but
even more significantly, the range of accessible vibrational (and continuum
translational) states and the corresponding internuclear distances sampled
are limited (Franck–Condon Principle). The use of multiple wavelength
laser photons greatly reduces these constraints.

We have demonstrated these advantages in studies of the homonuclear
alkali metal diatomic molecules (e.g., Na_2[1,2] and K_2[3–5]), but the advantages
are still significant in less symmetric molecules.

For a homonuclear diatomic molecule characterized by a term symbol
$^{2S+1}\Lambda_{\Omega\binom{u}{g}}^{(\pm)}$, the well known electronic selection rules are

$$\Delta\Lambda = 0, \pm 1$$

$$\Delta S = 0$$

$$\Delta\Omega = 0, \pm 1 = \Delta\Lambda$$

$$\Sigma^+ \leftrightarrow \Sigma^+, \qquad \Sigma^- \leftrightarrow \Sigma^-, \qquad \Sigma^+ \nleftrightarrow \Sigma^-$$

$$g \leftrightarrow u, \qquad g \nleftrightarrow g, \qquad u \nleftrightarrow u \,.$$

For vibration (and continuum translation), the Franck–Condon requirement
for significant transition probability is

$$|\langle v''|v'\rangle| > 0 \,.$$

For rotation, the selection rule is $\Delta J = 0, \pm 1$ per photon.

In Table I we illustrate the accessible electronic state symmetries using one to five photons starting in a totally symmetric ground state $(X^1\Sigma_g^+)$.

Table I. Singlet states accessible by multiple resonance from an $X^1\Sigma_g^+$ ground state.

ground state	one photon	two photons	three photons	four photons	five photons
					^1H_u
				$^1\Gamma_g$	$^1\Gamma_u$
			$^1\Phi_u$	$^1\Phi_g$	$^1\Phi_u$
		$^1\Delta_g$	$^1\Delta_u$	$^1\Delta_g$	$^1\Delta_u$
	$^1\Pi_u$	$^1\Pi_g$	$^1\Pi_u$	$^1\Pi_g$	$^1\Pi_u$
$X^1\Sigma_g^+$	$^1\Sigma_u^+$	$^1\Sigma_g^+$	$^1\Sigma_u^+$	$^1\Sigma_g^+$	$^1\Sigma_u^+$
		$^1\Sigma_g^-$	$^1\Sigma_u^-$	$^1\Sigma_g^-$	$^1\Sigma_u^-$

Table II. Triplet states accessible by perturbation-facilitated multiple resonance from an $X^1\Sigma_g^+$ ground state through $A^1\Sigma_g^+ \sim b^3\Pi_u$ "window" levels.

ground state	one photon ("window")	two photons	three photons	four photons	five photons
					^3H_u
				$^3\Gamma_g$	$^3\Gamma_u$
			$^3\Phi_u$	$^3\Phi_g$	$^3\Phi_u$
		$^3\Delta_g$	$^3\Delta_u$	$^3\Delta_g$	$^3\Delta_u$
$X^1\Sigma_g^+$	$A^1\Sigma_g^+ \sim b^3\Pi_u$	$^3\Pi_g$	$^3\Pi_u$	$^3\Pi_g$	$^3\Pi_u$
		$^3\Sigma_g^+$	$^3\Sigma_u^+$	$^3\Sigma_g^+$	$^3\Sigma_u^+$
		$^3\Sigma_g^-$	$^3\Sigma_u^-$	$^3\Sigma_g^-$	$^3\Sigma_u^-$

Using multiple tunable visible and IR photons with energies in the range $11\,000$–$25\,000$ cm^{-1}, virtually any important singlet state of Na$_2$ (IP = $39\,478.7$ cm^{-1}) and K$_2$ (IP = $32\,775.5$ cm^{-1}) can be accessed. If one takes advantage of perturbed levels of singlet–triplet mixed spin symmetry (e.g., the $A^1\Sigma_u^+ \sim b^3\Pi_u$ mixed levels in Na$_2$ and K$_2$), one can access even more states (e.g., triplets), as shown in Table II. Tables of electronic states correlating with asymptotes up to $6p$ in sodium[6] and $7d$ in potassium[7] have recently been presented.

Diatomic molecular electronic spectroscopy, based on single photon absorption, continues to emphasize observation of new and more highly excited electronic states. However, most information is restricted to the region near the equilibrium internuclear distance of the ground electronic state, corresponding to the Franck–Condon region for single photon absorption (Fig. 1). With the exception of a few states (e.g., the $A^1\Sigma_u^+$ state determined by modulated gain spectroscopy[8]), most of the states in Fig. 1 are unobserved at larger internuclear distances. In Fig. 2, we present recent data for the gerade states near the $3s+4s$, $3s+3d$, and $3s+4p$ limits, showing five states characterized in regions outside 10 Å ($3^1\Sigma_g^+$, $1^3\Delta_g$, $4^1\Sigma_g^+$, $2^1\Pi_g$, and $5^1\Sigma_g^+$). Analogous data is now being analyzed for the $2^3\Pi_g$ state. These states have been observed using double resonance excitation in combination with a very sensitive space charge diode ion detection system.[9]

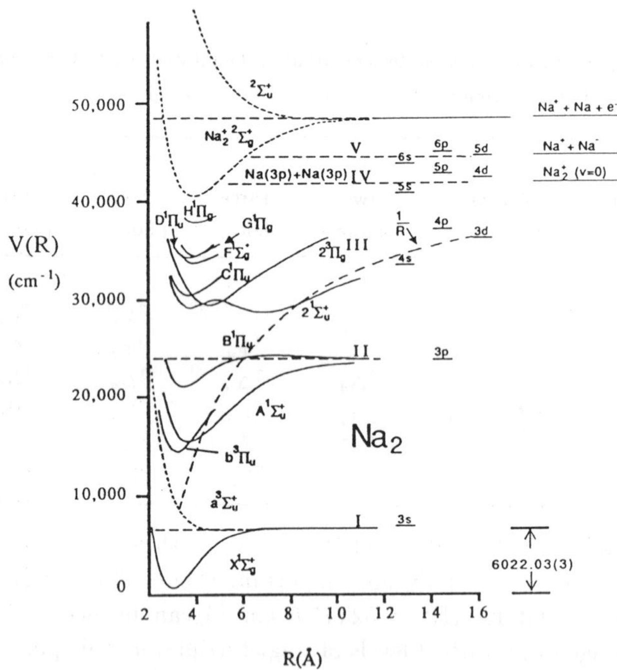

Fig. 1. Selected potential energy curves and asymptotic regions (I–V) for the Na$_2$ molecule.

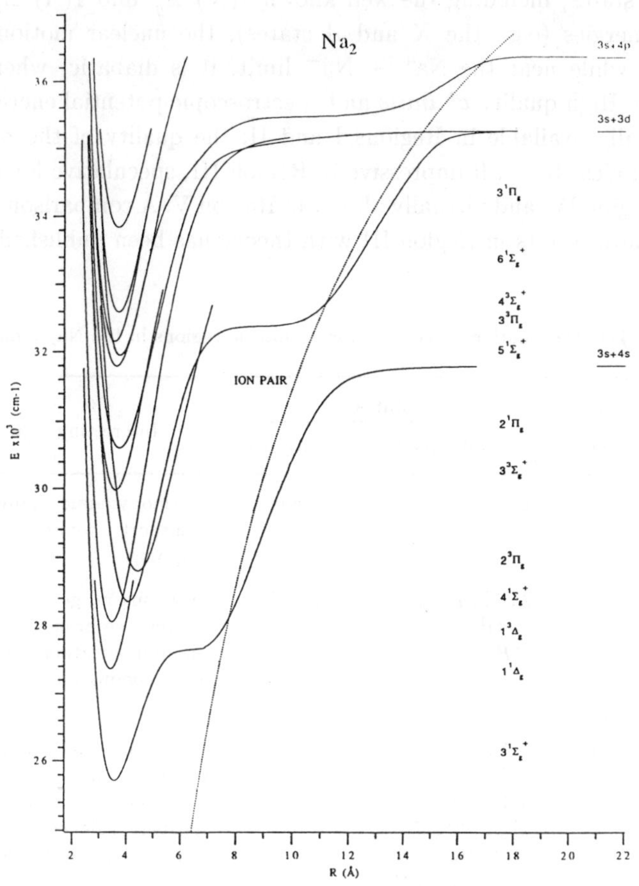

Fig. 2. Selected gerade potential energy curves (labeled by their minima at right) for Na$_2$ in Region III (the $3s + 4s$, $3s + 3d$, and $3s + 4p$ limits). Most of these curves were not well known in 1989. Also shown is the ion pair curve $-e^2/R$ with respect to the Na$^+$ + Na$^-$ asymptote.

To further pursue a concrete example, the various asymptotic regions of Na$_2$, shown in Fig. 1, are summarized in Table III. It is clear that for increasing energy there is decreasing information, particularly at large internuclear distances. This decreasing information is apparent from a theoretical as well as a spectroscopic perspective. For example, the Region V ion pair limit (Na$^+$ + Na$^-$) gives rise asymptotically to the $13^1\Sigma_g^+$ and $12^1\Sigma_u^+$ states,[6] yet ion pair configurations are important in all lower $^1\Sigma_g^+$

and $^1\Sigma_u^+$ states, including the well known $1(X)^1\Sigma_g^+$ and $1(A)^1\Sigma_u^+$ states. At low energies (e.g., the X and A states), the nuclear motion is fully adiabatic while near the $Na^+ + Na^-$ limit, it is diabatic; where is the crossover? High quality *ab initio* and spectroscopic potential energy curves are generally available in Regions I and II; the quality of the potentials is more modest but still impressive in Region III, speculative for the most part in Region IV, and virtually absent in Region V. A comparison of recent experimental results in Region III with theory has been published.[6]

Table III. Designations used here for asymtotic regions in the Na_2 molecule.

Region	Atomic Limits	Observed	Unobserved	Interesting Features
		States		
I	$3s + 3s$	$1(X)^1\Sigma_g^+, 1(a)^3\Sigma_u^+$	none	thermodynamic equilibrium; transport; atom cooling and trapping
II	$3s + 3p$	$2^1\Sigma_g^+, 1^1\Pi_g,$ $1(A)^1\Sigma_u^+,$ $1(B)^1\Pi_u,$ $1(c)^3\Sigma_g^+, 1(d)^3\Pi_g,$ $1(b)^3\Pi_u$	$2^3\Sigma_u^+$	pure long range molecules; atomic line broadening; photodissociation dynamics; atom cooling and trapping
III	$3s + 4s,$ $3d, 4p$	$3^1\Sigma_g^+, 4^1\Sigma_g^+,$ $5^1\Sigma_g^+, 2^1\Pi_g,$ $3^1\Pi_g, 1^1\Delta_g,$ $2^1\Sigma_u^+, 3^1\Sigma_u^+,$ $2(C)^1\Pi_u,$ $3(D)^1\Pi_u, 2^3\Sigma_g^+,$ $3^3\Sigma_g^+, 4^3\Sigma_g^+,$ $2^3\Pi_g, 3^3\Pi_g, 1^3\Delta_g$	8	exotic potential curves; predissociation and Feshbach resonances; Rydberg/valence/ion pair configuration interation
IV	$3s + 5s,$ $4d,$ $4f, 5p;$ $3p + 3p$	$6^1\Sigma_g^+, 7^1\Sigma_g^+,$ $8^1\Sigma_g^+, 9^1\Sigma_g^+,$ $5^1\Pi_g, 4(E)^1\Pi_u,$ $2^3\Delta_g$	57	association ionization/ dissociative recombination; atomic energy transfer; pure long range molecules; autoionization; atom cooling and trapping
V	$3s + 6s,$ $5d, 5f,$ $5g, 6p;$ $Na^+ +$ Na^-	$12^1\Sigma_g^+, 14^1\Sigma_g^+,$ $8^1\Pi_g, 5^1\Delta_g$	58	ion pair formation/ recombination; atomic energy transfer; associative ionization/dissociative recombination; autoionization

A sequence of three single photon transitions (AOTR) can reach long range ungerade states readily, for example, the $A^1\Sigma_u^+$, $a^3\Sigma_u^+$, and $b^3\Pi_u$ states of Na_2. Note that the ion pair states, such as the $4^1\Sigma_g^+$ state, are particularly important for reaching large internuclear distance outer turning points; this is due to the relatively large amplitude of the radial wavefunction at the inner turning point since the outer wall is anomalously steep. In ordinary long range levels close to dissociation (R^{-3}, R^{-5}, or R^{-6} long range attraction), normally > 99% of the wavefunction amplitude is at the outer turning point,[10] leading to very poor Franck–Condon factors with most vibrational levels in the lower states. Furthermore, multiple resonance experiments often exploit the outer turning points of vibration to reach long range levels in other states.

In Na_2, a great deal of information is now available at long-range for the $X^1\Sigma_g^+$, $a^3\Sigma_u^+$, $A^1\Sigma_u^+$, $b^3\Pi_u$, $1^1\Pi_g$, $3^1\Sigma_g^+$, $4^1\Sigma_g^+$, $5^1\Sigma_g^+$, $2^1\Pi_g$, $2^1\Sigma_u^+$, $2^3\Pi_g$, and $1^3\Delta_g$ states. Near dissociation vibrational levels, with outer classical turning points $R > 10$ Å, have been observed (up to ~ 200 Å for the $A^1\Sigma_u^+$ state[11]). The analysis is incomplete, however, and the accurate (~ 0.03 cm^{-1}) D_0 value from the $A^1\Sigma_u^+$ state is being used to fix the other D_0 values in long range fits to obtain improved values of the highly correlated parameters representing dispersion coefficients and exchange. Nevertheless, it is clear that a new era in long range analysis is at hand.

1.2. *All-Optical Triple Resonance*

Multiple resonance techniques such as optical–optical double resonance (OODR), perturbation-facilitated OODR (PFOODR) (see the chapter by Li and Field[12]), and stimulated emission pumping (SEP) can utilize two CW narrowband tunable lasers to achieve sub-Doppler resolution and spectral simplification. In addition, unambiguous assignment of all observed rotational levels is a major advantage offered by these techniques. In the following, we summarize briefly the relevant characteristics of these techniques, which have been described in Demtroder,[13] Hamilton *et al.*[14] and in several articles in the literature (e.g., Refs. 12, 15, 16). Stimulated emission pumping has revolutionized long range spectroscopy and dynamics studies as has been demonstrated beautifully in pulsed laser SEP work.[14,17] The strength of the SEP step is very much dependent on the number of photons in the single longitudinal mode of the laser field. It should be noted that even if CW lasers have on the average a factor of 10^7

fewer photons in a single longitudinal mode than a typical pulsed laser, CW SEP experiments are possible. Furthermore, because of the lower powers, CW SEP experiments do not suffer from competitive multiphoton processes such as multiphoton ionization (which leads to depletion of the population to the ionization continuum). These complications are typical in pulsed laser SEP experiments as unavoidable background processes. Peak-to-peak fluctuations in laser power are another problem in pulsed laser SEP experiments, whereas the CW laser output power is generally very stable. In the following examples we outline some of the advantages of the CW stimulated emission pumping step. These advantages include accurate knowledge of the laser frequency, high resolution, and very good signal-to-noise ratio. We have demonstrated these capabilities already in our recent CW triple resonance work.[1,2,5]

In a typical OODR experiment involving SEP, a PUMP laser excites molecules from a low level of the ground state to a specific vibration–rotation level of the chosen excited state. The DUMP laser drives some excited state molecules back to some unpopulated excited vibration–rotation level of the ground state via stimulated emission. This stimulated emission is a new decay channel for the PUMP final level, in competition with the spontaneous fluorescence channels. Therefore, the SEP spectrum can be recorded as decrement signals in the undispersed spontaneous fluorescence signal as the DUMP laser is scanned. Using this technique, vibrational levels near the dissociation limit have been excited in polyatomic molecules in a state-selected manner.[14,17]

Our recent research[1–5] involves extending these ideas to CW all-optical triple resonance (AOTR) laser experiments. Figure 3 gives the various triple resonance excitation schemes we have studied starting in the $X^1\Sigma_g$ state.[1] For example, in scheme (b) of Fig. 3b, the PUMP laser (normally modulated) is followed by a PROBE laser. The initial level of the transition stimulated by this PROBE laser is the final level of the PUMP laser transition. The final level of this PROBE laser transition is then used as the initial level of the DUMP laser transition. The PROBE signals are normally assigned and assignments confirmed by a dual scan of the PROBE frequency, while the PUMP is alternately set to an R or to a P line for these scans. Only signals that occur in both PROBE scans with the same frequency and intensity are true signals involving the chosen intermediate vibration–rotation level. If the PROBE laser frequency is then set to such a confirmed vibration–rotation level, the subsequent DUMP scan signals can again be confirmed using the same technique (two DUMP laser scans with R and P PUMP settings). This "all-optical triple resonance" (AOTR) scheme offers a variety of opportunities for overcoming selection rule constraints, e.g., the opportunity of observing long range levels in electronically excited states. Population transfer from the ground state initial level to the PROBE final level can be enhanced by saturation of the PUMP and PROBE transitions and by choosing favorable Franck–Condon factors for these transitions. We describe experiments of this kind below in Sec. 3 to examine low levels of the K_2 $A^1\Sigma_u^+$ state and unperturbed levels of the Na_2 and K_2 $b^3\Pi_u$ state.

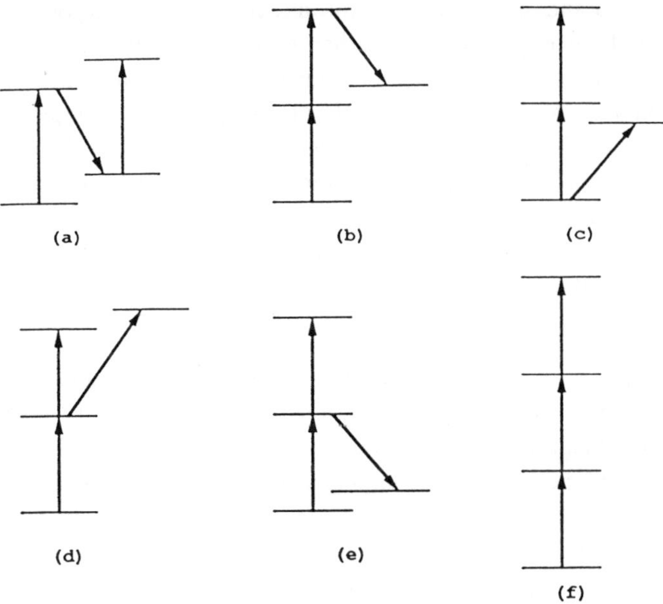

Fig. 3. All-optical triple resonance excitation schemes studied in our laboratories. Here we focus on the scheme (b) in Sec. 3 and scheme (a) in Sec. 4.

Alternatively, for scheme (a) as shown in Fig. 3a, the PUMP laser (normally modulated) is followed by a DUMP laser. The initial level of the DUMP laser transition is the final level of the PUMP laser transition. The DUMP signals are normally assigned and assignments confirmed by a dual scan of the DUMP frequency, while the PUMP is alternately set to a P or to an R line for these scans. Only signals that occur in both DUMP scans with the same frequency and intensity are true signals involving the same intermediate vibrational–rotational level. If the DUMP frequency is then set, the subsequent PROBE scan signals can then again be confirmed using the same technique (two PROBE laser scans with R and P PUMP settings). This AOTR scheme also offers many opportunities for overcoming selection rule constraints, e.g., state-selective photodissociation of excited vibrational levels of the ground state when thermally populated vibrational levels have extremely small photodissociation cross-sections. The population transfer of the first two steps (PUMP–DUMP) can be made 100% efficient in a molecular beam (the population can also be significantly aligned).[18-20] We describe experiments of this kind to examine high levels of the Na_2 $A^1\Sigma_u^+$ state in Sec. 3 and state-selected photodissociation of K_2 in Sec. 4.

It should be noted that the other schemes in Figs. 3c–f all involve stimulated emission and absorption (but only Fig. 3e has a net downward population transfer, i.e., a SEP (DUMP) step). For example Tsai and Bahns at Iowa have recently used scheme (f) to observe high Rydberg states of Na_2 (including ℓ uncoupling and autoionization) through the $3^1\Sigma_g^+$ ($v = 0, J = 0, 1, 2, 3$) levels.

2. Experimental

2.1. *Fluorescence Detection*

Figure 4 illustrates a typical AOTR scheme (b) experimental setup using fluorescence. Using this excitation scheme both in its normal and perturbation-facilitated versions, we have been able to determine with high resolution both perturbed and unperturbed levels of the $A^1\Sigma_u^+$ and $b^3\Pi_u$ states in Na_2 and K_2[2,5] (see Figs. 5 and 6 and Sec. 3). In addition to detecting the decrement signal caused by SEP in the side fluorescence double resonance signal (Fig. 7, top trace), the fluorescence signal following the SEP step has been used as well (Fig. 7, middle trace) as a means of detecting the triple resonance signal.

In general, the best signal-to-noise ratios are obtained from decrements in the short wavelength fluorescence detection (e.g., violet-blue in Na_2[21] and green-yellow in K_2[22]), since the corresponding background in these spectral regions (scattered laser light, laser-induced fluorescence, etc.) is low for orange-red-near infrared exciting lasers. Both triplet and singlet states can give rise to this relatively short wavelength emission. This is because collisional energy transfer among the highly excited states is

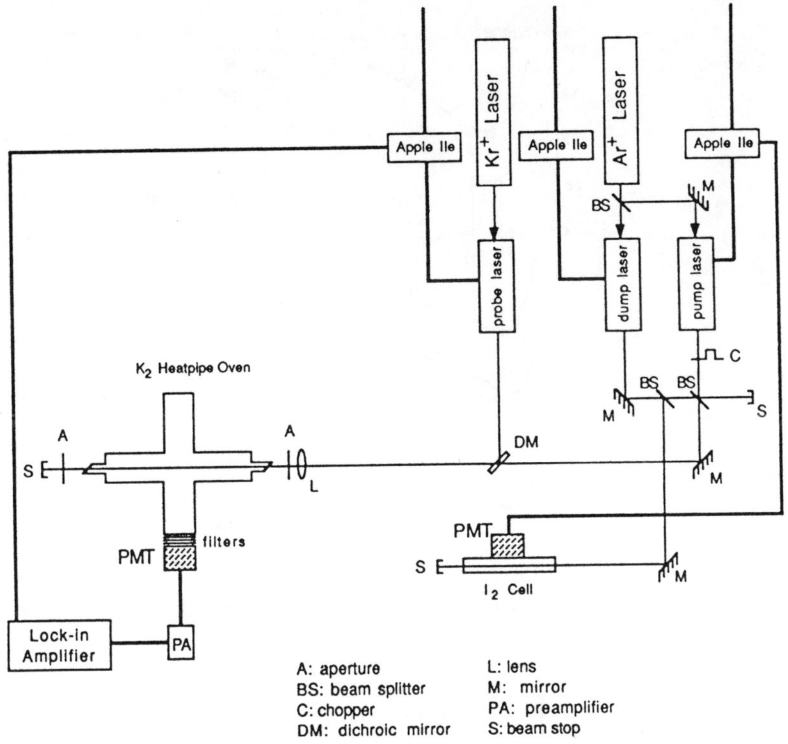

Fig. 4. An example of the experimental setup for all-optical triple resonance spectroscopy.

sufficiently rapid at a few torr pressure to give strong signals. Note that Demtröder's group[23] has observed very large pressure broadening cross-sections (~ 6000 Å^2) for comparable states in Li$_2$ broadened by Li. We suggest here that these high rates are associated with alkali trimer Rydberg states, i.e.,

$$\mathrm{Na_2^{**} + Na \rightarrow Na_3^{**} \rightarrow Na_2^{**\prime} + Na} \; ,$$

which are all clearly in the same energy region.[24]

Of course, for the scheme (a) in Fig. 3, one must detect fluorescence at wavelengths where there is often significant background unless careful baffling, etc., is used.

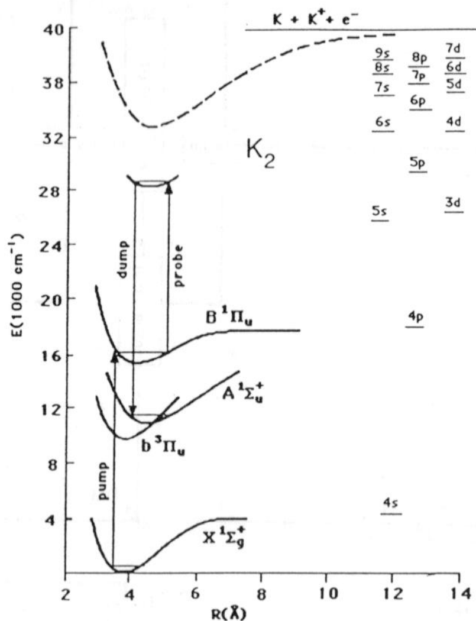

Fig. 5. All-optical triple resonance scheme (type b) for observing the low vibrational levels of the K_2 $A^1\Sigma_u^+$ state.

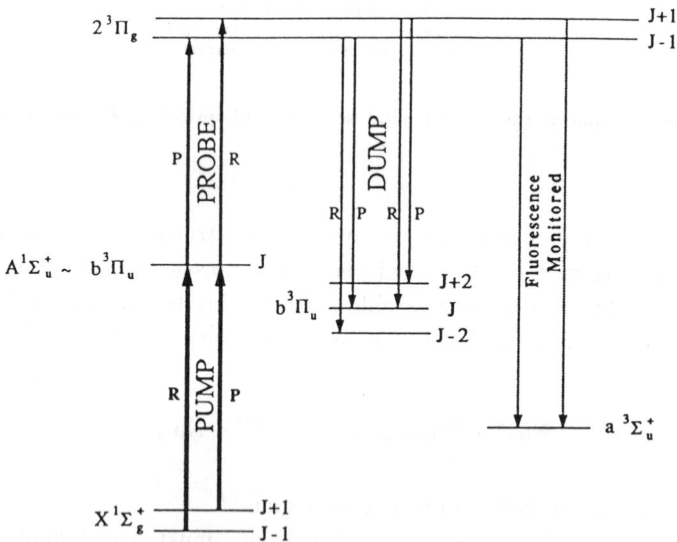

Fig. 6. Perturbation-facilitated all-optical triple resonance scheme (type (b)) for observing the $b^3\Pi_u$ states of Na₂ and K₂.

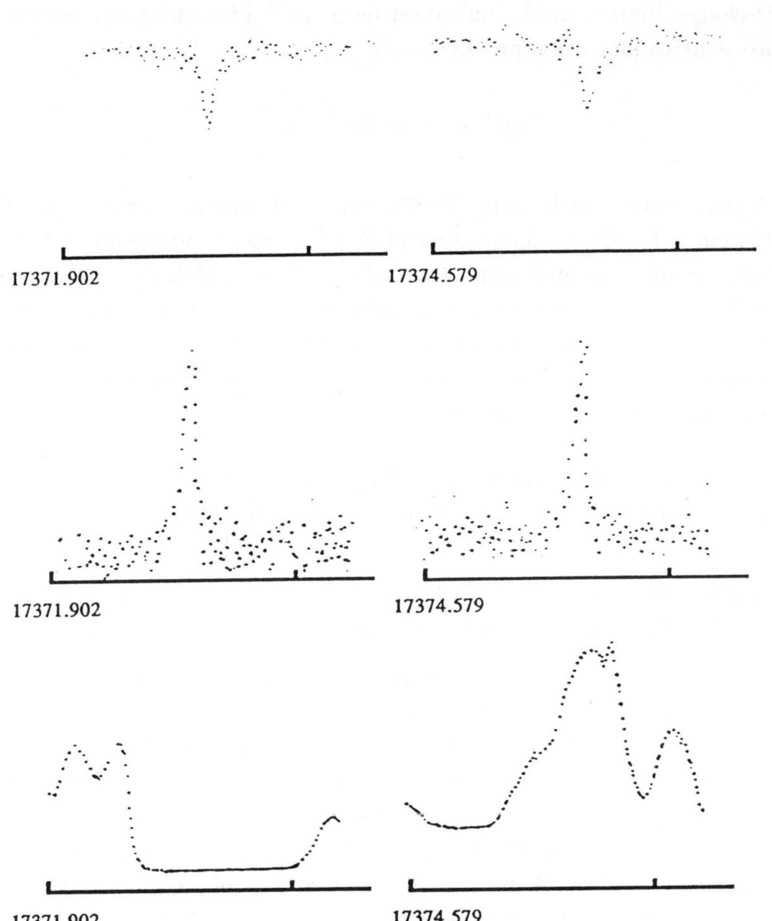

Fig. 7. Examples of AOTR signals in this work (set DUMP laser frequencies near the (5d) $^1\Pi_g(v^* = 7, J^* = 16) \to A^1\Sigma_u^+(v_A = 7, J_A = 17$ and $15)$ transitions. The left and right sides are the $P(17)$ and $R(15)$ lines, respectively. The top and the middle traces represent the decreased and the increased AOTR signals, respectively, and the bottom trace is the I_2 reference signal. In each case, the marks on the frequency axis are 5 GHz apart.

2.2. *Ionization Detection*

For both OODR (as in Ref. 12) and AOTR as described here, highly excited electronic states of Na_2 and K_2 can be very sensitively detected using a

space-charge limited diode ionization detector.[9] The ionization occurs by
the associative ionization process:

$$Na_2^{**} + Na \rightarrow Na_3^+ + e^- \ ,$$

which can occur with only $28\,003$ cm^{-1} of internal energy in Na_2
($24\,124$ cm^{-1} for the analogous case of K_2).[24] A space-charge-limited diode
provides an internal preamplification of $\sim 10^5$ (e.g., Ref. 25). Thus levels
above $28\,000$ cm^{-1} with Franck–Condon factors $\sim 10\text{--}10^3$ smaller than
those observed by fluorescence have recently been detected in OODR
experiments at Iowa. A more quantitative comparison of ionization and
fluorescence detection is planned.

3. All-Optical Triple Resonance Spectroscopy
of the Mutually Perturbed $A^1\Sigma_u^+$ and $b^3\Pi_u$ States
of Na$_2$ and K$_2$

3.1. *Observations of the $A^1\Sigma_u^+$ State*
and $A^1\Sigma_u^+ \sim b^3\Pi_u$ Perturbations

One classic example of the perturbations between singlet and triplet states
is that of the $A^1\Sigma_u^+ \sim b^3\Pi_u$ perturbations in Na_2 observed first by
R. W. Wood using magnetic rotation spectroscopy in 1907.[26] (The "bright"
$A^1\Sigma_u^+$ states of Na_2 and K_2 were first observed in 1874![27]).

Complete observations of the $A^1\Sigma_u^+$ states of Na_2 and K_2 were severely
restricted by the Franck–Condon principle. In particular, high vibrational
levels (up to 105) of the $A^1\Sigma_u^+$ state of Na_2 were first observed by modulated
gain spectroscopy[8] (the first all-optical triple resonance spectroscopy of
scheme (a), but where an optically pumped laser had to be stabilized
for each DUMP transition). Even higher levels were then observed by
fluorescence enhanced OODR spectroscopy.[11] Likewise, low vibrational
levels ($v = 0\text{--}12$) of the $A^1\Sigma_u^+$ state of K_2 were first observed by resolved
fluorescence[28] and then with higher resolution by AOTR scheme (b),[5] as
previously shown in Figs. 5 and 7.

Simultaneously, the $A^1\Sigma_u^+ \sim b^3\Pi_u$ mixed levels with significant singlet
oscillator strength have been observed by a variety of workers (see Refs. 2,5
and references therein). However, the "dark" unperturbed $b^3\Pi_u$ levels (with
negligible singlet oscillator strength) have remained unobserved until the
recent AOTR experiments described below.

3.2. *Observations of Unperturbed Levels of the $b^3\Pi_u$ State*

Sub-Doppler perturbation-facilitated all-optical triple resonance spectroscopy (PFAOTR) is a powerful technique for the study of perturbations. The high resolution of this technique can resolve even very small perturbations (< 0.05 cm^{-1} shifts) which cannot be detected in Doppler-limited spectra. More importantly, the unobserved "dark" unperturbed levels of the perturbing state can now be observed directly and characterized through the "window" perturbed levels. In the case discussed here, the "dark" $b^3\Pi_u$ state is unobservable in single photon spectroscopy except when there is a significant admixture of $A^1\Sigma_u^+$ character via the spin-orbit interaction. These few mixed $A^1\Sigma_u^+ \sim b^3\Pi_u$ levels then serve as "windows" into the triplet manifold by perturbation-facilitated multiple resonance spectroscopy[1,2,5,12] (as well as "gateways" for electronic energy transfer[12,29]). Figure 6 shows the scheme (b) AOTR approach we have used for both Na_2[2] and K_2.[5]

This perturbation-facilitated all-optical triple resonance spectroscopy has made it possible for the first time to observe with high resolution both unperturbed and perturbed levels in the $b^3\Pi_u$ state of the alkali dimers. The initial deperturbation effort of Whang *et al.*[2] has led to deperturbed molecular constants for both of these states below $v = 57$ of the $b^3\Pi_u$ state of Na_2. The deperturbation effort will resume after both of these states have been observed close to dissociation (there are also additional possibilities for perturbations (e.g., $B^1\Pi_u \sim b^3\Pi_u$)). The $A^1\Sigma_u^+$ state of Na_2 is known systematically to within 0.025 cm^{-1} of the dissociation limit.[11] Recently, we have obtained data (not yet analyzed) for all levels $v \leq 119$ in the $b^3\Pi_u$ state of Na_2 ($v = 119$ corresponding to a very weakly bound level with an outer turning point of ~ 20 Å). Similarly, initial analyses now exist allowing us to establish deperturbed molecular constants for the lower part of the $A^1\Sigma_u^+$ state of K_2.[5] Tables IVa and IVb summarize these preliminary results for Na_2 and K_2, respectively.

A further example of the interrelated sequential development of spectroscopic information by increasingly higher order all-optical multiple resonance spectroscopy is illustrated in Fig. 8 for K_2, here perturbation-facilitated (PF). At the left (Fig. 8a), a PFOODR experiment similar to our recent OODR experiments[7,16,28,30–32] (but at lower energy) is illustrated, using long wavelength Ti:Sapphire laser and LD700 dye laser photons to excite low-lying triplet gerade Rydberg states. Such experiments (currently

Table IVa. Deperturbed spectroscopic constants (in cm^{-1} except R_e in Å) for the Na_2 $A^1\Sigma_g^+$ and $b^3\Pi_u$ states. The $A^1\Sigma_u^+$ constants are still preliminary (based mainly on Ref. 8) and have not been merged with recent results.[11]

	$A^1\Sigma_u^+$	$b^3\Pi_u$
T_e	14680.58(5)	13520.946(9)
Y_{10}	117.315(6)	154.209(4)
Y_{20}	$-0.3546(12)$	$-0.47682(4)$
Y_{30}	$-0.415(88) \times 10^{-3}$	$-0.77(15) \times 10^{-4}$
Y_{40}	$3.848(31) \times 10^{-5}$	$-0.57(29) \times 10^{-6}$
$Y_{50}{}^{\text{a}}$	$-1.167(63) \times 10^{-6}$	$-0.142(2) \times 10^{-6}$
Y_{01}	$0.110745(8)$	$0.15195(4)$
Y_{11}	$-0.5288(29) \times 10^{-3}$	$-0.652(7) \times 10^{-3}$
Y_{21}	$-0.294(35) \times 10^{-5}$	$0.91(29) \times 10^{-6}$
$Y_{31}{}^{\text{a}}$	$0.213(16) \times 10^{-6}$	$-0.692(33) \times 10^{-7}$
$Y_{02}{}^{\text{b}}$	-0.386807×10^{-6}	$0.58676(6) \times 10^{-6}$
Y_{12}	—	$0.2199(34) \times 10^{-8}$
$Y_{03}{}^{\text{b}}$	0.100554×10^{-11}	—
Y_{00}	$0.0054(10)$	-0.0062
$A_e{}^{\text{c}}$	—	$7.106(3)$
$\alpha_A{}^{\text{c}}$	—	$-0.0183(3)$
$\gamma_A{}^{\text{c}}$	—	$0.33(5) \times 10^{-4}$
R_e	3.63900(63)	3.10667(41)
D_e	8297.63(5)	9474.46(4)

[a]For the $A^1\Sigma_u^+$ state ($v' = 0$ to 105) higher order constants ($Y_{60} = 0.1748(71) \times 10^{-7}$, $Y_{70} = -0.1411(41) \times 10^{-9}$, $Y_{80} = 0.4511(99) \times 10^{-12}$; $Y_{41} = -0.683(39) \times 10^{-8}$, $Y_{51} = 0.1127(53) \times 10^{-9}$, $Y_{61} = -0.981(36) \times 10^{-12}$, $Y_{71} = 0.3340(96) \times 10^{-14}$) are required for calculations of vibrational–rotational energies.
[b]Centrifugal distortion constants Y_{02} and Y_{03} for the $A^1\Sigma_u^+$ state have been determined using the program CDIST (J. M. Hutson, *J. Phys.* **B14**, 851 (1981)).
[c]Based on the equation $A_v = A_e + \alpha_A(v + 1/2) + \gamma_A(v + 1/2)^2$.

underway) will provide valuable information on the connection between low lying states and Rydberg states.

In the middle (Fig. 8b) is a PFAOTR experiment similar to our study of the unperturbed $b^3\Pi_u$ state of K_2;[5] however, a shorter wavelength laser stimulates emission (at high resolution) into the $a^3\Sigma_u^+$ state (previously known only from low resolution OODR-excited fluorescence[30]). This would provide a precise comparison with the $X^1\Sigma_g^+$ state of K_2[33] and could then

Table IVb. Deperturbed spectroscopic constants (in cm^{-1} except R_e in Å) for the $A^1\Sigma_g^+$ and $b^3\Pi_u$ states of $^{39}K_2$.[5,a]

Y_{ij}	$A^1\Sigma_u^+$	$b^3\Pi_u$
T_e	11107.92(11)	9911.8(18)
Y_{10}	70.545(17)	91.54(17)
Y_{20}	$-0.15656(64)$	$-0.1894(42)$
Y_{30}	$-0.1089(67) \times 10^{-3}$	—
Y_{01}	0.041801(10)	0.057635(76)
Y_{11}	$-0.14081(86) \times 10^{-3}$	$-0.1753(36) \times 10^{-3}$
Y_{00}	0.0002	-0.0004
A(spin-orbit)	—	21.75(8)
R_e(Å)	4.5498(5)	3.875(3)
D_e	6328.3(15)	7524.4(15)

[a]The constants (in cm^{-1}) are obtained for $v_A \le 62$ and $v_b \le 27$. Values in parentheses are one standard error, in units of the last significant digit of the corresponding constant.

be used to precisely determine the dispersion energy $V_D = ((V_a + V_x)/2)$ and the exchange energy $V_{EX} = ((V_a - V_x)/2)$ experimentally as functions of internuclear distance, as in recent work on Li_2,[34] and thereby test long range theory. The dispersion potential, V_D, is often approximated $-C_6 R^{-6} - C_8 R^{-8} - C_{10} R^{-10}$. However, at smaller internuclear distance, the coefficients C_n are "damped" (i.e., monotonically decreasing to zero as R decreases). Likewise, the exchange potential, V_{EX}, is often approximated as $Ae^{-\alpha R}$, with less justification as to the form than for V_D. Here, however, we can determine V_D and V_{EX} directly with no form dependent assumption. Similar arguments, of course, apply to other asymptotes, e.g., the $3s + 3p$ and $3s + 3d$ asymptotes, where we now have significant long range data (Fig. 2). Here, a shorter wavelength UV-pumped DUMP laser (e.g., Stilbene) must be used, but the experiment and analogous ones for other alkalis, e.g., Li_2,[35] are clearly feasible.

The $a^3\Sigma_u^+$ state governs the interatomic behavior of ultracold spin-polarized atoms.[36-39] For example, effective range theory can be used to show that the binding energy of the last rotationless level in this potential is inversely proportional to the low temperature total collision cross-section.[40] This cross-section has been recently measured for ultracold Cs, since it

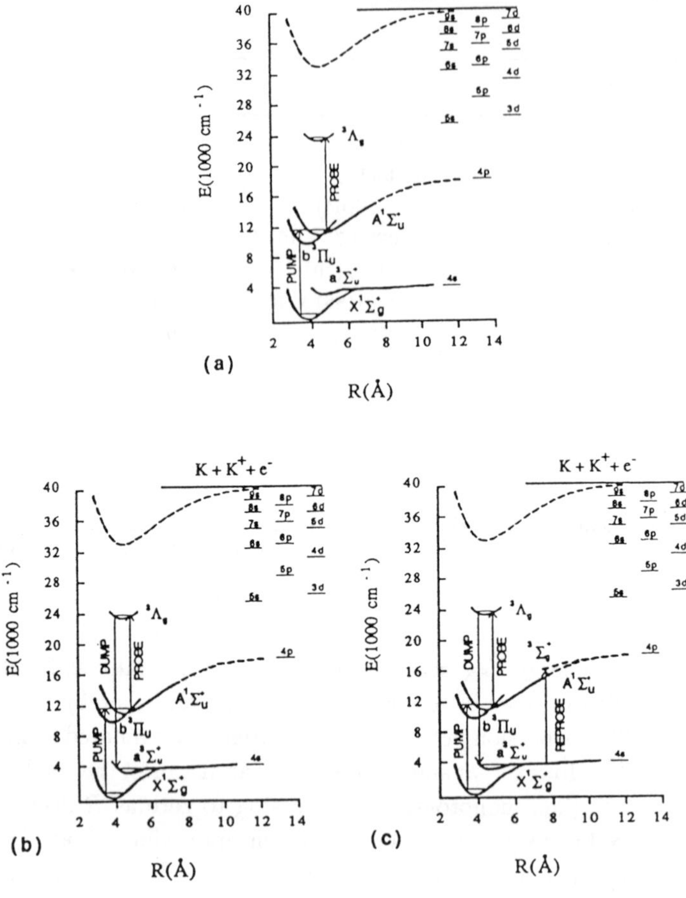

Fig. 8. An illustration of the sequential nature of all-optical multiple resonance experiments for K_2: (a) PFOODR studies of $^3\Lambda_g$ Rydberg states; (b) PFAOTR studies of the $a^3\Sigma_u^+$ "spin-polarized" state (previously studied by OODR-spectrally resolved fluorescence[30]); (c) PFAOQR studies of the long range region of the $1^3\Sigma_g^+$ state.

is an important loss mechanism in atom traps.[41,42] The $a^3\Sigma_u^+$ rho-type splitting (spin–spin splitting) is unobserved in low resolution (~ 0.1 cm^{-1}) spectrally resolved fluorescence experiments on Li_2,[35] Na_2,[15] and K_2,[30] but could probably be resolved in analogous AOTR experiments, at least for K_2. The magnitude of this spin-spin term in the molecular Hamiltonian is needed to estimate an important trap loss process.[42]

At the right (Fig. 8c) is a proposed perturbation-facilitated all-optical quadruple resonance (PFAOQR) experiment, with the final "REPROBE" laser starting in a long range level and examining the long range behavior of many excited states (e.g., the $1^3\Sigma_g^+$ (currently unknown) excited state with $-C_3/R_3$ long range behavior).

4. State-Selected Photodissociation by All-Optical Triple Resonance

4.1. *Photodissociation as a "Half Collision"*

State-selective laser spectroscopic techniques provide a powerful tool for investigating molecular dynamics, as well as molecular structure. These methods allow detailed state-resolved measurements of energy partitioning and collisional kinetics on a "state-to-state" basis,[43,44] and are particularly important for studying multichannel excited state collisions.

Excited state processes often involve the collisional interaction and mixing of a number of closely spaced molecular energy levels which become degenerate or nearly degenerate in the asymptotic separated atom limit. The details of excited state collision dynamics are determined by the adiabatic correlations of, and by the nonadiabatic mixing between, the Born–Oppenheimer potential curves.[43–46] The most important nonadiabatic corrections to the adiabatic Born–Oppenheimer Hamiltonian include the spin-orbit and Coriolis coupling terms. In bound-state molecular spectroscopy these nonadiabatic interactions result in spectroscopic perturbations.[47] In molecular dynamics they can have a profound influence on the final-state branching in multichannel collisions.[43–46] Nonadiabatic couplings can be important both at localized short range curve crossings, and in the long range limit between nearly degenerate asymptotic states (in both entrance and exit channels).

Our most detailed understanding of excited state collisions comes from highly state-selective molecular beam scattering experiments. In the ideal case, the experimenter would precisely select and control the initial state of the system, i.e., the electronic state symmetry, the internal rovibrational state of all molecular reagents, the relative collision velocity, and even the impact parameter for the scattering event. Of course, this ideal has never been realized. Stimulated emission pumping (SEP) or stimulated Raman adiabatic passage (STIRAP) techniques can be used to precisely select the internal rovibrational state of molecular reagents.[19,20] Polarized laser excitation techniques may be used to prepare the initial electronic state symmetry in some cases,[48] but long range Coriolis coupling in the entrance channel can readily mix near degenerate states of different symmetry, making it difficult to reliably specify the molecular symmetry at short range.[48] Standard molecular beam velocity selection techniques usually require a trade-off between resolution and beam intensity. Control of the impact parameter in a "full-collision" crossed beam scattering experiment cannot be accomplished.

Bound-free molecular photodissociation, as illustrated in Fig. 9, can be viewed as a "half-collision" analog to the "full-collision" processes discussed above. Photodissociation studies, however, can have distinct advantages in some cases. The excitation occurs at short range where the states of different molecular electronic symmetry are usually well-separated by electrostatic interactions. Hence, it is often possible to select a specific molecular symmetry state from which to initiate the "half-collision". The excited-state "half-collision" begins from a relatively narrow range of impact parameters, determined by the Franck–Condon principle and the equilibrium rovibrational distribution in the ground state. Some selectivity in the relative collision energy can be obtained by tuning the photodissociation laser above threshold, although the resolution is again limited by the ground state thermal distribution. This distribution can be narrowed by cooling the molecules in a supersonic expansion but this has the disadvantage of limiting the Franck–Condon accessible region in the excited state.

However, if supersonic molecular beam techniques are combined with optical multiple resonance excitation techniques (SEP or STIRAP), then the photodissociation can be studied with individual quantum state selectivity. This process is illustrated in Fig. 10. The subsequent excited-state "half-collision" begins with precisely defined angular momentum (to within 1 $\hbar$), and relative collision energy (to within the laser bandwidth). The angular momentum state can be selected by the "PUMP–DUMP" resonance excitation scheme, and the collision energy may by "tuned" by tuning the laser frequency above threshold. In the absence of barriers, the collision energy can be tuned arbitrarily close to threshold, allowing studies of very low energy "ultra-cold" collision dynamics.[49] The initial molecular orientation or alignment can also be selected through the multiple resonance excitation process. Furthermore, by multiple steps, one can achieve higher moments and larger alignments.

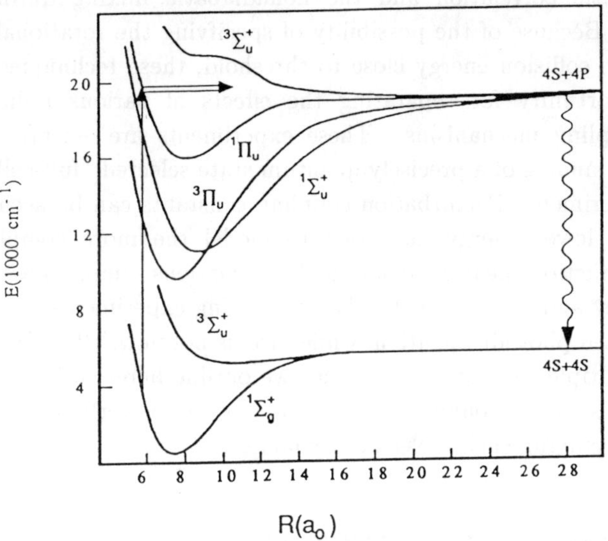

Fig. 9. Selected potential energy curves of K_2.

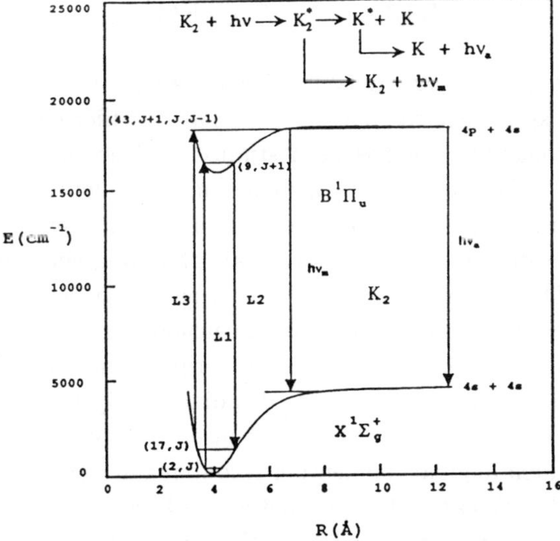

Fig. 10. State-selected photodissociation of K_2 by all-optical triple resonance scheme (a).

The final asymptotic distribution of fragment states will be determined by the adiabatic correlation and the nonadiabatic mixing during the dissociation.[50] Because of the possibility of specifying the rotational state and tuning the collision energy close to threshold, these techniques allow a unique opportunity for separating the effects of various radial and rotational coupling mechanisms. These experiments are essentially the "half-collision" analog of a precisely quantum state-selected "full-collision" scattering experiment. Perturbation coupling constants can be accurately determined at lower energy and used to model the more complicated multistate interactions near dissociation. Perturbations sample these terms more accurately and systematically than scattering experiments.

State-selected photodissociation, while rare, is not new.[4,51–54] However, the optical multiple resonance techniques we outline here, variants of all-optical triple resonance bound state spectroscopy, are powerful and flexible new tools for studying such molecular dynamics.

4.2. State-Selected Photodissociation of K_2

The photodissociation of K_2 via the $B^1\Pi_u$ state offers an excellent example:

$$K_2(X^1\Sigma_g^+) + h\nu \rightarrow (K_2(B^1\Pi_u))^*$$
$$(K_2(B^1\Pi_u))^* \rightarrow K^*(4^2P_J) + K(4^2S_{1/2}) \ .$$

The potential energy curves are well known through bound and quasi-bound molecular state spectroscopy.[4,53,54] The direct (thermally averaged) one-photon photodissociation has been studied both experimentally and theoretically.[55,56] The experiments included measurements of the total bound-free absorption spectrum, the fine-structure branching, and the final state alignment as determined from the atomic photofragment fluorescence polarization.[55] The agreement between experimental measurements and both semiclassical and quantum mechanical calculations is quite good.[55,56] These theoretical works give a complete description of the process and demonstrate that the dissociation proceeds adiabatically from the Hund's case (a) limit at short range to the Hund's case (c) limit at longer range, followed by a sudden angular momentum recoupling at very long range to the asymptotic Hund's case (e) (separated atom) basis.

One interesting result of the semiclassical theory is that the polarization of photodissociation through a specific $B^1\Pi_u$ continuum level characterized by E and J is a function only of the rotation angle $\alpha(E, J)$ during dissociation. The rotation angle $\alpha(E, J)$ can be expressed[57]:

$$\alpha(E, J)$$

$$= \int_{R_0}^{R_{\mathrm{L}}} \left[\frac{\hbar^2 (J(J+1) - \Omega^2)}{2\mu R^2} \right]^{1/2} \left[E - V(R) - \frac{\hbar^2 (J(J+1) - \Omega^2)}{2\mu R^2} \right]^{-1/2} \frac{dR}{R}$$

$$= \int_{R_0}^{R_{\mathrm{L}}} \left[\frac{E_{\mathrm{ROT}}(R)}{E_{\mathrm{KIN}}(R)} \right]^{1/2} \frac{dR}{R} \, , \tag{1}$$

where E is the continuum state energy, J is the total angular momentum (rotational plus electronic, excluding nuclear spin), $\Omega = 1$ is the magnitude of the projection of the electronic angular momentum on the diatomic axis, $V(R)$ is the $B^1\Pi_u$ rotationless potential energy curve, and $E_{\mathrm{ROT}}(R)$ and $E_{\mathrm{KIN}}(R)$ are the local rotational and kinetic energies (from the brackets on the preceding line). The integration is carried out from the inner classical turning point R_0 to a final "locking" radius R_{L}, beyond which the electronic cloud no longer follows the molecular axis.[58] Dynamical calculations demonstrate that rotational mixing is negligible, consistent with the choice $R_{\mathrm{L}} \to \infty$. Because of the barrier of 298.0 ± 0.6 cm^{-1} in the $B^1\Pi_u$ state,[4,54] the asymptotic kinetic energy is considerable for most states involved in the thermally averaged photodissociation. Thus, for these states, α is typically small ($\ll 2\pi$). Exceptions are the quasibound states/orbiting resonances described below.

Given a value of α, the resultant polarization of atomic fluorescence is given by the analytical semiclassical expression[55,56]:

$$P(\alpha) = \frac{3\cos^2 \alpha - 6\cos \alpha - 5}{\cos^2 \alpha - 2\cos \alpha + \left[\dfrac{40\sqrt{7} - 5}{3} \right]} \, . \tag{2}$$

For small α, $P(\alpha)$ approaches the limit of -0.245. The polarization (for all α) is also further reduced to $\sim 37\%$ of its Eq. (2) value (e.g., $P(0) \to -0.091$) by hyperfine depolarization.[55,59] When this effect is included, the final calculated polarization values agree with experimental values, which vary nearly linearly from $-7(\pm 3)\%$ at 605 nm to $-3(\pm 3)\%$ at 650 nm.[55] However, the small values of the polarization and especially the extensive thermal averaging reduce the quality of comparison between theory and experiment.

To more reliably test our understanding of the "half-collision" dynamics, we have studied vibrational and rotational state-selected photodissociation.[4] These measurements involve triply resonant state-selected photodissociation via quasibound resonances. Such resonances, called shape or orbiting resonances in the scattering literature, correspond classically to the orbiting points of metastable equilibrium at the maxima of effective potential barriers. This orbiting corresponds to a long scattering time delay and a large semiclassical rotation angle. Moreover, the time delay, rotation angle, line width, etc., all vary strongly with J and E. Thus adjacent quasibound levels (say $J' = 27, 28, 29$ of $v' = 43$ in the $B^1\Pi_u$ state of K_2 accessed from the single level $v'' = 17$, $J'' = 28$ in the $X^1\Sigma_g^+$ state (Fig. 10)) are expected to give three distinct polarizations for P, Q, and R branch photodissociation.

The experiments reported here were carried out with a molecular beam apparatus with three colinear laser beams at right angles to the molecular beam and atomic or molecular fluorescence detection in a third direction perpendicular to the laser and molecular beams (Fig. 11).

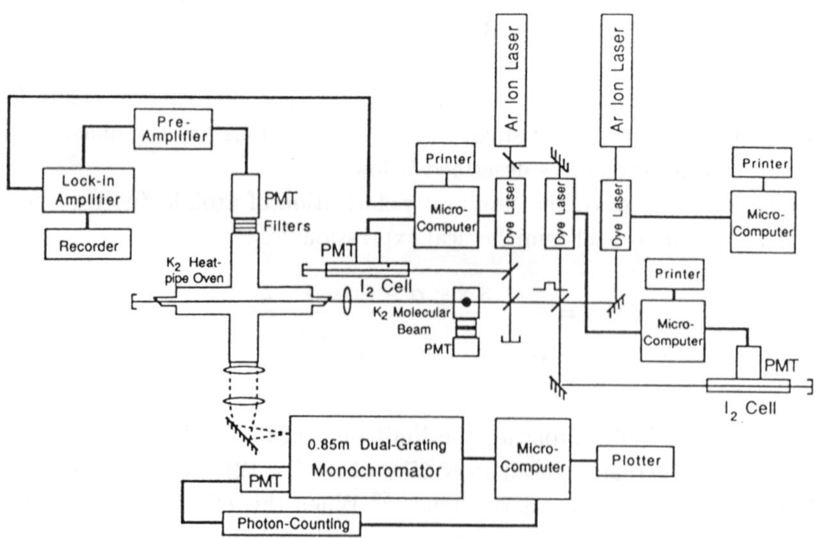

Fig. 11. Experimental setup for state-selected photodissociation.

In these state-selected triple resonance experiments, the lasers involved were three CW single mode scanning tunable dye lasers (Coherent Inc. 699–29) pumped by Ar^+ lasers. The dye laser beams were superimposed both in the molecular beam and in a heat pipe oven. The strong molecular fluorescence from the heat pipe oven was used to set the frequencies of the pump (L1) and dump (L2) lasers and then the frequency of the third probe laser (L3) was scanned continuously. The molecular fluorescence was detected using a double monochromator set to the wavelength of an appropriate $B^1\Pi_u(v' = 43, J') \rightarrow X^1\Sigma_g^+(v'' = 60, J'')$ transition (at ~ 732 nm). The atomic fluorescence was observed in a direction mutually perpendicular to the laser beam and the molecular beam. The fluorescence was detected with a bandpass filtered photomultiplier. The total atomic fluorescence was then determined as a function of probe laser (L3) wavelength.

In order to state-selectively photodissociate an individual vibration–rotation level, a particular vibrationally excited level was prepared by a two-laser (L1, L2) double resonance technique and then photodissociated with a third laser (L3) (Fig. 10). Such all-optical triple resonance (AOTR) spectroscopy is a direct extension of purely bound state techniques, both ordinary and perturbation-facilitated. The initial experiments have involved quasibound levels/orbiting resonances in the $B^1\Pi_u$ state of K_2. Excitation to such levels can produce either ordinary molecular fluorescence or photodissociation by tunneling to $K^* + K$ (with subsequent polarized atomic fluorescence). In addition to the increased linewidth of the molecular fluorescence, the decreased intensity of molecular fluorescence for higher, more rapidly tunneling levels, is evident in Fig. 12. Tunneling is thus observable at $J' \approx 18$, becomes comparable to molecular radiative decay at $J' \approx 24$, and extinguishes molecular fluorescence for $J' \gtrsim 30$.

More significantly, the direct production of K^* has been detected in all-optical triple resonance state-selected photodissociation experiments (Fig. 10). Selected triple resonance signals are shown in Fig. 13, which clearly show increased linewidths (detected by atomic fluorescence as laser L3 is scanned) for the more rapidly tunneling higher J' levels. At lower J', these linewidths agree (Fig. 14) with those of Heinze and Engelke[54] (detected by broadening of molecular fluorescence lines), but many high J' levels are observed for the first time. Energy and width calculations, based on a $B^1\Pi_u$ potential energy curve derived from Ref. 54, give very good agreement with observed energy levels and widths as shown in Table V

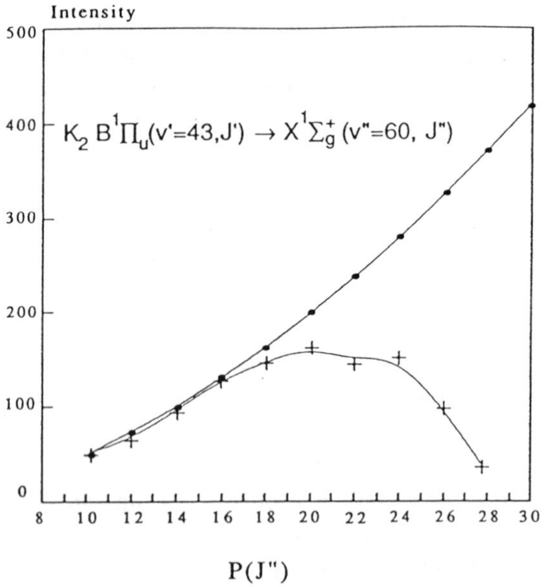

Fig. 12. Intensities of molecular fluorescence from $B^1\Pi_u(v' = 43, J')$ levels: experiment (+) and theoretical (assuming no tunneling and photodissociation) ($\bullet$). The experimental intensities are uncertain by $\pm 10\%$ and normalized to theory at $J' = 10$.

and in Fig. 14. We find agreement between theory and experiment for model potentials with barrier heights of 298.0 ± 0.6 cm^{-1}. Note that the radiative lifetime and AC Stark effect contributions to the linewidth (which are critically important in extracting the tunneling contributions in the lower J' work of Heinze and Engelke[54]) are much smaller than the tunneling contribution in our new, broader, higher J' results. Since the potential barrier is most sensitive to the broadest levels, we made no attempt to carefully characterize the AC Stark effect which is significant at low J'. Thus our low J' widths differ from those of Ref. 54. Previous linewidth calculations[54] for low J' agree with our calculations to within 30%, probably due to differences in interpolation. The Doppler effect is also small for this sub-Doppler triple resonance technique and again unimportant for high J' states. Note also that the $v' = 43$, $J' = 43$ level is apparently the very last J' level in $v' = 43$ below the barrier maximum (Fig. 15).

Heinze, Kowalczyk, and Engelke have used similar methods (relying, however, on less selective spontaneous emission (Franck–Condon) pumping rather than SEP for the DUMP step) to observe and characterize the

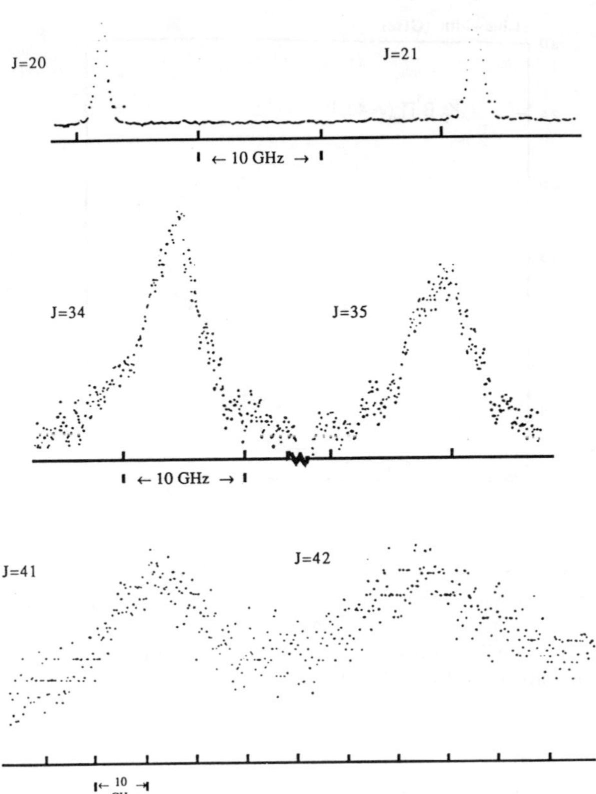

Fig. 13. Selected all-optical triple resonance signals for $B^1\Pi_u(v'=43, J')$ levels of K_2 as a function of probe laser (L3) frequency detected by observation of the intensity of filtered atomic fluorescence.

continuum shape resonances and the underlying resonant photodissociative background.[53]

These AOTR experiments have given detailed spectroscopic information on the $B^1\Pi_u$ barrier region in K_2. Quantum state-selective studies of the photodissociation or "half-collision" dynamics will require measurement of the final atomic fragment alignment through the $K^*(4^2P_{3/2} - 4^2S_{1/2})D_2$ line fluorescence polarization. These experiments are currently underway in our laboratory.

These techniques can be applied to other systems of dynamical interest. Another diatomic molecule of particular interest is the heteronuclear molecule NaK.[60]

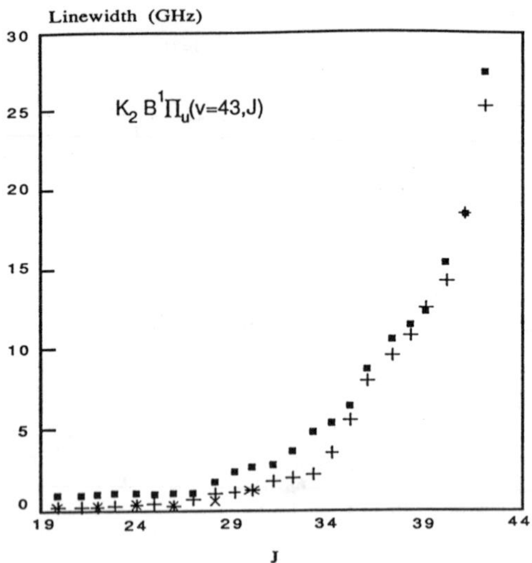

Fig. 14. Linewidths (FWHM in GHz) for quasibound resonances $B^1\Pi_u(v = 43, J)$ of K_2 from earlier molecular fluorescence results ($\times$), from the triple resonance atomic fluorescence experiments discussed here ($\blacksquare$), and from calculations which assume the linewidth is due to tunneling only ($+$). The linewidths are uncertain by $\pm 7\%$ for $J = 42$ and by lesser amounts down to $\pm 3\%$ at low J.

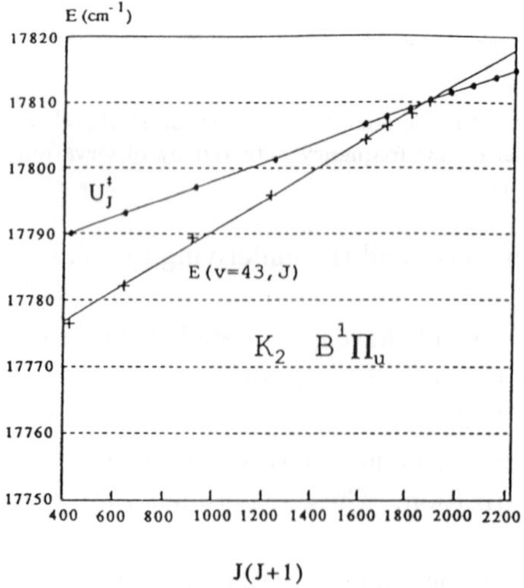

Fig. 15. The energies $E(v = 43, J)$ (in cm^{-1}) of the $B^1\Pi_u$ quasibound resonances in K_2 ($+$) and the maxima of the corresponding effective potential barriers $U_J^{\ddagger}$ (in cm^{-1}).

Preliminary NaK results are more complex and the theoretical and experimental results are much less complete. First, unlike K_2 where only the $^2P_{3/2}$ state (which adiabatically correlates with the $B^1\Pi_u$ excited state) is produced, both $^2P_{3/2}$ and $^2P_{1/2}$ K^* atoms are produced in NaK photodissociation through the $B^1\Pi$ state, in a ratio which varies strongly with wavelength. Experimental studies of the polarization have not yet been completed. In addition, the experimental potential energy curves needed for theoretical calculations are currently poorly determined. Furthermore, there is no barrier in the $B^1\Pi$ upper state potential, so the continuum energy can approach zero and an ultracold photodissociative "half-collision" at an asymptotic kinetic energy as low as $\sim 10^{-3}$ K is possible. For this reason, the rotation angle α can again become very large, as in the quasibound levels described here, and the polarization of the atomic fluorescence should likewise vary strongly with E' and J'.

Similar techniques could also be applied to the photodissociation of van der Waals complexes and polyatomic systems.

Finally, it should be noted that studies of other dynamical processes (e.g., predissociation, autoionization, state-to-state energy transfer) can be carried out using related extensions of AOTR spectroscopy.

Acknowledgments

We wish to particularly thank Professors Li Li and Robert W. Field for collaborations and discussions which led to the AOTR technique described here; and to an exceptional group of former Ph.D. students: Drs. He Wang, Thou-Jen Whang, Jiaxiang Wang, and Gwo Jong, who obtained most of the data reported here. Helpful discussions of aspects of this work with Professors Ken Sando and Mark Young, with Dr. John T. Bahns, and with current Ph.D. students, Chin-Chun Tsai, Bing Ji, Jin-Tae Kim, Alexandra Yiannopoulu, and Ken Urbanski are also gratefully acknowledged. Support from the National Science Foundation, from the Temple University Summer Research Fellowship Program and from the Donors of the Petroleum Research Fund, administered by the American Chemical Society, is gratefully acknowledged. The loan of a third 699–29 single mode dye laser by Coherent Inc. for our initial experiments is also gratefully acknowledged.

References

1. A. M. Lyyra, H. Wang, T.-J. Whang, L. Li, and W. C. Stwalley, *Phys. Rev. Lett.* **66**, 2724 (1991).
2. T.-J. Whang, W. C. Stwalley, A. M. Lyyra, and L. Li, *J. Chem. Phys.* **97**, 7211 (1992).
3. W. C. Stwalley, P. D. Kleiber, K. M. Sando, A. M. Lyyra, L. Li, S. Ananthamurthy, S. Bililign, H. Wang, J.-X. Wang, and V. Zafiropulos, *Faraday Disc. Chem. Soc.* **91**, 97 (1991).
4. J. X. Wang, H. Wang, P. D. Kleiber, A. M. Lyyra, and W. C. Stwalley *J. Phys. Chem.* **95**, 8040 (1991).
5. G. Jong, L. Li, T. J. Whang, A. M. Lyyra, W. C. Stwalley, M. Li, and J. Coxon, *J. Mol. Spectrosc.* **155**, 115 (1992).
6. T. J. Whang, W. C. Stwalley, L. Li, and A. M. Lyyra, *J. Mol. Spectrosc.* **155**, 184 (1992).
7. H. Wang, W. C. Stwalley, and A. M. Lyyra, *J. Chem. Phys.* **96**, 7965 (1992).
8. G. Chawla, H. S. Schweda, H. J. Vedder, R. W. Field, S. Churassy, A. M. Lyyra, W. T. Luh, and W. C. Stwalley, *Advan. in Laser Science I*, ed. W. C. Stwalley and M. Lapp (American Institute of Physics, New York, 1986), p. 466.
9. C. C. Tsai, J. T. Bahns, and W. C. Stwalley, *Rev. Sci. Instrum.* **63**, 5576 (1992).
10. W. C. Stwalley, *Contemp. Phys.* **19**, 65 (1978).
11. H. Knockel, T. Johr, H. Richter, and E. Tiemann, *Chem. Phys.* **152**, 399 (1991).
12. L. Li and R. W. Field, "Continuous Wave Perturbation-Facilitated Optical–Optical Double Resonance Spectroscopy of Na_2 and Li_2", Chapter 7 of this volume (1995).
13. W. Demtröder, *Laser Spectroscopy* (Springer-Verlag, New York, 1982).
14. C. E. Hamilton, J. L. Kinsey, and R. W. Field, *Ann. Rev. Phys. Chem.* **37**, 493 (1986) and references therein.
15. L. Li, S. F. Rice, and R. W. Field, *J. Chem. Phys.* **82**, 1178 (1985).
16. H. Wang, L. Li, A. M. Lyyra, and W. C. Stwalley, *J. Mol. Spectrosc.* **137**, 304 (1989).
17. H.-L. Dai, in *Advances in Multiphoton Processes and Spectroscopy*, ed. S. H. Lin (World Scientific, Teaneck, NJ, 1991), vol. 7.
18. U. Gaubatz, P. Rudecki, S. Schiemann, and K. Bergmann, *J. Chem. Phys.* **92**, 5363 (1990).
19. B. W. Shore, K. Bergmann, and J. Oreg, *Z. Phys.* **D23**, 33 (1992).
20. K. Bergmann and B. W. Shore, "Coherent Population Transfer", Chapter 9 of this volume (1995).
21. G. Pichler, J. T. Bahns, K. M. Sando, W. C. Stwalley, D. D. Konowalow, L. Li, R. W. Field, and W. Muller, *Chem. Phys. Lett.* **129**, 425 (1986).
22. W. T. Luh, J. T. Bahns, K. M. Sando, W. C. Stwalley, S. P. Heneghan, and K. P. Chakravorty, *Chem. Phys. Lett.* **131**, 335 (1986).

23. R. Bombach, B. Hemmerling, and W. Demtroder, *Chem. Phys.* **121**, 439 (1988).

24. W. C. Stwalley and J. T. Bahns, "Atomic, Molecular and Photonic Processes in Metal Vapor Laser-Induced Plasmas", in *Physics of High Power Laser Matter Interactions*, S. Nakai and G. H. Miley, eds. (World Scientific, Singapore, 1992), p. 185–191.

25. D. C. Thompson and B. P. Stoicheff, *Rev. Sci. Instrum.* **53**, 822 (1982).

26. R. W. Wood, *Phil. Mag.* **14**, 145 (1907).

27. H. E. Roscoe and A. Schuster, *Proc. Roy. Soc.* **22**, 362 (1874).

28. A. M. Lyyra, W. T. Luh, L. Li, H. Wang, and W. C. Stwalley, *J. Chem. Phys.* **92**, 43 (1990).

29. L. Li, Q. Zhu, A. M. Lyyra, T. J. Whang, W. C. Stwalley, R. W. Field, and M. H. Alexander, *J. Chem. Phys.* **97**, 8835 (1992).

30. L. Li, A. M. Lyyra, W. T. Luh, and W. C. Stwalley, *J. Chem. Phys.* **93**, 8452 (1990).

31. G. Jong and W. C. Stwalley, *J. Mol. Spectrosc.* **154**, 229 (1992).

32. G. Jong, H. Wang, C. C. Tsai, W. C. Stwalley, and A. M. Lyyra, *J. Mol. Spectrosc.* **154**, 324 (1992).

33. C. Amiot, *J. Molecular Spectroscopy* **146**, 370 (1991).

34. W. T. Zemke and W. C. Stwalley, "Analysis of Long Range Dispersion and Exchange Interactions of Two Li Atoms", *J. Phys. Chem.* **97**, 2053 (1993).

35. X. Xie and R. W. Field, *J. Chem. Phys.* **83**, 6193 (1985).

36. W. C. Stwalley in *Laser-Cooled and Trapped Atoms*, ed. W. D. Phillips, (NBS Special Publication No. 653, 1983), p. 95.

37. P. D. Lett, P. S. Jessen, W. D. Phillips, S. L. Rolston, C. I. Westbrook, and P. L. Gould, *Phys. Rev. Lett.* **67**, 2139 (1991).

38. P. S. Julienne and R. Heather, *Phys. Rev. Lett.* **67**, 2135 (1991).

39. R. W. Heather and P. S. Julienne, "Theory of Laser-Induced Associative Ionization of Ultracold Na", *Phys. Rev.* **A47**, 1887 (1992).

40. Y. H. Uang and W. C. Stwalley, *J. Chem. Phys.* **76**, 5069 (1982).

41. C. R. Monroe, E. A. Cornell, C. A. Sackett, C. J. Myatt, and C. E. Wieman, *Phys. Rev. Lett.* **70**, 414 (1993).

42. F. Mies, private communication (1992).

43. R. D. Levine and R. B. Bernstein, *Molecular Reaction Dynamics and Chemical Reactivity* (Oxford University Press, New York, 1987).

44. S. R. Leone, *Ann. Rev. Phys. Chem.* **35**, 109 (1984).

45. E. E. Nikitin and S. Ya. Umanskii, *Theory of Slow Atom Collisions* (Springer-Verlag, New York, 1984).

46. J. B. Delos, *Rev. Mod. Phys.* **53**, 287 (1981).

47. H. Lefebvre-Brion and R. W. Field, *Perturbations in the Spectra of Diatomic Molecules* (Academic Press, New York, 1986).

48. S. R. Leone, in *Selectivity in Chemical Reactions*, ed. J. C. Whitehead (Kluwer, Dordrecht) NATO ASI Ser. **C245**, 245 (1988).

49. P. S. Julienne, A. M. Smith, and K. Burnett, *Adv. At. Molec. Opt. Phys.* **30**, 141 (1993).

50. S. J. Singer, K. F. Freed, and Y. B. Band, *Adv. Chem. Phys.* **61**, 1 (1985).
51. G. Herzberg, *J. Mol. Spectrosc.* **33**, 151 (1970).
52. G. E. Hall and P. L. Houston, *Ann. Rev. Phys. Chem.* **40**, 375 (1989).
53. J. Heinze, F. Kowalczyk, and F. Engelke, *J. Chem. Phys.* **89**, 3428 (1988).
54. J. Heinze and F. Engelke, *J. Chem. Phys.* **89**, 42 (1988).
55. P. D. Kleiber, J. X. Wang, K. M. Sando, V. Zafiropulos, and W. C. Stwalley, *J. Chem. Phys.* **95**, 4168 (1991).
56. R. Dubs and P. S. Julienne, *J. Chem. Phys.* **95**, 4177 (1991).
57. M. S. Child, *Molecular Collision Theory* (Academic Press, New York, 1974), p. 9.
58. E. E. B. Campbell, H. Schmidt, and I. V. Hertel, *Adv. Chem. Phys.* **72**, 37 (1988).
59. U. Fano and J. H. Macek, *Rev. Mod. Phys.* **45**, 553 (1973).
60. J. X. Wang, P. D. Kleiber, K. M. Sando, and W. C. Stwalley, *Phys. Rev.* **A42**, 5352 (1990).

CHAPTER 13

**INTRAMOLECULAR VIBRATIONAL DYNAMICS
OF HIGHLY EXCITED SO$_2$ AND ACETYLENE:
COMBINATION OF SEP AND DF SPECTROSCPOY**

Kaoru Yamanouchi

*Department of Pure and Applied Sciences,
College of Arts and Sciences, The University of Tokyo,
Komaba, Meguro-ku, Tokyo 153 Japan*

Contents

1. Introduction

The recent development of stimulated emission pumping (SEP) spectroscopy has revealed characteristic features of the vibrational dynamics

491

of highly excited polyatomic molecules.[1] This novel spectroscopic method, which is a folded variant of optical–optical double resonance spectroscopy, enabled us to uncover the rovibrational level structure in the vibrationally highly excited region and made us realize the importance of the spectroscopic investigation of the vibrationally highly excited region for describing the most important dynamical processes in the chemically significant energy region, such as intramolecular vibrational energy redistribution (IVR) and unimolecular reactions.[2,3]

Recently, the SEP method was combined with conventional dispersed fluorescence (DF) spectroscopy for elucidation of the vibrational dynamics of highly excited SO_2 (sulfur dioxide)[4,6] and acetylene.[7,8] The simultaneous analysis of SEP and DF spectra amounts to an important new spectroscopic approach for a fundamental understanding of two mutually correlated problems, i.e., the IVR mechanism which distributes vibrational energy among individual vibrational modes, and the quantum manifestation of vibrational chaos in the classical system.

In this chapter, recent investigations of vibrationally highly excited SO_2 and acetylene by SEP and DF spectroscopy are introduced. In Sec. 2, the level structure of vibrationally highly excited SO_2 revealed by SEP and DF spectroscopy is interpreted in relation to vibrational chaos emerging in classical trajectory calculations, and the significance of the assignment of vibrational quantum numbers to peaks in an observed spectrum is discussed. In Sec. 3, a resonance tier mechanism, which forms the substructure of a feature state, is explained based on DF and SEP spectra of highly excited acetylene and the observed hierarchical level structure is related to IVR. Finally in Sec. 4, the prospects of SEP and DF measurements towards understanding of dynamical processes are briefly mentioned.

2. Vibrationally Highly Excited SO_2

2.1. *Classical Dynamics of SO_2 and Vibrational Chaos*

When treating vibrational motion of a polyatomic molecule in classical mechanics, its vibrational behavior changes rather drastically as a function of the total vibrational energy of the system; i.e., above a certain energy the vibrational motion starts to become formidably complex, in other words, "chaos" appears in the vibrational motion.[3,9,10] For example, this transition to chaos is visually depicted in the classical trajectory calculation of SO_2

performed by Farantos and Murrell.[11] They calculated the vibrational motion of SO$_2$ and showed that trajectories of the vibrational motion represented by symmetric stretching (Q_1) and antisymmetric stretching (Q_3) modes in the low energy region (below $12\,000$ cm^{-1}) are quasiperiodic, similar to a typical example shown in Fig. 1(a), representing a regular trajectory in the Q_1–Q_3 plane. However, above $12\,000$ cm^{-1}, chaotic trajectories, like the complex trajectory shown in Fig. 1(b), appear and the distance between two initially adjacent trajectories, whose initial conditions are only slightly different from each other, diverge exponentially as a function of time. In accordance with the well-known KAM theorem,[3,9,10] Farantos and Murrell clearly demonstrated that there is a threshold energy at which the Lyapunov exponent starts to increase and that this critical threshold energy was located at around $12\,000$ cm^{-1}. In other words, at the threshold energy, classical vibrational dynamics of SO$_2$ exhibits a transition from quasiperiodic motion to chaos. The classical trajectory calculation by Frederick, McClelland, and Brumer[12] also showed that 20% of trajectories are chaotic at $15\,000$ cm^{-1}.

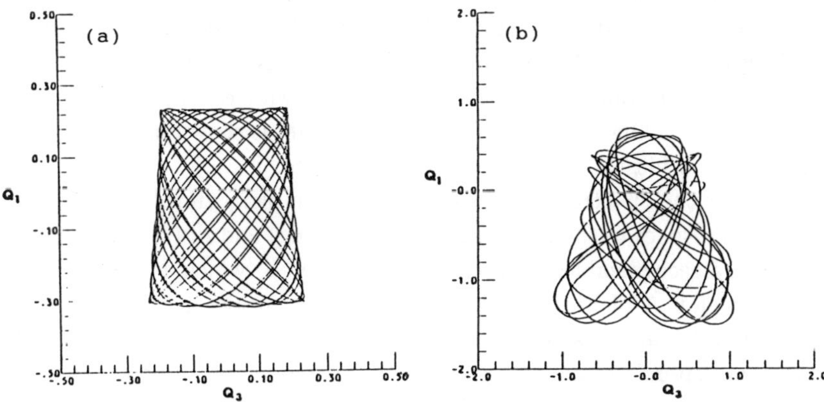

Fig. 1. A quasiperiodic trajectory at 0.335 eV representing the (1,0,0) state (a) and a chaotic trajectory at 3.013 eV (b) of SO$_2$ in the (Q_1, Q_3) plane. (Ref. 11).

In the low energy region where classical dynamics of a nonlinearly coupled oscillator is quasiperiodic, semiclassical quantization is possible and a set of quantum numbers can be assigned to a quantized trajectory. However, when chaos appears in the classical system in its high energy region, it is impossible to find quantum numbers via semiclassical quantization because there is no constant of motion other than the total energy of the system.[3,9,10] Thus, the possibility of assigning a set of quantum numbers to individual peaks in an observed spectrum may be closely related to the appearance of chaos in a corresponding classical system. A polyatomic molecule is a rare example of systems for which we can experimentally show a quantum level structure in the energy region where a corresponding classical system exhibits chaos. Thus, it is expected that an insight into the quantum correspondence of classical chaos may be obtained through spectroscopic investigation of vibrationally highly excited polyatomic molecules, i.e., through assignments of quantum numbers to an observed spectral structure.

2.2. *SEP and DF Spectroscopy of SO_2*

Among conventional spectroscopic methods, DF spectroscopy is the most suitable one to uncover vibrational level structure in a wide energy region including the vibrationally highly excited region. The DF spectra of SO_2 were recorded by exciting a single rotational level in eight vibrational levels, (0,0,0), (0,1,0), (0,2,0), (0,0,2), (1,0,0), (1,1,0), (1,2,2), and (3,1,0) of the electronically excited $\tilde{C}^1B_2$ state.[5] By using all of these eight DF spectra, which cover slightly different energy regions, it was possible to observe the $\tilde{X}^1A_1$ state vibrational level structure over the wide energy region of 4300 $\sim 21\,600$ cm^{-1}. One example of the DF spectrum is shown in Fig. 2, which was measured via the (1,2,2) level of the $\tilde{C}$ state.

Though the spectral structure is rather complex, vibrational progressions with respect to the ν_1 (symmetric stretching) and ν_2 (bending) modes are clearly recognizable. For example, in Fig. 3, which is an expanded view of region A of Fig. 2, the group of intense peaks distributed within 400–500 cm^{-1} appears repeatedly with an interval of around 500 cm^{-1}, which corresponds to one quantum of the ν_2 mode, and nine progressions of ν_2 mode account for almost all the observed peaks in this energy region. It was also noticed that there is no significant change in the level structure around 12 000 cm^{-1} where the onset of chaos was identified in the classical trajectory calculations.

Assignment of vibrational quantum numbers was also performed for seven other DF spectra, and a total of 1388 peaks were assigned and 484 vibrational levels were identified. The derived vibrational term values of these 484 levels were then fitted using a conventional anharmonic expansion,

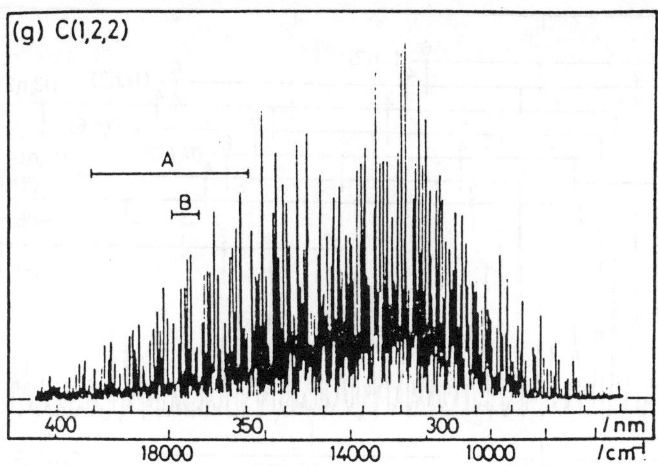

Fig. 2. The observed DF spectrum of SO$_2$ recorded via the $J_{K_a,K_c} = 3_{0,3}$ rotational level of the $\tilde{C}(1,2,2)$ state. The ordinate represents fluorescence intensity in arbitrary units. The abscissa shows the wavelength of the dispersed fluorescence (upper scale) and the vibrational energy measured from the ground vibrational level in the electronic ground $\tilde{X}$ state (lower scale). The region denoted as A is expanded in Fig. 3 and the region B is the one investigated by SEP spectroscopy and is expanded in Fig. 4. (Ref.5)

$$G(v_1, v_2, v_3) = \sum_i \omega_i(v_i + 1/2) + \sum_{i \leq j} x_{ij}(v_i + 1/2)(v_j + 1/2)$$

$$+ \sum_{i \leq j \leq k} y_{ijk}(v_i + 1/2)(v_j + 1/2)(v_k + 1/2) , \qquad (1)$$

by a least-squares procedure. It was found that, by the anharmonic expansion of Eq. (1), the vibrational level energies are well represented in the observed energy region within the resolution (± 10 cm^{-1}) of the DF spectra.

In order to expand further the energy region I in Fig. 3, SEP spectra were recorded. In Fig. 4, the observed SEP spectrum is shown and is compared with the DF spectrum. In the stick spectrum below the SEP spectrum, vibrational level positions predicted by Eq. (1) with optimized coefficients are shown by vertical bars, whose length represents the estimated Franck–Condon (FC) bright character. Though the distribution of bright levels reproduces the SEP spectrum, some of the peaks in the SEP spectrum have no correspondence in the calculated stick spectrum. It seems that *each*

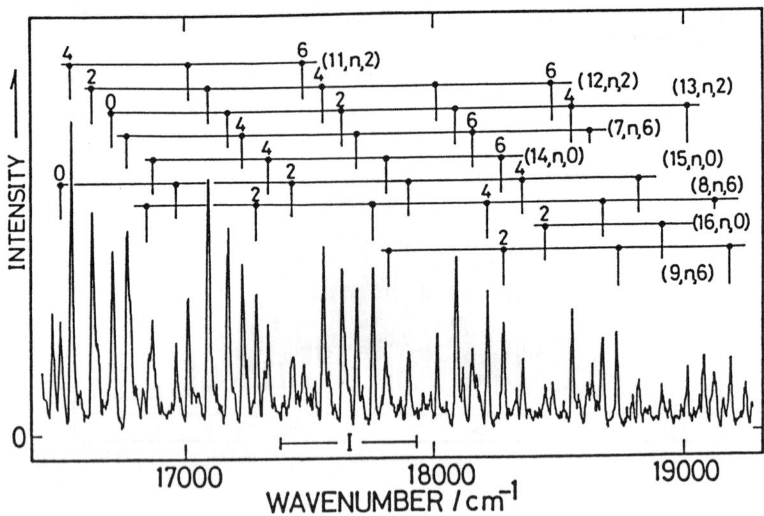

Fig. 3. An expanded DF spectrum recorded via the $\tilde{C}(1,2,2)$ state showing levels of the $\tilde{X}$ state in the region between $16\,500$ and $19\,200$ cm^{-1} of vibrational excitation. Nine identified progressions are shown with the assignments of the vibrational quantum numbers. The region denoted as "I" is that investigated by SEP spectroscopy shown in Fig. 4. (Ref. 5)

peak in the DF spectrum, which was assigned using a set of vibrational quantum numbers, splits into a *few* peaks in the SEP spectrum. It then becomes impossible to assign one set of vibrational quantum numbers to each one of the resolved peaks in the SEP spectrum.

The appearance of a few peaks in the SEP spectrum underneath a single peak in the DF spectrum can be ascribed to anharmonic couplings, which mix the FC bright (v_1, v_2, v_3) level and FC dark levels. In other words, FC dark levels which have originally only weak intensity can show up in the spectrum by borrowing character from FC bright levels. Taking account of the known force field of SO_2,[12] the 1:2 Fermi resonance and the 2:2 Darling–Dennison (DD) resonance between ν_1 (symmetric stretching) and ν_3 (antisymmetric stretching) modes were proposed as candidates for the anharmonic couplings. Since the off-diagonal element of the anharmonic coupling in which the quantum numbers v_j change by n_j units scale roughly as $\prod_j (v_j)^{(n_j/2)}$, anharmonic couplings will increase its influence

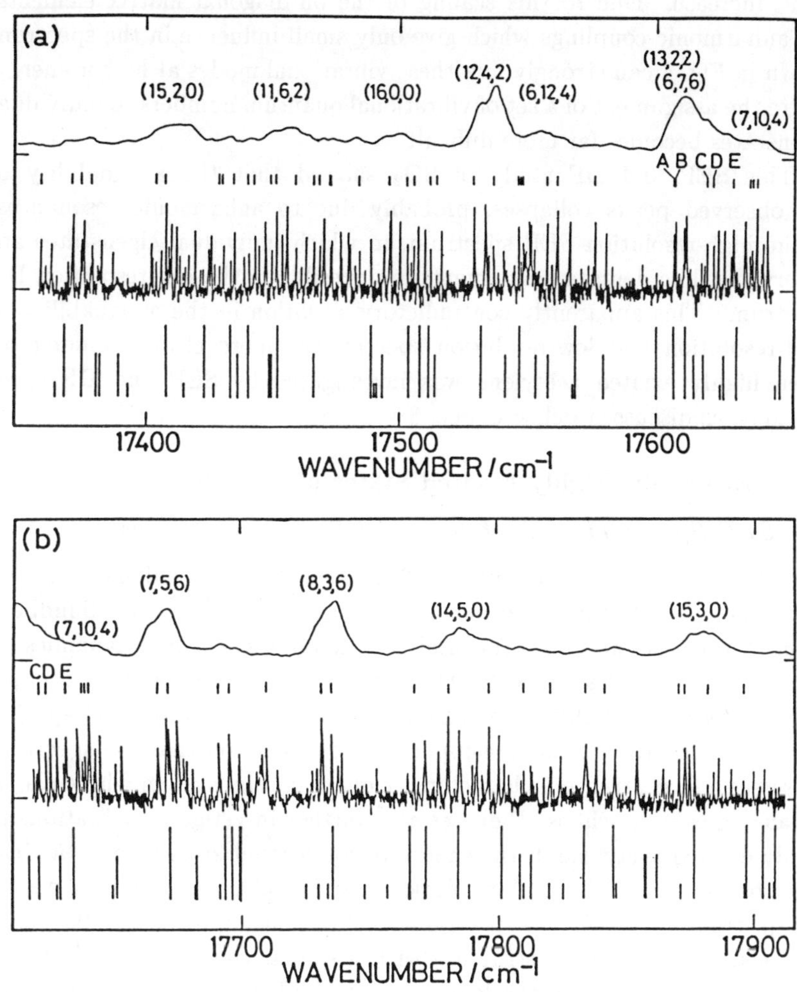

Fig. 4. Comparison of DF (upper trace) and SEP (middle trace) spectra of SO$_2$ for the $E_{\mathrm{vib}} = 17\,350\text{–}17\,910$ cm^{-1} region obtained via the $\tilde{C}(1,2,2)$ excitation. The ordinate of the DF spectrum represents the fluorescence intensity and that of the SEP spectrum the depth of the fluorescence dip with the maximum depth in the observed region being around 40% of the total fluorescence intensity. In the SEP spectrum, a pair of dips represents one vibrational level in the electronic ground state and a short vertical bar above the SEP spectrum indicates the energy position for the $J_{K_a,K_c} = 2_{0,2}$ rotational level. The stick spectrum (lower trace) represents positions of levels predicted by Eq. (1) and the length of each stick represents the estimated FC bright character. (Ref. 5)

as v_j increase. Due to this scaling of the off-diagonal matrix elements, the anharmonic couplings which give only small influence in the spectrum shown in Fig. 4 can strongly mix these vibrational modes at higher energy, where the assignment of a set of vibrational quantum numbers to individual eigenstates becomes far more difficult.

This SEP and DF study of SO_2 showed that the assignability of the observed peaks collapses, probably due to anharmonic resonances, in the high-resolution SEP spectrum, in which individual eigenstates are observed, while definite assignments are possible in the low-resolution DF spectrum. This apparently contradictory situation in the assignability of high-resolution and low-resolution spectra was more clearly understood when highly excited acetylene was investigated by SEP and DF spectroscopy, as discussed below in Sec. 3.

3. Vibrationally Highly Excited States of Acetylene

3.1. *SEP Spectrum of Acetylene and Statistical Analysis*

The SEP spectrum of acetylene near $27\,900$ cm^{-1} vibrational energy in the electronic ground state reported by Abramson *et al.*[14,15] and Sundberg *et al.*[16] was the first experimental attack on the chaotic dynamics of polyatomic molecules by SEP. The spectrum was composed of clumps, and the second and third moments of the nearest neighbor level spacing (NNLS) distribution strongly supported the conclusion that the spectrum possesses a level structure characteristic of a quantum system whose classical analog exhibits chaos. Pique *et al.*[17] further investigated vibrationally highly excited acetylene in the energy region around $26\,500$ cm^{-1} by SEP spectroscopy with 0.3 cm^{-1} resolution and found a very complex nested clump structure. When the spectral resolution was improved to 0.05 cm^{-1}, they also found a clump structure with a similar complexity underneath each peak in a clump in the lower resolution SEP spectrum. In place of assignments of individual peaks in observed clumps, an analysis using a statistical Fourier transform (SFT)[19] procedure was performed. The SFT analysis showed that there are two time scales (3 and 45 ps) representing certain dynamics corresponding to the two stages of the clump structure and that acetylene in that energy region is partly chaotic on both of these time scales.

By using this statistical procedure, short range spectral correlations, which can also be extracted by a NNLS distribution, as well as the spectral intensity distribution, are simultaneously treated and reinterpreted as dynamical behavior. The SFT of the SEP spectrum was further used to propose the existence of a large amplitude vibrational motion,[19] i.e., rotational motion of one H atom around a CCH core, which is related to the isomerization reaction from acetylene to vinylidene. The isomerization to vinylidene was further inferred from cross-correlation analysis of SEP spectra by Chen *et al.*[20–22]

These statistical analyses of SEP spectra of highly excited acetylene strongly stimulated both experimental[7,8] and theoretical efforts[23] towards an understanding of the dynamics in highly excited acetylene. Holme and Levine[23] performed classical trajectory calculations using a schematic potential surface and initial conditions designed to replicate an SEP prepared state and suggested that energy flows first from the CH stretches to the CC stretch and then to the *trans*-bend in the 26 500 cm^{-1} region and that the two time scales identified in the SFT analysis correspond to the stepwise energy flow via these two couplings. Though the initial conditions Holme and Levine adopted are incorrect, because there is no excitation of the CH stretch in the initial stage of the Franck–Condon projection onto the electronic ground state potential, their study demonstrated that the stepwise energy flow is reflected as a nested clump structure in the SEP spectrum.

The deviation from fully chaotic behavior revealed in the SFT analysis of highly excited acetylene in the 26 500 cm^{-1} region indicates the existence of at least one good quantum number, even in this highly excited region. Thus, it is expected that the assignment of at least one vibrational quantum number is possible and the origin of the complex SEP spectrum and the characteristic nested clump structure can be extracted in the course of the assignment of vibrational quantum numbers. The DF spectroscopy, which enabled the vibrational assignment of highly excited SO$_2$, was expected also to be powerful for the vibrational assignment of highly excited acetylene.

3.2. *DF Spectroscopy of Acetylene and Darling–Dennison Resonance Tier*

As demonstrated in previous absorption[24] and DF spectra[15,25] of acetylene, vibrational assignments in the low vibrational energy region are routine and unambiguous. Secure vibrational assignments may be extended towards the highly excited region once a wide vibrational energy region of a DF spectrum is measured. By exciting the $V_0^2 K_0^1$ and $V_0^3 K_0^1$ vibrational bands of the $\tilde{A}^1 A_u$–$\tilde{X}^1 \Sigma_g^+$ transition, two DF spectra spanning a wide energy region (5000–24 000 cm^{-1}) were recorded, where V represents the *trans*-bending mode (ν_3' in the $\tilde{A}$ state and ν_4 in the $\tilde{X}$ state). Due to the large change in geometry along the ν_2 (the CC stretch) and the ν_4 (the *trans*-bend) coordinates in the $\tilde{A}$–$\tilde{X}$ transition, long FC bright progressions in combinations of these two normal modes are expected.

The observed DF spectrum of acetylene below 14 000 cm^{-1} is shown in Fig. 5, where six progressions in even quanta of ν_4 ($\omega_4 = 624$ cm^{-1}) in combination with ν_2 ($\omega_2 = 2012$ cm^{-1}) are easily identified. The assigned progressions are marked with respect to the number of quanta in the ν_4 mode. In the high energy region above 14 000 cm^{-1}, the assignment of individual peaks becomes difficult due to the increasing density of peaks, but it is still possible to account for the gross distribution of intensity by extrapolating the progressions from the low energy region. Shown in Fig. 6 is the DF spectra above 14 000 cm^{-1} with the extrapolated progressions in the ν_2 mode.

Based on predictions of the vibrational term values from the anharmonic expansion coefficients tabulated by Strey and Mills,[26] 26 peaks in six progressions in v_4 were assigned to $(0, v_2, 0, v_4, 0)$ with $v_2 = 0$–5. A least-squares fit of the observed vibrational term values of the levels in these progressions was performed using an anharmonic expansion,

$$G(v_1, v_2, v_3, v_4, v_5) = \sum_i \omega_i (v_i + 1/2) + \sum_{i \leq j} x_{ij}(v_i + 1/2)(v_j + 1/2)$$

$$+ \sum_{i \leq j} x_{l_i l_j} l_i l_j \, , \tag{2}$$

where l_i ($i = 4$ or 5) represents the vibrational angular momentum of the degenerate bending mode, ν_4 or ν_5. The preliminary least-squares fit showed reasonably small residuals within the ~ 20 cm^{-1} experimental error except for relatively large (~ 25 cm^{-1}) residuals for levels with $v_4 = 14$.

Below 14 000 cm^{-1}, at least three additional *trans*-bend progressions built on excitations in a third vibrational mode were identified. It was found that term values of these additional progressions are well explained when they are assigned to $(0, v_2, 0, v_4, 2)$. However, the ν_5 mode is intrinsically FC dark and should not be detected in DF spectra without an interaction with FC bright levels having excitation in the ν_2 and/or ν_4 modes.

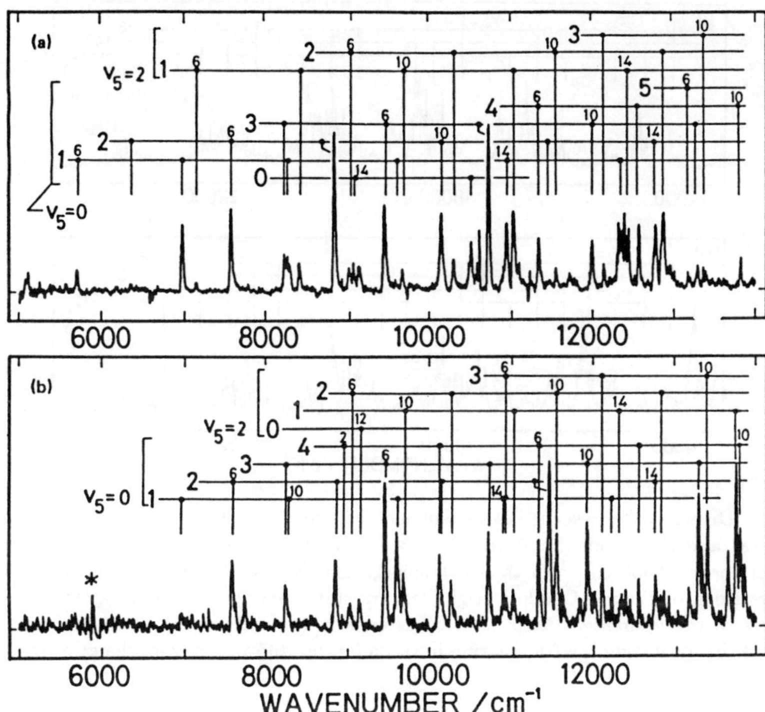

Fig. 5. Dispersed fluorescence spectra of acetylene from the $\tilde{A}^1A_u$ $v'_3 = 2$ $J_{Ka,K_c} = 3_{1,3}$ level (a) and that from the $\tilde{A}^1A_u$ $v'_3 = 3$ $J_{Ka,K_c} = 3_{1,3}$ level (b) in the low energy region below 14 000 cm^{-1} in the electronic ground $\tilde{X}$ state. The vibrational assignments are presented as progressions with respect to the ν_4 mode. In (b), a peak with an asterisk at 5910 cm^{-1} represents an Hg transition. (Ref. 5)

Taking account of (i) the intensity enhancement observed in these additional progressions at energies just above the $(0, v_2, 0, 14, 0)$ peaks, whose term values exhibit significant negative deviation from Eq. (2), and (ii) the prediction from the spectroscopic constants that the $(0, v_2, 0, v_4, 0)$ and $(0, v_2, 0, v_4, 2)$ levels cross between $v_4 = 16$ and 18, the Darling–Dennison (DD) resonance between the two degenerate bending modes was proposed as the perturbation which distributes FC bright character to the v_5 (*cis*-bending) mode. For C$_2$D$_2$, the importance of the DD resonance was shown by Huet *et al.*[27] based on far-infrared and infrared absorption spectra.

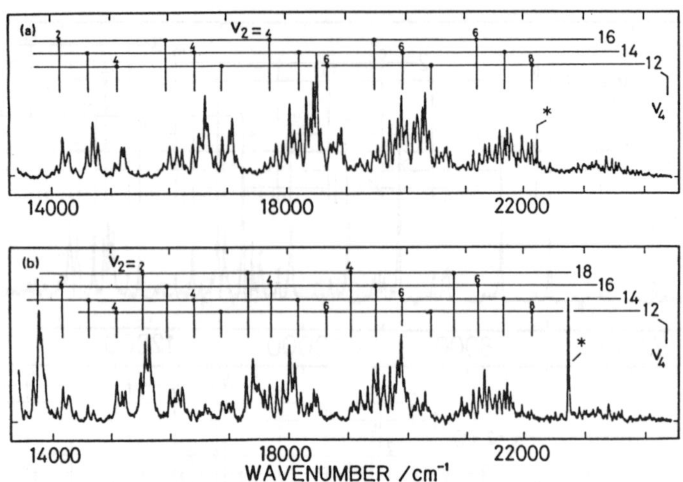

Fig. 6. Dispersed fluorescence spectra of acetylene from the $\tilde{A}^1 A_u$ $v'_3 = 2$ $J_{K_a,K_c} = 3_{1,3}$ level (a) and that from the $\tilde{A}^1 A_u$ $v'_3 = 3$ $J_{K_a,K_c} = 3_{1,3}$ level (b) in the high energy region above $14\,000$ cm^{-1} in the electronic ground $\tilde{X}$ state. The predicted vibrational progressions of Franck–Condon bright feature states are labelled as progressions with respect to the ν_2 mode. Peaks with an asterisk in (a) and (b) represent the scattered light of the excitation laser appearing in second order diffraction from the grating. (Ref. 7)

The DD interaction which occurs at the lowest energy is that between two levels; $(v_4, v_5) = (2, 0)$ and $(0, 2)$. However, due to the relatively large energy separation between these two levels, the DD interaction results in only a small perturbation for these levels. As the vibrational energy increases (as v_4 increases), due to the self- and cross-anharmonicities of the ν_4 and ν_5 modes, the separation between the two $(v, 0) \sim (v{-}2, 2)$ interacting levels gradually becomes smaller. At the same time the magnitude of the off-diagonal DD matrix elements becomes larger, roughly proportional to $v_4 v_5$. Therefore, in a certain energy region, energetically adjacent levels forming a polyad merge with each other and the DD resonance tier by which $(v_{\text{bend}}/2) + 1$ levels are connected by $(v_{\text{bend}}/2)$ resonances would be produced, where v_{bend} represents the total bending quanta, $v_4 + v_5$. Within this tier, each of quantum numbers v_4 and v_5 representing individual degenerate bending modes loses its significance, and only the sum of the two values, v_{bend}, can be used to label the entire $(v_{\text{bend}}/2 + 1)$ levels in a tier.

3.3. *Feature State*

As explained above, due to the mixing within the tier caused by the DD resonance, it becomes impossible to assign v_4 and v_5 vibrational quantum numbers to the peaks appearing around the energy position where the bright (or unperturbed) $v_4 = 14$ level is located. In other words, the bright character of the unperturbed state is shared by the levels within the tier, and many peaks, each of which represents a transition to one of the levels in the tier, are observed. When anharmonic couplings statistically mix the FC bright character into nearby FC dark levels, the intensity maximum of the group of eigenstates will coincide with the energy position of the zero-order FC bright level. By using the anharmonic expansion coefficients, which do not take account of the DD resonance tier, the vibrational energies of the FC bright levels were predicted in the energy region above $14\,000$ cm^{-1} and are shown in Fig. 6. This figure clearly shows that the predicted energy positions are almost equal to the centers of a clump whose full-width at half maximum (FWHM) is around 150 cm^{-1}. The $3 \sim 5$ peaks forming each clump represent a group of levels which share the bright character of the unperturbed zero-order state via the DD resonance. Thus, we can assign a set of vibrational quantum numbers representing a zero-order FC bright level to a group of levels. When a set of vibrational quantum numbers can be assigned to a *group* of eigenstates rather than to a *single* level, we call the group of eigenstates as a *feature state* (or a feature) which is represented by the set of vibrational quantum numbers. This type of assignment can be regarded as a generalization of the previous assignment, where one set of quantum numbers (i.e., a zero-order level) is assigned to each peak in a spectrum.

Shown in Fig. 7 is the $14\,000 \sim 16\,500$ cm^{-1} region of the DF spectra recorded via excitation of the two different vibronic levels. Due to the difference in the vibrational wave functions in the electronically excited state, the relative intensities of each feature, representing one bright state, in the two spectra are different. However, it can be noticed that the relative intensities of peaks within each feature in these two spectra are very similar. For example, the similarity of the intensity patterns within the groups of features assigned to $(v_2, v_4) = (2, 16)$, $(3, 16)$, and $(4, 14)$ is apparent. This observation strongly supports the explanation that the clump structure is formed by the DD resonance tier mechanism.

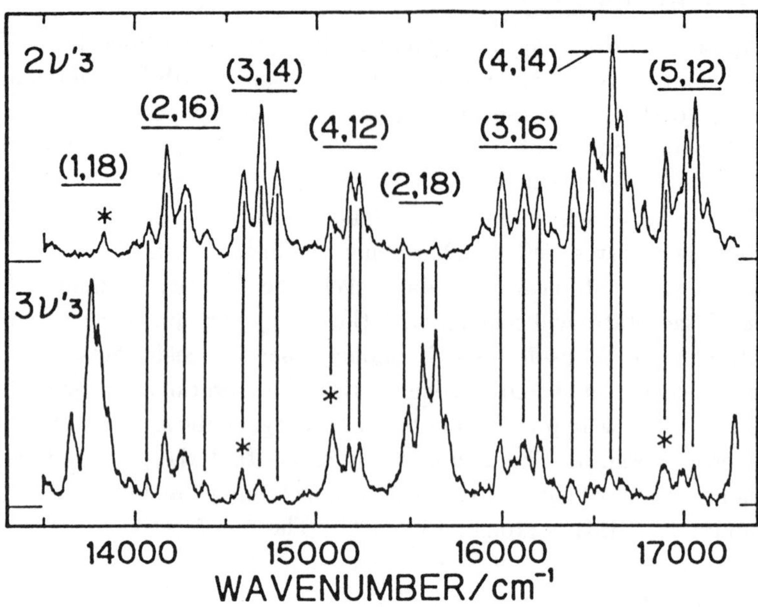

Fig. 7. Groups of acetylene DF features with nearly constant intensity ratios in $v'_3 = 2$ (upper trace) and $v'_3 = 3$ (lower trace) DF spectra are identified in the $E_{\mathrm{vib}} = 13\,500\text{–}17\,300$ cm^{-1} region. Vertical bars connect corresponding features in the two spectra. A horizontal bar with (v_2, v_4) above a group of features represents the energy region over which the character of a FC bright state is distributed. In the $v'_3 = 3$ DF spectrum, the intensities of a feature labelled with an asterisk in (3,14), (4,12), and (5,12) features may be affected by overlapping with the (0,22), (1,20), and (2,20) features, respectively. The peak with an asterisk at 13 820 cm^{-1} in the $v'_3 = 2$ DF spectrum is assigned to (4,10). (Ref. 7)

3.4. *Hierarchical Level Structure and IVR*

A comparison of the SEP and DF spectra in the 16 500 cm^{-1} region revealed that each peak in the DF spectrum (30 cm^{-1} resolution) is composed of $10 \sim 20$ peaks in the SEP spectrum (0.5 cm^{-1} resolution). Taking account of the rotational transitions, $5 \sim 10$ vibrational levels are contained within each peak of the DF spectrum. From the density of $\Sigma_g^+(l = 0)$ levels,[22] 50–70 vibrational eigenstates are expected within an interval of 70 cm^{-1}, which is the width of one peak in the DF spectrum at this energy. Thus, each vibrational "level" observed in the SEP spectrum may contain more than one vibrational eigenstate and this too can be regarded as a vibrational feature state. In other words, each DF feature at this energy contains $5 \sim 10$ subfeatures and each subfeatures contains a few eigenstates.

A comparison of DF and SEP spectra shown in Fig. 8 indicate that there are at least two stages in the hierarchical spectral structure in the highly excited region of acetylene. Due to the scaling of the off-diagonal matrix elements for anharmonic resonances mentioned in Sec. 2.2, the DD resonance $2\nu_4 \leftrightarrow 2\nu_5$, the vibrational l-resonance $2l_4 \leftrightarrow 2l_5$, and the Fermi $\nu_2 + \nu_4 + \nu_5 \leftrightarrow \nu_3$ (ν_3 is the antisymmetric C–H stretch) should all have large off-diagonal elements in the higher energy region above $14\,000$ cm^{-1}. Therefore, the splitting of the three groups of 300 cm^{-1} wide features in Fig. 8(b) into individual DF features was ascribed to the three main anharmonic resonances. The splitting of each single DF feature into the subfeatures observed in the 0.5 cm^{-1} resolution SEP spectrum was ascribed to weaker anharmonic and Coriolis resonances. The "giant clump" structure with the gaps within repeat intervals in 1900 cm^{-1} Fig. 8(a) shows that the coupling of the FC bright CC-stretch/*trans*-bend excitation is too weak to spread intensity over energy ranges larger than 1000 cm^{-1}.

Considering the results from the SFT analysis[17] and the classical trajectory calculations[23] together with the fact that shorter time dynamics is reflected in a lower resolution spectrum,[28] the hierarchical level structure implies the stepwise energy flow. That is, the energy flows from the initially localized CC-stretch/*trans*-bend excitation into the other vibrational modes in at least two distinct timescales in the highly excited region. Thus, the hierarchical structure in Fig. 8 can be interpreted as a stepwise energy transfer from the initially excited CC-stretch and *trans*-bend combination levels into other modes. The lifetime required for the evolution from the FC bright state of a DF feature is estimated to be 35 fs from the 150 cm^{-1} FWHM of each group of DF features. The further splitting observed in SEP reflects weaker resonances and a lifetime for this second stage of evolution is estimated to be 150 fs, corresponding to the 35 cm^{-1} FWHM of a DF feature. Here, we are able to describe directly from a spectrum where and how fast the energy flows. The hierarchy of subfeatures within features[29] may be found in a spectrum whenever there is an approximate separation of timescales and can be regarded as a general feature of highly excited polyatomic molecules. This explicitly described and characterized energy

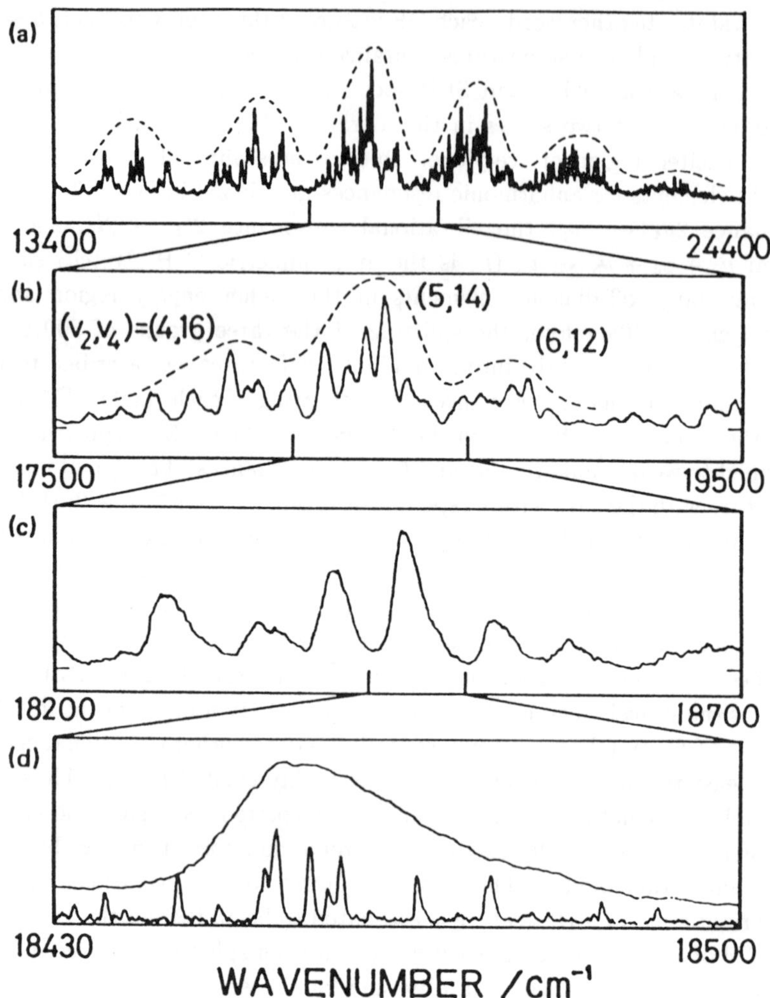

Fig. 8. A schematic diagram of the nested vibrational clump structure of highly excited acetylene; (a) giant clumps composed of several progressions of Franck–Condon (FC) bright groups; (b) three FC bright groups under the giant clump; (c) DF sub-features under a FC bright group; and (d) resolved SEP sub-subfeatures under a single DF feature. The work of Abramson *et al.* (Refs. 14,15) suggests the existence of further substructure under the features observed in SEP. (Ref. 7)

flow process is exactly the phenomenon of intramolecular vibrational energy redistribution (IVR).

It may be said that a feature state encodes the short time behavior of the evolution from the initially localized FC projection into vibrational eigenstates, and thus, the assignment of the feature states in the DF spectra can be viewed as an assignment of the short time IVR dynamics. By combining the DF and SEP spectrum, which has higher resolution, the time range of the dynamics can be extended towards longer time scale.

Quantum numbers which are good for early time may provide a theoretical framework for understanding the mechanism which produces chaotic level structure in the high resolution SEP spectra of acetylene at higher energies and may bear a close relationship to the scars of periodic orbits[28] for systems whose classical limit is chaotic.

3.5. *SEP Clarification of Anharmonic Resonances*

When both *trans*-bend and *cis*-bend are simultaneously excited, the individual vibrational angular momenta l_4 and l_5 couple to produce the total vibrational angular momentum $l = l_4 + l_5$, and vibrational l-resonance of $\Delta l_4 = -\Delta l_5 = \pm 2$ connects levels with common point group symmetry, differing only in the quantum numbers of l_4 and l_5. The constants determined by Pliva[30] allow prediction of the pattern of observed levels with total vibrational angular momenta $l = 0$ and 2. Therefore, even in the low energy region, a high resolution SEP experiment is expected to provide a definitive spectroscopic manifestation of the anharmonic resonances which play a central role in the vibrational energy flow at high energy.

By using the $V_0^3 K_0^1$ transition, the high-resolution (0.1 cm^{-1}) SEP spectrum was recorded in the 7000 cm^{-1} region of the electronic ground state, where the (0,1,0,8,0) and (0,1,0,6,2) DF peaks were observed.[8] The assignment of the (0,1,0,8,0) level is confirmed by both the rotational constant and the magnitude of the strong rotational l-resonance between $l = 0$ and the e parity component of $l = 2$. The high sensitivity of the SEP method also allowed us to identify the (0,3,0,2,0) vibrational state.

Taking account of the fact that two more $l = 2$ levels than $l = 0$ levels were identified in the observed energy region, the (0,1,0,6,2) and (0,0,1,5,1) levels were firmly assigned, based on vibrational band origins, rotational constants, and l-doubling constants. The observation of the (0,0,1,5,1) levels clearly showed the importance of the "2345 Fermi resonance",

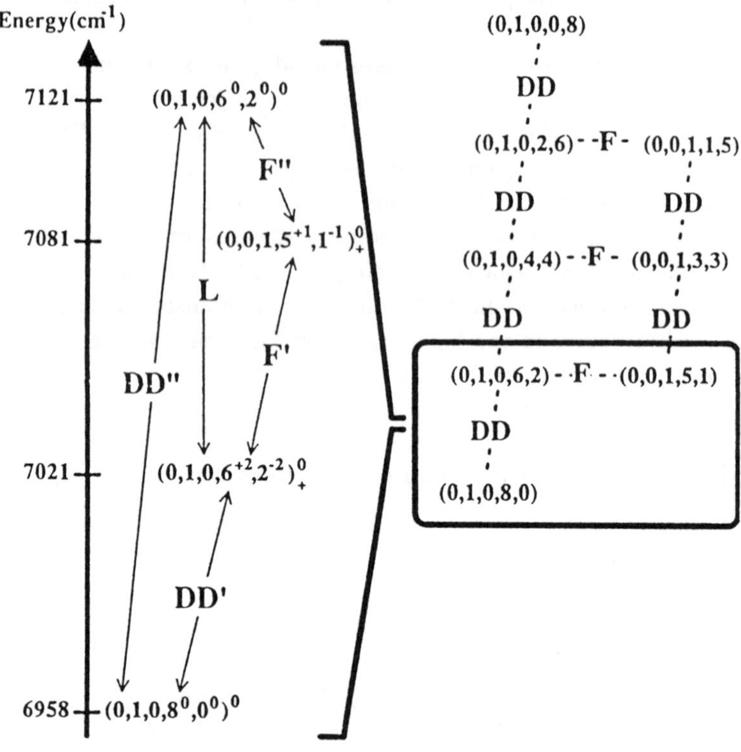

Fig. 9. Interaction diagram for observed Σ_g^+, $J = 0$ levels showing the anharmonic coupling due to the Fermi, Darling–Dennison, and vibrational l-resonance mechanisms. The observed levels are further coupled to unseen sets of levels at higher energy by the same interactions as specified in the inset. The matrix elements coupling the observed levels are $DD' = \sqrt{3}\,(r_{45} + 2g_{45})$, $DD'' = 4s_{45}$, $L = 4\sqrt{3}\,r_{45}$, $F' = -(1/\sqrt{2})K_{2345}$; $F'' = -(\sqrt{6}/4)K_{2345}$. The spectroscopic constants s_{45}, r_{45}, g_{45}, and K_{2345} are defined in Ref. 30. (Ref. 8)

which distributes intensity from $(0,1,0,6,2)$ to $(0,0,1,5,1)$ via the selection rules, $\Delta v_3 = -\Delta v_2 = -\Delta v_4 = -\Delta v_5 = \pm 1$ and $\Delta l_4 = -\Delta l_5 = \pm 1$. The identified resonances among the observed $J = 0$ Σ_g^+ levels are illustrated in Fig. 9.

High-resolution SEP spectroscopy in the low vibrational energy regions clearly proved the existence of the DD resonance between the two degenerate bending modes, which plays a crucial role in the earliest time

IVR in the highly excited region, and also successfully identified the vibrational *l*-resonance and the 2345 Fermi resonance, which were the predicted interactions that promote IVR towards the ν_3 mode from the ν_2, ν_4, and ν_5 modes. Due to this network of anharmonic resonances identified in the high resolution SEP spectrum at low energy, acetylene is destined to be strongly and specifiably perturbed at high energy. In other words, though their effects can be treated as small perturbations at low energy, the characteristic dynamics dominates at high energy by the large off-diagonal matrix elements of the anharmonic resonances.

4. SEP and DF Spectroscopy for Dynamical Processes

The DF spectrum of SO$_2$ revealed that the assignment of a set of vibrational quantum numbers to individual peaks in a low resolution spectrum is possible even when SO$_2$ is excited to the energy region where classical trajectories exhibit vibrational chaos. However, the comparison of the SEP and DF spectra at high energy showed that each peak in the low resolution DF spectrum splits into several peaks in the high resolution SEP spectrum, and the difficulty of the assignment of individual peaks in the SEP spectrum was ascribed to anharmonic resonances among three vibrational modes. The complementary DF and SEP spectroscopic measurements of acetylene more clearly uncovered the role of anharmonic resonances. It was shown that anharmonic resonances result in the characteristic nested vibrational level structure in the highly excited region and the mechanism of the stepwise intramolecular energy flow can be extracted by the assignment of the hierarchical structure of the feature states. Both the chaotic behavior of highly excited SO$_2$ in classical trajectory calculations[11,12] and the signature of chaos identified in the SFT analysis of highly excited acetylene[17] may be understood in terms of the network of anharmonic resonances which explicitly describe the time evolution of feature states. Each feature state identified in the highly excited region of acetylene is a mixed state in which the *trans*-bending and *cis*-bending modes are extensively mixed. This mixed state may have a character of a local bending mode of one hydrogen atom with respect to the CCH core. The local bending character was suggested by the SFT analysis of SEP spectra,[19] and the direction of the motion is in accord with the reaction coordinate for acetylene $\leftrightarrow$ vinylidene isomerization. It can be said that the DF and SEP measurements of highly excited acetylene clarified the mechanism of stepwise IVR prior to the isomerization reaction. Thus, in general, the SEP and DF measurements above the threshold for unimolecular dissociation and isomerization reactions may play a crucial role in characterizing the IVR mechanism prior to the unimolecular reactions.

Acknowledgments

The SEP and DF studies of SO_2 cited in Subsec. 2.2 were performed in collaboration with H. Yamada, S. Takeuchi, and S. Tsuchiya (Univ. of Tokyo) and those of acetylene in Subsecs. 3.2–3.5 were performed in collaboration with N. Ikeda and S. Tsuchiya (Univ. of Tokyo) and D. M. Jonas, J. K. Lundberg, G. W. Adamson, S. A. B. Solina, B. Rajaram, R. J. Silbey, and R. W. Field (MIT). The author is grateful for financial support from the Morino Foundation for Molecular Science which launched the Tokyo–MIT joint work on highly excited acetylene in 1987. The author is also grateful to R. W. Field and C. D. Pibel for their critical reading of the manuscript.

References

1. C. E. Hamilton, J. L. Kinsey, and R. W. Field, *Ann. Rev. Phys. Chem.* **37**, 393 (1986); For an overview of recent progress on SEP spectroscopy, see *J. Opt. Soc. Am.* **B7**, (1990).
2. R. A. Marcus, *Ber. Bunsenges. Phys. Chem.* **92**, 209 (1988).
3. D. W. Noid, M. L. Koszykowski, and R. A. Marcus, *Ann. Rev. Phys. Chem.* **32**, 267 (1981).
4. K. Yamanouchi, H. Yamada, and S. Tsuchiya, *J. Chem. Phys.* **88**, 4664 (1988).
5. K. Yamanouchi, S. Takeuchi, and S. Tsuchiya, *J. Chem. Phys.* **92**, 4044 (1990).
6. K. Yamanouchi, S. Takeuchi, and S. Tsuchiya, *Prog. Theoret. Phys. Suppl.* **98**, 420 (1989).
7. K. Yamanouchi, N. Ikeda, S. Tsuchiya, D. M. Jonas, J. K. Lundberg, G. A. Adamson, and R. W. Field, *J. Chem. Phys.* **95**, 6330 (1991).
8. D. M. Jonas, S. A. B. Solina, B. Rajaram, R. J. Silbey, R. W. Field, K. Yamanouchi, and S. Tsuchiya, *J. Chem. Phys.* **97**, 2813 (1992).
9. M. C. Gutzwiller, "Chaos in Classical and Quantum Mechanics" (Springer–Verlag, 1990).
10. L. E. Reichl, "The Transition to Chaos in Conservative Classical Systems: Quantum Manifestations" (Springer–Verlag, 1992).
11. S. C. Farantos and J. N. Murrell, *Chem. Phys.* **55**, 205 (1981).

12. J. H. Frederick, G. M. McClelland, and P. Brumer, *J. Chem. Phys.* **83**, 190 (1985).

13. Y. Morino, M. Tanimoto, and S. Saito, *Acta. Chem. Scand. Ser. A* **42**, 346 (1988).

14. E. Abramson, R. W. Field, D. Imre, K. K. Innes, and J. L. Kinsey, *J. Chem. Phys.* **80**, 2298 (1984).

15. E. Abramson, R. W. Field, D. Imre, K. K. Innes, and J. L. Kinsey, *J. Chem. Phys.* **83**, 453 (1985).

16. R. L. Sundberg, E. Abramson, J. L. Kinsey, and R. W. Field, *J. Chem. Phys.* **83**, 466 (1985).

17. J. P. Pique, Y. Chen, R. W. Field, and J. L. Kinsey, *Phys. Rev. Lett.* **58**, 475 (1987).

18. L. Leviandier, M. Lombardi, R. Jost, and J. P. Pique, *Phys. Rev. Lett.* **56**, 2449 (1986).

19. J. P. Pique, M. Lombardi, Y. Chen, R. W. Field, and J. L. Kinsey, *Ber. Bunsenges. Phys. Chem.* **92**, 422 (1988).

20. Y. Chen, D. M. Jonas, C. E. Hamilton, P. G. Green, J. L. Kinsey, and R. W. Field, *Ber. Bunsenges. Phys. Chem.* **92**, 329 (1988).

21. Y. Chen, D. M. Jonas, J. L. Kinsey, and R. W. Field, *J. Chem. Phys.* **91**, 3976 (1989).

22. Y. Chen, Ph.D. Thesis, MIT, 1988.

23. T. A. Holme and R. D. Levine, *J. Chem. Phys.* **89**, 3379 (1988); (b) T. A. Holme and R. D. Levine, *Chem. Phys. Lett.* **150**, 393 (1988); (c) T. A. Holme and R. D. Levine, *Chem. Phys.* **131**, 169 (1989).

24. J. K. G. Watson, M. Herman, J. C. Van Craen, and R. Colin, *J. Mol. Spectrosc.* **95**, 101 (1982).

25. J. C. Stephenson, J. A. Blazy, and D. S. King, *Chem. Phys.* **85**, 31 (1984).

26. G. Strey and I. M. Mills, *J. Mol. Spectrosc.* **59**, 103 (1976).

27. T. R. Huet, M. Herman, and J. W. C. Johns, *J. Chem. Phys.* **94**, 3407 (1991).

28. E. J. Heller, in *Chaos and Quantum Physics, NATO Les Houches Lecture Notes,* eds. A. Voros, M. Gianonni, and O. Bohigas (North-Holland, Amsterdam, 1990).

29. K. K. Lehmann and G. J. Scherer, *Adv. Laser Spectrosc.* **3**, 107 (1986).

30. J. Plíva, *J. Mol. Spectrosc.* **44**, 145 (1972); (b) **44**, 165 (1972).

CHAPTER 14

HIGH RESOLUTION SPECTROSCOPY OF CHEMICAL ISOMERIZATION: STIMULATED EMISSION PUMPING OF HCN

David M. Jonas*

*Department of Chemistry
Massachusetts Institute of Technology,
Cambridge, MA 02139, USA*

and

C. A. Rogaski, Alec M. Wodtke[†‡§††] and Xueming Yang[‡‡]

*Department of Chemistry
University of California, Santa Barbara
Santa Barbara, CA 93106, USA*

---

*present address: 5735 South Ellis Ave., Department of Chemistry, University of Chicago, IL 60637

[†]Author to whom correspondence should be addressed

[‡]Camille and Henry Dreyfus Teacher-Scholar

[§]Alfred P. Sloan Research Fellowsip

[††]National Science Foundation Presidential Young Investigator

[‡‡]present address: Department of Chemistry, Princeton University, Princeton, NJ 08544

513

Contents

Abstract

The simple and well understood models of rotational and vibrational motion which include "the rigid rotor", "the simple harmonic oscillator" and "normal coordinates", are fully applicable to the analysis of the spectra of highly vibrationally excited HCN, even when the molecule contains enough vibrational energy to isomerize to HNC. Further consideration of more sophisticated but fully understood concepts such as axis-switching, rotational-ℓ-doubling and local anharmonic perturbations allows the assignment of every vibrational level observed in stimulated emission pumping experiments. This allows the reduction of the experimentally derived spectra to a set of well defined molecular constants that can be directly compared to theoretical calculations on HCN. This paper reports on the present status of an on-going effort to determine the potential energy surface for the isomerization reaction, "Nature's own Hamiltonian" for a chemical reaction.

1. Introduction

The beginning student of spectroscopy may quickly obtain the impression that the field concerns itself only with the quantum description of microscopic *rigid* structures. Many fundamental models such as the "rigid rotor", the "symmetric top", or the "simple harmonic oscillator" are fairly simple (and, therefore, taught first) and indeed, accurately describe a great deal of the microscopic molecular universe. When such models are accurate, one is overwhelmed by the information content of high resolution spectra. Bond lengths, bond angles, bond strengths, electronic excitation energies, ionization potentials, and spin-orbit parameters are just the "meat and potatoes" of the multi-course menu of information that can and has been obtained from five decades of research in this field.[1] As Physical Chemists interested in the fundamental nature of molecules and their reactivity, we sense intuitively that spectroscopy must be an invaluable aid in our quest for understanding. On the other hand, we are troubled by the certain knowledge that chemical transformation has little or nothing to do with rigid structures. Indeed, the destruction of one chemical bond and the formation of a new one is essentially dynamic, not static. The conceptual link between chemical transformation and a quantum theory of nearly rigid molecular structures is not at all obvious.

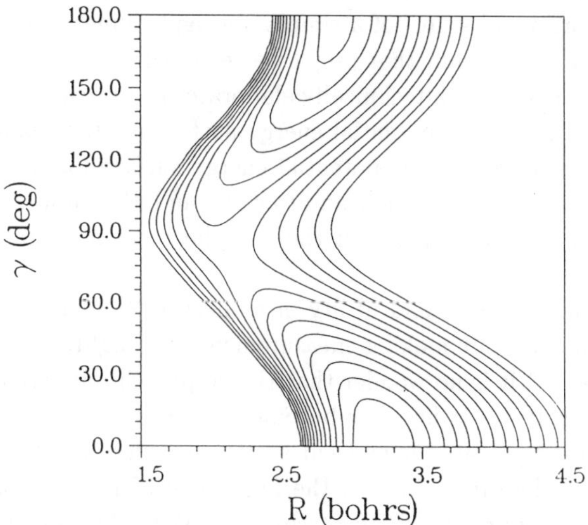

Fig. 1. A contour plot for the interaction of a H-, a C-, and an N- atom: The CN bond length is frozen at the ground electronic states equilibrium distance. R is the distance of the H atom to the center-of-mass of the CN. γ is the angle between the vector **R** and the CN bond axis ($\gamma = 0°$ is HCN, $\gamma = 180°$ is HNC). Contour intervals are 2000 cm^{-1} and the highest energy contour is 26 000 cm^{-1}.

One way to begin thinking about this problem is to restrict our attention to the consideration of isomerization reactions. Figure 1 shows a cut through the Born–Oppenheimer potential energy surface for interactions between a Hydrogen, a Carbon, and a Nitrogen atom.[2] The two wells represent the stable HCN and HNC isomers. The saddle point between them is the transition state for the chemical reaction. Although one might not immediately know how it is done, it is clear that one can solve for the vibrational wavefunctions that exist on this potential surface. Furthermore, the properties of these vibrational states, e.g., their band origins, rotational constants, and electric dipole moments, must be sensitive to the precise nature of the potential surface. For starters, we can classify the vibrational states into three categories: those that are localized on the HCN side of the isomerization barrier, those that are localized on the HNC side, and those that are *delocalized* with probability amplitude in both wells. This suggests a quantum picture of isomerization, in which collisions induce transitions from HNC states on one side of the isomerization barrier to delocalized states which can be collisionally stabilized to vibrational states localized on *either* side of the activation barrier. These vibrational energy levels represent the vibrational states of a simple nonrigid isomerizing molecule. They will exhibit a spectrum that might be recorded and by undertaking conceptually simple although technically elaborate theoretical analysis, one would be able to derive the potential energy surface for the isomerization reaction. Once we have the true potential energy surface for the reaction, we can construct "Nature's own Hamiltonian" and theoretically predict from first principles any property of the gas phase chemical reaction we may desire.

With the growing power and utility of modern laser methods that are capable of producing and making measurements on highly vibrationally excited molecules,[3–8] it is no wonder that this approach has recently come to some level of fruition. It has now become quite common to record high resolution spectra of polyatomic molecules with very high internal energies. Indeed, this book describes some of the best work of this sort using the stimulated emission pumping (SEP) technique. Why then has the above described strategy for investigating chemical reactions not been more widely employed? Furthermore, why is it that there is not a single experimentally derived potential energy surface for any chemical reaction?[a]

The answer lies in a more subtle analysis of the problem. First, imagine what one would like to do. Ideally, we wish to record high resolution spectra of a molecule's rovibrational quantum states at an internal energy where

---

[a]The $H + H_2 \rightarrow H_2 + H$ reaction is a possible exception.

chemical transformation is possible. Then we hope to reduce the data to a set of well-defined, spectroscopic constants, such as rotational constants and vibrational band origins. Finally, we plan to compare theoretically and experimentally determined molecular constants and investigate how the agreement is influenced by adjustments to the potential energy surface used in the theoretical calculations.

Such an approach would be greatly simplified if there was a separation of rotational and vibrational variables in the quantum mechanical rotation–vibration Schrödinger equation, that is, if

$$\Psi_{\text{rovibrational}} = \Psi_{\text{rotational}} \, \Psi_{\text{vibrational}} \; . \tag{1}$$

At low vibrational excitation, such assumptions can normally be made since vibration is fast in comparison to rotation and, as a result, vibrational state spacings are much larger than rotational state spacings. However, for polyatomics, this is no longer true at high levels of vibrational excitation. Consequently, the assumed separation of time scales between vibration and rotation is much less likely for highly vibrationally excited polyatomics. This fundamental characteristic of highly vibrationally excited molecules will significantly complicate our approach to most molecules, which will exhibit extreme spectral complexity. Despite the spectral complexity problem, some progress has been made on the HCCH $\leftrightarrows$ CCH$_2$ isomerization.[9]

Figure 2 shows a simple harmonic-count, vibrational density of states calculation for the HCN molecule. The upper curve is for all vibrational states, while the lower curve is for vibrational states with no vibrational angular momentum. One can see that even at internal energies as high as 20 000 cm^{-1}, which is thought to be more than 3000 cm^{-1} above the barrier to isomerization, the density of zero-angular momentum vibrational states is still less than 0.1 state per cm^{-1}. Consequently, the HCN $\leftrightarrows$ HNC system may be a rare example where the separation of vibrational and rotational motion persists even above the isomerization energy threshold. This would lead to a tractable analysis of the high resolution spectra and a simplified comparison between experiment and theory.

In the molecular orbital picture, the HCN $\tilde{A} \leftarrow \tilde{X}$ transition promotes one electron from the a'' component of the π bonding orbital to the a' component of the π^* antibonding orbital. Bending stabilizes the a' orbital by increasing bonding overlap with the hydrogenic 1s orbital. Consequently, the $\tilde{A}^1A''$-state is strongly bent (C_s point group) with an elongated CN

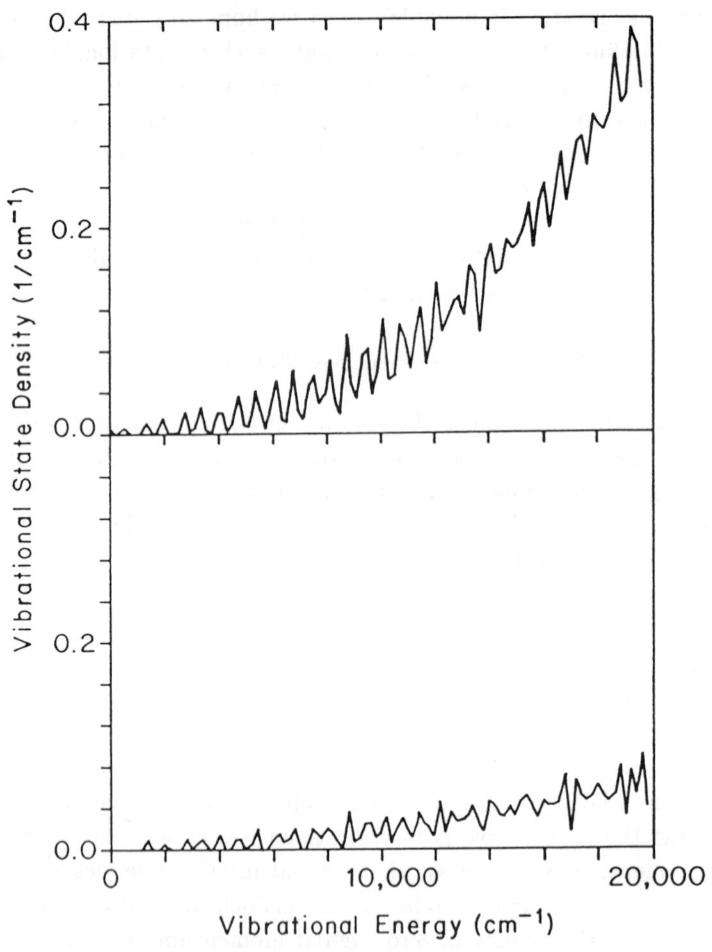

Fig. 2. Density of vibrational states for HCN as a function of Energy.

bond, in sharp contrast to the linear $\tilde{X}^1\Sigma^+$-state ($C_{\infty v}$ point group). From the $\tilde{A}$-state rotational constants for HCN and DCN, Herzberg and Innes deduced an HCN bond angle of 125° and an elongated 1.297 Å CN bond length (1.15324 Å in the ground state).[10] The CH bond length changes only slightly. From this information we can conclude that SEP spectroscopy using the intermediate $\tilde{A}^1 A''$ state would lead to highly excited *bending* states of HCN which might have a great deal to do with the isomerization reaction. This chapter describes the present state of affairs in research concerning this interesting and unique molecule and the on-going efforts at finding the potential energy surface (PES) for HCN $\leftrightarrows$ HNC isomerization.

2. SEP with a Tunable Argon Fluoride Laser

In 1981, Field and co-workers,[11] followed shortly thereafter by Knight and co-workers,[12] began systematically applying SEP to highly vibrationally excited molecules.[5] This method, which is a logical extension of the method of double resonance spectroscopy, uses two lasers to perform a special kind of two-photon spectroscopy on highly vibrationally excited molecules. The first laser, tuned to ω_{PUMP}, excites molecules out of a single, thermally-populated quantum state into an excited electronic state with a different equilibrium structure than the initial state. A second laser, tuned to ω_{DUMP} , induces emission of a photon transferring population back to the electronic ground state and leaves the molecule with $\hbar(\omega_{PUMP} - \omega_{DUMP})$ more internal energy than the initial state.

Since the transition probability of each of the two steps depends on a Franck–Condon factor for an allowed electronic transition, it is commonly found that the two-photon SEP transition can be saturated with conventional Excimer or Yag pumped dye lasers, even after one or two steps of nonlinear optical sum-frequency generation. Of course, SEP works best for the study of high vibrational states when the geometric structure of the excited and ground electronic states are as different as possible.

Figure 3 shows a schematic diagram of how our SEP spectroscopy experiment is carried out: the so-called "fluorescence-dip method". The PUMP beam from a tunable Argon Fluoride laser producing one-cm^{-1}-bandwidth light at 193 nm[13] is split into two equal pieces by a beam splitter (BS). Laser induced fluorescence is collected on two optically isolated photomultiplier tubes (PMT's), whose difference-signal is amplified for signal averaging. One of these two PUMP beams is spatially and temporally overlapped with the output of the DUMP dye laser. When a stimulated emission resonance is found by scanning the DUMP laser's frequency, a dip in the side-fluorescence seen by the PMT is observed, since the stimulated emission photons emerge from the sample cell with the DUMP laser beam. The shot-by-shot normalization to the PUMP laser accomplished with the dual beam geometry allows small fluorescence dips to be observed in the presence of relatively large PUMP laser intensity fluctuations.

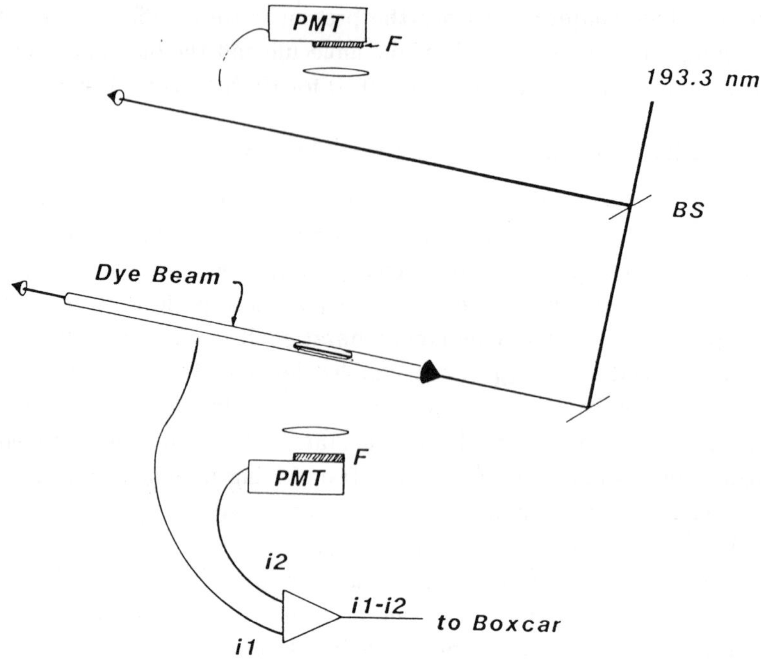

Fig. 3. Schematic of the experimental arrangment for SEP on HCN : See text.

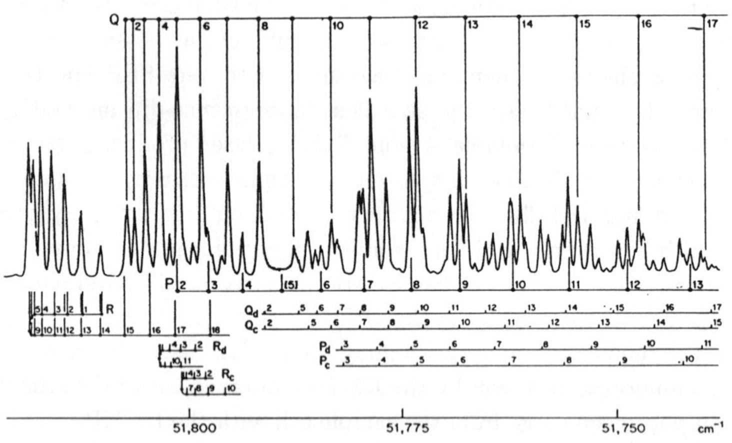

Fig. 4. Laser Induced Fluorescence Spectrum of the HCN $\tilde{A}^1 A''(0,1,0)^{K'=1} \leftarrow \tilde{X}^1\Sigma^+(0,2^{|\ell''|=0,2},0)$ band. The subscripts 'c' and 'd' label transitions reaching the lower and upper components of the $K'=1$ level from $\ell''=2$.

Figure 4 shows the results of scanning the tunable Argon Fluoride laser over its one nanometer wavelength scanning range, while collecting the HCN laser induced fluorescence signal on a single PMT. The spectrum is easily assigned to the $(0,1,0)^{K'=1} \leftarrow (0,2^{|\ell''|=0,2},0)$ hot-band transition of the bent $\leftarrow$ linear, $\tilde{A} \leftarrow \tilde{X}$ system first analyzed by Herzberg and Innes.[14] In this notation, $(v_1',v_2',v_3')^{K'} \leftarrow (v_1'',v_2''^{|\ell''|},v_3'')$, normal modes 1, 2, and 3 are approximately CH stretch, bend, and CN stretch, respectively. ℓ'' indicates the vibrational angular momentum accompanying the doubly degenerate bending motion of the linear ground state, which can take on the values $\ell'' = v_2, v_2-2, v_2-4, \ldots -v_2$. Anharmonicity lifts the degeneracy between states with different absolute values of ℓ'', but vibrational states with $\ell'' \neq 0$ are still doubly degenerate, corresponding to the two equal and opposite possible directions for the vibrational angular momentum. The excited $\tilde{A}$-state is nearly a prolate symmetric top. Therefore, we speak of a projection of the total angular momentum, K', along the top axis, which is approximately the CN bond.

Figure 5 shows examples of the first fluorescence dip SEP spectra obtained for HCN.[15,16] The five spectra shown are recorded for five different PUMP transitions of the $(0,1,0)^{K'=1} \leftarrow (0,2^{|\ell''|=0},0)$ Q-branch. The simplicity of the spectra is a noteworthy feature of the SEP technique, which provides emission spectra of the single PUMP-laser-prepared quantum state. An approximate selection rule may be derived from the direction of the transition dipole. The $\tilde{A}$-state has electronic symmetry A'' in the C_s point group, meaning that the electronic wavefunction changes sign upon reflection through the molecular plane. The $\tilde{X}$-state electronic symmetry is Σ^+ in the $C_{\infty v}$ point group, so that it is unchanged by reflection. A superposition of these two electronic wavefunctions will have an oscillating electric dipole moment which is perpendicular to the molecular plane of the bent A-state, *i.e.*, parallel to the c-inertial axis. For a rigid *symmetric* top, this direction of the transition moment implies the rotational selection rule $K_a' - \ell'' = 1$. Since ℓ'' is even or odd as v_2'' is even or odd, this "rotational" selection rule also restricts the vibrational quantum number, v_2''. In other words for the two-photon SEP transition, $|\ell'''|^{\text{b}}$ can only differ from $|\ell''|$ by 0 or 2. Since we have PUMPed out of $|\ell''| = 0, |\ell'''|$ must be 0 or 2. Because only vibrational states with an even number of bending quanta can have $|\ell'''|$ of 0 or 2, we immediately conclude that the SEP experiment will not be sensitive to the vibrational states with an odd number of bending

---

[b]The superscript, $'''$, is often used in this manuscript to distinguish the quantum numbers of the highly vibrationally excited vibrational states or SEP-final states, from those of the thermally populated levels or SEP-initial states.

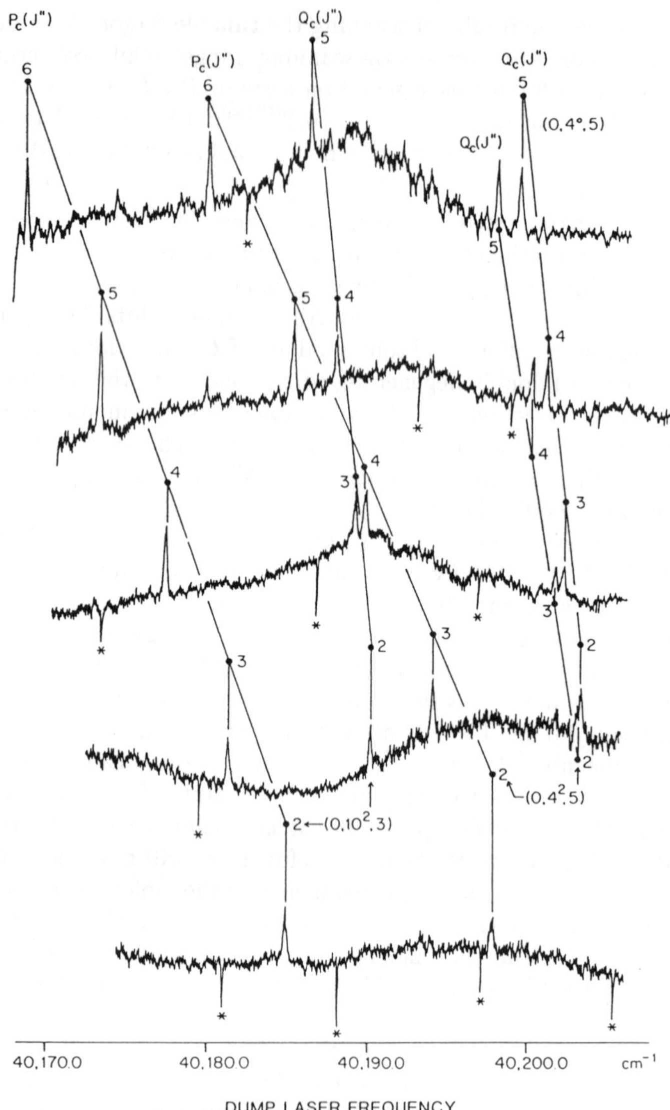

Fig. 5. Fluorescence Dip Spectra of highly vibrationally excited HCN : From top to bottom, respectively, the fluorescence dip spectra correspond to Q(5) through Q(1) PUMP transitions. Asterisks mark experimental checks of the overlap between PUMP and DUMP laser beams. The 'c' subscript is a reminder that the transitions labelled simply Q in Fig. 4 reach the lower component of $K' = 1$.

quanta, v_2'''. This conclusion turns out to be only approximately correct as will be discussed further below. Figures 6 and 7 graphically represents the rigorous rotational selection rules for Q-branch-PUMPed SEP transitions terminating in $|\ell'''| = 0$ and 2, respectively. The electric dipole parity selection rule is $+ \leftrightarrows -$. Inspection of Figs. 6 and 7 reveals that because we have exclusively used the $|\ell''| = 0$ Q-branch for the PUMP transition, stimulated emission transitions to $|\ell'''| = 0$ will only exhibit a Q-branch while those to $|\ell'''| = 2$ will exhibit P, Q, and R branches. This allows the determination of the ℓ''' quantum number for many vibrational states.

It is a simple matter to construct rotational term energy plots[c] for the highly vibrationally excited states since one knows the energy of the initial state as well as $\hbar(\omega_{\text{PUMP}} - \omega_{\text{DUMP}})$. Such plots derived from systematically obtained SEP spectra such as those of Fig. 5 reveal the identities of each state's rotational quantum numbers and disclose that the rotational energy is given by the surprisingly simple relation $B(v_1, v_2, v_3)J(J + 1)$ to nearly within the experimental error. Here, $B(v_1, v_2, v_3)$ refers to a vibrationally state specific rigid rotor rotational constant. The reader is referred to Refs. 15 and 16 for more details of the analysis. The fact that each of the observed vibrational states possessed a series of rigid rotor rotational states with an easily derived rotational constant specific to that vibrational state means that one of the above-mentioned hopes was indeed fulfilled and a simple reduction of the data to a set of vibrational band origins and rotational constants was possible, even at the highest experimentally achieved energies.

Between 8900 and 18 900 cm^{-1}, 67 vibrational bands were observed in the SEP spectrum of HCN. It was found that the band origins and the rotational constants for 30 of the $\ell''' = 0$ states could be predicted by two simple equations that have their roots in the simple harmonic oscillator, the rigid rotor and perturbation theory.

$$G(v_1''' = 0, v_2''', v_3''') = 710.77v_2''' - 2.57v_2'''^{\,2} - 2.71v_2'''v_3'''$$
$$+ 2107.06v_3''' - 10.74v_3'''^{\,2} \, . \tag{2}$$
$$B(v_1''' = 0, v_2''', v_3''') = 1.4782 + 0.00395v_2''' - 0.0109v_3''' \, . \tag{3}$$

23 more states could then be identified as the $|\ell'''| = 2$ "partners" of some of the $\ell''' = 0$ states since they had rotational constants which were nearly identical to their $\ell''' = 0$ "partners". In addition, the $|\ell'''| = 2$ states were

---

[c] Each observed SEP transition corresponds to observation of a given highly vibrationally excited quantum state with energy, $E_{\text{initial}} + \hbar(\omega_{\text{PUMP}} - \omega_{\text{DUMP}}) \equiv E_{\text{rot. term}}$.

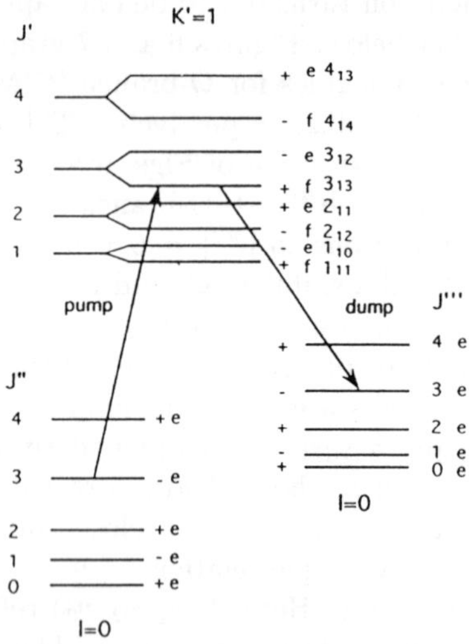

Fig. 6. Selection rules for QQ branches to an $\ell = 0$ level: All levels are indicated by their parity ($\pm$) as well as e/f symmetry. The $K' = 1$ levels are also denoted with their asymmetric rotor labels, $J'_{K_{a'}K_{c'}}$. The energy ordering labels 'c' and 'd' used for $K' = 1$ in Fig. 4 correspond to the symmetry labels 'f' and 'e', respectively. See text.

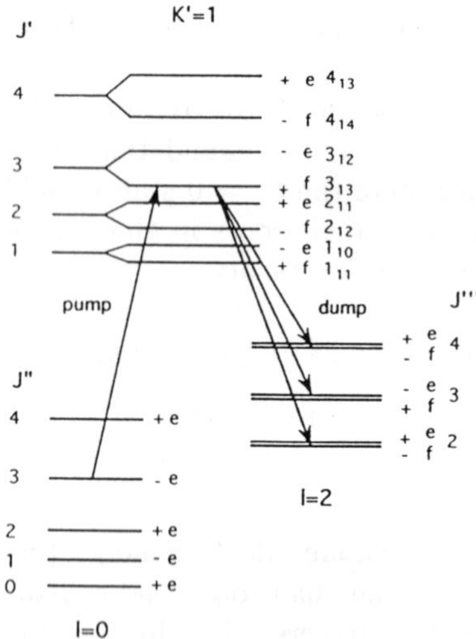

Fig. 7. Selection rules for QP, QQ, and QR branches to an $|\ell| = 2$ level: Quantum state labels are identical to Fig. 6.

always about 15–20 cm^{-1} higher in energy than the corresponding $\ell''' = 0$ states.

Both the rotational constants and vibrational energies for the assigned SEP transitions could be combined with the infrared and overtone data[17] and satisfactorily fit to an anharmonic normal mode expansion. This was used as a "prediction function" and it was believed to predict the band origins and rotational constants of all unobserved levels that were not influenced by perturbations or isomerization. From this a list of candidate states was constructed in the hope of finding the identities of the fourteen remaining unassigned states. See Table 13 of Ref. 16.

It was realized at that time that many vibrational states with odd numbers of bending quanta were close in energy to unassigned states, suggesting a violation of the approximate $K_a' - \ell$ angular momentum selection rule mentioned above. However, the observed rotational constants for these unassigned vibrational states disagreed with the values expected from simple interpolation between other observed and assigned vibrational states using Eq. 3. For example, it was known that the rotational constant for $(0, 12^0, 4)$ was 1.486 cm^{-1} and similarly that of $(0, 8^0, 4)$ was 1.464 cm^{-1}. Although an unassigned state at 15651.8 cm^{-1} appeared to be at the correct energy for the expected vibrational band origin of the $(0, 11^1, 4)$ {15651.6 cm^{-1}}, how could it be that $(0, 11^1, 4)$ had an observed rotational constant of 1.454 cm^{-1}, which is less than both $B(0, 8, 4)$ and $B(0, 12, 4)$? Equation 3 suggested that it *should* have a value of 1.477 cm^{-1}.

We found that the simple approach explains a great deal; however, since the unassigned states might be due to the isomerization, it is very important to consider all other possibilities. Let us start at the beginning and try to find any invalid assumptions that we might have made.

3. HCN SEP: A Closer Look[18]

The excited electronic state is actually an asymmetric top. Therefore, rotational levels of the bent $\tilde{A}$-state are characterized by three rotational quantum numbers $J'_{K_a' K_c'}$. J' is the total angular momentum; K_a' and K_c' are projections on the a-inertial axis (roughly the CN bond) and the c-inertial axis (perpendicular to the plane of the bent molecule), respectively. These labels are also indicated in Figs. 6, 7, and 8.

In the linear ground state vibrational states are accordingly labelled $(v_1'', v_2''^{|\ell''|}, v_3'')$ where again the normal modes v_1'' and v_3'' are roughly CH and CN bond stretches, respectively. The angular momentum, J'', must be greater than or equal to the vibrational angular momentum, ℓ''. Rotational states are labelled by J'', ℓ'', and a symmetry label e or f related to (but not) the parity.[d] HCN $\tilde{X}$-state $\ell'' = 0$ levels have rovibronic symmetry e for all J''; one linear combination of the two degenerate $\ell'' \neq 0$ levels has rovibronic symmetry e and one symmetry f for each J''. $\tilde{A}$-state levels with $K_c = J'[K_c = J' - 1]$ have symmetry $f[e]$.

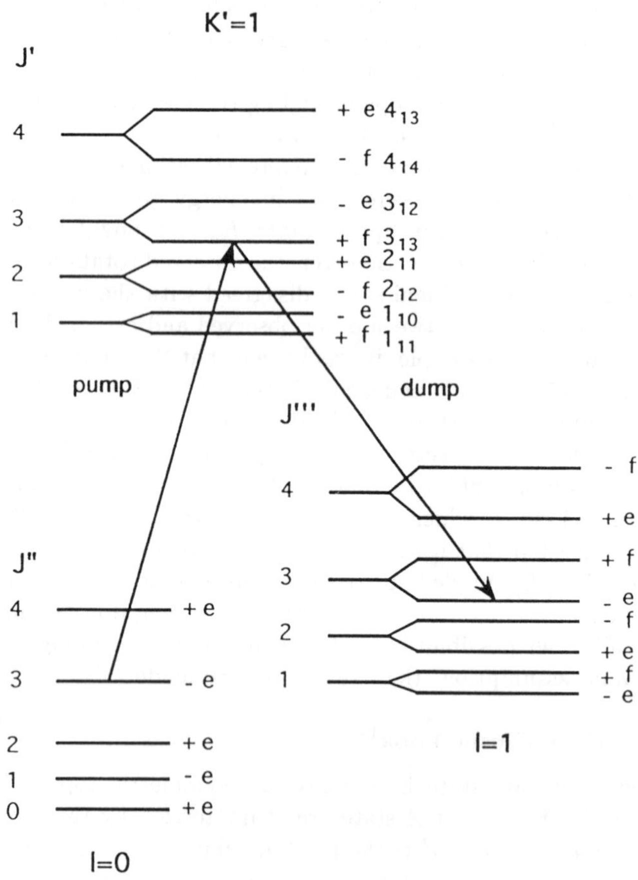

Fig. 8. Selection rules for the QQ branch of axis switching transtions to an $|\ell| = 1$ level: Quantum state lablels are the same as in Figs. 6 and 7. Note that P- and R- branch transitions are to $|\ell| = 1$; f levels are allowed but predicted to be more than an order of magnitude weaker than the Q-branch transitions shown in the figure.[18]

---

[d] J. M. Brown, *et al.*, *J. Mol. Spectrosc.* **55**, 500 (1975). For a state of parity, π, if $\pi(-1)^J$ is positive then the state is e and if $\pi(-1)^J$ negative then the state is f.

The electric dipole selection rules are: $e \leftrightarrow f$ for Q branch transitions with the additional restriction that $J' = 0 \leftrightarrow J'' = 0$ transitions are forbidden; and either $e \leftrightarrow e$ or $f \leftrightarrow f$ for P- and R-branch transitions. Since we are using the $\ell = 0$ Q-branch PUMP transition and since ten of the unassigned states were only observed as Q-branches, we can conclude *unambiguously* that the prepared rovibrational levels of these ten unassigned states are of e symmetry. This will become important in a moment.

Coriolis interactions also affect the rotational energy levels, especially at high J''. Classically, the Coriolis force due to end-over-end rotation will couple bending motion in a plane perpendicular to J'' with the stretching vibrations. Because the stretch vibrational frequencies are usually greater than bending frequencies, Coriolis coupling will decrease the effective rotational constant for bending perpendicular to J'' and increase the effective rotational constant for stretching; the rotational constant for bending parallel to J'' is unaffected by the Coriolis force. The manifestations of the Coriolis transition from symmetrically equivalent bending vibrations perpendicular to the linear axis at $J'' = 0$ to polarized bending vibrations parallel and perpendicular to the plane of rotation[19] at high J'' are called rotational-ℓ-doubling ($|\ell| = 1$ levels) or rotational-ℓ-resonance (for $|\ell| \neq 1$ levels).[20]

Quantum mechanically, rotational-ℓ-resonance and doubling split the $\ell \neq 0$ degeneracy between e and f symmetry levels.[e] In the signed ℓ basis *both* rotational-ℓ-resonance and doubling can be fit by off-diagonal matrix elements;

---

[e]T. Oka, *J. Chem. Phys.* **47**, 5410 (1967). Certain intermediate steps in a quantum mechanical calculation (e.g., the sign of off diagonal matrix elements such as q_2 or how one wavefunction transforms into another under a symmetry operation) depend on the arbitrary complex number (phase factor) by which any wavefunction can be multiplied without affecting the final result. Although many choices are possible, it is necessary to use the same phase choice in all aspects of a calculations to get correct results, and spectroscopists use various "phase conventions" for this purpose. Note that the phase convention used by Oka differs from that used here (compare Oka's Eq. 6 with Eq. 2 of Ref. 18) and that the phase convention adopted by Oka leads to a negative q_2 constant for HCN (see Oka's Eq. 32). Both phase conventions, when consistently applied, lead to the correct experimental result that the e component of the bending fundamental (bending in the plane perpendicular to J) lies below the f component (bending parallel to J).

$$\langle v_1, v_2^\ell, v_3; J, \ell | \mathcal{H} | v_1, v_2^{\ell \pm 2}, v_3; J, \ell \pm 2 \rangle = \frac{1}{4} q_2$$

$$\times \sqrt{(v_2 \mp \ell)(v_2 \pm \ell + 2)[J(J+1) - \ell(\ell \pm 1)][J(J+1) - (\ell \pm 1)(\ell \pm 2)]} \ . \tag{4}$$

The signed ℓ basis can be somewhat confusing because these basis states do not have a definite parity and one might incorrectly surmise from Eq. (4) that there is a physical interation between states of different parity. By taking symmetric and antisymmetric linear combinations of the signed ℓ basis states with $\ell \neq 0$, one obtains a new basis of states with definite parity. The matrix elements in the parity basis can be derived in a straightforward manner[18] from Eq. (4). For the largest splitting case of $|\ell| = 1$, referred to as rotational-ℓ-doubling, the energy corrections are given by Eq. (5).

$$\langle v_1, v_2^1, v_3; J, |\ell| = 1, {}_e^f | \mathcal{H} | v_1, v_2^1, v_3; J, |\ell| = 1, {}_e^f \rangle = \pm \frac{1}{4} q_2 (v_2 + 1) J(J+1) \ . \tag{5}$$

Of course, in the absence of external fields,[g] there is not and cannot be a physical interaction between states of differing parity or opposite e/f symmetry. Notice that for a given vibrational state the $|\ell| = 1$ rotational levels are perturbed in proportion to $J(J+1)$. Consequently, the ℓ-doubling acts to perturb the observed *rotational constants*. For the specific e/f symmetry of the unassigned levels (e-states) the effective rotational constant is lowered.

To first order, the constant q_2 appearing in Eqs. 4 and 5 depends on the form of the normal coordinates and the vibrational frequencies but is *independent of the vibrational quantum numbers*. The q_2 constant has in fact been determined from microwave and infrared spectroscopy of low vibrational states to be $+7.27 \times 10^{-3}$ cm^{-1}.[21,22] Let us consider, once again, our apparent paradox surrounding the $(0,11,{}^14)$ candidate state, which exhibited a reasonable vibrational band origin and an apparently anomalous rotational constant. From Eq. (5), the correction to the rotational constant is

$$(0.25)(7.27 \times 10^{-3})(11 + 1) = 0.0218 \text{ cm}^{-1} \ .$$

Adding this to the observed rotational constant (1.454 cm^{-1}) gives an expected "rigid-rotor rotational constant" of 1.476 cm^{-1} in near perfect

---

[g]This follows from the invariance of Maxwell's equations (which quantum mechanically govern all chemical forces) under the inversion of the coordinate system (parity operator).

agreement with the "prediction function" of Ref. 17 (1.477 cm^{-1}) that did not consider rotational-ℓ-doubling. This approach reduced the fitting error to approximately the experimental error for six of the fourteen unassigned states, which appear indeed to be $|\ell| = 1$ states with an odd number of bending quanta.[18]

A close inspection of the rotational constants for $|\ell'''| = 0$ and 2 in Table 7 of Ref. 16 shows that they are not always equal within the experimental uncertainty. This is also a result of the rotational-ℓ-resonance, which causes deviations from the simple rigid rotor energy formula. Subsequent rotational fits were performed which: i) included the rotational-ℓ-resonance interaction; ii) constrained the rotational constants, B_v, for $\ell''' = 0$ and $\ell''' = 2$ levels with the same normal mode quantum numbers to be equal;[h] and iii) constrained the rotational-ℓ-resonance constant q_2 to the literature value. These more highly constrained fits actually improved the global fit to the data for the assigned $|\ell'''| = 0$ and 2 states. See Tables III and V of Ref. 18.

Since the literature value of q_2 is derived from microwave and infrared data, this extrapolation is an impressive success for the anharmonic normal mode model. Similar extrapolations of rotational-ℓ-resonance have been observed in the iso-electronic molecule acetylene[26] and the isovalent molecule HCP.[23] In all of these molecules the effects of rotational-ℓ-resonances are "amplified" by the high bending vibrational quantum numbers. Consequently, these effects become quite large and easy to observe in an SEP experiment. One of the most dramatic results came in the C_2H_2 molecule where the $|\ell| = 2$ splittings were observed to be as large as 1 cm^{-1} for highly vibrationally excited states with J as low as 10.

It appears quite clear that some of the unassigned states possess the properties of $|\ell| = 1$ vibrational levels. How is it that there is significant transition probability despite the fact the the $K'_a - \ell'' = \pm 1$ selection rule is violated? The answer lies in the fact that HCN is not rigid *during* the electronic transition between the linear $\tilde{X}$-state and the bent $\tilde{A}$-state, so "axis-switching" transitions are weakly allowed. Hougen and Watson[24] showed that the rotational intensity of axis-switching transitions

---

[h]In the microwave and infrared data, a weak dependence of the rotational constants on ℓ was observed [G. Winnewisser, A. G. Maki and D. R. Johnson, *J. Mol. Spectrosc.* **39**, 149 (1971)]. The extrapolated differences are experimentally insignificant in the HCN SEP spectrum.

can be calculated from the sudden change in the Eckart axes on electronic excitation. The rotating Eckart axis system,[25] the body fixed frame within which all wavefunctions of a given electronic state are normally defined, sets the angular momentum of the rigid molecular equilibrium geometry (not the total angular momentum)[i] to zero, and thus depends on the equilibrium structure of that electronic state. For a given instantaneous geometry, the Eckart axes computed for the two different equilibrium geometries of each electronic state do not always coincide. For example, at the linear geometry, one $\tilde{X}$-state Eckart axis lies along the HCN symmetry axis; whereas, if the bent $\tilde{A}$-state is distorted to the linear geometry of the $\tilde{X}$-state, the closest $\tilde{A}$-state Eckart axis is tipped about 1.5° off the linear HCN symmetry axis. This 1.5° angle between the $\tilde{A}$- state Eckart axes and the $\tilde{X}$-state Eckart axes is called the axis switching angle. The axis switching angle is nearly constant for the range of geometries at which the electronic transition takes place.[24]

To calculate axis-switching transition intensities, one expresses the wavefunction(s) of one electronic state in terms of the Eckart coordinates of the other. Written in terms of symmetric top wavefunctions about the $\tilde{X}$-state Eckart axes, the HCN $\tilde{A}$-state wavefunction $J'_{K'_a=1,K'_c=J'}$ has some $K = 0$ and $K = 2$ character, allowing weak $\Delta K = \pm 1$ transitions to $\ell''' = 1$ and 3. In a transition to a bent electronic state, axis-switching causes forbidden rotational subbands, but in a transition terminating on a linear state, axis-switching can produce new vibrational bands. If we express the bent $\tilde{A}$-state rotational wavefunction in the $\tilde{X}$-state Eckart axes, the intensity of these axis-switching transitions can be formally separated into a rotational line-strength and a Franck–Condon factor. (Note that the Franck–Condon factors depend, in principle, on ℓ'''.)[j] These vibrational

---

[i]See L. D. Landau and E. M. Liftschitz, *Mechanics, third edition* (Pergamon Press, New York, 1976) Section 24 for a brief explanation on why the total angular momentum of rotating nonrigid body cannot be set to zero.

[j]The especially fuzzy dividing line between vibrational and rotational intensities in a bent-linear transition is illustrated by considering the $(v - l)/2$ radial nodes of the two-dimensional harmonic oscillator wavefunctions. [C. Cohen-Tannoudji, B. Diu, and F. Laloë, *Quantum Mechanics, Volume I,* (Wiley Interscience, New York, 1977) p. 737] The relative intensities for transitions from $K'_a = 1$ to $\ell'' = 0, 2$ are therefore *not* given simply by a rotational intensity. Since the difference in radial wavefunctions is greatest at the origin, relative $\ell'' = 0, 2$ intensities differing drastically from the rotational line-strengths are not expected (in the absence of perturbations) for the HCN $\tilde{A}-\tilde{X}$ transition, which is electronically forbidden at the linear geometry.

bands are, in effect, forbidden only by the rotational selection rules. An intensity interference effect enhances the Q-branch DUMP transition terminating on $\ell''' = 1\ e$ levels by a factor of nine in the high J limit. This forbidden axis-switching QQ transition is also enhanced by the SEP polarization geometry used in the experiments, and is actually predicted to be slightly more intense than the nominally allowed QR transition for equal Franck–Condon factors: both are approximately a factor of 100 weaker than the allowed Q-branch transition to $\ell''' = 0$ at low J and both are observed in the experiments.[18] In the SEP spectrum of isoelectronic C_2H2, axis-switching transitions from $K_a = 1$ to $|\ell'''| = 1$ and 3 have also been identified by their characteristic rotational intensity pattern.[26]

There are still eight vibrational states that are not explained even by the elegantly simple axis-switching mechanism. Let us start from the assumption that these are perturbed $|\ell| = 0$ and 2 states. This provides a source of oscillator strength for the unassigned states if some of the perturbed basis states were to have good Franck–Condon factors. Let us consider this point in more detail.

The SEP spectrum of HCN provides a particularly clear example of the Franck–Condon reflection of the excited state wavefunction onto the ground potential energy surface.[k] We discuss here a simple kind of plot used by Jonas to analyze emission spectra of acetylene.[27] For a single level of the excited state, we represent the emission intensity to each ground state level by a dot, with larger dots denoting greater intensity. We then plot the dots as a function of the ground state quantum numbers. If the spectra have "reflection character",[k] as will often be the case for emission from low vibrational levels of strongly displaced electronic states, we obtain a map of the nuclear probability density for the emitting level. This Franck–Condon map is extremely useful in identifying missing or misassigned levels and accidental resonances in the ground electronic state.

A Franck–Condon map of the HCN $|\ell'''| = 0$ and 2 levels assigned from the vibrational band origins and rotational constants is shown in Fig. 9.

---

[k]If, for any given photon energy, an electronic transition is classically allowed (according to the Franck principle) at only one instantaneous geometry, the strong Franck–Condon factors simply reflect the nodal pattern of the initial vibrational state. This is because the transition between the two vibrational states of the two electronic states can be viewed as taking place in a small region around the one classically allowed instantaneous molecular geometry. Therefore, interference effects do not contribute significantly to the overlap integral. See J. Tellinghuisen, *J. Mol. Spectrosc.* **103**, 455 (1984)

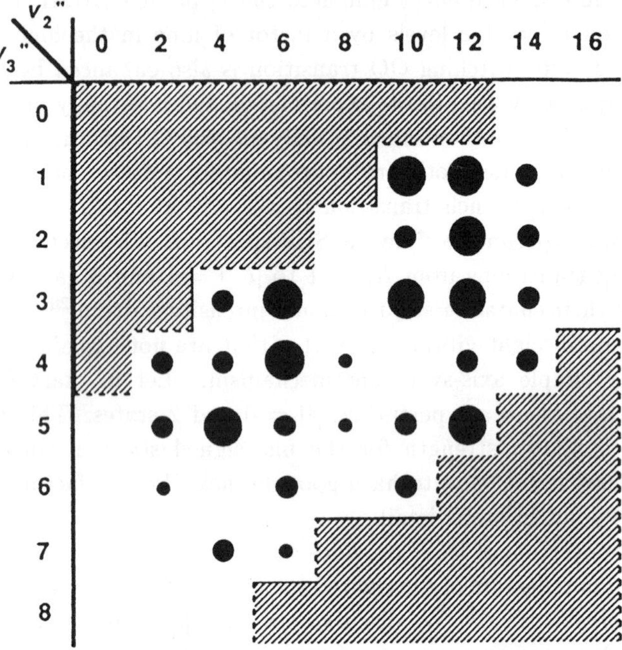

Fig. 9. Franck–Condon map of experimentally observed emission from the $\tilde{A}(010)$ level of HCN: The size of the dots indicates the approximate relative intensities of each observed vibronic band. Transitions to $v_2 = 8$ were weak or missing, while transitions to $v_2 = 6$ or 10 were strong, indicating that $v_2 = 8$ is near the Franck–Condon node in the bending progression due to the node in the A-state's (0,1,0) vibrational wavefunction. Transitions to excited CN stretching states show a single maximum near $v_3'' = 3$, as expected. Notice that (0,4,6) and (0,10,4) do not appear on the map despite residing in favorable regions of the map. The Franck–Condon analysis shown in this figure implies unambiguously that the transition intensity associated with the basis states (0,4,6) and (0,10,4) must be experimentally detectable. The fact that they are not assigned is strong evidence that they participate in a perturbation with a "dark" state.

This plot can be viewed as a map of the HCN $\tilde{A}$-state (0,1,0) wavefunction probability density in the $\tilde{X}$-state normal coordinates for bending, mode 2, and CN stretching, mode 3.[1] The probability density has a single maximum

---

[1] Note that the mapping from vibrational quantum numbers to normal coordinates is nonlinear.

in the "CN" stretch coordinate and the intensity gap along $v_2'' = 8$ reflects the (0,1,0) nodal line in the bending coordinate.

This map of intensities allows us to further verify that the axis-switching transitions to $|\ell| = 1$ states discussed above would have significant Franck–Condon factors. To see this compare the list of vibrational states assigned by the axis switching mechanism (Table III of Ref. 18) to the Franck–Condon map. These states all reside at favorable positions on the Franck–Condon map.

Note also that the Franck–Condon map implies transition strength to the empty squares for the states with quantum numbers $(0,10^{0,2},4)$ and $(0,4^{0,2},6)$. Why were these states not assigned? Referring to the list of candidate states (Table 13 of Ref. 16), we see that each of these states was a possible candidate for a series of three $|\ell| = 0, 2$ unassigned pairs of states near 15 000 cm^{-1}. If we were to assume that these four states, i.e., $(0,10^0,4)$, $(0,10^2,4)$, $(0,4^0,6)$, and $(0,4^2,6)$, are mixed up in some sort of perturbation with a third $|\ell| = 0, 2$ pair we could account for transition probability to all six unassigned states in this energy region. In fact, since no other state predicted near this energy will fit into the Franck–Condon map, the third $\ell''' = 0, 2$ pair *must* appear through a perturbation.

The visible absorption spectrum of HCN is dominated by $(n, 0, 0) \leftarrow (0, 0, 0)$ transitions at approximate multiples of the CH stretch frequency v_1''. The frequencies v_1'' (3444 cm^{-1}) and v_3'' (2130 cm^{-1}) are accidentally nearly in the ratio 2:3. Classically, oscillators with near integral frequency ratios will resonantly exchange energy when coupled. Quantum mechanically, the vibrational state (v_1'', v_2'', v_3'') can be mixed with the state $((v_1''+2), v_2'', (v_3'' - 3))$ by anharmonic coupling. Rank *et al.*[28] deperturbed $2v_1'' : 3v_3''$ resonances in the HCN overtone spectrum by assuming that the normal coordinates q_1 and q_3 were coupled by a term in the potential. The coupling then has the form

$$\langle v_1, v_2^\ell, v_3 | \mathcal{H} | (v_1 - 2), v_2^\ell, (v_3 + 3) \rangle$$
$$= K_{2:0:3} \sqrt{v_1(v_1 - 1)(v_3 + 1)(v_3 + 2)(v_3 + 3)} \, . \tag{6}$$

Note that this coupling increases with the stretch quantum numbers.

The level (2,10,1) is coupled to (0,10,4) by this resonance and has a predicted band origin within 15 cm^{-1} of the observed levels. This coupling can mix the two vibrational levels, transferring some of the (0,10,4) intensity to (2,10,1). Anharmonic mixing of the vibrational character also alters the individual rotational constants and vibrational energies, but the average rotational constant and average vibrational energy of the interacting levels are unaffected. The average rotational constant of the 6 observed states is 1.465(2). The average of the predicted zero order rotation constants from Table 13 of Ref. 16 is 1.463(1) if we assume the remaining $|\ell| = 0, 2$ pair is $(2, 10^{0,2}, 1)$. The $(3, 2^{0,2}, 2)$ state is also another possible candidate lying close in energy to these states. If we were to assume that the $(3, 2^{0,2}, 2)$ states were the fifth and sixth states involved in the perturbation the average of the six predicted rotation constants would then be 1.445(1), which disagrees significantly with experiment.[m] It appears quite clear that the six unassigned states near $15\,000$ cm^{-1} are a complex perturbation of the six states: $(0, 10^{0,2}, 4)$, $(0, 4^{0,2}, 6)$, and $(2, 10^{0,2}, 1)$. In addition to the $2v_1''$: $3v_3''$ resonance, the observed rotational constants indicated a previously undetected $6v_2''$: $2v_3''$ resonance between (0,10,4) and (0,4,6).

This leaves two unassigned states near $18\,260$ cm^{-1}. From the averaged rotational constants, it would seem the same set of resonances also occurs at $18\,260$ cm^{-1} between (2,12,2), (0,12,5), and (0,6,7). The last two unassigned levels in the SEP spectrum are, therefore, assigned as $(2, 12^{0,2}, 2)$. Every transition in the SEP spectrum of HCN has been assigned.

4. Two More Interesting Points

4.1. *The "CH" Stretch is absent*

The change in CH bond length is small, so the $\tilde{A}(0, 1, 0)$ emission intensity should be maximum at $v_1'' = 0$. The slight elongation of the CH bond in the $\tilde{A}$-state causes combination bands built on $v_1' = 1$ excitation to appear weakly in the $\tilde{A} \leftarrow \tilde{X}$ absorption spectrum, so one might expect weak combination bands with $v_1'' = 1$ in emission, but they do not appear. Chemical intuition suggests that bending vibrations take place at nearly constant bond length and stretching vibrations are nearly changes in bond length alone. One-dimensional harmonic Franck–Condon calculations using only the harmonic frequencies, the excited state geometry, and the form of the ground state normal coordinates can be quite instructive if this chemical intuition is used to define normal coordinates that are nearly separable.

---

[m]Dennison [*Rev. Mod. Phys.* **12**, 175 (1940)] has discussed why predictions of the unperturbed rotational constants are a more reliable perturbation diagnostic than predicted unperturbed vibrational energies.

The normal mode v_1'' contracts the CN bond while stretching the CH bond. At the $\tilde{A}$-state geometry, both the CH and CN bonds are elongated. If we define the $\tilde{X}$-state normal coordinate q_1'' in terms of changes in the CH and CN bond lengths, we find that the displacement along q_1'' at the $\tilde{A}$-state equilibrium geometry is very small; less than one tenth that at the classical turning point of the zero-point level. The increases in CN and CH bond length cancel. A simple harmonic calculation of the Franck–Condon factors[18] indicates transitions from $\tilde{A}(0,1,0)$ into $\tilde{X}$-state levels with $v_1'' = 1$ have about 300 times less intensity than $v_1'' = 0$. Transitions to these states could only be observed with a much more sensitive SEP spectrometer.

4.2. *The Duschinsky Effect in HCN is small*

The normal coordinates are found by a rotation of the mass-weighted Cartesian coordinates to diagonalize the force constant matrix.[n] Since the force constant matrix depends on the electronic state, the normal coordinates in one electronic state are related to those in another by a product of two rotations, which is also a rotation. Therefore, the normal coordinates of one electronic state are related to those of another by a rotation. When one calculates the vibrational overlap integrals for the Franck–Condon factors one must first represent the vibrational wavefunctions of one electronic state in terms of the normal coordinates of the other. This can cause anomalously large and nonintuitive Franck–Condon factors even for purely harmonic vibrational motion and is known as the Duschinsky effect.

In the Franck–Condon map of Fig. 9, a large Duschinsky rotation between the $\tilde{X}$ and $\tilde{A}$ state normal coordinates would yield a diagonal, rather than a vertical, nodal line. Since the mapping from normal coordinate to quantum number is nonlinear, a diagonal nodal line may

---

[n] In this discussion, it is assumed that the angular momentum is zero in an inertial frame. Then the Euler angles which describe the orientation of the "molecule fixed" coordinate axes relative to the laboratory axes may be taken as zero. For rotating molecules, it is customary to define rotation and vibration by the Eckart condition. Özkan has pointed out that the Eckart condition can make the transformation of the vibrational variables alone nonorthogonal, with important consequences for rotating molecules [Özkan, *J. Mol. Spectrosc.* **139**, 147 (1990)]. We emphasize that the Eckart frame is not inertial and is, therefore, inappropriate for molecules with zero angular momentum: for nonrotating molecules, the transformation between normal coordinates of two electronic states is orthogonal in an inertial frame. Note that the inertial coordinate axes are determined by objects in the laboratory and not by the instantaneous geometry.

even be curved on the Franck–Condon map. The observed vertical nodal line at $v_2'' = 8$ indicates that the Duschinsky effect is small for the bend and CN stretch in HCN.[o]

5. A Chemically Accurate Isomerization Potential

The vibrational levels of HCN provide valuable information about the HCN–HNC potential energy surface (PES) for isomerization. Most theoretical studies of HCN–HNC isomerization have used the empirical potential energy surface of Murrell *et al.*[31] Figure 5 of Ref. 16 compares the observed energy levels of highly excited HCN with the theoretically calculated energy levels[32] on the Carter–Murrell surface. The predicted energies are all too low, becoming systematically lower with increasing bend quantum number.

The simplest interpretation of the discrepancy between experiment and theory is that the 35 kcal/mol isomerization barrier on the Carter–Murrell potential energy surface is too low. A global potential obtained by Gazdy and Bowman[33] from empirical fits which included the SEP data supports this suggestion. Lee and Rendell, going to great lengths to account for electron correlation, have recently calculated an *ab initio* isomerization barrier 44.6 ± 1.0 kcal/mol above the HCN potential minimum.[34] Using the same level of theory, Bentley *et al.* have calculated a global *ab initio* PES and fit the computed points to a function, from which the vibrational Schrödinger equation could be solved.[35] This work has been extended to include a larger number of *ab initio* points.[36] The calculated vibrational energy levels on this potential energy surface reproduce the observed HCN vibrational energies to within 80 cm^{-1} wherever comparison is possible, as is shown in Fig. 10. The agreement is very encouraging and it appears that by using experiment as a control of the theory, we are rapidly converging on the true PES for isomerization. It is expected that in the very near future, slight empirical manipulation of the PES of Ref. 36 in order to optimize the agreement with experiment will yield a near spectroscopically accurate potential energy surface.

---

[o]It is possible although unlikely that a Duschinsky effect is just cancelled by an adiabatic change in the $\tilde{X}$-state normal coordinate q_2 with excitation of v_3''.

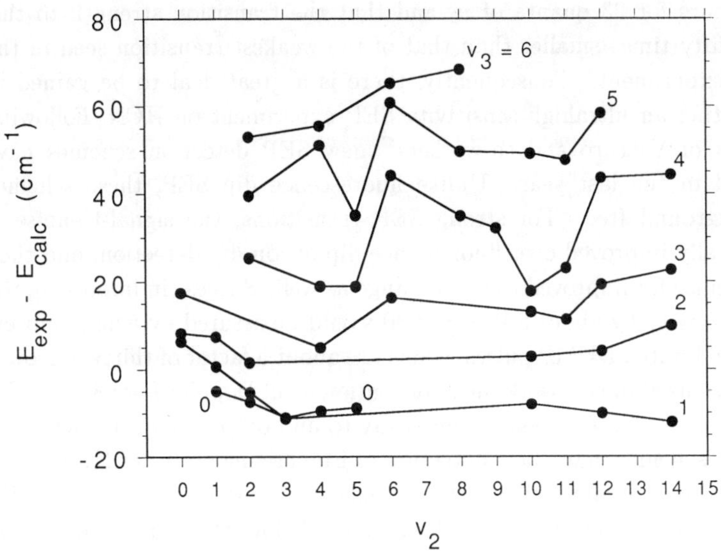

Fig. 10. A direct comparison between the predicted *ab initio* and the experimentally derived vibrational band origins: The potential energy surface of Ref. 35 has been used as input data to solve the vibrational Schrödinger equation. The vibrational wavefunction's nodal patterns were then analyzed to determine the vibrational quantum numbers of each state. A one-to-one comparison between assigned experimental and theoretical states was then carried out.

6. Future Work

The most important outstanding question regarding HCN is whether or not an experiment can be designed that will detect delocalized, "isomerization states". These states would be sensitive to the nature of the potential energy surface in the region of the transition state and the HNC structure. Furthermore, it is not at all obvious that the simple nature of the HCN rovibrational spectrum will be maintained for states that cross the isomerization barrier. It has even been reported that the calculated HCN/HNC vibrational spectrum displays the quantum signatures of classical chaos above the isomerization barrier, suggesting a connection between isomerization and chaos.

Joel Bowman's group has recently extended the efforts of Refs. 35 to calculate the Franck–Condon factors relevant to the SEP experiment:

$A(0, 1, 0) \rightarrow X(v_1, v_2, v_3)$. They have concluded that the first delocalized state occurs for 23 quanta of v_2 and that the transition strength to this state is fifty times smaller than that of the weakest transition seen in the present experiment. Consequently, there is a great deal to be gained in constructing an ultrahigh sensitivity SEP experiment on HCN. Following the work of Vaccaro and co-workers,[7] new SEP detection schemes have appeared in the last year. Unlike fluorescence dip SEP, these schemes are background free. For strong SEP transitions, the signal-to-noise is dramatically improved over fluorescence dip or ion dip detection, and they hold promise for improved dynamic range as well. Success in improving the SEP sensitivity by about a factor of 30 would be assured by using Fourier-transform limited DUMP pulses, which are about a factor of thirty narrower than those used in this work, in a conventional subDoppler fluorescence dip experiment. Work will soon be underway to find out if one of the new SEP detection schemes may allow detection of isomerizing states of HCN.

According to calculations, the intensity of SEP transitions to "CH" stretch excited levels is about three times below the current detection threshold, so a factor of 10 improvement in signal-to-noise would allow a whole new class of vibrational levels involving simultaneous excitation of all three normal modes to be detected. Since the CH stretch excited levels and the isomerizing states have been predicted to have roughly the same intensity in a number of calculations, it may be that some CH stretch excitation is required for isomerization: according to Lee and Rendell, the transition state has an elongated CH bond. Consequently, it may be possible to observe the delocalized states through this approach.

A final open question is the precise nature of the anharmonically interacting states which were analyzed with the average-rotational constant method above. There are at least three possible ways to clarify this. The most promising is the experimental method. If much more data, especially at high J''', were available so that both rotational constants and rotational-ℓ-resonances could be observed, a full deperturbation would be possible. Secondly, it may be possible to use a global *ab initio* potential which has then been optimized to experimental data, to identify the interacting levels and calculate the matrix elements. Thirdly, a direct inversion of the assigned vibrational levels has been used to obtain an empirical potential surface, which accurately reproduces the rotational constants of all of the unperturbed states.[37] If states with simultaneous excitation of all three vibrational modes could be accurately predicted, analysis of this potential may also shed light on the perturbed states.

HCN is an extremely interesting molecule since one can assign and interpret its rovibrational quantum structure even at chemically interesting energies. Astoundingly good agreement between experiment and *ab initio* theory has already been obtained and it is all but certain that quantitative agreement between experiment and theoretical predictions based on a semi-empirical potential energy surface will soon be achieved. In other words, we expect that soon the first chemical potential energy surface will have been derived through a pragmatic interaction between experiment and theory.

This suggests an enormous number of computer experiments that would be possible for HCN to investigate its quantum dynamics. Theoretical predictions regarding wave packet motion, quantum chaotic motion, collisional energy transfer, and collision induced isomerization are a few of the areas that are open to investigation on a chemically accurate PES. Many others possibilities will also arise to the imaginative theoretician. HCN certainly is a wonderful example of how high level theory and experiment can work hand in hand to resolve chemical problems at the most fundamental level.

Acknowledgments

The authors thank Professors Joel Bowman and Ian Mills for providing us with results prior to publication. DMJ thanks A.T. & T. Bell laboratories for a doctoral scholarship. XY wishes to thank Professors Giacinto Scoles and Kevin Lehmann for their support during the preparation of this article. AMW wishes to thank the Alfred P. Sloan Foundation and the Camille and Henry Dreyfus foundation. This research was partially supported by the following grants: Petroleum Research Fund Grant No. 21762 G6 and No. 24031-AC6, National Science Foundation Presidential Young Investigator Award CHE-8957978, and National Science Foundation Atmospheric Chemistry Division Grant ATM-8922214. In addition, a small grant from the University-wide Energy Research Group from the University of California is gratefully acknowledged. This work was also made possible by the Santa Barbara Laser Pool under NSF Grant No. CHE-8411302.

References

1. The classic monograph is G. Herzberg, *Molecular Spectra And Molecular Structure vols. I–III* (Krieger, Malabar, FL 1989–1991). A highly recommended, compact introduction to molecular spectroscopy may be found in G. Herzberg, *The Spectra and Structures of Simple Free Radicals: an Introduction to Molecular Spectroscopy* (Dover, New York, 1988).

2. This figure was graciously provided by Professor Joel Bowman (Emory University).

3. A representative collection of recent work can be found in a special issue devoted to stimulated emission pumping; *J. Opt. Soc. Am.* **B7**, (1990).

4. F. F. Crim, *Ann. Rev. Phys. Chem.* **35**, 657 (1984).

5. C. E. Hamilton, J. L. Kinsey, and R. W. Field, *Ann. Rev. Phys. Chem.* **37**, 493 (1986).

6. For a recent review see R. E. Weston Jr. and G. W. Flynn, *Ann. Rev. Phys. Chem.* **43**, 559 (1992).

7. Q. Zhang, S. A. Kandel, T. A. Wasserman, and P. H. Vaccaro, *J. Chem. Phys.* **96**, 1640 (1992).

8. H.-G. Rubahn and K. Bergmann, *Ann. Rev. Phys. Chem.* **41**, 753 (1990)

9. Y. Chen, D. M. Jonas, J. L. Kinsey, and R. W. Field, *J. Chem. Phys.* **91**, 3976 (1989)

10. S. Carter, I. M. Mills, and N. C. Handy, *J. Chem. Phys.* **97**, 1606 (1992).

11. C. Kittrell, E. Abramson, J. L. Kinsey, S. A. McDonald, D. E. Reisner, R. W. Field, and D. H. Katayama, *J. Chem. Phys.* **75**, 2056 (1981).

12. W. D. Lawrance and A. E. W. Knight, *J. Chem. Phys.* **76**, 5637 (1982).

13. A. M. Wodtke, L. Hüwel, H. Schlüter, and P. Andresen, *Rev. Sci. Instrum.* **60**, 801 (1989).

14. G. Herzberg and K. K. Innes, *Can. J. Phys.* **35**, 842 (1957). In this chapter, we follow the normal mode numbering established by the Joint Commission for Spectroscopy (R. S. Mulliken, *J. Chem. Phys.* **23**, 1997 (1955); erratum *J. Chem. Phys.* **24**, 1118 (1956)). The paper by Herzberg and Innes, vol. II of Ref. 1, and many older papers on infrared spectroscopy employ an older numbering. The recommended numbering which we use is used in most papers on electronic spectroscopy and Herzberg's book on electronic spectroscopy (vol. III of Ref. 1).

15. X. Yang, C. A. Rogaski, and A. M. Wodtke, *J. Opt. Soc. Am.* **B7**, 1835 (1990).

16. Xueming Yang, Ph.D. thesis, University of California, Santa Barbara, (1991).

17. A. M. Smith, S. L. Coy, W. Klemperer, and K. K. Lehmann, *J. Mol. Spectrosc.* **134**, 134 (1989).

18. D. M. Jonas, X. Yang, and A. M. Wodtke, *J. Chem. Phys.* **97**, 2284 (1992).

19. J. K. G. Watson, *J. Mol. Spectrosc.* **101**, 83 (1983).

20. G. Amat and H. H. Nielsen, *J. Mol. Spectrosc.* **23**, 359 (1967).

21. A. G. Maki and D. R. Lide Jr., *J. Chem. Phys.* **47**, 3206 (1967).

22. G. Strey and I. M. Mills, *Mol. Phys.* **26**, 129 (1972).

23. Y. T. Chen, D. M. Watt, R. W. Field, and K. K. Lehmann, *J. Chem. Phys.* **93**, 2149 (1990); It should be noted that the definition of q_2 in the parity basis used in this reference does not follow the accepted convention, and the value of q_2 should be divided by $\sqrt{2}$.

24. J. T. Hougen and J. K. G. Watson, *Can. J. Phys.* **43**, 298 (1965).

25. P. R. Bunker, *Molecular Symmetry and Spectroscopy* (Academic Press, Orlando, FL, 1979) Ch. 7; L. D. Landau and E. M. Lifschitz, *Quantum Mechanics, third edition* (Pergamon Press, New York, 1977) Section 104.

26. D. M. Jonas, S. Solina, B. Rajaram, R. W. Field, R. J. Silbey, K. Yamanouchi, and S. Tsuchiya, *J. Chem. Phys.* **97**, 2813 (1992); D. M. Jonas, Ph.D. thesis, (MIT, 1992); D. M. Jonas, S. Solina, B. Rajaram, R. W. Field, and R. J. Silbey (in preparation).

27. K. Yamanouchi, N. Ikeda, S. Tsuchiya, D. M. Jonas, J. K. Lundberg, G. W. Adamson, and R. W. Field, *J. Chem. Phys.* **95**, 6330 (1991).

28. D. H. Rank, G. Skorinko, D. P. Eastman, and T. A. Wiggins, *J. Opt. Soc. Am.* **50**, 421 (1960).

29. A. Meenakshi, K. K. Innes, and G. A. Bickel, *Mol. Phys.* **68**, 1179 (1989).

30. W. L. Smith and P. A. Warsop, *Trans. Faraday Society* **65**, 1164 (1968).

31. J. N. Murrell, S. Carter, and L. O. Halonen, *J. Mol. Spectrosc.* **93**, 307 (1982).

32. J. A. Bently, J. P. Brunet, R. E. Wyatt, R. A. Friesner, and C. LeForestier, *Chem. Phys. Lett.* **161**, 393 (1989).

33. B. Gazdy and J. M. Bowman, *J. Chem. Phys.* **95**, 6309 (1991).

34. T. J. Lee and A. P. Rendell, *Chem. Phys. Lett.* **177**, 491 (1991).

35. J. A. Bentley, J. M. Bowman, B. Gazdy, T. J. Lee, and C. E. Dateo, *Chem. Phys. Lett.* **198**, 563 (1992).

36. J. M. Bowman, B. Gazdy, J. A. Bently, T. J. Lee, and C. E. Dateo, *J. Chem. Phys.* **99**, 308 (1993).

37. S. Carter, I. M. Mills, and N. C. Handy, *J. Chem. Phys.* **99**, 4379 (1993).

CHAPTER 15

**STIMULATED EMISSION ION-DIP SPECTROSCOPY
OF JET-COOLED MOLECULES AND COMPLEXES —
VIBRATIONAL SPECTROSCOPY AND
INTRAMOLECULAR VIBRATIONAL REDISTRIBUTION**

Takayuki Ebata and Mitsuo Ito

*Department of Chemistry, Faculty of Science
Tohoku University, Sendai 980, Japan*

Contents

1. Introduction

The vibrational levels of ground-state molecules were traditionally studied
by vibrational spectroscopy such as infrared and Raman spectroscopy.

However, because of a small transition cross-section, vibrational spectroscopy is very hard to apply to molecules at very low concentration such as those in a supersonic free jet. An alternative way is to use electronic transition which has in general a large cross-section. A simple method is to excite molecules to an electronically excited state, say the S_1 state, and then measure the dispersed fluorescence spectrum. The dispersed fluorescence spectrum provides us with the information about the ground-state vibrational levels. However, in the spontaneous fluorescence process, the excited state molecules are distributed among transitions to many vibrational levels of the ground-state which are allowed by Franck–Condon factors. Therefore, the efficiency of producing the vibrationally excited molecules is low especially for a high vibrational level because of a small Franck–Condon factor. Also, it is not possible to produce selectively the molecules in a specific vibrational level.

The situation is greatly improved by the use of stimulated emission spectroscopy. In this spectroscopy, molecules are first pumped to a specific vibrational level of the S_1 state by the first laser (ν_1). Then the electronically excited state molecules are dumped to a specific ground-state vibrational level by stimulated emission with the second laser (ν_2). By choosing an appropriate ν_2 laser frequency, the electronically excited state molecules are efficiently transferred to the specific ground-state vibrational level. Even a level having a small transition probability can be detected with a very high sensitivity. The high sensitivity enables us to apply this spectroscopy for the molecules at very low concentration such as those in a supersonic free jet.

When the stimulated emission occurs, a fraction of the S_1 state molecules is dumped to the S_0 vibrational level, causing a decrease in the population of the S_1 state molecules. Therefore the stimulated emission can be probed by the population decrease of the S_1 state molecules. The most popular stimulated emission spectroscopy is stimulated emission pumping (SEP) spectroscopy (or fluorescence dip spectroscopy, see Fig. 1a).[1-8] In this spectroscopy, the stimulated emission to the ground-state vibrational level is monitored as a dip in the fluorescence. The fluorescence dip spectrum is obtained by scanning ν_2 while monitoring the fluorescence from the S_1 state. This method is very powerful for the detection of the ground-state vibrational levels over a wide energy range with a high resolution and a high sensitivity.

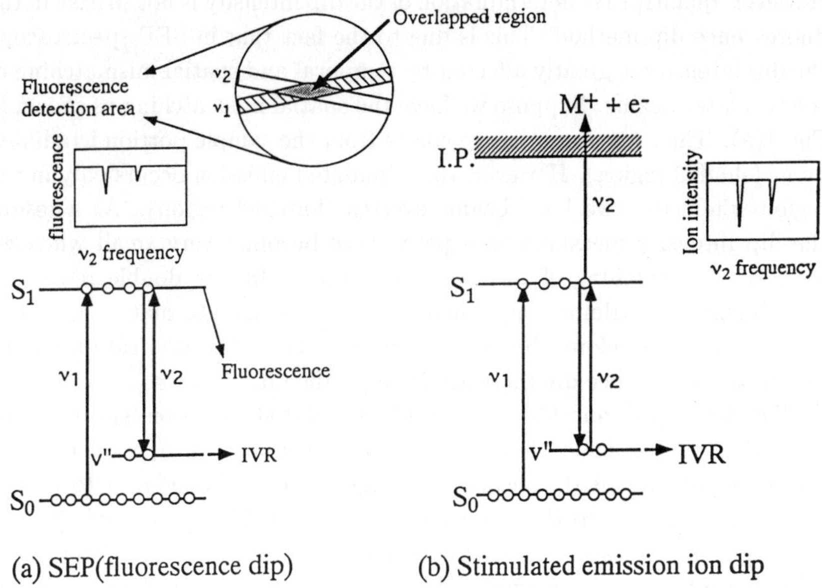

Fig. 1. (a) SEP (fluorescence dip) spectroscopy and (b) stimulated emission ion dip (SEID) spectroscopy.

In addition to the vibrational energies, the fluorescence dip gives us important information about the dynamics of the vibrational levels. The depth of the fluorescence dip reflects the decrease in the population of the S_1 state molecules. The population decreases in proportion to the transition probability. However, the molecules in the ground-state vibrational level reached by the stimulated emission will be reexcited by reabsorption of the same laser light of ν_2. The chance of the reabsorption is determined by the lifetime of the vibrational level, which is determined by the rate of relaxation by processes such as IVR and dissociation. If the relaxation is faster than the reabsorption, the molecules in the vibrational level decay very rapidly and have little chance to be pumped again to the S_1 state by reabsorption. Therefore, the S_1 state molecules are efficiently transferred to a ground-state vibrational level having a large relaxation rate, resulting in a deep dip. On the other hand, the dip depth is small for a vibrational level having a small relaxation rate. Therefore, it is possible to determine the relaxation rate from the observed fluorescence dip intensity.

However, quantitative determination of the dip intensity is not so easy in the fluorescence dip method. This is due to the fact that in SEP spectroscopy the dip intensity is greatly affected by temporal and spatial mismatching of the two laser beams. Suppose we have the spatial mismatching as shown in Fig. 1(a). Then, the fluorescence comes from the sample portion irradiated by ν_1 (shaded region). However, the stimulated emission occurs only in the region where the two laser beams overlap (hatched region). As a result, the dip intensity measured as a percentage becomes very small when we have the mismatching of the two laser beams. In any double resonance experiment, it is almost impossible to have a complete matching of two laser beams. Therefore, the determination of the relaxation rate from the observed fluorescence dip intensity is very difficult.

We developed another version of stimulated emission spectroscopy (Fig. 1b).[9,10] The molecule in S_0 is excited to a level in S_1 with the first laser light of ν_1 and the second laser light of ν_2 is introduced to induce stimulated emission to the ground-state vibrational level as usual. The ν_2 laser light is also used for ionization of the S_1 state molecule. The power of ν_1 is kept as low as possible. Therefore, the ionization does not occur with ν_1 alone. Under such a condition, the ions are produced only by the $\nu_1 + \nu_2$ ionization. The intensity of the ion signal is proportional to the population of the S_1 state molecules. By scanning the ν_2 frequency while monitoring the ion signal, the ionization efficiency curve from the S_1 state is obtained. When ν_2 is resonant to a ground-state vibrational level and stimulated emission occurs, the population of the S_1 state molecules decreases, resulting in a dip in the ion signal. The spectrum thus obtained is called "stimulated emission ion dip (SEID) spectrum". This spectroscopy has a great advantage in the quantitative determination of the dip intensity. In this method, both the ionization and the stimulated emission occur in the same spatial area where the two laser beams overlap. Therefore, the dip intensity measured as a percentage is not affected by mismatching of the two laser beams and gives us a reliable value. The measured reliable dip intensity is very useful for the study of dynamics of the ground-state vibrational levels of jet cooled molecules and complexes.

Two typical examples will be shown here for the application of SEID spectroscopy. Figure 2 shows the SEID spectrum of jet-cooled aniline.[9c] Here, the molecule is first excited to the zero point level of S_1 with ν_1 and the stimulated emission to the ground-state vibrational levels is observed

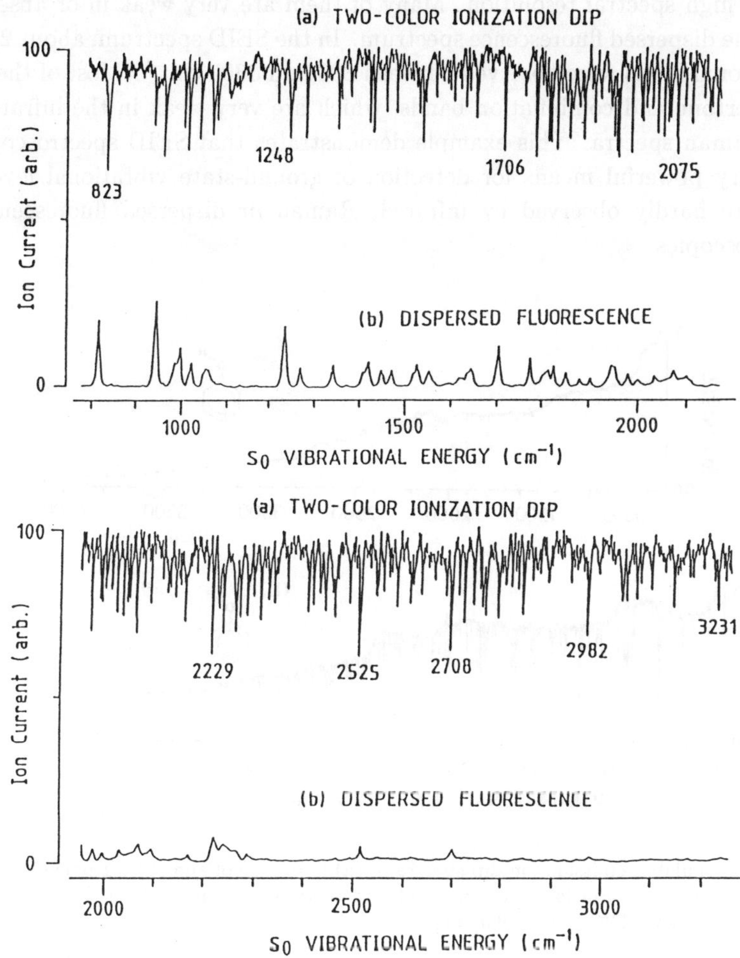

Fig. 2. (a) Stimulated emission ion dip spectrum (upper) and (b) dispersed fluorescence spectrum (lower) from the zero-point level of the S_1 state of object-cooled aniline. (*J. Phys. Chem.* **92**, 3774 (1988) Fig. 4)

as dips in the $\nu_1 + \nu_2$ ionization signal. The energy scale in the figure is given by $h(\nu_1 - \nu_2)$, which is the ground-state vibrational energy. In the lower part of the figure, the dispersed fluorescence spectrum from the zero point level of the S_1 state is also shown for comparison. It is clear from the comparison that many more bands are observed in the SEID spectrum with a high spectral resolution. Many of them are very weak in or absent from the dispersed fluorescence spectrum. In the SEID spectrum, about 200 vibrational levels were observed between 800 and 3300 cm^{-1}. Most of them are overtone and combination bands, which are very weak in the infrared and Raman spectra. This example demonstrates that SEID spectroscopy is a very powerful means for detection of ground-state vibrational levels that are hardly observed by infrared, Raman or dispersed fluorescence spectroscopies.

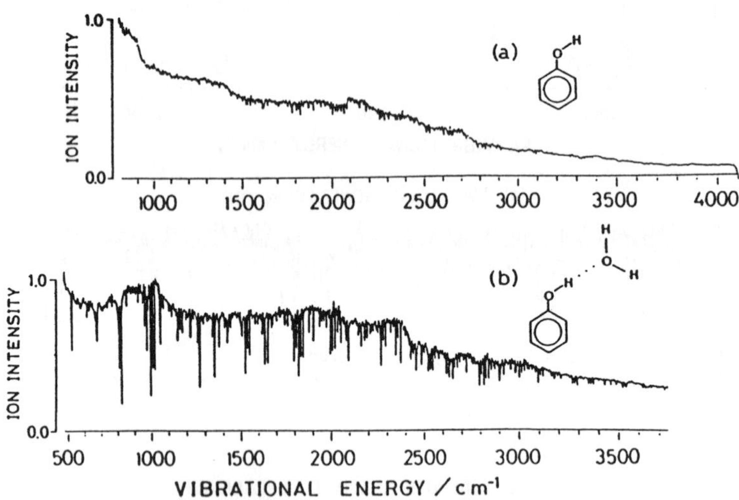

Fig. 3. Stimulated emission ion dip spectra of (a) bare phenol and (b) 1:1 phenol–H$_2$O complex after the excitation of the molecules to their zero-point levels in the S_1 state. (*J. Opt. Soc. Am.* **B7**, 1890 (1990) Fig. 2)

The second example is the SEID spectrum of bare phenol and that of its hydrogen bonded complex.[10a] Figure 3(a) shows the SEID spectrum of jet-cooled bare phenol obtained after exciting the molecule to the zero point level in S_1 with ν_1. There is no prominent dip over the entire spectral region. No prominent dip means that the vibrational levels of the ground-state of phenol have no efficient relaxation processes. Figure 3(b) shows the SEID spectrum of 1:1 hydrogen bonded complex of phenol with water in a supersonic jet. The spectrum was obtained after exciting the complex to its zero point level in S_1 and was recorded at the same laser powers as those in Fig. 3(a). In contrast to the case of bare phenol, many dips appeared strongly as seen in the figure. These dips are assigned to the vibrations of the phenol ring and intermolecular vibrations of the complex. As will be discussed later, the drastic enhancement in the dip intensity is due to the fact that the relaxation process in the ground-state vibrational levels is greatly accelerated by the complex formation. This example clearly demonstrates that SEID spectroscopy is also very useful for the study of dynamics of the ground-state vibrational levels of molecules and clusters.

Since the ground-state vibrational levels can be observed with a high resolution, even a small change in the vibrational frequencies of a molecule induced, for example, by weak intermolecular interaction like van der Waals interaction or hydrogen bonding can be studied in detail. In the section of "Experimental Results and Discussion", the effect of hydrogen bonding on the ground-state vibrational levels of phenol will be discussed in detail. The quantitative estimation of IVR rate constants will be also given for three cases: (1) the IVR rate of the hydrogen bonded complex of phenol, (2) the effect of alkyl chain length on the IVR rates of p-alkyl phenol and p-alkyl aniline, and (3) large amplitude motion and IVR of *trans*-stilbene. In the last section, it is shown that, together with stimulated emission spectroscopy, stimulated Raman spectroscopy is also a powerful means for the study of energies and dynamics of the ground-state vibrational levels of molecules and clusters.

2. Experimental: Method of SEID Spectroscopy

Figure 4 shows the experimental setup. The output of an excimer laser is split into two to pump two tunable dye lasers. The second harmonic of the first laser (Molectron DL 14) beam (ν_1) is introduced to a vacuum chamber to pump jet-cooled molecules to the zero point level of the S_1 state. The second harmonic beam (ν_2) of the second dye laser (Lambda Physik FL2002) is introduced to counterpropagate relative to the first laser beam and then it ionizes the S_1 state molecules. When we scan the frequency of the second dye laser while detecting the ion, the ionization efficiency curve from the zero point level of the S_1 state is obtained. The ionization efficiency curve of a large molecule is generally smooth and the population of the S_1 state molecules is monitored by this ion signal.

The second laser beam (ν_2) is also used for the stimulated emission of the S_1 state molecules to the vibrational levels in S_1. When the ν_2 frequency

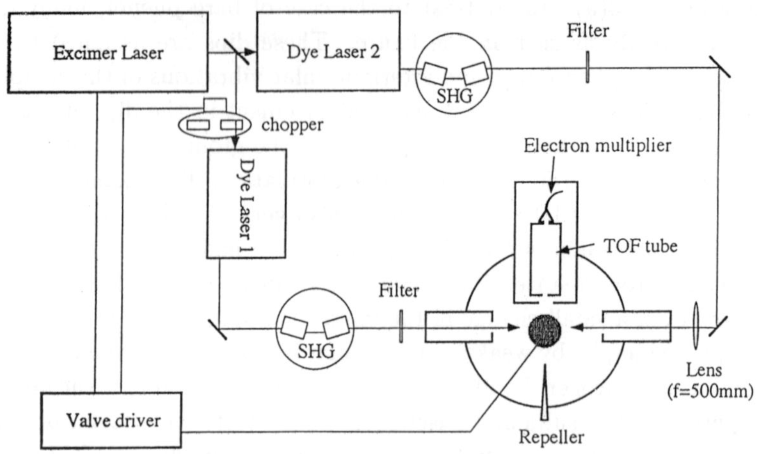

Fig. 4. Experimental setup of stimulated emission ion dip spectroscopy.

coincides with the transition to a vibrational level in S_0, stimulated emission occurs and the population of the S_1 state molecules decreases. The energy of the S_0 vibrational level reached by the stimulated emission is given by $h(\nu_1 - \nu_2)$. Since we monitor the S_1 state molecules by the ion signal due to the $\nu_1 + \nu_2$ ionization, the depopulation of the S_1 state molecules appears as a dip in the ionization efficiency curve and we obtain a SEID spectrum. If the relaxation of the ground-state vibrational level is fast, the population transferred into this level decays very rapidly and has little chance to be pumped again to the S_1 state by the reabsorption of the same ν_2 laser light within the laser pulse duration. Therefore, the S_1 state molecules are efficiently depopulated and a large dip appears in the spectrum. We will discuss this subject in more detail later.

Jet-cooled molecules are obtained by expanding a mixture of sample vapor and He into vacuum through a nozzle having a 400 μm orifice. To produce a molecular complex with a solvent molecule, the mixture of the solvent vapor and He passed through a sample vapor chamber and the gaseous mixture obtained was expanded into vacuum. The pressure of the solvent vapor was controlled by a thermoregulator. The jet-cooled molecules were irradiated by laser beams 15 mm downstream. The power of the first laser beam (ν_1) was kept as low as possible to avoid the ionization of the molecule by ν_1 alone. The power was ~ 200 μJ/cm^2 and it was used only for pumping the molecules to the S_1 state. The second laser beam (ν_2) was focused by an $f = 500$ mm focal length lens with its maximum power of 4 J/cm^2. Under such conditions, the ν_2 laser beam contributes to both the ionization and stimulated emission of the S_1 state molecules. Both processes occur in the same sample space where the two laser beams overlap. The dip intensity measured as a percentage is not affected by spatial mismatching of the two laser beams. Therefore, it is not necessary to worry about the change in the mismatching as the ν_2 frequency is scanned. The photoionized molecules are extracted to a sub-chamber by a repeller of 12 V/cm field. Either the total ion signal or the mass-separated ion signal obtained by using a 15 cm time-of-flight tube is detected by an electron multiplier (Murata Ceratron). The ion signals are amplified by an amplifier (NF Model BX-31) and processed by a boxcar integrator (PAR Model 4402/4400) connected to a microcomputer. Since the ν_2 laser power is very strong, nonresonant $1 + 1$ ionization by ν_2 alone occurs especially when the ν_2 frequency is close to the $S_1 - S_0$ band origin. This nonresonant ionization signal forms an undesired background in the SEID spectrum. To remove this background signal, a chopper was placed in the ν_1 laser beam and operated to block alternative pulses. For each ν_2 frequency, the alternating signals due to (i) the sum of $\nu_1 + \nu_2$ (SEID) and $2\nu_2$ (background signal) and (ii) $2\nu_2$ (background signal) alone are stored separately in two channels of a boxcar integrator. Then the signal (ii) was subtracted from signal (i), and the SEID spectrum without background was obtained. The dispersed fluorescence spectra of jet-cooled molecules were also measured to compare with the SEID spectra.

3. Results and Discussion

3.1. *Vibrational Frequencies of Hydrogen Bonded Complex of Phenol*

As was demonstrated previously for the hydrogen bonded complex of phenol with water, SEID spectroscopy is very useful for the study of very low density molecular complexes formed in supersonic jets. In this section, we will examine the ground-state vibrational levels of the various hydrogen bonded complexes of phenol in more detail and the effect of the hydrogen bonding on the vibrational frequencies of phenol will be discussed. Phenol is a prototype molecule for the study of hydrogen bonding and its hydrogen bonded complexes have been well studied by infrared and Raman spectroscopy. It is well known that the frequency of the OH stretching vibration of phenol decreases considerably upon complex formation.[11]

The OH stretching vibration is well isolated in frequency from other vibrations and it exhibits a great frequency change induced by hydrogen bonding. Therefore, this vibrational mode is easily studied by infrared and Raman spectroscopy. However, it is almost impossible to obtain the other vibrational frequencies of the individual hydrogen bonded complexes because infrared and Raman spectra will be heavily congested by serious overlapping of the bands due to bare phenol and phenol complex coexisting in the sample system. In this sense, a great advantage of SEID spectroscopy is the ability of selecting of a particular species. Figure 5 shows the $S_1 - S_0$ fluorescence excitation spectra of the gaseous mixtures of phenol with water (Fig. 5a) and with 1,4-dioxane (Fig. 5b). The band origin of the bare phenol is located at $36\,348$ cm^{-1}, while that of the 1:1 hydrogen bonded complex with water is located at $35\,995$ cm^{-1} and that with 1,4-dioxane at $35\,935$ cm^{-1}, being red shifted by 353 cm^{-1} and 413 cm^{-1}, respectively, from that of the bare phenol. The red shift means that the energy of stabilization upon complex formation increases in going from the S_0 state to the S_1 state. When we fix the frequency of the ν_1 laser at $35\,935$ cm^{-1} ($35\,995$ cm^{-1}), only the 1:1 complex of phenol with 1,4-dioxane (water) is excited to its zero-point level in S_1. Then the ground-state vibrational levels of this selected complex can be detected by stimulated emission with ν_2. The mass selection of ions by a time-of-flight mass spectrometer is also useful for identification of the complex.

Figure 6 shows the SEID spectra of 1:1 complexes of phenol–water and phenol-1,4-dioxane measured from the zero-point levels of their S_1 states. Many dips appear in the spectra and they are mainly assigned to the vibrations of the phenol ring. Moreover, the bands involving low frequency intermolecular stretching and bending vibrations can clearly be observed. The vibrational frequencies of various hydrogen bonded complexes of phenol are accurately obtained from their SEID spectra and they are listed in Table I. In the table, the frequency of the band origin of the $S_1 - S_0$ transition, red shift from that of phenol, intermolecular vibrations and vibrations of phenol ring in S_0 are listed. From the comparison of the vibrational frequencies of the bare phenol with its 1:1 complex, the changes of the individual vibrational frequencies upon complex formation can be studied. As can be seen in the table, the frequency change is generally

Table I. Band origins, red shifts from bare phenol and vibrational frequencies of several 1:1 hydrogen bonded complexes of phenol and deuterated phenol (Ph-OD).

	bare Ph-OH[a]	Benzene	dimer	H_2O	1,4-Dioxane	TMA	TEA	bare Ph-OD	D_2O
band origin	36348.2	36202.4	36043.9	35995.2	35934.8	35524.7	35448.5	36345.4	35999.5
red shift		145.8	304.3	353.0	413.4	823.5	899.7		345.9
inter-molecular vibration									
bending		24.3	18.1		18.6				
stretching			109.6	149.6	130.7	131.2	120.7		147.2
Phenol site									
$6a_1$	526	529.1	529.6	527.4	529.1	529.5	530.4	521.0	522.3
$10a_1$	817	818.5	814.2	812.5	814.2	813.2	813.8	806.9	806.6
12_1	823	827.7	828.8	825.1	827.9	828.0	827.5	821.7	822.6
1_1	999	1000.8	1003.5	999.5	1001.1	999.8	999.2	998.5	998.0
$16b_2$	1006	1012.3	1015.5	1011.9	1015.8	1013.5	1014.3	1008.2	1011.5
$18a_1$	1026	1025.0	1022.1	1024.7	1026.6	1023.7	1023.8	1025.9	1024.4
15_1	1070	1069.0	1074.1	1070.1			1069.8	1044.8	1044.7
$7a_1$	1261	1268.2	1273.3	1272.0	1273.5	1280.0	1280.5	1259.3	1266.0

[a]Vibrational frequencies of bare phenol (Ph-OH) was taken from Ref. 19.

small and it is within a few cm^{-1} for all the ring modes. Therefore, the effect of hydrogen bonding on the frequency of the ring localized modes are generally not large. An exception is mode $7a$. As can be seen in Table I, the frequency of the mode $7a$ (ν_{7a}) increases nearly in parallel with the red shift and the increase of its frequency from that of bare phenol amounts to as large as 20 cm^{-1} for the phenol-trimethyl amine (TMA) complex. Recently, Hartland *et al.* also reported the same frequency shift of ν_{7a} upon hydrogen bond formation by using the stimulated Raman technique.[12] It is known that the red shift of the $S_1 - S_0$ electronic transition is a good measure of the enthalpy of hydrogen bonding.[13] Therefore, the above results suggest a good correlation between the frequency shift of ν_{7a} and the enthalpy of the hydrogen bonding. In Fig. 7, the frequency of the mode $7a$ of the various hydrogen bonded complexes is plotted versus the enthalpy of the complex. As can be seen in Fig. 7, there exists a good linear relationship between them. Figure 8 shows the motion of the mode $7a$. It is essentially

 Molecular Dynamics and Spectroscopy

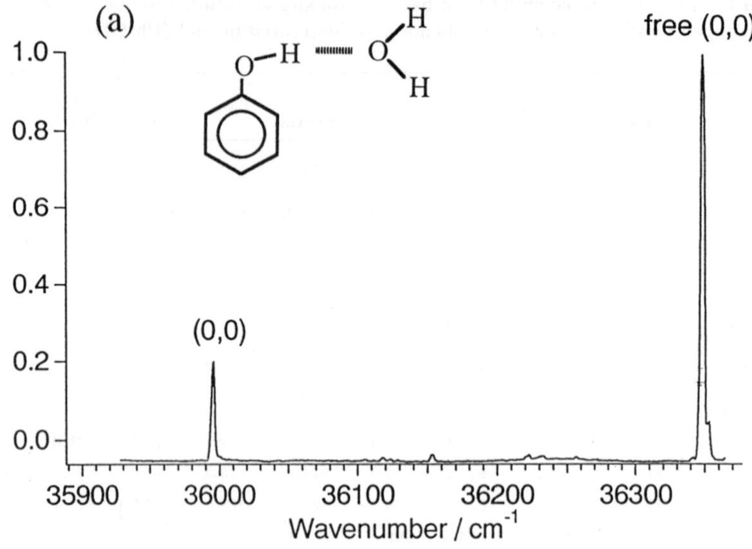

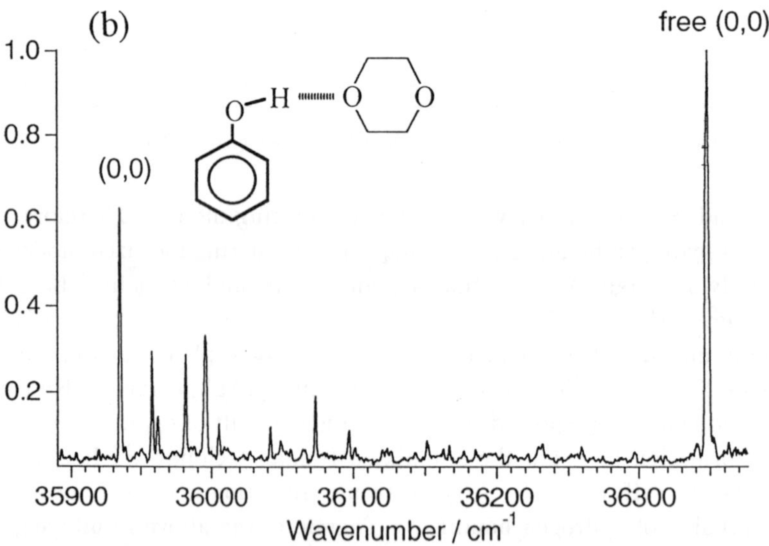

Fig. 5. Fluorescence excitation spectra of gaseous mixtures of (a) phenol and water and (b) phenol and 1,4-dioxane. (0,0) band of bare phenol is located at $36\,348$ cm^{-1} and those of the phenol–water (1:1) complex and the phenol–1,4-dioxane (1:1) complex are shifted to the red by 353 cm^{-1} and 413 cm^{-1}, respectively.

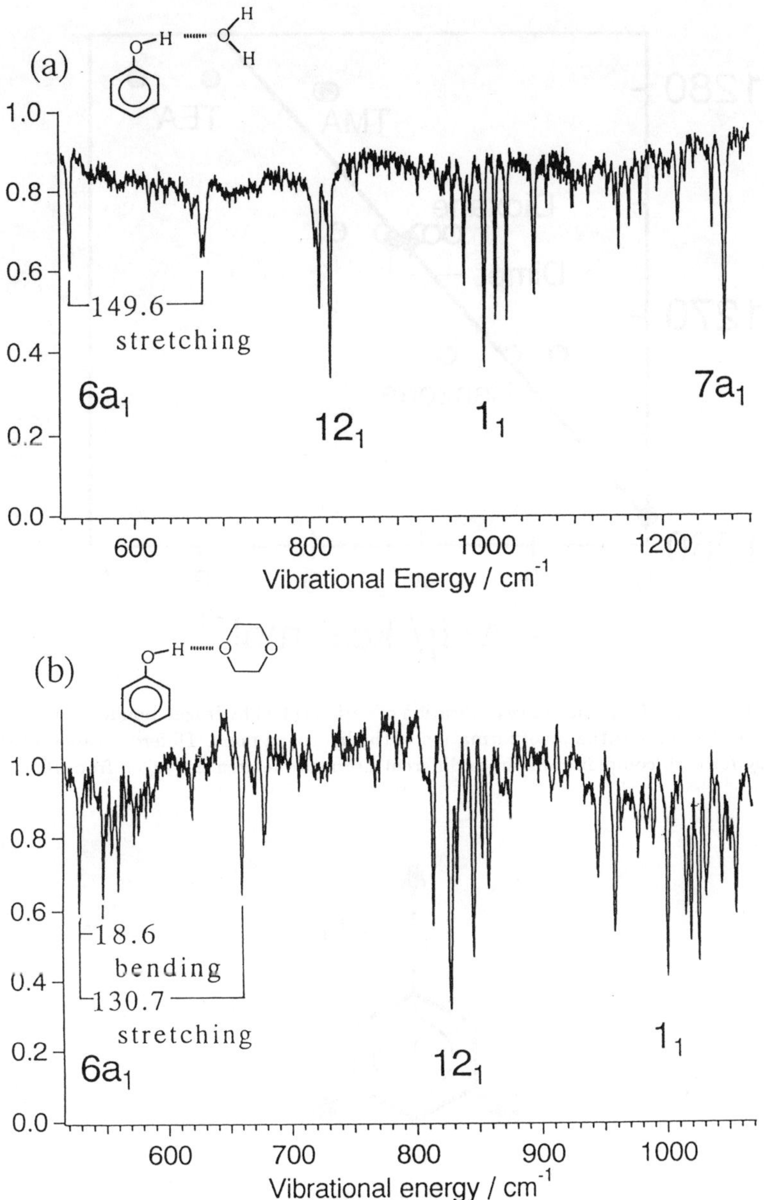

Fig. 6. Stimulated emission ion dip spectra of (a) phenol–water (1:1) complex and (b) phenol–1,4-dioxane (1:1) complex.

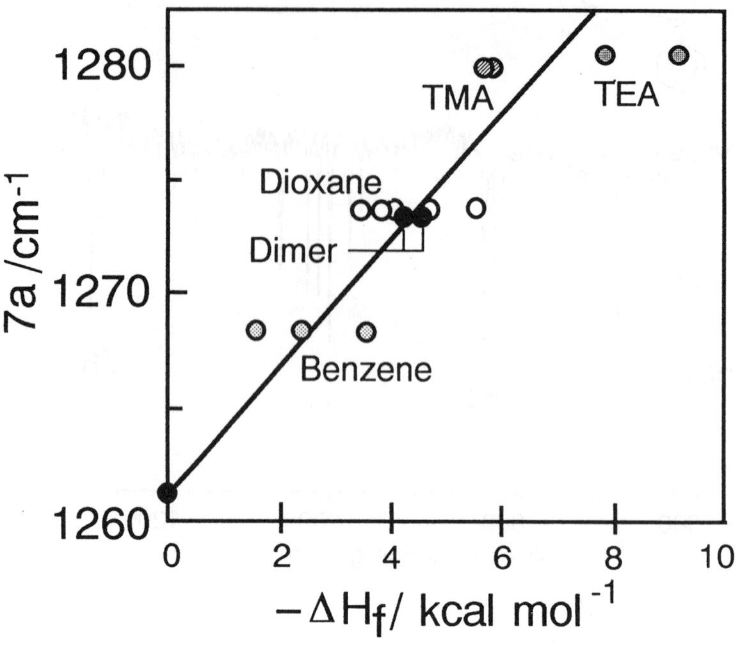

Fig. 7. Plot of the frequency of ν_{7a} vibration of various (1:1) hydrogen bonded complexes of phenol versus the enthalpy of formation of the hydrogen bond. Different values of the enthalpy for each complex come from different literature references (taken from Refs. 11 and 20).

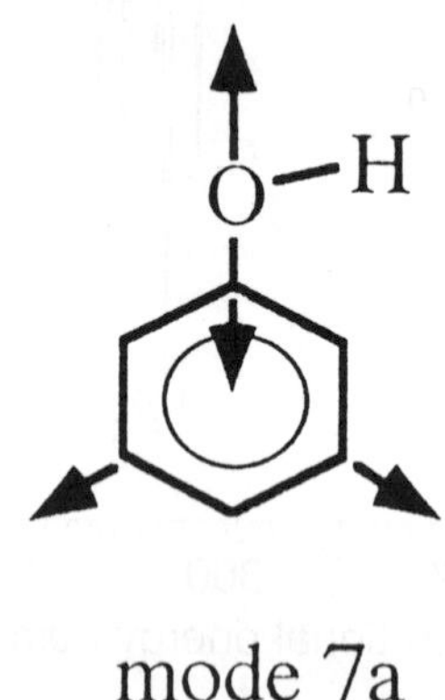

mode 7a

Fig. 8. Vibrational motion of mode 7a.

a CO stretching vibration. Therefore, the result indicates that the CO bond becomes stronger as the hydrogen bonding increases in strength. The increase in the bonding of CO is in contrast to the fact that the OH bond becomes weaker as the hydrogen bonding becomes stronger. When the hydrogen bond is formed between H atom of the OH group with a proton acceptor, the H atom becomes positively charged. This leads to a weakening of the OH bond. On the other hand, the O atom becomes negatively charged and this increases a double bond character in the CO bond and the CO bond strength will increase. These results have demonstrated that even a slight change in the vibrational frequency by complex formation can be accurately measured by SEID spectroscopy.

3.2. *Intramolecular Vibrational Redistribution (IVR) in the Ground-State Vibrational Levels of Isolated Molecules*

As mentioned previously, though bare phenol does not exhibit any appreciable dips in the SEID spectrum, a dramatic enhancement of the dip intensity occurs by forming a hydrogen bonded complex. In this section, more quantitative discussion will be given on how the dip intensity is related to IVR rate by using a model based on rate equations and the IVR rate will be discussed for three interesting cases: (1) IVR rate of the hydrogen bonded complex of phenol, (2) the effect of alkyl chain length on the IVR rates of p-alkyl phenol and p-alkyl aniline, and (3) large amplitude motion and IVR of *trans*-stilbene.

The rate equation for the change of the population of molecules in each state can be expressed as follows when the stimulated emission is occurring (resonant condition, see Fig. 9),

$$d[S_0]/dt = -\sigma_1 I_1[S_0] + \sigma_1 I_1[S_1] \tag{1}$$

$$d[S_1]/dt = -(\sigma_1 I_1 + \sigma_2 I_2 + \sigma_3 I_2 + k_T)[S_1]$$
$$+ \sigma_1 I_1[S_0] + \sigma_2 I_2[S_0^v] \tag{2}$$

$$d[S_0^v]/dt = -(\sigma_2 I_2 + k_v)[S_0^v] + \sigma_2 I_2[S_1] \tag{3}$$

$$d[\text{Ion}]/dt = \sigma_3 I_2[S_1] \ . \tag{4}$$

Here, $[S_0]$, $[S_1]$, $[S_0^v]$ and $[\text{Ion}]$ are the populations of the molecules in S_0, S_1, ground-state vibrational level (v), and ionic state, respectively. σ_1, σ_2, σ_3 are the cross-sections for the $S_1 - S_0(0,0)$, $S_1 - S_0^v$ and ionization transitions, respectively. σ_2 is proportional to the Franck–Condon factor of the transitions $S_1 - S_0^v$. I_1 and I_2 are the intensities of the ν_1 and ν_2 lasers. k_T is the decay rate of the vibronic level (zero point level) of the S_1 state that includes both radiative and nonradiative decay rates. k_v is the decay rate of the ground-state vibrational level and IVR can be assumed to be the only decay process when the vibrational energy is less than the dissociation limit. When stimulated emission does not occur under off-resonant condition, σ_2 is set to zero in Eqs. (2) and (3).

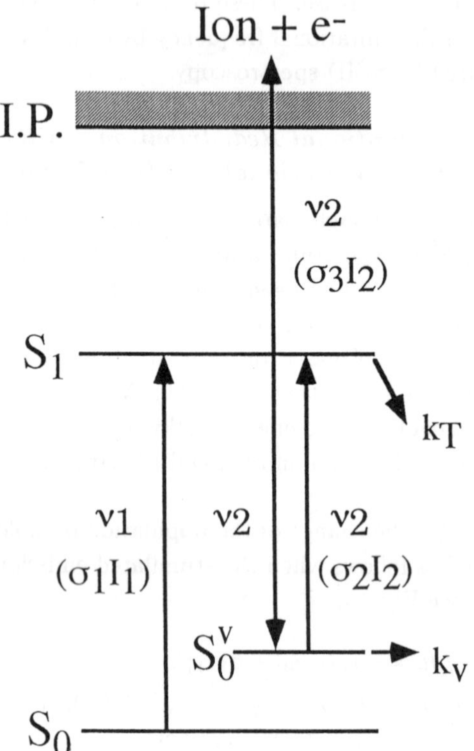

Fig. 9. Excitation scheme of SEID spectroscopy. σ_1, σ_2, and σ_3 are the cross-section for the S_1–S_0 $(0,0)$ and S_1–S_0^v and S_1–ion transition. I_1 and I_2 are laser powers. k_T and k_v are decay rate constants of the S_1 state and the S_0^v state.

The dip intensity is obtained by integrating [Ion] expressed in Eq. (4) over the laser pulse duration under resonant and off-resonant conditions. $[S_0(t = 0)] = S_0$ and $[S_1(t = 0)] = [S_0^v(t = 0)] = [\text{Ion}(t = 0)] = 0$. First, steady-state conditions for S_1 and S_0^v states are assumed and we will see roughly which term is important to affect the dip intensity. The laser pulse is assumed to be a rectangular shape with width of Δt. Then,

$$[S_1]_{\text{off}} = \frac{\sigma_1 I_1}{\sigma_3 I_2 + k_T}[S_0] \tag{5}$$

$$[\text{Ion}]_{\text{off}} = [S_1]_{\text{off}}\sigma_3 I_2 \Delta t , \tag{6}$$

for the off-resonant condition, and

$$[S_1]_{\text{on}} = [S_0]\sigma_1 I_1 \left[(\sigma_2 + \sigma_3)I_2 + k_T - \frac{\sigma_2^2 I_2^2}{\sigma_2 I_2 + k_v}\right]^{-1} \tag{7}$$

$$[\text{Ion}]_{\text{on}} = [S_1]_{\text{on}} \sigma_3 I_2 \Delta t , \tag{8}$$

for the resonant condition. The dip intensity (depth) is expressed as

$$I_{\text{DIP}} = ([\text{Ion}]_{\text{off}} - [\text{Ion}]_{\text{on}})/[\text{Ion}]_{\text{off}} . \tag{9}$$

From Eqs. (6) and (8) we can obtain

$$\frac{[\text{Ion}]_{\text{off}}}{[\text{Ion}]_{\text{on}}} = 1 + \frac{k_v \sigma_2 I_2}{(\sigma_3 I_2 + k_T)(\sigma_2 I_2 + k_v)} , \tag{10}$$

and I_{DIP} can be expressed as

$$I_{\text{DIP}} = \frac{k_v \sigma_2 I_2}{(\sigma_3 I_2 + k_T)(\sigma_2 I_2 + k_v) + k_v \sigma_2 I_2} . \tag{11}$$

Let's obtain I_{DIP} in Eq. (11) for three different cases of k_v, that is, (i) $k_v = \sigma_2 I_2/10$, (ii) $k_v = \sigma_2 I_2$, and (iii) $k_v = 10\sigma_2 I_2$. The results are

$$\text{(i)} \quad k_v = \sigma_2 I_2/10 \qquad : I_{\text{DIP}} = \frac{\sigma_2 I_2}{11(\sigma_3 I_2 + k_T) + \sigma_2 I_2} , \tag{12}$$

$$\text{(ii)} \quad k_v = \sigma_2 I_2 \qquad : I_{\text{DIP}} = \frac{\sigma_2 I_2}{2(\sigma_3 I_2 + k_T) + \sigma_2 I_2} , \tag{13}$$

and

$$\text{(iii)} \quad k_v = 10\sigma_2 I_2 \qquad : I_{\text{DIP}} = \frac{\sigma_2 I_2}{1.1(\sigma_3 I_2 + k_T) + \sigma_2 I_2} . \tag{14}$$

It is concluded from these results that the dip intensity, I_{DIP}, increases as the IVR rate, k_v, in the ground-state vibrational level increases. This can be explained as follows. The molecules transferred into the ground-state vibrational level either decay with a rate of k_v or are pumped back to the S_1 state by the reabsorption of ν_2 within the laser pulse duration Δt with a rate of $\sigma_2 I_2$. Therefore, if the IVR rate is much larger than the reabsorption rate, the stimulated emission process occurs predominantly. Then the S_1 state molecules are efficiently dumped to the ground-state vibrational levels, resulting in a large dip. Therefore, for two vibrational levels having the same transition cross-section σ_2, a large dip will be observed for the level having the larger IVR rate. The decay rate of the S_1 state, k_T, also affects the dip intensity, as can be seen in Eq. (11). Since the dip intensity is inversely proportional to k_T, a larger dip can be expected when the molecule has a small S_1 decay rate.

The steady-state approximation is not valid in actual cases and we have to solve Eqs. (1)–(4) numerically to obtain k_v. The decay rate of the S_1 state, k_T, can be obtained from the fluorescence lifetime, $\tau_f (= k_T^{-1})$. First, it is necessary to evaluate the cross-sections; σ_1, σ_2 and σ_3. Under our experimental condition, the first unfocused pump laser is quite weak and its photon density is estimated to be 10^{22} photons $\mathrm{cm}^{-2}\mathrm{s}^{-1}$. The second laser is focused by an $f = 500$ mm lens and its photon density is varied from 10^{25} to 10^{27} photons $\mathrm{cm}^{-2}\mathrm{s}^{-1}$. The ionization cross-section σ_3 is determined by the ν_2 laser power dependence of the two-color $1 + 1'$ ionization efficiency under the condition that stimulated emission does not occur.[10a] Figure 10 shows the observed power dependence of the ion intensity of jet-cooled bare phenol–d_1 from the $S_1(v = 0)$ state and the calculated power dependence by using the best fitted value of $\sigma_3 = 4 \times 10^{-18}$ cm^2. The cross-section of $S_1 - S_0(0,0)$ transition, σ_1, was obtained by measuring the ν_1 power dependence of the one-color $1+1$ ionization signal. Figure 11 shows the laser power dependences of the ion intensities of one-color $1+1$ ionization through the $S_1(v = 0)$ states of bare phenol–d_1 and the phenol–H_2O complex. The solid curve is the calculated power dependence with $\sigma_1 = 5 \times 10^{-17}$ cm^2. The stimulated emission cross-section, σ_2, can be obtained from the value of σ_1 and the $S_1 - S_0(0,v)$ Franck–Condon factor, which can be obtained from the relative intensities in the dispersed fluorescence spectrum. By using these parameters, the dependence of the dip intensity on ν_2 laser power was calculated by varying k_v to reproduce the observed one.

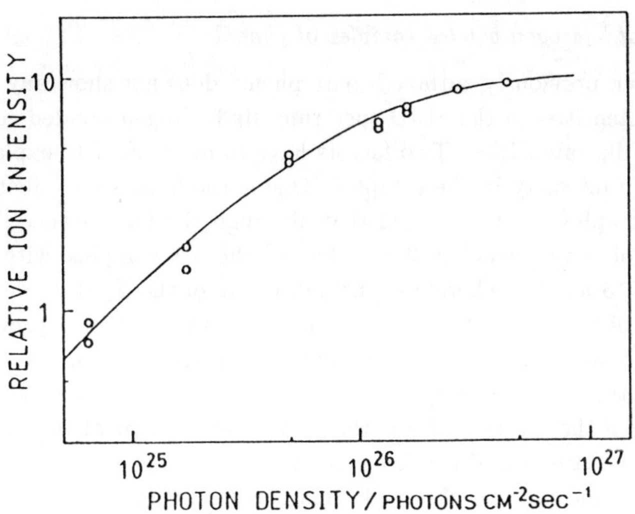

Fig. 10. ν_2 laser power dependence for the two-color $1 + 1'$ ionization signal of bare phenol–d_1 from the $S_1(v = 0)$ level. The solid curve is the calculated power dependence with $\sigma_3 = 4 \times 10^{-18}$ cm^2. (*J. Opt. Soc. Am.* **B7**, 1890 (1990) Fig. 5)

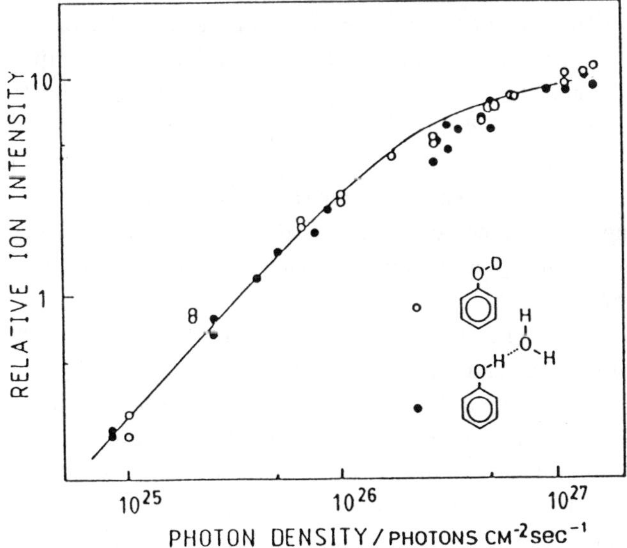

Fig. 11. ν_1 Laser power dependence for the one-color $1 + 1$ ionization signal of bare phenol–d_1 and phenol–H_2O complex. The solid curve is the calculated power dependence with $\sigma_1 = 5 \times 10^{-17}$ cm^2. (*J. Opt. Soc. Am.* **B7**, 1890 (1990) Fig. 6)

3.2.1. *IVR of hydrogen bonded complex of phenol*

As was shown previously, although bare phenol does not show any appreciable dip intensities in the SEID spectrum, its hydrogen bonded complex shows large dip intensities. Two factors have to be checked to explain the increased dip intensity in the complex. One is the fluorescence lifetime of S_1 in the complex. It is reported that although the fluorescence lifetime $\tau_f(=k_T^{-1})$ of bare phenol is 2 ns, that of the 1:1 complex with water increases to 15 ns.[14,15] Therefore, the decay rate of the S_1 state decreases by a factor of eight by complex formation. As was explained previously, even if the cross-section, σ_2, were the same, the dip intensity would become larger when the S_1 state decay rate becomes smaller. The other reason for the increase of the dip intensity is the great enhancement of the IVR rate in the ground-state vibrational levels of the complex. The SEID spectra of Fig. 12 answer the question about which factor is more important. Figure 12 shows the SEID spectra of (a) bare phenol, (b) phenol–H_2O (1:1) complex, (c) bare phenol–d_1, and (d) phenol–d_1–D_2O (1:1) complex. It is seen in the figure that the dip intensities of bare phenol and bare phenol–d_1 are quite weak. The fluorescence lifetime of bare phenol–d_1 is 16 ns,[16] which is almost equal to that of the phenol–H_2O complex. Therefore, if the S_1 state lifetime is primarily responsible for the dip intensity, a dip intensity similar to that found for the phenol–H_2O complex is expected for the SEID spectrum of bare phenol–d_1. However, the dip intensity of bare phenol–d_1 is much weaker than that of the phenol–H_2O complex. This shows that the large IVR rate is mostly responsible for the large dip intensity.

A next question is how large is the IVR rate constant in the complex? Figure 13 shows the ν_2 laser power dependences of the dip intensities of the $6a_1^0$, $6a_1^0\sigma_1^0$ and 12_1^0 bands of the SEID spectrum of the phenol–H_2O (1:1) complex (σ is the intermolecular stretching vibrational mode).[10a] Also shown are the calculated ν_2 power dependences of the dip intensities. Table II shows the obtained IVR rate constants for several vibrational levels of the complex. As can be seen in the table, the IVR rate constant of the 12_1 level is 10 times larger than that of the $6a_1$ level. The IVR rate is in the order of $10^9 \sim 10^{10} \mathrm{s}^{-1}$ for the levels in the vibrational energy range from 500 to 1000 cm^{-1} and it roughly increases with the density of vibrational states. In bare phenol, the IVR rate is very small and the populations decay very slowly with their infrared radiative decay rate. On the other hand,

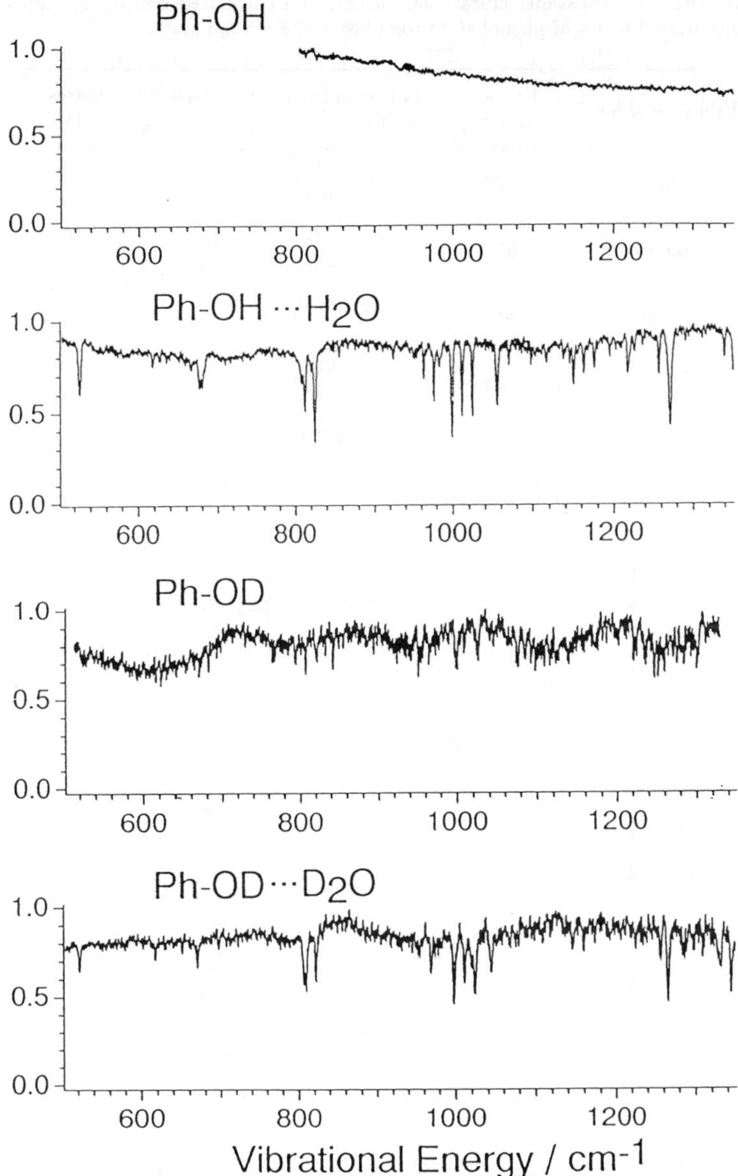

Fig. 12. SEID spectra of bare phenol, phenol–H_2O complex, bare phenol–d_1, and phenol–d_1–D_2O complex measured after exciting each species to the $S_1(v = 0)$ level.

Table II. IVR rate constant, vibrational energy, and calculated density of states for several vibrational levels of phenol-H_2O complex in the ground state.

Vibrational level	Energy (cm^{-1})	IVR rate constant $(\times 10^9 s^{-1})$	Density of states $(states/cm^{-1})$
$6a_1$	527	$1 \left(\begin{smallmatrix}+1\\-0.2\end{smallmatrix}\right)$	7
$6a_1\sigma_1$	677	$5 \left(\begin{smallmatrix}+3\\-2\end{smallmatrix}\right)$	15
$10a_1$	813	$10 \left(\begin{smallmatrix}+10\\-2\end{smallmatrix}\right)$	40
12_1	825	$10 \left(\begin{smallmatrix}+10\\-2\end{smallmatrix}\right)$	44
1_1	999	$8 \left(\begin{smallmatrix}+10\\-2\end{smallmatrix}\right)$	110

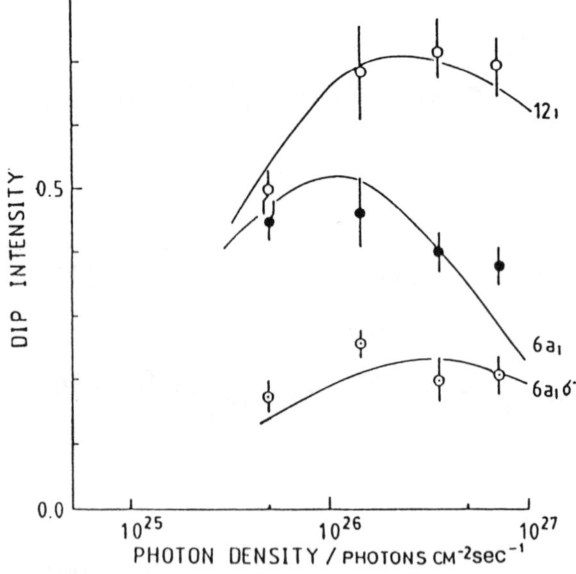

Fig. 13. ν_2 laser power dependences of the intensities of the $6a_1^0$, $6a_1^0\sigma''$ and 12_1^0 bands of the SEID spectrum of the phenol–H_2O complex. Here, $\sigma''(= \sigma_1^0)$ denotes the intermolecular stretching vibration. Solid curves are the calculated ones with the following parameters: $6a_1^0$ ($\sigma_2 = 2.0 \times 10^{-17}$ cm², $k_v = 1.0 \times 10^9$ s⁻¹), $6a_1^0\sigma_1^0$ ($\sigma_2 = 0.2 \times 10^{-17}$ cm², $k_v = 5.0 \times 10^9$ s⁻¹), and 12_1^0 ($\sigma_2 = 1.0 \times 10^{-17}$ cm², $k_v = 1.0 \times 10^{10}$ s⁻¹). (*J. Opt. Soc. Am.* **B7**, 1890 (1990) Fig. 7)

the IVR rate increases by more than several orders of magnitude when the complex is formed.

The results indicate that the energy put into an initial vibrational level of a molecule is immediately distributed to many isoenergetic vibrational levels involving low frequency intermolecular vibrations when a complex is formed. It should be noted that SEID spectroscopy is very useful to reveal such a great accelaration of the IVR rate upon complex formation and to obtain the IVR rate constant.

3.2.2. *IVR of p-alkyl phenols and p-alkyl anilines*

A similar enhancement of the IVR rate is also expected when we put an alkyl chain at the *para* position of phenol. The alkyl chain has low frequency vibrations such as C–C torsion and the density of states of phenol will increase very much by the addition of the alkyl chain. Also, the density of states will increase rapidly as the chain length is increased. We expect, therefore, that the IVR rate will greatly increase with the increase in the chain length. A similar situation is also expected for alkylanilines.

Figure 14 shows the SEID spectra of p-methylphenol (p-cresol), p-ethylphenol, p-propylphenol, and p-hexylphenol in supersonic jets.[10b] All the spectra were obtained by exciting the molecules to their zero-point levels in S_1 with ν_1. The experimental conditions such as laser powers and so on are the same as that for phenol. As mentioned before, no dip was observed for phenol, but dips appear in p-methylphenol (p-cresol) and the dip intensity progressively increases with the increase of the chain length. For each molecule, the laser power dependence of the dip intensity was observed for several vibrational modes and the IVR rates of these vibrational levels were evaluated using the rate equations given in a previous section. The results are summarized in Table III for the vibrational levels of modes 1 and $7a_1$. Both mode 1 and $7a_1$ are vibrations essentially localized in the phenyl ring and their frequencies are nearly constant for all the molecules. As seen from the table, the rate is very small for p-methylphenol and it increases rapidly with the increase of the chain length. However, the increase tends to saturate for the molecules of $n > 4$.

The results indicate that the IVR rate is primarily determined by the density of vibrational states, which increases rapidly with the increase in the chain length. But the tendency of saturation suggests that vibrational modes effectively coupled with the ring mode are restricted to those in

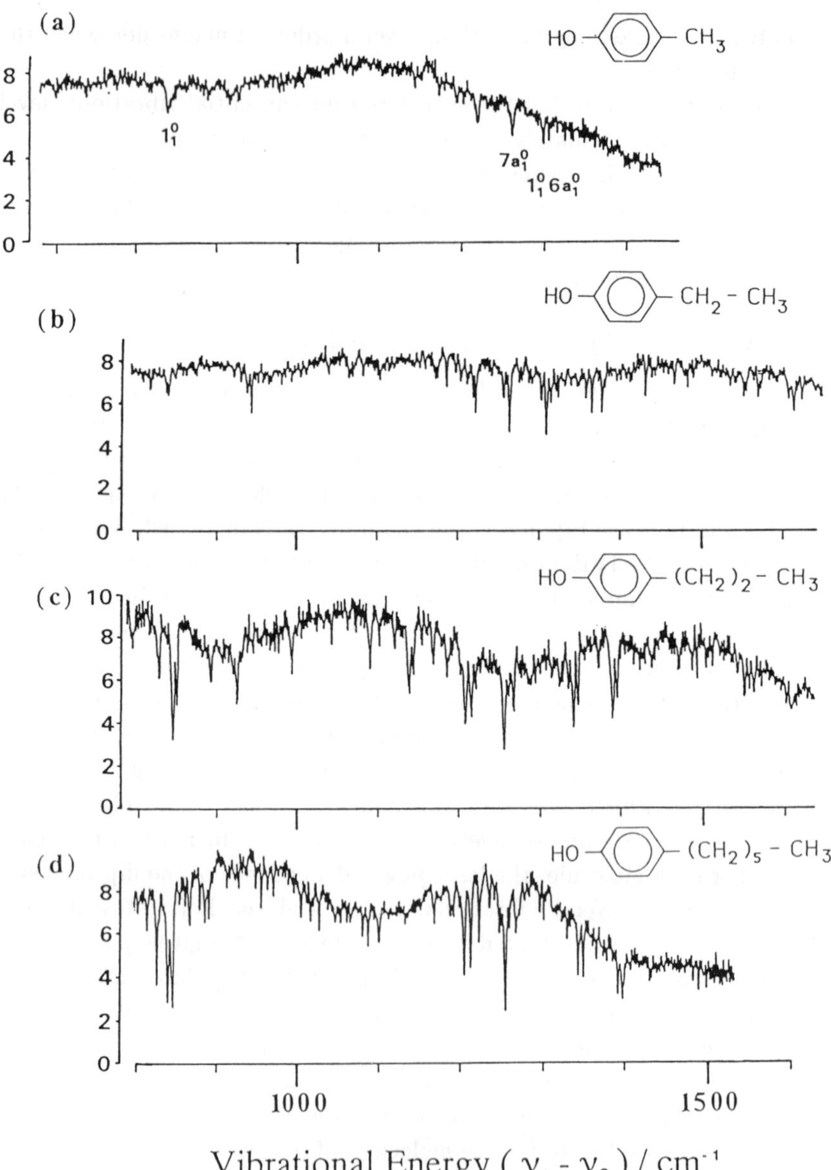

Fig. 14. SEID spectra of (a) *p*-methyl phenol, (b) *p*-ethyl phenol, (c) *p*-propyl phenol, and (d) *p*-hexyl phenol measured after exciting the molecules to the $S_1(v = 0)$ level. (*J. Phys. Chem.* **96**, 3224 (1992) Fig. 3)

Table III. IVR rate constants of p-alkyl phenols in the S_0 and S_1 states[a].

	p-methyl		p-ethyl		*trans-* p-n-propyl		*trans-* p-n-butyl		*trans-* p-n-hexyl	
	1	7a	1	7a	1	7a	1	7a	1	7a
S_1	0.25		3.3		5.1					
	(822)[b]		(817)		(815)		(809)		(816)	
S_0		< 1	2.0	3.0	3.3	10	5.0	10	5.0	> 10
	(839)	(1257)	(836)	(1260)	(844)	(1261)	(849)	(1259)	(843)	(1253)

[a]All values are in unit of $10^9 s^{-1}$.
[b]Vibrational frequency (cm^{-1}).

which the motions are localized in the chain near the ring and the modes involving motions far from the ring are not effective in the coupling.

In the same table, the IVR rate of mode 1 in the S_1 state is listed for three alkylphenols. The rates were determined from the measurements of the dispersed fluorescence spectra. The rates in S_1 are roughly comparable with those in S_0. At least, a large difference in the order of magnitude is not seen between S_0 and S_1. This suggests that the IVR rate is primarily determined by the density of vibrational states of the relevant electronic state and is not greatly different for different electronic states. A similar study of the IVR rate was also done for p-alkylaniline by SEID spectroscopy.

3.2.3. *Large amplitude motion and IVR of trans-stilbene*

Trans-Stilbene has a large amplitude torsional mode involving the phenyl group (mode 37). The vibration and the potential of this mode are shown in Fig. 15.[9b] As can be seen in the figure, the potential curve of ν_{37} is very anharmonic. It was also found that mode 37 couples strongly with other modes. Such anharmonicities play an important role in IVR processes. Figure 16 shows the comparison of the dispersed fluorescence spectrum of jet-cooled *trans*-stilbene, measured after exciting the molecule to the zero point level of the S_1 state, and the corresponding SEID spectrum. In the figure, the bands indicated by vertical lines are the progressions of mode 37 (37_n) starting from the third and forth overtones of the in-plane bending mode 25.

In the fluorescence spectrum, the intensity of the band in the progression decreases rapidly with the increase in the quantum number n. However, in

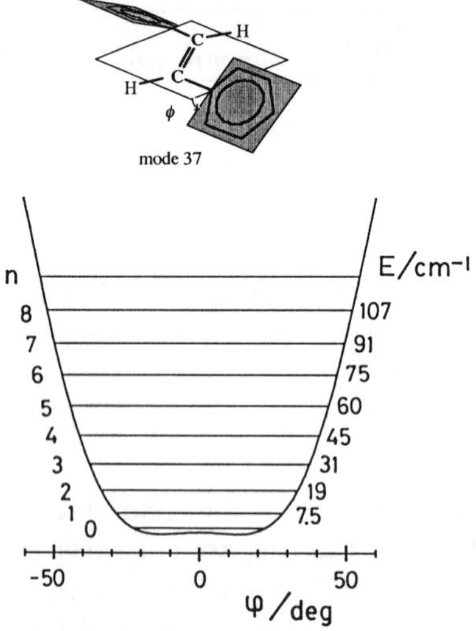

Fig. 15. Potential curve and energy levels for the out-of-plane torsional motion (mode 37) of *trans*-stilbene in the ground-state.

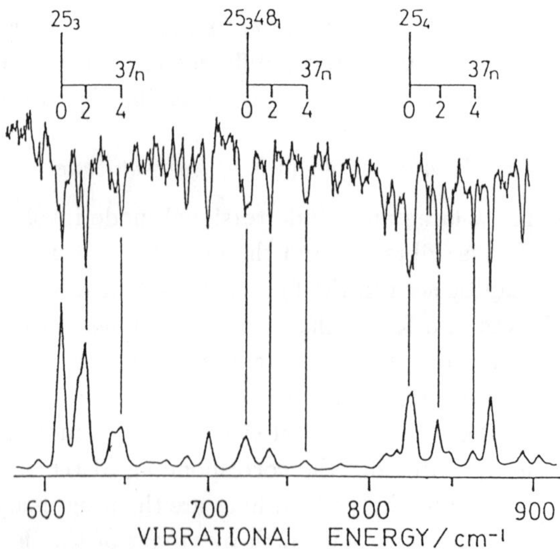

Fig. 16. SEID spectrum (upper) and dispersed fluorescence spectrum (lower) of *trans*-stilbene measured after exciting the molecules to the $S_1(v = 0)$ level. (T. Suzuki, Ph.D. Thesis, Tohoku University)

the SEID spectrum, the dip intensity of the band in the progression does not decrease very much as n increases, indicating that the dip intensity becomes large when mode 37 couples with other modes. This is a very clear demonstration that a large amplitude vibration with a great anharmonicity significantly accelerates the IVR process.

4. Stimulated Raman Spectroscopy

Stimulated emission spectroscopy including stimulated emission pumping and stimulated emission ion dip has been proven to be a very powerful method for the study of the ground-state vibrational levels of molecules and clusters in supersonic jets. It is a new type of vibrational spectroscopy which is complementary or very often superior to the usual vibrational spectroscopies such as Raman and infrared. As was shown in previous sections, stimulated emission spectroscopy has a great advantage in selecting a particular molecular species and studying its vibrational energies and dynamics. The selection of a particular species can be achieved by selective electronic excitation of the species with radiation from the first laser of ν_1. This advantage was actually used in SEID spectroscopy for the selective spectral measurement of a particular hydrogen-bonded phenol among many complexes formed in a supersonic jet.

In contrast to stimulated emission spectroscopy, all species contained in a sample are simultaneously excited to their vibrational states in ordinary vibrational spectroscopies such as Raman and infrared. As a result, the observed spectrum is degraded by overlap between spectra of many species and becomes very complicated. However, it is still possible to selectively pick out a particular species and observe only its spectrum. The selection of a particular species among many species can be achieved by resonance enhanced multiphoton ionization (REMPI) using the electronic excited state of this particular species as a resonance state. The ions produced by REMPI may be mass-selected. This is also very powerful in species selection. Therefore, the usual vibrational spectroscopies can also be used to select a particular species by combining with REMPI. However, the efficiency of vibrational excitation by Raman and infrared transitions is very small and it is not practical to use the usual Raman and infrared excitations. The situation can be greatly improved by use of the stimulated Raman process, which produces very effectively the vibrationally excited molecules. Therefore, the combination of stimulated Raman and REMPI will be very useful for the study of the ground-state vibrational state and a new spectroscopy based on this combination was actually developed by Owyoung *et. al.*[16]

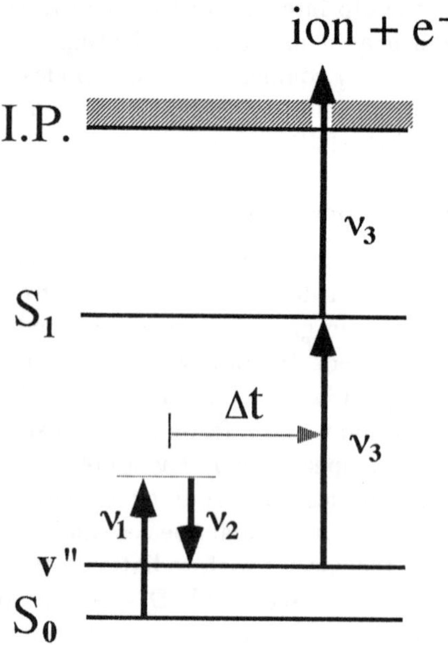

Fig. 17. Excitation scheme for stimulated Raman–UV optical double resonance spectroscopy of a jet-cooled molecule.

As shown in Fig. 17, an efficient excitation of molecules to a particular vibrational level can be accomplished by stimulated Raman pumping using two laser beams of ν_1 and ν_2. The vibrationally excited molecules are then probed with a third laser beam of ν_3 by resonance enhanced multiphoton ionization using the S_1 state as a resonance state. The ions generated by REMPI may be mass-selected. Since ν_3 used for electronic excitation is usually an ultraviolet laser beam, this spectroscopy is often called "stimulated Raman-UV optical double resonance spectroscopy." When the frequency of ν_3 is fixed to the electronic transition and the ν_2 frequency is scanned, we obtain the Raman spectrum. On the other hand, when the ν_1 and ν_2 frequencies are fixed to a particular Raman active vibrational level and the frequency of ν_3 is scanned, the electronic spectrum due to the transition from this Raman pumped level to S_1 can be obtained.

The application of this spectroscopy to benzene and its dimer in a supersonic jet will be shown as an example. Figure 18(a) shows the $(1+1)$ REMPI spectrum of benzene in a supersonic jet obtained after stimulated

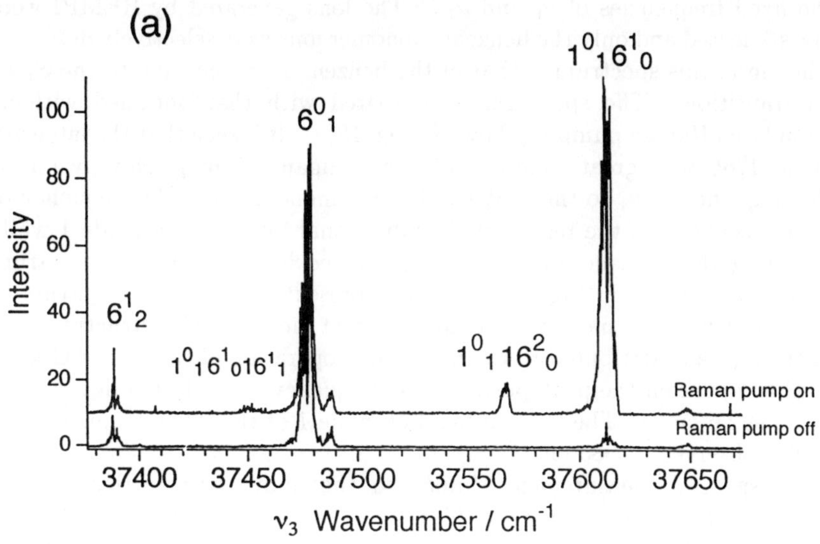

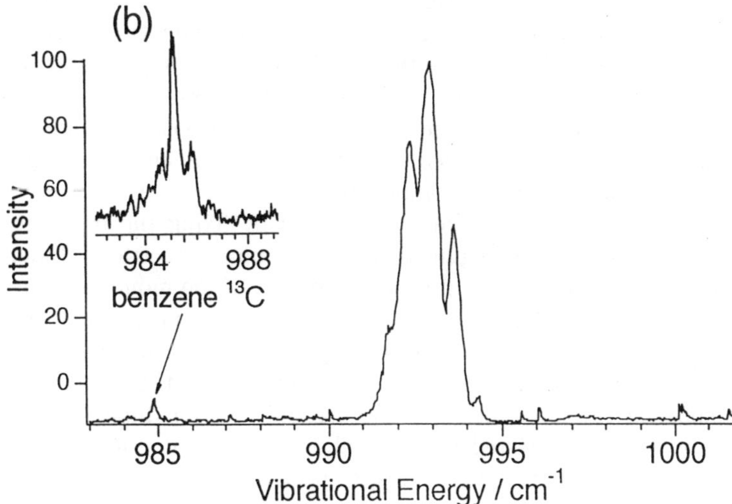

Fig. 18. (a) Mass-selected $1+1$ REMPI spectra of jet-cooled benzene monomer; (lower) measured without Raman pumping to the 1_1 level; (upper) measured after the Raman pumping to the 1_1 level. (b) Stimulated Raman spectrum of mode 1 of benzene monomer. ν_3 is fixed to the $1^0_16^1_0$ transition. (*Chem. Phys. Lett.* **199**, 33 (1992) Fig. 1)

Raman pumping to the ring breathing (mode 1) vibrational level with the fixed frequencies of ν_1 and ν_2.[17] The ions generated by REMPI were mass-selected and only the benzene monomer ions were selectively detected. Therefore, this spectrum is that of the benzene monomer due to the $S_1 \leftarrow S_0$ transition. The spectrum is compared with that obtained without stimulated Raman pumping shown in Fig. 18(a). It is seen that the intensity of the $1_1^0 6_0^1$ band greatly increases by the Raman pumping. Next, we fixed the frequency of ν_3 to the $1_1^0 6_0^1$ band of the monomer and the frequency of ν_2 was scanned in the region of the vibrational frequency of mode 1 while detecting the benzene monomer ion produced by REMPI. The spectrum obtained is shown in Fig. 18(b). This represents the Raman spectrum of jet-cooled benzene monomer in the region of mode 1. The structure seen in the figure is attributed to the multimode operation of the Nd:YAG laser used as ν_1. From the peak position, the frequency of mode 1 is determined to be 992.6 cm^{-1}. The Raman band of mode 1 of the ^{13}C-isotopic species is also observed at 985 cm^{-1}.

A spectral measurement similar to the above was carried out by detecting the benzene dimer ion this time. The mass-selected (1+1) REMPI spectrum obtained after Raman pumping is shown in Fig. 19. The bands observed in the spectrum are assigned to the $1_1^0 6_0^1$ transition of the benzene dimer. It was shown from the analysis of the spectrum that there exist two kinds of benzene dimers. The two bands of lowest frequency at 37 567 and 37 570.5 cm^{-1} belong to one of the benzene dimers (A) and the other higher frequency bands to another benzene dimer (B). When the frequency of ν_3 is fixed to one of the bands belonging to dimer A and the frequency of ν_2 is scanned, we obtain the Raman spectrum of dimer A. Similarly, the Raman spectrum of dimer B can be obtained by fixing the frequency of ν_3 to one of the bands belonging to B. The Raman spectra of dimers A and B in the region of mode 1 are similar to that of the monomer, but the frequency of mode 1 is slightly different among the monomer, dimer A, and dimer B. It was also found from Raman polarization measurements that the depolarization ratio of mode 1 is greatly different for three species, that is, $\rho = 0.008$ for the monomer and $\rho = 0.1$ and 0.02 for dimers A and B, respectively. It was suggested from the above result that dimers A and B are sandwich and T-shaped structures, respectively.[21]

As was demonstrated from the above result, stimulated Raman-UV optical double resonance spectroscopy is also very useful for selective observation of the ground-state vibrational levels of a particular molecular species. Felker *et al.* was successful in observing the individual Raman spectra of carbazole–(Ar)$_n$ clusters of $n = 1 \sim 30$ by using this spectroscopy.[18]

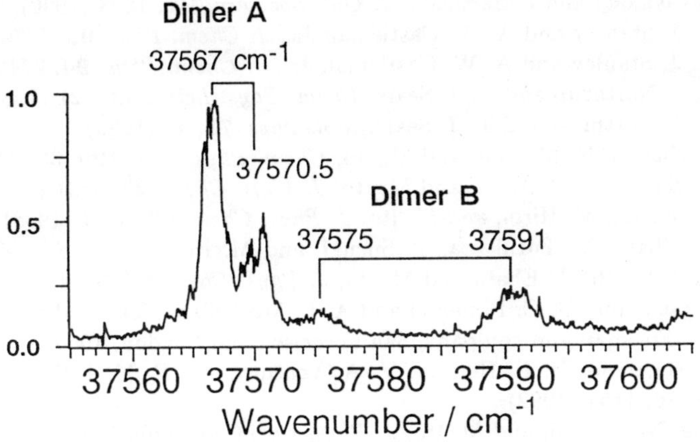

Fig. 19. Mass selected $1+1$ REMPI spectrum of benzene dimer measured after Raman pumping to the 1_1 level.

This spectroscopy also has a great advantage for the study of the dynamics of the Raman pumped level. By taking a suitable time delay between (ν_1, ν_2) and ν_3, we can directly follow the decay of the Raman populated vibrational level.[17]

In addition to stimulated emission spectroscopy, stimulated Raman spectroscopy will become a powerful vibrational spectroscopy in the near future.

References

1. C. Kittrell, E. Abramson, J. L. Kinsey, S. A. McDonald, D. E. Reisner, R. W. Field, and D. H. Katayama, *J. Chem. Phys.* **75**, 2056 (1981).
2. (a) W. D. Lawrance and A. E. W. Knight, *J. Chem. Phys.* **76**, 5637 (1982); (b) W. D. Lawrance and A. E. W. Knight, *J. Chem. Phys.* **77**, 570 (1982); (c) S. H. Kable and A. E. W. Knight, *J. Chem. Phys.* **86**, 4709 (1987); (d) S. H. Kable and A. E. W. Knight, *J. Chem. Phys.* **93**, 3159 (1990).
3. (a) H. L. Dai, C. L. Korpa, J. L. Kinsey, and R. W. Field, *J. Chem. Phys.* **82**, 1688 (1985); (b) C. E. Hamilton, J. L. Kinsey, and R. W. Field, *Ann. Rev. Phys. Chem.* **37**, 493 (1986); (c) Y. Chen, D. W. Watt, and R. W. Field, *J. Chem. Phys.* **93**, 2149 (1990).
4. H. L. Kim, S. Reid, and J. D. McDonald, *Chem. Phys. Lett.* **139**, 525 (1987).
5. (a) K. Yamanouchi, H. Yamada, and S. Tsuchiya, *J. Chem. Phys.* **88**, 4664 (1988); (b) K. Yamanouchi, S. Takeuchi, and S. Tsuchiya, *J. Chem. Phys.* **92**, 4044 (1990).

6. M. Takayanagi and I. Hanazaki, *J. Opt. Soc. Am.* **B7**, 1898 (1990).

7. (a) R. J. Stanley and A. W. Castleman Jr., *J. Chem. Phys.* **92**, 5770 (1990);
   (b) R. J. Stanley and A. W. Castleman Jr., *J. Chem. Phys.* **94**, 7744 (1991)

8. (a) F. J. Northrup and T. J. Sears, *Chem. Phys. Lett.* **159**, 421 (1989);
   (b) F. J. Northrup and T. J. Sears, *Mol. Phys.* **71**, 45 (1990).

9. (a) T. Suzuki, N. Mikami, and M. Ito, *Chem. Phys. Lett.* **120**, 333 (1985);
   (b) T. Suzuki, N. Mikami, and M. Ito, *J. Phys. Chem.* **20**, 6431 (1986);
   (c) T. Suzuki, M. Hiroi, and M. Ito, *J. Phys. Chem.* **92**, 3774 (1988).

10. (a) T. Ebata, M. Furukawa, T. Suzuki, and M. Ito, *J. Opt. Soc. Am.* **B7**, 1890 (1990); (b) T. Ebata and M. Ito, *J. Phys. Chem.* **96**, 3224 (1992).

11. See for example G. C. Pimentel and A. L. McClellan, "The hydrogen bond" (W. H. Freeman and company, San Francisco and London, 1960).

12. G. V. Hartland, B. F. Henson, V. A. Venturo, and P. M. Felker, *J. Phys. Chem.* **96**, 1164 (1992).

13. H. Abe, N. Mikami, and M. Ito, *J. Phys. Chem.* **86**, 1768 (1982).

14. (a) A. Sur and P. M. Johnson, *J. Chem. Phys.* **84**, 1206 (1986).

15. R. J. Lipert, G. Bermudez, and S. D. Colson, *J. Phys. Chem.* **92**, 3801 (1988).

16. (a) P. Esherick and A. Owyoung, *Chem. Phys. Lett.* **103**, 235 (1983);
    (b) P. Esherick, A. Owyoung, and J. Pliva, *J. Chem. Phys.* **83**, 3311 (1985);
    (c) J. Pliva, P. Esherick, and A. Owyoung. *J. Mol. Spectrosc.* **1125**, 393 (1987).

17. T. Ebata, M. Hamakado, S. Moriyama, Y. Morioka, and M. Ito, *Chem. Phys. Lett.* **199**, 33 (1992).

18. V. A. Venturo, P. M. Maxton, B. F. Henson, and P. M. Felker, *J. Chem. Phys.* **96**, 7855 (1992).

19. H. D. Bist, J. C. D. Brand, and D. R. Williams, *J. Mol. Spectrosc.* **24**, 402 (1967).

20. G. C. Pimentel and A. L. McClellan, *Ann. Rev. Phys. Chem.* **22**, 347 (1971).

21. Recently, these two species were reassigned to the benzenes in the different sites of the T-shaped dimer.
    (a) T. Ebata, S. Ishikawa, M. Ito, and S. Hyodo, *Chem.* **14**, 85 (1994);
    (b) B. F. Henson, G. V. Hartland, V. A. Venturo, and P. M. Felker, *J. Chem. Phys.* **97**, 2189 (1992).

Part III. Intermolecular Interactions

CHAPTER 16

PROBING VIBRATIONAL RELAXATION WITH STIMULATED EMISSION PUMPING SPECTROSCOPY

Scott H. Kable

School of Chemistry,
University of Sydney, NSW, 2006, Australia.

Warren D. Lawrance

School of Physical Sciences,
Flinders University, GPO Box 2100, Adelaide, SA, 5001, Australia.

Alan E. W. Knight

Faculty of Science and Technology,
Griffith University, Brisbane, QLD, 4111, Australia.

Contents

1. Introduction

This chapter is concerned with the application of stimulated emission
pumping (SEP) to investigate vibrational relaxation (also referred to as
vibration energy transfer) in the ground electronic state (S_0) of "large"
polyatomic molecules. In this context "large" molecules are considered to
be those with considerable vibrational complexity, e.g., molecules of the
size of benzene and other aromatics. Why is the study of vibrational
relaxation (VR) of interest in gas phase dynamics? Collisional energy
exchange controls the internal energy of molecules and, as such, determines
ultimately the fate of reactive encounters between molecules. Reaction
rates can vary dramatically among different vibrational levels of a molecule
and energy must often be provided to molecules in order for them to react
at all. The products of reactions can be internally hot, and their thermal
equilibration is brought about by collisions.

In order to understand how vibrational relaxation proceeds, it is
preferable to begin with a molecule whose vibrational quantum state is well
defined. Accordingly, the dissipation of this internal energy, due to collisions
with other atoms or molecules, has been a subject of intensive experimental
and theoretical investigation fro several decades. Several excellent reviews
are available on the subject.[1-5]

Experimental studies of the rate of collisional energy exchange may be divided into two broad classes of measurement. The total rate of redistribution of internal energy from the initially prepared state to the field of available destination state (a *state-to-field*, or deactivation rate) may be monitored as a decrease in population of the initial state. *State-to-state* rates are obtained if it is possible to measure the increase in population of the individual destination states that are reached following the collisional interaction. Data of both kinds are required for comparison with predictions derived from theoretical models for vibrational and rotational relaxation.

Another division in the available experimental data sets has been a consequence of the experimental techniques available to study vibrational relaxation. Infrared (IR) excitation methods have in the past provided the principal route for probing vibrational levels in the *ground* electronic state. Data are available for a large number of mostly *small* polyatomics for which infrared spectra are sufficiently discrete to permit state-selective excitation and probing.[2,4] However, IR selection rules are somewhat restrictive, and until recently, there have been few intense monochromatic, readily tunable IR light sources. Hence, most of the existing data refer to low-lying vibrational levels accessible through strongly allowed infrared transitions.

Visible or ultraviolet (UV) excitation provides a means of circumventing the restrictions of IR selection rules. Electronic transitions permit excitation of a wider variety of vibrational levels, albeit in *excited* electronic states, with limitations imposed only by Frank–Condon factors. Dispersed fluorescence from the initially prepared level, and from the state accessed through vibrational relaxation, is generally easy to detect, thus providing a convenient way of exploring vibrational relaxation pathways. As a consequence of these favourable experimental factors, the dispersed fluorescence technique has found application in state-selected studies of vibrational relaxation in electronically excited states of a variety of polyatomics[7–18] ranging as large as naphthalene in size.[20–22] It offers the advantages of state-selective excitation as well as relative simplicity in monitoring state-to-state kinetics. The technique is restricted to fluorescing species, but it has permitted a range of measurements to be made concerning the collision energy or temperature dependence of vibrational relaxation. Data are available in a few cases for a substantial temperature range. For example, in *p*-difluorobenzene, absolute rate coefficients are now available for *state-to-field* vibrational relaxation occurring at room temperature (from bulb experiments)[13,15,17] down to temperatures of a few degrees Kelvin (in a supersonic free jet expansion).[17,23,24] Similarly, in benzene, *state-to-state* rate constants have been measured over a similar temperature range.[7,18,19]

Crossed molecular beam experiments can potentially yield much more specific information about inelastic scattering processes by virtue of the experimentally controllable scattering energy (beam velocity). These crossed beam experiments are confined, generally, to ground electronic state species, and more particularly to investigating upward relaxation from the ground vibrational state. Indeed, such experiments have been carried out for a wide variety of diatomic and small molecules (see, for example, references in Krajnovich *et al.*[5]). Data for larger polyatomics are harder to find, the only two available at the present time being for aniline and p-difluorobenzene from the Gentry group.[25,26]

Crossed beam studies of vibrational energy transfer in excited electronic state are necessarily more restricted. The low densities in the crossed beam collision region mean that the molecule must have a long excited state lifetime in order for there to be the opportunity for collision transfer to occur. Consequently, only the molecules I_2[27–30] and glyoxal,[31] both of which have long lifetimes (> 1 μs) have been studied using this technique. The beauty of these S_1 studies is that they provide information on vibrational energy transfer from a variety of vibrational levels.

These experimental limitations have dictated that the data referring to polyatomic vibrational relaxation are divided into two relatively non-overlapping sets, with qualitative differences in the results. The bulk of the data available for small polyatomics refer to relaxation among low-lying vibrational levels in the ground electronic state. With few exceptions, vibrational relaxation efficiencies for these smaller polyatomics (e.g., triatomic, halogenated methanes) are found to be substantially less than unity when compared with the hard sphere collision rate.[2] For the handful of larger polyatomics studied so far (glyoxal, benzene, substituted benzenes), the data refer almost exclusively to the first excited electronic singlet state. Efficiencies, typically on the order of unity, are much larger than those observed for the ground state.[32] This contrast in efficiencies for vibrational relaxation in ground and excited states led earlier to postulates that ground and excited states may be fundamentally different with respect to the vibrational relaxation process.[32]

Clearly, there has been a need for developing experimental techniques to permit the exploration of vibrational relaxation in the ground and excited states of the same molecule. There have been a variety of techniques for preparing molecules in selected vibrationally excited levels in the ground electronic state, including direct IR pumping, Raman pumping, overtone pumping, and IR multiple photon pumping.[4] The first two of these techniques are generally restricted to relatively low levels of internal energy due to the stringent IR and Raman selection rules and these limitations have inhibited studies of higher levels where densities of vibrational states become appreciable. Infrared multiple photon pumping provided access to high internal energy states, but at the cost of a lack of state specificity: excitation is spread over a range of internal energies depending on the exact number of photons absorbed, which will not be the same for all molecules in the sample. Direct overtone pumping offers the advantages of initial state specificity and access to high vibrational energies (e.g., 6500–13 000 cm^{-1} in acetylene[33]). It is, however, limited to states with significant overtone intensities, which means that at high vibrational energies one is limited to excitation of C–H stretch modes. In this sense it is complementary to SEP (see below) since C–H modes rarely have Franck–Condon intensity, and hence cannot be prepared efficiently using SEP.

The need for vibrational relaxation data relating to chemically interesting regions of vibrational manifolds, where vibrational state densities are large, has stimulated development of other techniques for preparing ground electronic state molecules with high internal vibrational–rotational energies. For example, Barker and co-workers[34] have capitalised on the efficient internal conversion from excited electronic states to populate high-lying S_0 levels. The technique was pioneered using azulene, for which it was possible to prepare levels which vibrational energy in the region 10 000–30 000 cm^{-1}. It has since been applied to other systems including benzene and toluene.[35,36] Troe and co-workers[37,38] have used laser-induced isomerization to prepare toluene molecules with some 52 000 cm^{-1} of excess energy. With both of these techniques the definition of the initially prepared state is necessarily a little blurred: an ensemble of vibrational levels is populated, with a controllable range of internal energies, but with unknown and perhaps undefinable vibrational quantum numbers.

A more general technique is desirable for accessing selectively a variety of vibrationally excited states with precisely definable vibrational identities, at a wide range of energies. Stimulated emission pumping has been shown to be a suitable technique.[39,40] It has the capacity for accessing a wide range of S_0 levels at vibrational energies ranging from a few hundred to tens of thousands of wave numbers. In cases where the molecular manifold is still definable in terms of vibrational quantum numbers it provides the necessary state selection and energy specification, limited only by laser bandwidths.

The flexibility of SEP as a technique for S_0 state preparation has been demonstrated in past applications. Field, Kinsey, and co-workers have established that SEP may be used to prepare small molecules (e.g., acetylene[41] and formaldehyde[42]) in high-lying rovibrational levels and to characterise intramolecular state mixing amongst these levels. We ourselves have demonstrated that SEP may be used to selectively populate moderately high-lying S_0 vibrational levels in a large polyatomic (p-difluorobenzene[40,43–46]). Moreover, we have combined the SEP state preparation method with single vibronic level fluorescence (SVLF) (i.e., dispersed fluorescence) spectroscopy to provide for viewing state-resolved collision-induced relaxation among high-lying *ground state* vibrational levels in large polyatomics. The SEP–SVLF technique has now opened the opportunity for exploring collision-free and collision-induced phenomena in ground electronic states of the same molecules that have, hitherto, been studied in the *excited* electronic state with dispersed fluorescence methods. We have also demonstrated that SEP–SVLF may be used to explore vibrational relaxation induced by low energy collisions in supersonic free jets.[23,24] SEP has also been used in recent years to probe vibrational relaxation in the ground electronic state of smaller molecules.[47] In particular, energy transfer has been probed from high vibrational levels in diatomics.[48]

In this paper we review some of these applications of the SEP technique that have extended our view of S_0 vibrational relaxation in polyatomics. We focus on measurements of state-to-field rate constants for collisional relaxation from a range of selected levels in p-difluorobenzene. In addition, we describe a semiempirical model that we have developed for estimating rates of total vibrational deactivation from selected levels in polyatomics. The model is successful in reproducing, with an accuracy of $\pm 30\%$, state-to-field vibrational deactivation rates for a wide range of polyatomics, with a variety of collision partners, and thereby may prove useful to others in the field for making approximate predictions.

2. Experimental Methods and Data Analysis

The exploration of vibrational relaxation from S_0 vibrational levels requires state selection, measurement of level population, and kinetic analysis. In today's context, the SEP–LIF or SEP–SVLF method for studying S_0 vibrational relaxation may be considered as a fairly straightforward pump–probe experiment with the only added complication that the pump step is the folded double resonance technique that has been dubbed "stimulated emission pumping". Following state preparation by SEP, the time resolved population in this prepared state is probed by using laser induced fluorescence (LIF) or dispersed fluorescence. The latter probe (defined more specifically as single vibronic level fluorescence (SVLF) spectroscopy) helps to ensure a clean probe of the S_0 population since the detection of total fluorescence may contain contributions from levels other than one desired in cases where spectral congestion is appreciable. This issue is more than semantic in large polyatomics. It is therefore wise, before embarking on kinetic studies, to confirm that clean preparation has been achieved and that the integrity of the spectral signature of the level(s) probed is secure. We shall describe the spectroscopic procedure that we have used with the SEP–SVLF method and give some illustrative examples. The kinetic analyses used to extract state-to-field rate coefficients for vibrational relaxation from SEP–SVLF experiments are also summarised.

It is appropriate to comment here on why the molecule p-difluorobenzene (pDFB) was chosen for the polyatomic vibrational relaxation experiments that we have conducted. pDFB possesses several characteristics that both simplified and enhanced the experiment in the early days. The most important criterion was the presence of a strong $S_1 \leftarrow S_0$ absorption feature coincident with the 266 nm fourth harmonic of a Nd:YAG laser. This permitted us to transfer population to high-lying S_0 levels via SEP using a single Nd:YAG pumped dye laser system. Other favourable features of pDFB include the fact that the S_1–S_0 transition is strong ($\varepsilon \sim 6000$), facilitating transfer of population via SEP; that the quantum yield for fluorescence is almost unity; and, finally, that pDFB is an aromatic molecule, comparable with the type of molecules for which extensive data have been compiled for vibrational relaxation in the excited electronic state. The methodology we have implemented has shown that the combination of SEP state preparation with SVL fluorescence probing is an excellent method for measuring vibrational relaxation rates from selected vibrational levels of the ground electronic state.

2.1. *Stimulated Emission Pumping of Ground State Vibrational Levels in Polyatomics*

Stimulated emission pumping of S_0 vibrational states in larger polyatomics is, in principle, no different from nor more difficult than SEP in smaller molecules. Of course, the spectroscopic analysis for larger molecules, especially those containing many low frequency modes, may tend to become more demanding because of the myriad of vibronic transitions that are present in absorption and dispersed fluorescence. At room temperature, excitation of a single S_1 vibrational level in a polyatomic molecule may often be particularly difficult because of the crowding of each vibrational bandhead with sequences. Jet cooling can largely eliminate this complication. However, there are many incentives to make measurements at the higher kinetic temperatures that are readily achieved in bulb experiments, hence room temperature experiments in bulbs remain of interest.

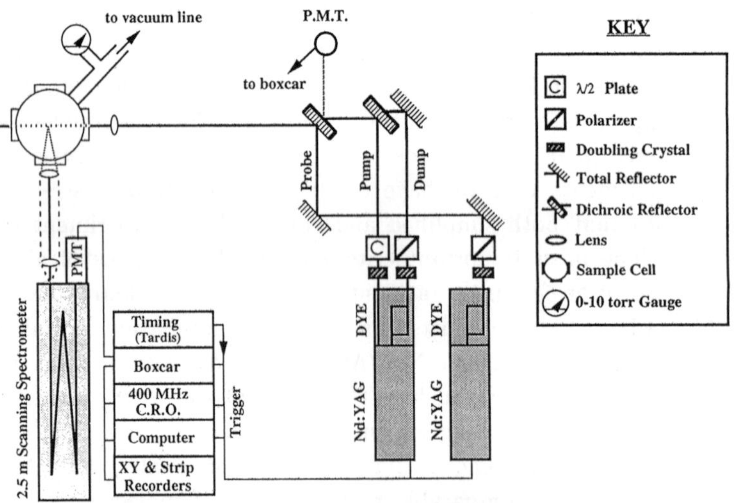

Fig. 1. Configuration of the SEP–SVLF experimental apparatus for the bulb experiments. The experiment is much the same for a free jet experiment where the bulb is replaced by the free jet chamber and the pump light is provided by a separate Nd:YAG pumped dye laser.

A schematic of the SEP–SVLF experimental apparatus used for our experiments with pDFB is shown in Fig. 1. One Nd:YAG laser, pumping either one or two lasers, provides two temporally near-coincident light pulses (266 nm + tunable dye, or two tunable dye pulses) that are used as the pump and dump steps for state preparation via SEP. A second Nd:YAG-pumped dye laser provides tunable radiation that is used to probe the population of the SEP-prepared state.

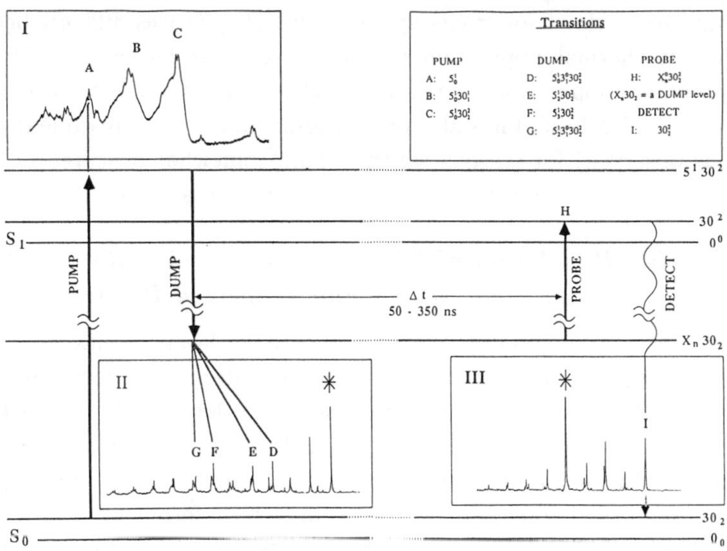

Fig. 2. Schematic representation of the temporal and spectral characteristics of the SEP–SVLF bulb experiment. On the left of the figure are shown the synchronous SEP lasers pumping a feature in the pDFB absorption spectrum (panel I) and dumping via one of four fluorescence transitions (panel II). On the right of the figure is shown the probing SVLF process, after a time delay of 50–350 ns. The probe reexcites the S_1 electronic state and subsequent fluorescence is detected through a monochromator. Panel III shows the triple resonance dispersed SVL fluorescence spectrum obtained after SEP and probing via the 30^2 state.

Figure 2 shows some of the details of spectroscopy and timing in the pump and dump steps. As noted above, a fortuitous resonance with the 266 nm fourth harmonic of the Nd:YAG laser is used to pump the hot-band transition $5_0^1 30_2^2$ in pDFB (panel I in Fig. 2). The residual 532 nm output following the 266 nm doubling crystal is used to pump a dye laser which provides the radiation for the dump transition (see Fig. 1). Any of the spectral features in the ensuing dispersed fluorescence spectrum (panel II, Fig. 2) can, in principle, be chosen as a dump transition: the transitions D–G (assigned in the figure) are used here to prepare four vibrational states of quite different energy. The SEP technique for state preparation is efficient and relatively easy to set up providing due care is taken with spatial alignment. Temporal coincidence for the pump and dump steps is assured by using the same Nd:YAG laser for pumping and dumping. Their relative timing can be optimised easily with short optical delays. Dumping to levels at lower energy than $D(3_1 30_2)$ proves difficult because of inherent thermal population of these levels interfering strongly with SEP-prepared population of the level. Dumping to states with higher energy than $G(3_1 5_2 30_2)$ has also proven problematic (no subsequent probe signal is observed) for reasons which remain somewhat obscure, although we allude to this issue in the discussion of the results below.

2.2. *Probing Population of SEP-Prepared Levels with Laser Induced Fluorescence and Dispersed Fluorescence*

The time-varying population of the prepared S_0 level v'' may be probed by re-exciting the molecule, after some predetermined delay, from its SEP-prepared S_0 level into a specific vibrational level in the S_1 electronic state. The fluorescence following this *probe* laser excitation serves as a monitor for the population in v''. If total fluorescence generated by the *probe* is monitored, then the technique may be referred to as SEP–LIF. Alternatively, added selectivity may often be necessary to discriminate against fluorescence resulting from the *pump* excitation, or against fluorescence arising as a result of overlapping bands accessed by the probe laser. In these instances, the *probe* fluorescence may be dispersed through a monochromator thus allowing a chosen fluorescence transition to be monitored selectively. This four-step wavelength selective procedure is what we refer to as SEP–SVLF. A particular advantage of SEP–SVLF over SEP–LIF in vibrational relaxation experiments relates to the fact that the *dispersed* fluorescence signal arising from probe excitation of the SEP-prepared S_0 level is likely to be contaminated by fluorescence generated by probe excitation of levels reached as a result of vibrational relaxation (VR), or by scattered probe laser light (shown by an asterisk in panel III, Fig. 2). Thereby, in a large polyatomic, SEP–SVLF provides a relatively secure technique for measuring accurately state-to-field VR rates. Moreover, it may be used for state-to-state VR measurements. Of course, the loss of signal through a dispersing monochromator may preclude SEP–SVLF from being practical in some circumstances. The use of dispersed fluorescence provides added confirmation of the SEP-prepared level identity, since it firmly identifies the emitting S_1 level. An illustration of this is provided in Subsec. 2.4 below.

Figure 1 shows the second Nd:YAG pumped dye system used as the probe, with the detection system configured for dispersed fluorescence. Figure 2 shows the spectral and temporal features of the collisional energy transfer experiment. A variable delay between the probe laser and the pump/dump state preparation lasers provides the time axis for rate measurements, i.e., we probe the population of the prepared state as a function of time after preparation, the population being continuously depleted as a result of collisions with the bath gas. A variable delay in the range of 50–350 ns is required in the pDFB VR experiments. This delay range covers the time scale whereby single collision events dominate the kinetics of depopulation. A set of fixed delays, using a series of different length 50 Ω cables, was used in our earliest experiments. A more convenient procedure, used in the majority of our studies, is to employ a digital delay generator where the delay between preparation and probe lasers may be varied continuously.

In order to probe a SEP-prepared S_0 vibrational level, one may in principle choose any favourable $S_1 \leftarrow S_0$ transition that re-excites v'' to some level v' in S_1. Any securely assigned band in the dispersed fluorescence from v' can serve as the monitor for the population in v''. For convenience, in the bulb experiments, where the pump transition is always $5_0^1 30_2^2$, we have chosen a common S_1 level, namely 30^2, as the level from which probe fluorescence is detected (see Fig. 2). We also choose the 30_0^2 fluorescence transition for detection in all cases (Fig. 2, panel III). The 30_2^2 transition is exceptionally bright in 30^2 fluorescence and it stands well clear of other fluorescence structure, including that due to the residual 266 nm excited (pump) fluorescence. Hence, the detection monochromator (slits set to give a relatively broad 20 cm^{-1} bandwidth) is normally tuned to accept 30_2^2 fluorescence. It transpires that there is a broad, unstructured component associated with $5^1 30^2$ fluorescence.[49] Part of this unstructured background appears in the region of the 30_2^2 band, and at short delay times can contribute to the fluorescence signal. To overcome this problem, an

A–B difference experiment is performed. This problem would be much more severe if the probe fluorescence was not dispersed.

A non-Boltzmann rotational population is prepared in the selected S_0 vibrational state by the SEP process, governed by the rotational selection rules operating for electronic transitions. We estimate[50,51] that our initial ensemble ($t = 0$) consists of pDFB molecules with low K_a, and a wide spread of J. After 50 ns, however, this population appears to have a near Boltzmann rotational distribution. For this reason we normally commence our analysis of VR 70 ns after initial state preparation.

2.3. *Application of the SEP–SVLF Technique to Vibrational Relaxation Studies in Supersonic Free Jets*

In experiments where the SEP–SVLF technique has been used to measure VR rates induced by low energy collisions occurring in a supersonic free jet, we cannot take advantage of the coincidence of the Nd:YAG fourth harmonic and a hot band in the S_1–S_0 transition since the intensity of the hot band structure is reduced drastically in the free jet environment. Consequently, a different set of pump, dump, and probe transitions are utilised. We have used two pulsed Nd:YAG lasers (532 and 532 + 355 nm output) to drive three frequency-doubled tunable dye lasers. The three dye laser beams (pump, dump, and probe, respectively) are combined spatially and temporally, and crossed with a supersonic free jet expansion of $< 1\%$ pDFB seeded in argon. LIF excitation spectra are obtained by collecting fluorescence perpendicular to both the laser and nozzle axes via $f/2$ optics and passing the collected light through a 0.25 m Spex double monochromator. The monochromator, in subtractive dispersion, serves as a tunable broad band filter with approximately 3000 cm^{-1} bandwidth. Dispersed SVL fluorescence spectra may also be obtained using the detection configuration shown in Fig. 1.

Temporal information concerning the population in the initially prepared vibrational state is obtained as in the bulb experiments by incrementing the delay between the pump/dump lasers and the probe laser.

2.4. *Verification of the Preparation and Probe Steps*

The pump–probe experiments described here are multilaser experiments in which population is transferred several times. It is important to ensure the integrity of these processes at each stage. It is perhaps instructive at this point to illustrate the type of evidence that can be used for this purpose. We do this for the case of vibrational relaxation from 3_1 in a supersonic expansion. Similar evidence has been presented for the case of state preparation in the room temperature experiments. The reader is referred to these papers for further details.[40,46]

The strongest and least congested band in the S_1–S_0 transition, namely the origin band at $36\,838$ cm^{-1} (vac),[46,52] is chosen as the *pump* transition. Figure 3 shows the dispersed fluorescence spectrum following pumping of the 0_0^0 transition. In principle, any transition in this spectrum can be selected to prepare population in the appropriate S_0 vibrational state via SEP. The *dump* laser was tuned to coincide with the strong emission band 3_0^1 at $35\,580$ cm^{-1}. This combination of two strong transitions (*pump* and *dump*) provides optimum conditions for preparing a significant number of molecules in a ground state vibrational level in pDFB.

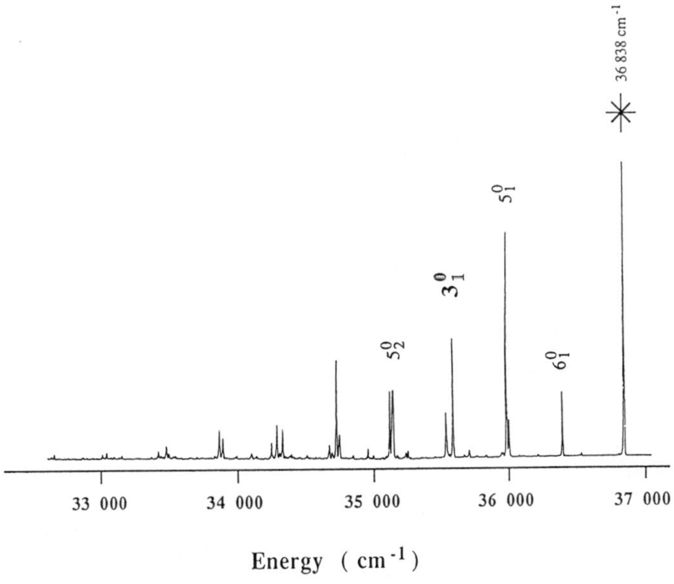

Fig. 3. Dispersed fluorescence spectrum following laser excitation of the 0^0 level in pDFB. The *dump* laser may, in principle, be tuned to any of the transitions that appear in the spectrum. In this work the strong 3_1^0 transition indicated was selected to dump population to the 3_1 level. (For other experimental details see Ref. 28.)

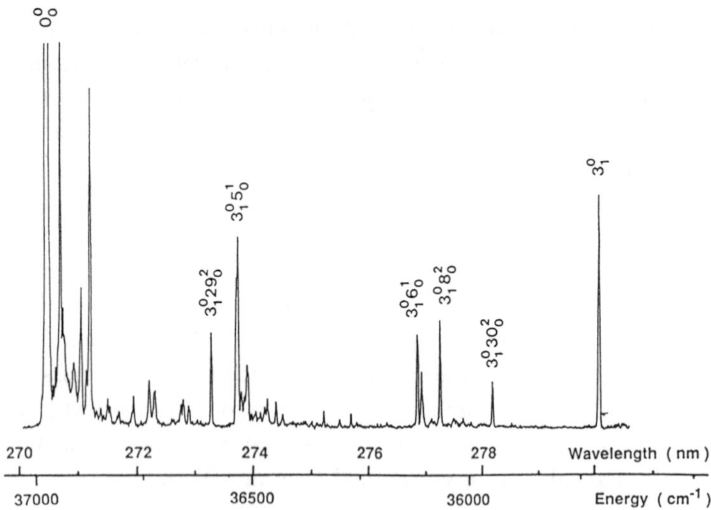

Fig. 4. The SEP–LIF spectrum of jet-cooled *p*DFB. The transition to longer wavelength than 273 nm can be positively assigned as originating from the *dump* level 3_1. To probe the time dependent population of 3_1 we tune the probe laser to the $3_1^0 5_0^1$ transition.

Figure 4 illustrates that state preparation of the level 3_1 is achieved successfully. The SEP–LIF spectrum, measured at a delay of 100 ns after state preparation, contains structure in the region $35\,600$–$36\,800$ cm^{-1} that may be identified exclusively with absorption originating from the S_0 level 3_1. Observed transitions, with assignments indicated in Fig. 4, are mainly of the type $3_0^1 X_0^n$, i.e., totally symmetric additions to the "false origin" 3_1^0. This confirms the assignment of the SEP populated level as 3_1. When the dump laser is blocked, the intensity of residual LIF features arising due to 3_1 population is found to be $< 5\%$ of the original intensity. This residual intensity may be ascribed to population of the 3_1 level via spontaneous emission from the *pump* level.

Population in the level 3_1 is monitored as a function of preparation – probe delay by probing the transition $3_1^0 5_0^1$, displaced by 818 cm^{-1} (v_5') from the *dump* transition. This transition is selected since it is the strongest transition in the SEP–LIF spectrum (Fig. 4) that is not overlapped with either of the *pump* or *dump* lasers. The stronger 3_1^0 band is of course resonant with the *dump* transition.

We may probe the population of the target level 3_1 by monitoring total SEP–LIF fluorescence. However, as discussed above, there are advantages

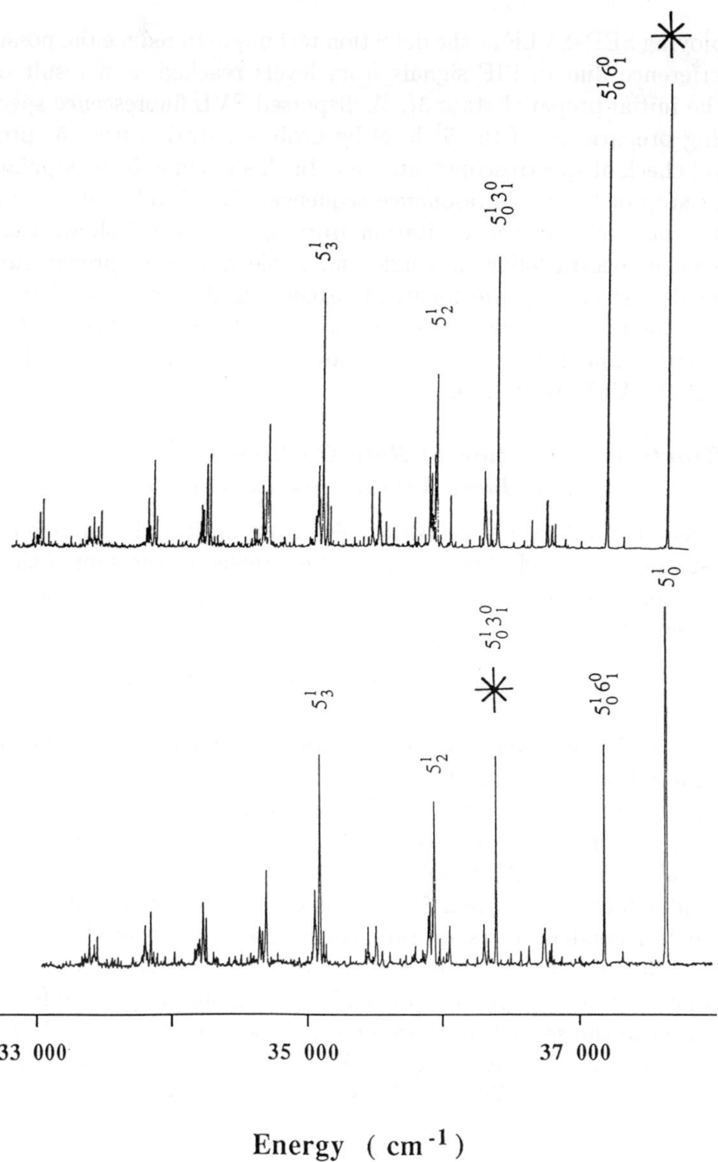

Energy (cm^{-1})

Fig. 5. Dispersed SVL fluorescence spectra from the 5^1 level of jet-cooled pDFB. The upper spectrum is generated following direct, single laser preparation of 5^1 via the 5_0^1 transition. The lower spectrum is generated via the $3_1^0 5_0^1$ transition following SEP preparation of the 3_1 level. The spectra are essential identical, thereby demonstrating the spectroscopic integrity of the three laser preparation.

in employing SEP–SVLF as the detection technique to reduce the possibility of interference due to LIF signals from levels reached as a result of VR from the initial prepared state 3_1. A dispersed SVL fluorescence spectrum following preparation of the 5^1 level by probe excitation from 3_1 provides the final check of spectroscopic integrity. In this scheme, 5^1 is populated as the last step in the triple resonance sequence. This level is also accessible directly via single photon excitation from 0_0. Figure 5 shows the SVL fluorescence spectra following single and triple resonance preparation. In all regards the spectra are identical (except in the position of the laser scattered light peak shown by asterisks in the figure) confirming that we are preparing and probing pDFB in known vibrational states at all stages of the SEP–SVLF procedure.

2.5. *Kinetics: Extraction of Rate Coefficients for Collision-Induced Vibrational Relaxation*

The observed total rate of relaxation ξ_T, from an initially prepared state into a surrounding field of states, may be expressed as the sum of the rates of all the contributing processes deactivating the initial state. Under single collision conditions,

$$\xi_T = k_0 + k_v^M[M] + k_v^{p\mathrm{DFB}}[p\mathrm{DFB}] + k_v^{p\mathrm{DFB}}[p\mathrm{DFB}^*]^2 \,, \tag{1}$$

where ξ_T is the observed total decay rate, k_v is a rate constant for the particular vibrational relaxation process, k_0 is the collision-free rate and is specific to each initial state, [   ] represents a concentration (pressure or density in our case), M is some foreign gas, and pDFB* represents the vibrationally excited pDFB molecules.

We introduce the notation ξ, which is a pressure dependent rate, while pressure independent rates or rate coefficients are denoted by k which may be first order (e.g., k_0) or second order (e.g., k_v^M). Under constant pressure conditions the pressure dependent rates become pseudo first order and related to the second order rate coefficients, e.g., $\xi^M = k^M[M]$.

The term in Eq. (1) involving $[p\mathrm{DFB}^*]^2$ may be neglected in the analysis since $[p\mathrm{DFB}^*] \sim 10^{-5}\,[p\mathrm{DFB}]$.[46] Thus Eq. (1) reduces to:

$$\xi_T = k_0 + k_v^M[M] + k_v^{p\mathrm{DFB}}[p\mathrm{DFB}] \,. \tag{2}$$

Provided one is careful to perform the experiment under single collision conditions, repopulation of the initial state is insignificant. In this case, the population of the initially prepared pDFB level decays exponentially with rate ξ_T and so a plot of $\ln[p\mathrm{DFB}^*]$ versus time is linear with slope

$-\xi_T$. One does not need to measure the absolute magnitude of $[p\text{DFB}^*]$ but only something proportional to it as the constant of proportionality will only affect the intercept, not the slope which provides ξ_T. The intensity of fluorescence generated by the probe is, for fixed laser intensity, directly proportional to the vibrationally excited pDFB concentration. Consequently, ξ_T is extracted by plotting $\ln[I(t)]$ versus t, where $I(t)$ is the probe-induced fluorescence intensity.

There are two contributions to collision-induced vibrational relaxation: that due to unexcited pDFB and that due to a foreign gas M. In the case of pDFB self-relaxation, the experiment is done in the absence of any foreign gas, and $[p\text{DFB}]$ is varied. For this situation it can be seen that Eq. (2) becomes:

$$\xi_T = k_0 + k_v^{p\text{DFB}}[p\text{DFB}] \ . \tag{3}$$

The slope of a plot of observed rate ξ_T versus $[p\text{DFB}]$ yields $k_v^{p\text{DFB}}$, the rate constant for vibrational relaxation of the SEP-prepared S_0 state due to collisions with the Boltzmann bath of pDFB molecules, while the intercept provides an estimate for k_0, the zero pressure rate.

To extract relaxation rate constants for a species M, one simply keeps $[p\text{DFB}]$ constant and varies $[M]$, obtaining ξ_T for each $[M]$. A plot of ξ_T versus $[M]$ (a so-called Stern–Volmer plot) then provides k_v^M as the slope in accordance with Eq. (2). The intercept is now of course $k_0 + k_v^{p\text{DFB}} \, [p\text{DFB}]$ where the second term has been kept constant. It can be seen from Eq. (2) that k_v^M can also be estimated directly from ξ_T provided that k_0 and $k_v^{p\text{DFB}}$ are known. We have used both methods (the direct and the Stern–Volmer) for estimating rate coefficients in the different studies of VR in S_0 pDFB described below.

2.5.1. *Low energy collision experiments*

For measurements carried out in a free jet expansion (Ar carrier gas in this case), the observed depletion rate ξ_T of population out of an initially prepared S_0 state is of the same form as that shown in Eq. (2). In this case, however, the collision-free rate includes an additional component which we label as k_{loss}. This arises from loss of excited pDFB molecules from the detection volume due to flow in the free jet. The density-dependent and density-independent contributions to ξ_T are evaluated experimentally.

In the bulb experiment discussed above, the pressure-independent rate is estimated by extrapolating a Stern–Volmer plot of observed rate versus pressure to zero pressure. In the free jet experiments, the rates of depletion are measured for a range of local molecular densities ($\rho \propto P_0$, i.e., density is determined by the nozzle backing pressure) and an extrapolation to zero density yields the density-independent contribution. From Eq. (2) we see that this intercept comprises the sum of all density-independent contributions. We have shown elsewhere [24,51] that in the free jet this density-independent term is dominated by k_{loss}.

$\xi_v^{p\text{DFB}}$ is a density-dependent rate. Since the seed ratio of pDFB in the carrier gas (Ar) is $< 1\%$, and since results obtained at room temperature provide the estimate $k_v^{\text{Ar}} \approx 0.5 k_v^{p\text{DFB}}$, we are justified in neglecting the vibrational relaxation rate due to pDFB*–pDFB collisions. Hence, for the supersonic expansion conditions, Eq. (2) reduces to

$$\xi_T = \xi_v^{\text{Ar}} + k_{\text{loss}} \ . \tag{4}$$

The final estimates for k_v^{Ar} are obtained from the relationship $k_v^{\text{Ar}} = \xi_v^{\text{Ar}}/\rho$, where ρ, the density of the argon carrier gas, is a function of distance X/D from the nozzle orifice. The value of ρ is determined from the well established relationship for nozzle expansions, as discussed in Sec. 3 below.

3. The Lennard–Jones Elastic Encounter Rate as a Means for Scaling Vibrational Relaxation Rates

Vibrational relaxation is viewed conventionally as a relatively inefficient inelastic process which may be characterised in terms of a rate coefficient, k_v. This may be written as a product of a rate coefficient for binary elastic collisions and a probability factor that expresses the likelihood of a relaxation event taking place during the elastic collision.[53] For a hard sphere (hs) elastic collision rate,

$$k_v = P_{\text{hs}} k_{\text{hs}} \ . \tag{5}$$

The probability, P_{hs} is the ratio of the observed vibrational relaxation rate to the hard sphere elastic rate. It is thus often couched in the language of a relative efficiency for VR.

The hard sphere rate constant for collisions between A and M is defined by

$$k_{\text{hs}} = \sqrt{\frac{8kT}{\pi \mu_{AM}}} \ \sigma_{\text{hs}} \ . \tag{6}$$

Here, $\sigma_{\text{hs}} = \pi d_{AM}^2 = \pi/4(d_A + d_M)^2$ is the hard sphere cross-section expressed in terms of the hard sphere diameters d_A and d_M, and μ_{AM} is the reduced mass for the collision pair $A \ldots M$.

Studies of vibrational relaxation in the ground electronic state of polyatomic molecules, induced by collisions with monatomic and small polyatomic collision partners, have shown[1,2,53] that the probability of vibrational relaxation occurring during a hard sphere collisional encounter is generally small, i.e., the quantity k_v/k_{hs} is less than unity. For example, efficiencies of less than 0.01 are typical for V–T relaxation in molecules such as the methyl halides, SF_6, and CO_2.[1,53] These observations are consistent with the notion that the steep repulsive wall of the intermolecular potential must be accessed for V–T transfer to take place. The use of the hard sphere elastic rate coefficient has, therefore, proved convenient for scaling the efficiencies for vibrational relaxation in these systems.

However, there are now ample data to show that vibrational relaxation in larger polyatomic molecules such as benzene and pDFB occurs with considerably greater efficiency than that observed for the vast majority of smaller polyatomics. $P_{hs} = k_v/k_{hs}$ may even exceed unity![14,44] In such cases the hard sphere collision rate is no longer a physically meaningful elastic collision rate against which to scale an observed inelastic vibrational relaxation rate. In essence, we are encountering a situation where the hard sphere potential is an unsatisfactory description of the elastic interaction between the target molecule and its collision partner. The problem is further highlighted in temperature dependent studies of vibrational relaxation. The hard sphere elastic rate is based on a temperature independent cross-section, hence it cannot provide a useful scale for vibrational relaxation that may be a strong function of temperature, as is the case at very low temperatures.[21,24]

We seek a better intermolecular potential to replace the hard sphere potential as a means for scaling vibrational relaxation efficiencies. While recognising that an accurate inelastic potential is what is really required, it must be remembered that it remains a challenge to determine accurate inelastic potentials for complex systems from theory alone. The philosophy proposed is to use, for scaling purposes, an elastic potential that has most of the isotropic dynamical features of the true inelastic potential, thus providing a suitable estimate of the elastic *encounter* rate for the collision partners. We have shown previously[44,46] that the Lennard–Jones (LJ) 6–12 potential serves this purpose well for foreign gas induced relaxation of S_0 pDFB and for explaining the temperature dependence of vibrational relaxation in the range 2–300 K.[24]

The rate coefficient k_{LJ} for collisions between particles A and M whose interaction conforms to the Lennard–Jones pair potential is given by

$$k_{LJ} = \sqrt{\frac{8kT}{\pi\mu_{AM}}}\, \pi d^2_{AM} \Omega^{(2,2)^*}_{AM} = k_{hs}\Omega^{(2,2)^*}_{AM} \ . \tag{7}$$

The collision integrals $\Omega_{AM}^{(2,2)^*}$ are exactly the factor by which the rate constant k_{LJ}^M differs from the hard sphere rate. The magnitude of $\Omega_{AM}^{(2,2)^*}$ is dependent on the reduced temperature $T^* = (kT/\varepsilon_{AM})$, hence, for a given temperature, its magnitude will be controlled by the intermolecular potential well depth ε_{AM}. Since the well depths for pDFB–M are not generally known experimentally, we employ the geometric mean approximation:

$$\varepsilon_{AM} = (\varepsilon_{AA} \cdot \varepsilon_{MM})^{1/2} \, , \tag{8}$$

which is reasonable for comparison of collision rates for a single target gas with a range of collision partners.

The $\Omega_{AM}^{(2,2)^*}$ collision integrals can be determined from published tables[54]; however, these tables do not go sufficiently low in T^* to enable comparison with very low temperature data of the type extracted in free jet experiments. In a previous study we have computed directly the magnitudes of the $\Omega_{AM}^{(1,s)^*}$ integrals for the reduced temperature range $0.0001 \leq T^* \leq 100$.[55] In this chapter the results of these calculations of the $\Omega_{AM}^{(2,2)^*}$ are used in calculating k_{LJ}.

Using the LJ potential, vibrational relaxation efficiencies, or *probabilities* P_{LJ}, are expressed, exactly analogous to the hard sphere case, as the ratio of the vibrational relaxation rate coefficient and the Lennard–Jones elastic encounter rate coefficient:

$$P_{\text{LJ}} = k_v/k_{\text{LJ}} \, . \tag{9}$$

A question that we have addressed in our experimental studies is whether either (or both) of these classical encounter rates can provide a useful measure of the magnitudes and temperature dependencies of the vibrational relaxation rates reported here and, if so, what significance may be associated with the magnitude of the efficiency P. We reiterate that the hard sphere cross-section σ_{hs} is independent of temperature, whereas σ_{LJ} contains a temperature term in the $\Omega^{(2,2)^*}$.

4. Experimental Results for S_0 p-Difluorobenzene

The experimental result on vibrational relaxation (VR) in S_0 (SEP-prepared) vibrational states of pDFB that are the subject of this chapter encompass a six year period from 1982 to 1987. The initial experiments were aimed at measuring the efficiencies of VR in both the ground and excited electronic states of the same molecule. As has been discussed earlier (Sec. 1), at the time that these experiments were undertaken the existing evidence could be construed to imply that vibrational relaxation was considerably more efficient in excited electronic states than in ground electronic states. In discussing the results of our experiments here, we shall address this question last. We have chosen to do this because it transpires that a comparison between VR in S_0 and S_1 is best accomplished by comparing the full gamut of data collected in our own experiments with the extensive VR data for the S_1 state of pDFB that has become available since then.

In presenting the results of our work, we shall take a somewhat historical approach. This has been done to illustrate how our understanding of the VR relaxation process in this relatively large molecule has evolved, and how the experiments evolved in sophistication to probe the hypotheses being developed. The experiments can be divided into three investigations probing different facets of vibrational relaxation: the influence of collision partner, the effect of the initial state, and the temperature dependence of the VR rate coefficients. The first two investigations were performed at 300 K in a static cell, while the temperature dependence study was undertaken in the low energy collision environment of a supersonic free jet expansion. Finally, we shall compare our data with VR data available in the S_1 states of the same molecule; this was, of course, the initial reason for undertaking these experiments and for utilising SEP as a state preparation technique.

4.1. *Extraction of Rate Coefficients*

4.1.1. *Bulb experiments*

Rate coefficients for VR from a specific S_0 state of pDFB are obtained from the experimental data by use of the kinetic equations described above, in particular Eqs. (2) and (3). Some typical decay curves plotted as $\ln(I)$ versus preparation-probe delay time are shown for three pressures of pDFB in Fig. 6. Each datum consists of an average of 150 measurements. The data show that, as predicted by Eq. (3), higher pressures of pDFB cause faster depopulation of the prepared state. The relaxation rates ξ_T for each pressure are obtained from the linear least squares fits as displayed in the figure.

Figure 7 shows self-relaxation decay rates, ξ_T, as a function of pDFB pressure (a so-called Stern–Volmer plot) for the levels $3_1 30_2$ and $3_1 5_2 30_2$ (E_{vib} = 1577 and 3276 cm^{-1}, respectively). A linear dependence of relaxation rate on pDFB pressure is observed in accord with Eq. (3). The slope of this line provides the experimental estimate of the second order

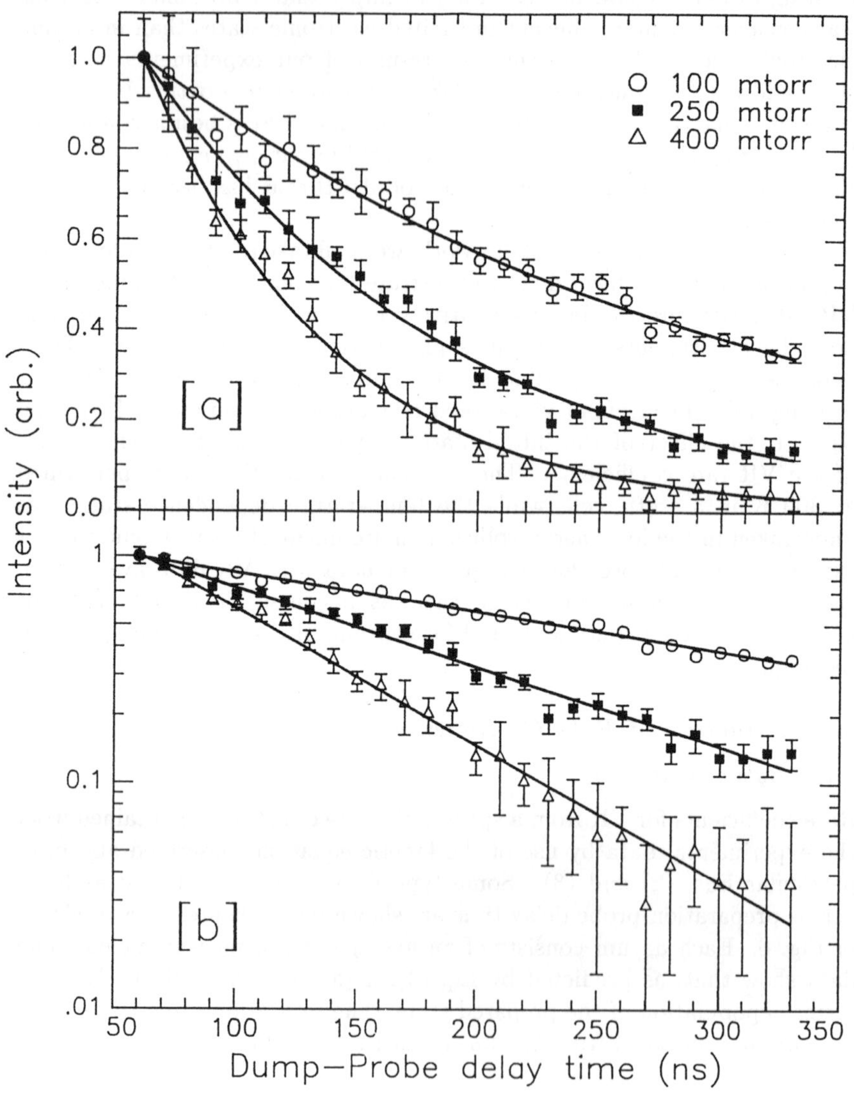

Fig. 6. Experimental decay curves for three pressures of neat *p*DFB. The upper panel (a) shows the single exponential decay of the fluorescence and the lower panel (b) shows the same data plotted with a logarithmic ordinate. In (b), the solid lines are linear least square fits to the data which are transformed back to the linear ordinate for plotting in (a). Vertical error bars represent the standard deviation of 150 measurements at each time point.

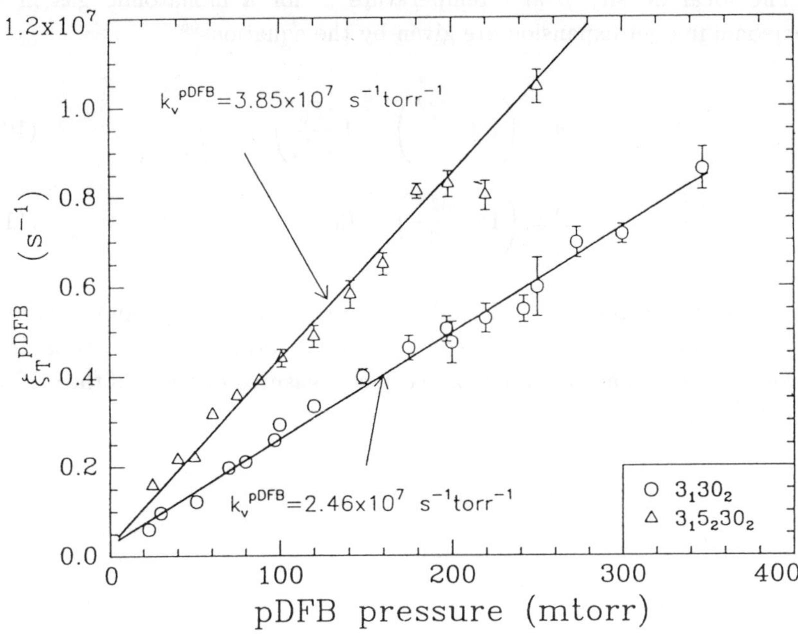

Fig. 7. Stern–Volmer plots for pDFB self-relaxation shown for two different vibrational levels. The rate coefficient indicated are calculated from the weighted (σ^2) linear least squares fit of each line.

rate coefficient $k_v^{p\mathrm{DFB}}$. The linearity of the log plots of Fig. 6 and the Stern–Volmer plots of Fig. 7 confirm that we have interpreted the kinetics correctly and that we have remained within the single-collision regime for the relaxation process. It is also apparent in Fig. 7 that self-relaxation from the higher level ($3_15_230_2$) is faster than from the lower level (3_130_2); an issue that is dealt with later in this section.

4.1.2. *Free jet experiments*

Measurements of ξ_T in a free jet experiment are as straightforward as the bulb experiment. The data look the same, as portrayed in Fig. 6. Extraction of the rate coefficients, however, requires knowledge of the density and temperature in the jet and an estimation of the k_{loss} term in Eq. (4).

The local density ρ and temperature T for a monatomic gas in a supersonic free jet expansion are given by the equations[56]

$$\rho = \left(1 + \frac{M^2}{3}\right)^{-\frac{3}{2}} \left(\frac{P_0}{kT_0}\right) , \tag{10}$$

$$T = \left(1 + \frac{M^2}{3}\right)^{-1} T_0 , \tag{11}$$

where $M \approx 3.26(X/D)^{2/3}$, T_0 and P_0 are the temperature and pressure in the stagnation chamber behind the nozzle orifice, and X/D is the distance downstream from the nozzle orifice measured in terms of the orifice diameter, D.

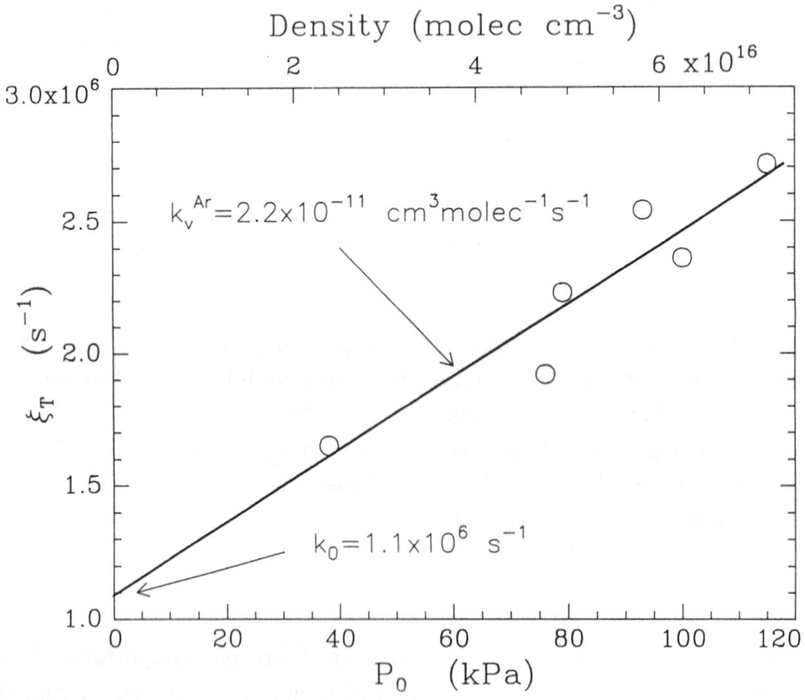

Fig. 8. Stern–Volmer type plot of the measured decay rate versus nozzle backing pressure P_0 and molecular density ρ, which is calculated from P_0 and X/D. The intercept at zero density provides a measure of the collision-free rate k_0.

Equation (10) shows that the density in the jet is directly proportional to P_0 (for fixed X/D and T_0) and so the density independent term in Eq. (5) can be extracted by plotting ξ_T versus P_0. This is shown in Fig. 8 for a variety of P_0 at a constant X/D of the probe laser from the nozzle orifice. The data conform reasonably to a straight line, intercepting the $\rho = 0$ axis at 1.1×10^6 s^{-1} and hence $k_{\text{loss}} = 1.1 \times 10^6$ s^{-1}. Using Eq. (10) to calculate the local density (upper abscissa), the rate coefficient, k_v^{Ar}, is simply the slope of the line, provided that pDFB*–pDFB collisions are negligible, as discussed in the kinetics section.

4.2. Dependence of Vibrational Relaxation Rates on the Nature of the Collision Partner

One of our first SEP experiments was to investigate VR in S_0 pDFB induced by collisions with a variety of collision partners.[44] Those experiments complemented the study of *state-to-field* vibrational relaxation in the S_1 state of benzene.[9,14] The $5_2 30_2$ level was chosen as the initial state in this study. It was populated by pumping the $5_0^1 30_0^2$ transition using the 266 nm fourth harmonic of a Nd:YAG laser, as described earlier, and dumping via the $5_0^1 30_0^2$ transition at $\sim$ 278 nm. The vibrational state density surrounding this level is about 6–8 per cm^{-1}. There were (and indeed still are) relatively few studies of vibrational relaxation of a polyatomic in the region of intermediate state density. This region is particularly important, as it bridges the low state density region, where state-to-state studies can chart the flow of vibrational energy with collision, and the high density region, where experiments of the type pioneered by Barker[34] and by Troe[37,38] extract properties such as the average energy transferred per collision. In all, a total of 23 collision partners were studied, including a variety of monatomics, diatomics, and a diverse mixture of polyatomics including pDFB itself.

The results of this study are given in Table I. It can be seen that the second order rate constants for depopulation of the $5_2 30_2$ level by all gases are extremely rapid, and for all the three lightest collision partners actually *exceed* the hard sphere elastic collision rate constant (i.e., their VR "efficiency" exceeds one). For this level in S_0 pDFB at least it is apparent that vibrational relaxation is extremely efficient, in contrast with small molecule behavior. While a comparison with the hard sphere encounter rate might provide a benchmark it is clearly of little use in rationalising the data with the hope of obtaining some physical insight into the process of vibrational relaxation. Relaxation induced by all but the three lightest collision partners occurs with a rate that exceeds the elastic hard sphere rate, often by a factor of ~ 2. It was this observation that led us to investigate a more realistic elastic potential as described in the theory section above. Consequently, we ask how well the vibrational relaxation rates are matched by elastic encounter rates determined using a potential that incorporates a more realistic long range attractive component than does the hard sphere potential (which of course has no attractive term).

The Lennard–Jones 6–12 potential (hereafter referred to as L–J potential) was chosen for this comparison as it is based on induced-dipole-induced-dipole interactions between molecules and this is most appropriate for the bulk of collision partners studied. Furthermore, values of the elastic collision rate are readily calculated from tabulated values for this potential.[54] The key quantity in the calculation of the L–J rate, k_{LJ}, is the intermolecular potential well depth. As described above (Eq. (8)), the geometric mean approximation is invoked in our comparison of data for different collision partners. The ratio k_v/k_{LJ} is also reported in Table I. This time, the efficiencies are generally close to unity (again with the exception of the three light collision partners). An efficiency of 1 of course indicates that VR occurs after every "encounter" of the Lennard–Jones spheres. Indeed the surprising result is not so much that He, H_2, and D_2 are so inefficient (there are many studies showing these light collisions partners to be inefficient relaxers) but that collision partners such as Ar, N_2, and O_2 are so efficient. They are comparable in efficiency to the more complex polyatomics such as c-hexane and even pDFB itself!

Tang and Parmenter's experiments probing VR rates from a variety of S_1 vibrational levels in benzene[14] provide a basis for understanding this behavior. These authors, using the collision partners CO and isopentane, measured VR rates from vibrational levels with increasing amounts of energy, and hence an increasing density of destination states. They found that isopentane, a complex polyatomic, has a high VR efficiency even at low vibrational energies, and does not show a significant increase in efficiency with increasing vibrational energy. CO, on the other hand, is inefficient at low energy but increases in efficiency significantly with increasing vibrational energy.

Table I. Vibrational relaxation rate constants for the deactivation of the $5_2 30_2$ level in pDFB induced by collisions with a variety of collision partners.

M	k_v^M		k_v/k_{hs}	k_v/k_{LJ}
	$(10^7 \mathrm{s}^{-1}\mathrm{Torr}^{-1})$	$(10^{-10}\mathrm{cm}^3\mathrm{molec}^{-1}\mathrm{s}^{-1})$		
He	0.76	2.5	0.26	0.26
H_2	2.0	6.5	0.45	0.38
D_2	2.2	7.1	0.68	0.58
N_2	1.8	5.8	1.19	0.88
O_2	2.6	8.4	1.83	1.30
Ar	1.8	5.8	1.42	1.00
CO_2	2.7	8.7	2.00	1.29
SF_6	2.5	8.0	2.02	1.30
C_2H_4	2.7	8.7	1.63	1.04
c-C_6H_{12}	3.1	10.0	1.96	1.15
i-C_5H_{12}	2.6	8.4	1.58	0.91
$(C_2H_5)_2O$	4.1	13.2	2.75	1.58
n-C_5H_{12}	2.6	8.4	1.57	0.90
CH_3OH	2.9	9.4	2.00	1.12
C_6H_{14}	2.6	8.4	1.71	0.96
C_2H_5OH	3.0	9.7	2.10	1.17
2-Propanol	2.9	9.4	2.15	1.20
t-Butanol	2.3	7.4	1.56	0.87
pDFB	3.5	11.3	2.11	1.17
1-Propanol	3.1	10.0	2.06	1.14
2-Butanol	2.9	9.4	1.82	1.00
1-Butanol	3.1	10.0	1.95	1.06
2-Pentanol	3.2	10.3	2.05	1.12
1-Pentanol	3.1	10.0	2.01	1.08

These observations by Tang and Parmenter, coupled with our measurements, led us to hypothesize at this time that the Lennard–Jones encounter rate might be forming some upper limit on the rate of vibrational energy transfer, i.e., the vibrational relaxation rates approach the collisional encounter rate as an upper bound when the relaxation is not limited by the availability of destination channels. This means that the VR rates for heavy and vibrationally complex collision partners are likely to be close to the collision-limited value (i.e., L–J rate) even at low vibrational energies and will increase only marginally with vibrational energy until they plateau near the L–J rate. However, the lighter collision partners will display VR rates well below the L–J rate at low vibrational energies and will increase in VR rate as the density of destination levels increases (i.e., vibrational energy increases) until they too will, at sufficiently high energy, also plateau near the L–J rate. This plateau will take longer to reach for the simplest collision partners as they are generally least efficient.

The attractive part of the potential is, of course, responsible for the colliding molecules influencing each other at distances greater than the sum of the hard sphere radii, and hence causing the vibrational relaxation to proceed at a rate greater than the hard sphere elastic collision rate. Given that the success of the L–J potential arises from its incorporation of an attractive component, it is likely that any similarly realistic potential will also match the observed trends.

The hypothesis above led to a set of experiments devised to test its validity. This study involved measuring vibrational relaxation rates from a number of levels, in the range from low to high vibrational state density, using a smaller subset of collision partners that still spanned the range from light atoms and diatomics to heavy polyatomics. It was planned to investigate relaxation from levels below $5_2 30_2$ to observe whether any of the medium-sized collision partners began to fall off in efficiency, and levels higher than $5_2 30_2$ to investigate whether the smaller collision partners would also approach the L–J limits, and indeed to test whether the L–J elastic rate remains a limit for the more complex species.

4.3. *Dependence of Vibrational Relaxation Rates on Initial State*

This series of experiments investigated VR from four vibrational levels in S_0 pDFB.[46] These levels cover the vibrational energy range from 1577 cm^{-1} to 3276 cm^{-1}. Over this energy range the background density of states rises from ~ 2 states/cm^{-1} to 200 states/cm^{-1}. The states are the $3_1 30_2$ (1577 cm^{-1}; $E_{\rm vib} \sim 2$ states/cm^{-1}), $5_2 30_2$ (2027 cm^{-1}; $E_{\rm vib} \sim 6$–8 states/cm^{-1}), $5_3 30_2$ (2884 cm^{-1}; $E_{\rm vib} \sim 80$ states/cm^{-1}), and $3_1 5_2 30_2$ (3276 cm^{-1}; $E_{\rm vib} \sim 200$ states/cm^{-1}) levels. VR was probed using a common set of eight collision partners for each of these four levels. These eight collision partners were chosen to span the regimes of light (He, H_2, and D_2), medium (Ar, N_2), and heavy/complex (SF$_6$, cyclohexane and pDFB) collision partners. The experimentally determined bimolecular vibrational

Table II. Vibrational relaxation rates and efficiencies for different levels in S_0 pDFB for a variety of collision partners.

Initial Level	Collision Partner	k_v^M (10^{-10}cm^{3}molec^{-1}s^{-1})	k_v/k_{hs}	k_v/k_{LJ}
$3_1 30_2$	He	1.49	0.16	0.17
	Ar	2.36	0.60	0.43
	N$_2$	2.89	0.60	0.46
	SF$_6$	2.76	0.72	0.46
	c-C$_6$H$_{12}$	4.97	1.02	0.58
	pDFB	7.64	1.48	0.79
$5_3 30_2$	He	1.89	0.20	0.22
	Ar	2.45	0.62	0.45
	N$_2$	3.60	0.75	0.57
	SF$_6$	3.48	0.90	0.57
	c-C$_6$H$_{12}$	6.15	1.26	0.72
	pDFB	8.32	1.62	0.86
$5_3 30_2$	He	3.38	0.37	0.39
	Ar	6.21	0.44	0.40
	D$_2$	5.12	0.51	0.47
	Ar	3.48	0.88	0.64
	N$_2$	4.81	1.01	0.77
	SF$_6$	4.78	1.24	0.79
	c-C$_6$H$_{12}$	7.79	1.60	0.91
	pDFB	9.31	1.81	0.97
$3_1 5_2 30_2$	He	4.94	0.54	0.57
	H$_2$	8.94	0.64	0.58
	Ar	5.03	1.28	0.93
	N$_2$	7.23	1.51	1.15
	SF$_6$	5.81	1.51	0.96
	c-C$_6$H$_{12}$	8.88	1.82	1.03
	pDFB	11.95	2.32	1.24

relaxation rate constants are shown in Table II. (It might be noted here that the rate coefficients in this study are typically 30–40% smaller than in the previous study. This is a consequence of the improved technology in the experiment over the four intervening years, providing the ability to take measurements at many pressure of added gas and hence to derive the rate coefficient from a Stern–Volmer plot rather than the direct single measurement method described in the kinetics section above.)

Recall that our previous hypothesis led us to predict that when the circumstances conspire to ensure that vibrational relaxation is facile, the relaxation rate will approach the elastic collision frequency given by a potential (such as the L–J 6–12 potential) that incorporates a realistic attractive component. Consequently, the heavy vibrational complex collision partners are expected to show relaxation rates that are close to the L–J elastic rate even at the $3_1 30_2$ level, and increase by a modest amount until they plateau at the L–J rate, by the $3_1 5_2 30_2$ level. In contrast, the light collision partners should display significantly smaller rate constants than the L–J elastic collision rate at the $3_1 30_2$ level (low E_{vib}), but gradually approach the L–J value as we progress in E_{vib} from $3_1 30_2$ to $5_2 30_2$, then to $5_3 30_2$, and finally to $3_1 5_2 30_2$.

The experimental data in Table II show that this prediction is realised. The vibrational relaxation rate for the light collision partner He is only 17% of the L–J elastic collision rate at the $3_1 30_2$ level, whereas at the $3_1 5_2 30_2$ level it has reached 57%, an increase by a factor of 3.4. The increase for the medium mass collision partner argon is less (as expected): it goes from 43% of k_{LJ} at $3_1 30_2$ to 93% of k_{LJ} at $3_1 5_2 30_2$ (a factor of 2.2). In contrast, the rate for relaxation by pDFB begins at 79% of the L–J rate at the $3_1 30_2$ level and climbs to 124% of the L–J rate at the $3_1 5_2 30_2$ level (an increase of a factor of only 1.6).

It is important to note, however, that the rate for the most efficient collider exceeds the L–J elastic encounter rate at the highest level studied. This suggests that the functional form of the L–J potential may underestimate the attractive part of the potential, i.e., the attractive part may be more long range than is allowed for in the L–J potential. The L–J parameters themselves may also be in question, the geometric mean approximation for the intermolecular well depth may also be inappropriate in some instances, and anisotropy of the real elastic potential may play a role. Nevertheless, for an intermolecular potential to be meaningful in its usage, the elastic encounter rate determined by that potential must normally serve as an upper limit to the rates for any inelastic process governed by the same potential. The observation that VR rates may surpass the elastic collision rate for the L–J 6–12 potential is not the only evidence from collisional studies that the L–J potential is not providing an adequate representation of the long range attractive part of the intermolecular potential. Measurements of rotational relaxation occurring within the 6^1 level of S_1 benzene by four collision partners have found relaxation rates in excess of the L–J elastic collision rate.[57,58] In the case of rotational relaxation of a large polyatomic like benzene there are many destination levels accessible. The large number of destination levels ensures that the relaxation is facile. The observation that relaxation can occur within the rotational manifold with a rate in excess of the L–J rate suggests that as the number of destination levels available in vibrational relaxation approaches those in rotational relaxation their efficiencies will be comparable, and exceed the L–J encounter rate. This suggests that the L–J rate may be surpassed by other collision partners in addition to pDFB for levels lying above $3_1 5_2 30_2$.

It is interesting to point out here that we attempted to study VR from one higher level: $5_4 30_2$ (with $E_{\mathrm{vib}} = 3745$ cm^{-1} and $\rho_{\mathrm{vib}} > 1000$ states/cm^{-1}) but could find no probe signal even after short delays. Although no clear explanation was forthcoming, we speculate that IVR may have been responsible for depleting our SEP-prepared level before collisional relaxation could compete.

These experiments have provided evidence that VR rates increase as the background density of accepting states increases. An approximate upper bound to the rate was found. This upper bound is approximately the elastic collision rate for two Lennard–Jones spheres. The data indicate, however, that the attractive part of the inelastic potential may be effective at longer ranges than the L–J 6–12 potential and, moreover, that this attractive part plays a significant role in determining the magnitude of VR rates. Consequently, our next set of experiments was designed to further probe the role of the attractive part of the intermolecular potential in VR.

4.4. *Vibrational Relaxation in S_0 p-Difluorobenzene Induced by Very Low Energy Collisions*

The influence of the long range attractive part of the intermolecular potential increases as the temperature is lowered because of the reduced relative velocity of the collision pair. The experiments discussed above were all undertaken in conventional gas bulbs at ambient temperature. In order to explore vibrational relaxation under conditions where the influence of the long range part of the intermolecular potential will be enhanced, experiments were undertaken in the collision region of a supersonic expansion, where the temperatures are very low, typically 1–10 K. The collision partner in these experiments is the gas in which pDFB is seeded, in this case argon.[23,24]

The S_0 level studied in the low energy pDFB collision experiment is 3_1. The SEP–LIF spectrum that establishes the integrity of SEP state preparation of 3_1 has been discussed above. Decay curves were collected and ξ_T evaluated for the lasers intersecting the free jet at $X/D = 7.5$ ($T \approx 6$ K from Eq. (14)) and for a variety of backing pressure, P_0. The data in Fig. 8, plotted against P_0, are also shown with an alternate abscissa above which is the local density at $X/D = 7.5$ calculated from Eq. (13). The slope of this plot provides the estimate of $k_v^{\mathrm{Ar}} = 2.2 \times 10^{11}$ cm^{3}molec^{-1}s^{-1} at 6 K.

The vibrational relaxation efficiency P was defined above as the probability of a vibrational relaxation event occurring relative to a benchmark encounter rate. As established above, an appropriate reference encounter rate is the Lennard–Jones elastic rate. It is interesting, however, to compare this measure of vibrational relaxation efficiency, P_{LJ}, with the efficiency relative to the hard sphere collision rate, i.e., P_{hs}. At room temperature, P_{hs} and P_{LJ} differ by approximately a factor of two, whereas in the low temperature regime, e.g., at 6 K, P_{hs} is 0.4 whereas P_{LJ} is 0.06. In other words, the efficiency of vibrational relaxation scaled with respect to the Lennard–Jones elastic rate is nearly an order of magnitude less than if scaled with respect to the hard sphere rate. This result is of course a ramification of the fact that the pDFB–Ar Lennard–Jones *elastic* encounter rate is considerably greater at very low temperatures than the hard sphere rate because the differential cross-section for the Lennard–Jones potential is contributed to very significantly by the attractive portion of the potential. The hard sphere differential cross-section is of course independent of temperature since, by definition, the potential has no attractive component. One can thus be misled into believing that VR remains efficient at low temperature if one uses the hard sphere potential to provide a reference collision rate.

The efficiency of VR has clearly fallen off significantly at low temperature compared with room temperature. There is no direct measurement of 3_1 relaxation at room temperature; however, the VR rate coefficient has been measured for the corresponding 3^1 level,[15] which provides an efficiency

$P_{LJ} = 0.31$. The efficiencies of VR for pDFB in the S_0 and S_1 states are shown below to be not substantially different. What has caused this falling off in efficiency? This is discussed in Sec. 5, where a semiempirical model is presented that is capable of reproducing the observed rate constants and efficiencies, including this temperature dependence.

4.5. *The Role of Electronic Excitation in Determining Vibrational Relaxation Rates in Polyatomics*

The initial experiments were aimed at measuring VR efficiencies in both the ground and excited electronic states of the same molecules. As has been discussed earlier, at the time that these experiments were undertaken the existing evidence could be construed to imply that vibrational relaxation was considerably more efficient in an excited electronic state than in the ground electronic state. We have presented above an appreciable compendium of data from S_0 pDFB and now it is time to return to our original question.

The data above indicate that vibrational relaxation is rapid from each of the levels studied; there is no reluctance on the part of the S_0 levels to undergo vibrational relaxation. To close the book, however, we should compare the data with rates for relaxation in the S_1 state of the same molecule. The first such comparison was obtained using data from this lab.[13] However, since then a more extensive survey of vibrational relaxation from a variety of S_1 vibrational states has been carried out by Parmenter and co-workers.[15,17] The data are reproduced in Table III. The magnitude of the rate coefficients are the same. In fact, upon closer inspection of k_v from states with a similar character (e.g., $3^1 30^1$ and $3_1 30_2$ or $5^2 30^1$ and $5_2 30_2$) the rate coefficients are in excellent agreement. It is clear that the apparent discrepancy between vibrational relaxation rates observed for S_1 polyatomics and those observed for molecules in their ground electronic state is non-existent, at least for this system. The question cannot be considered closed, however, until a compendium of data has been built up covering a range of systems. In particular, there exists limited data for VR in benzene that suggest that relaxation in S_0 may be inefficient compared with relaxation in S_1 for this molecule.[59] The method used to obtain these data was not as direct as that used here. We suggest that this system warrants fresh examination.

Table III. Experimental rate coefficients for VR from a variety of initial levels in both S_0 and S_1 pDFB for argon as a collision partner.

	Initial State	$E_{\mathrm{vib}}(\mathrm{cm}^{-1})$	$k_v^{\mathrm{Ar}}\ (10^{-10}\mathrm{cm}^3\mathrm{molec}^{-1}\mathrm{s}^{-1})$
S_0	$3_1 30_2$	1577	2.36
	$5_2 30_2$	2027	2.45
	$5_3 50_2$	2884	3.47
	$3_1 5_2 30_2$	3276	5.02
$S_1{}^{\mathrm{a}}$	5^1	818	1.21
	$5^1 30^1$	938	2.20
	$5^1 30^2$	1058	3.17
	3^1	1252	1.71
	$3^1 30^1$	1372	2.92
	5^2	1633	2.33
	$5^2 30^1$	1753	3.42
	3^2	2500	3.88

[a]Data from Catlett *et. al.*[15]

Given that vibrational relaxation rates are of similar magnitude in both the ground and excited electronic states of pDFB, we ask what feature of those molecules used in previous studies had led to the suggestion that vibrational relaxation in excited electronic states was enhanced relative to that in the ground electronic state. First, there had been no previous study of vibrational relaxation in the S_1 and S_0 states of the same molecule. Previous studies of relaxation in ground electronic states were concerned with small polyatomics, of which methane is typical, while S_1 studies had focused on aromatics because of the presence of favourable experimental factors. The vibrational state densities are considerably higher in aromatics compared to these similar molecules, and so in the case of aromatic there are more destination levels available. Furthermore, these destination levels tend to be closer to the original level than they are in the smaller polyatomics. We suggest that the inefficiency observed in previous studies of vibrational relaxation in S_0 is a consequence of the sparsity of readily accessible destination channels in the molecules studied.

5. A Semiempirical Model for Estimating Vibrational Relaxation Rates in Polyatomics

5.1. *Development of the Model*

The results summarised above illustrate how, using SEP to prepare selective S_0 states of pDFB, we have explored the dependence of vibrational relaxation rates on electronic state, vibrational state, collision partner, and temperature. It has been shown that VR efficiencies increase as the density of states increases with rising vibrational energy, reaching a plateau close to the L–J elastic encounter rate. Furthermore, it has been shown that VR has similar efficiencies in both S_0 and S_1 electronic states. With these trends as a guide, the challenge is to devise a model that can quantitatively reproduce the data, and moreover, provide a means of predicting VR efficiencies in other molecular systems.

A detailed derivation of the model devised is presented in a separate publication;[51,60] in this context the basis of the model lies in the preceding description of the Lennard–Jones collision efficiencies. We propose a relationship of the form

$$k_v = S_f \mu^{1/3} \langle v \rangle \sigma_{\text{LJ}} , \tag{12}$$

or, following the previous section,

$$k_v = S_f \mu^{1/3} \langle v \rangle \sigma_{\text{hs}} \Omega^{(2,2)^*} , \tag{13}$$

where σ_{LJ} and σ_{hs} are the Lennard–Jones and hard sphere cross-sections, respectively, $\Omega^{(2,2)^*}$ is the omega integral linking σ_{LJ} and σ_{hs}, $\langle v \rangle$ is the mean relative speed of the two collision partners which for a Maxwell–Boltzmann distribution is equal to $(8kT/\pi\mu)^{1/2}$ where each symbol has its usual definition, μ is the reduced mass of the collision pair, and S_f is a scaling factor which is described below. This relationship is found to describe the rate coefficients for vibrational relaxation for a large number of molecules when the collision partners are involved in V–T transfer alone. Comparing Eqs. (7) and (9) with Eq. (13) it is apparent that the product of terms $S_f \mu^{1/3} = P_{\text{LJ}}$, the efficiency of VR with respect to the L–J elastic encounter rate.

The $\mu^{1/3}$ factor that appears in Eq. (16) is entirely empirical. Terms of the form of $\mu^{1/x}$ appear quite regularly in theories of vibrational relaxation.[53] For this model it was found that an exponent of 1/3 provided a reasonable correlation with a large body of data. The data for S_0 pDFB for two different vibrational levels are shown in Fig. 9 demonstrating this $\mu^{1/3}$ correlation. Obviously the correlation is different for the two initial states, the state to higher energy ($3_1 5_2 30_2$) displaying higher efficiency. It is this different VR efficiency for different initial states which leads to the necessity for the inclusion of the state dependent factor, S_f, in Eq. (13).

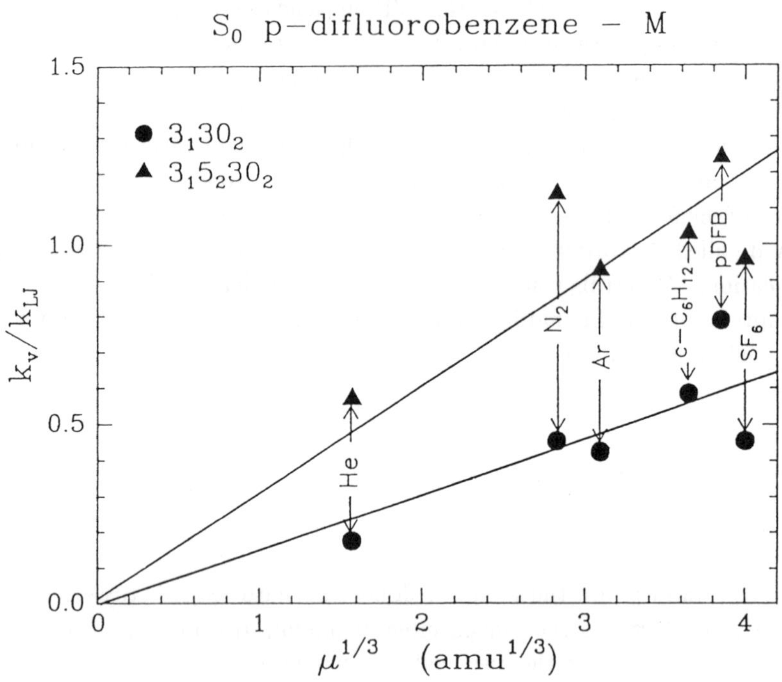

Fig. 9. Correlation between vibrational relaxation efficiency measured relative to the Lennard–Jones elastic encounter rate and the cube root of the reduced mass, μ, of the collision pair. The lines are linear least squares fits, constrained to pass through the origin.

The factor S_f represents a sum of energy transfer probabilities over all states, f, which act as the acceptor states for the energy transfer. It is this factor that accounts for the increase in VR efficiencies with increasing vibrational energy or state density. The factor is closely related to the relative propensity rules derived by Parmenter and Tang[7] based on the SSH semiempirical theory of vibrational relaxation.[61,62] Tang and Parmenter[14] extended their formalism to model state-to-field vibrational relaxation rates by summing their state-to-state propensities over all final states. This is, in principle, the approach taken to derive S_f. The precise formula used by Parmenter and Tang has been modified slightly to allow for the vibrational quantum number dependence of the harmonic oscillator matrix elements, H_{if}, in Eq. (17) below. The final formula for calculating S_f is provided by the following equation:

$$S_f = \sum_{i \geq f} H_{if}^2 10^{-\Delta v} e^{-0.01 \Delta E} + \sum_{i < f} H_{if}^2 10^{-\Delta v} e^{-0.01 \Delta E} e^{\frac{-\Delta E}{kT}} . \tag{14}$$

The two terms separate contributions to the total energy transfer propensity into downward energy transfer (the first term) and upward energy transfer (the second term). The second term only differs from the first by the inclusion of the Boltzmann factor which is required by microscopic reversibility. Obviously, then, S_f is explicitly temperature dependent which is important in later discussion of the temperature dependence of the vibrational relaxation process. The factors in each term of the S_f expression are an exponential energy gap $e^{-0.01 \Delta E}$, an inhibiting factor for changing vibrational quanta in the energy exchange $10^{-\Delta v}$, and a factor H_{if} based on the harmonic oscillator matrix elements which provides for the different coupling of the initial and final states for different numbers of quanta in any mode. These matrix elements are tabulated in Appendix 3 of the book of Wilson *et al.*[63] For our purpose, the H_{if} factors are normalised to the $v = 0 \rightarrow v = \Delta v$ channel. For example, for $\Delta v = 1$, the matrix elements contain a $\sqrt{v}$ dependence and a $v = 2 \leftrightarrow v = 1$ transfer would have $H_{if} = \sqrt{2}$, relative to the $v = 1 \leftrightarrow v = 0$ process for which $H_{if} \equiv 1$. The quantum number dependence of other Δv pathways can be found in the aforementioned table. As pointed out above, the utility of S_f is that it provides a dependence of k_v on the initially prepared state that is sensitive to the background density of states.

5.2. *Predictions of the Model for p-Difluorobenzene Vibrational Relaxation*

All the factors from Eq. (13) above can be calculated on the "back of an envelope" except for S_f. For even the highest lying state investigated in S_0 pDFB above ($3_1 5_2 30_2$), where the background density of vibrational states is about 200 per cm^{-1}, a simple counting algorithm can be programmed to count over these states evaluating the propensity for energy transfer via Eq. (14).

Figure 10 shows the experimental and predicted rate coefficients for pDFB under a variety of conditions including the S_0 data presented here and the S_1 data from Catlett *et al.*[15] The model predicts the rate coefficients for both electronic states and all vibrational states in a quantitative fashion. The similarity in the rates between each electronic state mentioned previously is reproduced, as is the nonuniform vibrational state dependence especially obvious for the S_1 data. Clearly, the primary scaling term in the formula is S_f. Its success strongly suggests that this term has captured one of the essential features of VR efficiency: viz., how VR is influenced by the structure of the surrounding destination states.

Also shown in Fig. 10 is the datum for low energy collisional relaxation from the 3_1 level of pDFB prepared by SEP corresponding to a temperature of approximately 6 K. Equation (16) contains an explicit temperature dependence from various factors, including the mean speed, the Omega integrals, and the S_f factor via the Boltzmann factor in Eq. (14). The predicted rate coefficient for this level and temperature is again in excellent agreement with the data. The largest contributor to the reduced efficiency at low temperature is the loss of endoergic channels due to the changed Boltzmann factor in Eq. (14).

Equation (13) uses the Lennard–Jones collision rate. The equation can be reworked in terms of the hard sphere collision rate which does provide an adequate fit to experimental data for 300 K. However, it cannot reproduce data at both 300 K and low temperature. This indicates that the Lennard–Jones cross-section captures the spirit of the temperature dependence of VR much better than does the hard sphere.

When the collision partners are more strongly polar (e.g., H_2O) the observed rate is in excess of that predicted by Eq. (13). The equation uses explicitly the (6–12) form of the Lennard–Jones potential which is appropriate for dispersion interactions. For interactions which involve dipole–dipole or dipole–induced-dipole forces, which are longer range than the dispersion forces, it is then expected that the L–J (6–12) elastic encounter rate would be exceeded. The more appropriate L–J potential could be applied to Eq. (16) in these cases. However, we have refrained from doing so because the experimental data are more scarce for these

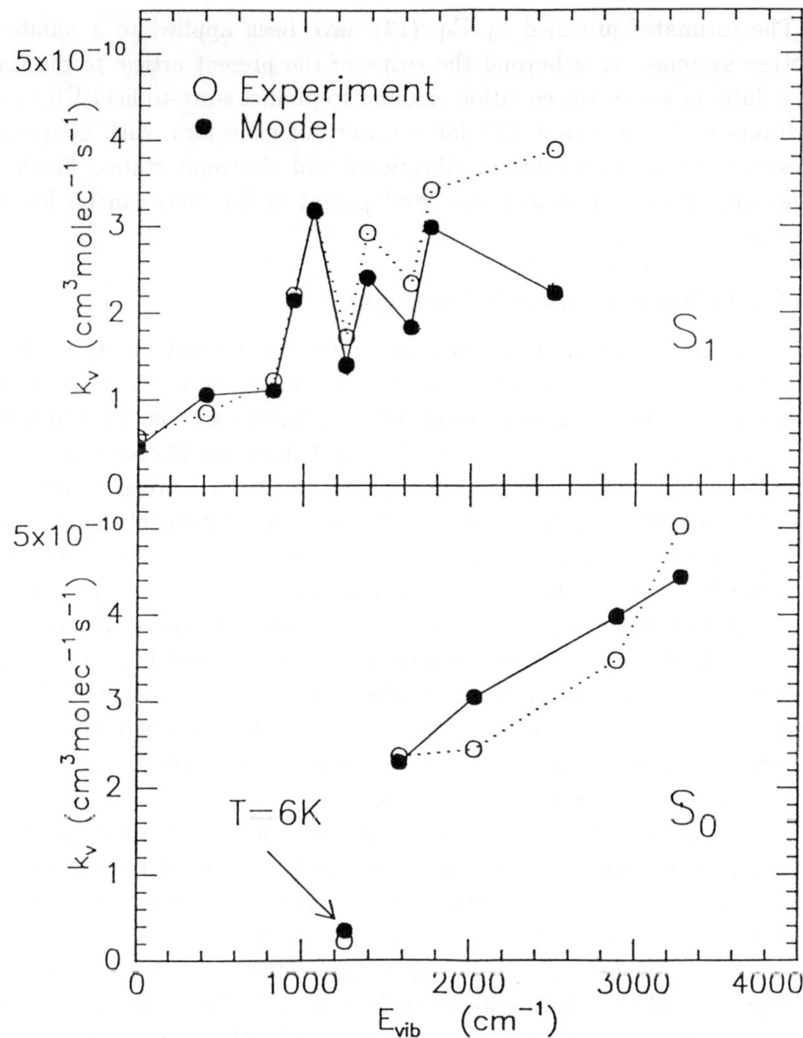

Fig. 10. Dependence of the vibrational relaxation rate on vibrational energy of the initial state for both S_0 and S_1 pDFB. All data are for room temperature (300 K) collisions with Ar except for the datum for the 3_1 state which is for 6 K. The experimental data are shown by open symbols; the rate calculated by Eq. (13) in the text by solid symbols and joined by lines.

systems, and the true intermolecular potential more complex, and so we have doubts about the worth of such a modification.

The estimates provided by Eq. (13) have been applied to a number of other systems. It is beyond the scope of the present article to discuss these data; however, the equation is found to predict state-to-field VR rate constants typically within 30% for a variety of molecules, with nonpolar collision partners, in a range of vibrational and electronic states. Further discussion of these data and the development of Eq. (36) can be found elsewhere,[51,60]

6. Conclusion and Future Directions

SEP has allowed us to access regions of the vibrational potential that have not been so cleanly accessible before. In doing so what have we learned about vibrational relaxation? The conclusions are that for collisions involving nonpolar species the elastic Lennard–Jones (6–12) collisions rate seems to provide a reasonable upper bound to the vibrational relaxation rate. The magnitude of the VR rate is influenced in an explicable manner by the mass and velocity of the collision partner and by the background density of "accepting states". The nature of the electronic state seems to play little role for p-difluorobenzene and presumably for similar systems. The caveat must be added that electronic excitation in such systems does not affect substantially many molecular parameters including bond lengths, force constants, intermolecular well depths, etc. In molecules where electronic excitation severely perturbs the system a significant change in vibrational relaxation behavior might be expected.

In the study of collision-induced vibrational relaxation (VR) in medium to large polyatomics, SEP remains a little-used technique for state preparation. One obvious reason for this is the relative experimental complexity: to probe VR in the ground electronic state using SEP requires three lasers; two to prepare the initial S_0 level and a further laser to probe the collision-induced transfer of population from the initially prepared level. In contrast, a single laser in combination with a monochromator can be used to probe vibrational relaxation in the S_1 state. It is this experimental ease of study that has made S_1 the favoured state in studies of VR in medium to large polyatomics. The S_0 SEP studies discussed in this chapter have shown that relaxation in the S_0 and S_1 states is similar in pDFB. While we suggest that this is likely to be a general feature of large molecule VR, it cannot be simply assumed that this is the case and more experiments probing VR in S_0 and S_1 of the same molecule are needed.

Theoreticians, of course, are more comfortable with calculations of dynamical processes in ground electronic states. However, the experimental evidence suggests that excited and ground electronic state VR is sufficiently similar for calculations using ground state parameters (but excited state frequencies) to be sufficiently accurate for comparison with electronic state data. (The caveat here, of course, is that lack of agreement can be put down to the different electronic states!) Consequently, we suggest that the long term future of SEP as a tool for probing VR lies in studies that rely on properties that are favorable in S_0 but unfavorable in S_1.

One problem encountered in studies of VR in S_1 states of large molecules is the onset of rapid nonradiative, intermolecular processes as one climbs the vibrational manifold. The onset of the "channel 3" region in S_1 benzene above 3000 cm^{-1} is a well-known, somewhat extreme example.[64] It is by no means unique, however. Thus one role for SEP is in the study of VR from high-lying levels in S_0. Understanding the distribution function describing how the energy is transferred in collisions at these high energies remains a considerable challenge. Studies using SEP may play a vital role in bridging the gap between the low energy state-to-state studies and the very high energy relaxation experiments typified by those of Barker's group.[34-36]

The second advantage of S_0 over S_1 is the considerably longer lifetimes of the vibrational levels. Due in part to nonradiative processes, it is rare for medium to large polyatomics to have S_1 lifetimes longer than 1 μs, and indeed it is common for them to be < 100 ns. In contrast, S_0 vibrational lifetimes are very long, due to the absence of internal conversion and intersystem crossing and the long radiative lifetime of infrared fluorescence. Experiments where it is an advantage to have intrinsically long lifetimes thus comprise another category likely to benefit from the use of SEP to prepare S_0 levels.

As a consequence, we suggest that a particularly fruitful use of SEP as a means of probing vibrational energy transfer in the ground electronic state of large molecules will come from molecular beam studies, particularly crossed molecular beams. In a conventional sample cell ("bulb") experiment the short S_1 lifetimes are not a problem, since the total pressure can be increased to ensure that collisions occur during the S_1 fluorescence lifetime. However, in crossed molecular beam studies the molecular density is low, and long lifetimes are necessary in order to obtain observable relaxation. Parmenter and co-workers have discussed the problem of short

S_1 lifetimes in relation to crossed beam measurements of VR in the S_1 state of polyatomics.[5] The only crossed beam data concerning VR in the S_1 state comes from studies of glyoxal[31] and iodine,[27–30] both of which have lifetimes on the order of 1 μs. The advantages of crossed beam studies of VR have been detailed in a review by Krajnovich *et al.*[5] The main advantage is the ability to specify the collision energy, rather than being forced to observe the average of a range of collision energies, as occurs in conventional bulb experiments. With the recent advances in computational methods for calculating VR,[65] it is clear that further advances in our understanding of VR will come from experiments that are more incisive than the conventional bulb studies that dominate the literature on VR.

Crossed beam studies of VR in S_0 polyatomics have been largely confined to up-pumping from the 0_0 level. SEP offers the opportunity to dramatically extend these studies to encompass a range of initial vibrational levels. The recent demonstration of this technique on I_2 by Gentry's group ushers in this new era.[66]

References

1. E. Weitz and G. W. Flynn, *Ann. Rev. Phys. Chem.* **25**, 275 (1974).
2. E. Weitz and G. W. Flynn, *Adv. Chem. Phys.* **47**, Part 2, 185 (1981).
3. S. A. Rice, *J. Phys. Chem.* **90**, 3063 (1986).
4. B. J. Orr and I. W. M. Smith, *J. Phys. Chem.* **91**, 6106 (1987).
5. D. J. Krajnovich, C. S. Parmenter, and D. L. Catlett, Jr., *Chem. Rev.* **87**, 237 (1987).
6. J. I. Steinfeld and W. Klemperer, *J. Chem. Phys.* **42**, 3475 (1965).
7. C. S. Parmenter and K. Y. Tang, *Chem. Phys.* **27**, 127 (1978).
8. D. A. Chernoff and S. A. Rice, *J. Chem. Phys.* **70**, 2521 (1979).
9. L. M. Logan, I. Buduls, A. E. W. Knight, and I. G. Ross, *J. Chem. Phys.* **72**, 5667 (1980).
10. D. B. McDonald and S. A. Rice, *J. Chem. Phys.* **74**, 4907 (1981).
11. M. Vandersall, D. A. Chernoff, and S. A. Rice, *J. Chem. Phys.* **74**, 4888 (1981).
12. M. Vandersall and S. A. Rice, *J. Chem. Phys.* **79**, 4845 (1983).
13. D. J. Muller, W. D. Lawrance, and A. E. W. Knight, *J. Phys. Chem.* **87**, 4952 (1983).
14. K. Y. Tang and C. S. Parmenter, *J. Chem. Phys.* **78**, 3922 (1983).
15. D. L. Catlett, Jr., K. W. Holzclaw, D. Krajnovich, D. B. Moss, C. S. Parmenter, W. D. Lawrance, and A. E. W. Knight, *J. Phys. Chem.* **89**, 4709 (1985).
16. M. W. Rainbird, B. S. Webb, and A. E. W. Knight, *J. Chem. Phys.* **88**, 2416 (1988).

17. C. J. Pursell and C. S. Parmenter, *J. Phys. Chem.* **97**, 1615 (1993).

18. R. A. J. Borg, PhD Thesis, Flinders University (1993).

19. E. R. Waclawik, W. D. Lawrance, and R. A. J. Borg, *J. Phys. Chem.* **97**, 5798 (1993).

20. D. B. Moss, S. H. Kable, and A. E. W. Knight, *J. Chem. Phys.* **79**, 2869 (1983).

21. D. J. Muller, R. I. McKay, G. B. Edwards, W. D. Lawrance, J. P. Hardy, A. B. Rock, K. J. Selway, S. H. Kable, and A. E. W. Knight, *J. Phys. Chem.* **92**, 3751 (1988).

22. S. H. Kable and A. E. W. Knight, *J. Phys. Chem.* **93**, 4766 (1990).

23. S. H. Kable and A. E. W. Knight, *J. Phys. Chem.* **86**, 4709 (1987).

24. S. H. Kable and A. E. W. Knight, *J. Phys. Chem.* **93**, 3151 (1990).

25. K. Liu, G. Hall, M. J. McAuliffe, C. F. Giese, and W. R. Gentry, *J. Chem. Phys.* **80**, 3494 (1984).

26. G. Hall, C. F. Giese, and W. R. Gentry, *J. Chem. Phys.* **83**, 5343 (1985).

27. D. J. Krajnovich, K. W. Butz, H. Du, and C. S. Parmenter, *J. Phys. Chem.* **92**, 1388 (1988).

28. D. J. Krajnovich, K. W. Butz, H. Du, and C. S. Parmenter, *J. Chem. Phys.* **91**, 7705 (1989).

29. D. J. Krajnovich, K. W. Butz, H. Du, and C. S. Parmenter, *J. Chem. Phys.* **91**, 7725 (1989).

30. H. Du, D. J. Krajnovich, and C. S. Parmenter, *J. Phys. Chem.* **95**, 2104 (1991).

31. K. W. Butz, H. Du, D. J. Krajnovich, and C. S. Parmenter, *J. Chem. Phys.* **89**, 4680 (1988).

32. V. E. Bondebey, *Ann. Rev. Phys. Chem.* **35**, 591 (1984).

33. A. L. Utz, J. D. Tobiason, E. Carrasquillo M., M. D. Fritz, and F. F. Crim, *J. Chem. Phys.* **97**, 389 (1992).

34. J. R. Barker, *J. Phys. Chem.* **88**, 11 (1984).

35. M. L. Yerram, J. D. Brenner, K. D. King, and J. R. Barker, *J. Chem. Phys.* **94**, 6341 (1990).

36. B. M. Toselli, J. D. Brenner, M. L. Yerram, W. E. Chin, K. D. King, and J. R. Barker, *J. Chem. Phys.* **95**, 176 (1991).

37. H. Hippler, J. Troe, and H. J. Wendelken, *J. Chem. Phys.* **84**, 257 (1981).

38. H. Hippler, L. Lindeman, and J. Troe, *J. Chem. Phys.* **83**, 3906 (1985).

39. C. Kitrell, E. Abramson, J. L. Kinsey, S. A. McDonald, D. E. Reisner, R. A. Field, and D. H. Katayama, *J. Chem. Phys.* **75**, 2056 (1981).

40. W. D. Lawrance and A. E. W. Knight, *J. Chem. Phys.* **76**, 5637 (1982).

41. D. E. Reisner, R. W. Field, J. L. Kinsey, and H.-L. Dai, *J. Chem. Phys.* **80**, 5968 (1984).

42. E. Abramson, R. W. Field, D. Imre, K. K. Innes, and J. L. Kinsey, *J. Chem. Phys.* **83**, 452 (1985).

43. W. D. Lawrance and A. E. W. Knight, *J. Chem. Phys.* **77**, 570 (1982).

44. W. D. Lawrance and A. E. W. Knight, *J. Chem. Phys.* **79**, 6030 (1983).

45. J. W. Thoman, Jr., S. H. Kable, A. B. Rock, and A. E. W. Knight, *J. Chem. Phys.* **85**, 6234 (1986).

46. S. H. Kable, J. W. Thoman, Jr., and A. E. W. Knight, *J. Chem. Phys.* **88**, 4748 (1988).

47. F. Temps, S. Halle, P. H. Vaccaro, R. W. Field, and J. L. Kinsey, *J. Chem. Phys.* **87**, 1895 (1987).

48. X. Yang, J. M. Price, J. A. Mack, C. G. Morgan, C. A. Rogaski, D. McGuire, E. H. Kim, and A. M. Wodke, *J. Phys. Chem.* **97**, 3944 (1993).

49. S. H. Kable, W. D. Lawrance, and A. E. W. Knight, *J. Phys. Chem.* **86**, 1244 (1982).

50. T. Cvitas and J. M. Hollas, *Mol. Phys.* **18**, 793 (1970).

51. S. H. Kable, PhD Thesis, Griffith University (1988).

52. R. A. Coveleskie and C. S. Parmenter, *J. Molec. Spectrosc.* **110**, 312 (1985).

53. J. T. Yardley, *Introduction to Molecular Energy Transfer* (Academic, NY, 1980).

54. J. O. Hirschfelder, C. F. Curtiss, and R. B. Bird, *Molecular Theory of Gases and Liquids*, 2nd ed. (Wiley, NY. 1964).

55. D. J. Muller, PhD Thesis, Griffith University (1988).

56. D. H. Levy, *Ann. Rev. Phys.Chem.* **31**, 197 (1980).

57. R. A. Coveleskie and C. S. Parmenter, *J. Phys. Chem.* **69**, 1044 (1978).

58. N. T. Whetten and W. D. Lawrance, *J. Phys. Chem.* **96**, 3717 (1992).

59. J. L. Lyman, G. Muller, P. L. Houston, M. Piltch, W. E. Schmid, and K. L. Kompa, *J. Chem. Phys.* **82**, 810 (1985).

60. S. H. Kable and A. E. W. Knight, *J. Phys. Chem.* to be submitted.

61. R. N. Schwartz, Z. I. Slawsky, and K. F. Herzfeld, *J. Chem. Phys.* **20**, 1591 (1952).

62. F. I. Tanzos, *J. Chem. Phys.* **25**, 439 (1956).

63. E. B. Wilson, Jr., J. C. Decius, and P. C. Cross, *Molecular Vibrations. The Theory of Infrared and Raman Vibrational Spectra*, (Dover, NY, 1980), Appendix 3.

64. J. H. Callomon, J. E. Perkin, and R. Lopez-Delgado, *Chem. Phys. Lett.* **13**, 125 (1972).

65. D. C. Clary, *J. Chem. Phys.* **86**, 813 (1987).

66. Z. Ma, S. D. Jons, C. F. Giese, and W. R. Gentry, *J. Chem. Phys.* **96**, 3717 (1992).

CHAPTER 17

**HOW DOES COLLISIONAL ENERGY TRANSFER DEPEND
ON VIBRATIONAL EXCITATION?
TRANSIENT CHEMICAL BOND FORMATION
VERSUS CHEMICAL REACTION**

X. Yang[*], J. M. Price, J. A. Mack,
C. G. Morgan, C. A. Rogaski and A. M. Wodtke[†],[‡],[§]

*Department of Chemistry, University of California
Santa Barbara, CA 93106, USA*

Contents

---

[*]Present Address: Department of Chemistry and Chemical Sciences Division, Lawrence Berkeley Laboratory, University of California, Berkeley, CA 94720.

[†]NSF Presidential Young Investigator.

[‡]Alfred P. Sloan Research Fellow.

[§]Camille and Henry Dreyfus Teaching Scholar.

619

Abstract

The stimulated emission pumping (SEP) technique has made the study of highly vibrationally excited molecules increasingly appealing, enabling sophisticated spectroscopic probes to be converted to preparation techniques. With such an approach even relatively improbable processes, such as vibrational energy transfer and chemical reactivity, may be studied. This article describes studies of the vibrational quantum number dependence of the vibrational relaxation of highly vibrationally excited Nitric Oxide and Oxygen. The NO experiments are one of the first data sets on bimolecular energy transfer for molecules with several hundreds of kJ/mol of internal energy. Information is obtained showing to what extent energy transfer theories designed for low vibrational energy can be applied in this case. Strong evidence is also obtained suggesting that qualitatively different mechanisms such as "transient chemical bond formation" can influence the rate of vibrational energy transfer at high vibrational energy. Vibrational energy transfer of highly vibrationally excited O_2 is of current interest to the full understanding of the stratospheric ozone problem. This problem and initial experimental results are also presented.

1. Introduction

During the last fifty years, the field of high resolution spectroscopy has made an immense contribution to the understanding of molecular structure and interaction, evolving into one of the most mature and sophisticated fields in physical chemistry.[1] Resolved rotational spectra routinely provide the most precise measure of molecular bond-lengths for ground and excited electronic states.[2] The understanding of atomic bonding within molecules has been profoundly influenced by Infrared and Raman spectroscopy.[3] High resolution spectroscopy of molecular clusters is even revealing the detailed nature of solute–solvent interactions. The understanding of vibrational energy flow within polyatomic molecules, which is at the root of chemical reaction rate theory, has been and continues to be strongly influenced by high resolution spectroscopy. Furthermore, the theoretical roots of molecular spectroscopy are deep and secure, linking experiment to the most fundamental level of molecular understanding, quantum mechanics.

In recent years, new and important ways of using lasers have been developed that have enabled significant progress to be made in the study of molecules containing large quantities of internal energy.[4-7] Methodologically, this is a spectroscopic field; but conceptually, this field mixes the purest mathematical rigor of high resolution spectroscopy with more intuitive "chemical thinking" and promises to deepen and intensify the precision with which we describe chemical reactions. Confinement of our attention to the arena of chemical issues highlights the importance of the study of highly vibrationally excited molecules, since chemical reactions often involve either the consumption or production of such species. For example, an exothermic reaction like $F + H_2 \rightarrow HF(v) + H$,[8] which is the driving force for the first infrared chemical laser,[9,10] efficiently produces highly vibrationally excited products. Photochemistry, which in one sense, is the process of converting ultraviolet photon energy into photo-product internal (as well as kinetic) energy, is another example of the role of highly vibrationally excited molecules in Chemistry. Specifically, it now appears clear that stratospheric ozone photodissociation, $O_3 + h\nu \rightarrow O_2(v) + O$, is an efficient means for producing vibrationally excited O_2 in quantum levels as high as $v = 22$.[11] Relaxation rates may be slow enough that a significant steady-state population builds up in the stratosphere.[12]

The role of vibrationally excited molecules is also important to endothermic chemical reactions, which are often strongly enhanced by vibrational excitation. In fact, it has been shown experimentally that control of vibrational excitation can be equivalent to control of chemical and photochemical reaction rates and branching ratio's.[13-22] Another area of chemical interest is isomerization. Completely stable, closed-shell molecules may undergo chemical isomerization when enough vibrational energy is available. Consequently, even the *chemical identity* of a molecule might be controlled by state specific control of a molecule's vibrational excitation. It has been demonstrated that vibrational structures of HCN at high vibrational excitation are very sensitive to the potential of the isomerization reaction $HCN \rightarrow CNH$.[23,24] The same is true for $HCCH \rightarrow H_2CC$:[25] and $HCP \rightarrow CPH$.[26] Such studies are leading to the experimental determination of chemical reaction potential surfaces, in essence Nature's Hamiltonian for chemical transformation.[27]

One of the most important technical advances in Physical Chemistry is the Stimulated Emission Pumping (SEP) method.[6] This method, which was a logical extension of the field of double resonance spectroscopy, uses two lasers to efficiently prepare highly vibrationally excited molecules. The first laser, tuned to ω_{PUMP}, excites molecules out of a single, thermally-populated quantum state into an excited electronic state with a different structure than the initial state. A second laser, tuned to ω_{DUMP}, induces emission of a photon transferring population back to the electronic ground state but leaving the molecule with $\hbar(\omega_{PUMP}-\omega_{DUMP})$ more energy than the initial state.

Most SEP experiments have been concerned with recording the positions and intensities of the stimulated emission transitions revealing the quantum level structure of highly vibrationally excited molecules. A far smaller number of investigations have attempted to use SEP as a means of preparing highly vibrationally excited molecules for study in bimolecular or gas/surface collisions. This is one of the most exciting new developments in the field of highly vibrationally excited molecules. Our technological ability is advancing rapidly and we are now entering a period where it will become quite standard to prepare molecules in the "Maxwell Boltzmann energy tail" with complete quantum state specificity and high enough efficiency to observe the chemistry and collision dynamics of these highly energized species!

Vibrational energy transfer has long been an important topic in chemical physics raising many fundamental questions. How long does it take for an energized species to disappear through collisional relaxation? What forms of motion soak up the energy? What mechanisms are important for energy transfer when the molecule's internal energy is comparable to its bond strengths? Laser preparation of highly vibrationally excited molecules for study in vibrational energy transfer is the specific topic of this article. SEP has been used to prepare large quantities of highly vibrationally excited NO as well as O_2. The bimolecular energy transfer properties of these species could then be studied using laser induced fluorescence as a probe. The work on NO has given one of the most complete data sets concerning the influence of vibrational excitation on molecular energy transfer. Initial results on O_2 contrast with that of NO and give valuable data for use in *nonlocal thermodynamical equilibrium* atmospheric models of stratospheric ozone.

2. Vibrational Energy Transfer in Highly Vibrationally Excited NO

NO is one of the most favorable molecules to which SEP can be applied.[28,29] Both the $B \leftrightarrow X$ and $A \leftrightarrow X$ electronic systems exhibit long Franck–Condon progressions in absorption and emission, meaning that high vibrational states can be prepared. An energy level diagram is shown in Fig. 1 which demonstrates the principle of the experiment. Using a tunable ArF laser, the $v' = 7 \leftarrow v'' = 0$ band can be pumped at 193 nm. A tunable XeCl pumped dye laser can then be used as the DUMP laser to prepare the NO molecules in vibrational states as high as $v'' = 25$, containing 4.6 eV of vibrational energy. NO B–X transitions are rather strong so usually a few mJ of dye laser power can saturate the dump transition. A precise description of the pulse energies necessary for efficient SEP of NO can be found in our previous work.[28] For preparation of NO in vibrational levels lower than $v'' = 12$ of the ground electronic state, the DUMP laser needed to be doubled by a BBO crystal.

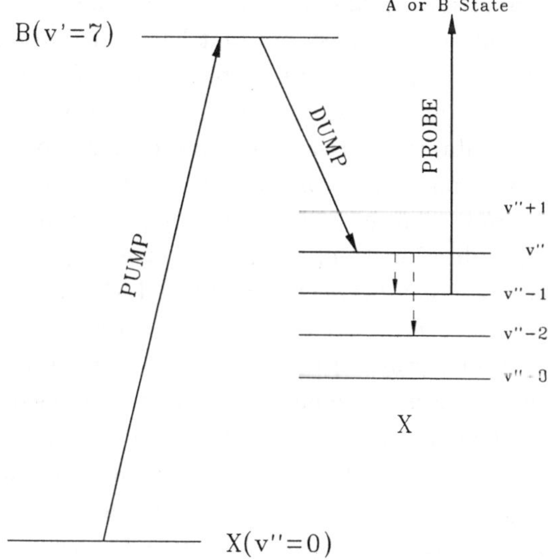

Fig. 1. Energy level diagram of the NO molecule: NO was pumped from the ground state ($v'' = 0$) to a high vibrational level v'' via the intermediate electronic state B ($v' = 7$) using SEP. After a variable delay, collisional quenched NO molecules were detected by LIF through either the A or B electronic state.

Vibrational relaxation from a prepared vibrational level is a first order kinetic process with an exponential decay. An example of such behavior is shown in Fig. 2.[30] Here the triangles indicate the measured LIF signal intensity resulting from $v'' = 19$ as a function of the DUMP–PROBE delay. The solid line is the best exponential fit to the data. Such measurements were made at several pressures for a specific vibrational state. The inverse of the pressure dependent exponential time constant was plotted versus pressure. Such a plot is similar to a Stern–Volmer plot. Several Stern–Volmer-like plots are shown in Fig. 3. The slopes of the curves are the vibrational relaxation rate constants. The vibrational relaxation rate constants for specific vibrational states of NO ($^{14}N^{16}O$, $^{15}N^{18}O$) between $v'' = 8$ and 25 have been measured, giving an uncommonly complete picture of the vibrational dependence of the energy transfer. These are tabulated in Ref. 30. Figure 4 shows the vibrational quantum number dependence of the measured NO self-vibrational relaxation rates, including the $v = 1$[31] and 2[32] data points from the previous experimental measurements. Figure 5 shows the vibrational energy dependence of the vibrational relaxation rate constant. The vibrational state specific rate constants reported here are an average deactivation rate constants that has contributions from all the populated rotational and spin-orbit states. The effect of the radiative loss is negligible.[33–35] From Figs. 4 and 5, it is readily apparent that the dependence of the vibrational relaxation rate constant on the vibrational excitation is very strong, increasing by about 200 times for the highest vibrational excitation achieved (4.6 eV) in comparison to the $v = 1$ level. Furthermore, the vibrational enhancement between $v = 1$ and $v \approx 14$ is much smaller than that between $v = 14$ and 25, already suggesting a switch over from low energy behavior to high energy behavior. Figure 6 shows this in a different way. In this figure, the function $k(v)/v$ is plotted versus vibrational quantum number. The function is nearly constant up to about $v = 14$. This means that the vibrational rate constant is linearly dependent on the vibrational quantum number over this range. Using the Schwartz–Slawsky–Herzfeld (SSH) theory[36–38] we interpret this as meaning the single quantum relaxation ($\Delta v = 1$) dominates below $v = 14$. Above $v = 14$ it is more strongly than linearly dependent on vibrational quantum number. Finally one can see that the heavier isotope $^{15}N^{18}O$ relaxes slightly more slowly than the normal isotope of NO.

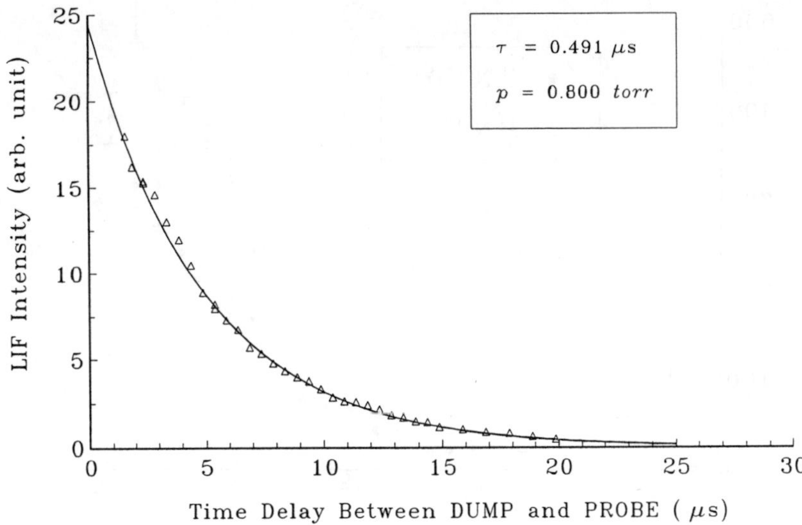

Fig. 2. Typical experimental measurement of collisional deactivation lifetime for $v'' = 19$ of $^{14}N^{16}O$ at a pressure of 0.80 torr. The triangles are the experimental data and the solid line is the fitted result using a single exponential.

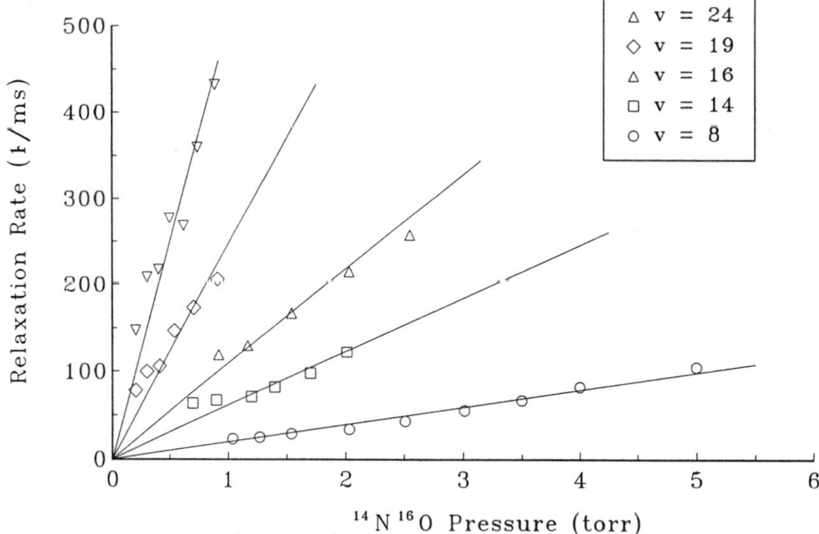

Fig. 3. Typical Stern–Volmer plots for collisional relaxation of $v'' = 8, 14, 16, 19, 24$ of $^{14}N^{16}O$.

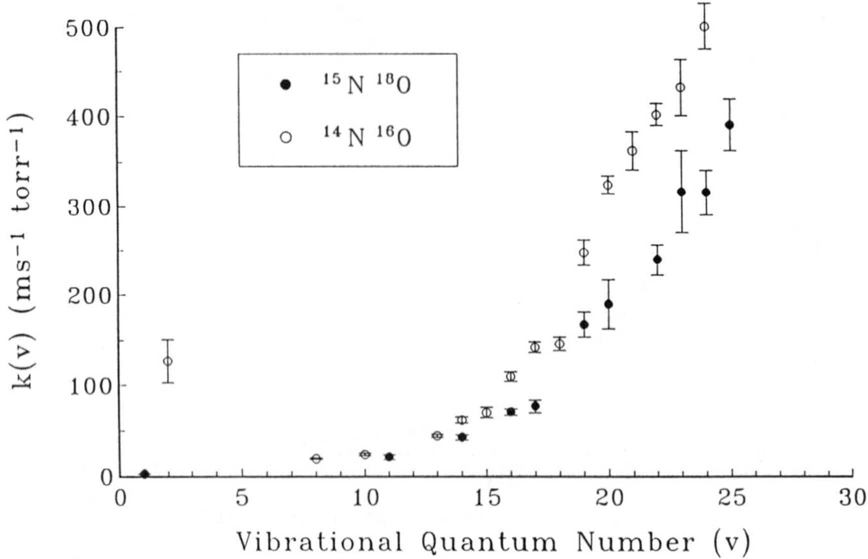

Fig. 4. Vibrational quantum number dependence of the vibrational relaxation rate constants of NO.

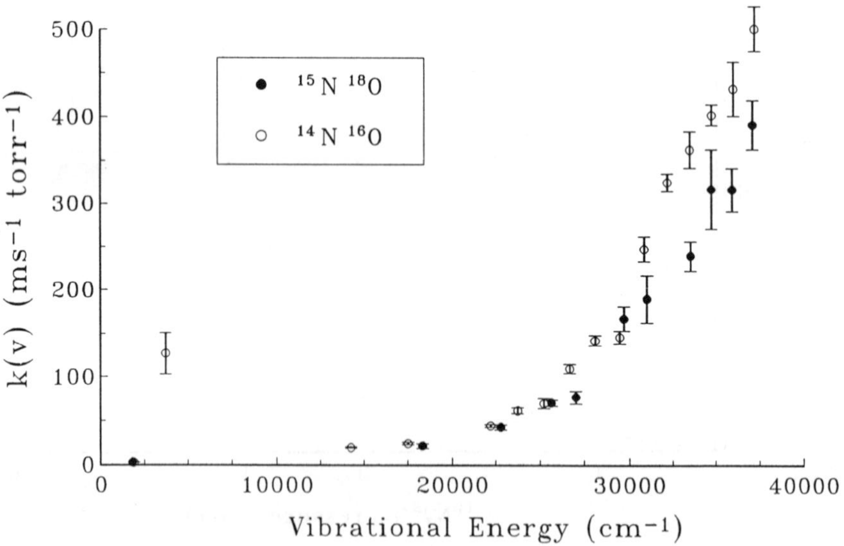

Fig. 5. Vibrational energy dependence of the vibrational relaxation rate constant of NO.

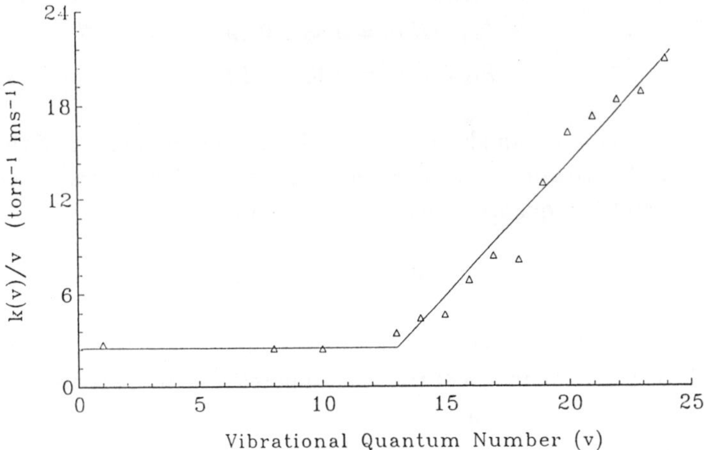

Fig. 6. The region of linear dependence on vibrational relaxation rate on vibrational quantum number.

The importance of multiple quantum relaxation was determined in the high energy region. In order to measure the relative importance of multiquantum relaxation, it is necessary to determine the relative population of both the prepared state v and the collisionally populated states $v - 1$ and $v - 2$. The rate constant from state v to $v - 1$ can be labeled as $k_{v,v-1}$, from v to $v - 2$ as $k_{v,v-2}$, and from $v - 1$ to $v - 2$ as $k_{v-1,v-2}$. The time integrated populations A_v, A_{v-1}, A_{v-2} of state v, $v - 1$ and $v - 2$ are directly related to the rate constants by the following relations[39]:

$$A_{v-1}/A_v = k_{v,v-1}/k_{v-1} \, , \tag{1}$$

$$A_{v-2}/A_v = (k_{v-1,v-2}k_{v,v-1} + k_{v-1}k_{v,v-2})/(k_{v-2}k_{v,v-1}) \, , \tag{2}$$

where k_v, k_{v-1} and k_{v-2} are the total relaxation rates from state v, $v - 1$ and $v - 2$. (See Figs. 4 and 5). Once the above ratio is determined, we can determine the relative importance of the $\Delta v = 2$ relaxation versus the $\Delta v = 1$ relaxation. Analyzing the experimental data for preparation of $v = 19$ yields the experimental evidence of multiquantum relaxation for highly vibrationally excited NO.

$$k_{19,18}/k_{19} = 0.47 \pm 0.10 \ ,$$

$$k_{19,17}/k_{19} = 0.33 \pm 0.19 \ ,$$

$$k_{18,17}/k_{18} = 0.48 \pm 0.12 \ .$$

These rate constants can also be used to determine the fraction of the total vibrational deactivation process we have accounted for in this experiment in the one and two quantum relaxation, which is

$$k_{19,18}/k_{19} + k_{19,17}/k_{19} = 0.80 \pm 0.29 \ .$$

From this result, one can say that the majority of the deactivation process is either through one-quantum or two-quantum relaxation. This also indicates that multiquantum (two-quantum) relaxation plays a significant role in the total vibrational relaxation at high vibrational excitation.

It is obvious that relaxation of NO from $v = 1$ is pure vibration–translation energy transfer, while the $v = 2$ relaxation is predominantly vibration–vibration relaxation since the V–V channel

$$NO(v = 2) + NO(v = 0) \rightarrow 2NO(v = 1) - 28 \text{ cm}^{-1}$$

is almost resonant. This is also evident in the 50 fold increase of the vibrational relaxation rate constant between $v = 1$ and $v = 2$, which is due to the efficient V–V energy transfer for $v = 2$. It is interesting to see that the measured total relaxation rate at $v = 8$ is much smaller than the $v = 2$ relaxation. This is not surprising since the V–V energy transfer is basically a resonant process, and the V–V rate constant is expected to drop very quickly as the energy defect grows with increasing v. For example, previous studies of V–V and V–T energy transfer in hydrogen fluoride showed that as vibrational energy increases, V–V transfer becomes less and less important and disappears after $v = 5$,[40–42] while V– T becomes the dominant channel for the vibrational relaxation.

This region where V–V energy transfer might be important is exactly within the gap of our data where we were unable to obtain high quality data. However it is possible to model this region of the vibrational quantum number dependence. Since the mechanism and the rate constant for NO $(v = 2)$ relaxation is known, the V–V rate constants for higher vibrational states of NO can be estimated using a semiclassical V–V energy transfer theory developed by Rapp and Englander–Golden[43,44] and a modified version of the SSH theory. V–V rate constants predicted by this theory essentially follow an exponential energy gap law. Figure 7 shows the predictions of the two models of the V–V energy transfer in the region between $v = 2$ and 11, covering the gap in our data. The open circles are the results of the Rapp–Englander–Golden model which is valid for the so called short range V–V energy transfer and the open squares are the results of the modified SSH theory of V–V energy transfer which uses the Keck–Carrier energy defect function.[45] The modified SSH theory is accurate even for relatively large energy defects, unlike the Rapp model which should be restricted to use with energy defect less than 200 cm^{-1}. In any case the V–V rates of NO decrease rapidly as vibrational energy increases. For vibrational states $v = 2$ to 7 V–V energy transfer should dominate the vibrational relaxation according to the calculated and experimental results, while above $v = 8$, V–V energy transfer makes a negligible contribution to the vibrational energy transfer, and V–T energy transfer dominates. The gap in our set of data actually "removes" the V–V vibrational energy transfer mechanism and our conclusions about the linearity of the vibrational relaxation rate constant apply only to the V–T component of the energy transfer.

Some early investigation of the influence of collision partners was also undertaken. The relaxation of NO ($v = 22$) with H_2 was measured in the same way as the NO self-relaxation except that two flow controllers were used to regulate the flow of NO and H_2. The collisional relaxation rate constant was found to be

$$k_{22}(H_2) = 2.3 \times 10^{-12} \ cm^3/s$$

at room temperature. Measurements of relaxation rates of NO ($v = 22$) with N_2, Ar, and Ne were also attempted. Unfortunately, the experimental setup was not suitable to measure these much slower processes. The rate constants for these processes are estimated to be smaller than 10^{-12} cm^3/s. The fact that the relaxation of NO ($v = 22$) with H_2 is much faster than that with other gases may be due to a mass effect. This refers to the simple prediction from SSH theory (see below) that when the reduced mass of the collision pair is smaller, the collision occurs with a higher velocity and thus the NO bond experiences a higher frequency variation of the external force imparted by the collision. This frequency is closer to the vibrational frequency of the bond and enhances vibration to translational energy transfer. The possibility of chemical reactions of highly excited NO with H_2 also exists. For example, the reaction

$$NO + H_2 \Longrightarrow H_2O + N$$

is only endothermic by 33 kcal/mol, indicating that the reaction can be accessed by NO with $v > 7$. However, how this reaction channel can effect the vibrational relaxation is unclear. More experiments are clearly needed in order to better understand the dynamics of this relaxation system.

As was seen above, the vibrational energy transfer at high vibrational excitation distinguishes itself through its high rate and multiquantum relaxation from that at low energy. How the vibrational excitation will affect the rotational and spin-orbit relaxation remains an open question. A study of rotational and spin-orbit relaxation for $v = 8$ and 19 has been conducted in order to answer this specific question.[46] The experimental set-up was exactly the same as that for measuring the vibrational relaxation rates. State-specific relaxation rates were measured by monitoring the time evolution of the prepared rotational and spin-orbit state. State-to-state relaxation rates were also measured by taking a spectrum of partially rotationally relaxed NO and determining the relative populations of the collisionally induced states and the SEP prepared state. Typical measurements were done at a few μs delay between the SEP step and the Probe step at a pressure of a few mtorr. The measured self-relaxation rates for $v = 8$ and 19 are shown in Fig. 8 and the state-to-state rates between the same spin-orbit manifold are shown in Fig. 9. In comparison with the $v = 2$ data of previous experimental studies,[47] the rotational relaxation rates for $v = 8$ and 19 are essentially the same, indicating that vibrational excitation has virtually no effect on the rotational and spin-orbit energy transfer. This is not too surprising since rotational energy transfer is normally dominated by long range forces important at very large impact parameters, and one would expect the process to be independent of the relatively small amplitude motion associated with vibration.

2.1. *Theoretical Background*

Vibrational energy transfer of NO at high vibrational excitation has been shown to have very different properties from that at low vibrational excitation. Molecular collisions at high vibrational excitation may be able to access completely different parts of the four-atom potential surface than those with little energy. Consider especially the possibility of traversing parts of the potential surface near to activated complexes of chemical reactions. For example, the energy threshold for the following reaction:

$$NO + NO \rightarrow N_2O + O \ , \tag{3}$$

is about 24 000 cm^{-1}. This means that above $v = 13$, NO can potentially traverse regions of the potential energy surface close to the transitions state for this reaction. This could have a dramatic influence on vibrational relaxation. Therefore, it is of fundamental importance here to determine to what extent vibrational energy transfer theories which are applicable at low vibrational excitation can be used in the "chemical energy regime". One of the most successful theories of vibrational energy transfer is the SSH theory[36–38] which has successfully predicted many vibrational relaxation rates to a quantitative level from intermolecular potentials obtained from transport properties. Before going into a discussion of the details of the NO relaxation, a simple introduction of the SSH theory is necessary. The SSH theory assumes that for the purposes of vibrational energy transfer, the intermolecular potential can be approximated by an exponential repulsion as follows:

$$V = V_o \exp(\alpha(r + A_1 s_1 + A_2 s_2 + \dots)) + \Phi_0 \ , \tag{4}$$

where r is the distance between the centers of mass of the two colliding molecules, the $s_i's$ are the vibrational normal coordinates of the two interacting molecules and $A's$ are the constant conversion factors, Φ_0 is the minimum value of the intermolecular potential function and it is normally negative, and α is a parameter related to the steepness of the repulsive wall. The rationale for this approximation was the recognition that the vibrational energy transfer processes are predominantly controlled by the short range forces, and thus mainly the repulsive part of the intermolecular potential. This is also reasonable in the sense that vibrational energy transfer occurs most efficiently at that point where the interaction potential is changing most rapidly with distance, and the high potential gradient of the repulsive wall produces a high frequency component to the impulsive collision that can efficiently couple the high frequency harmonic oscillator of the diatomic bond to translational motion. Using this approximation, the interaction between the two diatoms is modeled as an exponential repulsive wall and the problem is reduced to a single dimension. The probability of vibrational energy transfer of a diatomic molecule from initial vibrational state i to final vibrational state f is given by the SSH theory as follows:

$$P(i \to f) = P_0 P_c V^2(i \to f) 8(\pi/3)^{1/2} [8\pi^3 \mu \Delta E(\alpha^* h)^{-2}]^2 x^{1/2}$$
$$\cdot \exp[-(3x - \Delta E + \Phi_0)/(kT)] , \tag{5}$$

where

$$x = [2\pi^4 \Delta E^2 \mu(\alpha^*)^{-2} h^{-2}(kT)^{-1}]^{1/3} , \tag{6}$$

P_0 is a steric factor. P_c is a cross-section reference factor equal to $(r_c/r_0)^2$, where r_c is the separation of molecular centers at the point of closest approach and r_0 is the corresponding value at zero potential energy. V is the interaction coupling matrix element between the initial and final states. ΔE is the translational energy difference before and after collisional energy transfer. μ is the reduced mass of the collision pair, and α^* is closely related to the steepness of the repulsive wall for interaction between the collision partners.

The coupling matrix element V can be calculated from simple harmonic oscillator wavefunctions and has the following form for $\Delta v = -1, -2$:

$$V(v \to v - 1) = [\alpha A/(8\pi^2 \mu f)^{1/2}](hv)^{1/2} \tag{7}$$

and

$$V(v \to v - 2) = [\alpha^2 A^2/(8\pi^2 \mu f)](h^2 v(v - 1))^{1/2} \tag{8}$$

where

f is the vibrational frequency.

From this, it is apparent that if the vibration–translation energy transfer is mainly one-quantum relaxation, one would expect that the V–T rate be linearly dependent on the vibrational quantum number as was observed above for vibrational states below about $v = 14$.

So one can see that the SSH theory provides a closed form solution to the vibrational energy transfer problem. It makes specific and simple predictions regarding: 1) the vibrational quantum number dependence, 2) the influence of the collision pairs reduced mass, 3) the influence of the steepness of the repulsive interaction potential as well as 4) the temperature dependence of the relaxation rate. This allows us to perform many checks between experiment and theory in order to determine to what extent and to what level of vibrational excitation the SSH theory may be applicable.

2.2. *The NO Anomaly*

Despite the remarkable success of the SSH theory for many systems, the vibrational relaxation of NO ($v = 1$) by NO has long been one of the least well described systems by any theory of vibrational energy transfer. We will first discuss why we think this is and why the so-called "NO anomaly" does not totally discredit the SSH theory. The vibrational energy transfer of NO ($v = 1$) has long been known as an anomalous example of vibrational energy transfer in small molecules, since it is about 10^5 times faster than "similar" molecules such CO, N_2, O_2 etc.[31,48] Many explanations have been offered involving "attractive forces" as well as electronic nonadiabatic vibrational energy transfer. However most of these explanations lack fundamental theoretical support, and thus remain only suggestive. Here, we would like to provide a semiquantitative explanation of the NO anomaly based on the SSH theory.

The SSH theory has been very successful in calculating the vibrational relaxation rates of many molecular systems. The crucial part of the calculation is based on the knowledge of the interaction potential between the two collision partners. For most of the simple systems, the interaction potentials can be obtained from the transport properties and, in fact, all of the previous SSH calculations were done in this manner. But for the NO molecule, the interaction potential estimated from transport properties is wrong since $(NO)_2$ turns out to be bound by a few kcal/mol,[31] a much stronger interaction than for $(O_2)_2$ or $(CO)_2$, for example. Therefore, the normal way to calculate the SSH vibrational energy transfer rate is certainly doomed to fail. Spectroscopic studies have provided an accurate molecular structure for the NO dimer.[49,50] Even though the exact interaction potential is not known, the binding energy and the equilibrium bond lengths are clear at this point. A Lennard–Jones potential for NO–NO was constructed using the formulae from Atkins,[51] and the N–N bond length of 2.24 Å[49,50] and the dimer bond energy of 560 cm^{-1}.[31] A comparison of the two Lennard–Jones potentials is given in Fig. 10 for CO–CO and NO–NO. Here is the key point: Figure 10 shows that due to the stronger *attractive* forces and the deeper minimum of the NO–NO potential, the *repulsive* part of the potential is much steeper than for the case of CO–CO.

The implications of this on the vibrational relaxation rate can be estimated using the SSH theory. In the SSH theory, the steepness of the repulsive part of the potential is represented by the α^* parameter. The

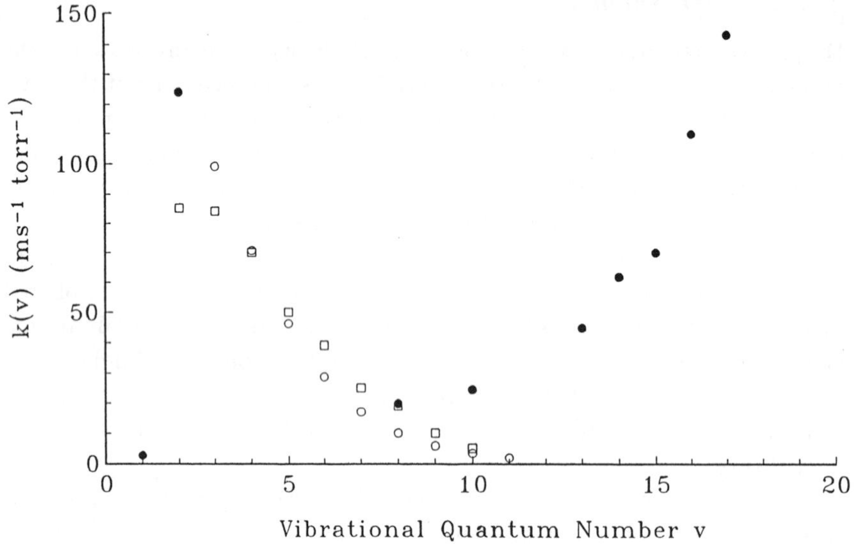

Fig. 7. Estimated V–V energy transfer rate constants for NO $v = 2$ to 10. The open circles are from the Rapp–Englander–Golden model, the open squares are from the modified SSH theory, and the closed circles are the experimental data points.

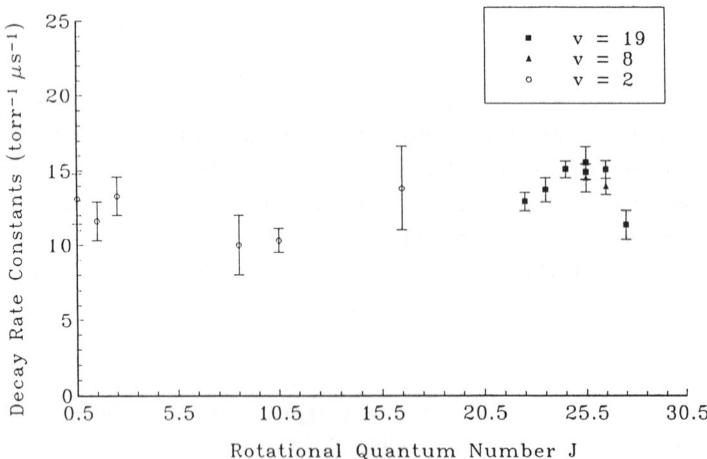

Fig. 8. Total rotational decay rate constants of $^{14}N^{16}O$ for $v'' = 8$ and 19 in comparison with previous $v'' = 2$ results.

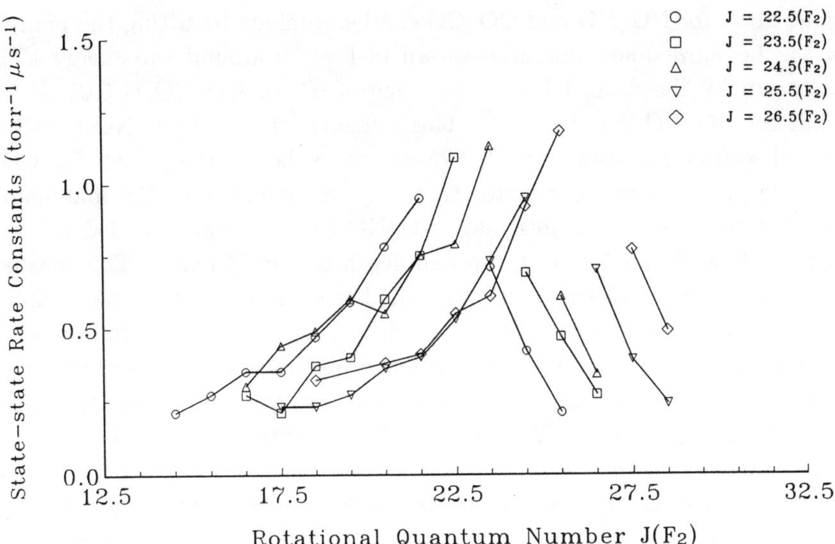

Fig. 9. State-to-state rotational energy transfer rate constants among the F_2 spin-orbit manifold, when $J = 22.5$, 23.5, 24.5, 25.5 and 26.5 levels are prepared using the SEP technique.

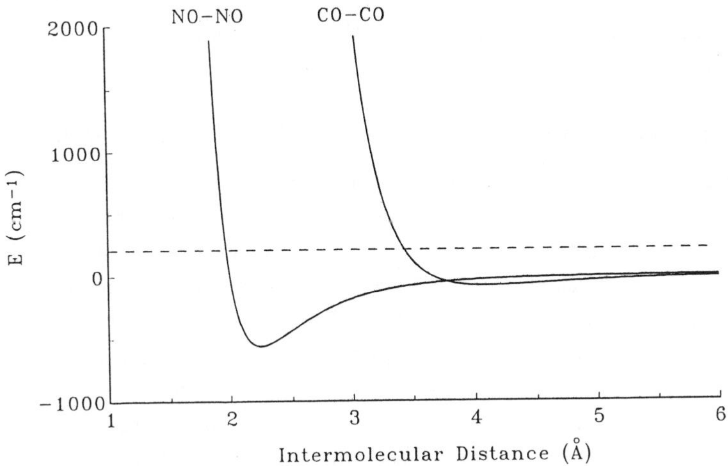

Fig. 10. Lennard–Jones potentials for NO–NO and CO–CO. The dashed line represents the average thermal energy, kT where $T = 300$ K.

value of α^* for NO–NO and CO–CO can be obtained by fitting the points of the Lennard–Jones potential shown in Fig. 10 around the energy kT, as shown by the dashed line. The value of α^* for CO–CO is 5.05 Å^{-1}, while for NO–NO it is 11.5 Å^{-1}. This single effect causes the NO ($v = 1$) self-relaxation probability to be 25 000 times larger than that for CO ($v = 1$). There is another factor that plays a smaller role: the minimum value of the interaction potential. NO–NO has a potential well depth of 560 cm^{-1}, while for CO–CO, the well depth is only 75 cm^{-1}. This means that the two NO molecules are accelerated toward one another and collide on the repulsive wall with a higher velocity. At room temperature, this will increase the NO ($v = 1$) self-relaxation probability another tenfold over CO ($v = 1$). The combined influence of these two factors suggests that NO ($v = 1$) self-relaxation is 250 000 times faster than that of CO ($v = 1$), approximately the observed difference.

This simple calculation using the SSH theory is consistent with previous experimental results, indicating that the attractive force is the most important factor in determining the NO vibrational relaxation, albeit, in a subtle way. The deeper attractive well in the case of NO–NO, leads to a significantly steeper repulsive wall. Similar arguments have been made independently by Tanner and Maricq[52] for ion-neutral vibrational relaxation, in which the potentials are qualitatively similar to that of the open shell NO system. Obviously, this analysis cries out for full-fledged quantum dynamical calculations. Nevertheless, we anticipate that the basic ideas discussed here would be confirmed by such an effort.

2.3. *The Temperature Dependence of NO Vibrational Relaxation*

The SSH theory predicts a very specific and very strong temperature dependence. Unfortunately, it is also well-known that NO vibrational self-relaxation exhibits only a very weak temperature dependence. We have increased the sophistication of the SSH theory to include the possibility of long range trapping in molecular collisions. This will be especially important for NO–NO at low temperature. An analogous but much more sophisticated approach[53] employing classical trajectory analysis on an empirical potential surface also indicated the importance of orbiting collisions at low collision energy for the vibrational relaxation of $O_2^+(v = 1)$ by Kr,[54,55] which has a well depth comparable to the NO–NO case. For a given collision energy, there will be a maximum value of the impact parameter, b_{max}, that creates a centrifugal barrier and prevents the forming collision complex from feeling the attraction between the NO monomers. All collisions at impact parameters within this value should be trapped. Therefore, we can use this maximum impact parameter as an estimate of the trapping cross-section.

$$\sigma_{TRAP} = \pi b_{max}^2 \; .$$

Next we imagine that the vibrational relaxation probability is proportional to the product of the trapping probability and the SSH relaxation probability that occurs on the repulsive wall.

$$P_{VET} = P_{TRAP} * P_{SSH} \; .$$

This includes two processes that have opposite temperature dependence. Trapping becomes more important at low temperature while SSH relaxation becomes less important. The expressions can easily be derived as a function of collision energy. The thermal averaging is done by a Monte–Carlo sampling program. Figure 11 shows the results of such calculations for CO–CO and NO–NO. One can see that this model indeed reproduces the qualitative difference between the temperature dependencies of the NO–NO with a strong attractive well and important trapping at low temperature and the CO–CO which is the conventional SSH molecule.

2.4. *Mass versus Energy Defect*

In the measurement of the vibrational relaxation rate constants for NO, it can be seen that the relaxation rate constants of $^{15}N^{18}O$ are always smaller than those of $^{14}N^{16}O$ (see Figs. 4 and 5). We believe this is due to the mass effect in the vibrational relaxation, predicted by the SSH theory. Another manifestation of this effect is the observation that relaxation of NO ($v = 22$) by H_2 is much faster than that with other gases such as Ar, N_2 and He. These observations imply that the heavier the molecule, the lower the vibrational relaxation rate. This phenomenon in vibrational relaxation has been observed previously[56] and is predicted by the SSH theory. In that theory, the vibrational relaxation probability has an exponential factor which is $\exp(-3x)$, where x is related to the reduced mass and, therefore, the average relative velocity of the collision pair. As the reduced mass increases, x increases, and the relaxation probability goes down. However, there is another effect due to the energy defect. Inspection of Eq. (6)

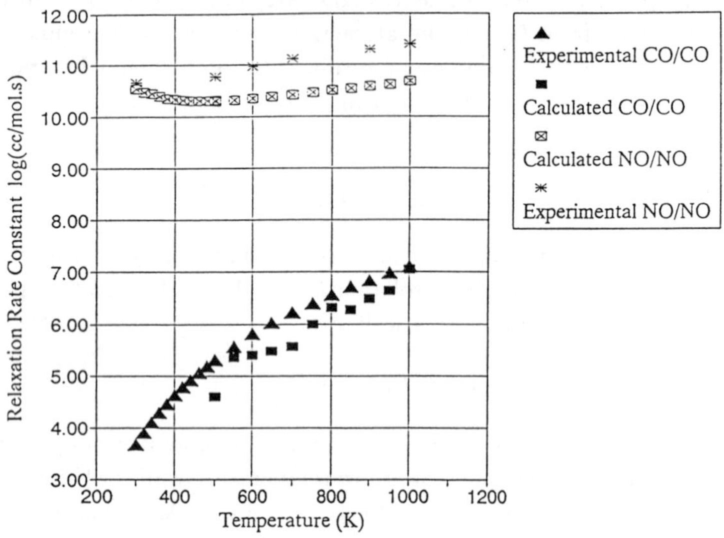

Fig. 11. The calculated and observed temperature dependencies for self-relaxation of NO ($v = 1$) and CO ($v = 1$). The calculated values were obtained using a model including long range trapping as well as SSH V–T energy transfer (see text).

also shows that x depends on the energy defect. That is, the lower the energy gap, the faster the vibrational relaxation. In the case of NO, the vibrational relaxation of $^{15}N^{18}O$ would be faster than that of $^{14}N^{16}O$ if one only considered the energy gap part of the problem. However, if both factors are included the vibrational rate constant of $^{15}N^{18}O$ is actually 20% lower, indicating the mass effect is slightly more important than the effect of energy defect. This is in agreement with our observed results and it is clear that the general SSH theory can account for this effect.

In summary, we have discussed several important reasons why it appears the basic picture given by the SSH theory is correct for low levels of vibrational excitation in the self vibrational relaxation of NO, perhaps for vibrational levels as high as $v = 14$. First, it appears that the steep repulsive wall, resulting from the presence of a relatively deep attractive well, increases the magnitude of NO–NO vibration-to-translation energy transfer by about five orders of magnitude compared to comparable closed-shell vibrational energy transfer systems. Secondly, the temperature dependence can be explained if one accounts for the long range trapping that is possible at low temperature for collision pairs that experience a significant attractive force. Thirdly, the vibrational quantum number dependence of the vibrational energy transfer is linear up to $v = 14$ which is also predicted by the SSH theory. Finally, the observed mass effect for heavy versus light NO is in agreement with the predictions of the SSH theory.

2.5. *Vibrational Dependence of Vibrational Relaxation at High Vibrational Excitation*

From the above discussion and analysis, the mechanisms of the vibrational relaxation up to $v = 13$ with $E_v = 22\,000$ cm^{-1} are rather clear. In summary, $v = 1$ is a pure V–T case, while relaxation of $v = 2$ is dominated by V–V energy transfer. From $v = 2$ to 7, V–V relaxation dominates, while above $v = 8$, the relaxation is again V–T energy transfer. The V–T component of the NO energy transfer between $v = 1$ and 13 is believed to be dominated by one-quantum relaxation because of the linear dependence of the V–T rate constant on v and the fact that this region appears to be consistent with SSH theory.

Above $v \approx 13$, or $22\,000$ cm^{-1} in energy, however, the quantum number dependence of the vibrational relaxation is clearly not linear. One interesting piece of evidence is the fact that the activation energy of the reaction, NO + NO $\rightarrow$ N$_2$O + O, is coincidental with the point ($v \approx 14$) where the vibrational relaxation starts to accelerate. Recent calculations of Gordon[57] also show that extremely unique interactions are possible for high energy (NO)$_2$ and that covalently bound complexes could be formed at these high energies (see Ref. 30). This has inspired us to suggest that a new mechanism for vibrational energy transfer may be possible at higher energy and we have dubbed this mechanism "transient chemical bond formation". The basic idea is simply that if the collision samples a region of the potential surface in the vicinity of the transition state to complex formation or bimolecular chemical reaction, (our experiment does not allow us to distinguish) it is quite likely that the interaction will be abnormally anharmonic, enhancing vibrational energy transfer.

This mechanism implies a significantly stronger and less harmonic interaction potential than in the low energy region. Consequently, one might expect to see the change in mechanism manifested, not only in the magnitude of the rate, but also in the vibrational quantum number changing propensity rules. As discussed above, it does appear that $\Delta v = 2$ is as important as $\Delta v = 1$ relaxation in the high energy region above $v'' \approx 14$. This is to be contrasted with the observed linear dependence of relaxation rate constant on vibrational quantum number, which implies the dominance of single quantum relaxation in the low energy region. We believe this is consistent with the idea of "transient chemical bond formation". It should be pointed out that efficient multiquantum relaxation has been observed when the perturbation approximation inherent to the SSH theory is invalid, for example, in I_2 where the vibrational spacing is significantly less than kT.[58,59] However, for NO, even at these high energies the vibrational spacing is still much bigger than kT. For example, the vibrational spacing in $^{14}N^{16}O$ between $v = 18$ and $v = 19$ is 1350 cm^{-1} while kT is 208 cm^{-1}.

On the other hand, the idea of "transient chemical bond formation" suggests that at high vibrational energy, the electronic structure of NO is much different than that of NO at its equilibrium geometry. *Ab initio* calculations of the vibrational state specific dipole moments show this conclusively.[60,61] All diatomics that separate to neutral atoms must asymptotically have a zero dipole moment. This means that NO's dipole moment, which is essentially zero at its equilibrium geometry, rises slowly and turns over going back to zero at infinite separation. Although this interesting effect has not yet been observed experimentally, this turn-over apparently occurs near the outer classical turning point of $v = 20$. One might, therefore, expect that the interaction potential of the tetra-atomic system is also significantly different than that of the equilibrium NO-dimer.

Let us now consider another example of highly vibrationally excited molecules which are thought to be present in the Earth's upper atmosphere.

3. Vibrational Energy Transfer of Highly Vibrationally Excited O_2: The Stratospheric Importance of Highly Vibrationally Excited Molecules

One of the most fundamental simplifications used in atmospheric modelling is the assumption of *local thermodynamical equilibrium* (LTE).[62] According to this approach *the internal quantum state distribution of all atmospheric constituents can be described by a Maxwell–Boltzmann distribution characterized by a local temperature*. In order to see how important the LTE assumption is, imagine the modelling complexities if it were not applicable. It is well-known that individual quantum states of molecules, especially excited electronic and vibrational states, can react with vastly different rate constants than the lowest quantum states that would be expected to be populated in a Maxwell–Boltzmann distribution at atmospheric temperatures (circa 200–300 K). Consequently if one cannot assume LTE, each quantum state of an atmospheric constituent would have to be treated as if it were a separate and unique atmospheric species with its own rate constants for chemical and photochemical production and destruction. Similarly, if one were to attempt a concentration retrieval from infrared emission data collected from a satellite, it would be necessary to know the steady-state distribution of emitting vibrational levels, a distribution governed by many state-to-state collisional processes.

On the other hand if the LTE assumption holds, all of this detailed microscopic information is contained, in a statistically averaged sense, within a chemical reaction's activation energy, a photochemical reactions temperature dependence or the Planck law of radiation. In essence, for the cases where the LTE assumption holds, quantum states of molecules "don't count". Traditional chemical kinetic's will be a sufficient conceptual framework for atmospheric modelling.

A priori, if the production rate of nonequilibrium quantum-state distributions of a given atmospheric constituent is large in comparison to the rate of return to equilibrium, the atmosphere will establish a nonequilibrium, steady-state distribution of molecular quantum states. If this nonequilibrium distribution of quantum states were to exhibit special chemical or photochemical behavior, no amount of modelling within the LTE assumption would accurately reproduce the actual atmospheric events. Indeed, "fiddling" with an LTE model to fit an inherently non-LTE set of data, might yield misleading results.

It has been recognized recently that the mid-latitude stratosphere is a likely region to find non-LTE phenomena. Two key factors are thought to be important to the production of the non-LTE species in the stratosphere: 1) the dominance of photochemical processes and 2) the low total pressure. One of the most important examples was suggested by Slanger *et al.*[11] to help explain the inability of LTE models of the stratosphere to reproduce measured ozone altitude profiles.[63] After considering the unexpected laboratory observation that 248 nm irradiation of pure O_2 lead to ozone formation, Slanger was able to show that UV photodissociation of O_3 produces highly vibrationally excited O_2 which can absorb light at much longer wavelengths than LTE distributions of O_2 vibrational states, leading to ozone formation. Steps (1)–(4) characterize the "Slanger mechanism" for non-LTE ozone formation.

$$(1) \qquad O_3 + h\nu(\lambda @ 250 \text{ nm}) \rightarrow O_2(^1\Delta) + O(^1D) \qquad 87\% \ .$$

$$(2) \qquad\qquad\qquad \rightarrow O_2(^3\Sigma, v'') + O(^3P) \qquad 13\% \ .$$

$$(3) \quad O_2(^3\Sigma, v'') + h\nu(\lambda > 300 \text{ nm}) \rightarrow 2O(^3P) \ .$$

$$(4) \qquad\qquad 2O(^3P) + 2O_2 + 2M \rightarrow 2O_3 + 2M \ .$$

Since there is much more Solar photon flux at wavelengths longer than 300 nm in the stratosphere, this mechanism allows $O_2(^3\Sigma, v'')$ to become an ozone precursor. Photofragmentation translational energy measurements of $O(^3P)$ atoms resulting from ozone photodissociation at 226 nm gave strong albeit low resolution evidence for efficient production of highly vibrationally excited O_2 in vibrational levels as high as $v'' = 28$.[64] Recently, the distribution of vibrationally excited O_2 resulting from ozone photolysis at 248 nm has been measured in Slanger's laboratory using high resolution LIF methods, showing efficient production of $O_2(v'')$ up to and including $v'' = 22$.[65] This measurement confirms the hypothesized *production* mechanism for high vibrational levels of O_2. A photochemical model of the upper atmosphere[66,67] was used to include the Slanger mechanism and it was found that the presence of highly vibrationally excited O_2 was a possible way to improve the altitude profiles of O_3 as well as a means to model recent 6.9μ infrared emission measurements from the LIMS.[68] A very recent model calculation[69] identified the important areas for experimental measurements to be 1) the vibrational distribution of $O_2(v'')$ from ozone photodissociation as a function of photolysis wavelength and 2) the determination of vibrational relaxation rate constants.

The SEP technique is clearly suited to investigation of the rate of return to equilibrium, as was seen for the case of the NO molecule.[28–30,46,70,71] One complication arises when considering application of the same methodology to O_2. Figure 12 shows a potential curve diagram for O_2. Unlike NO the intermediate "stepping stone" state used for the SEP is far above the ground state dissociation limit and predissociation is much more important for O_2 than for NO. Indeed the predissociation rates are about 10^{11}s^{-1}. Therefore, SEP on O_2 poses an exciting technical question in addition to the atmospheric ones: Is it possible to use the SEP method when the intermediate state predissociates so rapidly?

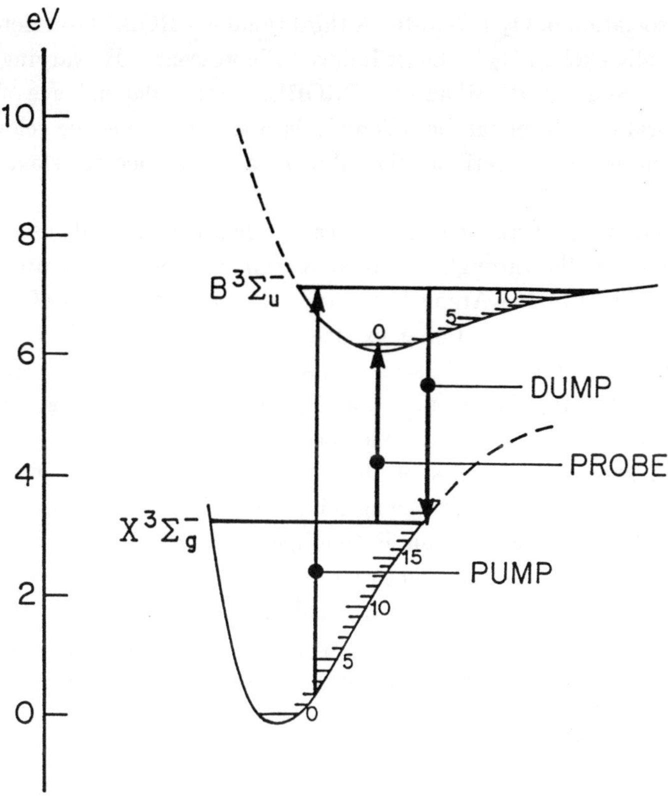

Fig. 12. Electronic states of O_2 and the PUMP–DUMP–PROBE scheme: This figure shows the stimulated emission pumping scheme for measuring the collisional properties of highly vibrationally excited O_2. The PUMP laser induces population transfer from the $X^3\Sigma_g^-$ to the $B^3\Sigma_u^-$ state. The DUMP laser induces emission back to high vibrational levels of the $X^3\Sigma_g^-$ state. A third PROBE laser observes the time evolution of the prepared state. The experimental approach is quite effective, despite the facts that nanosecond pulsed lasers are used and that all of the vibrational levels of the $B^3\Sigma_u^-$ state predissociate on the picosecond time scale.

Figure 12 also shows the experimental approach that was applied. A pulsed tunable Argon Fluoride PUMP laser is used to pump O_2 from $v'' = 1$ or 2 of the $X^3\Sigma_u^-$ ground electronic state to $v' = 7$ or 10, respectively, of the $B^3\Sigma_g^-$ excited electronic state by way of the well-known Schumann–Runge system. A powerful DUMP laser stimulates emission, inducing population transfer back to a specific vibrational level of the ground electronic state. As will be seen, stimulated emission is efficient enough to compete with the rapid predissociation of O_2's B-state. A third tunable PROBE laser detects the vibrationally excited O_2 by Laser Induced Fluorescence. By varying the time delay between the DUMP and the PROBE, the time dependence of the prepared vibrational level can be followed, data that contains the relevant information necessary to deriving the vibrational state specific relaxation rate constants.

The experimental arrangement is nearly identical to that used for NO. O_2 flows rapidly through a cell at a well controlled pressure and temperature. The tunable Argon Fluoride laser is overlapped carefully in space and time with the output of an excimer pumped dye laser, the DUMP. A second excimer pumped dye laser is spatially overlapped but variably delayed to give the PROBE signal. The PROBE signal is averaged on a boxcar and plotted out on a digital plotter or collected by PC-286 clone.

Figure 13 shows the results of the one laser experiment where the tunable Argon Fluoride laser's wavelength is scanned over its one nanometer tuning range. Many O_2 Schumann–Runge bands are readily observed and assigned, mainly to rotational structure in the $10 \leftarrow 2$ and $7 \leftarrow 1$ vibronic bands.[72] In order to enhance the hot band signal, the cell is heated to 460 K. Under these conditions very high quality data is obtained. In order to observe the magnitude of the temperature dependence some experiments were also carried out at room temperature, which was possible for a few vibrational levels. Figure 14 shows the results of the next most complicated experiment, where the PROBE laser's wavelength is scanned at a fixed delay with respect to the PUMP laser, while the PUMP laser excites one of the rovibronic transitions of Fig. 13. In this figure, one sees the results of the experiment are a high signal-to-noise spectrum of the $0 \leftarrow 21$ LIF band. The $v'' = 21$ population has been created in this experiment by the small fraction of O_2 B-state which has spontaneously emitted to $v'' \geq 21$, rather than undergone predissociation — the fate of the vast majority of molecules that were excited to the B-state. This creates a time evolving population in $v'' = 21$ as the molecules cascades back down the vibrational ladder. This experiment is performed in order to find the correct laser frequency at which the $v'' = 21$ target state can be detected with the PROBE laser.

Once this has been accomplished the PROBE laser is tuned to a strong feature of the $0 \leftarrow 21$ spectrum. Figure 15 shows the results of the three laser experiments, scanning the wavelength of the DUMP laser while observing the LIF signal from the PROBE. One can see that when the DUMP laser hits a resonance transferring population from the

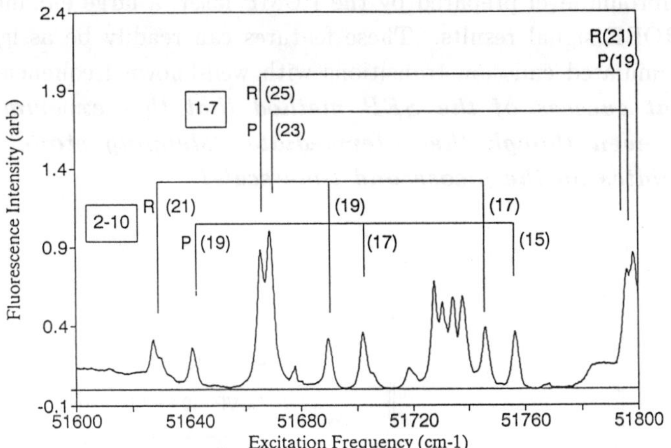

Fig. 13. O_2 Schumann Runge excitation spectrum: When the tunable Argon Fluoride laser is scanned while observing the gated fluorescence from a cell containing 100 torr of O_2 at 460 K, the spectrum shown is recorded. Hot band transitions are easily assigned from well established O_2 spectral studies as shown in the figure.

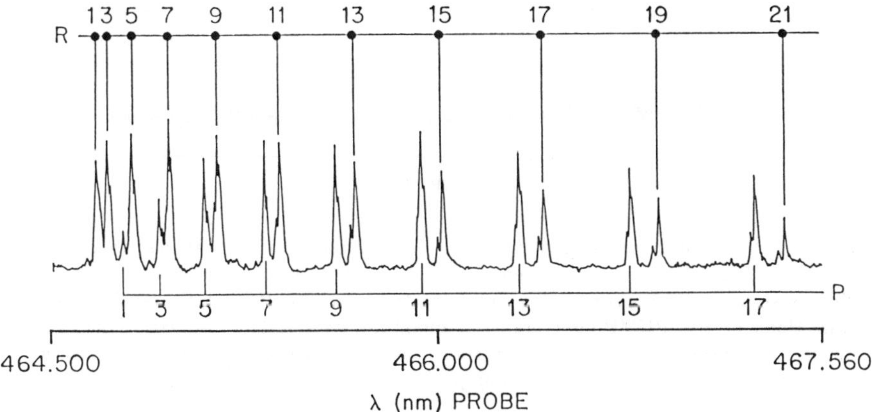

Fig. 14. Franck–Condon pumping of O_2: This figure shows the $0 \leftarrow 21$ vibronic band of the Schumann Runge system. The population in the $v'' = 21$ state has been induced by the small amount of spontaneous emission corresponding to production of vibrational levels higher than or equal to the $v'' = 21$. The presence of this state is detected using an excimer pumped dye laser. The spectral features can easily be assigned to a P- and R-branch. The fine structure is due to the well-known spin–spin and spin–rotation interactions of this triplet molecule.

single rovibronic level prepared by the PUMP laser, a large enhancement
of the PROBE signal results. These features can readily be assigned to
specific stimulated emission transitions with well-known frequencies.[73] *It
is a great success of the SEP method that this experiment is
possible, even though the intermediate "stepping stone state"
predissociates on the picosecond time scale!*

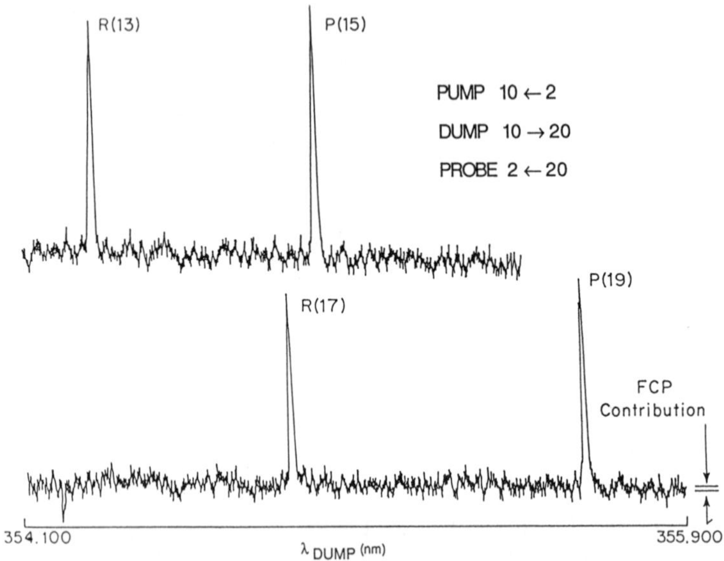

Fig. 15. Stimulated emission pumping of O_2: The PROBE signal described in the
caption of Fig. 14 is dramatically enhanced at specific frequencies of the DUMP laser.
This is shown in this figure. The two transitions are the *P*- and *R*-branch emission
transitions of the PUMP-laser-prepared quantum state. The two traces show the
dependence of the positions of the SEP transitions on the choice of the PUMP transition.
The contribution of the Franck–Condon pumping is indicated.

In Fig. 16 the collisional lifetime of the prepared vibrational state is
measured. This is accomplished by repeatedly scanning the DUMP laser
over the stimulated emission transition and recording the magnitude of this
signal as a function of DUMP–PROBE delay. This method of acquiring
the data guarantees that the time dependence of the contribution to the
prepared vibrational state's population which is due to SEP is independent
of the time evolution of the population due to Franck–Condon pumping.

This data is fit to an exponential function and the exponential time constant is derived. Similar measurements are made at a series of pressures. In Fig. 17 the inverse of the exponential time constant for a representative vibrational state was plotted against the pressure and linear regression analysis was carried out to determine the fitted straight line. One observes the expected linear dependence on pressure. The slope of this straight line is used to derive the vibrational state specific rate constant for disappearance from the prepared state.

Table 1. Relaxation rate constants for specific vibrational levels of O_2.[a]

v''	$k(v'')$ 295 K	$k(v'')$ 460 K
19	$4.7(0.3)\times10^{-15}$	$2.3(0.1)\times10^{-14}$
20	$3.2(0.3)\times10^{-15}$	$3.1(0.08)\times10^{-14}$
21	$5.8(1.2)\times10^{-15}$	$2.2(0.9)\times10^{-14}$
22	$5.4(0.8)\times10^{-15}$	$3.7(0.3)\times10^{-14}$
23	$1.2(0.4)\times10^{-14}$	$4.1(0.6)\times10^{-14}$
24	$0.84(0.04)\times10^{-14}$	$6.9(0.5)\times10^{-14}$
25	$1.8(0.05)\times10^{-14}$	$11.7(0.2)\times10^{-14}$
26	$4.7(0.2)\times10^{-14}$	$16.4(2.0)\times10^{-14}$
27		$> 83.0\times10^{-14}$

[a]The numbers in the parentheses are the error estimates in the measurements.

Table 1 as well as Fig. 18 show initial experimentally derived vibrational relaxation rate constants for a number of high vibrational states of O_2, which are thought to be produced by atmospheric ozone photodissociation. These experiments were carried out at two temperatures 460 K and 295 K. This work has not yet reached the stage where a clear explanation of the behavior can be offered although recent theoretical calculations give reasonable agreement with experiment for $v \leq 20$ and suggest the importance of single quantum relaxation with both V–T and V–V energy transfer.[74] The observation of single quantum relaxation as in the low energy region of NO appears to hold for higher vibrational states of O_2 as well. Figure 19 shows the measured time dependence of $v = 26$ and 25 obtained from preparing the 27th excited vibrational state, the last vibrational state whose vibrational energy is below the bimolecular reaction threshold[75]

$$O_2 + O_2 \rightarrow O_3 + O \ .$$

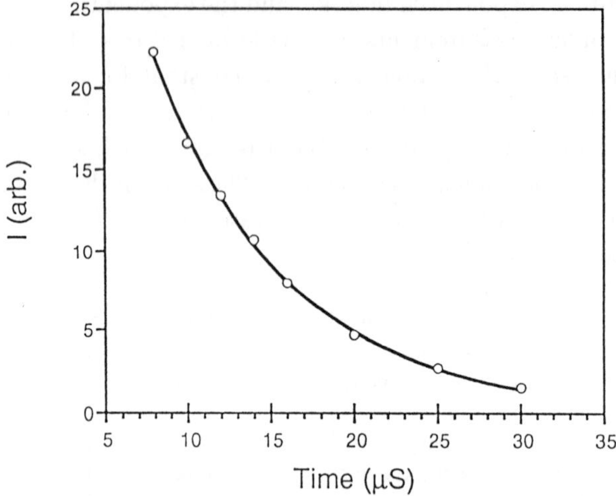

Fig. 16. Time dependence of the SEP signal: The SEP induced population of $v'' = 23$ is plotted versus delay time between the DUMP and PROBE lasers. The data are shown as circles. The solid line is the best fit to the data from which a pressure dependent lifetime can be derived. This experiment was carried out at 460 K and 170 torr

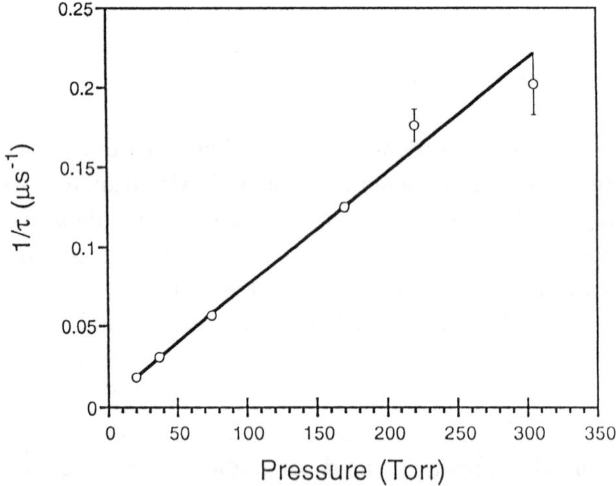

Fig. 17. The derived vibrational relaxation rate constant: the inverse of the derived pressure dependent collisional lifetime for $v'' = 20$ (prepared by SEP) is plotted versus pressure. This experiment temperature is 460 K. The slope of the line of the best fit gives the relaxation rate constant.

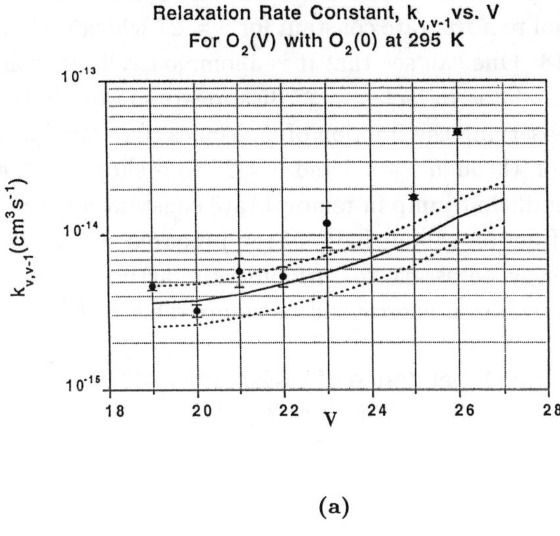

(a)

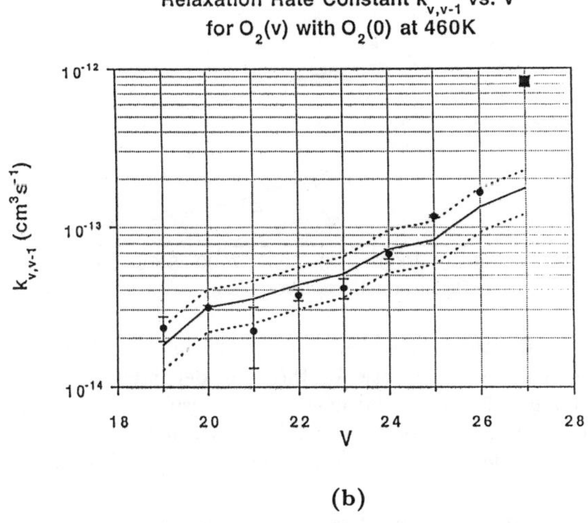

(b)

Fig. 18. Vibrational state specific relaxation rate constants for highly vibrationally excited O$_2$ being relaxed by collisions with O$_2$: at (a) 295 K, (b) 460 K. The plots show the results of the current study (circles) and extrapolations from the theoretical calculation of Billing[74] (solid line). Estimated errors in the calculation are shown by dashed lined. The deviation from theory increases as one approaches energies near the threshold to bimolecular chemical reaction.

The fit to the data is based on a model which only allows single quantum relaxation, that is: $27 \to 26 \to 25 \to 24$. From this analysis it is possible to derive the collisional removal rate constant for $v = 27$ which is also shown in Table 1 and Fig. 18. One can see that it is anomalously large compared to the rest of the set of data. In fact, $v = 27$ disappears so rapidly that we can only detect it by observing the fraction of molecules that "trickle-down" by collisional relaxation through $v = 26$ and $v = 25$ as is shown in Fig. 19. We believe that the significant jump in removal rate constant for $v = 27$ means that a significant fraction (about half) of the prepared $v = 27$ is reacting. The remaining energy necessary to surmount the activation barrier must be provided by the collision energy and be comparable to $kT = 319$ cm^{-1}.

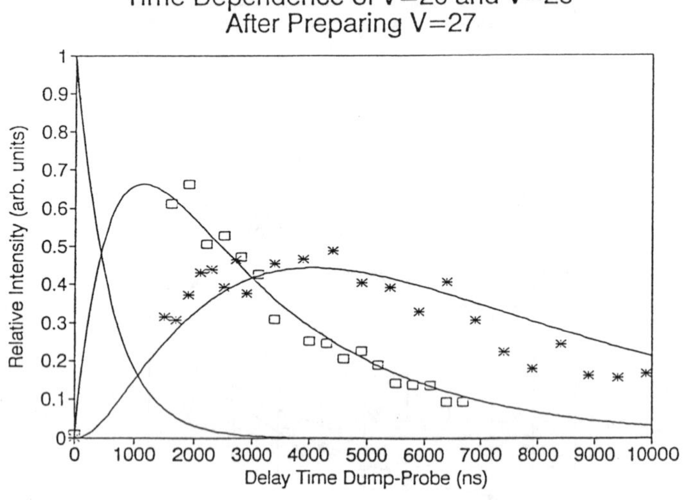

Fig. 19. Observation of the time dependence of $v = 26$ and 25 after preparation of $v = 27$: The open squares are the data for the time dependence of $v = 26$ while the asteriks are that for $v = 25$. The solid lines are the results of a model calculation that assumes single quantum relaxation. The solid line which appears as an early time exponential fall-off is the model calculation results for the prepared state, $v = 27$.

Even more interesting is the observation made when we apply the method of "trickledown spectroscopy" to $v = 28$. Figure 20 makes a comparison between two experiments where the DUMP laser is tuned over the resonances for preparation of $v = 28$ and 27. The transfer of population to either $v = 27$ or 28 is detected by probing the longer lived $v = 26$ vibrational state with the PROBE laser set at a 2–3 μs delay. It is necessary

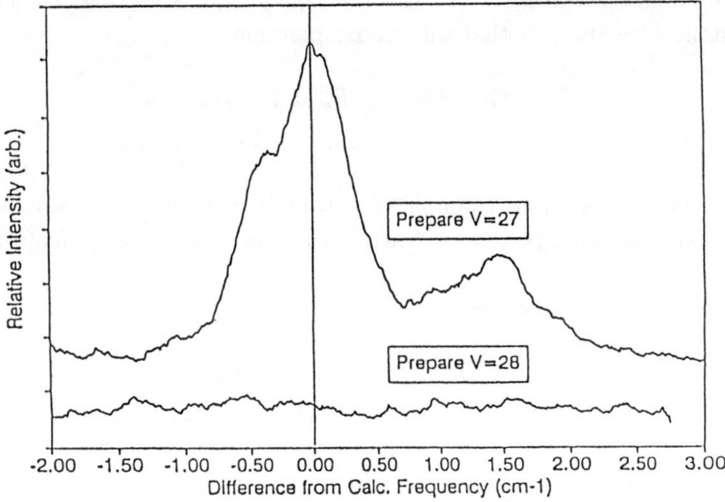

Fig. 20. "Trickle Down Spectroscopy" of $v = 27$ and $v = 28$: The PROBE laser is tuned to detect the presence of $v = 26$ which was observed through the $2 \leftarrow 26$ band, 2 μs after the SEP step. The DUMP laser is scanned over the DUMP transition producing $v = 27$ (upper trace) which is easily detected after the $v = 27$ "trickles down" to $v = 26$. The identical experiment for production of $v = 28$ (lower trace) from which one observes almost no "trickle-down" signal in $v = 26$. The $v = 27$ data is at least 10x larger than that of $v = 28$. This is strong evidence that $v = 28$ reacts with O_2 to form O and O_3. See text.

to point out here that the Franck–Condon factor to $v = 28$ in the SEP step is comparable to that of the $v = 27$ preparation step. As can easily be seen, in comparison to $v = 27$, $v = 28$ doesn't trickle down. Being the first vibrational state whose vibrational energy is above the reaction threshold, it appears that $v = 28$ reacts at least twenty-three times faster than $v = 27$ when collisionally relaxed. This allows us to estimate the reaction rate constant for $v = 28$ to be about 2×10^{-11} cm^3/s. In light of this new result, it is now of great atmospheric interest to know to what extent photolysis of O_3 leads to O_2 vibrational states where $v \geq 27$ — vibrational states that appear to react.

Before concluding, one is curious to ask why a reaction was observed in the Oxygen case, while no reaction, but instead an enhancement of vibrational relaxation due to transient chemical bond formation, was observed in the case of NO. We believe that this observation is telling us that for the case of Oxygen, the transition state for the chemical reaction is very similar in geometry to the reaction products O_3 and O. This is consistent with the fact that the reverse reaction,

$$O_3 + O \rightarrow O_2 + O_2, \qquad E_a = 4.6 \text{ kcal/mol} ,$$
$$\Delta H = -93.7 \text{ kcal/mol} ,$$

is believed to be a classic example of a strongly exothermic reaction with a small activation energy and an "early" transition state analogous to the F + H_2 reaction.

In contrast, the reaction

$$N_2O + O \rightarrow NO + NO, \qquad E_a = 28 \text{ kcal/mol} ,$$
$$\Delta H = -36 \text{ kcal/mol} ,$$

is significantly less exothermic and has a much larger activation energy. This suggests that the transition state could be much "later" than in the Oxygen case, meaning that the transitions state's geometry does not resemble N_2O + O. Consequently in the NO case, sampling the strong coupling region of the transition state leads to enhanced vibrational energy transfer instead of reaction.

There is a great deal of work to do in this area and the direction is now quite clear. One can carry out PUMP–DUMP–PROBE experiments on highly vibrationally excited O_2. One can investigate the role of all possible atmospheric collision partners. In particular, N_2, O_2 and O-atom are thought to be important. The temperature dependence can also be obtained by these types of measurements. The sum total of this data can go into atmospheric models to determine the role of highly vibrationally excited O_2 in the upper atmosphere. Work in this direction is now underway.

4. Conclusions

Vibrational energy transfer dynamics of very highly vibrationally excited NO and O_2 were studied by the PUMP–DUMP–PROBE method. These studies show two of the best examples of how modern laser methods can be used to transform spectroscopic techniques into preparative methods. Vibrational self-relaxation of NO was studied for vibrational energies as high as 4.6 eV. Several important observations came out of these experiments. First, it was found that below $v'' \approx 14$ the V–T energy transfer exhibited a linear dependence on vibrational quantum number.

This was interpreted as being consistent with low energy vibrational energy transfer theories and gave some indication of the range of applicability of such theories for highly vibrationally excited molecules. Second, the onset of the higher than linear dependence of vibrational energy transfer came at the energy threshold of the lowest accessible endothermic bimolecular reaction: $NO + NO \rightarrow N_2O + O$. This, as well as calculations by Gordon *et al.*, which showed the presence of stable high energy collision complexes, suggested that a qualitatively different relaxation mechanism may be important in these studies at high vibrational energy.

It was hypothesized that "transient chemical bond formation" was responsible for enhanced vibrational relaxation rates at high vibrational energy. The high vibrational states are influenced by the character of the outer turning point of the vibration. For such high vibrational levels as were studied here, the electronic structure may be significantly different than for NO near its equilibrium bond length. This presumably also has a large influence on the interaction potential important to the vibrational relaxation. Transient chemical bond formation means simply that trajectories, which pass near to the transition states of chemical reactions or collision complex formation channels, may sample a much more anharmonic part of the potential energy surface and, therefore, preferentially experience vibrational energy transfer. This hypothesis indicates a fundamentally stronger and less harmonic interaction at high vibrational excitation. Further evidence for this line of reasoning was found in the third important experimental observation. It was found that for high vibrational levels, $\Delta v = 2$ relaxation was about as important as $\Delta v = 1$ relaxation. This result was obtained despite the fact that the energy separation between vibrational levels was five to ten times kT.

The same method has also been applied to the important atmospheric problem of highly vibrationally excited O_2. The experimental approach works well even though the intermediate state of the SEP preparation step predissociated on a picosecond time scale. In this example, single quantum relaxation was observed even for relaxation of the last vibrational state below the bimolecular reaction threshold. Above the reaction threshold a dramatic increase in the collisional removal rate constant was observed, similar to the NO experiments. However, in contrast to NO, in the Oxygen experiments direct evidence of the dominance of bimolecular reaction for vibrational states above the reaction threshold was observed. The simlarities and diffrerences between NO and O_2 are consistent with simple concepts of early and late barriers to reaction and a comparison of the two experiments reinforces the concept of "transient chemical bond formation" in the NO relaxation system. The role, if any, of tetratomic Oxygen complexes as has been calculated from *ab initio* theory[76–78] remains an open question.

Acknowledgments

XY wishes to thank Professor Giacinto Scoles and Professor Kevin Lehmann for their support during the preparation of this article. AMW wishes to thank the Alfred P. Sloan Foundation and the Camille and Henry Dreyfus foundation. This research was partially supported by the following grants: Petroleum Research Fund Grant No. 21762-G6 and No. 24031-AC6, National Science Foundation Presidential Young Investigator Award CHE-8957978 and National Science Foundation Atmospheric Chemistry Division Grant ATM-8922214. In addition, a small grant from the University wide Energy Research Group from the University of California is gratefully acknowledged. This work was also made possible by the Santa Barbara Laser Pool under NSF Grant No. CHE-8411302. Special thanks go to Professor J. William Rich at the Ohio State University for his help with models of vibrational energy transfer as well as many useful discussions.

References

1. The number of volumes dedicated to high resolution spectroscopy is far too large to reference here. Two excellent and representative publications are *Molecules and Radiation,* by Jeffrey I. Steinfeld, (M. I. T. Press, Cambridge Massachusetts, 1985) and *Molecular Structure and Dynamics,* ed. W. H. Flygare, (Prentice Hall Inc., Englewood Cliffs, New Jersey, 1978).

2. *Molecular Spectra And Molecular Structure vol.'s I–IV,* eds. K. P. Huber and G. Herzberg, (Van Nostrand Reinhold Company, New York, 1979).

3. *Molecular Vibrations — The Theory of Infrared and Raman Vibrational Spectra,* eds. E. B. Wilson Jr., J. C. Decius, P. C. Cross, (McGraw Hill Book Company Inc., New York, 1955).

4. A representative collection of recent work can be found in a special issue devoted to stimulated emission pumping; *J. Opt. Soc. Am.* **B7**, (1990).

5. F. F. Crim, *Ann. Rev. Phys. Chem.* **35**, 657 (1984).

6. C. E. Hamilton, J. L. Kinsey, and R. W. Field, *Ann. Rev. Phys. Chem.* **37**, 493 (1986).

7. For a recent review see R. E. Weston Jr. and G. W. Flynn, *Ann. Rev. Phys. Chem.* **43**, 559 (1992).

8. D. M. Neumark, A. M. Wodtke, G. N. Robinson, C. C. Hayden, and Y. T. Lee, ACS Symposium Series No. 263, *Resonances in Electron Molecule Scattering, Van der Waals Complexes and Reactive Chemical Dynamics*, ed. D. G. Truhlar (1984) and references therein.

9. J. H. Parker and G. C. Pimentel, *J. Chem. Phys.* **51**, 91 (1969).

10. R. D. Coombe and G. C. Pimentel, *J. Chem. Phys.* **59**, 251 (1973).

11. T. G. Slanger, L. E. Jusinski, G. Black, and G. E. Gadd, *Science* **241**, 945 (1988).

12. X. Yang, J. M. Price, J. A. Mack, C. G. Morgan, C. A. Rogaski, D. McGuire, E. H. Kim, and A. M. Wodtke, *J. Phys. Chem.* (in Press).

13. F. F. Crim, M. C. Hsiao, J. L. Scott, A. Sinha, and R. L. Vander Wal, *Phil. Trans. R. Soc. Lond. A.* **332**, 259 (1990).

14. F. F. Crim, A. Sinha, M. C. Hsiao, and J. D. Thoemke, *Proceedings of the 24th Jerusalem Symposium in Quantum Chemistry and Biochemistry* (1992).

15. M. C. Hsiao, A. Sinha, and F. F. Crim, *J. Phys. Chem.* **95**, 8263 (1991).

16. A. Sinha, *J. Phys. Chem.* **94**, 4391 (1990).

17. A. Sinha, M. C. Hsiao, and F. F. Crim, *J. Chem. Phys.* **92**, 6333 (1990).

18. A. Sinha, J. D. Thoemke, and F. F. Crim, *J. Chem. Phys.* **96**, 372 (1992).

19. R. L. Vander Wal and F. F. Crim, *J. Phys. Chem.* **93**, 5331 (1989).

20. R. L. Vander Wal, J. L. Scott, and F. F. Crim, *J. Chem. Phys.* **94**, 3548–3555 (1991).

21. R. L. Vander Wal, J. L. Scott, F. F. Crim, K. Weide, and R. Schinke, *J. Chem. Phys.* **95**, 3548 (1991).

22. M. J. Bronikowski, W. R. Simpson, B. Girard, and R. N. Zare, *J. Chem. Phys.* **95**, 8467 (1991).

23. X. Yang, C. A. Rogaski, and A. M. Wodtke, *J. Chem. Phys.* **92**, 2111 (1990).

24. X. Yang, C. A. Rogaski, and A. M. Wodtke, *J. Opt. Soc. Am.* **B7**, 1835 (1990).

25. Y. Chen, D. M. Jonas, J. L. Kinsey, and R. W. Field, *J. Chem. Phys.* **91**, 3976–3987 (1989).

26. Y.-T. Chen, D. M. Watt, and R. W. Field, *J. Chem. Phys.* **93**, 2149–2151 (1990).

27. B. Gazdy and J. M. Bowman, *J. Chem. Phys.* **95**, 6309 (1991).

28. X. Yang and A. M. Wodtke, *J. Chem. Phys.* **92**, 116 (1990).

29. X. Yang, D. McGuire, and A. M. Wodtke, *J. Molec. Spectrosc.* **154**, 361 (1992).

30. X. Yang, E. H. Kim, and A. M. Wodtke, *J. Chem. Phys.* **96**, 5111 (1992).

31. J. T. Yardley, *Introduction to Molecular Energy Transfer* (Academic, New York, 1980).

32. J. C. Stephenson, *J. Chem. Phys.* **59**, 1523 (1973).

33. C. H. Kuo, C. G. Beggs, P. R. Kemper, M. T. Bowers, D. J. Leahy, and R. N. Zare, *Chem. Phys. Lett.* **163**, 291 (1989).

34. Henniger, S. Fenistein, M. Durup-Ferguson, E. E. Ferguson, R. Marx, and G. Mauclaire, *Chem. Phys. Lett.* **131**, 439 (1986).

35. G. Chambaud and P. Rosmus, *Chem. Phys. Lett.* **165**, 429 (1990).

36. R. N. Schwartz, Z. I. Slawsky, and K. F. Hertzfeld, *J. Chem. Phys.* **20**, 1591 (1952).
37. R. N. Schwartz and K. F. Hertzfeld, *J. Chem. Phys.* **22**, 767 (1954).
38. F. I. Tanczos, *J. Chem. Phys.* **25**, 439 (1956).
39. G. M. Jurisch and F. F. Crim, *J. Chem. Phys.* **74**, 4455 (1981).
40. J. M. Robinson, K. J. Rensberger, and F. F. Crim, *J. Chem. Phys.* **84**, 220 (1986).
41. L. S. Dzelzkalns and F. Kaufman, *J. Chem. Phys.* **79**, 3363 (1983).
42. G. D. Billing and L. L. Poulsen, *J. Chem. Phys.* **68**, 5128 (1977).
43. D. Rapp and P. Englander–Golden, *J. Chem. Phys.* **40**, 573 (1964)
44. D. Rapp, *J. Chem. Phys.* **43**, 316 (1965).
45. J. Keck and G. Carrier, *J. Chem. Phys.* **43**, 2284 (1965).
46. X. Yang and A. M. Wodtke, *J. Chem. Phys.* **96**, 5123 (1992).
47. Aa. S. Sudbo and M. M. T. Loy, *J. Chem. Phys.* **76**, 3646 (1983).
48. E. G. Nikitins, *Opt. Spectrosc.* **9**, 8 (1960).
49. C. M. Western, P. R. R. Langridge-Smith, B. J. Howard, and S. E. Novick, *Mol. Phys.* **44**, 145 (1981).
50. S. G. Kukolich, *J. Mol. Spectrosc.* **98**, 80 (1983).
51. P. W. Atkins, *Physical Chemistry* (Oxford University, Oxford, 1978).
52. J. J. Tanner and M. M. Maricq, submitted.
53. G. Ramachandran and G. S. Ezra, *J. Chem. Phys.* **97**, 6322 (1992).
54. E. E. Ferguson, *Comm. At. Mol. Phys.* **24**, 327 (1990).
55. E. E. Ferguson, *Adv. At. Mol. Phys.* **25**, 61 (1988).
56. H. Eyring, S. H. Lin, and S. M. Lin, *Basic Chemical Kinetics* (John Wiley & Sons, 1980).
57. M. Gordon, North Dakota State University (private communication).
58. J. I. Steinfeld and W. Klemperer, *J. Chem. Phys.* **42**, 3475 (1965).
59. R. B. Kerzel and J. I. Steinfeld, *J. Chem. Phys.* **53**, 3293 (1970).
60. F. P. Billingsley II, *J. Chem. Phys.* **62**, 864 (1975).
61. F. P. Billingsley II, *J. Chem. Phys.* **63**, 2267 (1975).
62. *Aeronomy of the Middle Atmosphere*, p. 134, eds. G. Brasseur and S. Solomon, (D. Reidel Publishing Co., Boston, 1984).
63. *Atmos. Ozone Rep.* No. 16 (World Meteorological Organization, Geneva, 1986).
64. T. Kinugawa, T. Sato, T. Arikawa, Y. Matsumi, and M. Kawasaki, *J. Chem. Phys.* **93**, 3289 (1990).
65. H. Park and T. G. Slanger, *J. Chem. Phys.* (in press).
66. P. Fabian, J. A. Pyle, and R. J. Wells, *J. Geophys. Res.* **87**, 4981 (1982).
67. R. Toumi, B. J. Kerridge, and J. A. Pyle, *Nature* **351**, 217 (1991).
68. B. J. Kerridge and E. E. Remsberg, "Satellite measurements from the Limb Infrared Monitor of the Stratosphere" *J. Geophys. Res.* **94**, 16323 (1989).
69. R. Toumi, *J. Atmos. Chemistry* **15**, 69 (1992).
70. X. Yang and A. M. Wodtke, *SPIE Conference Proceedings* (Los Angeles California, 1992).

71. X. Yang, C. A. Rogaski, E. H. Kim, D. McGuire, and A. M. Wodtke, *Lambda Physics Highlights* **24**, 1 (1990).
72. A. M. Wodtke, L. Hüwel, H. Schlüter, H. Voges, G. Meijer, and P. Andresen, *J. Chem. Phys.* **89**, 1929 (1988).
73. D. M. Creek and R. W. Nichols, *Proc. R. Soc. London Ser.* **A341**, 517 (1975).
74. G. D. Billing and R. E. Kolesnick, *Chem. Phys. Lett.* (in press) 1992.
75. *Chemical Kinetic and Photochemical Data Sheets for Atmospheric Reactions* (1980).
76. E. T. Seidl and H. F. Schaefer III, *J. Chem. Phys.* **88**, 7043 (1988).
77. V. Adamantides, D. Neisus, and G. Verhaegen, *Chem. Phys.* **48**, 215 (1980).
78. W. L. Feng and O. Novaro, *Intl. J. Quantum Chem.* **XXVI**, 521 (1984).

CHAPTER 18

STIMULATED EMISSION PUMPING AS A PROBE OF THE OH $(X^2\Pi)$ + Ar INTERMOLECULAR POTENTIAL ENERGY SURFACE

Marsha I. Lester and William H. Green, Jr.[*]

*Department of Chemistry,
University of Pennsylvania,
Philadelphia, Pennsylvania 19104-6323, USA*

Charusita Chakravarty[†] and David C. Clary

*Department of Chemistry,
University of Cambridge,
Lensfield Road, Cambridge CB2 1EW, UK*

Contents

---

[*]Current address: Exxon Research and Engineering Company, P.O. Box 998, Annandale, NJ 08801

[†]Current address: Dept. Chemistry, Indian Institute of Technology-Delhi, Hauz Khas, New Delhi, India 110016

659

1. Introduction

Over the last decade, much has been learned about intermolecular potentials from analysis of the spectroscopy of van der Waals complexes.[1-4] Microwave studies have probed the rotational degrees of freedom in these complexes, yielding primarily structural information. Since these studies are usually limited to the ground vibrational state of the complex, they characterize the potential energy surface only in the vicinity of the potential minimum. More global information on the potential has been gleaned by analyzing the rovibrational spectra (both infrared and far-infrared) of binary complexes. The rovibrational spectra access *excited inter*molecular vibrations, thereby sampling regions of the potential surface which lie far from the equilibrium position. The radial and angular dependence of the potentials is obtained by examining the intermolecular stretching and bending vibrations (in some cases, in combination with intramolecular vibrations). Realistic potential energy surfaces have been derived from this kind of spectroscopic data for systems such as Ar–HCl,[5] Ar–H$_2$,[6] Ne–HF,[7] and Ar–H$_2$O.[8] A compilation of information on intermolecular forces and intramolecular dynamics obtained by spectroscopic probes of intermolecular vibrations in binary complexes has recently been published.[9]

This laboratory has demonstrated that stimulated emission pumping (SEP) can also be used to access the intermolecular vibrational modes in the ground electronic state of the weakly bound complex OH–Ar ($X^2\Pi$). The experimental results have been published[10,11] and are summarized in this chapter. Virtually all of the bound vibrational levels of the complex from the zero-point level to the dissociation limit have been identified·using this technique. In addition, many metastable levels of the OH–Ar complex have been detected as much as 200 cm^{-1} above the OH $(X^2\Pi)$ + Ar dissociation limit. This data can be used to characterize a significant portion of the intermolecular potential energy surface at both long and short range. In this chapter, we derive a global potential energy surface for the average OH $(X^2\Pi)$ + Ar interaction based on the experimental SEP data.

Most of the earlier experimental work on OH–Ar complexes utilized electronic spectroscopy in the region of the OH $A^2\Sigma^+$–$X^2\Pi$ transition to probe the intermolecular potentials between Ar and OH in the ground $X^2\Pi$ and excited $A^2\Sigma^+$ electronic states.[12-22] These studies revealed an enormous change in the OH–Ar potential upon electronic excitation of OH: the binding energy of Ar to OH increases by nearly an order of magnitude and the OH (center-of-mass) to Ar distance decreases by 0.7 Å.[13] The intermolecular bond length change yielded spectroscopic access to many of the excited intermolecular stretching and bending levels supported by the OH $(A^2\Sigma^+)$ + Ar $(^1S_0)$ potential. This experimental data has provided extensive information on the radial and angular dependencies of the intermolecular potential correlating with OH $(A^2\Sigma^+)$ + Ar in the region of the O–H–Ar attractive well.[13,23]

These laser-induced fluorescence measurements provided only limited information about the ground state potential correlating with OH $(X^2\Pi)$ + Ar. All of the OH–Ar features observed in the OH A–X 0–0 and 1–0 spectral regions can be attributed to transitions originating from the lowest intermolecular vibrational level of the ground state potential.[16] Hot band transitions originating from excited intermolecular vibrational levels are not detected, since only the lowest OH–Ar level is populated at the ultralow temperatures achieved in the supersonic expansion. From this data, the ground state potential could be characterized only in the vicinity of the potential minimum. Spectroscopic analysis yielded the vibrationally averaged OH–Ar bond length of the vibrationless level, $r_0 =$ 3.7 Å,[16,18] and a lower limit for the ground state binding energy, $D_0 \geq$ 93 cm^{-1}.[16] Recent high resolution measurements, using electronic[20-22] and microwave[24] spectroscopies, have yielded more precise rotational constants for the lowest vibrational level of the ground electronic state. More global information about the ground state potential has been obtained from the SEP measurements which access excited intermolecular vibrations.[10,11]

Nearly all of the complexes studied in detail have been closed shell ($^1\Sigma$). Open-shell complexes, such as OH–Ar ($X^2\Pi$), are expected to exhibit a variety of new phenomena resulting from the unpaired electron, including first-order Stark effects, Renner–Teller interactions, and spin-orbit predissociation.[25–28] These phenomena provide a new way to examine the poorly understood spin-orbit relaxation process as well as degeneracy-splitting interactions which may influence the orientation of free radical attack in inelastic and reactive encounters. The OH–Ar system provides a test case for investigating the nature of open-shell complexes, as it is both experimentally observable[10–22,24] and theoretically tractable from first principles.[25–31]

Aside from OH–Ar, other open-shell systems studied in depth include Ar–NO, whose rotational spectrum was measured by Howard and coworkers.[32] In addition, the rovibrational spectrum of HF–NO has been examined in the HF stretching region[33] and electronic spectra of NeCN,[34] NeOH,[34] ArNH,[35] and KrOH[36,37] have been reported.[a] A number of theoreticians have been active in developing models capable of predicting the novel behavior of these sorts of complexes.[11,25–31,38] These theoretical models and predictions have proven to be important in understanding the experimental spectra of open-shell complexes, and in extracting the underlying potential energy surface and vibrational dynamics from the spectroscopic observables. Perturbation theory has been useful in elucidating the physical origin of various features observed in OH–Ar ($X^2\Pi$) stimulated emission spectra.[28]

This chapter is organized as follows. In Sec. 2, we review the relevant theoretical work on open-shell complexes which is needed to understand the experimentally observed stimulated emission spectra of OH–Ar ($X^2\Pi$). In Sec. 3, we describe the experimental method used to collect SEP data. Section 4 presents the experimental results on bound states supported by the ground state potential, the OH–Ar $\rightarrow$ OH ($X^2\Pi_{3/2}$) $j = 3/2 + $ Ar ($^1S_0$) dissociation limit, and metastable levels which lie above the dissociation limit and their predissociation dynamics. In Sec. 5, we derive an empirically adjusted intermolecular potential energy surface for the average OH ($X^2\Pi$)

---

[a]The references and discussion were current at the time this chapter was submitted (February 1993). Since that time, a number of additional open-shell complexes have been investigated. For entry into this more recent literature, see *Faraday Discuss. Chem. Soc.* **97** (1994).

+ Ar ($^1S_0$) interaction based on the experimental SEP results and compare it with the *ab initio* potential energy surface.[25] In addition, we summarize the effects which arise from the orientation of the OH radical's unpaired electron with respect to the OH–Ar plane.

2. Theoretical Background

2.1. *Ab Initio Potential*

A high-level *ab initio* calculation of the potential energy surfaces for OH ($X^2\Pi$) + Ar has been carried out by Degli Esposti and Werner.[25] Two different potentials correlate with OH ($X^2\Pi$) + Ar, depending on whether the orbital containing the unpaired electron of the OH radical lies in or out of the OH–Ar plane: the A' and A'' surfaces, respectively. The potentials are degenerate in collinear geometries, but the degeneracy is lifted in bent geometries as the $C_{\infty v}$ symmetry is lowered to C_s. In nonlinear configurations, the A' surface lies lower in energy than the A'' surface. The average ground state potential, $V_\Pi = (V_{A'} + V_{A''})/2$, exhibits a minimum at a linear hydrogen-bonded geometry O–H–Ar with an equilibrium bond length of 3.6 Å and a well depth, D_e, of about 100 cm^{-1}. A shallow secondary minimum occurs at the linear H–O–Ar configuration but has a low barrier for conversion to the O–H–Ar isomer. The average potential for the ground electronic state of OH–Ar is only weakly anisotropic. The difference potential, $V_2 = (V_{A''} - V_{A'})/2$, varies from exactly zero at linear configurations to a maximum of ~ 10 cm^{-1} at angles near 90° for the intermolecular separation distances accessed by the bound levels of the OH–Ar complex. The difference potential is highly anisotropic and increases exponentially as the intermolecular separation distance is reduced.

2.2. *Energy Level Pattern*

The bound rovibrational levels supported by the *ab initio* potential surfaces have been evaluated by Chakravarty and Clary using a variational method.[26,29] Dubernet, Flower, and Hutson have also predicted the bending level pattern for OH–Ar ($X^2\Pi$), based on their general treatment of open-shell complexes with various angular momentum coupling schemes and potential anisotropies.[27]

In OH–Ar ($X^2\Pi$), as in other HX-rare gas complexes (X = F, Cl), the bending states of the complex are best treated as those of a nearly free OH rotor. The anisotropy of the OH–Ar potential is small compared to the spacings between OH rotational levels, $2b_{OH}j$, and does not significantly mix the free rotor states. Therefore, the quantum numbers associated with the OH angular momentum, j, and the projection of the OH angular momentum on the OH axis, ω, are retained as nearly good labels for OH–Ar.[26-28] The presence of the Ar atom, however, does induce a weak perturbation on the OH molecule (a first-order Stark effect), removing the $2j + 1$ orientational degeneracy of each OH rotational level, j, ω.[26,27] This results in a manifold of $2j + 1$ bending levels in the complex, correlating with each j, ω level of OH $(X^2\Pi)$. (See Fig. 2 of Ref. 26 and Fig. 11 of Ref. 27.) This bending level pattern is unique to complexes which have nonzero electronic angular momentum along the diatom axis.[26,27] The bending levels can be described by another nearly good quantum number, P, which is the projection of the OH angular momentum j (as well as the total angular momentum) on the OH (center-of-mass) to Ar axis, R. (P was called K in previous work from this laboratory; here, we adopt the P notation of Dubernet *et al.*[27] for open-shell complexes.) In OH–Ar, P takes on half-integral values constrained by $|P| \leq j$. The energy spread across the manifold of bending levels (or P levels) is directly related to the anisotropy of the average potential V_Π.[26,27] The anisotropy of the OH–Ar potential is sufficiently weak, however, that the spacing between manifolds of bending levels derived from different OH j, ω levels is approximately the same as the energy difference between the j, ω rotational levels in free OH.

As noted by Dubernet, Flower, and Hutson,[27] the relative sign of ω and P yields information on the orientation of the Ar atom with respect to OH. If ω and P, the projections of j on the OH internuclear axis (r) and the OH–Ar axis (R), respectively, have the same sign, then the angle θ between r and R will likely be small, $\theta \leq 90°$, corresponding to OH–Ar configurations. If ω and P have opposite signs, then the most probable angles will be $\theta \geq 90°$ and correspond to HO–Ar geometries.[26,27] Since the *ab initio* calculation has shown the OH–Ar orientation to be more energetically favorable than HO–Ar,[25] the states with $\omega P > 0$ are expected to have lower energies than the states with $\omega P < 0$.

A stack of intermolecular stretching levels $v_s = 0, 1, 2, \ldots$ is also built on each j, ω level of OH. In OH–Ar, the intermolecular stretching interval is greater than the anisotropic splitting of the P levels.[10,11,26,28] Therefore, the bend–stretch combination states of OH–Ar derived from a given j, ω

level of OH ($X^2\Pi$) follow a fairly simple pattern: a manifold of $2j + 1$ bending levels is associated with each degree of stretching excitation.

The approximate quantum number labels j, ω, P, v_s are useful for understanding the energy level pattern of OH–Ar ($X^2\Pi$) and assessing which states may be spectroscopically observable. Only the total angular momentum J, however, is a rigorously good quantum number for the OH–Ar ($X^2\Pi$) system. Each j, ω, P, v_s state is composed of rotational levels with total angular momentum $J = |P|, |P|+1, |P|+2, \ldots$. The rotational levels appear as parity doublets of the form $1/\sqrt{2}\{|J, j, \omega, P\rangle + \epsilon|J, j, -\omega, -P\rangle\}$ where the parity is $\epsilon(-1)^{J-1/2}$. For simplicity in notation, we use the labels J, j, ω, P to refer to the parity doublets, with ω taken to be a positive number.

2.3. *Infrared Selection Rules*

Far-infrared spectra accessing excited intermolecular vibrations and infrared spectra involving the simultaneous excitation of the OH intramolecular and intermolecular vibrational modes of OH–Ar ($X^2\Pi$) have been predicted based on the *ab initio* potential energy surface.[9,10] The transition moment responsible for the infrared transitions in OH–Ar arises primarily from the dipole moment of OH and is therefore assumed to lie along the OH axis. The simulated infrared spectrum shows strong intensity to pure bending levels, following the selection rules $\Delta j = 0, \pm 1$ and $\Delta P = 0, \pm 1$.[26,27] Strongly allowed transitions occur from the lowest intermolecular level of OH–Ar, $j = 3/2$, $\omega = 3/2$, $P = 3/2$, to the $j = 3/2$, $\omega = 3/2$, $P = 1/2$ level at ~ 10 cm^{-1} and to the $j = 5/2$, $\omega = 3/2$, $P = 5/2$, $3/2$, and $1/2$ levels near 90 cm^{-1}. The latter are not predicted to be bound based on the *ab initio* potential energy surface,[25] although the experimental SEP results indicate that at least some of the $j = 5/2$ levels will lie below the dissociation limit.[11] Transitions to bending levels with $j = 3/2$, $\omega = 3/2$, $P = -1/2$ character are expected to be much weaker, as they are allowed only by Coriolis mixing, but should still be strong enough to be experimentally observable.[26] The transition dipole should have a weak dependence on the OH–Ar intermolecular bond length, so intermolecular stretching excitations gain intensity only indirectly through bend–stretch coupling. Neither far-infrared nor near-infrared spectra of OH–Ar ($X^2\Pi$) have been reported to date.

2.4. *Intensities of Stimulated Emission Pumping Transitions*

Stimulated emission pumping is governed by much less strict selection rules than infrared or far-infrared absorption. A theoretical simulation of the stimulated emission spectrum from the $v_s = 4$, $v_b = 0$ level of the OH–Ar $(A^2\Sigma^+)$ state to the bound rovibrational levels of the ground state reveals that SEP can access all of the intermolecular vibrational levels of OH–Ar $(X^2\Pi)$ from the zero-point level to the dissociation limit.[10] The bound levels of OH–Ar were calculated using *ab initio* potentials for both the ground and excited electronic states. The transition intensities were evaluated by assuming that the electronic transition dipole moment lies perpendicular to the OH internuclear axis, as is the case for a Σ–Π transition in free OH. Transitions are predicted to all four of the bending levels correlating with the $j = 3/2$ level of OH $(v_b = 0$–$3)$, as well as bend–stretch combination bands with 1–4 quanta of intermolecular stretch. The relative intensity of intermolecular stretching levels (for a given j, ω, P) in SEP spectra is determined by the Franck–Condon overlap between the stretching wavefunctions of the initial and final states involved in the downward electronic transition. The relative intensity of bending vibrations for a fixed value of v_s depends on a rotation matrix as well as the angular part of the OH–Ar wavefunction in the ground and excited electronic states.[27,29] The calculated spectrum terminates at 65 cm^{-1}, the dissociation limit (D_0) of the *ab initio* potential.[29]

3. Experimental

The experimental method used to produce OH–Ar complexes has been described previously.[11,16] Briefly, OH radicals are created by photolyzing gaseous nitric acid in Ar carrier gas (60 psi) using the 193 nm output of an ArF excimer laser. Photolysis occurs either within or just downstream from a quartz capillary (1.2 cm length; 0.05 mm bore) which is affixed to a pulsed valve. This produces a ~ 17.5 μs pulse of OH radicals within the gas pulse. OH–Ar complexes are formed as the gas mixture undergoes supersonic expansion.

The experimental setup for SEP experiments is shown in Fig. 1. A single Nd:YAG laser pumps two dye lasers denoted as the pump and dump lasers. The dump laser is frequency-doubled in a KDP crystal and delayed ~ 50 ns relative to the pump laser with a multipass optical delay line. The laser beams are then combined using a dichroic mirror, which reflects red light

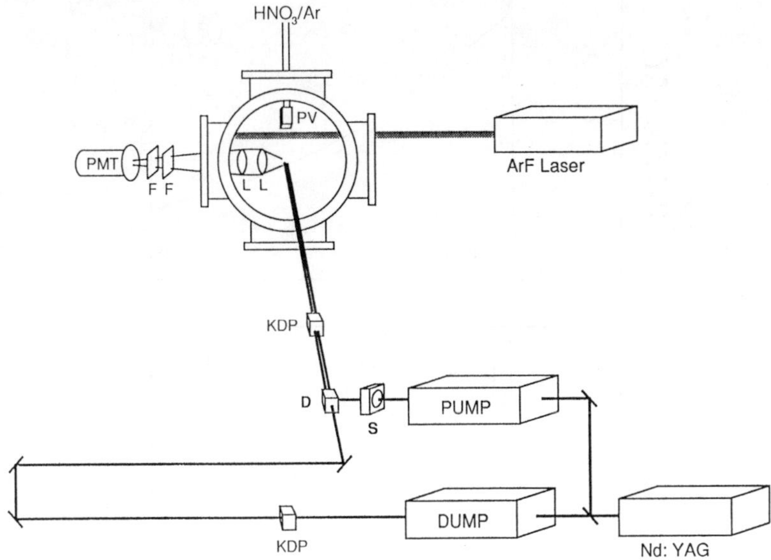

Fig. 1. Experimental apparatus for stimulated emission pumping of OH–Ar. OH–Ar complexes are produced in a pulsed supersonic expansion (PV = pulsed valve) by 193 nm photolysis (ArF laser) of HNO$_3$ entrained in Ar carrier gas. The frequency-doubled (KDP) output of the pump dye laser (Nd:YAG pumped) is blocked on alternate shots by a shutter (S). The output of the dump dye laser, which is pumped by the same Nd:YAG laser, is frequency-doubled (KDP), delayed in an optical delay line, combined with the pump laser in a dichroic mirror (D), and propagated along the same path as the pump laser. The induced fluorescence is collected by lenses (L), wavelength selected by filters (F), and detected with a photomultiplier tube (PMT). Figure has been reproduced from Ref. 11 with permission from the American Institute of Physics.

(pump laser) and passes blue light (dump laser). The copropagating laser beams are sent through an additional KDP crystal to frequency-double the pump laser. The laser beams are mildly focused (2 m focal length) into the vacuum chamber and intersect the gas pulse approximately 1.5 cm downstream of the exit of the quartz capillary tube. A blue-sensitive photomultiplier tube (EMI 9813Q) detects the fluorescence induced by the pump and/or dump lasers after filtering out background light using UG-5 (visible) and WG295 (193 nm) filters. The photomultiplier tube is gated off during the photolysis laser pulse.

The $\sim$ 50 ns delay between the pump and dump laser pulses allows the OH–Ar fluorescence signal ($\tau_{\mathrm{rad}} \sim$ 700 ns)[39,40] to be captured in two

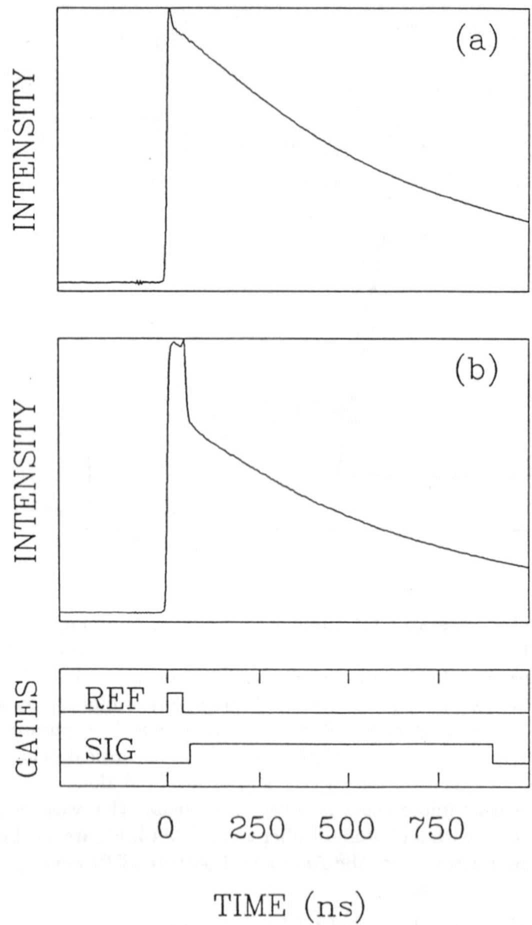

Fig. 2. Illustration of stimulated emission pumping experiment on OH–Ar in the time-domain. The fluorescence decay curves were captured with a digital signal analyzer and are signal averaged over 1024 laser shots. Panel (a) is the time profile of the fluorescence decay when pumping OH–Ar to the $v_s = 5$, $v_b = 0$ level of the excited $A^2\Sigma^+$ state with the dump laser blocked. Panel (b) shows the depletion of fluorescence due to the arrival of the dump laser at 50 ns delay when the dump laser is tuned into resonance with a downward transition. The spikes at 0 and 50 ns are due to laser scatter. The OH–Ar fluorescence signal is captured in two parts. Fluorescence induced solely by the pump laser is collected with a gated integrator at early times (REF), while signals due to the dump laser are collected with a second gated integrator at longer times (SIG), as shown in the lowest panel. Figure has been reproduced from Ref. 11 with permission from the American Institute of Physics.

parts as illustrated in Fig. 2: a reference signal ($t < 50$ ns) which is induced solely by the pump laser (REF) and the signal (SIG1) arising from both the pump and dump lasers ($t > 50$ ns). In addition, a mechanical shutter blocks every other pump laser pulse to provide a signal (SIG2) originating solely from the probe laser. Data is recorded by evaluating the ratio [(SIG1–SIG2)/REF]·100%; the REF signal is scaled to give a 100% baseline value when the probe laser is blocked.

4. Results

4.1. *Bound States*

Stimulated emission spectroscopy has been used to identify the intermolecular vibrational levels supported by the ground state potential, as well as many metastable levels which lie above the dissociation limit. A spectrum obtained when the pump laser prepares OH–Ar in the $v_s = 5$, $v_b = 0$ level of the $A^2\Sigma^+$ state with zero quanta of OH stretch ($\lambda_1 = 32\ 371.4$ cm^{-1}) and the dump laser (λ_2) is scanned from the wave number of the pump laser towards lower wave numbers is shown in Fig. 3. SEP spectra are displayed as a function of the difference in the wave numbers of the pump and dump lasers, which corresponds to the vibrational energy of OH–Ar ($X^2\Pi$). The energy region from the zero-point level (0 cm^{-1}) through 80 cm^{-1} is displayed in Fig. 3. SEP spectra have also been recorded with the pump laser exciting the OH–Ar complex to intermolecular levels with 3 or 4 quanta of stretch ($v_b = 0$) as well as bend–stretch combination levels in the excited electronic state.[10,11]

In addition to the fluorescence dips arising from stimulated emission, dips may also be observed due to hole-burning. As illustrated in Fig. 3, the pump (or "hole-burning") laser at λ_1 induces a transition from the lowest intermolecular level of the ground electronic state, $|a''\rangle$, to a specific intermolecular vibrational level of the excited electronic state, $|b'\rangle$, thereby significantly depleting the population of state $|a''\rangle$. When the dump laser (λ_2) is resonant with an upward transition from $|a''\rangle$ to $|a'\rangle$, as shown with a dashed arrow in Fig. 3, it will detect the depleted population in the $|a''\rangle$ state. The dump laser (λ_2) can excite fewer molecules from $|a''\rangle$ to $|a'\rangle$, and therefore induce less fluorescence, when λ_2 is preceded by the pump laser (λ_1) than when λ_1 is blocked. The fluorescence signal induced by λ_2 alone, SIG2, is greater than the *increase* in the fluorescence signal due to λ_2 when preceded by λ_1, SIG1–REF. Therefore, the ratio [(SIG1–SIG2)/REF]·100%

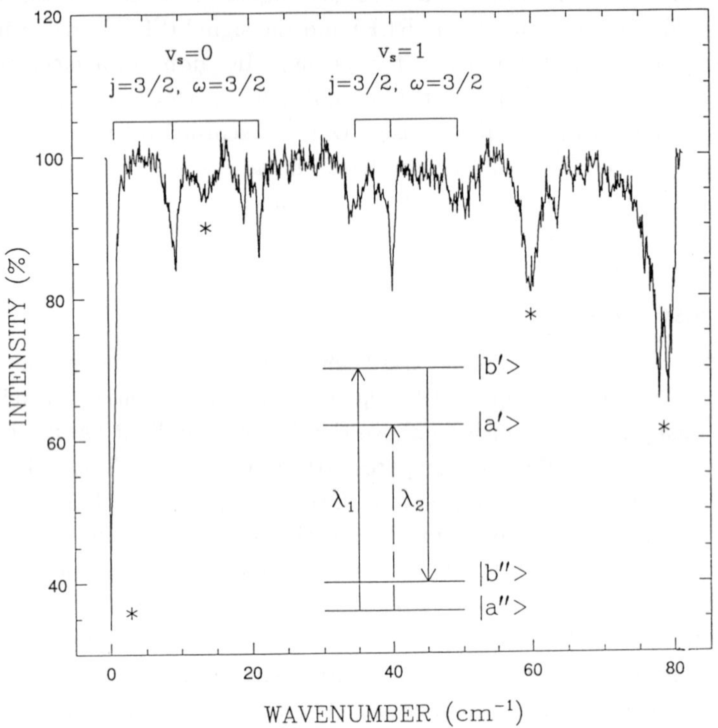

Fig. 3. Stimulated emission pumping spectrum of OH–Ar $(X^2\Pi)$ recorded over the 0–80 cm^{-1} energy range. Spectra are plotted as a function of the difference in the wave numbers of the pump and dump lasers. The pump laser $(\lambda_1 = 32\ 371.4.\ cm^{-1})$ promotes the OH–Ar complex from the lowest intermolecular level in the ground $X^2\Pi$ electronic state to the $v_s = 5$, $v_b = 0$ level correlating with the OH $A^2\Sigma^+$ $(v = 0)$ + Ar as the dump laser λ_2 is scanned. As shown schematically in the energy level diagram, the dump laser λ_2 can induce downward (solid arrow) or upward (dashed arrow) transitions, corresponding to stimulated emission and absorption, respectively. As explained in the text, these absorptions give rise to dips in the spectra due to hole-burning effects (labeled with asterisks). Intermolecular bending levels of OH–Ar derived from the lowest rotational level of OH $(X^2\Pi_{3/2})$, $j = 3/2$, $\omega = 3/2$, are observed with zero and one quanta of intermolecular stretching excitation $(v_s = 0\ ,1)$.

will decrease from its 100% baseline level when the dump laser is scanned over an upward transition originating from $|a''\rangle$.

Dips due to SEP can be readily distinguished from those due to hole-burning by changing the upper state excited by the pump laser (λ_1). The hole-burning dips will remain at the same absolute wave number of the dump laser, while the dips due to stimulated emission will shift to a dump laser wave number which keeps the *difference* between the pump and dump laser wave numbers constant. The four dips in Fig. 3 labeled with asterisks (*) are attributed to hole-burning.[41] These features obscure portions of the SEP spectra; by pumping different excited state levels, the SEP features can be shifted in absolute wave numbers to avoid overlap with hole-burning features.[10,11]

A catalog of the OH–Ar vibrational energy levels from 0 to 89 cm^{-1} and their j, ω, v_s assignments are given in Table I. Assignments were made by comparison with the energy level pattern predicted by bound state calculations based on the *ab initio* potential for the ground electronic state[26] and an empirically adjusted potential described in Sec. 5. The first group of levels at 0, 9.7, 19.2, and 21.3 cm^{-1} (see Fig. 3) correlate with the lowest rotational level of OH, $j = 3/2$, $\omega = 3/2$. Using the nearly good quantum numbers described in Sec. 2, these levels are assigned as the $P = 3/2, 1/2, -1/2$, and $-3/2$ states. The manifold of $j = 3/2$, $\omega = 3/2$ bending levels is repeated at higher energy with one quantum of intermolecular stretching excitation, $v_s = 1$ (Fig. 3). The $j = 3/2$, $\omega = 3/2$ manifolds with $v_s = 2$ and 3, which are obscured by hole-burning features in Fig. 3, are readily identified in SEP spectra recorded when pumping the $v_s = 4$, $v_b = 0$ level of the $A^2\Sigma^+$ excited electronic state at 32 291.8 cm^{-1}.[11] The spacings between the manifolds of $j = 3/2$, $\omega = 3/2$ bending levels decreases with v_s due to anharmonicity in the intermolecular stretching vibration. The energy spread across the manifold of bending levels also decreases with increasing v_s as a result of bend–stretch coupling. The nearly good quantum numbers become less exact labels with increasing energy due to a variety of perturbations which couple levels of different j, ω, P, v_s.[11]

Given the weak anisotropy of the OH–Ar ($X^2\Pi$) average potential, the spacing between the manifold of $j = 5/2$, $\omega = 3/2$ bending levels ($v_s = 0$) and the manifold of $j = 3/2$, $\omega = 3/2$ bending levels ($v_s = 0$) was expected to be approximately equal to 83.7 cm^{-1}, the energy separation between the $j = 3/2$ and 5/2 rotational levels in free OH ($^2\Pi_{3/2}$). The baricenter of the pure bending levels which correlate with $j = 3/2$, $\omega = 3/2$ occurs at 12.6 cm^{-1}, so the baricenter of the $j = 5/2$, $\omega = 3/2$ bending manifold was anticipated at 96.3 cm^{-1}. The features which appear just above and

Table I. Comparison of intermolecular vibrational energy levels of OH–Ar ($X^2\Pi$) observed experimentally by stimulated emission pumping with bound states computed variationally based on *ab initio*[a] and empirically adjusted potentials for OH ($X^2\Pi$) + Ar.

Assignment ($\omega = 3/2$)		Experimental[b]	*Ab Initio* Potential[c]	Adjusted Potential[d]
j	v_s	(cm^{-1})	(cm^{-1})	(cm^{-1})
3/2	0	0.0	0.0	0.0
		9.7±0.2	9.2	9.5
		19.2±0.2	16.6	19.3
		21.3±0.2	17.2	21.3
3/2	1	34.9±0.2	27.9	34.7
		41 ±1	33.9	42.1
		50 ±3[e]	39.1	48.9
		50 ±3[e]	39.6	50.6
3/2	2	60 ±1	47.2	60.2
		64 ±1	51.0	65.9
		71 ±1[e]	54.1	70.1
		71 ±1[e]	54.5	71.4
3/2	3	78 ±1	58.7	77.3
		81 ±1	61.1	81.2
		85 ±1[e]	62.2	83.3
		85 ±1[e]	62.6	84.2
5/2	0	89 ±1	–[f]	87.5

[a]Ref. 25.
[b]Refs. 10, 11.
[c]Refs. 10, 26.
[d]Computed for $J = 2.5$ with positive parity.
[e]Two of the $2j + 1$ bending levels of OH–Ar have not been resolved due to overlapping rotational structure of the bands.
[f]$j = 5/2$ levels are not predicted to be bound.

below this energy at 89, 94, 97, and 103 cm^{-1} (Fig. 4) are likely part of the manifold of six bending levels correlating with $j = 5/2$, $\omega = 3/2$ ($v_s = 0$).

The breadth of most of the SEP features appearing below the dissociation limit are ~ 1 cm^{-1}, due to unresolved rotational structure. Some features, such as those in the 50 cm^{-1} region, also have overlapping P

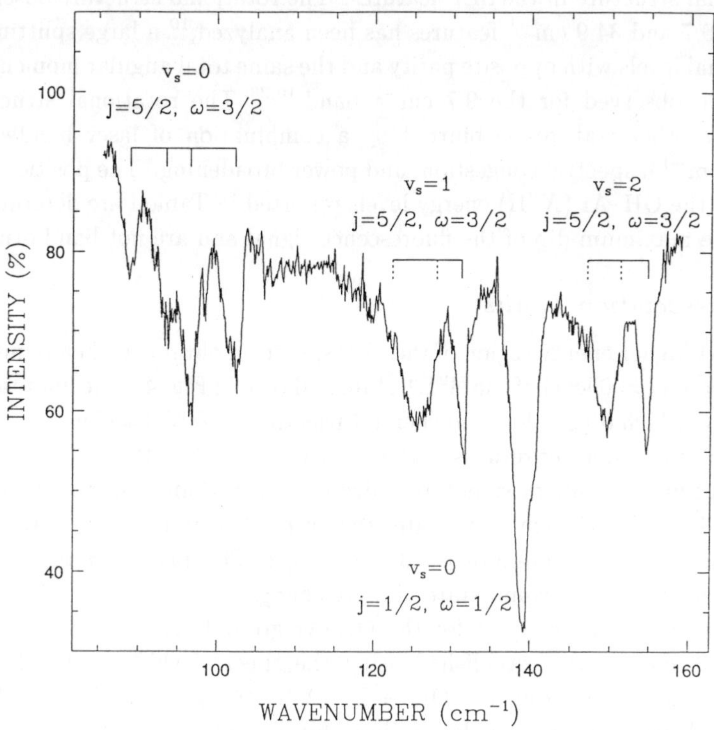

Fig. 4. Stimulated emission pumping spectrum of OH–Ar ($X^2\Pi$) in the 80–160 cm^{-1} energy range following preparation of OH–Ar complexes in the excited $A^2\Sigma^+$ electronic state with five quanta of intermolecular stretching excitation at 32 371.4 cm^{-1}. The dump laser induces downward transitions to intermolecular levels which lie beyond the dissociation limit (100 cm^{-1}). OH–Ar complexes prepared in these levels are metastable and may undergo predissociation by utilizing OH rotational or spin-orbit energy to break the OH–Ar bond. Bending levels correlating with the $j = 5/2, \omega = 3/2$ rotational level of OH ($X^2\Pi_{3/2}$) and containing 0–2 quanta of intermolecular stretch are identified. Some of these dips are broadened (dashed labels indicate energy range) due to rapid rotational predissociation. The intense dip at 139 cm^{-1} is assigned as a $j = 1/2, \omega = 1/2$ level ($v_s = 0$) correlating with the lowest rotational level of the spin-orbit excited state of OH ($^2\Pi_{1/2}$).

bands. In most experiments, the pump laser is fixed on the Q_1 bandhead of an upward transition, exciting several rotational levels (typically $N = 1$–5). The allowed downward transitions (following $\Delta J = 0, \pm 1$ selection rules) from the cluster of rotational levels prepared by the pump laser yield rotational structure in the SEP features. The rotational structure observed for the 9.7 and 34.9 cm^{-1} features has been analyzed;[10] a large splitting of rotational levels with opposite parity and the same total angular momentum has been observed for the 9.7 cm^{-1} band.[10,28] The rotational structure of many other features is blurred by a combination of laser bandwidth (~ 0.1 cm^{-1}), spectral congestion, and power broadening. The positions for most of the OH–Ar ($X^2\Pi$) energy levels reported in Table I are determined from the maximum dip of the fluorescence signal and are not band origins.

4.2. *Dissociation Limits*

The next higher energy region of the SEP spectrum obtained when pumping the $v_s = 5$, $v_b = 0$ level of the $A^2\Sigma^+$ state is shown in Fig. 4. The fluoresence intensity, which dips below and then returns to the 100% baseline from 0 to 89 cm^{-1} (Fig. 3), never returns to the 100% baseline after 103 cm^{-1} (Fig. 4). The continuous, though structured, decrease in the fluorescence intensity observed beyond 103 cm^{-1} indicates the onset of bound-free transitions to the dissociative continuum of $j = 3/2$, $\omega = 3/2$. This places an upper limit of 103 cm^{-1} on the ground state binding energy.

Previously, a lower limit for the OH–Ar ground state binding energy, $D_0 \geq 93$ cm^{-1}, was established using the spectral shift of the OH–Ar electronic spectrum from the OH $A^2\Sigma^+$–$X^2\Pi$ transition and the OH–Ar ($A^2\Sigma^+$) binding energy; a lower limit for the OH–Ar ($A^2\Sigma^+$) binding energy of greater than or equal to 742 cm^{-1} was determined from the position of the last observed bound level (denoted as feature D in previous work) above the zero-point level in the excited electronic state.[16] The SEP data now brackets the OH–Ar ($X^2\Pi$) binding energy D_0 between 93 and 103 cm^{-1}. Thus, the $j = 5/2$, $\omega = 3/2$ level at 89 cm^{-1} should be bound, while the 94, 97, and 103 cm^{-1} members of the $j = 5/2$, $\omega = 3/2$ bending manifold ($v_s = 0$) may lie above the dissociation limit. Rotationally resolved spectra of the $j = 5/2$, $\omega = 5/2$ bands ($v_s = 0$) may permit a more precise determination of D_0, particularly if homogeneous lifetime broadening arising from rotational predissociation can be observed for those rovibrational levels which lie above the dissociation limit.

4.3. *Metastable Levels*

Much structure appears in the continuum region of the SEP spectrum above the dissociation limit. The features appearing beyond 103 cm^{-1}

are metastable levels of OH–Ar. A progression in the intermolecular stretching mode built on the $j = 5/2$, $\omega = 3/2$ bending manifold is identified in Fig. 4. Broadened bands (FWHM ~ 10 cm^{-1}) centered at 125 and 149 cm^{-1} contain one and two quanta of intermolecular stretch, respectively. The discrete bending level structure observed for $j = 5/2$, $\omega = 3/2$, $v_s = 0$ (baricenter at 96 cm^{-1}) is lost upon v_s excitation. The $j = 5/2$, $\omega = 3/2$ levels with intermolecular stretching excitation lie above the dissociation threshold (93 cm$^{-1} \leq D_0 \leq 103$ cm^{-1}) and may undergo rotational predissociation to OH ($^2\Pi_{3/2}$) $j = 3/2$ + Ar. Direct rotational predissociation is induced by the anisotropy in the average interaction potential which couples states with different OH angular momenta, j, in a given spin-orbit state, ω.[5,6] Rapid rotational predissociation could cause sufficient line broadening to merge together several members of the bending manifold. This suggests that the rotational levels in these bands are homogeneously lifetime broadened, beyond the laser bandwidth of ~ 0.1 cm^{-1}, corresponding to rotational predissociation lifetimes of 50 ps or faster.

The direct coupling to the $j = 3/2$ continuum vanishes for some of the $j = 5/2$, $\omega = 3/2$ bending levels.[28] $|P| = 5/2$ states may only dissociate through second-order coupling mediated by the Coriolis term which mixes different P states. Therefore, the $|P| = 5/2$ states are predicted to predissociate at least 100 times more slowly.[28] This could leave $|P| = 5/2$ levels sharp, while the other bending levels ($|P| = 3/2$ and $1/2$) exhibit noticeable linebroadening in stimulated emission pumping (SEP) spectra. Such P-dependent rotational predissociation can explain the sharp dips observed at 131 and 154 cm^{-1} on the high energy side of the broadened $j = 5/2$, $\omega = 3/2$ features in the experimental SEP spectra (Fig. 4).[11]

The physical picture for P-dependent rotational predissociation in OH–Ar is much the same as that in closed-shell complexes,[42,43] although the additional angular momenta in the open-shell system due to the unpaired electron make the problem more complicated. Roughly speaking, when $|P| = j$, the OH radical (actually the H atom) rotates about the O–Ar axis. When $|P| = 1/2$, the OH radical rotates in the OH–Ar plane. Other values of $|P|$ lie between these two extremes. OH rotation in the OH–Ar plane is much more strongly coupled to the OH–Ar dissociation coordinate than when the OH radical rotates about the O–Ar axis ($|P| = 5/2$).

The intense dip at 139 cm^{-1} (FWHM 3.5 cm^{-1}) is assigned as one of the $j = 1/2$, $\omega = 1/2$ levels ($v_s = 0$) correlating with the lowest rotational level of the spin-orbit excited state of OH ($^2\Pi_{1/2}$). If the spin-orbit coupling constant of OH ($a = -139$ cm^{-1}) is not changed upon complexation with Ar, the lowest level of the OH–Ar spin-orbit excited state would be expected to appear at 139 cm^{-1}.[10,44] The fluorescence signal is depleted by more than 50% when the dump laser stimulates emission to the 139 cm^{-1} level of the spin-orbit excited state. The magnitude of the fluorescence dip shows that more than 50% of the OH–Ar complexes in the initially prepared level have been transfered to the 139 cm^{-1} level by the dump laser. This can only occur if the lower level is short-lived, with a decay time less than or equal to the laser pulse width, 6 ns. OH–Ar complexes prepared in the 139 cm^{-1} level and all other levels with $\omega = 1/2$ lie above the threshold for dissociation to OH ($^2\Pi_{3/2}$) $j = 3/2$ + Ar. The $\omega = 1/2$ levels may predissociate via spin-orbit relaxation, yielding OH fragments in the lower spin-orbit state. The difference potential (V_2) provides the direct coupling between the metastable $j = 1/2$, $\omega = 1/2$ spin-orbit excited levels and the dissociative continuum of the lower spin-orbit manifold ($\omega = 3/2$).[28,45]

5. Discussion

5.1. *Intermolecular Vibrational Levels of OH–Ar ($X^2\Pi$)*

The intermolecular bending and stretching vibrational levels of OH–Ar correlating with OH ($X^2\Pi_{3/2}$) + Ar which have been observed by SEP are shown on an energy level diagram in Fig. 5. The energy levels are displayed as discrete lines when experimentally resolved and as shaded regions when broad dips are observed. The dissociation threshold for producing OH fragments in the lowest $j = 3/2$ rotational level of the $^2\Pi_{3/2}$ ($\omega = 3/2$) state is indicated. The bound states lie between 0 and approximately 100 cm^{-1}. OH–Ar complexes prepared in rovibrational levels which lie beyond the dissociation limit are metastable, as they contain sufficient energy to dissociate to OH ($^2\Pi_{3/2}$) $j = 3/2$ + Ar fragments. The rate of predissociation, however, depends on the efficiency with which the excess energy is coupled to the OH–Ar dissociation coordinate. Fortunately, rotational and spin-orbit predissociation of OH–Ar is sufficiently slow to allow observation (at a laser resolution of ~ 0.1 cm^{-1}) of many metastable levels in the energy range from 100 to 300 cm^{-1} via stimulated emission spectroscopy.[11]

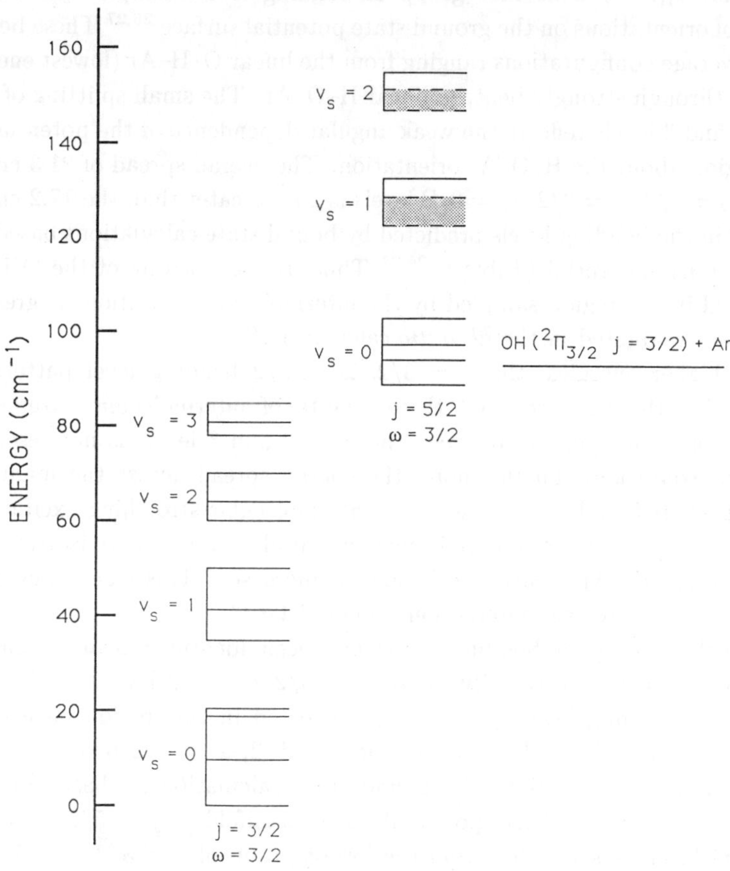

Fig. 5. Intermolecular vibrational energy levels of OH–Ar correlating with OH $^2\Pi_{3/2}$ ($\omega = 3/2$) which have been observed via stimulated emission spectroscopy from 0–160 cm^{-1}. The energy levels are stacked according to their j, ω assignment. The bending level structure is identified by discrete lines when experimentally resolved and by shaded regions when broadened dips have been observed. The OH–Ar dissociation limit, producing OH $X^2\Pi_{3/2}$ ($v = 0$) $j = 3/2$ + Ar fragments, occurs between 93 and 103 cm^{-1}.

The lowest four levels are a manifold of bending levels ($v_b = 0$–3) derived from the $j = 3/2$, $\omega = 3/2$ rotational level of OH ($^2\Pi_{3/2}$) with zero quanta of intermolecular stretch, $v_s = 0$. The energy spread across the bending levels is determined by the anisotropy of the average OH–Ar intermolecular potential V_Π.[26,27] This first group of bending levels samples the entire range of orientations on the ground state potential surface.[26,27] These bends have average configurations ranging from the linear O–H–Ar (lowest energy state), through strongly bent, to linear H–O–Ar. The small splitting of the $v_b = 2$ and 3 levels reflects the weak angular dependence of the potential in the region about the H–O–Ar orientation. The overall spread of 21.3 cm^{-1} in the $j = 3/2$, $\omega = 3/2$, $v_s = 0$ P levels is 20% greater than the 17.2 cm^{-1} spread in the bending levels predicted by bound state calculations based on the *ab initio* potential (Table I).[26,27] Thus, the anisotropy of the OH–Ar potential in the region sampled by the intermolecular vibrations is greater than that evaluated in the *ab initio* calculation.[25]

At higher energies, the $j = 3/2$, $\omega = 3/2$ bending level pattern is repeated with one, two, and three quanta of intermolecular stretching excitation. The splitting between the $v_b = 2$ and 3 levels is not resolved with v_s excitation. Furthermore, the energy spread across the manifold of P levels is found to decrease with intermolecular stretching excitation. This indicates that the intermolecular potential becomes more isotropic as the average OH–Ar separation distance is increased. This trend is evident in the *ab initio* average interaction potential V_Π.[25]

Another group of bending levels has been identified near 96 cm^{-1}, correlating to the rotationally excited $j = 5/2$, $\omega = 3/2$ level of OH. Four of the six bending levels ($v_b = 4$–10) expected in this region have been observed. The 14 cm^{-1} spread in the $j = 5/2$, $\omega = 3/2$ bending levels is similar to that predicted by bound state calculations,[26] based on the anisotropy of the *ab initio* potential. The manifold of $j = 5/2, \omega = 3/2$ bending levels is separated from the lowest group of $j = 3/2, \omega = 3/2$ P states by approximately the $j = 3/2 \to j = 5/2$ energy spacing in free OH ($^2\Pi_{3/2}$), which is 83.7 cm^{-1}.[44] Thus, the barrier to OH rotation in the OH–Ar potential is much less than the OH rotational spacing, leaving j a nearly good quantum number label for OH–Ar.

The intermolecular stretching levels v_s are observed at $\sim 20\%$ higher vibrational energies than predicted by the *ab initio* potential (Table I). For example, the first excited intermolecular stretch appears at 34.9 cm^{-1} versus the 27.9 cm^{-1} calculated value and the $v_s = 2$, $v_b = 0$ level is found at 60 cm^{-1} as compared to the 47.2 cm^{-1} theoretical value. Furthermore, the bound state calculations place D_0 at 65 cm^{-1},[25,26] more than 30% below the experimental binding energy[11] which has been bracketed between 93 and 103 cm^{-1}. The underestimation of both the intermolecular stretching frequency and the binding energy indicates that OH–Ar potential is changing more strongly in the radial coordinate (OH center-of-mass to Ar separation distance) over the bound state region than predicted by the *ab initio* calculation.[25]

5.2. *Refinement of the Average OH ($X^2\Pi$) + Ar Interaction Potential*

The intermolecular energy levels of OH–Ar ($X^2\Pi$) observed by SEP are compared with the bound states calculated based on the *ab initio* potential[25,26] in Table I. The differences between the calculated and experimental values suggested that the potential could be adjusted to better represent the experimentally observed energy levels.[b] This task is more complicated for OH–Ar ($X^2\Pi$) than for closed-shell systems[1,5–8] since two potential energy surfaces correlate with OH ($X^2\Pi$) + Ar. The A' and A'' surfaces differ in the orientation of the unpaired electron of the OH radical with respect to the OH–Ar plane. Previous theoretical calculations have shown that the energies of the intermolecular vibrational levels were determined essentially by the average potential, $V_\Pi = (V_{A'} + V_{A''})/2$, and the effect of the difference potential, $V_2 = (V_{A''} - V_{A'})/2$, was to induce parity splittings in the rotational level structure.[26,28] Since our goal was to improve the fit of calculated to experimental values for the bound intermolecular bending and stretching levels, we have varied only the average V_Π potential and held the difference potential fixed at its *ab initio* value.

A nonlinear least-squares fitting procedure has been carried out to construct a more accurate average intermolecular potential for OH ($X^2\Pi$) + Ar. An iterative procedure is required since no direct algorithm exists for potential inversion.[1] In each iteration, the bound rovibrational levels of

---

[b]Since submission of this paper (February 1993), Dubernet and Hutson have also developed potential energy surfaces for the ground state of Ar–OH, M. L. Dubernet and J. M. Hutson, *J. Chem. Phys.* **99**, 7477 (1993), based on the stimulated emission pumping results from this laboratory.[10,11] The fitting procedure of Dubernet and Hutson differs from that described in this chapter since they attempted to determine V_Π and V_2 simultaneously. The V_Π potential derived by Dubernet and Hutson, which has a well depth of -126 cm^{-1} in a near-linear Ar–O-H geometry and a barrier to internal rotation of 44 cm^{-1}, is in good accord with out fitted potential. They obtain binding energy of 95.5 cm^{-1} from their fit, which is slightly greater than the 93.2 cm^{-1} binding energy determined here.

OH–Ar were calculated using a variational method; the theoretical methods have been thoroughly described elsewhere.[26,29] The bound states computed for the trial potential were then compared with the experimental data and the potential was adjusted accordingly. The *ab initio* average potential was used as the initial guess for the fitting procedure.

The *ab initio* potential was originally parameterized in a functional form that was found to be too unwieldy for a least-squares fitting procedure. Therefore, the *ab initio* data points were refit to a simpler and more theoretically tractable form.[1,46] The average OH $(X^2\Pi)$ + Ar potential $V_\Pi(R,\theta)$ was expressed in a functional form composed of repulsion $(V_{\rm rep})$, dispersion $(V_{\rm disp})$, and induction $(V_{\rm ind})$ terms:

$$V_\Pi(R,\theta) = V_{\rm rep}(R,\theta) + V_{\rm disp}(R,\theta) + V_{\rm ind}(R,\theta) \,, \tag{1}$$

with

$$V_{\rm rep}(R,\theta) = \sum_{l=0}^{3} A_l P_l(\cos\theta) e^{-\beta R} \,, \tag{2}$$

$$V_{\rm disp}(R,\theta) = -U \sum_{n=6}^{8} \frac{G_n(\theta) D_n(R,\beta)}{R^n} \,, \tag{3}$$

$$V_{\rm ind}(R,\theta) = -\sum_{n=6}^{8} \frac{F_n(\theta)}{R^n} \,. \tag{4}$$

The angular dependence of $V_{\rm rep}$ is represented in terms of Legendre polynomials, $P_l(\cos\theta)$. The induction and dispersion coefficients, $F_n(\theta)$ and $G_n(\theta)$, are related to the electrical properties of OH and Ar. Expressions for $F_n(\theta)$ and $G_n(\theta)$ have been derived following Buckingham.[46] All of the electrical properties necessary to evaluate these constants were calculated by Degli Esposti and Werner;[25] the dipole moment of OH, however, was taken from the experiment.[47] The damping function $D_n(R,\beta)$ ensures that the dispersion energy does not overwhelm the repulsion term at short range; the Tang and Toennies damping function $D_n(R,\beta)$ was used.[48] Following Hutson,[1] no damping function was used for the induction terms.

The parameters A_l, β, and U were varied in the least squares fits. At least four terms were needed in the expansion for the repulsion, A_0–A_3, in order to fit the *ab initio* potential satisfactorily. The most important attractive term was found to be the isotropic dispersion. Induction was signficant only for the higher order anisotropic terms. The potential parameters determined in the least squares fit of the *ab initio* data were used as the starting point in deriving a more accurate potential energy surface which incorporates all of the new spectroscopic information obtained from experiment.

The same six potential parameters were then varied in order to derive an empirically adjusted OH ($X^2\Pi$) + Ar average potential which gives better agreement between the calculated and experimentally observed energy levels. In this least-squares fitting procedure, the variational calculation was inserted in a loop which varied the potential parameters. The data utilized for the fit were the vibrational energies of the assigned intermolecular energy levels from 0–89 cm^{-1} (Table I). In addition, the rotor constants for the 0.0, 9.7, and 34.9 cm^{-1} bands, 0.1023 ± 0.0010 cm^{-1},[21,24] 0.105 ± 0.010 cm^{-1},[10] and 0.099 ± 0.006 cm^{-1},[10] respectively, were included in the fit. The rotor constants were evaluated by taking the expectation value of $1/(2\mu R^2)$. The experimental data were weighted in the least-squares fit by the inverse of their experimental uncertainties. The potential parameters determined in the least-squares fit are listed in Table II. Further information is available from the authors upon request.

Table II. Potential Parameters for OH–Ar ($X^2\Pi$).

A_0, Hartree	515.6
A_1, Hartree	122.8
A_2, Hartree	91.2
A_3, Hartree	172.3
β, a_0^{-1}	2.054
U, Hartree	0.427

The average intermolecular potential for OH ($X^2\Pi$) + Ar obtained in the least-square fitting procedure is shown as a contour plot in Fig. 6. Contours are displayed at 10 cm^{-1} intervals relative to the absolute minimum at -127 cm^{-1}. The potential minimum occurs at an OH (center-of-mass) to Ar separation (R) of 3.7 Å in the linear O–H–Ar geometry ($\theta = 0°$). The barrier to internal rotation is 45 cm^{-1} and is located at $R = 3.7$ Å and $\theta \sim 120°$. A secondary minimum with a well depth of -113 cm^{-1} is found at the linear H–O–Ar configuration

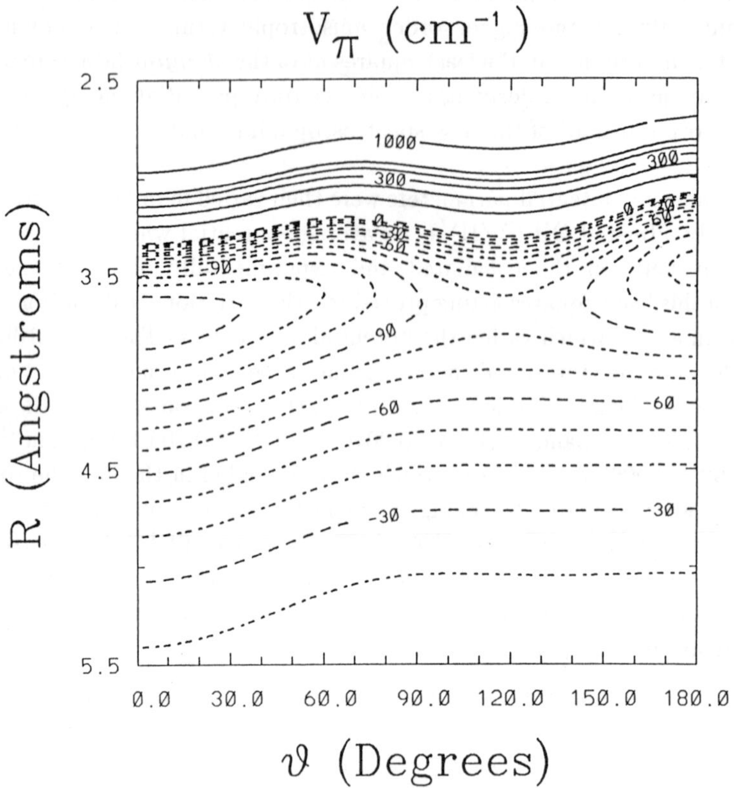

Fig. 6. Contour plot of the average intermolecular potential for OH $(X^2\Pi)$ + Ar derived in a least-square fitting procedure from the experimentally observed vibrational energy levels from 0–89 cm^{-1}. In the attractive well region (dashed contour lines), contours are shown at 10 cm^{-1} intervals relative to the absolute minimum at -127 cm^{-1}. Along the repulsive wall (solid contour lines), contours are drawn at 100 cm^{-1} intervals from 0–500 cm^{-1}. $\theta = 0°$ corresponds the linear O–H–Ar configuration.

$(\theta = 180°)$ at a slightly shorter bond length of 3.5 Å. The bound states supported by this empirically adjusted potential are listed in Table I. The energies of the bound states are in very good agreement with the positions of the experimentally observed features (Table I). The calculated vibrational energies agree within $\sim 2\%$ of the experimental values.

The lowest intermolecular energy level is computed to have a binding energy of 93.2 cm^{-1} with respect to the OH $(X^2\Pi_{3/2})$ $j = 3/2 +$

Ar asymptote. It is characterized by a rotor constant of 0.1021 cm^{-1}, $v_s = 0$ (0.99), $v_b = 0$ (0.99), and a wide amplitude bending motion with $\cos^{-1}\langle\cos\theta\rangle = 51°$. The calculated rotor constant is in good agreement with that obtained by high resolution electronic[20,21] and microwave[22] spectroscopies.

The magnitude of the calculated binding energy is within the experimental limits, 93 cm^{-1} $\leq D_0 \leq$ 103 cm^{-1}, deduced from electronic spectroscopy[16] and the onset of bound-free transitions in stimulated emission pumping spectra,[11] but it is much greater than the 65 cm^{-1} predicted from an *ab initio* calculation of the OH($X^2\Pi$) + Ar potential.[26] The well depth of the empirically adjusted potential is $\sim$ 25% greater than the *ab initio* potential. A closer comparison of the *ab initio* and least-squares derived potentials reveals that the *ab initio* isotropic attraction is too weak by $\sim$ 50%. Smaller, but significant, discrepancies are found in the anisotropic terms. The fitted potential exhibits a deeper secondary minimum at $\theta = 180°$ than the *ab initio* surface.

5.3. *OH ($X^2\Pi$) + Ar Difference Potential*

The electronic state degeneracy in OH ($^2\Pi$) is lifted in nonlinear configurations of OH–Ar due to an electrostatic interaction with the argon atom. The intermolecular potential splits into two components, the A' and A'' surfaces, depending on the whether the orbital containing the unpaired electron on OH lies in or out of the OH–Ar plane. This degeneracy-splitting interaction and subsequent vibronic coupling is known as the Renner–Teller effect in chemically bound species.[49–51] The difference potential, $V_2 = (V_{A''} - V_{A'})/2$, is clearly implicated in two important phenomena in OH–Ar ($X^2\Pi$): parity splitting of rotational levels and spin-orbit predissociation. Both of these effects have been experimentally observed by stimulated emission spectroscopy.

5.3.1. *Parity splitting of rotational levels*

In OH–Ar ($X^2\Pi$), the difference potential (V_2) was predicted to induce a splitting between rotational levels with the same total angular momentum but opposite parity.[26,27] Theoretical calculations based on the *ab initio* potential predicted the largest parity splitting for bending levels with $|P| =$ 1/2. Experimentally, the first excited bending level at 9.7 cm^{-1}, assigned as $j = 3/2, \omega = 3/2, P = 1/2, v_s = 0$, exhibited a noticeable parity splitting.[10]

The spacings between parity doublets are quite large, comparable to the energy spacings between rotational levels of the complex. This leads to a gross shift of the Q_1 branch relative to the P_1 and R_1 branches in stimulated emission spectra (see Fig. 3 of Ref. 11) since Q_1 transitions always terminate on the lower energy component of the parity doublet while P_1 and R_1 transitions always terminate on the higher energy component. The parity splitting at $J = 3/2$ was determined to be 0.23 ± 0.03 cm^{-1}. The parity splitting is predicted to be extremely small for the lowest intermolecular level, $j = 3/2$, $\omega = 3/2$, $P = 3/2$, $v_s = 0$, [c,26,28,52] but has been observed by microwave spectroscopy.[24]

Using perturbation theory, Green and Lester have deduced that the parity doublets in $|P| = 1/2$ bending levels are split apart in third order by a term proportional to the product of the Coriolis ($J \cdot j$), difference potential (V_2), and rotational decoupling ($j \cdot s$) terms divided by the appropriate energy denominators.[28] Using the approximations discussed in Ref. 28, an expression for the splitting between parity doublets in the $j = 3/2$, $\omega = 3/2$, $P = 1/2$ band has been derived:

$$\text{splitting} = -0.009 \text{ cm}^{-1} \left(J + \frac{1}{2} \right) \langle \nu_s | V_{22}(R) | \nu_s \rangle \,, \tag{5}$$

where $V_{22}(R)$ is the leading term in the expansion of the difference potential in terms of reduced rotation matrix elements (see Eq. (9) of Ref. 28). From the experimentally determined parity splitting for the 9.7 cm^{-1} band, the expectation value $\langle v_s | V_{22}(R) | v_s \rangle$ is found to be ~ 12 cm^{-1}. The difference potential is then evaluated as $V_2 \approx 6$ cm^{-1} in the region sampled by the

---

[c]Using perturbation theory, Dubernet *et al.*[52] have derived a theoretical relationship between the magnitude of the parity splitting in the lowest $P = 3/2$ level and that for the first excited $P = 1/2$ level. Their theoretical ratio differs substantially from experimental findings.[10,24] In light of this work, we have reanalyzed our experimental SEP spectra for the $P = 1/2$ band at 9.7 cm^{-1} [see *Faraday Discuss. Chem. Soc.* **97**, (1994)] using the following form for the parity splitting

$$\Delta\nu_{P=1/2} = p(J + 1/2)$$

and find a best fit with $p = -0.155$ cm^{-1} and a rotor constant of 0.105 cm^{-1}. The negative sign on the parity splitting indicates that the $+$ parity state is lower in energy. We emphasize that the SEP spectra were recorded using a Nd:YAG pumped dye laser with a laser resolution of 0.1 cm^{-1}.[10,11] Clearly, higher resolution spectra of the $P = 1/2$ band at 9.7 cm^{-1} would be desirable. This band should be observable by far-infrared or near-infrared spectroscopy.[26,27]

first excited bend (centered at $R \approx 3.6$ Å, $\theta \approx 60°$).[11] This is about a factor of 2 smaller than expected based on the *ab initio* difference potential.[25]

5.3.2. *Spin-orbit predissociation*

OH–Ar complexes prepared in the spin-orbit excited state ($\omega = 1/2$), resulting from the coupling of the spin and electronic orbital angular momenta of the unpaired electron, may undergo spin-orbit predissociation. Spin-orbit predissociation of OH–Ar complexes prepared in $j = 1/2$, $\omega = 1/2$ intermolecular levels, like the parity-splitting of rotational levels discussed above, is induced by the difference potential (V_2). The experimental measurements bracket the spin-orbit predissociation lifetime of the $j = 1/2$, $\omega = 1/2$ level at 139 cm^{-1} between approximately 6 ns, the temporal pulse width of the laser, and 1.5 ps, the transform of the observed linewidth in the time domain.[d,11] In principle, measured spin-orbit relaxation rates can be used to obtain the magnitude of the difference potential at short-range,[28] where the continuum wave function has its largest amplitude. This information would be complementary to that derived from the parity splitting of rotational levels in the 9.7 cm^{-1} band, which yields the magnitude of the difference potential in the region sampled by the first excited bend.[28] The magnitude of the difference potential is expected to increase exponentially as the intermolecular separation distance is reduced.[25]

6. Conclusions

The intermolecular bending and stretching vibrational levels of the ground electronic state of the weakly bound OH–Ar ($X^2\Pi$) complex have been characterized by stimulated emission pumping. Virtually all of the bound vibrational levels of the complex from the zero-point level to the dissociation

---

[d]The 139 cm^{-1} SEP feature has been reanalyzed taking into account the underlying rotational structure of the band which arises from the pump laser preparing OH–Ar ($A^2\Sigma^+$) is several rotational levels (pumping Q_1 bandhead). The rotational contour was simulated using a Voigt line profile and iteratively varying the transition type, lower state rotational constant, and the homogenous component of the linewidth until the calculated rotational band contour agreed well with the experimental feature. The feature can be characterized as a perpendicular transition with a lower state rotor constant of 0.095 cm^{-1}. The homogenous contribution to the linewidth of the 139 cm^{-1} features is found to be 1.2 cm^{-1}, corresponding to a spin-orbit predissociation lifetime of 4.5 ps (C. C. Chuang, P. M. Andrews and M. I. Lester, to be published).

limit have been accessed. The bending level pattern characteristic of open-shell complexes, in which $2j + 1$ bending levels correlate with each j, ω rotational level of OH $(X^2\Pi)$, has been experimentally observed. The dissociation limit to OH $(X^2\Pi)$ $v = 0$, $j = 3/2 + $ Ar $(^1S_0)$ occurs between 93 and 103 cm^{-1} above the zero-point level, as determined by the onset of bound-free transitions. In addition, metastable levels which lie as much as 200 cm^{-1} above the dissociation limit have been identified. OH–Ar complexes prepared in these levels may undergo rotational predissociation induced by anisotropic terms in the average interaction potential V_Π. Complexes prepared in intermolecular levels correlating with the spin-orbit excited state of OH $^2\Pi_{1/2}$ $(\omega = 1/2)$ may predissociate via spin-orbit relaxation which is induced by the change in the intermolecular potential when the odd electron in the free radical lies in or out of the O–H–Ar plane (V_2).

An empirically-adjusted potential energy surface has been derived for the average OH $(X^2\Pi)$ + Ar interaction. The bound states supported by this potential reproduce the positions of the intermolecular vibrational levels identified by stimulated emission pumping experiments which lie up to 89 cm^{-1} above the zero-point level as well as the rotational constants determined for several of these levels. The adjusted potential exhibits a stronger isotropic attraction than the *ab initio* potential of Degli Esposti and Werner[25] as well as increased anisotropy. The potential minimum is located at an OH (center-of-mass) to Ar separation of 3.7 Å with a well depth of -127 cm^{-1} in the linear hydrogen-bonded O–H–Ar configuration. The zero-point level is calculated to have a binding energy of -93.2 cm^{-1}.

Acknowledgments

This research has been sponsored by the Division of Chemical Sciences, Office of Basic Energy Sciences of the the Department of Energy. Acknowledgement is made to the donors of The Petroleum Research Fund, administered by the ACS, for partial support of the theoretical portion of this research. Theoretical calculations performed at the University of Cambridge were supported by the Science and Engineering Research Council. The experimental part of this research was performed by M. T. Berry, R. A. Loomis, L. C. Giancarlo, and M. R. Brustein at the University of Pennsylvania. Their contributions to this research have been essential.

References

1. J. M. Hutson, *Ann. Rev. Phys. Chem.* **41**, 123 (1990).
2. See articles in *Dynamics of Polyatomic van der Waals Complexes*, eds. N. Halberstadt and K. C. Janda, (Plenum Press, New York, 1990).
3. R. E. Miller, *Science* **240**, 447 (1988).
4. D. J. Nesbitt, *Chem. Rev.* **88**, 843 (1988).
5. J. M. Hutson, *J. Chem. Phys.* **89**, 4550 (1988).
6. R. J. LeRoy and J. S. Carley, *Adv. Chem. Phys.* **42**, 353 (1980).
7. C. M. Lovejoy and D. J. Nesbitt, *J. Chem. Phys.* **94**, 208 (1991).
8. R. C. Cohen and R.J. Saykally, *J. Phys. Chem.* **94**, 7991 (1990).
9. R. C. Cohen and R.J. Saykally, *J. Phys. Chem.* **96**, 1024 (1992).
10. M. T. Berry, M. R. Brustein, M. I. Lester, C. Chakravarty, and D. C. Clary, *Chem. Phys. Lett.* **178**, 301 (1991).
11. M. T. Berry, R. A. Loomis, L. C. Giancarlo, and M. I. Lester, *J. Chem. Phys.* **96**, 7890 (1992).
12. M. T. Berry, M. R. Brustein, J. R. Adamo, and M. I. Lester, *J. Phys. Chem.* **92**, 5551 (1988)
13. M. T. Berry, M. R. Brustein, and M. I. Lester, *Chem. Phys. Lett.* **153**, 17 (1988).
14. K. M. Beck, M. T. Berry, M. R. Brustein, and M. I. Lester, *Chem. Phys. Lett.* **162**, 203 (1989).
15. M. T. Berry, M. R. Brustein, and M. I. Lester, *J. Chem. Phys.* **90**, 5878 (1989).
16. M. T. Berry, M. R. Brustein, and M. I. Lester, *J. Chem. Phys.* **92**, 6469 (1990).
17. W. M. Fawzy and M. C. Heaven, *J. Chem. Phys.* **89**, 7030 (1988).
18. W. M. Fawzy and M. C. Heaven, *J. Chem. Phys.* **92**, 909 (1990).
19. S. K. Bramble and P. A. Hamilton, *Chem. Phys. Lett.* **170**, 107 (1990).
20. J. Schleipen, L. Nemes, J. Heinze, and J. J. ter Meulen, *Chem. Phys. Lett.* **175**, 561 (1990).
21. B.-C. Chang, L. Yu, D. Cullin, B. Rehfuss, J. Williamson, T. A. Miller, W. M. Fawzy, X. Zheng, S. Fei, and M. Heaven, *J. Chem. Phys.* **95**, 7086 (1991).
22. B.-C. Chang, J. M. Williamson, D. W. Cullin, J. R. Dunlop, and T. A. Miller, *J. Chem. Phys.* **97**, 7999 (1992).
23. J. M. Bowman, B. Gazdy, P. Schafer, and M. C. Heaven, *J. Phys. Chem.* **94**, 2226 (1990); *ibid. J. Phys. Chem.* **94**, 8858 (1990); U. Schnupf, J. M. Bowman, and M. C. Heaven, *Chem. Phys. Lett.* **189**, 487 (1992).
24. Y. Ohshima, M. Iida, and Y. Endo, *J. Chem. Phys.* **95**, 7001 (1991).
25. A. Degli Esposti and H.-J. Werner, *J. Chem. Phys.* **93**, 3351 (1990).
26. C. Chakravarty and D. C. Clary, *J. Chem. Phys.* **94**, 4149 (1991).
27. M.-L. Dubernet, D. Flower, and J. M. Hutson, *J. Chem. Phys.* **94**, 7602 (1991).
28. W. H. Green, Jr. and M. I. Lester, *J. Chem. Phys.* **96**, 2573 (1992).

29. C. Chakravarty, D. C. Clary, A. Degli Esposti, and H.-J Werner, *J. Chem. Phys.* **93**, 3367 (1990).

30. C. Chakravarty and D. C. Clary, *Chem. Phys. Lett.* **173**, 541 (1990).

31. C. Chakravarty, D. C. Clary, A. Degli Esposti, and H.-J. Werner, *J. Chem. Phys.* **95**, 8149 (1991).

32. P. D. A. Mills, C. M. Western, and B. J. Howard, *J. Phys. Chem.* **90**, 3331 (1986); *ibid.* 4961 (1986).

33. W. M. Fawzy, G. T. Fraser, J. T. Hougen, and A. S. Pine, *J. Chem. Phys.* **93**, 2992 (1990).

34. M. C. Heaven, *Ann. Rev. Phys. Chem.* **43**, 283 (1992).

35. R. W. Randall, C.-C. Chuang, and M. I. Lester, *Chem. Phys. Lett.* **200**, 113 (1992).

36. S. Fei, X. Zheng, and M. C. Heaven, *J. Chem. Phys.* **97**, 1655 (1992).

37. G. W. Lemire and R. C. Sausa, *J. Phys. Chem.* **96**, 4821 (1992).

38. W. M. Fawzy and J. T. Hougen, *J. Mol. Spectrosc.* **137**, 154 (1989).

39. S. K. Kulkarni, Y. Lin, and M. C. Heaven, *Chem. Phys. Lett.* **167**, 597 (1990).

40. K. R. German, *J. Chem. Phys.* **63**, 5252 (1975).

41. M. I. Lester, R. A. Loomis, L. C. Giancarlo, M. T. Berry, C. Chakravarty, and D. C. Clary, *J. Chem. Phys.* **98**, 9320 (1993).

42. C. J. Ashton, M. S. Child, and J. M. Hutson, *J. Chem. Phys.* **78**, 4025 (1983).

43. D. C. Clary, C. M. Lovejoy, S. V. O'Neil, and D. J. Nesbitt, *Phys. Rev. Lett.* **61**, 1576 (1988); D. J. Nesbitt, C. M. Lovejoy, T. G. Lindeman, S. V. O'Neil, and D. C. Clary, *J. Chem. Phys.* **91**, 722 (1989).

44. G. H. Dieke and H. M. Crosswhite, *J. Quant. Spectrosc. Radiat. Transfer* **2**, 97 (1961).

45. M. H. Alexander, *J. Chem. Phys.* **76**, 5974 (1982).

46. A. D. Buckingham, *Adv. Chem. Phys.* **12**, 107 (1967).

47. S. Chu, M. Yoshimine, and B. Liu, *J. Chem. Phys.* **61**, 5390 (1974).

48. K. T. Tang and J. P. Toennies, *J. Chem. Phys.* **80**, 3726 (1984).

49. Ch. Jungen and A. J. Merer, *Mol. Phys.* **40**, 1 (1980).

50. G. Duxbury and R. N. Dixon, *Mol. Phys.* **43**, 255 (1981).

51. J. A. Pople and H. C. Longuet-Higgins, *Mol. Phys.* **1**, 372 (1958).

52. M.-L. Dubernet, P. A. Tuckey, and J. M. Hutson, *Chem. Phys. Lett.* **193**, 355 (1992).

CHAPTER 19

CLUSTER ION-DIP SPECTROSCOPY

R. J. Stanley

Department of Chemistry
Stanford University
Stanford, CA 94305

and

A. W. Castleman, Jr.

Department of Chemistry
Pennsylvania State University
University Park, PA 16802

Contents

1. Molecular Clusters and Spectroscopy

The study of molecular clusters provides considerable insight into a variety of subjects, including solvation effects on reaction dynamics, radiationless transitions, supramolecular structure, and on the evolution of the electronic properties of isolated molecules from their behavior in the gas phase to the bulk condensed phase. The term "cluster" is usually taken to mean aggregates of molecules or atoms which are held together by forces much weaker than those found in a chemical bond. These forces lead to cluster bond energies ranging in strength from a few tens to hundreds of wave numbers for those involving the very weak van der Waals (vdW) interactions, to ones of several thousands of wave numbers where they are stabilized through hydrogen bonding (H-bond). Because of these weak interactions, clusters are usually unstable with respect to dissociation or further aggregation and hence must be studied in an isolated environment.

The method of choice for the creation and isolation of clusters is the supersonic expansion-molecular beam technique. By the adiabatic expansion of a gas mixture from a high pressure reservoir into vacuum, even weakly bonded clusters, such as $I_2 \cdot He$, have been formed and characterized spectroscopically.[1] The supersonic expansion technique has several extremely desirable characteristics. The clusters which are formed can be made extremely cold, with rotational temperatures down to several degrees Kelvin being achieved fairly easily, and vibrational temperatures about an order of magnitude higher, which considerably simplifies the optical spectra. The collimated molecular beam can be probed optically such that the Doppler linewidth is usually negligible, making possible very high resolution spectroscopy of molecules and clusters which would be difficult to obtain otherwise. Conditions can be varied to yield a large size distribution, making a comprehensive study of aggregation effects possible.

Generally, the study of the spectral perturbations of a bare chromophore molecule in contact with another "solvent" species which can vdW or hydrogen bond forms the beginning of the investigation, followed by a determination of the evolution of the spectra with ever increasing numbers of "solvent" molecules and the appearance of new vibronic structure. Normally, the first such addition results in the largest spectral change, such as a red or blue shift in the absorption spectrum of the "solute" chromophore. Because of this shift it is relatively easy to isolate the 1:1 complex spectroscopically, permitting its characterization without interference from larger aggregates. However, as the cluster size increases, the spectral perturbation becomes smaller, with the result that the overlapping absorption bands of different species requires another dimension of discrimination. Mass spectrometry provides the species selectivity necessary to work with larger clusters.

Dispersed fluorescence,[2] FTIR,[3] direct infrared absorption,[4] molecular beam resonance studies,[5] and most recently vibration rotation tunneling spectroscopy[6] techniques have been used to obtain ground state information on clusters, typically for dimeric or trimeric species. However, for large molecules with significant spectral congestion, and for larger clusters, a mass resolved ground state spectroscopic method is clearly warranted, one that can serve to identify transitions with specific species. A promising approach for isolated small molecules, called ion-dip spectroscopy (IDS), was first developed in 1981 by Cooper *et al.*[7] It was thereafter taken up in the mid 80s by Ito's group[8] for the study of intramolecular vibrational relaxation in aromatic molecules. The use of a mass spectrometer for an additional level of discrimination was worked out simultaneously by Schlag's group in Germany[9] and in our laboratory.[10]

2. Ion-Dip Spectroscopy of Molecular Clusters

Stimulated Emission Pumping (SEP), pioneered by Kinsey and Field,[11] is a double resonance technique which involves the selective preparation of an excited electronic state such that selection rules for the optical transition from this intermediate level to the final state(s) lead to a simplification of the resultant molecular spectra.[12] This selectivity is especially useful if the sample is comprised of a mixture in which the absorption bands of the species in the sample do not overlap. Ion-Dip Spectroscopy (IDS) is a variant of SEP. In the case where there is spectral overlap of the different species (e.g., clusters in a molecular beam) a third species-selective dimension is needed to resolve the spectra. Utilizing a mass spectrometer as this third selective filter makes ion-dip spectroscopy, in effect, a *triple* resonance technique. The basic idea is shown in Fig. 1. Consider an ensemble of atoms or molecules which are all in the same initial state, S_0 (ground state singlet). A laser beam of frequency ω_1 is used to couple the ensemble to some excited electronic state, S_1, by resonant absorption of light. There are a number of photophysical processes which may occur in the excited state. These include fluorescence, intersystem crossing (ISC) to the triplet manifold, and internal conversion (IC) to highly excited vibrational states of the S_0 state, as well as inter- and intramolecular vibrational relaxation processes (IVR) within the S_1 manifold itself.

A second laser beam of frequency ω_2 (the "dump" beam) generates an ion current from the intermediate state if the sum of the energies of the pump and dump photons exceeds the ionization potential (I.P.) of the system. If the frequency of the dump beam is scanned, and the energy difference between the pump and dump lasers ($\omega_1 - \omega_2 = \omega_{\text{dip}}$) comes into resonance with a vibronic level in S_0, then stimulated emission pumping can result in a transfer of the population from the intermediate state to the final state, and will appear as a dip in the ion current as a function of the dump laser frequency, ω_2. The modulation of the ion current will depend on the stimulated emission rate, the ionization rate, the inverse lifetime of the intermediate state, γ, and the relaxation rate of the final state, δ. The spectral resolution of IDS is limited primarily by the bandwidth of the pump laser, as was so convincingly demonstrated by Weber *et al.* for the rotationally resolved IDS of benzene.[13]

The dip intensity can also provide information on energy transfer in the final state. The implementation of this approach can be understood by first considering a system where the ionization rate from S_1 is slow compared to the rate of SEP and there is no energy transfer possible in the final state (i.e., $\delta = 0$). In the high fluence limit of the dump laser, the rates of up-pumping from S_0 to S_1 and stimulated emission pumping from S_1 to S_0 become equal so that the maximum observable dip is 50% (two-level system). However, in the case where $\delta > 0$ (e.g., a cluster), energy

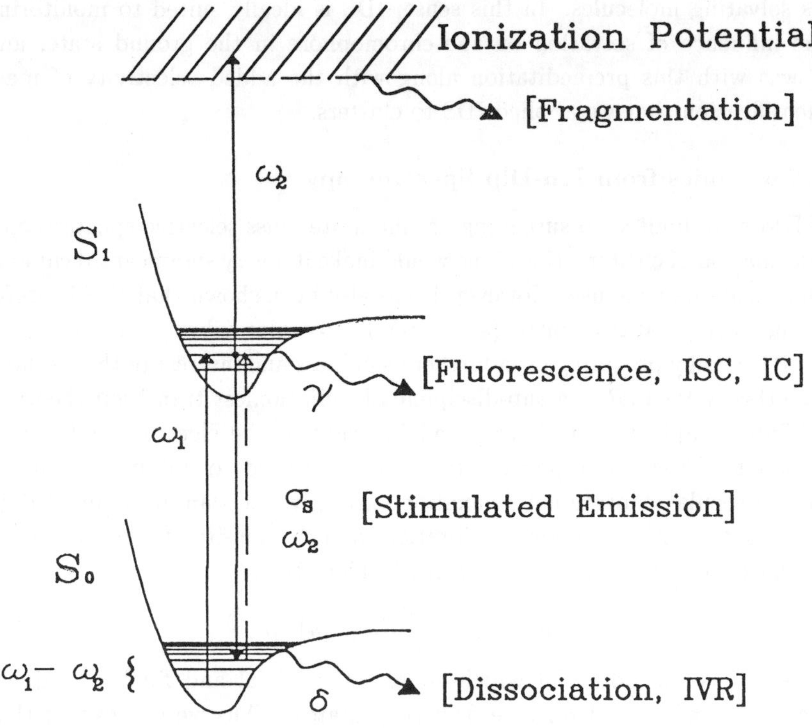

Fig. 1. Schematic representation of ion-dip spectroscopy. Note that the final state may be depleted through IVR or dissociation at a rate δ.

transfer can occur from the chromophore into bath states of the complex on a time scale shorter than the dump laser pulse width ($\delta^{-1} < \Delta t$). Hence, the final state will become depopulated, allowing more population from S_1 to be dumped into S_0. This may result in a dip modulation of $> 50\%$, if relaxation can compete kinetically with up-pumping from the state in resonance with the dump laser. Such mechanisms include IVR and dissociation of the cluster bond, if the final state energy is sufficient. In the particular case of a chromophore solvated by one or more ligands, the low frequency intermolecular modes of the cluster system act as a "bath" to sink population from the down-pumped state. Thus, a dip modulation of greater than 50% is the signature that either or both a non-negligible coupling and large density of states exists between the chromophore and its solvating molecules. In this sense, IDS is ideally suited to monitoring the influence of solvation on the chromophore in the ground state, and it was with this premeditation along with the added selectivity of mass identification that we applied IDS to clusters.

3. Dynamics from Ion-Dip Spectroscopy

If IDS were limited to supplying ground state mass selective spectroscopic information of clusters, this alone would make it a very significant technique and ensure its wide use. However, it has also been shown that IDS is useful in measuring rates of intra (and inter-) molecular vibrational relaxation (IVR) thereby providing even further usefulness and interest in the method. The theory for IVR is a sub-discipline of radiationless transition theory,[14] originally applied by Robinson and Frosch[15] to the electronic relaxation in solids. First order perturbation theory is used to obtain the rate of intramolecular vibrational relaxation, k_{IVR}, which can be expressed in terms of the coupling between vibrations and the density of states available to the system, also known as Fermi's Golden Rule:[16]

$$k_{\mathrm{IVR}} = \delta = \frac{2\pi}{\hbar} |\langle m|V|n \rangle|^2 \rho \, , \tag{1}$$

where V is the potential which couples the initial and final vibrational states, m and n, and ρ is the density of states. Thus we can expect that k_{IVR} scales from low values in systems comprised of pure van der Waals interactions, to larger values for hydrogen bonded complexes and clusters with more covalent character. Particularly important in the understanding of energy redistribution in clusters is that the density of states is usually quite large, due to the large number of degrees of freedom from the very low frequency "bath" modes of the cluster bonds. Implicit in Eq. (1) is that the density of states is continuous over the energy interval being examined. However, for the lowest levels of the final state manifold, the continuum model for the bath states of the cluster may break down.

The measurement of k_{IVR} is performed on the basis of a straightforward kinetic rate equation model for the IDS process, as developed by Suzuki, Mikami, and Ito.[8] If the pump intensity, I_p, is kept constant for both the on-resonant and off-resonant experiments, then the ratio of the off-resonant to on-resonant ion currents, $S = M_{\mathrm{off}}^+/M_{\mathrm{on}}^+$ can be represented by,

$$S = \frac{M_{\mathrm{off}}^+}{M_{\mathrm{on}}^+} = \left[1 + \frac{\sigma_s I_d \delta}{(\sigma_s I_d + \delta)(\sigma_i I_d + \gamma)} \right] \, , \tag{2}$$

IDS Kinetic Model

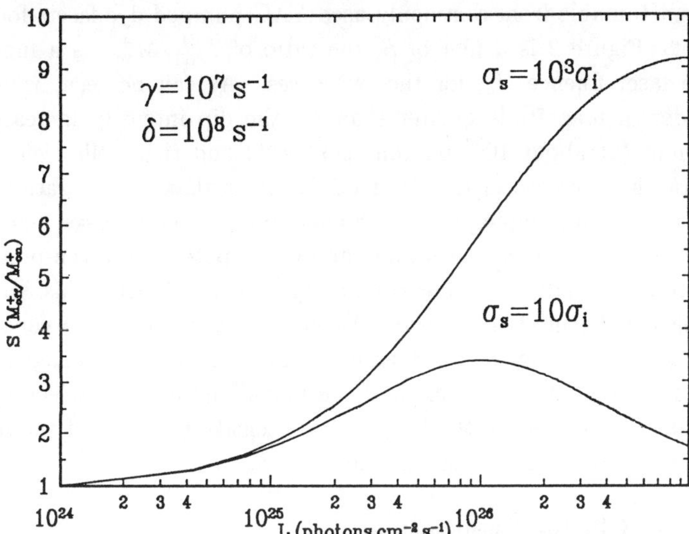

Fig. 2. IDS Kinetic Model Calculations for two different rates of ionization, $10\sigma_s$ and $1000\sigma_s$. In the two level system ($10\sigma_s$), the signal gain is high because of low losses from ionization. When the rate of ionization is higher the curve saturates at a lower dump laser fluence and the IDS signal decreases with a further increase in the dump laser power.

where I_d is the intensity of the dump laser, and σ denotes cross-sections for stimulated emission (s), ionization (i), and pumping (p). In the limit of low dump laser fluence, and where the lifetime of the intermediate state is such that $\gamma \gg \sigma_i I_d$, we find that S becomes

$$S = 1 + \frac{\sigma_s}{\gamma} I_d \ . \tag{3}$$

Thus, in the limit of low I_d fluence, S is linear with slope σ_s/γ. It is instructive to model two limiting cases where IDS spectroscopy could be used to measure IVR rates: where $\sigma_s \gg \sigma_i$, and where $\sigma_s \sim \sigma_i$. In the first case, stimulated emission controls the dip modulation, but in the second case, the ionization rate competes effectively with down-pumping. We will consider a molecule for which $\sigma_s = 1000\sigma_i$ and $10\sigma_i$. The model parameters are chosen to be: $\sigma_p = 10^{-17}$ cm^2, $\delta = 10^8$s^{-1}, $\gamma = 10^7$s^{-1}, and $\sigma_i = 10^{-20}$cm^2. The laser pulse width is 5 ns. We apply Eq. (2) and let I_d vary from 10^{24} to 10^{27} photons cm^{-2}s^{-1}, which is what is typically attainable from commonly used YAG pumped dye lasers focused to a spot. Figure 2 is a plot of S, the ratio of $M_{\text{off}}^+/M_{\text{on}}^+$, as a function of dump laser fluence, I_d, for the two cases. As can be seen, when σ_s is an order of magnitude greater than σ_i, the dip intensity increases to a maximum (at about 10^{26} photons cm^{-2}s^{-1}) and then rolls over. The reason for the decrease in the dip modulation is that the ionization rate exceeds the down-pumping rate at higher dump laser fluences. For the case where $\sigma_s = 1000\sigma_i$, ionization cannot compete with down-pumping until extremely high laser fluences are attained. Clearly, this kind of modelling can be used to measure the cross-sections and rate constants of a system undergoing similar dynamics. Another practical consideration is that knowledge of parameters and the modelling assist in selecting the best conditions to achieve the largest dip intensity in terms of the dump laser fluence which leads to the maximum in S.

4. IDS and Cluster Dissociation

Cluster dissociation plays an important part in the interpretation of cluster spectra and this is particularly true for IDS. The internal energy of both the final state of the neutral which has undergone SEP, and that of the ion, is a function of the dump wavelength; the internal energy may be great enough to cause fragmentation of the weak cluster bond in either the ground state neutral or in the ion which is detected. As such, there are three scenarios involving cluster dissociation: (1) the cluster ion dissociates due to excess energy above the ionization limit ($\tau_{\text{diss}}^{\text{ion}} < \Delta t_{\text{dump}}$); (2) the intermediate state decays before stimulated emission can occur; and (3) the final ground state *neutral* cluster dissociates rapidly compared to the rate of down-pumping in the IDS process ($\delta_{\text{diss}}^{-1} < \Delta t_{\text{dump}}$).

Case 1: The cluster ion dissociates

If cluster ion dissociation is fast, then the ion-dip signal in the cluster ion mass channel should be greatly enhanced. In other words, if the final energy of the ion, $E_{\text{ion}} = [(\omega_1 + \omega_2) - \text{I.P.}]$ is greater than the cluster binding energy, part of the ion signal will appear in a fragment ion mass channel. To obtain the true ion-dip spectrum it is necessary to sum the parent and fragment mass channels together. An example of this can be found for the case of phenylacetylene–Ar, NH_3 and for the case of phenol–H_2O, where the ion is prepared with $> 6000 cm^{-1}$ of excess energy. This presents a problem only if the fragment ion mass channel is contaminated by dissociative ionization from larger clusters (a nontrivial problem). Furthermore, structure in the ion-dip spectrum may result if the cross-section for ionization, σ_i, is not constant with wavelength.

Case 2: The cluster intermediate (S_1) state decays rapidly

If the cluster is prepared in an intermediate state which decays on a time scale faster than the stimulated emission process, then stimulated emission could occur from lower vibronic levels of the excited state manifold of the *chromophore*. If fluorescence, internal conversion, and intersystem crossing compete with stimulated emission, the signal-to-noise ratio can be severely impacted. IVR and internal conversion can lead to the population of a number of lower vibronic levels, some of which may give rise to stimulated emission as well. This would probably manifest itself in a very broad line width, where a multiplicity of closely lying intermediate states are made available through IVR. Similar results in dispersed fluorescence were obtained by Levy and coworkers for the *s*-tetrazine–argon system,[2] and the benzene–He, Ar system as observed by Stephenson and Rice.[17] Thus, the excited state binding energy of the cluster restricts the intermediate vibronic states which are accessible. This is not a severe limitation as there are usually strong transitions, both intramolecular and intermolecular, which can be used. A more subtle complication arises when the intermediate state undergoes rapid energy transfer, either radiative or nonradiative. In this case, randomization of the prepared intermediate state leads to a diffuse ion-dip signal. An example of this is seen for the case of the IDS of phenol, where the S_1 lifetime is much less than the pulse width of the pump laser.[13]

Case 3: The cluster dissociates rapidly in the final state

The situation is somewhat different for dissociation in the final state. If the final state energy lies above the dissociation energy for the cluster, then it is entirely possible that the cluster bond may break. If the dissociation occurs faster than the stimulated emission rate then the down-pumping step can never saturate, which allows a greater percentage of the intermediate state population to be transferred into the ground state manifold. In fact, it should be possible to achieve a 100% dip signal if the cluster dissociation is rapid enough compared to the ionization rate. *IDS should prove to be a very sensitive technique for detecting cluster dissociation.* It should be noted, however, that it would be very difficult to distinguish dissociation from rapid inter (inter-) molecular vibration relaxation since the signal is identical in either case. However, at lower energies in the vibronic manifold, dissociation can usually be discounted.

5. Limitations of Ion-Dip Spectroscopy

IDS has its limitations, of course. Anomalous dips may appear in an ion-dip spectrum due to the selection rules in the double resonance scheme if the rotational states cannot be resolved by the pump laser. Krätzschmar, Selzle, and Schlag[18] have studied the effect of slight offsets in the pump wavelength from the peak of the intermediate state transition. Using the $6_0^1(S_1 \leftarrow S_0)$ transition in jet-cooled benzene as the intermediate state, they were able to obtain a single dip of $\Delta\omega = 1.7$ cm^{-1} for the 6_2 state when the pump laser was centered on the 6_0^1 transition. As they introduced a 1 cm^{-1} offset into the pump laser, the single dip split into two bands, giving three dips when the offset was increased to 3 cm^{-1}. This leads to the interesting case when IDS is implemented using ultrafast lasers to examine directly the dynamics of energy transfer in cluster systems. We can expect that time resolved IDS will produce a complicated dependence of the final state spectra on the broad spectral bandwidth which is inherent in femtosecond spectroscopy.

The sensitivity of IDS is limited by several considerations. IDS is a nonzero background technique; the signal is a small modulation of a large background signal. In practice, baseline fluctuations are in the order of 5%, using non-injection seeded lasers. Furthermore, it is very difficult to normalize the data to the laser intensities, as IDS (see below) is inherently nonlinear. Several schemes used in SEP might have application to IDS. Frye and coworkers[19] have used a scheme where two separate pulsed molecular jets, a signal and reference jet, are irradiated by the same pump laser, but only the reference jet is intersected by the dump laser. Because the shot-to-shot fluctuations of the pulsed jets is much less ($\sim 1\%$) than the fluctuations in the pump laser, the variations in the pump laser can be nullified. In this case, the signal of interest was fluorescence, not an ion current, which poses problems of its own.

A similar scheme might be possible for a TOFMS based experiment using the technique of Page, Shen, and Lee[20] in performing infrared–UV double resonance experiments. The idea is that two spatially separated focal spots are formed in the jet, one consisting of the pump and dump lasers (signal), and one consisting of the pump beam alone (reference). The resulting ion signal consists of two ion peaks, separated in time, the ratio of which can be used to normalize the data to the fluctuations of both the pump laser and the pulsed jet expansion on a shot-to-shot basis. One problem with this idea is that it assumes that the pump laser fluence is large enough to produce an ion signal which is not desirable in the case of two color experiments described here. In these experiments, the pump laser fluence is attenuated until no one-color ionization occurs. Also, it is not always advantageous to focus either pump or dump beam, which would make application of the above technique much more difficult. Another normalization method consists of measuring the total fluorescence due to the pump beam just before and after the dump pulse. These two data points can be fitted to an exponential decay function which is thereby used to normalize the ion-dip data.[21]

6. IDS in the Weak Electronic Coupling Limit: Phenylacetylene–NH_3,Ar

The first major attempt at ion-dip spectroscopy in our laboratory[10] was made on phenylacetylene (hereafter referred to as PA: C_6H_5CCH), PA–Ar, and PA–NH_3. The 0_0^0 transition in PA is just over half-way to its ionization potential (I.P.). This energy level scheme is highly desirable for the study of van der Waals clusters (e.g., PA clustered with NH_3 and Ar) because the low excess energy upon resonant (1+1) ionization of PA is not likely to fragment the cluster ion. Ironically, this attribute actually limits the usefulness of using the $S_1 \leftarrow S_0$ transition for the preparation of the intermediate state in an IDS experiment, because the dump photon has a rather limited working range. However, PA has an interesting and intense S_1 vibronic manifold, which contains several levels that are useful as intermediate states.

The most intense transition, 35_0^1, was used as the intermediate state for the IDS experiments due to the relationship of this transition relative to the ground state and the ionization potential, which is 8.814 eV, or 71 089 cm^{-1}.[22] The maximum energy level that can be probed in the ground state well is set by I.P. $-E_{\mathrm{pump}} = E_{\mathrm{max}}$, and the minimum energy, E_{min}, is usually the energy of the 0_0^0 transition. For $\omega_{pump} = 0_0^0$, ω_{min} and ω_{max} are 0 and 669 cm^{-1}, respectively. For the $\omega_{pump} = 35_0^1$ transition, ω_{min} and ω_{max} are 492 and 1652 cm^{-1}. Thus, 35_0^1 affords a wider scan range, as well as having a larger oscillator strength.

The ion-dip spectrum of phenylacetylene, PA–Ar, and PA–NH$_3$ at 1000 torr stagnation pressure and $\sim$ 1.5 cm^{-1} resolution is presented in Fig. 3. The intermediate state is the ν_{35} vibration in the S_1 state (35_0^1). The x-axis is the energy above the vibrationless ground state, just the difference $(\omega_1-\omega_2)$ between the pump and dump photons. The strongest features in PA are at 978, 1002, 1028, and 1038 cm^{-1}. The maximum dip intensity, as determined from the ratio of the dip to the ion current before the dip feature, is less than 40%.

The assignment of the strongest (and broadest) feature at 1028.0 cm^{-1} is somewhat confused by the accidental overlap in energy of 35_2 and 10_1, which are the acetylenic carbon in-plane bend (2×514 cm^{-1} (ν_{35}) = 1028 cm^{-1}, assuming a harmonic potential for the 35_2 vibration) and the phenyl C–H in-plane bending mode (1028 cm^{-1}), respectively. While both are allowed by symmetry, the 35_2 vibrational wave function should have two nodes ($v = 2$), while the 10_1 wave function has only one node ($v = 1$). From a Franck–Condon point of view, the overlap of the 35^1 wave function, which also has one node, should be greatest for transitions to vibrations with one node as well. This is true only if the internuclear distance between the ground and excited electronic states is similar. The studies of King and So[23] suggest that there is very little change in the geometry in the $S_1 \leftarrow S_0$ transition. However, the comparison of the IDS data with those obtained from the single vibronic level fluorescence study[24] suggests that intensity stealing is a more important factor in deciding which of the allowed transitions will carry more intensity. On this basis, the 1028 cm^{-1} band has been tentatively assigned to 35_2, though the 10_1 band could not be ruled out given the resolution.

A power study of ion-dip depth versus dump laser fluence (data not shown) indicates that the spectra were obtained close to saturation, as borne out by the 40% dip for the PA 1028 cm^{-1} band. The dip modulation of the cluster spectra is similar to that of the PA ion-dip spectrum which indicates that dissociation in the ground state does not make a dominant contribution to the SEP kinetics. This is because the dissociation of the cluster on the time scale of the dump step would lead to depletion of

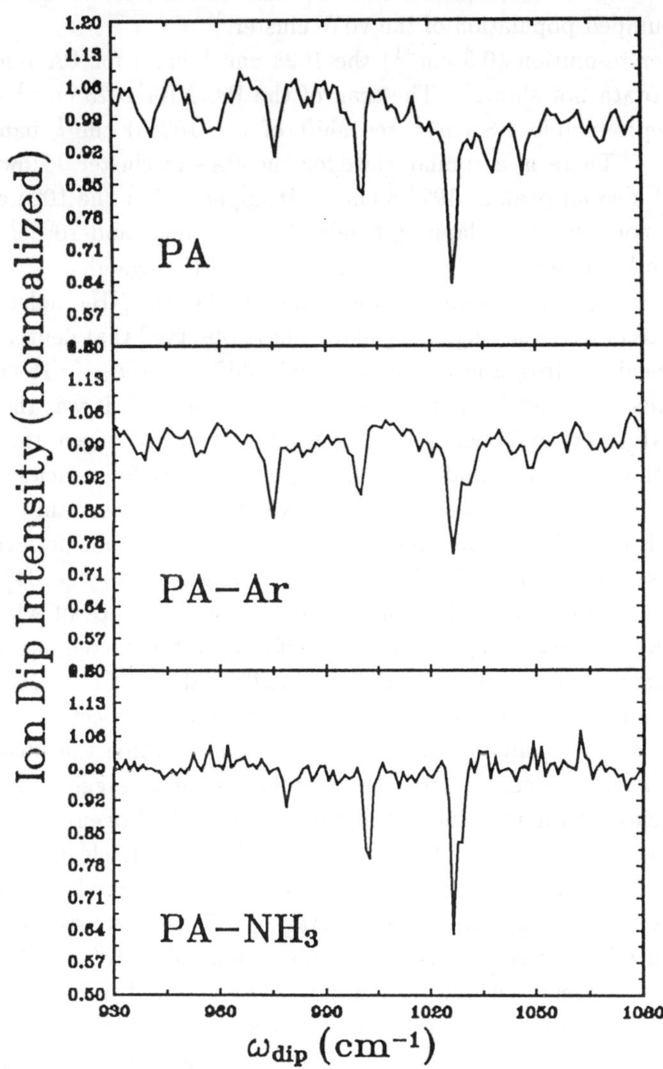

Fig. 3. Ion-dip spectra of phenylacetylene and its van der Waals clusters, PA–Ar and PA–NH$_3$ in the 930–1060 cm^{-1} region. Note that the dip modulation is approximately the same for all three species. In both vdW species there is a splitting of the 1028 cm^{-1} band (see text).

population in the dump state, and hence a larger ion-dip signal. Thus, the dissociation of the complex is not the main channel for relaxation of the down pumped population of the vdW cluster.[25]

At higher resolution (0.3 cm^{-1}) the 1028 cm^{-1} band for PA remains unresolved (data not shown). The scan of the PA–NH$_3$ 1028 cm^{-1} band reveals a reproducible 1.8 cm^{-1} red shift of the 1028.0 cm^{-1} band to 1026.2 cm^{-1}. There is a similar shift for the PA–Ar cluster. However, there is still a small peak at 1028.8 cm^{-1}. It appears that the 1028 cm^{-1} band of PA consists of overlapping bends of 35_2^1 (strong) and $10_1^0 35_0^1$, and that this overlap is removed by the perturbation of the ligand.

A subsequent study was done to ascertain whether the NH$_3$ molecule is on the C$_2$ axis or over the ring. King and So have observed that deuteration of the acetylenic hydrogen leads to an isotopic shift in the ν_{35} (+18 cm^{-1}) and ν_{34} bands (+10 cm^{-1}) in the ground electronic state.[26] It was thought that if the NH$_3$ were hydrogen bonded to the acetylenic hydrogen, then the presence of this molecule would have a similar effect on the frequencies of the non-deuterated PA chromophore. This idea was explored by exciting the 0_0^0 band in both the PA molecule and the PA–NH$_3$ complex and performing IDS on these species over the 480–650 cm^{-1} range. Both the 34_1^0 and 35_1^0 modes are observed at 514 and 623 cm^{-1}, the difference of which is 109 cm^{-1} and within the experimental uncertainty of the uncomplexed PA results. In analogy to the deuterated case described above, if the NH$_3$ were hydrogen bonded to the acetylenic hydrogen, one might expect to see the difference in the ν_{34} and ν_{35} shifts to be $\approx$ 8 cm^{-1}, while the observed difference is at most 1 cm^{-1}. Thus, the evidence seems to suggest that the NH$_3$ is attached to the ring and not at the acetylenic hydrogen.

In summary, from the moderate dip modulations it is clear that the rate of IVR from the chromophore modes into the vdW modes must be slow relative to Δt_{dump}. Thus $k_{\text{IVR}} \leq \Delta t_{\text{dump}}^{-1}$. This is consistent with IVR rates in vdW clusters measured by the groups of Levy[2] and Rice.[17] Qualitatively, this relatively low rate cannot depend strongly on the density of states available to the chromophore, as PA–Ar has fewer degrees of freedom than PA–NH$_3$. The similarity of the spectra undoubtably comes from the quadratic dependence of k_{IVR} on the electronic coupling term, which arises from the same type of intermolecular force in both cases.

7. IDS in the Strong Electronic Coupling Limit: Phenol–H_2O

When electronic (or vibronic) coupling between the chromophore and the ligand is stronger, new features are observed in the ion-dip spectrum. This is indicative of a highly coupled system.[16] We now examine a hydrogen-bonded cluster system where the coupling between chromophore and ligand is much larger than that of the van der Waals systems: phenol–H_2O (PhOH$(H_2O)_n$, $n = 0$–4) and PhOD$(D_2O)_m$, $m = 0, 1$.[37,38] From a fundamental point of view, the water molecule is one of the simplest polyatomic molecules which exhibits H-bonding, and consequently considerable experimental and theoretical work has been done on water clusters and other molecules H-bonded to H_2O.[27]

Much spectroscopic work has been done on phenol in the gas phase, extending from the earlier IR and UV absorption studies of Bist, Brand, and Williams,[28–30] to studies of the decay dynamics of the S_1 state of the molecule solvated by one water molecule, as demonstrated by Lipert, Bermudez, and Colson.[31] Most reported studies on water complexes of phenol provide information on the S_1 excited state,[32–34] though some dispersive LIF work has been done on phenol with other solvent molecules.[35] Higher complexes of phenol with water ($n > 3$) had not been studied prior to the present work and only very recently has there been any information on the ground state spectroscopy of any of the phenol–water complexes.[36–38]

7.1. *Phenol and PhOH$(H_2O)_n$, $n = 1$–4 Excitation Spectra*

The excited state frequencies were measured using a $1 + 1$ two-color multiphoton ionization scheme. The purpose for using a two-color ionization scheme is to limit, as much as possible, the amount of excess energy that is deposited in the resulting photoion, with the aim of reducing fragmentation of the relatively weak cluster bond. The necessity of using a threshold energy photon for the ionization step is dramatically demonstrated in the PhOH$(H_2O)_{0-4}$ system. Figures 4a and 4b were taken with the ionizing laser set at 287 nm and 347.5 nm, respectively. Note that the fragmentation which occurs when the ionization laser is set to $\omega_2 = 287$ nm is almost completely eliminated when $\omega_2 = 347.5$ nm. The intensity scale for each scan has been normalized to unity for comparison purposes. Previous measurements are available from the work of Fuke and Kaya[32] (one-color MPI) and more recent two-color MPI results have been reported by Lipert and Colson[34] for PhOH$(H_2O)_n$, $n = 0$–3. Excitation spectra for the $n = 4$

cluster have not been reported before and the findings reported herein are new.

The $PhOH(H_2O)_1$ spectrum shows structure that has been assigned[39] to intermolecular vibrations of the H_2O H-bonded to the phenol. The origin is found to be red shifted -354 cm^{-1} relative to phenol 0_0^0, in excellent agreement with findings of other workers (a closeup of this region can be seen in Fig. 5 where the energy scale is set relative to the 0_0^0 of the $PhOH(H_2O)_1$, $PhOD(D_2O)_1$ clusters).[34] A weak feature at 120 cm^{-1} has been assigned by Lipert and Colson[34] as an in-plane "wag" of the H_2O ligand, although Oikawa *et al.*,[33] using LIF excitation spectroscopy, designated this peak to be an in-plane bend. For convenience we use the label β''^1 when referring to this mode. Both groups agree that the moderately strong peak at 156 cm^{-1} is the H-bond stretch, σ^1. Lipert and Colson[34] have tentatively assigned weak features at 279 cm^{-1} as a combination band of the wag and stretch modes $(156+120)$; however, bands at 96, 336, and 356 are unassigned. A peak observed at $482(\pm2)$ cm^{-1} (not shown) is assigned as the $6a_0^1$ transition for the cluster, as compared to the uncomplexed $6a_0^1$ vibration, which lies 475.0 cm^{-1} above the 0_0^0 transition for phenol.[28] Similar features are observed for the $PhOD(D_2O)_1$ system.

The $PhOH(H_2O)_2$ channel in Fig. 4a shows only a broad peak at about -121 cm^{-1}, also seen by Lipert and Colson.[34] The feature at -91 cm^{-1} is due to fragmentation of $PhOH(H_2O)_3^+$, as can be seen from the figure. Note that when ionized with 287 nm radiation, all of the $PhOH(H_2O)_2$ cluster fragments into the $PhOH(H_2O)_1$ channel, and the features in the $PhOH(H_2O)_2$ channel correspond to the $PhOH(H_2O)_3$ cluster at -91 cm^{-1}, which has fragmented, losing one H_2O molecule. Softer ionization reveals a broad, structured band for the $PhOH(H_2O)_2$ cluster (Fig. 4b), which consists of ten closely spaced (3 cm^{-1}) peaks, has been shown to originate from only one isomer, using hole burning techniques.[34]

The $PhOH(H_2O)_3$ scan with $\omega_2 = 287$ nm shows three prominent peaks at -50, -91, and -180 cm^{-1} (Fig. 4b). All three peaks were observed by Fuke and Kaya,[32] and we agree with their assignment of the -91 cm^{-1} band as being the most stable of the $PhOH(H_2O)_3$ conformers observed. However, the assignment of the peak at -49 cm^{-1} as a third conformer of $PhOH(H_2O)_3$ is in contradiction with the evidence presented here. It is seen clearly that this peak is due to fragmentation of the $PhOH(H_2O)_4^+$ cluster (-50 cm^{-1}). It is instructive to note that Oikawa

Phenol–Water Complexes

Hard and Soft R2PI S$_1$ Spectra

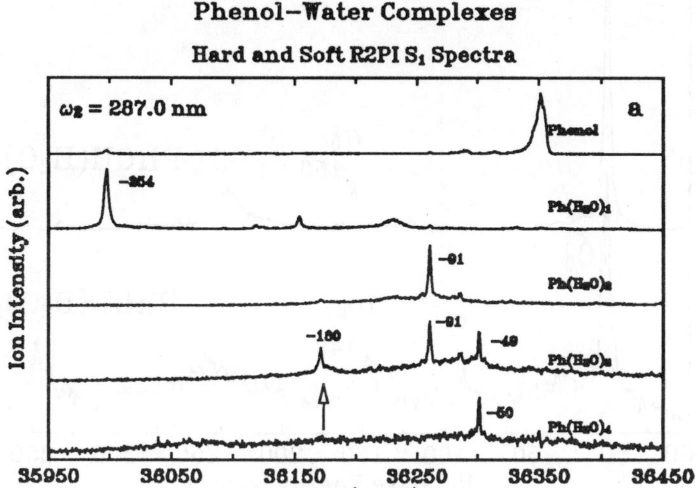

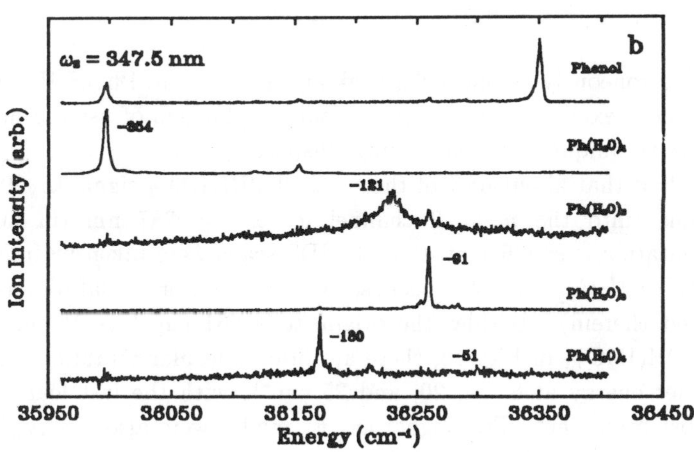

Fig. 4. PhOH(H$_2$O)$_n$, n = 0–4 R2PI Excitation Spectra. a) Hard ionization with ω_2 = 287 nm, which gives (for example) PhOH(H$_2$O)$_1$ 6879 cm^{-1} of energy above its I.P. of 63 980.[34] b) Soft ionization with only 813 cm^{-1} of energy above the I.P. of PhOH(H$_2$O)$_1$.

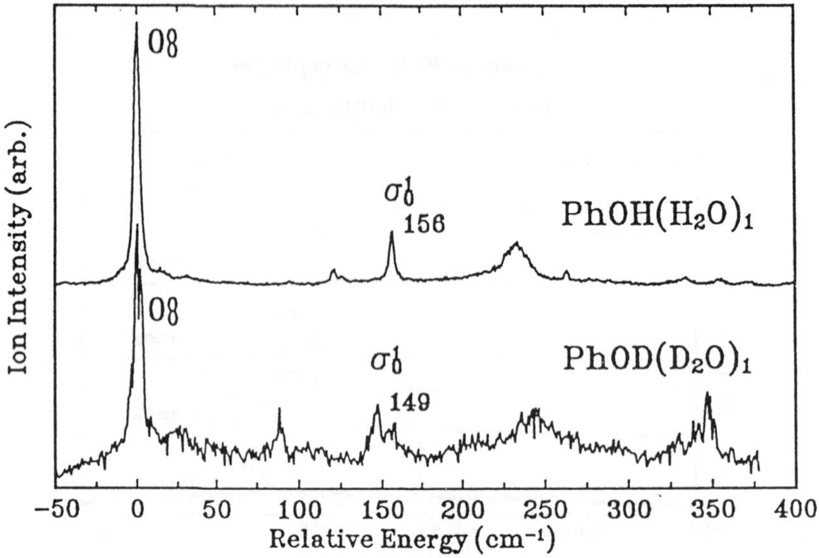

Fig. 5. Detail of the one-color $1 + 1$ excitation spectrum of PhOH(H$_2$O)$_1$ and PhOD(D$_2$O)$_1$. The broad peak at 248 cm^{-1} is due to dissociative ionization of the PhOH(H$_2$O)$_2$ (PhOD(D$_2$O)$_2$ cluster.

et al.[33] erroneously assigned the peak at -91 cm^{-1} to PhOH(H$_2$O)$_2$ using fluorescence excitation, which points out the dangers of using a non-mass selective technique in deconvoluting cluster spectra.

We find that about 68% of the total PhOH(H$_2$O)$_3$ signal at -91 cm^{-1} fragments into the $n = 2$ channel for $\omega_2 = 287$ nm (in fact, the fragmentation is so efficient, that the IDS scans were taken by integrating the PhOH(H$_2$O)$_2^+$ channel because of the superior signal-to-noise ratio obtained therein). Besides the origin at $36\,261$ cm^{-1} (-91 cm^{-1}) seen for PhOH(H$_2$O)$_3$ in Fig. 4b, there are intermolecular vibrational features to higher energy at 8, 14, 20, and 25 cm^{-1}, with the last feature being the most prominent. Two higher energy modes were also observed at 136 and 188 cm^{-1} (to the blue of the origin and not shown). We assign the latter mode on the basis of similar features in the p-cresol–water system, to the σ^1 intermolecular mode, analogous to the PhOH(H$_2$O)$_1$ hydrogen bond donor stretch at 156 cm^{-1},[32,34] and the 136 cm^{-1} band corresponds to β'''^1 although the motion is probably not a simple bend or wag as in the case of the 1:1 complex.

The $PhOH(H_2O)_4$ cluster absorption maximum is nearly lost in the 287 nm ionization spectrum (see arrow, Fig. 4a), but is seen clearly in Fig. 4b, where softer ionizing conditions prevailed. The origin for this complex is at -180 cm^{-1} relative to the phenol origin, which is red shifted from the $PhOH(H_2O)_3$ origin by 89 cm^{-1}. The peak at -51 cm^{-1} in this channel may be an isomer, as it appears only weakly under these expansion conditions. From spectra for $PhOH(H_2O)_{5,6}$ it was determined that the -51 cm^{-1} peak is not the result of fragmentation of higher clusters into the $PhOH(H_2O)_4$ channel.

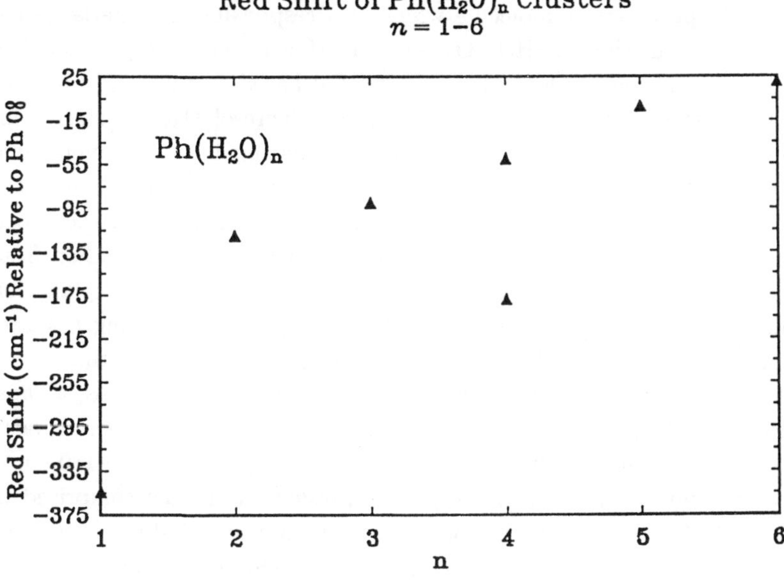

Fig. 6. Electronic red shifts observed for the $PhOH(H_2O)_n$, $n = 1-6$. The $PhOH(H_2O)_4$ cluster seems to have two isomers, one of which may be undergoing proton transfer.

A plot of the red shift versus cluster size is shown in Fig. 6. With the exception of the -180 cm^{-1} red shifted $PhOH(H_2O)_4$ cluster, there is a smoothly decreasing red shift until the $PhOH(H_2O)_6$ cluster, at which point the shift becomes blue relative to phenol. The red shift for $PhOH(H_2O)_1$ is quite large when compared to aromatic-van der Waals systems (where the binding energy of the 1:1 cluster may be 5–10 times weaker) and can be attributed to the change in the inductive effect of the phenolic oxygen atom on the $\pi^* \leftarrow \pi$ transition when phenol hydrogen bonds to water.[40] The proposed structure of $PhOH(H_2O)_1$ is based on phenol acting as the hydrogen donor (acid) with the oxygen on the water as the acceptor (base). It is likely that the geometry of this complex is "*trans*-linear", as is the case of water dimer,[5] and p-cresol$(H_2O)_1$.[41] In this structure, the hydrogen atoms of the water are out of the plane of the aromatic ring, while the $O_{water}-HO_{phenol}$ bond is linear. Although there are apparently two sites to which the second water might bond, *ab initio* calculations done by Pohl and coworkers[41] indicate that the second water molecule acts as a proton donor to the phenolic oxygen. This somewhat reduces the inductive effect of the proton accepting water molecule which was responsible for the large red shift observed in the $PhOH(H_2O)_1$ cluster. Hence, the shifts measured in Fig. 6 are consistent with the trends expected based on these calculations.

These same calculations suggest that the p-cresol–$(H_2O)_3$ cluster involves the OH group on the p-cresol participating in a four-membered ring in which the water opposite the OH group is out of the plane of the aromatic ring. Pohl and coworkers justify the transition from the broad p-cresol$(H_2O)_2$ spectrum to the sharp p-cresol$(H_2O)_3$ as a transition from a flexible open structure to a rigid ring structure.[40] Such an explanation should also be applicable to the $PhOH(H_2O)_{2,3}$ system. Using a Watts potential, Vernon *et al.*[42] has shown that $(H_2O)_{3,4,5}$ form rings with increasingly linear hydrogen bonds as the cluster size increases. Thus, the $PhOH(H_2O)_4$ system might be analogous to the four membered water cluster, also having a fairly rigid structure. In this picture, the -180 cm^{-1} red shift from the $PhOH(H_2O)_4$ peak represents a species comprised of a $PhOH(H_2O)_3$ ring structure with the fourth water H-bonded to one of the H_2O molecules. This would be consistent with a shift back to the red from an otherwise slowly progressive trend towards the blue for the larger clusters. Likewise, an isomer comprised of a full ring of four waters bound to phenol would be consistent with the continued trend of shifts to the blue, and is suggested for the structure of the $PhOH(H_2O)_4$ observed at -51 cm^{-1} with respect to the Ph 0_0^0 origin. This interpretation is also substantiated by the ion-dip spectrum for $PhOH(H_2O)_4$ (see below). Unfortunately, Pohl *et al.* did not perform calculations for the p-cresol–$(H_2O)_4$ cluster, as they did not observe this species under their experimental conditions.[40] The lack of a well defined spectrum for $PhOH(H_2O)_2$, an analogue for the $(H_2O)_3$ cluster, is in stark contrast to recent vibration–rotation tunneling results from Saykally's work.[6] The strained geometry of the trimer must be incompatible with a bulky substituent like the phenyl group in phenol.

7.2. *Ion-Dip Spectroscopy of Phenol: Pumping the 0_0^0 Band*

The ion-dip spectra of phenol–OH and phenol–OD in the range of 500–1300 cm^{-1} are presented in Figs. 7a and 7b, respectively. The temporal delay between the pump and dump pulses was ≈ 5 ns. The x-axis is the energy above the vibrationless (zero point) ground state ($\omega_1 - \omega_2 \equiv \omega_{dip}$), and the ion intensity plotted along the ordinate is normalized to unity. Because of variations in the dye laser gain curve, and due to the inherent nonlinear dependence of the IDS signal on the dump laser fluence, it is difficult to correct the spectra for laser power fluctuations. Since these spectra are composites of several scans which cover the entire region shown, the separate spectra were baseline corrected and normalized, and then "spliced" together to give the figures presented. Thus, the graphed dip intensities are only approximately correct. However, all dip positions, widths, and modulations were taken from the raw spectra and are free from any artifact of the splicing process. The discernible features are tabulated in Table I, along with assignments from the work of Bist, Brand, and Williams.[28–30] The most prominent features are those associated with symmetric ring modes and vibrations involving in-plane motions of the OH group, and combinations thereof (Wilson notation is used throughout[43]).

The vibrational line widths were hard to determine because of the poor signal-to-noise (S/N) of the spectrum, but their apparent excessive width is due to power broadening from the dump laser. Using a much lower dump fluence, we were able to obtain dips of 0.6 cm^{-1} full width at half maximum, which is roughly the bandwidth of the dye laser (0.3 cm^{-1}). A similar power dependence was observed for benzene by Weber *et al.*[13] where using lower dump power, the vibrational bands were resolved into sub-bands, with the more intense component blue-shaded and a band separation of 1.5 cm^{-1}.

The PhOH dip modulation is $< 20\%$, which is far below the maximum of 50% even though the power broadening of the vibrational lines suggests that the stimulated emission pumping step must be saturated. Furthermore, the S/N ratio is inferior to that achieved in our previous results for phenylacetylene (see Sec. 6). The underlying reason for the poor S/N ratio may be

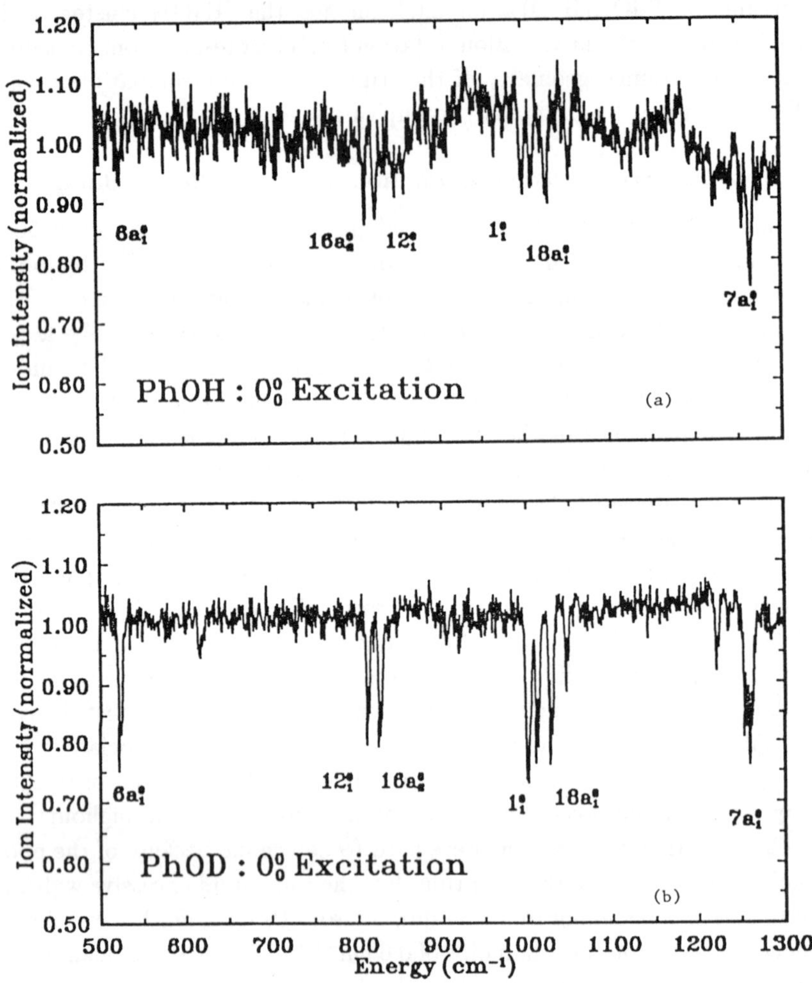

Fig. 7. IDS of PhOH and PhOD through the 0^0 intermediate state. The difference in modulation is due to the longer S_1 lifetime (7×) of PhOD.

found by considering the dynamics of the S_1 state of phenol, in particular the lifetime of the singlet S_1 state relative to radiationless processes that would deplete its population. Using a multiphoton ionization, time resolved pump–probe technique, Sur and Johnson[44] have measured the singlet state lifetime of phenol to be 2 ns. They found that the major pathway for

Table. I. PhOH and PhOD Ground State Assignments.

PhOD[PhOH] (cm^{-1})	Literature (cm^{-1})	Assignment[a]	Error (Obs-Lit)
521[526]	522[526]	$6a_1^0$	-1
616[N/A]	617[619]	$6b_1^0$	-1
824[813]	820[811]	$16a_2^0$	$+4$
810[826]	806[823]	12_1^0	$+4$
903[N/A]	*	N/A	N/A
917[N/A]	917[1176]	$\beta(OH)_1^0$	0
1000[1000]	997[999]	1_1^0	$+3$
1010[1009]	1009[1010]	$16b_2^0$	$+1$
1025[1026]	1025[1026]	$18a_1^0$	0
1043[1054]	1044[1053]	$6a_2^0$	-1
1221[N/A]	*	N/A	N/A
1234[N/A]	1234	$6b_2^0$	0
1259[1266]	1258[1261]	$7a_1^0$	$+1$

[a]harmonic value assumed.

decay of the S_1 state was through internal conversion (IC) to high-lying vibrational states in the S_0 manifold. The rate constant for this process was reported to be $k_{IC} = 27 \times 10^7 \text{s}^{-1}$ and the quantum yield, $Q_{IC} = 0.54$. Also important was intersystem crossing (ISC) to vibrationally "hot" levels of the excited triplet state, T_1, where $k_{ISC} = 19 \times 10^7 \text{ s}^{-1}$ and $Q_{ISC} = 0.38$. Interestingly, the fluorescence lifetime of phenol in the gas phase has not been measured, but Sur and Johnson used values from the liquid phase measurements in cyclohexane in order to make their analyses. However, the quantum yield for fluorescence is $Q_f = 0.08$ and is, therefore, not a major factor in the decay of the S_1 state of phenol.

Lipert, Bermudez, and Colson[31] remeasured these rate constants and quantum yields for phenol, having taken into account their instrument response function. They found that the values for ISC were significantly different; $k_{ISC} = 5 \times 10^7 \text{ s}^{-1}$ and $Q_{ISC} = 0.10$. This means that most of the decay of the S_1 state proceeds via internal conversion, with a rate of $k_{IC} = 42 \times 10^7 \text{ s}^{-1}$, about twice that of Sur and Johnson. Thus, the lifetime of the phenol S_1 state was found to be ≤ 2 ns, which represents an upper limit set by their instrumental response function. Lipert and Colson confirmed their measurements with isotopically substituted PhOD,[45] which showed that

IC was effectively quenched by the heavier mass of the deuterium atom, leading to an S_1 lifetime of 16 ns.

How does this fast decay of the S_1 state impact the ion-dip technique? Since the time delay between pump and dump lasers is in the order of the singlet state lifetime, it must be that the population in S_1 is drastically depleted. Adjusting the temporal delay so that the pump and dump pulses coincided to within an uncertainty of 2 ns produced only slightly better (30%) dip modulation. If we assume an exponential decay for the S_1 state with a lifetime of 2 ns and a time delay of 5 ns between pump and dump pulses, then only 8% of the intermediate population in S_1 would remain for stimulated emission pumping, 92% of the S_1 state having undergone internal conversion in the intervening time delay. In terms of the process of internal conversion, one would expect that high-lying vibrational modes of S_0 would be populated by the time the dump laser interrogates the system. Thus, the IDS spectrum of phenol represents a convolution of the ion signal modulated by stimulated emission pumping with ionization from many highly excited vibrational states of the phenol S_0 manifold. To improve the situation, shorter (picosecond) pulse lasers could be employed. It appears that if one uses only nanosecond lasers, phenol is quite unsuitable for study with IDS due to the short-lived S_1 state.

7.3. *Ion-Dip Spectrum of PhOD: Pumping the* 0_0^0 *Band*

The ion-dip spectrum of PhOD through the 0^0 state was obtained under the identical conditions as the PhOH studies and can be compared directly (see Fig. 7). Immediately apparent is the greater dip modulation and signal-to-noise ratio of the PhOD spectrum. This was expected as a consequence of the longer S_1 lifetime of PhOD (16 ns) relative to PhOH (2 ns) as discussed above. The dip intensity has increased from a maximum of 20% for the $7a_1^0$ band in PhOH to 35% in PhOD, which is suggestive that the rate of ionization is not negligible compared to the rate of stimulated emission.

The observed isotopic frequency shifts for all bands are in accord with the previous IR and UV studies of Bist, Brand, and Williams[28-30] (see Table I). Note that the effect of deuteration on the frequencies of the Fermi doublet at ~ 820 cm^{-1} leads to a crossing of the $16a_2^0$ and 12_1^0 bands. Also, the intensity ratio of the Fermi doublet upon deuteration is much the same as the PhOH doublet, 0.998 vs. 0.989 for PhOH vs. PhOD. This is contrasted with the results for the monohydrated clusters discussed below. The OH bend appears as a weak band at 917 cm^{-1}. This band occurs at 1176 cm^{-1} in PhOH but was not observed due to the poor signal-to-noise. Of interest is the complex band structure around the $7a_1^0$ band at ~ 1250 cm^{-1}.

7.4. *Cluster Ion-Dip Spectroscopy of PhOH(H$_2$O)$_1$*

The IDS of PhOH(H$_2$O)$_1$ was performed through three intermediate vibronic states, as seen in Fig. 8 . These are the 0^0, stretch ($0^0 + 156$ cm^{-1}), and "wag" ($0^0 + 120$ cm^{-1}) modes, as assigned from previous studies. We will deal with each intermediate state IDS spectrum separately; however, all scans were performed under identical conditions. The excited state of PhOD(D$_2$O)$_1$ showed only a moderately intense 0_0^0 transition but very weak intermolecular bands, in contrast to the monohydrated cluster. Hence, ion-dip spectra of PhOD(D$_2$O)$_1$ were not obtained for the stretch or wag intermediate states.

7.4.1. *PhOH(H$_2$O)$_1$: Pumping the 0_0^0 Band*

The addition of one water to phenol makes a dramatic difference in the lifetime of the S_1 state and hence a difference in the IDS spectrum for PhOH(H$_2$O)$_1$ and PhOD(D$_2$O)$_1$ as seen from Figs. 7 and 8. Immediately obvious are the following points:

(i) the major dips are those belonging to fundamentals observed in the bare phenol spectrum, though greatly enhanced in S/N and dip modulation,

(ii) several of the dips now exceed 50%, and

(iii) new features appear throughout the spectrum that are not assignable to phenol fundamental or combination bands.

The increased S/N and the greater modulation of the dips in the cluster spectra compared to the phenol spectrum have the same basis, namely an increase (by a factor of 7) in the lifetime of the S_1 state relative to internal conversion. Both Sur and Johnson,[44] and Lipert and Colson[31,45] measured a huge suppression of the k_{IC} and Q_{IC} in PhOH(H$_2$O)$_1$, which increased the lifetime of the S_1 state to 15 ns. In line with the reasoning presented above for phenol, the S_1 population is now reduced by only 28% relative to the time scale of stimulated emission pumping in this experiment; hence, the increased S/N and dip modulation.

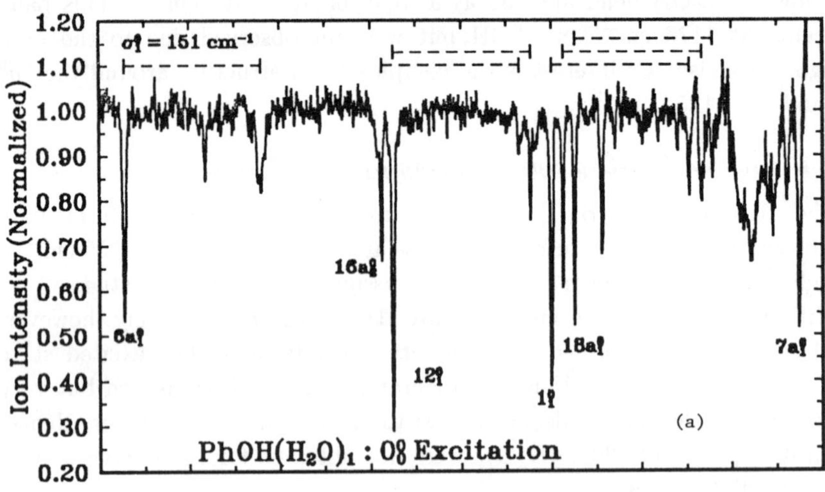

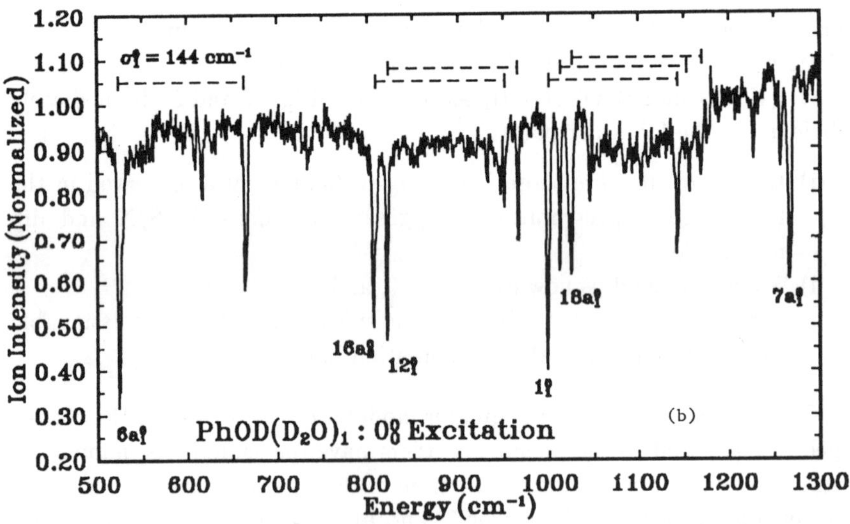

Fig. 8. IDS of the (a) PhOH(H$_2$O)$_1$ and (b) PhOD(D$_2$O)$_1$ clusters through the 0^0 intermediate level. The profound modulation of both spectra is the result of ground state IVR from the phenol chromophore modes into the H$_2$O "bath" modes. Brackets indicate the phenol–water intermolecular stretch frequency.

However, as mentioned above, the maximum dip modulation is 50% as long as the rate of relaxation from the down-pumped state is slower than stimulated emission. This can be either dissociation or IVR. We believe that dissociation is unlikely; H-bond energies are typically larger than 1300 cm^{-1}, although no estimates for this system are evidently known. However, Pohl and coworkers have calculated that the hydrogen bond energy for the p-cresol/H_2O system is approximately 2600 cm^{-1} using an STO-3G basis set.[41] It seems unlikely that the p-cresol methyl group, being in the *para*-position, would have much effect on the strength of the hydrogen bond of that system which allows for a valid comparison to be made between the two systems. Modes with energies less than 1000 cm^{-1} still exhibit $> 50\%$ dip modulation and, thus, it is reasonable to expect that IVR is responsible for the increased magnitude of the dips.

What is driving the IVR process? We believe that it is the influence of the low frequency modes of the attached H_2O, i.e., the *intermolecular* modes between the water and the phenol that lead to depopulation of the various ground state vibrational modes. It should follow that those intramolecular modes which are most highly coupled to the H-bond intermolecular modes will have large IVR rates. Qualitatively, this is what is observed. The "X-Sensitive" (substituent sensitive) modes $6a_1$, 12_1, 1_1, and $18a_1$ (at 528, 825, 1000, and 1026, respectively) all show pronounced enhancement of their dip depth. We are seeing the effects of solvation on vibrational energy transfer, after the addition of only one solvent molecule to the bare chromophore. It is instructive to compare the results for the $PhOH(H_2O)_1$ system with the $PA–NH_3$ cluster. Both clusters have roughly the same density of states available for IVR to occur, but the large difference in dip intensities is more likely a consequence of the profound difference in the intermolecular potential, i.e., a vdW bond as compared to the much stronger hydrogen bond. This is manifested in a much larger electronic coupling term in Eq. (1), leading to a much faster rate for IVR ($> 100\times$). It is possible and valuable to measure the ground state IVR rate ($\delta = k_{IVR}$) using IDS, as will be demonstrated in the next section.

7.5. *Measurement of the Relaxation Rate Constant,* δ, *for the* 1_1^0 *Dip*

The dip modulation can give quantitative information on the kinetics of intermolecular (or intramolecular) vibrational relaxation, even though it may not be a state-to-state rate in the sense that the bath states are unknown. The ratio of M_{off}^{+} to M_{on}^{+} vs. I_2 for the 1_1^0 band in $PhOH(H_2O)_1$ is shown in Fig. 9.

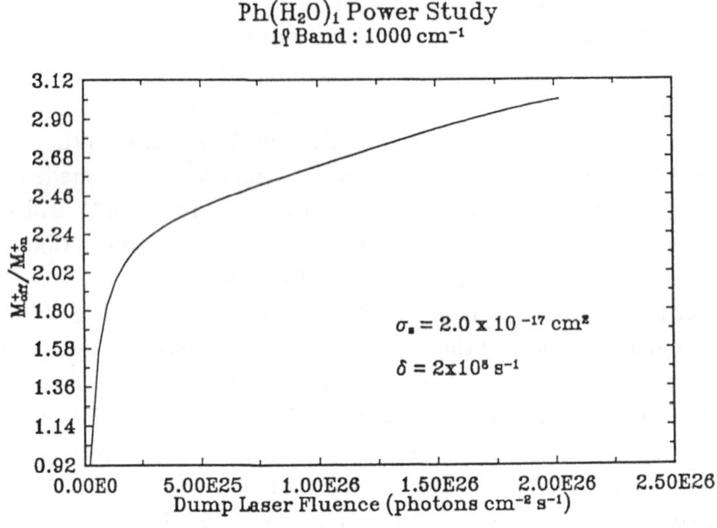

Fig. 9. IVR rate for the 1_1 mode of $PhOH(H_2O)_1$ from a study varying the fluence of the dump laser.

The fluence measurements for the 1_1^0 mode of the $PhOH(H_2O)_1$ system pumped through 0_0^0 give $\sigma_s = 2.0 \times 10^{-17}$ cm^2 and $\delta = 2 \times 10^8$ s^{-1}. The data do not increase linearly with the ionizing laser fluence, as would be expected if $\sigma_i I_2 > \delta$. The stimulated emission cross-section is quite reasonable for an aromatic system. However, we believe the relaxation rate represents a lower limit. By way of comparison, IVR studies on the s-tetrazine–Ar van der Waals molecule give rates[2] of $k_{\text{IVR}} \sim 10^8$ s^{-1}. Given the larger coupling constants for a H-bonded system, we would expect the rate in the $PhOH$–H_2O system to be significantly faster than the preliminary result stated above. Ebata *et al.*[36] have published results for the IDS of $PhOH(H_2O)_1$, pumped through the 0^0 intermediate state. They also measured δ and σ_s for the 1_1^0 dip band. We are in excellent agreement with their measurement of $\sigma_s = 2.0 \times 10^{-17}$ cm^2 but slower than their value of $\delta = 8 \times 10^9$ s^{-1}. This is probably due to a breakdown in the assumption that the rate of ionization is less than the rate of relaxation for high dump laser fluence.

7.6. *PhOH(H$_2$O)$_1$ IDS Line Assignments*

The most prominent of the new features are the bands at 679, 963, and 977 cm^{-1}, with many less intense dips that are dealt with in more detail in the next section. These bands appear to be phenol fundamentals in combination with a 151($\pm$1) cm^{-1} intermolecular band, possibly the intermolecular stretch between the water and phenol molecules. A weaker series is observed involving a 141($\pm$2) cm^{-1} intermolecular mode, which is seen more clearly when the σ^1 state is used as the pump level. The average line width is 3 cm^{-1}, which appears to be independent of the dump laser power over 2 orders of magnitude and may be due to the IVR process.[16] A compilation of the spectral features is found in Table II. The broad band at 1222 cm^{-1} is not seen when pumping the σ_0^1 or $\beta_0''^1$ bands. This leads us to believe that we are hitting a dissociative vibronic state in the *ion* with the absorption of a third photon. This would manifest itself as a broad dip, especially if the S_1 and ion states have different geometries over a large energy gap, as is likely for aromatic systems.

Ebata *et al.*[36] have also observed these two progressions using IDS. Ebata and coworkers attribute the 151 cm^{-1} band to the phenol–H$_2$O stretch. The 141 cm^{-1} band is assigned as an overtone of an intermolecular bend, designated as $\beta_2''^0$. The $\beta_1''^0$ band was not observed. The anomalous transition strength of the $\beta_2''^0$ band (in combination with phenol modes) might arise through interaction with the σ_1^0 band, because overtone transitions are usually much less intense than the fundamentals. The possibility of a Fermi resonance between σ_1^0 and $\beta_2''^0$ exists, but the data cannot directly support this speculation.

7.7. *PhOD(D$_2$O)$_1$ Ion-Dip Spectra: Pumping the 0_0^0 Band*

The ion-dip spectrum for the PhOD(D$_2$O)$_1$ cluster through the 0^0 intermediate state is shown in Fig. 8. The measured frequencies of the ground state intramolecular vibrations are in good agreement with previous IR work on the PhOD molecule,[30] with the exception of the assignment of the 7a band; this is discussed below. The dip modulation of the PhOD(D$_2$O)$_1$ complex is large, which is indicative that the low frequency modes of the hydrogen bonded D$_2$O molecule are responsible for increased IVR. As in

Table II. PhOD(D$_2$O)$_1$/PhOH(H$_2$O)$_1$ Assignments.

PhOD(D$_2$O)$_1$ (cm^{-1})	PhOH(H$_2$O)$_1$ (cm^{-1})	Assignment[a]
523	528	$6a_1^0$
615	618	$6b_1^0$
664	679	$6a_1^0 + \sigma_1^0$
808	825	12_1^0
822	813	$16a_2^0$
951	977	$12_1^0 + \sigma_1^0$
966	964	$16a_2^0 + \sigma_1^0$
999	1000	1_1^0
1013	1012	$16b_2^0$
1026	1025	$18a_1^0$
1056	1045	$6a_2^0$
1095	1103	N/A
1143	1153	$1_1^0 + \sigma_1^0$
1157	1167	$16b_2^0 + \sigma_1^0$
1170	1179	$18a_1^0 + \sigma_1^0$
1227	*	$6b_2^0$
*	1246	N/A
1256	1261	N/A[b]
1267	1274	$7a_1^0$

[a]assumes $\sigma_1^0 = 151$ (144) cm^{-1} for σ_1^0 in PhOH(H$_2$O)$_1$ [PhOD(D$_2$O)$_1$].
[b]see discussion.

the PhOH(H$_2$O)$_1$ case, ground state cluster vibrational activity is observed in combination with the PhOD bands.

The ground state intermolecular stretch is 144 cm^{-1}, observed in combination with the fundamental and overtone bands of the phenol–OD chromophore. This can be compared with the 151 cm^{-1} band assigned as the intermolecular hydrogen bond stretch for the undeuterated complex. The frequency observed in the isotopically substituted complex can be used to calculate whether the above assignment is corroborated by the new data. Thus, the ratio of the undeuterated to deuterated complexes is related to the reduced masses, where the phenol and water molecules are considered to be vibrating relative to each other (point-mass approximation),

$$\frac{\omega_D}{\omega_H} = \sqrt{\frac{\mu_H}{\mu_D}} \, ,$$

where $\mu_D = (m_{D_2O} \cdot m_{PhOD})/(m_{D_2O} + m_{PhOD})$ and $\mu_H = (m_{H_2O} \cdot m_{PhOH})/(m_{H_2O} + m_{PhOH})$. Both ground state and excited state values agree to within 0.3% of the expected harmonic value. This provides satisfactory proof that the 151 [144] cm^{-1} band represents the stretching vibration of the H$_2$O [D$_2$O] molecule against the PhOH [PhOD], and that the phenolic hydrogen is not appreciably transferred to the hydrogen bonded water molecule.

Both our work and that of Ebata *et al.*[36] lead to the assignment that the strong 1277 [1267] cm^{-1} band in PhOH(H$_2$O)$_1$ [PhOD(D$_2$O)$_1$] is the $7a_1^0$ band, an OH[OD] sensitive mode. This is at variance with the band position published by Bist *et al.*, 1262 [1258] for PhOH [PhOD].[30] However, the appearance of a significantly weaker but reproducible band at 1261 [1256] brings into question the accuracy of the assignment of the $7a$ band. While a shift due to hydrogen bonding is expected, we would normally expect this shift to be to a lower rather than a higher energy because the hydrogen bond can support or stabilize non-bonding or anti-bonding electron density. However, a shift to higher frequency would be consistent with a shortening of the C–O bond. This is what is observed for the $7a_0^1$ band in the S_1 (B_2) electronic state of phenol due to the mixing of charge transfer states into the $\pi \rightarrow \pi^*$ transition through greater conjugation in the excited state. In fact, the frequencies of the $7a_0^1$ PhOH and PhOD bands are 1273 cm^{-1} and 1269 cm^{-1}, respectively,[28] which agree well with the ground state monohydrated cluster $7a_1^0$ assignments at 1274 cm^{-1} and 1267 cm^{-1} for PhOH(H$_2$O)$_1$ and PhOD(D$_2$O)$_1$, respectively. Interestingly enough, the $7a_1^0$ band reverts back to 1268 cm^{-1} in the PhOH(H$_2$O)$_3$ cluster system, which is within 2 cm^{-1} of the *ground state* phenol $7a_1^0$ band value at 1266 cm^{-1} (see below). This observation is consistent with the idea that the three water molecules along with the phenol form a four-membered ring, as suggested by the *ab initio* calculations of Pohl *et al.* for the p-cresol–(H$_2$O)$_3$ cluster.[41] The fact that there now exists both proton acceptor and donor would effectively nullify the effect of charge transfer from the primary water acceptor molecule to the C–O bond, thereby leaving the C–O bond length (and frequency) relatively unchanged from its ground state value. An indication of the decreasing degree of charge transfer for the ring system PhOH(H$_2$O)$_3$ is also manifested by the smaller redshift of $PhOH(H_2O)_3$, 91 cm^{-1}, than that of the monohydrated cluster (354 cm^{-1}), as observed previously.

7.8. Intermolecular Intermediate States: the σ_0^1 and $\beta_0'^1$ Bands

When the $PhOH(H_2O)_1$ complex is prepared through the stretch and wag modes in the S_1 manifold the Franck–Condon Principle provides us with a spectrum that is composed primarily of bands of the intermolecular modes of the phenol–water complex (see Fig. 10). Two different low frequency modes are evident, one with a frequency of 141($\pm$2) cm^{-1} and one with 151($\pm$1) cm^{-1}. This can be compared to the water dimer and p-cresol$(H_2O)_1$, which have stretch frequencies of 144 and 146 cm^{-1} respectively. In both cases the H-bond is thought to be "trans-linear".[42,41]

The intermolecular bands now dominate the spectroscopy, and the phenol fundamentals and combination bands that were most prominent in the 0^0 prepared state are much less intense. The most prominent intermolecular activity is seen from 1100–1250 cm^{-1}, where the strong 1_1^0, $16b_2^0$, and $18a_1^0$ bands appear in combination with σ_1^0 and $\beta_2''^0$. Generally, a maximum of two quanta are seen in combination with the observed phenol bands, with the σ_1 mode being the most intense. This would suggest that the ground and excited state potentials of the cluster are slightly displaced.

Pumping the wag mode of the $PhOH(H_2O)_1$ system gives rise to the lower spectrum in Fig. 10. Now, almost all of the intramolecular intensity is gone, with only the intermolecular modes having any appreciable intensity. The S/N is much worse in this spectrum due to the low oscillator strength of the $\beta'^1 S_1 \leftarrow S_0$ transition. However, it is clear that the bands at 669 and 680 cm^{-1} are intermolecular in nature, as are the bands between 950–980 cm^{-1}. Once again, the combination of the intermolecular ground state stretch and the fundamentals (and combinations) of the phenol are seen above 1100 cm^{-1}. These features are tabulated in Table II (see above).

In analogy to the S_1 stretch, we might expect to see activity in the ground state manifold that implicates the bend or wag mode, which is being used as the intermediate state for the spectrum in Fig. 10. Based on the percent reduction for the stretch mode in going from S_1 to S_0, a reasonable range for the ground state wag frequency is 105–115 cm^{-1}. However, no evidence is found for this intermolecular mode from the spectrum. This is odd because the Franck–Condon Principle would suggest that if we pump the wag, we ought to see the wag band in the ground state spectrum, just

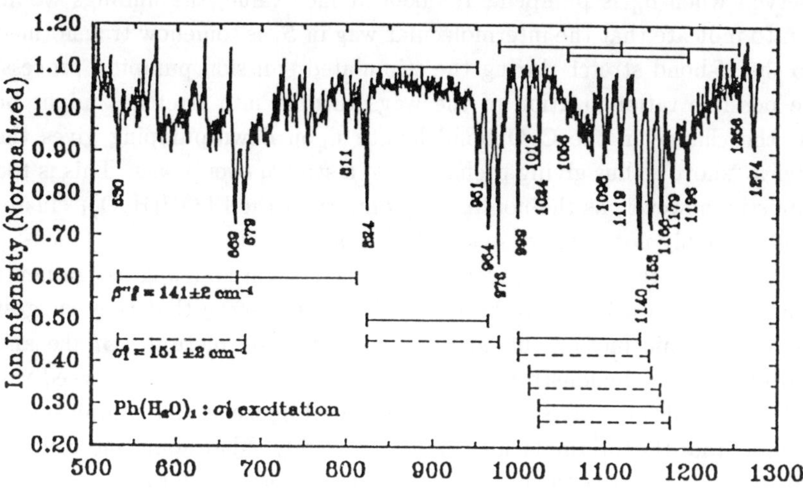

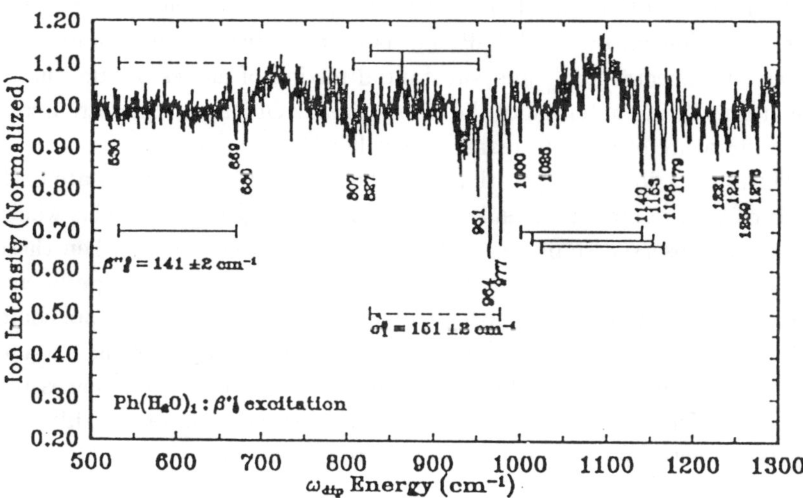

Fig. 10. IDS of the PhOH(H$_2$O)$_1$ and PhOD(D$_2$O)$_1$ clusters through the $S_1\sigma^1$ and β'^1 stretch and wag intermolecular modes. Note that the intramolecular transition intensity decreases relative to the intermolecular transition intensity as the intermediate level takes on more intermolecular mode character.

as was the case when combinations of σ_1^0 and $\beta_2''^0$ with phenol bands were observed when σ_0^1 is pumped. If taken at face value, the findings would seem to indicate that the intermolecular wag in S_1 is somehow transformed into the H-bond stretch during the stimulated emission pumping process. One possible rationalization of the wags turning into stretches might be that the change in the C–O bond length upon down-pumping gives the water a "shove", thus giving preference to a stretch over a wag. This is too simplistic, however, as the nonlinear geometry of the $PhOH(H_2O)_1$ cluster would probably not support so simple a motion.

Another more intriguing idea is that the stretch and wag assignments of the previous workers may be misleading in the sense that they suggest a pure harmonic motion, based on a normal mode picture. Bieske and coworkers[47] have demonstrated that it was possible to assign observed van der Waals modes in the S_1 state of substituted benzene–Ar_1 complexes by assuming that the intermolecular modes are highly anharmonic and coupled through cubic terms in the potential. The wave functions for any particular intermolecular mode are therefore a superposition of several modes, with coefficients that allow for the degree of character of that particular mode. Another example[48] is the fluorescence spectrum of benzene–$[^2H_6]$ where, after excitation into 6_0^1 and $1_0^1 6_0^1$, ν_{10} is quite active in the ground state fluorescence spectrum. Again, substantial mixing of modes occurs due to cubic terms in the potential of the Hamiltonian. Other examples are found in the literature of the s-tetrazine molecule.[2,46]

The implication for this work is that the σ_0^1 and $\beta_0'^1$ bands in S_1 are both superpositions of stretching and wagging (or bending) motions. This would explain the observed similarity in the stretch and wag spectra when these bands are used for intermediate state preparation in the ion-dip experiment. The intramolecular charge transfer character of the S_1 state in phenol could lead to substantially different anharmonic coupling constants than found in the ground electronic state. Given that the H_2O ligand influences the charge transfer nature of the phenol, it is not surprising that quite different vibrational activity is observed between ground and excited states.

7.9. *IDS of PhOH(H_2O)$_2$: Pumping the 0_0^0 Band*

Attempts to obtain a IDS spectrum of $PhOH(H_2O)_2$ were thwarted by the very poor S/N for this cluster. Recent time-resolved measurements of the S_1 lifetime of $PhOH(H_2O)_2$ have been performed by Lipert and

Colson,[49] giving a value of 6 ($\pm$1) ns. Given that the time delay between pump and dump lasers was 7 ($\pm$2) ns for these scans, we would expect less than 32% of the $PhOH(H_2O)_2$ S_1 population to be available for down-pumping. We have not tried to rescan the region with a smaller time delay, though we might expect somewhat better S/N if the lasers were overlapped temporally. Furthermore, the S_1 transition is broad and weak for this cluster, compromising the experiment.

7.10. *IDS of PhOH(H$_2$O)$_3$: Pumping the 0_0^0 Band*

The -91 cm^{-1} origin band in the $PhOH(H_2O)_3$ excitation spectrum (Fig. 4) was used as the intermediate state for the IDS of $PhOH(H_2O)_3$. The results are shown in Fig. 11(a). Again, the major features correspond to vibrations of the phenol chromophore (see Table III). As can be seen from the figure, prominent new features are found at 725, 1191, and 1217 cm^{-1}, which we have assigned as the σ_1 intermolecular mode built off of the major phenol bands, just as in the case of $PhOH(H_2O)_1$. The 1247 cm^{-1} band also appears in the $PhOH(H_2O)_1$ ground state spectrum. We find the S_0 intermolecular stretch (σ_1) for $PhOH(H_2O)_3$ to be 189 ± 1 cm^{-1}, as compared to 188 cm^{-1} for the S_1 state. The stretching motion would be analogous to the phenol–water stretch in the $PhOH(H_2O)_1$ system. For comparison, the value for $PhOH(H_2O)_3$ is nearly the same as the 185 cm^{-1} value for p-cresol$(H_2O)_3$,[41] which was measured using dispersed fluorescence. No evidence could be found for an intermolecular vibration due to an acceptor stretch, which should lie to higher energy around 230 cm^{-1}.

There is also substantial vibrational activity to the high energy side of the Fermi doublet[28] at 816/825 cm^{-1} ($16a_2^0/12_1^0$) and the 1025 cm^{-1} ($18a_1^0$) band, similar to the fine structure observed in the $PhOH(H_2O)_3$ S_1 spectrum. Especially interesting are the two groups of bands centered around 850 and 900 cm^{-1}. With the exception of the 914 cm^{-1} band (assigned as $6a_1^0 + \sigma_2^0$), these bands seem to be very low frequency modes built off of the strongest observed transitions. Interestingly enough, Lipert and Colson[50] have reported similar low frequency structure in the hole burning spectrum of 0_0^0 transition of the phenol–ethanol complex. Such very low frequency structure might be indicative of torsional or constrained "wag" mode behavior in the water ring.

Table III. Assignments for $Ph(H_2O)_3$ and $Ph(H_2O)_4$.

$Ph(H_2O)_3$ $(0_0^0)^a$	$Ph(H_2O)_4$ $(0_0^0)^b$	Assignment
534	535	$6a_1^0$
625	–	–
–	673	–
725	722	$6a_1^0 + \sigma_1^0$
816	815	$16a_2^0$
828	829	12_1^0
843	841	–
852	853	–
860	–	–
868	868	–
893	–	–
907	–	–
914	–	$6a_1^0 + \sigma_2^0$
926	–	–
940	–	–
958	961	–
981	–	–
1004	1004	1_1^0
1017	1013	$16b_2^0, \ 12_1^0 + \sigma_1^0$
1029	1030	$18a_1^0$
1038	–	–
1046	1043	–
1056	–	$6a_2^0$
1064	1065	–
1070	–	–
1083	–	–
1107	–	–
1151	–	–
1191	1189	$1_1^0 + \sigma_2^0$
1205	–	$12_1^0 + \sigma_2^0$
1217	1215	$18a_1^0 + \sigma_1^0$
1247	N/A	–
1268	N/A	$7a_1^0$

[a] $\sigma_1^0 = 189 \pm 1$ cm^{-1}.
[b] $\sigma_1^0 = 185 \pm 1$ cm^{-1}.

7.11. *IDS of PhOH(H₂O)₄: Two Isomers*

Unlike the case for $PhOH(H_2O)_3$ which allows comparison with the results for the p-cresol$(H_2O)_3$ system, no comparisons can be made between the $PhOH(H_2O)_4$ and p-cresol$(H_2O)_4$ because the latter was not observed to have a sharp absorption spectrum under the experimental conditions of Pohl *et al.*[40] As suggested in the discussion of the two isomers for the $PhOH(H_2O)_4$ excitation spectra, it is possible that the -180 cm^{-1} state has the same ring-like structure as $PhOH(H_2O)_3$ but has the fourth water hydrogen bonded (possibly bridging) the cyclic ring formed by the other four molecules. The difference in the ability to obtain a IDS spectrum for these two isomers may be the result of a proton transfer reaction in S_1 for the $\omega_1 = 36\,302$ cm^{-1} state discussed below.

7.11.1. Pumping the 0_0^0 Band: $\omega_{\text{pump}} = 36\,172$ cm^{-1}

The IDS of $PhOH(H_2O)_4$ is shown in Fig. 11(b). There is a striking similarity between the $PhOH(H_2O)_3$ spectrum and the $PhOH(H_2O)_4$ system. Again, an intermolecular stretch has been identified at 185 ± 1 cm^{-1}. The stretch bands are built off of the stronger phenol bands, in particular the $6a_1$, 12_1, 1_1, and $18a_1$ modes (at 535, 829, 1004, and 1030 cm^{-1}) and are seen at 722, 1013, 1189, and 1215 cm^{-1}, respectively. Also, there is significant activity arising to the blue of the Fermi doublet at $815/829$ cm^{-1}, similar to the $PhOH(H_2O)_3$ spectrum. Interestingly enough, the second group of bands observed at 900 cm^{-1} in the $PhOH(H_2O)_3$ spectrum are not present here. Although this might be a consequence of the slightly inferior signal-to-noise ratio for the tetramer system, it is more likely that the absence of the 900 cm^{-1} grouping is the result of a difference in geometry between the $PhOH(H_2O)_3$ and $PhOH(H_2O)_4$ clusters. In particular, if the fourth water was not participating directly in the ring, we might expect that the $PhOH(H_2O)_4$ IDS spectrum would look qualitatively similar to the $PhOH(H_2O)_3$ spectrum; the fourth water would affect only the finer structure of the spectrum.

7.11.2. Pumping the 0_0^0 Band: $\omega_{\text{pump}} = 36\,302$ cm^{-1}

An IDS scan for this state over the 980–1100 cm^{-1} range (not shown) gave only one dip of any intensity, at 1095 cm^{-1}. No features were observed in the 1000–1030 cm^{-1} "fingerprint" region, where we might expect to find the prominent 1_1^0 and $18a_1^0$ bands as in the $PhOH(H_2O)_3$ IDS spectrum in Fig. 11. It is possible that this isomer undergoes a proton transfer reaction in S_1 resulting in a large Stokes shift, which would move the fingerprint dips

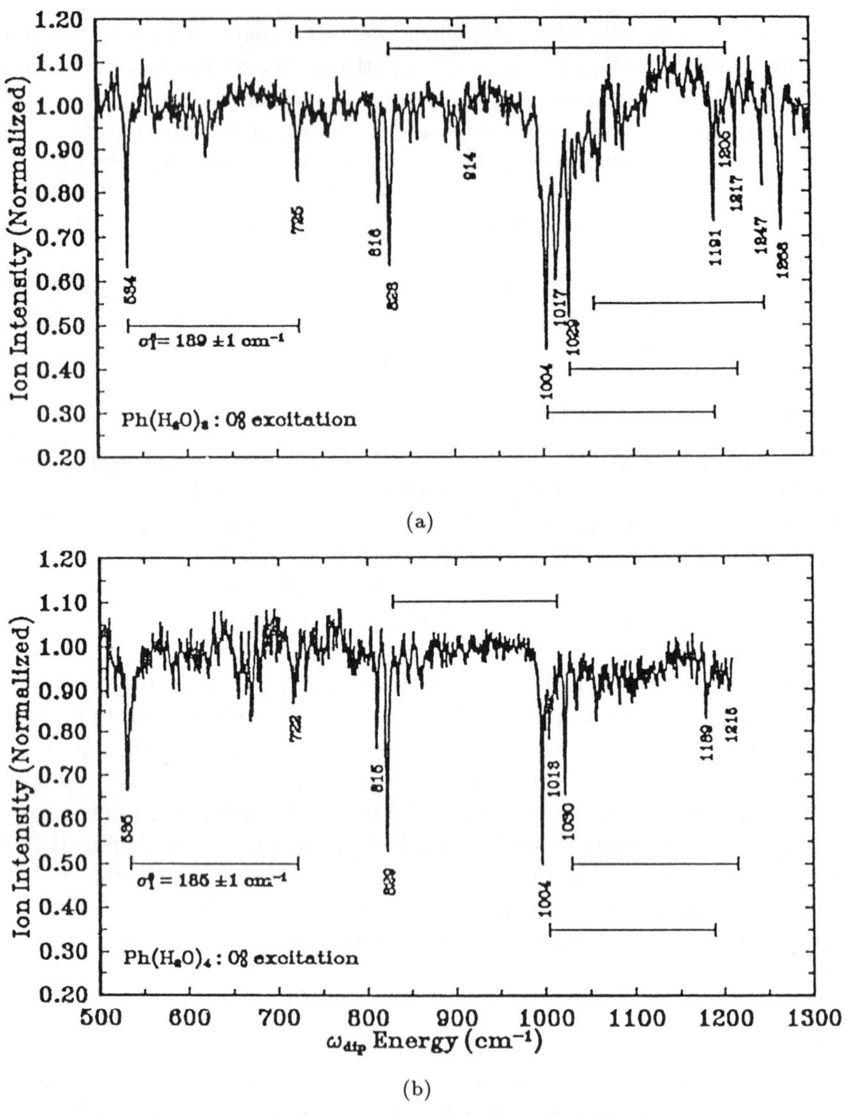

Fig. 11. IDS of (a) $PhOH(H_2O)_3$ and (b) $PhOH(H_2O)_4$ pumping through 0^0. The stretch frequency has moved to 189 cm^{-1} and 185 cm^{-1}, respectively. Note carefully the very low frequency vibrational activity to slightly higher energy of the phenol modes. These intermolecular features are probably bends and constrained torsional vibrations of a four and five membered hydrogen bonded ring structure.

out of the range of the scanned region. Jouvet and coworkers[51] observed that proton transfer occurred in the $Ph(NH_3)_m$ system for $m > 3$. On the other hand, Steadman and Syage have performed picosecond mass-selective pump–probe experiments on the phenol–water system that shows no proton transfer product for $PhOH(H_2O)_m$, $m \leq 5$ when pumping at 266 nm,[52] but it may be that detection of proton transfer via dissociation is wavelength (energy) dependent.

8. Summary

It is shown herein that ion-dip spectroscopy is peculiarly well suited to obtaining well resolved spectra of intermolecular mode structure in cluster systems. The mass selectivity gained by ion detection is extremely important in separating out contamination of the spectra from other species. We can expect that various technical enhancements will expand the usefulness of IDS. Use of injection seeded lasers will improve the signal-to-noise of the data significantly. The use of high resolution lasers can lead to resolved rovibronic spectra of large polyatomics. The possibility for time resolved IDS will involve a trade-off between spectral and temporal resolution, but this is certainly a profitable avenue for new research.

The advantage to using different intermediate states to bring out intermolecular spectral features has been demonstrated above. This can be contrasted with Raman techniques which have not demonstrated this selectivity.[53] However, the most important aspect of IDS that pertains to cluster spectroscopy is that the ion-dip signal *will be enhanced in the presence of intermolecular coupling*, making it exquisitely sensitive to studying the transition from the gas to the condensed phase.

Acknowledgments

Financial support by the U.S. Department of Energy, Grant No. DE-FGO2-88ER-60648, is gratefully acknowledged.

References

1. R.E. Smalley, D.H. Levy, and L. Wharton, *J. Chem. Phys.* **64**, 3266 (1976).
2. D. V. Brumbaugh, J.E. Kenny, and D.H. Levy, *J. Chem. Phys.* **78**, 3415 (1983).
3. D.J. Nesbitt, *Chem. Rev.* **88**, 843 (1988).
4. Z.S. Huang and R.E. Miller, *J. Chem. Phys.* **91**, 6613 (1989).
5. T.R. Dyke, K.M. Mack, and J.S. Muenter, *J. Chem. Phys.* **66**, 498 (1977).
6. N. Pugliano and R.J. Saykally, *Science* **257**, 1937 (1992).
7. D.E. Cooper, C.M. Klimcak, and J.E. Wessel, *Phys. Rev. Lett.* **46**, 324 (1981).
8. T. Suzuki, N. Mikami, and M. Ito, *J. Phys. Chem.* **90**, 6431 (1986).
9. H. Bader, O. Krätzscmar, H.L. Selzle, and E.W. Schlag, *Z. Naturforsch.* **44A**, 1215 (1989).
10. R.J. Stanley and A.W. Castleman, Jr., *J. Chem. Phys.* **93**, 3215 (1990).
11. C.E. Hamilton, J.L. Kinsey, and R.W. Field, *Ann. Rev. Phys. Chem.* **37**, 493 (1986).
12. J.A. Syage and J.E. Wessel, *Appl. Opt.* **26**, 3573 (1987).
13. T. Weber, E. Riedle, and H.J. Neusser, *J. Opt. Soc. Am.* **B7**, 1875 (1990).
14. M. Bixon and J. Jortner, *J. Chem. Phys.* **48**, 715 (1968).
15. G.W. Robinson and R.P. Frosch, *J. Chem. Phys.* **38**, 1187 (1963).
16. K.F. Freed, *Acc. Chem. Res.* **11**, 74 (1978).
17. T.A. Stephenson and S.A. Rice, *J. Chem. Phys.* **81**, 1083 (1984).
18. O. Krätzschmar, H. L. Selzle, and E. W. Schlag, *Z. Naturforsch.* **43A**, 765 (1988).
19. D. Frye, L. Lapierre, and H.-L. Dai, *J. Opt. Soc. Am.* **B7**, 1905 (1990).
20. R.H. Page, Y.R. Shen, and Y.T. Lee, *J. Chem. Phys.* **88**, 4621 (1988).
21. E.A. Rohlfing and J.E.M. Goldsmith, *J. Opt. Soc. Am.* **B7**, 1915 (1990).
22. Unpublished results obtained from two color ionization in our laboratory.
23. G.W. King and S.P. So, *J. Mol. Spectrosc.* **37**, 543 (1971).
24. A.R. Bacon, J.M. Hollas, and T. Ridley, *Can. J. Phys.* **62**, 1254 (1984).
25. Further evidence is supplied by the relative stability of the $PA-NH_3$ complex in S_1 as observed in the parent-to-fragment ion ratio in the excitation scans.
26. G.W. King and S.P. So, *J. Mol. Spectrosc.* **36**, 468 (1970).
27. A. Beyer, A. Karpfen, and P. Schuster, *Topics in Current Chemistry, Vol. 120*, Ed. P. Shuster, (Springer-Verlag, Berlin, 1984) p. 1.
28. H.D. Bist, J.C.D. Brand, and D.R. Williams, *J. Mol. Spectrosc.* **21**, 76, (1966).
29. H.D. Bist, J.C.D. Brand, and D.R. Williams, *J. Mol. Spectrosc.* **24**, 402, (1967).
30. H.D. Bist, J.C.D. Brand, and D.R. Williams, *J. Mol. Spectrosc.* **24**, 413, (1967).
31. R.J. Lipert, G. Bermudez, and S. D. Colson, *J. Phys. Chem.* **92**, 3801 (1988).
32. K. Fuke and K. Kaya, *Chem. Phys. Lett.* **94**, 97 (1983).
33. A. Oikawa, H. Abe, N. Mikami, and M. Ito, *J. Phys. Chem.* **87**, 5083 (1983).
34. R.J. Lipert and S.D. Colson, *J. Chem. Phys.* **89**, 4579 (1988).

35. H. Abe, N. Mikami, M. Ito, and Y. Udagawa, *J. Phys. Chem.* **86**, 2567 (1982).

36. T. Ebata, M. Furukawa, T. Suzuki, and M. Ito, *J. Opt. Soc. Am.* **B7**, 1890 (1990).

37. R.J. Stanley and A.W. Castleman, Jr., *J. Chem. Phys.* **94**, 7744 (1991).

38. R.J. Stanley and A.W. Castleman, Jr., *J. Chem. Phys.* **98**, 796 (1993).

39. H. Abe, N. Mikami, and M. Ito, *J. Phys. Chem.* **86**, 1768 (1982).

40. M. Pohl, M. Schmitt, and K. Kleinermanns, *J. Chem. Phys.* **94**, 1717 (1991).

41. M. Pohl, M. Schmitt, and K. Kleinermanns, *Chem. Phys. Lett.* **177**, 252 (1991).

42. M.F. Vernon, D.J. Krajnovich, H.S. Kwok, J.M. Lisy, Y.-R. Shen, and Y.T. Lee, *J. Chem. Phys.* **77**, 47 (1982).

43. E.B. Wilson, Jr., *Phys. Rev. Lett.* **45**, 706 (1934).

44. A. Sur and P.M. Johnson, *J. Chem. Phys.* **84**, 1206 (1986).

45. R.J. Lipert and S.D. Colson, *J. Phys. Chem.* **93**, 135 (1989).

46. P.M. Weber and S.A. Rice, *J. Chem. Phys.* **88**, 6120 (1988).

47. E.J. Bieske, M.W. Rainbird, I.M. Atkinson, and A.E.W. Knight, *J. Chem. Phys.* **91**, 762 (1989).

48. G. Fischer, *Vibronic Coupling*, (Academic Press, Inc., Orlando, 1984) pp. 108–119.

49. R. J. Lipert and S.D. Colson, *J. Phys. Chem.* **94**, 2358 (1990).

50. R. J. Lipert and S.D. Colson, *J. Phys. Chem.* **93**, 3894 (1989).

51. C. Jouvet, C. Lardeux-Dedonder, M. Richard-Viard, D. Solgardi, and A. Tramer, *J. Phys. Chem.* **94**, 5041 (1990).

52. J. Steadman and J. Syage, *J. Chem. Phys.* **92**, 4630 (1990).

53. V.A. Venturo, P.M. Maxton, B.F. Henson, and P.F. Felker, *J. Chem. Phys.* **96**, 7855 (1992).

Part IV. Theoretical Methods for Extracting Vibrational Dynamics

CHAPTER 20

SPECTROSCOPY AND DYNAMICS IN THE WINGS

E. J. Heller

Department of Chemistry and Department of Physics,
University of Washington, Seattle, WA 98195, USA

Contents

Abstract

Assigning spectra and inferring dynamics from c.w. data can be difficult, and sometimes ambiguous and controversial. The ambiguity can be due to a poor understanding of the nature of the "feature" state whose Franck–Condon amplitudes determine the spectral intensities. Such difficulties can arise when the vibrational overlaps are small, corresponding to the wings of an absorption or emission band. This is the case in SEP at $28\,000$ cm^{-1} in acetylene or in typical radiationless processes.

1. Introduction

Behind every interesting spectrum there is a nonstationary state, one that is "here" and may go "there" during the dynamics. This nonstationary state gives the spectral intensities as the squared overlaps with the eigenstates. Terms like "zeroth order state", "bright state", or "initial wave packet", are used to describe it. The term "feature state" is often appropriate. It has the advantage of focusing attention on the known or knowable nonstationary state associated with a spectral feature, such as identifiable bumps or clumps in spectra. As we discuss below, the feature state may arise naturally out of the built-in spectral intensities, or it may be produced artificially through manipulation of the laser field. A feature state can exist in the laboratory as the state of the system at a given time, by using pulsed laser fields, or it may exist only virtually. In either case, it is vital to decoding the information in a spectrum.

We can obtain the spectrum from the feature state $\phi(0)$ and the eigenstates ψ_n:

$$\phi(0) = \sum_i a_i \psi_i \,,$$

$$\epsilon(\omega) = \sum_i |a_i|^2 \delta(\omega - E_i/\hbar) \,, \tag{1}$$

where ω is the frequency and E_i are the energy eigenvalues of the states ψ_i. Alternatively, the feature state $\phi(0)$ and its dynamics $\phi(t)$ give the spectrum as the Fourier Transform of the overlap[1] as

$$\epsilon(\omega) = \omega(2\pi)^{-1} \int_{-\infty}^{\infty} e^{i\omega t} \langle \phi(0)|\phi(t)\rangle \, dt \,. \tag{2}$$

This form is more directly relevant to the issue of the *dynamics* of the molecule. One of the confusions in this field concerns the two ways, Eq. (1) and Eq. (2), of writing the spectrum. They are equivalent. One does not have to create the wave packet $\phi(t)$ literally in the laboratory to use the form, Eq. (2). The time dependent form also makes clear the potential for interpreting the spectrum at low resolution, since short time dynamics is sufficient to determine the low resolution spectrum.

Uncertainty about the feature state $\phi(0)$ leads necessarily to vagueness about the meaning and assignment of the spectrum. There is a residue of suspicion that some aspects of the spectrum might not be attributable to the dynamics, but rather to the feature state. The spectrum is thus suspected of having a type of nature/nurture problem. ("Nature" is $\phi(0)$; "nurture" is the dynamics.) This problem may be more widespread than is generally believed.

Normally, in spectroscopic measurements the "natural" spectral intensities are allowed to determine the feature state. This means that the amplitudes a_i are determined from matrix elements of an initial eigenstate ψ_0 and final eigenstates ψ_i with the transition moment μ:

$$a_i = \langle \psi_i | \mu | \psi_0 \rangle \equiv \langle \psi_i | \phi(0) \rangle . \tag{3}$$

There is a problem when the matrix elements get very small. This happens for Franck–Condon factors in the wings of electronic absorption or emission bands, in radiationless transitions with moderate to large energy gaps, or for high overtone local mode spectroscopy. All these cases have chronically poor vibrational overlap integrals. In this circumstance, the wings of the feature state $\phi(0)$ can make a large relative difference in the overlap. When considering a single bond coordinate, as in a diatomic molecule, this means some uncertainty in intensity of a transition. However, when several modes are involved, there are much larger issues at stake. A somewhat imprecise statement of the biggest issue is: which mode or combination of modes will have the largest of the small Franck–Condon factors? Another statement of the issue is: in a radiative or radiationless transition, where (in coordinate space, or perhaps phase space) does the amplitude "arrive" on the new potential energy surface?

Normally the geometry differences of the two electronic states are thought to play the crucial role in determining the propensity rules for the excitation of various modes. For example, if a bending mode is the most displaced coordinate in one electronic state relative to another, it is assumed that states with bending excitation will be produced in an electronic transition between the Born–Oppenheimer potential energy surfaces. As we show here, this is not necessarily the case if the transition is governed by small Franck–Condon factors. Chronically small Franck–Condon factors occur in the wings of electronic absorption bands. Large Franck–Condon factors result from the two wavefunctions involved in the integral "doing the same thing in the same place"; i.e., oscillating similarly in some region. Such concerted oscillation can only occur where the momenta are the same in the same region the two potential energy surfaces involved. This is clear from the simplest of semiclassical ideas, e.g., de Broglie's relation $p\lambda = h$, where p is the classical momentum at a given place and λ is the local wavelength there. Consider first the case of a radiationless transition. The two wavefunctions involved in the Franck–Condon factor are necessarily of approximately equal energy, if the coupling is assumed to be moderate to weak. In order for the momenta to be nearly the same at a given position the two *potential energy surfaces must be degenerate.* That is, the region of potential energy surface crossing contributes most to the Franck–Condon factor, provided it occurs where both wavefunctions are reasonably large (classically allowed region). This crossing condition is also well known from the stationary phase approximation to the Franck–Condon integral. Again, the crossing must take place in a region where both the donor and the acceptor wavefunctions are in their classically allowed zones. Otherwise, the integral will be small due to the tunneling attenuation of the states.

The radiative case is very similar to the radiationless transition if we think of raising (in absorption) or lowering (in emission) the initial "donor" potential energy surface by the photon energy. This is nothing fundamental, just an energy bookeeping tool. We use it in Fig. 1 to understand the distinction between the allowed (band center) and forbidden (wing) parts of an absorption spectrum.

2. Vertical and Nonvertical Electronic Transitions

An absorption spectrum for a bound state going to a steep part of an excited electronic state is shown in Fig. 1.[2] As the incident light frequency is increased, we go from the "red" wing, through the band center, to the "blue" wing. A horizontal line between the two classical turning points of the "donor" or initial surface at the quantum energy of the donor state on the donor surface marks the classically allowed region of the initial wavefunction. This surface is progressively raised by increasing $\hbar\omega$. When the surfaces cross in the classically allowed region, the overlap is good and the spectrum is near its maximum. When they cross outside the classically allowed region, the overlap is poor and corresponds to the nonclassical wings.

There are two clearly defined regimes: (i) the "classical regime" corresponds to the realm of favorable conditions for the Franck–Condon overlap, and (ii) the "nonclassical regime" of small Franck–Condon factors. The distinction between the two is simply whether the potential energy

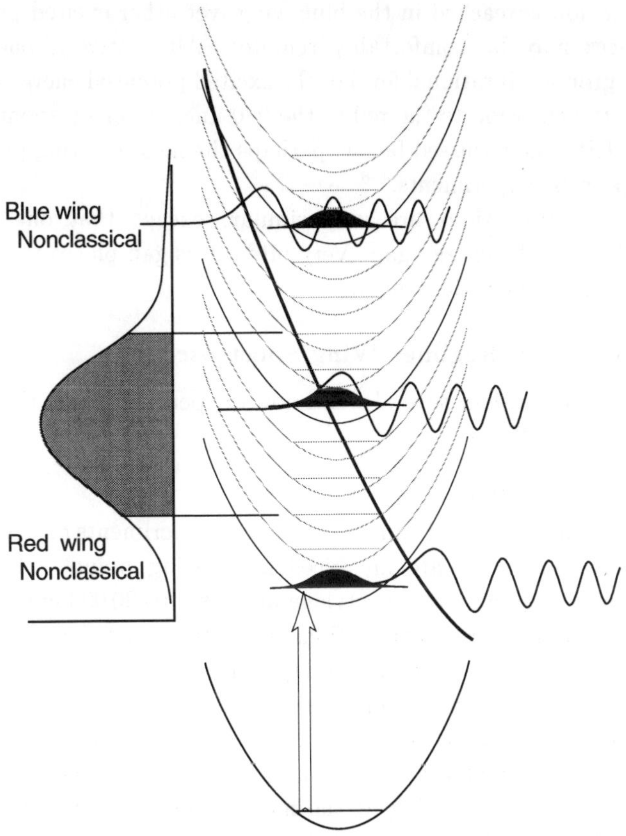

Fig. 1. The construction of the nonclassical wings of an absorption band from the crossing of the donor and acceptor Born–Oppenheimer potential energy surfaces.

surface crossing takes place in the classically allowed region of both surfaces, as we now discuss. This distinction extends to any number of dimensions. Loosely speaking, when the surfaces cross in classically allowed regions, the electronic state change can occur without any instantaneous change or jump in position or momentum. Otherwise, in the nonclassical wings, a jump is required.

A common situation finds the relative horizontal displacement of the two surfaces small. There is no red wing in absorption, and the blue wing starts just above the strong "0–0" transition. This case has interesting experimental implications, because by tuning to the blue of the 0–0 line the nonclassical region is reached in the blue wing, yet other excited potential energy surfaces may be comfortably remote. Alternatively, one could populate the ground vibrational level of the excited potential energy surface and examine the emission to the red of the 0–0 line, either by spontaneous emission or SEP.[3] As discussed below, perhaps the most exciting prospects are for pump-probe experiments.[4–6]

Radiationless transitions are very similar, except that one cannot systematically vary the energy gap. Very often, this gap places the system in question in the nonclassical regime.

3. The Nonclassical Regime: Wing Spectroscopy

As motivation, we consider several well-known experiments and the potential for unexpected effects.

Acetylene SEP in the Wings

The Field *et al.* stimulated emission pumping experiments on acetylene[3] in the red emission wing are an excellent starting point. Red wing stimulated emission pumping in acetylene at 28 000 to 30 000 cm^{-1} excess energy depends on nonvertical transitions in which all the Franck–Condon factors are small because of poor overlap. In the electronically excited S_1 state, most of the excess energy is locked up in the electrons; but 28 000 cm^{-1} of this energy is suddenly delivered to the vibrational degrees of freedom in the stimulated emission pumping transition. Where are the nucleii and what are they doing at the moment of electronic transition? This is again the issue of the nature of the feature state in the wing region.

Considering the electronic transition from S_1 to S_0 as a whole, the corresponding "feature" is the entire electronic emission band. Figure 2 shows an analogous situation. The broad envelope of the band has the feature state shown at the upper left. This is the state whose short time autocorrelation function would Fourier Transform to give the broad envelope. The feature state $\phi(0)$ is at rest on S_0 in the S_1 geometry. Of course, it is on a slope and will move downhill for $t > 0$.

Next, consider the narrower bands within the broad envelope. These have an envelope of their own, and the darkened one has the feature state shown. In acetylene, this would correspond to one of the C–H *trans*-bend, C–C stretch combination bands. It would have been a vibrational

FEATURE STATES

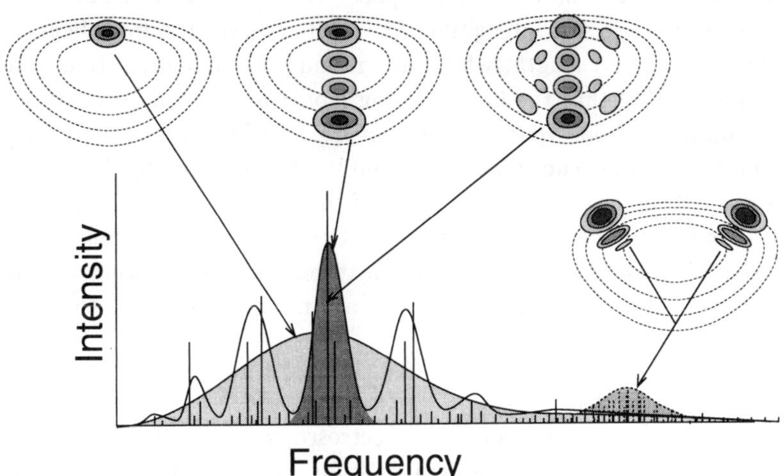

Fig. 2. Natural and artificial features in an absorption, emission, or SEP spectrum. The natural features are the overall envelope (shaded broad band) and the vibrational features, one of which is shaded more darkly. A wing feature is shown, created by a laser pulse with a Gaussian profile; the artificial feature state then takes its character from both the natural Franck–Condon factors and the spectral density of the laser pulse.

eigenstate, but for the couplings to other modes. It decays on a time scale given by the inverse spectral width.

For a resolved line, the feature state is simply the corresponding eigenstate.

Finally, we address the question posed above about what the nucleii are doing as they make the transition in the "red" emission wing to the right in the spectrum. This question is a bit different from the others. What the nucleii are doing at the moment of the transition should show up in the wing feature state. We need to identify a separate feature state for the wing, because the expectation value of the energy of the donor or feature state $\phi(0)$ on the acceptor potential energy surface lies in the middle of the band, and not in the wings. Significant displacement from the Franck–Condon geometry in one or more coordinates, and/or possibly a boost in momentum is necessary to bring the energy, as measured on the accepting potential energy surface, into the wing region; this displacement need not be in the mode most displaced in the Franck–Condon sense.

Only if the feature state is well understood (usually this means localized in a known region of coordinate space) can *dynamical* significance be attached to the spectral intensities. The excited S_1 state is *trans*-bent, with a longer C–C bond length, while the ground state is linear. It is almost taken for granted that the C–H bending modes are initially excited in the transition down from S_1 to S_0, but *which* kind of bend is excited? Three possibilities are *cis*-bend, *trans*-bend, and local mode bending, in which one C–H gets all the energy. In the middle of the band, it has been established that as expected the *trans*-bending and C–C stretching combinations are responsible for the medium resolution features.[7] In the red wing, one must reassess the situation. We take this up again below, after the theory and examples have been presented.

Formaldehyde Photochemistry

Another good example of wing spectroscopy is provided by formaldehyde, which has received considerable study, notably by Moore's group.[8] The lowest vibrational states in S_1 lie at approximately the dissociation energy of S_0. The S_1 surface lies above the S_0 and is "nested" inside it. The radiationless decay from the lowest vibrational states of S_1 must find quasidegenerate states on S_0. These have poor Franck–Condon overlap; the transition is nonclassical.

The couplings to individual levels of S_0 have been cleverly investigated experimentally, Stark tuning them relative to the S_1 level. Statements about the ergodic-like behavior of the eigenstates on S_0 have been made based on these spectral intensities. The lifetimes measured from individual resonance peaks are rather long, spanning nanoseconds to microseconds. However, to interpret spectra, or make statements about ergodicity, one should know the nature of the initially prepared state, i.e., the wavefunction initially prepared on the S_0 potential energy surface.

Suppose the molecule starts in the ground vibrational state ψ_0^1 of S_1. Assuming that the transition moment connecting S_1 and S_0 is only weakly coordinate dependent, the state ψ_0^1 is (apart from an overall constant) just $\phi(0)$, the donor state.

Although the energy of ψ_0^1 is quite sharp on the S_1 surface, it spans a broad range of energies on the S_0 surface. The expectation value of the energy of ψ_0^1 measured on S_0 can be estimated by

$$\langle H_0 \rangle = \langle \psi_0^1 | H_0 | \psi_0^1 \rangle \approx H_0(\langle \mathbf{p} \rangle, \langle \mathbf{x} \rangle) \,, \tag{4}$$

where p represents the momenta and x the positions of the nucleii, and H_0 is the Hamiltonian for the S_0 Born–Oppenheimer potential energy surface. Now, since $\langle \mathbf{p} \rangle = 0$, we have $\langle H_0 \rangle \approx V(\langle \mathbf{x} \rangle)$. That is, the expectation value of the energy of ψ_0^1 measured on S_0 is roughly the energy of the S_0 potential evaluated at the equilibrium geometry of S_1. For nested surfaces, this energy is far below the narrow band of energy eigenstates coupled nonradiatively to ψ_0^1. Clearly, to visualize the quasidegenerate nonradiative transition, the relevant amplitude arriving on S_0 from S_1 due to the weak nonadiabatic coupling cannot be thought of as starting at rest at the geometry of the S_1 minimum. This would correspond to far too low an energy. There must, therefore, be some large displacement in position or momentum on S_0, bringing the energy up to the quasidegenerate requirement. What is this displacement, and what factors control it? May we view the wing-feature state as still localized, or is it somehow already spread out on the S_0 Born–Oppenheimer potential energy surface?

The state ψ_0^1 of S_1 has a probability density ${\psi_0^1}^2(\mathbf{x})$ in coordinate space, and ${\psi_0^1}^2(\mathbf{p})$ in momentum space. Most of the time the nucleii do not have the right position or momentum to arrive on the acceptor potential energy surface with the necessary energy to remain quasidegenerate with the initial energy. But the wavefunction has tails which extend in every direction in position and momentum, implying that the molecule spends some time at large excursions from equilibrium. Occasionally, an excursion will happen which gives it the correct quasidegenerate energy, and then a transition $S_1 \rightarrow S_0$ is possible. The probability of such an excursion is governed by the form of ${\psi_0^1}^2$ and thus the excited potential energy surface, but *which* excursions permit the degeneracy is a function also of the ground state potential energy surface.

Singlet–Triplet High Resolution Spectra

The high resolution singlet–triplet spectra due to spin-orbit radiationless transitions, as for example in the work of Pratt and co-workers,[9] are another important class of experiments. One measures features, which are fragmented into clusters of lines of lower and fluctuating intensity due to the spin-orbit interactions between a lowest triplet T_1 Born–Oppenheimer surface and S_1, or possibly between T_1 and S_0. The spectra are again controlled by Franck–Condon factors that are very often quite small. How do we interpret such spectra, not knowing much about the feature state that produced them? What do they tell us about dynamics on T_1?

Van der Waals Clusters

Another example comes from the decay of excited quasibound van der Waals complexes. High frequency vibrations are playing the role of electronic states. The essential story is the "momentum gap", as elucidated by Ewing.[10] Again, nonvertical, nonclassical Franck–Condon factors (in the sense defined above) are responsible for the decay of the clusters that have energy above the threshold for dissociation. They decay slowly because of the small Franck–Condon factors. The cluster fragments are launched, so to speak, on their separate ways with a large boost in momentum.

Isotope and Promoting Mode Effects in Photodissociation

The work of Imre and co-workers[11] pointed to a way to exploit the wings of the Franck–Condon feature state to enhance isotope selectivity in certain photodissociation experiments. For example, branching ratios in the dissociation of HOD between hydrogen and deuterium production depend dramatically on wavelength and the initially excited (local) mode of HOD. These effects can be understood in terms of the theory given below.

Pulsed Transient Feature State in the Wing

It is easy to produce certain kinds of artificial feature states. By applying a laser pulse centered in frequency on a part of the wing, and of a duration such that some range of states is excited, an artificial feature is produced (Fig. 2, right). The population in each of the eigenstates will be governed by two factors: the natural Franck–Condon factor, and the energy density of the pulse at that frequency, which is controlled by the pulse duration and center frequency. There is an associated transient wave packet or feature state that can be defined as

$$|\phi_{\omega_0,\sigma}\rangle = \sum_n e^{-(E_n - \hbar\omega_0)^2/(2\sigma^2)}|n\rangle\langle n|\phi\rangle \; . \tag{5}$$

Clearly, $|\phi_{\omega_0,\sigma}\rangle$ is a nonstationary wave packet that can be prepared in the laboratory.

The corresponding feature state shown in Fig. 2, right, is surprising. What causes it to move out to the sides? The displaced mode is the vertical one; why wouldn't this mode still carry all the excitation, as it does in the middle of the spectrum? Clearly, to understand the spectral intensities in this region we should be thinking of the strange feature state in which both modes are excited (displaced) initially. The reason for this and for even stranger behavior will become clear below.

4. Position versus Momentum Jumps

The Franck–Condon factors in the wings will be small, but in the blue wing there are two possible sources of the small residual: Either it results from an incomplete oscillatory cancellation in the region of large amplitude, or it could come from the overlapping region of the tails of the two states. This distinction leads to important effects in two or more dimensions. If the tail region is the major contributor, the integral is coming from a region not within the classically allowed region in coordinate space; we call this a "position jump".

On the other hand, suppose the major part of the (small) integral comes from the region of large amplitude. Then the rapid oscillations of the final (acceptor) state imply a sudden change of momentum as the transition from donor to acceptor is made. We call this a momentum jump case.

Consider the Golden Rule expression for the rate of the electronic transition,

$$k = \frac{2\pi}{\hbar} \langle V^2 \rangle \rho(E_0),\tag{6}$$

where $\rho(E_0)$ is the density of states at energy E_0, and $\langle V^2 \rangle$ is the mean squared matrix element. Making the Condon approximation, we write this as

$$k = \frac{2\pi}{\hbar} V_0^2 \ \mathrm{Tr}\ [\delta(E_0 - H_a)|d\rangle\langle d|]$$

$$\equiv \frac{2\pi}{\hbar} V_0^2 \ \mathrm{Tr}\ [\delta(E_0 - H_a)\rho_d]\tag{7}$$

where $|d\rangle$ is the donor wavefunction, V_0^2 is the constant coupling strength, and $\rho_d = |d\rangle\langle d|$ is the density matrix for the donor wavefunction. We simplify this further here by approximating Eq. (7) as

$$k \approx \frac{2\pi}{\hbar} V_0^2 \int d\mathbf{p}\, d\mathbf{q}\ \delta[(E_0 - H_a(\mathbf{p}, \mathbf{q}))]\, \rho_d^W(\mathbf{p}, \mathbf{q})\ .\tag{8}$$

In Eq. (8) the rate k is proportional to the integral of the Wigner density of the donor, $\rho_d^W(\mathbf{p}, \mathbf{q})$, over the energy hypersurface $\delta[E_0 - H_a(\mathbf{p}, \mathbf{q})]$ of the acceptor. We want to ask *where* on the acceptor energy surface the bulk of the rate integral comes from. This is a matter of investigating the *integrand* of Eq. (8). By understanding the phase space structure of the donor wavefunction and the energy hypersurface of the acceptor Hamiltonian, we understand where amplitude appears on the acceptor potential energy surface. The rate k is proportional to the average Franck–Condon factor; it is a non state specific quantity. By interpreting the integrand of Eq. (8), we are putting back the specificity in a very useful way, by saying where (and with what velocity) the amplitude appears on the acceptor surface.

Considered as a Wigner phase space distribution, the donor wavefunction is a smooth distribution peaked at zero momentum. In coordinate space it is peaked at the minimum of the donor potential energy surface. The bulk of such a phase space density *has too low an energy* on the *acceptor* potential energy surface to be germane to the donor $\rightarrow$ acceptor transition. The transition has to be quasidegenerate with the donor state. To find the relevant regions in phase space we must examine those parts of the "tail" of the Wigner density that have the right energy on the acceptor potential energy surface.

We are treating the donor and acceptor wavefunctions differently. The donor is given a full fledged Wigner density, whereas the acceptor is treated as a classical energy hypersurface. The inequivalent treatment requires justification. This was done before in a treatment of the *rate* of radiative transitions, where Eq. (8) was used directly, in the classical and nonclassical regime.[12] The results were very satisfactory, and an improvement over the reflection approximation. However, we did not attempt to go deep into the nonclassical regime, nor did we analyze the rate in terms of the integrand of the rate expression.

We examine a one dimensional system in Fig. 3 to get started. Two cases are shown corresponding to blue wing absorption (or red wing emission). The wavefunctions are shown on the left. On the right, the phase space pictures are shown. The Wigner phase space distribution for the donor state is a smeared distribution, and the quasidegenerate contour interval on the acceptor potential energy surface is the parabolic curve. In the top case, the donor wavefunction is narrow, so the oscillations of the acceptor wavefunction do not completely kill the integral. Furthermore, the tail of the donor wavefunction is very small near the tail of the acceptor wavefunction. There is no doubt that the overlap comes from the region where the donor wavefunction is large. This is a momentum jump case, because the amplitude leaving the donor wavefunction may be thought of as appearing near the donor wavefunction, with considerable momentum on the repulsive acceptor potential. The phase space picture reflects the increased momentum uncertainty and decreased position uncertainty associated with the narrow coordinate space distribution. Clearly a shift in momentum at (nearly) constant position gives the shortest path between the two distributions.

At the bottom of Fig. 3 we see a much wider donor wavefunction, centered as in the previous case. A glance at the wavefunction plots shows there is now good overlap of the tails, and more nearly complete cancellation of the integral in the oscillatory region. The integral is coming mostly from the region of the tails, and the phase space plot shows that a position jump (again at zero momentum) gives the shortest path between the two states.

We have just seen that position jumps and momentum jumps can compete to be the major contribution to the Franck–Condon integral in the nonclassical regime. In several dimensions, each position and each momentum become competitors. The winning coordinates can be quite surprising. Thus, for nonclassical transitions, the usual Franck–Condon propensity rules can be very misleading. For example, the mode with the largest equilibrium position displacement need not be excited in a nonclassical Franck–Condon transition in the wings.

Teeter-totter Effect

Some examples of this kind are well-known.[14] Radiationless transitions in aromatic hydrocarbons can be dominated by the carbon–hydrogen stretching motion (they are the accepting modes on the accepting surface) even though the carbon–carbon backbone is much more displaced. The competition between two modes for Franck–Condon intensity can be understood by a simple picture, which we call the "teeter–totter" model (Fig. 4). It illustrates a two mode case, with both modes displaced. There is a low frequency mode and a high frequency mode, but the low frequency mode is more displaced. The modes are assumed to be separable normal coordinates. The solid curves show potentials and states from the ground potential energy surface, and the dashed curves show potentials and

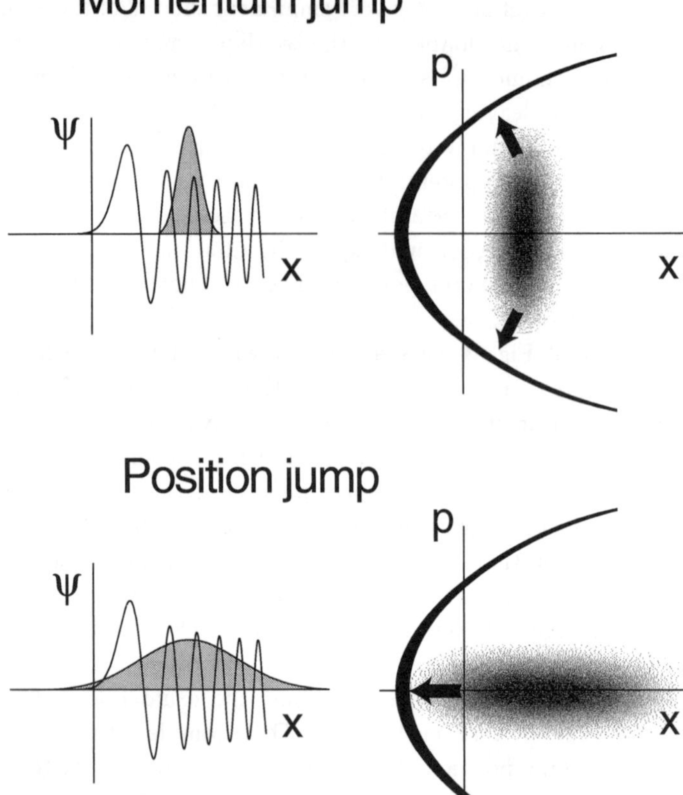

Fig. 3. Two blue absorption wing cases, a momentum jump and a position jump, are shown in coordinate and phase space representations.

states from the excited potential energy surface. The teeter-totter has the property that the sum of the heights of the two ends is a constant. This corresponds to the necessity that the energy of the two ground state modes add up to the fixed energy of the initial, donor state. The top scenario wins, with the accepting mode being the high frequency mode. The small Franck–Condon factor of the high frequency mode in that case is still much larger than if the same energy is taken up by the low frequency mode, as in the lower scenario. This is the situation in the case of the aromatic hydrocarbons.

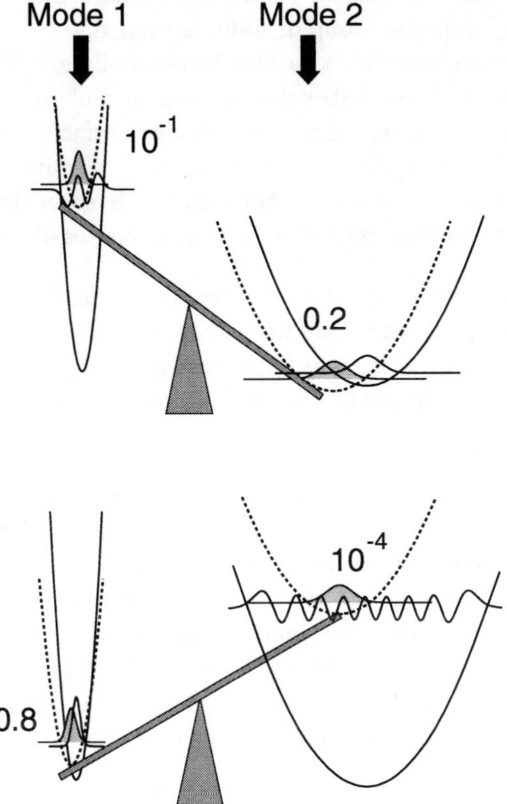

Fig. 4. This figure shows why the high frequency mode will become the acceptor mode in the two mode competition illustrated. This is true even though the high frequency mode is much less displaced. A fixed amount of energy must be accomodated by the two modes, and the teeter-totter accounts for this. The Franck Condon factor is much smaller if the low frequency mode accepts the energy. The individual Franck–Condon factors are shown; the product of the two upper Franck–Condon factors is larger than the two lower Franck–Condon factors; therefore, the upper scenario has the faster rate.

5. C–H Oscillators

A pulsed laser (typically of 50 femtoseconds to a few picoseconds in duration) may prepare a nonstationary wave packet on the acceptor potential energy surface. We want to create a "nonclassical" wave packet by tuning the laser frequency to the blue of the absorption band (or analogously the red wing in emission) and adjusting the pulse duration so that an appropriate group of states is excited.

Consider two identical, separable Morse oscillators. They may represent two equivalent C–H bond stretches in the same molecule, for example. Both the donor and the acceptor potential energy surfaces are Morse potentials but the equilibrium positions, well depths, and effective force constants will vary between the donor and acceptor states. Even in this simplified donor $\rightarrow$ acceptor transition, several interesting scenarios are possible:

- Either position jumps or momentum jumps can be dominant.
- The C–H bonds may share the energy in the acceptor nonclassical wave packet equally, or one may be excited much more than the other.
- The two previous items, each with two extremes, combine to make four scenarios.

In what follows, we shall see two of the four scenarios arise with slight adjustments of the *donor* potential only; the acceptor potential will remain the same. (Indeed, it is only the differences in the two potentials that matter.) We will examine the cases with two tools: first, we create the wing state with a transform limited pulse bandwidth. By inspecting the location and nodal structure of the wing state, the various possibilities for position and momentum jumps appear as the parameters are varied. Second, we plot the phase space density of the function

$$\delta[E_0 - H_a(\mathbf{p}, \mathbf{q})]\, \rho_d^W(\mathbf{p}, \mathbf{q}) \,, \tag{9}$$

where $\delta[E - H_a(\mathbf{p}, \mathbf{q})]$ is the energy hypersurface ("energy contour") of the accepting Born–Oppenheimer Hamiltonian, and $\rho_d^W(\mathbf{p}, \mathbf{q})$ is the Wigner density of the donor wavefunction. The Wigner method is fast and applicable to several nonseparable degrees of freedom, making it possible to explore the effects of small changes in the potential energy surfaces rather easily.

The parameters for the two cases are as follows. Both have the acceptor potential energy surface

$$V_a(x, y) = D_a[1 - \exp(-\lambda_a x)]^2 + D_a[1 - \exp(-\lambda_a y)]^2 \,. \tag{10}$$

with $D_a = 20, \lambda_a = 0.2$.

The kinetic energy is simply

$$T = \frac{1}{2}(p_x^2 + p_y^2) \, . \tag{11}$$

The two donor potential energy surfaces are of the form

$$V_d(x,y) = D_d[1 - \exp(-\lambda_d(x - x_0))]^2 + D_d[1 - \exp(-\lambda_d(y - y_0))]^2 \, . \tag{12}$$

The parameters for the two examples are:

Case	D_d	λ_d	$x_0 = y_0$	E_0	σ
1 (momentum jump, unshared)	28	0.20	0.55	16	2.24
2 (position jump, shared)	17	0.25	−0.5	16	1.73

The examples nicely illustrate competition for the role of accepting mode, or more properly accepting coordinate. The donor potential in the first example is tighter and has its equilibrium shifted to a larger distance, as compared to the accepting potential. Both differences tend to favor momentum jumps. The donor potential in the second example is looser and has its equilibrium shifted to a smaller distance, as compared to the accepting potential. In this case, both differences tend to favor position jumps.

These trends are exactly what is seen in Fig. 5 and Fig. 6. Both figures show the accepting potential contours, with the contour for $E = 16$ darkened; this is the average final energy on the acceptor potential energy surface. This is the center energy E_0 of the Gaussian energy envelope, Eq. (5). It is also the energy E_0 of the classical energy surface (see Eq. (9)) on which the Wigner density of the donor wavefunction is evaluated.

The phase space density of the Wigner function $\rho_d^W(\mathbf{p}, \mathbf{q})$ is plotted as a series of line segments representing the momentum distribution at selected positions. The direction of a line segment shows the direction of the momentum, and its length is proportional to the magnitude of ρ_d^W as a function of momentum for the given position, and not the momentum itself.

In case 1 (Fig. 5), note that the wave packet and the phase space distribution agree on the predominantly *independent* bond momentum jump in this case (the wave packet is a linear combination of two symmetrically related wave packets, each with energy mostly in one bond). Further, the energy is stored mostly as momentum, as seen from the nodal structure of the wave packet and the fact that the wave packet is very small near the classical contour $V(x,y) = E_0$.

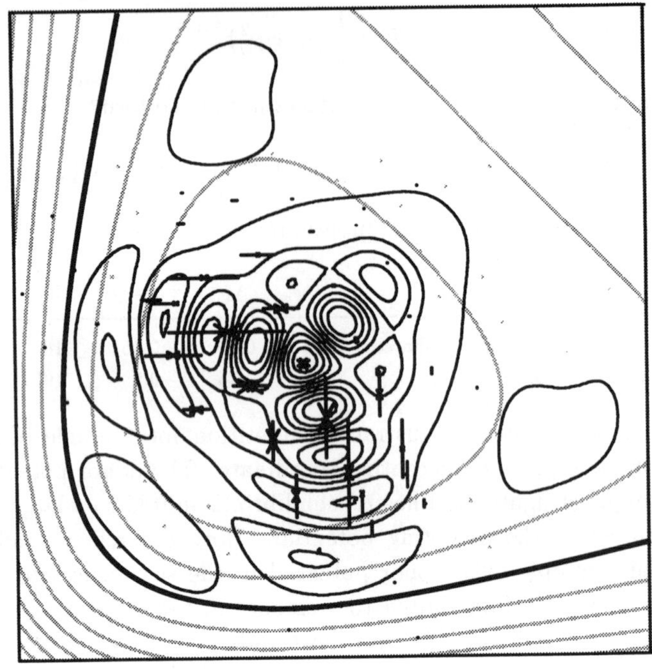

Fig. 5. The acceptor potential energy surface contours ($E = 16$ darkened) together with the pulse excited wave packet and the phase space density of the Wigner function of the donor ground state, ρ_d^W at $E = 16$. The latter is plotted as a phase space distribution, as a series of line segments representing the momentum distribution at selected positions. The direction of a line segment shows the direction of the momentum, and its length is proportional to the magnitude of ρ_d^W as a function of momentum for the given position, not the momentum itself. Note that the wave packet and the phase space distribution agree on the predominantly independent bond momentum jump in this case (the wave packet is a linear combination of two symmetrically related wave packets, each with energy mostly in one bond.

In case 2 (Fig. 6), the wave packet and the phase space distribution agree on the *correlated* bond position jump (the wing state has both bonds sharing the energy simultaneously). The energy is stored mostly as potential energy, as seen from the fact that the wing state wave packet is very near the classical contour $V(x, y) = E_0$.

The wave packet has energy uncertainty, given by the energy width σ of the envelope used. There is a reciprocal time uncertainty implied; one effect of this is that the wing state has some amplitude and nodes leading away from the region in phase space where the amplitude is largest according to the Wigner phase space density.

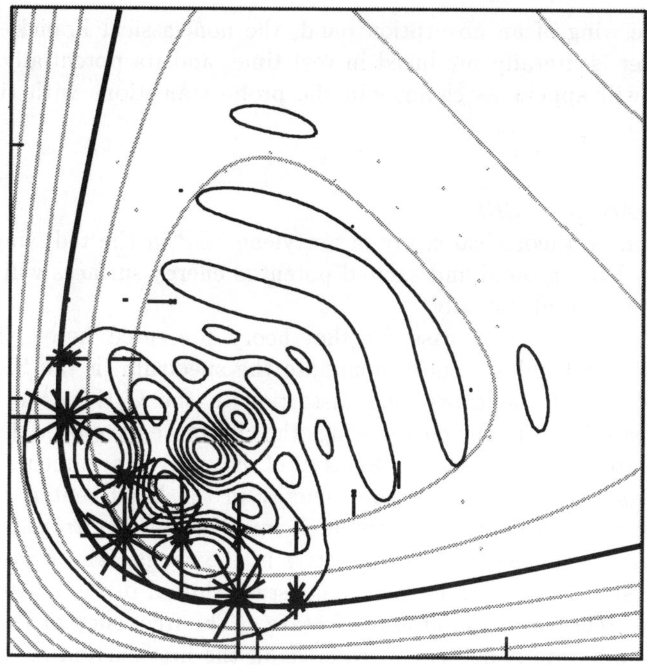

Fig. 6. Same as Fig. 5, for case 2. The wave packet and the phase space distribution agree on the correlated bond position jump (the wave packet has both bonds sharing the energy simultaneously).

The wing state wave packets and the Wigner phase space density agree remarkably well. This raises the hope that predictions of the acceptor coordinates in general circumstances will be quite feasible. The wing state wave packet is quite sensitive to the relative shape of the potentials.

Our simple results have already revealed a case where the electronic energy shows up primarily in one of two nominally equivalent bond modes (case 1), and another case where the energy is shared equally (case 2). One case (case 1) was primarily a momentum jump, and the other (case 2), a position jump.

Direct or indirect experiments depend critically on the nature of the nonclassical wave packet. In direct measurement, the effects of putting energy into very specific modes (and indeed with specific initial momenta or positions) can have a dramatic effect; e.g., the breaking of a specific bond. In a pulsed laser experiment, the accepting modes can dramatically change as the energy gap is increased by tuning the center frequency to the blue. Pump-probe experiments[4-6] would be ideal. By exciting just a part of the blue wing of an absorption band, the nonclassical Franck–Condon wave packet is literally produced in real time, and its potentially bizarre character will appear as changes in the probe transitions with time and detuning.

Return to Acetylene SEP

A definitive theoretical study of acetylene SEP in the red wing lies in the future. Both ground and excited potential energy surfaces will have to be explored computationally.

This is a fascinating case for the theory presented here. The low frequency C–H bending modes dominate the spectrum in the SEP band center because of the *trans*-bend distortion in the C_2H_2 $\tilde{A}^1 A_u$ excited electronic state, but in the far red wings the effects we have been discussing can dominate the displacement. Jonas *et al.* have given arguments to show that the *cis*-bend is active at lower energy in the band center, but only by resonance with the initially populated *trans*-bend.[7] The *trans*-bend also stands a good chance of remaining active in the wings: The *trans*-bend is strongly displaced *and* strongly force constant shifted in the excited C_2H_2 $\tilde{A}^1 A_u$ state relative to the ground X state. (The frequency of *trans*-bend in the ground state is 609 cm^{-1}, whereas in the first excited A state it is 1048 cm^{-1}).[15] This is a large shift that suggests momentum jumping in the *trans*-bend coordinate, since the momentum uncertainty of the donor wing state is so relatively high because of the high force constant.

Force constant changes between the ground and excited states, or high frequency modes with relatively small displacements, can dominate the most displaced modes in wing transitions. In this case the implication is that the C–C and C–H stretches could enter the picture in the red emission wing. Will the C–H stretch, which Jonas has shown to be inactive in the 7000 cm^{-1} region, rear its head in the wing?

The recent report of six sparse, narrow features in the red emission SEP wing near 27 500 cm^{-1} C_2D_2 is especially intriguing.[16] One plausible explanation is that the wings of the initially populated vibrational level in the C_2D_2 $\tilde{A}^1 A_u$ electronic state have found a region of stable, quasiperiodic motion in phase space, leading to the sharp, sparse features.

Less evident but just as important is the competition between types of C–H motion. Is the *trans*-bend still the accepting mode in the red wing?

Many existing experiments could benefit from an understanding of the effects of the wing states discussed here. For example, given the extensive work on formaldehyde $S_1 \rightarrow S_0$ predissociation,[8] it would be good to know whether the hydrogens are "launched" on the S_0 potential energy surface with equal or unequal energies (this is the question treated here for a model potential).

6. Dynamical Information in a Spectrum

Until now, the concern has been about the nature of the initial state that is "launched" on the accepting potential energy surface, and not about the subsequent dynamics of that state or the resulting spectrum. We discuss these issues now.

It appears that the wing states are typically rather well localized in coordinate space. More generally, they are localized in phase space. The potential for delocalization exists, should the countervailing forces affecting one type of excitation over another happen to balance approximately.

It is important to recall the inherent limitations of absorption or emission spectroscopy. We would like to know the full dynamics that ensues from the initial localized wing state. All we get to see, however, is the view through the "keyhole" of the initial state itself, as is evident from the formula for absorption (Eq. (2)). (See Fig. 7.) However, this is a dynamic view; the full time history of the overlap $\langle \phi | \phi(t) \rangle$ is available as

$$\langle \phi | \phi(t) \rangle = \int_{-\infty}^{\infty} e^{-i\omega t} \frac{\epsilon(\omega)}{\omega} \, d\omega \ . \tag{13}$$

By peering through the keyhole long enough, much about the dynamics of the molecule can often be inferred. However, an unambiguous sequence of events is difficult to infer in the wing region, because the spectrum there can be much less well organized hierarchically into clumps and sub-clumps, which translate into definite sequential recurrences in the time domain.

The difficulty with a typical spectrum is that the evidence for dynamics is indirect. Photodissociation offers a more straightforward view of the dynamics, especially if the dissociation is direct. Figure 8 shows some possibilities and serves to illustrate the effects we have been discussing. In case A the ground electronic, ground vibrational state eigenstate of the donor potential energy surface (not shown) is seen as a hatched ellipse. This ellipse represents the feature state corresponding to the whole electronic band.

In case B, the result of tuning near the band center is shown. This is an artificial feature state which is produced by a pulsed laser with center frequency at the band center, and a width somewhat less than the bandwidth of the electronic transition. The wavefunction in B starts at the Franck–Condon region and heads downhill (along X) from there, just

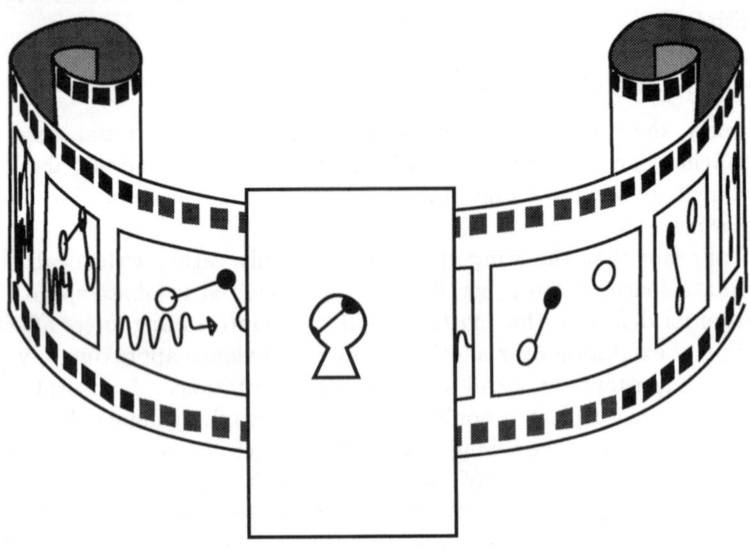

Fig. 7. Cartoon showing the limits to the information available in an absorption (or emission) experiment. The projection of the dynamcs $\phi(t)$ onto $\phi(0)$ is "seeing" the dynamics through just one state in Hilbert space.

as did the time dependent wave packet whose Fourier Transform produced the wavefunction.

The result of preparing a wing state by tuning off resonance can be quite different. In case C we see the result of tuning to the blue of the absorption maximum for the same donor and acceptor wavefunctions and potential energy surfaces shown in A and B. The blue (higher energy) wing state has position tunneled (jumped) along the Y direction, the long axis of the ellipse representing the ground state. The state starts from this displaced position and heads downhill from there. The potential energy surface is soft (and the wavefunction tails relatively large and slowly decaying) in the Y direction, corresponding to the low frequency mode. Since the acceptor potential energy surface is also fairly steep in that direction, the easiest path is to position jump in the Y direction.

The two donor normal modes are reversed in case D. With the "hard" and "soft" modes exchanging roles, a possible outcome is shown. A position jump has occurred along the "soft" direction again, this time along X. In

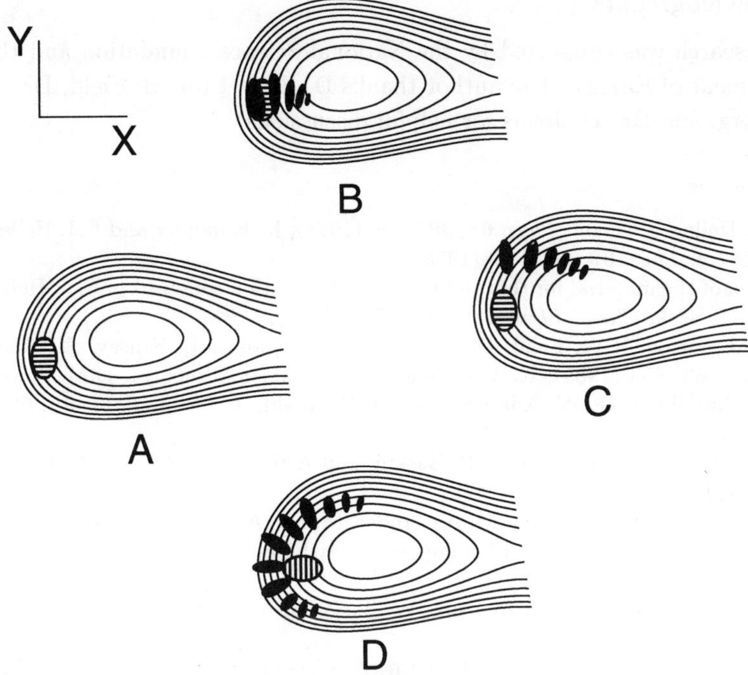

Fig. 8. Some natural and pulse prepared feature states for a dissociative potential energy surface. See text for details.

addition, momentum jumping has taken place in the second mode, causing the nodal structure seen in the wavefunction corresponding to movement in the Y direction. The narrowness of the donor wave packet in that direction leads to large momentum uncertainty on the acceptor potential energy surface.

The final state product distributions leading from the three cases B, C, D will differ from one another and be distinct from the naive expectation of amplitude leaving from the Franck–Condon region A. Even if a pulse preparation is not made, the dissociation should be viewed as arising from a dispaced location. The product states that correspond to the wings are distinguished easily by the total energy, and may be produced simply by tuning the laser well to the blue of the absorption maximum. The final state product distributions are the "smoking gun," revealing something about the region from which the amplitude is "launched" before dissociation.

Acknowledgments

This research was supported by the National Science Foundation and the Department of Energy. The author thanks D. Beck, Prof. B. Field, Dr. J. Lundberg, and Dr. D. Jonas for helpful discussions.

References

1. E.J. Heller, *J. Chem. Phys.* **68**, 3891–96 (1978); K. Kulander and E.J. Heller, *J. Chem. Phys.* **69**, 2439–49 (1978).
2. Much of the material for this and the following section comes from E. J. Heller and D. Beck, *Chem. Phys. Lett.* **202**, 350 (1993).
3. E. Abramson, R.W. Field, D. Imre, K.K. Innes, and J.L. Kinsey, *J. Chem. Phys.* **83**, 453 (1985); K. Yamanouchi, N. Ikeda, S. Tsuchiya, D.M. Jonas, J.K. Lundberg, G. W. Adamson, and R.W. Field, *J. Chem. Phys.* **95**, 6330 (1991).
4. M. Dantus, R.M. Bowman, J.S. Baskin, and A.H. Zewail, *Chem. Phys. Lett.* **159**, 406 (1989).
5. N.F. Scherer, R.J. Carlson, M. Alexander, M. Du, A.J. Ruggiero, V. Romero-Rochin, J. A. Cina, G.R. Fleming, and S.A. Rice, *J. Chem. Phys.* **95**, 1487 (1991).
6. R. A. Mathies, C. H. Brito Cruz, W. T. Pollard, and C.V. Shank, *Science* **240**, 777 (1988).
7. D. Jonas, Ph.D. Thesis, MIT Department of Chemistry, 1992; *J. Chem. Phys.* **97**, 2813 (1992).
8. W. F. Polik, D. R. Guyer, W. H. Miller, and C. Bradley Moore , *J. Chem. Phys.* **92**, 3471 (1990), and references therein.
9. W. Siebrand, W. Leo Meerts, and D. W. Pratt, *J. Chem. Phys.* **90**, 1313 (1989).
10. G. E. Ewing, *J. Chem. Phys.* **71**, 3143 (1979).
11. J. Zhang, D. G. Imre, and J. H. Frederick, *J. Phys. Chem.* **93**, 1840 (1989); D. G. Imre and J. Zhang, *Chem. Phys.* **139**, 89 (1989).
12. E.J. Heller, *J. Chem. Phys.* **68**, 2066 (1978).
13. D. Beck and E. J. Heller, unpublished.
14. P. Avouris, W.M. Gelbart, and M.A. El-Sayed, *Chem. Rev.* **77**, 793, (1977).
15. K.K. Innes, *J. Chem. Phys.* **22**, 863 (1954).
16. J.K. Lundberg, R.W. Field, C.D. Sherrill, E.T. Seidl, Y. Xie and H.F. Schaefer III, *J. Chem. Phys.* **98**, 8384 (1993).

CHAPTER 21

COMPUTATION OF SEP SPECTRA

Claude Leforestier

Laboratoire de Chimie Théorique
Université de Paris-Sud
91405 Orsay, France

Robert E. Wyatt

Department of Chemistry and Biochemistry
University of Texas
Austin, Texas 78712, USA

Contents

1. Introduction

1.1. *Computing Spectra*

This chapter is concerned with the efficient computation of the spectral lineshape function,

$$I(\omega) = \sum_{\alpha} |\langle \Psi_0 | \mu | \Psi_\alpha \rangle|^2 \delta(\omega - \omega_\alpha) , \qquad (1)$$

where $|\langle \Psi_0 | \mu | \Psi_\alpha \rangle|^2$ is the intensity at frequency ω_α for the dipole transition linking the initial or prepared state $|\Psi_0\rangle$ to the eigenstate $|\Psi_\alpha\rangle$. In applications to SEP (stimulated emission pumping) spectroscopy,[1] $|\Psi_0\rangle$ represents a vibrational eigenstate on the excited electronic surface, $|\Psi_\alpha\rangle$ is one of a number of probed eigenstates on the ground electronic surface, and μ is the geometry dependent electronic transition dipole linking the excited and ground electronic states.

Although there is at present only one example where the SEP lineshape function has been computed (see Sec. 1.2.), the theoretical and computational methodologies are now in place so that complex spectra can be computed and analyzed. This statement is correct to the extent that the necessary electronic potential surface information is available. The computational approach that is reviewed and developed in this chapter is an outgrowth of the recursive residue generation method, RRGM, developed by Wyatt and Nauts[2,3] in 1983. The aim of this method is the efficient computation of just those quantities, intensities and frequencies, that are needed to construct the lineshape function.

In the following section, we will review the only example where an SEP spectrum has been computed and measured — in fact, this is one of those infrequent cases where the theoretical prediction came before the experimental measurement. Then, in Sec. 2, we will review the original formulation of the RRGM and see how intensities and frequencies can be computed, even for situations involving very many states. Section 3 is devoted to an application of this method to the SEP spectrum of the HCN molecule. In the last section, we discuss its extension to tetratomic systems.

1.2. *A Case Study: HCN*

With respect to both theory and experiment, the HCN molecule has served as the focus of considerable effort. For the SEP spectrum in particular, HCN is the only example where the spectrum has been computed theoretically and measured in the laboratory. However, the story is not over; the results of new calculations and experiments should provide unprecedented opportunities for further detailed comparisons. In the remainder of this section, we will first review the energetics and potential surfaces, and then the calculations of vibrational states and SEP spectra and, finally, we will comment on the experimental results. Later, in Sec. 3, new computational results on the SEP spectrum for HCN will be presented.

The history of *ab initio* and semiempirical calculations of both the energetics and potential surfaces for HCN were nicely reviewed by Lee and Rendell[4] in 1991. Before commenting on this history, we will let ΔE denote the energy difference (without zero-point corrections) between equilibrium linear HCN and equilibrium linear HNC. In addition, let $\Delta E^{\neq}$ denote the energy difference between the bent HCN transition state and equilibrium linear HCN, again without zero-point corrections. The relatively early (1975) *ab initio* calculations of Schaefer and coworkers[5] (at the TZP/CISD level) gave $\Delta E = 5106 \text{ cm}^{-1}$ and $\Delta E^{\neq} = 17\,313 \text{ cm}^{-1}$. Five years later, the TZ2P/MP4 calculations of Bartlett and coworkers[6] gave a similar result of $\Delta E^{\neq} = 16\,928 \text{ cm}^{-1}$. Calculations in the early 1980s at the 6-311G*/MP3 or MP4 levels[7,8] gave $\Delta E^{\neq} - 16\,963 \text{ cm}^{-1}$ and $16\,018 \text{ cm}^{-1}$, respectively (at the SCF optimized geometries). Believing the Schaefer *et. al.* barrier height to be too high, Murrell, Carter, and Halonen[9] (MCH) developed a global semiempirical potential surface with the barrier height $\Delta E^{\neq} = 12\,179 \text{ cm}^{-1}$. Until recently, the MCH surface was used in virtually all calculations of vibrational levels and spectra for this system.[10–20]

Quite recently, Lee and Rendell[4] reported a new series of *ab initio* calculations of ΔE and $\Delta E^{\neq}$. Using DZP and TZ2P basis sets at the CCSD and CCSD(T) levels, they performed a series of HCN, HNC, and transition state calculations leading to the "best estimates" $\Delta E = 5036 \text{ cm}^{-1}$ and

$\Delta E^{\neq} = 15\,599$ cm^{-1}, including the zero-point corrections. In a second study,[21] energies were computed at 1124 geometries. These points were then fit in Jacobi coordinates to build a global potential surface.

Using input from the above electronic structure calculations, there have been a number of calculations of vibrational energy levels[10–22] in the HCN system. Most of these calculations, which were reported between 1986 and 1991, used the MCH potential because it was thought to be the most complete and (possibly) the best available potential surface. In spite of the now well understood deficiencies in the MCH potential, the vibrational calculations that have been done using this potential are very valuable, for at least several reasons: new computational methods (such as the discrete variable representation employing distributed Gaussian bases,[10–11] the truncation recoupling scheme,[15] and the RRGM[20]) were developed and applied; the transition in the nature of the vibrational eigenfunctions as the energy was increased through and above the isomerization barrier could be charted.[22] A detailed study of the vibrational eigenfunctions on the MCH potential has recently been reported.[22] Also, vibrational eigenvalues have been computed for total angular momentum $J = 0$ and 1 on a new global fit to the latest set of *ab initio* energies.[21] Agreement between the calculated and experimental energies over the energy range 740–18 500 cm^{-1} was generally quite good: 27 out of 47 reported energies had $|\text{error}| < 50$ cm^{-1}, while 15 out of 47 states had energies with errors in the range 50–100 cm^{-1}. This new PES leads to a closer agreement between the computed and experimental energies, particularly for the higher bending overtones.

In 1989, the RRGM was used to compute the first SEP spectrum for HCN.[20] The MCH potential surface was used and the calculations were performed in Jacobi coordinates (R, r, θ) (see Sec. 3). The RRGM calculations then used basis sets of dimension up to 8619 terms consisting of the product of up to 221 radial functions $F_j(R, r)$ times up to 39 Legendre polynomials $P_j(\theta)$. The initial ground vibrational state, (0,0,0), on the upper $\tilde{A}^1 A'$ electronic surface was expanded in this basis set and up to 6000 Lanczos recursion steps were used to generate the SEP spectrum.

Following calculation of the SEP spectrum, Wodtke and coworkers[23,24] reported the first experimental observation of the SEP spectrum for HCN. In the experimental study, a pulsed tunable ArF laser was used to pump ground electronic state HCN up to the $(0, 1, 0)^{K=1}$ level of the $\tilde{A}^1 A'$ excited electronic surface. A frequency doubled pulsed dye laser was then used to dump the molecule to a state lying between 8900 cm^{-1} and 18900 cm^{-1} on the lower electronic surface. The usual fluorescence depletion spectrum was recorded as a function of the dump laser frequency. Because of the selection rule $K - l = \pm 1$ (where K and l refer to the components of angular momentum along the principal axis in the upper or lower electronic surface, respectively), the allowed l values are $l = 0, 2$. After assigning the lines, comparisons with calculated line positions based upon the MCH surface showed good agreement, except for higher bend excitations, $v_2 > 4$. The systematic deviations became progressively worse, with the error $>$ 100 cm^{-1} for $v_2 > 8$. The low isomerization barrier on the MCH potential surface is the cause of this deviation. As mentioned earlier, calculated vibrational frequencies on the recent *ab initio* ground electronic surface show much better agreement with the experimental results.[21]

Although there are similarities between the computed and experimental SEP spectra, the assumed initial state used for computation of the theoretical spectrum, (0,0,0), was different from the initial state used in the experiment, $(0,1,0)^{K=1}$. Therefore, the intensities are not directly comparable. However, a computed spectrum using the experimental initial state and the recent *ab initio* potential surface should be available in the near future.

2. The Recursive Residue Generation Method

2.1. *Overview*

In this section, application of the recursive residue generation method to the computation of spectra is presented. However, in order to establish some of the notation, the standard matrix diagonalization approach is discussed first in Sec. 2.2. Then, Secs. 2.3–2.7 present the recursive approach, followed by a summary in Sec. 2.8. The central core of Sec. 2 is concerned with the Lanczos algorithm for transforming a matrix into tridiagonal form.[25–29]

As mentioned in Sec. 1.2, the RRGM has been used[20] to compute SEP spectra in HCN. In other spectroscopic applications, the RRGM has recently been used to compute: (a) the photodissociation spectrum and survival probability[30] for O_3; (b) spectrum and survival probabilities for CH overtones in benzene;[31,32] and (c) spectra and survival probabilities for CH overtones[33,34] in CD_3H.

The first application of the Lanczos algorithm was in 1951; a "difficult" 8×8 eigenproblem was solved in 100 hrs. on a mechanical calculator.[26] Since then, many large-scale eigenproblems have been solved with this algorithm. The algorithm has been described in detail in a number of textbooks, monographs, and conference proceedings.[27–29,35–37] A comprehensive history of the Lanczos algorithm covering the period 1948–1976, including both formal developments and applications, has been presented by Golub and O'Leary.[38] One of the best references continues to be Volume 35 of the series *Solid State Physics*.[39] Recent (1985 and thereafter) applications have been made in a number of fields, including the following: molecular electronic structure,[40] resolvent operator calculations,[41] resonance eigenvalues,[42] atom-molecular reactive scattering,[43,44] molecular-surface scattering,[45] bound state eigenvalues,[46] vibronic coupling,[47] intramolecular dynamics,[48] survival probabilities,[49] ESR relaxation,[50,51] and lattice gauge theory.[52,53] The RRGM itself was reviewed by Wyatt[3] in 1989.

2.2. *The Standard Approach: Matrix Diagonalization*

The time honored way to compute a spectrum is worth a brief review. Assume that we are given an orthonormal basis set

$$B = \{|i\rangle, \quad i = 1, 2, \ldots N\}, \tag{2}$$

where each member could represent a multi-mode function, $|i\rangle = \phi_i (r_1, \ldots, r_N)$. We assume that this basis is sufficient to represent both the initial state $|0\rangle$ and the eigenstates $|\Psi_\alpha\rangle$ that make a major contribution to the spectrum. The expansion coefficients of the initial state in the basis,

$$|0\rangle = \sum_{i=1}^{N} d_i |i\rangle \tag{3}$$

can be represented as the column vector $\mathbf{d}$,

$$\mathbf{d} = \begin{bmatrix} d_1 \\ d_2 \\ \vdots \\ d_N \end{bmatrix} . \tag{4}$$

In order to generate eigenvectors of the Hamiltonian, we follow the two-step procedure: (1) Compute the matrix elements of the Hamiltonian operator in the basis, $H_{ij} = \langle i|H|j\rangle$; (2) Using standard direct diagonalization algorithms, find the eigenvalues and eigenvectors of the $N \times N$ matrix $\mathbf{H}$,

$$\mathbf{HC} = \mathbf{CE} , \tag{5}$$

where $\mathbf{E}$ is the diagonal matrix containing the eigenvalues $\{\omega_\alpha\}$. Column number "α" in the $N \times N$ eigenvector matrix $\mathbf{C}$ is then denoted $\mathbf{c}_\alpha$,

$$\mathbf{c}_\alpha = \begin{bmatrix} c_{1\alpha} \\ c_{2\alpha} \\ \vdots \\ c_{N\alpha} \end{bmatrix} . \tag{6}$$

The amplitude for finding the initial state localized on eigenvector α is then given by the inner product,

$$\langle 0 | \Psi_\alpha \rangle = \mathbf{d}^t \mathbf{c}_\alpha . \tag{7}$$

Finally, from Eq. (1), the lineshape function is given by

$$I(\omega) = \sum_\alpha (\mathbf{d}^t \mathbf{c}_\alpha)^2 \delta(\omega - \omega_\alpha) . \tag{8}$$

In spite of its apparent simplicity, this method has several major shortcomings. First, the computational effort involved in solving the matrix eigenproblem in Eq. (5) scales as $O(N^3)$. This feature, along with the storage requirements for $\mathbf{H}, \mathbf{C}$, and the required workspace, practically limit N to no more than a few thousands. The approaches described below are designed to overcome these unfavorable characteristics.

2.3. *General Aspects of the Recursion Method*

The recursive residue generation method[2,3] provides a computationally efficient way to compute spectra in large multistate systems. Before presenting any details, we will first review some of the basic features of the method. In order to make a link between the lineshape function and the term "residue", we will rewrite Eq. (1) as

$$I(\omega) = \sum_\alpha r_\alpha \delta(\omega - \omega_\alpha) , \tag{9}$$

where the residues are defined by $r_\alpha = \langle 0 | \Psi_\alpha \rangle^2$. The primary output from an RRGM calculation is the set of residues $\{r_\alpha\}$ and the eigenvalues $\{\omega_\alpha\}$. The eigenvectors $|\Psi_\alpha\rangle$ are not computed at all. The recursion sequence that eventually leads to the residues and the eigenvalues is initiated with the normalized starting vector $|0\rangle$. The idea is then, in stepwise fashion, to develop an $M \times M$ tridiagonal matrix $\mathbf{T}$. This process is accomplished by the Lanczos algorithm,[25-29] which involves the construction of the sequence of vectors $\mathbf{U}_0, \mathbf{U}_1, \ldots, \mathbf{U}_{M-1}$, in which $\mathbf{U}_0$ represents $|0\rangle$, the starting vector. All of these vectors are not stored; it is possible to develop $\mathbf{T}$ by storing only three vectors. We will return to this point when the Lanczos algorithm is described.

As a result of performing M Lanczos recursion steps, we have formed the $M \times M$ tridiagonal matrix $\mathbf{T}$. From $\mathbf{T}$, we can then quickly compute M eigenvalues and residues. This, in turn, permits an M term approximation to the lineshape function,

$$I^M(\omega) = \sum_{\alpha=1}^{M} r_\alpha \delta(\omega - \omega_\alpha) \,. \tag{10}$$

Of course, it is necessary to verify convergence by examining the spectra for several values of M. It is an important characteristic of the method that approximations to the largest residues are developed quickly in the recursion sequence. Many more recursion steps may be required in order to develop the small residues.

2.4. *What is a Residue?*

A major goal of the RRGM is the "direct" computation of the residues $\{r_\alpha\}$, without computing the eigenvectors. The term residue arises in the following way. *If* we did know the eigenvalues and eigenvectors of H, then the spectral resolution of the identity, Hamiltonian, and Green ("inverse Hamiltonian") operators could be written

$$\hat{1} = \sum_{\alpha} |\Psi_\alpha\rangle \langle\Psi_\alpha|$$

$$\hat{H} = \sum_{\alpha} |\Psi_\alpha\rangle E_\alpha \langle\Psi_\alpha|$$

$$\hat{G}(E) = (E - \hat{H})^{-1} = \sum_{\alpha} |\Psi_\alpha\rangle (E - E_\alpha)^{-1} \langle\Psi_\alpha| \,. \tag{11}$$

The matrix element of the Green operator (referred to as the Green function) over the initial state is then

$$G_{00}(E) = \langle 0|\hat{G}(E)|0\rangle = \sum_\alpha \langle 0|\Psi_\alpha\rangle (E - E_\alpha)^{-1}\langle \Psi_\alpha|0\rangle$$

$$= \sum_\alpha r_\alpha (E - E_\alpha)^{-1} . \tag{12}$$

From the last expression, it is apparent that r_α is the "strength" of the Green function at the pole E_α on the energy axis. The residue is thus the spectral strength (intensity) at this energy.

2.5. *Lanczos Recursion: How it Works*

The Lanczos recursion algorithm plays a key role in the RRGM. This algorithm, developed by C. Lanczos[25] in 1950, has the following goal: given the starting vector $\mathbf{U}_0$, and the $N \times N$ Hamiltonian matrix $\mathbf{H}$, construct the $M \times M$ tridiagonal representation of the Hamiltonian, denoted $\mathbf{T}$. In order to illustrate how the Lanczos algorithm is used, we will first state the general algorithm, and then write out the first two recursion steps. First, in order to establish the notation, the recursion steps are labeled, $0, 1, \dots, M-1$. At each step, we generate a new recursion vector, along with a pair of diagonal and off-diagonal elements of $\mathbf{T}$. These diagonal and off-diagonal elements are denoted $\{a_0, a_1, \dots\}$ and $\{b_1, b_2, \dots\}$, respectively. At the start of step n, we have $\mathbf{U}_n$, a_{n-1}, and b_n. We then compute the matrix–vector product $\mathbf{HU}_n$ and use the three-term recurrence equation,

$$\mathbf{HU}_n = a_n \mathbf{U}_n + b_{n+1}\mathbf{U}_{n+1} + b_n \mathbf{U}_{n-1} , \tag{13}$$

where a_n, b_{n+1}, and $\mathbf{U}_{n+1}$ are to be found. The recursion vectors are orthonormal by design:

$$\mathbf{U}_m^t \mathbf{U}_n = \delta_{mn} . \tag{14}$$

So multiplying Eq. (13) by $\mathbf{U}_n^t$ gives the new diagonal element,

$$a_n = \mathbf{U}_n^t \mathbf{HU}_n . \tag{15}$$

We now write Eq. (13) as

$$U_{n+1} = \frac{1}{b_{n+1}}\{\mathbf{HU}_n - a_n \mathbf{U}_n - b_n \mathbf{U}_{n-1}\} \tag{16}$$

where each term within the brackets is known. Forcing normalization then gives

$$1 = \mathbf{U}_{n+1}^t \mathbf{U}_{n+1} = \frac{1}{b_{n+1}^2} , \tag{17}$$

so that the new off-diagonal element is given by

$$b_{n+1} = \|\mathbf{H}\mathbf{U}_n - a_n\mathbf{U}_n - b_n\mathbf{U}_{n-1}\|^{1/2} . \tag{18}$$

This completes step n; we now have $\mathbf{U}_{n+1}$, a_n, and b_{n+1}.

For the first two steps, we have

Step 0: Given $\mathbf{U}_0$, with $\mathbf{U}_0^t\mathbf{U}_0 = 1$ (note that $\mathbf{U}_{-1} = 0$),

$$\mathbf{H}\mathbf{U}_0 = a_0\mathbf{U}_0 + b_1\mathbf{U}_1 \tag{19}$$

$$a_0 = \mathbf{U}_0^t\mathbf{H}\mathbf{U}_0 \tag{20}$$

$$b_1 = \|\mathbf{H}\mathbf{U}_0 - a_0\mathbf{U}_0\|^{1/2} \tag{21}$$

$$\mathbf{U}_1 = \frac{1}{b_1}\{\mathbf{H}\mathbf{U}_0 - a_0\mathbf{U}_0\} \tag{22}$$

$$\mathbf{U}_0^t\mathbf{U}_1 = 1, \qquad \mathbf{U}_0^t\mathbf{U}_0 = 0 . \tag{23}$$

Step 1: We now have $\mathbf{U}_1$, a_0, and b_1.

$$\mathbf{H}\mathbf{U}_1 = a_1\mathbf{U}_1 + b_2\mathbf{U}_2 + b_1\mathbf{U}_0 \tag{24}$$

$$a_1 = \mathbf{U}_1^t\mathbf{H}\mathbf{U}_1; \qquad b_2 = \|\mathbf{H}\mathbf{U}_1 - a_1\mathbf{U}_1 - b_1\mathbf{U}_0\|^{1/2} \tag{25}$$

$$\mathbf{U}_2 = \frac{1}{b_2}\{\mathbf{H}\mathbf{U}_1 - a_1\mathbf{U}_1 - b_1\mathbf{U}_0\}; \qquad \mathbf{U}_2^t\mathbf{U}_2 = 1, \quad \mathbf{U}_2^t\mathbf{U}_1 = 0 . \tag{26}$$

From this point onward, the recursion method just loops over the same steps. After M recursion steps, we have the $M \times M$ tridiagonal matrix $\mathbf{T}$,

$$\mathbf{T} = \begin{bmatrix} a_0 & b_1 & 0 & 0 & & 0 \\ b_1 & a_1 & b_2 & 0 & & 0 \\ 0 & b_2 & a_2 & b_3 & & 0 \\ & & & & & \\ & & & & & a_{M-2}\,b_{M-1} \\ 0 & 0 & 0 & 0 & \cdots & b_{M-1}\,a_{M-1} \end{bmatrix} . \tag{27}$$

In the next step, we will extract the residues and eigenvalues from $\mathbf{T}$.

Finally, we note that the Hamiltonian matrix enters only through the matrix–vector products, $\mathbf{H}\mathbf{U}_{\text{old}} = \mathbf{U}_{\text{new}}$, so that $\mathbf{H}$ is not modified. For large basis sets where the storage of $\mathbf{H}$ in central memory would be prohibitive, it may be possible to compute the matrix–vector product "on-the-fly," using properties of the Hamiltonian. Alternatively, blocks of

H may be read into central memory from secondary storage (e.g., the Cray SSD). This is also a place where parallelization would be useful: parts of the matrix–vector multiplication could be executed simultaneously on multiple processors.

2.6. *Residues from the Tridiagonal Matrix*

In order to compute residues and eigenvalues from the tridiagonal matrix **T**, we will carry out some formal manipulations on the Green function in Eq. (12). Some readers will likely want to skip to the end, where the main result is given by Eq. (40).

In matrix notation, we can rewrite Eq. (12) for the Green function as

$$G(E) = \mathsf{U}_0^t (\mathbf{H} - E\mathbf{1}_N)^{-1}\mathbf{U}_0 , \qquad (28)$$

where **H** and $\mathbf{1}_N$ are both $N \times N$ matrices. Next, imagine that we stack the recursion vectors (each $N \times 1$) side-by-side to build the $N \times M$ matrix **Q**,

$$\mathbf{Q} = [\mathbf{U}_0, \mathbf{U}_1, \ldots, \mathbf{U}_{M-1}] . \qquad (29)$$

The matrix **Q** formally brings **H** into tridiagonal form,

$$\mathbf{Q}^t \mathbf{H} \mathbf{Q} = \mathbf{T} , \qquad (30)$$

where **T** is an $N \times N$ matrix. Also, **Q** is orthogonal since the recursion vectors themselves are orthonormal,

$$\mathbf{Q}\mathbf{Q}^t = \mathbf{1}_N . \qquad (31)$$

As a result,

$$\mathbf{H} - E\mathbf{1}_N = \mathbf{Q}\mathbf{T}\mathbf{Q}^t - E\mathbf{Q}\mathbf{Q}^t = \mathbf{Q}[\mathbf{T} - E\mathbf{1}_M]\mathbf{Q}^t , \qquad (32)$$

where $\mathbf{1}_M$ is an $M \times M$ unit matrix and the inverse matrix is

$$[\mathbf{H} - E\mathbf{1}_N]^{-1} = \mathbf{Q}[\mathbf{T} - E\mathbf{1}_M]^{-1}\mathbf{Q}^t . \qquad (33)$$

With these preliminaries now completed, we are ready to return to Eq. (28),

$$G(E) = \mathsf{U}_0^t \mathbf{Q}[\mathbf{T} - E\mathbf{1}_M]^{-1}\mathbf{Q}^t \mathbf{U}_0 . \qquad (34)$$

Now, because the recursion vectors are orthonormal, the matrix product $\mathsf{U}_0^t\mathbf{Q}$ has a very interesting form:

$$\mathsf{U}_0^t\mathbf{Q} = \mathsf{U}_0^t[\mathbf{U}_0, \mathbf{U}_1, \ldots, \mathbf{U}_{M-1}] = [1, 0, 0, \ldots, 0] = \mathbf{e}_1^t \ . \tag{35}$$

The result is just the unit row vector, $\mathbf{e}_1^t$, a very simple result indeed. Consequently,

$$G(E) = \mathbf{e}_1^t[\mathbf{T} - E\mathbf{1}_M]^{-1}\mathbf{e}_1 = [\mathbf{T} - E\mathbf{1}_M]_{11}^{-1} \ . \tag{36}$$

The final step arises from the matrix product $\mathbf{e}_1^t\mathbf{A}\mathbf{e}_1 = A_{11}$; we select the 1,1 upper-left element from matrix $\mathbf{A}$. In other words, $G(E)$ is the 1,1 element of the inverse of $(\mathbf{T} - E\mathbf{1}_M)$.

The quantity $[\mathbf{T} - E\mathbf{1}_M]^{-1}$ can be evaluated directly in terms of the a's and b's using continued fraction techniques. However, it is more convenient for our purposes to consider the diagonalization of $\mathbf{T}$, where S and $\mathbf{E}$ are both $M \times M$ matrices

$$S^t\mathbf{T}S = \mathbf{e} \ . \tag{37}$$

The 1,1 element of $[\mathbf{T} - E\mathbf{1}_M]^{-1}$ is then

$$\mathbf{e}_1^t S[\mathbf{e} - E\mathbf{1}_M]^{-1}S^t\mathbf{e}_1 \ , \tag{38}$$

but since $[\mathbf{e} - E\mathbf{1}_M]$ is a diagonal matrix, we have

$$G(E) = \sum_{\alpha=1}^{M} S_{1\alpha}^2(e_\alpha - E)^{-1} \ . \tag{39}$$

This is the result that we have been striving for: *only the elements in the first row of the eigenvector matrix S are needed to compute $G(E)$.* This equation shows that the residues are given by $r_a = S_{1\alpha}^2$, so that the lineshape function is (where $\omega_\alpha = e_\alpha$)

$$I(\omega) = \sum_{\alpha=1}^{M} S_{1\alpha}^2\delta(\omega - \omega_\alpha) \ . \tag{40}$$

It is rather astounding that all spectral intensity information is contained within the first row of the eigenvector matrix of $\mathbf{T}$. In the next section, we will examine an efficient algorithm for generating the residues $\{S_{1\alpha}^2\}$ and eigenvalues $\{\omega_\alpha\}$.

2.7. *Using the QL Algorithm*

The standard technique for computing all of the eigenvalues of a symmetric matrix is to use a finite sequence of orthogonal similarity transforms to tridiagonalize the matrix, followed by use of the QL algorithm.[35,54–56] This algorithm uses a theoretically infinite sequence of orthogonal similarity transformations which preserves the tridiagonal form and converges to a diagonal matrix. The resulting eigenvector matrix is the product of all of the transformation matrices. If the eigenvectors are desired, then these transformations can be accumulated *as they are applied* to the matrix.[57] The QL algorithm costs $\mathrm{O}(M^2)$ to compute the eigenvalues but also costs $\mathrm{O}(M_3)$ if *all* of the eigenvectors are computed.

In our context, the matrix is already tridiagonal and only the first row of the eigenvector matrix ($\mathbf{S}$ in Eq. (37)) is needed. Fortunately, it is possible to obtain the first row of the eigenvector matrix of $\mathbf{T}$ for $\mathrm{O}(M_2)$ operations. To initiate the QL algorithm, we first factor $\mathbf{T}$ (now relabeled $\mathbf{T}_1$) into an orthogonal matrix and a lower triangular matrix (with non-negative diagonal elements):

$$\mathbf{T}_1 = \mathbf{Q}_1 \mathbf{L}_1 \rightarrow \mathbf{Q}_1^t \mathbf{T}_1 = \mathbf{L}_1 \ . \tag{41}$$

When these factors are multiplied in reverse order, the next tridiagonal matrix in the sequence is formed:

$$\mathbf{T}_2 = \mathbf{L}_1 \mathbf{Q}_1 = \mathbf{Q}_1^t \mathbf{T}_1 \mathbf{Q}_1 \ . \tag{42}$$

The process is continued by "doggedly iterating".[29] At step k, we have

$$\mathbf{T}_k = \mathbf{Q}_k \mathbf{L}_k \tag{43}$$

and

$$\begin{aligned}
\mathbf{T}_{k+1} &= \mathbf{L}_k \mathbf{Q}_k = \mathbf{Q}_k^t \mathbf{T}_k \mathbf{Q}_k \\
&= (\mathbf{Q}_k^t \ldots \mathbf{Q}_2^t \mathbf{Q}_1^t) \mathbf{T}_1 (\mathbf{Q}_1 \mathbf{Q}_2 \ldots \mathbf{Q}_k) \\
&= \mathbf{S}_k^t \mathbf{T}_1 \mathbf{S}_k \ .
\end{aligned} \tag{44}$$

There are several significant features about this iterative process:

(i) The tridiagonal form is preserved in the sequence of $\mathbf{T}$ matrices.

(ii) The sequence $\mathbf{T}_1, \mathbf{T}_2, \ldots \mathbf{T}_k$ converges to a *diagonal* (eigenvalue) matrix

$$\mathbf{T}_1 \to \mathbf{T}_2 \to \ldots \mathbf{T}_k = \mathrm{diag}(\mathbf{e}_1, \mathbf{e}_2, \ldots, \mathbf{e}_M) \tag{45}$$

(iii) As $\mathbf{T}_k$ becomes diagonal, $\mathbf{S}_k$ simultaneously converges to the eigenvector matrix of $\mathbf{T}$

$$\mathbf{S}_k^t \mathbf{T}_1 \mathbf{S}_k = \mathbf{E} = \mathrm{diag}(\mathbf{e}_1, \ldots, \mathbf{e}_M) \ . \tag{46}$$

In the preceding section, we emphasized that only the top row $(S_{11}, S_{12}, \ldots, S_{1M})$ of the eigenvector matrix of $\mathbf{T}$ is needed in order to compute the lineshape function. The QL algorithm, as normally implemented, computes *all eigenvectors* of $\mathbf{T}$; the converged $M \times M$ matrix $\mathbf{S}_k$ is generated. However, the algorithm may be modified to produce *only the first row* of $\mathbf{T}$.[57] The key observation is the fact that each new transformation is applied *on the right* of the current approximation to the eigenvector matrix. That is, if $\mathbf{Q}_1, \mathbf{Q}_2, \ldots, \mathbf{Q}_k$ are the transformations used so far, then the approximation to the eigenvector matrix is

$$\mathbf{S}_k = \mathbf{Q}_1 \mathbf{Q}_2 \ldots \mathbf{Q}_k \tag{47}$$

with the $\mathbf{Q}$'s multiplied in that order. Formally, the first row of the eigenvector matrix is

$$\mathbf{e}_1^t \mathbf{S}_k = \mathbf{e}_1^t \mathbf{Q}_1 \mathbf{Q}_2 \ldots \mathbf{Q}_k \ , \tag{48}$$

where $\mathbf{e}_1^t = (1, 0, 0, \ldots, 0)$. Thus, to obtain the desired row, it is only necessary to obtain a *single vector* by operating on the right by each tranformation as it is generated. The only extra storage needed is the vector of length M which will hold the final result.

In practice, we have modified the efficient EISPACK routine IMTQL2 to compute only the first row of $\mathbf{S}$ (which is named $\mathbf{Z}$ in the routine). Input to the routine includes:

diagonal of $\mathbf{T}$: $\mathrm{A}(1), \ldots, \mathrm{A}(M)$ (in array D)

subdiagonal of $\mathbf{T}$: $\mathrm{B}(1), \ldots, \mathrm{B}(M-1)$ (in array E)

order of matrix: M

Output from the routine includes:

> eigenvalues of T : $\mathbf{e}(1), \ldots, \mathbf{e}(M)$ (in array D)
>
> first row of eigenvector matrix: $\mathbf{S}(1), \ldots, \mathbf{S}(M)$ (in array Z)

2.8. *Concluding Remarks*

In Sec. 2, we have reviewed the recursive residue generation method for the computation of the lineshape function. The RRGM aims at computation of the set of residues and eigenvalues $\{r_\alpha, \omega_\alpha\}$. The steps are summarized as follows:

a. The normalized initial vibrational state on the upper electronic surface (A) is expanded in the basis set used to represent the Hamiltonian on the lower electronic surface. This expansion provides the starting vector $\mathbf{U}_0$.

b. Given $\mathbf{U}_0$ and a subroutine for computing the matrix–vector product $\mathbf{H}\mathbf{U}_{\text{old}} = \mathbf{U}_{\text{new}}$, we perform M Lanczos recursion steps. This generates the set of diagonal $\{a\}_0^{M-1}$ and off-diagonal $\{b\}_1^{M}$ elements in the $M \times M$ tridiagonal matrix $\mathbf{T}$.

c. Given the a's and b's, the QL algorithm is used to compute the eigenvalues $\{\omega_\alpha\}$ of $\mathbf{T}$. In addition, the elements $\{S_{1\alpha}\}_1^{M}$ in the first row of the eigenvector matrix $\mathbf{S}$ associated with $\mathbf{T}$ are also computed.

d. The lineshape function is computed from the eigenvalues and the squares of the elements in the first row of $\mathbf{S}$.

e. Convergence is checked by recurring to a higher value of M, then steps (c) and (d) are repeated.

In most applications, the most time consuming step in this procedure is the series of matrix–vector products involved in step (b).

3. Application to HCN

3.1. *Description of the Molecular System*

The HCN molecule is the prototype of a system undergoing large amplitude motion. This is especially true when considering the SEP spectrum. As a consequence, one must depart from the standard approach in spectroscopic calculations.[58] The system description should not make reference to some equilibrium geometry, which has lost all significance in the floppy regime.

The basis set associated with some particular geometry will grow exponentially in size as the energy increases, due to the larger deformations involved. When most of the configuration space is available, general coordinates, such as those used in scattering calculations, prove more useful. In the case of HCN, Bačić and Light[10] have shown that Jacobi coordinates allow one to very efficiently describe the system, when used in conjunction with the Adiabatic Discrete Variable Representation (ADVR) method. As the method is quite general, and appears to be the key to future calculations involving larger systems, we will present below the basic concepts in the case of a non-rotating ($J = 0$) molecule, as initially given by Bačić and Light. We will then extend the formulation to the rotating case ($J \neq 0$), as required for the SEP spectrum calculations.

The geometry of the HCN molecule is described in the body fixed (BF) frame by the set of Jacobi coordinates (R, r, θ), as shown in Fig.1. The Hamiltonian operator of the rotating triatomic is

$$\mathbf{H}^J = -\frac{\hbar^2}{2M}\frac{\partial^2}{\partial R^2} - \frac{\hbar^2}{2\mu}\frac{\partial^2}{\partial r^2} - \frac{1}{2I}\left\{ \frac{\hbar^2}{\sin\theta}\frac{\partial}{\partial\theta}\sin\theta\frac{\partial}{\partial\theta} - \frac{\ell_z^2}{\sin^2\theta} \right\}$$

$$+ \frac{1}{2\mu r^2}\left\{ \mathbf{J}^2 - 2\mathbf{J}_z\ell_z - \mathbf{J}_+\ell_- - \mathbf{J}_-\ell_+ \right\} + V(R, r, \theta) \qquad (49)$$

with $\frac{1}{M} = \frac{1}{m_H} + \frac{1}{m_{CN}}$, $\frac{1}{\mu} = \frac{1}{m_C} + \frac{1}{m_N}$, and $\frac{1}{I} = \frac{1}{MR^2} + \frac{1}{\mu r^2}$, associated with the volume element $d\tau = \sin\theta dR dr d\theta d^3\Omega$. In Eq. 49, $\mathbf{J}$ stands for the total angular momentum, and l corresponds to the relative (orbital) angular momentum. In the above formulation, we have defined the CN axis as the BF $\mathbf{z}$ axis. This allows us to reduce the Coriolis coupling terms,[59] due to the light mass of the hydrogen atom. The rotational motion is described in terms of the symmetric top basis set $\{|J, K, M\rangle\}$[60]

$$\langle\Omega|J, K, M\rangle = \sqrt{\frac{2J+1}{8\pi^2}}D^J_{MK}(\Omega) \qquad (50)$$

where Ω stands for the three Euler angles which specify the orientation of the Body Fixed (BF) frame in the Space Fixed (SF) frame. We will also make use of the symmetrized basis set $\{|J, K, M, \pm\rangle, K = 0, J\}$,

$$|J, K, M, \pm\rangle = N_K\{|J, K, M\rangle \pm (-1)^{J-K}|J, -K, M\rangle\}, \qquad (51)$$

$$N_K = \{2(1 + \delta_{0K})\}^{-1/2},$$

which displays the $\pm$ parity with respect to inversion.

The overall wave function $|\Psi^{JM\sigma}\rangle(\sigma = \pm)$ can be expanded as

$$|\Psi^{JM\sigma}\rangle = \sum_K |\Phi_K^{JM\sigma}\rangle.|J, K, M, \sigma\rangle , \qquad (52)$$

where the $|\Phi_K^{JM\sigma}\rangle$ functions depend only on the three Jacobi coordinates.

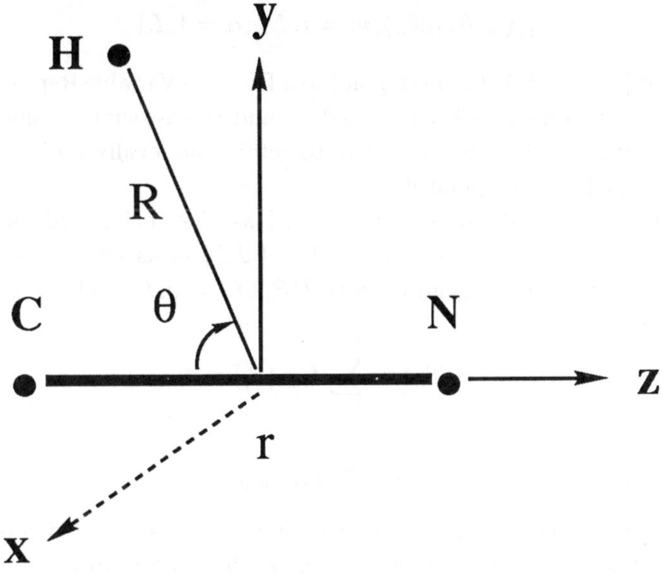

Fig. 1. Definition of the (x, y, z) BF axes and associated (R, r, θ) Jacobi coordinates used to describe the triatomic system.

We now present the ADVR method in the case $J = 0$, i.e., when only one component Φ_0 enters the expansion (Eq. 52). It consists in defining an adiabatic basis set $\{|\chi_m(\theta)\rangle\}$ with respect to the angular coordinate θ, as eigenfunctions of the $J = 0$ Hamiltonian operator $H^{J=0}(\theta)$ corresponding to fixed values θ_α of the angle:

$$\left\{ -\frac{\hbar^2}{2M}\frac{\partial^2}{\partial R^2} - \frac{\hbar^2}{2\mu}\frac{\partial^2}{\partial r^2} + V(R, r, \theta_\alpha) \right\} |\chi_m(\theta_\alpha)\rangle = E_m(\theta_\alpha)|\chi_m(\theta_\alpha)\rangle . \quad (53)$$

Each *local* basis set $\{|\chi_m(\theta_\alpha)\rangle\}$ is then truncated according to some energy threshold E_{th}, $\{|\chi_m(\theta_\alpha)\rangle, m = 1, N_\alpha\}$. The justification for such a truncation is that the procedure constitutes a very efficient prediagonalization of the (R,r)-basis set adapted to the local potential $V(R,r,\theta_\alpha)$. One thus reduces the off-diagonal coupling terms in the total Hamiltonian $H^{J=0}$. Furthermore, Bačić and Light have shown how the scheme could be implemented recursively on the different variables.[61] One now has an optimal basis set for each value of the θ variable. The overall basis set is then defined as the *non direct* product

$$\{|\chi_m(\theta_\alpha)\rangle|\theta_\alpha\rangle, m = 1, N_\alpha; \alpha = 1, L\} \ , \tag{54}$$

where the $\{|\theta_\alpha\rangle, \alpha = 1, L\}$ correspond to a Discrete Variable Representation (DVR), i.e., functions highly localized around the associated value $\theta = \theta_\alpha$. This constitutes a very efficient way to retain the locally optimized basis set $\{|\chi_m(\theta_\alpha)\rangle\}$ at each point θ_α.

In practice, the θ_α values are defined as the abcissae of an L-point Gauss–Legendre quadrature scheme. The $\{|\theta_\alpha\rangle\}$ basis set is then deduced from the *normalized* Legendre basis $\{|P_\ell\rangle, \ell = 0, L - 1\}$ by a unitary transform $\mathbf{U}$

$$|\theta_\alpha\rangle = \sum_{\ell=0}^{L-1} \mathbf{U}_{\alpha\ell}|P_\ell\rangle \ , \tag{55}$$

where

$$U_{\alpha\ell} = \sqrt{\omega_\alpha}\ P_\ell(\cos\theta_\alpha) \ . \tag{56}$$

In the above equation, the ω's correspond to the associated quadrature weights. The unitarity of the $\mathbf{U}$ matrix can be verified from the orthogonality relation between Legendre polynomials

$$\sum_{\alpha=1}^{L} P_\ell(\cos\theta_\alpha)\omega_\alpha P_{\ell'}(\cos\theta_\alpha) = \delta_{\ell\ell'} \ . \tag{57}$$

Although in principle a similar formulation can be cast without the use of any Gaussian quadrature,[62,63] an effective underlying quadrature scheme is required in order to achieve a reasonable accuracy.

In the $\{|\chi_m(\theta_\alpha)\rangle|\theta_\alpha\rangle\}$ ADVR basis set, the Hamiltonian matrix has the form

$$\langle\theta_\beta|\langle\chi_{m'}(\theta_\beta)|H^{J=0}|\chi_m(\theta_\alpha)\rangle|\theta_\alpha\rangle = E_m(\theta_\alpha)\delta_{mm'}.\delta_{\beta\alpha} + \hbar^2 \mathbf{L}_{\beta\alpha}\mathbf{G}^{\beta\alpha}_{m'm} \tag{58}$$

where

$$\mathbf{G}^{\beta\alpha}_{m'm} = \langle\chi_{m'}(\theta_\beta)|\frac{1}{I}|\chi_m(\theta_\alpha)\rangle \tag{59}$$

and

$$\mathbf{L}_{\beta\alpha} = \sum_{\ell=0}^{L-1} \mathbf{U}_{\beta\ell}\ell(\ell+1)\mathbf{U}_{\alpha\ell} \ . \tag{60}$$

The second term on the right hand side of Eq. (58), which corresponds to the bend kinetic energy operator, also contains the nonadiabatic coupling terms originating from the dependence of the $\{|\chi_m(\theta)\rangle\}$ basis set on θ.

The specifications of the $J = 0$ ADVR basis set are as follows. The so-called ray eigenvectors $\{|\chi_m(\theta)\rangle\}$ are calculated for 45 θ-values, spanning the range $[0,\pi]$. At each of these values, all the states lying below an absolute energy threshold of $24\,000$ cm^{-1} were kept, resulting in a total number N of 1113 ray eigenvectors.

The $\{|\chi_m(\theta_\alpha)\rangle|\theta_\alpha\rangle\}$ ADVR basis set, as defined above, cannot directly be used in the $J \neq 0$ case. The presence of a $\ell_z^2/\sin^2\theta$ term in the Hamiltonian operator precludes the use of the $\{|P_\ell\rangle\}$ basis set to describe the internal motion of a *linear* molecule. In the associated DVR $\{|\theta_\alpha\rangle\}$, this term will lead to singularities at the $\theta = 0$ and $\theta = \pi$ configurations, as the $|\theta_\alpha\rangle$ states satisfy

$$f(\theta)|\theta_\alpha\rangle = f(\theta_\alpha)|\theta_\alpha\rangle \ . \tag{61}$$

This singularity can be removed by using instead the associated Legendre basis set $\{|P_\ell^K\rangle\}$, eigenfunctions of the $\ell^2(K)$ operator

$$\left\{\frac{-1}{\sin\theta}\frac{\partial}{\partial\theta}\sin\theta\frac{\partial}{\partial\theta} + \frac{K^2}{\sin^2\theta}\right\}|P_\ell^K\rangle = \ell(\ell+1)|P_\ell^K\rangle \ . \tag{62}$$

Consequently, two different ways of extending the ADVR method to the $J \neq 0$ case have been proposed. The first method, as proposed by Tennyson and Henderson,[64] and Choi and Light,[65] consists in computing the adiabatic ray-eigenvectors $\{|\chi_m(\theta_\alpha^K)\rangle\}$, corresponding to the different $\{\theta_\alpha^K\}$-grids $(K = 0, 1, \ldots, J)$, where the $\{\theta_\alpha^K\}$ are defined as the abcissae of a Gauss-associated Legendre quadrature sheme. For each K value entering the expansion of $|\Psi^{JM\sigma}\rangle$ (Eq. 54), one computes the adiabatic solutions

$$\left\{-\frac{\hbar^2}{2M}\frac{\partial^2}{\partial R^2} - \frac{\hbar^2}{2\mu}\frac{\partial^2}{\partial r^2} - \frac{\hbar^2}{2I}\mathbf{L}_{\alpha\alpha}^K + \frac{\hbar^2}{2\mu R^2}\{J(J+1) - 2K^2\}\right.$$

$$\left. + V(R,r,\theta_\alpha^K)\right\}|\chi_m(\theta_\alpha^K)\rangle = |E_m(\theta_\alpha^K)\rangle|\chi_m(\theta_\alpha^K)\rangle \ . \tag{63}$$

As a straight generalization of the $J = 0$ case, one can define the K-dependent DVR's $\{|\theta_\alpha^K\rangle\}$ by the unitary transforms $\mathbf{U}^K$

$$|\theta_\alpha^K\rangle = \sum_{\ell=K}^{L+K-1} \mathbf{U}_{\alpha\ell}^K |P_\ell^K\rangle \,, \tag{64}$$

where

$$\mathbf{U}_{\alpha\ell}^K = \sqrt{\omega_\alpha^K}\, P_\ell^K(\cos\theta_\alpha^K)\,. \tag{65}$$

The overall basis set is then defined as the *non direct* product

$$\{|\chi_m(\theta_\alpha^K)\rangle|\theta_\alpha^K\rangle|J, K, M, \sigma\rangle, \quad m = 1, N_\alpha^K; a = 1, L; K = 0, J\}\,. \tag{66}$$

The matrix elements of the Hamiltonian operator $\mathbf{H}^J$ in this basis set are given in Ref. 65. This formulation thus requires construction of the corresponding Hamiltonian matrix of size approximatively $[N(J + 1) \times N(J + 1)]$. For the case of the HCN SEP spectrum considered here, where $N \approx 1100$ and J values up to 4 have to be considered, it would result in a matrix too large to fit in core memory.

A second approach has been proposed by Leforestier[17] which allows one to circumvent this problem. It consists in using a *unique* set of θ-values for the different K-components of the wave function. The $\ell_z^2/\sin^2\theta$ singularity is then handled through the Generalized Discrete Variable Representation of Light, Hamilton and Lill[62] as briefly recalled now.

In this approach, one defines a unique set of eigenvectors, as given by Eq. (53), which corresponds to the $\theta_\alpha^{K=0}$ quadrature points. In the following, we drop the $K = 0$ explicit notation. The total wave function $|\Psi^{JM\sigma}\rangle$ is thus written as

$$|\Psi^{JM\sigma}\rangle = \sum_{m\alpha K} C_{m\alpha K}^{JM\sigma} |\chi_m(\theta_\alpha)\rangle|\theta_\alpha\rangle|J, K, M, \sigma\rangle\,. \tag{67}$$

In order to handle the $J \neq 0$ terms in the Hamiltonian operator, one then makes use of the GDVR scheme which allows one to associate a unique set of points $\{\theta_\alpha, \alpha = 1, L\}$ to different basis sets $\{|P_\ell^K\rangle, \ell = K, K + L - 1\}$ through unitary transforms $U^{(K)}$:

$$|P_\ell^K\rangle = \sum_{\alpha=1}^{L} [\mathbf{U}_{\ell\alpha}^{(K)}]^+ |\theta_\alpha\rangle \tag{68}$$

where

$$\mathbf{U}^{(K)} = \mathbf{Y}^{(K)}[\mathbf{S}^{(K)}]^{-1/2} , \tag{69}$$

$$\mathbf{Y}_{\alpha\ell}^{(K)} = P_\ell^K(\theta_\alpha) , \tag{70a}$$

and

$$\mathbf{S}^{(K)} = [\mathbf{Y}^{(K)}]^+ \mathbf{Y}^{(K)} . \tag{70b}$$

The above relations (Eq. 68) allow one to switch from the $\{|\theta_\alpha\rangle\}$ representation to the $\{|P_\ell^K\rangle\}$ ones when evaluating the effect of the angular momentum operators, left over in the calculation of the ray-eigenvector (Eq. 53), and considered now :

$$\left[-\frac{1}{2I} \left\{ \frac{\hbar^2}{\sin\theta} \frac{\partial}{\partial\theta} \sin\theta \frac{\partial}{\partial\theta} - \frac{\ell_z^2}{\sin^2\theta} \right\} + \frac{1}{2\mu r^2} \{ J^2 - 2J_z\ell_z \} \right] |P_\ell^K\rangle|J,K,M,\sigma\rangle$$

$$= \left[\frac{\ell(\ell+1)\hbar^2}{2I} + \frac{\hbar^2}{2\mu r^2} \{ J(J+1) - 2K^2 \} \right] |P_\ell^K\rangle|J,K,M,\sigma\rangle , \tag{71}$$

$$\frac{1}{2\mu r^2} \{ J_+\ell_- + J_-\ell_+ \} |P_\ell^K\rangle|J,K,M,\sigma\rangle$$

$$= \frac{1}{2\mu r^2} \{ c_-(J,K)c_-(\ell,K)|P_\ell^{K-1}\rangle|J,K-1,M,\sigma\rangle$$

$$+ c_+(J,K)c_+(\ell,K)|P_\ell^{K+1}\rangle|J,K+1,M,\sigma\rangle \} , \tag{72}$$

where $c_\pm(a,b) = \sqrt{a(a+1) - b(b\pm 1)}$.

Two matrices are required in this formulation, $\mathbf{G}$ as defined in Eq. (59) and

$$g_{m'm}^{(\beta,\alpha)} = \langle \chi_{m'}(\theta_\beta)|\frac{1}{2\mu r^2}|\chi_m(\theta_\alpha)\rangle \tag{73}$$

of size $N \times N$ each. This formulation thus requires a storage of $2N^2$ words, as compared to $(J+1)^2 N^2$ in the former case.

3.2. *Excited State Description, Dipole Function, and Selection Rules*

In accord with the experiments of Wodtke and coworkers,[23,24] we used as initial rovibrational state for the $\tilde{A} \to \tilde{X}$ DUMP transition the same state $(0,1,0)^{K'=1}$ (See also in this book the chapter "High Resolution Spectroscopy of Chemical Isomerization: Stimulated Emission Pumping of HCN″ by Jonas, Rogaski, Wodtke, and Yang). The (v_1, v_2, v_3) quantum numbers correspond respectively to the C–N stretching, HCN bending, and C–H stretching motions. In the absence of any potential energy surface description for the excited $\tilde{A}$ electronic state, we have approximated the vibrational wave function Φ_v^A in the following way

$$\Phi_v^A = H_{v_1}(q_1)H_{v_2}(q_2)H_{v_3}(q_3)\exp\left\{-\frac{1}{2}[q_1^2 + q_2^2 + q_3^2]\right\}, \qquad (74)$$

where

$$q_1 = \sqrt{\frac{\mu\omega_1}{2\hbar}}\ (r - r_{\text{eq}})\ ,$$

$$q_2 = \sqrt{\frac{I^0\omega_2}{2\hbar}}\ (\theta - \theta_{\text{eq}}), \qquad \frac{1}{I^0} = \frac{1}{MR_{\text{eq}}^2} + \frac{1}{\mu r_{\text{eq}}^2}\ ,$$

$$q_3 = \sqrt{\frac{M\omega_3}{2\hbar}}\ (R - R_{\text{eq}})\ ,$$

and $H_v(q)$ corresponds to a Hermite polynomial. The above expression corresponds to a normal mode description, with the (R, r, θ) Jacobi coordinates playing the roles of the Q_1 (C–N stretch), Q_3 (C–H stretch), and Q_2 (HCN bend), respectively. The equilibrium geometry and vibrational frequencies have been taken from experiments.[66,67]

Jonas, Yang, and Wodtke[24] recently gave a detailed description of the rotational distribution of the excited $\tilde{A}$ state, corresponding to their experiments. They first approximated the molecule as a symmetric top, due to the $\kappa = -0.992$ value of its asymmetry parameter. Consequently, the associated rovibrational function displayed a single K-component, corresponding to the value $K' = 1$

$$|\Psi_A^{JM\sigma}\rangle = |\Phi_v^A\rangle|J', K' = 1, M, \sigma\rangle_A\ , \qquad (75)$$

where the subscript A indicates that the $\tilde{A}$ rotational wave function is written in terms of symmetric top wave functions associated to the Eckart axes of the $\tilde{A}$ state. During the bent $\rightarrow$ linear transition, there is a sudden change in Eckart axes characterized by a single rotation (β_T) about the common axis of greatest moment of inertia. As shown by these authors, the correct rotational wave function, expressed in terms of the symmetric top basis associated to the Eckart axes of the $\tilde{X}$ state, reads as

$$|J', K' = 1, f\rangle_X$$

$$= \rho\sqrt{2J'(J'+1)}\,|J', k = 0\rangle_X$$

$$+ \{1 - \rho^2[J'(J'-1) - 1]\}\frac{1}{\sqrt{2}}\{|J', k = 1\rangle_X - |J', k = -1\rangle_X\}$$

$$- \rho\sqrt{(J'-1)(J'+2)}\frac{1}{\sqrt{2}}\{|J', k = 2\rangle_X + |J', k = -2\rangle_X\}\,, \qquad (76)$$

where $\rho = \tan(\frac{1}{2}\beta_T)$. As done by Wodtke and collaborators, the SEP spectrum is usually dealt with by separating the calculation into vibrational Franck–Condon factors and rotational line strengths. By supposing a ℓ-dependent vibrational factor, the spectrum can be reconstructed by summing up over all the different possibilities ($\Delta J = 0, \pm 1$; $\Delta K = \pm 1$; $M = -J', J'$; $M'' = -J', J''$).

Our concern in this study is to demonstrate the efficiency of the overall method previously developed in this chapter. We will not attempt a direct comparison to the experimental results as the MCH surface[9] is known to present too low a barrier height to isomerization.[23] Such a comparison has to wait for an accurate description in order to be meaningful, as the SEP experiments probe the high energy region of the spectrum. As an illustration of the method, we will consider only DUMP transitions originating from the single $|k'| = 1$ component of Eq. (76):

$$|\Psi_A^{JM\sigma}\rangle = |\Phi_v^A\rangle|J', K' = 1, f\rangle_X\,. \qquad (77)$$

The X-state wave function produced by the DUMP laser can be written[68]

$$\Psi_X = \sum_a \Phi_{Za}\mu_a|\Psi_A^{JM\sigma}\rangle \qquad (a = x, y, z)\,, \qquad (78)$$

where we have assumed a laser polarized along the space fixed $\mathbf{Z}$ axis. In the above equation, x, y and z stand for the BF axes. Herzberg and Innes[69] showed that the $\tilde{A} \to \tilde{X}$ electronic transition moment is perpendicular to the plane of the bent molecule, which corresponds to the x axis (see Fig. 1) in our conventions. We used the $\mu = \mu \sin\theta x$ dipole function of Ross and Bunker.[70]

We now consider the selection rules associated with the Φ_{Zx} term. According to accepted spectroscopic conventions,[71] levels with total parity $(-1)^J$ and $-(-1)^J$ will be denoted as e and f, respectively. For Q-branch

PUMP transitions, the SEP intermediate state $|\Psi_A^{JM\sigma}\rangle$ has f parity. As a consequence, the parity for the DUMP transitions will be e for the Q-branch ($J'' = J'$), and f for the $P(J'' > J')$ and $R(J'' < J')$ branches. The SEP spectrum corresponds to all the different possibilities ($\Delta J = 0, \pm 1; \Delta K = \pm 1$) of these DUMP transitions *expressed in the X symmetric top basis set.* For our calculations, we will consider only one of these transitions at a time, i.e., corresponding to a well-defined DUMP transition ($J' \to J'', K' \to K''$). In the above notation, K'' makes reference to the quantum number of the rotational component used to dress the BF wave function $|\Phi_v^A\rangle$:

$$|\Psi_X^{J'' M \sigma''}\rangle = \mu \sin\theta |\Phi_v^A\rangle |J'', K'', M, \sigma''\rangle . \tag{79}$$

[It should be clear to the reader that this procedure does not mean that K'' is a good quantum number. It only reflects the dipole selection rules in the symmetric top basis set]. This wave function $|\Psi_X^{J'' M \sigma''}\rangle$ is then used to compute the line intensities in terms of its projections onto the exact rovibrational eigenstates of the ground electronic state. Details of the calculations are given in the next section.

3.3. *Results*

3.3.1. *The $^r R_0(0)$ DUMP transition*
$$[(J' = 1, K' = 1, f) \to (J'' = 0, K'' = 0, e)]$$

As a first example, let us consider the transition $(J' = 1, K' = 1, f) \to (J'' = 0, K'' = 0, e)$. This transition produces the simplest SEP spectrum in terms of the number of lines, as shown by a stick diagram in Fig. 2. We present in Table I the convergence properties as a function of the intensity range considered and of the number of Lanczos recursions. Each row in Table I corresponds to a given intensity range, as different ranges display different convergence rates. As discussed in Sec. 2, this behavior reflects the property that large residues converge faster. The sum of all the intensities I_α has been normalized to 1 for convenience. The convergence of any given line ε_α is defined according to the following two criteria :
i) the line positions $\varepsilon_\alpha^{(n)}$ and $\varepsilon_\alpha^{(n')}$ corresponding to two different numbers n and n' of Lanczos recursions have to agree within 1 cm^{-1}. Typically, $n' = n + 200$. Such an intrinsic criterion is valid as our method employs the same basis set as would have been used in a standard diagonalization approach.

ii) In order to avoid fortuitous coincidence in the dense part of the spectrum, the $I_\alpha^{(n)}$ and $I_\alpha^{(n')}$ intensities also have to correspond within 0.1%. For very small intensities ($< 10^{-4}$), such a criterion is too stringent. Instead, the two absolute intensities are required to correspond within 10^{-5}.

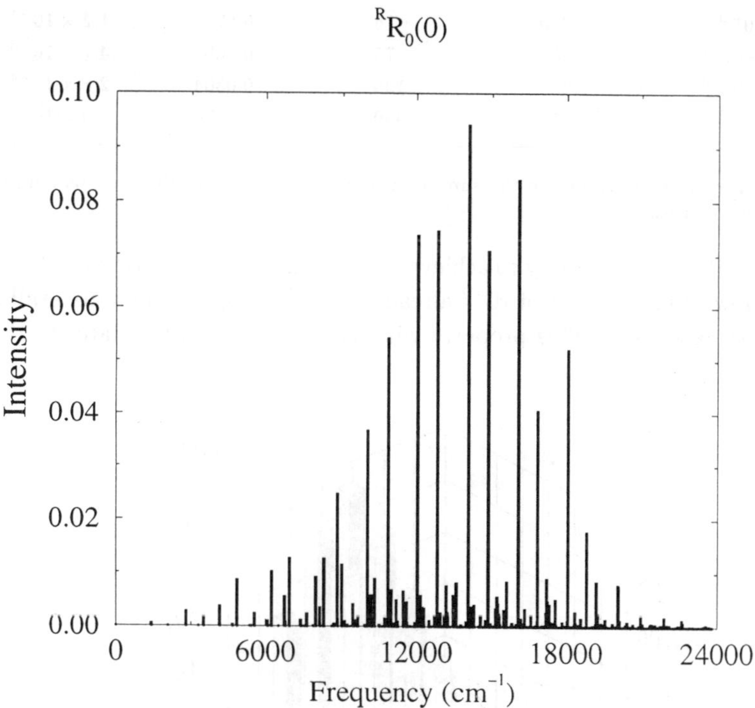

Fig. 2. $^{r}R_0(0)$ DUMP transition spectrum, represented as a stick diagram. This spectrum corresponds to the initial $(0, 1^1, 0)$ level of the excited A electronic state and to a dipole function $\mu_{A \to X} = \sin\theta x$.

For a given number n of Lanczos recursions, one then defines the error of the spectrum as

$$\text{error} = 1 - \sum_{\alpha}^{\text{converged}} I_\alpha^{(n)} . \tag{80}$$

Table I shows that, by increasing the number of recursions, one can eventually reach any desired accuracy on the computed spectrum. In

Table I. Convergence properties for the $^r R_0(0)$ DUMP transition as a function of the intensity range. For each range, the number of recursions corresponds to a convergence of 1% on the contribution to the spectrum.

Intensity Range	Number of Lanczos recursions	Number of Lines	Contribution to the Spectrum	Error[a]
$> 10^{-2}$	1250	15	0.679	1.2×10^{-1}
10^{-2}–10^{-3}	3500	75	0.269	4.7×10^{-3}
10^{-3}–10^{-4}	4000	135	0.0504	2.8×10^{-4}
$< 10^{-4}$	4250	350	0.0076	1.0×10^{-6}

[a] The error makes reference to the sum of unconverged line intensities for this number of Lanczos recursions.

practice, it is not necessary to achieve such a high accuracy. Friesner *et al.*[72] have shown that an error of 10^{-2} already produces a spectrum indiscernible from the exact one. This property will also be demonstrated later in this chapter.

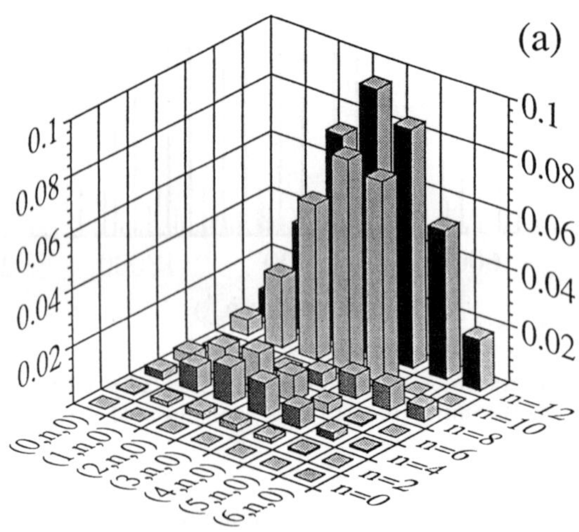

Fig. 3a. Populations of the different (v_1, v_2, v_3) states reached in the $^r R_0(0)$ transition: a) $v_3 = 0$; b) $v_3 = 1$; c) $v_3 = 2$. v_1, n and v_3 correspond to the C–N stretching, H–C–N bending and C–H stretching modes respectively.

An important tool, required for the analysis of simulated SEP spectra, concerns the labelling of energy levels singled out in the calculation. Two

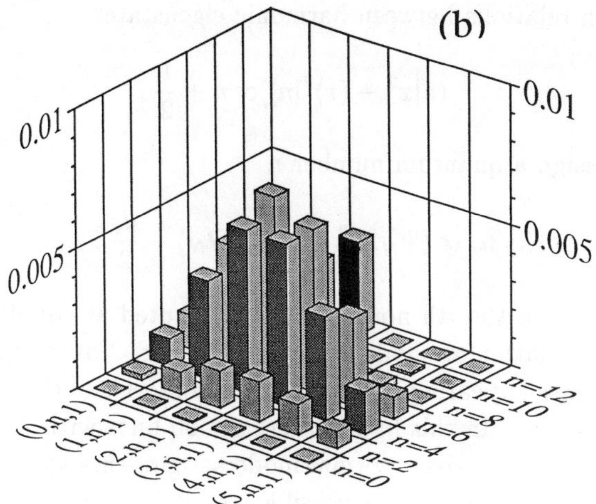

Fig. 3b.

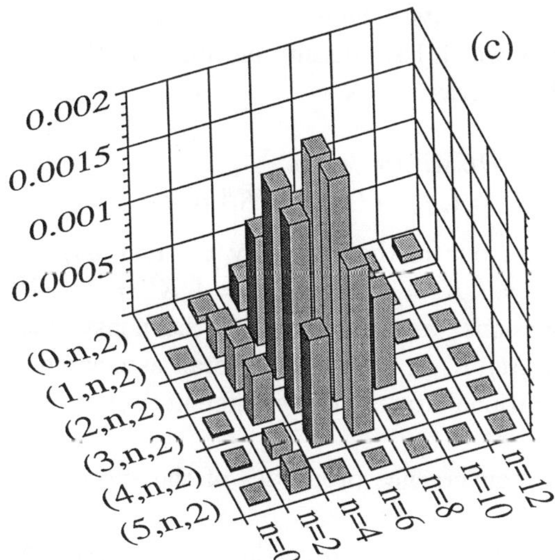

Fig. 3c.

different methods have recently been proposed for this purpose, and applied to the highly excited levels of HCN. The first one, due to Bačić,[73] consists in projecting the exact eigenstates onto a zero order adiabatic basis set. The second method, developed by Menou and Leforestier,[74] makes use of the dispersion relations between harmonic eigenstates

$$\langle n|x^2 - \langle x \rangle^2|n\rangle \propto n + \frac{1}{2}\,, \tag{81}$$

in order to assign a quantum number n

$$n_i \propto \langle \Psi_N|Q_i^2 - \langle Q_i \rangle^2|\Psi_N\rangle - \frac{d_i}{2}\,. \tag{82}$$

(Q_i corresponds to the ith normal mode computed at equilibrium geometry). Using the latter method, we present in Fig. 3 an assignment of the most intense lines appearing in the spectrum, in terms of the three quantum numbers (v_1, v_2, v_3). In this figure, there are 90 lines with $I > 0.001$. Most of these lines could be given normal mode assignments as shown in Fig. 3. The fact that assignments are possible at all indicates very little mode mixing. However, as the energy increases above about $11\,000$ cm^{-1}, an increasing number of lines develop, many of which could not be given HCN normal mode assignments. These "unassignable" lines are associated with delocalized states which have density on both the HCN and HNC sides of the isomerization barrier.

3.3.2. The $^P P_2(4)$ DUMP transition
$$[(J' = 3, K' = 1, f) \to (J'' = 4, K'' = 2, e)]$$

As an example of our method's ability to treat large systems, we now consider the $(J' = 3, K' = 1, f) \to (J'' = 4, K'' = 2, e)$ DUMP transition. As these calculations were conducted in the nonsymmetrized rotational basis set $\{|J, K, M\rangle, K = -J, J\}$, the overall associated basis set involves over 9900 rovibrational states. Three different properties will be addressed in this study:

i) The absolute convergence properties of the spectrum for such a large system.

ii) Differences of this spectrum from the simple $^r R_0(0)$ case.

iii) The accuracy required as a function of the resolution of the spectrum.

We present in Table II the convergence properties of the Lanczos algorithm for this $^P P_2(4)$ transition. Each row corresponds to a specified intensity range. The first remark concerns the number of intense lines ($I > 10^{-2}$). This number (16) is very close to the number (15) of intense lines for the $^r R_0(0)$ transition (see Table I). However, the number of

Table II. Same as Table I for the $^PP_2(4)$ transition

Intensity Range	Number of Lanczos recursions	Number of Lines	Contribution to the Spectrum	Error[a]
$> 10^{-2}$	10 000	16	0.523	6.5×10^{-1}
10^{-2}–10^{-3}	23 000	102	0.348	5.4×10^{-3}
10^{-3}–10^{-4}	25 000	311	0.099	5.5×10^{-5}
$< 10^{-4}$	27 000	1948	0.031	3.0×10^{-6}

[a]The error makes reference to the sum of unconverged line intensities for this number of Lanczos recursions.

recursions required to achieve convergence in this case is much larger (10 000 as compared to 1250). Such an increase in the number of recursions reflects the fact that the Lanczos algorithm is sensitive to the density of states. In fact, the spectrum associated with the $H^{J=4}$ Hamiltonian is 9 times denser than the $H^{J=0}$ one, due to the use of the nonsymmetrized basis set. The moderate intensity range ($10^{-2} > I > 10^{-3}$) basically displays the same behavior, i.e., a similar number of lines (102 instead of 75), but a much larger number of recursions (23 000 vs. 3500). Finally, the number of lines significantly increases (1948) for the low intensity range. These extraneous lines correspond to $K \neq 2$ values, which can be reached only through Coriolis coupling.

As could be anticipated from the above discussion, the stick diagram for the $^PP_2(4)$ transition, presented in Fig. 4, looks very similar to the previous one. The main changes can be summarized as follows:

i) a blue shift and slight intensity reduction of the high and moderate intensity lines ($I > 10^{-3}$) , which correspond to the approximate quantum number $K'' = 2$.

ii) an increase in the number of small intensity lines ($I < 10^{-3}$), resulting from Coriolis coupling and associated with $K'' = 2$ values. It should be noted that these changes are barely noticable on the scale of this figure.

Having established above the absolute convergence properties of the method, we now consider the convergence with respect to a given spectral resolution. For this purpose,we define the simulated spectrum as

$$I(\omega) = \sum_\alpha I_\alpha L(\omega, \varepsilon_\alpha, \Gamma) , \tag{83}$$

with $L(\omega, \varepsilon_\alpha, \Gamma)$ being a Lorentzian of width Γ:

$$^P\text{P}_2(4)$$

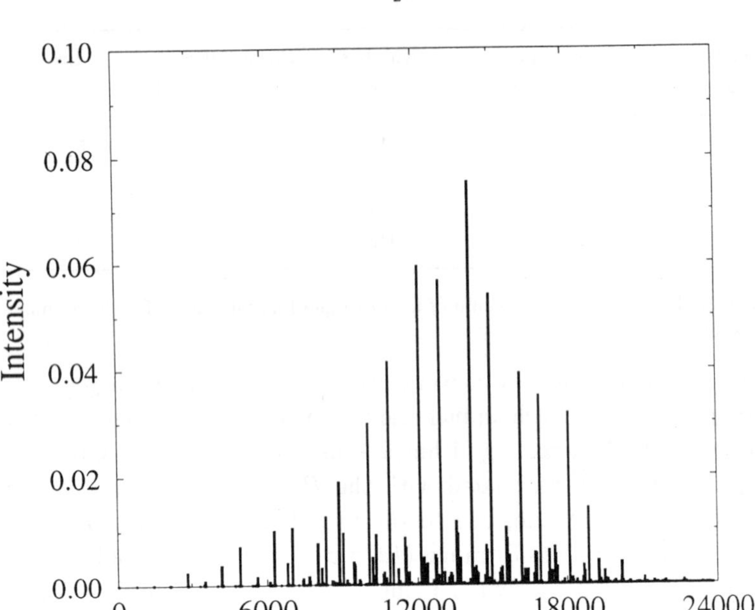

Fig. 4. Same as Fig. 2 for the $^P P_2(4)$ DUMP transition.

$$L(\omega, \varepsilon_\alpha, \Gamma) = \frac{\Gamma^2}{(\hbar\omega - \varepsilon_\alpha)^2 + \Gamma^2} \, . \tag{84}$$

By varying the width Γ, one can generate spectra corresponding to different resolutions, as can be required in order to compare with experimental spectra. Figures 5, 6, and 7 display spectra corresponding to three different resolutions ($\Gamma = 1, 10,$ and 100 cm^{-1}, respectively). In each figure, we represent the exact simulated spectrum $I_{\text{ex}}(\omega)$, as computed from the results obtained for the maximal number of Lanczos iterations (27 000), and the error $|I_{\text{ex}}(\omega) - I^{(N)}(\omega)|$. The $I^{(N)}(\omega)$ spectrum has been computed using results corresponding to a smaller number N of Lanczos recursions. In each case, the number N of recursions has been chosen such that the relative error is about 1%. At high resolution ($\Gamma = 1$ cm^{-1}, Fig. 5), 20 000 recursions are necessary to achieve such an accuracy. When the resolution is lowered to $\Gamma = 10$ cm^{-1} (Fig. 6), a smaller number of recursions (4000) is sufficient to converge the calculated spectrum. Finally, when the

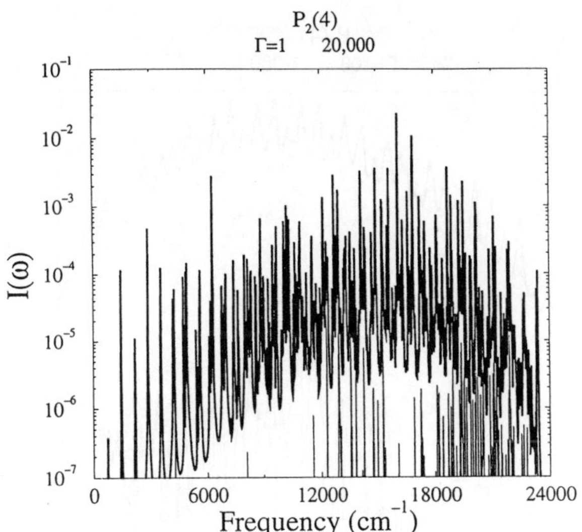

Fig. 5. High resolution SEP spectrum $I_{\mathrm{ex}}(\omega)$ (thick line),as computed from Eq. (79), and error $|I_{\mathrm{ex}}(\omega) - I^{(N)}(\omega)|$ (lower curve) for the $^P P_2(4)$ DUMP transition. A width $\Gamma = 1$ cm^{-1}, and eigenvalues and residues corresponding to 20 000 recursions, have been used.

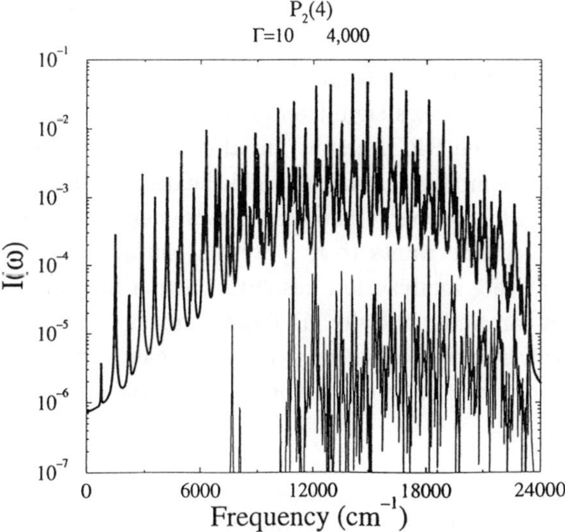

Fig. 6. Same as Fig. 5 for a medium resolution spectrum corresponding to $\Gamma = 10$ cm^{-1} and 4000 recursions.

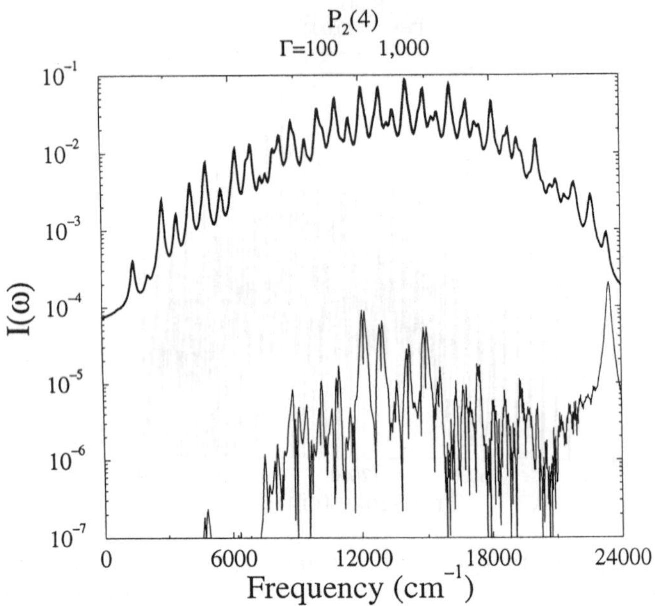

Fig. 7. Same as Fig. 5 for a low resolution spectrum corresponding to $\Gamma = 100$ cm^{-1} and 1000 recursions.

resolution is lowered to $\Gamma = 100$ cm^{-1}, as shown in Fig. 7, the number of recursions further decreases significantly to $N = 1000$. This illustrates a major advantage of the RRGM; substantial computational effort can be saved by tuning the recursion number to yield the desired level of resolution in the spectrum. This property should be extremely useful in practical applications in which one wants to fit potential surfaces to experimental data[76] (or qualitatively analyze low resolution experimental data). Low resolution spectra can be fit initially via large searches of parameter space; once a qualitatively reasonable set of parameters has been found, further refinements to reproduce details of the spectral patterns can be carried out.

4. Discussion

The method discussed in this chapter represents a new formulation developed to compute the SEP spectrum of a triatomic molecule. It combines two very efficient approaches aimed at overcoming the traditional bottlenecks associated with large systems:

i) The Recursive Residue Generation method, based on the Lanczos algorithm, which allows one to specifically compute only the relevant eigenvalues of the molecular system (i.e., those associated with non-negligible line intensities). One can thus avoid the N^3 scaling law of standard diagonalization procedures. Furthermore, it has been shown that it constitutes a progressive method, which gradually improves the resolution of the spectrum as the number of recursion steps increases.

ii) As the RRGM does not require explicit knowledge of the Hamiltonian matrix, but only its action on any given vector, one can use alternative representations, such as grid methods, specifically tailored to treat large systems. However, a straight grid description leads to a very slow convergence of the Lanczos algorithm.[72] The Adiabatic Generalized DVR method allows one to define a compact, very efficient representation of the Hamiltonian operator. In the examples treated here, the central memory occupancy was dictated by the size of the vibrational subsystem only. (All calculations reported here have been carried out on an IBM 550 RISC workstation, using at most 32 MB of core memory). For tetratomic systems, the large memory required would constitute a bottleneck for the calculation. Friesner *et al.*[72] have recently proposed the Adiabatic Pseudo Spectral (APS) method as a way to greatly reduce the memory requirements.

The recursive residue generation method has been applied to the SEP spectrum of the HCN molecule. Although HCN constitutes one of the simplest physical systems for which detailed experimental results are available, it still presents a challenge to the theoretician. Basis sets containing over 9000 rovibrational states had to be considered in the calculations. Our results show that highly excited bending states (assigned up to $v_2 = 12$) are reached in the DUMP transition. The maximum of intensity peaks at the $v_2 = 12$ value, close to the HCN $\leftrightarrow$ CNH isomerization barrier maximum. The shape of the intensity distribution indicates that higher energy states can be significantly populated, for which no assignment could be made. These non-assignable states account for about 15% of the line intensities. In accordance with the analysis of Carter, Mills and Handy,[75] transitions in the $(v_1 = 3, v_2, v_3 = 0)$ progression display the highest intensity. A comparative study of theoretical and experimental results should be performed on a recently improved potential energy surface of HCN.[21]

Acknowledgments

The authors gratefully acknowledge Prof. Robert Field for very helpful discussions. This research was supported in part by the Robert Welch Foundation and the National Science Foundation.

References

1. C. E. Hamilton, J. L. Kinsey, and R. W. Field, *Ann. Rev. Phys. Chem.* **37**, 493 (1986).
2. A. Nauts and R. E. Wyatt, *Phys. Rev. Lett.* **51**, 2238 (1983).
3. R. E. Wyatt, *Adv. Chem. Phys.* **73**, 231 (1989).
4. T. J. Lee and A. P. Rendell, *Chem. Phys. Lett.* **177**, 491 (1991).
5. P. K. Pearson, H. F. Schaefer III, and U. Wahlgren, *J. Chem. Phys.* **62**, 350 (1975).
6. L. T. Redmon, G. D. Purvis, and R. J. Bartlett, *J. Chem. Phys.* **72**, 986 (1980).
7. C. Glidewell and C. Thomson, *J. Comp. Chem.* **5**, 1 (1984).
8. J. A. Pople, K. Raghavachari, M. J. Frisch, J. S. Binkley, and P. von R. Schleyer, *J. Am. Chem. Soc.* **105**, 6389 (1983).
9. J. N. Murrell, S. Carter, and L. O. Halonen, *J. Mol. Spect.* **93**, 307 (1982).
10. Z. Bačić and J. C. Light, *J. Chem. Phys.* **86**, 3065 (1987); J. C. Light and Z. Bačić, *J. Chem. Phys.* **87**, 4008 (1987).
11. M. Mladenovic and Z. Bačić, *J. Chem. Phys.* **93**, 3039 (1990).
12. Z. Bačić, *J. Chem. Phys.* **95**, 3456 (1991).
13. P. R. Fleming and J. S. Hutchinson, *J. Chem. Phys.* **90**, 1735 (1989).
14. K. M. Dunn, J. E. Boggs, and P. Pulay, *J. Chem. Phys.* **85**, 5838 (1986).
15. B. Gazdy and J. M. Bowman, *Chem. Phys. Lett.* **175**, 434 (1990); *J. Chem. Phys.* **95**, 6309 (1991).
16. J. M. Bowman and B. Gazdy, *Theor. Chim. Acta.* **79**, 215 (1991).
17. C. Leforestier, *J. Chem. Phys.* **94**, 6388 (1991).
18. B. H. Chang and D. Secrest, *J. Chem. Phys.* **94**, 1196 (1991).
19. J. P. Brunet, R. A. Friesner, R. E. Wyatt, and C. Leforestier, *Chem. Phys. Lett.* **153**, 425 (1988).
20. J. A. Bentley, J. P. Brunet, R. E. Wyatt, R. A. Friesner, and C. Leforestier, *Chem. Phys. Lett.* **161**, 393 (1989).
21. J. A. Bentley, J. M. Bowman, B. Gazdy, T. J. Lee, and C. E. Dateo, *Chem. Phys. Lett.* **198**, 563 (1992).
22. J. Bentley, C. M. Huang, and R. E. Wyatt, to be published.
23. X. Yang, C. A. Rogaski, and A. M. Wodtke, *J. Chem. Phys.* **92**, 2111 (1990); X. Yang, C. A. Rogaski, and A. M. Wodtke, *J. Opt. Soc. Am.* **B7**, 1835 (1990).
24. D. M. Jonas, X. Yang, and A. M. Wodtke, *J. Chem. Phys.* **97**, 2284 (1992).
25. C. Lanczos, *J. Res. Nat. Bur. Stand.* **45**, 58 (1950).

26. J. B. Rosser, C. Lanczos, M. R. Hestenes, and W. Karush, *J. Res. Nat. Bur. Stand.* **47**, 291 (1951).

27. J. K. Cullum and R. A. Willoughby, "Lanczos Algorithms for Large Symmetric Eigenvalue Computations" (Birkhäuser, Boston, 1985).

28. G. H. Golub and C. F. Van Loon, "Matrix Computations" (Johns Hopkins, Baltimore, 1989) Ch. 9.

29. B. N. Parlett, "The Symmetric Eigenvalue Problem" (Prentice-Hall, Englewood Cliffs, N. J., 1980), Ch. 13.

30. F. LeQuere and C. Leforestier, *J. Chem. Phys.* **92**, 247 (1990); *Chem. Phys. Lett.* **189**, 537 (1992).

31. R. E. Wyatt, C. Iung, and C. Leforestier, *J. Chem. Phys.* **97**, 3458 (1992).

32. R. E. Wyatt, C. Iung, and C. Leforestier, *J. Chem. Phys.* **97**, 3477 (1992).

33. C. Iung and C. Leforestier, *J. Chem. Phys.* **90**, 3198 (1989).

34. C. Iung and C. Leforestier, *J. Chem. Phys.* **97**, 2481 (1992).

35. G. Strang, "Introduction to Applied Mathematics" (Wellesley–Cambridge, Wellesley, 1986), pp. 389–392.

36. D. G. Pettifor and D. L. Weaire (eds.), "The Recursion Method and Its Applications" (Springer, Berlin, 1985).

37. N. S. Sehmi, "Large Scale Structural Eigenanalysis Techniques" (Wiley, New York, 1989).

38. G. H. Golub and D. P. O'Leary, *SIAM Review* **31**, 50 (1989).

39. H. Ehrenreich, F. Seitz, and D. Turnbull (eds.), "Solid State Physics," Vol. 35 (Academic, New York, 1980).

40. R. Riedinger and M. Benard, *J. Chem. Phys.* **94**, 1222 (1991).

41. H. D. Meyer and S. Pal, *J. Chem. Phys.* **91**, 6195 (1989); T. J. Godin and R. Haydock, *Comp. Phys. Comm.* **64**, 123 (1991).

42. K. F. Milfeld and N. Moiseyev, *Chem. Phys. Lett.* **130**, 145 (1986).

43. M. D'Mello, C. Dunczky, and R. E. Wyatt, *Chem. Phys. Lett.* **148**, 169 (1988).

44. C. Duneczky and R. E. Wyatt, *J. Phys.* **B21**, 3727 (1988).

45. O. Kolin, C. Leforestier, and N. Moiseyev, *J. Chem. Phys.* **89**, 6836 (1988).

46. G. C. Groenenboom and H. M. Buck, *J. Chem. Phys.* **92**, 4374 (1990).

47. E. Haller, H. Koppel, and L. S. Cederbaum, *J. Mol. Spec.* **111**, 377 (1985).

48. K. Marshall and J. Hutchinson, *J. Phys. Chem.* **91**, 3219 (1987); *J. Chem. Phys.* **95**, 3232 (1991).

49. N. Ben-Tal and N. Moiseyev, *J. Phys.* **A24**, 3593 (1991).

50. K. V. Vasavada and J. H. Freed, *Comp. in Phys.* Sept/Oct, 61 (1989).

51. D. J. Schneider and J. H. Freed, *Adv. Chem. Phys.* **73**, 387 (1989).

52. I. M. Barbour, N. E. Behilil, P. E. Gibbs, M. Rafig, K. J. M. Moriarty, and G. Schierholz, *J. Comp. Phys.* **68**, 227 (1987).

53. A. N. Burkitt and A. C. Irving, *Comp. Phys. Comm.* **59**, 447 (1990).

54. J. G. F. Francis, *Comp. Journal* **4**, 265 (1961/62).

55. B. Parlett, *SIAM Review* **6**, 275 (1964).

56. G. W. Stewart, "Introduction to Matrix Computations" (Academic, New York, 1973), Ch. 7.

57. R. E. Wyatt and D. Scott, in "Large Eigenvalue Problems", eds. J. Cullum and R. Willoughby (North-Holland, Amsterdam, 1986), p. 67.

58. E. B. Wilson, J. C. Decius, and P. C. Cross, "Molecular Vibrations" (McGraw-Hill, New York).

59. J. Tennyson and B. T. Sutcliffe, *J. Chem. Phys.* **77**, 4061 (1982); **79**, 43 (1983).

60. A. R. Edmonds, "Angular Momentum in Quantum Mechanics" (Princeton University Press, 1974).

61. Z. Bačić and J. C. Light, *J. Chem. Phys.* **85**, 4594 (1986).

62. J. C. Light, I. P. Hamilton, and J. V. Lill, *J. Chem. Phys.* **82**, 1400 (1985).

63. W. Yang and A. Peet, *Chem. Phys. Lett.* **153**, 98 (1988).

64. J. Tennyson and J. R. Henderson, *J. Chem. Phys.* **91**, 3815 (1989).

65. S. E. Choi and J. C. Light, *J. Chem. Phys.* **92**, 2129 (1990).

66. G. Herzberg, "Molecular Spectra and Molecular Structure, III" (Van Nostrand Rheinhold, Princeton, New Jersey).

67. G. A. Bickel and K. K. Innes, *Can. J. Phys.* **62**, 1763 (1984).

68. G. Herzberg, "Molecular Spectra and Molecular Structure, II", (Van Nostrand Rheinhold, Princeton, New Jersey).

69. G. Herzberg and K. K. Innes, *Can. J. Phys.* **35**, 842 (1957).

70. S. C. Ross and P. R. Bunker, *J. Mol. Spectrosc.* **105**, 369 (1984).

71. J. M. Brown *et al.*, *J. Mol. Spectrosc.* **55**, 500 (1975).

72. R. A. Friesner, J. Bentley, M. Menou, and C. Leforestier, *J. Chem. Phys.*, in press.

73. M. Mladenovic and Z. Bačić, *J. Chem. Phys.* **93**, 3039 (1990).

74. M. Menou and C. Leforestier, *Chem. Phys. Lett.*, in press.

75. S. Carter, I. M. Mills, and N. C. Handy, *J. Chem. Phys.* **93**, 3722 (1990).

76. R. C. Cohen and R. J. Saykally, *Annu. Rev. Phys. Chem.* **42**, 369 (1991).

CHAPTER 22

THE SEP SPECTRUM OF ACETYLENE: SYMMETRY PROPERTIES AND ISOMERIZATION

James K. Lundberg

Joint Institute for Laboratory Astrophysics
and National Institute of Standards and Technology
Boulder, Colorado 80309-0440, USA

Contents

1. Introduction

Spectroscopy is a powerful method of determining molecular structure. Decoding the patterns in molecular spectra provides us with detailed information about their electronic structure, the geometrical arrangement of the nuclei, and the forces between these atoms.

Molecular spectroscopy has its roots in the early days of quantum mechanics. Since that time the vast body of information collected about polyatomic molecules concerns the properties of molecules near the equilibrium geometry of the ground electronic state. With the advent of tunable lasers it has become possible to study highly vibrationally excited molecules with techniques such as direct single photon overtone absorption spectroscopy,[1-3] multiphoton absorption,[4] and cavity ringdown spectroscopy.[5]

The overtone absorption technique overcomes the $\Delta v = \pm 1$ selection rules by exciting the high frequency molecular vibrations which have large anharmonicity. In polyatomic molecules, vibrations with large anharmonicity are usually subject to strong coupling of the vibrational and rotational motions. Strong anharmonic coupling muddles the already complicated absorption spectrum, making the extraction of useful information from a one-photon absorption spectrum difficult.

However, this strong coupling leads to very interesting and complex dynamical behavior,[1,6,7] including isomerization.[8] Studying the complex dynamics of highly vibrationally excited molecules presents major challenges for both the spectroscopist and the dynamicist.

An extremely powerful experimental technique for studying highly excited molecules is the *double resonance* technique of Stimulated Emission Pumping (SEP).[9-11] Since the intermediate state in an SEP experiment typically belongs to an electronic state with an equilibrium structure different from the ground state, the $\Delta v = \pm 1$ single photon propensity rule for vibrational spectroscopy is not even approximately valid. Due to the prior state selection of double resonance spectroscopy, and the restrictive ΔJ and ΔK_a optical selection rules, the resulting spectrum is devoid of the congestion inherent in one photon and multiphoton absorption techniques. Typically, only several J, K_a levels per vibrational state appear in the SEP spectrum.

The conventional method of decoding the spectrum to derive structural information is based on the assumptions that the electronic, vibrational, rotational, nuclear spin, and electron spin parts of the molecular wavefunction are separable. These approximations are a useful starting point for the analysis of molecular spectra. This is especially so when an electronic state is isolated in energy from other electronic states, and when the molecule contains a small amount of vibrational energy (i.e., the nuclei remain near their equilibrium positions). Even under such ideal conditions, this separation is not rigorous. For example, Huet and coworkers[12] have shown that in acetylene-d_2, extreme mixing between the bending modes of vibration occurs at only ~ 1000 cm^{-1} above the zero point vibrational level of the ground state.

At sufficiently high vibrational excitation, where exceedingly large amplitude motions occur, the approximations which were useful at low energy may no longer be beneficial. The separation of the electronic and vibration–rotation parts of the wavefunction, i.e., $\psi_{\mathrm{evr}} = \psi_{\mathrm{e}}\psi_{\mathrm{rv}}$ (the Born–Oppenheimer approximation) which at low energy is a good approximation, may no longer be valid. Coriolis interactions between states which differ in both V (vibrational excitation of all normal modes) and K_a degrade the validity of these two quantum numbers.[13] In such cases, the labeling of levels with the standard vibrational and rotational quantum labels, i.e., $|V_n, J, K_a\rangle$ becomes either misleading or infeasible. Indeed, due to extreme mixing, all of the $(3n - 6)$ vibrational modes may no longer be approximately separable and the quantum numbers associated with these motions may be completely lost.[14]

The issue of vibrational identity can be clouded even further when the internal energy of the molecule is sufficient to promote tunneling between isomeric structures. Clearly, in such cases, the traditional vibrational analysis must fail.

How then are we to proceed with our goal of identifying the molecular eigenstates (i.e., assigning vibrational quantum numbers), and what if any eigenstate classification scheme can be valid? The answer is found in group theory. It can be shown that under all circumstances, molecular energy levels can be rigorously labeled according to the irreducible representations of a symmetry group of the molecular Hamiltonian.[15] Even in extreme cases, when the quantum numbers associated with the traditional factorization of the wavefunction must fail, the eigenstates can still be labeled using their symmetry properties and the rigorously good quantum number J. This labeling scheme provides us with a basis for understanding complex dynamics and spectroscopic transition selection rules.

1.1. *Symmetry Properties of the Molecule*

What is a "symmetry group of a Hamiltonian"? A symmetry group of the molecular Hamiltonian is a group of symmetry operations that convert an eigenfunction Ψ_n of the Hamiltonian to a new function Ψ'_n having the same eigenvalue, or energy, as the original eigenfunction Ψ_n. In other words, the molecular Hamiltonian is *invariant* to a symmetry operation and, therefore, the energy of any eigenstate of the molecule remains unaffected.

What kinds of symmetry does a molecule possess? Typically, one thinks of the symmetry of the rigid structure of the molecule in terms of reflections and rotations. These *point group* operations are widely used when the molecule possesses a unique equilibrium structure. These widely used and useful symmetry properties of the molecule are not valid when considering weakly bound (nonrigid) systems, or even when considering a rotating molecule.

Molecules possess another type of symmetry that is universally applicable. These symmetries consist of the exchange of identical nuclei in combination with the inversion of the laboratory fixed reference frame.[16] Clearly, exchanging identical nuclei in a molecule does not affect the Hamiltonian. Likewise, the molecule remains unaware of our choice of axis systems, and an inversion of the lab fixed reference frame cannot affect the energy of the molecule.

These permutation-inversion symmetry elements remain rigorous in all cases, including rotating *and* nonrigid molecules, and provide a fundamental and powerfully descriptive means for understanding complex molecular dynamics. It is this set of symmetry properties which provide the framework for labeling the molecular eigenfunctions and understanding the complex spectra of highly excited molecules.

1.2. *Structures of Acetylene*

The fact that molecules can undergo qualitative changes in conformation upon electronic excitation is well established.[17] These changes in geometry are exhibited by spectroscopic patterns or "fingerprints" which are diagnostic of the difference between the structures of the states involved in the transition. A classic analysis of such "fingerprints", which appeared in the near ultraviolet spectrum of acetylene, enabled Ingold and King[18] and Innes[19] to conclude that the first excited singlet electronic state of C_2H_2 was stable in a planar *trans*-bent structure qualitatively different from that of the linear ground state.

Acetylene is a prime example of a molecule possessing several qualitatively different stable structures. Many electronically excited states of acetylene possess structures distorted from the linear geometry of the ground state. In addition, many potential energy surfaces of acetylene possess several deep potential minima corresponding to different isomeric structures on the *same* electronic potential energy surface. It has been shown in several *ab initio* studies,[20–23] that acetylene has stable isomeric minima not only in the linear $D_{\infty h}$ and *trans*-bent C_{2h} conformations of the well studied ground and first excited singlet states, but also in the *cis*-bent and *vinylidene* structures associated with C_{2v} point group symmetries, and *nonplanar* C_2 point group structures. While there has been no direct spectroscopic evidence of a singlet state of C_2H_2 with the *cis*-bent structure, there are recent experimental observations of the vinylidene and *nonplanar* conformations in singlet C_2H_2.[8,24–26]

It is interesting that although both the *vinylidene* and *cis*-bent structures of C_2H_2 possess C_{2V} symmetry properties, the symmetry elements of the two structures are not identical. For example, in *vinylidene* where both hydrogens are bonded to the same carbon atom, the C_2 point group operation is a rotation about the a-inertial axis and permutes the two H's. In the *cis*-bent structure, where one hydrogen is bonded to each carbon atom, the C_2 operation is a rotation about the b-inertial axis, and permutes both the H's and the C's. There are also two possible, and nonequivalent, *nonplanar* structures of acetylene which belong to different C_2 point groups. One has the C_2 symmetry element corresponding to a rotation about the *b*-inertial axis and the other has a C_2 rotation about the *c*-inertial axis.

Rigid molecule point group theory provides no consistent framework for dealing with transitions between any of the large number of possible stable structures of acetylene. A group theoretical treatment is required which focuses on symmetry operations which are common to all imaginable configurations. This is especially important when one electronic surface supports multiple minima. The limitation of the point group treatment for acetylene, which is most widely used in the literature, arises from the use of *near symmetry* operations such as rotations and reflections about *molecule fixed* axes. These *near symmetry* operations are *not* common to the point groups corresponding to the stable structures of C_2H_2, and are only rigorously applicable to nonrotating molecules. The more generally applicable and useful Complete Nuclear Permutation Inversion (CNPI) groups and their Molecular Symmetry (MS) subgroups are composed of *true symmetry* operations, such as nuclear permutations and laboratory-fixed frame inversions, which *are* common to all of the possible isomeric forms of C_2H_2.

When the symmetries of the rotational wavefunctions are expressed in the CNPI or MS framework, a simplification of notation for irreducible representation labels can be made which replaces the 'Mulliken symbol' labels (i.e., A_1, A_2, ..., A_g, A_u, ...)[27] corresponding to the behavior of the wavefunctions under the *near symmetry* operations of the various point groups, by "true" symmetry labels which are rigorous and point group independent properties of the wavefunction itself. The explicitly defined and rigorous CNPI symmetry elements are common to all the relevant acetylene MS subgroups isomorphic to any given point group geometry, and allow us to label the levels directly with their true symmetry properties rather than the usual but more obscure irreducible representation labels of point groups.

This procedure, essentially a generalization of the procedure utilized by Ingold and King[18] for *cis*-bent and *trans*-bent acetylene, has the desirable consequence that the selection rules for rovibronic transitions and perturbations between states belonging to different electronic configurations and isomeric forms are much clarified. This allows the peculiarities of the patterns manifest in the spectrum to be readily identified.

In this discussion, the *point group independent* CNPI–MS group treatment[28] is used to identify the rotational energy level patterns of all predicted and observed structures of acetylene. After identifying the patterns unique to each structure, the CNPI–MS treatment is used to identify the pattern of perturbations basis states associated with the vinylidene minimum on the ground state electronic surface produce in the dense manifold of linear acetylene basis states. In addition, the CNPI–MS group theory is used to illustrate the unique patterns these perturbations leave in a Stimulated Emission Pumping (SEP) spectrum.

2. CNPI–MS Group Theory of Acetylene

The CNPI–MS treatment we will develop applies to all parts (i.e., nuclear, electronic, vibrational, rotational, and electron spin) of the wavefunctions for symmetric and asymmetric tops. We will concentrate here on the symmetry labels for the rotational wavefunctions, including nuclear permutation symmetry, keeping in mind that we must ultimately account for the symmetries of the electronic, vibrational and electron-spin parts of the wavefunction when deriving the exact overall symmetry labels for the C_2H_2 rovibronic levels.

Many different notations exist for labeling the rotational wavefunctions of asymmetric top molecules. Here, we will combine the asymmetric rotor designation as given in Townes and Schawlow[29] with the CNPI and MS group picture of Hougen[30] and Longuet–Higgins,[16] as described by Bunker.[28] This combination of the two notations serves a useful purpose in that the labels derived using the CNPI and MS groups explicitly display the symmetry properties of the wavefunctions. The more widely used asymmetric rotor designation is retained here in order to explicitly display the orientation of a transition moment, relative to the inertial axes of a rigid reference configuration, and to maintain a link with standard notation.

Acetylene is composed of two carbons and 2 hydrogens. We will label the two carbons with the letters a and b, and the two hydrogens with the numbers 1 and 2. The group of symmetry operations involving exchange of identical nuclei with and without a laboratory fixed reference frame inversion are the exchange of the two carbons designated with the notation (ab). We designate the symmetry operation of exchanging the two hydrogens as (12). The identity operation, which leaves the molecule unaffected, is designated E, and the operation that inverts the laboratory fixed reference frame E^* is called the parity operator. The parity operator either transforms a wavefunction into itself (the wavefunction has + parity), or it transforms the wavefunction into its negative ($-$ parity). These operations, alone or in combination, constitute the eight membered Complete Nuclear Permutation Group of the C_2H_2 molecule, generally called G_8.[31] The operations are, E, E^*, (12), $(12)^*$, (ab), $(ab)^*$ $(12)(ab)$, and $(12)(ab)^*$. $((12)^* \equiv (12) \otimes E^*$, etc.$)$.

This group of operations forms a complete set of symmetry labels with which to identify the molecular eigenfunctions of acetylene. Because the molecular Hamiltonian for C_2H_2 and C_2D_2 is invariant to these symmetry operations, these labels are rigorously good at any level of treatment. Table I is the G_8 character table for acetylene.

Table. I. The G_8 group character table for acetylene. The full permutation-inversion group (G_8) must be used when the internal energy of the molecule is sufficient to allow the breaking and reforming of C–H bonds.

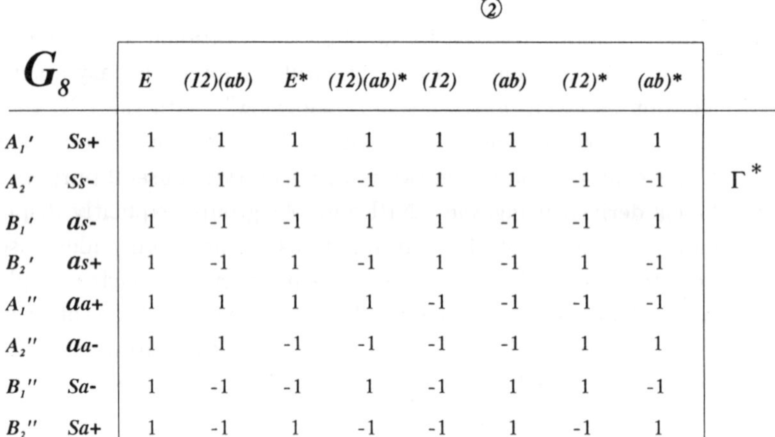

G_8		E	$(12)(ab)$	E^*	$(12)(ab)^*$	(12)	(ab)	$(12)^*$	$(ab)^*$	
A_1'	$Ss+$	1	1	1	1	1	1	1	1	
A_2'	$Ss-$	1	1	-1	-1	1	1	-1	-1	Γ^*
B_1'	$as-$	1	-1	-1	1	1	-1	-1	1	
B_2'	$as+$	1	-1	1	-1	1	-1	1	-1	
A_1''	$aa+$	1	1	1	1	-1	-1	-1	-1	
A_2''	$aa-$	1	1	-1	-1	-1	-1	1	1	
B_1''	$Sa-$	1	-1	-1	1	-1	1	1	-1	
B_2''	$Sa+$	1	-1	1	-1	-1	1	-1	1	

2.1. *Asymmetric Rotor Labels*

We can now begin labeling the rotational eigenstates using their CNPI symmetry. The behavior of the asymmetric rotor functions with respect to

180° rotations about the inertial axes of a rigid molecule serves as a useful starting point for this discussion.

The rotational wavefunctions either transform into themselves after a C_2 rotation about the corresponding inertial axis $(+)$, or they transform into their negative $(-)$. Performing a C_2 rotation about any one of the three inertial axes is equivalent to carrying out C_2 rotations about both of the other two inertial axes. Therefore, indicating the behavior of these functions with respect to C_2 rotations about two of the inertial axes uniquely displays the behavior of the wavefunction with respect to a C_2 rotation about the remaining inertial axis.[32] Consequently, all symmetry information about the asymmetric rotor functions can be displayed by two labels. Labels describing the behavior of the rotational wavefunction about the a- and c-inertial axes are usually used. Here we use the conventional order K_a, K_c,[29] where K_a and K_c are the rotational quantum numbers for rotation about the a- and c-inertial axes. Table II summarizes the asymmetric rotor designations of the rotational levels.

Table II. The $+-$ symbols represent the *even* or *odd* behavior of the rotational wavefunctions with respect to a C_2 rotation about the relevant inertial axis. Column 4 relates Dennison's $(+, -)$ notation with the commonly used K_a and K_c rotational wavefunction labels.

$C_2^{\,a}$	$C_2^{\,b}$	$C_2^{\,c}$	K_a	K_c
$+$	$+$	$+$	even	even
$-$	$-$	$+$	odd	even
$-$	$+$	$-$	odd	odd
$+$	$-$	$-$	even	odd

For all asymmetric rotors, whatever the point group or MS group of the molecule, these symmetry properties of the rotational wavefunctions are rigorously valid. The particular ordering of the labels that is used here is specific to all near prolate tops such as acetylene. With these invariant labels for asymmetric rotor functions one can readily construct an energy level "template" for the nontunneling asymmetric top molecule (Fig. 1).

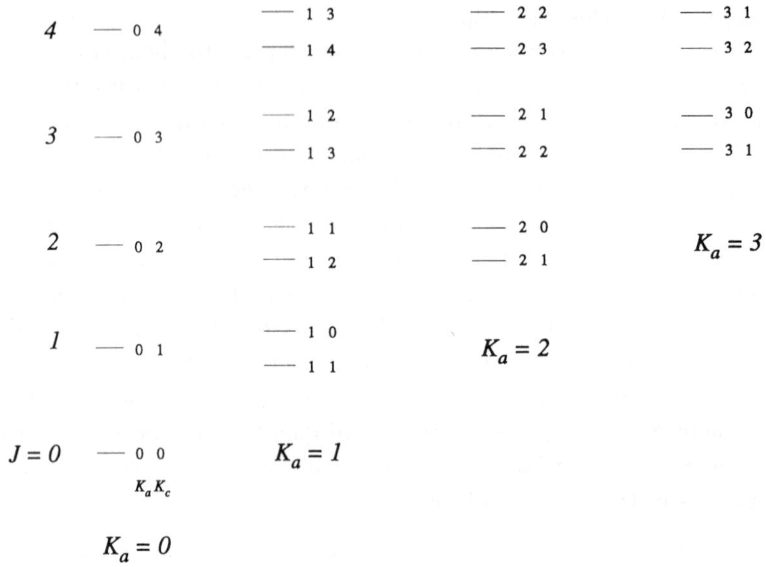

Fig. 1. Rotational energy level "template" for nontunneling near prolate asymmetric top molecules. The (K_a, K_c) notation indicates the *even* or *odd* behavior of the rotational wavefunctions about the **a**- and **c**-inertial axes, respectively. This pattern of the rotational wavefunctions is common to all near prolate asymmetric tops regardless of the symmetry group of the molecule.

2.2. CNPI–MS Operations

2.2.1. Parity and particle exchange

In acetylene, the rigorously good rovibronic quantum labels in a field free environment are: J, parity $(+, -)$, and for all pairs of identical nuclei, nuclear permutation symmetry (ss/aa). The large letter refers to the symmetry (s) or antisymmetry (a) with respect to exchange of the carbons, and the small letter refers to the exchange symmetry (s or a) of the hydrogens.

It is useful to retain the approximately good quantum label, K_a, because the rotational energy levels for a near prolate top arrange themselves in groups of like K_a, as can be seen from the energy expression for a near prolate top:

$$E = (B + C)/2J(J + 1) + [A - (B + C)/2]K_a^2 \ldots . \tag{1}$$

The basis set of "quantum labels" we will use throughout is then $|J, K_a, ss/aa, +/-\rangle$.

Using these basis state labels we can label the rotational energy levels with the G_8 symmetry properties of the wavefunctions, suppressing the standard character representation of a particular group.

While quantum mechanics tells us that to treat rigorously the acetylene problem we must use the full set of G_8 symmetry operations, it is often convenient to consider a reduced set of symmetry elements to label the energy levels of the molecule. For instance, at low levels of molecular excitation the probability for the hydrogens or carbons to tunnel to an equivalent minimum on the potential energy surface is insignificant. The time scale of these motions is long and the corresponding level splitting is small and may not be resolvable. In such cases, it is acceptable to ignore the *'infeasible'* operations (12), (ab), $(12)^*$, and $(ab)^*$ which correspond to these tunneling motions. The remaining group of *'feasible'* operations comprise the Molecular Symmetry (MS) group of the molecule.

For *rigid* nonlinear molecules the MS group is isomorphic to the molecular point group. The permutation–inversion operations of the MS group have a direct correlation to the reflections and rotations of the rigid molecule point group. This isomorphism links the familiar near symmetry operations of the point group with the rigorous true symmetry operations of the CNPI group. It is then a simple process for those familiar with point groups to use the MS subgroup to label the rovibronic levels of the molecule and subsequently incorporate the doubling introduced when considering the (12), (ab), $(12)^*$, and $(ab)^*$ operations of the full G_8 symmetry group.

2.2.2. *Trans-Bent C_{2h} Structure*

In C_2H_2 the equilibrium structure of the $\tilde{A}^1 A_u$ state is known to be *trans*-bent (C_{2h} symmetry). This state is used as an intermediate in many double resonance experiments including ours, therefore, it serves as a useful starting point from which to proceed with the discussion of the rigorous symmetries and spectroscopic selection rules for C_2H_2.

Three Euler angles relate the nuclear and electronic coordinates in the molecule fixed axis system to those in the laboratory fixed axis system.[33] Operations of a symmetry group can cause a change in the Euler angles of the molecule and, therefore, a change in the rovibronic wavefunction which is defined in terms of these angles. The change of the Euler angles caused by *any* group operation is equivalent to that caused by a rotation about one of the principal axes of inertia. Therefore, when working with rotational wavefunctions it is convenient to substitute the symmetry elements of the group with their *equivalent rotations.*

In Table III the Molecular Symmetry subgroup of G_8, which is isomorphic to the C_{2h} point group of the nontunneling molecule, is given. The point group operations (i.e., reflections and rotations etc.) are included, along with those of their corresponding CNPI–MS group operations, and their *equivalent rotations* are indicated.

Table III. The MS group table for *trans*-bent C_2H_2. The electric dipole operator Γ^* belongs to the representation which changes the parity (-1 under E^*), while leaving the nuclear permutation symmetry unaffected ($+1$ under $(12)(ab)$). The *equivalent rotations* of the permutation-inversion operations are indicated, and the point group operations (i.e., reflections and rotations etc.) are included. The Mulliken symbols for the relevant groups (first three columns) are correlated with the true symmetry properties of the wavefunction. The symmetry of the rovibronic tunneling doublets is indicated in column 4.

Equivalent Rotation				R_0	R_c^{π}	R_c^{π}	R_0		
CNPI-MS				E	$(12)(ab)$	E^*	$(12)(ab)^*$		
$D_{\infty h}$	G_8	C_{2h}	True Symmetry Labels	E	$C_2^{\,c}$	σ_{ab}	i		
Σ_g^+	$A_1'\ A_1''$	A_g	$ss+\ \ aa+$	1	1	1	1		J_c
Σ_u^-	$A_2'\ A_2''$	A_u	$ss-\ \ aa-$	1	1	-1	-1	Γ^*	c
Σ_g^-	$B_1''\ B_1'$	B_g	$sa-\ \ as-$	1	-1	-1	1		$J_a\ \ J_b$
Σ_u^+	$B_2''\ B_2'$	B_u	$sa+\ \ as+$	1	-1	1	-1	a	b

The parity of the wavefunction, given by the operation E^* in the MS group, corresponds to a reflection through the plane of the molecule (σ_{ab} in the point group notation). From the *equivalent rotation* of this reflection it can be seen that parity labels can be affixed to the C_{2h} rotational energy diagram by using the K_c label. *In fact, for all planar molecules the parity operation E^* will correspond to a C_2 rotation about the **c**-inertial axis.*[34]

Because the permutation of the carbons and hydrogens correlates with the C_2^c rotation, the nuclear permutation symmetry can be affixed to the rotational energy level "template" by, again, using the K_c labels.

It is now possible to assign rigorous symmetry labels to the rotational levels of *trans*-bent acetylene, keeping in mind that we have so far only considered the rotational symmetries. The electronic and vibrational symmetries will ultimately have to be included.

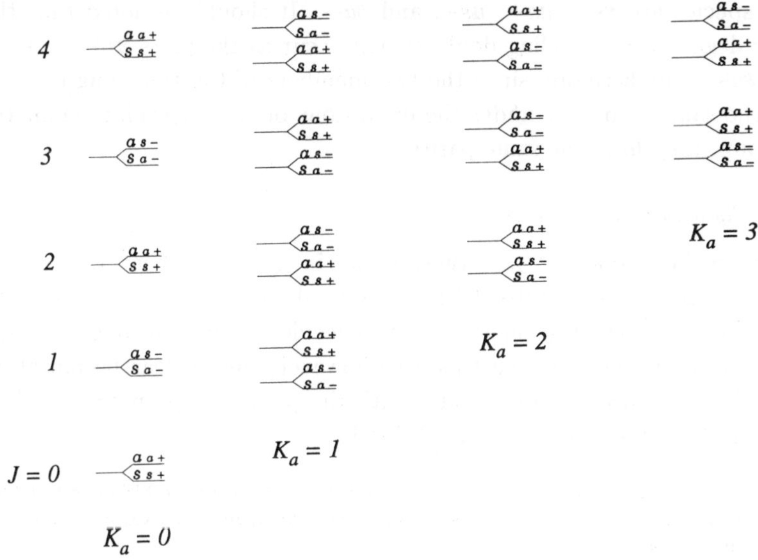

Fig. 2. The parity $(+/-)$ and nuclear permutation symmetry (Ss/aa) of the rotational wavefunctions in *trans*-bent acetylene. The G_8 species wavefunctions that have the same transformation properties in the MS subgroup of the G_8 group comprise the tunneling doublets. In the rigid molecule, the doublets are exactly degenerate. When the internal energy of the molecule is sufficient to allow nuclear permutation tunneling, this degeneracy is lifted. Only four of the 8 possible rovibronic species appear in the *trans*-bent structure.

When we chose to use the reduced symmetry MS subgroup, the G_8 operations (12), $(12)^*$, (ab) and $(ab)^*$, were ignored. These operations correspond to the breaking and forming of C–H bonds and lead to doubling of the rovibronic levels of the rigid molecule. The pairs of G_8 species which transform identically under the reduced symmetry MS group describe the doublets. The pairs of species that have the same transformation properties in the MS group are listed together in the group table. We can then affix the rigorous symmetry properties of both components of the doublet to the rotational energy level "template" (Fig. 2). We shall see later that this doubling of levels may or may not manifest itself in the molecule, depending on whether both components of the doublet have nonzero statistical weights.

In the rotational manifold of the rigidly *trans*-bent structure, only 4 of the 8 possible symmetries occur. In totally symmetric vibronic states the 4 species are $ss+$, $aa+$, $as-$, and $sa-$. It should be noted that the nearly degenerate tunneling doublets will occur as the pairs $ss+$, $aa+$, or $as-$, $sa-$. Furthermore, since the two members of the tunneling doublet have the same character under the operations of the molecular symmetry subgroup, they have the same parity.

2.2.3. *Cis-bent* C_{2v} *structure*

In the *cis*-bent b-axis C_{2v} structure, as in the *trans*-bent c-axis C_{2h} structure, the parity operator E^* in the MS subgroup group correlates with a reflection through the plane of the molecule or an *equivalent rotation* about the c-inertial axis. In this case however, the nuclear permutation $(12)(ab)$ in the MS group correlates with the point group operation C_2^b, a rotation about the b-inertial axis (Table IV).

Table IV. The MS group table for *cis*-bent acetylene. Although the MS operations for the *cis*-bent and *trans*-bent structures are identical, the symmetry axes correspond to different inertial axes.

Equivalent Rotation				R_0	R_b^{π}	R_c^{π}	R_a^{π}		
CNPI-MS				E	$(12)(ab)$	E^*	$(12)(ab)^*$		
$D_{\infty h}$	G_8	C_{2V}	True Symmetry Labels	E	$C_2^{\,b}$	σ_{ab}	σ_{bc}		
Σ_g^+	$A_1'\ A_1''$	A_1	$ss+\ aa+$	1	1	1	1	b	
Σ_u^-	$A_2'\ A_2''$	A_2	$ss-\ aa-$	1	1	-1	-1	Γ^*	J_b
Σ_g^-	$B_1''\ B_1'$	B_1	$sa-\ as-$	1	-1	-1	1	c	J_a
Σ_u^+	$B_2''\ B_2'$	B_2	$sa+\ as+$	1	-1	1	-1	a	J_c

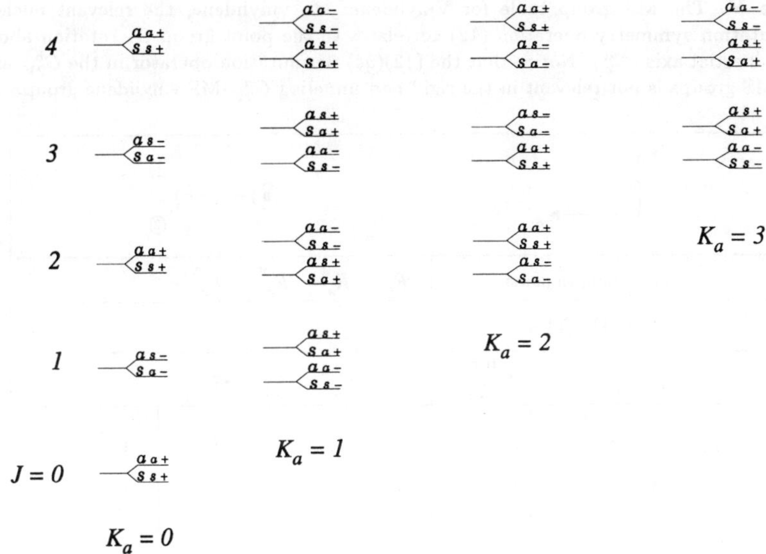

Fig. 3. The parity $(+, -)$, and nuclear permutation symmetry (Ss/Aa) of the rotational wavefunctions in *cis*-bent acetylene. Because the symmetry axis in the *cis*-bent structure lies along the *b*-inertial axis, this structure supports rotational-nuclear permutation tunneling wavefunctions with all eight possible combinations of nuclear permutation symmetry and parity.

The parity ordering of the rotational levels in this C_{2V} group is identical with that of the C_{2h} group because the equivalent rotation for the parity operation E^* is, again, a *c*-axis rotation. However, the nuclear permutation symmetry alternates between K_a stacks as a result of the *b*-axis *equivalent rotation* of the (12) (ab) MS operation. These different *equivalent rotations* for the two MS operations have the effect that the *cis*-bent structure of acetylene supports rotational wavefunctions with all eight possible combinations of nuclear permutation symmetry and parity (Fig. 3). The two quite different patterns of symmetry labels on the K_a structures of the C_{2h} and C_{2V} isomers give rise to different selection rules and perturbations.

2.2.4. *Vinylidene* **a**-*axis* C_{2v} *structure*

In the *a*-axis C_{2v} (vinylidene) structure, (Table V), the parity of each rotational level is given by a reflection through the plane of the molecule or

Table V. The MS group table for Vinylidene. In vinylidene, the relevant nuclear permutation symmetry operation (12) correlates to the point group C_2 rotation about the **a**-inertial axis (C_2^a). Notice that the (12)(ab) permutation operator in the C_{2V}^b and C_{2h}^c MS groups is not relevant in the rigid nontunneling C_{2V}^a MS vinylidene group.

				Equivalent Rotation				R_0	R_a^{π}	R_c^{π}	R_a^{π}	
				CNPI-MS				E	(12)	E^*	$(12)^*$	
$D_{\infty h}$		G_8		C_{2V}	True Symmetry Labels			E	$C_2^{\,a}$	σ_{ab}	σ_{ac}	
Σ_g^+	Σ_u^+	A_1'	B_2'	A_1	$ss+$	$as+$		1	1	1	1	a
Σ_u^-	Σ_g^-	A_2'	B_1'	A_2	$ss-$	$as-$		1	1	-1	-1	Γ^* J_a
Σ_g^-	Σ_u^-	B_1''	A_2''	B_1	$sa-$	$aa-$		1	-1	-1	1	c J_b
Σ_u^+	Σ_g^+	B_2''	A_1''	B_2	$sa+$	$aa+$		1	-1	1	-1	b J_c

by the *equivalent rotation* (C_2^c) about the **c**-axis, consequently, the parity ordering of the levels is the same as in the *cis-* and *trans*-bent planar states. In this case, however, the relevant nuclear permutation symmetry operation (12) correlates to the point group C_2 rotation about the **a**-inertial axis (C_2^a), and in vinylidene the nuclear permutation symmetry alternates with K_a (i.e., all the levels of the same value of K_a have identical (12) permutation symmetry). Notice that the (12) (ab) permutation operator in the C_{2v}^b and C_{2h}^b MS groups, is not relevant in the rigid nontunneling C_{2v}^a MS vinylidene group. In the C_{2v}^a group the relevant permutation operator, (12), does not permute the carbons. The energy level diagram with the exact and point group independent symmetry labels is given in Fig. 4.

It is important to stress that, in the three different isomeric structures considered so far, each has a unique rotational energy level structure, and all three structures can be present on the same electronic potential energy surface.[22] Interactions (nonzero matrix elements) between wavefunctions approximately localized in different isomeric regions of configuration space can *only* occur between basis states with the *same* exact symmetry properties. All electric dipole transitions *must* follow the rigorous selection rules $ss \leftrightarrow ss$, $sa \leftrightarrow sa$, ... $aa \leftrightarrow aa$, and $+ \leftrightarrow -$.

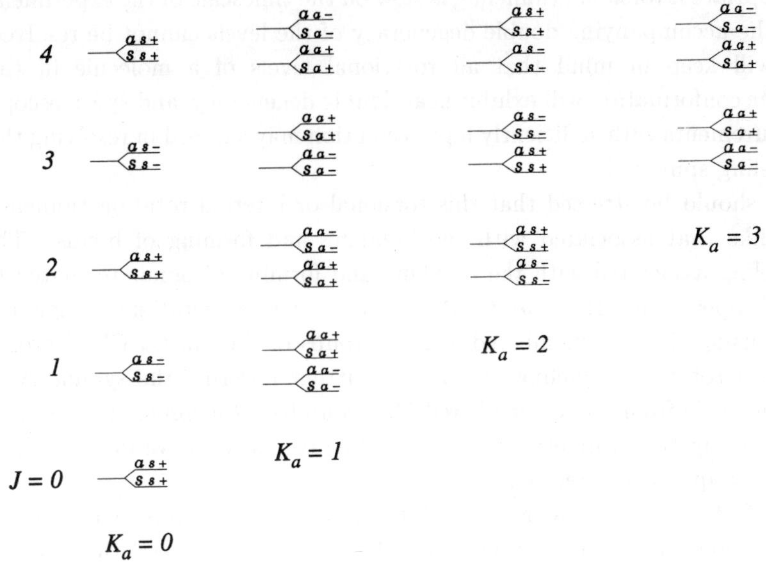

$$K_a = 3$$

$$K_a = 2$$

$$K_a = 1$$

$$K_a = 0$$

Fig. 4. The parity $(+, -)$ and nuclear permutation symmetry (Ss/aa) of the rotational and tunneling levels in the vinylidene structure of acetylene. As indicated in Table V, the vinylidene structure supports a unique pairing of tunneling doublets due to the a-inertial symmetries axis.

2.2.5. *Nonplanar C_2 acetylene*

For nonplanar structures, in the limit of an infinite barrier to internal rotation, the parity label is usually considered to be undefined. This leads to the effective MS group with only two operations. This does not mean, however, that the wavefunctions cannot have a unique parity; it can be simply considered a result of there being two exactly degenerate levels corresponding to the symmetric and antisymmetric linear combinations of the basis states localized in the two energetically equivalent regions of the symmetric double minimum potential surface. In reality, however, barriers to internal rotation are not infinite, and the parity label is both rigorous and relevant. Hougen discussed a related problem concerning hydrogen peroxide, and considered the case of torsional tunneling occurring through both the *cis-* and *trans*-barriers.[35]

When considering a double well potential for C_2H_2 in a *nonplanar gauche* structure, it is convenient to use a C_2 group. Using a C_2 group is appropriate if torsional tunneling is slow on the timescale of the experiment, and the accompanying double degeneracy of the levels cannot be resolved. We will keep in mind that all rotational levels of a molecule in this *gauche* conformation will exhibit near double degeneracy, and spectroscopic measurements with sufficiently high resolution may succeed in resolving this tunneling splitting.

It should be stressed that this torsional or internal rotation tunneling is unlike that associated with the breaking and forming of bonds. The tunneling associated with the breaking and forming of bonds is related to the G_8 operations (12), (ab), $(12)^*$, and $(ab)^*$. These operations are ignored when using the four membered MS subgroups of the full G_8 CNPI group. Internal rotation tunneling arises from further lowering the symmetry of the molecule from a four membered MS group to a 2 membered MS group. By lowering the symmetry from 8 to 2 operations, we have introduced in essence a quadruple degeneracy.

In C_2H_2 there exists the possibility of two C_2 structures, one in which the C_2 operation is a rotation about the b-inertial axis, and the other in which the C_2 operation is about the c-inertial axis. We will first consider the c-axis C_2^c case.

2.2.5.1. *The c-axis C_2 structure*

It can be seen in the group table for this rigid C_2^c gauche near-*trans*-bent structure (Table VI) that the parity operation E^* is absent. Because the 2-element C_2^c group is introduced after distorting the molecule from the 4-element *trans*-bent C_{2h} group, a degeneracy has been introduced which is related to the "lost" symmetry operation E^*. It should be stressed that there can be no real decrease in the number of states when the symmetry of a molecule is lowered. When the planar *trans*-bent molecule is deformed into the rigidly nonplanar near-*trans*-bent structure, two vibrational levels, which in the C_{2h} MS group have the same nuclear permutation symmetry but opposite parity, become *effectively* degenerate. For example, the C_{2h} even and odd quanta torsional vibrations of A_g and A_u species ($ss+$ and $ss-$ nuclear permutation symmetry and parity) *effectively* become one C_2^c A (ss symmetry) degenerate level.

We can label the rotational levels of this structure as before. The operation which permutes the nuclei $(12)(ab)$ has an *equivalent rotation* about the c-inertial axis. Therefore, the (s, a) labels are equivalent to the

Table VI. The C_2^c MS group table for rigid *nonplanar* C_2H_2. The operation which permutes the nuclei, $(12)(ab)$, has an *equivalent rotation* about the **c**-inertial axis.

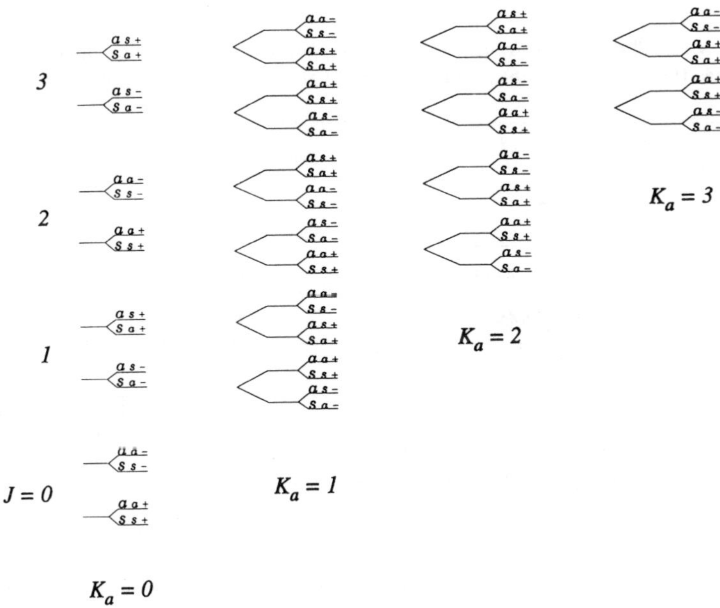

Fig. 5. The parity $(+, -)$, and nuclear permutation symmetry (Ss/aa) of the rotational wavefunctions in *nonplanar* C_2^c C_2H_2. This structure supports a pattern of levels similar to the rigid planar *trans*-bent molecule. An additional doubling of levels has been introduced due to internal rotation tunneling.

K_c labels, and the near degenerate torsional tunneling levels can be given a parity $(+/-)$ label. This has been done in Fig. 5.

The symmetry pattern in the *near-trans*-bent structure closely resembles that of the planar *trans*-bent conformation. Clearly, it would be difficult to differentiate experimentally the rotational structures of states with these similar geometries without directly observing the splitting arising from torsional tunneling .

2.2.5.2. *The b-axis C_2 structure*

As for the c-axis C_2^c group, the b-axis C_2^b group ("near *cis*-bent") seemingly lacks the parity operation E^* (Table VII). The two symmetry operations in this group appear identical to those in the C_2^c group. However, the (12)(ab) operation in the two groups has *different* effects on the Euler angles of the rotational wavefunctions. This is seen by noting that the *equivalent rotations* of the two groups are rotations about different inertial axes.

Table VII. The C_2^b MS group table for *nonplanar* acetylene. The operation which permutes the nuclei, (12)(ab), has an *equivalent rotation* about the b-inertial axis.

Equivalent Rotation				R_0	R_b^{π}	
CNPI-MS				E	(12)(ab)	
$D_{\infty h}$	G_8	C_2^b	True Symmetry Labels	E	C_2^b	
Σ_g^+ Σ_u^-	A_1' A_2' A_1'' A_2''	A	$ss+$ $ss-$ $aa+$ $aa-$	1	1	Γ^* b J_b
Σ_g^- Σ_u^+	B_1'' B_2'' B_1' B_2'	B	$sa-$ $sa+$ $as-$ $as+$	1	-1	a c J_a J_c

In this C_2^b group the rotation equivalent to the nuclear permutation operation, (12)(ab), is about the b-axis. This subtle difference between the C_2^c and C_2^b group properties has dramatic consequences in the energy level patterns of these two structures.

The nuclear permutation symmetry and parity of the rotational energy levels in this nonplanar *near-cis* conformation closely follow those in the

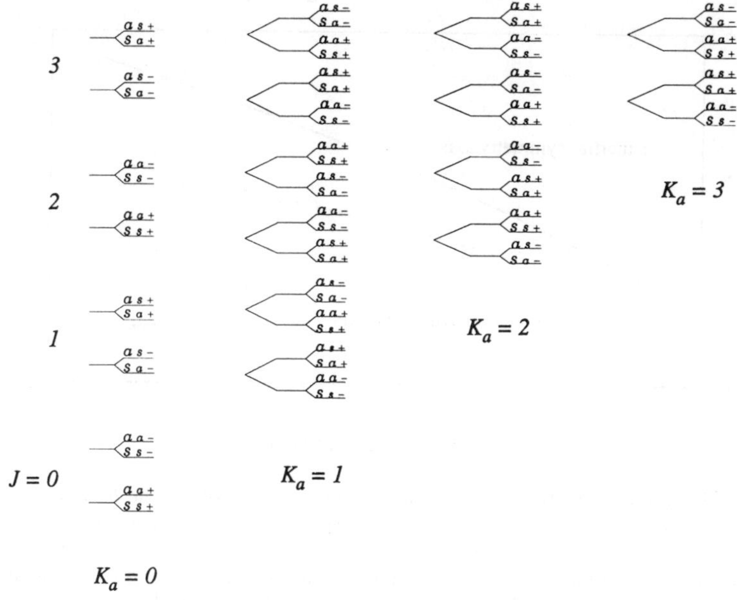

Fig. 6. The parity $(+, -)$ and nuclear permutation symmetry (Ss/aa) of the rotational wavefunctions in the *nonplanar* C_2^b conformation of acetylene.

planar *cis*-bent structure, and are shown in Fig. 6. As the barrier to internal torsional tunneling becomes finite the effective degeneracy introduced by the omission of the parity label is broken. The parity ordering of the levels then reflects that of the planar *cis*- or *trans*-bent structure at to the top of the lowest torsional barrier.[35]

2.2.5.3. *Isotopic substitution effects*

There is an interesting effect which may add to the complexity of the spectroscopy of acetylene. If a rigidly stable conformation of a *nonplanar* (C_2) symmetry exists at the proper configuration, replacing the hydrogens by deuterium atoms may cause the *c*- and *b*-inertial axes of the molecule to exchange places with respect to the external laboratory fixed reference frame. This effect is profoundly different from the *axis switching* effect discussed by Hougen and Watson,[36] which causes anomalous ΔK_a branches to appear in an electronic spectrum.

     Molecular Dynamics and Spectroscopy

$$b \leftrightarrow c \text{ axis exchange angle}$$

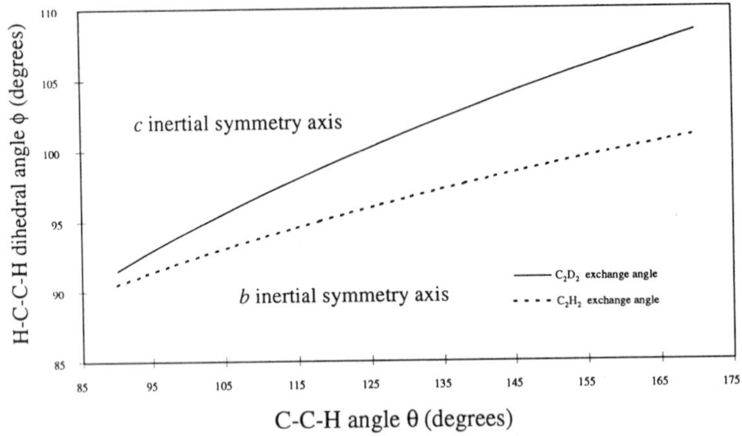

Fig. 7. The graph indicates the bond angle parameters at which the **b**- and **c**-inertial axes instantaneously switch orientation with respect to the laboratory fixed reference frame. The calculation was done at $r_{CC} = 1.40$ Å and $r_{CH} = 1.10$ Å. Above the curves, the molecular C_2 symmetry axis corresponds to the **c**-inertial axis. Below the curves, the symmetry axis corresponds to the **b**-inertial axis. If acetylene has a rigid structure with values of the angles ϕ_{HCCH} and θ_{CCH} lying between the two curves, it will have the C_2^c rotational energy structure in C_2H2, and the C_2^b rotational structure in C_2D_2.

It can be shown that at a dihedral angle slightly larger than $90°$, acetylene could have a C_2^c rotational energy level structure for the C_2H_2 species and a C_2^b rotational energy level structure for the C_2D_2 isotopomer. This unusual isotope effect can be visualized if you start with acetylene in the planar *trans*-bent configuration in which the inertial axes are aligned with a laboratory fixed reference frame, the **c**-inertial axis being perpendicular to the plane of the molecule. If the two hydrogens are then *simultaneously* rotated out of the *trans*-bent plane and into a *cis*-bent configuration, the **b** and **c**-inertial axes must switch places as the **c**-inertial axis is *always* the out-of-plane axis in a planar molecule. This switching of the inertial axes will occur instantaneously and at different values of the dihedral angle, ϕ_{HCCH}, for the C_2H_2 and C_2D_2 isotopic species.

The bond angle parameters at which the **b**- and **c**-inertial axes exchange places with respect to the laboratory fixed reference frame have been calculated for values of the C–C and C–H bond lengths indicated in Fig. 7. If acetylene has a stable potential energy minimum at a value of ϕ_{HCCH}

which lies between the two curves in Fig. 7, the C_2H_2 species will have the *near trans-bent* nonplanar (C_2^c) rotational energy level structure shown in Fig. 5. The isotopically substituted C_2D_2 species will have the *near-cis* (C_2^b) rotational energy level structure of Fig. 6. Such an effect may be present in at least one of the nonplanar states of acetylene predicted in the *ab initio* studies, all of which have stable potential energy minima near $\phi_{HCCH} = 90°$.[20,23,37,38] Furthermore, a molecule with a structure in or near this configuration may experience this exchange of the *b*- and *c*-inertial axes while undergoing vibrational motion. It would be interesting to see how this phenomenon might affect the spectrum of the molecule.

2.2.6. *The linear $D_{\infty h}$ structure*

The linear molecule is somewhat of a special case. The rotational variables for linear molecules consist of only two Eulerian angles.[30] It is the Euler angles which specify the orientation of the molecule fixed axes in the laboratory reference frame. For linear molecules, the undefined or "missing" third Euler angle introduces complications in describing the vibronic wavefunction (Γ^{ev}). The consequences are that while the rovibronic wavefunctions (with which we are here concerned) can be described by the MS group, the vibronic wavefunctions cannot. To describe the vibronic wavefunction in a manner consistent with the MS picture, there is the need to introduce the Extended Molecular Symmetry group (EMS), which is isomorphic to the molecular point group ($D_{\infty h}$ for linear acetylene). The EMS group[39] allows the classification of all parts of the wavefunction including the vibronic part. The use of the EMS group to classify the vibronic wavefunction is explained in Ref. 40.

To classify the rovibronic wavefunctions for linear acetylene, we use the $D_{\infty h}$ MS group. The $D_{\infty h}$ MS group table is given in Table VIII. The CNPI (G_8 group) symmetry species correlate to the MS group classifications as follows.

$$^1\Sigma_g^+ \Rightarrow \left\langle \begin{matrix} ss+ \\ aa+ \end{matrix} \right. \qquad ^1\Sigma_g^- \Rightarrow \left\langle \begin{matrix} as- \\ sa- \end{matrix} \right. \tag{2}$$

$$^1\Sigma_u^+ \Rightarrow \left\langle \begin{matrix} sa+ \\ as+ \end{matrix} \right. \qquad ^1\Sigma_u^- \Rightarrow \left\langle \begin{matrix} ss- \\ aa- \end{matrix} \right. . \tag{3}$$

In linear acetylene, the *trans*-bending and the *cis*-bending vibrational modes are doubly degenerate. When there is excitation in either or both of

Table VIII. The MS group table for *linear* C_2H_2. For the linear molecule, the direction of the two axes perpendicular to the molecular axis (a) are not uniquely defined. This is reflected in the notation corresponding to the $D_{\infty h}$ point group operations where the ∞ indicates that there are infinitely many mirror planes (σ^{∞}) and rotational symmetry axes C_2^{∞}. J_b and J_c belong to the doubly degenerate Π_g symmetry species of the extended molecular symmetry (EMS) group.

Equivalent Rotation (EMS)			R_0	R_{∞}^{π}	R_{∞}^{π}	R_0		
CNPI-MS			E	(12)(ab)	E^*	(12)(ab)*		
$D_{\infty h}$	G_8	True Symmetry Labels	E	C_2^{∞}	σ_{ab}^{∞}	i		
Σ_g^+	$A_1'\ A_1''$	$ss+\ \ aa+$	1	1	1	1	Γ^*	
Σ_u^-	$A_2'\ A_2''$	$ss-\ \ aa-$	1	1	-1	-1		
Σ_g^-	$B_1''\ B_1'$	$sa-\ \ as-$	1	-1	-1	1		J_a
Σ_u^+	$B_2''\ B_2'$	$sa+\ \ as+$	1	-1	1	-1		

these bending modes, the molecule can possess angular momentum along the symmetry axis (coincident with the C–C bond), which arises from this vibration. The vibrational angular momentum generated by each bending mode is $\ell_i(h/2\pi)$ where $\ell_i = v_i, v_i - 2, \ldots, 1$ or 0.[41] The total vibrational angular momentum (designated by the quantum number ℓ) is then

$$\ell = \left| \sum_i (\pm \ell_i) \right| , \tag{4}$$

where the $\pm$ indicates the direction of the vibrational angular momentum with respect to the internuclear axis. The values $\ell = 0, 1, 2, \ldots$ correspond to $\Sigma, \Pi, \Delta, \Phi, \ldots$ vibrational levels.

Linear molecules can also possess nonzero electronic angular momentum. This electronic angular momentum Λ arises from circulation of the electrons about the internuclear axis. The total vibronic angular momentum

$$K = |\pm \Lambda \pm \ell| \tag{5}$$

Table IX. Correlation of the vibronic symmetries of linear acetylene $D_{\infty h}$ species with the true CNPI symmetries.

$D_{\infty h}$ Symmetry species	CNPI "true" symmetry			
Σ_g^+	$\left\{ \begin{array}{l} Ss+ \\ aa+ \end{array} \right.$			
Σ_g^-	$\left\{ \begin{array}{l} Sa- \\ as- \end{array} \right.$			
Σ_u^+	$\left\{ \begin{array}{l} Sa+ \\ as+ \end{array} \right.$			
Σ_u^-	$\left\{ \begin{array}{l} Ss- \\ aa- \end{array} \right.$			
Π_g	$\left\{ \begin{array}{l} Ss+ \\ aa+ \end{array} \right.$	$\oplus$	$\left\{ \begin{array}{l} Sa- \\ as- \end{array} \right.$	
Π_u	$\left\{ \begin{array}{l} Ss- \\ aa- \end{array} \right.$	$\oplus$	$\left\{ \begin{array}{l} Sa+ \\ as+ \end{array} \right.$	
Δ_g	$\left\{ \begin{array}{l} Ss+ \\ aa+ \end{array} \right.$	$\oplus$	$\left\{ \begin{array}{l} Sa- \\ as- \end{array} \right.$	
Δ_u	$\left\{ \begin{array}{l} Ss- \\ aa- \end{array} \right.$	$\oplus$	$\left\{ \begin{array}{l} Sa+ \\ as+ \end{array} \right.$	
$\vdots$	$\vdots$		$\vdots$	

is then the sum of the electronic and vibrational angular momentum. For linear molecules with nonzero vibronic angular momentum (K), $K = 0, 1, 2, \ldots$ levels correspond to $\Sigma, \Pi, \Delta, \Phi, \ldots$.

A correlation can be made between the linear molecule vibronic angular momentum K and the *nonlinear* molecule K_a structure. In the linear molecule, however, only even *or* odd K levels can appear in the same vibrational state. Table IX correlates the vibronic symmetries of linear acetylene $D_{\infty h}$ species (i.e., $\Sigma_g, \Sigma_u, \Pi_g, \Pi_u, \Delta_g, \Delta_u, \ldots$) with the true CNPI symmetries of the molecule.

The rotational levels in a totally symmetric vibronic state of linear acetylene have symmetry properties as given in the energy level template in Fig. 8.

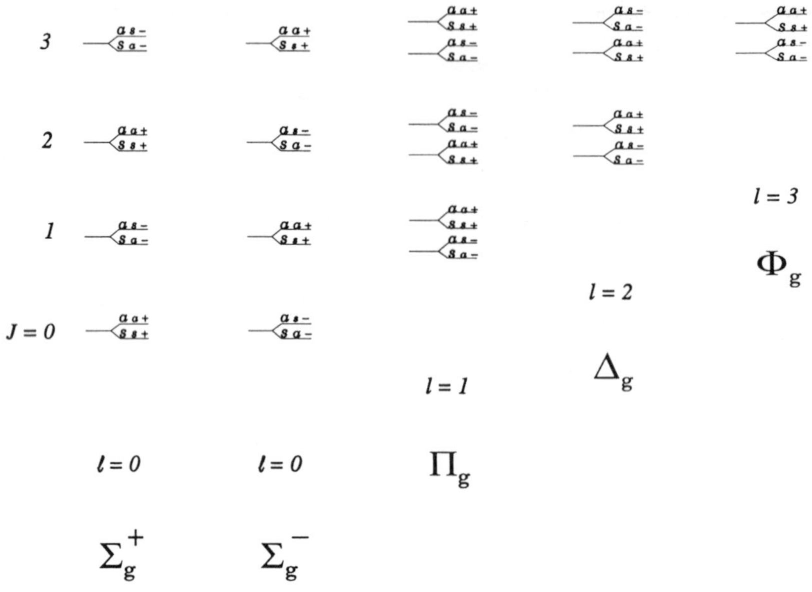

Fig. 8. The energy level template for totally symmetric vibronic states of the linear molecule. Only even *or* odd $\ell(K)$ manifolds can appear in each vibrational state.

3. Statistical Weights

Knowledge of the nuclear spin statistical weights of the rovibronic levels in the molecule is instrumental in identifying the patterns that a local minimum may impose on the spectrum. The intensities of lines appearing in the spectrum are directly related to the statistical weights of the rovibronic states, and only levels with nonzero statistical weight will be present in the spectrum.

The interaction of nuclear spin with the other molecular coordinates is extremely weak, and it is a very good approximation to separate the total molecular wavefunction into nuclear spin and rotation–vibration–electronic–electron spin parts.

$$|\Psi^{\text{total}}\rangle = |\Psi^{\text{ns}}\rangle \otimes |\Psi^{\text{r,v,e,es}}\rangle \, . \tag{6}$$

We know that the total wavefunction must be an eigenfunction of all operators of the CNPI symmetry group G_8, which includes the exchange of identical nuclei. This leads us to the requirement[42] that the wavefunction must be antisymmetric for exchange of fermions (particles of half integer spin), and symmetric for exchange of bosons (particles with integer spin). Because hydrogen is a fermion (nuclear spin 1/2), and carbon-12 is a boson (nuclear spin 0), the total wavefunction $|\Psi^{total}\rangle$ for acetylene $^{12}C_2H_2$ must be changed in sign (antisymmetric) by a permutation of the hydrogens (the (12) CNPI operation), and $|\Psi^{total}\rangle$ must be unchanged (symmetric) by the operation (ab) which exchanges the identical carbon atoms. Inspection of the G_8 group table (Table I) shows that for $^{12}C_2H_2$, $|\Psi^{total}\rangle$ must have symmetry $sa+$ or $sa-$. Knowing the symmetry of $|\Psi^{total}\rangle$ and $|\Psi^{ns}\rangle$ restricts the symmetry of the remaining part of the wavefunction $|\Psi^{r,v,e,es}\rangle$.

We determine the symmetry of the nuclear part of the wavefunction ($|\Psi^{ns}\rangle$), under the G_8 group by generating the symmetry adapted nuclear spin wavefunctions:

$$\alpha_1\alpha_2\theta_a\theta_b \qquad \beta_1\alpha_2\theta_a\theta_b$$
$$\alpha_1\beta_2\theta_a\theta_b \qquad \beta_1\beta_2\theta_a\theta_b \; .$$

Here, α and β represent hydrogens with spin 1/2 and $-1/2$, respectively. θ represents the spin 0 carbons. These basis functions form a reducible representation of the G_8 symmetry group of the molecule. By reducing this representation to its irreducible components, we can determine the statistical weight of $|\Psi^{ns}\rangle$, or the number of times each symmetry species can occur. These basis functions generate a reducible representation of the G_8 group as follows:

G_8	E	$(12)(ab)$	E^*	$(12)(ab)^*$	(12)	(ab)	$(12)^*$	$(ab)^*$
χ^{ns}	4	2	4	2	2	4	2	4

Using the orthogonality relation[43] we can project the characters of this reducible representation of $|\Psi^{ns}\rangle$ onto the irreducible representations of the G_8 group as follows. Note that the numbers in parentheses are the characters of the irreducible representations of the group which are given in the G_8 group table, their coefficients are the number of the basis functions that transform as the corresponding operator, and the sum is divided by the order (number of operations) of the group. The commonly used Mulliken symmetry species can be obtained from Table 1.

$$\chi^{ns} = 1/8[4(1) + 2(1) + 4(1) + 2(1) + 2(1) + 4(1) + 2(1) + 4(1)]$$
$$= 3 \; ss+ \tag{7}$$
$$= 1/8[4(1) + 2(1) + 4(-1) + 2(-1) + 2(1) + 4(1) + 2(-1) + 4(-1)]$$
$$= 0 \; ss- \tag{8}$$
$$= 1/8[4(1) + 2(-1) + 4(-1) + 2(1) + 2(1) + 4(-1) + 2(-1) + 4(1)]$$
$$= 0 \; as- \tag{9}$$
$$= 1/8[4(1) + 2(-1) + 4(1) + 2(-1) + 2(1) + 4(-1) + 2(1) + 4(-1)]$$
$$= 0 \; as+ \tag{10}$$
$$= 1/8[4(1) + 2(1) + 4(1) + 2(1) + 2(-1) + 4(-1) + 2(-1) + 4(-1)]$$
$$= 0 \; aa+ \tag{11}$$
$$= 1/8[4(1) + 2(1) + 4(-1) + 2(-1) + 2(-1) + 4(-1) + 2(1) + 4(1)]$$
$$= 0 \; aa- \tag{12}$$
$$= 1/8[4(1) + 2(-1) + 4(-1) + 2(1) + 2(-1) + 4(1) + 2(1) + 4(-1)]$$
$$= 0 \; sa- \tag{13}$$
$$= 1/8[4(1) + 2(-1) + 4(1) + 2(-1) + 2(-1) + 4(1) + 2(-1) + 4(1)]$$
$$= 1 \; sa + \; . \tag{14}$$

This process reveals that there are 3 nuclear spin wavefunctions with the symmetry $ss+$ and 1 nuclear spin wavefunction of $sa+$ symmetry.

$$\Gamma^{ns} = 3ss + \oplus sa + \; . \tag{15}$$

The permutation symmetry restrictions imposed on the total wavefunction limit $|\Psi^{total}\rangle$ to either $sa-$ or $sa+$ symmetry when $\Gamma^{ns} = ss+$. These levels have a statistical weight of 3. When $\Gamma^{ns} = sa+$, $|\Psi^{total}\rangle$ can have either $ss-$ or $ss+$ symmetry with a statistical weight of 1. Wavefunctions of the remaining symmetries ($aa+$, $aa-$, $as-$, and $as-$) have statistical weight zero and, therefore, are not present in the $^{12}C_2H_2$ isotopomer.

For a state in which the electronic, vibrational and electron spin functions are totally symmetric ($ss+$), the symmetry of $|\Psi^{total}\rangle$ will be given by the product $|\Psi^{ns}\rangle \otimes |\Psi^{r}\rangle$. In this case the rotational part of the wavefunctions can have $sa-$ or $sa+$ symmetry with statistical weight 3, and $ss-$ or $ss+$ symmetry with statistical weight 1.

$$\Gamma^{r,v,e,es} = sa + \text{ and } sa-, \quad SW = 3 \; . \tag{16}$$
$$\Gamma^{r,v,e,es} = ss - \text{ and } ss+, \quad SW = 1 \; . \tag{17}$$

To determine the statistical weights of the C_2D_2 rovibronic levels we must include representations for the deuterium spin projection $+1$, 0, and -1 wavefunctions. We choose ϕ, η and ρ, respectively. As for C_2H_2, θ represents the spin 0 carbon. The nuclear spin basis functions are then:

$$\phi_1\phi_2\theta_a\theta_b \qquad \rho_1\phi_2\theta_a\theta_b \qquad \eta_1\phi_2\theta_a\theta_b$$
$$\phi_1\rho_2\theta_a\theta_b \qquad \rho_1\rho_2\theta_a\theta_b \qquad \eta_1\rho_2\theta_a\theta_b$$
$$\phi_1\eta_2\theta_a\theta_b \qquad \rho_1\eta_2\theta_a\theta_b \qquad \eta_1\eta_2\theta_a\theta_b \; .$$

These functions form a reducible representation of the G_8 group with the following character:

G_8	E	$(12)(ab)$	E^*	$(12)(ab)^*$	(12)	(ab)	$(12)^*$	$(ab)^*$
χ^{ns}	9	3	9	3	3	9	3	9

This reducible representation of the nuclear spin wavefunctions reduces to:

$$\Gamma^{ns} = 6ss + \oplus 3sa + \; . \tag{18}$$

Because deuterium is a boson, as is ^{12}C, $|\Psi^{total}\rangle$ for C_2D_2 must be symmetric under both (12) and (ab). When $\Gamma^{ns} = ss+$, rovibronic levels of $ss-$, $ss+$ symmetry have a statistical weight of 6. When $\Gamma^{ns} = sa+$, the rovibronic levels of $sa-$ or $sa+$ symmetry have a nuclear spin statistical weight of 3. The statistical weights now favor $|\Psi^{total}\rangle$ wavefunctions that are symmetric under exchange of the deuterium atoms.

$$\Gamma^{r,v,e,es} = sa + \text{ and } sa-, \qquad SW = 3 \; . \tag{19}$$
$$\Gamma^{r,v,e,es} = ss - \text{ and } ss+, \qquad SW = 6 \; . \tag{20}$$

Wavefunctions of the remaining symmetries ($aa+$, $aa-$, $as-$, and $as-$) again have statistical weight zero, and are not present in the C_2D_2 isotopomer.

When the ^{13}C isotope of carbon is used, a very different rotational energy level structure is present. Wavefunctions of all eight G_8 symmetry species have nonzero statistical weight. To see this, we construct the nuclear spin basis functions as before. First we consider the $^{13}C_2H_2$ molecule. We use θ and τ to represent the nuclear spin $1/2$, and $-1/2$ carbons, and α and β to represent the hydrogens with spin $1/2$, and $-1/2$. The basis functions are:

$$\alpha_1\alpha_2\theta_a\theta_b \qquad \alpha_1\beta_2\theta_a\theta_b \qquad \beta_1\alpha_2\theta_a\theta_b \qquad \beta_1\beta_2\theta_a\theta_b$$
$$\alpha_1\alpha_2\tau_a\tau_b \qquad \alpha_1\beta_2\tau_a\tau_b \qquad \beta_1\alpha_2\tau_a\tau_b \qquad \beta_1\beta_2\tau_a\tau_b$$
$$\alpha_1\alpha_2\theta_a\tau_b \qquad \alpha_1\beta_2\theta_a\tau_b \qquad \beta_1\alpha_2\theta_a\tau_b \qquad \beta_1\beta_2\theta_a\tau_b$$
$$\alpha_1\alpha_2\tau_a\theta_b \qquad \alpha_1\beta_2\tau_a\theta_b \qquad \beta_1\alpha_2\tau_a\theta_b \qquad \beta_1\beta_2\tau_a\theta_b \, .$$

These functions form a reducible representation of the G_8 group with the following characters:

G_8	E	$(12)(ab)$	E^*	$(12)(ab)^*$	(12)	(ab)	$(12)^*$	$(ab)^*$
χ^{ns}	16	4	16	4	8	8	8	8

This reducible representation of the nuclear spin wavefunctions projects as

$$\Gamma^{\mathrm{ns}} = 9ss + \oplus 3as + \oplus 1aa + \oplus 3sa + \, . \tag{21}$$

Permutation symmetry requirements then limit the symmetry of $|\Psi^{\mathrm{total}}\rangle$ to $aa+$ or $aa-$ (antisymmetric with respect to both exchange of the hydrogens and carbons). This produces rovibronic wavefunctions with the following statistical weights:

$$\Gamma^{\mathrm{r,v,e,es}} = aa+ \text{ and } aa-, \qquad \mathrm{SW} = 9 \tag{22}$$
$$\Gamma^{\mathrm{r,v,e,es}} = sa+ \text{ and } sa-, \qquad \mathrm{SW} = 3 \tag{23}$$
$$\Gamma^{\mathrm{r,v,e,es}} = ss- \text{ and } ss+, \qquad \mathrm{SW} = 1 \tag{24}$$
$$\Gamma^{\mathrm{r,v,e,es}} = as+ \text{ and } as-, \qquad \mathrm{SW} = 3 \, . \tag{25}$$

A slightly different pattern of statistical weights appears in the $^{13}\mathrm{C}_2\mathrm{D}_2$ isotopomer. Using θ and τ to represent the nuclear spin $1/2$, and $-1/2$ carbons, and ϕ, η and ρ to represent the deuterium atoms with spin 1, 0 and -1, the basis functions are:

$$\phi_1\phi_2\theta_a\theta_b \qquad \phi_1\phi_2\theta_a\tau_b \qquad \phi_1\phi_2\tau_a\tau_b \qquad \phi_1\phi_2\tau_a\theta_b$$
$$\phi_1\eta_2\theta_a\theta_b \qquad \phi_1\eta_2\theta_a\tau_b \qquad \phi_1\eta_2\tau_a\tau_b \qquad \phi_1\eta_2\tau_a\theta_b$$
$$\phi_1\rho_2\theta_a\theta_b \qquad \phi_1\rho_2\theta_a\tau_b \qquad \phi_1\rho_2\tau_a\tau_b \qquad \phi_1\rho_2\tau_a\theta_b$$
$$\eta_1\eta_2\theta_a\theta_b \qquad \eta_1\eta_2\theta_a\tau_b \qquad \eta_1\eta_2\tau_a\tau_b \qquad \eta_1\eta_2\tau_a\theta_b$$
$$\eta_1\phi_2\theta_a\theta_b \qquad \eta_1\phi_2\theta_a\tau_b \qquad \eta_1\phi_2\tau_a\tau_b \qquad \eta_1\phi_2\tau_a\theta_b$$
$$\eta_1\rho_2\theta_a\theta_b \qquad \eta_1\rho_2\theta_a\tau_b \qquad \eta_1\rho_2\tau_a\tau_b \qquad \eta_1\rho_2\tau_a\theta_b$$
$$\rho_1\rho_2\theta_a\theta_b \qquad \rho_1\rho_2\theta_a\tau_b \qquad \rho_1\rho_2\tau_a\tau_b \qquad \rho_1\rho_2\tau_a\theta_b$$
$$\rho_1\phi_2\theta_a\theta_b \qquad \rho_1\phi_2\theta_a\tau_b \qquad \rho_1\phi_2\tau_a\tau_b \qquad \rho_1\phi_2\tau_a\theta_b$$
$$\rho_1\eta_2\theta_a\theta_b \qquad \rho_1\eta_2\theta_a\tau_b \qquad \rho_1\eta_2\tau_a\tau_b \qquad \rho_1\eta_2\tau_a\theta_b \, .$$

Table X. The statistical weights of the rovibronic levels in the acetylene isotopomers.

Rovibronic Statistical Weights

G_8		$^{12}C_2H_2$	$^{12}C_2D_2$	$^{13}C_2H_2$	$^{13}C_2D_2$
A_1'	$Ss+$	1	6	1	6
A_2'	$Ss-$	1	6	1	6
B_1'	$as-$	0	0	3	18
B_2'	$as+$	0	0	3	18
A_1''	$aa+$	0	0	9	9
A_2''	$aa-$	0	0	9	9
B_1''	$Sa-$	3	3	3	3
B_2''	$Sa+$	3	3	3	3

These nuclear spin wavefunctions form a reducible representation with the following characters:

G_8	E	$(12)(ab)$	E^*	$(12)(ab)^*$	(12)	(ab)	$(12)^*$	$(ab)^*$
χ^{ns}	36	6	36	6	12	18	12	18

This representation reduces to

$$\Gamma^{ns} = 18\,ss + \oplus 6\,as + \oplus 3\,aa + \oplus 9\,sa + . \tag{26}$$

In $^{13}C_2D_2$ the permutation symmetry requirements applying to the D and ^{13}C atoms restrict $|\Psi^{total}\rangle$ to $as+$ or $as-$ (antisymmetric under (ab) and symmetric under (12)). The rovibronic wavefunctions then have the following statistical weights:

$$\Gamma^{r,v,e,es} = aa + \text{ and } aa-, \qquad SW = 9 \tag{27}$$

$$\Gamma^{r,v,e,es} = sa + \text{ and } sa-, \qquad SW = 3 \tag{28}$$

$$\Gamma^{r,v,e,es} = ss - \text{ and } ss+, \qquad SW = 6 \tag{29}$$

$$\Gamma^{r,v,e,es} = as + \text{ and } as-, \qquad SW = 18 . \tag{30}$$

Again, rovibronic levels of every symmetry species have a nonzero statistical weight.

The intensities of the lines appearing in the spectrum are fundamentally dependent on the nuclear spin statistical weights of the rovibronic levels. Knowledge of the statistical weights of the rovibronic levels is vitally important for identifying the spectral patterns associated with the possible molecular configurations. The rovibronic statistical weights of the acetylene isotopomers are compiled in Table X.

4. Vibrational and Electronic Symmetry

After labeling the rotational energy levels with the rigorous symmetry properties of the rotational wavefunctions, it is necessary to account for the vibrational and electronic wavefunctions. Classification of the vibrational and electronic part of the wavefunction must be done in a manner slightly different than that used to classify the rotational and spin parts of the wavefunction. By dealing with each part of the wavefunction independently, we have in effect performed a separation of the coordinates used to define the wavefunction. To determine the symmetry properties of the rotational and nuclear spin parts of the wavefunction, we considered the relative orientation of the molecule fixed and laboratory fixed axis systems. The orientation, (relative to the laboratory fixed axis system) of the molecule fixed axis system, which is rotating and translating with the molecule, is established by the Euler angles.

By separating the wavefunction, we eliminated the Euler angle dependence of the vibrational and electronic motions. Equivalently, the vibrational and electronic coordinates (and symmetry properties) are tied to the molecule fixed axis system, and thus unaffected by the orientation of the molecule in the lab fixed reference frame (Euler angles). To determine the symmetry properties of the electronic and vibrational parts of the wavefunction we can then neglect the effects the MS group permutation operations have on the Euler angles and permutation of the nuclear spins associated with the Euler angle coordinates. Classifying the vibronic part of the wavefunction in this manner is equivalent to using the molecular point group, and the molecular point group symmetry operations are included in the character tables.

4.1. *Symmetry of the Vibrational Wavefunction*

Labeling the vibrational part of the wavefunction with its symmetry properties takes on special significance in the analysis of the spectra of vibronic transitions and perturbations between states with different stable geometries. The strengths of interactions and transitions is proportional to the square of the vibrational overlap integrals between the states involved.

$$R_{v'v''} = \int \psi_v'^* \psi_v'' d\tau_v \ . \tag{31}$$

These overlap integrals are the essence of the Franck–Condon principle.[44] When analyzing vibronic transitions, it is useful to know which vibrational modes will be Franck–Condon active, or have a high probability of appearing in the spectrum. This is particularly true in double resonance spectra where levels in the final state of the two-step process are accessed via a single selected intermediate level of well known vibrational character.

In order to evaluate the vibrational contribution to the transition moment integral it is essential that ψ_v' and ψ_v'' be expressed in a common basis. The vibrational wavefunctions in the final state can be expressed as linear combinations of the initial state vibrational wavefunctions.[45]

$$\psi_v'' = \sum_i a_i \psi_{v_i}' \ . \tag{32}$$

It follows that only vibrational modes belonging to the same rigorous symmetry species can contribute to the linear combination. It is clear that the product

$$\psi_v'^* \psi_v'' = \psi_v'^* \left[\sum_i a_i \psi_{v_i}' \right] \tag{33}$$

must be totally symmetric for the integral to be nonzero. Therefore, the symmetry and parity $(+, -)$ of the final vibrational states appearing in an electronic spectrum must be identical to those of the initial state. It should be stressed that the CNPI–MS symmetry labels (i.e., ss/aa and parity $(+/-)$) are structure independent, and provide a consistent group theoretical 'symmetry basis' to aid in evaluating the integrals.

The symmetry properties of the vibrational wavefunctions can be determined by considering the transformation properties of the Cartesian displacement coordinates under the operations of the proper MS group (ignoring the permutation of nuclear spins as stated above). Since it is not our purpose to repeat the derivation of the symmetry properties for the vibrational normal modes, the reader is referred to an excellent discussion of the formalism for this process given in Ref. 46.

It is easiest to determine the symmetry properties of the nuclear
Cartesian displacements by examining figures of the normal mode displacement vectors. The symmetry properties of the vibrational normal modes
corresponding to the isomeric structures of acetylene, for which we have
already discussed the rotational structure, are given below. We will begin
with the *trans*-bent C_{2h} configuration of acetylene.

4.1.1. *The trans-bent normal modes*

In Fig. 9 the nuclear displacement coordinates corresponding to the normal
modes of vibration in *trans*-bent acetylene are given along with the symmetry properties of these displacements under the C_{2h} MS group. It should be
apparent from the group table for this structure (Table 3) that only nuclear
motions that have motion perpendicular to the plane of the molecule can
have negative parity $(-)$. This is understood by noting that the parity
operation (an inversion of the laboratory fixed axis system) correlates with
a reflection through the plane of the molecule. This operation will *only*
affect those motions that have vector components perpendicular to the

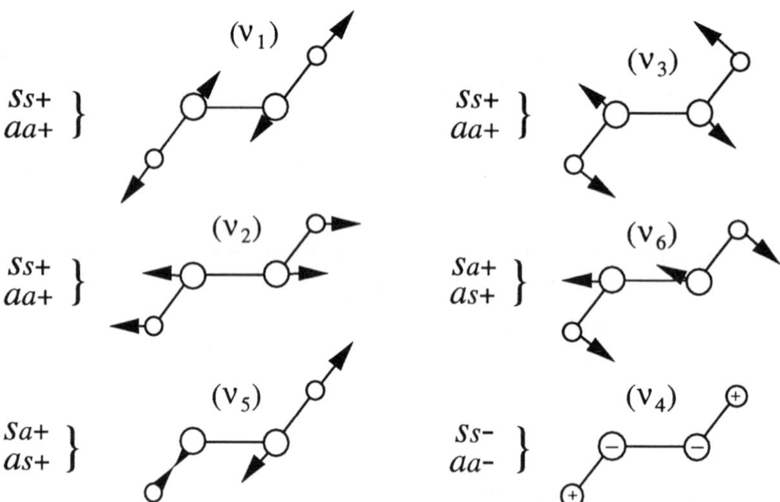

Fig. 9. The displacements and the MS group symmetry of the normal modes in *trans*-
bent C_{2h} acetylene. The conventional numbering of the normal modes is given in
parentheses and the correlated symmetry properties of the vibrational wavefunction is
indicated.

plane of the molecule. The behavior of the vibrational motion with respect to the relevant nuclear permutation operator $C_2 \leftrightarrow (12)(ab)$, is readily apparent from the displacements depicted in Fig. 9. When the vibrational wavefunction has character (1) under the MS $(12)(ab)$ operation, the wavefunction must transform identically under the G_8 (12) and (ab) operations. The point group species A_g and A_u, therefore correlate to $ss+$, $aa+$ and $ss-$, $aa-$ symmetry species, respectively. The B_g and B_u species, with character (-1) under $C_2 \leftrightarrow (12)(ab)$, have opposite character under the G_8 (12) and (ab) operations. These species correlate to $sa-$, $as-$ and $sa+$, $as+$, symmetry species, respectively. The correlation of the point group 'Mulliken species' with the CNPI true symmetry properties is provided in the character tables.

4.1.2. *The cis-bent normal modes*

The vibrational normal modes in the *cis*-bent conformation of acetylene and their symmetries are shown in Fig. 10. The parity of the vibrational wavefunctions can be determined as in the C_{2h} planar conformation by observing which normal mode has motion perpendicular to the plane of the molecule. Again, there is only one normal mode with negative $(-)$ parity. It is readily apparent which normal modes will be symmetric under the operation $C_2 \leftrightarrow (12)(ab)$. Readers should convince themselves that the parity and nuclear permutation symmetries of the normal modes are as indicated in Fig. 10. The correlation of the point group $C_2 \leftrightarrow (12)(ab)$ operation with the G_8 (12) and (ab) operations parallels that for the *trans*-bent structure.

4.1.3. *The vinylidene normal modes*

The normal modes of the vinylidene structure are shown in Fig. 11. Using the group table for the vinylidene structure of C_2H_2 (Table 5), readers should convince themselves of the symmetry properties of these vibrational motions. As for all planar molecules, only normal mode vibrations with motion perpendicular to the plane of the molecule can possess $(-)$ parity.

In the a-axis C_{2v} vinylidene structure, the C_2 operation corresponds to the MS (12) operation. The CNPI operation (ab) is not relevant to this group. Under the $C_2 \leftrightarrow (12)$ operation the vibrational wavefunction transforms either as (1) or (-1). Therefore, the A_1 and A_2 species wavefunctions, which are symmetric under exchange of the hydrogens, correlate to the positive parity $ss+$, $as+$, and negative parity $ss-$,

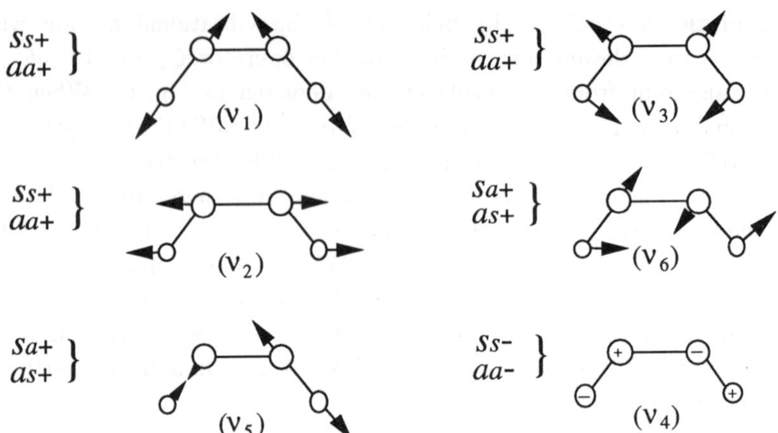

Fig. 10. The displacements and CNPI–MS symmetry species of the normal modes in C_{2v} acetylene.

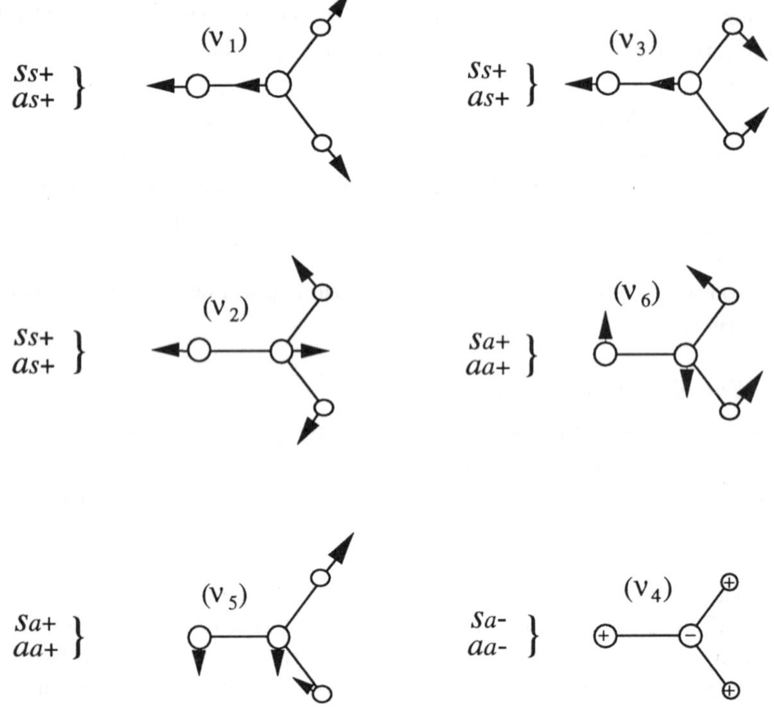

Fig. 11. The displacements and CNPI–MS symmetry species of the vibrational normal modes of the vinylidene structure.

$as-$ species, respectively. The B_2 symmetry vibrational wavefunction is antisymmetric (-1) under the $C_2 \leftrightarrow (12)$ operation and correlates to $ss+$, $aa+$ symmetry under the G_8 group.

4.1.4. *Vibrations in nonplanar C_2 acetylene*

Since the parity label has 'seemingly' lost significance in the C_2^b and C_2^c groups, the vibrational motions of the *nonplanar* structures present a slightly different picture. It should be realized that the 'doubling of the rotational levels' discussed earlier, when considering the rotational structure of *nonplanar* acetylene, has arisen as a result of torsional tunneling through the (*cis* or *trans*) barriers to planarity and not the breaking and reforming of bonds. If tunneling through one or both of these barriers occurs, as in reality must always be the case, the symmetry properties of the *planar* configuration at the top of the lowest barrier provides all the necessary group theoretical results. In this case, the tunneling motion can be considered a part of the vibrational wavefunction, and the parity label will be significant. Hougen has considered the cases when tunneling through the *cis, trans,* or both barriers is present in the HOOH molecule.[35]

If tunneling through either the *cis* or *trans* barrier is considered, as is the case here, the slow nuclear tunneling motion accompanying the fast normal modes of vibration would show up as symmetric and antisymmetric linear combinations of the vibrational basis functions localized on opposite sides of the potential barrier separating the two equivalent equilibrium conformations.

Since the tunneling motion is relatively slow, on the timescale of a rotation, the tunneling can be accounted for with the rotational wavefunctions, and is responsible for the parity doubling indicated in the rotational energy level diagrams for the *nonplanar* structures. This is reassuring as the rotational wavefunctions, which depend only on the ellipsoid of rotation, must in reality be of single well defined parity. The normal modes of vibration for the *nonplanar* C_2^c and C_2^b structures are shown in Fig. 12 and Fig. 13. The symmetry properties of these vibrational motions can be obtained from the relevant group table. (Table VI and VII).

4.1.5. *The linear normal modes*

Determining the symmetry of the normal modes of linear acetylene is somewhat more complicated than the nonlinear cases. In linear acetylene the two bending modes are doubly degenerate. The *trans*-bending mode is symmetric with respect to the $(12)(ab)^*$ operator, while the *cis*-bend

     Molecular Dynamics and Spectroscopy

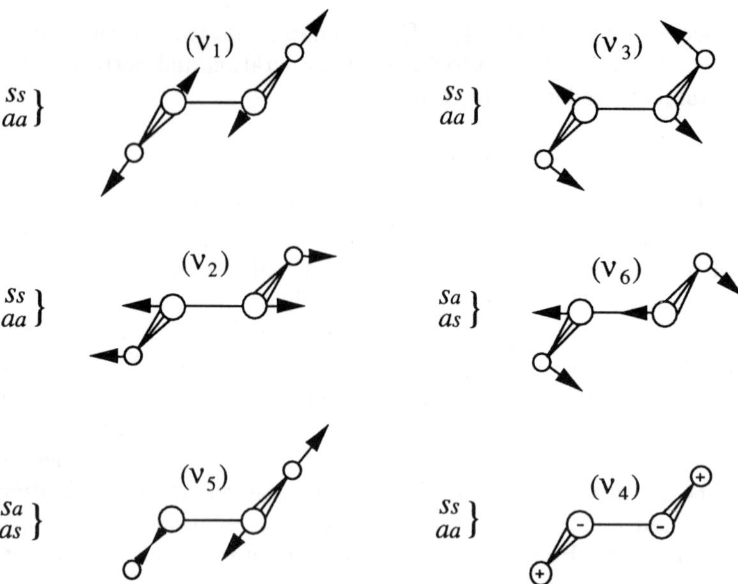

Fig. 12. The displacements and MS symmetry species of the normal modes in *nonplanar* C_2^c acetylene.

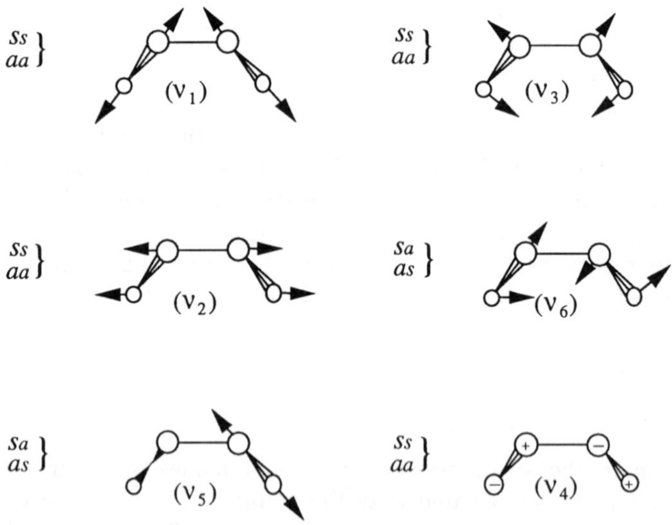

Fig. 13. The displacements and the MS group symmetry of the normal modes in *nonplanar* C_2^b acetylene.

is antisymmetric under $(12)(ab)^*$. Both doubly degenerate vibrations have components symmetric and antisymmetric with respect to the parity ($E^* \leftrightarrow \sigma_{ab}$) operation. The positive parity component of the vibrations lies in the σ_{ab} plane, the negative parity component is perpendicular to the σ_{ab} plane. The stretches are of positive parity ($+$) and are either symmetric or antisymmetric with respect to the $(12)(ab)^* \leftrightarrow i$ (molecule fixed inversion) operation. Figure 14 depicts the nuclear displacements and MS symmetry species of the normal modes of linear acetylene.

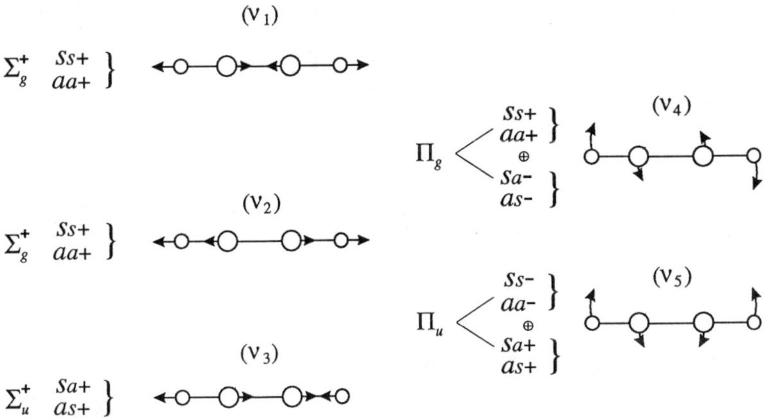

Fig. 14. The displacements and symmetry properties of the normal modes in *linear* $D_{\infty h}$ acetylene. The conventional numbering of the normal modes is given in parentheses.

By observing the normal mode displacement vectors it is possible to judge which modes may be Franck–Condon active in electronic transitions between states with the chosen structures. Figure 15 correlates the nuclear displacements, symmetry properties and accepted numbering of the vibrational normal modes of the acetylene structures with the E, $(12)(ab)$, E^*, and $(12)(ab)^*$ MS operations. The symmetry properties, parity ($+/-$), and when relevant, a-axis rotations of the normal modes are given. Because the MS group for the vinylidene structure does not contain the $(12)(ab)$, and $(12)(ab)^*$ permutation–inversion operations, correlation of the vinylidene vibrational normal modes does not fit into this diagram. However, using the rigorous symmetry properties of the vibrational wavefunctions the normal modes of vinylidene can be individually correlated with those of the other MS groups.

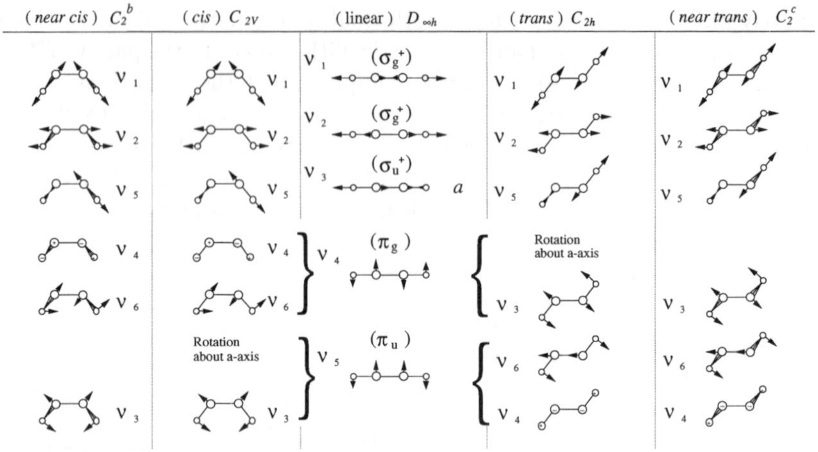

Fig. 15. The correlation diagram for the normal mode displacements of the closely related structures of acetylene. Franck–Condon activity in vibronic transitions may be judged by observing the displacement vectors and symmetry species of vibrations in the appropriate structures.

4.2. *Symmetry of the Electronic Wavefunction*

Molecular orbitals can be expressed as linear combinations of atomic orbitals (LCAO). If we start with separated atoms and bring the atoms together to form the molecule, the atomic orbitals will mix to create the molecular orbitals. For example, in the ground state of acetylene, the triple bond between the carbon atoms consists of two π (pi) bonds arising from the p_x and p_y orbitals of the two carbon atoms, and a σ (sigma) bond which is primarily a linear combination of the carbon, the p_z orbitals, and the carbon valence s orbitals (Fig. 16). As the molecule is deformed, the atomic orbital contribution, symmetry properties, and energy of the molecular orbitals can change.

The Walsh diagrams[47] indicate qualitatively the variation of the energies of the individual molecular electronic orbitals with changing geometry, and explicitly display the correlation of the orbital symmetries in the different symmetry groups. A concise and comprehensive description of the energetic aspects of the Walsh diagrams has been given by Buenker and Peyerimhoff,[48] but it is the symmetry properties of the electronic wavefunctions in which we are interested. The symmetry of the electronic orbitals is easily determined using the relevant group table.

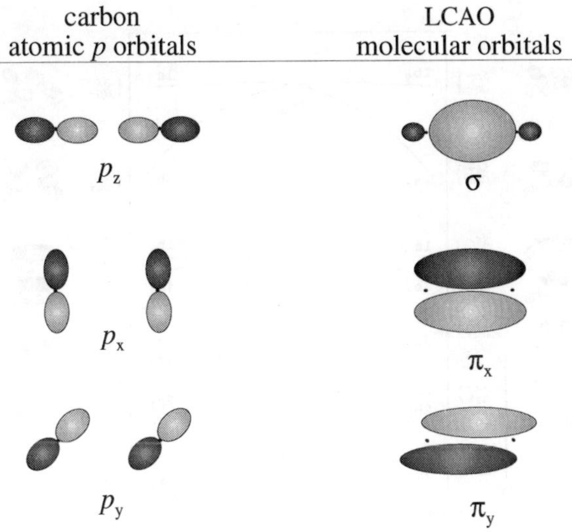

Fig. 16. The atomic orbitals of the separated atoms combine to form the molecular orbitals. The separated carbon atom p orbitals combine to form σ, π_x and π_y molecular orbitals. The triple bond of the $\tilde{X}$ state of acetylene is composed primarily of the separated carbon atom p_x, p_y the p_z orbitals.

There are two Walsh diagrams which are relevant to this discussion of the symmetry of the electronic wavefunctions of acetylene. These cover the *trans* (C_{2h}) $\leftrightarrow$ *linear* $(D_{\infty h})$ $\leftrightarrow$ *cis* (C_{2v}) and *trans* (C_{2h}) $\leftrightarrow$ (C_2^c) *gauche* (C_2^b) $\leftrightarrow$ *cis* (C_{2v}) distortions, depicted in Fig. 17 and Fig. 18, respectively.

Acetylene is a 14 electron molecule. In its ground electronic state, the orbitals are doubly occupied through the doubly degenerate $(1\pi_u)$ bonding orbital (the doubly degenerate $(1\pi_u)$ orbital contains 4 electrons). Note that the Walsh diagram displays only the valence orbitals; the $1\sigma_g$ and $1\sigma_u$ core orbitals are not shown. The symmetries of the individual electron orbitals for the linear configuration of acetylene are given in the center of the diagram. The correlated bent molecule 'Mulliken symbols' and CNPI–MS symmetry properties of the wavefunction are given on the left and right sides of the figure. Notice that when the linear molecule is distorted, the orbitals assume the symmetry properties of the group on the left or right side of the diagram, and the degeneracy of the π orbitals is removed. The contribution of the individual electron orbitals to the overall binding energy of the molecule changes as the geometry changes, as indicated on the vertical axis in the diagrams.

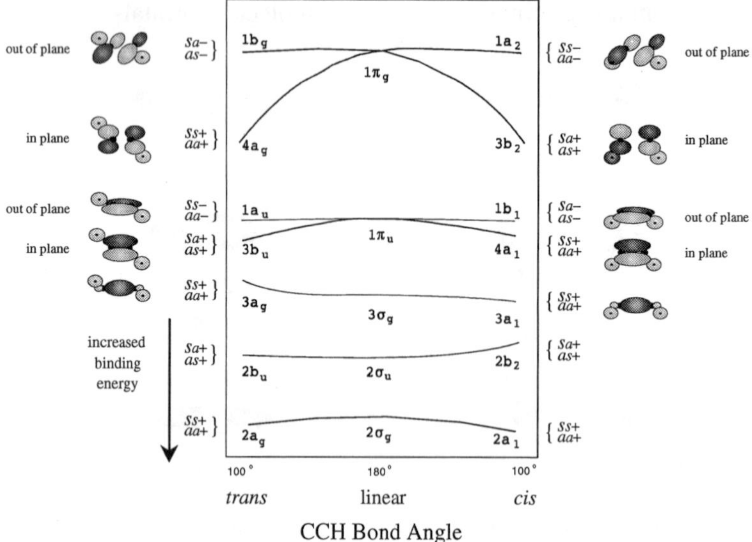

Fig. 17. The Walsh diagram for the HCC bending coordinate. This bending coordinate corresponds to distorting the *trans*-bent configuration on the left of the diagram by symmetrically moving the hydrogen atoms into the linear configuration and then into the *cis*-bent configuration on the right. The contribution of the molecular orbitals to the binding energy is indicated by the curves connecting the orbitals on the left to those on the right.

The electronic wavefunction ψ_e can be approximated by the product of the individual orbital wavefunctions ϕ_e.

$$\psi_e = \phi_1 \phi_2 \phi_3 \phi_4 \phi_5 \cdots \tag{34}$$

The electronic symmetry of the molecule is determined by taking the *direct product* of the species of the individual orbital functions;[49]

$$\Gamma_e(\psi_e) = \Pi_n(\Gamma_n(\phi_n)) \ . \tag{35}$$

These are single electron orbital functions, and the doubly occupied orbitals must be included in the product twice. Using Walsh diagrams, and the direct product rules for symmetry species given in Table XI with the group tables for the relevant structures, the symmetry of the electronic part of the molecular wavefunction (Γ_e) for any electronic configuration is readily determined.

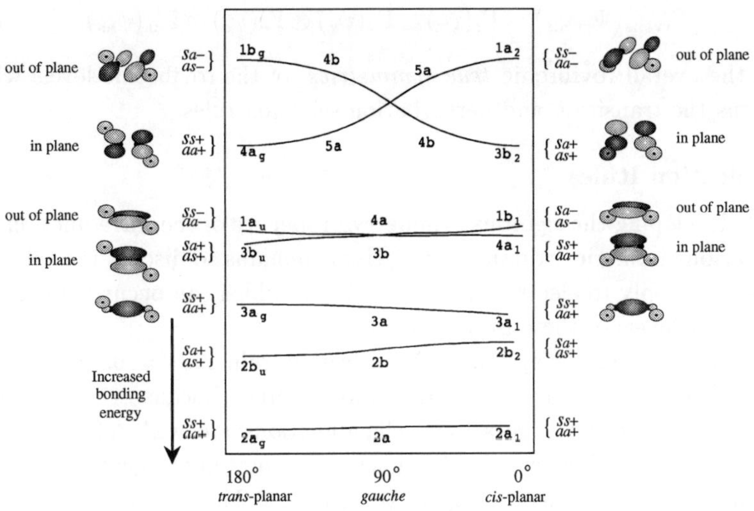

HCCH Torsional Bond Angle

Fig. 18. The Walsh diagram for the torsional coordinate of acetylene. The torsional coordinate of this diagram corresponds to distorting the *trans-bent* structure on the left by simultaneously moving the hydrogen atoms out of the plane of the page into the plane perpendicular to the page and containing the carbon atoms.

Table XI. The direct product table for the permutation-inversion symmetry species.

	Ss	Sa	as	aa	$+$	$-$
Ss	Ss	Sa	as	aa		
Sa		Ss	aa	as		
as			Ss	Sa		
aa				Ss		
$+$					$+$	$-$
$-$						$+$

After establishing the symmetry species for the electronic (ψ_e), vibrational (ψ_v), rotational (ψ_r) and nuclear spin (ψ_{ns}) parts of the acetylene wavefunction, the overall symmetry of the individual rotational levels is obtained by the direct product:

$$\Gamma_{\mathrm{rvens}}(\Psi_{\mathrm{rvens}}) = \Gamma_{\mathrm{r}}(\psi_{\mathrm{r}}) \otimes \Gamma_{\mathrm{v}}(\psi_{\mathrm{v}}) \otimes \Gamma_{\mathrm{e}}(\psi_{\mathrm{e}}) \otimes \Gamma_{\mathrm{ns}}(\psi_{\mathrm{ns}}) \,. \qquad (36)$$

It is the overall rovibronic *true symmetries* of the rovibronic levels which governs the transition and perturbation selection rules.

5. Selection Rules

After developing the rigorous symmetry properties of the wavefunctions for the various isomeric structures of C_2H_2, it remains to discuss the selection rules that apply to electric dipole transitions which can occur between the rovibronic levels of these isomers.

As for all atomic systems, the selection rule $\Delta J = 0, \pm1$ is valid (the transition $J = 0 \leftrightarrow J = 0$ cannot occur). Because the properties of the asymmetric top rotational eigenfunctions are well defined, strong restrictions control which combinations of the rotational eigenfunctions can be connected by an electric dipole transition.[50] The symmetry properties developed in the previous sections show rigorously which levels can be connected by electric dipole transitions or by spectroscopic "perturbations". The K_a, K_c designation of the rotational levels participating in a transition display the direction of the transition moment with respect to the inertial axes of the molecule.

The pairs of asymmetric rotor functions connected by allowed transitions is dependent on the direction of the transition moment $(\vec{m})$ with respect to the inertial axes of the molecule. If the transition moment lies in the direction of the a-inertial axis, the transitions

$$\Delta K_a = 0, \pm2, \ldots, \qquad \Delta K_c = \pm1, \pm3, \ldots \qquad (37)$$

can occur. If the transition moment lies in the direction of the b-inertial axis, the transitions

$$\Delta K_a = \pm1, \pm3, \ldots, \qquad \Delta K_c = \pm1, \pm3, \ldots \qquad (38)$$

can occur. When the transition moment lies along the c-inertial axis, the transitions

$$\Delta K_a = \pm1, \pm3, \ldots, \qquad \Delta K_c = 0, \pm2, \ldots \qquad (39)$$

can occur.

Because the rotational eigenfunctions are well defined, these restrictions will be valid even when the transition moment has components that lie along more than one of the three inertial axes.

Accompanying these selection rules for electric dipole transitions are the rigorous selection rules for parity and nuclear permutation symmetry. These restrictions arise from the symmetry of the electric dipole operator (Γ^*). As indicated in the group tables of the previous sections, Γ^* has the CNPI–MS symmetry properties ($ss-$). Therefore, states that can be connected by the electric dipole operator must have the same nuclear permutation symmetry *and* opposite parity.

$$ss \leftrightarrow ss, \quad sa \leftrightarrow sa, \quad \ldots aa \leftrightarrow aa \ . \tag{40}$$

$$+ \leftrightarrow - \ . \tag{41}$$

It must be emphasized that, because all near prolate asymmetric tops have identical K_a, K_c rotational symmetry patterns, transitions between states with the same structure (or between rotation–vibration states within the same electronic isomer) are completely described by the asymmetric rotor function selection rules. *However, when transitions between states which have different isomeric structures are considered, the point group independent CNPI–MS paradigm must be introduced.* Extra restrictions resulting from the subtle interplay between the rotational wavefunctions and the nuclear permutation symmetry in the different isomeric structures will give rise to unique patterns in the spectrum. Identification of these patterns will be crucial in determining the isomeric identity of the states under investigation. These selection rules are compiled in Table XII.

Table XII. The electric dipole selection rules for the asymmetric rotor functions display the direction of the transition dipole moment ($\vec{m}$) with respect to the inertial axes. The nuclear spin species and parity selection rules apply rigorously to all structures of acetylene.

if $\vec{m} \parallel a$	$\Delta K_a = 0, \pm 2, \ldots$	$\Delta K_c = \pm 1, \pm 3, \ldots$
if $\vec{m} \parallel b$	$\Delta K_a = \pm 1, \pm 3, \ldots$	$\Delta K_c = \pm 1, \pm 3, \ldots$
if $\vec{m} \parallel c$	$\Delta K_a = \pm 1, \pm 3, \ldots$	$\Delta K_c = 0, \pm 2, \ldots$

$$\text{parity} \quad + \longleftrightarrow -$$

$$\text{nuclear permutation} \quad ss \longleftrightarrow ss \ \ldots\ldots\ aa \longleftrightarrow aa$$

6. Isomerization in Acetylene

Understanding the acetylene $\leftrightarrow$ vinylidene isomerization process on the ground electronic state potential energy surface is an excellent example of the necessity to invoke the CNPI–MS paradigm.

The vinylidene radical ($H_2C{=}C$) is probably the most interesting of the isomers of acetylene. Isomerization of the ground electronic state of acetylene to vinylidene has been implicated as the first step in both homogeneous pyrolysis of acetylene and carbene addition to aromatic rings.[51-54]

Several of the electronic surfaces of acetylene support deep vinylidene minima.[22,55] However, the electronically excited states with deep vinylidene minima are high in energy ($\sim$ 90 kcal/mol)[56] and are not relevant to thermal reaction processes.[54] The vinylidene minimum on the ground electronic state lies at a more interesting energy ($\sim$ 45 kcal/mol above the vibrationless level), but this vinylidene well is known to be very shallow ($\sim$ 2 kcal/mol).[57]

The ground state vinylidene minimum lies far below the first dissociation energy of acetylene ($\sim$ 46 000) and there can be no lifetime broadening of a vinylidene spectral line. There is, however, a width associated with the vinylidene isomerization lifetime. This vinylidene linewidth (Γ) is the width of a band of predominantly acetylene eigenstates carrying the character of one vinylidene basis state.

$$\Gamma(|\psi_{\text{vinyl}}\rangle) = \Gamma(|\Psi_{\text{acetylene}}\rangle)$$

$$\text{where} \qquad |\Psi_{\text{acetylene}}\rangle = a|\psi_{\text{vinyl}}\rangle + \sum_{i=1}^{n} b_i|\psi_{i\ \text{linear}}\rangle \qquad (42)$$

$$\text{and} \qquad\qquad a \neq 0 \,.$$

The spread of this vinylidene basis state character into the acetylene eigenstates is analogous to that proposed in the theory of radiationless transitions where an isolated spectroscopically "bright" state is coupled to a dense manifold of "dark" states.[58]

Using Fermi's golden rule,

$$\Gamma = 2\pi\rho\langle V^2\rangle \,, \qquad (43)$$

the width (Γ) of the isolated vinylidene basis state with respect to decay via isomerization is given in terms of the density of 'background' linear acetylene basis states, ρ, and the average square modulus of the coupling matrix element ($\langle V^2 \rangle$).

Because the short isomerization lifetime causes a dilution of the vinylidene basis state character into many acetylene eigenstates, it is difficult to identify the ground state vinylidene species using high resolution spectroscopy. In spite of these limitations, two interesting methods of system preparation and observation have been utilized to study this species.

The study of Ervin *et al.*[56] exploited the fact that the structure of certain anionic isomers correspond to interesting regions of the potential energy surface of the neutral. Because the anion of C_2H_2 possesses a stable vinylidene structure, photodetachment of $H_2C{=}C^-$ yields the transient neutral vinylidene species. Kinetic energy analysis of the photodetached electrons yielded information on the vibrational frequencies of neutral vinylidene. Vibrational states up to 3400 cm^{-1} above the origin were observed, including the CH_2 rocking mode. The CH_2 rocking mode corresponds closely to the isomerization reaction coordinate and should be the most short-lived of the vinylidene vibrational states. The observation of the CH_2 rock $2 \leftarrow 0$ transition suggests that the vinylidene well depth on the singlet ground state potential energy surface is at least 450 cm^{-1} (1.3 kcal/mol), with an estimated lifetime for the vinylidene origin level of approximately 0.2 ps (~ 25 cm^{-1}).

Chen and coworkers observed the vinylidene radical using the technique of Stimulated Emission Pumping — Spectral Cross Correlation spectroscopy SEP-SCC.[59,60] Spectral cross correlation measures the similarity of two double resonance spectra recorded from intermediate states of maximally different vibrational character. The similarity of the two spectra is then assessed over some energy window of the spectrum using the cross correlation index:

$$C(E) = \frac{\langle I_1'(E') I_2'(E') \rangle}{\{\langle I_1'^2(E') \rangle \langle I_2'^2(E') \rangle\}^{1/2}} , \qquad (44)$$

with $I'(E)$ the derivative of the SEP spectrum and $\langle\ \rangle$ meaning integration over an energy window centered at E, $E - \delta < E' < E + \delta$.

The SEP spectra that Chen *et al.* cross correlated were recorded via acetylene S_1 vibrational intermediate levels of different vibrational character, and were expected to access distinctly different vibrational levels of the electronic ground state surface. SEP spectra recorded from such different character intermediates would normally produce a low level of spectral cross correlation. In regions where the potential energy surface is distorted, (i.e., near a secondary potential minimum), the normal mode description of vibrations will weaken and mixing of the vibrational modes will be promoted. In such strongly mixed regions a high spectral cross correlation is expected.

Using the SCC method, Chen and coworkers located an S_0 vinylidene vibrational state at $15\,617$ cm^{-1} above the $v'' = 0$ level of acetylene. This measurement, in conjunction with those of Ervin *et al.*, who observed both the singlet and triplet vinylidene species but could not tie these to the acetylene $v'' = 0$ level, place the vinylidene minima on the T_1 and T_2 electronic surfaces near $32\,100$ cm^{-1} and $37\,600$ cm^{-1}, respectively.

6.1. *State Specific Coupling and Isomerization*

The theory of the cross correlation method provides a good introduction to the spectroscopic characterization and identification of the state specific perturbations and isomerization induced by secondary minima on a potential energy surface.

Due to Franck–Condon restrictions, single and double resonance experiments are able to access only certain regions of the potential energy surface. In acetylene, the spectra are usually limited to the linear and *trans*-bent regions. In spite of the Franck–Condon restrictions, the presence of secondary minima on the potential energy surface will leave identifiable marks on the rotationally resolved spectrum.[59]

To first order, the Hamiltonian for acetylene can be written in block diagonal form with the blocks corresponding to the different isomeric structures on the potential energy surface. At higher levels of approximation, coupling between the various isomeric structures must be included. The Hamiltonian matrix then includes off-diagonal matrix elements. The general Hamiltonian matrix is

$$\begin{pmatrix} E_{\text{linear}} & H_{l-t} & H_{l-c} & H_{l-v} \\ H_{l-t} & E_{trans-\text{bent}} & H_{t-c} & H_{t-v} \\ H_{l-c} & H_{t-c} & E_{cis\text{-bent}} & H_{c-v} \\ H_{l-v} & H_{t-v} & H_{c-v} & E_{\text{vinylidene}} \end{pmatrix}. \tag{45}$$

Identifying the effects of the off-diagonal coupling matrix elements is the essence of the Spectral Cross Correlation (SCC) method.

Due to the comparatively large mass of the carbon "backbone" in acetylene, all of its isomeric structures are near prolate tops. The heavy carbon "backbone" induces a propensity to conserve the orientation, in the lab fixed reference frame, of the a-inertial axis, and hence the quantum number K_a, during isomerization. This propensity to conserve $K_a \equiv \ell$ (at low J) during isomerization accompanies conservation of the rigorously good symmetry properties of the wavefunction, parity $(+/-)$ and nuclear permutation symmetry (s/a), and the rotational quantum number J.

As can be seen in the rotational energy templates for the different structures of acetylene (Sec. 2), the J, $K_a \equiv \ell$ rotational levels in the various conformations have different rigorous symmetry properties. During isomerization these symmetry properties and J must be conserved. The conservation of the rigorous symmetry properties and quantum numbers, in conjunction with the different rotational structures of the various isomers, will prohibit some of the numerous "background" acetylene rovibronic states from interacting with the isolated vinylidene vibrational levels. These restrictions affect the structure of the Hamiltonian matrix.

The H_{n-m} matrix elements connecting different zero-order structures are strongly J and K_a dependent. In *all* levels of approximation, only those rovibronic levels with the same true molecular symmetries and good quantum numbers will be connected. For example, for acetylene (linear structure) and vinylidene basis levels of totally symmetric vibrational character in $^{12}C_2H_2$ (where one member of each tunneling doublet has statistical weight zero), the acetylene-vinylidene $K_a \equiv \ell = 0$ submatrix will only connect levels of even-J. In other words, the linear acetylene totally symmetric $\ell = 0$ vibrational states of even-J are of $ss+$ symmetry, and will contain $K_a \equiv \ell = 0$ vinylidene character and isomerize. The odd-J $sa-$ symmetry levels of the totally symmetric acetylene $\ell = 0$ vibrational states will, by symmetry, contain no $K_a \equiv \ell = 0$ vinylidene character and be prohibited from isomerizing. The odd-J levels of the totally symmetric vinylidene $\ell = 0$ vibrational states can isomerize to the linear structure. However, because they are of $ss-$ and $as-$ symmetry, they interact with vibrationally nontotally symmetric acetylene levels which are spectroscopically unobservable in the cold band PUMPed $\tilde{X} \leftarrow \tilde{A} \leftarrow \tilde{X}$ SEP.

The submatrix for this interaction can be written:

$$
\begin{array}{c}
\begin{array}{cc}
\text{Linear} & \text{Vinylidene} \\
\ell = 0 & K \equiv \ell = 0 \\
\begin{array}{cccc} J=0 & J=1 & J=2 & J=3 \end{array} \;\cdots & \begin{array}{cccc} J=0 & J=1 & J=2 & J=3 \end{array} \;\cdots \\
\begin{array}{cccc} Ss+ & Sa- & Ss+ & Sa- \end{array} & \begin{array}{cccc} Ss+ & Ss- & Ss+ & Ss- \end{array}
\end{array}
\end{array}
$$

$$
\begin{array}{l}
\text{Linear} \\
\ell=0
\end{array}
\;
\begin{array}{l}
J=0\; Ss+ \\
J=1\; Sa- \\
J=2\; Ss+ \\
J=3\; Sa-
\end{array}
\left(
\begin{array}{cccccccccc}
E_{\text{lin}0} & 0 & 0 & 0 & \cdots & H_{0-0} & 0 & 0 & 0 & \cdots \\
0 & E_{\text{lin}1} & 0 & 0 & \cdots & 0 & 0 & 0 & 0 & \cdots \\
0 & 0 & E_{\text{lin}2} & 0 & \cdots & 0 & 0 & H_{2-2} & 0 & \cdots \\
0 & 0 & 0 & E_{\text{lin}3} & \cdots & 0 & 0 & 0 & 0 & \cdots \\
\vdots & \vdots & \vdots & \vdots & \ddots & \vdots & \vdots & \vdots & \vdots & \ddots \\
H_{0-0} & 0 & 0 & 0 & \cdots & E_{\text{vinyl}0} & 0 & 0 & 0 & \cdots \\
0 & 0 & 0 & 0 & \cdots & 0 & E_{\text{vinyl}1} & 0 & 0 & \cdots \\
0 & 0 & H_{2-2} & 0 & \cdots & 0 & 0 & E_{\text{vinyl}2} & 0 & \cdots \\
0 & 0 & 0 & 0 & \cdots & 0 & 0 & 0 & E_{\text{vinyl}3} & \cdots \\
\vdots & \vdots & \vdots & \vdots & \ddots & \vdots & \vdots & \vdots & \vdots & \ddots
\end{array}
\right)
\begin{array}{l}
\\ \\
J=0\; Ss+ \\
J=1\; Ss- \\
J=2\; Ss+ \\
J=3\; Ss-
\end{array}
\tag{46}
$$

(Row labels for the lower block: Vinylidene, $K \equiv \ell = 0$.)

This J and K_a dependent interaction can be seen in the energy level templates for the two structures (Fig. 19).

When one of the $^{13}C_2H_2$ isotopomers is used, the statistical weights of *both* tunneling doublets are nonzero. The nonzero weights of the tunneling doublets then allow coupling between levels of even-J and between those of odd-J. In this case, both the even-J and the odd-J coupling will be spectroscopically observable using SEP. The coupling matrix will display the following pattern:

$$
\begin{array}{c}
\begin{array}{cc}
\text{Linear} & \text{Vinylidene} \\
\ell = 0 & K \equiv \ell = 0 \\
\begin{array}{cccc} J=0 & J=0 & J=1 & J=1 \end{array} \;\cdots & \begin{array}{cccc} J=0 & J=0 & J=1 & J=1 \end{array} \;\cdots \\
\begin{array}{cccc} Ss+ & Aa+ & Sa- & As- \end{array} & \begin{array}{cccc} Ss+ & As+ & Ss- & As- \end{array}
\end{array}
\end{array}
$$

$$
\begin{array}{l}
\text{Linear} \\
\ell=0
\end{array}
\;
\begin{array}{l}
J=0\; Ss+ \\
J=0\; Aa+ \\
J=1\; Sa- \\
J=1\; As-
\end{array}
\left(
\begin{array}{cccccccccc}
E_{\text{lin}0} & 0 & 0 & 0 & \cdots & H_{0-0} & 0 & 0 & 0 & \cdots \\
0 & E_{\text{lin}1} & 0 & 0 & \cdots & 0 & 0 & 0 & 0 & \cdots \\
0 & 0 & E_{\text{lin}2} & 0 & \cdots & 0 & 0 & 0 & 0 & \cdots \\
0 & 0 & 0 & E_{\text{lin}3} & \cdots & 0 & 0 & 0 & H_{1-1} & \cdots \\
\vdots & \vdots & \vdots & \vdots & \ddots & \vdots & \vdots & \vdots & \vdots & \ddots \\
H_{0-0} & 0 & 0 & 0 & \cdots & E_{\text{vinyl}0} & 0 & 0 & 0 & \cdots \\
0 & 0 & 0 & 0 & \cdots & 0 & E_{\text{vinyl}1} & 0 & 0 & \cdots \\
0 & 0 & 0 & 0 & \cdots & 0 & 0 & E_{\text{vinyl}2} & 0 & \cdots \\
0 & 0 & 0 & H_{1-1} & \cdots & 0 & 0 & 0 & E_{\text{vinyl}3} & \cdots \\
\vdots & \vdots & \vdots & \vdots & \ddots & \vdots & \vdots & \vdots & \vdots & \ddots
\end{array}
\right)
\begin{array}{l}
\\ \\
J=0\; Ss+ \\
J=0\; As+ \\
J=1\; Ss- \\
J=1\; As-
\end{array}
\tag{47}
$$

(Row labels for the lower block: Vinylidene, $K \equiv \ell = 0$.)

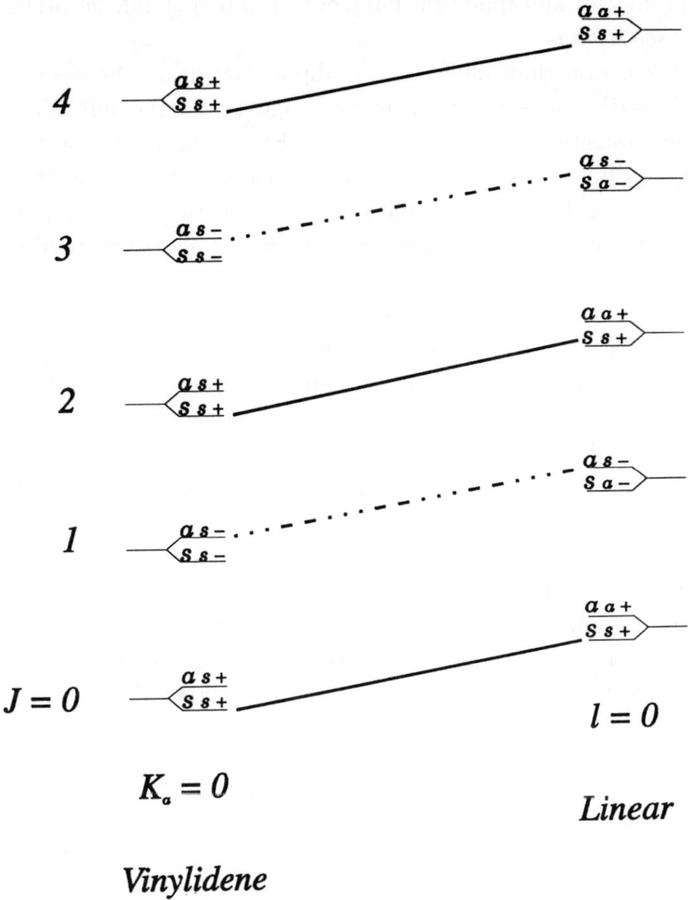

Fig. 19. The J and K_a dependent interaction between the linear and vinylidene structures arises from the different symmetries of the rotational energy levels. The solid lines indicate the allowed isomerization pathways for the $^{12}C_2$ acetylene $K_a \equiv \ell = 0$ rotational manifold of states. When a $^{13}C_2$ isotopomer is used, the statistical weights of both tunneling doublets are nonzero. The dotted lines indicate the additional allowed isomerization pathway for the $^{13}C_2$ isotopomers.

6.2. *Identifying Vinylidene States*

Due to the unique features of the SEP method it is possible to identify the acetylene $\leftrightarrow$ vinylidene J-specific couplings in the spectrum. In double resonance schemes, such as SEP, intermediate states with precisely defined symmetry properties and quantum numbers (J and K_a) can be prepared with *perfect selectivity*.

Rigorous selection rules allow electric dipole transitions between rovibronic states with the same nuclear exchange symmetry but different parity. These transitions obey the $\Delta J = 0$, ± 1 selection rules, and have a propensity to follow the $\Delta K_a = 0$, ± 1 selection rule. Effects such as axis-switching and Coriolis coupling will weaken the ΔK_a propensity rule[36,61] but have no effect on the symmetries of the accessible rovibronic states.

The overall symmetries of the $S_0(\tilde{X}^1\Sigma_g^+)$ and $S_1(\tilde{A}^1 A_u)$ electronic states of acetylene restrict the rovibronic levels accessible by the PUMP photon in the SEP (or double resonance) $\tilde{A} \leftarrow \tilde{X}$ process to $ss-$, $aa-$, $sa+$, $as+$. At the levels of vibrational excitation accessible in the $\tilde{A}^1 A_u$ state, these rovibronic states will appear in the unresolvably degenerate pairs $ss-$, $aa-$, or $sa+$, $as+$. For $^{12}C_2$ acetylene molecules, the $aa-$ and $as+$ species have statistical weight 0, and only intermediate rovibronic states of $ss-$, and $sa+$ symmetry species are reached.

Restricting the symmetry characteristics of the PUMP prepared intermediate in the double resonance process in turn limits the symmetry of the rovibronic states accessible by the DUMP photon. In the acetylene $\tilde{X}(\ell = 0, 2) \leftarrow \tilde{A}(K_a = 1) \leftarrow \tilde{X}(\ell = 0)$ system final states of $ss+$ and $aa+$ symmetry can be reached utilizing a P- or R-branch PUMP transition (Fig. 20). Final states of $sa-$, $as-$ symmetry can be accessed using a Q-branch PUMP transition (Fig. 21). At sufficiently high levels of excitation in the ground state of acetylene, however, the levels will necessarily appear in resolvably degenerate pairs. This is due to the feasibility of the (12) or (ab) permutations. Furthermore, in totally symmetric vibrational states, rovibronic levels localized in the vinylidene region of the potential will appear in the pairs $ss+$ $as+$, $ss-$ $as-$, $sa+$ $aa+$, and $sa-$ $aa-$, while those localized in the linear acetylene region appear in the pairs $ss+$ $aa+$, and $sa - as-$. Note that all eight symmetry species are supported in totally symmetric vibrational levels of vinylidene while only 4 are supported by the totally symmetric vibrational levels of linear acetylene. In $^{12}C_2H_2$ and $^{12}C_2D_2$ the $aa+$, $aa-$, $as+$, $as-$ symmetry rovibronic levels have statistical weight zero and are not present. Due to the rigorous symmetry restrictions imposed on electric dipole transitions only four of the 8 symmetry species of vinylidene levels are accessible from the $\tilde{A}^1 A_u$ state SEP intermediates.

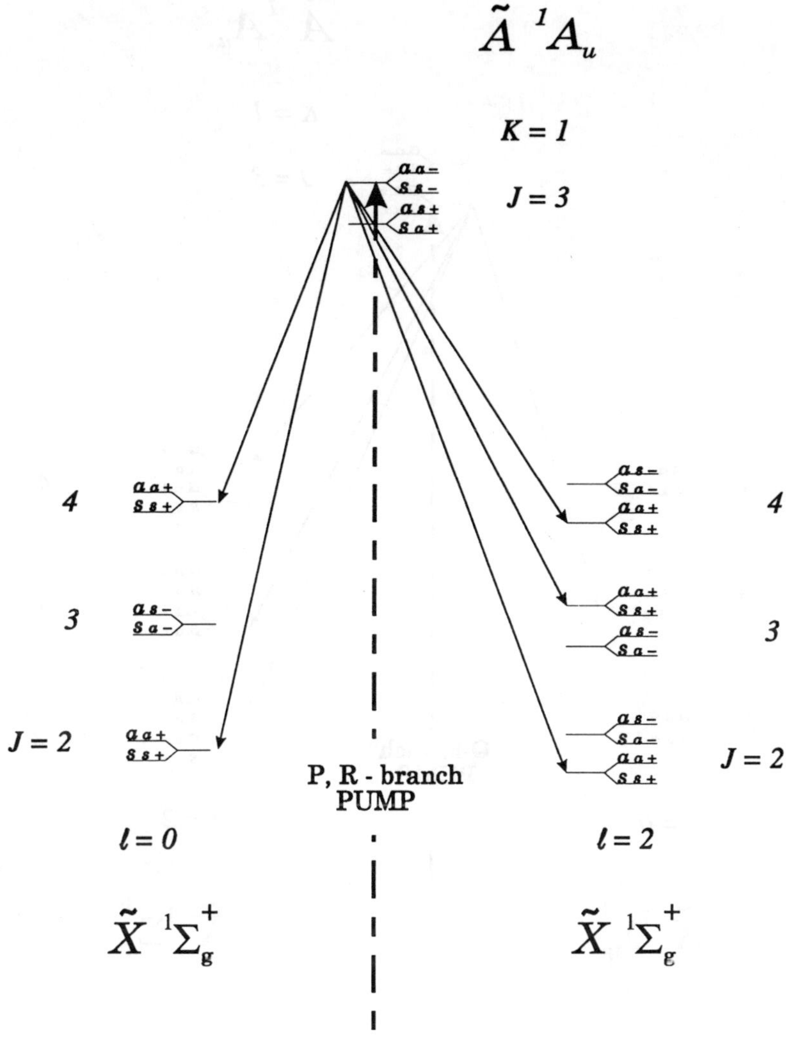

Fig. 20. In the acetylene $\tilde{X}(\ell = 0, 2) \leftarrow \tilde{A}(K_a = 1) \leftarrow \tilde{X}(\ell = 0)$ system, final states of $Ss+$ and $aa+$ can be reached utilizing a P- or R-branch PUMP transition.

The relatively well separated vinylidene vibrational basis states ($\sim$ 15 600 cm^{-1} above the S_0 vibrationless level) lie in a very dense region of linear acetylene vibrational basis states. The true acetylene eigenstates

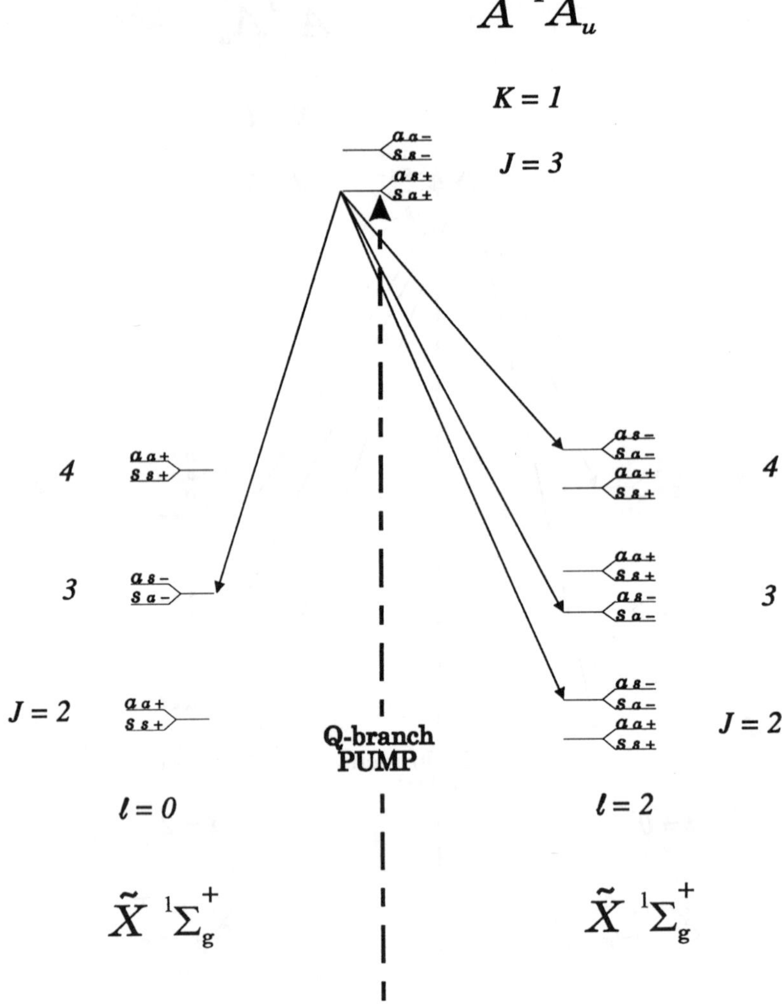

Fig. 21. SEP can access final states of $Sa-$, $as-$ symmetry using a Q-branch PUMP transition in the $\tilde{X}(\ell = 0, 2) \leftarrow \tilde{A}(K_a = 1) \leftarrow \tilde{X}(\ell = 0)$ system.

$|\Psi_{\text{acetylene}}\rangle$ are linear combinations of the one vinylidene and $N(N \gg 1)$ linear acetylene basis states.

$$|\Psi_{\text{acetylene}}\rangle = v|\psi_{\text{vinyl}}\rangle + \sum_{i=1}^{n} a_i |\psi_{i\ \text{linear}}\rangle \ . \tag{48}$$

At excitation energy far away from that of the vinylidene zero point basis level the coefficient (v) of the contributing vinylidene basis function is very small. When the internal energy of the acetylene molecule approaches that of the vinylidene zero point level, the coefficient, v, reaches its maximum (still very small) value. The eigenstate then acquires appreciable vinylidene character (Fig. 22). Of course, by symmetry all basis states contributing to an eigenstate must have the same true symmetry properties (i.e., good quantum numbers).

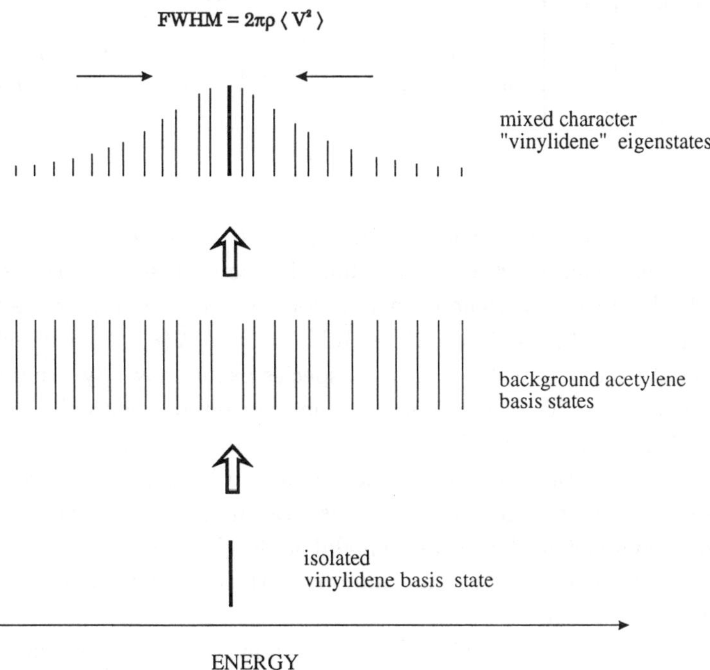

Fig. 22. The isolated vinylidene basis state couples to the dense manifold of acetylene background states. Near an isolated vinylidene basis state the acetylene eigenstates carry appreciable vinylidene character.

If we consider levels of *low J*, the B- or C-axis Coriolis coupling that would mix $K_a \equiv \ell$ levels and is proportional to $\sqrt{J(J+1) - K_a(K_a+1)}$, leaves $K_a \equiv \ell$ a nearly good quantum number. In this case, the Hamiltonian matrix for vibrationally totally symmetric $J = l \equiv K_a$ levels of the coupled system will be of the form:

848

$$
\begin{array}{cc}
 & \begin{array}{cc} \text{Vinylidene} \\ J = 0 \\ Ss+ \quad as+ \end{array} \quad\quad \begin{array}{cccccc} \text{Linear} \\ J = 0 \\ Ss+ \quad aa+ \quad Ss+ \quad aa+ \quad Ss+ \quad aa+ \end{array} \\
\begin{array}{l}
\text{Vinylidene} \quad Ss+ \\
J = 0 \qquad\quad as+ \\[2ex]
\\
\qquad\qquad\quad Ss+ \\
\qquad\qquad\quad aa+ \\
\text{Linear} \quad\;\; Ss+ \\
J = 0 \qquad\quad aa+ \\
\qquad\qquad\quad Ss+ \\
\qquad\qquad\quad aa+ \\[1ex]
\;\vdots \quad\;\; Ss+
\end{array}
&
\left(
\begin{array}{ccccccccc}
E_V & 0 & & h_1 & 0 & h_1 & 0 & h_1 & 0 & \cdots \\
0 & E_V & & 0 & 0 & 0 & 0 & 0 & 0 & \cdots \\
 & & & \vdots & \vdots & \vdots & \vdots & \vdots & \vdots & \cdots \\
h_1 & 0 & \cdots & E_{A1}-E & & & & & & \cdots \\
0 & 0 & \cdots & & E_{A1} & & & & & \ddots \\
h_1 & 0 & \cdots & & & E_{A2}-E & & & & \cdots \\
0 & 0 & \cdots & & & & E_{A2} & & & \cdots \\
h_1 & 0 & \cdots & & & & & E_{A3}-E & & \cdots \\
0 & 0 & \cdots & & & & & & E_{A3} & \cdots \\
\vdots & \vdots & \vdots & \vdots & & & \vdots & \vdots & \vdots & \ddots
\end{array}
\right)
\end{array}
\tag{49}
$$

Note that the vinylidene type $J = 0$ states of totally symmetric vibrational character have tunneling doublets of $ss+$ or $as+$ symmetry, while the linear acetylene tunneling doublets are $ss+$ and $aa+$ symmetric. Furthermore, the $\tilde{X}(\ell = 0, 2) \leftarrow \tilde{A}(K_a = 1) \leftarrow \tilde{X}(\ell = 0)$ SEP scheme can populate the $J = 0$ $ss+$ and $aa+$ species exclusively. By symmetry, only the vinylidene $K_a = \ell = 0$ $ss+$ species can interact with the $\ell = 0$ linear acetylene basis states. The $J = 0$, $ss+$ symmetry acetylene states will be perturbed by the presence of the vinylidene basis state and shifted with respect to their zero order positions. Because there are no vibrationally totally symmetric $J = 0$ $aa+$ vinylidene levels, the $J = 0$ $aa+$ symmetry linear acetylene states can not interact with the vinylidene basis state, and remain unshifted.

Due to the nuclear spin statistical weights of $^{13}C_2$ isotopomers, the $ss+$ and $aa+$ nuclear permutation doublets will appear in the spectrum with intensities given by their nuclear spin statistical weights. For example, for $K \equiv \ell = 0$ $^{13}C_2H_2$ ($^{13}C_2D_2$) with totally symmetric vibrational character, even J levels will appear in 9/1 (9/6) pairs with the lower intensity component shifted. Odd J levels in the same manifold will appear in 3/3 (1/6) pairs, with the lower intensity (in C_2D_2) shifted.

There are two important characteristics of this type of level shift. First, the shift is dependent on whether the acetylene basis level lies energetically above or below the zero-order vinylidene basis state. Levels higher in energy than the vinylidene state will be pushed to higher energy while levels lower in energy will be pushed to lower energy.[62]

Using this diagnostic it will be possible to locate the zero-order position of the vinylidene state to within the average level spacing of the coupled background acetylene basis states ($\sim \pm 1$ cm^{-1}). This will be true even when rapid tunneling causes the vinylidene character to be diluted among *many* states spread over a wide energy region.

Second, regardless of the strength of the interaction, the levels can not be shifted past the zero-order position of the adjacent interacting vibrational levels (Fig. 23). In pure sequence spectra (spectra containing rovibronic states of only one $J, K_a \equiv \ell$ value) this assures that, if the experimental resolution is sufficient to resolve the permutation doublets, adjacent levels in the spectrum will be tunneling doublet pairs with the unperturbed component marking the zero-order position of the perturbed doublet partner. This is of great practical importance since identifying the resolvable splittings in a dense spectrum requires certainty that the levels indeed belong to doublet pairs. The intensities of the rovibrational levels appearing in the spectrum will be random, reflecting the Franck–Condon factors of the individual vibrational levels. However, the permutation doublet levels belong to the same vibrational state, and when resolvable, their intensities necessarily reflect their relative nuclear spin statistical weights.

6.3. *Isomerization and Isomer Specific Spin-Orbit Coupling*

Other electronic states of acetylene will also display state specific isomerization. The lowest triplet (T_1) surface has deep minima in the *cis*-bent, *trans*-bent and vinylidene conformations[22,55,63] (Fig. 24). The electronic symmetries of the T_1 minima are $^3B_2$, ^3B_u, and $^3B_2$ respectively. There will be symmetry specific couplings between states localized in the different minima on the T_1 surface, and isomer specific couplings between the T_1 and the ground state (S_0) surfaces.

If two totally symmetric vibrational levels (one localized in the *cis*-bent and one in the *trans*-bent minimum) are energetically resonant, the pattern of couplings will alternate with K_a. The rotational energy level symmetry templates for these structures, Fig. 2 and 3, explicitly show this pattern. Note that due to the rigorously good symmetry properties of these totally symmetric vibrational levels, states of even K_a can undergo *cis* $\leftrightarrow$ *trans* isomerization while states of odd K_a cannot.

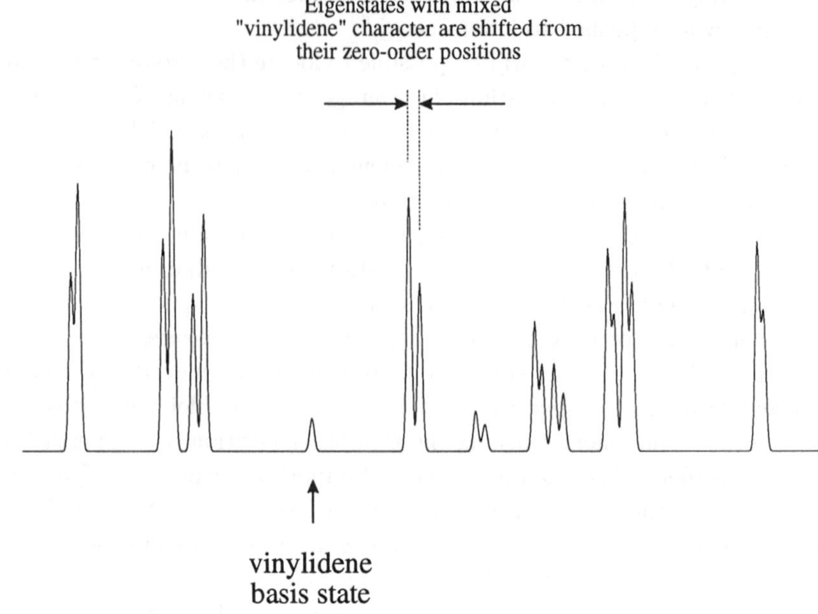

Fig. 23. Near an isolated vinylidene basis state, linear acetylene basis states with the same symmetry properties as the vinylidene level acquire appreciable vinylidene character. This interaction causes the mixed eigenstates to be shifted. The direction of this shift is dependent on whether the acetylene basis level is energetically above or below the zero-order vinylidene basis state. Levels higher in energy than the zero-order vinylidene basis state are pushed to higher energy while levels lower in energy will be pushed to lower energy. Regardless of the strength of the interaction, the levels can not be shifted past the zero-order position of the adjacent interacting vibrational levels. If the experimental resolution is sufficient to resolve the permutation doublets, adjacent levels in the spectrum will be doublet pairs, with the unperturbed component marking their zero-order position. The intensity of these doublet pairs reflect their relative nuclear spin statistical weights. In this synthetic SEP spectrum, of $^{13}C_2D_2$ with totally symmetric vibrational character, and $K \equiv \ell = 0$, even J levels will appear in 9/6 pairs with the lower intensity component shifted.

While such *cis* $\leftrightarrow$ *trans intra*-system coupling has yet to be detected, strong evidence for isomer specific spin-orbit coupling (*inter*-system crossing) between the acetylene ground state (S_0) and the *cis* structure on the first excited triplet surface ($T_{1\,cis}$) has been obtained.

Lisy and Klemperer[64] impacted a beam of electronically excited acetylene (~ 4.5 eV) onto a Auger effect detector. The energy of their excited

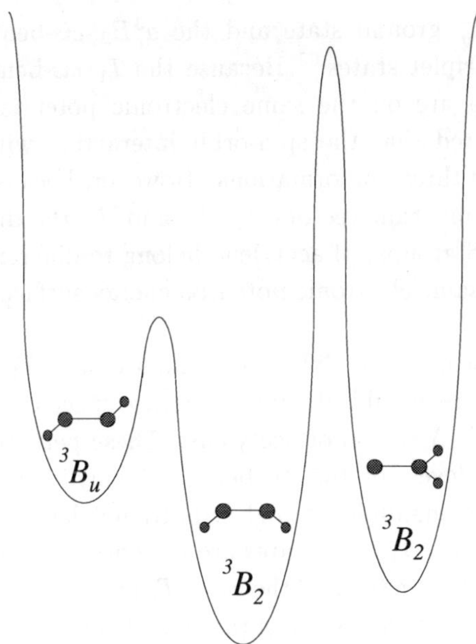

Fig. 24. The T_1 surface has deep minima in the *cis*-bent, *trans*-bent and vinylidene conformations. The electronic symmetries of the T_1 minima are $^3B_2$, ^3B_u, and $^3B_2$, respectively. Level specific couplings analogous to the vinylidene $\leftrightarrow$ linear tunneling in the ground state will be present when multiple minima exist on the same electronic potential energy surface.

acetylene was sufficient to generate T_1 acetylene in its *cis*-bent, *trans*-bent and vinylidene conformations. This beam of metastable C_2H_2 molecules (produced by low energy electron bombardment of C_2H_2) was observed to contain an undetectable fraction of species deflectable in a quadrupole electric field. The polar *cis*-bent and vinylidene C_2H_2 species, with the A, B, and C rotational constants and electric dipole moments computed *ab initio* for the T_1 $\tilde{a}^3B_2$ states, would have been observably deflectable at the electric field strengths employed.

The conclusion that the metastable C_2H_2 beam exclusively contained the nonpolar (undeflectable) *trans*-bent T_1 isomer is especially surprising since the *cis*-bent and vinylidene structures are calculated to be 0.35 eV and 0.25 eV more stable than the *trans*-bent conformation.[22,63]

The explanation of this experimental result lies in the very different strengths of spin-orbit perturbations between the high lying vibrational levels of the $\tilde{X}^1\Sigma_g^+$ ground state and the $\tilde{a}^3 B_2$ *cis*-bent, vinylidene and *trans*-bent $\tilde{a}^3 B_u$ triplet states.[65] Because the T_1 *cis*-bent, *trans*-bent and vinylidene minima are on the same electronic potential energy surface, it might be expected that the spin-orbit interaction with the $\tilde{X}$ state is the same for these three conformations. However, because Ψ^{es} transforms as the angular momentum vectors J_a, J_b, and J_c, the different structural conformations (MS groups) of acetylene belong to different Γ^{es}, even when they occur on the same electronic potential energy surface (Tables 3, 4, and 5).

Basis states localized in different regions of the T_1 potential energy surface will, therefore, exhibit distinct structure specific spin-orbit coupling to the 'background' $\tilde{X}$ states of acetylene. These perturbations cause the lowest vibrational levels of the *cis*-bent and vinylidene triplet states to mix into the dense manifold of high vibrational levels of the $\tilde{X}^1\Sigma_g^+$ S_0 state. The mixed $S_0 \sim T_1$ eigenstates are no longer Auger-detectable. The electronically forbidden nature of the $S_0 \sim T_1$ (*trans*) spin-orbit interaction causes the lowest vibrational levels of the *trans*-bent triplet state to be much less diluted into the high vibrational levels of the $\tilde{X}^1\Sigma_g^+$ state. Therefore, the *trans*-bent T_1 molecules retain their Auger-detectability.

6.4. *Torsional Tunneling in Nonplanar Acetylene*

While the excited electronic potential energy surfaces of acetylene have minima corresponding to many isomeric structures, spectroscopic observations of acetylene have largely been limited to the linear and *trans*-bent conformations. Transitions within, and between, the rigid *linear* and *trans*-bent conformations of acetylene can be completely described by the usual ΔK_a, ΔK_c selection rules. However, when transitions between these planar structures and one of the *nonplanar* structures is considered, more complicated patterns appear in the spectrum. Use of the rigorous and point group independent CNPI group theory is of paramount importance to properly assign these spectra.

A prime example of the need to use the CNPI paradigm is the assignment of the acetylene $\tilde{E}$-state. The vibrational progressions observed in the one-photon vacuum ultraviolet (vuv) spectrum above $74\,000$ cm^{-1}, and assigned to the $\tilde{E}$-state,[66,67] also appear in the ultraviolet–ultraviolet double resonance (UVUVDR) spectrum.[26] While the $\tilde{E}$-state is strongly predissociated, and the vuv spectra can only be classified as diffuse, the UVUVDR double resonance technique permitted rotationally resolved spectra to be recorded. The rotationally resolved spectra of the E-state exhibit a-type ($\Delta K_a = 0$, $\Delta K_c = \pm 1$) electric dipole transitions from odd-K_a intermediates, and b-type ($\Delta K_a = \pm 1$, $\Delta K_c \pm 1$) electric dipole transitions from even-K_a intermediates.

A CNPI group theoretical analysis of the unprecedented rotational energy level patterns in the UVUVDR spectrum, taken in conjunction with the one photon vuv spectrum, reveal that the $\tilde{E}$-state has a *nonplanar* near *cis*-bent (C_2^b) structure.[26]

This *nonplanar* near *cis*-bent (C_2^b) structure may also be *nonrigid*. Torsional tunneling through both the *cis* and *trans* barriers to planarity may explain why the $\tilde{E}$-state term values reported from the one-photon $\tilde{E} \leftarrow \tilde{X}$ transition are systematically 25 ± 8 cm^{-1} above the term values reported for the two photon $\tilde{E} \leftarrow \tilde{A} \leftarrow \tilde{X}$ transition, as well as the significant Franck–Condon overlap between the structurally quite different $\tilde{E}$ and $\tilde{A}$ states. Because a ± 20 cm^{-1} error was ascribed to the measured term values of the vuv experiment, this discrepancy is not experimentally significant. However, it is systematic. If this discrepancy is real and not due to calibration error in the vuv spectra, then the $\tilde{E} \leftarrow \tilde{X}$ transition accesses the upper torsional tunneling component while the UVUVDR $\tilde{E} \leftarrow \tilde{A} \leftarrow \tilde{X}$ spectra accesses the lower torsional tunneling component. This 25 cm^{-1} difference then indicates the $\tilde{E}$-state is *nonrigid*, with a torsional splitting of the same order of magnitude as that in hydrogen peroxide.[68]

7. Conclusions

Molecular eigenstates can be classified according to a symmetry group of the Hamiltonian. Molecular point groups are most commonly used in the literature to provide a descriptive framework. However, when a molecule possesses distinct isomeric structures, the near symmetry operations (i.e., reflections and rotations) of the point group do not provide a consistent or convenient basis with which to classify rovibronic states. The Complete Nuclear Permutation Inversion (CNPI) groups and their Molecular Symmetry (MS) subgroups are composed of *true symmetry* operations which are common to all possible isomers of the molecule. Labeling the molecular eigenstates with their true symmetry properties provides a rigorous and point group independent description of the molecular eigenstates.

Labeling the eigenstates using the CNPI symmetry properties of the molecule greatly clarifies the selection rules for spectroscopic transitions and perturbations involving states localized in different regions of the potential energy surface. Indeed, as has been shown, the CNPI–MS group theoretical description is instrumental in understanding and characterizing the isomerization processes in the structurally complex acetylene molecule.

Acknowledgments

The author is indebted to Professor R. W. Field and Dr. D. M. Jonas for clarifying discussions during the course of this work, and R. M. Williams for valuable criticism of the manuscript.

References

1. G. J. Scherer, K. K. Lehmann, and W. Klemperer, *J. Chem. Phys.* **78**, 2817–2832 (1983).
2. F. F. Crim, *Ann. Rev. Phys. Chem.* **35**, 675 (1984).
3. T. R. Rizzo, C. C. Hayden, and F. F. Crim, *J. Chem. Phys.* **81**, 4501 (1984).
4. D. S. King, *Adv. Chem. Phys.* **50**, 105 (1982).
5. J. J. Scherer, A. L. Cooksey, R. Sheeks, J. Heath, and R. J. Saykally, *Chem. Phys. Lett.* **172**, 3, 4 (1990).
6. S. L. Coy, R. Hernandez, and K. K. Lehmann, *Phys. Rev. A*, **40**, 10 (1989).
7. K. Yamanouchi, N. Ikeda, S. Tsuchiya, D. M. Jonas, J. K. Lundberg, G. W. Adamson, and R. W. Field, *J. Chem. Phys.* **95**, 9, 6330–6342 (1991).
8. Y. Chen, D. M. Jonas, C. E. Hamilton, P. G. Green, J. L. Kinsey, and R. W. Field, *Ber. Bunsenges. Phys. Chem.* **92**, 329 (1988).
9. C. Hamilton, J. L. Kinsey, and R. W. Field, *Ann. Rev. Phys. Chem.* **37**, 493–524, 1986.
10. C. Kittrell, E. Abramson, J. L. Kinsey, S. A. McDonald, D. E. Reisner, and R. W. Field, *J. Chem. Phys.* **75**, 2056 (1981).
11. D. E. Reisner, R. W. Field, J. L. Kinsey, and H.-L. Dai, *J. Chem. Phys.* **80**, 5968–5978 (1984).
12. T. R. Huet, M. Herman, and J. W. C. Johns, *J. Chem. Phys.* **94**, 3407 (1991).
13. H. L. Dai, R. W. Field, and J. L. Kinsey, *J. Chem. Phys.* **82**, 2161 (1985).
14. M. E. Kellman and G. Chen, *J. Chem. Phys.* **95**, 8671 (1991).
15. P. R. Bunker, *Molecular Symmetry and Spectroscopy 75 ff* (Academic Press, New York, NY, 1979).
16. H. C. Longuet-Higgins, *Mol. Phys.* **6**, 445 (1963).
17. A. D. Walsh, *J. Chem. Soc.* 2260–2331 (1953).
18. C. K. Ingold and G. W. King, *Nature* **169**, 1101 (1952).
19. K. Keith Innes, *J. Chem. Phys.* **22**, 5, 863–876 (1954).
20. D. Demoulin, *Chem. Phys.* **11**, 329–341 (1975).

21. S. P. So, R. W. Wetmore, and H. F. Schaefer *J. Chem. Phys.* **73**, 5706–5710 (1980).
22. M. P. Conrad and H. F. Schaefer, *J. Am. Chem. Soc.* **100**, 7820–7823 (1978).
23. W. E. Kammer, *Chem. Phys. Letts.* **6**, 529–532 (1970).
24. Yongqin Chen, Ph.D. Thesis, Massachusetts Institute of Technology, 1988.
25. K. M. Ervin, J. Ho, and W. C. Lineberger, *J. Chem. Phys.* **91**, 5974 (1989).
26. J. K. Lundberg, D. M. Jonas, B. Rajaram, Y. Chen, and R. W. Field, *J. Chem. Phys.* **97**, 7180–7195 (1992).
27. F. Albert Cotton, "Chemical Applications of Group Theory" (Wiley-Interscience, 1971) pp. 86–92.
28. P. R. Bunker, *Molecular Symmetry and Spectroscopy* (Academic Press, New York, NY, 1979).
29. C. H. Townes and A. L. Schawlow, "Microwave Spectroscopy" (Dover, New York, NY, 1975) pp. 83–114.
30. J. T. Hougen, *J. Chem. Phys.* **39**, 358 (1962).
31. P. R. Bunker, *Molecular Symmetry and Spectroscopy* (Academic Press, New York, NY, 1979) 374.
32. D. M. Dennison, *Rev. Mod. Phys.* **3**, 280 (1931).
33. P. R. Bunker, *Molecular Symmetry and Spectroscopy* (Academic Press, New York, NY, 1979) 94.
34. J. T. Hougen, *J. Chem. Phys.* **36**, 519 (1962).
35. J. T. Hougen, *Can. J. Phys.* **62**, 1392–1402 (1984).
36. J. T. Hougen and J. K. G. Watson, *Can. J. of Phys.* **43**, 298–320 (1965).
37. H. Lischka and A. Karpfen, *Chem. Phys.* **102**, 77–99 (1986).
38. M. Peric, S. D. Peyerimhoff, and R. J. Buenker, *Mol. Phys.* **62**, 6, 1339–1356 (1987).
39. P. R. Bunker and D. Papousek, *J. Mol. Spectrosc.* **32**, 419 (1969).
40. P. R. Bunker, *Molecular Symmetry and Spectroscopy* (Academic Press, New York, NY, 1979) pp. 330–344.
41. G. Herzberg, "Electronic Spectra of Polyatomic Molecules" (Van Nostrand Reinhold, New York, 1966) pp. 24–26.
42. W. Pauli, *Phys. Rev.* **58**, 716 (1940).
43. H. Eyring, J. Walter, and G. E. Kimball, "Quantum Chemistry" (John Wiley, New York, 1944), p. 371.
44. E. U. Condon, *Phys. Rev.* **32**, 858–872 (1928).
45. F. Duschinsky, *Acta Physicochim. URSS* **7**, 551 (1937).
46. P. R. Bunker, *Molecular Symmetry and Spectroscopy* (Academic Press, New York, NY, 1979) pp. 160–165.
47. A. D. Walsh, *J. Chem. Soc.* 2260–2331 (1953).
48. R. J. Buenker and S. D. Peyerimhoff, *Chem. Rev.* **74**, 127–188 (1974).
49. G. Herzberg, "The Spectra and Structure of Simple Free Radicals" (Dover Publications Inc. New York, NY, 1988) pp. 128–137.
50. G. Herzberg, "Electronic Spectra of Polyatomic Molecules" (Van Nostrand Reinhold, New York, 1966) pp. 244–249.

51. R. P. Durán, V. T. Amorebieta, and A. J. Colussi, *J. Am. Chem. Soc.* **109**, 3154–3155 (1987).

52. A. H. Laufer, *J. Chem. Phys.* **76**, 945–948 (1982).

53. R. P. Durán, V. T. Amorebieta, and A. J. Colussi, *J. Phys. Chem. Soc.* **92**, 636–640 (1988).

54. T. Tanazawa and W. C. Gardiner Jr, *J. Phys. Chem.* **84**, 236–239 (1980).

55. J. H. Davis, W. A. Goddard III, and L. B. Harding, *J. Am. Chem. Soc.* **99**, 2919–2925 (1977).

56. K. M. Ervin, J. Ho, and W. C. Lineberger, *J. Chem. Phys.* **91**, 5974 (1989).

57. T. Carrington Jr, L. M. Hubbard, H. F. Schaefer III, and W. H. Miller, *J. Chem. Phys.* **80**, 4347 (1984).

58. M. Bixon and J. Jortner, *J. Chem. Phys.* **48**, 715 (1968).

59. Y. Chen, D. M. Jonas, C. E. Hamilton, P. G. Green, J. L. Kinsey, and R. W. Field, *Ber. Bunsenges. Phys. Chem.* **92**, 329 (1988).

60. Yongqin Chen, Ph.D. Thesis, Massachusetts Institute of Technology, 1988.

61. C. L. Brummel, S. W. Mork, and L. A. Philips, *J. Chem. Phys.* **95**, 7041–7053 (1991).

62. Yongqin Chen, Ph.D. Thesis, Massachusetts Institute of Technology, appendix A, 85–88 (1988).

63. R. W. Wetmore and H. F. Schaefer III, *J. Chem. Phys.* **69**, 1648–1654 (1978).

64. J. M. Lisy and W. Klemperer, *J. Chem. Phys.* **72**, 3880 (1980).

65. J. K. Lundberg, R. W. Field, C. D. Sherril, E. T. Seidl, Y. Xie, and H. F. Schaefer III, *J. Chem. Phys.* **98**, 8384–8391 (1993).

66. P. G. Wilkinson, *J. Mol. Spectrosc.* **2**, 387–404 (1958).

67. M. Herman and R. Colin, *J. Mol. Spectrosc.* **85**, 449–461 (1981).

68. P. Helminger, W. C. Bowman, and F. C. DeLucia, *J. Mol. Spectrosc.* **85**, 120 (1981).

CHAPTER 23

TREES FROM SPECTRA: GENERATION, ANALYSIS,
AND ENERGY TRANSFER INFORMATION

Michael J. Davis

Chemistry Division,
Argonne National Laboratory,
Argonne, IL 60439, USA

Contents

1. Introduction

Stimulated emission pumping spectroscopy[1] has allowed an unprecedented view of molecules excited to the upper regions of their ground state potential energy surface. The complexity of the spectra of such highly excited molecules has led to the application of several approaches different than the traditional assignment procedures of molecular spectroscopy.[2] These include statistical methods, which investigate distributions of energy levels or intensities. Reference 3 lists two recent applications with extensive bibliographies, and Ref. 4 has an application to an experimental spectrum. Another approach involves the investigation of spectra at lower levels of resolution. It has been found that even when high resolution spectra of highly excited molecules are complicated, patterns may emerge at lower levels of resolution. This smoothing has been done by degrading otherwise higher resolution spectra,[5,6] or by comparing spectra generated from higher resolution techniques with ones generated by inherently lower resolution techniques.[7] The smoothing has aided in the understanding of complicated spectra. For example, spectral patterns have been associated with periodic orbits,[6] and spectral features have been assigned.[7] Such assignments have aided in the understanding of energy transfer in highly excited molecules.[8]

Although the smoothing techniques have proven very enlightening, they generally discard high resolution information. In addition, the smoothing has not been done in a particularly systematic fashion, in our opinion. These two drawbacks have been part of the motivation for the development of a hierarchical approach to the study of molecular spectra.[9–11] This approach involves viewing the spectra at all levels of resolution in a systematic fashion, addressing the two drawbacks. At the same time, the hierarchical approach retains any simplicity which might be present in low resolution versions of spectra. A hierarchical approach is consistent with many well-known views of IVR (intramolecular vibrational redistribution),[12] because many of these are inherently hierarchical. For example, one common approach models the energy redistribution as the coupling of manifolds of states, a hierarchy.

Another approach used to understand complicated spectra involves the generation of the exact eigenstates for the potential surface of a molecule.[13–15] This approach suffers from several problems, including the size of computations and the availability of potential energy surfaces which are accurate at high energy for strongly bound molecules. However, it generates the exact answer within these constraints, and thus has motivated a considerable amount of effort. Other problems may arise which involve the analysis of the eigenstates. First, given the resolution of hundreds, or perhaps a few thousand eigenstates, it may be difficult to process the information contained in the states. In addition, it may be difficult to understand states at high energy, where there may be strong mixing. It is here that a hierarchical tree once again can be an aid in understanding and decomposing complicated information. In particular, a hierarchical tree in conjunction with the eigenstates can offer some very specific information concerning assignability of complicated molecular spectra and energy transfer time scales and pathways.

The outline of this paper is as follows. Section 2 has an overview of the motivation, generation, and some initial discussions of hierarchical trees. Section 3 reviews several analysis tools for the study of the trees. An experimental spectrum of NO_2 is studied in Sec. 4, which also includes a comparison of results for the experimental spectrum with that of the spectrum of the model system of Secs. 2 and 3. Section 5 has a discussion of how the hierarchical analysis can be employed to address issues concerning assignability and energy transfer time scales and pathways, given a theoretical model of a spectrum generated from the eigenstates. Section 6 has a brief conclusion.

2. Spectral Trees

Figure 1 motivates the idea of a hierarchical description of vibrational spectra. The stick spectrum of Fig. 1(a) is the same one described in Ref. 9. One may be able to observe a sequence of tall lines in Fig. 1(a), but these lines are not necessarily monotonic, as can be observed for the two lines of nearly equal height which are at the energies of -0.038 and -0.037 (these are the 55th and 57th lines). There does not appear to be any other obvious sequences in the plot. However, if the spectrum is smoothed, new patterns may emerge and other more obvious patterns, such as the previously noted tall lines, can be confirmed.

The simplicity obtained from the smoothing can be observed in Figs. 1(b)–1(d). First, if the spectrum is smoothed enough, all structure is lost and a single peak is observed (Fig. 1(b)). When the resolution is increased patterns may emerge, and Figs. 1(c) and 1(d) show such patterns. These plots, like the other smooth spectra, are for a middle portion of the spectrum of Fig. 1(a), as described in the caption of Fig. 1. Past work on smoothed spectra has generally concentrated on the study of smoothed spectra such as those of Figs. 1(c) and 1(d), particularly 1(c). As noted in the Introduction, this tends to throw out other information available in the spectrum at higher levels of resolution (e.g., Figs. 1(e) and 1(f)), and, in general, the smoothing has not been done in a particular systematic fashion. The algorithm for the generation of trees, described in Refs. 9 and

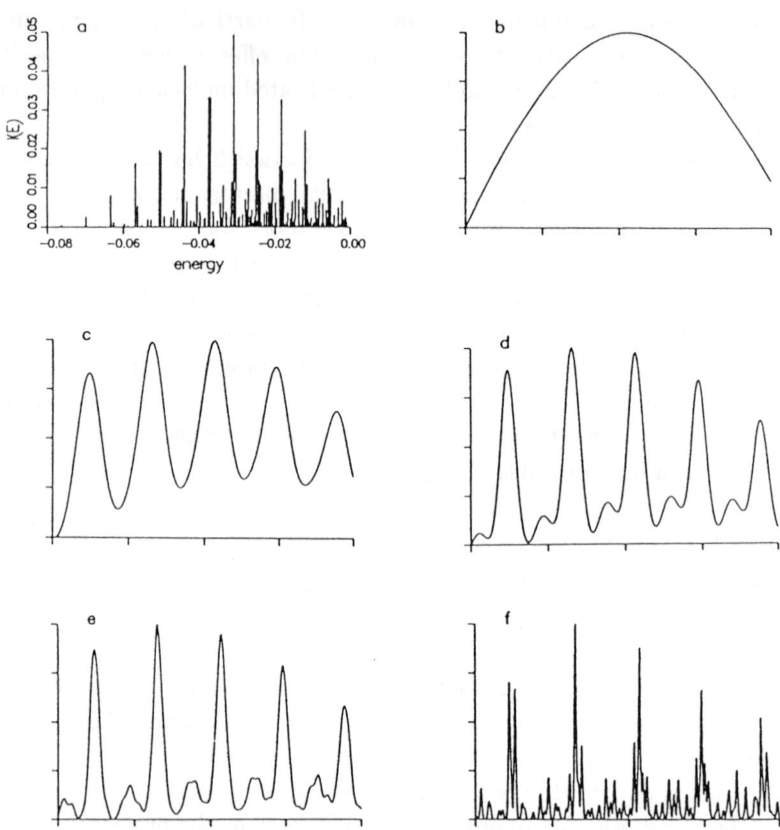

Fig. 1. A stick spectrum is shown in 1(a) and smoothed versions of the middle portion ($E = -0.0412$ to $E = -0.0103$) of the spectrum are shown in 1(b)–1(f). The model system, the generation of the spectrum, and the smoothing are all described in Ref. 9.

10, can be used to study a spectrum at all levels of resolution in a systematic fashion. Indeed, the plots shown in Figs. 1(c) and 1(d) were chosen from a hierarchical tree described below, and, in fact, the generation of the tree preceded the plotting of the smoothed spectra. Figure 1(d) demonstrates that patterns which may not be obvious at high resolution emerge with systematic smoothing. The small side peaks attached to the larger peaks in Fig. 1(d) are not obvious in the fully resolved stick spectrum of Fig. 1(a).

The motivation for a hierarchical tree can be understood by studying the smoothed spectra in Fig. 1, particularly Figs. 1(c) and 1(d). As resolution is increased, smaller peaks tend to split off larger peaks, suggesting parent/child relationships among the peaks of a spectrum. In general, a peak splits into two peaks rather than three or more, and the splitting usually occurs along the side of the parent peak. By carefully following the splitting of peaks as resolution is increased, one can generate a hierarchical tree like the one in Fig. 2(a) for the spectrum of Fig. 1(a).

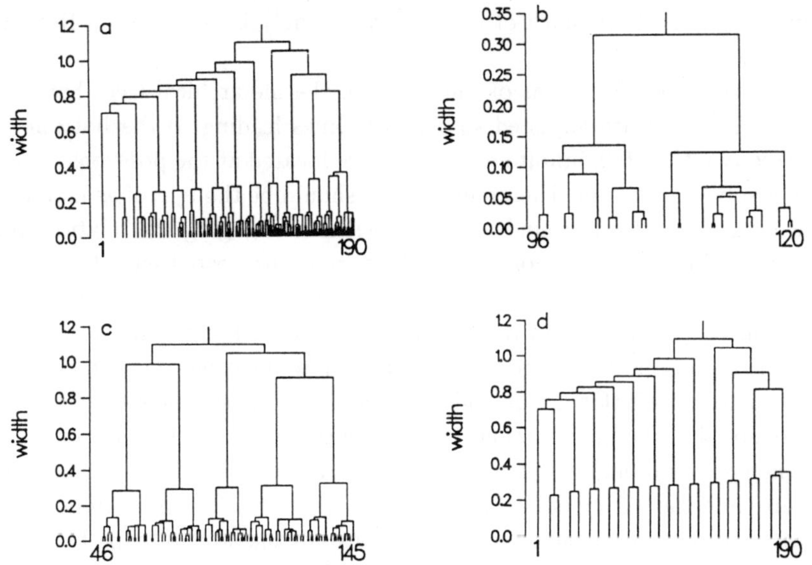

Fig. 2. The tree generated from the spectrum of Fig. 1(a) is shown in 2(a) and several manipulations of the tree are shown in Figs. 2(b)–2(d). Widths have been scaled as described in Ref. 9. See text for further details.

One can observe the previously noted sequences of peaks in the tree of Fig. 2(a). The set of nodes descending from the top of the tree down to a width of 0.71 point to a set of large, broad peaks, five of which are shown in Fig. 1(c). The splittings which occur between 0.23 and 0.36 are the set of side peaks splitting off the large peaks, which can be observed in Fig. 1(d). This demonstrates that the tree retains the simplicity of the smoothed spectra of Figs. 1(c) and 1(d), but there is also information concerning the spectra at higher levels of resolution, as indicated by the structure of the tree below a width of 0.23.

The number of peaks at a given level of resolution can be read off the tree by counting the number of vertical lines at that resolution (for example, 13 peaks at 0.4). In addition, the tree defines a relationship among all lines of the spectrum. One can define the distance between two lines as the height of their most recent common ancestor. For example, the first line splits from the group containing lines 1–4 at a width of 0.71, and this split defines the distance between the first line and the next two as 0.71. The calculation of all split widths generates a distance matrix, and such matrices are used for much of the statistical analysis described in Ref. 10 and Sec. 3.

The horizontal locations of the nodes in the hierarchical trees have no meaning,[9,10] but are adjusted so that the lines leading to the bottom of the tree (width = 0.0, infinite resolution) are located at the positions of the fully resolved spectrum. Hierarchical trees should be thought of as mobiles, which can be rotated around any node without changing the relationship among the lines. Furthermore, hierarchical trees are nested sets of subsets. Thus, one can use the trees to "divide and conquer" a spectrum, either by studying various pieces of a tree separately or by studying a tree at a given level of resolution. There are several simple manipulations one can do to a tree to accomplish this divide and conquer strategy (see below).

As noted in the Introduction, two important and related issues concerning the spectra of highly excited molecules are the assignability of spectral features and the nature of IVR implied by a spectrum. The most important features of the trees concerning these issues are the sizes of gaps between nodes and the nature of the branching, regardless of any apparent regularity in the smoothed spectra at the level of resolution indicated by particular gaps. Large gaps between successive nodes point to strong bottlenecks to intramolecular energy transfer. When large gaps occur across a portion of a spectrum they point to assignability of groups below them. For example, there is a large gap across the whole spectrum between 0.36 and 0.71 and there is another sizable gap between the node at 0.23 (far left) and lower nodes which runs across the whole spectrum, except for the first spectral line. As demonstrated in Ref. 9, the first large gap leads to assignability of the features in Fig. 1(c) and the second gap to assignability of the features in Fig. 1(d).

Much of the previous discussion is summarized and explained more fully in Figs. 2 and 3. Figure 2(a) shows the previously discussed tree, and the three other panels illustrate three manipulations of trees we have found useful (for a fourth, see Fig. 13(c) in Sec. 5). The one shown in Fig. 2(b) involves cutting off a portion of the tree to generate a so-called subtree. While the original tree visualizes information concerning the whole spectrum (lines 1–190), the subtree contains information concerning a subset of the spectrum (lines 96–120). One can also study portions of a tree which are not strictly subtrees (one cut of the tree). This involves cutting off the ends of a tree, and Fig. 2(c) shows a tree generated from this procedure. A third (Fig. 2(d)) involves replacing all existing structure below a given level of resolution by single lines.

Figure 3 further summarizes the information contained in a hierarchical tree. The left-hand column is a subset of various levels of resolution of the spectrum of Fig. 1(a). The first plot shows a portion of Fig. 1(c), the next two plots show a portion of Fig. 1(d), and the bottom two plots show a portion of the fully resolved spectrum. The top three plots in the first column indicate that the tree algorithm assigns a group of lines to every smooth spectral feature. Because the tree algorithm follows all the splittings of peaks as resolution is increased (see Fig. 1), it assigns a parent peak to every splitting. This leads to the assignment of a parent line to every group of lines, at the fully resolved level (the bottom of the tree). The bottom two plots demonstrate this. The 83rd line is the parent of the group 78–87 and the 91st line is the parent of lines 88–95. The parent line is often the largest line in a group, but need not be, although it is generally one of the largest.[10] Identification of a parent is useful for understanding assignability and energy transfer (see below and Ref 9).

Column 2 of Fig. 3 shows a series of subtrees which describe the evolution of the portions of smooth spectra shown in the first column. At the top is shown the subtree associated with lines 78–95. The next two panels show subtrees cut from this subtree. These smaller subtrees were generated by cutting all the way across the larger subtree above its second highest node. The bottom two subtrees of column 2 show the splitting at the bottom of the tree, where the lines of the spectrum become fully resolved.

Plots like the ones in the first two columns of Fig. 3 can be generated for a typical spectrum, whether it is experimental or theoretical, and we have studied several cases. If the eigenstates (not just the eigenvalues) of the system are available one can go well beyond this analysis. The tree construction allows for the generation of "smoothed eigenstates" associated with smooth spectral features[9] (for other approaches, see Refs. 16 and 17). The first three rows of column 3 show smoothed states for each

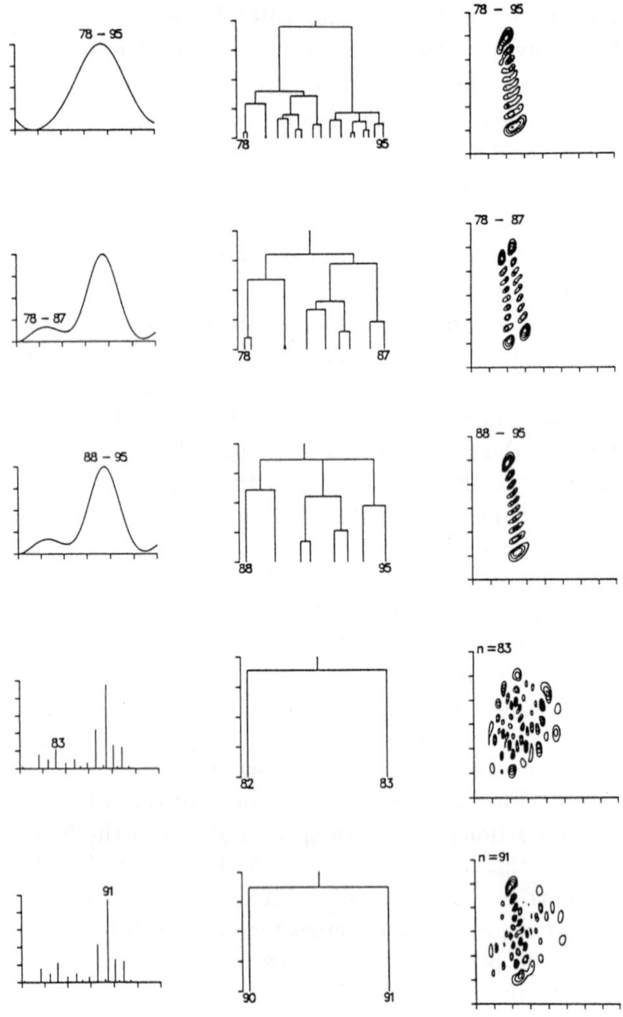

Fig. 3. The first column shows pieces of smoothed spectra (top three) and the fully resolved spectrum (bottom 2). The spectra are portions of the one in Fig. 1. The numbers above the smoothed spectra refer to lines associated with the smoothed spectra as calculated from the tree algorithm. The middle column shows a series of subtrees associated with the features in column 1 and the column on the right shows smoothed states calculated for the ranges associated with each subtree.

subtree. These are generated by summing the eigenstates with the correct coefficients.[9] These coefficients are the amplitudes, that is the square roots of the intensities with the correct phases ($+1$ or -1 for real functions). The last two rows of column 3 show the 83rd and 91st eigenstates of the system, which are considerably more complicated than the smoothed states.

The first three subtrees in column 2 and the trees and subtrees of Figs. 2(a)–2(c) illustrate several important points. By comparing the subtree at the top of column 2 of Fig. 3 with the next two, one can observe how a tree summarizes information at all levels of resolution below it. The subtrees plotted in rows 2 and 3 of column 2 can easily by observed in the subtree at the top of column 2. In addition, one can appreciate the utility of the divide and conquer approach, by observing that there are many large gaps in the subtrees of these figures which do not range the whole spectrum. A good example of such a gap can be observed in the first two subtrees of Fig. 3, where there is a large gap between the node splitting off the group 82–87 and subsequent nodes below it (this node is the second highest of the second subtree). Due to anharmonicity and change of molecular dynamics with energy, there can be considerable differences in time scales in different parts of a spectrum and the divide and conquer approach of a hierarchical tree allows one to probe for this. It provides a systematic means of investigating local energy transfer properties which may change or disappear as the energy of a molecule changes.

The hierarchical approach not only provides a divide and conquer prescription, but it does this in a systematic fashion. For example, one can generate a series of subtrees from a cut across the whole tree at a given level of resolution. The subtree at the top of column 2 of Fig. 3 and the subtree of Fig. 2(b) were generated from a cut across the full tree of Fig. 2(a) between a width of 0.36 and 0.71. These subtrees provide information related to the "decay" of "overtone" states like the one on the top right of Fig. 3. A hierarchical approach also provides a systematic method for the study of portions of a spectrum which are not subtrees. These can be generated by cutting off the ends of a spectrum at a natural boundary (above nodes at the ends of the tree), or by combining a set of consecutive subtrees which were cut off at a given level of resolution. For example, the tree containing line 78–120 can be formed by combining the subtrees for lines 78–95 and 96–120.

The tree and resulting smoothed states give information concerning assignability and energy transfer. For example, the state in the middle of column 3 can easily be assigned a set of quantum numbers. There are 9 nodes along the longer part of the state and no nodes along the narrower part. The utility of smoothed states can be appreciated by comparing this particular state to the eigenstate ($n = 91$) which is the parent line of the group 88–95 (the smoothed state), shown at the bottom of column 3. It is much more difficult to assign the eigenstate a set of quantum numbers. Also, it is clear that the nodal pattern of the state in the middle of column 3 is not strictly Cartesian and the density pattern is noticeably different from a normal mode. Thus further illustrates the utility of the present approach over other procedures, for example, the overlap of the eigenstates with a normal mode basis state, to judge degree of normal character, or assignability. Finally, compare the state in the second row with the eigenstate in the fourth row, the parent of the group 78–87. While the previously noted states (88–95 and 91) have a similar density build-up, these two states are quite different and there is no hint of such a simple state as 78–87 embedded in the 83rd eigenstate.

The smoothed state on the top right of Fig. 3 is part of a progression of states in which additional nodes appear, mostly along the vertical, as energy is increased.[9] In the same way that this state is part of a progression, so are the two states in rows 2 and 3 of column 3, although this latter progression is much shorter, involving only two states. One can observe that this latter progression involves the substitution of a node along the longer part with one along the narrower part. The length of progressions like this depend on the sizes of various regions in phase space, as emphasized often in the past (see Ref. 11 for a discussion of this particular system). The idea of progressions within progressions is a familiar one in molecular spectroscopy,[2] and the generation of a tree provides a convenient way to explore such progressions. An important aspect of this exploration is that it does not depend on choice of coordinate system or basis set (although one needs to know the eigenstates). The previous discussion of the states in column 3 emphasizes this. Section 5 shows additional progressions of states which are descendants of combination states similar to the one in column 3 (78–87).

While assignability can be investigated through sequences of subtrees cut from an ancestral subtree, insights into energy flow can result from following a particular path down a tree. The sizes of the gaps between nodes indicate time scales for energy flow, and the smoothed states generated from a given subtree under a node demonstrate energy flow pathways. One can observe energy flow in the plots of column 3 of Fig. 3 by comparing the 88–95 smoothed state with 91st eigenstate. Although the eigenstate has most of its density along the same direction as the smoothed state, there is noticeable density along a direction which is roughly perpendicular to this vertically oriented coordinate, and this indicates that significant energy flow has taken place.

3. Analysis of Trees

One can make a more quantitative analysis of trees[10] than what was done in the previous section. The main goal of this analysis is to generate a systematic means to divide a spectrum into sets of groups to aid investigation of assignability and energy flow discussed in the previous section and in Secs. 4 and 5. The analysis is meant to be free from any particular model of a spectrum and is meant to provide a systematic procedure to reach our stated goals. The key to this procedure is the ability to divide a spectrum into a set of groups at various levels of resolution, and to discern what levels of resolution (time scales) are important. It is clear from the previous section that this can be done to some degree by merely viewing a particular tree, but what we put forth here is a more systematic and quantitative approach. In this section, we describe three types of statistics which describe three related aspects of the trees: (1) dimensionality, (2) clustering, and (3) the number of time scales implied by a particular tree. Although these statistics are generated for a particular tree, they reflect the underlying spectrum from which the tree is generated.

The reader should be warned ahead of time that the following section is very detailed and condensed and probably can be skipped without impairing the understanding of subsequent sections too much (Ref. 10 has a more thorough discussion of the techniques outlined here). The key point to be taken from the section is that one can study shapes of curves generated from the various statistics which highlight important time scales and groupings of complicated spectra. The significant features of the shapes are inflection points and sudden drops or rises, with these features further indicated by studying instantaneous slopes of the curves. One final aspect of the analysis is that several different statistics are used so that significant features of energy flow and assignability will become obvious even on a cursory study of the statistics.

Figure 4 summarizes the analysis tools we have found useful for a quantitative analysis of trees. These techniques have been taken from the classification and multivariate analysis literature.[18–21] The top of Fig. 4 shows another subtree cut from the tree of Fig. 2(a). This subtree is part of the sequence of subtrees generated by cutting between 0.36 an 0.71 (see above), and it also has the characteristic combination and overtone subtrees embedded in it (lines 121–131 and 132–145). The remaining plots in Fig. 4 pictorially summarize the analysis tools outlined in Ref. 10. Figure 4(b) shows results for a geometrical analysis of the tree which leads to an understanding of the inherent dimensionality of a spectrum. Figure 4(c)

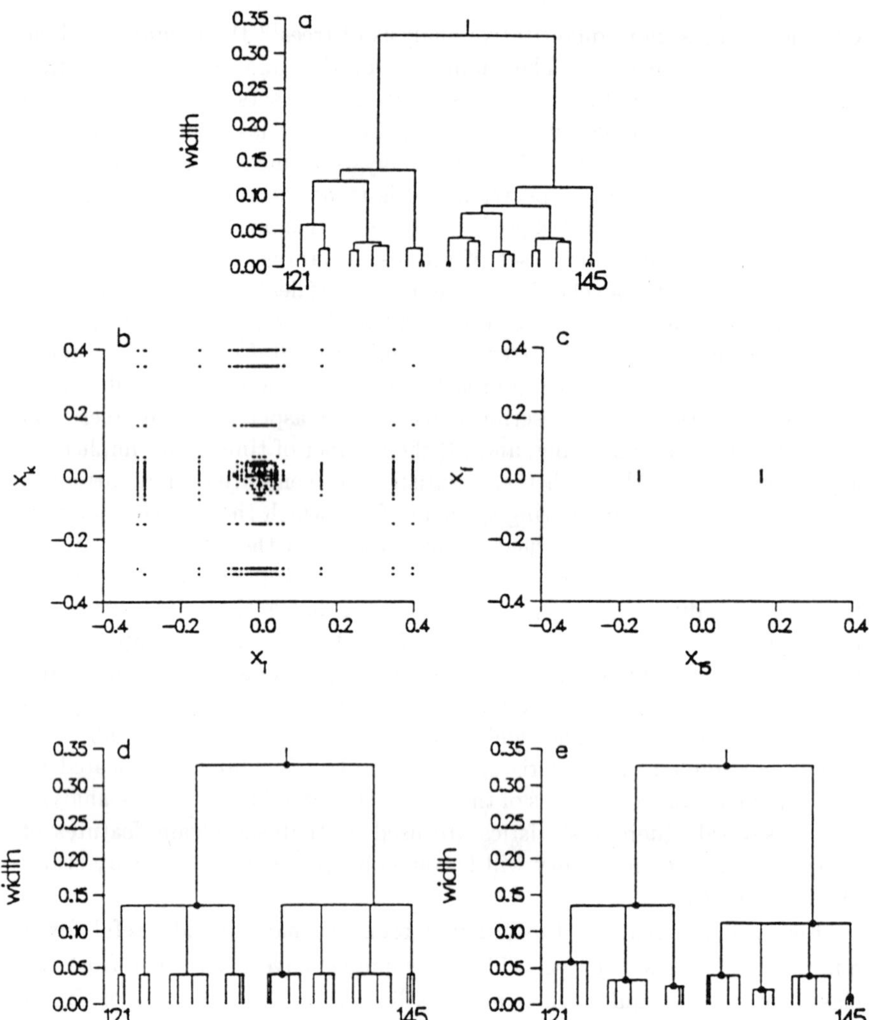

Fig. 4. This figure shows a series of plots (b)–(e) which visualize the various measures which can be applied to trees, for the subtree of Fig. 4(a), which is cut from the tree of Fig. 2(a). Figure 4(b) presents results for Critchley's geometrical representation, 4(c) for the cluster analysis, and Figs. 4(d) and 4(e) show results for Gordon's global (d) and local (e) parsimonious tree analysis.

shows results for a cluster analysis and Figs 4(d) and 4(e) shows results for the parsimonious tree analysis of Sec. III(c) of Ref. 10. These latter two plots relate to point (3) above, that is they reflect the number of distinct time scales inherent in a given spectrum.

Using a method due to Critchley,[22] one can embed a hierarchical tree with $n - 1$ nodes into an $(n - 1)$-dimensional Euclidean space. Each dimension of the geometrical representation is associated with a node of the tree and each spectral line is a point in this space. The coordinates in a given dimension for a spectral line are zero unless that node is part of the path from the top of the tree to the line. In general, the higher the node the greater the magnitude of the coordinates associated with it, but the importance of the node may be reduced if the branching at the node is not significant. For example, a node which splits m lines into two groups of $m/2$ will give rise to noticeably higher coordinate values than a node of the same relative height which splits a group of m lines into groups of $m - 1$ and 1.

Results for Critchley's geometrical representation of the subtree of Fig. 4(a) are presented in Fig. 4(b). This plot shows all unique pairwise projections of the geometrical representations of the 25 spectral lines. There is a large number of possible pairwise projections, but it is severely reduced by all the zero coordinate values noted above and the hierarchical nature of the tree (there are only 24 unique distances). Figure 4(b) shows that the pairwise projections come in groups which are arranged roughly radially around the center of the plot. There are noticeable gaps in the nodes of the subtree of Fig. 4(a) and the branching is fairly even, which leads to this result.

The reason why Critchley's method is useful for the evaluation of dimensionality is because once a geometrical representation of the tree has been found, one can discern to what extent a given number of dimensions contribute to the overall variance in the data (see Fig. 5(b)). For example, Fig. 5(b) demonstrates that approximately 99% of the variance of the data is found in the first 5 dimensions (that is, the first 5 nodes of the tree).

Figure 4(c) shows results which emphasize the nature of cluster analysis.[21] The top node of the subtree is the 15th node of the full tree, and thus all coordinates of lines 121–145 are the same for the first fourteen nodes of the tree. The first node which differs is 15 and this splits the group in two, a combination (121–131) and an overtone (132–145) group. The

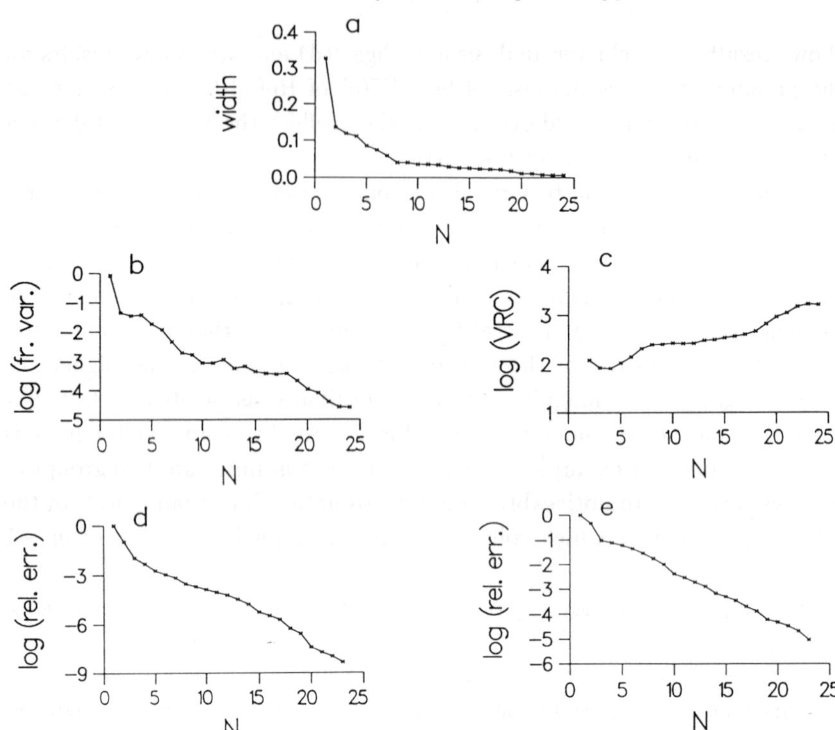

Fig. 5. A series of results for the various analysis tools of Fig. 4 are presented in this figure. The top plot (a) shows the width at which each node splits for the subtree of Fig. 4(a) and the plots (b)–(e) show results for the analysis tools of Figs. 4(b)–4(e), in the same order as that earlier figure. See text for further details.

abscissa of Fig. 4(c) shows the values of the coordinates for the 25 spectral lines in the 15th dimension, which take on two different values, -0.15 and 0.15, depending on whether the spectral lines are in the combination group on the left or the overtone group on the right. The ordinate of Fig. 4(c) is the coordinate of the highest dimension which is nonzero for each of the lines. There is strong clustering in this plot, which is defined as the ratio of the squared distance between the two groups divided by the average of the squares of the distances within the two groups.[10,23,24]

Figures 4(d) and 4(e) show results for Gordon's global and local parsimonious tree analyses,[10,25] respectively. The parsimonious trees are generated by replacing the full set of heights with a subset of these heights, called, "chosen nodes". The global method pictured in Fig. 4(d) takes these chosen nodes and acts all the way across the tree, replacing all heights between two consecutive chosen ones with the higher of the two. The local method replaces a particular node of the tree with the height of a chosen node which is closest to that node and lies on a path from the node to the top of the tree. For a given number of chosen nodes (three in Fig. 4(d) and ten in Fig. 4(e)), the particular chosen nodes are picked which minimize an error function between the parsimonious tree and the original tree.

The techniques pictured in Figs. 4(a)–4(e) can be used to analyze the trees at all levels, and Figs. 5 and 6 present such results. Figure 5(a) is a plot of the width of the nodes of the tree of Fig. 4(a) ordered in terms of node number. This plot shows the large gap apparent in the tree below the top node, but also shows some additional interesting structure (for example, a distinct inflection point at $N = 8$). As noted in Ref. 10, there are two important aspects of plots such as Fig 5(a) and those below it. The overall range of the data is important, because it indicates the range of time scales inherent in a spectrum, and the shapes of the curves are important, including large drops and inflection points, because they indicate hierarchical structure in the spectrum.

The other plots in Figs. 5 and 6 relate to those of Figs. 4(b)–4(e) in the same order, with Fig. 6 included to make the study of the shapes of the curves more apparent. This figure shows the ratios between consecutive points of the data in Figs. 5(b)–5(f). The geometrical representation of the spectrum can be employed to discern the importance of particular nodes, by judging what fraction of the variance in the data is contained in each dimension, and Fig. 5(b) shows results for this analysis. The plot presents the fraction of variance versus node number. Note that the fraction of variance is not monotonic with node number because, as noted above, coordinate values are not strictly monotonic with node height. This plot shows results which are analogous to those of Fig. 5(a), there being a large drop after the first node and some changes in slope at higher values of N. This is shown more explicitly in the ratio plot of Fig. 6(a).

The results of a complete cluster analysis are shown in Figs. 5(c) and 6(b), with the y-axis showing a clustering statistic.[10,24] Note that $N = 2$ (the first point on the plot) is a relative maximum and that there are several rapid rises to plateaus (starting at $N = 7$ or $N = 8$, and $N = 22$). These observations indicate, for example, that there are two groups of lines and within these two groups there are additional good groupings yielding a division into seven or eight groups. The data may not be sufficient to distinguish between seven or eight groups, but a study of the two subtrees

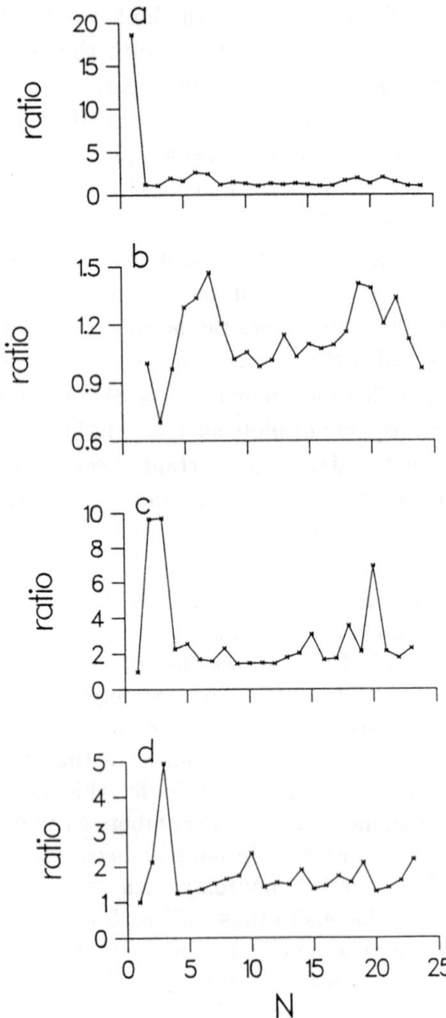

Fig. 6. This figure shows a series of plots meant to elucidate the information in Fig. 5 more fully. The plots here [(a)–(d)] refer to (b)–(e) of Fig. 5. These are plots of the changes in the various measures. Figure 6(a) shows the ratio of the ordinates of Fig. 5(b) at the current value of N to their value at $N + 1$, and the other plots show the current value to previous values ratios. The first numbers in (b)–(d) are set to one, as is the last one of Fig. 6(a).

from the $N = 2$ grouping (121–131 and 132–145) indicates that 121–131 is best viewed as three groups and 132–145 as four groups, indicating seven groups for 121–145.

Results for Gordon's global parsimonious trees are summarized in Figs. 5(d) and 6(c). There is a rapid drop to $N = 3$, which indicates three distinct time scales. The parsimonious tree generated at this point is pictured in Fig. 4(d). Like several of the plots shown above, this parsimonious tree indicates 8 groups of lines (one can count eight groups under the lowest height in 4(d)). Although the cluster analysis points to seven groups when the subtree is divided into two smaller subtrees, the global parsimonious trees still point to eight groups under this condition.

The local parsimonious analysis is presented in Figs. 5(e) and 6(d). The implications of this plot can be understood by studying the local parsimonious tree of Fig. 4(e). This tree corresponds to the $N = 10$ points of Figs. 5(e) and 6(d), which are considered important based on the shape of the curves there. One can observe in Fig. 4(e) that $N = 10$ indicates seven distinct groups. Similar comparisons would show that the $N = 3$ points of the plots indicate two distinct groups, because the three chosen nodes are the top node of the tree and the next two, which delineate two distinct groups.

The discussions of the previous few paragraphs show some of the strengths and weaknesses of the analysis. They should also make clear why we prefer several types of analyses. The analyses are exploratory and definite conclusions may be difficult to make, indicating the need for several different types of information. However, the dynamics of a particular system as reflected in a spectrum may not be particularly simple, and it is often difficult to make definitive conclusions under any circumstance. It also might seem to the reader at this point that the previous discussions were an exercise in minutiae. However, the assignment of groups of spectral features depends on the resolution of the groups and interesting groupings are important for assignments, although the difference between seven and eight groups is usually not critical. Section 5 will revisit the subtree studied here, and the assignment of some of the groups will be discussed.

Figure 2 presented three ways of accomplishing the "divide and conquer" strategy of the hierarchical analysis. Figure 7 shows results of Critchley's analysis for these manipulations, with the same ordering as Fig. 2. Figures 7(a) and 7(b) once again emphasize the utility of studying subtrees. The plot of the full tree shows evidence of the splitting of the overtones (between $N = 10$ and $N = 11$) and the combinations (between $N = 23$ and $N = 24$), but the structure is not particularly distinct. However, Fig. 7(b)

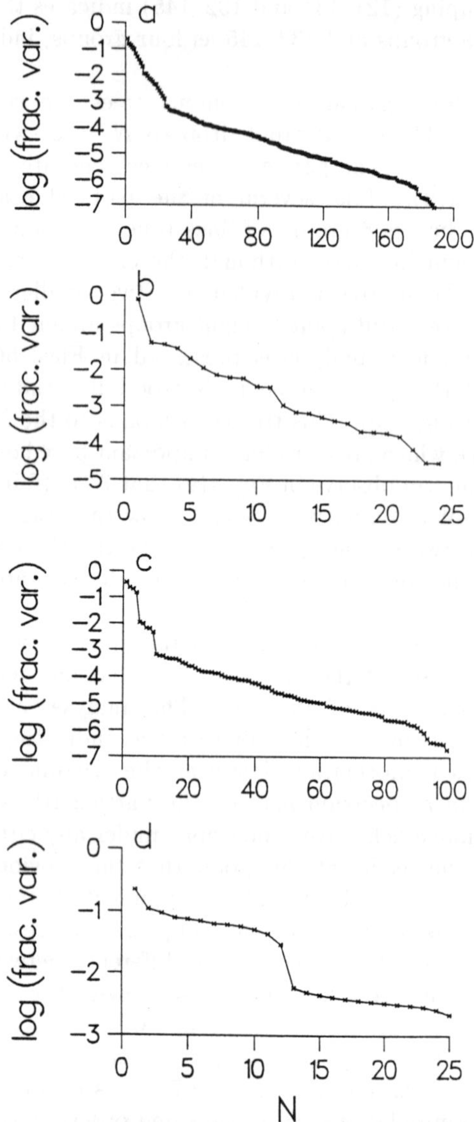

Fig. 7. These plots show the value of Critchley's fraction of variance of each node for the trees of Fig. 2 in the same order as the trees of that previous figure. The x-axis labels (N) refer to the ordering of the variances with their value (from high to low), and not to node number.

shows a large drop below $N = 1$, indicative of the splitting of the initial overtone into a combination and an overtone, and this is much more distinct than what is observed in Fig. 7(a).

Figure 7(a) is another example of the difference between the direct study of nodal gaps and the investigation of results generated from Critchley's geometrical representation. The gaps between the nodes in Fig. 2(a) are much more distinct than what is observed in Fig. 7(a). This occurs because the full spectrum encompasses a large change in local density of states. The number of states under an overtone range from one for the ground vibrational state to 28 for the last full overtone (146–173, see Sec. 5). This change in density of states is reflected in the branching, which effects the values of the coordinates of the lines and is ultimately reflected in the statistics in Fig. 7(a).

The effect of change in density of states can be compensated for in several ways, and the manipulations of Fig. 2 are meant, in part, to do this. Those in Figs. 2(b) and 2(c) generate smaller portions of the spectrum and the one in Fig. 2(d) eliminates a large portion of the states. Other reasons for using these manipulations include the fact that time scales may change with energy, regardless of the change of density of states, because the dynamics may change with energy. The first two manipulations generate trees for narrower ranges of energy, and the changes in the dynamics may be less severe over these narrower ranges. The third manipulation views dynamics over a reduced time scale, and the dynamics is more likely to be similar over a larger energy range under this condition. Results for the last two manipulations are presented in Figs. 7(c) and 7(d). Both these plots show very distinct drops. The first (7(c)) shows drops due to both the initial overtone and the subsequent combination/overtone. The second (7(d)) shows only the initial overtone split, because that is all that is left after higher resolution information has been eliminated. Hierarchical structure in these two plots is much more distinct than it is in Fig. 7(a).

Figure 8 shows a series of plots that we have found useful for summarizing all the statistics. We usually generate such "summary" sheets for all spectra we investigate for the full tree and all subtrees with more than a minimum number of lines, usually nine. Similar summary sheets can also be generated for the types of manipulations illustrated in Figs. 2(c) and 2(d).

The first column of Fig. 8 shows plots of width (top) or data from Critchley's analysis. The middle column shows clustering statistics and the third column shows results for the parsimonious trees. These plots are

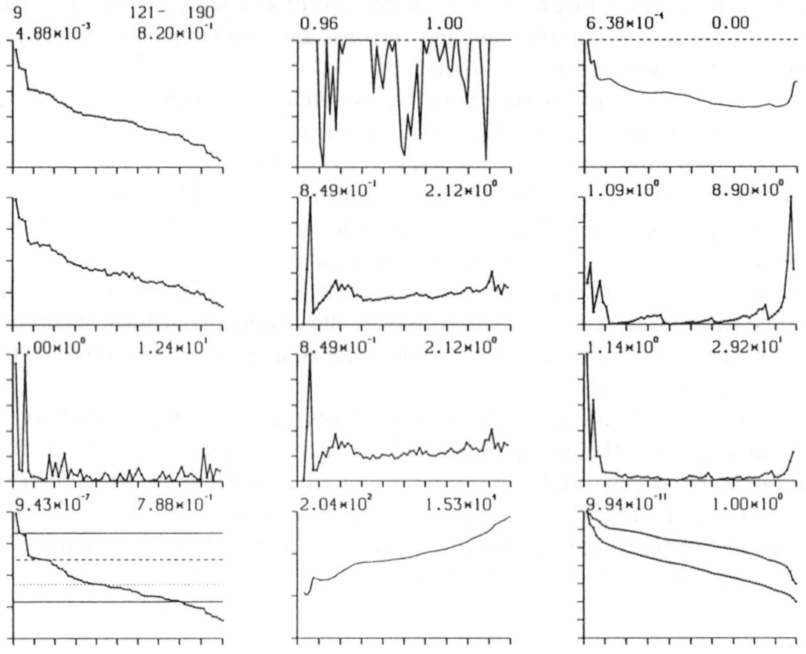

Fig. 8. A summary sheet for the measures of Ref. 10 for a subtree (lines 121–190) of the tree of Fig. 2(a). The text has an extensive discussion of this figure.

for a subtree of the tree of Fig. 2. The numbers on the top of column 1 indicate that this subtree is cut above the ninth node of the tree and includes the lines 121–190. All plots in Fig. 8 have their minimum and maximum ordinates printed on the top, except for one (first column, second from the top), which has the same range as the plot at the bottom of the column. All plots have logs along the y-axis, except for the top plot of column 2. The x-axes all have the same range, 0.0 to 1.0, which is the fraction of node or group number. For example, the value at the first tick mark is one-tenth of 68, the number of nodes included in the calculation (see Ref. 11; the 174th and 175th lines are not resolved, giving 68 nodes out of a possible 69).

In column 1 the top plot shows width, the second plot from the top shows fraction of variance, and the bottom plot shows an ordered version of the second plot. In addition, the bottom plot has horizontal lines across it which mark cumulative variance (the sum of the variances up to that point). The third plot of column 1 shows the ratio of the current variance to that of the next one for the bottom plot.

The second column shows results for the clustering criterion outlined above and the optimal clustering presented in Sec. III.B. of Ref. 10. In the bottom plot of the column, the values of both are plotted on top of each other with different line types, but have nearly the same value for all N (a dashed line is used for the non-optimal clustering). The top plot of column 2 shows the ratio of the two, which again indicates they have nearly the same value, 0.96 being the minimum non-optimal to optimal ratio. The other two plots in column 2 show ratios similar to what was plotted in Fig. 6(b). These are for the optimal clustering in the plot second from the top and for non-optimal clustering in the other plot. Both plots show the ratio of current to next, whereas Fig. 6(b) shows current to previous, with $N = 2$ set to one there. The first points on these plots have values less than one, which indicates that $N = 2$ is a relative maximum (the plots start at $N = 3$).

Column 3 shows results for the parsimonious tree analysis. The bottom plot shows both the relative error of the global (bottom curve) and local (top curve) analyses. The top column shows the log of the ratio of the local error to the global error and the middle two plots show the drop in error between consecutive points (ratios of error of current to previous, starting at $N = 2$). Results for the global parsimonious trees are in the lower of these two curves.

The data presented in Fig. 8 point to the three types of groups noted previously for the subtree containing lines 121–145 (Figs. 4–6). All the plots in Fig. 8 show these to some degree, but there are small differences in the third grouping as implied by the different measures. For example, the cluster and geometrical analyses point to 18 groups, whereas Gordon's local method picks 17. This difference is similar to the one for the 121–145 group (7 versus 8, see the above discussion).

Because the results for the three columns of Fig. 8 are similar, we only discuss those for the cluster analysis of column 2 in more detail. In the bottom plot of column 2 there is a local maximum at $N = 2$, indicative of two overtone states (121–145 and 146–173). There is another local maximum at $N = 5$ for two combination/overtone groups (121–131 and 132–145, as well as 146–160 and 161–173) and a combination state without an overtone (174–190). There is also a weak ledge at $N = 18$, indicative of the previously described grouping. This grouping is supported by the 121–145 overtone, which was judged to be divided into seven groups, because the 18 groups include the division of 121–145 into seven groups. We will investigate the assignment of some of these groups in Sec. 5, and this will demonstrate that the third sequence of groups is significant (see Figs. 15 and 16).

4. A Comparison to an Experimental Spectrum

A vibronic spectrum of NO_2[26] was studied in Ref. 10. The tree generated from that spectrum is presented in Fig. 9(a), and various subtrees are plotted in the other panels of the figure. The subtrees include a larger one (9(b)), and four smaller subtrees generated by cutting all the way across the larger subtree above its fourth highest node. One can observe that there is a large gap below the second highest node and another one below the ninth node for the tree in Fig. 9(a). These gaps indicate that there are first three distinct groups and then 10. Reference 10 suggested that the first grouping of three peaks is indicative of short time motion on the electronically excited B_2 surface before there is sufficient time[27] for coupling with the ground A_1 surface from which the molecule is originally excited. It is the coupling between the two surfaces which is generally accepted to cause much of the complexity in the excitation spectrum of NO_2 (see Ref. 26 for a discussion and a listing of previous studies). The three peak structure is most likely connected to symmetric stretch/bend motion on the electronically excited surface, based on the spacing of the peaks. This type of progression is expected, based on the displacement of the two electronic surfaces from each other.[27]

The ten peak case is resolved on a time scale which is longer than the previously noted time scale for internal conversion, so this case probably involves coupling between the two surfaces. At first glance, the spacings of the ten peaks do not appear to be related to any type of progression.[10] However, as noted above, we have found secure assignments for peaks in our theoretical studies of both model and experimental spectra even when there was no perceived regularity in peak spacings, particularly if the gaps between nodes are large.

Information concerning the subtree of Fig. 9(b) is summarized in Fig. 10. Most of the panels in Fig. 10 show evidence of the previously noted 10 peak case. For example, there is a rather distinct maximum at $N = 4$ in the

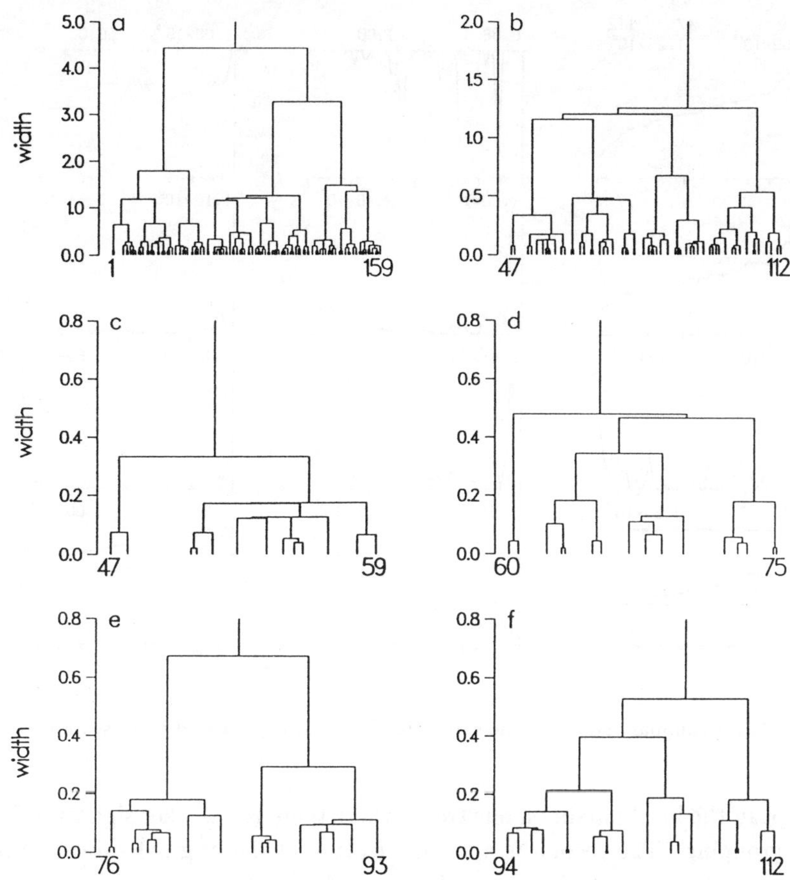

Fig. 9. A tree for the NO$_2$ vibronic spectrum is shown in Fig. 9(a) and subtrees are shown in Figs. 9(b)–9(f). The subtree of Fig. 9(b) was cut from the tree of 9(a) above its third highest node, and the subtrees of 9(c)–9(f) are a complete set of subtrees cut from the subtree of 9(b) above its fourth highest node. The widths are scaled as described in Ref. 10.

ratio plots of the cluster analysis in the middle panels of column 2 (the first point is once again at $N = 3$). These four peaks are part of the 10 peak case for the complete spectrum. Most of the plots in Fig. 10 show two other distinct instances of strong clustering at relatively low values of N. The first is a set of 13 groups which appears strongly in all measures,

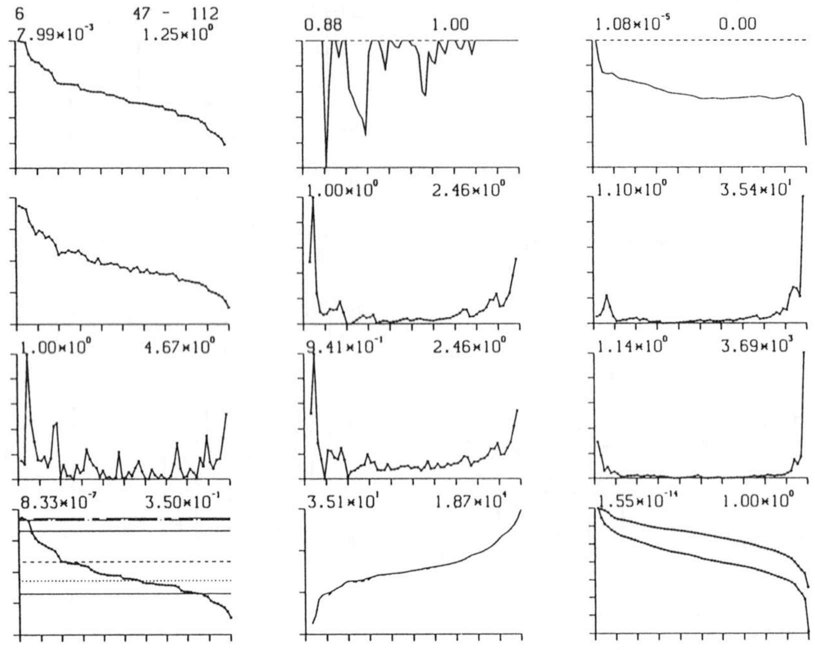

Fig. 10. A summary sheet for the subtree of Fig. 9(b) is presented here (see Fig. 8).

except in the local parsimonious trees, where there is a weaker signature for this grouping. The second is a weaker feature of the original subtree, the geometrical representation, and cluster analysis. In the measures which do not take into account branching (the width plot and the sub-optimal clustering) there is a signature of 20 groups and in the two other measures, which do take into account branching (Critchley's measure and the optimal clustering), this shows up more strongly as 22 groups. However, both the width plot and the sub-optimal clustering show weaker features associated with 22 groups. Also, when the smaller subtrees of Figs. 9(c)–9(f) are investigated, it appears that 22 groups is a better way to view the data than 20. Therefore, we conclude that 22 is the better grouping, although 20 is close and should yield very similar results concerning assignability and energy transfer. Like the 10 peak case for the full spectrum, we are unable at the present to find any definitive assignment for the 22 peak case. However, this grouping suggests to us that there are distinct bottlenecks to energy flow in this portion of the spectrum and assignability of the features if the eigenstates of NO_2 could be used to generate smoothed states as they are for the model system studied in Secs. 2 and 5. In fact, Sec. 5 demonstrates that smoothed states generated from as few as three eigenstates lead to regular progressions, although the smoothed states are complicated (see Figs. 15 and 16).

Comparisons between the NO_2 spectrum and the one of Fig. 1(a) are made in Figs. 11 and 12. Several important caveats need to be made here before the comparisons are discussed. First, the NO_2 spectrum is affected strongly by the coupling between two potential surfaces, compared to the model system which involves a single surface. In addition, the model system involves fewer degrees of freedom than NO_2 and the NO_2 spectrum is estimated to be missing many lines[26] (approximately 50 out of 210). These lines are assumed to be missing because they are too weak to be detected and their presence would cause all the statistics studied here to have a larger range (the weaker lines would extend the time scale range for resolution of all the lines and result in all statistics having a larger range).

The top plots in Figs. 11 and 12 show results for the full tree and below them are results for various of the manipulations of Fig. 2. Figure 11 presents results for Critchley's geometrical representation and Fig. 12 shows results for the cluster analysis. Comments concerning these plots are centered on one of the aspects noted in the discussion of Fig. 5 in Sec. III; the overall range in a given measure. For Critchley's measure this relates to the dimensionality inherent in the data. The idea behind viewing the data in this manner is to determine how much of the variance in the data is contained in a reduced number of dimensions.

Figure 11(a) shows that the fraction of variance for the tree of the model system is spread out more evenly than that of the NO_2 spectrum, indicating that the NO_2 spectrum is a lower dimensional data set than the spectrum of the model system. The cluster analysis of Fig. 12 further indicates that the tree for the NO_2 spectrum displays more clustering than the tree of the model system. Both these results indicate that the NO_2 spectrum is, in some sense, simpler than the spectrum of the model system. Since the eigenstates of the model system show significant localization,[11] this suggests to us[10] that "chaotic" is an inappropriate description of the NO_2 spectrum.[26(b)]

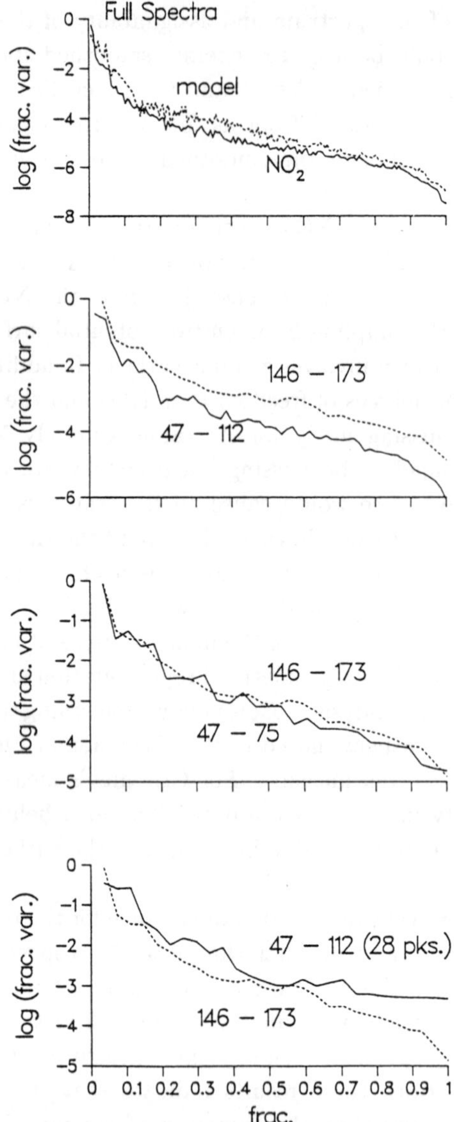

Fig. 11. Various comparisons between the spectrum of the model system and that of NO_2 are presented here, for fraction of variance. The text has further details.

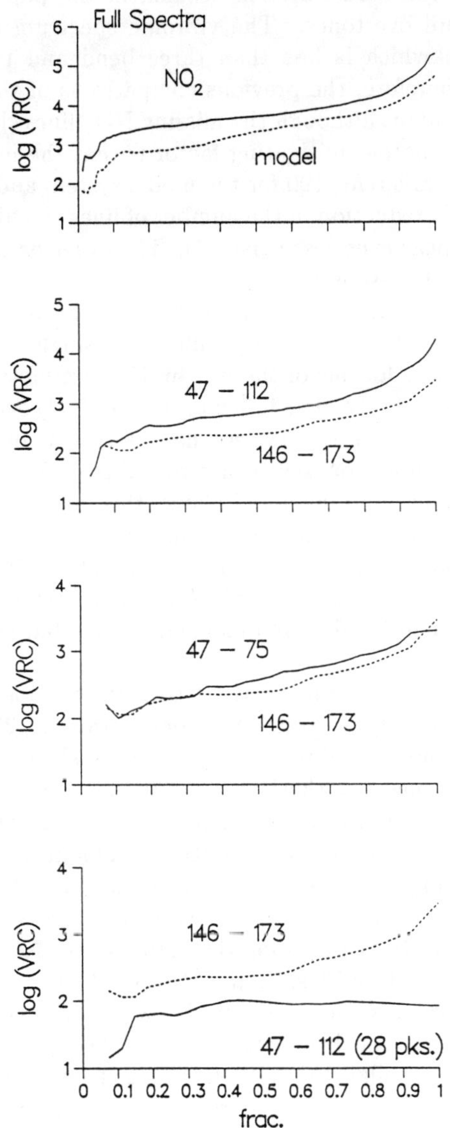

Fig. 12. Comparisons of the same sets as those of Fig. 11 are made here, for the clustering measure.

The spectrum of the model system studied in the paper has a large range, covering 13 full overtones. The vibronic spectrum of NO_2 has a range of 2000 cm^{-1}, which is less than three bend and two symmetric stretch overtones. Therefore, the previous comparisons of the two spectra might be deemed as unfair, although the missing NO_2 lines should mitigate such an assessment. Another mitigating factor is that the number of lines in the two spectra are different; 190 for the model system and 159 for NO_2. From our experience, a reduction in the number of lines would tend to make the model system appear even less clustered. The other panels in Figs. 11 and 12 address some of these issues.

The second plots in Figs. 11 and 12 show results for a subtree of NO_2 (47–112). This subtree is taken to be similar to a single overtone of the model system and results for one of these (146–173) are plotted along with the NO_2 subtree results. These plots indicate that the NO_2 overtone subtree is of lower dimensionality (Fig. 11) and has more clustering (Fig. 12). However, this is a comparison between two subtrees with significantly different numbers of lines, and, as noted above, this may cause difficulties. Therefore, a third comparison is made which involves a smaller NO_2 subtree (47–75). This new comparison is shown in the plots third from the top in both Figs. 11 and 12. Once again these plots show that a piece of NO_2 spectrum has lower dimensionality and more clustering, but now the results are much more similar.

A final comparison is made using the tree manipulation of Fig. 2(d). The 47–112 subtree of NO_2 is replaced by one with only 28 lines at the bottom, representing the level of resolution needed to show 28 peaks. This is the same number of peaks as there are lines in the subtree of the model system. The results are shown on the bottoms of Figs. 11 and 12. Under these conditions the subtree of the NO_2 spectrum is of higher dimensionality and shows less clustering than the subtree of the model system.

Figures 11 and 12, along with other results we have studied, do not change our basic conclusions concerning the nature of the NO_2 spectrum.[10] However, the bottom panels of Figs. 11 and 12 suggest some caution. Based on experience with systems for which we know the eigenstates (Secs. 2 and 5, for example), our conclusion is that the NO_2 spectrum indicates a level of localization similar to that of the model system, which has significant mixing within its original coordinate system, but shows a high degree of localization along new sets of coordinates (see Figs. 15 and 16, for example). This statement is made with the caveats presented in the previous discussions.

5. Energy Transfer and Assignability

Figure 13 presents a series of overtone subtrees for the model system. Results for all these subtrees have been presented in previous sections of the paper. There are markings on two of the trees; dots on Fig. 13(b) and short dashed lines on Fig. 13(c). The dots show a specific path down the tree. They lie on nodes which define groups of states which are summed to form "smoothed eigenstates" (Sec. 2 and Ref. 9). For example, the third dot from the top defines the group 96–101.

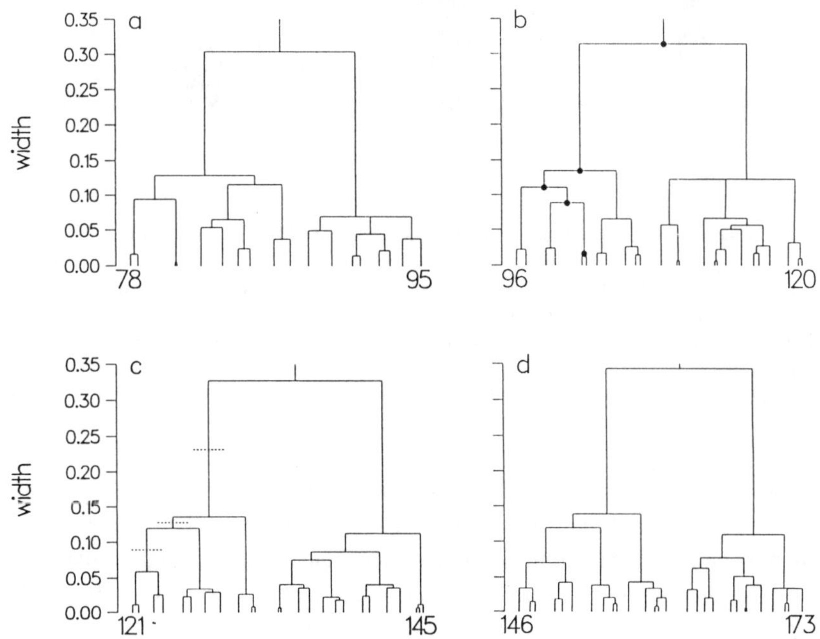

Fig. 13. A set of overtone subtrees for the spectrum of Fig. 2(a) is shown. The dots in 13(b) show a path down the tree and the short, dashed lines on 13(c) show a set of local cuts of the subtree.

The smoothed states for the path of Fig. 13(b) are shown in Fig. 14. These plots illustrate the energy transfer which occurs down this particular path of the tree. One way of thinking about these plots is that they are showing one of the decay pathways of an overtone state (top left in the figure). There are several significant features of energy transfer indicated by the states in Fig. 14. The first is evident in the plot in the middle of

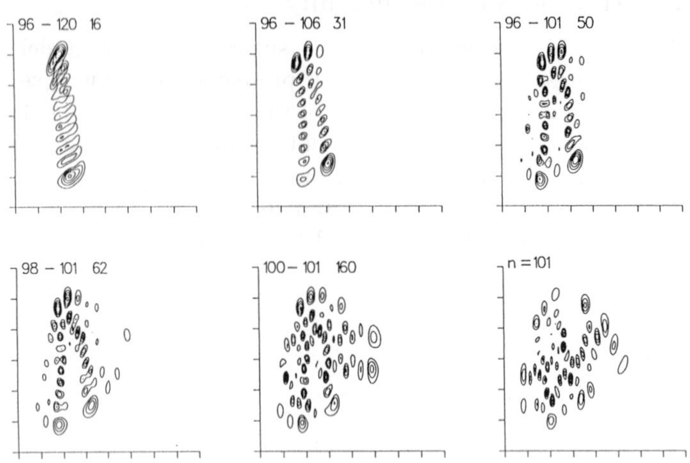

Fig. 14. A set of smoothed states down the path indicated by the dots in Fig. 13(b). The order of the plots is left to right on the top and then the same direction on the bottom. The numbers separated by the dashes show ranges of lines used to generate the states and the isolated numbers show the node number as counted from the top of the full tree of Fig. 2(a). An exception to this numbering is the plot on the bottom right, which shows the 101st eigenstate. These plots demonstrate an energy flow pathway present in this system, which can be described alternatively as a decay pathway of the overtone state of Fig. 14(a).

the top row (96–106), which is the familiar combination state. This state represents significant energy transfer, because it involves the formation of a node perpendicular to the major part of the density of the original overtone state. Additional energy transfer is indicated in the next two plots by the movement of density further from the original overtone axes, and by a significant distortion of the nodal patterns.

One way to interpret[11] the first four plots in Fig. 14 is the following. An initial wave packet is excited and begins to move on the potential surface in a region of phase space dominated by overtone motion. At sufficiently long times, portions of the wave packet explore other parts of the overtone region near its boundary, and evidence for this is the state in the middle of the top row (the region is large enough to contain one or two states, depending on energy). At still longer times, a portion of the wave packet has moved into a region dominated by a 2:1 Fermi resonance, and evidence of this is the morphology of the state on the lower left of Fig. 14.

The final two plots in Fig. 14 show the following. The plot in the middle of the bottom row shows evidence of exploration of a subregion of phase space within the region of phase space encompassing the 2:1 resonance. This conclusion is based partly on the study of the phase space representations of the states.[11] The final plot in Fig. 14 on the bottom right is the 101st eigenstate. The eigenstate lies outside of both the 2:1 region and the overtone region, indicating that a piece of the wave packet has entered a third significant region of phase space (there are four significant regions of phase space in this system: overtone, 2:1 resonance zone, 5:3 resonance zone, and a 3:2 resonance zone).

One can also address issues of assignability by studying sequences of subtrees. We have generally done this by making a cut above a particular node of the tree all the way across the tree, but an alternative is suggested in Fig. 13(c). The dashed lines in this plot give a series of groups: 121–131 and 132–145; 121–128, 129–131, and 132–145; 121–124, 125–128, 129–131, and 132–145. These sets are different from those generated by making cuts all the way across the subtree. One can make an approximate assignment of the three groups within the combination portion of the subtree (121–131). This progression is in addition to the first two progressions studied previously in this paper (overtone progression and combination progression). Evidence for these have been presented in several other plots in the paper (e.g., Fig. 5), and have been discussed in previous sections. They are most obvious under the last two subtrees of Fig. 13, for reasons discussed below.

Figure 15 shows a set of 12 smoothed states chosen from the combination portions of the subtrees of Fig. 13. These states were generated from groups resolved by cutting all the way across the subtree above the third highest nodes of the combination subtrees. It is clear from inspection of the combination halves of the subtrees of Fig. 13 that three groups represent strong groupings for only the last two subtrees, and this is borne out by also studying summary sheets for the combination subtrees. Note that the first combination subtree (78–87) shows a weak tendency for three groups, but a stronger one for four.

The states in Fig. 15 are arranged in two types of progressions. As one goes down a column the states have an increase in one type of node (along a hyperbola or parabola) and as one goes across a row there is an increase in the number of nodes perpendicular to the first coordinate. However, the top two rows do not show a complete progression of three states, there being only two states in the progression at these energies. An important aspect

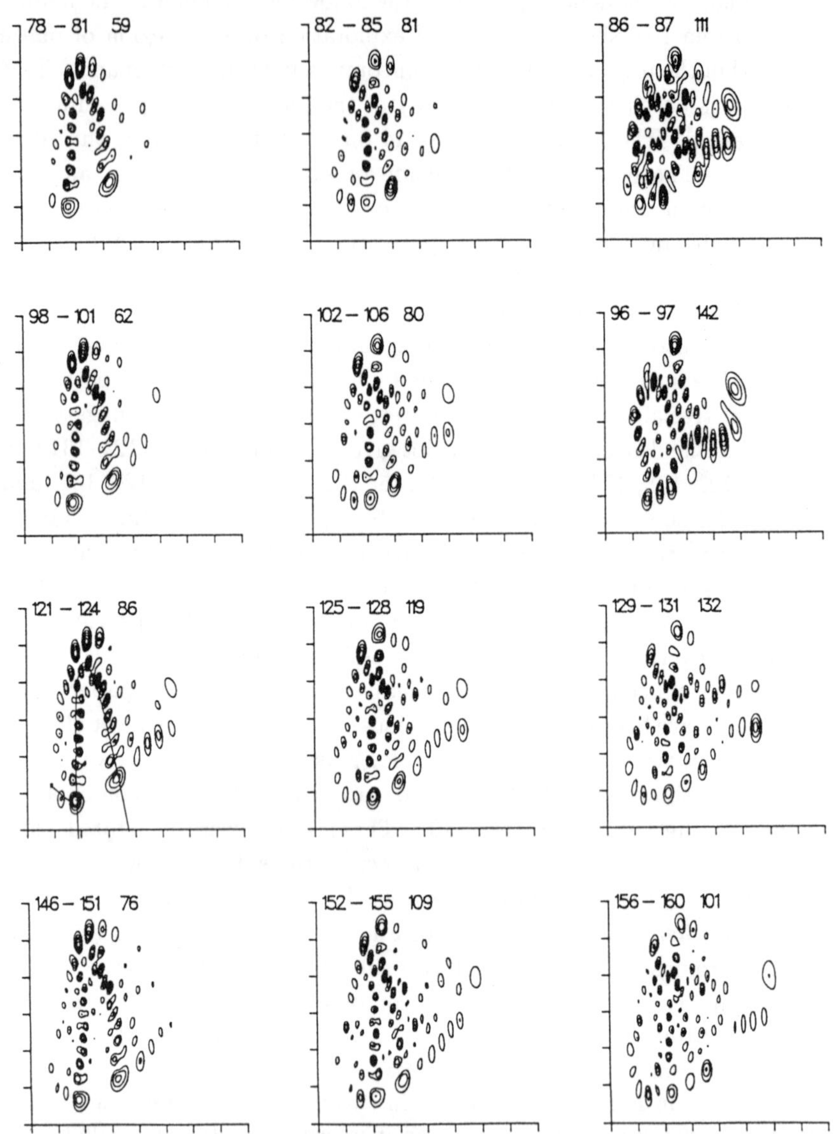

Fig. 15. A series of plots generated for the combination portion of the subtrees of Fig. 13 (the left half of each of these subtree). The text has further details.

of the first two rows is that the additional state is the last one in terms
of spectral line numbers in the first row (86–87 out of the group 78–87),
but is the first one in the next row (96–97 out of 96–106). This indicates
to us a crossing between smoothed states, something which often happens
for individual eigenstates due to anharmonicity, and generally results in
an avoided crossing in parameter space. This crossing in energy space is
interesting, because it means that a spectrum may become locally very
complicated, but may be much simpler on either side of the crossing. The
crossing can also be observed in the subtrees of Fig. 13. In the first subtree
(a), the first node of the combination subtree splits off to the right, but the
first nodes of the other combination subtrees (b)–(d) split to the left.

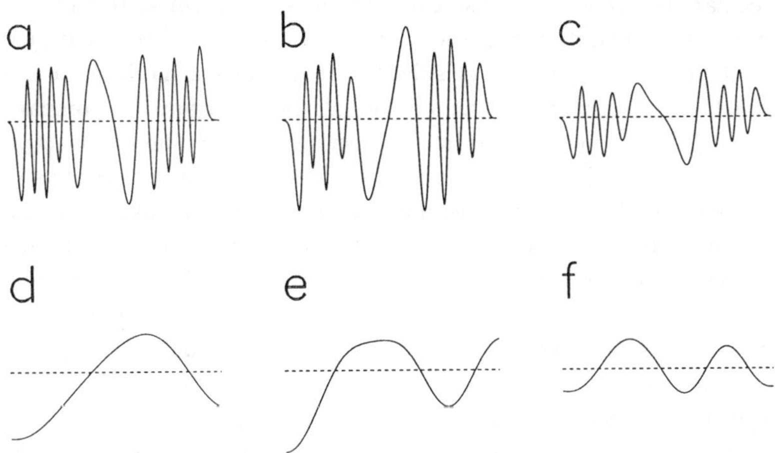

Fig. 16. Values of the wave functions in the third row of Fig. 15 along paths like the one
plotted on the state in the first column of the third row. The top row (a)–(c) shows the
wave functions along the parabolic paths and the bottom row along straight line paths
like the one drawn on the state. The plots in this figure correspond to moving left to
right on the third row of Fig. 15, and demonstrate approximate assignments of these
states as: (19,2), (17,3), and (15,4), respectively.

Approximate assignments of the three states in the third row of Fig. 15
can be made by choosing curves like those for the 86th smoothed state
(numbered by node height), which contains the 121st–124th lines, and these
curves are shown in Fig. 15. We use a parabola which is adjusted to cut
through one type of nodal path and a straight line to cut through the other
type node. This procedure is used for all states in the third row and plots

of the results are presented in Fig. 16. These plots show the wave function along the paths for the parabolas in the top row and the straight lines in the bottom row. The plots in Fig. 16 have the same ordering as the states of the third row of Fig. 15. Each state of the progression has two fewer nodes in the top row and one more node in the bottom row, moving from left to right. The 2:1 resonance nature of these states is evident from the replacement of two nodes along one coordinate with only one node in the other coordinate.

6. Conclusion

The hierarchical analysis reviewed here allows for a systematic study of molecular spectra at all levels of resolution. A hierarchical tree provides a convenient method to "divide and conquer" a spectrum, because it pictures a spectrum as a nested set of subsets. It provides a method for both a detailed characterization of an individual spectrum and for comparisons between different pieces of the same spectrum or two different spectra.

We view the hierarchical approach as a very useful way to study spectra which may be difficult to understand using traditional assignment procedures. In addition, we believe that a hierarchical tree provides a framework consistent with many approaches to the study of IVR. For example, it is consistent with the approach common to the spectroscopy literature, where energy flow is modeled via the coupling between manifolds of states.[12] It is also consistent with another model; energy flow occurring via the movement between regions of phase space separated by intramolecular bottlenecks.[28,29] The hierarchical approach provides information concerning energy transfer and assignability, by extracting time scales and by pointing to important, distinct features of smoothed spectra at specific levels of resolution. Significantly, the approach is nonparametric, it does not depend on a particular model of a spectrum (normal mode, local mode, etc.).

In addition to the information available for interpretation of experimental spectra, the hierarchical analysis of theoretical spectra generated form the eigenstates of a system can be employed to get a detailed understanding of energy flow properties. Information concerning energy transfer pathways can be extracted from the study of the smoothed states associated with portions of the hierarchical tree. The hierarchical analysis can also be used to study the eigenstates of a system without reference to a particular experimental spectrum.[11]

There are several ongoing studies using the hierarchical analysis. The study of a theoretical photodetachment spectrum of OHCl generated by Koizumi and Schatz[30] is underway,[31] as well as the study of a theoretical infrared spectrum of HO_2,[32] generated by Gazdy, Chapman, and Bowman.[33] Also, the study of a new set of SEP spectra[34] of acetylene is planned and it is anticipated that comparisons will be made to recent analyses of previous SEP spectra of acetylene.[8] All these studies are focused on vibrational spectroscopy, while an ongoing study involves the analysis of Perry's model[35] of IVR in his group's experimental spectra,[36] which involves the effect of vibration–rotation interactions.

Acknowledgment

This work was supported by the Office of Basic Energy Sciences, Division of Chemical Sciences, U.S. Department of Energy, under Contract No. W-31-109-ENG-38.

References

1. C. E. Hamilton, J. L. Kinsey, and R.W. Field, *Ann. Rev. Phys. Chem.* **37**, 493 (1986).
2. See, for example, J. M. Hollas, *High Resolution Spectroscopy* (Butterworths, London, 1982).
3. R. D. Levine, *Adv. Chem. Phys.* **70**, 53 (1988), and references cited therein; W. Karrlein, *J. Chem. Phys.* **94**, 3293 (1991), and references therein.
4. Y. Chen, S. Halle, D. M. Jonas, J. L. Kinsey, and R. W. Field, *J. Opt. Soc. Am.* **B7**, 1805 (1990), and references therein.
5. J. P. Pique, *J. Opt. Soc. Am.* **B7**, 1816 (1990), and references therein.
6. J. M. Gomez Llorente, S. C. Farantos, O. Hahn, and H. S. Taylor, *J. Opt. Soc. Am.* **B7**, 1851 (1990) and references therein.
7. (a) K. Yamanouchi, S. Takeuchi, and S. Tsuchiya, *J. Chem. Phys.* **92**, 4044 (1990); (b) K. Yamanouchi, N. Ikeda, S. Tsuchiya, D. M. Jonas, J. K. Lundberg, G. W. Adamson, and R. W. Field, *J. Chem. Phys.* **95**, 6330 (1991), and references therein.
8. D. M. Jonas, S. A. B. Solina, B. Rajaram, R. J. Silbey, R. W. Field, K. Yamanouchi, and S. Tsuchiya, *J. Chem. Phys.* **97**, 2813 (1992).
9. M. J. Davis, *Chem. Phys. Lett.* **192**, 479 (1992).
10. M. J. Davis, *J. Chem. Phys.* **98**, 2614 (1993).
11. M. J. Davis, "Hierarchical analysis of molecular spectra: Further considerations and a study of energy transfer", *J. Chem. Phys.* (to be submitted).
12. See, for example, T. Uzer, *Phys. Rep.* **199**, 73 (1991), and references therein.
13. Z. Bacic and J. C. Light, *Ann. Rev. Phys. Chem.* **40**, 469 (1989).

14. E. L. Sibert, *Int. Rev. Phys. Chem.* **9**, 1 (1990).
15. J. M. Bowman, ed., *Advance in Molecular Vibrations, Vols. IA and IB* (JAI Press, Greenwich, CT, 1991).
16. J. Zhang and D. G. Imre, *J. Chem. Phys.* **90**, 1666 (1989); S. L. Tang, E. H. Abramson, and D. G. Imre, *J. Phys. Chem.* **95**, 4969 (1991).
17. J. M. Gomez Llorente, F. Borondo, N. Berenguer, and R. M. Benito, *Chem. Phys. Lett.* **192**, 430 (1992).
18. A. D. Gordon, *Classification* (Chapman and Hall, London, 1981).
19. A. D. Gordon, *J. R. Statist. Soc.* **A150**, 119 (1987).
20. K. V. Mardia, J. T. Kent, and J. M. Bibby, *Multivariate Analysis* (Academic Press, London, 1979).
21. A. K. Jain and R. C. Dubes, *Algorithms for Clustering Data* (Prentice-Hall, New Jersey, 1988).
22. F. Critchley, in *Classification and Related Methods of Data Analysis*, ed. H. H. Bock (Elsevier Science Publishers, The Netherlands, 1988).
23. H. P. Friedman and J. Rubin, *J. Am. Statist. Assoc.* **62**, 1159 (1967).
24. T. Calinski and J. Harabasz, *Comm. in Stat.* **3**, 1 (1974).
25. A. D. Gordon, *J. Classification* **4**, 85 (1987).
26. (a) A. Delon and R. Jost, *J. Chem. Phys.* **95**, 5686 (1991); (b) A. Delon, R. Jost, and M. Lombardi, *ibid.* **95**, 5701 (1991).
27. K. K. Lehmann and S. L. Coy, *Ber. Bunsenges. Phys. Chem.* **92**, 306 (1988).
28. See, for example, R. A. Marcus, W. L. Hase, and K. N. Swamy, *J. Phys. Chem.* **84**, 73 (1984), and references therein.
29. M. J. Davis, *J. Chem. Phys.* **83**, 1016 (1985).
30. H. Koizumi, Ph.D, Thesis, Northwestern University, 1991; H. Koizumi and G. C. Schatz, to the submitted.
31. H. Koizumi, G. C. Schatz, and M. J. Davis, to be submitted.
32. D. Chapman, B. Gazdy, J. M. Bowman, and M.J. Davis, to be submitted.
33. D. Chapman, B. Gazdy, and J. M. Bowman, to be submitted; for earlier work on HO_2 by this group, see D. Chapman, J. M. Bowman, and B. Gazdy, *J. Chem. Phys.* **96**, 1919 (1992), and references therein.
34. S. A. B. Solina, R. W. Field, and R. J. Silbey (in preparation).
35. D. S. Perry, *J. Chem. Phys.* **98**, 6665 (1993), and "Random matrix treatment of intramolecular vibrational redistribution II. Coriolis interactions in 1-Butyne and Ethanol", to be submitted.
36. A. M. Kouza, D. Kaur, and D. S. Perry, *J. Chem. Phys.* **88**, 4569 (1988); J. S. Go, G. A. Bethardy, and D. S. Perry, *J. Phys. Chem.* **94**, 6153 (1990); G. A. Bethardy and D. S. Perry, *J. Mol. Spectrosc.* **144**, 304 (1990).

CHAPTER 24

EXTRACTING DYNAMICAL INFORMATION FROM COMPLEX AND CONGESTED SPECTRA: STATISTICS, PATTERN RECOGNITION, AND PARSIMONIOUS TREES

Stephen L. Coy, David Chasman[1] and Robert W. Field

Massachusetts Institute of Technology

Department of Chemistry and G. B. Harrison Spectroscopy Laboratory

Cambridge, Massachusetts 02139, USA

Contents

---

[1]Current Address: Dept. of Chemistry, Columbia University, New York, N.Y.

Abstract

Stimulated Emission Pumping (SEP) and other double resonance techniques provide very large quantities of high-resolution, wide-dynamic-range, quantum-number-sorted spectral data. Deriving information about intramolecular dynamics without detailed line-by-line assignment of fully-resolved transitions between individual eigenstates requires the development of new methods of recognizing patterns and of identifying hierarchical levels of coupling in the spectrum. Power spectrum analysis, and methods using level statistics have been the only alternative to traditional eigenstate assignment and modeling methods for extracting dynamical behavior. We illustrate the limitations of these methods and describe two promising new techniques for spectral pattern recognition. The first, the extended autocorrelation function (XAC), allows complex patterns parameterized in a multi-dimensional way to be located in a spectrum in the presence of interfering data. We describe the motivation for this method, and illustrate its application to synthetic systems, and to C_2H_2 dispersed fluorescence data. The second, parsimonious trees, allows direct model-free identification of multiple clusters of coupling matrix element values from a spectrum in which the underlying dynamics has a hierarchical or sequential structure. For this method, we also describe the motivation, apply it to synthetic systems, and to both band origin and rovibrational data in the optical spectrum of NO_2.

1. Introduction

Stimulated Emission Pumping (SEP) has made possible the rapid accumulation of very large quantities of high quality spectral data. The data provided by SEP and a few other double resonance techniques is especially valuable because the observed spectrum originates from a purposefully selected and fully specified intermediate level and illuminates many or all of the levels in the sampled energy region of a target manifold. The intermediate and target levels may be single eigenstates in a fully resolved spectrum, or groups of eigenstates when the transitions conceal unresolved interactions. SEP is able to exploit differences between electronic state equilibrium geometries and to use intermediate state perturbations to allow systematic Franck–Condon exploration of most of the target manifold. These double resonance datasets often have effective dynamic ranges after correction for saturation of 500:1 for SEP[1,2] and up to 5000:1 for MODR.[3] Using double resonance spectra exploiting access via different intermediate levels, it is possible to obtain pure-sequence spectra, that is, sequences of levels in which observed features have common sets of the rigorously-conserved quantum numbers for all resolved interactions. An understanding of the interactions within sets of these pure-sequence levels, levels which may interact only with each other, may be translated to a detailed time-domain picture of intramolecular dynamics, as has been done for NH_3 overtone polyads.[4]

SEP spectra now cover the complete range of internal energies: from low energies previously studied by IR spectroscopy, to intermediate energies where some approximate quantum numbers are destroyed, to regions, unanalyzable in detail, which appear ergodic. No approximate or qualitative method can hope to replace the definiteness and certainty of a complete analysis of an energy region where treatment of a limited group of states as isolated is still appropriate. The analysis in progress of C_2H_2 near 7000 cm^{-1},[2,5,6] which completely analyses a group of 12 interacting vibrational states and identifies the first stage of energy flow as *trans* $\rightarrow$ *cis* bend energy transfer, is an example. The information on vibrational interactions obtained in this way may be scaled to higher energies, but the extent to which such extrapolation remains useful and valid is an important open question. The answer to this question lies in better methods for detecting the fragments of order, or the traces of early time localization, left in spectra at ever-increasing energies. Achieving a greater understanding requires the development of more rapid methods of assignment, fitting, and analysis, methods which will be alternatives to the exquisitely detailed but highly laborious eigenstate modeling approach of traditional high-resolution spectroscopy.

Polyatomic molecular dynamics for a wide range of internal energies is strongly hierarchical. Electronic, vibrational, vibronic, rovibronic, rovibrational, rotational, spin, and hyperfine interactions occur with matrix elements which range from thousands of cm^{-1} to tens of kilohertz. This hierarchy of couplings leaves traces in the spectrum of the surviving approximately conserved quantities, and of the energy ranges (or delocalization widths) over which basis states for the mostly-broken quantum numbers are distributed. The extended autocorrelation function (XAC) is designed to detect and locate those remnants of order. It allows complex patterns parameterized in a multi-dimensional way to be located in a spectrum in the presence of interspersed unrelated patterns. We describe the motivation for the XAC method, and illustrate its application to synthetic systems, and to C_2H_2 dispersed fluorescence data. The parsimonious trees method of analysis can estimate the delocalization widths. It allows direct identification of multiple average coupling matrix element values from a spectrum in which the underlying dynamics has a hierarchical structure of initially unspecified nature. For this method, we also describe the motivation, apply it to synthetic systems, and to both band origin and rovibrational data from the NO_2 optical spectrum.

2. Statistical and Fourier Transform Methods

2.1. *Eigenstate Statistics and their Limitations*

The first prototypes for strongly coupled pure sequence eigenspectra chosen by chemical physicists in attempting to understand complex double resonance spectra were developed for systems so much simpler than molecular systems that it is not surprising that it has been difficult to apply these oversimplified prototypes to molecular spectra. Statistical tools such as the level-spacing distribution, the staircase function, cluster functions, Δ_3, Σ^2, and others, and the ergodic reference system, the Gaussian orthogonal ensemble (GOE), were developed by Dyson and Mehta and others[7-9] for the analysis of energy levels of atomic nuclei.

An atomic nucleus is free of the hierarchical structure of polyatomic systems. Nonetheless, limiting cases of some molecular systems may be analyzed using classical statistical tools because they are sufficiently close to the case represented by the prototypes, where a single type of coupling and a simple matrix-element distribution predominates. For a model like the strong coupling of multiple bright states to a bath of non-interacting dark states,[10] which is useful in understanding the band origin data from the NO_2 optical spectrum, comparison with GOE and Poisson statistical measures is appropriate. The behavior of Δ_3 and other measures can quantify level repulsion, extract a single timescale for intramolecular dynamics and the corresponding delocalization range, and can characterize the disappearance of rigidity in the spectrum as a function of range.

When there are multiple levels of coupling, eigenstate statistics are difficult to interprete. Examples of multiple levels of coupling are almost as numerous and various as small polyatomic molecules themselves. As a schematic example of an additional level of coupling, we can consider a simple extension of the multiple-bright-state (MBS) model[10] (strong bright-state $\longleftrightarrow$ dark-state coupling to Poisson distributed dark states), by the addition of dark–dark coupling. The additional dark–dark coupling can be viewed as modifying the initial distribution of the dark states to be intermediate between Poisson and GOE, before the coupling to the bright states is taken into account. Increasing the interaction among the dark states causes the final spectrum to be closer to the GOE limit, and increases the apparent delocalization width of the bright state, in all of the statistical measures. The statistical measures do not successfully distinguish this case, with two types of states, and two magnitudes of coupling matrix elements, from a single group of states coupled among themselves by a single mechanism. As a second example we can consider a similar model in which bright states and dark states alike are identified by an approximate quantum number, K, which is good in the bright states but weakly broken down in the dark states. This model, the K-breakdown model, is described in more detail as part of our discussion of parsimonious trees. All of the dark states appear in the spectrum accessed by a single bright state because of the weak K breakdown in the dark manifold, and yet dark states with different K values interact so weakly that they do not repel each other. Level repulsion is greatly reduced, and all statistics appear Poisson in cases where the MBS model is close to GOE. Both effects occur in the optical spectrum of NO_2,[11] the first for the band origin data, and the second for intermediate J pure-sequence data generated by double resonance.

We have also found that the sharp distinction between a regular spectrum with Poisson level-spacing statistics and an ergodic spectrum with GOE statistics is blurred by limited resolution, even with arbitrarily high dynamic range. Peakfinding a spectrum with varying intensities, and with limited resolution, causes the statistics of a GOE spectrum to become more Poisson, and those of a spectrum with Poisson statistics to become more GOE. In the case of strongly overlapped lines, a peakfound Poisson

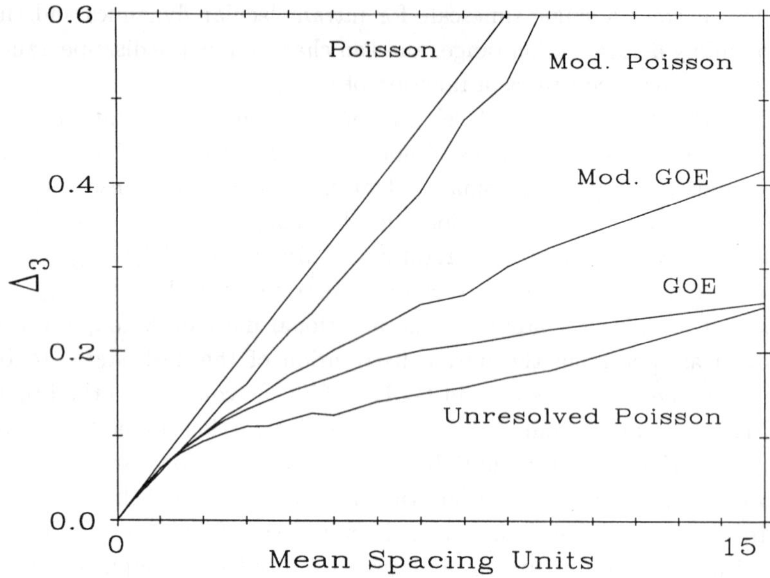

Fig. 1. Δ_3 modified by limited spectral resolution and dynamic range. A spectrum with Poisson distributed spacings and $\chi^2_{\nu=1}$ distributed intensities was broadened with a width of 10% (Modified Poisson) and of 300% (Unresolved Poisson) of the mean spacing. A GOE spectrum was broadened with a width of 3% of the initial mean spacing. Both were peakfound with an effective dynamic range of about 300:1. Meaningful statistics requires very high resolution and dynamic range.

spectrum can appear more rigid than GOE over significant lengths, with the length scale based on the average spacing of observed spectral features.

This blurring effect is illustrated by Fig. 1 for Δ_3, based on the following calculation:

1. Both Poisson and GOE spectra were generated with intensities chosen from a Porter-Thomas $(\chi^2_{\nu=1})$ distribution, where intensities were uncorrelated with the eigenvalues. A Poisson spectrum has energies chosen randomly from the uniform distribution, or equivalently, energies generated by using uncorrelated spacings from the negative-exponential distribution. The GOE eigenvalues were generated by diagonalizing a Gaussian orthogonal matrix of dimension 1000, and then transforming by semi-circular unfolding[7] the central 800 eigenvalues. Semi- circular unfolding is extremely effective at extending GOE statistics very close to the energy boundaries of the matrix eigenspectrum.

2. Both datasets were processed as if they were level sequences experimentally observed by some unspecified technique. The Poisson levels were interpreted as a spectrum by assigning a Lorentzian linewidth equal to 10% of the average line spacing, and by peakfinding with threshold intensity of 1/300 of the mean intensity. The level statistics of the peakfound data were calculated. Similarly, the GOE data were given a linewidth equal to 3% of the average line spacing, peakfound with the same threshold, and then level statistics calculated.

These are relatively high-resolution datasets; nonetheless their Δ_3 curves on Fig. 1 show that the distinction between Poisson and GOE has been significantly reduced. An unresolved Poisson spectrum was also generated by assigning a linewidth = 300% of the average spacing, and the spectrum peakfound. The number of peaks is reduced by a factor of 12 so that the linewidth appears to be only 25% of the line spacing, and, surprisingly, the statistics mimic the GOE for lengths a factor of 15 greater than the new apparent level spacing, with Δ_3 increasing significantly only beyond $L = 15$.

2.2. *Fourier Transform Analysis of Spectra and its Limitations*

The squared Fourier transform (SFT) (power spectrum or survival probability) of the raw spectrum has also been developed as a statistical tool.[12–15] It is closely related to the other eigenspectrum statistical measures through the two-level cluster function, Y_2, and its Fourier transform, b_2. Although other statistical measures contain similar information, the Fourier transform or power spectrum expresses dynamical information in the time domain: at short times, the delocalization of the bright state causes an initial decay; at intermediate times, two-level correlations result in a correlation hole; at longer times, dynamical recurrences may identify characteristic intramolecular time scales. It has been suggested that the Fourier transform applied directly to the raw spectrum would be expected to be less sensitive to inadequate resolution and other experimental problems than analysis of peakfound data. Fig. 2 illustrates how the Fourier transform of peakfound data from an unresolved Poisson spectrum, can show a spurious correlation hole, just as the Δ_3 values of Fig. 1 appear to be nearly GOE. Since convolution of a stick spectrum with a Lorentzian or Gaussian lineshape is

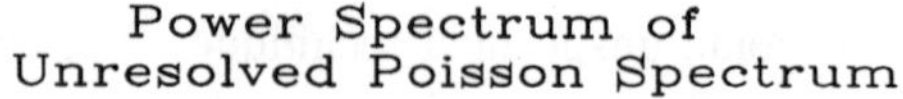

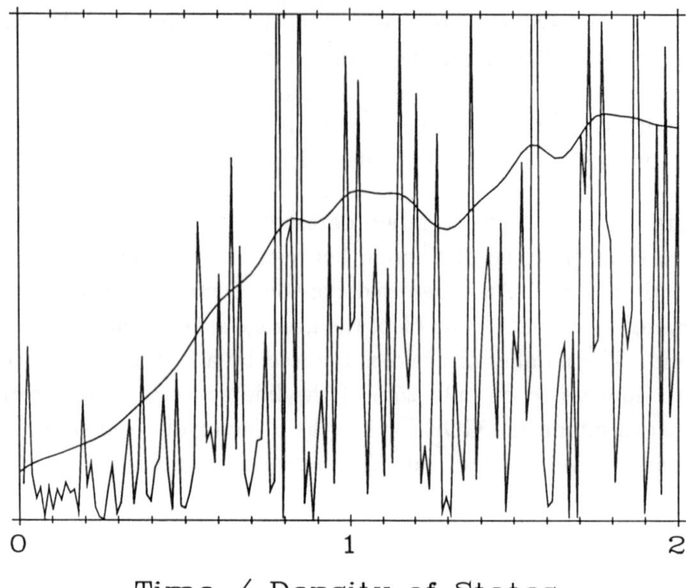

Fig. 2. Squared Fourier transform or power spectrum of the Unresolved Poisson spectrum from Fig. 1, calculated from peakfound data. The smoothed power spectrum shows a spurious correlation hole, and the unsmoothed power spectrum illustrates 100% speckle noise.

equivalent to multiplication by a decaying exponential or Gaussian in the time domain, it seemed reasonable to expect that a correlation hole at early time might appear to be immune to the effects of inadequate resolution.

Several effects cause this to be not quite true. The correlation hole is reduced in depth, narrowed, pushed toward earlier time, and obscured by speckle in much the same way that other statistical measures are modified by experimental limitations:

- The position of the correlation hole is moved toward $t = 0$ by the fact that experimental spectra, except for a few special cases, do not consist of a single pure sequence. When N pure sequences are present in the data, the width of the correlation hole is reduced by a factor of $1/N$.

Peakfound data, with the attendant risk of missing and spurious levels,[16] is required to generate a pure sequence. This is the case because the peaks common to two double resonance spectra, with an intermediate-level energy offset (i.e., intermediate-level combination-differences), are those which form the pure sequence. In addition, underestimation of the true density of levels due to limited dynamic range or resolution further reduces the width of the correlation hole.

- There is no correlation hole at $t = 0$. In time domain language, the rate of appearance of the correlation hole is related to the rate of delocalization of the bright state (Ref. 10, Figs. 18 and 19) through Fermi's Golden Rule.

- The depth of a correlation hole in a power spectrum calculated from the raw spectrum is at most $\frac{1}{3}$. When positions and intensities are uncorrelated, and when the two-level cluster function $Y_2(f_1, f_2)$ is independent of position in the spectrum (i.e., $= Y_2(f_1 - f_2)$), the depth of the correlation hole is multiplied by the F_{aa} factor of Heller and Sundberg.[10,12,17] This factor, computed as the squared mean intensity over the mean squared intensity ($\langle I \rangle^2 / \langle I^2 \rangle$), varies from zero in the regular limit to $\frac{1}{3}$ in the ergodic limit. Only when the power spectrum is calculated from peakfound data with all lines assigned unit intensities, can the correlation hole have a depth approaching 1 (Fig. 2).

- Since the power spectrum shows 100% speckle noise (Fig. 2), it must be smoothed to reveal a correlation hole or recurrence which is necessarily of much smaller amplitude than the noise. In addition to speckle noise, baseline noise also contributes to the power spectrum. It is quite difficult to smooth the SFT with such delicacy that "noise" is eliminated and a "real" correlation hole or "real" recurrences are revealed, rather than smoothed away. A specialized smoothing procedure has been recommended (the dichotomic convoluted window (DCW); Ref. 15), but this scheme has not prevented disagreement in the literature over the meaning of the SFT of experimental data.[18] We have examined the limitations of the SFT rather than extensively evaluating the DCW method. Nonetheless, regardless of the procedure used, we have found that features which remain after smoothing may either be noise remnants or "real" dynamical features.

- When the spectral dataset covers such a wide frequency range that "unfolding" (correction for a smooth variation in the expected density of states) is necessary, the details of the unfolding method can cause peaks in the power spectrum to appear or disappear, and cause some of the statistical measures, especially Σ^2, to become unreliable. Delon, Jost, and Lombardi[19] discuss this effect in their App. C.

The common basis of these classical statistical methods, including the SFT, in cluster functions (usually 2-level) is especially debilitating to these methods when they are applied to polyatomic systems with hierarchical dynamics. Because increasing internal energy often opens up new rotational, vibrational, and finally electronic coupling mechanisms, each bringing a new bottleneck to restrict energy flow, most molecular systems will never truly reach the "bag of atoms" limit in which no approximate quantum numbers remain. The extended autocorrelation function and parsimonious trees methods are initial steps toward providing more flexible tools, suited to hierarchical dynamics in polyatomic systems, to supplement existing statistical methods.

3. The Extended Autocorrelation Function (XAC)

3.1. *Description of the XAC Function*

Searching for fragments of order in imperfect spectroscopic data, where some but not all approximate quantum numbers are broken, is a very different problem from the measurement of ergodicity that results from a single type of coupling. We expect regularity to be most evident in SEP or other double resonance data which provide a partial presorting by rigorous quantum numbers; still, these spectra are often too complex and congested to assign. The fragments of order remaining in these spectra will take the form of multiple, interfering, patterns which repeat with regular and irregular variations within a spectrum, or patterns which are common to two double resonance spectra recorded in different ways. The most familiar methods for the recognition of patterns in a spectrum or other data stream, or the recognition of patterns in common between two spectra, are the autocorrelation function and the cross-correlation function. The autocorrelation function is closely related to the Fourier transform technique, since the SFT and the autocorrelation are a Fourier transform pair. A form of the cross-correlation function has been applied to locate a breakdown in selection rules associated with the vinylidene to acetylene

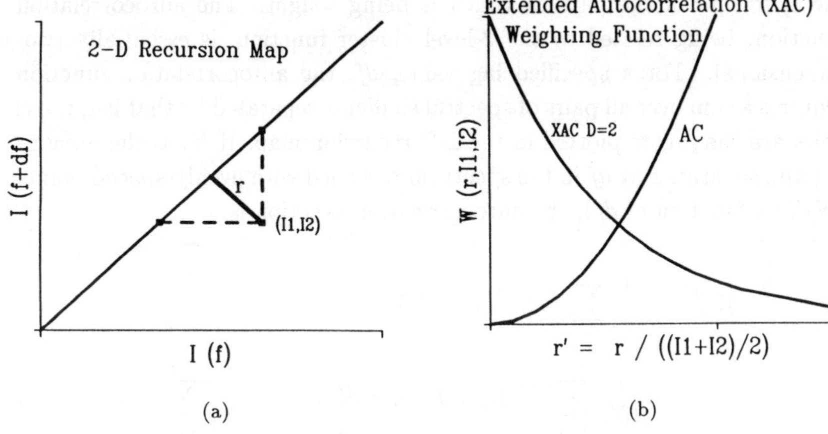

Fig. 3. (a) Two-dimensional recursion map of a spectrum. Intensities from spectral elements *df* apart locate a point on the recursion map. Paired spectral features correspond to points on the line $I_1 = I_2$. Baseline–baseline pairs appear near the origin. Peak–baseline pairs appear along either axis. The distance, r, from the equal-intensity line to recursion point appears in the autocorrelation and extended autocorrelation function (XAC) weighting functions, as discussed in the text. (b) Weighting functions for pairs of points in the spectrum in the autocorrelation function and the XAC. The autocorrelation function amplifies the contributions of mismatched intensities, while the XAC minimizes them.

isomerization.[20] In spite of some success with these techniques, they have generally been difficult to apply: the kinds of patterns that can be detected are limited, and the contrast ratio is poor, especially in spectra which contain spectral congestion, or multiple types of patterns.

The reason for poor contrast ratio in the autocorrelation function can be understood from the *recursion map* of the spectrum (Fig. 3(a)). For a spectrum recorded with evenly-spaced points across its range, the recursion map in two dimensions is a scatter plot of (I_1, I_2) points where I_1 is the intensity at frequency f_i, and I_2 is the intensity at frequency $f_i + df$. The point (I_1, I_2) falls a distance r from the $I_1 = I_2$ line, where $r = |I_1 - I_2|/\sqrt{2}$. The number of points is the number of possible locations in the spectrum for the frequency pattern, $(f_i, f_i + df)$, at a specified value of df. The recursion map can be generalized to N dimensions as a scatter plot of N-tuples of spectral elements. The individual elements in each N-tuple are the intensities at frequencies spaced in a way described by some specifiable pattern. The number of N-tuples is the number of possible locations in the spectrum of the pattern which is being sought. The autocorrelation function, being related to the 2-level cluster function, is essentially two-dimensional. For a specified lag value, df, the autocorrelation function requires a sum over all pairs of spectral elements separated by that lag; those pairs are the points plotted in the 2-D recursion map. If N_f is the number of pairs separated by df in the spectrum recorded with evenly-spaced points (N_f is a function of df), the autocorrelation function is

$$A(df) = \frac{1}{N_f} \sum_{i=1}^{N_f} I(f_i)I(f_i + df)$$

$$= \frac{1}{2N_f} \sum_{i=1}^{N_f} (I^2(f_i) + I^2(f_i + df)) - \frac{1}{\sqrt{2}\,N_f} \sum_{i=1}^{N_f} r_i^2 \; . \tag{1}$$

Of these two terms, the first term is, when df is much smaller than the spectral width, approximately equal to the spectral average value of the squared intensity. Thus this term is nearly independent of df, regardless of the structure of the spectrum. The remaining variation in that term is due to the variation in the few intensity values at the extreme edges of the spectrum which are omitted from the sum as df is increased. The second term is the sum of squares of distances, r, from the equal-intensities line, and is a *minimum* when many lines in the spectrum are spaced by df. The points on the recursion map with large r are the pairs in which lines are matched with baseline or noise elements. The autocorrelation function heavily weights baseline points, and fluctuations in the baseline are, therefore, amplified.

The extended autocorrelation function for two dimensions, X_2, for a pattern of equal intensity doublets, weights most heavily the points on the recursion map near the equal-intensities line which have significant values of average intensity:

$$X_2(df) = \frac{1}{N_f} \sum_{i=1}^{N_f} \pm \sqrt{|I(f_i)I(f_i + df)|}\; e^{-\alpha(r_1/\langle I \rangle)^2} \; . \tag{2}$$

The exponential factor, α, sets the required accuracy of match between line intensities; $\langle I \rangle$ is the average intensity value for the pair. The sign of the radical is the sign of the product of the two intensities. The intensity product may have either sign when one or both of the points is near the baseline. This factor discriminates against pairs with low intensity at both frequencies, regardless of how well matched. Only points on the recurrence map with significant intensity in both pattern positions, and with matched intensities, contribute to X_2; baseline points contribute very little. Unlike the autocorrelation function itself, the extended autocorrelation function is dependent on the location of the spectral baseline. The XAC is intended to find splittings and patterns of intensity in the spectrum, and requires the intensities to be represented as accurately as possible. Inaccuracy in the baseline affects the XAC by making it more difficult to distinguish between spectral features and noise.

The general form for the extended autocorrelation function for N_p lines in a pattern specified by pattern parameters, $\{\lambda\}$, for a particular set of pattern parameters, is a sum over all possible occurrences of the pattern in the spectrum ($i = 1 \ldots N_f$), of a pattern weight function, W_X, which is a function of the set of intensities at the frequency positions of the pattern, $\{I\}_i$:

$$X_{N_p}(\{\lambda\}) = \frac{1}{N_X(\{\lambda\}_0)} \frac{1}{N_f} \sum_{i=1}^{N_f} W_r(r_i, \langle I \rangle) W_I(\{I\}_i)$$

$$= \frac{1}{N_X(\{\lambda\}_0)} \frac{1}{N_f} \sum_{i=1}^{N_f} W_X(\{I\}_i) \, . \tag{3}$$

The total pattern weight, W_X, has been partitioned into a radial weight factor, W_r, and an intensity weight factor, W_I. We have explored a number of forms for these weight factors. The number of possible locations for patterns of a particular type in the spectrum, N_f, is a function of the pattern parameters $\{\lambda\}$. We can write the N_p frequencies forming the pattern at which the intensities are evaluated in Eq. (5), and which appear in a single term of the sum in Eq. (3) schematically as:

$$\{f_{i,j}\} = \{f_{i,0} = f_i, \quad f_{i,1} = f_i + s_{\{\lambda\},1}, \ldots, \quad f_{i,N_p-1} = f_i + s_{\{\lambda\},N_p-1}\} \, , \tag{4}$$

where f_i is the initial frequency for the i'th placement of the pattern, and the $s_{\{\lambda\},j}$ are the $N_p - 1$ spacings that define the pattern for a particular set of pattern parameters $\{\lambda\}$. The average intensity $\langle I \rangle$ and the intensity set $\{I\}_i$ are then written as:

$$\langle I \rangle = \frac{1}{N_p} \sum_{j=0}^{N_p-1} I(f_{i,j}) \, ,$$

$$\{I\}_i = \{I(f_{i,0}), \ldots , I(f_{i,N_p-1})\} \, . \tag{5}$$

The normalization factor, $N_X(\{\lambda\}_0)$ is chosen to make the value of the XAC function (Eq. (3)) equal to one for a reference set of pattern parameters, $\{\lambda\}_0$. Those parameters are chosen to be the set which makes all the spacings, $s_{\{\lambda\},j}$, zero when that set is allowed by the parameterization of the pattern; otherwise, an arbitrary set is chosen.

The radial weight function, W_r, reduces the weight of points on the recurrence map as the distance from the equal-intensity line increases. For the radial weight function, we have chosen to use a normal distribution for weights with a variance which is based on known experimental uncertainties in intensity:

$$W_r(r, \langle I \rangle) = e^{-0.5 r^2 / V_r(r|\langle I \rangle)} \, ,$$

$$r = \mathrm{Max}(\{I - \langle I \rangle\}) \, . \tag{6}$$

Although the most straightforward generalization of the 2-D case for the distance, r_2, used in Eq. (6), is the variance of the intensities $\{I\}_i$, the simple maximum deviation from the mean intensity is proportional to the distance r in 2-D, and is quicker to compute for larger N_p. The weights for r, W_r, are from a normal distribution with a variance calculated from estimates of the experimental error in intensities. For the variance of r for a particular average intensity, $V_r(r|\langle I \rangle)$ in Eq. (6), we use the expected variance of $I - \langle I \rangle$, assuming that the mismatch in intensities which causes r to be non-zero is due solely to experimental errors in intensity. The variance is calculated allowing experimental errors in intensity to consist of two types of contributions: a constant or baseline noise level, σ_0, given in the units in which the spectrum is recorded, and a fractional error, σ_1, which is proportional to the observed intensity. This yields:

$$V_r(r|\langle I \rangle) = (\sigma_0^2 + \sigma_1^2 \langle I \rangle^2) \frac{(N_p + 1)}{N_p} \, . \tag{7}$$

For the intensity weight factor, W_I, we use the geometric mean of all intensities in the pattern:

$$W_I(\{I\}) = \pm \bar{I} \, ,$$

$$\bar{I} = \left[\prod_{i=1}^{N_I} |I_i| \right]^{\frac{1}{N_I}} \, .$$

The sign of W_I is the sign of the product of the intensities, which may be negative when baseline points are selected. As in the 2-D case, this factor discriminates against pattern placements, $\{I\}_i$, where some of the intensity values in the group are small, and emphasizes points with all intensities large. Other choices for W_I include the arithmetic mean. In most cases, very comparable results are obtained regardless of the detailed choices for the weight functions, as long as the falloff of the weight values away from the equal-intensity line (W_r), and the intensity dependence, W_I, are similar.

The extended autocorrelation function has a number of advantages over the autocorrelation function:

- The XAC weighting function strongly discriminates against interspersed unrelated patterns. Patterns formed by subsets of the lines in a spectrum may be located. For instance, a subset of lines may have splittings, such as l-doubling, or tunneling to another conformation.
- Complex patterns, parameterized in a multi-dimensional way by the parameter set $\{\lambda\}$, may be used. Peaks in the extended autocorrelation function in parameter space correspond to the molecular parameter set $\{\lambda\}$ present in the spectrum.
- The pattern may include any number of lines with unequal spacings. These may include vibrational progressions, Fermi resonance intervals, fluxional motions, etc.
- The frequency locations of the lines making up the patterns may be obtained by plotting the XAC weights, the individual terms of the sum in Eq. (3), as a function of the initial frequency of the pattern, f_i.
- The expected intensities do not need to be equal; the intensities sampled by the pattern from the spectrum can be normalized before computing the weights. These intensities may represent P, Q, and R branch lines parameterized by rotational constants, Hönl-London relative linestrength factors, Franck–Condon factors, etc.
- Experimental errors in intensity can be correctly modeled as including separate contributions independent of, and proportional to, the observed intensity.
- A variable resolution element may be used. This can compensate for random frequency perturbations. The width of XAC signals can be used as a measure of dispersion of rotational constants, $l = 0, 2$ splittings, etc.

Although we do not discuss it in detail here, the extended autocorrelation function may also be generalized to a spectral cross-correlation form. The spectral cross-correlation can be used to find the fragmented parts of a single bright state, which appear in two different double resonance spectra with different total intensities, in the presence of different sets of interfering features. This case is illustrated in Fig. 7 of Yamanouchi *et. al.*,[21] where similar intensity patterns appear in dispersed fluorescence from two different upper states. The spectral cross-correlation may also be used to locate and quantify the appearance of similarity between two different spectra in a specific energy range, as was done in Ref. 20, where access to the vinylidene isomeric structure was expected to break down selection rules which caused two SEP experiments normally to access disjoint sets of levels.

3.2. *Application to Synthetic Spectra*

The simple synthetic spectra that we have used to test the extended autocorrelation function emphasize its ability to locate specific patterns in the presence of interfering features, and its adaptability to perturbed spectra, and to experimental errors.

The first synthetic spectrum, illustrated in Fig. 4, is an example of a simple doublet pattern in the presence of much more numerous interfering lines. The spectrum contains 21 pairs of lines among 151 additional unrelated lines over a spectral interval of 235 units. For both the unrelated lines, and the doublets, the positions were chosen randomly from the uniform distribution (spacings from a negative-exponential distribution, Poisson statistics). Each doublet consists of two lines of equal intensity 5 units apart. All intensities are taken from the Porter–Thomas $(\chi^2_{\nu=1})$ distribution, with an average intensity of 5. The spectrum was then convolved with a Gaussian of 0.1 units width, and sampled at every 0.05 units. The presence of doublets is not at all obvious in the spectrum. Figure 5 shows the autocorrelation of that spectrum. There is a peak at the doublet separation of 5 units, but many other peaks of similar intensity are present, and the largest peak, between 14 and 15, is not related to the doublets. The high noise level in the autocorrelation function is related to the speckle noise in the power spectrum (Fig. 2), since the autocorrelation function and the power spectrum form a Fourier transform pair. Figure 6 shows the extended autocorrelation function (XAC) of the spectrum of

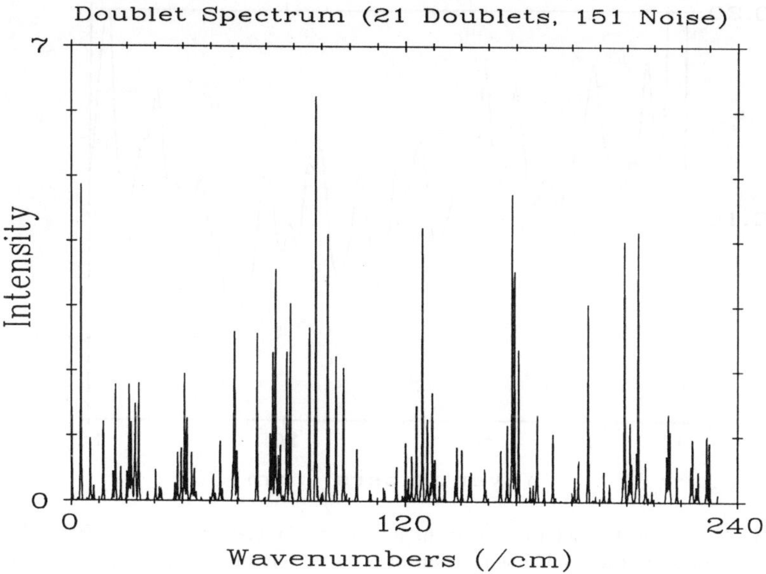

Fig. 4. Spectrum containing 21 pairs of lines among 151 unrelated lines over a range of 235 units. For both sets, the positions are uniformly distributed. Each pair consists of two lines of equal intensity 5 units apart. For both sets, the intensities are taken from the Porter–Thomas $(\chi^2_{\nu=1})$ distribution, with an average intensity of 5. The spectrum was then convolved with a Gaussian of 0.1 units width, and sampled every 0.05 units. The presence of doublets is not at all obvious.

Fig. 4. The XAC was computed with a doublet pattern, and an uncertainty in the intensity of 7% of the observed intensity ($\sigma_1 = 0.07$, Eq. (7)). The abscissa is the doublet spacing. The doublet peak is clearly evident at the correct location, and the spurious peaks present in the autocorrelation are suppressed.

The second synthetic spectrum is similar to that of Fig. 4, except that the signal consists of 21 triplets with two equal spacings of 4.5 units. The features are again uniformly distributed, and the intensities chosen from the Porter–Thomas distribution. Figure 7 shows the XAC of this spectrum. The triplet peak is much more prominent against the background level than was the doublet peak of Fig. 6. Increasing the number of lines in the pattern makes the XAC a sharper measure because it decreases the background signal owing to accidental overlaps with lines that are not a

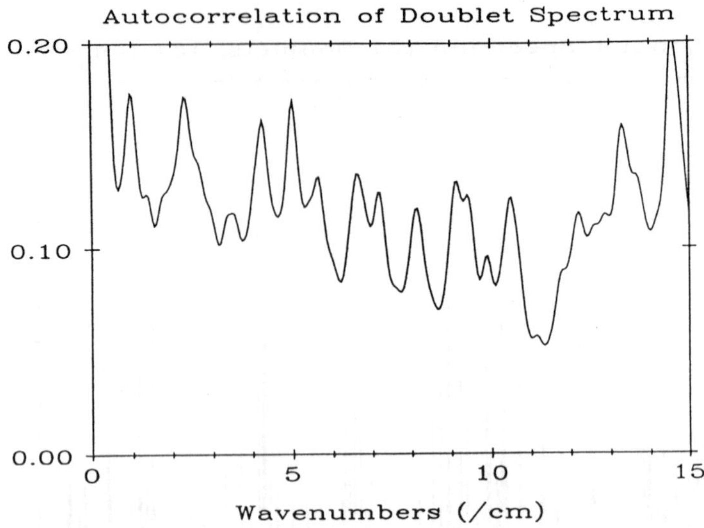

Fig. 5. Autocorrelation of the spectrum in Fig. 4, doublets among interfering lines. There is a peak at the double separation of 5 units, but many other peaks of similar intensity are present, and the largest peak, between 14 and 15, is not related to the doublets.

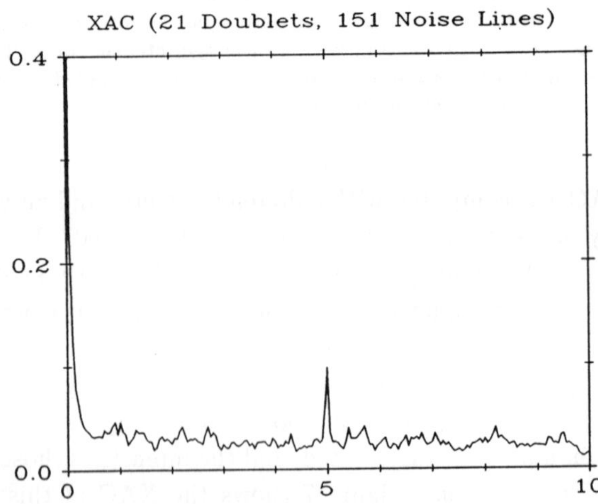

Fig. 6. Extended autocorrelation function (XAC) of the spectrum of Fig. 4. The XAC was computed with a doublet pattern, and an uncertainty in the intensity of 7% of the observed intensity. The abscissa is the doublet spacing. The doublet peak is clearly evident, and conflicting signals are suppressed.

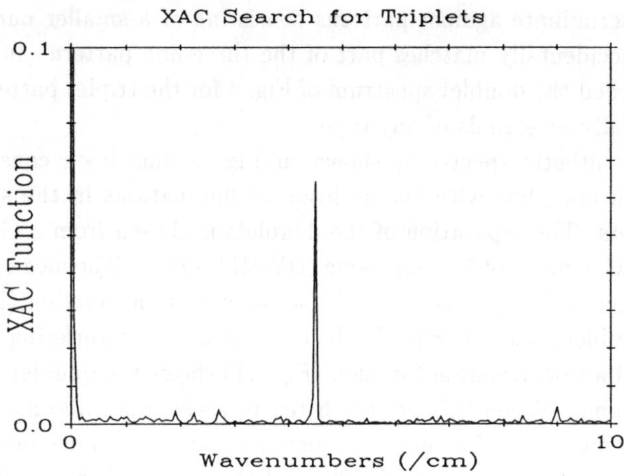

Fig. 7. Extended autocorrelation function (XAC) of a spectrum of 21 triplets among 151 interfering lines. The triplet spacing was 4.5 units, with other characteristics similar to that of Fig. 4. Increasing the number of lines in the pattern sought reduces the background from interfering lines.

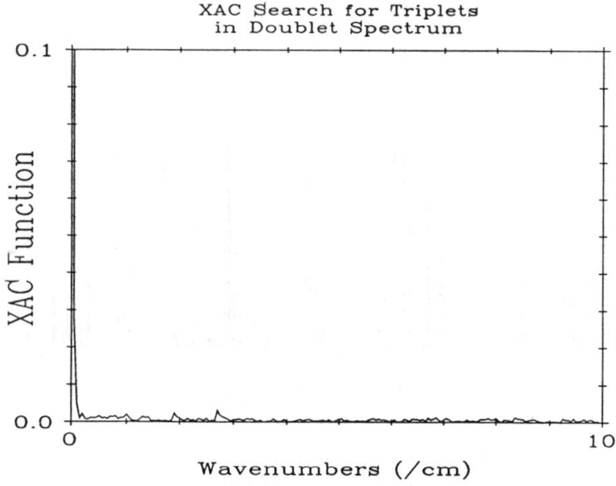

Fig. 8. Searching for triplets with the XAC in a spectrum containing only doublets (Fig. 4), gives essentially no signal, and a lower background signal than that in the doublet XAC (Fig. 4, spectrum, and Fig. 6, XAC).

part of the pattern. Figure 8 illustrates that the XAC for the 3-line pattern is able to discriminate against patterns consisting of a smaller number of lines which accidentally matches part of the three-line pattern. In Fig. 8, we have searched the doublet spectrum of Fig. 4 for the triplet pattern, and found essentially no signals of any type.

A third synthetic spectrum, shown in Fig. 9, has been constructed similar to the first, but with the addition of fluctuations in the spacings of the doublets. The separation of the doublets is chosen from a Gaussian distribution with mean of 5.0 and sigma (HWHM) of 0.2. The mean spacing of the 21 selected doublets is 4.82. The autocorrelation function (Fig. 10) shows the doublet as a minor peak which is lost among interfering signals. The extended autocorrelation function (Fig. 11) shows the doublet peak as the largest signal. When this peak is fitted to a Gaussian lineshape using a peakfinding program, the center frequency agrees to two decimal places with the correct mean spacing of 4.82, and the width of the peak agrees with the distribution width to within 0.02.

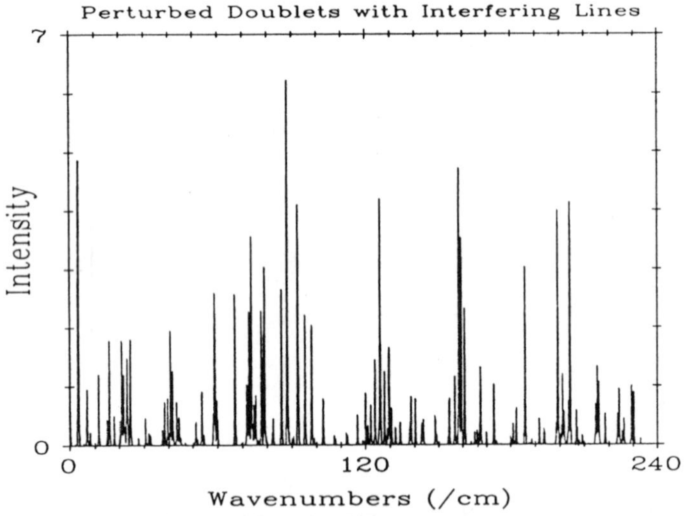

Fig. 9. Spectrum similar to that of Fig. 4, but in which the separation of the doublets is chosen from a Gaussian distribution with mean 5.0 and variance 0.2. The actual mean spacing of the 21 selected doublets is 4.82.

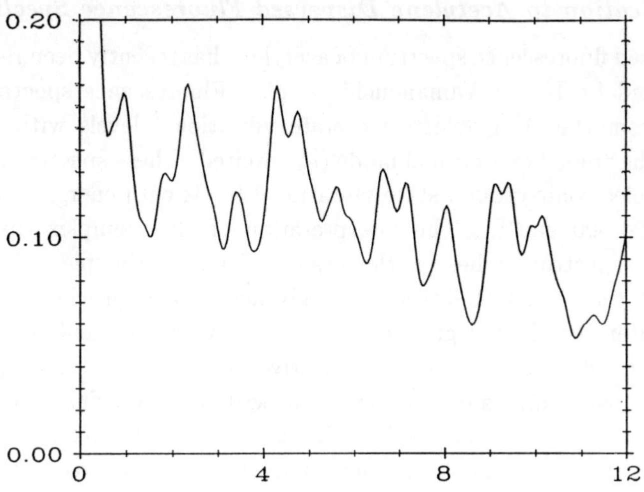

Fig. 10. Autocorrelation of the spectrum of Fig. 9. Although the doublet peak is present at 4.82, it is lost among the signals from the interfering lines.

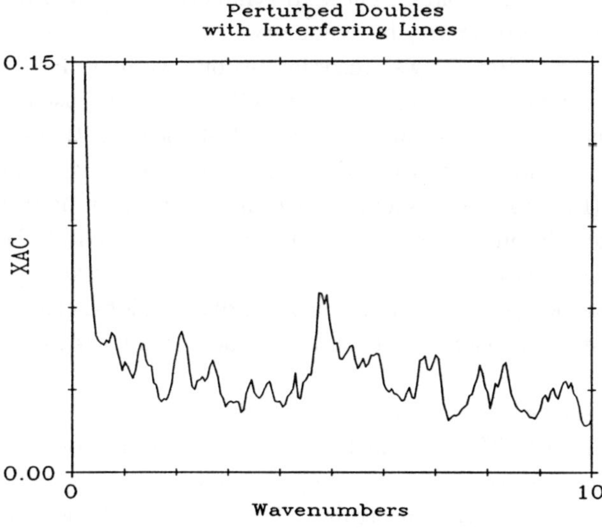

Fig. 11. The extended autocorrelation function of Fig. 9 shows the doublet peak clearly. When fit to a Gaussian lineshape, this peak has the position and width corresponding to the 21 selected doublets.

3.3. *Application to Acetylene Dispersed Fluorescence Spectra*

The dispersed fluorescence spectrum of acetylene has recently been recorded and analyzed by Kaoru Yamanouchi *et. al.*[21] Fluorescence spectra were recorded from the $\tilde{A}^1 A_u$ electronic state vibrational levels with 2 or 3 quanta of the *trans*-bend normal mode (ν_3') excited. These spectra sampled the $\tilde{X}^1 \Sigma_g^+$ electronic ground state vibrational levels with energies between 4000 and 24 000 cm^{-1}. The interpretation of this complex dispersed fluorescence spectrum relies on the Franck–Condon principle. The C–C stretch and *trans*-bend modes are strongly active, and provide all of the observed intensity. In the ground state, the C–C stretch and *trans*-bend modes are labeled as ν_2 and ν_4, respectively. We will refer to ground state levels with unprimed, rather than with double-primed quantum numbers, since we will not need to discuss the initial level in the SEP double resonance experiment. $\tilde{A}$ vibrational levels are indicated with single primes. Although the symmetric C–H stretch (vibration 1 (ν_1, ν_1') in both $\tilde{X}$ and $\tilde{A}$) was previously believed to be active, it has been noted[5,20] that the displacement in that mode is small, and that the spectrum is most convincingly interpreted in terms of Franck–Condon displacement along $\tilde{X}^1 \Sigma_g^+$ modes ν_2 and ν_4 only. The anti-symmetric stretch, ν_3, and the *cis*-bend, ν_5, have zero activity by symmetry.

The spectrum of Ref. 21 was recorded in four sections; for our application of the extended autocorrelation function, we have used only one of the sections, i.e., fluorescence from the $\tilde{A}^1 A_u$ level with two quanta of *trans*-bend, to lower states with energies between 4200 cm^{-1} and 18 000 cm^{-1} (Fig. 12). The spectrum which we currently have available does not reflect the final frequency calibratation used in Ref. 21, but is sufficiently linear over limited ranges for our current purposes. We find that the extended autocorrelation function can identify the global characteristics of the spectrum, and can also be used to locate and assign individual features.

Figure 13 shows a series of extended autocorrelation results for the spectrum of Fig. 12. The bottom trace shows the results of a search for doublets in the spectrum. The center trace identifies patterns of three equally spaced lines in the spectrum. The top trace finds patterns of three lines in which the second spacing is twice the first. The effective resolution element used in generating these traces was about 20 cm^{-1}, and the intensity uncertainty used was 30% with all expected intensities equal.

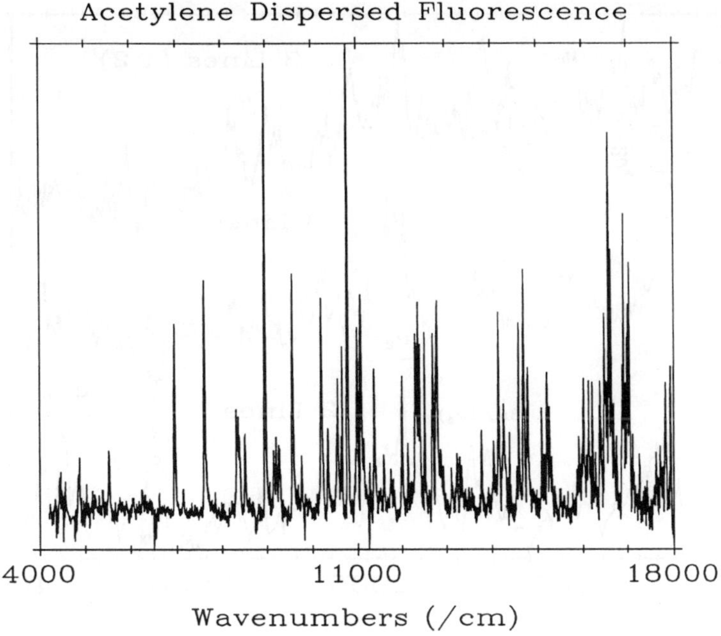

Fig. 12. One section of the C_2H_2 dispersed fluorescence spectrum of Yamanouchi *et. al.*[21] Emission from the $J_{K_a,K_c} = 3_{1,3}$ rotational level of the $\tilde{A}^1A_u$ ν_3' (*trans*-bend) $= 2$ vibrational state to $\tilde{X}^1\Sigma_g^+$ electonic ground state energies from 4200 cm^{-1} to 18 000 cm^{-1}.

The intensity uncertainty was increased from the experimental accuracy of 10% in order to allow for more variation in the Franck–Condon pattern.

The most prominent feature in each Fig. 13 XAC result is the ν_2 progression which varies in spacing from about 1900 cm^{-1} to below 1800 cm^{-1}, as the number of ν_4 quanta excited increases (see Refs. 21, and Ref. 5 for anharmonic constants). The three line, 1:1 spacing and 1:2 spacing, ν_2 results show substructure which results from the Franck–Condon nodal pattern of the ν_4 progression. The two line result identifies the global peak of the Franck–Condon intensity pattern in this energy range. Additional peaks appear with spacing $2\nu_2$ in the doublet and 3 line (1:1) results. The presence of $2\nu_2$ patterns indicate that the ν_2 progression is very long (and known to cover the range from 0 to 5 quanta) since the first 3 line (1:1) $2\nu_2$ pattern would contain 0, 2, and 4 quanta of ν_2, and the pattern containing

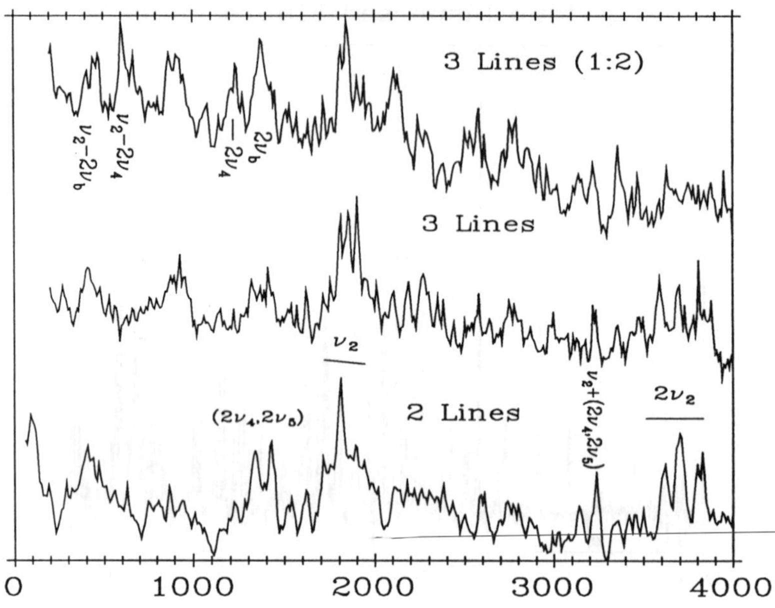

Fig. 13. Extended autocorrelation results on the spectrum of Fig. 12 for three different patterns: a 2 line or doublet pattern, a 3 line pattern with 1:1 spacings, and a 3 line pattern with 1:2 spacings. The features in these traces are interpreted in Sec. 3.3 of the text.

1, 2, and 5 quanta would also contribute. For the 3 line (1:2) pattern the first $2\nu_2$ pattern would contain 0, 2, and 6 quanta of ν_2; the absence of those patterns in the 3 line (1:2) result indicate that the ν_2 progression does not extend over $7\nu_2$ quanta with significant intensity. Because of the 30% intensity uncertainty used in calculating the XAC, both of the 3 line results show weaker peaks at one half the ν_2 frequency, which results from accidental matches with a line at the nominally incorrect third frequency in the pattern. This peak can be recognized as spurious because it is absent in the doublet XAC.

One of the reasons for running the 3 line (1:2) pattern was the report in Ref. 21 that, for the $2\nu_4$ progressions present in the spectrum which we reproduce as Fig. 12, weak or missing transitions typically occur after 2 strong transitions. In Fig. 13, the $2\nu_4$ peak just above 1200 cm^{-1} appears strong only in the (1:2) XAC result, and all $2\nu_b$ (bending levels ν_4 or ν_5)

peaks are quite weak in the 3 line (1:1) result. It is now known[5,6] that the first stage of energy delocalization in ground state acetylene involves transfer from $2\nu_4$ to $2\nu_5$ via a Darling–Dennison resonance. An additional peak occurs midway between 1300 and 1400 cm^{-1} in the doublet and 3 line (1:2) XAC which corresponds to the bending interval in the energy range where the two bending frequencies become approximately equal. Because of this strong $2\nu_4$–$2\nu_5$ resonance, additional progressions in the dispersed fluorescence built on levels with $2\nu_5$ excited were definitely identified in the range below 14 000 cm^{-1}. The doublet XAC clearly identifies the presence of these progressions through the peak above 1400 cm^{-1} which corresponds to the energy difference between 0 and 2 quanta of ν_5.

A number of additional ν_2, ν_4 combination peaks also appear in the XAC. Because of the similarity between the ν_4 frequency and the ν_2–$2\nu_4$ interval, the dispersed fluorescence and low-resolution SEP spectra have some similarity to a progression in ν_4. The ν_2–$2\nu_4$ peak occurs in the 3 line (1:2) XAC, and the ν_2–$2\nu_b$ peak occurs in all three traces. In the doublet XAC, there is an additonal peak between 3200 and 3300 cm^{-1} which we identify as $\nu_2 + 2\nu_5$. Although this peak might be associated with excitation of ν_3 via the "2345 resonance" ($\nu_3 \approx \nu_2 + \nu_4 + \nu_5$), the sharpness of the peak and its frequency support the other assignment. In the part of the dispersed fluorescence spectrum that we have analyzed, the XAC reveals no evidence for either the "2345" or "1234" ($2\nu_3 \approx \nu_1 + \nu_2 + 2\nu_4$)[22] resonances.

Figure 14 illustrates how extended autocorrelation weights may be used to assign individual features in the dispersed fluorescence spectrum. We have selected the two XAC peaks from the ν_2 region which are common to both the 3 line (1:1) and 3 line (1:2) traces in Fig. 13. These peaks correspond to patterns with first spacings of 1800 cm^{-1} and 1850 cm^{-1}, respectively. The next spacings were given a negative anharmonicity of -10 cm^{-1} per quantum. The XAC weights shown in Fig. 14 are the individual terms, $W_X(\{I\}_i)$, in Eq. (3) that contributed to those two XAC peaks in the 3 line (1:1) result as a function of pattern placement in the spectrum. Each peak in the weight function identifies the frequency position of the initial peak in the three line pattern. The other two lines in the pattern can be identified by replotting the same weight curve shifted successively by the spacings. The three replications of the same feature in the three shifted weight curves align with the 3 peaks of the pattern in the dispersed fluorescence spectrum. The weights aligned with the first and second lines of the pattern are plotted in Fig. 14. The weights aligned with the third line of the pattern are omitted for clarity. Franck–Condon peaks in the ν_4 progression occur for about 6–8 and about 14–16 quanta of ν_4. ν_2 progressions built on these Franck–Condon peaks have different ν_2 intervals. The XAC weights of Fig. 14 show that the 1850 cm^{-1} and 1800 cm^{-1} peaks in the 3-line XAC (ν_2 progressions) correspond to the two ν_4 Franck–Condon peaks in the ν_2 separation as a function of ν_4 excitation, 6–8 ν_4 and 14–16 ν_4, respectively. For example, the most prominent peaks in the 1850 cm^{-1} trace align with peaks assigned to $(\nu_2, \nu_4) = (3, 4)$, $(2,6)$, $(3,6)$ and others with low ν_4 excitation, while the prominent peaks in the 1800 cm^{-1} trace align with peaks assigned to $(\nu_2, \nu_4) = (0, 16)$, $(1,14)$, $(2,14)$, and others with high ν_4 excitation.

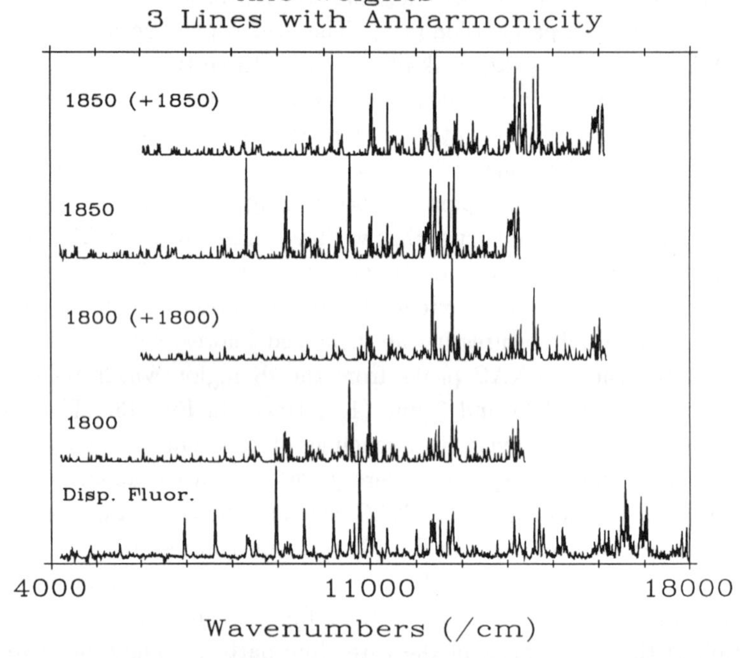

Fig. 14. Extended autocorrelation weights which may be used to assign individual features in the dispersed fluorescence spectrum. XAC weights for the 3 line (1:1) pattern are plotted aligned with the first and second lines of the pattern. The weights aligned with the third line of the pattern are omitted for clarity. Corresponding XAC weights identify the positions of the lines that form the pattern.

Because of the multi-dimensional capabilities of the XAC, it is also possible to use contour plots, mesh surface plots, or 3D visualization techniques, to discover relationships between spacings in a pattern in the spectrum. We have made considerable use of these methods, with other dimensions assigned to anharmonicity or to other spacings, to find concentrations of patterns in the spectrum as a global function of more than one variable. The anharmonicity used in Fig. 14 corresponds to a peak in the anharmonicity coordinate for a 3-line pattern.

Both applications of the XAC, to synthetic spectra and to C_2H_2 dispersed fluorescence spectra, indicate that the technique has the ability to locate patterns in the presence of interfering features. In addition, we have seen that the XAC has the ability to suggest assignments for spectral features. The use to suggest assignments is similar to the classical technique of combination-differences, but our formulation of the weight functions provides a sharp quality measure which was absent in the traditional implementation of the combination-difference method, and which made that technique impossible to use in congested spectra. We have also pointed out that the extended autocorrelation function can be generalized to a spectral cross-correlation function which has a number of potential applications. The spectral cross-correlation could be used to find the fragmented parts of a single bright state, which appear in two different double resonance spectra with different total intensities, and in the presence of different sets of interfering features. The spectral cross-correlation might also be used to locate and quantify the appearance of similarity between two different spectra in a specific energy range, as when energy levels of other fluxional structures tune into resonance, introducing new types of coupling, as in Ref 20.

4. Parsimonious Trees

4.1. *Description of Parsimonious Trees*

The generation of trees from spectra, and their interpretation, as well the use of a number of other classification methods, are also discussed by Michael Davis, in this volume,[23] and in earlier publications.[24] Our modification of the parsimonious trees technique, and our interpretation of the results, is driven by the experimentalist's need to understand intramolecular dynamics in terms of a "natural" zero-order basis set of Poisson-distributed levels, a Hamiltonian containing a hierarchy of coupling matrix elements or timescales, and quantum numbers which may be rigorously, approximately, or weakly conserved in the fully coupled system. As we have noted, polyatomic molecular dynamics for a wide range of internal energies is strongly hierarchical, with off-diagonal matrix elements which range from thousands of cm^{-1} to tens of kilohertz. This hierarchy of coupling strengths leaves traces in the spectrum of the remaining approximately conserved quantities, and of delocalization widths for the mostly-broken quantum numbers. That energies of the assumed zero-order basis states should have Poisson statistics is based on the result that the levels of a regular, classically integrable system follow Poisson statistics at high energy.[25] For a perfectly harmonic system, the energy range over which Poisson statistics are followed is limited; nonetheless, Poisson-distributed levels are the appropriate reference point.

We will show how the parsimonious trees technique allows direct identification of multiple average coupling matrix element values from a spectrum in which the underlying dynamics has a hierarchical structure of initially unspecified nature. Equivalently, knowledge of off-diagonal matrix element distributions enables the delocalization width of a Franck–Condon bright state over the coupled bath of dark states to be estimated.

The motivation for our implementation of parsimonious trees is quite simple and rather qualitative. Two interacting basis states, with unperturbed energies separated by a gap not much larger than the interaction matrix element, $\mathbf{V}$, have perturbed energies separated by slightly more than $2\mathbf{V}$. If a two-line spectrum, with equal intensities and separation $2\mathbf{V}$, is convolved with a Gaussian lineshape, the two lines are resolved only for Gaussian HWHM less than $\mathbf{V}\sqrt{2\ln(2)}$. When there is only one contributing bright state in a local region of the spectrum the resolving HWHM is expected to approximate the coupling matrix element even when the matrix element is so small as to be comparable to the spacing, and when the intensities are are unequal. This can be understood by noting that, if the lines have unequal intensities, a smaller Gaussian HWHM is required to resolve the lines, but the unequal intensities imply a smaller value of the off-diagonal matrix element between the two basis states.

Application of the parsimonious trees method to a spectrum involves three steps:

1. Determining the "branching widths" of the spectrum. The branching widths in a tree structure are the widths (HWHM) of a convolving Gaussian lineshape for which the number of peaks in a spectrum changes by one. For a spectrum of N lines at maximum resolution, there are $N - 1$ branching or split widths.
2. Translating the branching widths and energy levels into a connected tree structure. The connected tree structure is a pictorial representation of the evolution of a spectrum as the width of the convolving Gaussian lineshape decreases.
3. Determining a *reduced* form of the tree using "parsimonious widths." The parsimonious widths characterize clusters of branching widths, and thus, clusters of matrix elements in the spectrum. The principal widths determined by this procedure are the centers of width ranges where the number of lines in the spectrum changes most rapidly.

The first step, determination of the branching widths, is a straight-forward, if time-consuming, bounded-search computational procedure which must be done with sufficient accuracy to resolve the $N - 1$ values. Determining the two coalescing peaks at each split-width allows a tree structure to be drawn; however, this procedure must be done with some care. Associated with each split width are two sets of line positions, the positions of peaks in the spectrum convolved with a Gaussian with the two widths bracketing the split-width. Only a very few lines will show any difference in frequency position between the two bracketing branching widths. Matching the peak positions in the tree which move very little allows the line that disappears as the Gaussian width is increased to be identified by a process of elimination. Completing a tree structure requires knowing whether the disappearing line coalesces with the line to lower or higher frequency. When the disappearing line is similarly distant from strong and weak lines, it is better described as coalescing with the stronger line, because its peak position moves toward the stronger peak as the convolving width is increased. Thus, identifying the two coalescing lines requires considering both the energy spacing to the nearby peaks, and the intensity contributed to the convolved spectrum by lines below and above the disappearing line. Figures 15 and 16, respectively, show a stick spectrum and its corresponding connected tree, which we will discuss in detail below. We have, so far, limited our application of parsimonious widths to stick spectra, although the method could be applied

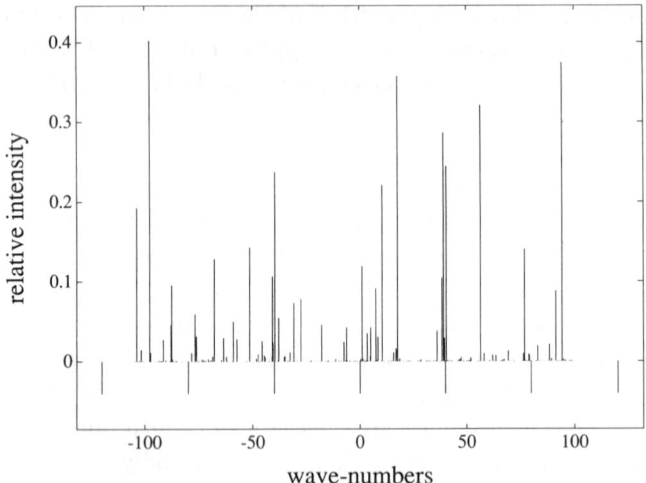

Fig. 15. Synthetic spectrum generated from the K-breakdown model (discussed in Sec. 4.2 and illustrated in Fig. 17). Two synthetic quantum numbers are used in the model: "brightness" (or darkness), and "K". The 7 evenly-spaced zero-order "bright" basis state positions are marked below the spectrum. Each bright basis state has a five-fold K degeneracy.

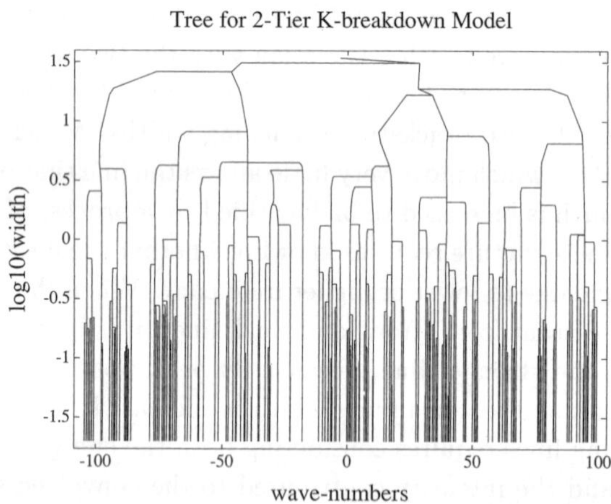

Fig. 16. Tree for the K-breakdown spectrum of Fig. 15. The tree traces the evolution of peak positions in the spectrum as resolution is reduced. Although clumping is not very obvious in Fig. 15, the tree does reveal a weak grouping near each of the 5 bright states in the model within the range of the spectrum.

to complete, unprocessed, spectra. A complete spectrum might contain many spurious peaks owing to baseline noise for very small convolving widths, but those spurious peaks would disappear as the convolving width is increased.

The distribution of branching widths results from the distribution of bright and dark basis state energies, the distributions of coupling matrix elements, and the coupling scheme between or among the bright and dark states. In order to extract information about the distributions of matrix elements from a spectrum, we would like to reduce the $N - 1$ branching widths to a very considerably smaller number of *principal widths*, and in addition, we would like to devise a criterion for the "optimum" reduction. These principal widths will be the centers of width ranges where the number of lines in the spectrum changes most rapidly. We expect these widths to identify interactions and clusters of matrix elements in the spectrum. The optimum reduction would correctly identify the number of distinct interactions operating among the levels observed in the spectrum. These interactions cause an initial basis set with Poisson level statistics to be transformed into a system with the observed spectrum. This reduction would simplify the tree of Fig. 16 by clustering the branch points at the principal widths, and would identify clusters of matrix element values in the spectrum.

To find the optimum set of principal widths for a particular number of principal widths, we define a distance function, $d(w, W)$, for the distance between two widths, where w is a branching width, and W is a principal width, as:

$$d(w, W) = \left| \log \left(\frac{w}{W} \right) \right| . \tag{9}$$

This distance measure was chosen because of the scale-invariant way in which it divides the branching width space between two adjacent principal widths. In comparison with $|w - W/w|$ or $w - W/W$, it treats widths above and below a principal width more reasonably if we are looking for interactions related by multiplicative scale factors. For instance, 0.17 is closer to 0.08 than to 0.73 using Eq. (9), but the order is reversed using $|w - W/W|$, and $|w - W/w|$ puts the dividing point at the arithmetic mean of the two principal widths, while Eq. (9) divides the space at the geometric mean of two adjacent principal widths.

Using this distance function, we define a stress function for the deformation of the tree:

$$S_{N_W}(\{W\}) = \sum_{i=1}^{N_W} \sum_{\substack{j \\ w_j \approx W_i}} w_j [d(w_j, W_i)]^2 \; . \tag{10}$$

The widths w_j are the complete set of branching widths, while the widths W_i are the principal widths. The stress must be minimized over all possible sets of N_W principal widths. The extra weighting factor w_j equalizes the expected contribution to the stress from different principal widths because the expected number of peaks in a limited spectral width is inversely proportional to the width of the peaks. We have also explored the use of the nearest principal width, W_i, as the extra factor, but our optimization procedure is more stable when the non-varying w_j is used, and the results from both schemes are similar. Optimizing the stress function requires a simulated annealing procedure[26] since there may be local minima in the stress function. In effect, this procedure applies parsimony to the logarithmically-scaled tree illustrated in Fig. 16, with a weight correction equalizing the expected number of peaks.

4.2. *Application to Synthetic Spectra*

The stress in the simplified tree will continue to decrease as the number of principal widths increases until it becomes zero when $N_W = N - 1$. Because of the variability of basis-state level spacings (Poisson distributed), and of coupling matrix elements (perhaps Gaussian distributed), the optimum set of principal widths (which characterize the interactions which transformed the basis-state levels into the observed levels) will correspond to some intermediate stress level. A number of synthetic spectra have been examined which suggest an explicit procedure for selecting the optimum number of principal widths.

The spectrum of Fig. 15 was generated by diagonalizing a 2-tier bright state — dark state model called the K-breakdown model. The Hamiltonian for this model is illustrated in Fig. 17. Two synthetic quantum numbers are used in the model: "brightness" (or darkness), and "K". The brightness quantum number is strongly broken by bright–dark coupling, while K is broken at a weaker level by dark–dark interactions. The dark levels consist of N_k subgroups of interspersed levels with each subgroup having Poisson statistics, and a single value of K. The ensemble of all dark levels thus also have Poisson statistics. Although there is a schematic relationship, K is not identical to the K_a quantum number of NO_2, but is simply a counting

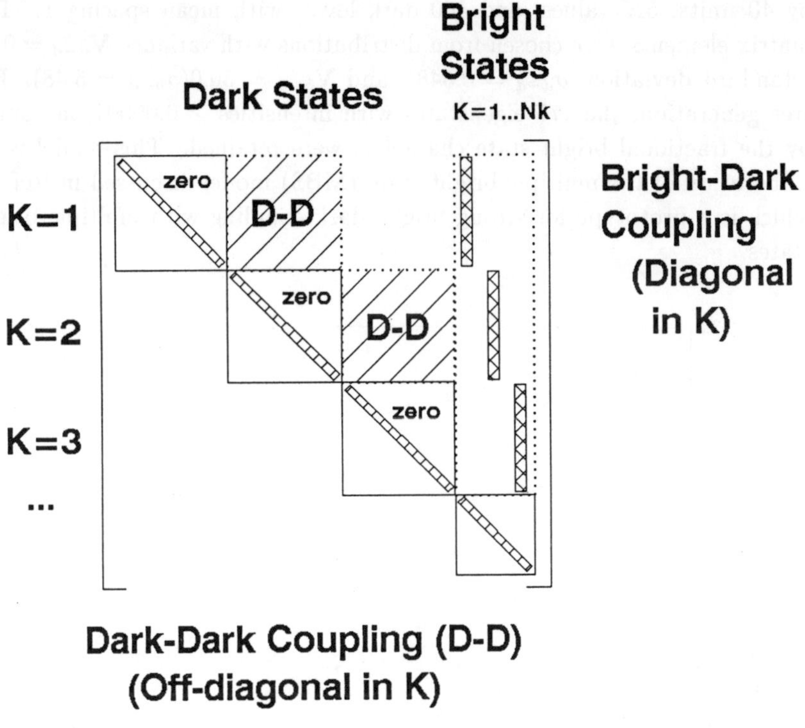

Fig. 17. Hamiltonian for the K-breakdown model. Two synthetic quantum numbers are used in the model: "brightness" (or darkness), and "K". The brightness quantum number is strongly broken by bright–dark coupling, while K is broken at a weaker level by dark–dark interactions. The dark levels consist of N_k groups of Poisson-distributed levels. The bright levels consist of N_b evenly spaced levels, each of which are N_k degenerate. K is simply a counting index.

index. The bright levels consist of N_b evenly spaced levels, each of which is N_k degenerate. The interactions between dark states are off-diagonal by one in K, with matrix elements taken from a normal distribution with mean zero, and variance $\mathbf{V}_{d-d}$. The interaction between the bright and dark levels is diagonal in K, with matrix elements from a normal distribution with mean zero, and variance $\mathbf{V}_{b-d}$. The spectrum of Fig. 15 consists of the central 220 eigenstates of a K-breakdown model with 7 bright levels spaced by 40 units, $5K$ values, and 400 dark levels with mean spacing 1. The matrix elements were chosen from distributions with variance $\mathbf{V}_{d-d} = 0.30$ (standard deviation, $\sigma_{d-d} = 0.548$), and $\mathbf{V}_{b-d} = 30.0(\sigma_{b-d} = 5.48)$. For tree generation, the 177 eigenstates with intensities ≥ 0.00001, measured by the fractional bright state character, were retained. This model is an elaboration of the multiple-bright-state (MBS) model discussed in Ref. 10 which is a prototype for strong bright–dark coupling with multiple bright states.

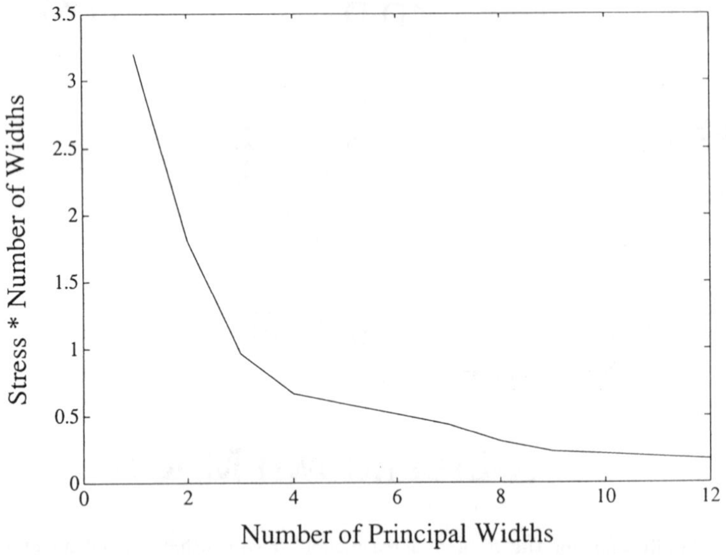

Fig. 18. Stress (Eq. (10)) for the synthetic K-breakdown spectrum of Figs. 15 and 16 as the number of principal widths is increased.

The stress as a function of the number of principal widths is shown in Fig. 18. Because of the non-uniformity of spacings and matrix elements, the two-tier structure appears at a level of significant residual stress. Based on a number of variations of level counts and matrix element distributions, in this model and others, we tentatively suggest a criterion for determining the characteristic, or optimum, number of principal widths. This criterion is based on the parsimonious width structure of Fig. 19, generated by using simulated annealing to minimize the stress in the tree, (Eq. (10)). The largest parsimonious width, as the number of widths increases, eventually

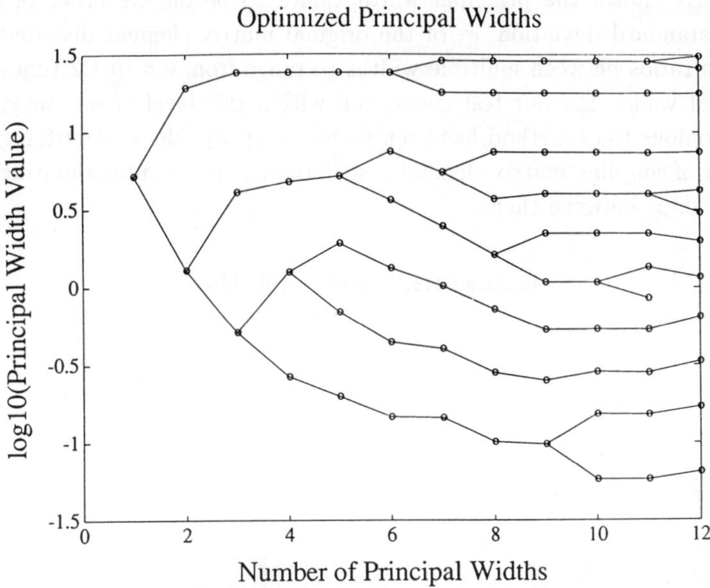

Fig. 19. Principal width values as a function of the number of principal widths for the K-breakdown spectrum of Fig. 15. The optimum number of principal widths is the number at which the principal width for the full spectral width becomes nearly constant. In this spectrum, this occurs for $N_W = 3$.

reaches a nearly constant value which corresponds to sampling the entire width of the spectrum. We have found that the most characteristic number of principal widths is usually the number at which the principal width for the full spectral width becomes nearly constant, and occasionally, one fewer than that number. In this spectrum, the largest principal width becomes nearly constant at $N_W = 3$. The smaller two principal widths occur at values of 4.1 and 0.51, corresponding to the distributions with $\sigma_{b-d} = 5.48$ and $\sigma_{d-d} = 0.548$.

Figure 20 shows the reduced tree for $N_W = 3$. The largest principal width corresponds to sampling the entire spectral width. The next two principal widths correspond approximately to the most probable matrix elements from the two distributions which were used in generating the spectrum, and the fragmentation of the spectrum into 5 bright states is evident. Thus, all of the elements which went into the construction of this spectrum have been revealed by the parsimonious trees method. Other

tests have shown the principal width values to be on the order of 80%
of the standard deviation, σ, of the original matrix element distribution,
and the ratios between multiple widths to range from 0.8 to 1.2 times the
expected value. For our test cases, and within this level of accuracy, the
parsimonious trees method has been found to be capable of identifying the
number of coupling matrix element distributions, and of approximating the
relationships between them.

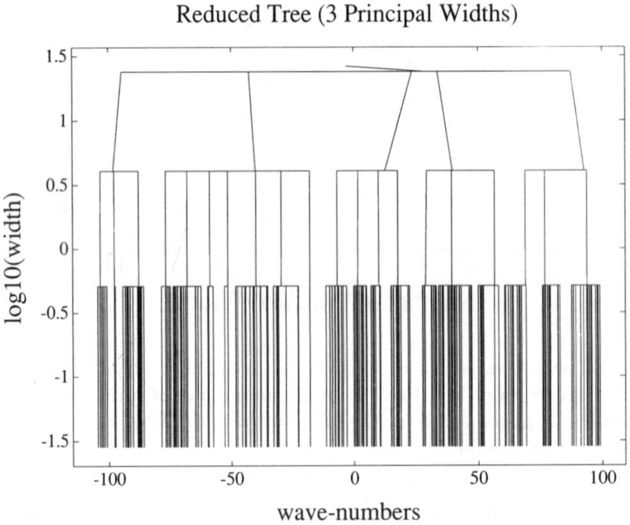

Fig. 20. Reduced tree for the K-breakdown spectrum of Fig. 15 corresponding to
the optimum number of widths, $N_W = 3$. The largest principal width corresponds
to sampling the entire spectral width. The next two principal widths correspond
approximately to the most probable matrix elements from the two distributions which
were used in generating the spectrum. The 5 bright states which are in the selected
spectral range, as marked in Fig. 15, are clearly identified by the reduced tree at the first
level of interaction.

4.3. *Application to the NO_2 Optical Spectrum*

4.3.1. *Coupling mechanisms in NO_2*

The NO_2 optical spectrum is an example of extreme spectral complexity
that continues to challenge experimentalists and theoreticians. Recent
results on the mid-optical region near 590 nm include the extensive analysis

of band origins and low-J data from jet-cooled fluorescence excitation spectra by Jost and others in Grenoble,[19,27] and microwave-detected microwave-optical double resonance (MD MODR) data at intermediate-J by Lehmann and Coy.[3] Although the results from Grenoble have indicated the previously unsuspected presence of collisionally-tranfered signals in the MD MODR datasets, a critical reevaluation[11] has shown that the fundamental results of Ref. 3, which show that the intermediate-J data exhibit dramatic differences from the low-J data, are unchanged. In order to illustrate the capabilities of the parsimonious trees technique, we will describe the possible sources of complexity in the NO_2 spectrum, describe the ambiguities that result from the application of eigenstate level-spacing statistical measures to the NO_2 data, and show how parsimonious trees can contribute to a unified understanding of the two datasets.

NO_2 has four electronic states with origins in the range 0 to 2 eV. In Table 1, we show the character table for NO_2, the symmetries and energies of the electronic states, as well as the symmetries of a number of operators useful in determining the selection rules which regulate the important coupling mechanisms.

The visible spectrum of NO_2 is dominated by the $\tilde{A}^2B_2 \leftarrow \tilde{X}^2A_1$ transition, which is μ_a electric-dipole allowed (parallel transition). Thus, rotational selection rules are determined by the direction cosine λ_a (ΔK_a even, ΔK_c odd), and the vibrational symmetry remains unchanged. The total symmetry (spin-rovibronic symmetry) of NO_2 eigenstates must be A_1 or A_2, and that symmetry must be changed by absorption in the $\tilde{A} \leftarrow \tilde{X}$ transition since $\mu_a\lambda_a$ is A_2. The primary Franck–Condon displacement is along the bend, ν_2, with a smaller displacement along the symmetric stretch, ν_1. Since NO_2 is very nearly an accidental symmetric top, with $\kappa = -0.9986$, transitions with ΔK_a nonzero are expected to be much weaker than $\Delta K_a = 0$, especially for higher J values. Centrifugal mixing is also small at the experimental values of N and K_a.

The potential number of coupling mechanisms affecting the $\tilde{A}^2B_2 \leftarrow \tilde{X}^2A_1$ spectrum is rather large. These mechanisms progressively mix all of the approximate quantum numbers of the molecule: electronic (Γ_{elect}), vibrational (n_1, n_2, n_3), rotational (N), electron spin ($J = N + S$), and nuclear spin ($F = J + I$). Although we will not discuss hyperfine coupling because its effects are only detectable at higher resolution than that of these two datasets, splittings due to the other quantum numbers are

Table I. Character table for NO_2, labeled by the principal inertial axes. For the allowed electronic transition, $\tilde{A} \leftarrow \tilde{X}$, the electronic transition is μ_a electric-dipole allowed. Thus, rotational selection rules are determined by the direction cosine λ_a, and vibrational symmetry does not change. The primary Franck–Condon displacement is along the bend, ν_2, with smaller displacement along the symmetric stretch, ν_1 (even-ν_3 might also be active). Since the $\tilde{}$ vibrational levels which are bright from the $\tilde{X}$ ground vibrational state are of A_1 vibrational symmetry, $\tilde{X}$ vibrational levels coupled to those levels vibronically are of B_2 vibrational symmetry. Because Hund's case (b) spin functions are of A_1 symmetry,[d] rovibronic and e^--spin rovibronic symmetry are identical.

				$N^{16}O_2$ e^--spin rovibronic symmetry	electronic states (T_e cm^{-1})	permanent dipole moments	vibrations	rotations and direction cosines	rigid rotor functions (K_a, K_c)
C_{2v}	E	σ_{bc}	C_{2b}						
A_1	1	1	1	allowed	$\tilde{X}^2A_1$ (0.0)	μ_b, T_b	ν_1, ν_2, even ν_3		$e\ e$
A_2	1	-1	1	allowed	$\tilde{C}^2A_2$ (2.028(9) eV)[a]			R_b, λ_b	$o\ o$
B_1	1	1	-1	forbidden	$\tilde{B}^2B_1$ (1.66 eV)[b]	μ_c, T_c		R_a, λ_a	$e\ o$
B_2	1	-1	-1	forbidden	$\tilde{A}^2B_2$ (9737. cm^{-1})[c]	μ_a, T_a	odd ν_3	R_c, λ_c	$o\ e$

[a] A Weaver, R. B. Metz, S. E. Bradforth and D. M. Neumark, *J. Chem. Phys.* **90**, 2070–2071 (1989).
[b] G. D. Gillespie, A. U. Khan, A. C. Wahl, P. R. Hosteny and M. Krauss, *J. Chem. Phys.* **63**, 3415–3444 (1975).
[c] A. Delon and R. Jost, *J. Chem. Phys.* **95**, 5686–5700 (1991).
[d] J. T. Hougen, *J. Chem. Phys.* **39**, 358 (1963).

fully resolved, and their breakdown contributes to the complexity of the spectrum. Predominant among those mechanisms is vibronic coupling:

1. Vibronic coupling. The importance of vibronic coupling is established by the large number of vibrational bands observed in the optical spectrum.[28] The strong vibronic coupling between the $\tilde{A}$ and $\tilde{X}$ electronic states allows $\tilde{X}$ vibrations of the proper symmetry (B_2 vibronic symmetry) to appear in the absorption spectrum. Allowed transitions from the $\tilde{X}$ ground vibrational state are to $\tilde{A}^2 B_2$ vibrational levels of A_1 vibrational symmetry; vibronic coupling mixes these bright levels with $\tilde{X}$ vibrational levels of B_2 vibrational symmetry (odd n_3). Because ν_3 is the only vibronically active mode, the observation in the spectrum of nearly all $\tilde{X}$ vibrations of the proper symmetry, indicates that additional coupling mechanisms are present.

The additional coupling mechanisms listed below contribute to the spectral complexity by mixing the vibronically-illuminated highly-vibrationally-excited electronic ground state levels with other, nominally dark, ground-state spin-rovibrational levels:

2. Anharmonic vibrational coupling (Fermi (cubic) and higher order coupling). Anharmonic coupling is expected to be significant between the highly vibrationally excited $\tilde{X}^2 A_1$ vibrational levels of B_2 vibrational symmetry, since anharmonic coupling strengths increase with vibrational quantum number. Anharmonic coupling does not cause additional vibrational levels to appear in the spectrum, but does modify the expected vibrational level-spacing statistics. The importance of this mechanism has not been directly established, but may be inferred, as discussed below, using evidence from the density and eigenvalue statistics of the observed spectrum, from analysis of model systems, and from parsimonious trees.

3. Rovibrational coupling (Coriolis and centrifugal distortion). Coriolis coupling in NO_2 may only be of perpendicular type ($\sim q_3 \lambda_c$), since only ν_3 can combine with a direction cosine, λ_c, to give a coupling term in the Hamiltonian with the full molecular symmetry (A_1). This coupling increases with ν_3 quanta, with rotational quantum number (N), and is maximum at low K_a. Centrifugal effects, although weak at the N and K_a values considered here, increase with both N and K_a.

4. Spin–rovibronic coupling. Both parallel (electronic μ_c, vibrational (ν_1, ν_2, or $2\nu_3$), rotational λ_a) and perpendicular (μ_c, ν_3, λ_b) spin–rovibronic coupling mechanisms can couple the dense $\tilde{X}^2 A_1$ levels to levels in $\tilde{B}^2 B_1$. Spin-orbit coupling is expected to be large between the $\tilde{X}$ and $\tilde{B}$ electronic states because they form a degenerate Renner–Teller doublet in the linear configuration. In second order, this electronically off-diagonal interaction causes effective spin–rotation matrix elements to appear between $\tilde{X}^2 A_1$ levels. These interactions cause the breakdown, in the $\tilde{X}^2 A_1$ state, of the body-frame rotation and spin quantum numbers, mixing $N = J - 1/2$ and $N = J + 1/2$ levels, as well as inducing vibrational and rotational mixing. Since the second-order interaction may result from sequential perpendicular or parallel steps, changes in K_a may be as large as 2, and levels of odd n_3 may interact with levels of even or odd n_3 (and conversely). The $\tilde{A}^2 B_2$ and $\tilde{C}^2 A_2$ levels (Table I) also form a Renner–Teller pair in the linear configuration. Shubiya *et. al.*[29] observed this interaction to follow $\Delta K_a = 0$ selection rules in the 19 500 cm^{-1} region. Thus, it does not contribute to K breakdown. The density of $\tilde{C}$ levels is so much reduced at lower energies, that the interaction can be neglected.

Without going into extensive detail, we can summarize the effect of these mechanisms by noting that only vibronic and anharmonic coupling affect the band origins of the bright B_2 vibronic levels, while the effects of the other two mechanisms increase with rotational quantum number, and contribute significantly to the complexity of the intermediate-J double resonance spectrum.

4.3.2. *Eigenstate statistics applied to NO$_2$*

The NO$_2$ band origin data in the optical region[19] covers a much wider spectral range than the MD MODR data. In order to compare statistics from similar parts of the spectrum, and in order to avoid the need for unfolding of the band origin spectrum,[7,19] we have selected the restricted range of band origins from 16 600 cm^{-1} to 17 600 cm^{-1} from Table II of Ref. 19. Two pure sequence datasets have been generated from the MD MODR data: eigenstates with $J = 8.5$ of A_1 total symmetry (accessed from $K_a = 1$ lower states), and eigenstates with $J = 9.5$ of A_2 total symmetry (accessed from $K_a = 0$ lower states). All three datasets should be largely free of spurious levels, although levels may be missing. Δ_3 for these three datasets are plotted in Fig. 21. The band origin data follows the GOE curve closely, while the higher-J data is nearly Poisson. This dramatic difference can be understood as a result of hierarchical levels of coupling, with supporting evidence coming from eigenstate statistics, nuclear dynamics simulations, and parsimonious trees.

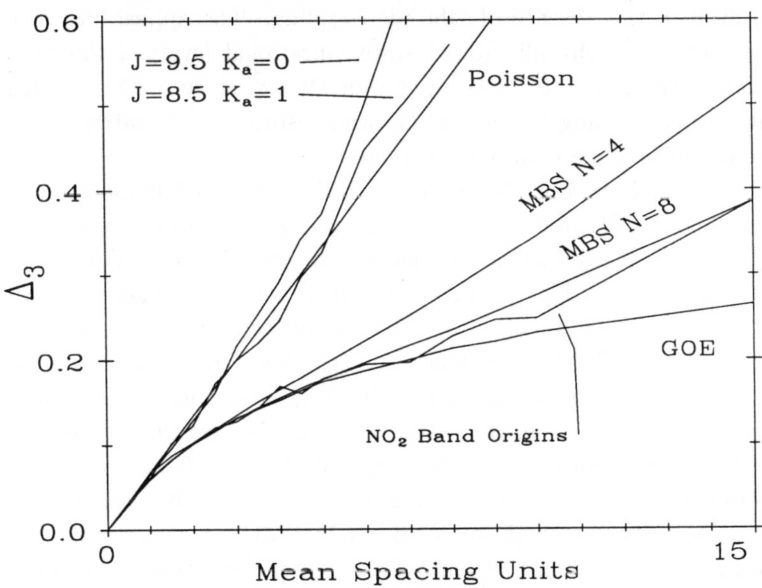

Fig. 21. Δ_3 for NO_2 datasets and for the multiple-bright-state (MBS) model[10] with 4 and 8 bright states. The band origin data[19] in the 16 600 cm^{-1} to 17 600 cm^{-1} range follow the GOE curve closely, while the $J = 8.5$ (from $K_a = 1$) and the $J = 9.5$ (from $K_a = 0$) data are nearly Poisson. The multiple-bright-state model with $N = 4$ (MBS $N = 4$) would be expected to agree with the band-origin data, using the nuclear dynamics results of Ref. 30.

Nuclear dynamics simulations[30] of NO_2 have suggested that the number of $\tilde{A}^2 B_2$ vibrational levels mixed into a typical eigenstate in this region is about 4. The multiple bright state (MBS) model,[10] which was intended to approximate the strong vibronic coupling in NO_2, is a statistical Hamiltonian simulation of a single-tier coupling in which multiple bright states couple very strongly to a bath of dark states with Poisson statistics. Based on the nuclear dynamics results, the MBS model for 4 bright states would be expected to approximate the NO_2 vibrational spectrum, as long as the underlying manifold of dark $\tilde{X}^2 A_1$ vibrational levels follows Poisson statistics.

Figure 21 indicates that the band origins are considerably more rigid than can be expected with 4, or even 8, coupled bright states in the MBS model. This excess rigidity is an indication that anharmonic interactions between the $\tilde{X}^2 A_1$ vibrational levels would significantly mix those basis states, even in the absence of vibronic coupling. The appearance in the spectrum of essentially all ground state vibrational levels of the proper symmetry also argues for mixing within the $\tilde{X}$ state. This evidence indicates that coupling exists, but does not estimate the matrix element values, as can be done with parsimonious trees.

The Δ_3 result in Fig. 21 for the $J = 8.5$ (A_1 total symmetry, from $K_a = 1$) MD MODR data is nearly Poisson, and contains approximately 4 to 5 times as many lines as the number of observed vibrational bands in the same region. This count makes allowance for a small number of bands expected to be missing from Table II of Ref. 19 because of limited dynamic range. The large number of extra lines, in combination with experimental results on breakdown of the N quantum number,[31] indicate that both N and K_a are broken to some extent. We see that the presence of small matrix elements for $\Delta N \neq 0$ and $\Delta K_a \neq 0$ interactions has completely disguised the rigidity of the band origins, resulting in Δ_3 values near Poisson. The very high sensitivity of MD MODR, which results in a dynamic range of between 1000 and 5000 to 1 after correction for saturation, may accentuate the tendency to appear Poisson: many weakly-coupled levels that experience little level-repulsion appear in the spectrum. Because these weak interactions involve the breakdown of N, K_a, and vibrational symmetry, and may include Coriolis, and several types of effective spin–rotation coupling, the 2-tier K-breakdown model is too simple to represent this system. Nonetheless, in results not presented here, we have found that the presence of weak coupling matrix elements can make a rigid spectrum appear much more Poisson, much as occurs in NO_2.

From statistical measures alone, the discrepancy between band-origin and higher-J data is difficult to interpret; the same system appears GOE on one hand, and Poisson on the other. Eigenstate statistics cannot easily be used to recognize the common vibrational structure of these two datasets, or to understand why they are so different. Knowledge of the probable interactions, and comparison with model systems can be combined with the following results from a parsimonious trees analysis to form a more complete picture of the complex NO_2 optical spectrum.

4.3.3. *Parsimonious trees applied to NO_2*

Figure 21 indicates that the NO_2 band origin data is more rigid than expected from the nuclear dynamics simulations. The band origin data, the corresponding tree, and the optimized principal width values are shown in Figs. 22, 23, and 24, respectively. As we have discussed, our criterion for the optimum, or characteristic, number of principal widths is the lowest number that allows the full width of the spectrum to be sampled, as indicated by the appearance of a nearly constant largest principal width. Figure 24 indicates that the optimum number of widths is 3. The largest width corresponds to a Gaussian HWHM which merges the entire spectrum into a single line. The next two smaller widths correspond to additional interactions (clusters of matrix elements) *required* to create a spectrum of the observed complexity. This result provides additional evidence that vibronic $\tilde{A} \leftrightarrow \tilde{X}$ coupling *alone* is unable to fully explain the rigidity observed in the band origin data. Based on the list of NO_2 interactions given above, the two remaining widths are believed to correspond to vibronic $\tilde{A} \leftrightarrow \tilde{X}$ coupling, with an estimated average matrix element of about 56 cm^{-1}, and anharmonic $\tilde{X} \leftrightarrow \tilde{X}$ coupling, with an average matrix element of 8.6 cm^{-1}. The tree reduced to 3 principal widths is shown in Fig. 25.

We have also used the parsimonious trees method to determine the number of levels of coupling required to produce the complexity observed in the $J = 8.5$ (A_1 total symmetry, from $K_a = 1$), and the $J = 9.5$ (A_2 total symmetry, from $K_a = 0$) MD MODR data. The $J = 9.5$ data will not be presented here in order to limit our discussion to the essential results. The $J = 8.5$ levels, with saturation-corrected intensities as observed from the $\tilde{X}^2 A_1$ $N_{K_a,K_c} = 7_{1,7}$, $J = N + S = 7.5$, $F = J + I = 8.5$, lower state, are shown in Fig. 26. The corresponding tree is shown in Fig. 27. Simulated annealing yields the optimized principal widths shown in Fig. 28, for which the optimum number of widths is 4. Fig. 29 is the reduced tree corresponding to that number of principal widths. The spectral range available in the MD MODR data is insufficient to resolve the structure due to vibronic coupling sampled in the band origin data, since that principal width in Figs. 24 and 25 is larger than

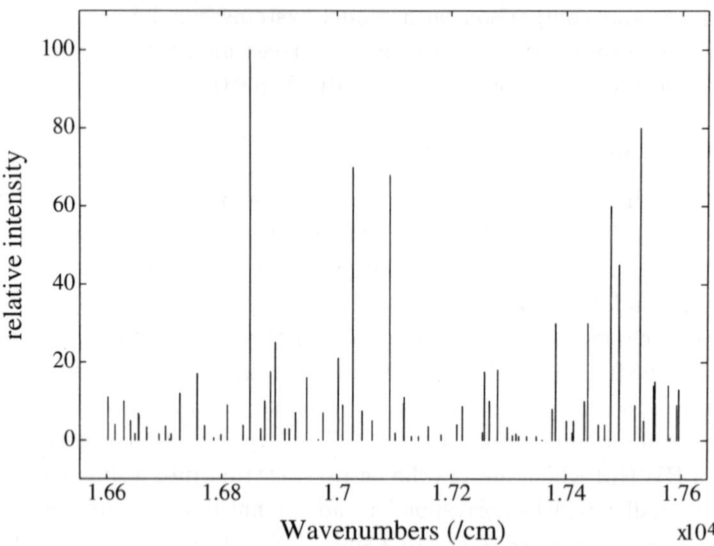

Fig. 22. NO_2 band origins between 16 600 cm^{-1} and 17 600 cm^{-1} identified by Delon *et. al.*[19] This region was selected to overlap the MD MODR datasets.[3,11]

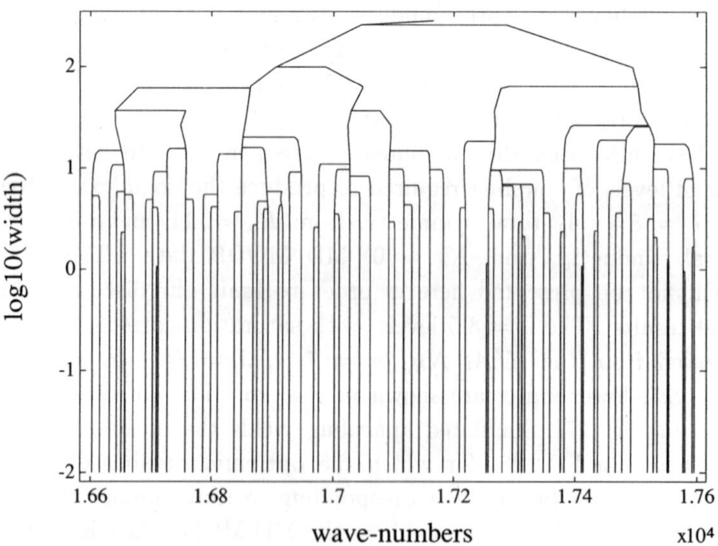

Fig. 23. Tree for the NO_2 band origins of Fig. 22. The tree traces the evolution of peak positions in the spectrum as resolution is reduced.

the full-spectrum width in this dataset. The width illustrated by Figs. 24 and 25, which we tentatively associate with anharmonic coupling among $\tilde{X}^2 A_1$ vibrational levels at 8.6 cm^{-1}, corresponds closely to the width of 9.6 cm^{-1} seen in the $J = 8.5$ data in Figs. 28 and 29. In addition, the parsimonious trees diagnostic shows that two additional clusters of matrix elements with smaller values are necessary to create a spectrum of the observed complexity. These values are 1.9 cm^{-1}, and 0.4 cm^{-1}. Without additional experimental information on the J and K_a dependence of N and K_a breakdown, it is not possible to identify these matrix element clusters with the interactions that we have listed.

When parsimonious trees are applied to the $J = 9.5$ (A_2 total symmetry, from $K_a = 0$) MD MODR data, using intensities observed from $\tilde{X}^2 A_1$ $N_{K_a, K_c} = 8_{0,8}$, $J = 8.5$, $F = 9.5$, the results are quite similar to those for the $J = 8.5$ data, in spite of the differences between the spectra: the $J = 9.5$ range of intensities is somewhat wider than in the $J = 8.5$ data, and the levels are more clustered, as indicated by the super-Poisson Δ_3 values of Fig. 21, which probably indicates that there are more levels missing from this dataset. Just as for $J = 8.5$, the calculation of principal widths shows the optimum number of principal widths to be 4. The matrix element cluster values identified for the the $J = 9.5$ data are similar to, but somewhat larger than, those from $J = 8.5$. The values are 12.3 cm^{-1}, 2.4 cm^{-1}, and 0.5 cm^{-1}. Whether or not these small differences are significant is unknown at this stage of development of the parsimonious trees diagnostic.

We find it encouraging that all three datasets identify the same cluster of matrix elements near 10 cm^{-1}, possibly associated with anharmonic coupling, even though the difference between the level statistics is as large as possible: from near GOE for the band origins, to Poisson, or hyper-Poisson, for the MD MODR data. We also consider it significant that both MD MODR datasets show that two additional levels of clustering are implicit in the data. It is necessary to further validate the parsimonious trees method on additional synthetic datasets, and on additional experimental data with known and unknown interactions. Nonetheless, the method has shown itself to be quantitatively reliable on synthetic spectra, and in accord with qualitative expectations on experimental data.

                    Molecular Dynamics and Spectroscopy

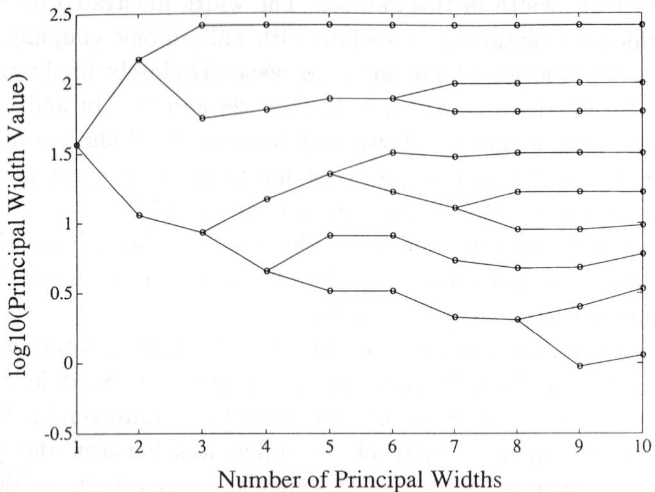

Fig. 24. Principal widths for the NO_2 band origin tree of Fig. 23 obtained by simulated annealing. Based on the criterion that the optimum number of principal widths is the lowest number that allows the full width of the spectrum to be sampled, as indicated by a nearly constant largest principal width, the optimum number of widths is 3. The smaller two widths are believed to correspond to vibronic $\tilde{A} \sim \tilde{X}$ coupling (~ 56 cm^{-1}) and anharmonic $\tilde{X} \sim \tilde{X}$ coupling (~ 8.6 cm^{-1}).

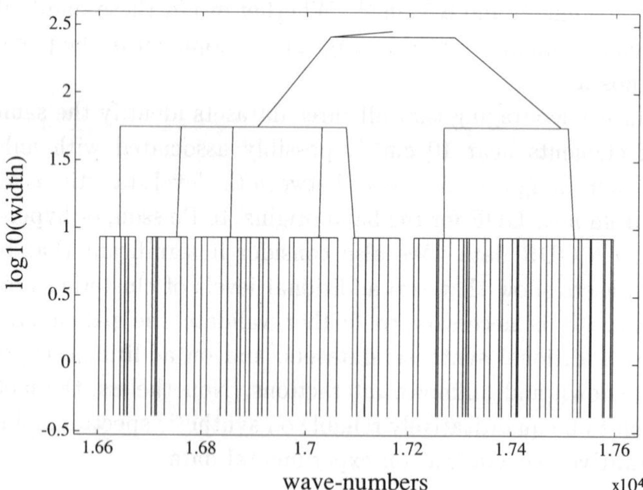

Fig. 25. Reduced tree for the NO_2 band origin data of Fig. 22 corresponding to the optimum number of widths, $N_W = 3$.

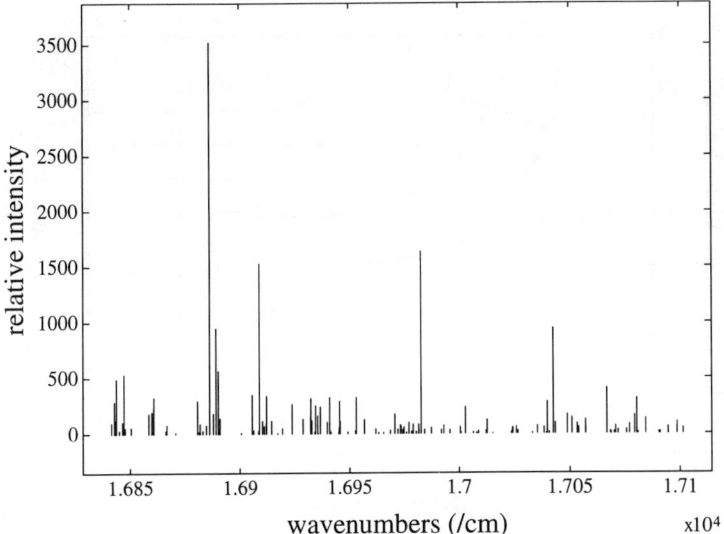

Fig. 26. Pure sequence of NO_2 energy levels with $J = 8.5$ (A_1 total symmetry, accessed from $K_a = 1$) derived from MD MODR data. The $J = 8.5$ levels are shown with saturation-corrected intensities as observed from the $\tilde{X}^2 A_1$, $N_{K_a,K_c} = 7_{1,7}$, $J = N+S = 7.5$, $F = J + I = 8.5$, lower state.

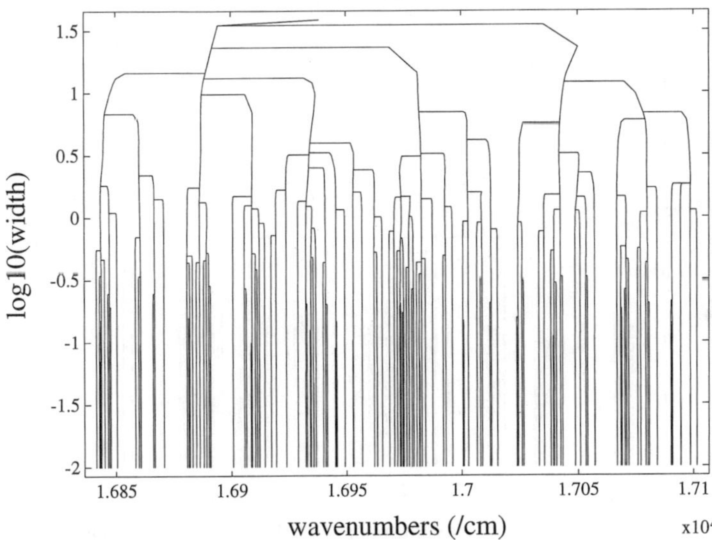

Fig. 27. Tree corresponding to the NO_2 $J = 8.5$ pure sequence of Fig. 26. The tree traces the evolution of peak positions in the spectrum as resolution is reduced.

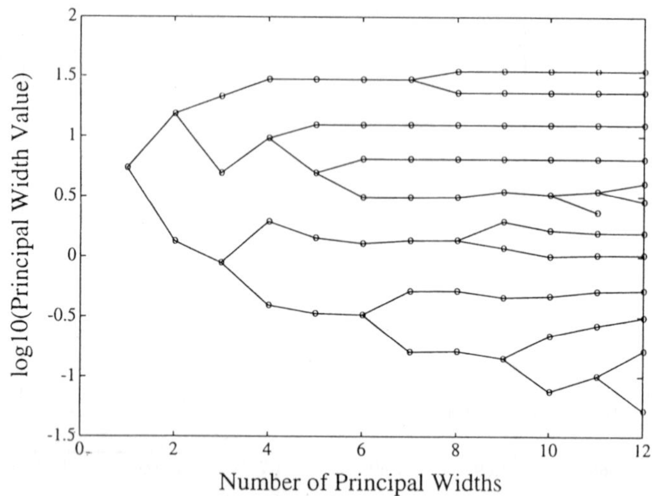

Fig. 28. Principal widths for the NO_2 $J = 8.5$ tree of Fig. 27 obtained by simulated annealing. Based on the criterion that the optimum number of principal widths is the lowest number that allows the full width of the spectrum to be sampled, as indicated by a nearly constant largest principal width, the optimum number of widths is 4. The next smallest width (9.6 cm^{-1}) is close to the smallest width determined for the band origin data (8.6 cm^{-1}). The remaining width values are 1.9 cm^{-1}, and 0.4 cm^{-1}.

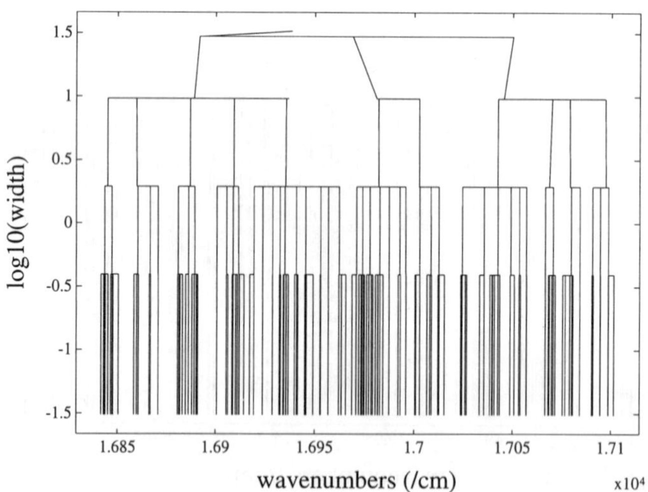

Fig. 29. Reduced tree for the NO_2 $J = 8.5$ tree of Fig. 27 corresponding to the optimum number of widths, $N_W = 4$.

5. Summary

We have described and tested two numerical methods which are designed to recognize patterns and to identify hierarchical levels of coupling in spectra: the extended autocorrelation function (XAC) and the method of parsimonious trees. These methods are especially valuable because other existing methods are too inflexible and too limited to apply to molecular spectra in regions with high vibrational excitation. The traditional spectroscopic techniques of line-by-line assignment and polyad fit models are virtually impossible in highly-congested, strongly-interacting regions of the spectrum. Methods using level statistics, and the related method of power spectrum analysis, have been the only alternative to traditional eigenstate assignment and modeling for extracting dynamical behavior. We have shown that these methods may be quite misleading when the underlying molecular dynamics involves several types of interactions with a hierarchy of strengths. The information encoded in the wealth of data provided by SEP and related double resonance techniques cannot be extracted effectively without rapid and reliable methods of discovering and interpreting the patterns which remain even in the spectra of nearly ergodic systems.

We have shown how the extended autocorrelation function (XAC), allows complex patterns parameterized in a multi-dimensional way to be located in a spectrum in the presence of interfering data. Because the pattern being searched for, as well as the resolution of the search, may be flexibly specified, this technique may be used to locate and assign either broad "feature" states which describe short-time dynamics, or individual eigenstates in a traditional spectroscopic analysis which describes the full dynamics of an entire polyad. We applied the XAC to synthetic spectra and to C_2H_2 dispersed fluorescence data. For the C_2H_2 dispersed fluorescence spectrum, we found that the XAC results were very effective at identifying Franck–Condon progressions, were able to identify individual features and suggest assignments, and were also able to provide strong support for interpretation of the spectrum in terms of $\nu_2 + 2\nu_5$ (*cis*-bend) instead of the alternative ν_1 (symmetric CH stretch).[2]

We have also described and tested a modification of the parsimonious trees method, constructed from the perspective of the experimentalist, which allows direct model-free identification of multiple clusters of coupling matrix element values. This method was initially applied to synthetic spectra in which the underlying dynamics had a hierarchical or sequential structure. That experience has led to a criterion for identifying the number of distinct matrix element clusters necessary to transform an assumed-Poission initial basis set into a spectrum with the observed complexity. When applied to the band origin data for NO_2, this method indicates that an additional weaker cluster of matrix elements, in addition to the established stong vibronic coupling, is necessary to explain the observed near-GOE statistics. Additional support to this conclusion was provided by previous nuclear dynamics and statistical Hamiltonian results. Applying this method to $J = 8.5$ and $J = 9.5$ pure sequences derived from MD MODR data also reveals the same weaker mechanism evident in the band origin data, and indicates the presence of two still weaker classes of interactions which may be rovibrational, or spin–rovibronic in character.

Both of these methods are promising additions to the existing techniques of spectral interpretation and analysis, especially in cases where line-by-line analysis is very difficult, and where existing statistical methods, such as the Fourier transform and Δ_3, are likely to be misleading.

Acknowledgments

Our interest in parsimonious trees was inspired by the work of Michael Davis. We thank him for providing information on his current work. This research has been supported by AFOSR grant 91-0079, and by the Donors to the Petroleum Research Fund (23239-AC6-C), administered by the American Chemical Society.

References

1. D. M. Jonas, S. A. B Solina, B. Rajaram, R. J. Silbey, R. W. Field, K. Yamanouchi, and S. Tsuchiya, *J. Chem. Phys.* **97**, 2813 (1992).
2. D. M. Jonas, S. A. B Solina, B. Rajaram, S. J. Cohen, R. J. Silbey, R. W. Field, K. Yamanouchi, and S. Tsuchiya, *J. Chem. Phys.* **99**, 7350 (1993).
3. Kevin K. Lehmann and Stephen L. Coy, *Ber. Bunsenges. Phys. Chem.*, **92**, 306–311 (1988) and references therein.
4. Stephen L. Coy and Kevin K. Lehmann, *Spectrochim. Acta*, **45A**, 47–56 (1989) and K. Lehmann, and Stephen L. Coy, *J. Chem. Soc. Faraday Trans. 2*, 84, 1389–1406 (1988).
5. D. M. Jonas, Ph.D Thesis, MIT, 1992.
6. Work in progress by S. A. B. Solina and D. M. Jonas.
7. *Statistical Theories of Spectra: Fluctuations*, edited by C. E. Porter (Academic, New York, 1965).

8. M. L. Mehta, *Random Matrices and the Statistical Theory of Energy Levels* (Academic, New York, 1967).

9. O. Bohigas and M. Giannoni, in "Mathematical and Computational Methods in Nuclear Physics", Vol. 209 *Lecture Notes in Physics*, eds. J. S. Dehesa, J. M. G. Gomez, and A. Polls (Springer-Verlag, New York, 1984).

10. S. L. Coy, R. Hernandez and K. K. Lehmann, *Phys. Rev.* **40**, 5935 (1989). The acronym FWHM (full width at half maximum) in this paper is a typographical error everywhere it appears and should read HWHM (half width at half maximum).

11. A paper revaluating the microwave-detected MODR data and more fully discussing the NO_2 optical spectrum in terms of statistical models and levels of coupling is in preparation.

12. L. Leviandier, M. Lombardi, R. Jost, and J. P. Pique, *Phys. Rev. Lett.* **56**, 2449 (1986).

13. R. Jost and M. Lombardi in *Quantum Chaos and Statistical Nuclear Physics*, Vol. 263 *Lecture Notes in Physics*, eds. T. H. Seligman and H. Nishioka (Springer-Verlag, New York, 1986), p. 72.

14. J. P. Pique, Y. Chen, R. W. Field, and J. L. Kinsey, *Phys. Rev. Lett.* **38**, 475 (1987).

15. J. P. Pique, *J. Opt. Soc. Am.* **B7**, 1816 (1990).

16. Kevin K. Lehmann and Stephen L. Coy, *J. Chem. Phys.*, **87**, 5415 (1987).

17. E. J. Heller and R. L. Sundberg, in *Chaotic Behavior in Quantum Systems*, ed. Giulio Casati (Plenum, New York, 1985).

18. J. Wilkie and P. Brumer, *Phys. Rev. Lett.* **69**, 2018 (1992) and the following reply by J. P. Pique, Y. Chen, R. W. Field, and J. L. Kinsey, *Phys. Rev. Lett.* **69**, 2019 (1992).

19. A. Delon, R. Jost and M. Lombardi, *J. Chem. Phys.* **95**, 5701-5718 (1991).

20. (a) Y. Chen, D. M. Jonas, J. L. Kinsey, and R. W. Field, *J. Chem. Phys.* **91**, 3976 (1989). (b) Y. Chen, D. Jonas, C. Hamilton, P. Green, J. L. Kinsey, and R. W. Field, *Ber. Bunsenges. Phys. Chem.* **92**, 329 (1988).

21. K. Yamanouchi, N. Ikeda, S Tsuchiya, D. M. Jonas, J. K. Lundberg, G. W. Adamson, and R. W. Field, *J. Chem. Phys.* **95**, 6330–6342 (1991).

22. (a) B. C. Smith and J. S. Winn, *J. Chem. Phys.* **89**, 4638 (1988). (b) B. C. Smith and J. S. Winn, *J. Chem. Phys.* **94**, 4120 (1991).

23. Michael J. Davis in *Molecular Dynamics and Spectroscopy by Stimulated Emission Pumping*, eds. Hai-Lung Dai and Robert W. Field (World Scientific, 1995).

24. Michael J. Davis, *J. Chem. Phys.*, **98**, 2614–2641 (1993), and references therein.

25. M. V. Berry and M. Tabor, *Proc. R. Soc. London* **356**, 375 (1977).

26. F. Aluffi-Pentini, V. Parisi and F. Zirilli, *ACM Transactions on Mathematical Software* **14**, 345–365 and 366–380 (1988), and *J. Optim. Theo. Appl.* **47**, 1–16 (1986).

27. A. Delon and R. Jost, *J. Chem. Phys.* **95**, 5686–5700 (1991).

28. R. E. Smalley, L. Wharton, and D. H. Levy, *J. Chem. Phys.* **63**, 4977 (1975).
29. (a) Kazuhiko Shibuya, Tadashi Kasumoto, Hidekazu Nagai, and Kinichi Obi, *J. Chem. Phys.* **95**, 720–721 (1991). (b) Hidekazu Nagai, Kazuhiko Shibuya, and Kinichi Obi, *J. Chem. Phys.* **93**, 7656–7665 (1990). (c) Koichi Tsukiyama, Kazuhiko Shibuya, Kinichi Obi, and Ikuzo Tanaka, *J. Chem. Phys.* **82**, 1147–1152 (1995).
30. E. Häller, H. Köppel, and L. S. Cederbaum, *J. Mol. Spectrosc.*, **111**, 377 (1985).
31. Stephen L. Coy, Kevin K. Lehmann, and Frank C. DeLucia, *J. Chem. Phys.* **85**, 4297–4303 (1986).

CHAPTER 25

DYNAMICAL ANALYSIS OF HIGHLY EXCITED VIBRATIONAL SPECTRA: PROGRESS AND PROSPECTS

Michael E. Kellman

Department of Chemistry
University of Oregon
Eugene, Oregon 97403 USA

Contents

1. Introduction

This chapter reviews work toward a program whose ultimate goal is a comprehensive and systematic theoretical framework for analysis of highly excited vibrational states of molecules. The basic motivation arises from the view that the traditional spectroscopic picture of states defined in terms of normal mode quantum number assignments is fundamentally inadequate for the highly excited, strongly coupled, anharmonic regime. Underlying all the work described here is the general outlook that the road to progress lies in harnessing recent developments in nonlinear dynamical systems to problems of highly excited spectra. While I will attempt to give a reasonably self-contained summary of progress achieved in prior work, the greater emphasis is to outline the central challenges facing completion of the program. I will offer a number of suggestions and speculations, by no means conclusive, of possible routes toward the desired end.

The motivation to seek a new framework for spectral analysis arises from the explosive growth of experimental spectroscopy and dynamics at high levels of excitation. These highly excited systems give direct evidence for previously unknown or unobservable dynamical effects in molecules. Examples include:

(i) High overtone and combination levels of strongly coupled vibrations, both by direct absorption and methods such as stimulated emission pumping and resonance Raman scattering;

(ii) Signatures of chaotic classical dynamics in very highly excited spectra with high densities of states[1-4];

(iii) Spectra which give evidence of multiple interacting Fermi resonances[5–7];

(iv) Hierarchies of structure of features and subfeatures[8] in spectra of classically chaotic molecules, related to energy transfer pathways[7,9,10] and approximate dynamical constants of motion[9,10];

(v) Pronounced spectral complexity and intramolecular vibrational coupling induced by strong rotation–vibration interaction in spectra with high-J excitations[11,12];

(vi) Spectra of transition states in chemical reactions[13–15]; and

(vi) New spectral methods for highly excited states of radical species such as CH_2.[16]

These new phenomena offer possibilities for obtaining previously inaccessible information such as energy transfer pathways and reliable potential energy surfaces at high energies. At the same time, unlocking this information presents unprecedented challenges for the spectral interpretation. The strongly coupled and highly anharmonic nature of these detailed high energy spectra introduces fundamental new problems of molecular dynamics which cannot be handled by existing methods. Traditional methods[17] are predicated on assignment of eigenstates in terms of quantum numbers, such as the number of quanta in each normal mode, which are physically meaningful only in the low-energy, near-harmonic regime. In highly excited spectra, these zero-order quantum numbers are destroyed by the strongly coupled dynamics, which means the spectra become unassignable[4] in ordinary terms. What is needed is a new or at least much more general framework for molecular dynamics complicated by strong interactions among some or all of the degrees of freedom. I believe the best chance for obtaining such a framework lies in a systematic understanding of the complex underlying semiclassical dynamics of the molecule. A systematic procedure for classification of the dynamics would make possible a dynamically based assignment of high energy spectra. This framework should be designed to extract as easily as possible the dynamical information encoded in the newly rationalized spectra, including energy transfer processes, and eventually, molecular rearrangements. The main emphasis of this review is the assignment of highly excited spectra with quantum numbers appropriate to the dynamics. However, a brief discussion is presented in Sec. 5 on use of fits of highly excited spectra of chaotic systems to determine potential surfaces for high energy molecules.

Section 2 outlines the accomplishments of our prior work toward analysis of highly excited spectra. Our original goal was a method to extract dynamical information about molecules analyzed with standard spectroscopic fitting Hamiltonians. The result[18] was a way to visualize the dynamics associated with spectra on a phase space construct called the "polyad phase sphere". Success in this original endeavor of decoding and visualizing the dynamical information in spectra led to a more systematic analysis using tools from nonlinear dynamics, in particular, bifurcation theory.[19] This led to an essentially complete classification[20] of the spectroscopic fitting Hamiltonian, and application[21,22] of this analysis to assignment of spectra in terms of new quantum numbers appropriate to the underlying dynamics of the quantum states. This program is essentially complete for comparatively simple systems such as a pair of normal modes coupled by a single resonance interaction.

Sections 3 and 4 deal with the steps now needed for completion of the desired framework for analyzing highly excited spectra. The difficulties are considerable, and the greatest challenges undoubtedly still lie ahead. In contrast to the description of prior work, I will offer some frankly highly speculative ideas. There are two main questions of principle which are unanswered in our work to date; in fact, they touch at the heart of central problems in nonlinear Hamiltonian dynamics. (1) The first problem, discussed in Sec. 3, is the classification of the quantum spectra of chaotic systems. There are still fundamental questions regarding the relation between nonlinear classical dynamics and quantum theory. A body of beautiful and powerful results has been achieved in the last dozen years in dynamical systems theory toward understanding the structure of chaotic classical dynamics of two degree of freedom systems. I believe that important results are waiting to be achieved by establishing the connections of this work to quantum mechanics. (2) The second problem, discussed in Sec. 4 is that of understanding the phase space classification of systems with many degrees of freedom, and using this to classify the quantum spectrum. Unlike two degree of freedom systems, understanding of the classical dynamics of multidimensional systems is just beginning.

2. Dynamics and Assignment of Coupled Vibrations from Experimental Spectra of Systems with a Single Strong Resonance

This section describes prior work on dynamical analysis and quantum number assignment of systems of highly anharmonic vibrations coupled by a single strong resonance, for instance, the coupled stretches in an ABA triatomic. We begin with depiction of experimental spectra on an easily visualized phase space construct called the *polyad phase sphere*. This procedure starts with the fit of the spectrum to a phenomenological Hamiltonian, such as the Darling–Dennison Hamiltonian[23] for coupled anharmonic stretches, or a Fermi resonance Hamiltonian[24] for a coupled stretch and bend. The "Heisenberg correspondence principle" is then used to depict the energy levels in terms of their corresponding classical trajectories on the phase sphere. Subsection 2.1 describes the operational working of the phase sphere procedure. Subsection 2.2 discusses the systematic classification of the phase sphere structure using the tools of bifurcation theory. Subsection 2.3 uses the resulting classification of phase space structure to make new assignments for strongly coupled systems using quantum numbers based on the underlying semiclassical dynamics. Subsection 2.4 finishes with a much more complete description of the classical and semiclassical principles underlying the operational phase sphere procedure of Sec. 2.1. The exposition of Sec. 2.4 is important not only for full understanding of the work presented here in Sec. 2, but also for understanding the issues involved in later sections in extending the program to systems with a strong degree of classical chaos, and systems with many degrees of freedom.

2.1. *Polyad Phase Sphere Dynamics From Fits of Resonance Spectra*

The earliest aim of our work was to devise a systematic method to visualize and classify the dynamics of coupled molecular vibrations using experimental spectral data, for example, a fit with a Darling–Dennison resonance Hamiltonian for the stretches[25] of a symmetric triatomic like H_2O, or a Fermi resonance Hamiltonian for coupled bend and stretch[26] in a molecule like CO_2. We devised a method[18] to associate classical dynamics with the individual vibrational levels obtained from fits of overtone and combination spectra. Each energy level in the spectrum corresponds to a classical trajectory which is mapped on the polyad phase sphere, giving a convenient representation of the dynamics. (The representation of molecular vibrational dynamics on a spherical surface has been investigated by numerous workers[27-31] following work on "rotational energy surfaces" for rotational spectra.[32]) The phase sphere trajectories can be transformed into easily visualized trajectories in coordinate space. These coordinate space trajectories have distinctive patterns which are reflected in the quantum wave functions. The phase sphere procedure has led to a method[21] described in Sec. 2.3 to classify and assign spectra of ABA triatomics which display the transition from normal to local modes. This new assignment procedure has also been applied[22] to Fermi resonance spectra of an interacting stretch and bend in approximate 2:1 frequency ratio.

I present here a brief outline of the phase sphere procedure. (A more detailed treatment is found in Ref. 18.) One begins with a standard effective Hamiltonian for fitting an experimental spectrum. For instance, for a pair of identical coupled stretches such as the local mode O–H stretches in water, the Hamiltonian is[18]

$$\begin{aligned}
\mathbf{H} &= \mathbf{H}_0 + \mathbf{V}_{1:1} + \mathbf{V}_{2:2} \\
\mathbf{H}_0 &= \omega_0(N+1) + \alpha/2(n_1 + 1/2)^2 \\
&\quad + \alpha/2(n_2 + 1/2)^2 + \alpha_{12}(n_1 + 1/2)(n_2 + 1/2) \\
\mathbf{V}_{1:1} &= 1/2[\beta + \varepsilon/2(n_1 + n_2 + 1)(a_1^+ a_2 + a_2^+ a_1)] \\
\mathbf{V}_{2:2} &= \delta'(a_1^+ a_1^+ a_2 a_2 + a_2^+ a_2^+ a_1 a_1) .
\end{aligned} \tag{1}$$

(This local mode Hamiltonian is equivalent to the standard Darling–Dennison Hamiltonian in the normal mode representation[18,25,33].)

The action of the raising and lowering operators is defined for the anharmonic local modes by analogy to harmonic oscillators (see Ref. 18 for a more detailed discussion of this point):

$$\begin{aligned}
a_i^+ |n_i\rangle &= (n_i + 1)^{1/2}|n_i + 1\rangle \\
a_i |n_i\rangle &= n_i^{1/2}|n_i - 1\rangle .
\end{aligned} \tag{2}$$

The connection with classical mechanics is made by taking the Heisenberg correspondence[34,35] of the raising and lowering operators to expressions in terms of the actions and conjugate angles:

$$a_i^+ \to (n_i + 1/2)^{1/2}e^{i\phi_i}, \qquad a_i \to (n_i + 1/2)^{1/2}e^{-i\phi_i} . \tag{3}$$

A simple canonical transformation is now very useful, with actions

$$I = 1/2(n_1 + n_2 + 1), \qquad I_z = (n_1 - n_2) \tag{4a}$$

and conjugate angles

$$\theta = (\phi_1 + \phi_2), \qquad \psi = (\phi_1 - \phi_2) . \tag{4b}$$

Using the Heisenberg correspondence (3), in terms of the variables (4), the classical correspondent of the quantum Hamiltonian (1) is

$$H = 2\omega_0 I + (\beta + \epsilon I)(I^2 - I_z^2)\cos\psi + \alpha_1 I^2 + \alpha_2 I_z^2 + 2\delta'(I^2 - I_z^2)\cos 2\psi \tag{5}$$

with

$$\alpha_1 = 1/2(\alpha + \alpha_{12}), \qquad \alpha_2 = 1/2(\alpha - \alpha_{12}) .$$

The total action I is a very important quantity for the model fitting Hamiltonian (1) and its classical analog (5). Because the classical Hamiltonian is independent of the rapidly varying angle θ, the conjugate action I is a constant of the motion. Equivalently, I commutes with the quantum Hamiltonian (1) and is a conserved quantity. The quantum Hamiltonian is therefore diagonal in blocks of constant I. Diagonalizing these sub-blocks gives a set of $2I = (n_1 + n_2 + 1)$ levels. This set of levels is sometimes referred to as a "polyad" characterized by the total or polyad quantum number

$$P = (n_1 + n_2) = 2I - 1 . \tag{6}$$

From the classical Hamiltonian (5), phase space trajectories can be obtained corresponding to the individual energy levels of the spectrum. These trajectories are obtained from the fit to the experimental spectrum by the following procedure. The least-squares fit to the spectrum gives a set of optimized parameters for the Hamiltonian. From diagonalization of this Hamiltonian are obtained a set of quantum energy levels, presumably with a good match to the available experimental levels used as input for the fit. The Hamiltonian is block diagonalized into polyads of levels defined by their common value of the conserved total quantum number $P = 2I - 1$. Corresponding to these energy levels are the classical trajectories which we wish to obtain. Because of the existence of the extra conserved quantity I, the corresponding classical motion is quasiperiodic: the trajectories move

     Molecular Dynamics and Spectroscopy

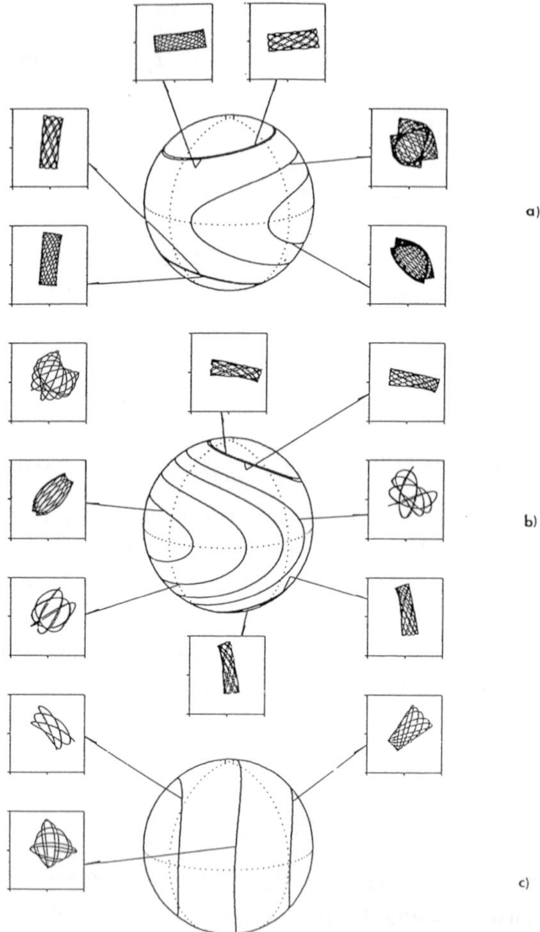

Fig. 1. Polyad phase spheres with trajectories corresponding to levels from nonlinear least squares fit to experimental spectra, and corresponding coordinate space trajectories. (a) H_2O polyad 3. (b) O_3 polyad 5. (c) SO_2 polyad 2. For each energy level, we solve for the action I_z as a function of ψ; the solution $I_z(\psi)$ is the trajectory corresponding to the given energy level. Each trajectory with a given value of the total quantum number P is plotted on a sphere with radius proportional to $2I = P + 1$. Points on a trajectory are parameterized on the sphere by the azimuthal angle ψ and a polar angle θ given by $\Theta = \cos^{-1}(I_z/I)$. The set of all such trajectories plotted on a sphere for the energy levels of a given polyad P constitutes the polyad phase sphere. There is one such sphere for each value in the spectrum of the polyad number P.

on surfaces of constant energy and action I. As discussed in detail below in Subsection 2.4, the motion of each trajectory is therefore confined to a surface which is an invariant torus. One degree of freedom on the invariant torus is the rapidly varying angle θ conjugate to the fixed action I. However, because the Hamiltonian (5) is independent of θ, this degree of freedom is trivial and plays no role in our visualization. Rather, the important phase space behavior is all contained in the nontrivial mutual dependence of I_z, ψ in the Hamiltonian (5). This is a one-dimensional problem solvable by quadrature, rather than by recourse to numerical integration of the full Hamilton's equations. For each energy level, we solve for I_z as a function of ψ; the solution $I_z(\psi)$ is the trajectory corresponding to the given energy level. Each trajectory with a given value of the total quantum number P is plotted on a sphere with radius proportional to $2I = P + 1$. Points on a trajectory are parameterized on the sphere by the azimuthal angle ψ and a polar angle θ given by $\Theta = \cos^{-1}(I_z/I)$. (The polar angle Θ on the sphere is not to be confused with the phase space variable θ.) The set of all such trajectories plotted on a sphere for the energy levels of a given polyad P constitutes the polyad phase sphere. There is one such sphere for each value in the spectrum of the polyad number P.

The phase sphere analysis for stretches of ABA molecules has been applied most extensively[18] to experimental spectra of H_2O, O_3, and SO_2. This analysis shows that H_2O is a molecule which displays the normal to local modes transition, while SO_2 is a pure normal modes molecule. Surprisingly, ozone also displays the normal-local transition. The phase sphere and coordinate trajectories for a representative experimental polyad of each of these molecules is shown in Fig. 1. Other examples analyzed in detail include 2:1 Fermi resonance between stretch and bend modes in a variety of molecules.[20]

2.2. *Bifurcation and Catastrophe Map Classification of Strongly Coupled Dynamics*

The phase sphere gives a very convenient representation of molecular dynamics extracted from the information inherent in the experimental spectrum. The phase sphere was first used[18] to analyze systems of coupled stretches which have the normal/local modes transition, as in Fig. 1. The next step[20,22] has been to investigate a systematic classification of the dynamical information in the phase spheres of spectra of molecular vibrations coupled by arbitrary Fermi resonance, such as systems in 2:1 frequency ratio. Essentially, what is desired is a *systematic generalization of the language of the normal-local transition.* This is a compelling goal for the following reasons.

The idea of the transition from normal to local modes[36–41] has been a landmark in thinking about dynamics of coupled molecular vibrations. It has stimulated new interpretations of spectra,[42] new formulations of the quantum mechanical description of vibrations,[43] and theoretical investigations of the transition's dynamical origin.[44,45] The dynamics of ABA triatomics depicted on the phase sphere in Fig. 1 are described in a completely satisfactory way in the language of the normal-local transition. However, we found[46] that the normal-local taxonomy was inadequate to characterize the behavior of trajectories on the phase sphere of systems with 2:1 Fermi resonance. To get beyond these limitations of the language of the normal-local transition, we have used bifurcation theory[19] to obtain the *phase space bifurcation structure* of resonant systems. In our view, the systematic taxonomy of phase space structure brought to light in the bifurcation analysis constitutes the desired classification of the dynamical information contained in fits of spectra.

The basic goal of a bifurcation analysis is to understand the variation in the structure of phase space as all the relevant *control parameters*[47] of the system are systematically varied. In the case of our molecular fitting Hamiltonian, the six fitting parameters can be reduced to two reduced control parameters.[19] One parameter μ measures the coupling strength between the two oscillators. (For example, these oscillators might be the two local stretch modes of an ABA molecule, or the coupled stretch and bend normal modes common in 2:1 Fermi resonance systems.) The other parameter β' measures the asymmetry between the oscillators, i.e., if the oscillators are in approximate $n : m$ frequency resonance, the asymmetry β' measures how far off exact resonance the zero-order harmonic frequencies are. For example, in ABA triatomics with identical local mode stretches, the asymmetry β' is zero.

Once the control parameters of a system are determined, the problem is to understand how the structure of the phase space changes as the parameters are varied. The structure of phase space is defined in terms of the structure of the *fixed points* and their *bifurcations.* It may be helpful at this point to illustrate both these terms by reference to a pictorial example.

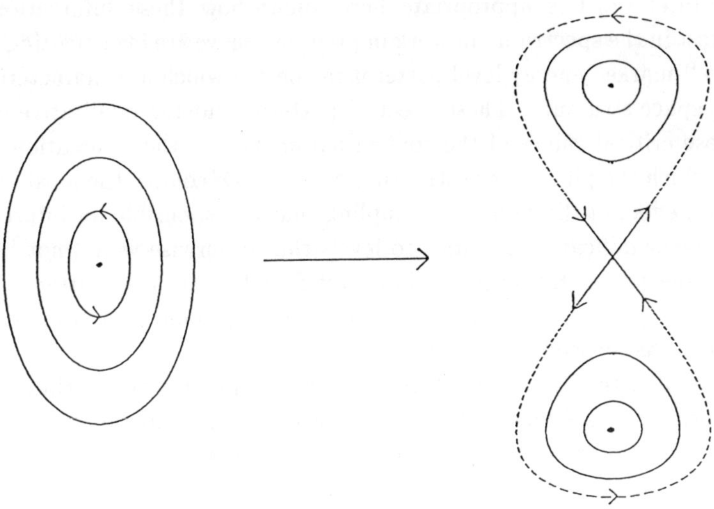

Fig. 2. Pitchfork bifurcation in which an elliptic fixed point (stable normal mode) bifurcates into an unstable fixed point and two elliptic fixed points (unstable normal mode and stable local modes). Elliptic fixed points are dots at centers; hyperbolic fixed point is at "X" of figure eight.

Figure 2, left, shows a phase space flow similar to that for one of the normal modes on the phase sphere of SO_2, a system with pure normal modes dynamics (See Fig. 1.) The phase space is characterized by flow around a *stable*, or *elliptic* fixed point (so-called because of the geometry of flow in phase space in the vicinity of the fixed point). Figure 2, (right), shows the same system, but with the coupling between modes reduced. The system has now bifurcated and a new fixed point structure has come into being. The elliptic point of Fig. 2, (left) has become unstable, turning into an unstable or hyperbolic fixed point (with "X" shaped flow) from which are spawned two new elliptic points (the local modes). As the coupling is reduced further from its value at the bifurcation, the new fixed point structure remains. The unstable fixed point persists (still as a normal mode), and the two stable local modes move away from the unstable point. In general, we say that the phase space structure is determined by the number, types, and organization of the fixed points. This fixed point structure changes at certain critical values of the control parameters at which bifurcations occur.

A brief word is appropriate here about how these bifurcations are manifest in the spectrum. In work in preparation we are investigating subtle but well-marked energy level patterns in spectra which are characteristic of phase space structure. These spectral patterns undergo qualitative change at those critical values of the control parameters — the bifurcation points — at which the phase space structure changes. Of course, the most obvious spectral effect of the resonance coupling that is responsible for bifurcations is to give significant intensities to levels that otherwise would not be seen in the spectrum. However, we have not found systematic patterns in the intensities characteristic of specific bifurcations, so our search for spectral patterns has focused on the energies.

Formally, the fixed point behavior for the spectral fitting Hamiltonian is determined as follows. The condition of a point on the sphere (I_z, ψ) being a fixed point of the Hamiltonian $H(I_z, \psi)$ is

$$\dot{\psi}(I_z, \psi) = d\psi/dt = 0$$
$$\dot{I}_z(I_z, \psi) = dI/dt = 0 \ . \tag{7}$$

These fixed point conditions are satisfied by those solutions of Hamilton's equations for which

$$\dot{\psi} = \frac{\partial H}{\partial I_z} = 0$$
$$\dot{I}_z = -\frac{\partial H}{\partial \psi} = 0 \ . \tag{8}$$

It can easily be shown[19] from the second equation that all of the fixed points for the Hamiltonian (5) lie on a *great circle on the sphere*. This great circle can be parameterized by an angle $0 \leq \alpha \leq 2\pi$ such that

$$\sin \alpha = I_z/I \ . \tag{9}$$

It is worth pausing at this point to summarize where the analysis has taken us so far. We are relating spectra of the quantum fitting Hamiltonian (1) to its corresponding classical Hamiltonian (5). These Hamiltonians have the special property that they have the polyad number I as a conserved quantity. As already discussed earlier and further discussed in Subsecs. 2.4 and 4.2, the existence of the polyad number yields enormous simplifications to several aspects of the problem of analyzing highly excited spectra. For the bifurcation analysis, it implies that all the fixed points on the phase sphere lie on one great circle. It has the further very important implication that the fixed point conditions (7) and (8) and their bifurcations are easily solved for, essentially analytically, without recourse to numerical integration of Hamilton's equations. The details of finding these solutions may be found in Ref. 19. Here, we proceed with the methods for depiction and analysis of the results for the fixed points and bifurcations.

Information about the fixed points and their bifurcations is plotted on a *bifurcation diagram*. One way of plotting such a diagram is to fix the value β' of the molecular asymmetry, and to plot the fixed points with the reduced coupling strength μ as x-axis and the angle α as y-axis. The resulting bifurcation diagram for $\beta' = 0$, i.e., for the stretches of an ABA triatomic, is shown in Fig. 3. This shows that the Darling–Dennison Hamiltonian with local to normal transition is characterized by a *pitchfork bifurcation*. The change in phase space flow as a pitchfork bifurcation occurs has been seen in Fig. 2. The behavior of the phase sphere under a pitchfork bifurcation is illustrated in Fig. 3. As in Fig. 2, at the bifurcation, a stable normal mode becomes unstable and two new stable fixed points representing local modes are born. Similar bifurcation diagrams can be plotted for the Darling–Dennison Hamiltonian for other values of the asymmetry parameter β'.

We call the Darling–Dennison Hamiltonian for coupled stretches the Hamiltonian for a 1:1 resonance system because of the frequency ratio of the stretching modes when they are in exact resonance. Bifurcation diagrams can be constructed for spectral fitting Hamiltonians satisfying arbitrary frequency ratios $n : m$. As discussed above, the behavior of the phase sphere as the molecular parameters are varied cannot be described[46] in terms of the local-normal transition, which we have seen is characterized by a pitchfork bifurcation. This is readily understood from the bifurcation diagrams for the 2:1 system. We found[19] that this system is characterized by a *transcritical*[47] rather than a pitchfork bifurcation. It is therefore not surprising that the 1:1 and 2:1 systems have very different qualitative behavior in phase space with variation of the parameters. (As noted in paragraph 5 of this subsection, work is in progress on characteristic spectral patterns associated with these bifurcations.)

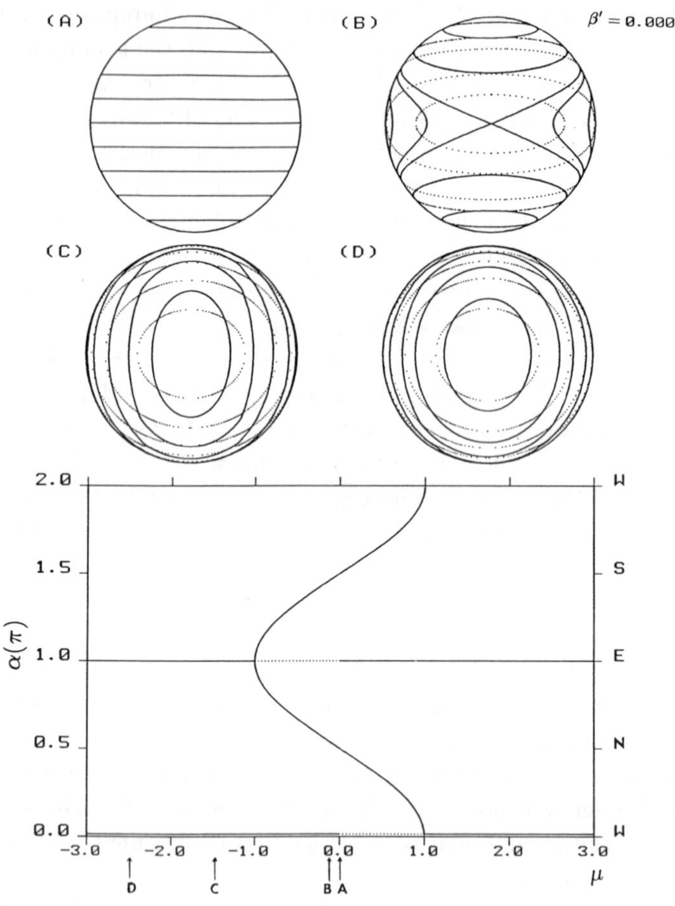

Fig. 3. Bifurcation diagram for the normal-local transition of the stretches of a symmetric triatomic such as H_2O, described mathematically as a pitchfork bifurcation. Phase spheres, illustrated with local modes at the poles and normal modes on the equator, are shown for the four values of the reduced coupling parameter μ indicated at the bottom of the bifurcation diagram. The value β' of the molecular asymmetry is fixed, in this case at $\beta' = 0$, and the fixed points are then plotted with the reduced coupling strength μ as x-axis and the angle α as y-axis. Stable and unstable fixed points are represented by solid and dotted curves, respectively. At the bifurcation point, a stable normal mode becomes unstable and two new stable fixed points representing local modes are born. Similar bifurcation diagrams can be plotted for the Darling–Dennison Hamiltonian for other values of the asymmetry parameter β'.

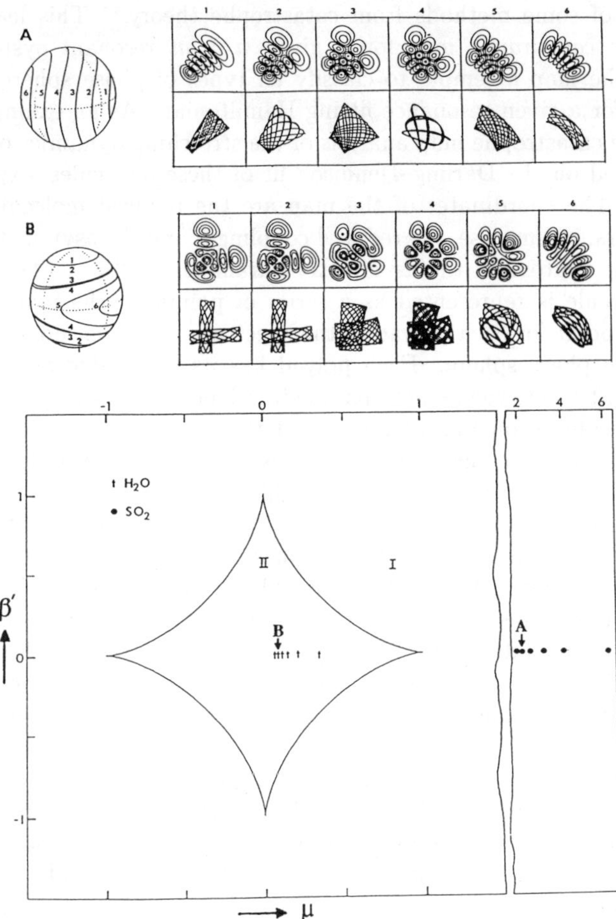

Fig. 4. Catastrophe map classification of spectra of 1:1 resonant systems with normal-local transition. The axes are labeled by the reduced coupling strength μ and the reduced asymmetry β'. The spectrum of a molecule is represented as a series of points on the map. There is one such point for each polyad of the spectrum, i.e., each point represents a separate phase sphere. Each polyad has its own value of the reduced coupling μ because the reduced parameters depend on the polyad number.[19] The map shows the locations of polyads 1–6 of SO_2 (A, zone I) and polyads 1–6 of H_2O (B, zone II). The phase spheres and corresponding trajectories and probability densities for all of the states belonging to polyad 5 of SO_2 (A) and H_2O (B), marked by bold downward arrow, are shown above the catastrophe map.

The bifurcation analysis becomes especially transparent and useful with the help of some methods from catastrophe theory.[48] This leads to the *molecular catastrophe map* for a given $n : m$ resonant system. The catastrophe map attempts to classify all types of phase sphere structure possible for a given resonance fitting Hamiltonian. As an example, Fig. 4 shows the catastrophe map analysis of the stretching dynamics of SO_2 and H_2O, based on the Darling–Dennison fit of these molecules' experimental spectra. The coordinates of the map are the reduced molecular control parameters, in this case, the reduced coupling μ and the asymmetry β'. For symmetric triatomics, the asymmetry parameter is $\beta' = 0$. The spectrum of a molecule is represented as a series of points on the map. There is one such point for each polyad of the spectrum, i.e., each point represents a separate phase sphere. Each polyad has its own value of the reduced coupling μ because the reduced parameters depend on the polyad number.[19] This dependence on polyad number of the position on the map of each polyad is related to the quite natural expectation that the dynamics will vary with polyad number, and so, roughly speaking, with the energy. The map is divided into zones, each of which has a distinct type of phase sphere structure. The boundaries between zones are the points in the parameter space μ, β' at which bifurcations take place. (For instance, the pitchfork bifurcation is represented on the catastrophe map as the two pointed cusps at the zone boundaries for $\beta' = 0$.) The classification of the map into zones is, therefore, tantamount to the classification of phase space structure types for the resonance Hamiltonian. The catastrophe map in Fig. 4 for the 1:1 system has two zones. Zone II is that for which phase spheres have the normal-local transitions, i.e., a mixture of normal and local mode regions. Zone I has no such transition. Figure 4 shows that SO_2 is a molecule for which all the polyads of the experimental spectrum have pure normal modes. H_2O is a molecule for which all the polyad spheres show the normal-local transition. The profound difference between zones is seen in the phase spheres, illustrated in Fig. 1 for a polyad of each molecule.

Figure 5 shows the catastrophe map analysis of the 2:1 Fermi resonance dynamics of CO_2 and the coupled C–H bend–stretch motion in a series of substituted methane molecules. It is readily seen that the catastrophe map classifications for the two kinds of resonance, 1:1 Darling–Dennison and 2:1 Fermi resonance, are quite different. The 1:1 system had only two zones. In contrast, the catastrophe map for 2:1 Fermi resonance has a very different structure, with four zones instead of two. The existence of different catastrophe map zone classifications for different resonance orders $n : m$ is intimately related to the occurrence of different types of bifurcations for

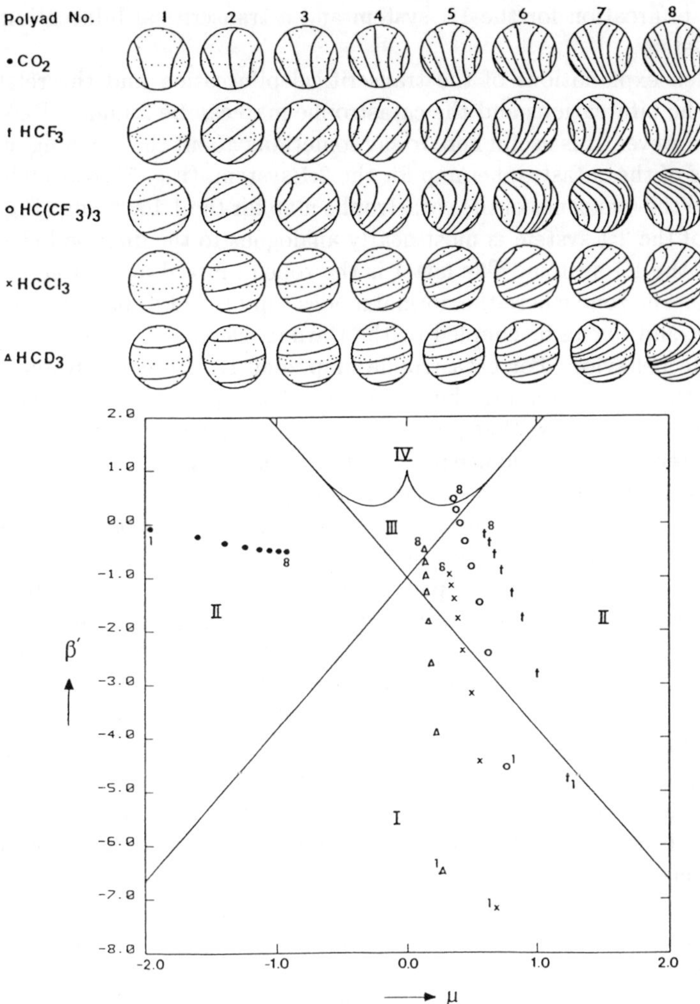

Fig. 5. Catastrophe map classification of polyads 1–8 of spectra of CO_2, HCF_3, $HC(CF_3)_3$, $HCCl_3$, and HCD_3 on the map for the 2:1 Fermi resonance system. Each molecule is represented by one of the symbol types. Each individual polyad of a molecule is marked at a point on the map by the symbol for that molecule. The polyads are partially numbered on the map by their polyad quantum number. The phase spheres for the polyads marked on the map are shown, with the polyad quantum numbers given at the top of the figure.

different resonance orders $n : m$, as discussed above — for instance, a pitchfork bifurcation for the 1:1 system and a transcritical bifurcation for the 2:1 system.

Detailed explanations of the transcritical bifurcation and the relation between the bifurcations and the catastrophe map can be found in Refs. 19 and 20. However, it is useful here to say something about the meaning of the four zones of the catastrophe map for the 2:1 system, (Fig. 7), especially in relation to the two zones of the catastrophe map for the 1:1 system, (Fig. 4). Zone III of the 2:1 system is most nearly analogous to the diamond-shaped zone II of the 1:1 system. The latter is the zone of mixed normal and local modes, i.e., the zone where the normal modes can be said to be in frequency resonance, and so can be coupled by the Darling–Dennison coupling to give local mode behavior. For catastrophe map zone II of the 1:1 system, we follow customary terminology by saying the phase sphere has a "resonance region". We say likewise that phase spheres in catastrophe map zone III of the 2:1 system have a resonance region. In contrast, zones I, II, and IV of the 2:1 system taken together are somewhat analogous to the 1:1 system zone I, which has an undivided phase space with no resonance region. Zones I and IV of the 2:1 system share this property of an undivided phase space. However, the analogy is not complete, because although the 2:1 zone II lacks a resonance region, it does have a cusp at the north pole. This cusp lies on a separatrix which divides two regions of "resonant collective modes", as discussed in detail below in connection with Fig. 7.

We end this subsection with two final remarks. First, as discussed briefly earlier in this subsection, we are working on energy level patterns associated with the bifurcation structure we are discussing. It must suffice to say here that there are very interesting energy level patterns characteristic of all of the types of phase space structure we have been discussing here. Second, the relationship between catastrophe maps for different resonances $n : m$ involves some interesting topological questions. It would be very interesting to give a unified treatment of *all* resonances. If this proved to be possible, it would potentially lead to a universal assignment scheme for all systems with a single resonance coupling. At present, we must be content with using the individual catastrophe maps for specific resonance orders $n : m$ to devise a new, dynamically-based assignment procedure — the subject we consider next.

2.3. *New Assignments of Spectra of Strongly Coupled Vibrations*

The information from the polyad phase sphere and catastrophe map analysis is the key to a new method for assigning spectra, rooted in a systematic understanding of the underlying semiclassical dynamics. As noted in paragraph 5 of the preceding subsection, the most obvious effect of resonant couplings is often to give significant intensities to levels that would otherwise be unobservable. Therefore, it is important to be able to use quantum number assignments to organize the spectra when the resonant couplings are producing increased spectral complexity. But the resonant couplings destroy the meaning of the usual normal mode quantum numbers precisely here, when they are most needed. The traditional assignment of normal mode quantum numbers has little meaning (except as a mere labeling device[49]) in a molecule, for example, with the normal-local transition where some levels correspond to normal mode dynamics, but others to local mode dynamics. Instead, it makes far more sense to make a *mixed assignment* of normal mode quantum numbers to levels with normal mode dynamics, and local mode quantum numbers to levels with local mode dynamics. An example of this new assignment procedure[21] for the normal-local transition of H_2O is given in Fig. 6.

In contrast to this assignment for the 1:1 resonance normal/local system, the new assignment[22] for a molecule with 2:1 Fermi resonance is more novel, since the phase space is much less familiar than the simple normal-local taxonomy, as shown by the existence of four zones in the catastrophe map of Fig. 5. The catastrophe map is thus the key to extending the new assignment for 2:1 systems beyond the normal-local classification of ABA triatomics. We illustrate the procedure for the experimental C–H bend–stretch spectrum of $CH(CF_3)_3$ with Fermi resonance, using the polyad with $P = (n_s + n_b/2) = 5$ as an example. (This assignment is for one of a series of spectra of substituted methanes classified[20] by the catastrophe map method.) From the catastrophe map, (Fig. 5), this polyad lies in zone II. In this zone, the meaning of an assignment with zero-order bend and stretch quantum numbers has been completely lost. Instead, the vibrations in zone II are better characterized as "resonant collective modes", to be described shortly. The structure of zone II is seen in the phase sphere for $P = 5$ in Fig. 7. There are two regions a and b on the sphere for two different types of collective modes, divided by a separatrix, shown as a dashed line. The new assignment of the energy levels on the basis of quantum numbers for these collective modes is given in Fig. 7. The division into regions characterized as resonant collective modes is reflected in the coordinate space trajectories and wave functions for the levels in the polyad. In particular, the overtone states of the collective modes assigned $[0\ 10]_{IIa}$ and $[10\ 0]_{IIb}$ have the characteristic "horseshoe" shape of a 2:1 resonant periodic orbit. This is especially evident for $[0\ 10]_{IIa}$. The fixed

	EXPERIMENT	FIT	ASSIGNMENT
$P = 1$	3657.05	3654.13	$(10)^+$
	3755.93	3754.92	$(10)^-$
$P = 2$	7201.54	7199.69	$(20)^+$
	7249.82	7249.67	$(20)^-$
	7445.05	7443.13	$[02]$
$P = 3$	10599.66	10601.62	$(30)^+$
	10613.41	10615.73	$(30)^-$
	10868.86	10866.40	$[12]$
	11032.40	11031.91	$[03]$
$P = 4$	13828.30	13832.41	$(40)^+$
	13830.92	13834.72	$(40)^-$
	14221.14	14219.34	$(31)^+$
	14318.80	14319.34	$[13]$
	14536.87	14537.50	$[04]$
$P = 5$	16898.40	16895.46	$(50)^+$
	16898.83	16895.73	$(50)^-$
	17458.20	17458.95	$(41)^+$
	17495.52	17495.61	$(41)^-$
	17748.07	17748.57	$[14]$
		17948.12	$[05]$

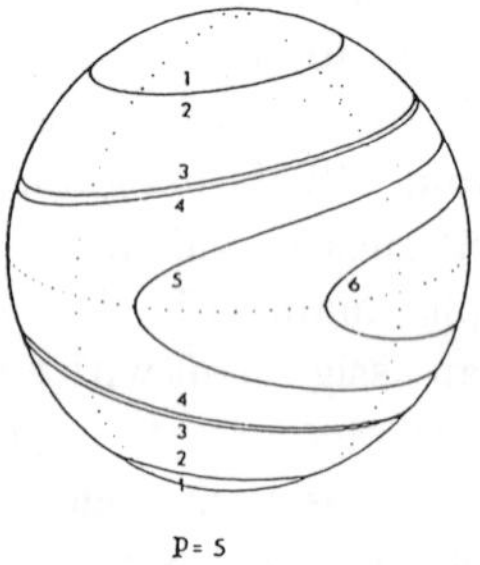

LEVEL	ASSIGNMENT
1	$(50)^+$
2	$(50)^-$
3	$(41)^+$
4	$(41)^-$
5	$[14]$
6	$[05]$

Fig. 6. Dynamically based assignment of H_2O. In the figure, polyad 5 is shown. The trajectories are numbered in order of increasing energy. Levels corresponding to trajectories in the local mode region are labeled with local mode quantum number n_1, n_2 as $(n_1\, n_2)^+$ or $(n_1\, n_2)^-$, depending on symmetry with respect to the C_2 axis. Levels corresponding to trajectories in the normal mode region are labeled with normal mode quantum numbers n_s, n_a as $[n_s\, n_a]$. The table gives the assignment of the stretching spectrum of H_2O for polyads 1–5. The levels are grouped into polyads P. The second column gives the experimental levels; the third column, the levels from the fit with the Darling–Dennison Hamiltonian. The right column gives the new mixed normal-local assignment.

point **a** in Fig. 7 associated with this overtone is very near the equator, so is a "pure" resonant collective mode. Note the profound difference of these 2:1 resonant modes from the "resonant modes" of the 1:1 system — the normal mode states 1 and 6 of polyad A in Fig. 4 for SO_2.

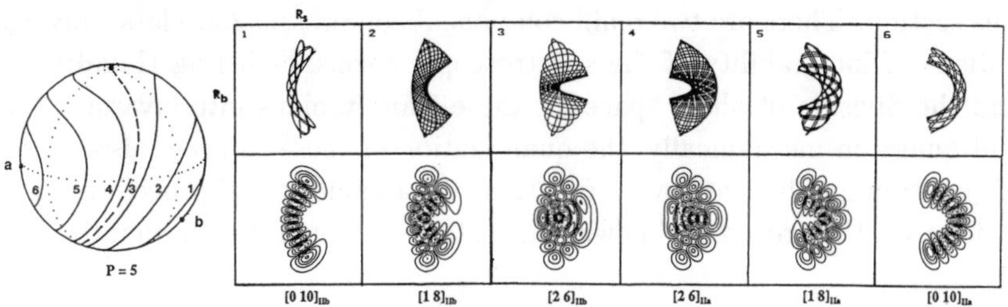

Fig. 7. Dynamically based quantum number assignment for 2:1 Fermi resonance between C–H stretch and bend of polyad 5 of $CH(CF_3)_3$, a polyad from zone II of the catastrophe map. Shown are the phase sphere, wave functions, and coordinate trajectories of levels 1–6 (in order of increasing energy). The sphere is divided into two regions defined by the elliptic fixed points **a** and **b**. The fixed points **a** and **b** correspond to "resonant collective modes" in the full phase space. The two regions are divided by a separatrix, represented by the dashed curve. The separatrix has a cusp at the north pole. The cusp is a fixed point on the sphere; in the full phase space, it corresponds to an unstable pure stretch periodic orbit. The new quantum number assignment for each level is given under the wave function.

We have seen that resonant systems are assignable in terms of new dynamically-based quantum numbers, as if in a sense the system is separable in terms of the new types of modes. More discussion on this point is presented in Ref. 21, in particular, the relation of the resonant collective modes and the striking nodal patterns evident in Figs. 4 and 7, to "nodal coordinates" introduced by DeLeon and Heller.[50] In terms of these nodal coordinates, the classical resonant dynamics is indeed approximately separable.

2.4. *Nonlinear Dynamics and Semiclassical Quantization of Integrable Resonance Fitting Hamiltonians*

We have seen how the phase sphere procedure leads to a new dynamically based assignment scheme based on a systematic classification of phase space structure using bifurcation theory. This subsection provides a much more complete discussion of the classical and semiclassical dynamics underlying the operational phase sphere procedure of Subsec. 2.1. This exposition is important for understanding the issues involved in later sections in extending the spectral assignment program to systems with many degrees of freedom and systems with a strong degree of classical chaos, but it is also very useful for a full understanding of the work presented up to now in this section. There are two main components of Subsec. 2.4: classically, the notions of integrability of the spectroscopic resonance fitting Hamiltonian and the division of phase space for these Hamiltonians into invariant tori; and quantum mechanically, the quantization of these tori, i.e., association of quantum numbers and wave functions via a semiclassical procedure. This is done via the procedure known as Einstein–Brillouin–Keller quantization.

2.4.1. *Integrable single resonance fitting Hamiltonians and invariant tori and actions*

We begin by describing how the phenomenological fitting Hamiltonian leads to a correspondence of the quantum energy levels with well-defined classical structures in phase space (invariant tori) and how a classical property of the tori (the invariant actions) leads to the new quantum number assignment, e.g., the mixed local/normal mode scheme.

The key point is that even though the classical version of the spectroscopic Hamiltonian, e.g., Eq. (5), has strong coupling between anharmonic modes, the system is nonetheless *integrable*, as explained shortly, and therefore has $N = 2$ invariant actions, corresponding quantum mechanically to assignable quantum numbers. To be integrable, a system must have $(N - 1)$ constants of the motion in addition to the energy. This means that for the two-mode spectroscopic Hamiltonian, it is necessary to have a second constant in addition to the energy. If this second constant exists, it is possible to find a set of two invariant actions to describe the system. It is easy to see for the Hamiltonian of Eq. (5) that this condition holds. The angular dependence of the resonance coupling involves only the single angle coordinate ψ. Since the Hamiltonian does not depend on the second angle θ, the action I conjugate to θ is a constant of motion for the Hamiltonian. This invariant action I, corresponding quantum mechanically to the polyad number $P = 2I - 1$, is the second constant needed to make the system integrable.

The existence of the second constant implies the existence of invariant tori, as follows. For any system which has the energy as a constant, the motion of an individual trajectory is confined to at most a $(2N - 1)$ dimensional submanifold of the full $2N$-dimensional phase space. Each additional constant reduces the dimensionality of the allowed submanifold by one. In our case of a two-mode system, the existence of the energy and the extra action I reduces the motion to a two-dimensional subspace. This subspace is a torus on which the energy and the action I are both constants.

(It may be useful here to clarify the relation between the phase sphere and the invariant tori. A trajectory on the torus is related to a trajectory on the phase sphere as follows. A phase sphere trajectory is a function of the action I_z and its conjugate angle ψ. This trajectory is a closed loop on the subspace (I_z, ψ) constituting the sphere defined at constant I. However, the trajectory is actually also moving along the rapidly-varying (in comparison with ψ) "fast angle" θ. The phase sphere loop residing in the spherical subspace (I_z, ψ) is, therefore, obtained as a surface-of-section of a two-dimensional torus residing in the space (I_z, ψ, θ).)

The invariant actions and conjugate angles defining the torus are described as follows. The torus is parameterized by two angles. Since the zero-order action I is conserved, it is one of the invariant actions, so its conjugate angle, the zero-order angle θ, is one of the angles appropriate to parameterize the torus. However, the second zero-order angle ψ is not suitable to parameterize the tori of the spectroscopic Hamiltonian, since its conjugate action I_z is not a constant of the motion. Instead, a new second angle $\bar\psi$ is needed. The second angle $\bar\psi$ is defined so as to parameterize the closed elliptic curves obtained from a section of the invariant tori (or equivalently, the closed loops on the phase sphere). The angle $\bar\psi$ is, therefore, defined with respect to that elliptic fixed point around which a particular loop flows. There is thus a separate angle $\bar\psi$ within each distinct phase space region defined by an elliptic point. For example, for the Darling–Dennison Hamiltonian there are two distinct types of angles: one type $\bar\psi_{\mathrm{normal}}$ for the trajectories in the normal mode region of the sphere, and a second type $\bar\psi_{\mathrm{local}}$ for each of the local mode regions.

It is easily shown that conjugate to each second angle $\bar\psi$ is a second invariant action, as follows. The energy is constant on each torus (or equivalently, on each phase-sphere loop when the fast angle is removed). Therefore, the energy does not depend on the angle $\bar\psi$ parameterizing the loop, so by Hamilton–Jacobi theory, there is an invariant action $\bar I_z$ conjugate to $\bar\psi$. There are different types of such actions $\bar I_z$, depending on the phase space region in which the torus resides: in the normal mode region, a normal-mode action $\bar{I}_{z\,\text{normal}}$; in the local regions, local-mode actions $\bar{I}_{z\,\text{local}}$.

2.4.2. *EBK quantization of the tori and dynamically based quantum number assignments corresponding to the invariant actions*

The existence of the invariant tori and associated invariant actions is the key to the assignment procedure based on the use of the spectroscopic fitting Hamiltonian. This is because there is a well-defined procedure for associating a semiclassical energy level and wave function with an invariant torus satisfying appropriate quantization conditions on the actions. This procedure is known as Einstein–Brillouin–Keller, or EBK quantization. The initial idea was developed by Einstein[51] in a paper of 1917. In 1958, Keller published a paper[52] which combined Einstein's ideas of 1917 with ideas of the WKB semiclassical procedure used to find wave functions for one-dimensional systems, to obtain semiclassical wave functions for multidimensional integrable systems.

Einstein's quantization was an attempt to get beyond the shortcomings of the Bohr–Sommerfeld quantization procedure. The latter had given a successful treatment of the hydrogen atom spectrum, but suffered from the shortcoming that it uses a quantization condition on the classical orbits of hydrogen which is not applicable to general systems, whether integrable or not. Because of the special form of the Coulomb potential, which gives the problem a special rotation symmetry in an abstract four-dimensional space,[53,54] the hydrogen atom orbits are all periodic (in fact, they are ellipses). The Bohr–Sommerfeld quantization condition is that the action along the quantizing periodic orbits be a multiple of Planck's constant. In the later de Broglie wave picture, the Bohr–Sommerfeld orbits are those whose quantization condition results in phase coherence of a wave attached to the orbit.

In contrast to hydrogen, a nonintegrable system such as the helium atom does not have the simple periodic orbit structure of the one-electron Coulomb potential. In contemporary parlance, there is every reason to expect the semiclassical quantization of such a system to be plagued by chaos. It is, therefore, no surprise that the Bohr–Sommerfeld quantization procedure failed for helium — giving impetus to the search for a fully quantum mechanics.

Einstein apparently was quite aware of the problem that classical nonintegrability entails for quantization. In the 1917 paper,[51] he postulated the existence in some circumstances of invariant tori even for nonintegrable systems, thus anticipating in some respects the KAM theorem[55-58] of the 1950's and '60's, which proved their existence. It is quantization of the invariant tori with which the EBK procedure is concerned, rather than quantization of families of periodic orbits associated with special symmetries such as that of the Coulomb potential.

Unlike a periodic orbit, which is a one-dimensional subspace of phase space, an invariant torus is a two-dimensional subspace. This immediately signals that the Bohr–Sommerfeld periodic orbit quantization procedure is problematic, despite the system being integrable. The orbits on a torus are only periodic if the two frequencies $\dot{\theta}, \dot{\psi}$ have an integral ratio $n : m$. But for the vast majority of tori, the frequency ratio is irrational, hence, the orbits are not periodic. Einstein postulated that if the action was quantized along two independent circuits parameterizing the torus, e.g., along the two angle coordinates θ, ψ, then the torus corresponds to a quantum state. The quantum numbers for the state are given by the invariant classical actions with, as was recognized later, a half-integral contribution due to the zero-point energy. It was Keller's contribution[52] to show how to obtain the wave function for a quantizing torus by extending the WKB procedure.

The EBK quantization procedure is related very directly to the phase sphere analysis of the spectroscopic fitting Hamiltonian described earlier. The spectrum is fit with the single resonance fitting Hamiltonian. Each energy level is associated with a trajectory on the phase sphere in the space (I_z, ψ), i.e., an invariant torus in the zero-order action-angle space (I_z, ψ, θ). The phase sphere is categorized according to the bifurcation and catastrophe map analysis of Subsec. 2.2. This classifies the division of the sphere into regions, defined by separatrices, with different types of dynamics in each region. The dynamics within a region are described by a distinct set of action-angle variables suitable to that region, as described in Subsec. 2.4.1. The tori corresponding to the levels of the quantum spectrum are assumed to be EBK tori, i.e., their actions satisfy EBK quantization conditions. Each level in a particular region of the sphere is therefore assigned quantum numbers corresponding to the invariant actions suitable for that region.

The only qualification is that the phase sphere trajectories correspond to energies obtained from a quantum mechanical matrix diagonalization, rather than a primitive semiclassical quantization. Therefore, the actions of the phase sphere trajectories obtained by our procedure deviate from exact EBK quantization conditions to the extent that the energy levels are influenced by effects of classically forbidden processes. Typically, significant deviation of energy levels and their trajectories from EBK conditions occurs near a separatrix. In such instances of the inadequacy of "primitive" semiclassical quantization methods, i.e., methods such as EBK which do not take cognizance of classically forbidden processes, it is necessary to employ a "uniform" semiclassical procedure, where explicit consideration of the classically forbidden region leads to quantization conditions different from those of EBK. This is closely related to the fact[46,59] that a resonance zone (defined by a separatrix) may not be large enough to support an EBK trajectory for each quantum level. In a certain sense, our phase sphere trajectories are "uniform" in that they correspond to energies obtained from a fully quantal procedure — diagonalization of a matrix Hamiltonian — rather than a semiclassical method. In our assignment procedure, we ascribe an integer quantum number to a state even when the phase sphere trajectory deviates significantly from primitive EBK conditions or when there is no corresponding primitive EBK state at all. This is perfectly sensible as part of a labeling scheme for ordering states within the resonance zone. For instance, consider level 5 of H_2O in Fig. 6. The trajectory for this level lies in the normal mode region fairly near the separatrix, so its classical actions deviate somewhat from integer values (plus zero-point actions). Nonetheless, we assign it as the normal mode combination state [1 4], since it is the next level in the progression starting with level 6, the overtone [0 5]. In the case where there is no primitive EBK state at all, our procedure for going from the quantum spectrum to trajectories on the sphere still gives a unique trajectory, which is assigned integral quantum numbers proper to its position in the phase sphere region in which it resides.

3. Dynamical Analysis of Spectra of Classically Chaotic Systems

3.1. *Introductory Remarks and Summary*

This section is perhaps the most difficult of this review chapter, as it unavoidably is involved with very complicated and technical issues, which nonetheless can only be surveyed. On a first exposure, the reader may wish to study only this introductory Subsec. 3.1, which summarizes the principal issues and conclusions, before proceeding to Secs. 4–6. The reader more deeply interested in chaotic quantization may then wish to return to the detailed discussion of Subsecs. 3.2–3.5.

Section 2 has described our previous work on bifurcation analysis of phase space structure for classification and assignment of spectra with a single strong Fermi resonance coupling. The first major obstacle to expanding this approach to a comprehensive framework for highly excited spectra is the problem of quantum spectra of classically chaotic systems. (The second problem, that of many-dimensional systems, is taken up in Sec. 4). The reason chaos is a major problem is that all of the bifurcation analysis described so far has been based on effective spectroscopic Hamiltonians with a single resonance coupling. As we have seen, such a Hamiltonian has nontrivial bifurcations, associated with phenomena such as the local-normal transition, but it nonetheless is integrable, or quasiperiodic. Integrability has two crucial simplifying implications for spectral analysis. As discussed above in Subsec. 2.3, in an integrable system the classical motion takes place on invariant tori in phase space. The tori can be classified with respect to invariant actions defined within different phase space regions, e.g., local mode actions in the local mode region, and normal mode actions in the normal mode region. A further implication of the invariant tori is that we can associate them and their actions with individual quantum states, i.e., assign the states with quantum numbers corresponding to the actions. The reason for this is that there is a well-defined semiclassical procedure, the Einstein–Brillouin–Keller (EBK) quantization, for associating quantum states with invariant tori.

The classification of the classical phase space for nonintegrable systems and assignment of their quantum states is a much more difficult problem. According to the KAM theorem,[55–58] when an integrable system is perturbed, there is a mixture of quasiperiodic and chaotic motion, if the perturbation is sufficiently weak. At first, the quantizing EBK tori with their invariant actions and associated quantum numbers still exist as the perturbation is turned on. Even so, it is not altogether clear what the effect of the developing chaos on the spectrum and eigenstates is. Eventually, each EBK torus itself is enveloped by the chaos and destroyed. When the torus corresponding to a state is destroyed, one is really cast adrift in trying to identify a classical structure to associate with the state. It does not seem likely on the face of it that the phase space structure that remains should have anything like the regularity of the invariant tori that has been destroyed. In particular, whatever structure may exist in the chaotic phase space, it is not at all evident a *priori* that one can associate invariant actions with it. Without the invariant actions associated with tori, the EBK quantization procedure is no longer available, so even if the classical phase space structure were well-understood, it is not clear how one would make the connection with quantum mechanics. And finally, even if the classical-quantum connection were understood outside the EBK framework, it is not evident how this would entail a meaningful assignment of the spectra, given the absence of invariant actions.

I will outline here some highly speculative ideas we are currently investigating for these problems of spectral classification of chaotic systems. These ideas make use of some recent developments in classical nonlinear dynamics concerning the phase space structure of chaotic systems, and some extensions of the very old notions of Bohr–Sommerfeld and WKB semiclassical quantization. Our point of view has some points of contact with the program of Gutzwiller[60] for semiclassical quantization of chaotic systems based on sums over periodic orbits, and presently we will examine the relevance of this work to our problem. However, the main thrust of our approach is an attempt to make direct use of some profound results of Mather and others[61–65] on the structure of phase space in chaotic systems. This mathematical work constitutes a highly nontrivial extension of the notion of invariant tori chaotic systems. Essentially, Mather proved the existence of chaotic remnants of tori which retain some of the useful properties of the tori. These invariant structures have come to be known as *cantori* because mathematically they are Cantor sets — a kind of fractal object described below in Subsec. 3.4. Of great importance for us, this work seems to bear on our surprising success with the simple single resonance spectroscopic fitting Hamiltonian even in situations in which the tori have probably been destroyed. For this reason, we believe a semiclassical approach which makes use of the work of Mather *et al.*[61–65] has a good chance of leading to the desired end, a quantization scheme allowing a natural classification and assignment of spectra. To our knowledge, the periodic orbit schemes, at least as implemented to date, do not make use of the results of Mather *et al.*[61–65] on the structure of chaotic phase space. To directly exploit these results, we are investigating a procedure, outlined below, to semiclassically quantize the cantorus, in a manner analogous in many respects to EBK quantization of invariant tori in quasiperiodic systems.

This section will begin with a discussion of the most widely studied existing approaches to the semiclassical quantization of chaotic systems, the periodic orbit theories of Gutzwiller and others. Next, we discuss the periodic orbit program in relation to molecular problems of spectral classification and assignment, and conclude that so far at least, some key elements necessary for the spectral problem are lacking in the periodic orbit approach. Then, we take up the theory of Mather and others[61-65] of cantori as the chaotic remnants of invariant tori. Finally, we discuss prospects for semiclassically quantizing the cantori in a WKB or EBK approach, which if successful would lead to a natural extension of our scheme for classifying and assigning spectra.

3.2. *Periodic Orbit Theory of Semiclassical Quantization*

In this subsection, I consider the work of other investigators on semiclassical quantization schemes for chaotic systems based on periodic orbits, an approach which has attracted great attention. The main conclusion is that while suggestive hints of assignability emerge from some work using this approach, the overall picture is too unclear at this point to offer the prospect of an assignment scheme of the generality we are seeking. This will motivate us to turn to an outline in Subsec. 3.4 of the much different approach of quantizing the chaotic remnants of invariant tori, the "cantori" of Mather *et al.*[61-65]

Gutzwiller and others have developed a program[60] which has attracted great interest as potentially offering a general solution to the semiclassical quantization problem for chaotic systems. This theory is based on the idea of a semiclassical approximation to Feynman's path integral approach[66] to quantum mechanics. The theory starts with Feynman's fully quantum mechanical expression for the propagator. The propagator contains all the information about the quantum mechanics of a system. The first simplification is to make a classical approximation to the propagator, as obtained originally by van Vleck.[67] It has been thought that in practice even the classical propagator would be far too difficult to calculate. (However, see some recent work of Tomsovic and Heller.[68]) It turns out, though, that the quantum mechanical density of states as a function of energy can be obtained by summing the propagator, leading to the *trace formula*. The paths that contribute to the sum in the trace formula are periodic orbits. In principle, calculation of the trace formula to sufficient accuracy should give spikes in the density of states corresponding to the (semiclassical approximation to) the quantum mechanical eigenvalues. In practice, this has proved to be a difficult program to carry out. A chaotic system has an infinite number of periodic orbits; moreover, there are questions concerning the basic validity of the whole procedure, such as problems of convergence. Although there has been considerable progress recently in implementing the trace program for simple systems,[69] it seems fair to say that a variety of problems probably rule out much hope in the foreseeable future of implementing it for systems as complicated as molecules.

Even apart from practical difficulties of implementation, there is a basic question about using the trace program as presently understood for the kind of analysis desired for highly excited molecular states. This has to do with the issue of classification and assignment of states, which of course is the entire point and goal of our work. There is no reason *a priori* to expect that assignable quantum numbers should exist for quantum states corresponding to classically chaotic systems, and this seems to be reflected in the periodic orbit (p.o.) formalism itself. Unlike the Bohr–Sommerfeld or EBK quantization procedures, the general p.o. trace formalism is not based on any general quantization condition which leads naturally to a set of quantum numbers assignable to a given state. But despite the absence of a manifest assignment scheme emerging automatically from the p.o. formalism, the situation is by no means so simple as to enable one to say that states of chaotic systems are without quantum numbers and intrinsically unassignable. In fact, there is circumstantial evidence from several of the p.o. studies themselves pointing at least in some cases to the existence of good or approximately good quantum numbers. I will attempt to outline some of the main highlights of this unsettled and still unfolding situation.

In the first place, in those *nonchaotic* cases where definite quantum numbers are known to exist, they should emerge in some way from the p.o. formalism as well. For example, consider a semiclassical state obtained from EBK quantization of an invariant torus. The EBK quantization procedure for the energies and wave functions[52] is a rigorous semiclassical procedure in the circumstances where it applies, i.e., when there are quantizing invariant tori. In order for the p.o. formalism to be a complete procedure in the sense of being able to capture a semiclassical approximation to an arbitrary quantum state, it must certainly be able to obtain the EBK states. It should therefore be possible to show that the p.o. and EBK formalisms are equivalent when the latter is applicable. This problem was investigated by Berry and Tabor.[70] They started with the EBK solution, then showed by doing a Poisson transform that the EBK solution is equivalent to a sum of periodic tori converging to the quasiperiodic quantizing torus. This was further investigated by Ozorio de Almeida,[71] who concluded that the periodic tori converging to the quasiperiodic quantizing torus make a phase-coherent contribution to the sum, leading to a singularity in the density of states.

Furthermore, there is evidence from investigations of the p.o. formalism in particular cases for the existence of at least approximate quantum numbers and assignments in *chaotic* systems. It has been found[69] that semiclassical energies in good agreement with fully quantum mechanical calculations can be successfully obtained by quantizing motion transverse to *stable* periodic orbits. This is not surprising on the basis of EBK theory if the excited states correspond to invariant tori surrounding the stable periodic orbit, which is certainly possible. But this success would be very surprising if the invariant tori have been destroyed by chaos, which is also very much possible. Furthermore, following earlier work of Heller[72] on "scarring" of quantum wavefunctions, there is evidence[73] that a *single unstable* periodic orbit can sometimes be quantized and excited states obtained by quantizing excitations transverse to the unstable orbit. Quantizability of a *single* orbit is not at all evident *a priori* from the trace formalism, and is very suggestive of some kind of Bohr–Sommerfeld quantization condition.

This was investigated by Ozorio de Almeida,[74] following his earlier investigation[71] of the equivalence of the p.o. and EBK formalisms. His idea was to account for a quantum state associated with an unstable periodic orbit in terms of a certain type of phase space structure incorporating the unstable orbit. The structure in question has certain properties in common with an invariant torus. This structure is obtained by considering the unstable orbit and its stable and unstable manifolds. When the surface of section is taken as shown in Fig. 8, the unstable orbit gives a point and the stable and unstable manifolds form curves emanating from this point. If the system were quasiperiodic, these curves would joint to form a closed loop pinched at the point of the unstable orbit, similar except for the pinched point to the circle obtained in the surface of section of an invariant torus. In a chaotic system, however, the stable and unstable manifolds intersect an infinite number of times, forming a *homoclinic tangle*, as shown in Fig. 8.

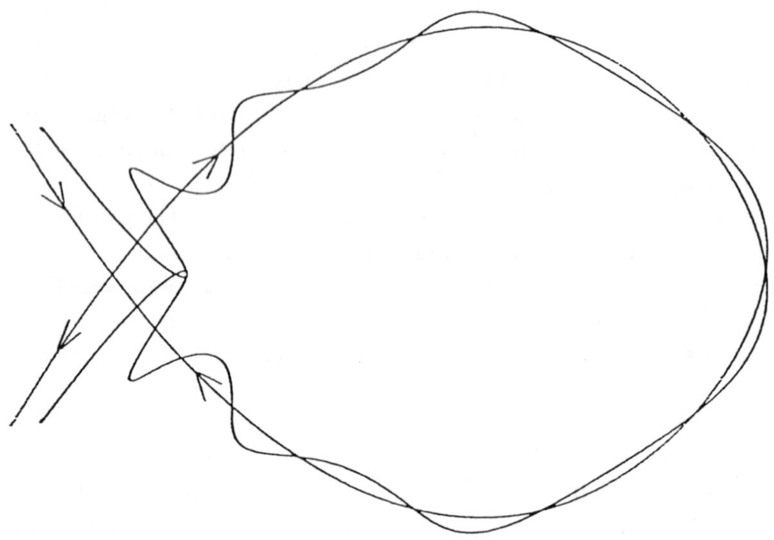

Fig. 8. Homoclinic tangle of manifolds of an unstable fixed point. The unstable fixed point is at the "X" from which the manifolds emanate. The flow away from and toward the fixed point along the manifolds is indicated by arrows. The stable manifold (flow from unstable fixed point) and unstable manifold (flow toward unstable fixed point) intersect an infinite number of times. Only the first few intersections are shown.

The tangle is certainly a chaotic structure, and has an infinite number of nearby periodic orbits associated with it. Nonetheless, it is an invariant structure. Furthermore, in a certain sense described by Ozorio de Almeida, it is possible to construct a "homoclinic torus" with associated actions. The tangle, therefore, has aspects related to both the p.o. and EBK theories.

Ozorio de Almeida[74] therefore investigated whether a homoclinic torus satisfying EBK quantization conditions supports a quantum state in the p.o. formalism in the sense of yielding a singularity in the density of states as a function of energy. The conclusion apparently is that for systems which are "not too chaotic", in a sense which is not yet precisely quantified, it is always possible to associate a state with an *unstable* orbit attached to a homoclinic torus satisfying approximate EBK conditions. This work, therefore, seems to be a step toward understanding the occurrence of states apparently assignable[73] to unstable periodic orbits.

Further insight into the nature of this near, but not exact, EBK assignability of chaotic systems comes from some recent work of Takatsuka.[75] He develops a "dynamical characteristic function" formalism and investigates some issues in the p.o. theory, including quantization conditions for chaotic systems. He first investigated integrable systems and concluded that to contribute, a periodic orbit must satisfy a quantization condition of the form

$$\sum_k^N r_k^R \{I_k^0 - (m_k + 1/2)\hbar\} = 0 \ . \tag{10}$$

Here, the m_k are integers ≥ 0 and r_k^R are the rotation numbers of periodic orbit R. For example, for a 2:3 orbit, i.e., one which winds two times around angle 1 in one complete period while winding three times around angle 2, one has $r_1^{2:3} = 2$ and $r_2^{2:3} = 3$.

Takatsuka calls (10) a "resonant-type quantization condition". It can be seen that this is similar to a Bohr–Sommerfeld quantization on the periodic orbit: to contribute, the *total* phase must go through an amount corresponding to an integral (*modulo* zero-point motion) phase contribution for each circuit of every degree of freedom. Note that (10) is not an EBK quantization condition on each of the actions, which would be

$$I_k^0 = (m_k + 1/2)\hbar \qquad (k = 1, \ldots, N) \ . \tag{11}$$

Rather, in (10) the action variables are not quantized individually as in EBK quantization, but only as the sum $\sum_k^N r_k^R I_k^0$, so the total phase in (10) is quantized along the periodic orbit. However, suppose one considers quantization of an invariant torus instead of a periodic orbit. In general, a torus has quasiperiodic rather than periodic motion, with an irrational relation between the frequencies. Takatsuka argued that a quasiperiodic orbit comes arbitrarily close to periodic orbits with values of the r_k^R which are linearly independent. A separate resonant quantization condition like (10) must be satisfied for each such periodic orbit and corresponding set of r_k^R. But this is equivalent to demanding that each of the actions be quantized individually as in the EBK conditions (11). Thus, the equivalence of the p.o. and EBK formalisms can be thought of as the convergence of Bohr–Sommerfeld quantization conditions on independent periodic orbits to EBK conditions on the quantizing torus.

Takatsuka then went on to consider two nonintegrable or chaotic cases: stably nonintegrable systems, and weakly unstable systems. The former have stable orbits in a sea of chaos; the latter, unstable orbits in a still weakly chaotic sea. In the case of stable orbits, the resonant-type quantization condition of (10) is found. In the case of unstable orbits, a set of quantum numbers is found which locates the center of a Lorentzian peak in the density of states. According to Takatsuka, however, this is not strictly a "quantization condition" because it does not lead to *singularities* in the density of states, and further investigation of quantization conditions for unstable orbits is warranted.

This leaves us in a situation as regards the periodic orbit theory which is suggestive but at this point by no means entirely clear. For stable orbits surrounded by chaos,[69] the "resonant" Bohr–Sommerfeld type quantization condition of (10) suggests a rigorous basis for the existence of approximate quantum numbers which, moreover, must converge to exact EBK conditions when the motion is on a torus. This may well be related to the success[69] in quantizing excited states transverse to stable orbits. For unstable orbits, there is the surprising success[73] in quantization when the orbits are not too unstable. Insight into this is provided by the work of Takatsuka[75] and the work of Ozorio de Almeida[74] on "homoclinic tori". However, the former approach does not yet appear to be completely successful in establishing quantization conditions, while the latter approach has not yet precisely defined just what it means for an orbit to be "not too chaotic" to support a quantum state.

3.3. *Relation of Chaotic Quantization Conditions to Molecular Spectral Classification and Assignment*

We have seen that the periodic orbit quantization investigations contain several hints for the existence of quantization conditions governing states of classically chaotic systems, though the picture at this time is very incomplete and unsettled. I now want to relate the notion of quantization conditions for chaotic systems to our procedure for classifying spectra using the fit to an integrable Hamiltonian, and the hoped-for extension to chaotic systems. I will argue that the simple spectral fitting Hamiltonian used so far is successful in regimes where there is probably a significant degree of classical chaos. This suggests that the fitting Hamiltonian is mimicking chaotic structures analogous to invariant tori. I will suggest that the purported quantizing structures in fact are the chaotic remnants of tori known as "cantori", which have been the object of much recent work in classical nonlinear dynamics. I will then describe some ideas we are currently investigating for quantizing these cantori, analogous to obtaining EBK quantization conditions and semiclassical wave functions for tori.

The single resonance fitting procedure gives an integrable approximation to the spectrum in surprisingly good agreement[25] with experiment for such a simple approximation. For the stretching spectrum of water, the r.m.s. deviation from experiment is about 2 cm^{-1} for levels up to 30 000 cm^{-1}. For the stretching spectrum of ozone, the r.m.s. deviation of 7 cm^{-1} is larger, but within experimental error. In the case of ozone, it is very likely that at least some of the levels cannot correspond to invariant KAM tori. Evidence for this comes from work of Standard and Kellman[76] who obtained a potential surface in coordinate space for the stretches of ozone. They obtained this surface by fitting the spectrum *directly to a coordinate potential*, as distinct from the phenomenological Darling–Dennison fit. Classical trajectories on this surface show a considerable degree of chaos, and at higher levels, it is unlikely that the quantizing structure for every state corresponds to an invariant torus. Nonetheless, the agreement of the fit with the integrable Darling–Dennison Hamiltonian is within experimental error. Moreover, when wave functions from the Darling–Dennison Hamiltonian are compared[77] with converged variational wave functions on the nonintegrable potential surface, the differences are minor, with little evidence of much effect of the classical chaos on the latter wave functions. If, as seems likely, there are quantum states on the potential surface that do not correspond to invariant tori, they nonetheless correspond to structures that in their energetics and semiclassical wave functions behave very much like tori.

Hints of such structures occur in some of the periodic orbit investigations described above. One example comes in the quantization of excited states in degrees of freedom transverse to a stable periodic orbit.[69] This is analogous to our spectral fits inasmuch as the tori of the classical spectroscopic

Hamiltonian are defined with respect to a stable orbit, e.g., the stable orbits defining local mode regions on the sphere. However, the quantization of Ref. 69 uses a harmonic approximation, which is not adequate to give energy levels in as good agreement with experiment as the anharmonic fitting Hamiltonian. More importantly, the procedure of Ref. 69 is an *ad hoc* one which does not necessarily yield insight regarding the phase space structures corresponding to the excited states — they could be chaotic, but they might simply be invariant tori embedded in a weakly chaotic sea.

The closest thing we are aware of in the p.o. literature explicitly resembling the type of structure we are after — a chaotic torus remnant — is the homoclinic torus investigated by Ozorio de Almeida. This has both a desired torus-like structure and an infinite number of periodic orbits convergent to it. However, it is formed from the manifolds of an *unstable* periodic orbit, with the unstable orbit at one "pinched" end of the homoclinic torus. What we are after is an unpinched torus defined in relation to a *stable* fixed point, or at least the chaotic remnant of such a torus.

A clue is contained in the chaotic quantization condition (10) of Takatsuka. This "resonant" Bohr–Sommerfeld condition is expected to hold for periodic orbits representing excitations transverse to a central stable orbit. It thus could apply to the ozone example where some of the states calculated on a potential surface look like EBK states from the Darling–Dennison fit, but correspond to classical dynamics with considerable chaos surrounding a central stable orbit. Suppose the quantum states do correspond to a structure like a chaotic remnant of a torus. By analogy to Takatsuka's analysis of the integrable case, one might think that there should be a series of p.o.'s converging to the torus-like structure such that the resonant quantization conditions on the p.o.'s of (10) converge to approximate or exact EBK conditions. In pursuit of a notion of this kind, a unified mathematical treatment of cantori as the torus remnants, accompanied by a series of periodic orbits convergent to the cantori, is outlined next.

3.4. *Invariant Cantori as Phase Space Remnants of Invariant Tori*

I believe that the phase space structures we are seeking may be the objects known as cantori, which have been extensively investigated in the classical nonlinear dynamics literature by Mather and others.[61–65] A cantorus is what remains of a torus which has been destroyed by the encroaching sea of chaos. The torus breaks up into a Cantor set — a totally disconnected set with properties like Cantor's "middle third" set[78] — which preserves many of the characteristics of an invariant torus, hence the name cantorus. Following some earlier work of Davis[79] on energy levels of a chaotic system, we are investigating the possibility that the cantori may be the phase space structures corresponding to quantum wave functions via a semiclassical quantization procedure. This would be the case if a quantization procedure could be found for cantori analogous in some respects to EBK quantization of invariant tori in integrable systems. If this proves to be possible, the correspondence would lend itself quite naturally to spectral classification and assignment.

Breakup into a Cantor set as the fate of a torus when it is destroyed by chaos was first suggested by the numerical investigations of Greene.[80] Proofs of the existence of cantori and some precise mathematical conditions for the breakup of tori are the results of investigations of Mather,[61,62] working from the point of view of a mathematician, and independent work by Aubry,[63] investigating a particular physical model in solid-state physics. Katok[64,65] has extended the work of Mather, obtaining a simplified proof, in the process formulating the problem in a way nicely suited to make contact with periodic orbit theory.

Cantori have properties which are useful for our purposes because they are analogous to important properties of invariant tori. These properties of cantori and their significance can be summarized as follows:

(1) The cantorus has a well-defined *winding number*. This is analogous to the frequency ratio of the two degrees of freedom for motion confined to an invariant torus. For a periodic orbit, the winding number is the ratio p/q of the number of circuits p completed around one angle as q circuits are completed around the other angle, with $q > p$ the period of the orbit. For a cantorus, as for most invariant tori, the winding number is an irrational number ω.

(2) There are manifolds emanating from (unstable manifold) and approaching (stable manifold) the points of the set making up the cantorus. *The manifolds together with the points of the cantorus can be used to define closed curves in phase space.* This affords the possibility of finding EBK-like quantization conditions for the cantorus, a possibility that

will be discussed further below. Orbits on the stable manifold approach points of the cantorus forward in time; orbits on the unstable manifold approach points of the cantorus backward in time.

(3) There is a set of p.o.'s with well-defined winding numbers p/q converging to a cantorus. For unstable p.o.'s, stable and unstable manifolds exist. The existence of a convergent set of p.o.'s means that, in principle, it is feasible to calculate approximations to the fractal cantorus to any desired degree of accuracy.

3.5. *Quantization of Cantori and Spectral Classification and Assignment*

If the cantori turn out to be phase space structures that correspond to quantum states, in a way that lends itself to a natural spectral classification and assignment, they may be the key to the goal of classifying states of chaotic molecular systems. We are currently investigating these ideas in relation to finding quantum energy levels, and the essential but even more difficult problem of finding semiclassical wave functions. Our investigations can be outlined as a series of hypotheses:

Hypothesis 1. Quantum states are associated with invariant cantori satisfying EBK quantization conditions. The action is defined as the area inside a closed curve constructed from the points of the cantorus and the invariant manifolds connecting these points. These closed curves are a feature of the cantori and their associated manifolds, as noted under property (1) above. Remark: From the work of Katok,[64,65] it is known that a cantorus and its manifolds can be approximated increasingly well by a series of periodic orbits converging to the cantorus, and the manifolds of these periodic orbits. These convergent p.o.'s are the key to finite approximate calculations of the cantori and their semiclassical properties.

Evidence that an EBK quantization condition on periodic orbits is associated with states of a chaotic system comes from work of Davis,[79] who investigated a model nonintegrable Hamiltonian for the normal-local transition. He found for several states that at energies near those of the exact quantum states of the system, there are low-order periodic orbits near cantori satisfying resonant quantization conditions with near-integer quantum numbers.

Hypothesis 2. A periodic orbit contributes to the density of states if it satisfies "resonant" Bohr–Sommerfeld quantization conditions like Eq. (10) of Takatsuka. In the case of periodic orbits converging to a torus satisfying EBK quantization conditions, the resonant quantization conditions converge to the EBK conditions.

Hypothesis 3. A semiclassical wave function can be obtained by a WKB procedure applied to the postulated EBK cantorus. In the WKB approach, a function depending on the action is attached to an orbit. The phase of the wave function is determined by the action along the orbit, with phase reflection factors (contained in the Maslov indices[58]) picked up at classical turning points. It is possible that most of the true quantum wave function may be given by WKB quantization of trajectories on the cantorus. Still more of the wave function will perhaps be obtained from trajectories on the manifolds connecting the points of the cantorus.

Hypothesis 3 states that semiclassical wave functions may be obtained which are associated with invariant cantori. If these wave functions satisfy EBK quantization conditions, they would have associated with them quantum numbers suited to classifying and assigning states of the chaotic system. We are in the process of numerical investigation of this possibility, including the assessment of the degree of chaos for which the procedure is effective. If this procedure works, I believe that when applicable it will essentially solve the problem of assigning spectra of two degree of freedom chaotic molecular systems.

A final important point, albeit rather technical, is that the main circumstance where application of this procedure is expected to be problematic is when cantori satisfying EBK conditions do not exist. This is expected to be a problem near a separatrix, where a resonance zone may destroy the initially sought-after nonresonant torus or cantorus. The next step is then to search for a quantizing cantorus within the resonance zone. However, there will undoubtedly be cases where the resonance zone is too small to support a primitive quantizing cantorus. This is well-known to occur for tori in quasiperiodic systems, as discussed in Subsec. 2.4.2, and has been observed for cantori in chaotic systems by Davis,[79] where the problem is essentially the same, but perhaps more common due to the proliferation of resonance zones. The problem of absent primitive EBK states arises in regions where classically forbidden processes are very important. As discussed in Subsec. 2.4.2 for integrable systems, the solution from a semiclassical standpoint is to use a uniform method which treats classically allowed and forbidden regions on a similar footing. Development of such a procedure for the cantorus quantization proposed here will undoubtedly be very challenging. As discussed in Subsec. 2.4.2, development of a successful uniform semiclassical method for such levels and wave functions is probably not necessary in practice for assigning states in cases where primitive semiclassical methods are problematic. However, this would be a very desirable logical development in a complete assignment procedure.

To summarize where this long and somewhat convoluted section has taken us: I began with a discussion of the most widely studied existing approaches to the semiclassical quantization of chaotic system, the theories of Gutzwiller and others based on periodic orbits. Next, I discussed the periodic orbit program in relation to molecular problems of spectral classification and assignment, and concluded that from our point of view, key elements are lacking at the present time. I then took up the theory of Mather and others[61-65] of cantori as the chaotic remnants of invariant tori, hoping to take advantage of the natural connections of this beautiful work to the problems of spectral classification and assignment. Finally, I discussed prospects for semiclassically quantizing the cantori in a WKB or EBK approach. If this is successful, it would lead to a natural extension of our scheme for classifying and assigning spectra, and might essentially solve the molecular spectral problem, at least for two degree of freedom chaotic systems.

4. Systems with Many Degrees of Freedom

All of the discussion so far, both of our past work on single resonance Hamiltonians and work in progress on chaotic systems, has been for two degree of freedom systems. Independent of the problems that arise from the presence of chaos in these systems, it is evident that a successful approach to highly excited spectra must confront directly the challenge of many interacting degrees of freedom. This divides naturally into two parts. Subsection 4.1 discusses the problem of identifying the dimensions of phase space actually affected by bifurcations. A very important byproduct is the identification of *approximate constants of motion and superpolyad quantum numbers* associated with the unbifurcated degrees of freedom. The superpolyad quantum numbers are proving very useful in accounting for some of the main features of complex spectra, in particular, energy transfer pathways[7,10] deduced from C_2H_2 absorption spectra, and hierarchies of splittings[8] in dispersed fluorescence and SEP spectra. Subsection 4.2 then discusses the problem of classifying the structure of the bifurcated degrees of freedom. It will be seen that the superpolyad quantum number permits an enormous simplification, similar to that in the two-mode system, in that it makes it possible to solve *analytically* for the *large-scale bifurcation structure* of spectroscopic fitting Hamiltonians of chaotic many degree of freedom systems.

4.1. *Approximate Dynamical Constants of Motion and Quantum Number Assignments of Polyatomics with Multiple Fermi Resonances*

The most challenging part of the multidimensional problem, discussed in Subsec. 4.2, is to carry out the bifurcation analysis for three or more coupled degrees of freedom. As a prelude, it is essential to determine the partitioning of phase space between those dimensions in which bifurcations take place, and those which are unaffected by bifurcations. It will turn out that determination of the degrees of freedom which remain unbifurcated leads to the inference of approximate constants of motion, which is very important information in its own right. These constants of motion in general are associated with a complex partitioning of molecular phase space between coupled and regular degrees of freedom, so they bear a nontrivial relationship to the zero-order normal modes, as will be seen. They have important ramifications for analysis of highly excited spectra. They are useful as quantum numbers which can be assigned to spectra of highly excited states. Furthermore, a system with one or more approximate constants has motion in a subset of phase space of reduced dimension, with implications for energy transfer pathways. In recent papers,[9,10] we presented a theory of these approximate constants. This theory, which is closely related to a method used by Fried and Ezra in a van Vleck perturbation theory study,[81] focused on spectral fitting Hamiltonians with multiple Fermi resonances. The formalism reduces to simple methods of vector algebra, so is very easy to apply. The constants of motion correspond to "superpolyad" and other quantum numbers very useful for assigning complex spectra. In ongoing studies following our earlier exploratory work,[9,10] we are analyzing these quantum numbers in fits of experimental and simulated spectra of C_2H_2 currently under investigation in several laboratories. These systems are interesting in part because they require multiple Fermi resonance couplings in the fits to get good agreement with experiment. Hence, they are both multidimensional and nonintegrable, which implies an interesting bifurcation problem. At the same time, at the level of detail to which they can currently be fit, they possess approximate constants and quantum numbers.

The inference of approximate dynamical constants from fits of spectra is readily understood. The crudest type of fitting Hamiltonian simply uses uncoupled normal modes, expressing the energy in terms of powers of the quantum numbers. This kind of Hamiltonian has N independent actions as constants for the uncoupled N-oscillator system, and the corresponding semiclassical motion is quasiperiodic on N-tori. Hamiltonians which incorporate oscillator coupling most often contain a single resonance coupling such as Darling–Dennison or Fermi resonance, as in the cases discussed in Sec. 2. The motion is still quasiperiodic, with $(N-1)$ global actions, or quantum numbers, plus actions defined in local regions of phase space (e.g., the local or normal mode actions $\bar{I}_z$ of Subsec. 2.4.1.) However, examples[5-7] now exist of systems where it is necessary to invoke several resonance couplings to obtain a good fit. These "multiresonance spectra" undoubtedly will proliferate as spectroscopists probe the transition from regular motion to global chaos. Intuitively, one might expect that spectra with multiple resonances would have some actions left intact as good quantum numbers, if the number of resonances is not too large. However, to the best of our knowledge no one before has tried to determine these constants of motion from analysis of experimental data.

The theory of Refs. 9 and 10 for finding these constants yields nontrivial good quantum numbers, yet is extremely simple to apply. The method can be stated as follows. A molecule with N vibrational degrees of freedom can be described in terms of N zero-order quantum numbers for the normal modes. Couplings of these normal modes are given by a set of m resonance coupling operators. The theory then represents these resonance couplings as a set of m *resonance vectors*. The resonance vectors may not all be independent; from the m "raw" resonance vectors of the fit, form a set of n independent resonance vectors. The independent vectors form a vector space V^n. Then the $(N-n)$ vectors $J^{(N-n)}$ orthogonal to V^n constitute a set of $(N-n)$ good actions, i.e., *good quantum numbers*, for the system.

To apply this method to finding new quantum numbers for a real experimental system, we analyzed the fit of the FTIR spectrum of acetylene up to $10\,000$ cm^{-1} of Smith and Winn.[6] These workers found that three resonance couplings, a Darling–Dennison and two Fermi resonances, were needed to obtain an adequate fit. These resonance couplings and their corresponding resonance vectors are:

$$V_1 = a_1^+ a_1^+ a_3 a_3 + \text{Hermitian conjugate} \longleftrightarrow (2, 0, -2, 0, 0)$$
$$V_2 = a_3^+ a_2 a_4 a_5 \longleftrightarrow (0, -1, 1, -1, -1)$$
$$V_3 = a_3^+ a_3^+ a_1 a_2 a_4 a_4 \longleftrightarrow (-1, -1, 2, -2, 0) \,. \quad (12)$$

These three resonance vectors are independent, so we expect two degrees of freedom to be left uncoupled in the system, corresponding to two good quantum numbers. However, the resonance operators couple all five vibrational modes. The good quantum numbers must therefore correspond to a nontrivial partitioning of the original zero-order normal mode space. Two good quantum numbers, chosen to be orthogonal in the same sense as the resonance vectors, turn out to be:

$$J_1 = (n_1 + n_2 + n_3)$$
$$J_2 = (2n_1 - 4n_2 + 2n_3 + 3n_4 + 3n_5) \,. \quad (13)$$

The second quantum number, in particular, is one which is very unlikely to be discovered using intuition, but the reader can easily verify that it is left invariant by all three resonance coupling operators. (It should be noted that Holme and Levine[82] found that the first action J_1 of Eq. (13) is approximately conserved in classical trajectory simulations.)

A second set of orthogonal quantum numbers P, Q can easily be constructed as linear combinations of J_1, J_2:

$$P = (5n_1 + 3n_2 + 5n_3 + n_4 + n_5) = \frac{1}{3}J_1 + \frac{13}{3}J_2$$
$$Q = (4n_1 - 22n_2 + 4n_3 + 13n_4 + 13n_5) = \frac{13}{3}J_1 - \frac{14}{3}J_2 \,. \quad (14)$$

The quantum number P is the superpolyad quantum number corresponding to the approximate frequency ratios of the zero-order normal modes. The second, orthogonal quantum number Q turns out be awkward to use in practice. Fortunately, it is not necessary that the second quantum number be orthogonal to P, only that it be a quantum number P' independent of P. For instance, one could choose $P' = J_1 = (n_1 + n_2 + n_3)$.

Both the superpolyad number P and the second constant P' are very useful for assigning the spectrum of C_2H_2 and identifying otherwise unrecognized patterns in the energy levels, and for describing pathways for intramolecular vibrational energy redistribution. Smith and Winn[7] observed that the resonance couplings used in the fit of the acetylene absorption spectrum obey relations that can be interpreted in terms of energy transfer pathways from the initially excited C–H stretch absorber mode. With the loss of each quantum from the C–H stretch, there is simultaneous flow of one quantum into the C–C stretch, and two quanta into the bends. It is easily shown[10] that the relations of Smith and Winn are equivalent to the existence of the two approximate quantum numbers P, P'. The resonance vector analysis has also been applied by Jonas *et al.*[83] to dispersed fluorescence and SEP spectra of acetylene in a model including vibrational angular momentum and vibrational l-resonance. They find that the superpolyad number is very useful for organizing information about energy transfer pathways, including the observation of hierarchies of time scales for the energy flow. It is noteworthy that McCoy and Sibert have found,[84] following the analysis of Ref. 9, that the constants P, P' are good approximate constants in a van Vleck perturbation theory calculation of C_2H_2 energy levels starting from a molecular potential energy surface. This is evidence that the superpolyad quantum number P and the second constant P' abstracted from spectral fits are real features of molecular Hamiltonians.

We have therefore seen that in the transition from regular motion to global chaos, there are easily determined quantum numbers such as P, P' constructed from the zero-order normal mode quantum numbers. The vector methodology to determine these quantum numbers from spectra is extremely simple to apply. If there are n independent resonance vectors and $(N - n)$ constants, then n is the number of bifurcated degrees of freedom: the bifurcated dimensions consist of the subspace V^n spanned by the independent resonance vectors. The method is extremely helpful for the more difficult bifurcation problem to come, in that it reduces the dimensionality of the subspace in which bifurcations occur. But more importantly, the superpolyad number makes the bifurcation problem itself enormously simpler, as we are about to see.

4.2. *Bifurcation Structure of the Many Degree of Freedom Phase Space*

We turn now to the very difficult problem of the degrees of freedom of the many-dimensional problem bifurcated by the resonance couplings. One of the main virtues of the single resonance Hamiltonians used in our earlier work is that they make possible a simple and exhaustive classification of the bifurcated phase space structure, thereby leading to a set of quantum numbers appropriate to the classical dynamics underlying a given spectrum. Thus, for example, the 1:1 coupled stretch systems had a two-zone catastrophe map, with a mixture of local and normal modes in one zone, and only normal modes in the second zone. For the single resonance Hamiltonian, it is possible to achieve this classification analytically, without numerical solution of Hamilton's equations. This capability of finding an analytic classification affords great insight into the molecular dynamics underlying the assignment. For chaotic two degree of freedom systems, the hope is that this classification will carry over to most physically important situations. This will happen if the *large-scale bifurcation structure* of the chaotic Hamiltonian is essentially that given by the single resonance approximate Hamiltonian. That this expectation may often be reasonable is indicated, for example, by the finding discussed in Subsec. 3.3 that wave functions of coupled stretches in ozone calculated on a potential surface are essentially similar to those of the single resonance fitting Hamiltonian, with the same local/normal dichotomy suitable for spectral assignment.[77] (One exception of relevance to molecular spectroscopy is chaotic systems where the primary resonance coupling, e.g., the coupling between local modes which gives rise to the symmetric–antisymmetric stretch splitting, is very large. For example, the work of Matsushita and Terasaka[85] showed that for very large coupling between Morse oscillators, there is a series of resonances between normal modes; these resonances fall outside the two-zone classification of the catastrophe map of the Darling–Dennison Hamiltonian. While this kind of situation is likely to be relatively rare in molecular spectroscopy, except possibly near dissociation or when there is isomerization, there are significant cases at intermediate energies where it is probably important, for example, CO_2 and CS_2. We are currently at work on extending the use of fitting Hamiltonians and their bifurcation analysis to these cases.)

We have therefore seen that one of the cardinal features of the single resonance Hamiltonian is that it gives an analytic solution to the large-scale bifurcation structure and spectral assignment. Insofar as possible, it is desirable that the analysis of the many degree of freedom problem incorporate and generalize this property. At first this might appear unlikely, because the multidimensional Hamiltonian is likely to have more than one resonance, hence be chaotic. For example, a fit of the spectrum of all three vibrational modes of water requires not only a stretch–stretch coupling, but also a 2:1 Fermi resonance between the symmetric stretch and the bend.[86] However, for the single resonance Hamiltonian it is not the lack of chaos *per se* that allows an analytic classification. Rather, the key factor is the existence of a conserved global action (polyad number). The integrability of the single resonance Hamiltonian that confines trajectories to invariant tori is for our purposes merely a secondary byproduct of this extra constant of motion. As I will show, the crucial implication of a conserved polyad number for classifying phase space is that it makes the large-scale bifurcation structure analytically solvable. Furthermore, this analytic solvability as a consequence of the polyad number extends to many degree of freedom chaotic systems.

The basic idea is that the large-scale bifurcation structure is defined by the *lowest-order periodic orbits* and their bifurcations. First consider a two-mode system, without loss of generality, the 1:1. From Hamilton's equations, if the action I is conserved, then the dependence of the Hamiltonian on the fast angle $\theta = (\phi_1 + \phi_2)$ is zero:

$$\dot{I} = \partial H/\partial \theta = 0 \, . \tag{15}$$

That is, the Hamiltonian is independent of θ. For a given value of θ, consider a point $(I, I_z, \psi)_0$ at which the following conditions hold:

$$\dot{I} = 0; \qquad \dot{I}_z = 0; \qquad \dot{\psi} = 0 \, . \tag{16}$$

Since the Hamiltonian does not depend on θ, then for any value of θ, the point $(I, I_z, \psi)_0$ satisfies the conditions of Eq. (16). These conditions define the *critical points* of the Hamiltonian written as a function $H(I, I_z, \psi)$. In the subspace (I, I_z, ψ), the point $(I, I_z, \psi)_0$ is therefore a fixed point. Since in the remaining coordinate θ the frequency $\dot{\theta} \neq 0$, in the full space (I, I_z, θ, ψ) the point $(I, I_z, \psi)_0$ is a periodic orbit of lowest period, i.e., with the period of the fast angle θ. We have therefore shown that for the two degree of freedom Hamiltonian with conserved action I, any orbit satisfying the critical point conditions of Eq. (16) is a lowest period orbit, i.e., an orbit defining what we call the large-scale structure of the phase space. The role of the conserved quantum number I is crucial: without it, a point $(I, I_z, \psi)_0$ which satisfies Eq. (16) for a certain value of θ need not satisfy Eq. (16) for another value θ'. Thus, a point satisfying Eq. (16) is guaranteed to define a periodic point in the full space (I, I_z, θ, ψ) if and only if there is a conserved polyad number.

This result for two degree of freedom systems is immediately generalized to the following statement for a system with n degrees of freedom and a conserved (super)polyad number. Consider a Hamiltonian defined in terms of n actions $J_1 \ldots J_n$ and n conjugate angles $\theta_1 \ldots \theta_n$. Let J_1 be a good superpolyad number, or classically a constant of motion. Then the Hamiltonian does not depend on θ_1 and is a function $H(J_1 \ldots J_n; \theta_2 \ldots \theta_n)$. And by the same argument as above, the critical points of H defined by

$$\dot{J}_1 = 0, \ldots, \dot{J}_n = 0$$
$$\dot{\theta}_2 = 0, \ldots, \dot{\theta}_n = 0 \tag{17}$$

are lowest period orbits of the Hamiltonian.

This is the basic result needed to find the lowest period orbits and their bifurcations, i.e., the sought-after large-scale bifurcation structure. The mapping out of this structure for systems with three and more degrees of freedom is presently being carried out with the aim of a classification analogous to the catastrophe map for two degree of freedom systems.

5. Bootstrap Fitting and Potential Surfaces of Highly Excited Polyatomics

The preceding sections have been concerned with analysis of phase space structure of molecules and assignment of quantum numbers from the resulting dynamical classification. We have proceeded from simpler to more complex systems, as regards both the transition from regular to chaotic motion, and the inclusion of many degrees of freedom. The point of view adopted throughout is the desirability of starting with experimental spectral data as input for the phase space analysis and dynamically based quantum number assignment. Since an effective resonance Hamiltonian is used as the starting point for the theoretical analysis, it is crucial for analysis of particular molecules to be able to use experimental data to define the Hamiltonian. In our approach, the resonance Hamiltonian is obtained directly from the fit to the spectrum. A less direct but perhaps more general approach would be to fit the experimental data first to a molecular potential surface, as in the fit of Standard and Kellman for ozone,[76] and then determine the phenomenological resonance Hamiltonian from the surface by a method such as van Vleck perturbation theory, as implemented by Sibert and coworkers.[84]

An important question therefore is whether it will be possible in practice to fit spectra of molecules with many degrees of freedom having classically chaotic dynamics in the same way as is customary for spectra which are nonresonant or have one, or at most a few, important resonance couplings. Spectroscopists are accustomed to detailed state-by-state fitting of spectra with assignment of quantum numbers to the eigenstates, or at least to broad features containing many individual molecular eigenstates. At the very least, the existence of chaos in highly excited states makes this traditional spectroscopic program somewhat problematic. Complete chaos results in the loss of all approximate zero-order quantum numbers, and a breakdown of all the usual approximate selection rules. One may then ask whether it is possible to analyze molecular spectra in the chaotic regime with anything at all like the accustomed detail.

On the other hand, we have argued[49] that there is no inherent reason why it should not be possible to fit spectra of classically chaotic systems which are unassignable in terms of the zero-order quantum numbers. We investigated a "bootstrap" method of fitting of spectra of highly chaotic systems, using simulated model spectra as "experimental" input. The basic idea of the bootstrap procedure is to start with the relatively straightforward fit of the quasiperiodic region, and to gradually work up through increasingly strongly coupled energy ranges by using what is learned at each previous, slightly simpler step. In this way, it was possible to label and fit with a potential surface the spectra of strongly coupled systems, even when there are "missing" levels in the experimental spectrum unknown to the fitting program. This test of the bootstrap method fit a spectrum directly with a global potential, rather than the usual phenomenological resonance Hamiltonian of the kind used in this chapter. However, the basic bootstrap method can be used equally well with the latter type of Hamiltonian.

The model investigation of the bootstrap fitting method in Ref. 49 shows that there is no barrier, in principle, to detailed level-by-level fitting of spectra of systems with a high degree of classical chaos. In practice, however, the density of states in such systems will become so large that a fit of each individual level will be impractical for the foreseeable future. Instead, one will aim to fit the hierarchy of feature and subfeature structure such as that seen in the acetylene spectra of Ref. 8, probably stopping before the level of individual eigenstates. The goal will be to extract an approximate effective Hamiltonian, and from this Hamiltonian, to follow the procedure of Sec. 4. Approximate constants of motion and global quantum numbers will be inferred for the unperturbed degrees of freedom as in Subsec. 4.1, and a phase space bifurcation analysis and assignment will then be attempted for the remaining, perturbed degrees of freedom as in Subsec. 4.2. As deeper feature and subfeature structure is accommodated in the fit, fewer approximate global constants remain; if the fit reaches the level of individual eigenstates, the entire phase space will have to be subjected to the bifurcation analysis.

This chapter is almost entirely about spectral analysis using ideas of phase space structure to classify and organize the spectra of highly excited vibrational states. One of the motivations for trying to analyze spectra at this level of sophistication is the promise of deducing high-energy molecular potential surfaces. As a test of some of the bootstrap fitting ideas, we have sought to obtain an improved ground state electronic potential for the stretching spectrum of ozone. The experiments of Imre *et al.*[87] obtained the overtone and combination spectrum of ozone almost up to dissociation, so they probe the highly anharmonic region of the potential. We therefore used[76] the ozone spectrum as an ideal test case to investigate the status of current potential energy surfaces and the use of the bootstrap method to obtain potential surface information from fitting of very highly excited spectra. The vibrational spectrum on the ground electronic state of ozone is a system where one might expect existing potential surfaces to fail badly in accounting for highly excited data. Usual methods of obtaining empirical surfaces essentially rely on a low-energy expansion of the surface around the equilibrium configuration. *Ab initio* potential calculations give a more "global" view of the potential surface than empirical surfaces, but as recent work on ozone by Schaeffer *et al.*[88] shows, a very large basis set and a high degree of correlation are needed just to get the correct ordering for the symmetric and antisymmetric stretch fundamentals. We performed accurate quantum mechanical calculations of the ozone vibrational spectrum to test two of the best current surfaces. The resulting comparison of these surfaces with experimental data showed gross qualitative disagreement between the calculated and experimental spectrum, pointing up the inadequacy of current empirical methods for surfaces for truly highly excited states. We compared spectra calculated on our new surface with experiment and found good qualitative agreement with the data, far better than the spectra we calculated on previous surfaces. We believe this new surface for the stretching modes of ozone is the best now in existence, and shows the promise of the bootstrap approach for obtaining molecular potentials directly from analysis of high energy spectra.

6. Summary and Conclusions

In this chapter, I have reviewed progress toward a systematic theoretical framework for analysis of highly excited vibrational states of molecules. The basic problem at which this work is aimed is the breakdown of the traditional normal modes spectral analysis in the highly excited, strongly coupled anharmonic regime. Our general point of view is that progress can be achieved by applying recent developments in nonlinear dynamics of Hamiltonian systems. I have attempted to give a reasonably self-contained summary of prior work for a pair of modes coupled by a single Fermi resonance. I have then outlined steps toward the central remaining challenges, the problems of chaos and of many coupled degrees of freedom.

The survey began with a dynamical analysis and quantum number assignment of systems of highly anharmonic vibrations coupled by a single strong resonance, for instance, the coupled stretches in an ABA triatomic. The starting point is the fit of the spectrum to a phenomenological Hamiltonian. This leads to semiclassical depiction of experimental spectra as trajectories on the polyad phase sphere. A systematic classification of the structure of the phase sphere is obtained using bifurcation theory. (It should be noted that a Russian school[89] has been using bifurcation theory from a rather different point of view.) The catastrophe map gives an especially transparent view of the resulting phase space classification. This is used in a new assignment scheme based on the underlying semiclassical dynamics.

This assignment scheme is essentially complete for the problem of two modes coupled by a single resonance interaction with quasiperiodic dynamics. I next outlined some ideas for the much more difficult problem of extending this scheme to molecules whose dynamics have a significant degree of chaos. These ideas make use of some recent developments in classical nonlinear dynamics which show that many of the properties of the invariant tori of quasiperiodic dynamics extend to chaotic remnants of tori, known as cantori. We are investigating whether quantum wave functions of classically chaotic systems can be obtained semiclassically using WKB or EBK quantization methods on the cantori. If this is successful, it should allow a dynamical classification and assignment of states of chaotic systems analogous to what we have earlier obtained for single resonance, quasiperiodic fitting Hamiltonians.

It is evident that in addition to the problems introduced by chaos, a successful comprehensive approach to highly excited spectra has to surmount the problem of bifurcations in systems with many interacting degrees of freedom. In this chapter, I first discussed the problem of identifying the dimensions of phase space actually affected by bifurcations. A very important byproduct is identification of approximate constants of motion associated with the unbifurcated degrees of freedom. These approximate constants are associated with energy transfer pathways and superpolyad quantum numbers. These quantum numbers are proving very useful in accounting for some of the main features of complex spectra, in particular, hierarchies of structure and energy transfer in spectra of acetylene. I then discussed the problem of classifying the structure of the bifurcated degrees of freedom. The superpolyad quantum number is crucial here since it permits a simplification similar to that in the two-mode system, in that it makes it possible to solve analytically for the large-scale bifurcation structure of spectroscopic fitting Hamiltonians for chaotic many degree of freedom systems.

I finished with an outline of investigations of a bootstrap method for fitting spectra of systems with a high degree of classical chaos. This is important because our entire approach is deliberately focused on phase space analysis of a molecular Hamiltonian tied as closely as possible to experimental data. The question is then whether it is possible in practice to obtain a Hamiltonian from spectral fits for molecules with many degrees of freedom and classically chaotic dynamics. We have argued that with the use of the bootstrap method, there is no inherent reason why it should not be possible to fit spectra of classically chaotic systems which are unassignable in terms of the zero-order quantum numbers. We successfully tested this method on model chaotic systems and found that it was possible successfully to label and fit a spectrum of a highly chaotic molecular model, even when there are "missing" levels in the experimental spectrum. We then applied the bootstrap method to obtain a new potential surface for the stretching spectrum of ozone. The new surface gives results for the spectrum in much better agreement with experiment than existing surfaces.

The model investigation of the bootstrap fitting method in Ref. 49 shows that there is no barrier, in principle, to fitting of spectra of systems with a high degree of classical chaos. This should enable the kind of analysis we foresee to be carried out for complex spectra with, for example, the hierarchy of structure seen in acetylene spectroscopy. Approximate constants of motion and global quantum numbers will be inferred for the unperturbed degrees of freedom. A phase space bifurcation analysis and assignment will then be implemented for the remaining, perturbed degrees of freedom. As deeper feature structure is fit, fewer approximate global constants remain, and the bifurcation analysis becomes more complicated. It is to be hoped that the only limitations in practice will be those of computational power and numerical technique.

Acknowledgment

Support of the Air Force Office of Scientific Research under Grant F49620-92-J-0035, the Department of Energy Basic Energy Sciences Program under Grant DE-FG06-92ER14236, and the National Science Foundation under Grant CHE-9120400 is gratefully acknowledged.

References

1. E. Abramson, R. W. Field, D. Imre, K. K. Innes, and J. L. Kinsey, *J. Chem. Phys.* **83**, 453 (1985).
2. J. L. Hardwick, *J. Mol. Spectrosc.* **109**, 85 (1985).
3. J. P. Pique, Y. Chen, R. W. Field, and J. L. Kinsey, *Phys. Rev. Lett.* **58**, 475 (1987).
4. J. P. Pique, Y. M. Engel, R. D. Levine, Y. Chen, R. W. Field, and J. L. Kinsey, *J. Chem. Phys.* **88**, 5972 (1988).
5. A. Amrein, H.-R. Dubal, and M. Quack, *Mol. Phys.* **56**, 727 (1985).
6. B. C. Smith and J. S. Winn, *J. Chem. Phys.* **89**, 4638 (1988).
7. B. C. Smith and J. S. Winn, *J. Chem. Phys.* **94**, 4120 (1991).

8. K. Yamanouchi, N. Ikeda, S. Tsuchiya, D. M. Jonas, J. K. Lundberg, G. W. Adamson, and R. W. Field, *J. Chem. Phys.* **95**, 6330 (1991).

9. M. E. Kellman, *J. Chem. Phys.* **93**, 6630 (1990).

10. M. E. Kellman and G. Chen, *J. Chem. Phys.* **95**, 8671 (1991).

11. H. L. Dai, C. C. Korpa, J. L. Kinsey, and R. W. Field, *J. Chem. Phys.* **82**, 1688 (1985).

12. N. L. Garland and E. K. C. Lee, *J. Chem. Phys.* **84**, 28 (1986).

13. P. Arrowsmith, F. E. Bartoszek, S. H. P. Bly, T. Carrington, Jr., P. E. Charters, and J. C. Polanyi, *J. Chem. Phys.* **73**, 5895 (1980).

14. S. E. Bradforth, D. W. Arnold, R. B. Metz, A. Weaver, and D. M. Neumark, *J. Phys. Chem.* **95**, 8066 (1991).

15. R. D. van Zee, Charles D. Pibel, Thomas J. Butenhoff, and C. B. Moore, *J. Chem. Phys.* **97**, 3235 (1992).

16. W. Xie, C. Harkin, H.-L. Dai, W. H. Green, Jr., Q. K. Zheng, and A. J. Mahoney, *J. Mol. Spectrosc.* **138**, 596 (1990).

17. G. Herzberg, *Infrared and Raman Spectra* (Van Nostrand, New York, 1950).

18. L. Xiao and M. E. Kellman, *J. Chem. Phys.* **90**, 6086 (1989).

19. Z. Li, L. Xiao and M. E. Kellman, *J. Chem. Phys.* **92**, 2251 (1990).

20. L. Xiao and M. E. Kellman, *J. Chem. Phys.* **93**, 5805 (1990).

21. M. E. Kellman and L. Xiao, *Chem. Phys. Lett.* **162**, 486 (1989).

22. M. E. Kellman and L. Xiao, *J. Chem. Phys.* **93**, 5821 (1990).

23. B. T. Darling and D. M. Dennison, *Phys. Rev.* **57**, 128 (1940).

24. E. Fermi, *Z. Phyzik* **71**, 250 (1931).

25. M. E. Kellman *J. Chem. Phys.* **83**, 3843 (1985).

26. M. E. Kellman and E. D. Lynch, *J. Chem. Phys.* **85**, 7216 (1986).

27. D. Farrelly, *J. Chem. Phys.* **85**, 2119 (1986).

28. W. G. Harter, *J. Chem. Phys.* **85**, 5560 (1986).

29. C. Jaffc, *J. Chcm. Phys.* **89**, 3395 (1988).

30. M. E. Kellman and E. D. Lynch, *J. Chem. Phys.* **89**, 3396 (1988).

31. D. K. Sahm, S. W. McWhorter, and T. Uzer, *J. Chem. Phys.* **91**, 219 (1989).

32. W. G. Harter and C. W. Patterson, *J. Chem. Phys.* **80**, 4241 (1984).

33. K. K. Lehmann, *J. Chem. Phys.* **79**, 1098 (1983).

34. W. Heisenberg, *Z. Phyz.* **33**, 879 (1925), translated in *Sources of Quantum Mechanics*, ed. B. L. van der Waerden (Dover, New York, 1967).

35. A. P. Clark, A. S. Dickinson, and D. Richards, *Adv. Chem. Phys.* **36**, 63 (1977).

36. W. Siebrand and D. F. Williams, *J. Chem. Phys.* **49**, 1860 (1968).

37. B. R. Henry, *Acc. Chem. Res.* **10**, 207 (1977).

38. M. E. Long, R. L. Swofford, and A. C. Albrecht, *Science* **191**, 183 (1976).

39. M. S. Burberry and A. C. Albrecht, *J. Chem. Phys.* **71**, 4631 (1979).

40. R. Wallace, *Chem. Phys.* **11**, 189 (1975).

41. R. Wallace and A. A. Wu, *Chem. Phys.* **39**, 221 (1979).

42. R. T. Lawton and M. S. Child, *Mol. Phys.* **40**, 773 (1980).

43. M. L. Sage, *Chem. Phys.* **83**, 1455 (1979).

44. C. Jaffe and P. Brumer, *J. Chem. Phys.* **73**, 5646 (1980).

45. E. L. Sibert III, J. T. Hynes, and W. P. Reinhardt, *J. Chem. Phys.* **77**, 3583 (1982).

46. M. E. Kellman and E. D. Lynch, *J. Chem. Phys.* **88**, 2205 (1988).

47. M. Golubitsky and D. G. Schaeffer, *Singularities and Groups in Bifurcation Theory, Vol. I* (Springer, New York, 1985).

48. T. Poston and L. Stewart, *Catastrophe Theory and its Applications* (Pitman, London, 1978).

49. J. M. Standard, E. D. Lynch, and M. E. Kellman, *J. Chem. Phys.* **93**, 159 (1990).

50. N. DeLeon and E. J. Heller, *Phys. Rev.* **A30**, 5 (1984).

51. A. Einstein, *Verh. Dtsch. Phys. Ges.* **19**, 82 (1917).

52. J. B. Keller, *Ann. Phys.* **4**, 180 (1958).

53. V. Fock, *Z. Phys.* **98**, 145 (1935).

54. V. Bargmann, *Z. Phys.* **99**, 576 (1936).

55. A. N. Kolmogorov in *Proceedings of the 1954 International Congress of Mathematics* (North-Holland, Amsterdam, 1957).

56. J. Moser, *Nachr. Akad. Wiss. Goettingen Math. Phys.* **K1**, 1 (1962).

57. V. I. Arnold, *Russ. Math. Surv.* **18**, 85–191 (1963).

58. M. Tabor, *Chaos and Integrability in Nonlinear Dynamics* (Wiley, New York, 1989).

59. C. C. Martens and G. S. Ezra, *J. Chem. Phys.* **86**, 279 (1987).

60. M. C. Gutzwiller, *Chaos in Classical and Quantum Mechanics* (Springer-Verlag, New York, 1990).

61. J. N. Mather, *Topology* **21**, 457 (1982).

62. J. N. Mather, *Publ. Math. I.H.E.S.* **63**, 153 (1986).

63. S. Aubry, in *Structures et Instabilites*, ed. C. Godreche, (Les Editions de Physique, Paris, 1986).

64. A. Katok, *Ergod. Theor. and Dynami. Sys.* **2**, 185 (1982).

65. A. Katok, in *Dynamical Systems and Chaos, Proceedings of the Sitges Conference on Statistical Mechanics, Springer Lecture Notes on Physics #179* (Springer, New York, 1983).

66. R. P. Feynman and A. R. Hibbs, *Quantum Mechanics and Path Integrals* (McGraw-Hill, New York, 1965).

67. J. H. van Vleck, *Proc. Nat. Acad. Sci. USA* **14**, 178 (1928).

68. S. Tomsovic and E. J. Heller, *Phys. Rev. Lett.* **70**, 1405 (1993).

69. D. Wintgen, *Phys. Rev. Lett.* **61**, 1803 (1988).

70. M. V. Berry and M. Tabor, *Proc. Roy. Soc. London* **1349**, 101 (1976).

71. A. M. Ozorio de Almedia, *Hamiltonian systems: chaos and quantization*, (Cambridge University, Cambridge, 1988).

72. E. J. Heller, *Phys. Rev. Lett.* **53**, 1515 (1984).

73. H. S. Taylor and J. Zakrzewski, *Phys. Rev.* **A38**, 3732 (1988).

74. A. M. Ozorio de Almeida, "On the quantization of homoclinic motion", *Nonlinearity* **2**, 519 (1989).

75. K. Takatsuka, *Phys. Rev.* **A45**, 4326 (1992).

76. J. M. Standard and M. E. Kellman, *J. Chem. Phys.* **94**, 4714 (1991).

77. J. M. Standard and M. E. Kellman, unpublished.

78. R. L. Devaney, *An Introduction to Chaotic Dynamical Systems* (Addison-Wesley, Redwood City, 1987).

79. M. J. Davis, *J. Phys. Chem.* **92**, 3124 (1988).

80. J. M. Greene, *J. Math. Phys.* **20**, 1183 (1979).

81. L. E. Fried and G. S. Ezra, *J. Chem. Phys.* **86**, 6270 (1987).

82. T. A. Holme and R. D. Levine, *Chem. Phys. Lett.* **150**, 393 (1988).

83. D. Jonas, MIT thesis.

84. A. B. McCoy and E. L. Sibert III, *J. Chem. Phys.* **95**, 3476 (1991).

85. T. Matsushita, A. Narita, and T. Terasaka, *Chem. Phys. Lett.* **95**, 129 (1983); T. Matsushita and T. Terasaka, *Chem. Phys. Lett.* **100**, 138 (1983); *ibid* **105**, 511 (1984); *Phys. Rev.* **A32**, 538 (1985).

86. J. E. Baggott, *Mol. Phys.* **65**, 739 (1988).

87. D. G. Imre, J. L. Kinsey, R. W. Field, and D. H. Katayama, *J. Phys. Chem.* **86**, 2564 (1982).

88. G. E. Scuseria, T. J. Lee, A. C. Scheiner, and H. F. Schaeffer III, *J. Chem. Phys.* **90**, 5635 (1989).

89. V. B. Pavlov-Verevkin and B. I. Zhilinskii, *Chem. Phys.* **128**, 429 (1988); B. I. Zhilinskii, *Chem. Phys.* **137**, 1 (1989); D. A. Sadovskii, B. I. Zhilinskii, J. P. Champion, and G. Pierre *J. Chem. Phys.* **92**, 3874 (1990).

CHAPTER 26

ROTATION–VIBRATION MIXING
OF HIGHLY EXCITED STATES

Darin C. Burleigh, Anne B. McCoy and Edwin L. Sibert III

Department of Chemistry and Theoretical Chemistry Institute

University of Wisconsin – Madison,

Madison, Wisconsin 53706, USA

Contents

1. Introduction

A flurry of recent experimental work has indicated that rotation–vibration
mixing plays an integral part in the theoretical description of the dynamics

of polyatomic molecules in the gas phase,[1-8] especially as vibrational energy E and rotational quantum number J are increased. These findings have spurred the development of many theoretical models of rotation–vibration interactions in an effort to understand more fully the origins of these coupling effects.[9-21] In this Chapter we describe three components of research in this area. The first area entails calculating observables associated with highly excited rotation–vibration states. The second involves generating schemes of data analysis, so that one can readily understand the general trends of rotation–vibration coupling as well as gain insights into the predominant coupling mechanisms. Finally, we introduce a random matrix approach to rotation–vibration mixing that allows one to extract from a complicated molecular Hamiltonian the essential ingredients controlling the breakdown of rotation–vibration separability. In presenting these ideas, the formaldehyde molecule will be used as a prototype.

In a series of papers, we have shown the utility of Canonical Van Vleck perturbation theory (CVPT) for describing highly excited vibrational states of polyatomic molecules.[18,22-25] Although low order CVPT has long been pivotal in analyzing molecular spectra, traditionally it has not been used to examine more highly excited vibrational states, where configuration interaction between nearly degenerate states leads to a rapid divergence of the perturbative expansions.[26] We have found, however, that even at high energies CVPT can be used to transform a Hamiltonian, if the final form of the Hamiltonian includes all the important resonance interactions.

Three important issues that arise in studying rotation–vibration interactions are the choice of coordinates, choice of embedding conditions, and the ensuing form of the Hamiltonian. Specific forms for the rotation–vibration Hamiltonian has been derived for rectilinear normal coordinates and Jacobi coordinates. There are, however, many molecules where other choices of internal coordinates are more convenient for variational calculations[27] and lead to more rapid convergences in the perturbative expansions.[28] We describe in this Chapter a method for obtaining rotation–vibration Hamiltonians that can be applied readily to many different coordinate systems and choices of embedding.

Given that one can calculate various observables from a rotation–vibration Hamiltonian at high energies, one quickly realizes that, due to the high density of states, the presentation of the data in an enlightening form is a challenge in its own right. We consider several formats for displaying the data that allow one to visualize the extent of rotation–vibration mixing and the role of Coriolis and centrifugal coupling. In developing these formats, our motivation is not to examine rotation–vibration mixing for specific zero-order states. Instead, we are more interested in elucidating general trends. For example, the participation number[29] η is a useful measure of rotation–vibration mixing of an individual zero-order state. For a given J and within an energy range $E \pm \Delta E$ there will be a distribution of η. Rather than focusing on an individual member of the distribution, one can determine the coupling mechanisms that most strongly influence the properties of the distribution.

We have recently introduced a random matrix approach[30] to answer these questions. The goal is to define an ensemble of Hamiltonians such that the ensemble yields the same distribution of η as is found for the original molecular Hamiltonian. In particular, the ensemble results should reproduce the trends observed for the actual η distributions as J and E are increased. One can then relate these trends to the properties of the random matrix ensemble in order to elucidate the mechanisms of mode mixing. Our choice of distribution functions comes from simple ideas about the correlation of diagonal and off-diagonal Hamiltonian matrix elements that do not depend on state-specific details.

2. Rotation–Vibration Hamiltonians

There are several possible routes to obtaining the rotation–vibration Hamiltonian of a polyatomic molecule expressed as a function of the internal coordinates. Handy and coworkers[27,31] started with the kinetic energy operator, expressed in terms of the Cartesian coordinates of the atoms. This operator is reexpressed in terms of the three Euler angles, a set of $3n - 6$ internal coordinates, and their conjugate momenta *via* the chain rule. Although the algebra is complicated, all that is required are the derivatives of internal vibrational coordinates and the three Euler angles with respect to the $3n$ Cartesian coordinates.

Although analytic expressions have been obtained using this approach, the resulting Hamiltonian does not take a simple form. Furthermore, one does not have the flexibility of changing readily the choice of embedding or redefining the internal coordinates. On the other hand, Pickett's derivation provides the Hamiltonian written in a form that is ideally suited for our purposes. Pickett,[32] following the work of Meyer and Günthard,[33] started with the classical rotation–vibration Hamiltonian expressed as a function of the angular momenta, $3n - 6$ internal coordinates and their conjugate momenta. This Hamiltonian is then transformed to its quantum mechanical form following standard procedures.[34] Pickett's treatment is general, so there is a great deal of flexibility in the choice of both the internal coordinates and the embedding of the body-fixed axis system. The form of the Hamiltonian is the same, regardless of the choice of embedding and choice of internal coordinates. Since we are using a perturbative approach, it is only necessary to expand the various coordinate dependent terms in the kinetic energy operator about the equilibrium configuration in a Taylor series.[18,23,24] In order to evaluate these expansion coefficients, one needs to transform between the curvilinear internal coordinates and the Cartesian coordinates. We now outline the central steps in this procedure.

Following the work of Pickett,[32] the quantum mechanical rotation–vibration Hamiltonian takes the form

$$\hat{H} = \hat{H}_v + \frac{1}{2}\mathbf{J}^T(\boldsymbol{\mu} + \boldsymbol{\Lambda})\mathbf{J} - \frac{1}{2}\left[\mathbf{P}^T\mathbf{A}^T\mathbf{J} + \mathbf{J}^T\mathbf{A}\mathbf{P}\right] . \tag{1}$$

The first contribution is the $(J = 0)$ vibrational Hamiltonian which takes the familiar form

$$\hat{H}_v = \frac{1}{2}\mathbf{P}^T\mathbf{G}\mathbf{P} + V + V' . \tag{2}$$

The kinetic energy contribution is expressed in terms of the internal coordinates, S_t, and their conjugate momenta, $P_t = -i\hbar\frac{\partial}{\partial S_t}$. If the internal coordinates are the usual bond stretching and bending extensions, then the **G**-matrix elements, which are functions of S_t, are those of Wilson, Decius, and Cross.[35] The potential type terms consist of the Born–Oppenheimer potential V and a mass dependent contribution V' which results from the transformation of the kinetic energy operator from Cartesian coordinates to internal coordinates.

We have demonstrated the accuracy and utility of numerical derivations of $\hat{H}_v$ for several nonlinear polyatomic molecules.[23,24] The initial step is to note that both the **G**-matrix elements and V' are functions of the $(3n - 6) \times 3n$ **B**-matrix, whose elements are

$$B_{tk} = \frac{\partial S_t}{\partial x_{\alpha j}} . \tag{3}$$

We use α to represent the Cartesian coordinates, x, y, z, so that $x_{\alpha j}$ is a Cartesian coordinate of the jth atom in the body-fixed coordinate system. The B_{tk} matrix elements can be readily constructed using a finite difference scheme. For a given configuration, one extends a Cartesian coordinate a small distance and then determines the corresponding changes in the internal coordinates. Although the B_{tk} matrix elements are dependent on the choice of rotating reference frame, $\hat{H}_v$ is independent of this choice. Once the **B**-matrix is constructed, expressions for G_{ij} and V' can be evaluated readily.

The remaining contributions to the Hamiltonian are more complicated. The matrices $\boldsymbol{\mu}$, $\boldsymbol{\Lambda}$, and $\mathbf{A}$ are all solely functions of the internal coordinates. In contrast to $\mathbf{G}$, however, the functional form depends on the criterion used for embedding the body-fixed reference frame. For example, $\boldsymbol{\mu} \equiv \mathbf{I}^{-1}$ is the instantaneous inverse moment of inertia tensor whose components are defined with respect to the body-fixed frame. A finite difference scheme can also be used here to determine the functional dependence of $\boldsymbol{\mu}$. However, for any value of the internal coordinates at which the components of $\boldsymbol{\mu}$ are evaluated, one must transform to the body-fixed coordinate system. For example, if one wants to minimize Coriolis interactions at the molecular equilibrium configuration, Eckart[36] has supplied a simple algorithm that enables one to transform to such a reference frame.

The remaining two terms, $\boldsymbol{\Lambda}$ and $\mathbf{A}$, are obtained in a similar fashion. Here, in addition to requiring knowledge of how the internal coordinates depend on the body-fixed Cartesian coordinates, we also require the inverse relationship, $\partial x_{\alpha j}/\partial S_t$. These derivatives are obtained by inversion of the $3n \times 3n$ **B**-matrix, where the additional six rows are used to define the origin and embedding of the body-fixed axis system.

The pure rotational contribution to the Hamiltonian is

$$\hat{H}_r = \frac{1}{2}\mathbf{J}^T(\boldsymbol{\mu}^0 + \boldsymbol{\Lambda}^0)\mathbf{J} \,, \tag{4}$$

where $\hat{J}_\alpha$ are the components of the total angular momentum along the body-fixed axes. They obey the commutation relations

$$\left[\hat{J}_\alpha, \hat{J}_\beta\right] = -i\hbar\epsilon_{\alpha\beta\gamma}\hat{J}_\gamma. \tag{5}$$

The matrices $\boldsymbol{\mu}^0$ and $\boldsymbol{\Lambda}^0$ are evaluated in the equilibrium configuration. It should be noted that in an Eckart frame $\boldsymbol{\Lambda}^{(0)} = \mathbf{0}$, and $\hat{H}_r$ takes the familiar form of the Hamiltonian for the rigid-body motion.

The vibrations and rotations are coupled via the remaining terms not included in $\hat{H}_r$ and $\hat{H}_v$. This coupling can be divided into two contributions. The first includes those terms that are proportional to $\hat{J}_\alpha \hat{J}_\beta$. These are referred to as centrifugal coupling. The second includes those terms that are linear in $\hat{J}_\alpha$. These contributions are referred to as α-axis Coriolis coupling.

In these calculations, it is computationally convenient to carry out the above derivations in terms of curvilinear normal coordinates q_t and their conjugate momenta p_t instead of S_t and P_t. These coordinates are defined to be the linear combination of the internal coordinates that leads to a separable harmonic oscillator Hamiltonian to lowest order. Since this transformation is invertible, one can readily determine the functional dependencies of the terms in the Eq. (1) using either the normal or internal coordinates. Interestingly, in our study of vibrational states of the well known local mode molecule H_2O and its deuterated analogs we found only minor differences between the results of CVPT in the internal and normal mode representations.[24] The normal mode calculations, however, were significantly faster, since many terms in the Hamiltonian are constrained to zero by symmetry. For this reason we have chosen to work with the normal modes.

Finally, the normal coordinates need not be linear combinations of the internal extension coordinates. In this work the stretch coordinates are the Simons–Parr–Finlan (SPF) coordinates.[37] The normal coordinates are then defined as the appropriate linear combination of these. When we expand the coordinate dependent terms of the Hamiltonian in a Taylor series to a given order, the normal coordinates based on the SPF coordinates lead to a more accurate representation of *both* the model potential and the **G**-matrix elements than do expansions based on the usual internal extension coordinates. Likewise, an appropriate choice of bending coordinates can also provide a more rapid convergence of these terms.[18]

With the above approach we can combine the use of curvilinear normal coordinates with the Eckart frame. When we do so, the harmonic oscillator, rigid rotor and, to lowest order, the Coriolis and centrifugal coupling contributions to $\hat{H}$ have exactly the same form as those found for the more commonly used Watson Hamiltonian. Therefore, the detailed discussion of Dai *et al.*[1] concerning the origin and magnitude of various low order rotation–vibration coupling terms in H_2CO applies to our Hamiltonian as well. The higher order contributions are different.

3. Canonical Van Vleck Perturbation Theory

CVPT is used to transform the Hamiltonian of Eq. (1) to a block-diagonal Hamiltonian $\hat{K}$. Our initial discussion describes the transformation of the $(J = 0)$ vibrational Hamiltonian. We will then include the effects of rotation. One begins by determing the dimensionless normal coordinates using standard methods[35] and then expanding the coordinate dependent terms of the Hamiltonian [cf. Eq. (1)] in a Taylor series in the dimensionless normal coordinates. Following Nielsen,[26] this Hamiltonian is then reexpressed in the form

$$\hat{H} = \hat{H}^{(0)} + \lambda\hat{H}^{(1)} + \lambda^2\hat{H}^{(2)} + \ldots + \lambda^n\hat{H}^{(n)} , \tag{6}$$

where λ is the perturbation parameter. We take $\hat{H}^{(0)}$ to be N uncoupled harmonic oscillators with frequencies ω_j. Terms that are cubic in coordinates and momenta are included in $\hat{H}^{(1)}$; quartic terms are included in $\hat{H}^{(2)}$, and so forth. The expansion includes terms up to nth order, the order to which you wish to carry out the perturbative analysis. An exception is made for the partitioning of V' [cf. Eq. (2)]. Here, the constant term is put in $H^{(2)}$, the linear term in $H^{(3)}$, etc. With this choice of ordering, all the terms in $\hat{H}^{(l)}$ are proportional to $\hbar^{\frac{l+2}{2}}$.

The transformations are accomplished via a succession of canonical transformations,

$$\exp\left[i\lambda^n[S^{(n)},]\right]\ldots\exp\left[i\lambda^2[S^{(2)},]\right]\exp\left[i\lambda[S^{(1)},]\right]\hat{H} = \hat{K} . \tag{7}$$

Other observables, such as the dipole moment function, are transformed likewise[38,39] as

$$\exp\left[i\lambda^n[S^{(n)},]\right]\ldots\exp\left[i\lambda^2[S^{(2)},]\right]\exp\left[i\lambda[S^{(1)},]\right]\hat{\mu} = \hat{M} . \tag{8}$$

As the form of these transformations implies, one first transforms $\hat{H}$ to an intermediate Hamiltonian

$$\exp\left[i\lambda[S^{(1)},]\right]\hat{H} = \hat{K}_1 \tag{9}$$

where

$$\exp\left[i\lambda[S^{(1)},]\right]\hat{H} = \hat{H} + i\lambda[S^{(1)}, \hat{H}] - \frac{\lambda^2}{2!}\left[S^{(1)}, [S^{(1)}, \hat{H}]\right] + \ldots . \tag{10}$$

Expanding $\hat{H}$ and $\hat{K}_1$ in powers of λ and then equating powers of λ one obtains

$$\hat{K}_1^{(0)} = \hat{H}^{(0)} \, ,$$
$$\hat{K}_1^{(1)} = \hat{H}^{(1)} + i[S^{(1)}, \hat{H}^{(0)}] \, ,$$
$$\hat{K}_1^{(2)} = \hat{H}^{(2)} + i[S^{(1)}, \hat{H}^{(1)}] - \frac{1}{2!}\left[S^{(1)}, [S^{(1)}, \hat{H}^{(0)}]\right] \, , \qquad (11)$$

and so forth. If $\hat{H}^{(1)}$ is written in the normal form

$$\hat{H}^{(1)} = \sum_{\mathbf{m}} \sum_{\mathbf{n}} C_{\mathbf{m},\mathbf{n}} \prod_{j=1}^{N} (a_j^{\dagger})^{m_j} (a_j)^{n_j} \, , \qquad (12)$$

then a choice of

$$\hat{S}^{(1)} = \sum_{\mathbf{m}}{}' \sum_{\mathbf{n}}{}' C_{\mathbf{m},\mathbf{n}} \frac{\prod_{j=1}^{N} (a_j^{\dagger})^{m_j} (a_j)^{n_j}}{i \sum_{j=1}^{N} (m_j - n_j) \hbar \omega_j} \qquad (13)$$

leads to

$$\hat{K}_1^{(1)} = \sum_{\mathbf{m}}{}'' \sum_{\mathbf{n}}{}'' C_{\mathbf{m},\mathbf{n}} \prod_{j=1}^{N} (a_j^{\dagger})^{m_j} (a_j)^{n_j} \, . \qquad (14)$$

The primes and double primes refer to restricted summations; a term in $H^{(1)}$ will not appear in $\hat{K}_1^{(1)}$ if the corresponding term is included in $S^{(1)}$. In this way, $\hat{K}_1^{(1)}$ can be designed to include just those terms which lead to significant resonant coupling. Although $\hat{K}_1$ is calculated through nth order, it is constructed to have the desired form only through first order.

In order to obtain a Hamiltonian which has the correct form through λ^2, a second transformation must be carried out. In analogy to the first

$$\exp\left[i\lambda[S^{(2)},]\right] \hat{K}_1 = \hat{K}_2 \, . \qquad (15)$$

In general, the operator $S^{(n)}$ can be chosen such that, through order λ^n, $\hat{K}_n$ contains only those terms that couple nearly degenerate states. In order to extend these ideas to include rotation we expand $\mathbf{A}$, $\boldsymbol{\mu}$, and $\boldsymbol{\Lambda}$ about the equilibrium configuration in a Taylor series. If we then express the full rotation–vibration Hamiltonian in the form of Eq. (12) and apply the transformations of Eq. (7), we run into difficulties. The coefficients $C_{\mathbf{m},\mathbf{n}}^{(l)}$

no longer commute, as they are now functions of the angular momentum operators. Analytical expansions for $\hat{K}_1$ and $\hat{K}_2$ through fourth order are given by Amat *et al.*,[40] but in order to implement high order CVPT additional modifications are necessary.

To remove the operator dependence of the $C^{(l)}_{\mathbf{m},\mathbf{n}}$, while maintaining the simple form of $\hat{H}$ given in Eq. (12), we reexpress the rotation operators in terms of the raising and lowering operators for two coupled, degenerate harmonic oscillators. Following Schwinger,[41] in a body-fixed coordinate system, the rotation operators are redefined as

$$
\begin{aligned}
\hat{J}_x &= \frac{1}{2}\left(a_{r1}^{\dagger}a_{r2} + a_{r1}a_{r2}^{\dagger}\right) \ , \\
\hat{J}_y &= \frac{i}{2}\left(a_{r1}^{\dagger}a_{r2} - a_{r1}a_{r2}^{\dagger}\right) \ , \\
\hat{J}_z &= \frac{1}{2}\left(a_{r1}^{\dagger}a_{r1} - a_{r2}^{\dagger}a_{r2}\right) \ , \\
\hat{J}^2 &= \hat{J}_x^2 + \hat{J}_y^2 + \hat{J}_z^2 \\
&= \frac{1}{2}\left(a_{r1}^{\dagger}a_{r1} + a_{r2}^{\dagger}a_{r2}\right) \times \left[\frac{1}{2}\left(a_{r1}^{\dagger}a_{r1} + a_{r2}^{\dagger}a_{r2}\right) + 1\right] \ .
\end{aligned}
\tag{16}
$$

These definitions have been used by several groups[12,42–44] to gain insights into the properties of two harmonically coupled degenerate anharmonic oscillators.

In contrast to the work of Nielsen,[26] we introduce the rotation terms at higher orders in the perturbative expansion. In expanding $\mathbf{A}$, $H^{(1)}$ includes constant terms, $H^{(2)}$ includes linear terms, and so forth. In expanding $\boldsymbol{\mu}$ and $\boldsymbol{\Lambda}$, $H^{(2)}$ includes constant terms, $H^{(3)}$ includes linear terms, and so forth. Once again, with this choice of ordering, all the terms in $\hat{H}^{(l)}$ are proportional to $\hbar^{\frac{l+2}{2}}$. When taking the perturbation theory to high orders, it is not crucial to introduce the rotational couplings in $H^{(0)}$ because these terms are much smaller than the corresponding contributions to the vibrational Hamiltonian.

In order to take advantage of the molecular symmetry, the rotational contribution to the full wave function must be symmetry adapted. This is accomplished by transforming to the Wang basis. Second, the Hamiltonian matrix representation of $\hat{K}$ is not, in general, real symmetric. Huber[45] has discussed a rotational basis which is both symmetry adapted and which leads to a real symmetric Hamiltonian matrix. Following his approach, the rotation basis is expressed as linear combinations of the harmonic oscillator basis vectors $|n_{r1}, n_{r2}\rangle$ as

$$|J, K^+\rangle = \frac{1}{\sqrt{2}} \left[|J + K, J - K\rangle + (-1)^K |J - K, J + K\rangle \right] , \qquad (17)$$

$$|J, K^-\rangle = \frac{1}{i\sqrt{2}} \left[|J + K, J - K\rangle - (-1)^K |J - K, J + K\rangle \right] ,$$

and

$$|J, 0\rangle = |J + 0, J - 0\rangle . \qquad (18)$$

The harmonic oscillator labels n_{ri} are related to J and K by $n_{r1} = J + K$ and $n_{r2} = J - K$, where J and K are the quantum numbers associated with the total angular momentum and the projection of the angular momentum onto the molecular z-axis, respectively. As shown by Huber,[45] these basis vectors lead to the following relations

$$\hat{J}^2 |J, K^\pm\rangle = J(J + 1)\hbar^2 |J, K^\pm\rangle \qquad (19)$$

and

$$\hat{J}_z |J, K^\pm\rangle = \pm iK\hbar |J, K^\mp\rangle . \qquad (20)$$

Combining these rotation functions with the usual vibrational basis functions provides a symmetry adapted rotation–vibration basis.

With this basis the Hamiltonian matrix for $\hat{K}$ is block-diagonal, where each block contains states of the same Γ, J, and N_t quantum numbers, where Γ refers to the molecular symmetry. Since CVPT is performed in an operator framework, the perturbative transformations to determine $\hat{K}$ are performed only once.

Having described the general method we now describe the 'exact' system of interest. H_2CO has C_{2v} symmetry and, therefore, four point group representations: a_1, a_2, b_1, b_2. In the energy regime of interest there is an additional approximate constant of motion, which we reveal by using Van Vleck perturbation theory to transform the Hamiltonian to a block-diagonal form. This block-diagonal structure is not an isolated case. Similar structure has been found for many triatomic molecules as well as for acetylene using Van Vleck perturbation theory.[24,25] This structure is also the basis of the construction of the polyad phase spheres of Kellman and co-workers.[46–48]

In order to understand how these blocks are defined one can inspect the fundamental frequencies of Table I. This table shows that there are two similar high frequency modes (ω_1 and ω_5) which are approximately twice the frequency of the other four. Consequently, there are several likely candidates for Fermi resonance interactions between a CH stretching degree of freedom and the lower frequency modes. In addition, there is a strong Darling–Dennison resonance between the two high frequency stretches. We expect that the lower frequency modes might also be coupled by Darling–Dennison resonances. Despite all these resonances there remains remarkably a single constant of the motion,

$$N_t = 2n_1 + n_2 + n_3 + n_4 + 2n_5 + n_6 , \qquad (21)$$

where n_i = number of quanta in the ith normal mode. Given that the states within each block are coupled by multiple resonance interactions, we will refer to N_t as the superpolyad quantum number as opposed to the polyad number.

Table I. Calculated harmonic frequencies[a], ω_j, of H_2CO and D_2CO given in units of cm^{-1}.

Mode	Symmetry	ω_j/cm^{-1}		motion
		H_2CO	D_2CO	
1	a_1	2937.4	2137.9	CH symmetric stretch
2	a_1	1777.8	1728.5	CO stretch
3	a_1	1544.0	1127.8	HCH bend
4	b_1	1188.3	952.5	out-of-plane bend
5	b_2	3012.0	2249.4	CH asymmetric stretch
6	b_2	1269.4	1001.2	in-plane wag

[a]Calculated from the potential of Romanowski *et al.*[50]

For ($J = 0$) and with this choice of N_t, the magnitude $|E_6 - E_4|$, where E_n is the result of nth order perturbation theory, is less than one cm^{-1} for all the states up to 10 000 cm^{-1} above the zero point energy (cf. Fig. 1). This difference, which we take to be roughly a measure of convergence, is about the same size as $|E_v - E_6|$ were E_v are the eigenvalues obtained from a variational calculation.[18] Higher J results were found to yield similar convergence.

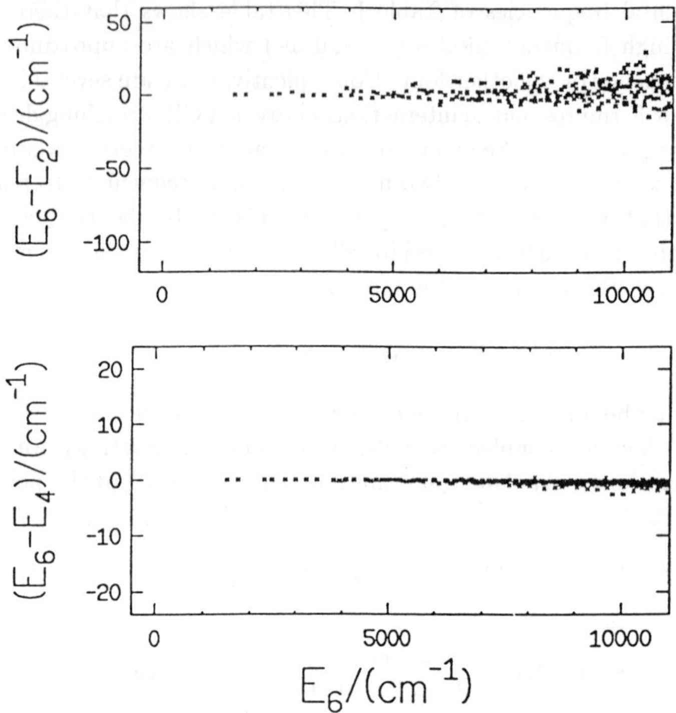

Fig. 1. Convergence plots of perturbative energies for $J = 0$ curvilinear normal coordinate Hamiltonian of H_2CO. Energy differences $E_n - E_6$ are plotted as a function of E_6 for eigenstates with a_1 symmetry and $N_t \leq 8$. E_n is the eigenvalue of the effective Hamiltonian transformed through nth order. N_t is the superpolyad quantum number describing the states of $\hat{K}$. Note the differences in scale of the y-axes.

The existence of the superpolyad quantum number is not simply based on energy considerations. We have found, in fact, that there is considerable overlap in the spectrum between states of differing superpolyad numbers. For example, many states with $N_t = 7$ have a greater energy than states with $N_t = 8$. The additional considerations for constructing the superpolyad quantum number involves the connectivity of states. The size of matrix elements between two vibrational normal mode zero-order states $|n\rangle$ and $|n'\rangle$ falls off rapidly as the difference in the quanta in the two states

$$\Delta = \sum_{i=1}^{6} |n_i - n_i'| \tag{22}$$

increases. At the vibrational energies considered in this paper, the magnitude of coupling terms with $\Delta \geq 5$ are typically much less than the inverse of the density of states. This is not true of the lower order resonances. Therefore, the consideration of *low order* resonances ($\Delta = 3$ or 4) are crucial for determining the superpolyad structure.

We have also carried out calculations on D_2CO.[30] Isotope effects reduce all the frequencies, maintaining the same 2:1 resonance for all modes except ω_2, the CO stretch (cf. Table I). The selection of N_t is not as straightforward as for H_2CO, since ω_2 is now about halfway between the high and low frequency modes. One possibility is to group ω_2 with the high-frequency modes, giving

$$N_t^a = 2n_1 + 2n_2 + n_3 + n_4 + 2n_5 + n_6 \ . \tag{23}$$

A more numerically exact condition is given by

$$N_t^b = 4n_1 + 3n_2 + 2n_3 + 2n_4 + 4n_5 + 2n_6 \ . \tag{24}$$

However, as discussed above, the 4:3:2 resonance condition implied by this constant is higher order than a 2:1 resonance, and the coupling elements are significantly smaller. We find that calculations carried out with either choice lead to $|E_6 - E_4| \leq 2$ cm^{-1} for states up to 10 000 cm^{-1}. This suggests that both N_t^a and N_t^b are good constants. Equivalently, one can take a linear combination of these constants to obtain

$$\begin{aligned} N_t^{c1} &= 2n_1 + n_3 + n_4 + 2n_5 + n_6 , \\ N_t^{c2} &= n_2 \ . \end{aligned} \tag{25}$$

The results obtained with these two constants is comparable to the above choices. This suggests that the CO stretch of D_2CO is not mixing significantly with other modes.

4. Rotation–Vibration Mixing in H_2CO

In this section we present results for rotation–vibration mixing in H_2CO. For all the calculations reported, coordinate dependent terms in the

Hamiltonian have been expanded through fourth order in the normal coordinates. The CVPT results are similar for the 4th and 6th order, so we have used 4th order. We use the potential of Harding and Ermler[49] as modified by Bowman and co-workers.[50] This potential is a quartic expansion in terms of bond-angle extension coordinates, where the stretch coordinates are Simons–Parr–Finlan coordinates. The Eckart conditions were used to embed the body-fixed axes.

4.1. *K-Mixing*

In order to quantify the extent of rotation–vibration mixing we examine two properties of the molecular eigenfunctions. The first property is a measure of the "goodness" of the rotational energy as a constant of the motion. To do this we rewrite Eq. (4) as

$$
\begin{aligned}
\hat{H}_r &= \frac{\hat{J}_x^2}{2I_{xx}^0} + \frac{\hat{J}_y^2}{2I_{yy}^0} + \frac{\hat{J}_z^2}{2I_{zz}^0} \\
&= \frac{\hat{J}^2 - \hat{J}_z^2}{4}\left[\frac{1}{I_{xx}^0} + \frac{1}{I_{yy}^0}\right] + \frac{\hat{J}_z^2}{2I_{zz}^0} + \frac{\hat{J}_x^2 - \hat{J}_y^2}{4}\left[\frac{1}{I_{xx}^0} - \frac{1}{I_{yy}^0}\right] .
\end{aligned}
\tag{26}
$$

The z-axis is chosen to lie along the CO bond axis when the molecule is in its equilibrium configuration. H_2CO is a near prolate top, consequently, the $\{x, y, z\}$-axes are referred to as the $\{c, b, a\}$-axes. The third term, the asymmetry term, found in the last line of the above equation is small. In the absence of this term the rotational Hamiltonian is just that of a symmetric top, and K is a good quantum number. Since the effect of this term is small we include it in the definition of $\hat{H}_{vr}$.

In Fig. 2 we have plotted the variance

$$
\Delta E_S = \left[\langle E_S^2 \rangle - \langle E_S \rangle^2\right]^{\frac{1}{2}}
\tag{27}
$$

as a function of $\langle E_S \rangle$, the expectation value of $\hat{H}_r$ with respect to the molecular eigenfunctions. The four panels (a–d) correspond to excitation energies of 0, 3000, 6000, and 9000 cm^{-1}, respectively. These panels correspond to $J = 4$. The little K-mixing that occurs for $N_t = 0$, in contrast to the higher N_t blocks (note the change in the scale of the y-axis), is a result of H_2CO being an asymmetric top. The separability of the rotational and vibrational degrees of freedom observed for $N_t = 0$ diminishes noticeably

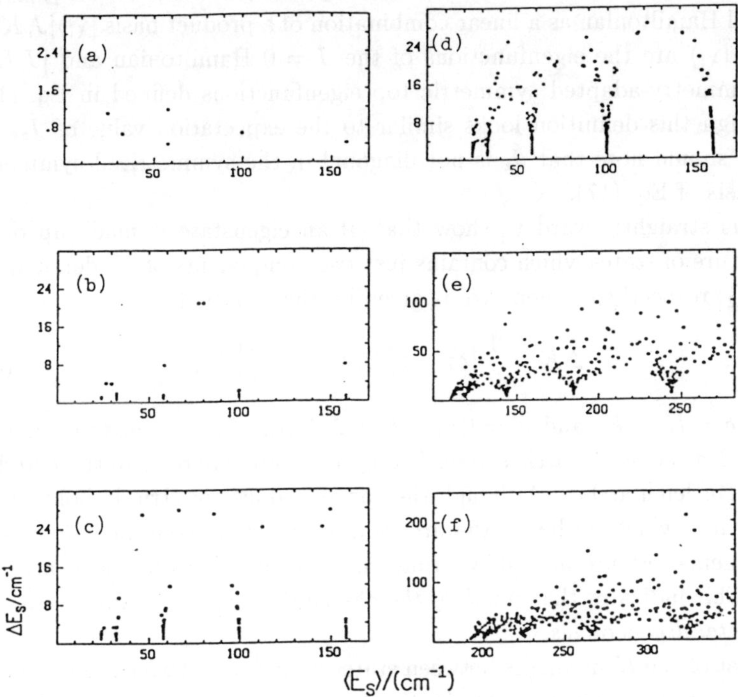

Fig. 2. Standard deviation of symmetric top energies ΔE_S (Eq. 27) plotted as a function of expectation values for $J = 4$ in (a–d). These panels contain states with $N_t = 0, 2, 4,$ and 6 respectively. The approximate energies corresponding to these blocks are 0, 3000, 6000, and 9000 cm^{-1}. Panels (d–f) show results corresponding to $J = 4, 9$ and 12, all with $N_t = 6$. Not all states are shown in panels (e) and (f).

with increasing vibrational energy. K-mixing increases more dramatically as J is increased. In Figs. 2(d–f) we show results of increasing J for states with about 9000 cm^{-1} of excitation.

Some insights into the nature of this mixing are obtained by plotting for each eigenstate the standard deviation ΔK

$$\Delta K = \left[\langle K^2 \rangle - \langle K \rangle^2 \right]^{\frac{1}{2}} \tag{28}$$

where the moments are given by

$$\langle K^q \rangle = \sum_l \sum_K K^q |d_m^{lJK\pm}|^2 . \tag{29}$$

The expansion coefficients are those for expanding the eigenstates $|\psi_{mJ}\rangle$ of the full Hamiltonian as a linear combination of a product basis $|\chi_l\rangle|J,K^{\pm}\rangle$, where $|\chi_l\rangle$ are the eigenfunctions of the $J=0$ Hamiltonian and $|J,K^{\pm}\rangle$ are symmetry-adapted symmetric top eigenfunctions defined in Eq. (17). Although this definition looks similar to the expectation value of $\hat{J}_z$, the reader should note that $\hat{J}_z$ is not diagonal in the symmetrized symmetric top basis of Eq. (17).

It is straightforward to show that, if an eigenstate is made up of an admixture of states which contains just two components of K, denoted K_1 and K_2, respectively, then ΔK is given by the equation

$$\Delta K = \frac{1}{2}\left[\delta^2 - 4(\langle K\rangle - \overline{K})^2\right]^{1/2}, \tag{30}$$

where $\delta = K_1 - K_2$ and $\overline{K} = (K_1 + K_2)/2$. Most of the eigenstates plotted in Fig. 3(a) lie on the arches described by Eq. (30). We refer to these arches as the Burleigh arches. Although the converse does not strictly hold, those eigenstates which do lie on the arches almost always contain *only* two K components. Figure 3(a) shows that many of the states lie on or near an arch. This indicates that *much of the observed K-mixing is due to isolated accidental degeneracies.*

Most of the K-mixing is between states with $\delta = 1$. Two coupling terms which are primarily responsible for this mixing are the b- and c-axis Coriolis coupling terms. To verify this, we set to zero the rotation–vibration terms that are linear in either $\hat{J}_x$ or $\hat{J}_y$ in Eq. (1). Since we are in an Eckart frame, these terms are, to lowest order identical to the corresponding terms $\mu_{xx}\hat{\pi}_x\hat{J}_x + \mu_{yy}\hat{\pi}_y\hat{J}_y$ of the Watson Hamiltonian.[51] The results, analogous to Fig. 3(a) but with the b- and c-axis Coriolis coupling terms omitted, are shown in Fig. 3(b). Most of the states which lay on the $\delta = 1$ arches in the fully coupled problem now have greatly diminished values of ΔK. Furthermore, all of the states now lie on the arches. Clearly the b- and c-axis Coriolis coupling terms play an important role in K-mixing for $J = 4$.

Given that the magnitude of these coupling terms increases with J, their contribution to the overall mixing becomes more pronounced as J increases. The centrifugal terms, however, are expected to increase more rapidly since these coupling terms increase quadratically with J. The enhanced mixing is evident in Fig. 4 where results for $J = 9$ are plotted. Here, there is substantially more mixing. Furthermore, a significant amount of mixing remains, even after the b- and c-axis Coriolis terms are set to zero. This

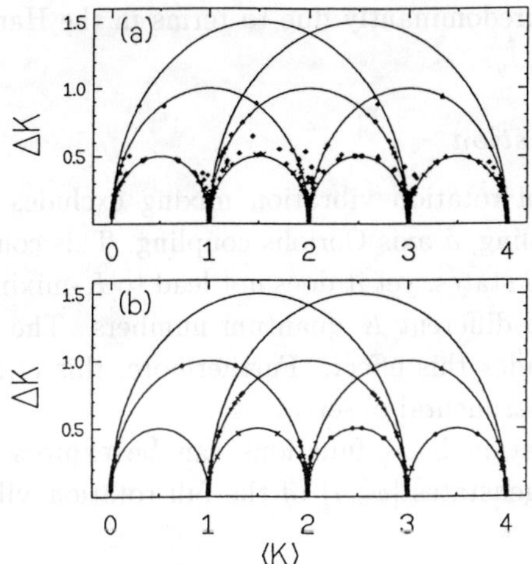

Fig. 3. Standard deviation ΔK (Eq. (28)) plotted as a function of the average value $\langle K \rangle$ for: (a) the full rotation–vibration problem with $J = 4$ and $N_t = 6$, and (b) same as (a) but b- and c-axis Coriolis coupling terms are set to zero. States which lie on the arches, defined by Eq. (29), result from mixing of states with just two K components. The different symbols correspond to rotation–vibration states of different symmetries.

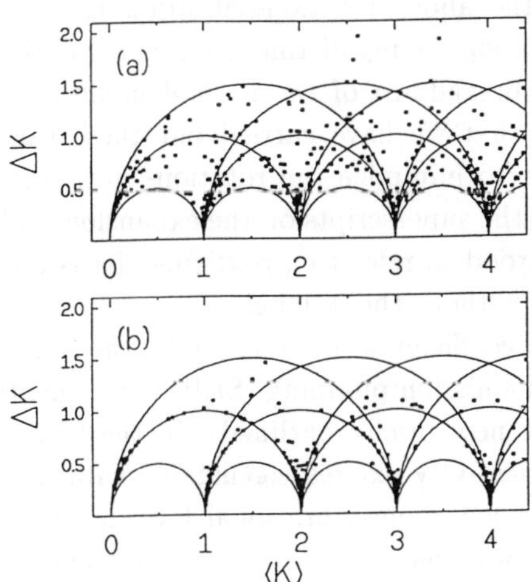

Fig. 4. Standard deviation ΔK (Eq. (28)) plotted as a function of the average value $\langle K \rangle$ for: (a) the full rotation–vibration problem with $J = 9$ and $N_t = 6$, and (b) same as (a) but b- and c-axis Coriolis coupling terms are set to zero. Only states with $\langle K \rangle \leq 4.5$ are shown.

$\Delta K = 2$ coupling is predominantly due to terms in the Hamiltonian that are proportional to $\hat{J}_x \hat{J}_y$.

4.2. *Spectral Congestion*

The above measure of rotation–vibration mixing excludes an important effect of rotation coupling, a-axis Coriolis coupling. This coupling leads to a mixing of vibrational states, yet it does not lead to K-mixing, i.e., it does not couple states with different K quantum numbers. The next criterion to be considered includes this effect. Furthermore, this criterion is more directly related to experimental observables.

The rotation–vibration basis functions can be expressed as a linear combination of the eigenstates $|\psi_{mJ}\rangle$ of the full rotation–vibration block-diagonal Hamiltonian as

$$|\chi_l\rangle|J, K^{\pm}\rangle = \sum_{m=1}^{N_s} d_m^{lJK^{\pm}} |\psi_{mJ}\rangle \, . \tag{31}$$

The summation is over the number of states N_s of the particular (J, Γ, N_t) block.

An advantage of the above rotation–vibration basis is that one can examine rotation–vibration mixing distinct from many of the complications that arise due to the breakdown of the normal mode approximation. It should also be noted that we have carried out these calculations in an Eckart frame, in order to minimize the rotation–vibration interactions.[35] For simplicity we drop the superscripts on the expansion coefficients $d_m^{lJK^{\pm}}$, and d_m will be understood to refer to a particular basis function. Also, in general, we refer to K without the $\pm$ label.

The above mixing coefficients are related to the spectral congestion observed in stimulated emission pumping (SEP) experiments of Dai *et al.*[1] These experiments are now briefly outlined. In the SEP experiments, a two photon process selectively excites eigenstates with a specific J. The first photon prepares a low lying vibrational level of an upper electronic state. Because of the low density of states at this energy in the upper electronic manifold, both K and J are nearly "good" quantum numbers for describing the intermediate state. Using this separability, Dai *et al.* were able to obtain spectra that were equivalent to those that would be obtained if only one of the zero order states $|\chi_l\rangle|J, K^{\pm}\rangle$ of Eq. (31) were a "bright" state. Consequently, we simulate the spectrum for such an experiment by plotting $|d_m|^2$ as a function the eigenvalues E_m. Moreover, as Dai *et al.* did, we obtain a measure of the rotation–vibration mixing by quantifying the change in the spectral congestion that occurs for a series of "bright" states $|\chi_l\rangle|J, K^{\pm}\rangle$, where this series corresponds to increasing J and K, but holding l constant.

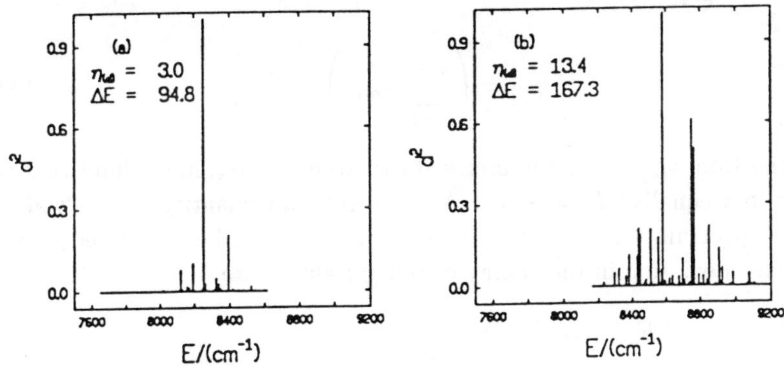

Fig. 5. Simulated spectra for two choices of "bright" states, from the $N_t = 6, J = 9$ block. All states have same vibrational character. Lines are scaled so the largest peak is 1. (a) $K = 4$; (b) $K = 8$.

Figures 5(a–b) are representative examples of simulated spectra for formaldehyde. The congestion shown in Figs. 5(a–b) is a consequence of extensive rotation–vibration mixing. These spectra differ only in the value of K. It is well known that a-axis Coriolis coupling is significant in H_2CO and that this coupling increases with $|K|$.[1,9,11] This coupling is an important factor in the increase in congestion shown in the figures. These results are consistent with the experimental results of Dai *et al.*[1]

Aside from these generalizations, however, a case by case discussion of the coupling mechanisms and assignment of peaks is not necessarily warranted. Furthermore, it is currently beyond our capability to make this close of a connection with experiments. Although the potential we have employed for these studies is good,[50] the spectral features shown in Fig. 5 are extremely sensitive to small changes in the potential surface. Consequently, we are unable to reproduce, level by level, the spectra of Dai *et al.* We, therefore, focus instead on the much simpler and more manageable goal of understanding global features of the intramolecular dynamics. To do this, we need to be able to parametrize the extent of mixing with just a few variables.

The extent of "spectral congestion" can be quantified with a minimum of two parameters. The first is the number of lines in the spectrum, and the second is the energy range over which the intensity is spread. The parameter we opt to use for the first measure is the participation number,[52] defined relative to Eq. (31) as

$$\eta_{(d)} = \left(\sum_{m=1}^{N_s} |d_m|^4 \right)^{-1}. \tag{32}$$

It varies from $\eta_{(d)} = 1$, for unmixed states, to $\eta_{(d)} = N_s$, in the limit that all states mix equally $(d_m = \frac{1}{\sqrt{N_s}})$. Another relevant quantity is the "width" of the spectrum. To lowest order this quantity can be calculated as the standard deviation in the energy of the "bright" state,

$$\Delta E = \langle (\hat{H} - \langle \hat{H} \rangle)^2 \rangle^{\frac{1}{2}}$$

$$= \left[\sum_{m=1}^{N_s} |d_m|^2 E_m^2 - \left(\sum_{m=1}^{N_s} |d_m|^2 E_m \right)^2 \right]^{\frac{1}{2}}. \tag{33}$$

The values of these quantities, calculated for the examples in Figs. 5(a–b), are displayed in the figures. They are in good qualitative agreement with the interpretation of the particpation number and spectral width given above.

Using these two measures of spectral congestion one can examine rotation–vibration mixing from a more global perspective. We do this by reducing the spectrum associated with each zero-order state of a (J, Γ, N_t) block to a point on a plot of $\eta_{(d)}$ versus ΔE. In Fig. 6 we have plotted results for those states with $J = 9, \Gamma = a_1$. The panels (a–c) correspond to states of increasing vibrational energy, i.e., different N_t blocks. These results show that spectral congestion increases rapidly with vibrational energy. Figure 6(c), contains the points for our two sample spectra; they are seen to be representative.

The rotation–vibration coupling that is primarily responsible for these results is a-axis Coriolis coupling. This coupling, which increases linearly with K, is known to play an important role in the description of the rotation–vibration states of formaldehyde. To highlight the role of this

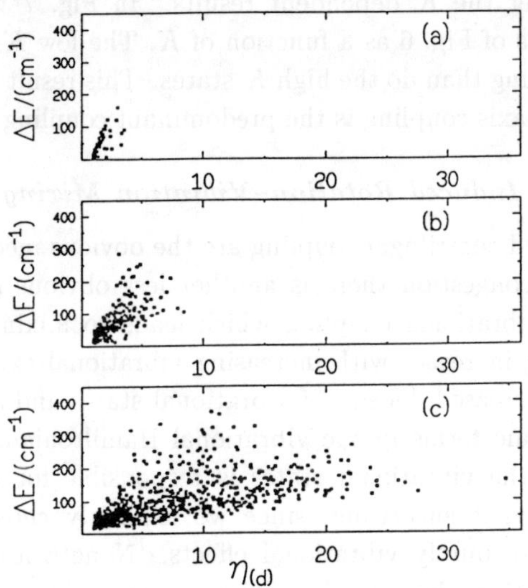

Fig. 6. Scatter plot of ΔE (Eq. (33)) versus the participation number $\eta_{(d)}$ (Eq. (32)), for a_1 states with $J = 9$, and (a) $N_t = 2$; (b) $N_t = 4$; (c) $N_t = 6$.

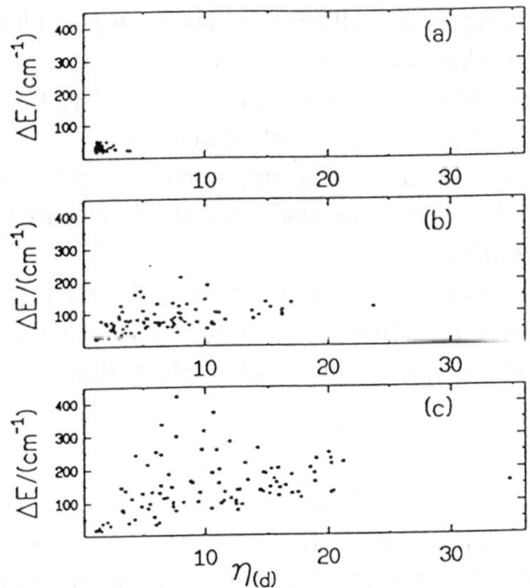

Fig. 7. Scatter plot of ΔE (Eq. (33)), versus the participation number $\eta_{(d)}$ (Eq. (32)), for a_1 states with $N_t = 6$, $J = 9$, and (a) $K = 0$; (b) $K = 4$; (c) $K = 8$.

coupling, we consider the K dependent results. In Fig. 7 we show the $J = 9$, $N_t = 6$ results of Fig. 6 as a function of K. The low K states show significantly less mixing than do the high K states. This result is consistent with the idea that a-axis coupling is the predominant coupling mechanism.

4.3. *Vibrationally Induced Rotation–Vibration Mixing*

Although Coriolis and centrifugal coupling are the obvious mechanisms for explaining spectral congestion there is another less obvious effect. This is the anharmonic vibrational coupling which leads to mixing of normal modes. Such mixing increases with increasing vibrational excitation as a consequence of an increased density of vibrational states and an increased role of the anharmonic terms in the vibrational Hamiltonian. This idea, that the mixing of the vibrational states is responsible for the spectral congestion, is somewhat surprising, since we explicitly chose our basis functions to minimize purely vibrational effects. Nonetheless, there are many isolated examples where mode mixing occurs only as a combined result of both rotation–vibration mixing and purely vibrational mixing. The $2\nu_1$ band of propyne provides a recently observed example.[53] The importance of this mixing, which we henceforth refer to as vibrationally-induced rotation–vibration mixing, is now illustrated via a simple calculation.

In order to explain this calculation, we introduce the quantity $\langle \eta_{(d)} \rangle$ that is obtained by averaging $\eta_{(d)}$ over all the zero-order states in a block of the Hamiltonian. In Fig. 8, $\langle \eta_{(d)} \rangle$, calculated for $N_t = 6$ and $\Gamma = a_1$, is plotted for several values of J. The upper curve, where the mixing is seen to increase rapidly with rotational excitation, is based on the full rotation–vibration Hamiltonian.

The role of vibrational mixing can be demonstrated by examining the importance of coupling terms in the vibrational Hamiltonian. This Hamiltonian can be written as a sum of two contributions

$$\hat{H}_v = \hat{H}^{(0)} + \hat{H}_{\text{anh}} \, . \tag{34}$$

Here $\hat{H}^{(0)}$ is the zero-order normal mode Hamiltonian, and $\hat{H}_{\text{anh}}$ includes the higher-order anharmonic terms. By applying CVPT to this Hamiltonian we obtain the transformed Hamiltonian. It takes the standard form

$$\hat{K}_v = \sum_{i=1}^{6} \hbar\omega_i \left(\hat{n}_i + \frac{1}{2} \right) + \sum_{i=1}^{6} \sum_{j \geq i}^{6} x_{ij} \left(\hat{n}_i + \frac{1}{2} \right) \left(\hat{n}_j + \frac{1}{2} \right) + \hat{K}_c \, . \tag{35}$$

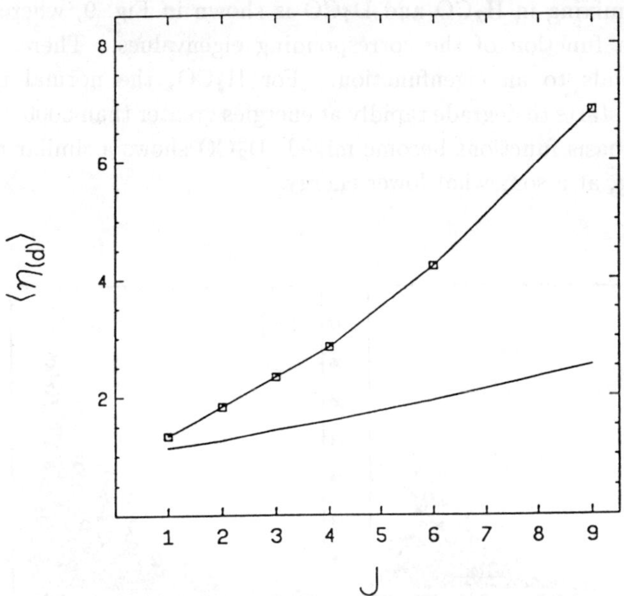

Fig. 8. Average participation number $\langle \eta_{(d)} \rangle$ for states with $N_t = 6$ as a function of total angular momentum J. Upper curve connects the results ($\square$) for the model Hamiltonian $\hat{K}$ of Eq. (7). The vibrational functions of Eq. (31) are eigenfunctions of $\hat{K}_v$ [cf. Eq. (35)]. Bottom curve is same as above but $\hat{K}_v$ was redefined by setting $\hat{K}_c$ was set to zero in Eq. (35).

The coupling term $\hat{K}_c$ contains all the off-diagonal coupling terms. These terms couple states within a superpolyad. In Eq. (35) the diagonal contributions include only terms up to second order. In reality, the Hamiltonian we have used includes terms through third order in the number operator, i.e., $n = 4$ in Eq. (7).

The eigenfunctions of $\hat{K}_v$ can be written as $|\chi_l\rangle = \sum_{j=1}^{N_s} c_l^j |\mathbf{n}\rangle$. Here, j is an index to the separable normal mode 'like' basis functions. These basis functions are eigenfunctions of $\hat{K}_v - \hat{K}_c$; the corresponding zero-order eigenvalues are denoted $K_{jj} = E_j^0$.

The participation number,[52] defined as

$$\eta_l^{(c)} = \left(\sum_{j=1}^{N_s} |c_l^j|^4 \right)^{-1} , \tag{36}$$

is used as measure of vibrational mixing. A comparison of the distribution of vibrational mixing in H_2CO and D_2CO is shown in Fig. 9, where $\eta_l^{(c)}$ is plotted as a function of the corresponding eigenvalues. There, each point corresponds to an eigenfunction. For H_2CO, the normal mode approximation starts to degrade rapidly at energies greater than $6000\ \mathrm{cm}^{-1}$, as many more basis functions become mixed. D_2CO shows a similar rapid onset of mixing, at a somewhat lower energy.

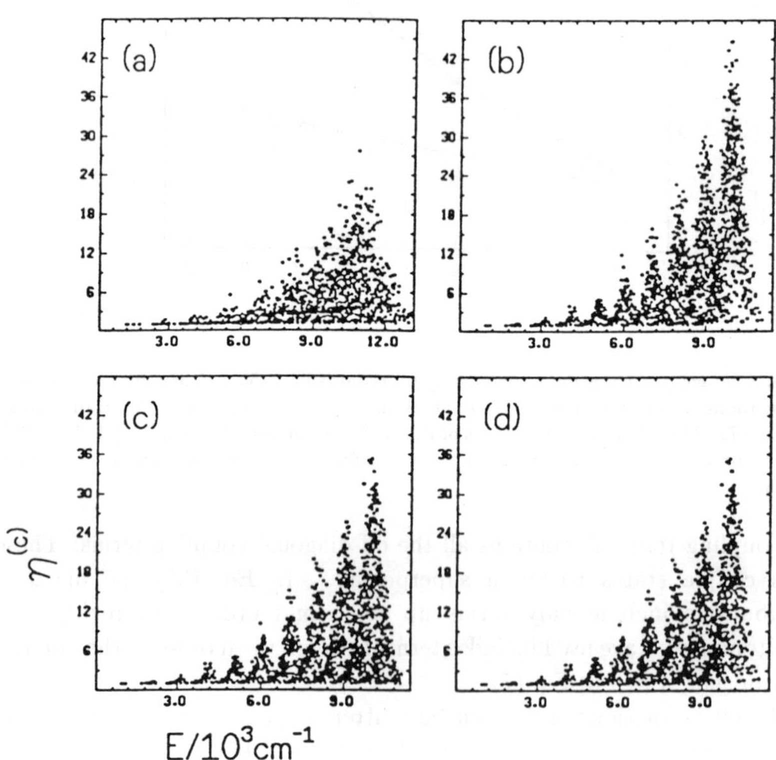

Fig. 9. Scatter plot of $\eta^{(c)}$ (Eq. (36)) versus eigenvalues E_v of all symmetries of the $J = 0$ Hamiltonian. The use of the superscript indicates that the participation number refers to the number of zero-order states contributing to the makeup of an eigenstate, as opposed to the number of eigenstates contributing to the makeup of a zero-order state. (a) Results for H_2CO with $N_t = 1 - 8$. (b–d) Results over same range of energies as (a) but for D_2CO, using three different resonance conditions for N_t. It should be noted that the extent of mixing is insensitive to the particular choice of resonance conditions.

To investigate the influence of this mixing on rotation–vibration mixing, we artificially remove the effects of the mixing of normal modes due to anharmonic terms in the vibrational Hamiltonian, i.e., we set $\hat{K}_c = 0$. We then repeated the calculation of $\langle \eta_{(d)} \rangle$. The results are displayed in the lower curve of Fig. 8. In sharp contrast to the full Hamiltonian, the lack of vibrational coupling results in only a slight increase of $\langle \eta_{(d)} \rangle$ with rotational excitation. We have also found that this result is insensitive to small variations in the values of E_j^0 used in the calculation. Clearly, the terms in $\hat{K}_c$ are playing a significant role in the rotation–vibration mixing of this molecule. In the next section, we describe a simple model that reproduces the central features of this effect.

5. Random Matrix Approach to State Mixing

The generic features of mixing are elucidated by introducing an ensemble of Hamiltonians whose matrix elements are randomly chosen from distribution functions. The choice of distribution functions comes from simple ideas about the correlation of diagonal and off-diagonal Hamiltonian matrix elements that do not depend on state-specific details.[30] Given these distribution functions, we will show results that indicate that the ensemble reproduces the extent of vibrational mixing and rotation–vibration mixing as a function of J and as a function of total internal energy.

5.1. *Vibration–Vibration Mixing*

The results of Fig. 8 demonstrate that rotation–vibration mixing in H_2CO increases rapidly as a function of J for states whose vibrational energy is greater than 6000 cm^{-1}. In the last section, we argued that a major contributor to the mixing is the mixing of the vibrational normal mode functions. This vibrational mixing is now examined using random matrices. The idea is to generate an ensemble of random Hamiltonian matrices $\mathbf{H}_{\mathrm{rm}}$, of which the model Hamiltonian matrix $\mathbf{H}$ is a member. In the present context, the matrix elements H_{ij} of the model Hamiltonian are given by the matrix elements of $\hat{K}_v$ defined with respect to the basis functions $|\mathbf{n}\rangle$ of the transformed representation.

Each matrix of the ensemble is generated from simple distribution functions. The essential difference from much of the earlier work using random matrices is that we determine the distribution functions that define the ensemble from the form of $\mathbf{H}$. This is in contrast to dealing with complicated systems where the actual distribution functions are unknown, and one therefore makes *ad hoc* assumptions concerning the distributions. The long-term goal of such an approach is to develop ideas of transferability from one molecular system to another.

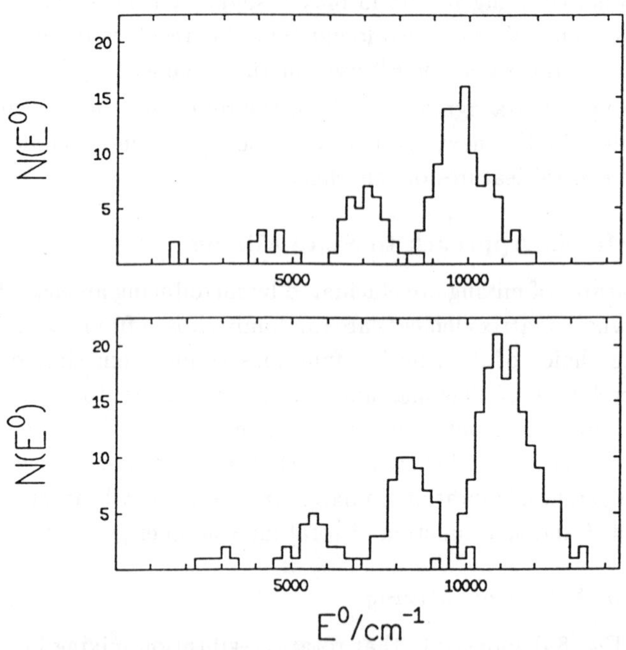

Fig. 10. Histogram of diagonal matrix elements of the transformed Hamiltonian $\hat{K}_v$ (Eq. (35)) corresponding to states of a_1 symmetry. Upper panel, $N_t = 1, 3, 5, 7$; lower panel $N_t = 2, 4, 6, 8$.

For each superpolyad there are at least three important components to the distribution functions. The first is the distribution of diagonal energies E_i^0. These are shown in Fig. 10. For $N_t \geq 5$, we find that the distributions *within* each block can be adequately described with a Gaussian function, where the width and center of Gaussian is chosen to match the first and second moments of the actual distribution. As an aside, if one takes a superposition of the histograms shown in Fig. 10 in order to determine the *total* density of states, then, as expected one finds that the total density of states is monotonically increasing. In other words, the superpolyad structure of the molecule cannot be discerned solely from energetic considerations.

The second component is the distribution of off-diagonal coupling elements, H_{ij}. These are shown in Fig. 11. Most of these terms are less than one cm^{-1}, while the rest are distributed evenly out to large values. The distribution used to model these couplings is as follows. The off-diagonal elements are either 0 or $\pm\overline{V}$. The ratio f_0 of zero and nonzero matrix elements is determined from the actual distribution in Fig. 11, where the fraction of zero elements corresponds to the fraction of elements whose magnitude is less than some arbitrary value $V_{\min}$. The quantity $\overline{V}$ is the average magnitude of the remaining elements. We have found that these results are insensitve to the particular value of $V_{\min}$ for a wide range of values.

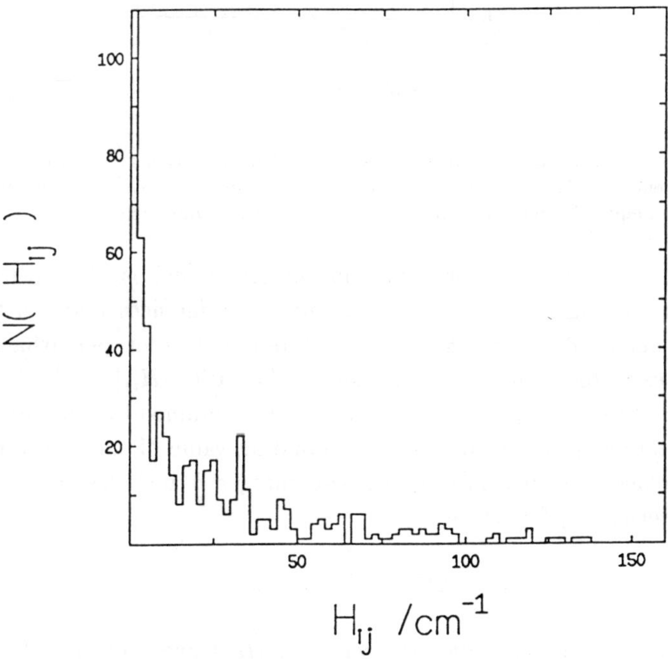

Fig. 11. Histogram of the number of off-diagonal coupling elements H_{ij} $(i > j)$ for states with a_1 symmetry, $J = 0, N_t = 6$. The plot was cut off at $N(H_{ij}) = 120$; the number of elements ≤ 2 cm^{-1} is 1769.

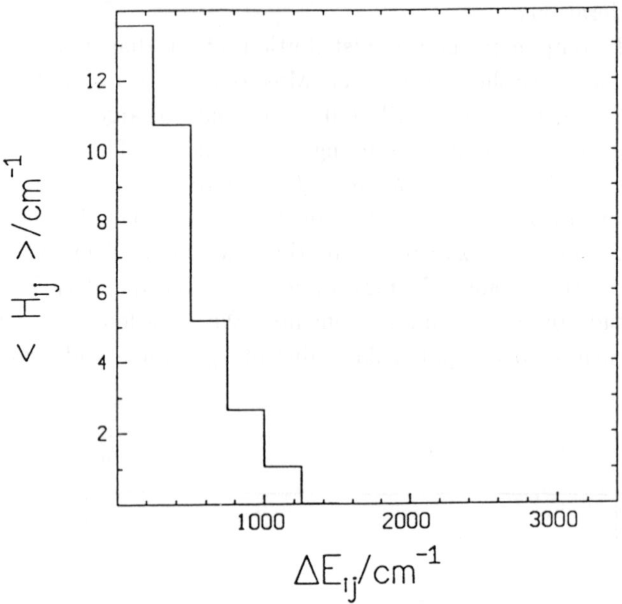

Fig. 12. Average magnitude of off-diagonal coupling elements $|H_{ij}|$, grouped by range of energy difference ΔE_{ij}^0, for a_1 states of $N_t = 6$. All terms with $\Delta E_{ij}^0 > 1300$ cm^{-1} are found to be identically zero. Results are similar for other symmetries.

The final component in our random matrix model is the correlation between off-diagonal elements, H_{ij}, and the differences between the diagonal elements, $\Delta E_{ij}^0 = |E_i^0 - E_j^0|$. From first-order perturbation theory, it is expected that the magnitude of the ratio, $|H_{ij}|/\Delta E_{ij}^0$, is an important measure of coupling. This correlation is obtained by binning the off-diagonal terms according to the corresponding value of the zero-order energy difference. A bin labeled by the integer b includes all those off-diagonal terms H_{ij} for which

$$bW_E \leq \Delta E_{ij}^0 < (b+1)W_E . \tag{37}$$

The bar graph of Fig. 12 shows the value of $|H_{ij}|$ averaged over 'bins' that are $W_E = 250$ cm^{-1} wide. The final results are insensitive to this particular choice of W_E. We incorporate this correlation into the random matrix model by making $\overline{V}$ a function of ΔE_{ij}^0. In preliminary work we use

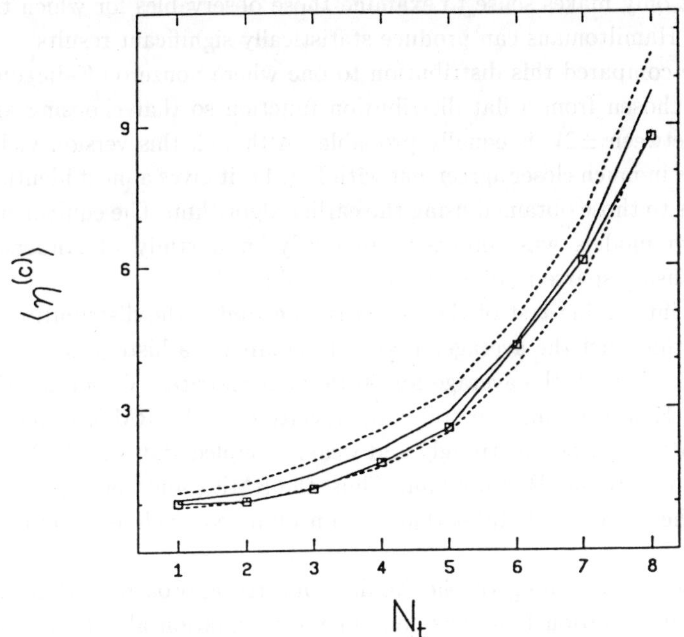

Fig. 13. Average participation number $\langle \eta^{(c)} \rangle$ (Eq. (36)) for $N_t = 1 - 8$ states of H_2CO. Solid curve connected by squares ($\square$) are exact results for $\hat{K}_v$ (Eq. (35)); second solid curve is the ensemble average over 50 realizations of the simple random matrix model; dotted lines are $\pm$ one standard deviation. $V_{\min}$ and W_E (Eq. (37)) were chosen as 1 cm^{-1} and 250 cm^{-1}, respectively, to generate this plot.

a binning procedure to do this. Both $\overline{V}$ and f_0 are determined for each of the bins shown in Fig. 12.

The results for this model are presented in Fig. 13 where an average of $\eta_l^{(c)}$ has been computed for each value of N_t (cf. Fig. 9). This average $\langle \eta^{(c)} \rangle$ is then compared to the average value obtained from the corresponding averaged results for 50 realizations of the random matrix ensemble (RME). The quantitative accuracy obtained in these results suggests that the particular RME we have proposed is a useful model for understanding vibrational mixing. This hypothesis is strengthened by the fact that equivalent accuracy is found for all four symmetry representations of H_2CO, as well as for D_2CO. A particularly important feature of the RME results is that the standard deviation of $\langle \eta^{(c)} \rangle$ is small. In using a random matrix approach, it only makes sense to examine those observables for which the ensemble of Hamiltonians can produce statistically significant results.

We have compared this distribution to one where nonzero off-diagonal terms were chosen from a flat distribution function so that choosing any H_{ij} value between $\pm 2\overline{V}$ is equally probable. Although this version yields distributions in much closer agreement with Fig. 11, it gives almost identical $\langle \eta^{(c)} \rangle$ results to those obtained using the earlier algorithm. The equivalence of these two models was reported previously in a study of Anderson localization using sparse random matrix ensembles.[54]

Another interesting test of the model is to examine the distribution of $\eta_l^{(c)}$, rather than just the average. Figure 14 compares a histogram of $\eta_l^{(c)}$ for the exact $\hat{K}_v$ with the average for 50 random matrices. Generally, the values from $\mathbf{H}$ are within one standard deviation of the RME, implying that the relative number of strongly and weakly coupled states in the RME is typical of our original Hamiltonian. Thus, the RME should be capable of predicting the presence of states that remain unmixed, such as some local mode states.

A possible shortcoming of the random matrix approach is that the above three distribution functions contain no information about the connectivity of states. The matrix representation of a superpolyad has a very rich structure. The superpolyad is comprised of many interconnecting polyads, and each polyad, by definition, is dominated by a single resonance interaction term. Consequently, the matrix representation of the polyad states can be represented as a tridiagonal matrix,[55] where this tridiagonal structure describes a specific connectivity of states. A possible route to improving the agreement between the random matrix approach and the model Hamiltonian results would be to incorporate some notion of connectivity into the random matrix approach.

One of the central motivations of pursuing a random matrix approach, is that it allows one to determine the sensitivity of mixing results to the parameters in the model. From a preliminary investigation it appears that the mixing, as measured by $\eta_l^{(c)}$, is a very simple function of the parameters that make up the RME. In fact, for the range of states considered here, $\langle \eta^{(c)} \rangle$ is roughly linearly related to these parameters. This was found by varying each parameter independently and holding the others constant. To highlight the functional dependencies we write

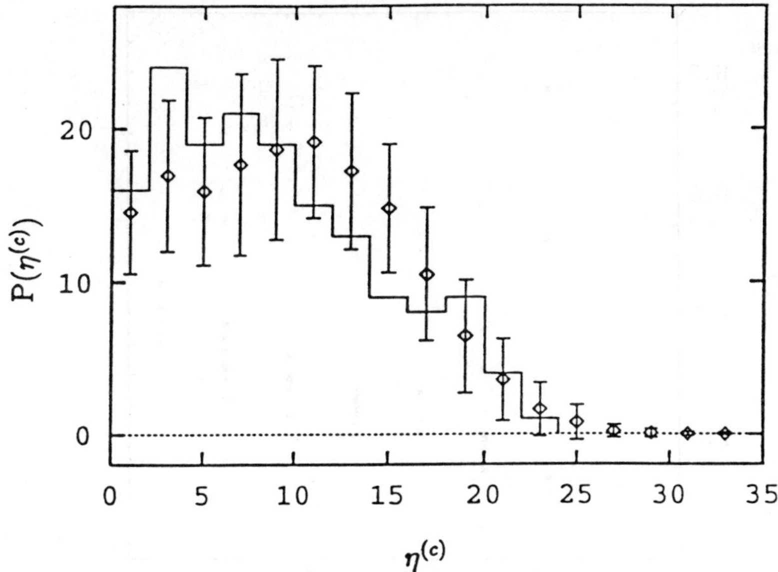

Fig. 14. Distribution of $\eta^{(c)}$ (Eq. (36)) for $N_t = 8$ is shown in the upper and lower panels respectively. The solid line is for $\hat{K}_v$ (Eq. (35)), the points ($\diamond$) are an ensemble average for 50 realizations of the RME, and the error bars are $\pm$ one standard deviation.

$$\langle \eta^{(c)} \rangle = S \frac{N_s \overline{V}(1)(1 - f_0(1))}{\Delta E^0} , \tag{38}$$

where $\overline{V}$ is the average coupling for the first bin, N_s is the number of states, f_0 is the fraction of zero terms in the first bin, ΔE^0 is the variance of the Gaussian used for determining the diagonal matrix elements, and S is a scaling factor that accounts for any remaining dependencies. We have included the most obvious functional dependencies of $\langle \eta^{(c)} \rangle$ explicitly so that S is not a sensitive function of the parameters of our model. The value of S for the four symmetry groups of H_2CO and D_2CO is plotted for $N_t = 1 - 8$ in Fig. 15. For values of N_t less than 4, it is evident that there is a great deal of scatter in the data. These blocks contain too few states for a statistical approach to be meaningful. For higher N_t values, however, the variation in S is small, and there is only a slight trend of S increasing. Assuming a constant S in Eq. 38, although qualitatively correct, is an oversimplification that deserves further scrutiny. Improving

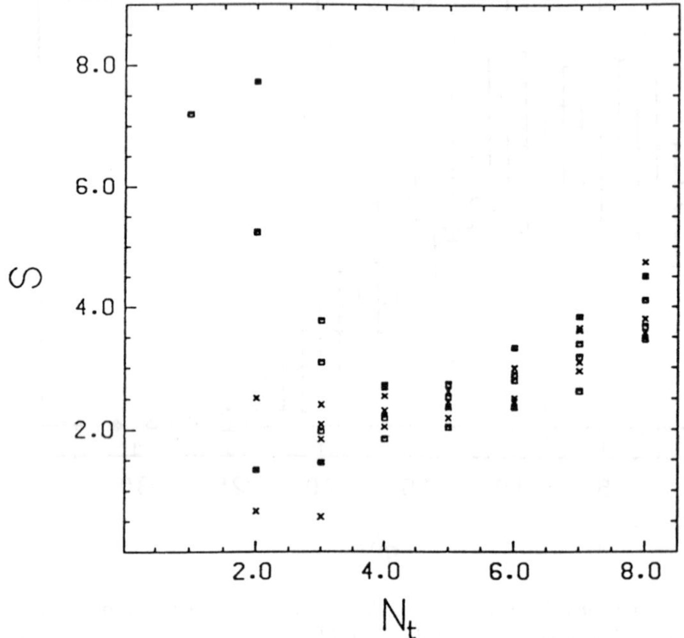

Fig. 15. Scaling factor S (cf. Eq. (38)), as a function of N_t for all four symmetry blocks. $\square$: H_2CO results; $\times$: D_2CO results.

this equation would give further insights into the nature of coupling, and this warrants future investigation.

5.2. *Rotation–Vibration Mixing*

We turn now to the goal of elucidating rotation–vibration mixing and, in particular, the role of vibrational mixing on rotation–vibration mixing. The random matrix approach generates both the eigenvalues $\mathbf{E}_v$ and the transformation matrix $\mathbf{T}$ between the normal mode like states $|\mathbf{n}\rangle$ and the vibrational states $|\chi_l\rangle$ of the $J = 0$ Hamiltonian. This allows for the calculation of the rotation–vibration matrix $\mathbf{K}'_{vr}$ in the new representation via

$$\mathbf{K}' = \mathbf{T}^{\dagger}\mathbf{K}\mathbf{T} \,,$$
$$= \mathbf{K}_r + \mathbf{T}^{\dagger}\mathbf{K}_{vr}\mathbf{T} + \mathbf{E}_v,$$
$$= \mathbf{K}_r + \mathbf{K}'_{vr} + \mathbf{E}_v \,. \tag{39}$$

Given the choice of embedding and choice of coordinates, $\hat{H}_r$ and $\hat{H}_{vr}$ can be calculated exactly, at least in principle. Thus, only the vibrational part of the problem has been treated using the RME. The goal is to examine the effects of vibrational mixing on rotation–vibration mixing. To do this one diagonalizes a random matrix, to obtain $\mathbf{T}$ and the vibrational eigenvalues; $\langle \eta_{(d)} \rangle$ can then be calculated as an ensemble average of the full rotation–vibration problem.

The results of these calculations are shown in Fig. 16 where $\langle \eta_{(d)} \rangle$ has been calculated for an ensemble of random matrices and compared to the results of the model H_2CO Hamiltonian. In order to retain the correlation between off-diagonal and diagonal matrix elements of $\mathbf{K}$ both the diagonal terms in the model Hamiltonian and the diagonal terms of H_{rm} have been ordered by energy. Figure 16 shows that the RME results for $\langle \eta_{(d)} \rangle$ match the values obtained for the model Hamiltonian closely. As for $\langle \eta^{(c)} \rangle$, the standard deviations for the random matrix results are relatively small; this indicates that $\langle \eta_{(d)} \rangle$ is a measure for which the random matrix approach is producing statistically meaningful results.

To gain insight into the consequences of vibrational mixing on rotation–vibration interactions, we multiplied the anharmonic coupling term $\hat{K}_c$ of Eq. (35) by a multiplicative factor F_v. This factor allows one to control the extent of vibrational mixing. In the limit that F_v is zero there is no vibrational mixing. For each value of F_v, the relevant quantities of the RME are scaled accordingly. The results for $J = 9$ and $N_t = 6$ are shown in Fig. 17. One can see that for small values of coupling the results are sensitive to F_v. In fact, the rotation–vibration coupling dramatically decreases as F_v approaches zero. For larger values, however, the results are surprisingly insensitve to F_v over a wide range of values. Similar trends hold for $J = 4$.

The F_v dependence should be investigated further, since these results imply that one may not need to obtain an accurate expression for $\hat{K}_c$ to predict general trends in rotation–vibration mixing. It will be very interesting to see whether similar results obtain for other systems. It should be noted that, as F_v continues to increase, the rotation–vibration mixing in our model must approach zero. This is an artificial result; for large values

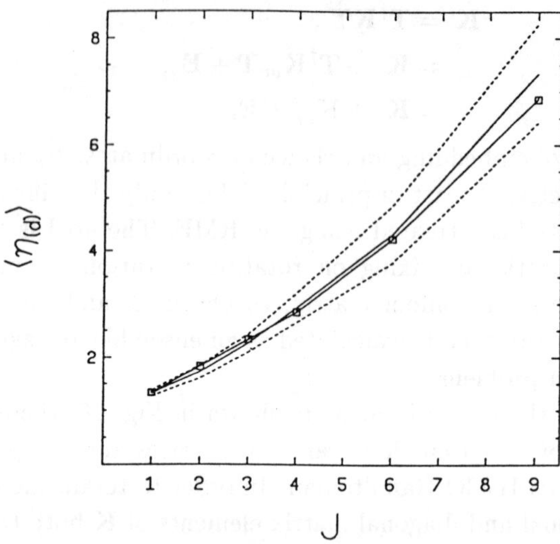

Fig. 16. The average participation number $\langle \eta_{(d)} \rangle$ versus J calculated for the full rotation–vibration Hamiltonian (points connected by solid curve) and random matrix analogue (solid curve). The dotted lines correspond to $\pm$ one standard deviation of random matrix results.

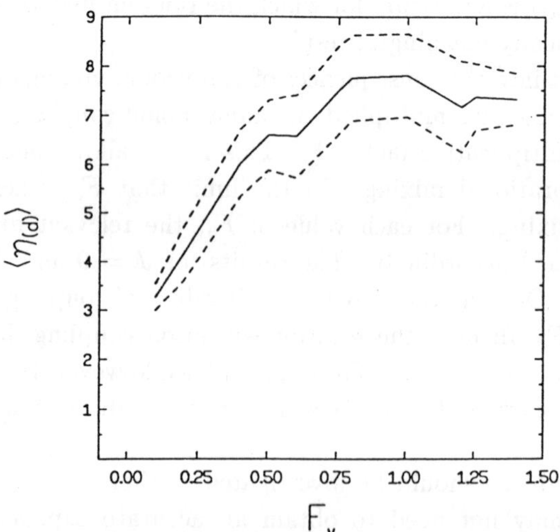

Fig. 17. The average participation number $\langle \eta_{(d)} \rangle$ for states with $J = 9$, $N_t = 6$, and a_1 symmetry plotted as a function of F_v. Here, F_v is a multiplicative factor multiplying $\hat{K}_c$ in Eq. (35).

of F_v the vibrational eigenvalues become so spread out that they cannot be coupled by the rotation–vibration contribution to the Hamiltonian.

6. Summary and Future Directions

In this Chapter we have discussed our recent efforts to elucidate rotation–vibration mixing in polyatomic molecules. There are several steps in this process. These include deriving the Hamiltonian in a suitable coordinate representation, finding its solutions, and then characterizing the nature of the highly excited eigenstates. We presented a perturbative approach for obtaining observables associated with these states. Although our implementation of the perturbative approach is limited to molecules under-going motion about a single equilibrium configuration, we have obtained accurate energies even where there is significant state mixing. We then examined two quantities that are measures of this state mixing. The first quantity measured the "goodness" of the K quantum number for various rotation–vibration eigenstates. The second quantity, the participation number, measured the spectral congestion associated with the preparation of zero-order states. An analysis of these measures revealed the relative importance of Coriolis and centrifugal coupling in H_2CO.

In energy regions where there is considerable state mixing, a detailed description of vibrational state mixing is very sensitive to small variations of the potential energy surface. Consequently, if one is asking questions at the state-to-state level, it is difficult to study one molecule in order to make predictions about another. We have, therefore, adopted a perspective that concentrates on describing the global features of state mixing. For example, we model the distribution of rotation–vibration mixing amongst an ensemble of rotation–vibration states with a given J and superpolyad quantum number. These ensemble properties are found to be insensitive functions of the potential energy surface. Furthermore, by examining the properties of an ensemble of states rather than the properties of each individual member of the ensemble, we have shown that one can tremendously reduce the parameter space needed to describe the mode mixing.

Our approach to studying ensemble properties is based on random matrices. We define an ensemble of Hamiltonian matrices of which the exact matrix is considered to be a member. These matrices are random to the extent that their elements are randomly chosen from distribution functions that are defined with the absolute minimum number of parameters. The distribution functions and their parameters are extracted from the properties of the exact matrix, such that the ensemble reproduces those observables associated with the exact matrix that are of interest. If one is interested in sufficiently averaged properties, then one can find simple forms for the parametrization of the RME. We have done so using three distributions: one describing the distribution of diagonal elements, a second describing the distribution of off-diagonal elements, and a final distribution describing the correlation between diagonal and off-diagonal elements. Having parametrized the ensemble, we then determined how mixing varies as a function of these parameters. In principle, the parameter space is now small enough that one can begin to explore rotation–vibration mixing as a function of all of the parameters.

The longer term goal of this research is to see whether these ideas can be extended to other molecules. In particular, if a molecule has a superpolyad structure, one can investigate whether the same distribution functions can be used to describe the Hamiltonian matrix as were used for H_2CO. If so, then the state-mixing found for this molecule can be compared in the same parameter space as was used for H_2CO. If different distribution functions are needed, one will need to generalize the results for H_2CO in order to find new distribution functions, described with a few additional parameters, that now describe state mixing for both systems. In this sense, we hope to develop ideas of transferability from one molecular system to another.

In a recent study of mode-mixing, Logan and Wolynes[56] mapped the IVR problem onto that of Anderson localization. One of the central features of their analytical treatment, which is missing in our treatment, are certain assumptions about the connectivity of the Hamiltonian matrix. In particular, they assumed that connectivity could be modeled using a Cayley tree. With this approximation, two states coupled by a particular resonance coupling cannot both be simultaneously coupled to a third state. For the large systems they were interested in, this may be a good approximation. For smaller molecules, this approximation of connectivity is less valid, i.e., three states are often all directly coupled to each other. Nevertheless, the three distributions we have used to determine the RME, discussed in this Chapter, do not take into consideration specifically any connectivity. It will be interesting to examine other molecular properties, such as nearest neighbor level spacing statistics, to see whether details such as connectivity become important factors in making accurate comparisons with actual Hamiltonians.

Acknowledgments

This work was supported by National Science Foundation Grant No. CHE–9013904 and CHE–895764. Calculations were carried out using the Department of Chemistry Computer Center Facilities, partially supported by the National Science Foundation Grant No. CHE–9007850.

References

1. H. L. Dai, C. L. Korpa, J. L. Kinsey, and R. W. Field, *J. Chem. Phys.* **82**, 1688 (1985).
2. G. M. Nathanson and G. M. McClelland, *J. Chem. Phys.* **81**, 629 (1984).
3. T. J. Kulp, R. W. Ruoff, and J. D. McDonald, *J. Chem. Phys.* **82**, 2175 (1985).
4. E. K. C. Lee, *J. Chem. Phys.* **85**, 1261 (1986).
5. C. L. Brummel, S. W. Mork, and L. A. Philips, *J. Chem. Phys.* **95**, 7041 (1991).
6. T. K. Minton, H. L. Kim, and J. D. McDonald, *J. Chem. Phys.* **88**, 1539 (1988).
7. D. J. Nesbitt and R. Lascola, *J. Chem. Phys.* **97**, 8096 (1992).
8. D. S. Perry, *J. Chem. Phys.* **97**, 6994 (1992).
9. D. C. Burleigh, R. C. Mayrhofer, and E. L. Sibert, *J. Chem. Phys.* **89**, 7201 (1988).
10. T. Uzer, G. A. Natanson, and J. T. Hynes, *Chem. Phys. Lett.* **122**, 12 (1985).
11. S. K. Gray and M. J. Davis, *J. Chem. Phys.* **90**, 5420 (1989).
12. D. K. Sahm, R. V. Weaver, and T. Uzer, *J. Opt. Soc. Am.* **B9**, 1865 (1990).
13. M. Aoyagi, S. K. Gray, and M. J. Davis, *J. Opt. Soc. Am.* **B7**, 1859 (1990).
14. J. H. Frederick and G. M. McClelland, *J. Chem. Phys.* **84**, 876 (1986).
15. J. H. Frederick and G. M. McClelland, *J. Chem. Phys.* **84**, 4347 (1986).
16. G. S. Ezra, *Chem. Phys. Lett.* **127**, 492 (1986).
17. W. B. Clodius and R. B. Shirts, *J. Chem. Phys.* **81**, 6244 (1984).
18. A. B. McCoy, D. C. Burleigh, and E. L. Sibert, *J. Chem. Phys.* **95**, 7449 (1991).
19. M. S. Krishnan and Tucker Carrington, Jr., *J. Chem. Phys.* **95**, 1884 (1991).
20. K. K. Lehmann, *J. Chem. Phys.* **95**, 2361 (1991).
21. H. Li, G. S. Ezra, and L. A. Philips, *J. Chem. Phys.* **97**, 5956 (1992).
22. E. L. Sibert, *J. Chem. Phys.* **88**, 4378 (1988).
23. E. L. Sibert, *J. Chem. Phys.* **90**, 2672 (1989).
24. A. B. McCoy and E. L. Sibert, *J. Chem. Phys.* **92**, 1893 (1990).
25. A. B. McCoy and E. L. Sibert, *J. Chem. Phys.* **95**, 3476 (1991).
26. H. H. Nielsen, *Rev. Mod. Phys.* **23**, 90 (1951).

27. M. J. Bramley, W. H. Green, and N. C. Handy, *Mol. Phys.* **73**, 1183 (1991).
28. E. L. Sibert, *Int. Rev. Phys. Chem.* **9**, 1 (1990).
29. R. J. Bell and P. Dean, *Faraday Discuss. Chem. Soc.* **50**, 55 (1970).
30. D. C. Burleigh and E. L. Sibert, *J. Chem. Phys.* (in press).
31. N. C. Handy, *Mol. Phys.* **61**, 207 (1987).
32. H. M. Pickett, *J. Chem. Phys.* **56**, 1715 (1972).
33. R. Meyer and Hs. H. Günthard, *J. Chem. Phys.* **49**, 1510 (1968).
34. E. C. Kemble, *The Fundamental Principles of Quantum Mechanics* (McGraw-Hill, New York, 1937) p. 394.
35. E. B. Wilson, J. C. Decius, and P. C. Cross, *Molecular Vibrations* (McGraw-Hill, New York, 1955).
36. C. Eckart, *Phys. Rev.* **47**, 552 (1935).
37. G. Simons, R. G. Parr, and J. M. Finlan, *J. Chem. Phys.* **59**, 3229 (1973).
38. A. B. McCoy and E. L. Sibert, *J. Chem. Phys.* **95**, 3488 (1991).
39. A. B. McCoy and E. L. Sibert, *Mol. Phys.* **77**, 697 (1992).
40. G. Amat, H. H. Nielsen, and G. Tarago, *Rotation–Vibration Spectra of Molecules* (Marcel Dekker, New York, 1971).
41. J. Schwinger, in eds. L. C. Biedenharn and H. van Dam, *Quantum Theory of Angular Momentum* (Academic, New York, 1965) p. 229.
42. D. Farrelly, *J. Chem. Phys.* **85**, 2119 (1986).
43. M. E. Kellman, *J. Chem. Phys.* **76**, 4528 (1982).
44. K. K. Lehmann, *J. Chem. Phys.* **79**, 1098 (1983).
45. D. Huber, *Int J. Quantum Chem.* **28**, 245 (1985).
46. L. Xiao and M. E. Kellman, *J. Chem. Phys.* **90**, 6086 (1989).
47. L. Xiao and M. E. Kellman, *J. Chem. Phys.* **93**, 5805 (1990).
48. M. E. Kellman, *J. Chem. Phys.* **93**, 6630 (1990).
49. L. B. Harding and W. C. Ermler, *J. Comput. Chem.* **6**, 13 (1985).
50. H. Romanowski, J. M. Bowman, and L. B. Harding, *J. Chem. Phys.* **82**, 4155 (1985).
51. J. K. G. Watson, *Mol. Phys.* **19**, 465 (1970).
52. H. L. Kim, T. J. Kulp, and J. D. McDonald, *J. Chem. Phys.* **87**, 4376 (1987).
53. J. Go and D. S. Perry, *J. Chem. Phys.* **97**, 6994 (1993).
54. S. N. Evangelou and E. N. Economou, *Phys. Rev. Lett.* **68**, 361 (1992).
55. J. Segall, R. N. Zare, H. R. Dübal, M. Lewerenz, and M. Quack, *J. Chem. Phys.* **86**, 634 (1987).
56. D. E. Logan and P. G. Wolynes, *J. Chem. Phys.* **93**, 4994 (1990).

CHAPTER 27

VIBRATIONAL ENERGY FLOW
IN ROTATING MOLECULES

T. Uzer

School of Physics
Georgia Institute of Technology
Atlanta, Georgia 30332-0430, USA

and

Tucker Carrington, Jr.
Université de Montréal
Départment de chimie, C.P. 6128, succursale A
Montréal (Québec) H3C 3J7, Canada

Contents

1. Introduction

The stimulated emission pumping (SEP) experiments of Dai, Korpa, Kinsey, and Field[1] on the ground electronic state of formaldehyde implicate rotation-induced vibration coupling as a major factor in the intramolecular energy flow dynamics[2] even for modest values of the total angular momentum. This detailed exploration constitutes a milestone in our understanding of rovibrational interactions in highly excited molecules,[3] and has inspired a number of theoretical descriptions of intramolecular dynamics in highly excited molecules where vibrations and rotations are far from separable.[4-14] The purpose of this chapter is to provide an elementary introduction to the subject of energy flow in vibration–rotation problems. In the spirit of this monograph the examples cited are selected on the basis of their didactic value; it is hoped that after reading this chapter, readers will be in a position to use the techniques described here in their own research.

Much of this chapter will be devoted to simple but fundamental models of energy transfer between vibrational modes. This subject is best introduced by recalling features of energy exchange between classical oscillators. The large-amplitude vibrations of the highly excited member of the pair act as an external force on the unexcited one, turning it into a forced oscillator. It is well-known that the "cold" oscillator can absorb energy very effectively if the forcing frequency is in "resonance" (i.e., it is the same or rationally related to the frequency of the cold oscillator). The subject of resonance is a recurring one throughout this chapter.

If the oscillators involved are harmonic, or "linear" (i.e., if their frequencies do not change with excitation), a relatively simple picture of energy transfer results: The two oscillators remain in resonance (or remain out of resonance) and the flow is plentiful (or modest) — in any case, a remarkably steady regime is maintained. In what follows, this regime will be called the "mode mixing" type of energy transfer because the dynamics of the overall system can be followed easily in terms of its normal modes, the superposition of which accounts for changes in the mode energies.

However, the energy transfer mechanism changes quite dramatically when the oscillators involved are anharmonic. The frequencies of these "nonlinear" oscillators depend on the excitation (as in the Morse oscillator), and are, consequently, dynamical quantities. Recall that energy flow depends on the existence of resonances — yet the resulting energy flow detunes these resonances that helped the flow in the first place. Therefore, energy flow through "nonlinear", excitation- dependent resonances has a self-regulating or feedback character: In contrast to the mode-mixing case, only a restricted amount of energy can be exchanged. Most oscillators in nature are nonlinear, as are the Fermi resonances of molecular systems.

It follows that in most realistic energy transfer scenarios, the amount of energy exchanged is determined by a delicate balance between how nonlinear the resonances are (i.e., how easy it is to detune them), and how strong the coupling is. In addition to the "push" and "pull" of coupling and anharmonicity, additional parameters like initial excitations may also influence the dynamics. In realistic models of molecular vibrations, many such resonances may be coupled and vibrational energy may proceed through a complicated sequence of resonances, each with its own detuning characteristics.

It is certainly possible to study numerically the energy flow characteristics of a molecular Hamiltonian given its mode frequencies, potential/kinetic couplings, anharmonicities, and initial excitations. But do the rather different roles of these factors fit into a unified, compact, geometrical picture? The answer turns out to be yes, and will be illustrated in this chapter. In particular, we will show that many aspects of energy transfer that we described above can be explained in terms of two well-known physical systems: the hindered rotor (or pendulum) and asymmetric top. In the vibration–rotation problems we are considering here, the latter is a generalization of the former.

The plan of our chapter is as follows: After examining the separation of vibrations and rotations, we first illustrate the conversion of molecular Hamiltonians into hindered rotors (or pendula) using action-angle variables. Because of its central role, the pendulum Hamiltonian is then treated in some detail. Next, we show the conversion of a strongly coupled Hamiltonian to an asymmetric top, again using tools from classical mechanics. Lastly, we illustrate the contrasting mode-mixing regime of energy transfer using a reduced dimensionality model of formaldehyde. Although the modes in our study are anharmonic and coupled, the expected nonlinear resonance disappears and instead we see the abundant energy flow characteristic of mode mixing. We conclude with a summary of recently developed vibration–rotation decoupling schemes.

2. Vibration–Rotation Separation and the Watson Hamiltonian

A molecule may be thought of as a collection of N nuclei and N_e electrons. In terms of space-fixed coordinates, it is straightforward to write down the Hamiltonian operator

$$H = -\frac{\hbar^2}{2m_e} \sum_{i=1}^{N_e} \left(\frac{\partial^2}{\partial X_i^2} + \frac{\partial^2}{\partial Y_i^2} + \frac{\partial^2}{\partial Z_i^2} \right)$$

$$- \frac{\hbar^2}{2} \sum_{\alpha=1}^{N} \frac{1}{m_\alpha} \left(\frac{\partial^2}{\partial X_\alpha^2} + \frac{\partial^2}{\partial Y_\alpha^2} + \frac{\partial^2}{\partial Z_\alpha^2} \right) + V , \qquad (1)$$

where m_e is the mass of an electron, m_α is the mass of the αth nucleus, X_k, Y_k, Z_k are space-fixed coordinates of the kth particle (either an electron or a nucleus), and V is the potential energy. In principle, it would be possible, using the Hamiltonian in this form, to solve the Schrödinger equation to study either the bound states of a molecule or, given an initial state, energy flow within a molecule. However, in practice the space-fixed Hamiltonian of Eq. (1) is not very useful. Its failings are due to both the computational complexity of the space-fixed equations and to the opaque character of its solutions. We know that for most molecules the nuclei are held fairly close to equilibrium configurations by electronic forces, and that electronic and nuclear degrees of freedom can be separated by invoking the Born–Oppenheimer approximation. In what follows, we consider the motion of the nuclei on one electronic potential surface. To facilitate solution of the equations which describe the motion of the nuclei and to understand the solutions we seek, we introduce a "molecule-fixed" axis system, embedded in the molecule, the orientation of whose axes changes as the nuclei move. It is only after having introduced molecule-fixed axes that we can describe the motion of the nuclei in terms of translation, rotation, and vibration. Introducing a molecule-fixed axis system necessarily complicates the Hamiltonian, but it enables one to think in terms of "vibration" and "rotation". Vibration and rotation are valuable concepts because for many polyatomic molecules a useful theory can be built upon a zeroth-order model of small-amplitude vibrations and rigid rotations of the nuclei about a fixed point on the potential energy surface.[15–20]

Associated with the molecule-fixed axis system are $3N$ coordinates which describe the configuration of the nuclei: three Euler angles (θ, ϕ, χ)

which specify the orientation of the molecule-fixed axes with respect to the space fixed axes, $3N - 6$ (vibrational) coordinates which determine the shape of the molecule (but not its orientation), and three coordinates for the position of the center of mass. To a first approximation, the Hamiltonian, in terms of the molecule-fixed coordinates, takes the form

$$H^0 = H_{\text{rot}} + H_{\text{vib}} \ . \tag{2}$$

To derive the general Hamiltonian it is simpler to start with a classical approach (from which a quantum mechanical operator may be obtained afterwards). Space-fixed and molecule-fixed coordinates are related by

$$\mathbf{R}_i = \mathbf{R}_0 + \mathbf{S}^{-1}(\theta, \phi, \chi) \cdot \mathbf{r}_i \ , \tag{3}$$

where $\mathbf{S}(\theta, \phi, \chi)$ is the 3×3 matrix of direction cosines which relates the orientations of the molecule- and space-fixed axis systems, $\mathbf{R}_0$ is the position vector of the center of mass of the nuclei (and the origin of the molecule-fixed axis system) in space-fixed coordinates, $\mathbf{r}_i$ is the position of the ith nucleus in the molecule-fixed axis system, and $\mathbf{R}_i$ is the position of the ith nucleus with respect to the space-fixed axis system. Six constraints must be introduced to reduce the number of independent variables on the right hand side of Eq. (3). We write

$$\mathbf{r}_i = \mathbf{a}_i + \mathbf{d}_i \qquad (i = 1, \ldots, N) \ , \tag{4}$$

where $\mathbf{a}_i$ represents the equilibrium geometry of the nuclei. The origin of the molecule-fixed axes is chosen so that

$$\sum_{i=1}^{N} m_i \mathbf{a}_i = \mathbf{0} \ . \tag{5}$$

The molecule-fixed position vectors are written

$$\mathbf{r}_i = \mathbf{a}_i + \mathbf{d}_i \ , \tag{6}$$

where $\mathbf{d}_i$ denotes a Cartesian displacement vector for the ith nucleus.

Three of the six constraints serve to maintain the origin of the molecule-fixed axis system at the nuclear center of mass (we ignore the small difference between the nuclear and molecular centers of mass):

$$\sum_{i=1}^{N} m_i \mathbf{d}_i = \mathbf{0} \ . \tag{7}$$

The remaining three constraints are chosen to minimize the Coriolis terms which couple rotation and vibration. This is achieved by imposing the condition

$$\sum_{i=1}^{N} m(\mathbf{a}_i \times \mathbf{d}_i) = 0 \ . \tag{8}$$

Together Eqs. (7) and (8) are known as the Eckart conditions.[17]

For our analysis, it is useful to write the Hamiltonian in terms of $3N-6$ normal coordinates. The normal coordinates are combinations of the Cartesian displacements $d_{i\alpha}$ $(i = 1, \ldots, N; \alpha = x, y, z)$ chosen so that the vibrational kinetic energy and the quadratic part of the potential energy do not couple normal coordinates. In terms of normal coordinates q_r and their conjugate momenta p_r the zeroth-order vibrational Hamiltonian takes the form,

$$H^0_{\text{vib}} = \sum_r \frac{1}{2}(p_r^2 + \lambda_r q_r^2) \ . \tag{9}$$

Often one writes the Hamiltonian in terms of dimensionless variables defined as

$$Q_r = \left(\frac{2\pi c \omega_r}{\hbar}\right)^{1/2} q_r \qquad \text{and} \qquad P_r = \frac{p_r}{(2\pi c \hbar \omega_r)^{1/2}} \ , \tag{10}$$

where ω_r is the rth harmonic normal mode frequency in cm^{-1} units. A particularly clear discussion of the molecular Hamiltonian and the Eckart conditions has been given by Bunker.[18]

Starting from the classical space-fixed Lagrangian, transforming to molecule-fixed coordinates, and applying the Eckart constraints, one obtains the complete normal coordinate rovibrational (classical) Hamiltonian in dimensions of wave numbers,[19]

$$H = \frac{\hbar^2}{2hc} \sum_{\alpha\beta} (J_\alpha - \pi_\alpha)\mu_{\alpha\beta}(J_\beta - \pi_\beta) + \frac{1}{2}\sum_k \omega_k P_k^2 + \frac{V}{hc} \ , \tag{11}$$

where

$$\pi_\alpha = \sum_{kl} \zeta_{kl}^\alpha \sqrt{\frac{\omega_l}{\omega_k}} \, Q_k P_l \ . \tag{12}$$

Latin letter subscripts denote normal modes, Greek letter subscripts refer to the molecule-fixed axes, $\mathbf{J}$ is the dimensionless total angular momentum in the molecule-fixed (Eckart) frame, Q_k and P_k are dimensionless normal coordinates and their conjugate momenta, V is the potential energy, and $\mu_{\alpha\beta}$ is essentially the inverse moment of inertia tensor. There are two kinds of vibration–rotation coupling: "Centrifugal coupling" is due to the coordinate dependence of the moment of inertia tensor, and "Coriolis coupling" is introduced by the rotation of the molecule-fixed axis system and appears in the cross terms $\pi\mu\mathbf{J}$ even if the coordinate dependence of μ is neglected. It would be possible to choose the orientation of the molecule-fixed axes so as to diagonalize μ and thereby minimize coupling due to the coordinate dependence of μ, but doing so engenders large Coriolis terms. It is preferable to adapt the Eckart condition which minimizes Coriolis coupling. It is not possible to choose molecule-fixed axes which completely eliminate all Coriolis terms. If one expands the elements of $\mu_{\alpha\beta}$ it is possible to write the general Hamiltonian as

$$
\begin{aligned}
H = \quad & H_{20}+H_{30}+H_{40} + \ldots \text{ (vibrational terms)} \\
& +H_{21}+H_{31}+H_{41} + \ldots \text{ (Coriolis terms)} \\
& +H_{12}+H_{22}+H_{32}+H_{42} + \ldots \text{ (rotational terms) ,}
\end{aligned}
\tag{13}
$$

the first index being the number of vibrational operators and the second index, the number of rotational operators. Explicit equations for many of these terms are given by Aliev and Watson:[20]

$$
H_{20} = \frac{1}{2} \sum_k \omega_k (Q_k^2 + P_k^2) ,
\tag{14}
$$

$$
H_{30} = \frac{1}{6hc} \sum_{klm} \left(\frac{\partial^3 V}{\partial Q_k \partial Q_l \partial Q_m} \right) Q_k Q_l Q_m ,
\tag{15}
$$

$$
H_{21} = -\hbar^2 \left(\sum_{\alpha kl} \mu_{\alpha\alpha}^e \zeta_{kl}^{\alpha} J_\alpha \right) \sqrt{\frac{\omega_l}{\omega_k}} \, Q_k P_l ,
\tag{16}
$$

$$
H_{02} = \frac{\hbar^2}{2} \sum_\alpha \mu_{\alpha\alpha}^e J_\alpha^2 ,
\tag{17}
$$

$$
H_{12} = \sum_{k\alpha\beta} B_k^{\alpha\beta} J_\alpha J_\beta Q_k ,
\tag{18}
$$

$$
H_{22} = \sum_{\alpha\beta\gamma kl} \frac{3}{4\hbar^2 \mu_{\gamma\gamma}^e} (B_k^{\alpha\gamma} B_l^{\gamma\beta} + B_k^{\gamma\beta} B_l^{\alpha\gamma}) J_\alpha J_\beta Q_k Q_l ,
\tag{19}
$$

where

$$B_k^{\alpha\beta} = -\sqrt{\frac{hc}{\hbar^2\omega_k}} \left(\frac{\partial I_{\alpha\beta}}{\partial Q_k}\right)_e \frac{\hbar^4 \mu_{\alpha\alpha}^e \mu_{\beta\beta}^e}{2} . \tag{20}$$

$I_{\alpha\beta}$ is the $(\alpha\beta)$-component of the instantaneous moment of inertia tensor, $\mu_{\alpha\alpha}^e$ is proportional to the inverse of the moment of inertia about the principal α-axis,

$$\mu_{\alpha\alpha}^e = \frac{1}{hcI_{\alpha\alpha}^e} , \tag{21}$$

and all derivatives are evaluated at the equilibrium configuration.

The Hamiltonian of Eq. (13) represents $3N - 6$ coupled vibrational modes. Often, only two of these modes are strongly coupled, usually because they are nearly degenerate. The remaining modes do not have an important effect on energy flow between the two strongly coupled modes. In later sections of this chapter, we shall illustrate classical theories of energy transfer using simple two mode Hamiltonians. There are no molecules with two vibrational modes — the two mode Hamiltonians we use are obtained by neglecting all but two modes. One could obtain a slightly more realistic effective Hamiltonian for the two strongly coupled modes by applying perturbation theory to the general Hamiltonian (Eq. (13)). The original Hamiltonian would be transformed to remove terms which couple modes which are not nearly degenerate. This is accomplished by using the perturbation theory treatment developed by spectroscopists.[17] New terms are added to the Hamiltonian by the transformation process. For example, eliminating the centrifugal terms linear in vibrational coordinates would give rise to terms in the transformed Hamiltonian with four powers of angular momentum operators. The effective Hamiltonian for the two strongly coupled degrees of freedom would be obtained by taking matrix elements of the transformed Hamiltonian for all but the two special degrees of freedom. In what follows, energy flow in formaldehyde is treated using a Hamiltonian which does not include centrifugal terms, but in principle their inclusion is straightforward.

3. Rotation-Induced Vibrational Energy Flow
— An Elementary Example

We examine the simplest possible model to illustrate our techniques, ideas, and results. Consider a model linear ABA type triatomic rotating in a plane, using classical mechanics.[5] Two normal modes, the bend and antisymmetric stretch, are Coriolis coupled,[21,22] and the importance of this coupling increases as the mode frequencies approach each other. To bring the phenomenon into clearest focus, we will assume that these frequencies are 1:1 resonant, i.e., equal at equal mode energies even in the presence of mode anharmonicity. Rather than examining energy flow between rotation and vibration,[23] or energy flow between bonds,[4] our focus is on the Coriolis-induced energy flow between the normal modes.

The Hamiltonian for our model ABA triatomic can be easily derived using two vectors:[24] the vector $\mathbf{R}$ between two A atoms, and the vector $\mathbf{r}$ from the AA center of mass to atom B. The result is

$$H_\mu = \frac{1}{2\mu}(P_b^2 + P_s^2) + \frac{(J - \pi_z)^2}{2\mu_R R^2} + V(Q_s, Q_b) , \tag{22}$$

in which we have ignored the symmetric stretch which is decoupled. Here,

$$\mu = 2m_\mathrm{A} m_\mathrm{B}/(2m_\mathrm{A} + m_\mathrm{B}), \quad \mu_R = m_\mathrm{A}/2, \quad \text{and} \quad R \equiv |\mathbf{R}| \approx 2R_\mathrm{AB}^e ,$$

with R_AB^e the equilibrium AB bond length. The antisymmetric stretching and bending normal mode coordinate definitions are (with z as the axis perpendicular to the molecule frame)

$$Q_s = \mathbf{r} \cdot \mathbf{R}/R; \qquad Q_b = (\mathbf{r} \times \mathbf{R})_z/R . \tag{23}$$

The momenta conjugate to Q_s and Q_b are P_s and P_b, respectively. The total conserved angular momentum is J, and π_z is the vibrational angular momentum, defined for this molecule as

$$\pi_z \equiv P_b Q_s - P_s Q_b . \tag{24}$$

The potential energy is assumed to be quadratic and quartic and identical for each normal mode,

$$V(Q_s, Q_b) = \frac{1}{2}f(Q_s^2 + Q_b^2) + B\mu^2(Q_s^4 + Q_b^4) \tag{25}$$

with positive anharmonicity. With this model, the stretching and bending modes have the same fundamental frequency, but because of anharmonicity are only 1:1 resonant when there are nearly equal energies in the modes.

The simplest route to conversion of Eq. (22) to a hindered rotor representation is via the approximate action-angle transformation

$$Q_i = (2I_i/\Omega_\mu)^{1/2} \sin \phi_i; \qquad P_i = (2I_i/\Omega_\mu)^{1/2} \cos \phi_i , \tag{26}$$

where (I_i, ϕ_i) are the action-angle variables and Ω is the (common) fundamental frequency $\Omega = (f/\mu)^{1/2}$. A double Fourier analysis of the transformed H gives $H = J^2/2\mu_R R_e^2 + H_{\rm res}$, with the "resonance Hamiltonian"

$$H_{\rm res} = \Omega(I_s + I_b) + \frac{3B}{2\Omega^2}(I_s^2 + I_b^2) + \frac{(I_s I_b)^{1/2}}{m_{\rm A}(R_{\rm AB}^e)^2} J\sin(\phi_s - \phi_b) \ . \qquad (27)$$

In obtaining $H_{\rm res}$, we have discarded all angular terms apart from those involving $\phi_s - \phi_b$, since those discarded terms will be rapidly varying when the system is nearly 1:1 resonant. We have also omitted an angular term involving $\cos[2(\phi_s - \phi_b)]$ that arises from the π_z^2 term in Eq. (22) which is unimportant for modest to large J values.

The variable frequencies for the anharmonic normal mode motions are

$$\omega_i(I_i) = \frac{\partial H_{\rm res}}{\partial I_i} \approx \Omega + \frac{3B}{\Omega^2} I_i \ . \qquad (28)$$

The critical last term in Eq. (27), which leads to energy transfer, can be approximately evaluated at the "resonance center" (I_s^r, I_b^r) where

$$\omega_s(I_s) = \omega_b(I_b) \ ,$$

i.e., when $I_s^r = I_b^r$. Since the total vibrational energy is $E \approx \Omega(I_s^r + I_b^r)$, we see that $I_s^r = I_b^r \approx E/2\Omega$. Thus the resonance Hamiltonian

$$H_{\rm res} = \Omega(I_s + I_b) = \frac{3B}{2\Omega^2}(I_s^2 + I_b^2) + \frac{(I_s^r I_b^r)^{1/2}}{m_{\rm A}(R_{\rm AB}^e)^2} J\sin(\phi_s - \phi_b) \qquad (29)$$

has an interaction term coupling the normal modes that scales approximately as $E/2\Omega$, and, of course, with J.

The final step in the derivation is the canonical transformation

$$P = I_s + I_b, \qquad p = I_s - I_b \ ;$$

$$\phi_s + \phi_b = 2\Phi, \qquad \phi_s - \phi_b - \frac{\pi}{2} = 2\psi$$

to give

$$H_{\rm res} = \Omega P + \frac{3B}{4\Omega^2} P^2 + \frac{3B}{4\Omega^2} p^2 + \frac{(I_s^r I_b^r)^{1/2}}{m_{\rm A}(R_{\rm AB}^2)^2} J\cos(2\psi) \ . \qquad (30)$$

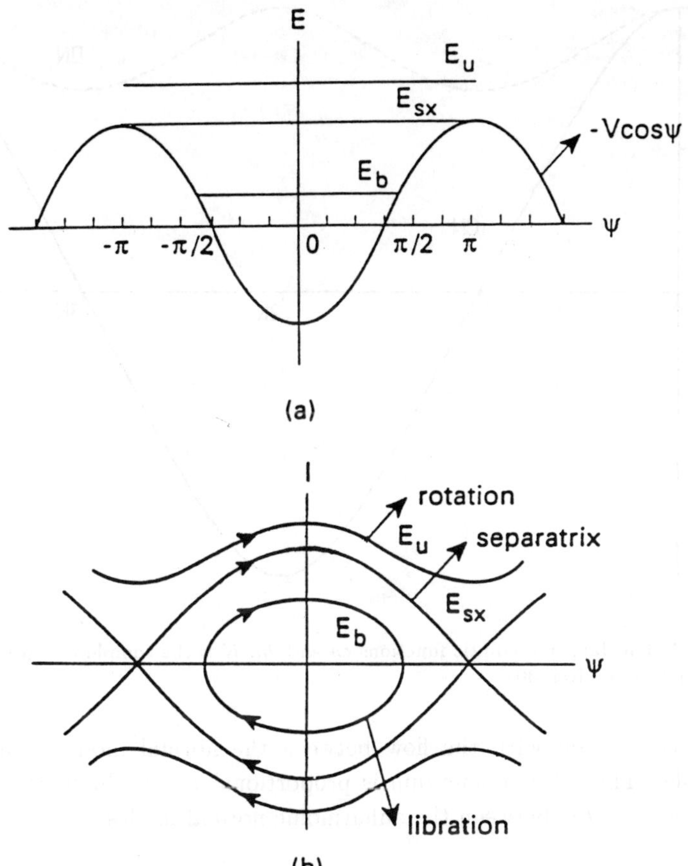

Fig. 1. Correspondence between (a) an energy diagram and (b) a phase space diagram for the pendulum Hamiltonian, Eq. (45). (From Lichtenberg and Lieberman, Ref. 26)

Since P is a constant of the motion, the anharmonic normal mode energy transfer induced by the Coriolis coupling is given by the last two terms of Eq. (30). These have precisely the hindered rotor form (see also Fig. 1):

$$H_r = \frac{p^2}{2m} + V_0 \cos 2\psi \, , \tag{31}$$

where

$$m = \frac{2\Omega^2}{3B}, \qquad V_0 = \frac{(I_s^r I_b^r)^{1/2}}{m_A (R_{AB}^2)^2} J \, . \tag{32}$$

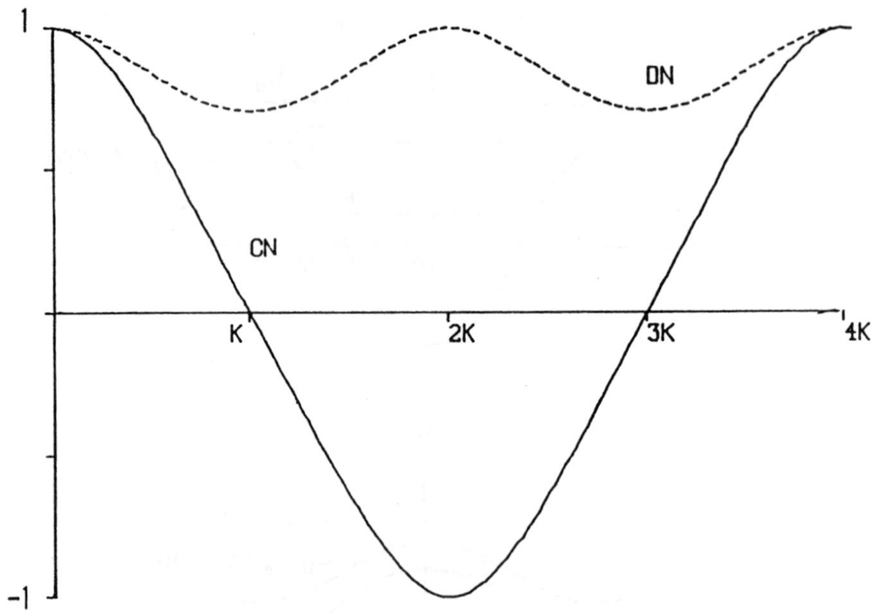

Fig. 2. The Jacobian elliptic functions *cn* and *dn*. K is the complete elliptic integral of the first kind. (Ref. 30)

The connection with the flow between the normal modes is direct and simple. The rotor momentum is proportional to the difference in energy, $\Delta E = E_s - E_b$, between the anharmonic normal modes

$$p = \frac{\Delta E}{(\partial H_{\text{res}}/\partial P)} = \frac{\Delta E}{\frac{1}{2}[\omega_s + \omega_b]} \approx \frac{\Delta E}{\Omega}, \tag{33}$$

and thus is a measure of Coriolis-induced intramolecular energy flow. If the rotor kinetic energy $K_r = p^2/2m$ far exceeds the barrier height $2V_0$, energy remains *localized* in a given normal mode (as in the "*dn*" in Fig. 2); the Coriolis coupling is ineffective. But if $K_r < 2V_0$, energy is periodically *exchanged* between the modes (as in the "*cn*" in Fig. 2). The latter is more easily attained the higher the molecular angular momentum J since $V_0 \propto J$, i.e., as the Coriolis coupling strength is increased. At this point, it is useful to examine the dynamics of energy transfer through analysis of the motion of a pendulum.

4. Nonlinear Resonances and the Classical Mechanics of the Pendulum

Like many other applications of nonlinear dynamics in molecular physics, the pendulum or the hindered rotor picture originated in plasma physics research. This paradigmatic picture was recognized and developed to understand the trapping of charged particles. Soon it was realized that this picture could be applied generally to a number of other trapping phenomena, including trapping and leakage in phase space. For an extensive review of the derivation of the pendulum picture the reader is referred to Chirikov,[25] as well as to standard texts.[26,27] For our purposes, the following brief derivation will suffice:[28] Suppose one is considering the dynamics of a coupled oscillator system, the zero-order frequencies of which have a "resonance", i.e., a commensurability relation like

$$\sum n_k \omega_k = 0, \text{integer } n_k\text{'s}\ .$$

It is convenient and customary to analyze the nonlinear coupling in terms of the action-angle variables of the unperturbed system, $(\mathbf{I}, \boldsymbol{\phi})$

$$H(\mathbf{I}, \boldsymbol{\phi}) = H_0(\mathbf{I}) + \sum_{\{m\}} V_{\{m\}}(\mathbf{I}) \exp[i(\mathbf{m} \cdot \boldsymbol{\phi})]\ , \tag{34}$$

where $\mathbf{m}$ is a multidimensional vector of integers. In particular, for two coupled oscillators

$$H(\mathbf{I}, \boldsymbol{\phi}) = H_0(\mathbf{I}) + \sum_{m_1, m_2} V_{m_1, m_2}(\mathbf{I}) \exp[i(m_1\phi_1 + m_2\phi_2)]\ , \tag{35}$$

or, concentrating on a single resonance,

$$H(\mathbf{I}, \boldsymbol{\phi}) \cong H_0(\mathbf{I}) + V_{m_1, m_2}(\mathbf{I}) \exp[i(m_1\phi_1 + m_2\phi_2)]\ . \tag{36}$$

If the system contains a resonance

$$m_1\omega_1(\mathbf{I}) + m_2\omega_2(\mathbf{I}) = 0 \tag{37}$$

for a particular set of actions $\mathbf{I}^r$, then terms like

$$\exp[i(m_1\phi_1 + m_2\phi_2)] \tag{38}$$

are slowly varying in time. Introducing the new independent angle variables

$$\psi_k = \sum_{i=1}^{2} \mu_{ki}\phi_i \qquad (k = 1, 2) \, , \tag{39}$$

where μ_{ki} are constants such that

$$\mu_{11} = m_1, \quad \mu_{12} = m_2, \quad \mu_{21} = m_1, \quad \mu_{22} = m_2 \, . \tag{40}$$

The new actions **I** are defined by the generating function

$$F_2(\mathbf{I}, \boldsymbol{\phi}) = \sum_{i=1}^{2} \left(I_r^i + \sum_{k=1}^{2} I_k \mu_{ki} \right) \phi_i \tag{41}$$

giving

$$I_k = \sum_i (I_i - I_i^r)\mu_{ik}^{-1} \, , \tag{42}$$

where μ_{ik}^{-1} are elements of the matrix $\boldsymbol{\mu}^{-1}$.

The new momenta **I** measure deviations from $\mathbf{I}^r$. Since ψ_2 does not occur in the resonance Hamiltonian (36), I_2 is a constant of the motion, and Eq. (36) can be rewritten

$$H_r(\mathbf{I}, \boldsymbol{\psi}) = \sum_{i,k} I_k \mu_{ki}\omega_i + \sum_{k,\ell} \frac{I_k I_\ell}{2M_{k\ell}} + V_{m_1,m_2}(\mathbf{I}) \cos \psi_1 \, . \tag{43}$$

Here, $M_{k\ell} = \sum_{i,j}^{2} \mu_{ki}(\partial \omega_i / \partial I_j)\mu_{\ell j}$ is a generalized mass.

At the resonance center, Eq. (37) can be applied, and Eq. (43) becomes

$$H_r(\mathbf{I}, \boldsymbol{\psi}) = \frac{1}{2M_{11}} I_1^2 + V_{m_1,m_2}(\mathbf{I}) \cos \psi_1 \, . \tag{44}$$

The approximation to the dynamics has been obtained by averaging over the fast motion, leaving only the slowly varying terms. Therefore, this process has resulted in a lower-dimensional "resonance Hamiltonian" H_r (Eq. (44)), which can be examined for some of the salient features of the original Hamiltonian.

Special cases arise when the anharmonicity disappears, i.e., the system becomes "intrinsically degenerate"[29] as opposed to accidentally degenerate, which is the case we treat. Often one can approximate the Fourier coefficients at the center of the resonance, i.e., when condition (37) is fulfilled.

The resulting approximate Hamiltonian (the subscripts are not needed)

$$H_r(I, \psi) = \frac{1}{2M} I^2 + V(I^r) \cos \psi = E_r \tag{45}$$

is that of a pendulum, or a hindered rotor. It is clearly an integrable system, and it has three types of motion: When the total energy is less than the potential barrier,

$$E_r < V(I^r) , \tag{46}$$

the motion is that of "vibration". When the energy is above the barrier, it is "rotation". These terms originate from the motion of a physical pendulum, but are commonly used when the pendulum variables are abstract quantities, as well. The third type of motion is the motion on the separatrix, i.e., when the energy is precisely equal to the barrier height. The phase space profiles of these three motions are shown in Fig. 1. The pendulum is such a useful system that we quote the results of its analysis in full. Its action-angle variables (I, ψ) are given by Lichtenberg and Lieberman.[26]

$$I = \frac{8}{\pi}(VM)^{1/2} \begin{cases} E(\kappa) - (1 - \kappa^2)K(\kappa) & \kappa < 1 \\ \frac{1}{2}\kappa E(\kappa^{-1}) & \kappa > 1 \end{cases} \tag{47}$$

$$\psi = \frac{\pi}{2} \begin{cases} [K(\kappa)]^{-1} F(\eta, \kappa) & \kappa < 1 \\ 2[K(\kappa^{-1})]^{-1} F(\psi/2, \kappa^{-1}) & \kappa > 1 \end{cases} \tag{48}$$

$$2\kappa^2 = 1 + E_r/V , \tag{49}$$

where K, E are complete elliptic integrals of the first and second kinds,[30] F is an incomplete elliptic integral of the first kind,[30] and

$$\kappa \sin \eta = \sin(\psi/2). \tag{50}$$

The approximate width of the resonance in action and angle, $(\Delta \mathbf{I})^r, (\Delta W)^r$, respectively, is given as[28] (in terms of the original variables of (44))

$$(\Delta \mathbf{I})^r = 2\mathbf{m}(M_{11} V_{m_1, m_2})^{1/2} ,$$

$$(\Delta W)^r = \frac{2}{|\mathbf{m}|} \left(\frac{V_{m_1, m_2}}{M_{11}} \right)^{1/2} , \tag{51}$$

where $\mathbf{m} = (m_1, m_2)$. The explicit time dependence of the variables of Eq. (45), on the other hand, are[31] (see Fig. 2)

$$I(t) = 2\omega_0 \kappa cn(\omega_0 t) \qquad \kappa < 1 , \tag{52}$$

$$I(t) = 2\omega_0 \kappa dn(\omega_0 t) \qquad \kappa < 1 , \tag{53}$$

where

$$\omega_0 = (V/M)^{1/2} \tag{54}$$

and the frequency of the pendulum is

$$\frac{\omega(\kappa)}{\omega_0} = \frac{\pi}{2} \left\{ \begin{array}{ll} [K(\kappa)]^{-1} & \kappa < 1 \\ 2\kappa/K(\kappa^{-1}) & \kappa > 1 \end{array} \right. . \tag{55}$$

Near the separatrix

$$\lim_{\kappa \to 1} \frac{\omega}{\omega_0} = \left\{ \begin{array}{ll} \frac{\pi}{2}/\ln[4(1-\kappa^2)^{-1/2}] & \kappa < 1 \\ \pi/\ln[4(\kappa^2-1)^{-1/2}] & \kappa > 1 \end{array} \right. . \tag{56}$$

The behavior of ω/ω_0 with varying κ is plotted in Lichtenberg and Lieberman,[26] (Fig. 2.2.)

5. Application of the Pendulum Picture to Rotation Mediated Vibrational Energy Flow

Comparing Eqs. (31) and (33) with Eqs. (45), (52), and (53), we see that the absence or presence of complete energy exchange between the Coriolis coupled anharmonic normal modes is dictated exclusively in the rotor picture by κ. This parameter can be expressed alternatively as the ratio of the initial rotor kinetic energy $K_r(0)$ (i.e., kinetic energy at the potential minimum) to the rotor barrier height V_0 (the $V(I^r)$ of Eq. (45)),

$$\kappa^2 = K_r(0)/2V_0 . \tag{57}$$

The initial K_r value is set by the initial energy disparity $\Delta E(0)$ between the modes. For a given J, κ is smaller for higher total vibrational energy $E(V_0 \propto E)$, and for less asymmetry in the initial mode energy distribution $(p \propto \Delta E)$. Since the barrier height $2V_0 \propto J$, this energy flow regime is more likely for larger J, i.e., for large Coriolis coupling.

For $\kappa < 1$, i.e., below the rotor barrier, flow between the normal modes is complete in that E_s and E_b are exchanged: ΔE is periodic in time (Eq. (52) and Fig. 2). The transfer time t_{tr} is:[32]

$$t_{\mathrm{tr}} = 2(m/4V_0)^{1/2}K(\kappa) , \tag{58}$$

where K is the complete elliptic integral of the second kind.[30] Since $V_0 \propto J$, the energy transfer time is predicted to decline with overall angular momentum J as $t_{\text{tr}} \propto J^{-1/2}$, in stark contrast to the harmonic normal mode case, where[33] $t_{\text{tr}} \propto J^{-1}$.

For $\kappa > 1$, i.e., above the rotor barrier, the Coriolis coupling is powerless to cause the periodic energy exchange between the normal modes. Rather, the energy exchange is only partial and is quenched in the limit $\kappa \gg 1$. The anharmonic normal modes in the regime $\kappa > 1$ are too far from resonance, due to the anharmonicity, to be effectively Coriolis coupled. This energy localization in a given mode is, of course, completely absent between coupled harmonic modes.

6. Generalizing the Pendulum — The Asymmetric Top Mapping

It was mentioned earlier that during the reduction of the full nonlinear Hamiltonian into a pendulum (Eqs. (44),(45)), it was necessary to approximate the Fourier coefficients by constants; for example, by evaluating them at the resonance center. This approximation is only good when the coupling between the degrees of freedom is weak. For strong couplings, the variation of the Fourier coefficients with the momentum needs to be taken into account. An example of this procedure is provided by the vibrational Hamiltonian

$$H_v = \frac{1}{2}\omega(P_1^2 + Q_1^2 + P_2^2 + Q_2^2) + k'(Q_1^4 + Q_2^4) + k'_{12}(Q_1 Q_2)^2 \ . \tag{59}$$

A single-resonance approximation to this Hamiltonian, when expressed in the action-angle variables of the unperturbed Hamiltonian, reads[11,13]

$$H = \frac{1}{4}(3k - k_{12})\Delta^2 + \frac{1}{8}k_{12}(I^2 - \Delta^2)\cos 4\phi_\Delta \ , \tag{60}$$

where (see also Eq. (64)),

$$I \equiv \frac{1}{2}(P_1^2 + Q_1^2 + P_2^2 + Q_2^2) \ ,$$

$$\Delta \equiv \frac{1}{2}(P_2^2 + Q_2^2 - P_1^2 - Q_1^2) \ ,$$

and ϕ_Δ is the phase angle conjugate to Δ. In deriving Eq. (60), Eqs. (26) have been used with the replacements $\Omega \to \omega$, $b \to 2$, and $s \to 1$. Therefore, I and Δ play the role of P and p in Eq. (30). Additionally,

$$k = k'\omega^{-2}; \qquad k_{12} = k'_{12}\omega^{-2} \ .$$

This Hamiltonian is similar to that of an asymmetric rotor in torque-free space. The relation between this and the pendulum Hamiltonian can be made more transparent by transforming the rotor Hamiltonian into conjugate variables (K, χ), where K is the z-component of the angular momentum in the body-fixed axes. This transformation,[34-36]

$$J_x = -(J^2 - K^2)^{1/2} \sin \chi \, ,$$
$$J_y = -(J^2 - K^2)^{1/2} \cos \chi \, ,$$
$$J_z = K \, , \tag{61}$$

converts the asymmetric top Hamiltonian

$$H_{AR} = C J_x^2 + B J_y^2 + A J_z^2 \tag{62}$$

into

$$H_{AR} = \frac{1}{2}(B+C)J^2 + (A-(B+C)/2)K^2 + \frac{1}{2}(B-C)(J^2-K^2)\cos 2\chi \, , \tag{63}$$

which approximates the pendulum Hamiltonian (45) for small variations of K.

Remarkably, any problem containing the same SU(2) symmetry group structure as the asymmetric top can be treated similarly, except that the components of J might not be physical angular momenta, but quantities with the same Poisson bracket relations as angular momentum components.[37-47] For example, the two-dimensional isotropic harmonic oscillator (in x, y coordinates) has four constants of motion ("generators")[43,47]

$$S_0 \equiv \frac{1}{4}(p_x^2 + x^2 + p_y^2 + y^2) \, ,$$
$$S_1 \equiv \frac{1}{2}(p_x p_y + xy) \, ,$$
$$S_2 \equiv \frac{1}{2}(y p_x - x p_y) \, ,$$
$$S_3 \equiv \frac{1}{4}(p_y^2 + y^2 - p_x^2 - x^2) \, , \tag{64}$$

where the third quantity is proportional to the vibrational angular momentum. The generators $\{S_i\}$, $i = 1$–3, behave like angular momentum components and are related by

$$S_1^2 + S_2^2 + S_3^2 = S_0^2 \, . \tag{65}$$

This sphere in $\{S_i\}$ coordinates has a radius equal to the "principal action".[43] Various authors[40,44-47] have identified spherical geometries as apt topologies for displaying the dynamics of vibrational systems. When the energy of the system is also expressed in terms of S_0 and $\{S_i\}$ ("constant energy surface"[37,40,41]) the motion of these vibrational generators can be followed in the same way as the (body-fixed) angular momentum components of the asymmetric top. In more complicated problems where the good constants are not obvious, this picture can suggest which of the three generators (if any), or which of their combinations, is an approximate constant of motion: Because all three are displayed in one figure through their connection, Eq. (65), their relative roles are at once evident.

Another, more direct and quantum mechanical conversion of Eq. (59) into a rotational Hamiltonian makes use of Schwinger's treatment of angular momentum,[48] in which the group theoretical connection between the two dimensional isotropic oscillator

$$H_0 \equiv \frac{1}{2m}(P_1^2 + P_2^2) + \frac{1}{2}m\omega^2(Q_1^2 + Q_2^2) \tag{66}$$

and rotations is used through the ladder (raising-lowering) operators

$$a_j \equiv \sqrt{\frac{m\omega}{2\hbar}}\left(q_j + i\frac{p_j}{m\omega}\right); \quad a_j^\dagger = \sqrt{\frac{m\omega}{2\hbar}}\left(q_j - i\frac{p_j}{m\omega}\right); \quad j = 1, 2 \ . \tag{67}$$

Using the Pauli spin matrices

$$\sigma_1 = \begin{pmatrix} 0 & 1 \\ 1 & 0 \end{pmatrix}; \quad \sigma_2 = \begin{pmatrix} 0 & -i \\ i & 0 \end{pmatrix}; \quad \text{and } \sigma_3 = \begin{pmatrix} 1 & 0 \\ 0 & -1 \end{pmatrix}, \tag{68}$$

the following operators

$$J_x \equiv \frac{1}{2}(a_1^\dagger, a_2^\dagger)\sigma_1\begin{pmatrix} a_1 \\ a_2 \end{pmatrix} = \frac{1}{2}(a_1^\dagger a_2 + a_1 a_2^\dagger) \ ,$$

$$J_y \equiv \frac{1}{2}(a_1^\dagger, a_2^\dagger)\sigma_2\begin{pmatrix} a_1 \\ a_2 \end{pmatrix} = \frac{1}{2i}(a_1^\dagger a_2 - a_1 a_2^\dagger) \ ,$$

$$J_z \equiv \frac{1}{2}(a_1^\dagger, a_2^\dagger)\sigma_3\begin{pmatrix} a_1 \\ a_2 \end{pmatrix} = \frac{1}{2}(a_1^\dagger a_1 - a_2^\dagger a_2) \ , \tag{69}$$

satisfy the commutation relations for angular momentum.

The subsequent transformations of

$$a_j = \sqrt{n_j + \frac{1}{2}}\exp(-i\phi_j); \quad a_j^\dagger = \sqrt{n_j + \frac{1}{2}}\exp(i\phi_j) \tag{70}$$

into a representation involving quantum numbers n_j and phase ϕ_j complete the connection to classical mechanics.

To appreciate the advantages of treating vibrational problems as rotations, let us take a brief look at the dynamics and energy level structure of an asymmetric top.

7. Dynamics and Quantization of the Asymmetric Top

The physics of the asymmetric top is essential for describing the rotational motion and quantized energy level structure of polyatomic molecules.[49] The asymmetric top also happens to be a paradigm for nonlinear physical systems with momentum-dependent periods.[50] A surprisingly large number of systems show asymmetric top behavior when examined in the appropriate action-angle variables. Moreover, the quantization of such systems can then be performed succinctly using the analogy to the asymmetric top.

The Hamiltonian of a torque-free asymmetric top is given by

$$H = CJ_x^2 + BJ_y^2 + AJ_z^2 \tag{71}$$

where A, B, C are rotational constants such that $A > B > C$ and J_x, J_y, and J_z are the components of the angular momentum $\mathbf{J}$ in the body-fixed principal axis system. The quantization of the asymmetric top was achieved by Kramers and Ittmann[51] (their results can be simplified considerably by using quantal counterparts of action-angle variables[50]). The classical-mechanical motion of the asymmetric top has the well-known property that only two of the axes (the ones with the largest and smallest moments of inertia) are stable axes of rotation, whereas rotations about the intermediate axis are unstable.[47,52] This is also reflected in the quantum-mechanical energy levels, which correspond to quantization along either of the two stable axes depending on the energy.

As usual, a semiclassical description is very useful in elucidating the physics of the asymmetric top. A superb way of visualizing the mechanics, which provides insight into the energy level structure, was invented by Harter and Patterson[37] and consists of constructing a rotational energy surface for the problem using the generators of the Lie algebra. A rotational energy surface (RES) is a plot of total energy as a function of the direction of an angular momentum vector (in a body fixed frame) for a constant value of $|\mathbf{J}|$. The components of angular momentum are interpreted as Cartesian coordinates of a position vector whose length is given by H which is plotted

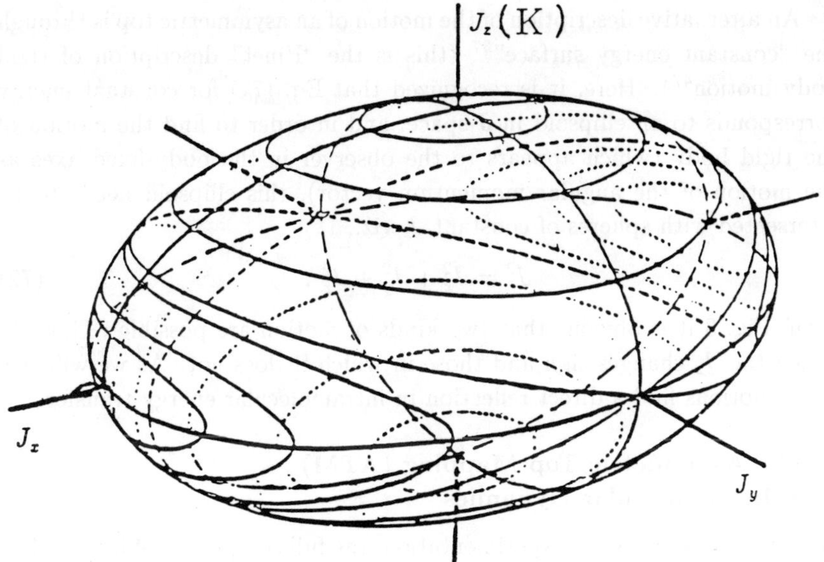

Fig. 3. The motion of the asymmetric top, as seen from the body-fixed frame. Two quantities are conserved: The energy

$$T = CJ_x^2 + BJ_y^2 + AJ_z^2$$

and the magnitude of the angular momentum,

$$J^2 = J_x^2 + J_y^2 + J_z^2 \, .$$

On the ellipsoid in the figure (representing T) energy is constant. The contours on this surface are possible trajectories for the tip of the J vector, and are obtained by intersecting this surface with the spherical J^2 surface. If the energy difference between the modes is defined as J_z (or K), trajectories encircling the J_z axis correspond to local modes (because J_z does not change sign), and those around the J_x axis correspond to normal modes because J_z changes sign as the J vector crosses the (J_x, J_y) plane.

radially outwards. The contours on the surface represent the intersection of spheres whose radii are the total energy with the rotational energy surface. The contours thus represent equipotentials on the rotational energy surface. The equal-energy contours on a RES contain a great deal of information about the semiclassical states of the asymmetric top; the emergence of two classes of states which are separated by a separatrix is particularly obvious in this description.[37]

An alternative description of the motion of an asymmetric top is through the "constant energy surface"[37] (this is the "Binet" description of rigid body motion[47]). Here, it is recognized that Eq. (71) for constant energy corresponds to an ellipsoid in $\mathbf{J}$-space, and in order to find the motion of the rigid body (which appears to the observer in the body-fixed axes as the motion of the angular momentum vector), this ellipsoid needs to be intersected with spheres of constant J, viz.,

$$J^2 = J_x^2 + J_y^2 + J_z^2 \ . \tag{72}$$

From Fig. 3 it is obvious that two kinds of motion are possible: Those in which the J_z changes sign and those in which it does not. As we will see, these motions find a direct reflection in intramolecular energy transfer.

8. The Asymmetric Top Mapping (ATM) in Intramolecular Dynamics

With recent advances in experimentation, the full complexity of intramolecular vibrational energy redistribution (IVR) is beginning to be probed in detail, and it has become desirable to understand the connections between the various microscopic factors that drive intramolecular energy flow. These are Fermi resonances, anharmonicities, vibrational and rotational couplings as well as initial excitations.[2] In certain cases, it is possible to obtain a compact picture of the interplay of all these factors in mediating intramolecular energy flow by mapping the vibrational problem onto a rotational one by means of the Asymmetric Top Mapping (ATM[53]). This procedure has been illustrated in a number of recent papers and relies on the homomorphism of the R(3) and SU(2) symmetry groups of the asymmetric top and the two-dimensional isotropic oscillator,[40] which means that the vibrational constants of motion in the case of the isotropic oscillator generate a Lie algebra just like the components of the angular momentum of the asymmetric top.[43]

Specifically, let us consider mode–mode vibrational energy transfer under the joint action of vibrational and rotational couplings on the basis of a model system consisting of two degenerate anharmonic modes, which are coupled to each other by means of vibrational and Coriolis couplings,[11,13] with the Hamiltonian

$$H = \omega(P_1^2 + Q_1^2 + P_2^2 + Q_2^2)/2 + k(Q_1^4 + Q_2^4) + k_{12}(Q_1 Q_2)^2 + B\pi_z^2 - 2BJ\pi_z \ , \tag{73}$$

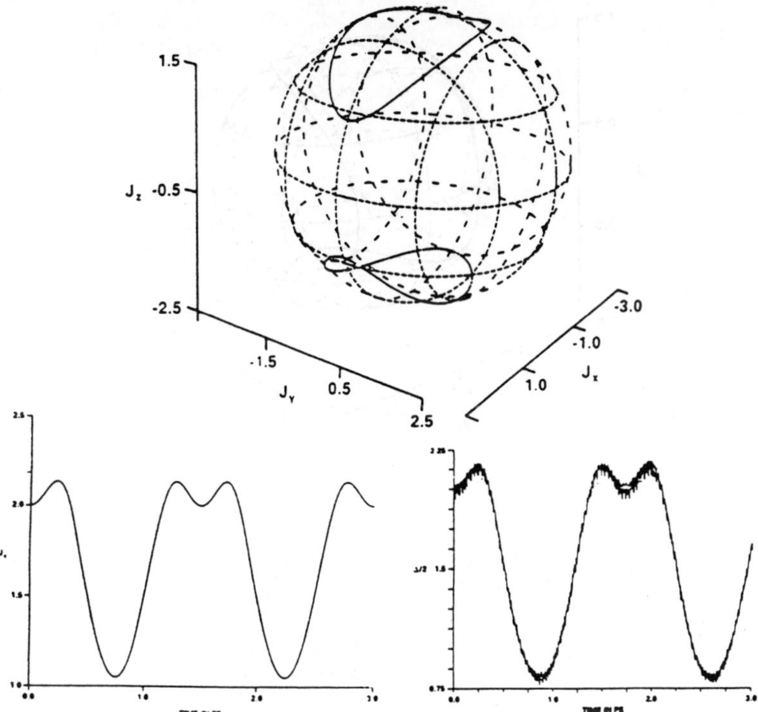

Fig. 4. A typical local mode. The top panel shows the constant energy surface, which is the ellipsoid

$$H = CJ_x^2 + BJ_y^2 + AJ_z^2 + cJ_x \ .$$

The solid curve is the trajectory corresponding to the local mode. The bottom two panels show the vibrational energy difference between modes (vertical axis) as a function of time. Left: Obtained from the constant energy surface; Right: Obtained by numerical integration of the full vibration–rotation Hamiltonian adapted from Ref. 13.

where the vibrational angular momentum π_z is given by

$$\pi_z = \zeta(P_2 Q_1 - P_1 Q_2) \ , \tag{74}$$

and where k_{12} here denotes a coupling constant, k an anharmonicity, J is the (conserved) angular momentum, ζ is the Coriolis coupling constant, and B is the rotational constant. To obtain this Hamiltonian, we have set $\hbar = 1$. The mapping of this problem onto an asymmetric top[11,13]

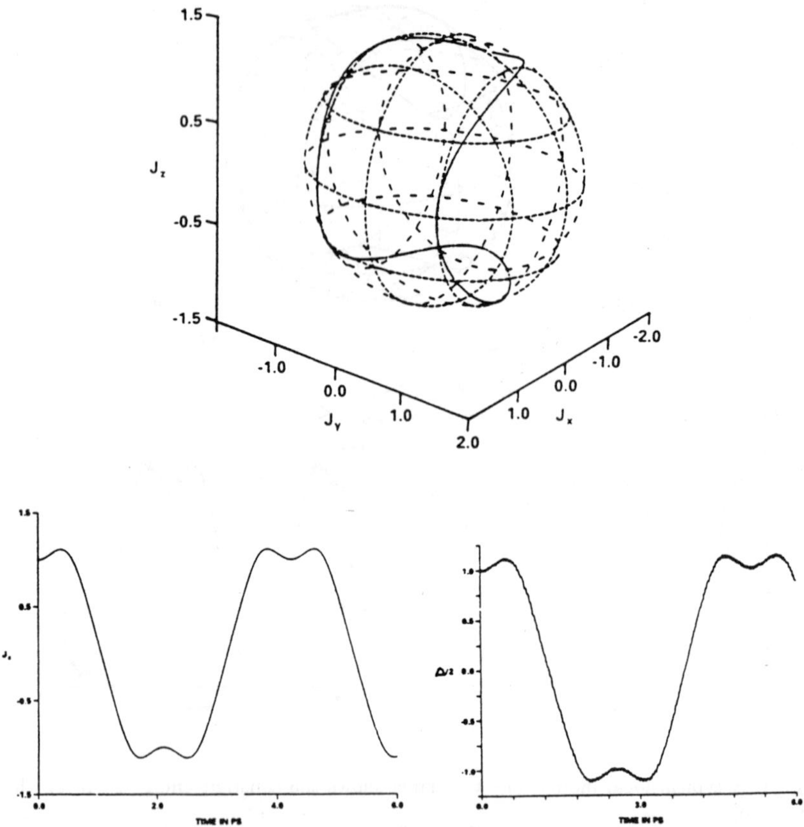

Fig. 5. A "precessional" normal mode. Compare with the preceding figure: the Coriolis coupling was increased, therefore, the pairs of local modes have joined to form a normal mode. The bottom panels again compare the energy difference between the two modes; the one on the left obtained from the constant energy surface, the one on the right from the numerical integration of the full rotational–vibrational Hamiltonian. Adapted from Ref. 13.

consists of identifying angular momentum components with vibrational quantities which are generators of the appropriate Lie algebra (in particular, the z-component can be mapped onto the mode–mode energy difference, an identification suggested by the expression for S_3 in Eq. (64)). Using these relations, the perturbations can be expressed in terms of angular momentum components, and a perturbed asymmetric top results, viz.,

$$H = C J_x^2 + B J_y^2 + A J_z^2 + c J_x \, . \tag{75}$$

The rotational constants A, B, C are constructed from the vibrational and rotational parameters of the original problem. In this case, the Binet picture differs from the one in the previous section in that the center of the ellipsoid is offset from the origin by an amount c proportional to the Coriolis coupling. Again, two kinds of motion result: The motion in which the z-component keeps its sign (Fig. 4) corresponds to restricted energy transfer ("local mode" behavior or energy localization, Eq. (53)), and a sign change in the z-component (Fig. 5) signals extensive energy transfer ("normal mode" behavior or energy delocalization, Eq. (52)). The variation of the energy difference in time is very well represented by motion of the angular momentum vector on the Binet ellipsoids.

The reverse problem, namely, the influence of molecular vibrations on rotational quantum numbers[9,12,54,55] (i.e., vibration–induced rotation mixing, or "K-mixing", from the component K of the angular momentum) has also been studied using the ATM.[54] In this case, it turns out that the breakdown of K as a good quantum number is due to its coupling to the vibrational angular momentum. Of course in molecules, both these phenomena are present at once, and the picture is considerably more complicated than in these models. However, a good example of how helpful these models are to the understanding of intramolecular dynamics is provided by the formaldehyde molecule,[1] which we examine next.

9. Application to Energy Flow in Formaldehyde

Despite its small size, formaldehyde possesses a number of strongly coupled motions and, in general, it is neither easy nor rigorous to concentrate on one particular motion, while ignoring the others. Vibration–rotation energy flow, vibration-induced rotational mixing ("K-mixing"), and Coriolis-induced vibrational energy flow take place simultaneously in formaldehyde to varying degrees. However, recent theoretical work shows[9,12] that in some energy regimes, the full dimensional intramolecular dynamics of formaldehyde can be approximated by two strongly interacting vibrational modes (the out-of-plane bending and the HCO wagging, called modes 4 and 6, respectively, in the standard mode numbering) in addition to molecular rotation. The validity of these classical-mechanical results has been established by parallel semiclassical and quantal calculations.[9,10,12] But even the motion of a three-degree of freedom system is not easily condensed and visualized. However, since the two vibrational frequencies are only 80 cm^{-1} apart, they are connected by an approximate adiabatic invariant (I_+ of Eq. (86)) which can be invoked to reduce the dimensionality by another degree of freedom.[9,12] In contrast with the predictable dynamics of the single-degree of freedom case, even this much simplified description contains some complicated (and fascinating) rovibrational dynamics.[53]

The purpose of this section is to introduce a picture which displays quasiclassical trajectory information on formaldehyde intramolecular dynamics in a way that is more informative than the "dynamical-variable-versus-time" plots of Figs. 2, 4, and 5. However, formaldehyde differs from the problems introduced in the preceding sections somewhat: The two modes are not exactly degenerate, their anharmonicities are slightly different, and the dynamics turns out to be driven not by nonlinear resonances, but by mode mixing. In practice, this means that a "local mode" regime, in which there is only incomplete energy flow between modes,[43] is the exception rather than the rule here: Vibrational energy flows freely between the modes even when the coupling is not particularly strong. The unrestricted energy flow is somewhat surprising, since the anharmonicities are small but nonzero. The following examination of the rovibrational dynamics shows the reason behind the inability of the anharmonicities to hinder energy flow.

The ATM is helpful in understanding and depicting rotational dynamics of formaldehyde because this molecule is a slightly asymmetric prolate top with Ray's asymmetry parameter[56] $\kappa = -0.961$, and when the Coriolis effect is included, a perturbed asymmetric top. On the other hand, since the frequencies of the two modes are accidentally degenerate, its vibrational dynamics can be regarded as that of a perturbed isotropic oscillator.

Using the notation of Burleigh et $al.$[9] the effective three-degree of freedom Hamiltonian of formaldehyde can be summarized as

$$H = H_v + H_{vr} + H_r \, , \tag{76}$$

where the rotational part is the asymmetric top

$$H_r = \frac{1}{4}(\mu_{xx} + \mu_{yy})\hbar^2 J^2 + \left[\frac{1}{2}\mu_{zz} - \frac{1}{4}(\mu_{xx} + \mu_{yy}) \right] \hbar^2 K^2$$

$$+ \frac{1}{4}(\mu_{xx} - \mu_{yy})\hbar^2(J^2 - K^2)\cos 2\phi \, , \tag{77}$$

which is connected by the vibration–rotation (Coriolis) coupling

$$H_{vr} = c_2 K(\lambda P_4 Q_6 - \lambda^{-1} P_6 Q_4) = -\mu_{zz}\hbar K \pi_z ,\qquad (78)$$

with

$$c_2 = -\mu_{zz}\hbar^2 \zeta_{4,6}^{(z)} ,\qquad (79)$$

to the vibrational Hamiltonian

$$H_v = \frac{1}{2}\Omega_4(P_4^2 + Q_4^2) + \frac{1}{2}\Omega_6(P_6^2 + Q_6^2) + k_{44}Q_4^4 + k_{66}Q_6^4$$

$$+ k_{46}Q_4^2 Q_6^2 + c_1(\lambda P_4 Q_6 - \lambda^{-1} P_6 Q_4)^2 ,\qquad (80)$$

with

$$c_1 = \frac{1}{2}\mu_{zz}(\hbar\zeta_{4,6}^{(z)})^2 .\qquad (81)$$

Here, P_i and Q_i are dimensionless normal mode variables, K is the component of the angular momentum along the body-fixed a-axis, which at equilibrium is aligned with the CO bond, and π_z is the vibrational angular momentum:

$$\pi_z = \hbar\zeta_{4,6}^{(z)}(\lambda P_4 Q_6 - \lambda^{-1} P_6 Q_4) .\qquad (82)$$

Furthermore,

$$\lambda \equiv (\Omega_4/\Omega_6)^{1/2} .\qquad (83)$$

It is known that the variation of the moments of inertia with Q_4 and Q_6 is negligible.[12] The asymmetry in the top is very small and, therefore, K is nearly always conserved. The combined Hamiltonian which describes the rotation (ignoring the vibrational Hamiltonian)

$$H_r^{\text{eff}} \equiv H_r + H_{vr}\qquad (84)$$

has the structure of Eq. (13). For the vibrational problem, there is a near-adiabatic invariant, and using it, an adiabatic Hamiltonian can be constructed by averaging over the fast oscillation,[12] giving

$$H_v^{\text{ad}} = (\Omega_4 + \Omega_6)I_+ + (\Omega_4 - \Omega_6)I_- + 3(k_{44} - k_{66})I_+ I_-$$

$$+ \frac{3}{2}(k_{44} + k_{66})(I_+^2 + I_-^2) + [k_{46} + c_1(\lambda + \lambda^{-1})(I_+^2 - I_-^2)]$$

$$+ \left\{\frac{1}{2}k_{46} - c_1\left[1 + \frac{1}{2}(\lambda^2 + \lambda^{-2})\right]\right\}(I_+^2 - I_-^2)\cos 2\theta_-$$

$$- c_2(I_+^2 - I_-^2)^{1/2}K(\lambda + \lambda^{-1})\sin\theta_- ,\qquad (85)$$

where we have renamed some of the generators to bring them in line with the literature,[9] namely,

$$I_+ = \frac{1}{4}(P_4^2 + Q_4^2 + P_6^2 + Q_6^2) = S_0 \;,$$

$$I_- = \frac{1}{4}(P_4^2 + Q_4^2 - P_6^2 - Q_6^2) = S_3$$

and

$$\theta_- = \tan^{-1}(Q_4/P_4) - \tan^{-1}(Q_6/P_6) \;. \tag{86}$$

The vibrational dynamics yields to the ATM only approximately because the two vibrations are connected by an approximate adiabatic invariant. However, this invariant is usually a very good constant of the motion, and therefore in many energy regimes, we can display the vibrational dynamics on a sphere of radius I_+. The variation of the energy difference

$$\Delta = \Omega_4 I_4 - \Omega_6 I_6 \tag{87}$$

can be approximated by the generator S_3, which is denoted by I_- in Eq. (86) and which is approximately proportional to one half of the energy difference. Similarly, the generator S_1 approximates (one half of) the vibrational angular momentum, Eq. (78). In most low-energy ranges, the adiabatic invariant is conserved very precisely and the ATM should be very good.

Since K does not change much, we can adopt its value for use in the adiabatic Hamiltonian (85). If we now draw traces on the vibrational sphere for a variety of energies, tilted latitudes appear (Fig. 6) — there is no separatrix, and the contours on the sphere do not resemble Fig. 1(b), which shows three kinds of motion resulting from a nonlinear resonance. We obtain these latitudes because in J-space the constant energy surfaces which determine the traces are tilted planes and do not show the ellipsoids of Fig. 1(b) as would be expected if the dynamics were dominated by Fermi resonances. This difference is intriguing because previously, when we examined the vibrational dynamics of the clearly related Hamiltonian of Eq. (73),

$$H_{\text{rel}} = \frac{1}{2}\omega(P_1^2 + Q_1^2 + P_2^2 + Q_2^2) + k(Q_1^4 + Q_2^4) + k_{12}(Q_1 Q_2)^2 + B\pi_z^2 - 2BJ\pi_z \;, \tag{88}$$

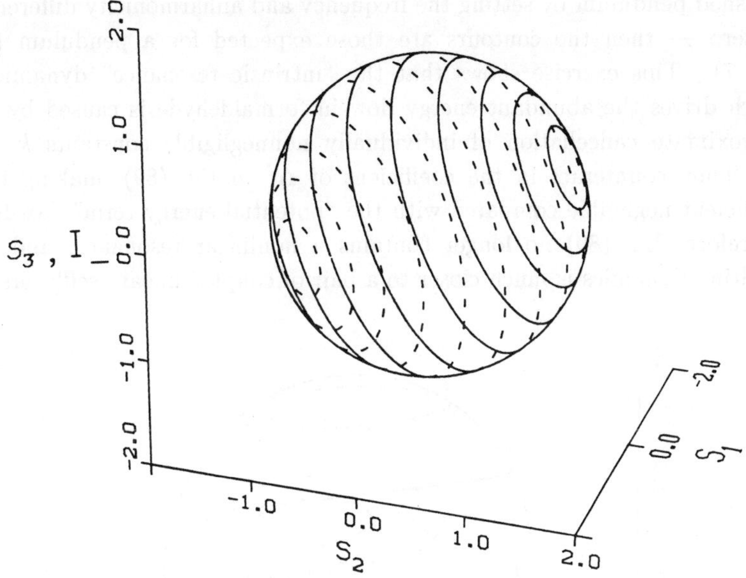

Fig. 6. These series of traces were obtained using the adiabatic Hamiltonian, Eq. (85), and $K = 8$ and $v_4 + v_6 = 2$. Each of the traces correspond to different v_4 and v_6 (allowed to be fractional in this case), and thus to different energies. Clearly, there is no separatrix.

we did obtain the characteristic pendulum contours (Fig. 1(b)) for values of the anharmonicity k, mode excitations and couplings k_{12} and BJ, which were not much different from the formaldehyde case. A clue to the whereabouts of the vanished pendulum is provided by the action-angle version of Hamiltonian H_{rel} ,

$$H = \frac{1}{4}(3k - k_{12} - 2B\zeta^2)\Delta^2 + \frac{1}{8}(k_{12} - 4B\zeta^2)(I^2 - \Delta^2)\cos 4\phi_\Delta$$

$$- 2BJ\zeta(I^2 - \Delta^2)^{1/2}\sin 2\phi_\Delta \ . \tag{89}$$

H_v^{ad} reduces to Eq. (89) for $k_{44} = k_{66} = k$, $k_{46} = k_{12}$, $\Omega_4 = \Omega_6 = \omega$, $c_1 = B\zeta^2$, $c_2 = -2B\zeta$, $K = J$, $I_- = -\Delta/2$, $I_+ = I/2$ and $\Theta_- = -2\phi_\Delta$. When we substitute representative numbers for formaldehyde coupling and anharmonicity constants into Eq. (89), we see that the pendulum (or "asymmetric top") part of the Hamiltonian (which supplies the separatrix) is vanishingly small compared with the Coriolis part, which therefore dominates the dynamics. However, it is still possible to uncover the vanished pendulum by setting the frequency and anharmonicity differences to zero — then the contours are those expected for a pendulum (see Fig. 7). This exercise shows that the "intrinsic resonance" dynamics[29] which drives the abundant energy flow in formaldehyde is caused by the approximate cancellation of individually nonnegligible constants k and k_{12}. They counteract in the coefficient of Δ^2 in Eq. (89), making that coefficient negligible compared with the "potential energy term", $\cos 4\phi_\Delta$. Therefore, Eq. (89) no longer contains a nonlinear resonance, and the resulting dynamics is much closer to a pair of coupled linear oscillators.

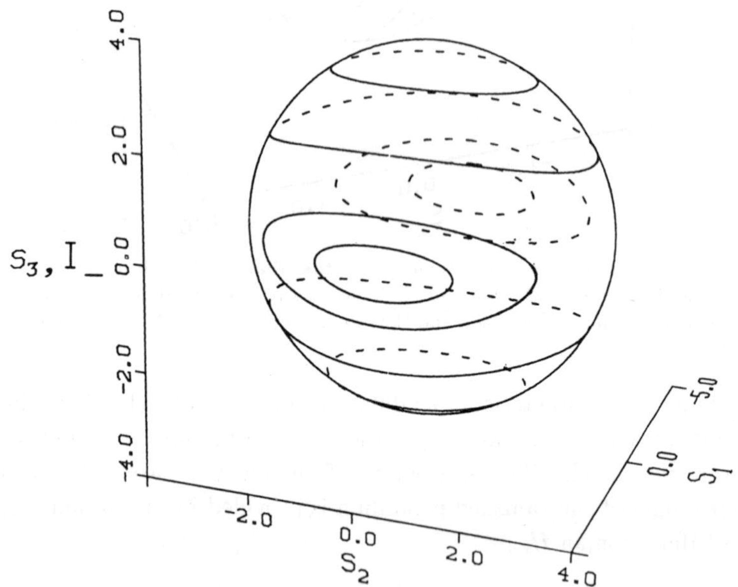

Fig. 7. The result of eliminating Coriolis coupling, taking $k_{44} = k_{66}$ and $v_4 + v_6 = 6$. Note that like in any other asymmetric top, there are two stable axes of rotation and one unstable axis.

10. New Decoupling Schemes for the Vibration–Rotation Hamiltonian — the BT Transformation

Using van Vleck perturbation theory[17] and an harmonic oscillator zeroth-order model, it is often possible to derive effective rotational Hamiltonians for each vibrational state. The two vibrational modes of the Hamiltonian of Eq. (76) are strongly coupled (because the zeroth-order frequencies are nearly degenerate). The coupling is not amenable to perturbation theory. Burleigh *et al.* have shown that if two normal modes are Coriolis coupled by the figure axis component of the angular momentum, the classical Hamiltonian may be rewritten in terms of what they call "Coriolis adapted normal modes" (CANMs).[9] The CANMs are not as strongly coupled as the original normal modes. It has recently been shown that if the quantum mechanical Hamiltonian is written in terms of "vibrational" operators which depend on the components of the angular momentum which couple the original normal mode variables, it is possible to obtain a good zeroth-order rotation–vibration Hamiltonian. Zeroth-order wavefunctions including Coriolis and centrifugal effects are easily obtained from the new Hamiltonian.[57–59] If only the figure-axis component of the angular momentum couples the vibrational degrees of freedom, and if only linear and quadratic terms in the expansion of $\mu_{\alpha\beta}$ are retained and anharmonic effects are neglected, one is able to determine an exact energy level expression for a symmetric top molecule. For asymmetric top molecules the transformation to the angular momentum dependent "vibrational" coordinates introduces small matrix elements which couple effective rotational Hamiltonians for the new vibrational states. These matrix elements may be treated by perturbation theory.

A general two mode effective Hamiltonian is obtained from Watson's general Hamiltonian by progressively eliminating coupling between the vibrational blocks which are not nearly degenerate. We shall illustrate the transformation to angular momentum dependent "vibrational operators" by applying it to a simple two mode Hamiltonian which does not include centrifugal terms. It is somewhat more difficult to transform more complicated effective Hamiltonians. We begin with a two mode effective Hamiltonian

$$H_{\mathrm{vr}} = \frac{1}{2}[\omega_1(P_1^2 + Q_1^2) + \omega_2(P_2^2 + Q_2^2)] + H_{\mathrm{cor}} + H_{\mathrm{rr}} \qquad (90)$$

where H_{rr} is a rigid rotor Hamiltonian and

$$H_{\mathrm{cor}} = -\hbar^2[\mu_{xx}\zeta_{12}^x J_x + \mu_{zz}\zeta_{12}^z J_z]\left(\sqrt{\frac{\omega_2}{\omega_1}}\,Q_1P_2 - \sqrt{\frac{\omega_1}{\omega_2}}\,Q_1P_2\right) . \qquad (91)$$

This Hamiltonian is slightly more complicated than that used to study formaldehyde because it contains two Coriolis terms. In terms of harmonic oscillator raising and lowering operators

$$B_\alpha^\dagger = \frac{1}{\sqrt{2}}(Q_\alpha - iP_\alpha) \,, \tag{92}$$

$$B_\alpha = \frac{1}{\sqrt{2}}(Q_\alpha + iP_\alpha) \,, \tag{93}$$

the Hamiltonian can be rewritten

$$\begin{aligned}
H = {} & \frac{\omega_1}{2}(B_1^\dagger B_1 + B_1 B_1^\dagger) + \frac{\omega_2}{2}(B_2^\dagger B_2 + B_2 B_2^\dagger) \\
& - \frac{1}{i}[X_1(J_z) + Y_1(J_x)] \left(\sqrt{\frac{\omega_1}{\omega_2}} - \sqrt{\frac{\omega_2}{\omega_1}} \right) (B_1^\dagger B_2^\dagger - B_1 B_2) \\
& - \frac{1}{i}[X_1(J_z) + Y_1(J_x)] \left(\sqrt{\frac{\omega_2}{\omega_1}} - \sqrt{\frac{\omega_1}{\omega_2}} \right) (B_1^\dagger B_2 - B_1 B_2^\dagger) + H_{\text{rr}}
\end{aligned} \tag{94}$$

where $X_1(J_z) = \frac{\hbar^2 \mu_{zz} \zeta_{12}^z J_z}{2}$ and $Y_1(J_x) = \frac{\hbar^2 \mu_{xx} \zeta_{12}^x J_x}{2}$.

The vibrational degrees of freedom can be decoupled by employing an operator dependent Bogoliubov–Tyablikov (BT) transformation. The BT transformation is described in Refs. 60 and 61. One introduces new raising and lowering operators A_α and $A_\alpha^\dagger$, which are linear combinations of B_α and $B_\alpha^\dagger$. The new raising and lowering operators are chosen so that they have the same commutation relations as the old ones and so that the Hamiltonian assumes the simple form

$$H = \sum_\mu E_\mu(J_z, J_x)(2A_\mu^\dagger A_\mu + 1) + H_{\text{rr}} \,. \tag{95}$$

The functions $E_1(J_z, J_x)$ and $E_2(J_z, J_x)$ are obtained by solving a 2×2 matrix equation,

$$E_1(J_z, J_x) = \frac{1}{2\sqrt{2}}\sqrt{\lambda_{11}(J_z, J_x) + 8[X_1(J_z) + Y_1(J_x)]^2 + \omega_1^2 + \omega_2^2} \tag{96}$$

$$E_2(J_z, J_x) = \frac{1}{2\sqrt{2}}\sqrt{-\lambda_{11}(J_z, J_x) + 8[X_1(J_z) + Y_1(J_x)]^2 + \omega_1^2 + \omega_2^2} \tag{97}$$

and

$$\lambda_{11}(J_z, J_x) = \sqrt{32(\omega_1^2 + \omega_2^2)[X_1(J_z) + Y_1(J_x)]^2 + (\omega_1^2 - \omega_2^2)^2} \,. \tag{98}$$

In order to calculate rotational matrix elements, to calculate energy levels, or to study energy flow, one chooses the orientation of the molecule-fixed axes at equilibrium so that the Coriolis coupling term, in the rotated coordinate system, is proportional to the projection of the angular momentum on the rotated z axis, $J_{z'}$. A matrix representation of the Hamiltonian is obtained using basis functions which are the products of symmetric top eigenfunctions and the eigenfunctions of $A_\mu^\dagger A_\mu$,

$$A_\mu^\dagger(J_{z'})A_\mu(J_{z'})|n_\mu(J_{z'})\rangle = n_\mu|n_\mu(J_{z'})\rangle \qquad \mu = 1,2 \, . \tag{99}$$

If the rigid rotor Hamiltonian contains terms which are not diagonal in the symmetric top basis (terms which do not commute with $J_{z'}$), the transformed Hamiltonian matrix is not block diagonal. The off-block matrix elements tend to be small and do not have a big effect on the energy levels. If one neglects terms in the Hamiltonian of Eq. (94) with two raising or two lowering operators, one is able to find fairly simple expressions for the off-block matrix elements.[62] Expressions for the off-block matrix elements can be determined from reduced rotation matrices when these are available.[63]

The BT transformed Hamiltonian enables one to decouple nearly degenerate vibrational states, and thereby provides a new zeroth-order picture which incorporates rotation–vibration coupling.

11. Conclusion

In this chapter, we illustrated the rich dynamics of vibrational energy flow generated by rotation–vibration Hamiltonians. The classical mechanics of coupled oscillators can be invoked to distinguish two energy transfer mechanisms: Harmonic modes exchange energy simply and abundantly because energy flow does not detune the resonances between them. In contrast to this mode-mixing mechanism, energy flow between coupled anharmonic modes is more complicated because frequencies of the oscillators are determined by their excitation, and resonances between modes are established and destroyed by a subtle feedback mechanism: Energy flows because there is a resonance, but the same energy transfer destroys the resonance, thus restricting further flow.

The factors driving the dynamics can be unified in a compact picture by converting a properly decoupled molecular Hamiltonian to a hindered rotor or an asymmetric top using action-angle variables. This picture shows how the energy imbalance [approximated by the hindered rotor momentum p in Eq. (33) or the quantity I_- in Eq. (86)] changes in time by the disparate factors like anharmonicities, couplings, and initial excitations. It is very difficult to obtain such insights from the analytical form of the Hamiltonian or from extensive numerical studies using it.

We presented examples for both energy transfer regimes: Energy flow through a nonlinear resonance was illustrated on the model triatomic ABA containing a 1:1 resonance, and mode mixing on two coupled low-frequency modes of formaldehyde. While the behavior of realistic, multimode molecular Hamiltonians is considerably more complicated because of the number and dynamical nature of Fermi resonances, the models introduced in this chapter remain useful qualitative guides to energy flow scenarios in vibration–rotation Hamiltonians.

References

1. H.-L. Dai, C. L. Korpa, J. L. Kinsey, and R. W. Field, *J. Chem. Phys.* **82**, 1688 (1985).
2. For a recent review, see T. Uzer, *Phys. Rep.* **199**, 73 (1991).
3. For example, A. E. W. Knight, *Excited States* **7**, 1 (1988).
4. W. B. Clodius and R. B. Shirts, *J. Chem. Phys.* **81**, 6224 (1984).
5. T. Uzer, G.A. Natanson, and J. T. Hynes, *Chem. Phys. Lett.* **122**, 12 (1985).
6. J. H. Frederick, G. M. McClelland, and P. Brumer, *J. Chem. Phys.* **84**, 190 (1985); J. H. Frederick and G. M. McClelland, *J. Chem. Phys.* **84**, 876; 4347 (1986).
7. G. S. Ezra, *Chem. Phys. Lett.* **127**, 492 (1986).
8. S. K. Gray and M. S. Child, *Mol. Phys.* **53**, 961 (1984).
9. D. C. Burleigh, R. C. Mayrhofer, and E. L. Sibert III, *J. Chem. Phys.* **89**, 7201 (1988).
10. E. L. Sibert III, *J. Chem. Phys.* **90**, 2672 (1989).
11. D. K. Sahm and T. Uzer, *J. Chem. Phys.* **90**, 3159 (1989).
12. S. K. Gray and M.J. Davis, *J. Chem. Phys.* **90**, 5420 (1989).
13. D. K. Sahm, S. W. McWhorter, and T. Uzer, *J. Chem. Phys.* **91**, 219 (1989).
14. R. Parson, *J. Chem. Phys.* **91**, 2206 (1989).
15. W. Gordy and R. L. Cook, *Microwave Molecular Spectra, Techniques of Chemistry Vol XVIII* (John Wiley & Sons, New York, 1984).
16. I. M. Mills in *Molecular Spectroscopy: Modern Research*, eds. K. Narahari Rao and C. W. Matthews (Academic Press, New York, 1972) p. 115.
17. D. Papousek and M. R. Aliev, *Molecular Vibrational–Rotational Spectra* (Elsevier, Amsterdam, 1982).
18. P. R. Bunker, *Molecular Symmetry and Spectroscopy* (Academic Press, Toronto, 1979).

19. E. B. Wilson Jr., J. C. Decius, and P. C. Cross, *Molecular Vibrations* (Dover, New York, 1980).

20. M. R. Aliev and J. K. G. Watson, in *Molecular Spectroscopy: Modern Research Vol III*, ed. K. Narahari Rao (Academic Press, New York, 1985) p. 1.

21. H. A. Jahn, *Phys. Rev.* **56**, 680 (1939).

22. G. Herzberg, *Infrared and Raman Spectra* (Van Nostrand Reinhold, New York, 1945) p. 374 ff.

23. For example, J. H. Frederick, G. M. McClelland, and P. Brumer, *J. Chem. Phys.* **83**, 190 (1985).

24. G. N. Natanson, *Mol. Phys.* **66**, 129 (1989).

25. B. V. Chirikov, *Phys. Rep.* **52**, 263 (1979).

26. A. J. Lichtenberg and M. A. Lieberman, *Regular and Stochastic Motion* (Springer, New York, 1983).

27. G. M. Zaslavskii, *Chaos in Dynamical Systems* (Harwood, New York, 1985).

28. P. Brumer, *Adv. Chem. Phys.* **47**, 201 (1981).

29. C. C. Martens and G. S. Ezra, *J. Chem. Phys.* **83**, 2990 (1985); *ibid.* **86**, 279 (1987).

30. L. M. Milne-Thomson, in *Handbook of mathematical functions with formulas, graphs and mathematical tables*, eds. M. Abramowitz and I. A. Stegun, National Bureau of Standards Applied Math Series No. 55 (U.S. Department of Commerce, Washington, DC, 1972) chapt. 17.

31. G. M. Zaslavskii and N. N. Filonenko, *Sov. Phys. JETP* **25**, 851 (1968).

32. E. L. Sibert III, W. P. Reinhardt, and J. T. Hynes, *J. Chem. Phys.* **77**, 3583 (1982).

33. J. A. Combs and W. G. Hoover, *J. Chem. Phys.* **80**, 2243 (1984).

34. S. D. Augustin and W. H. Miller, *J. Chem. Phys.* **61**, 3155 (1974).

35. S. D. Augustin and H. Rabitz, *J. Chem. Phys.* **71**, 4956 (1979).

36. G. S. Ezra, *Chem. Phys. Lett.* **127**, 492 (1986).

37. W. G. Harter and C. W. Patterson, *J. Chem. Phys.* **80**, 4241 (1984).

38. M. E. Kellman, *J. Chem. Phys.* **76**, 4528 (1982); *ibid.* **83**, 3843 (1985); *ibid.* **85**, 6542 (1986); K. K. Lehmann, *J. Chem. Phys.* **79**, 1098 (1983).

39. O. S. van Roosmalen, F. Iachello, R. D. Levine, and A. E. L. Dieperink, *J. Chem. Phys.* **79**, 2515 (1983); O. S. van Roosmalen, I. Benjamin, and R. D. Levine, *J. Chem. Phys.* **81**, 5986 (1984); R. D. Levine and J. L. Kinsey, *J. Chem. Phys.* **90**, 3653 (1986).

40. W. G. Harter, *J. Chem. Phys.* **85**, 5560 (1986).

41. W. G. Harter, *Comput. Phys. Rep.* **8**, 319 (1988).

42. H. Hopf, *Math. Ann.* **104**, 637 (1931); H. V. McIntosh, *Am. J. Phys.* **27**, 620 (1959); V. A. Dulock and H. V. McIntosh, *Am. J. Phys.* **33**, 109 (1965).

43. D. Farrelly, *J. Chem. Phys.* **85**, 2119 (1986).

44. M. E. Kellman and E. D. Lynch, *J. Chem. Phys.* **88**, 2205 (1986); *ibid.* **89**, 3396 (1988).

45. C. Jaffe, *J. Chem. Phys.* **89**, 3395 (1988).

46. L. Xiao and M. E. Kellman, *J. Chem. Phys.* **90**, 6086 (1989).

47. H. Goldstein, *Classical Mechanics*, Second Edition (Addison-Wesley, Reading, 1980).

48. J. Schwinger, in *Quantum Theory of Angular Momentum*, eds. L. C. Biedenharn and H. van Dam (Academic Press, New York, 1965), p. 229.

49. R. N. Zare, *Angular Momentum: Understanding Spatial Aspects in Chemistry and Physics* (Wiley, New York, 1988).

50. V. Postell and T. Uzer, *Phys. Rev.* **A41**, 4035 (1990).

51. H. A. Kramers and G. P. Ittman, *Z. Phys.* **53**, 553 (1929); **58**, 217 (1929); **60**, 663 (1930).

52. L. D. Landau and E. M. Lifschitz, *Mechanics* (Pergamon, New York, 1980).

53. D. K. Sahm, R. V. Weaver, and T. Uzer, *J. Opt. Soc. Am.* **7**, 1865 (1990).

54. D. K. Sahm and T. Uzer, *Chem. Phys. Lett.* **163**, 5 (1989).

55. M. Aoyagi and S. K. Gray, *J. Chem. Phys.* **94**, 195 (1991).

56. C. H. Townes and A. L. Schawlow, Microwave *Spectroscopy* (Dover, New York, 1975).

57. M. S. Krishnan and T. Carrington, Jr., *J. Chem. Phys.* **94**, 461 (1991).

58. M. S. Krishnan and T. Carrington, Jr., *J. Chem. Phys.* **95**, 1884 (1991).

59. M. S. Krishnan and T. Carrington, Jr., *J. Chem. Phys.* **98**, 83 (1993).

60. N. N. Bogoliubov, *Lectures on Quantum Statistics, Vol. 1. Quantum Statistics* (Gordon and Breach, New York, 1967); S. V. Tyablikov, *Methods in the Quantum Theory of Magnetism* (Plenum, 1967).

61. An alternative derivation of the Coriolis-adapted normal models originates from the "cranked oscillator" model of nuclear physics. For the solution of an anharmonic oscillator with a Coriolis term, see P. Ring and P. Schuck, *The Nuclear Many-body Problem* (Springer, New York, 1980) p. 133, as well as the article by M. A. Z. Habeeb, *J. Phys. G: Nucl. Phys.* **13**, 651 (1987).

62. Formulae are given in M. S. Krishnan and T. Carrington, Jr., *J. Chem. Phys.* **95**, 1884 (1991).

63. C. C. Martens, *J. Chem. Phys.* **96**, 8971 (1992).

CHAPTER 28

THE EXTRACTION OF DYNAMICS
FROM COMPLEX MOLECULAR SPECTRA IN CASES
WHERE THE CLASSICAL MOTION IS CHAOTIC

Howard S. Taylor

Department of Chemistry
University of Southern California
University Park Campus, Los Angeles, CA 90089-0482, USA

Contents

1. Introduction

A reader performing a quick perusal of lists of titles appearing in the
chemical, physical, mathematical, biological, and engineering scientific
literature cannot fail to note that in recent years there has been a virtual

explosion in the use of the word "chaos". Such diverse phenomena and fields as heartbeats, brain waves, weather patterns, turbulent flows, periodic chemical reactions, and particle accelerator design have all felt the impact of this relatively new concept. A book, *Chaos, the Making of a New Science*,[1] has been written for public consumption and recommended as a source. In this article we try to heuristically explain the connection between concepts underlying spectroscopy as used by chemists, chemical physicists, and molecular physicists with ideas from nonlinear dynamics. The result will be the beginning of an understanding of how chaotic motion impacts on atomic and molecular spectroscopy. As always in spectroscopy and scattering experiments, the problem will be to go from observations in the form of spectral lines and bands, to a model of the molecule and its motions. The model of the molecule and the types of molecular motions are what we aim to learn. The reason we believe in our model is that when the equations of quantum and/or classical mechanics are applied to the model, the experimentally measured spectrum, or features thereof, are reproduced. The common feature that a spectroscopic and scattering problem has with the other fields that bear the "chaos" label is that the differential equations describing the classical motion (in our case, the equation of classical mechanics) are nonlinear and have at least one class of solutions that are exquisitely sensitive to initial conditions. In what follows, we shall attempt to describe what a chaotic spectrum is and how we use qualitative ideas carried over from the mathematical theory of classical and quantum chaos[2] to extract a model of the molecular motion.

Immediately, because we are dealing with chaos, it is necessary to add a long caveat to our "promise" to supply a model of the motion. As is clear, classical chaotic motion is complicated and quantization, even with its averaging over regions of phase space of measure $\hbar$, never makes the motions simple. As such the models we shall supply will be low resolution models. The total motion shall be divided into repetitive regular parts which cause higher intensity low resolution spectral features and chaotic, erratic parts to whose quantization in a generic sense is found by observation to correspond to a region of the spectrum where one finds in high resolution studies a very dense, complex and difficult to interpret (i.e., unassignable) set of transitions. In fact, in the work of the USC group, some of which is covered in this review, the detailed chaotic motion and the nature of the individual eigenstates are generally ignored. Often, the actual work is purely classical and, therefore, intrinsically is a smoothing of the discreteness of all functions (as correlation functions) that depend on the quantum spectra. The loss of information is not as bad as it first sounds. Mandelbrot's[3] analogy to beaches and coastlines is appropriate here. One can describe beaches by specifying the coordinates (and even momenta) of each grain of sand. Clearly this is not desirable from the information storage and human conceptualization point of view. One can alternately specify low resolution and smoothed local features such as dunes and the slope and curvature of the coast. This latter view will be the preference of this review. This is not to say that more detailed studies are not useful. A large school of work recognizes the complexity of eigenvalues, spectra, and wavefunctions as an opportunity to apply statistical methods. The aim is to associate statistical features such as the distribution of spectral energy spacings, extracted from complex experimental or theoretically calculated spectra with certain types of motion and localization mechanisms.[4] As interesting as the results are, they do not yet give answers in a language translatable into a knowledge of molecular motions as required by chemists. Simply put, they never talk about specific motions of electrons or atoms in molecules. As such, this large school of work is not considered here. We also do not cover here periodic orbit theory.[2] This school is concerned with the semiclassical quantization of chaos. It yields powerful theoretical insights into fundamental questions on the relationship between classical and quantum dynamics but is not yet a practical computational tool to compute energy levels of spectra. Most of the applications are to models or at best to two electron atoms.

For us a necessary condition for having a fully chaotic spectrum is that it be congested and complex. The lines should be transitions between two groups of states, one or both of which cannot be assigned quantum numbers other than that for the energy and total angular momentum. As such, the spectral lines are intrinsically unassignable. It is also fair to say that the inability of spectrocopists to make an assignment only hints, but does not demonstrate, that a spectrum is chaotic. The final word comes when interpretative methods that assume that the spectrum is chaotic succeed in explaining features of the spectrum. Chaotic spectra have been seen in many systems, the most famous of which are the rovibrational spectrum of the ground state of acetylene around $28\,000$ cm^{-1},[5] the photodissociation spectrum[6] of excited predissociating H$_3^+$ and the rovibrational spectrum above 250 cm^{-1} of the sodium trimer[7] Na$_3$. In atomic physics the best known chaotic spectra are those involving transitions ending near the ionization threshold of the hydrogen atom in a 6-T magnetic field.[8] Generally, chaotic spectra appear when highly excited electronic and/or rovibrational states are probed. In floppy molecules like Na_3, in big molecules, and in nonrigid clusters, chaos and chaotic spectra appear at surprisingly low energies. In the process of spectroscopically identifying reactants, products, and transition states of chemical reactions, chaotic spectra appear.[6] The chemical dynamics then involves chaotic dynamics and, hence, our added interest in these cases.

The question addressed here is how information about the motions of the system's constituent particles can be extracted from a chaotic spectrum. To prepare to answer this question, Sec. 2 of this review gives a highly personal, physically motivated and heuristic view of regular and chaotic motion and its effect in quantum mechanics. [Note: its effect is as close as we will ever come to defining "quantum chaos".] After this section, follows Sec. 3 which uses the example of Na_3 as an illustration of our first mechanism that causes chaos and of the success of a purely classical method of interpretation. Section 4 will briefly discuss H_3^+ and will demonstrate how a periodic orbit can control dynamics. Section 5 is a theoretical exercise carried out by Gomez-Llorente *et al.*,[9] that simulates an LiCN–LiNC hypothetical SEP spectrum and shows how low resolution analyses can extract dynamics and information about IVR from such data. Section 6 will discuss SEP spectra of acetylene and will demonstrate another route to chaos and its connection to a traditional method of molecular spectral analysis that is applied in relatively highly excited, but not chaotic, regions. In short, all the methods will explain the low resolution spectra obtained by smoothing the high resolution spectra in the chaotic regime. A short summary and conclusion ends the review.

2. Regular and Chaotic Spectra

2.1. *Regular Spectroscopy, Regular States, and Regular Motions*[10]

Spectroscopy, as most chemists know it, is regular spectroscopy and refers to that type of atomic and molecular spectroscopy typically covered in the books of G. Herzberg.[11] Regular spectroscopy assumes that the initial and final states of a transition can be assigned N (equal to the number of degrees of freedom) quantum numbers. More fundamentally, it assumes that there exists a coordinate system, albeit convoluted or even non-analytically related to the usual textbook coordinates, in which the major part H_0 of the Hamiltonian H of the system, is essentially separable. For vibrational motions,[12] such coordinate systems are normal modes, local modes, mixed modes, etc. Since the system is essentially separable, quantum mechanics tells us that each degree of freedom (ith mode) has its own associated quantum levels n_i and its own one degree of freedom wavefunctions $\psi_{n_i}^0$ with $n_i - 1$ nodes. The states of H_0 have a wavefunction $\psi_{\bar{n}}^0$ which is the product of the N one-mode wavefunctions $\psi_{n_i}^0$. If n_i is the quantum number associated with the state of the ith degree of freedom, then $\bar{n} = (n_1, n_2, \ldots, n_N)$ is the assignment of the state of H_0. Regular spectroscopy assumes that $\bar{n}$ can be used to label (regular) states of the full system controlled by H in the sense that a perturbation theory starting with H_0 and $\psi_{\bar{n}}^0$ will converge to an exact unique eigenstate $\psi_{\bar{n}}$ of H. If two or more different H_0's and $\psi_{\bar{n}}^0$'s give convergent results, the one with the biggest integral overlap of $\psi_{\bar{n}}^0$ and $\psi_{\bar{n}}$ converges faster and is the better assignment. The pattern of states of the system, and therefore the pattern of lines in the spectrum, can then be related to changes in the various n_i quantum numbers.

As stated, chaotic molecular spectroscopy typically involves high densities of lines and, therefore, a high density of states. This latter means that we are working in the correspondence region and that quantum guided classical (and vice versa) ideas will be used to interpret chaotic spectra. For this reason, and because particle motions are thought of classically, it behooves us before discussing chaos to review first the semiclassical view of a regular quantum state $\bar{n}$ and to consider the associated classical motions. Semiclassical mechanics associates the state $\bar{n}$ with a very particular "quantizable" trajectory that moves in the $2N$-dimensional phase space of positions $\bar{q}$ and momenta $\bar{p}$ of the vibrating atoms. If the trajectory is "regular", N conserved momenta exist and the motion is restricted to the surface of an N-dimensional torus (the surface of a donut is a 2D torus) which, itself, lies in the hypersurface defined by the trajectory's constant energy, $E = H(\bar{p}, \bar{q})$. In the regular regions of phase space, most trajectories are "regular" as defined above.

Phase space is essentially filled with such trajectories on tori, but the $\bar{n}_i$ quantum state is associated with that particular torus (trajectory) that has N independent constants of the motion called actions, $\bar{J}$ (which all tori in the regular region have) that satisfy $J_i = (n_i + \frac{1}{2})$ where J_i is computed as in Eq. (1) from the trajectories motion and n_i is the "integer" mode quantum number:

$$J_i = \oint_{C_i} \bar{p} \cdot d\bar{q} \qquad i = 1, \dots, n \ . \tag{1}$$

Here, each C_i is a closed path around the torus in phase space. N independent C_i can be found for regular motion. All tori, including our quantizable ones, when projected into the coordinate plane, reveal a quasiperiodic (almost but never quite periodic) motion.

Figure 1a shows for the Na_3 trimer a projection onto a surface in coordinate space of a quasiperiodic orbit moving on the torus in phase space. Here, a time-stepped reading of the trajectory's coordinates would enable visualization of a simultaneous bending and symmetric stretch of this obtuse triangular molecule. In the figure, the usual "arrow" diagrams are given for the two modes in the usual displacement coordinates.[11] Our figure looks different because Fig. 1a is in internal Jacobi Coordinates. What is generic to regular motion is the almost woven thread pattern of the trajectory. In molecules, regular quantum states generally and roughly occur in the same energy and mode coupling ranges in which classical tori (quasiperiodic trajectories) exist. When there are differences almost always classical regularity will disappear before its quantum analog. This is caused by the fact that the $\hbar$ measure of quantum mechanics smoothes those features in phase space that "cause" the transition to chaos. Generally, these chaos causing features and regions in phase space increase in size as the energy (coupling among modes) increases and, thereby, become visible to the quantum dynamics. Regular wavefunctions have their regions of large amplitude in the same region of coordinate space as the quantizable quasiperiodic trajectories; off the torus, the wavefunctions damp exponentially.

2.2. *Chaotic Spectra, Chaotic States, and Chaotic Motions*

As the energy increases three mechanisms or combinations thereof cause the breakdown of regularity. Sometimes, as in the case of Na_3, the potential can change abruptly causing the "old" H_0 to no longer represent the motion. Unless a new H_0 exists with near separable properties in coordinates that describe local regions of phase space large enough to hold quantizable tori, regularity is lost and technically chaos sets in. This can also happen if the

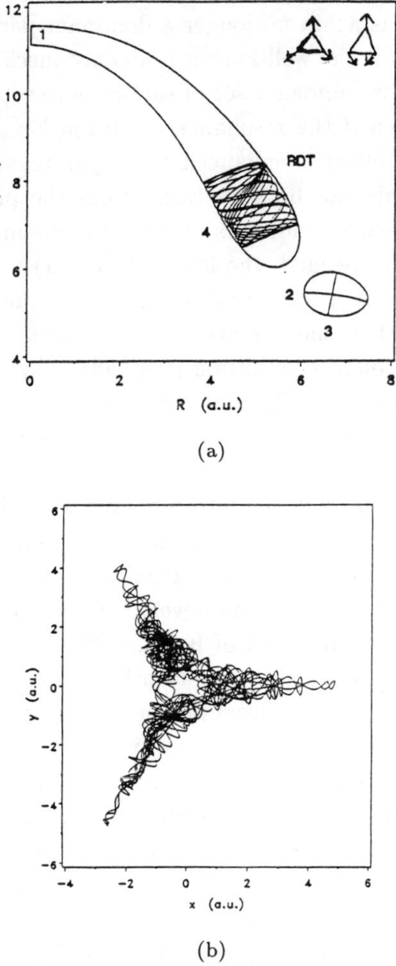

Fig. 1.(a) A quasiperiodic trajectory computed at $E = 575$ cm^{-1} and $J = 0$ for Na$_3$ represented by the potential surface of Ref. 43. R and r are mass scaled Jacobi coordinates, and the solid outer line is an equipotential for Na$_3$ held in C$_{2v}$ position. The dark line through the quasiperiodic trajectory is a symmetric stretch periodic orbit typical of many periodics orbits (tori in 1D) which have been tested and found not to contribute significantly to the spectra. In the upper-right corner are, in the more recognizable displacement coordinates, pictures of the coupled motions represented by the trajectory. (b) A chaotic trajectory for Na$_3$ run at $E = 575$ cm^{-1}. It is generic to the region of the experiment. Figures 1a and 1b differ in initial conditions.

coupling between the once separable coordinates increases sufficiently with energy such that H_0 is again no longer a dominant part of the dynamics. The third mechanism is the well-known resonance mechanism. Here, if the mode frequencies ω_i are among a set of say m, once separable modes, and if n_i are integers, then if the resonance condition $\Sigma n_i \omega_i = 0$ is satisfied, the mode interaction effect is enhanced and again regularity breaks down for the given H_0. This can be seen easily from the perturbation theory point of view where such conditions cause zero denominators for various order (the lower the n_i needed, the lower the order) terms and the series diverges. Again, chaos is the breakdown of any possible perturbation theory associated with a single state of any possible H_0. The chance of an occurrence of a resonance condition is enhanced by two considerations. First, as molecules get larger there are more ω_i, and as such their chance of resonating is therefore greater. Second, the ω_i, due to anharmonicity, can be considered functions of energy. As such, as the energy increases, the $\omega_i(E)$ decrease in size and fall into resonance. At this point, although regularity is lost in the sense that the original H_0 and its quantum numbers are no longer descriptive of the motions, regularity may not be lost because, as has been shown by Chirikov[13] and developed for molecular spectroscopy by the groups of Reinhardt[14] and of Kellman,[15] a transformation can be made to other coordinates (generically to other action-angle variables) for which a new H_0 can be found at least in a large enough part of phase space to hold quantum states. An example or this is the normal to local mode transformation and the well-known $2:1$ or $2:3$, etc., Fermi resonances.[15-17]

In such regions now, the quantum states are regular but different (often interspersed) states can be better represented by one or the other type of H_0 and its associated quantum numbers. Also, in such energy regions, different types of regular tori exist in phase space and the quantum states associated with the quantizable subset of these tori have wavefunctions that are localized over the region of configuration space that is the projection of the tori onto configuration space. The wavefunctions, again, are regular in the sense that along appropriate (usually now curvelinear) coordinates the wavefunctions have nodes that correlate with the quantum numbers. Hence, even in the regular region successive regular states can have different shaped regular wavefunctions.[18]

As energy increases, again, depending on the strength of the mode coupling, regularity can definitively be lost but now the mechanism is that of overlapping resonances;[13] that is, several different resonances at the same energy exist and are coupled in the same region of phase space and, therefore, one can no longer find a single H_0. Wherever any of these three above mechanisms dominate the dynamics, chaos sets in. When this happens, our regular tori and states begin to disappear and many things now happen simultaneously.

Most importantly, the density of states increases rapidly. That this will occur can be seen from several points of view. First, with increasing energy there are many more degenerate and near degenerate states of H_0 which will split upon strong perturbation. A second point of view is that as energy increases so does the size of the energy shell in phase space. As such, when its volume is divided by $\hbar^{2N}$ to give the number of quantum states, this number becomes large. The third view is the quantum chemists' point of view. At high energy not only is the potential larger and there are no effective constants of the motion (quantum number) to restrict the space, but the kinetic energy can increase greatly. The latter requires a high frequency oscillation in the wavefunction. Both considerations indicate that many very localized simple, say Gaussian-type, basis functions would be needed to span the configuration space. The strong coupling would mix these and give rise to a large multitude of states. In any case, the density of states becomes higher in the chaotic region. Moreover, the states themselves are no longer dominated by a single $\psi_{\bar{n}}^0$, and as such the wavefunctions can only be labeled by a few $(< N)$ such quantum numbers as energy and total angular momentum. The wavefunctions also become erratic in nodal pattern, highly oscillatory in amplitude, and generally delocalized, being confined only by the potential itself. This wavefunction complexity is easily anticipated from our above discussion on resonances. For example, when two resonance zones overlap, one expects the resulting wavefunctions to reflect the mixing of the respective resonance wavefunctions. Since these latter functions have different and, in some cases, already complicated nodal patterns the result is essentially no global nodal patterns.

In classical mechanics, irregularity starts with stochastic layers forming near the separatrix that borders the regular regions of phase space. The separatrix that surrounds the now ever smaller regular regions, as always, connects to unstable hyperbolic periodic orbits at the edges of the region. The stochastic layers get thicker as the chaos causing mechanisms get stronger. If two such layers overlap or if the layer seems to have broken and to have crossed itself, chaos begins to dominate in this region of phase space.

In this more chaotic region the quantum wavefunctions tend to localize about these hyperbolic periodic orbits as the regular region disappears. As chaos increases, the wavefunctions begin to spread or "loan amplitude" to regions in the stochastic layers about the separatrix. Further increases in chaos see the wavefunctions begin to spread off the single resonance region's periodic orbit following the growth of the broken separatrices. When the wavefunctions following the separatrices, whether from the same or different resonance zones, overlap any high degree of localization is truly lost and trajectories become delocalized and eventually "spill out" into phase space. The wavefunctions then achieve a very delocalized low amplitude nature and have a complex nodal pattern. In our model, mixing between two resonance structures is described although it may take several more of these mixing of resonances to untrap all modes and to see the types of spreading of chaos described above.

Figure 1b shows a chaotic trajectory in the case of Na_3.[19] The propeller shape is due to the trigonal nature of the bounding potential and reflects the possibility of sodium interchange. With chaos, the one-to-one correspondence between trajectories (tori) and states is lost, as is "assignability". Transitions into such a region yield congested, unassignable chaotic spectra. The trajectories and particle motions, the wavefunction's nodal patterns and amplitude gyrations, and the spectral lines become so complex that they defy systematic cataloging or categorizing. Statistical patterns and distributions of lines which are detached from dynamic interpretations are used for a minimal and relatively uninformative categorization of the spectra.

It would seem that nothing dynamically worth knowing can be learned from chaotic spectra. This would remain true were it not for two saving graces. First, in our description of the onset of chaos, via our first two mechanisms, we assumed that as the potential (coupling) grew larger in strength and size, it affected all coordinates at all their values at once. Fortunately, this does not happen for the cases studied so far until such high energies that dissociation and/or ionization has occurred. What does happen?[20-22] Simply put, in some cases the potential coupling and changes cause only some degrees of freedom to go chaotic. The mechanical results of this can be anticipated from our above discussion. Depending on the system, tori can still exist in $n = 1$ or $n = 2$ or $...n = N - 1$ reduced dimensions; we call these reduced-dimension tori. For $n = 1$ the reduced-dimension torus is a periodic orbit. Most N-dimensional trajectories act chaotically in all degrees of freedom except when they approach the region of phase space where the reduced-dimension tori exist. At this point, they move nearly quasiperiodically in the reduced-dimension space, e.g., they mimic the motion on the reduced-dimension tori for several picoseconds before they leave the region and again act chaotically. As such, a subset of almost regular motions exists for this short time. Correspondingly, the wavefunction looks chaotic (low amplitude, highly oscillatory, no simple model patterns) and is unassignable except in those regions of coordinate space and energy where the quantizable reduced dimension tori exist; there it looks regular and has much bigger amplitude (square root of the probability) than in other regions of space and energy. Since most of the trajectory and ψ is "chaotic", the state density is still high due to the coupling among the chaotic coordinates and the ranges of the regular coordinates not near the reduced-dimension tori. The high-resolution spectrum is, however, still uninterpretable. This is unfortunate as there exists a simply understood motion, namely, that almost-quasiperiodic motion that mimics the reduced-dimension tori. This regular motion *inserted into the chaotic trajectory* is itself a short-lived (here picoseconds) species.

Like motion in all short-lived species, e.g., transition states, resonant scattering states, etc., knowledge of the inserted regular motion is worth having. What can one do to go from the experimental spectrum to the knowledge of the motion of the reduced-dimension tori? The answer comes from the essence of chaos. Chaotic trajectories (or more precisely, the parts of a trajectory that are chaotic) are greatly different if the initial conditions change and, in particular, if the energy changes. Likewise, chaotic wavefunctions change greatly with energy in regions not located over configuration-space projections of the reduced dimension tori. Low-resolution experiments that average over energy should (because they average the different erratic motions or wavefunction oscillations) de-emphasize the chaotic motions, and emphasize the regular spectra of the regular inserted motions that mimic the reduced-dimension tori.

The second "saving grace" is most appropriate for chaos mechanisms two and three, i.e., increased coupling and mixing and overlapping resonances. Here, again, low resolution will come to the rescue from the point of view of retrieving from the chaotic situation some ability to extract dynamics. In essence, what happens here is that when conditions of coupling and anharmonically shifted frequency allow, two (or three or more) resonances (Fermi, Denison–Darling),[11] each involving a subset of not necessarily all different modes, come into play and overlap classically; chaos sets in. This occurs because generically no new set of even local action angle variables can be found in which the system is still quasiseparable. Now depending on the coupling, the chaotic trajectories, when near that part of phase space containing the resonance zones (i.e., regions of phase space where the resonance has a strong effect) imperfectly mimic the now defunct tori that were, at lower coupling and energy, typical of the zone. When not in the zones the trajectories are chaotic (i.e., sensitive to initial conditions) and if one waits long enough, fill phase space. The quantum analog can best be explained by describing how the wavefunction and final spectrum form from the simpler uncoupled spectra by mixing between tiers of states localized in separate regions of space. First, one imagines that when the configurations of H_0 are allowed to mix separately via the respective resonance, a wavefunction is produced in each zone whose Husimi transform[23] (the analogue of the wavefunction in phase space) resembles the topology of the resonant torus (not the H_0 torus). We are now at "stage" 1. Next, one imagines the mixing of the wavefunctions from each zone and at roughly the same energy to give new wavefunctions localized in the region of phase space corresponding to the two zones. This wavefunction is, as discussed above with reference to nodal structure, already quite complex albeit localized and of moderate amplitude. The resulting wavefunctions at this second stage are irregular in the modes that have undergone resonance but may still have some good quantum numbers in the modes that haven't mixed, i.e., fallen into resonance. The last mixing is with the large number of basis states that span the regions of phase space in which chaos exists. This third stage, as in our first mechanism, causes a large and dense number of unassignable states with wavefunctions that, inside the multiple resonance zone, still have the same features as in the previous stage and are relatively large in amplitude compared to the even more complex lower amplitude sections outside the overlapping resonance region. As such, and for the same reason as before, one sees here in a lowered resolution a spectrum representative of the resonances that are mixing with each other. Further lowering of the resolution may even untangle these mixings to spectral structures representative of single resonant zones.

As energy increases one can envision a third, then a fourth, etc., resonance coming into play. At each stage similar arguments say that fitting the low resolution spectra to resonant Hamiltonians (by, as usual with resonances, adjusting the linear strength parameter in front of the resonant term in the Hamiltonian, such that when a diagonalization is made the spectrum is reproduced) will allow one to reveal which resonances are coming sequentially into play.[17] This of course allows a mapping of mode mixing. If an experiment excites a particular subset of modes then a knowledge of the resonance shows how energy can transfer to other modes. This, therefore, shows how Intramolecular Vibrational Redistribution of energy (IVR) works.[24] Often, the excitation, like the classical trajectory, could remain trapped for a short period of time in the region of phase space occupied by the interacting strong resonances. Only after a delay (not unlike a scattering resonance lifetime) will the system really "feel" the quasicontinuum and mix with it. When it does, the system (wave packet) fills all of phase space as does the chaotic classical trajectory. If, at a given excitation, one assumes that the first resonance causes the biggest splitting then, under fortunate circumstances, studying the system at a succession of higher resolutions[9,25] will yield a succession of spectra that show the modes that are accessed successively in time. An example of some of these ideas will be seen to be the acetylene SEP and DF spectra that will be discussed briefly in Sec. 6. Here, not only will a flow into a vibrational mode be seen but other types of motional couplings will also show up in low resolution.

An added useful way to consider these processes is worth mentioning here. As stated, classical chaotic trajectories get trapped in quasiregular regions where they mimic tori (disintegrating tori, resonant, nonresonant and reduced dimensional) and periodic orbits. For simple cases[26] one can show, and in more complicated cases infer, that the trapping is due to local dynamic potentials that "cause" the tori and periodic orbits, etc. In quantum theory, evolving and spreading wave packets get trapped by these same potentials. Such trapping of packets for short periods of time can easily be shown to give rise to eigenfunctions that not only are localized in these local dynamic potentials but resemble locally the eigenfunctions of the local dynamic potential.

Another feature and complication must be addressed and added to our discussion. It was left out because it would complicate our already complicated "simplified" pedagogical and heuristic discussion. This feature, which will be illustrated in the discussion in Sec. 4 of the theoretical SEP spectra of LiCN,[9] is anticipated by our above discussion when we pointed out that in our "thought" picture mixing mixed not just modes, but effectively different regions of configuration of space. We saw that embedded in the very chaotic parts of phase space were structures that ranged from islands containing tori, to quasiregular (reduced-dimension tori), to coupled but localized and simply parametrized chaos due to overlap of a few resonances. In short, if there was an index that measured the position on a "regularity to chaos" axis at a given energy, it would depend on the position in phase or configuration space. Now, recalling that different spectra access different regions of phase space, one concludes that the same state may be judged locally "chaotic" or locally regular depending on where it is accessed. Several different regular features existing at different configurations, respectively, can show up in the chaotic wavefunction. Between such features the wavefunction could well be of low intensity and reminiscent of a fully mixed state. Such is the case in LiCN where over the LiCN and LiNC wells a local regularity exists up to an energy well above isomerization that is typical of the expected regular local modes. These same states, when probed in the region of the isomerization barrier in high resolution, appear chaotic and when taken together in low resolution with other states show a reduced dimension regular feature. This feature is characteristic of a trapping of the system in the transition state at the saddle point that leads to isomerization. In short, the degree of regularity or irregularity in a spectra depends not only on energy but on the configuration and phase space region the spectra accesses.

The examples of the next few sections are intended to illustrate some of these ideas and then some varied classical, quantum and mixed classical and quantum methods used to extract the dynamic information or contained in a chaotic or near chaotic spectra.

3. Low Resolution Spectra of the Sodium Trimer

In the experiment of interest, sodium trimer is produced in an optimally cooled beam assuring that it is in its ground vibronic state and low ($J < 20$) rotation state. It is then excited with a single photon to the lowest regular vibrational state of the electronically excited $C^2E''(0)$ state. From parallel regular-type spectroscopic experiments, the geometry of the C state and its frequencies are known, and from this a simple normal mode ground vibrational wavefunction ψ_{initial} can be constructed. A second laser is used to stimulate emission (stimulated emission pumping) down to the rovibration levels of the ground state. Since those molecules that do not emit are ionized by absorbing a second photon of the first laser, the ionization signal dips at the emission frequencies. The spectrum below 250 cm^{-1} is well assignable in terms of three normal modes for C$_{2v}$ geometry, which are symmetric stretch, bend, and asymmetric stretch modes with frequencies 139, 49, and 87 cm^{-1}, respectively. Figure 1a (upper-right corner) shows pictures of two of these motions.

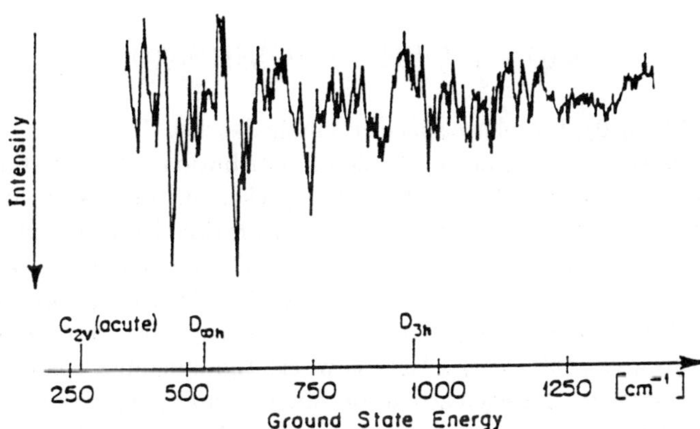

Fig. 2. The SEP spectrum of Na$_3$ for the $C^2E''(0)3/2$ intermediate state (Ref. 7). THe resolution is 5 cm^{-1}. The intensity scale, in arbitrary units, increases in a downwards direction and measures the depletion of the intermediate state by recording the multiphoton ionization signal for this state. Energy is measured with respect to the minimum of the potential. The activation energies of the different geometries are also indicated.

Figure 2 shows a low-resolution (5 cm^{-1}) spectrum in which the reproducible fine structure hints at a very congested, erratic, chaotic high-resolution spectra and which is also suggestive of a double oscillation. The frequencies of the latter double oscillation, which we anticipated in a general sense in the previous section, show up better in the magnitude square of the Fourier transform of the spectrum. This quantity, symbolized as $C(t)$, is plotted in Fig. 3a, and two frequencies are seen at 130 and 40 cm^{-1}. In quantum mechanics $C(t)$ measures the square magnitude of the time correlation function of the wave packet formed on the lower surface; it is large when the system returns to its initial region of space since trapping, if it is to be seen in this experiment, occurs in this region. The periods ($T = 2\pi/\omega$) seen are indicative of the vibrational frequencies of the "trapped species".

We now turn to classical mechanics to extract the dynamics and use the classical limit of $C(t)$. The potential used in the calculations was that of Ref. 44. It is worth noting that the spectrum itself has no meaningful classical limit as its information is in its discreteness of energies; a property which classical mechanics can never have. $C(t)$, on the other hand, measures the frequencies that represent energy differences and has a physically meaningful classical analog given as

$$C(t) = \int d\bar{p}\,d\bar{q}\,\rho(\bar{p}(0), \bar{q}(0))\rho(\bar{p}(t), \bar{q}(t))$$

where $\rho(\bar{p}(0), \bar{q}(0))$ is a phase space ensemble density. This density is, itself, obtained as the classical analog (Husimi distribution) of the initial packet created on the lower surface. The initial packet, in turn, is the transition dipole, here taken as 1, times the known vibrational wavefunction of the intermediate dump state, ψ_{initial}. Hence, $\rho(\bar{p}(0), \bar{q}(0))$ weights each point $(\bar{p}, \bar{q})$ with a weight characteristic of how the experiment produced the initial packet. For each point $(\bar{p}(0), \bar{q}(0))$ the trajectory $(\bar{p}(t), \bar{q}(t))$ is put in $\rho(\bar{p}, \bar{q})$ to give $\rho(\bar{p}(t), \bar{q}(t))$. Here is where the dynamics comes in. $C(t)$ is clearly a density–density correlation function and measures the classical analogue of the recurrences discussed above. As explained in Sec. 2, we expect the recurrences from the chaotic motions to average out as each contributing trajectory to $C(t)$ has different "chaotic" recurrences. What does not average out is the recurrences of the trajectory in the trapped region. Every trajectory "oscillates" similarly in the trapped region and the periods do not average out.

Since the experiment measures relative intensity, the computed $C(t)$ shown in Fig. 3b is matched at one value of t to the experiment. The agreement of theory and experiment is excellent. Figure 3a yields $\omega_1 = 130$ cm^{-1} and $\omega_2 = 40$ cm^{-1}, and Fig. 3b yields $\omega_1 = 128$ cm^{-1} and $\omega_2 = 44$ cm^{-1}. Since it was checked that all trajectories used in computing $C(t)$ were chaotic trajectories, the spectrum can be deemed "chaotic".

Having demonstrated the chaotic nature of the motion and the spectrum, we turn to the heart of this effort, namely, the problem of extracting

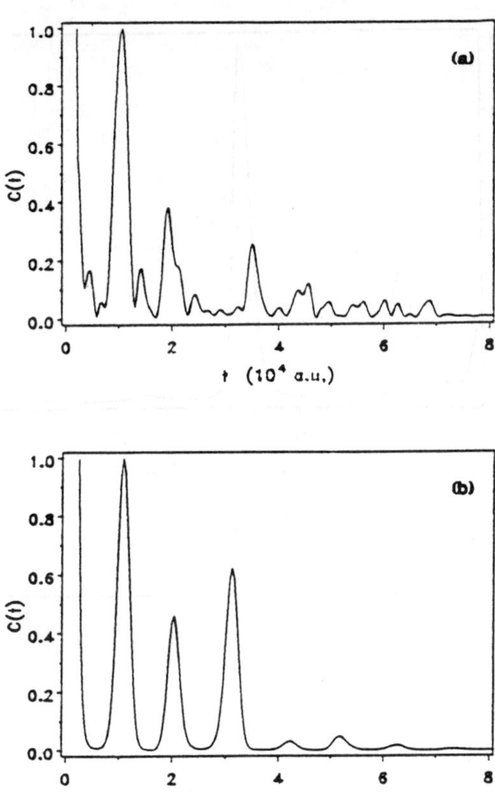

Fig. 3. The survival probability $C(t)$ of the SEP experiment. Intensities are in arbitrary units and relative to the first peak. Panel (a) shows the experimental survival probability defined [Eq. (2)] as the squared magnitude of the Fourier transform (FT) of the experiment. The FT was done after smoothing the spectrum up to a 15 cm^{-1} resolution and using a window function that maximizes slightly the 575 cm^{-1} spectral region. The smoothing was done because our purpose was to analyze the low resolution features. In any case, the graphical data available to us were not pictorially well defined below that resolution. The window function was used to reduce the contribution of both very low energies and energies close to and above the Jahn–Teller intersection. Three most intense peaks can be observed. The frequencies corresponding to the first and third peaks are $\omega_1 = 130$ cm^{-1} and $\omega_2 = 40$ cm^{-1}, respectively. The second peak is an overtone of the first one. Panel (b) shows the simulated survival probability for $J = 0$.

The window function mentioned above was also taken into account in the calculation of $C(t)$. The frequencies corresponding to the first and third peaks are $\omega_1 = 128$ cm^{-1} an $\omega_2 = 44$ cm^{-1}.

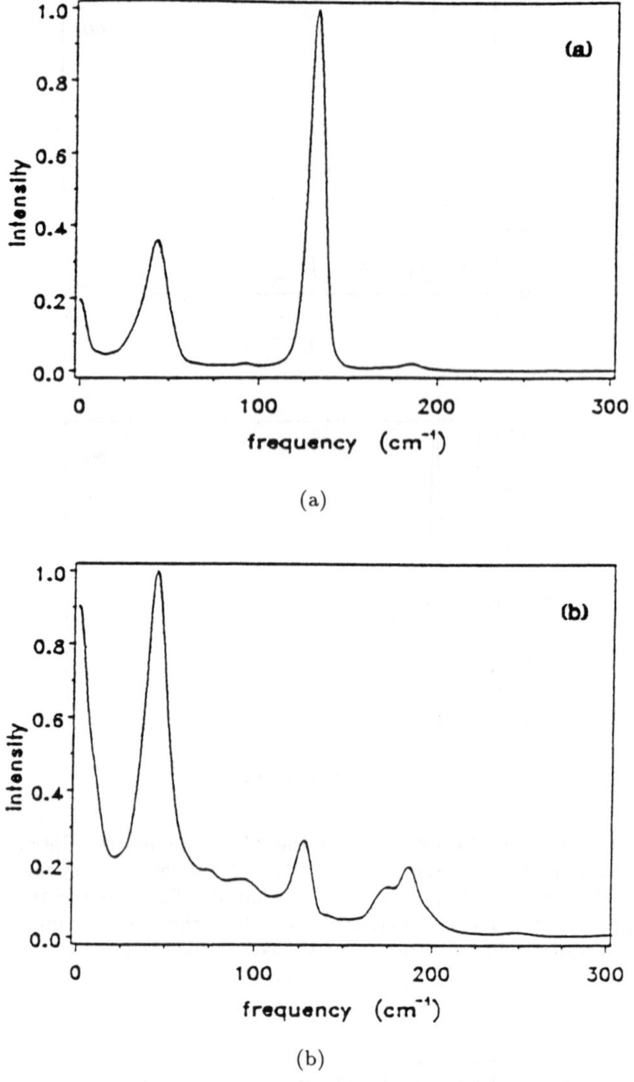

(a)

(b)

Fig. 4. Averaged power spectrum of two different dynamical variables $u(t)$ for a resolution of 15 cm^{-1}. $E = 575$ cm^{-1} and $J = 0$. In panel (a) $u = z$, the symmetric stretch coordinate of the D_{3h} symmetry group. In panel (b) u is the bending variable $u = \cos(3\alpha)$, where α is the polar angle in the plane defined by the two D_{3h} symmetry coordinates x and y of Fig. 1b.

and interpreting the regular motion that is argued to come from a reduced-dimension torus inserted in the chaotic motion. The inserted regular motion is in this case two dimensional, since two fundamental frequencies for $C(t)$ are seen in Fig. 3b. This motion can be analyzed by computing averaged power spectra of various coordinates evaluated along the three-dimensional chaotic trajectory. Power spectra are basically Fourier transforms of the time evolution of a coordinate. The chaotic regions of the trajectories, over which we average, contribute small almost never repeated peaks at the myriad of frequencies necessary to describe chaos. The regular region's insert, when visited and revisited, always adds to the intensity at the regular insert frequencies. At all energies in the spectral region, a result similar to Fig. 4 occurs; namely, if the symmetric stretch or the bend coordinate is used as the coordinate in the power spectra, one obtains two narrow peaks at frequencies 131 and 44 cm^{-1} near those of Fig. 3. Clearly, one can safely assume that in Fig. 4a (4b), which is for the symmetric stretch (bend) coordinate, the big peak is the frequency of this symmetric stretch (bend), and the small peak is that of the bend (symmetric stretch). Note that no asymmetric stretch appears and a power spectrum of the asymmetric stretch gives no sharp peaks. Conclusion: our torus is reduced to two dimensions and must be a two degree of freedom symmetric stretch-and-bend family of tori that exist in the transition region, and hence in the same region of position space, but at higher energy than the three-degree (symmetric stretch, bend, and symmetric) tori that give the regular levels (by emission from the same intermediate state). The lower energy three-dimensional tori have lost a degree of freedom (the asymmetric stretch), and trajectories that would have been on them have become chaotic trajectories which mimic a two-dimensional reduced dimension tori. Put in another way, the chaotic trajectories have inserts that act as the 2D reduced dimensional tori.

Now, having decided what one expects to see and where to see it, it is a simple matter to search the transition region for such an insert. We also need to check that other long-lived inserts do not exist. This latter is here the case except for a 1D torus (stable periodic orbit; see Fig. 1a) which does not contribute significantly to $C(t)$. Such an insert was found and can be demonstrated to be very similar to a two degree of freedom torus (symmetric stretch and bend) by the following procedure. First, near the insert, the asymmetric displacement coordinate and its conjugate momentum are set to zero, predestining the trajectory to lie in two dimensions. The resulting motion is the two-dimensional (symmetric stretch and bend) torus and this torus actually was used as our example in Fig. 1a. The frequencies of this torus are 135 and 46 cm^{-1} and are extremely close to those of Fig. 4 (obtained with the chaotic trajectory). Now the momentum and/or the position displacement coordinate of the asymmetric stretch is set near, but not equal, to zero. Motion similar to that of the torus is observed for picoseconds after which the trajectory moves ever farther away in the asymmetric stretch direction until it becomes chaotic in all variables; time reversal demonstrates the approach to the torus. The long-time power spectrum again becomes that of Fig. 4.

Recalling that the ground-state potential has a three-blade propeller shape with symmetrically equivalent wells (that hold the five regular states), one in each blade, and three saddle points at small radii between the blades, a pleasing physical picture arises from our results. At low energies, the four observable levels can be viewed as the result of a semiclassical quantization of tori with fundamental normal mode frequencies $\omega_{01} = 139$ cm^{-1} (symmetric stretch or breathing), $\omega_{02} = 49$ cm^{-1} (obtuse bend), and $\omega_{03} = 87$ cm^{-1} (asymmetric stretch). Such tori lie in each one of the three symmetrically equivalent wells of the potential surface. The normal-mode frequencies calculated quantum mechanically in Ref. 27 (142, 58, and 94 cm^{-1}, respectively) support this conclusion. (For completeness, we note that the fifth bound state (011) at roughly 136 cm^{-1} has not been observed experimentally due to selection rules.) Past the fourth level, the vibrational amplitudes are quite large and floppy, and the trajectory feels the confining wells of the potential. At about 250 cm^{-1} and along the asymmetric stretch coordinate (acute bend saddle point) direction, the walls fall away and open to the saddle point between two of the three equivalent regions. Initially, only motion in this direction becomes unstable, leaving a C_{2v} two degree of freedom family of tori, describable as the combination of a symmetric stretch mode (now because of anharmonicity $\omega_1 \approx 135$ cm^{-1}) and an obtuse bend mode (now $\omega_2 \approx 40$ cm^{-1}). Later, as we move far along the asymmetric stretch direction and far from the two-dimensional torus, complete chaos takes over until a reduced-dimension torus is again approached. Clearly, Na$_3$ is an example of how, when the potential changes rather suddenly to allow one of the "springs" to break and to permit motion in a much larger space, chaos sets in.

4. Application to H$_3^+$

Here, the influence on the spectrum of a single type of robust periodic orbit embedded in the chaos is demonstrated. Physically, the idea to remember is that any short level regular insert must be the result of the trajectory "feeling" a dynamic force (due to the potential and the motion) that is in some sense confining, albeit with places to enter and leave the confined localized region. This same potential then traps wave packets. The fact of wave packets trapped in a local region implies that the stationary wavefunctions in the energy range of the packet are localized, regular looking, and of relatively large amplitude in the trapping region and are small in amplitude outside the region. This is the picture discussed in Sec. 2.2. The strategy used to unravel the dynamics is then to find the periodic orbit and to see if its frequencies mirror the quantum low resolution frequencies. If so, the motion along the periodic orbit is the classical analogue of the quantum motion and, assuming tunneling is not dominant, the dynamics become evident from how the coordinates change along the "insert" which here is a periodic orbit. Parenthetically, wavefunctions in the "chaotic" region were observed[28] to be scarred by classical periodic orbits. Our just competed discussion on the nature of the stationary continuum wavefunction is our view of the cause of this phenomena.

When high rovibrationally excited H_3^+ produced in an ion source was irradiated with CO_2 laser photons (with Doppler shifted frequency between 870 and 1094 cm^{-1}) the collected H^+ signal contained about 27 000 narrow lines.[6] This assignable, dense and chaotic appearing spectrum yielded, under simulated low resolution, several clumps separated by roughly 53 cm^{-1}, a frequency not reminiscent of lower energy excitation spectra.

It was shown that the states leading to narrow lines in the high resolution spectrum were the resonant states of H_3^+ trapped behind the total angular momentum barrier.[29,30] The semiclassical approach[31] succeeded in showing that the tunneling lifetimes of these resonant states were consistent with the widths of the observed high resolution spectrum. Berblinger et al.[32] studied the classical mechanics of H_3^+ at energies above the dissociation limit. At a total angular momentum $J = 0$, they found that almost all phase space was chaotic apart from a small stability island surrounding the "horseshoe" stable periodic orbit. A "classical" estimate of the number of resonance states based on one state per $\hbar^N$ phase space volume ($\sim$ 80 000 states) verified that the density of resonance states trapped behind the total angular momentum barrier is sufficient to account for the number of lines observed experimentally.

Although these studies form a basis for understanding the formation of the experimentally observed high resolution spectrum, the regular structure observed in the low resolution remained a puzzle. Pfeiffer and Child[30] failed to simulate the spectrum quantum mechanically within the rigid rotor model. Moreover, a full exact quantum mechanical treatment is, at the present date, prohibitive. The frequency 50 cm^{-1} was observed in classical studies;[33] however, the authors were not able to relate it to the experimental spectrum.

The breakthrough occurred when Gomez-Llorente and Pollak,[34] after computing and analyzing the classical dipole spectrum for the chaotic manifold, concluded that the regularity seen in the low resolution spectrum could be associated with a stable manifold of trajectories. They found that the horseshoe periodic orbit[32] was stable in the three dimensional space for total angular momentum values J up to 25 and at energy values consistent with the experiment.

Perhaps the observed low resolution 53 cm^{-1} was the recurrence of a wave packet trapped in the same effective dynamic potential as the horseshoe periodic orbit? The horseshoe motion was called horseshoe because of the periodic's orbits appearance in Jacobi coordinates. It corresponds to a rapid (975 cm^{-1}) vibrational inversion of the bent H–H$^+$–H complex accompanied by a slow H$_2$ rotation in the plane perpendicular to the inversion axis. On the average, from the point of view of the rotation, the molecule is linear. One might expect to see low resolution features spaced at 975 cm^{-1}. Even if this does exist, the experiment, which for reasons of the laser's ability to scan only a less than 300 cm^{-1} range, obviously implies that such a 975 cm^{-1} feature cannot be seen. Question: what motions are the 53 cm^{-1} clump representing? The answer was to realize that the rotations had not yet been quantized. For rotations of "linear" systems one expects the rotation lines to be spaced equally at $2B$ where B is the rotation constant. Now it is known that B is given semiclassically as $\frac{\partial E}{\partial J} = 2B$ where E is the energy of the periodic orbit and J its total angular momentum, and where the derivative is computed along the periodic orbit. The horseshoe gave $2B = 60$ cm^{-1} for one available potential surface and $2B = 50$ cm^{-1} for another. The result suggests strongly that the observed clump was the localized complex's rotational spectra with each line broadened by mixing with, as discussed in Sec. 2.2, the chaotic high density of states quasicontinuum that was the heuristically expected quantum analog of the chaotic part of the motion.

This picture was verified quantum mechanically in Ref. 21 where the quantum stabilization method,[35] designed originally to compute resonances embedded in the scattering continuum, was used to calculate the low resolution spectrum. That this can be done is easily understood if one recalls that by our tier picture of Sec. 2 the bound (negative energy) states that show up in low resolution are localized ones adjacent to a quasicontinuum. If one now realizes that in low resolution the quasicontinuum appears as a continuum, then the described picture of the quantum wavefunctions and spectrum in Sec. 2 is exactly what one envisions when resonances are discussed. In fact, much of the work of the USC group started[36] with the resonance picture and then applied the analogy to the low resolution picture of chaotic spectra. Interestingly, in classical mechanics the explanation of localization is the same, and as given in Sec. 2, whether one is dealing with bound or open systems. Localization mechanisms are sensitive to local dynamic potentials and are not very sensitive to the existence of confining walls of the true potential since they are often "far away". This fact explains why that when the spectrum is taken in the chaotic high energy region on both sides of breakup, that the high resolution spectra changes dramatically from discrete to continuous, while the qualitative appearance of the low resolution spectra is the same.[8]

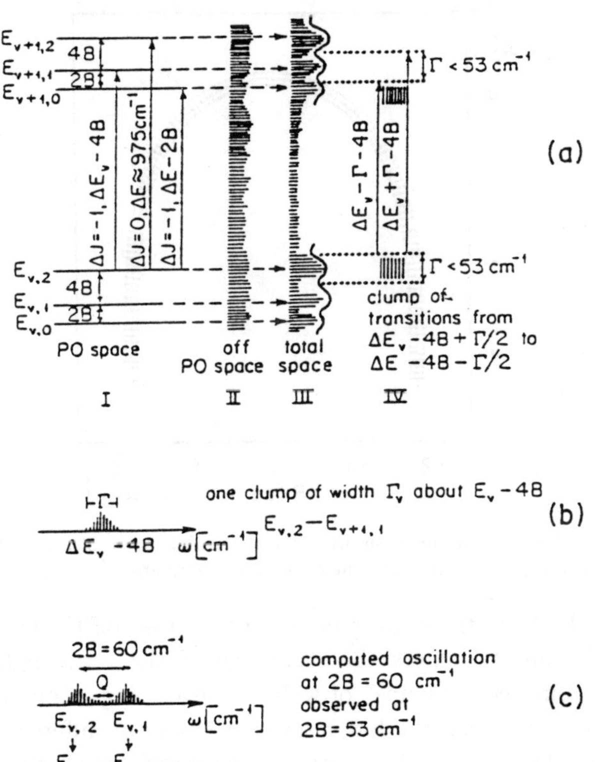

Fig. 5. Scheme of clump-to-clump transitions. For explanation see the text.

The result of the quantum stabilization calculation is shown in Fig. 5 where in Fig. 5a column I shows schematically the computed stable low resolution localized state levels. Column I could be right out of discussion on linear molecules explaining why the rotation spectra has P, Q, R branches and the rotation transitions are spaced by $2B$. The quantum calculations give the same values of B as did the classical result.

The quantum stabilization calculation which gave these levels placed a Gaussian basis set only inside the dynamic trapping potential. Since this latter is not known it used the ideas above and placed the basis around the horseshoe periodic orbit as shown in Fig. 6. Diagonalization of the Hamiltonian, for given J, gave the quoted results.

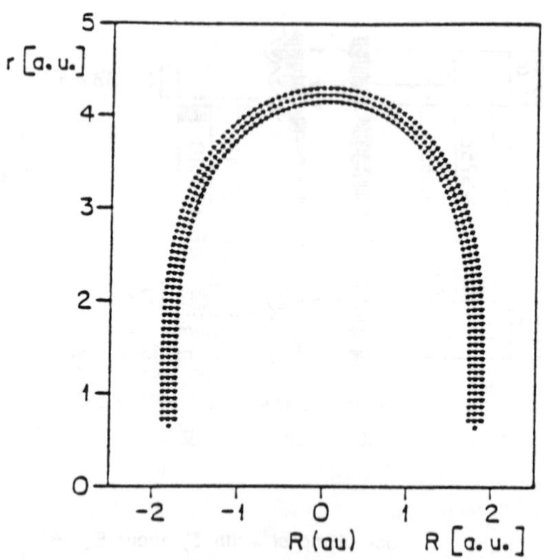

Fig. 6. The placement of the basis used for the stabilization study of the horseshoe periodic orbit in H_3^+. Stars indicate the centers of Gaussians.

Column II shows the chaotic space spectra (the "off the periodic orbit" space). Column III shows the schematic of the total high resolution spectra and one sees "clumps" of width Γ developing. Column IV shows the transitions between the low resolution clump of different vibrational quantum number. Figure 5b shows a mock up of one clump and Fig. 5c demonstrates how the spectra in low resolution appears and why it looks like an oscillation of 53 cm^{-1} in experiment and 50 cm^{-1} for the potential used in this work.

5. LiCN Spectra in the Isomerization Region

The purpose of this section is not only to again illustrate the power of low resolution methods in extracting dynamics, but to demonstrate how what type of spectra (regular, chaotic, mixed, etc.) is seen depends on the transition region of the process under study. The purely theoretical simulation studied here is taken from Ref. 9. Here, a two mode, Li ... CN stretch (coordinate R) + Li–CN bend, (coordinate + two dimensional model) is taken. First, a study is made of the nature of the classical trajectories in the energy region of the isomerization barrier and the result is seen in Fig. 7. Here, the phase space is cut so that one sees the trajectories

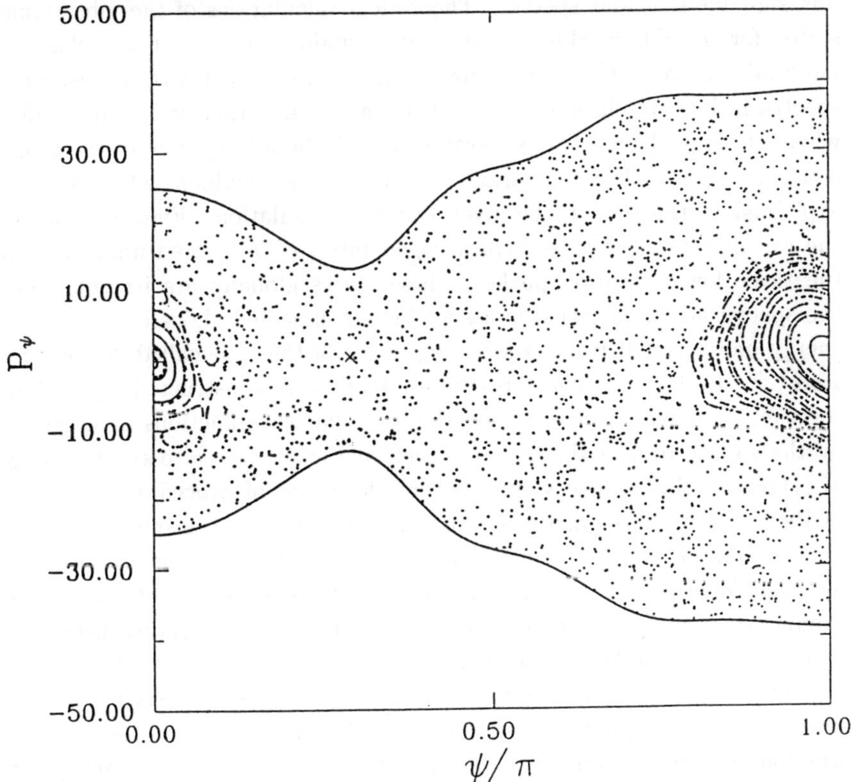

Fig. 7. Classical Poincaré surface of section for $E = 0.0172925$ a.u. The position of the cross indicates the location of the saddle point periodic orbit (SPPO).

coming through the page toward the reader. The trained eye notices that at $\psi/\pi = 0$ where the molecule is LiCN, or at $\psi/\pi = 1$ where it is CNLi (ψ/π measure a curvilinear motion passing through the saddle point at $\psi/\pi \approx 0.30$ of Li from one side of the CN to the other) that tori (the circles) exist when the geometry is LiCN or LiNC. This means that one expects that any SEP spectra taken into these regions will see some very strong lines and that they will be assignable as regular spectra. Under high resolution one may also see some very low intensity unassignable lines due to the obvious existence in these regions of a small amount of chaos. Now, a theoretically computed SEP spectra from a hypothetical upper electronic state over $\psi/\pi \approx .3$ will see chaos and a chaotic, generally unassignable irregular spectra. Theoretical calculations of the vibrational states for $J = 0$ in this region[37] yield qualitatively what the classical mechanics predicted. In particular, Fig. 8c shows such a computed SEP spectra in high resolution where it is irregular, and then in low resolution where three peaks are seen separated roughly by a frequency in the range $680\ \mathrm{cm}^{-1}$. A little imagination also sees some shoulders which will be explained in the soon to be given classical simulation. Figure 8a shows the wavefunctions of three typical eigenstates, each located under one of the three low resolution peaks. These states globally are irregular and (no special nodal patterns) unassignable, as expected. Locally, the lowest energy one at small ψ has a local regularity of the type called "a rotating state".[38] In Ref. 9, a low resolution state was constructed by adding together, under each broad peak, the irregular wavefunctions as weighted by the square root of the intensity seen in Fig. 8c (see also the paper by M. Davis in this review volume where the types of procedure suggested in Ref. 20 were developed contemporaneously with Ref. 9 but with much greater mathematical systemization). The result shown in Fig. 8b is exactly as outlined in Sec. 2. The result resembles three local bound states (one ground state and two excited states – count the nodes) representing the quantization of the oscillation of the Li against the CN at an angle of about 3.5 radians. These states are truly the transition states for the isomerization process. Not surprisingly, a periodic orbit (shown drawn under the packets) was found, whose frequencies was that of the low resolution spacing and whose quantization clearly would give a good approximation to the energy of the three low resolution peaks. The stability frequency of the periodic orbit is comparable to the widths of the low resolution peaks which further

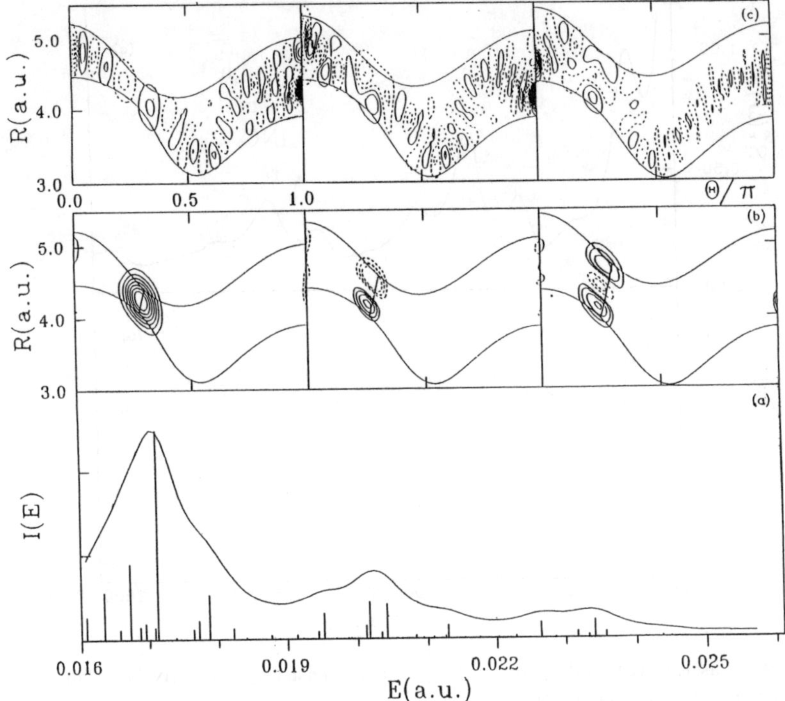

Fig. 8. (a) Simulated SEP stick spectrum and its low resolution version for LiCN. (b) Resonance functions corresponding to the three bands appearing in the low resolution spectrum of (a). The saddlepoint periodic orbit and the equipotentials corresponding to the energies of Table 1 of Ref. 9 are also indicated. (c) Three typical wavefunctions (levels 67, 95, 133) each one contributing to a different spectral band.

affirms the analogy. For further low resolution study, a classical correlation function which, as in the Na_3 study, launches from the transition region an ensemble of trajectories and counts the periods of their return to the region, was computed and its Fourier transform is shown in Fig. 9a. The quantum analogue of this result is the quantum intensity–intensity correlation function which by the Fourier transform convolution theorem is the square magnitude of the Fourier transform of a smoothed version of the stick spectra in Fig. 8c. The smoothing was chosen to make the quantum correlations look as nearly the classical as possible. The results in Fig. 9a show clearly the expected frequency of 680 cm^{-1} but reveal more clearly

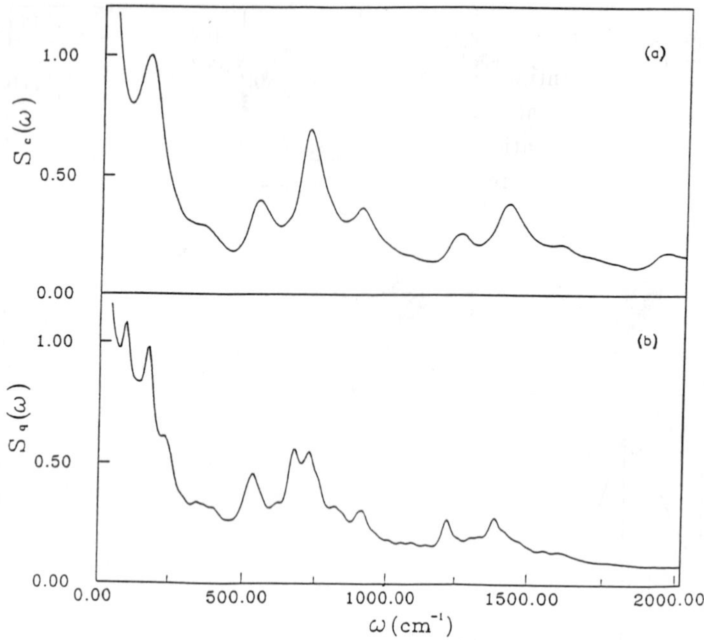

Fig. 9. Classical (a) and quantal (b) spectral densities, respectively. For LiCN: Transition to the barrier region.

shoulders (triplets) that are separated by a frequency ω_2 which was found to correspond to an average bending frequency of the classical trajectories used to calculate the classical correlation function. Several families of unstable periodic orbits were found connecting the TS region with each well region. Some of these orbits are of rotating type.[38] All have a similar return time to the transition state; the corresponding return frequency is ω_2 and the stability frequencies of the least unstable orbits were very close to the widths of the bands within the triplets. A mechanism for the energy relaxation of the resonances can, therefore, be given: energy put into the stretch mode of the TS is first transferred to the motion along the isomerization coordinate through coupling to the bending mode, and then uniformly distributed from here to throughout the chaotic phase space. In the quantum case, the discreteness of the spectrum prevents from a true relaxation, but this fact only shows up at much longer times; long enough for the finer structure in the quantum correlation to be resolved. Some of this finer structure can already be observed at the resolution of Fig. 9b. Quantum tunneling is also the origin of some of the discrepancies observed in Fig. 9 between the classical and quantum results. The behavior of the system is obviously quite far from statistical.

The nontrivial relaxation mechanism for the resonances of Fig. 8 involves two decay times which are the result of the interplay between the SPPO and a set of families of unstable periodic orbits connecting the saddle point region with both isomer wells. Many of the features associated to potential saddlepoints may now be understood in term of saddlepoint resonances. In particular, one could predict a nonstatistical, nonuniform behavior of quantum isomerization rates as a function of the energy, enhanced by the possibility of complex relaxation pathways.

6. SEP and DF Spectra in HCCH

Probably the cleanest and most developed study of the third mechanism for the onset of chaos, that is overlapping resonances and the subsequent use of low resolution spectra as a diagnostic, comes from the Stimulated Emission Pumping (SEP) and Dispersed Florescence (DF) work on acetylene. Here, emission to the electronic ground $\tilde{X}^1\Sigma_g$ state from a prepared regular vibration rotation state of the electronically excited $\tilde{A}^1A_u$ is monitored. High resolution (0.5 cm^{-1}) rovibronic spectra can be seen when the emission is stimulated (SEP) and low resolution spectra (30 cm^{-1}) have been recorded with spontaneous emission (DF). Emission occurs from the *trans*-bending (called v_3' in the $\tilde{A}$ state; "prime" is upper, "double prime" is lower state) $v_3' = 2$ or 3 states of the $\tilde{A}$ state where the transition moment is perpendicular to the plane of the initial state. This latter implies specific propensity rules $\Delta K = \pm 1$ and allows assignment of the lower spectra with respect to the total angular momentum J and the total vibrational angular momenta l. Specifically, since emission is from a $K_a' = 1$ rotational level the bends in the final state must have $l'' = 0$ or 2. The DF spectra showed a total of 140 features between 5700 and 21 200 cm^{-1} and long progressions in the *trans*-bend $(v_4' = 6\text{--}18)$ and C$\equiv$C stretch $(v_2' = 0\text{--}6)$ were identified. The high resolution SEP and estimation of the number of allowed and expected final (J'', l'') states showed that below 8000 cm^{-1} each DF structure was a single vibrational level, while above 16 500 cm^{-1} an observed DF structure contained as many as ten levels. It is the change of *trans*-bend angle from upper state to lower state and the fact that the upper state is electronically a double C=C bond, and therefore longer than the ground states triple C≡C bond, that makes the observed progressions "expected". Generally, because of the small Franck–Condon (FC) factors the symmetric CH stretch v_1'' was not excited; nor, because of small distortion and frequency change, was the excitation level of the v_3'' asymmetric stretch and the v_5'' *cis*-bend expected to be changed upon transition from whatever excitation it originally had (which for the most part was zero). Below $14\,000$ cm^{-1} a second order anharmonic expansion, slightly modified from the one used to explain levels observed in infrared and Raman studies, was used to "fit" and assign to within the 20 cm^{-1} experimental error many of the 50 observed DF features. The assignment was in terms of v_2'', v_4'' combinations and had an intensity pattern in accord with the Franck–Condon principle. As the spectra from the two possible ($v_3 = 2$ or 3) upper states had essentially the same relative intensities in the SEP spectra and the same energy positioning of the maxima in the DF spectra, one must conclude that where several SEP lines appeared under a DF feature that they were mixtures of zero order bright states and that the DF itself was to one zero order bright state (which was usually assignable). Since in the DF the same zero order ground state "localized packets" are formed from the different initial, upper states, the dynamics is expected to be controlled by the forces of the lower surface and the initial bright state accessed (i.e., the packet at $t = 0$). As such it can be safely said that the dynamics is independent of the upper state. The actual eigenstates seen in SEP are accessed because they have a component on the time zero low resolution bright state.

Let us consider the less than $14\,000$ cm^{-1} region. Now, besides the simple low resolution (v_3'', v_3'') progressions at least three other progressions built on excitation in a third vibrational mode were identified. Working at 7000 cm^{-1} where each DF feature is one state, an assignment scheme was constructed that explained all the unassigned "3rd mode" DF (low resolution) progressions below $14\,000$ cm^{-1}. Jonas *et al.* invoked[24] a stepwise pattern of resonances that is "proven" by parametrization of model Hamiltonians for bright states plus one, then plus two and then plus three resonances. The first and strongest of these resonances is one which mixes with the dark *cis*-bend (v_5'') by a Darling–Dennison (DD) resonance between it and the *trans*-bend; in particular $\Delta v_4'' = -\Delta v_5'' = \pm 2$. This "lights" up low resolution states with v_2'', v_4'' and now v_5'' excited. It was then realized that this resonance could not stand alone. This is because if both v_4'' and v_5'' were excited their individual particular bending vibrational angular momentum combination l_4'' and l_5'' would couple to produce a total vibrational angular momentum $l'' = l_4'' + l_5''$ combinations. It is known that such states of a common point group symmetry differing only in the quantum numbers l_4'' and l_5'' are mixed by a $\Delta l_4 = -\Delta l_5 = \pm 2$ (i.e., $\Delta l'' = 0$) vibrational-l-resonance. This "kills" l_4'' and l_5'' as good quantum numbers and spreads the intensity of the DF bright level with the original l_4'', l_5'' among levels of the same symmetry and same remaining quantum numbers. Hence, in higher resolution one expects extra levels whose observation would and did confirm that v_4'' and v_5'' combinations were present. The DD vibrational l-resonance will spread the intensity of levels as $(0, 1, 0, 8^\circ, 0^\circ)^\circ$ [notation: $v_1'', v_2'', v_3'', v_4''^{l_4''}, v_4''^{l_4''}$] and $(0, 1, 0, 8^2, 0^2)^\circ$ (seen as $(0, 1, 0, 8, 0)$ in the DF [30 cm^{-1} resolution] spectrum around 7000 cm^{-1}), respectively over two levels as $(0, 1, 0, 6^\circ, 2^\circ)^\circ$ and $(0, 1, 0, 6^{\pm 2}, 2^{\mp 2})^\circ_+$ and three levels as $(0, 1, 0, 6^2, 2^\circ)^2$, $(0, 1, 0, 6^\circ, 2^2)^2$ and $(0, 1, 0, 6^{\pm 4}, 2^{\pm 2})^2$. These latter levels which include the "extra" $l = 2$ levels could be seen in higher than DF resolution, albeit still low resolution (30 cm^{-1}) SEP spectra. This observation and assignment was further supported and confirmed by both the extraction of the rotation constant and by observation of expected rotational-l-resonances. The mechanism also explained the assignment of a DF spectral peak in the 7000 cm^{-1} region as $(0, 1, 0, 6, 2)$. It was the low resolution envelope of the two most intense among the above five levels. A similar discussion could be given for all low resolution states in the 9400 cm^{-1} to 9700 cm^{-1} region except that $(0, 1, 0, 12, 0)$ is the starting low resolution (v_2'', v_4'') combination. The DD resonance in particular becomes more prominent with increasing excitation of the *trans*-bend due to a catastrophic degeneracy of the entire $v_b'' = (v_4'' + v_5'')$ manifold near $v_b'' = 16$ which causes a massive perturbation of all DF features with $v_4'' = 14$.

The IVR and mode coupling story would now be finished except for the fact that the high resolution study of the $(0, 1, 0, 8, 0)$ and $(0, 1, 0, 6, 2)$ region showed within 100 cm^{-1} of the latter a set of levels assignable using band origins rotational constants and vibrational angular momentum considerations, as $(0, 0, 1, 5, 1)$. This indicates the presence of a resonance that allows $(0, 0, 1, 5, 1)$ to borrow intensity from $(0, 1, 0, 6, 2)$ and to take one phonon out of each of the *trans*-bend, *cis*-bend, and CC stretch and put them into the antisymmetric stretch. This resonance is denoted as a $\Delta v_3 = -\Delta v_2 = -\Delta v_4 = -\Delta v_5 = \pm 1, \Delta l_4 = -\Delta l_5 = \pm 1 ("2345")$ Fermi resonance and is familiar in that it causes the well-known perturbation of the fundamental v_3''.[39] At this stage, the resonances involved in the mechanism that governs the observed levels in the 7000 cm^{-1} region and which controls IVR flow at higher energies out of the v_2, v_4 combination is now known for the first few hundred femtoseconds. The *cis-trans* resonance acts immediately in conjunction with the vibrational-l-resonance and this allows "flow" into the antisymmetric stretch via the Fermi resonance. The timings of the flow are indicated by the resolution needed for the particular observation and the short time dynamics is clearly revealed.[40]

Our discussion so far has illustrated the earlier discussion of the 3rd mechanism in Sec. 2.2. It is useful to continue this discussion even a bit further now that the acetylene example is available to exemplify what otherwise would have been too abstract a presentation if given earlier. Using the vector notation $(\Delta v_1, \Delta v_2, \Delta v_3, \Delta v_4, \Delta l_4, \Delta v_5, \Delta l_5)$ to exemplify which modes are coupled in a resonance, it was noted[23] that a linearly independent set of the vectors used above to discuss the coupling are:

vibration-l-resonance	$(0,0,0,0,2,0,-2)$
DD bend resonance	$(0,0,0,2,0,-2,0)$
"2345" Fermi Resonance	$(0,1,-1,1,1,1,-1)$

If the strong, well characterized $\Delta v_1 = -\Delta v_3 = \pm 2$ DD resonance $[(2, 0, -2, 0, 0, 0, 0)]$ between the two CH stretches is also considered, then all modes are coupled and no separable part of our normal mode "torus" will still exist at energies where all these resonances exist. Does this mean that there are no approximate quantum numbers and no restrictions on the motion, i.e., that full chaos exists? The answer, as indicated in Sec. 2, is that, at least for short times, the ability to even fit low resolution spectra to a model (here one of a few interacting resonances) leads us to say that motion is localized for short times in a part of phase space. Now, using ideas from perturbation theory,[41,42] a vector model was developed that shows that to the degree that perturbation theory converges and on the time scale of the resolution used to observe the resonances, that if the above (here four) vector resonances (if need be reduced to a linearly independent set) are considered to span a part of a vector space of dimension (here 7) of the vector, then the basis of the orthogonal complement space [denoted as $(m_1^i), m_2^i, m_3^i)]$ are indicative of approximately, [here three $(7 - 4 = 3)$] conserved quantities of the form

$$n^i = m_1^i v_1 + m_2^i v_2 + m_3^i v_3 + m_4^i v_4 + m_5^i v_5 \qquad (i = 1, 2, 3) \; .$$

In proper linear combination these n_i show that there are here three conserved quantities. First is the expected and confirmed $l = l_4 + l_5$. Then, the $n_s = v_1 + v_2 + v_3$, which reflects why in the Fermi resonance when the CC stretch loses a phonon, another stretch must pick it up. Finally, the $n_{\text{res}} = 5v_1 + 3v_2 + 5v_3 + v_4 + v_5$ quasiconstant which, because the normal mode frequency ratios are 5:3:5:1:1, seems to reflect the idea (that is evident from perturbation theory) that strongly interacting levels (i.e., in resonance) need be of the same energy. The conserved quantities, or in a sense selection rules, clearly indicate that the trapping of the "packet" on the short time scale has some "good" quantum numbers. In a recent paper,[43] it has been shown and implemented for a very simple case, how under the assumption that perturbation theory converges (at least for short time processes) that the resonances $(\Delta v_1, \Delta v_2, \Delta v_3, \Delta v_4, \Delta v_5, \Delta v_6, \Delta v_7)$ can in principle be used to variationally find implied localized regions of configuration space. If this can be extended to more complicated systems, one might actually be able to see where the localization takes place.

Another "motional" fact can be obtained from contemplating the short time resonance pattern. The initial *trans*-bend to *cis*-bend mixing indicates the new bend combination allows one hydrogen to bend at a time. The CC stretch to antisymmetric CH stretch resonance that includes such bends in equal mix indicates that, when the molecule bends, one of its CH bonds bends and stretches while the other shortens and bends. This motion, which could eventually cause the extended hydrogen to swing to the other side of the molecule, suggests that the acetylene vinylidene isomerization barrier may well be the feature of the potential that causes the onset of these resonances. Hence mechanism I may be related to mechanism III.

From the splitting of the low resolution DF features into several vibrational states in high resolution SEP spectra[17] we know that after our "short time", new resonances break down even the "short time quantum numbers." Eventually, new methods of analysis will be needed to decide if chaos is achieved or if resonances between new distorted modes would simply explain the spectra and, hence, the motion. Even if the spectra could be fit by adding new resonances, after three or four stages of splittings, the physics would be lost as the fittings would hardly be unique.

In the $\sim 28\,000$ cm^{-1} wave number region[5] low resolution (0.3 cm^{-1}) SEP spectra showed, for final $l'' = 0$, many 1 to 2 cm^{-1} clump-like features separated by about 5 to 10 cm^{-1}. Higher resolution (0.03 cm^{-1}) resolved the clumps into many lines for which a statistical analysis suggested the presence of chaos (Wigner distribution of energy spacings). What the $l = 0$ trapping that causes clump is, is not known although many have suggested that since the system is above the acetylene–vinylidene isomerization barrier, which has already been suggested as the "cause" of mixing, and since the frequency spacings are suggestive of a rotation, the clumps may be the result of a trapped motion where one or even two hydrogens rotate about the carbon–carbon bond.

The above mechanisms hopefully illustrate ways of learning about how a molecule approaches full chaos. Often, the molecule will not get there as it will dissociate before such a state of excitations is achieved. The subject and the motions are complex as expected of the approach to the complete complexity that is chaos. The study of trappings in low resolution not only helps us understand this evolution to chaos but also supplies "states" where the system exists long enough as a simple species where predictable processes may be initiated by chemical or physical attack. As such, the trapped regions play the same role as scattering resonances do in scattering physics.

Acknowledgment

I thank the National Science Foundation, Grant #9120493, for financial support.

References

1. *Chaos*, ed. J. Gleick (Penguin Books, New York, NY, 1990).
2. M.C. Gutzwiller, *Chaos in Classical and Quantum Mechanics* (Springer-Verlag, NY, 1990).
3. B.B. Mandelbrot, *Fractal Geometry of Nature* (W.H. Freeman & Co., 1977).
4. O. Bohigas, *Random Matrix Theories and Chaotic Dynamics*, Proceedings of the Les 1989 Houches Summer School on "Chaos and Quantum Physics", eds. M.J. Giannoni, A. Voros, and J. Zinn-Justin (Elsevier Science Publishers, B.V., North Holland, 1991), p. 547.
5. E. Abramson, R.W. Field, D. Imre, and K.K. Innes, *J. Chem. Phys.* **83**, 453 (1985).
6. A. Carrington and R.A. Kennedy, *J. Chem. Phys.* **81**, 91 (1984).

7. Figure 2 is from a private communication. For similar data see M. Boyer, G. Delacretaz, G.Q. Ni, R.L. Whetten, J.P. Wolf, and L. Woste, *Phys. Rev. Lett.* **62**, 2100 (1989).

8. A. Holle, G, Wiebush, J. Main, B. Hager, H. Rottke, and K.H. Welge, *Phys. Rev. Lett.* **56**, 2594 (1986).

9. J.M. Gomez-Llorente, F. Borondo, N. Berenguer, and R.M. Benito, *Chem. Phys. Lett.* **192**, 430 (1992).

10. The following books are very helpful introductions to the subject: a) R. Abram and C. Shaw, *Dynamics, The Geometry of Behavior*, Part 1-4 (Aerial Press, Santa Cruz, CA 1982-1988); b) Edward Oh, *Chaos in Dynamical Systems* (Cambridge University Press, 1993).

11. G. Herzberg, *Molecular Spectra and Molecular Structure, Volume I, II, III*: Nostrand, NY, 1966: *Atomic Spectra and Atomic Structure;* (Dover Publications; 1944).

12. For simplicity and without loss of generality, our discussion will be restricted to vibrational motions.

13. B.V. Chirikov, *Phys. Rep.* **52**, 263 (1979).

14. E.L. Sibert and W.P. Reinhardt, *J. Chem. Phys.* **77**, 3583 (1982).

15. L. Xiao and M.E. Kellman, *J. Chem. Phys.* **90**, 6086 (1989).

16. H.R. Dubal and M. Quack, *J. Chem. Phys.* **81**, 3774 (1984).

17. K. Yamanouchi, N. Ikeda, S. Tsuchiya, D.M. Jonas, J.K. Lundberg, G.W. Adamson, and R.W. Field, *J. Chem. Phys.* **95**, 6330 (1991).

18. G. Hose and H.S. Taylor, *J. Chem. Phys.* **76**, 5356 (1982).

19. J.M. Gomez-Llorente and H.S. Taylor, *J. Chem. Phys.* **91**, 953 (1989).

20. H.S. Taylor and J. Zakrzewski, *Phys. Rev.* **A38**, 3732 (1988).

21. (a) J. M. Gomez-Llorente, J. Zakrzewski, H.S. Taylor, and K.C. Kulander, *J. Chem. Phys.* **89**, 5859 (1988); (b) *J. Chem. Phys.* **90**, 1505 (1989).

22. (a) J.M. Gomez-Llorente and E. Pollak, *Chem. Phys. Lett.* **138**, 125–130 (1987); (b) *J. Chem. Phys.* **89**, 1195 (1988); (c) *J. Chem. Phys.* **90**, 5406 (1989).

23. K. Husimi, *Proceedings of Physical Mathematical Society, JPM* **22**, 264 (1940).

24. D.B. Jonas, S.A. B. Solina, B. Rajaram, R.J. Silbey, R.W. Field, K. Yamanouchi, and S. Tsuchiya, *J. Chem. Phys.* **97**, 2813 (1992); D.M. Jonas, Ph.D. Thesis, MIT (1992).

25. M. J. Davis, *Chem. Phys. Lett.* **192**, 479 (1992).

26. (a) H.S. Taylor and K. Stefanski, *Phys. Rev.* **A31**, 2810 (1985); (b) Y. Y. Bai, G. Hose K. Stefanski, and H.S. Taylor, *Phys. Rev.* **A31**, 2821 (1985).

27. J. Tennyson, Private Communication.

28. E.J. Heller, *Phys. Rev. Lett.* **53**, 1515 (1984).

29. E. Pollak, *J. Chem. Phys.* **86**, 1645 (1987).

30. R. Pfeiffer and M.S. Child, *Molec. Phys.* **60**, 1367 (1987).

31. J.M. Gomez-Llorente and E. Pollak, *Chem. Phys.* **120**, 37 (1988).

32. M. Berblinger, E. Pollak, and C. Schlier, *J. Chem. Phys.* **88**, 15 (1988).

33. J.M. Gomez-Llorente and E. Pollak, *Chem. Phys. Lett.* **138**, 125 (1987).
34. J.M. Gomez-Llorente and E. Pollak, *J. Chem. Phys.* **89**, 1129 (1988).
35. A. Hazi and H.S. Taylor, *Phys. Rev.* **A1**, 1109 (1970).
36. H.S. Taylor, *Acc. Chem. Res.* **22**, 263 (1989).
37. R.M. Benito, F. Borondo, J.H. Kim, B.G Sumpter, and G.S. Ezra, *Chem. Phys. Lett.* **161**, 60 (1989).
38. M. Founargiotakis, S.C. Farantos, and J. Tennyson, *J. Chem. Phys.* **88**, 223 (1988).
39. B.C. Smith and J.S. Winn, *J. Chem. Phys.* **94**, 4120 (1991).
40. E.J. Heller, *Acc. Chem. Res.* **14**, 368 (1981).
41. L.E. Fried and G.S. Ezra, *J. Chem. Phys.* **86**, 6270 (1986).
42. M.E. Kellman, *J. Chem. Phys.* **93**, 6630 (1990).
43. C. Jung and H.S. Taylor, *Phys. Rev. E* (1993) submitted.
44. T.C. Thompson, G. Izmirlian Jr., S.J. Lemon, D.G. Truhlar, and C.A. Mead, *J. Chem. Phys.* **82**, 5597 (1985).

INDEX